ANNUAL REVIEW OF EARTH AND PLANETARY SCIENCES

ANNUAL REVIEW OF EARTH AND PLANETARY SCIENCES

VOLUME 22, 1994

GEORGE W. WETHERILL, *Editor*
Carnegie Institution of Washington

ARDEN L. ALBEE, *Associate Editor*
California Institute of Technology

KEVIN C. BURKE, *Associate Editor*
National Research Council

ANNUAL REVIEWS INC 4139 EL CAMINO WAY P.O. BOX 10139 PALO ALTO, CALIFORNIA 94303-0139

ANNUAL REVIEWS INC.
Palo Alto, California, USA

International Standard Serial Number: 0084-6597
International Standard Book Number: 0-8243-2022-0
Library of Congress Catalog Card Number: 72-82137

♾ The paper used in this publication meets the minimum requirements of American National Standard for Information Sciences—Permanence of Paper for Printed Library Materials, ANSI Z39.48-1984.

TYPESET BY BPC-AUP GLASGOW LTD., SCOTLAND
PRINTED AND BOUND IN THE UNITED STATES OF AMERICA

Annual Review of Earth and Planetary Sciences
Volume 22, 1994

CONTENTS

SOME RELATED ARTICLES IN OTHER *ANNUAL REVIEWS*

From the *Annual Review of Astronomy and Astrophysics*, Volume 31 (1993):

The Long-Term Dynamical Evolution of the Solar System, Martin J. Duncan and Thomas Quinn

Planet Formation, Jack J. Lissauer

Proto-Planetary Nebulae, Sun Kwok

The Atmospheres of Uranus and Neptune, Jonathan I. Lunine

Jupiter's Great Red Spot and Other Vortices, Philip S. Marcus

From the *Annual Review of Ecology and Systematics*, Volume 24 (1993):

Physical-Biological Interactions Influencing Marine Plankton Production, Kendra L. Daly and Walker O. Smith

From the *Annual Review of Energy and the Environment*, Volume 18 (1993):

Status of Stratospheric Ozone Depletion, Jonathan P. D. Abbatt and Mario J. Molina

From the *Annual Review of Fluid Mechanics*, Volume 26 (1994):

Three-Dimensional Long Water-Wave Phenomena, T. R. Akylas

Climate Dynamics and Global Change, R. S. Lindzen

Dynamics of Coupled Ocean-Atmosphere Models: The Tropical Problem, J. David Neelin, Mojib Latif, and Fei-Fei Jin

Double Diffusion in Oceanography, Raymond W. Schmitt

From the *Annual Review of Physical Chemistry*, Volume 44 (1993):

Theory of the Equation of State at High Pressure, Marvin Ross and David A. Young

ANNUAL REVIEWS INC. is a nonprofit scientific publisher established to promote the advancement of the sciences. Beginning in 1932 with the *Annual Review of Biochemistry*, the Company has pursued as its principal function the publication of high quality, reasonably priced *Annual Review* volumes. The volumes are organized by Editors and Editorial Committees who invite qualified authors to contribute critical articles reviewing significant developments within each major discipline. The Editor-in-Chief invites those interested in serving as future Editorial Committee members to communicate directly with him. Annual Reviews Inc. is administered by a Board of Directors, whose members serve without compensation.

ANNUAL REVIEWS OF
Anthropology
Astronomy and Astrophysics
Biochemistry
Biophysics and Biomolecular Structure
Cell Biology
Computer Science
Earth and Planetary Sciences
Ecology and Systematics
Energy and the Environment
Entomology
Fluid Mechanics
Genetics
Immunology
Materials Science
Medicine
Microbiology
Neuroscience
Nuclear and Particle Science
Nutrition
Pharmacology and Toxicology
Physical Chemistry
Physiology
Phytopathology
Plant Physiology and Plant Molecular Biology
Psychology
Public Health
Sociology

SPECIAL PUBLICATIONS

Excitement and Fascination of Science, Vols 1, 2, and 3

Intelligence and Affectivity, by Jean Piaget

For the convenience of readers, a detachable order form/envelope is bound into the back of this volume.

Robert Dietz

Annu. Rev. Earth Planet. Sci. 1994. 22:1–32

EARTH, SEA, AND SKY: Life and Times of a Journeyman Geologist

Robert S. Dietz

Professor Emeritus, Arizona State University, Tempe, Arizona 85287-1404

It is the practice and privilege of age to hark back to days of yore, and I cannot resist—especially after being asked to do so by the Editor. Apparently this practice of inviting an earth scientist to write a prefatory chapter for the *Annual Review of Earth and Planetary Sciences* has been going on for over two decades. The list of authors has included several fellow scientists who have been my close friends such as Roger Revelle, Walter Munk, Sir Edward Bullard, and J. Tuzo Wilson. It is a privilege to be associated with them and I will try to roughly use their autobiographical style. I will follow the advice of the king in *Alice in Wonderland* who said, "Begin at the beginning, and go on 'till you come to the end: then stop."

My career as a government and academic geologist has been both fascinating and rewarding. I have often felt like a child let loose in a candy store. I have been able to sail on all the seven seas and with echo soundings pierce the blue mirror that reflects the sky and reveals the deep abyss. I have also been able to do research on all seven continents, including Antarctica on the Navy-Byrd Expedition of 1946–1947. My chosen profession has permitted me to fly over a million miles and into the stratosphere and to dive more than a mile deep in the oceans. My life has spanned an era of remarkable change; however, if I compare my college geology textbook of 1933 with one of today, I find that few really new chapters have been added. I take a modicum of pride in having been involved early-on in three of these: marine geology, astrogeology, and plate tectonics. I am pleased that a mountain in Antarctica, a tablemount in the mid-Pacific Mountains and an asteroid (#4666-Dietz) have been named in my honor. I recommend a scientific career as the best of all possible pursuits. Understanding the rainbow as a phenomenon of light

0084–6597/94/0515–0001$05.00

refraction rather than a heavenly miracle makes it no less interesting. Similarly, knowing the geologic history and structure of a mountain makes it no less awesome. My life has spanned a fascinating era of remarkable change—the telephone, the automobile, and the airplane were invented only shortly before my birth. Geology has evolved from an observational and field science into largely a laboratory science with instrumental capabilities that have improved data collection and data processing by orders of magnitude.

I have tried to pursue a policy of lifelong learning zeroed in on the earth sciences. I have been chided for never reading works of fiction but I have always considered novels a waste of time especially for the likes of me—those of us not bright enough to spread our efforts too thinly. My mentor Francis Shepard had it right when he was questioned regarding his obvious lack of student status in a discount line at the Scripps Institution of Oceanography lunch kiosk. Replied he, "Professors are students all their lives." No doubt this tunnel vision has stunted my intellectual maturity.

One way that I differ from nearly all research scientists is that I have never had a government grant—but this is not for lack of applying. It is also not for lack of need as I am not wealthy and never have I inherited so much as a red cent. Twice I almost received grants but these never finally materialized. On the first occasion in the late 1960s, I was funded for three dives with the *Reynolds Aluminaut*, a remarkable deep research vehicle with a hull less compressible than water. The principal purpose was to investigate a presumed south-setting bottom current that ran counter to the Gulf Stream in the Strait of Florida. This current, disbelieved by physical oceanographers, was indicated by the asymmetry of bottom current ripples on photographs. I arrived at the Miami dock in 1967 only to be told that my program had been cancelled. The *Aluminaut* had been preempted to search for an H-bomb that had fallen from an Air Force bomber into the sea off Spain's Mediterranean coast.

The second lost grant was from NASA and occurred in 1976. A year earlier I had been involved in astronaut tutoring and photographic target selection of volcanos and astroblemes for the Apollo-Soyuz handshake-in-space mission. My accepted post-mission proposal was to investigate some circular features in central Brazil which the astronauts had photographed. Unfortunately the funding was six months late in materializing. This exactly coincided with my retirement from government service and the beginning of my professorship at Arizona State University. This necessitated turning the grant over to a colleague, John McHone. He was able to reconfirm the astrobleme status of Serra da Cangalha and to locate another probable example in Brazil. So the number of grants I ever received remains at zero. Thus, perforce over the years I have had to be largely content with utilizing data sets in the public domain.

I have received some private support, however, for my research from the Barringer Crater Company, the nonoperating partner in Meteor Crater Enterprises of Arizona which operates Meteor Crater as a touristic facility. Over six decades, this company has supported much research in meteoritics under the successive leadership of D. Moreau Barringer Jr., Brandon Barringer, and J. Paul Barringer. In 1968 Paul and Dorothy Barringer supported and accompanied me (along with Robert Fudali of the Smithsonian, and William Cassidy of the University of Pittsburgh) in an investigation of the remarkable Richat dome in central Mauritania—a giant bull's-eye feature photographed by astronauts on the early Gemini missions. I had hoped that Richat would prove to be an astrobleme par excellence, but unfortunately we found it to be endogenic and presumably related to a buried intrusion. Dorothy and Paul also accompanied me in 1972 on a study by Indian canoe and Eskimo boat of the Nastapoka Arc, a giant hemi-circle marking the southeastern shoreline of Hudson Bay. Again our search for shock evidence suggesting an astrobleme was negative. This was really not surprising as ground zero lies far offshore and deeply buried beneath the Belcher Islands. I personally believe, however, that it does indeed mark an Archean impact event which will eventually be demonstrated. In 1978 the Barringers supported a month-long expedition by myself, my son Rex, John McHone, and Philippe Lambert into central Algeria where we investigated with good success several probable meteorite craters and ancient impact scars (see below).

I have been fully satisfied with my career as a research scientist in governmental and university settings. Even trivial discoveries in science are fun and scientific discoveries are a unique way of adding to knowledge. If I were a musician, a writer, or a theologian, I would never know for sure if I was a star performer, good, bad, or mediocre. Fortunately the star system does not generally apply to science; we are all journeymen doing the best we can with the tools we have. It is good to have lived long enough to see some findings and concepts believed; e.g. Sudbury Basin as an astrobleme, which achieved a consensus only in 1992 by my reckoning. Athletes tend to peak out in their twenties but scientists can do useful work, even beyond their usual biblical allotment of three score and ten. If Leonardo da Vinci suddenly came back to life, he probably would be unimpressed by the progress in the fine arts. Isaac Newton, however, would be utterly amazed by modern science and technology. And, as Mark Twain wrote in *Life on the Mississippi*, "There is something fascinating about science—one gets such a wholesale return of conjecture out of such a trifling investment of fact."

EARLY YEARS/UNIVERSITY OF ILLINOIS (1914–1936)

I was born in Westfield, New Jersey, in 1914, a bedroom city for white-collar workers who commuted to New York City. My forebears were entirely Anglo-Germanic having mostly arrived in the USA about 1850. They were average middle-class people with conservative values; my father was a civil engineer. Although I normally disapprove of large families, I make an exception here as I was the second youngest of seven children: six boys and one girl, Helen, the eldest. I would rate my schooling through high school as average. Helen was virtually a surrogate mother and, as the assistant librarian in our town's library, she introduced me to the world of books. Financially, times were rough especially during the Great Depression (1928–1936). My mother was a devout Christian Scientist, a nineteenth century cult which became an established church. By high school, I had largely rejected all religions and developed a serious interest in science, devouring library books by James Jeans, Arthur Eddington, and H.G. Wells. A school friend (Ralph Hall) and I became ardent amateur naturalists. We were rockhounds before this word was even invented, hitchhiking to such world class mineral localities as Franklin Furnace, New Jersey. The police offered to put us up in the jail. It was a friendly gesture but we opted instead for a farmer's barn. Visits to the American Museum of Natural History stimulated our interest and were enhanced by a retired Yale professor who was present every Saturday to answer our questions. He assured us that our specimens were as good as those in the Museum's J. Pierpont Morgan Collection, simply because we found them ourselves. My mother died while I was in high school and my father several years later. With no advice from anyone, I decided to go to college and study geology. I worked two summers in a resort hotel on the New Jersey shore for a paltry recompense of $12 per month plus room and board and rare tips. I was urged by a coworker there to split to Miami Beach, Florida and work for big money at the famous Roney Plaza Resort Hotel where he had connections. It was a tough choice but, instead, I hitchhiked west to the University of Illinois choosing that school mainly because the Chicago World's Fair (1933) was in progress. Another significant factor was the out-of-state tuition of only $65 per semester.

My liberal leanings probably derive from the slim pickings during my childhood when there was hardly enough food on the table for seven hungry children. There was a lot of sibling rivalry. My parents, however, remained staunch Republicans—even convervatives—throughout the Great Depression. Two summers of work on the New Jersey coast and one in Yellowstone in tourist hotels made me recognize the need for organized labor. Later, part-time work at the University of Illinois for 25 cents per hour under the National Youth Administration was critical to my survival as a student and made Franklin D. Roosevelt a hero figure to me.

I arrived at the University of Illinois in September 1933, already planning to major in geology. The room I rented was actually a windowless closet but the price was right—$5 per month. I had saved a little money from my summer job and my father sent me $400 over the first two years; but beyond that I would earn my own way over eight years: BS, MS, PhD at Illinois. This was not the best way to go but it was a matter of survival. These were maturing years and I am forever grateful to my alma mater. Thanks, Illinois—the truth is, I never left you. These were lean years but full of the exhilaration and challenge of youth. A defect in my college education was my being hell-bent for high grades. This was successful in that I earned a Phi Beta Kappa key. But this also stunted my learning as I purposely avoided the advanced levels of physics, chemistry, and math and, hence, I became more qualitative than quantitative, a skewing that I have regretted. One's best learning years are over by age twenty-five.

One way life has changed is the ease of travel for Americans in the modern world of jet aircraft. My lifelong desire to see the world has been amply fulfilled, the only major gap being mainland China. My father was regarded as well-traveled having been as far west as Chicago, built a logwood dye factory in Haiti, and worked on the Panama Canal. My mother's life was confined to the limits of Boston, Cleveland, and Washington, D.C.

I chose my country, my hometown, and my parents well. Nonetheless life is a crapshoot—snake eyes, boxcars; once in a while you make your point or even roll a lucky seven. Life is a rocky road, but along its highways and byways I never met a rock I did not like. And rocks are composed of crystals, the glittering flowers of the inorganic world. It is a theorem of geometry that when lost in a maze, if you place your hand on one wall and never take it off, you will find your way out. It works!—because I am a survivor. But even a geologist must recognize that the awesome story of the rocks must yield to the mysteries of life on Earth.

MARINE GEOLOGY WITH FRANCIS SHEPARD (1936–1941)

In my junior year at Illinois, I struck up a friendship with fellow student, K. O. Emery whom I now recognize as the only true genius I have ever known—but defined in Edison's terms, "90% perspiration and 10% inspiration." (I have just recently written the preface for his latest treatise with Elazar Uchupi, *Morphology of the Rocky Planets of the Solar System*).

Working for Professor Francis ("Fran") P. Shepard, Emery was making contoured bathymetric maps using raw sounding data collected by the U.S. Coast and Geodetic Survey. The main effort was to define the many submarine canyons off California, but eventually the entire continental borderland off southern California was mapped. I asked to join the effort and was accepted, an association that lasted for five years and was a defining event in my career. These were the early days of marine geology. The seafloor was *terra incognito* and the sea surface simply a mirror reflecting the sky.

I hold no brief for the "good old days" but perhaps it adds to the perspective of marine sciences, and certainly to humor, to recall something of the beginnings of marine geology in the United States by citing some of our early experiences. This subdiscipline of geology commenced almost simultaneously in the mid-1930s on the East Coast at the Woods Hole Oceanographic Institution with the research of Henry C. Stetson and the West Coast with the studies of Francis P. Shepard. Emery and I were the first of Shepard's sixty or so marine geology students, shuttling with him between Illinois and California. We originally met at the University of Illinois, where we arrived via modes of transportation that were the norm for those days of the Great Depression. I arrived by hitchhiking from the East Coast; Emery came by train, riding boxcars from San Diego.

In 1936 Shepard received a grant from the Penrose Bequest of the Geological Society of America for studying submarine canyons and the seafloor generally off the coast of California. The amount of $10,000 was a handsome grant for those days—in fact, the largest ever given by the GSA in prewar years. With the money he was able to charter the 96-foot schooner *E. W. Scripps* of the Scripps Institution of Oceanography for six one-month cruises, build the necessary scientific equipment, and employ us as his assistants at a salary of $30 per month. The low funding at least required us to develop some ingenuity in devising simple, inexpensive instrumentation. For example, we used the two-meter-tall Roger Revelle, later director of Scripps Institution of Oceanography, as a wave staff by having him stand at various distances from shore in the buffeting surf. As another example, we organized a rock preparation and sedimentation laboratory for which a budget of $50 per year was arranged. This was considered a reasonable proportion of the Scripps overall budget of $125,000 per year. The road to this now famous marine laboratory was marked by a forlorn sign reading *Fish Lab*.

Life aboard *E. W. Scripps* was somewhat different from shipboard duty today. The ship's crew consisted of only four persons—captain, engineer, deck hand, and cook; the scientific party was seven—the number of bunks available. We generally worked around the clock, six hours on and six off.

Members of the scientific party were expected to be sailors and to run the ship; technicians would operate oceanographic winches, assemble and use the water and bottom samplers, and do various shipboard analyses for water chemistry. Among our duties while steering the ship was to tabulate by hand the water depth every two minutes. We did this with great enthusiasm since we had installed aboard the latest Submarine Signal Co. fathometer which indicated the depth on a revolving red-flashing neon light. Graphic recorders had not yet been invented, so this instrument represented to us a remarkable advance over the sounding lead. And, in fact, we continued to use the hand-powered, wire-and-lead sounding winch installed on a rowboat for making hydrographic surveys of the inner heads of several submarine canyons. Rather remarkably, it was possible to demonstrate that canyon heads were repeatly filling with sediment and then emptying out. Prior to the cruises, we built dredges, grab samplers, sediment traps, and corers. The best corer that we constructed was a 600-pound open-barrel gravity model that increased the weight of such devices over earlier models by a factor of ten. We purchased junk lead at 3 cents per pound, used scrap 2.5-inch pipe, and built two corers for about $50 each. It was not until after the war that we heard about Kullenberg's invention in Sweden of the piston corer. Nevertheless, we commonly obtained cores 12 feet long, and in one instance, a diatomaceous ooze core in the Gulf of California which at 17 feet long was a new record.

In 1939, we joined Shepard and Roger Revelle for a three-month cruise of the Gulf of California. Our additional compensation for this cruise was one dollar, which made us nominal, hence insurable, employees of the Scripps Institution of Oceanography. Over a five year period, the Shepards took us in as part of their extended family traveling back and forth from Illinois for academic studies and Scripps for research. Shepard would eventually attain the status of the Father of Marine Geology.

Emery and I were as poor as church mice but I count these times as among the best days of my life. Shepard displayed contagious enthusiasm for seagoing research but great disdain for other than simple mechanical instruments and for armchair geologists. He was a skeptic and an iconoclast bent upon collecting his own new data. He preferred to "count the lion's teeth rather than consult Aristotle." His passion was submarine canyons which he wrongly believed until late in life were river-cut during great secular drops in sea level. Fran's intuitive grasp of physical laws lacked a certain spin control. In formal exchanges of views with Reginald Daly and Philip Kuenen (two great geologists of those days in my opinion) he invariably came out second best. As Boston Brahmin, Fran had considerable wealth but, while at the same time generous, was penurious to a fault. He kept a close tally on all expenses but never bothered to add them up. Looking back, these were halcyon years; and in memory of Fran Shepard, I recall some lines from Coleridge "Where Alph the sacred river ran/Through caverns measureless to man/Down to a sunless sea."

Students expect to be financially supported nowadays and to be assigned research topics. I chose my own theses subject—the phosphorite deposits Shepard, Emery, and I discovered on the continental borderland off southern California for my Master's degree, and deep sea clays for my PhD. I hitchhiked from Illinois to Washington, D.C. to obtain some *Challenger* Expedition worldwide, deep-sea samples from the Smithsonian to supplement those we had collected off California. One finding was a considerable uptake of potassium by illitization of clays. At my thesis defense one examiner wanted to disqualify it on the basis that I had been paid by the Illinois State Geological Survey while doing some minor aspects of this work. A guideline in those days was that a student should not receive any pay for research that pertained to his own thesis. Yes, times have changed, and for the better.

In 1939 I attended my first scientific meeting: the Geological Society of America's annual convention in Minneapolis. Such meetings were always convened between Christmas and New Year's so as to not conflict with the academic classes. I recall that I seemed to know all of the few hundred participants, at least by reputation, and to understand all of the papers. I naively thought I was a jack-of-all-trades and a master of four or five. The papers were generalized and totally lacking in sophisticated mathematics or "flute music." It has been downhill every since. Scientists know more and more about less and less. The age of the generalist seems to have passed.

Why is the Earth such a special planet? Of its many universal aspects the unique presence of water in the liquid state must rank first as this makes life possible. And so we have the oceans whose volume is often called Inner Space. In a similar vein I like to consider the ocean floor as the Third Surface. Earth uniquely has three volumes or "spheres"—hydrosphere, atmosphere, and lithosphere (liquid, gaseous, and solid states of matter). Geometrically, volumes intersect forming planes; thus there is the First Surface (land/air), the Second Surface (air/water), and the Third Surface (water/ocean floor). It is in the nature of things that energy exchanges take place primarily along surfaces; hence their importance in the scheme of things. The Greek philosophers of ancient time believed the universe consisted of four things or "elements"—earth, air, water (the three states of matter), plus fire. But if we consider life to be fire (reasonably so because we live by burning calories) and add a fourth sphere, the biosphere, then the list agrees with the Greek's classification. Earth, as perhaps Planet Ocean is, indeed, a remarkable abode—and is everyone's favorite planet.

The cosmological paradigms from Copernicus to Newton were a blow to Man's pride. We live on an average planet circling an average star in an average universe: the principle of mediocrity applies. Planet Earth, however, is habitable because of two key criteria—temperature and size, sometimes called the Goldilock's paradox because, as in the children's tale, things are "just right." These insure the *sine qua non* for life, the presence of liquid surface water creating the oceans unique among the terrestrial planets and their satellites. The oceans are, of course, where life began. We must rank in order of increasingly stressful environments, the oceans, fresh water, and the land. Rigor is hard on the individual but is good for the species. Hence animals that now rule the sea are the mammals, all of which spent considerable time evolving on land. And even modern Man would not be here today were it not for the rigors of the Ice Age.

MILITARY SERVICE IN WORLD WAR II (1941–1945)

Yogi Berra once advised, "When you come to a fork in the road take it!" Beginning my junior year in 1935 at the University of Illinois I unknowingly did just that. All students were required to take two years of basic military training but accepting advanced training and earning a commission was optional. I had definitely decided not to go on but, as I passed through registration, the $16 per month stipend for advanced Reserve Officers' Training Corps was a windfall I could not forego. The thought that I would ever be called to active duty never crossed my mind. But, as is often said, the rest is history. Eventually, the U. of I. ROTC program would provide more officers for WWII than even West Point.

When I received my PhD in June 1941, I quickly realized I was not any smarter than the day before. There was no obvious employment for a marine geologist as this discipline had not been invented yet; nor had such things as post-doctoral fellowships. I painfully recalled a research paper in the early twenties in which Chester Wentworth questioned the need for more than a couple of new PhDs in the USA each year. I wistfully recognized that my co-graduating colleague Emery would have first choice of any position as his qualifications were truly outstanding: For example, while still an undergraduate, he was the coauthor with Shepard of the Geological Society of America Special Paper #31 concerning submarine canyons off California. Emery, a human dynamo and later a member of the National Academy of Sciences, sent out 137 job applications. I opted to hang back and accept one of his discards but he received not a single job offer. Accordingly, I returned to the Scripps Institution of Oceanography in La Jolla and applied for work with the Navy-sponsored newly organized University of California Division of War Research on Point

Loma in San Diego—the dogs of war already had been unleashed in Europe. They refused to hire me as I was a reserve First Lieutenant in the U.S. Army Reserve. And, indeed, on August 7, 1941 the traditional President's "Greeting" arrived. I was called to active duty as a ground officer in the U.S. Army Air Corps (probably because I already had a private flying license) with the 91st Observation Squadron in Fort Lewis, Washington. Enroute I climbed Mount Adams in Washington, my first and last high mountain although I was destined to eventually sail over and map many others from the deck of a rolling ship. Everyone of my vintage knows where they were and what they were doing on Pearl Harbor day and when President Kennedy was shot. On Dec. 7, 1941, I was the squadron adjutant at Pine Camp, New York, making up a duty roster when the news broke. Rumors ran rampant and we were led to believe German aircraft might even attack our snowbound airfield.

To play a more active role even before Pearl Harbor, I applied for flight training immediately after being called up, but failed the rigorous physical exam because, it was claimed, I had a slight muscular eye imbalance. Believing this to be nonsense, a month later I drove 200 miles overnight with no headlights in a driving rain to retake the exam. It was the day before my 27th birthday—when I would have been too old, disqualified because of age. This was the first time in my life when I would have been too old for anything rather than too young. I passed. In 1942 I went through the seven-month flying training (Kelly, Randolph, and Ellington fields) and graduated in Class 42-1. The class consisted mostly of cadets but included seven student officers of which I was one. Six of this group were ordered to B-26 (later regarded as a flying coffin) training and on to Europe from whence, I understand, few ever returned. Presumably because of my academic background, I was posted to an air navigator's school in Hondo, Texas, where I served as an instructor for about 18 months.

By emphasizing my geologic training, I was eventually able to be transferred to a photo mapping squadron. My wish, however, for duty either in Europe or the Pacific, was not accommodated. My principal overseas mission was commanding a four plane flight of B-25s (Billy Mitchells) modified for photo mapping in South America. Our major mission was to map a boundary dispute region in the Amazon claimed by both Equador and Chile. This required flying over the high Andes to the headwaters of the Amazon from which we could never return if we ever lost even ten percent of our engine power. Fortunately this never happened although our operational casualties were high elsewhere. We also mapped extensively in Chile where our official task was to provide the data for air navigation maps. In reality we were sent in as soon as Chile broke diplomatic relations with Germany, so I suppose our primary purpose was to establish an

American presence. We worked closely with the Chilean Air Force and in appreciation I was made an honorary pilot in that Air Force. Our squadron eventually did mapping in all South American countries except Bolivia and Argentina which never did break relations with Germany. My total flying time was nearly 3000 hours which was twice the normal life expectancy in terms of flying hours for a WWII pilot. As the war wound down, I was being retrained for the assault on Japan but the A-bomb ended that program. After World War II, I remained in the reserves for 15 more years and am now a retired Lt. Colonel. Not being a West Point graduate, I did not envision a bright future in the military service. It was my plan to establish a private photo mapping company but that was not to be.

NAVY CIVILIAN SCIENTIST (1946–1963)

After the War, in the early fall of 1945, I received a letter from Eugene LaFond, one of my prewar colleagues at Scripps who had served with the University of California Division of War Research and was then organizing an oceanographic research section at the new Navy Electronics Laboratory (NEL) in San Diego. He asked me to organize a seafloor studies group. It was a time for decision. The bait of being the oceanographer with Admiral Byrd on his Navy-sponsored fourth and last expedition to Antarctica (Operation Highjump) was too tempting to turn it down. I joined the Naval Electronics Laboratory as a civilian scientist. My tenure there for 17 years would be interrupted by a one-year (1953) Fulbright Fellowship in Japan and four years (1954–1958) with the Office of Naval Research in London, England.

These were productive years with much work at sea and numerous scientific publications. Science was simpler in those days so it was possible to make contributions outside of my specialty in marine geology; for example, on the origin of natural slicks on the ocean and the worldwide distribution of the deep sound scattering layers as zooplankton migrations. My best coup was the hiring of the brilliant H. W. (Bill) Menard—a new PhD from Harvard. We collaborated for several years on problems of the Pacific Ocean floor which was then little known. We published, for example, the first contoured bathymetric map of the Gulf of Alaska based on USC&GS sounding lines and also a paper on the discovery of the Mendocino Fracture Zone, the first such fracture zone to be described. We wrongly interpreted the nature of this prototype fracture which only became apparent some years later in terms of plate tectonics. Menard later joined Scripps Institution of Oceanography and under the Carter administration was appointed Director of the U.S. Geological Survey. With the coming of the Reagan years, Menard was fired from this position by James Watt with whom he had philosophical differences—as did practically all scientists. Watt, a Mormon, thought there was no need to conserve natural resources because the second coming of Christ was imminent. This was not a joke: It is recorded in the Congressional Record which reports official testimony. Watt, however, had the right to expunge such remarks, but he chose not to do so. Bill Menard returned to Scripps where he died an untimely death at the peak of his career in 1985.

In 1952 I co-led the joint Navy-Scripps Mid-Pacific Expedition, a two-ship operation with the major goal of seismic refraction profiling by Russel Raitt to determine the thickness of coral reef limestones atop Bikini Atoll—the recent site of atom bomb tests. A principal effort of mine, however, was to dredge the edges of some drowned ancient islands (tablemounts or guyots) to determine if these features were Precambrian, as Harry Hess supposed, or much younger. We were successful in obtaining an excellent fauna of both micro- and macro-fossils including, for example, rudists. The results, published by E. L. Hamilton, revealed a Cretaceous age for these guyots; hence, it was inferred that Pacific seamounts generally were not an ancient graveyard of geomorphology like mountains on the moon but were geologically young. This was one of several surprises in marine geology that conditioned some of us in the mid-1950s to become mobilists, eventually accepting continental drift. Later a Scripps expedition doing a more extensive survey of these Mid-Pacific Mountains honored me by giving one of the guyots my name. Work as a Navy scientist was not without its frustrations. There were great geographic and geologic discoveries to be made about the virtually unknown Pacific floor. But all our work required military justification and an "admiral's page"—something the "brass" could understand if they ever actually bothered to read it. In the early years at NEL our research had somehow to apply to submarine warfare and underwater sound. My small group of five persons churned out a considerable number of papers in the open scientific literature which gained media attention and so was a source of embarrassment. We were criticized for not being invisible. In the late 1950s, the so-called Mansfield amendment became law; it was intended to quash all basic research within Navy laboratories. About the same time oceanography suddenly became a high profile science and an OK study to pursue. Many acousticians changed their position description to oceanographer creating considerable confusion.

On leave from the Naval Electronics Laboratory, I spent my entire year of 1953 in Japan at the University of Tokyo and the Japanese Hydrographic Office as a Fulbright Fellow. Japan was still recovering from the ravages of World War II. I acquired great respect for the Japanese culture—the people's intelligence, skill, and diligence. Their modern economic power is no surprise to me. I have even considered writing a book about why this is so, which would emphasize their genetics, the analog nature of the writing, etc. I arrived with an underwater camera and scuba gear—the first such diving equipment in Japan. It created quite a sensation all around that sea-oriented country. I was able, for example, to obtain some good footage of the tai fish, sacred to the Nicherin Buddhist sect.

My principal effort was to study the marine geology of the northwest Pacific seafloor using soundings collected by the Imperial Navy during World War II. These had been organized by the Japanese Hydrographic Office into a bathymetric chart which I revised and reorganized for my own interpretation. One aspect was the description of a majestic range of drowned ancient islands (guyots) striking northwest from near Midway Island to the Kamchatka trench. I named these the Emperor Seamounts after ancient Japanese emperors (e.g. Jimmu, alleged founder of the royal lineage in 600 B.C.), and empresses (e.g. Suiko). My idea for this nomenclature came from the Presidential Range in New Hampshire and Vermont. Much objection was voiced against this nomenclature but it seems now to be accepted.

Another study concerned the undersea eruption of Myojin Reef volcano in the fall of 1952. By happenstance I searched the underwater sounds recorded on the Navy's newly installed SOFAR stations (two off California and one off Hawaii) for sounds coming from the erupting Barcena volcano off Mexico. Instead I found explosions arriving from Myojin. A research ship had mysteriously disappeared with all 31 hands including my intended Fulbright fellowship host, R. Tayama. A single plank was later found embedded with bits of fresh scoria. I was able to show from the SOFAR records that the vessel was blown up by a great explosion at 12: 33 hours on September 23, 1953. At the time, it was the most distant undersea sound transmission ever recorded.

A high point of this Fulbright fellowship was a visit to the royal palace's marine laboratory where nudibranchs (shell-less gastropods), the emperor's hobby, were studied. A 90-minute audience with Emperor Hirohito followed. I explained my underwater film of kelp beds off California which he had seen. I also spoke about the tragic disappearance of a research ship *Kaiyo Maru* 5. Then I discussed the marine geology of the northwest Pacific Ocean and the majestic chain of seamounts stretching from Midway to Kamchatka, the Emperor Seamounts. The Emperor seemed uninterested, so upon leaving I asked my escort what the Emperor had said—as my Japanese was limited. My escort replied, "Emperor he say, 'Ah so des ka?' (Is that so?) one hundred and thirty-seven times." I assess Hirohito as a very shy, average Japanese, and a bird in a gilded cage. It seemed evident to me that he probably had little to do with the glossy books on marine invertebrates printed with his nominal authorship, except to collect the nudibranchs at his seashore summer palace. During World War II, the Japanese people were told that the emperor was brighter than the sun and staring at his highness directly would be blinding. Such naive beliefs could only be foisted upon a populace before the advent of television. All heroic figures have been reduced to common people under the glare of this new medium.

From 1954–1958 I once again took leave from the NEL in San Diego to join the Office of Naval Research, London, a small group attached to the American Embassy. A principal task was to help rehabilitate war-ravaged research institutes in the United Kingdom and western Europe, administer contracts, and re-establish firm ties to the USA. Part of my work was also overt intelligence—learning about new technologies and scientific discoveries. A whole generation of scientists, especially in Germany, had been lost to World War II. The work was enjoyable and worthwhile but a critical turn occurred in 1955 when I met Jacaques Piccard at a deep-sea diving symposium in London. I offered to promote support for the further development and testing of his bathyscaph *Trieste*—essentially a deep-diving "blimp" with a gasoline-filled float for buoyancy and a high-pressure-resistant sphere for the pilot and one observer. The rationale was to get the Navy involved in oceanographic research using deep research vehicles (DRVs). The Navy, for example, spent enormous sums on aircraft, thus successfully invading the atmosphere, but zero on invading the hydrosphere (their principal realm) below the shallow depths penetrated by submarines. Military submarines of that time were surface vessels that could dive only to the upper several hundred feet of the ocean. As machines of war they were not useful as research vehicles.

After a protracted effort the Navy agreed to sponsor a series of dives off Italy in 1957. Twenty six dives were made off Naples to depths of a mile or more. Because a bathyscaph is uniquely capable of ultra-deep diving, I envisioned a plan I called Project Nekton to descend to the seven-mile bottom of the Marianas Trench, the world's deepest spot and a mile deeper than Mt. Everest is high. With the purchase of the *Trieste* by the Navy and its transfer to the NEL in San Diego, this dream became a potential reality and soon full sponsorship emerged. A new float and a stronger sphere was manufactured in Germany. By the fall of 1959 the *Trieste* was stationed in Guam ready for test dives preparatory to the ultimate plunge. By January 1960, it was finally ready to dive. The Navy, of course, wanted this to be a uniformed military event, and as the saying goes, "He who rules the waves, waives the rules." Although still a member of the team, my role had been reduced to being a consultant. As the zero day approached, push gave way to shove and an attempt was made to co-opt Piccard's role as pilot. This was saved only when Piccard invoked a clause in his Office of Naval Research contract that I had originally negotiated stating that Piccard had the option of being the pilot on "any special dives." A flurry of messages flashed back between Guam and Washington. The Office of Naval Research stood firm supporting Piccard and left the copilot choice to the Bureau of Ships. So Piccard piloted the dive to the nadir of the Earth, on January 20, 1960 with Lt. Don Walsh as copilot. As I was the most qualified scientist, Piccard requested my participation in a second dive. But major structural damage made further descents impossible. Irked with his treatment, Piccard asked me to coauthor the definitive documentary account. Our book *Seven Miles Down: Story of the Bathyscaph Trieste* was published by G. P. Putnam & Sons in 1961.

For the first two years after my return from London to the San Diego naval laboratory (by then renamed Navy Undersea Center), my major involvement remained with the bathyscaph *Trieste*, but after the ultimate dive the Navy created a deep submergence group within the submarine service on Point Loma, San Diego, allowing me to return to more scientific pursuits in the field of marine geology. Deep submergence was alive and well—a multi-million dollar program. With this program winding down I vowed to try to work alone: 1. eschewing team projects, 2. avoiding classified work which could not be published in open scientific journals, 3. avoiding contributing to the grey or internal Navy reports, and 4. selecting research on the leading edge of science. Although it was not in my position description to engage in scientific generalities, I chose to write about the marine geologic evidence for continental drift. A flurry of papers resulted concerning the origin and nature of continental slopes, the continental rise prism as nascent eugeoclines, and a mechanism for continental drift which I called seafloor spreading—a name I considered awkward but it has stuck.

In late 1960, I initially wrote a brief speculative paper entitled *Continent and Ocean Basin Evolution by Spreading of the Sea Floor* which eventually appeared in *Nature* in June 1961. It provided a reasonable mechanism for continental drift by implanting new ocean crust at mid-oceanic rift zones. It was a potboiler that boiled over attracting much attention, as continental drift was a hot subject. I also read this paper at the 1961 Pacific Science Congress in Hawaii where J. Tuzo Wilson was present. I recall that he immediately found the concept appealing which surprised me as he had recently written about the impossibility of continental drift. I believe it converted him to mobilism and he then became a major contributor to the plate tectonic revolution. Unbeknownst to me, Harry Hess, independently and earlier (1960 preprint published in 1962), had suggested almost the same idea. Bill Menard received a copy of the preprint in May 1961 and contacted me since he had earlier reviewed my manuscript already in press. I agreed that Harry Hess should be accorded priority and did so in a 1962 publication (*AGU Monogr.* 6: 11–12). A full and accurate account of these seminal days for plate tectonics is found in Menard's excellent 1986 book, *The Ocean of Truth*. Looking back, Hess and I were agents of the inevitable: It was an idea whose time had come. The data base had been largely generated by Maurice Ewing and the Lamont Geological Observatory. Ewing remained a fixist until the bitter end at the 1967 American Geophysical Union meeting in Washington. I believe I was sitting next to him when he finally accepted the reality of drifting continents. Acceptance of my paper by *Nature* probably involved a bit of luck, as soon they would turn down L. W. Morley's correct explanation of the ocean's magnetic reversal stripes leaving it for Vine and Matthews to rediscover, and thereby provide an important verification of seafloor spreading.

The merit of seafloor spreading to me was in its explanatary power. It provided a reasonable mechanism for continental drift. It also offered an explanation for continental slopes as rift scars (Atlantic-type) or accretionary prisms (Pacific-type). Seafloor spreading further offered a new understanding for ensimatic eugeoclines and ensialic miogeoclines. It explained the mystery of the thin pelagic ocean floor sedimentary cover and the apparent youth of tablemounts—the ocean floor was not a repository of ancient geomorphology like the moon. Most significantly, rifting mid-ocean ridges provided the logical couplet for trench underthrusting about which I was fully convinced. I was also earlier impressed by S. W. Carey's mobilism and appearance of new ocean floor but rejected his expanding Earth model. He also showed that the cratonic fits across the Atlantic were cartographically-precise hard data and not on a par with Italy looking like a boot. Plate tectonics has subsequently revolutionized our understanding of terrestrial and planetary tectonics.

Nearly three decades ago we learned how our world works—by plate tectonics and uniquely so. Alone among Solar System members, Earth has a carapace of eight major lithospheric plates and many lesser ones which drift and rotate a few centimeters per year interacting along their boundary by subducting, shearing, or spreading apart. Mountains, volcanoes, and earthquakes and even the continents and ocean basins are the result. Hypsographic bimodality is unique to Earth and is the hallmark of a highly evolved planet where basalt is extruded as a partial melt of mantle peridotite and is then remelted to create the sialic rocks of the granitoid continental plateaus. A convecting body like the sun is understandable and so is a solid rock in space that has gravitationally collapsed into a sphere like the Moon. But a planet with surficial plates drifting only as fast as one's fingernails grow is nonintuitive. This must be why we were so slow in understanding our own planet. Continental drift is simply part of a larger scenario: the drift of lithospheric plates in which the cratons are imbedded. Certainly plate tectonics ranks as a paradigm if not *the* paradigm of geology. It has proven to be a concept of broad explanatory power and prediction.

USC&GS AND NOAA (1963–1977)

In 1963 at the behest of Harris B. Stewart, one of Shepard's early students, I joined the U.S. Coast and Geodetic Survey in Washington, D.C., an ancient, honorable and traditional governmental establishment primarily involved in geodesy and constructing coastal navigation charts within the Department of Commerce. These were the heyday years of oceanography which unfortunately were short-lived. Other than making charts this organization had little scientific interest in oceanography. Sounding data were rarely contoured into bathymetric maps and there were almost no scientific interpretations of the ocean floor.

The Kennedy and Johnson years supported science, unlike the subsequent administrations. We organized an Indian Ocean expedition to take part in the International Decade of the Indian Ocean. An excellent survey was made of the *Swatch of No Ground*, a picturesque name for a remarkable submarine canyon off the Ganges delta down which a substantial percentage of the world's sediments are dumped to create a complex system of deep-sea channels in the Bay of Bengal. I had originally spotted these channels from the Swedish Albatross Expedition results where they had been wrongly interpreted as graben rifts.

With a new oceanographic research ship, the *Oceanographer*, we then mounted an around-the-world cruise with myself as chief scientist for the leg from Perth to Sydney, Australia. It was a voyage of exploration in the heroic tradition such as started by the British ship *Challenger* (1872–1876) and carried on after World War II by the Swedish *Albatross* and the Danish *Galathea*. This was the last of such voyages. NOAA had two ships capable of extended deep-sea voyages. They were used to some extent in running trans-Atlantic geophysical traverses between the east coast of the United States and the congruent bulge of Africa. My suggestion was that they should be used for studying plate boundaries especially along the mid-ocean ridges (even before the "black smokers" were discovered), but such cruises never materialized.

Another expedition on the newly launched *Discoverer* explored the South Atlantic. As a historical curiosity we resurveyed the location of the first deep-sea sounding made by James Ross in 1843 enroute to Antarctica.

Ross's depth measurement was too deep—presumably he did not recognize when his cannonbell sounding weight first touched bottom. Also, the sounding, still recorded on modern charts, had drifted in position and been mis-transcribed—a 5 had become a 3. I was reminded of an oceanographic study I made of the Bering Sea in 1949. I searched for two anomalous holes in this broad shelf sea on the standard charts only to find them nonexistent. Upon completing the search, I realized that some early cartographer had recorded two longitudes, 174°W and 175°W, as depths in fathoms.

Around the bulge of Africa I charted in detail two great submarine canyons, both of which create deep-sea channel systems: the *Trou sans Fond* ("Bottomless Hole") off the Ivory Coast and Cayar off Mauritania. Both canyons tap the shoreline and are of considerable economic importance. The *Trou sans Fond* makes the port of Abijan (second only in importance to Dakar along western Africa) possible. The wave refraction pattern associated with the Cayar canyon permit piroques to launch through a low surf to a rich fishing ground.

Our survey of the small Bijagos excrescence off Portuguese Guinea proved especially interesting since, like the large Bahama platform, it creates a continental drift overlap when the Atlantic Ocean is closed. Since seafloor spreading is usually a symmetrical process, it would appear that a Jurassic hot spot beneath the Bahamas caused asymmetrical spreading in the initial phases of the opening of the Atlantic Ocean. Thus these congruent excrescences are now of grossly unequal area. This is a matter resolved by hot spots which measure absolute drift while drifting plates measure relative motion.

Unfortunately, changes were in store for the USC&GS which did not bode well for deep-sea marine geologic research. Through a series of reorganizations USC&GS became the Environmental Sciences Administration (ESSA) and then the National Oceanographic and Atmospheric Administration (NOAA). The oceanographic effort was spun off to laboratories at Miami and Seattle where they languished. The "O" in NOAA became very small while the "A" became very large. Efforts in geology and geophysics were cut down to nearly zero. Disenchanted with prospects, I looked forward to retirement from federal service and joining academia.

Mention must be made here of extensive collaboration with John C. Holden, a fine scientist with unique graphic and cartooning ability. We wrote about a dozen or so papers together in the mid-1960s to the early 1970s. I wrote the words and he composed the graphics ("music"). Holden was unfortunately lost to NOAA by a reduction in force in 1973. Holden was an "almost PhD" from the University of California, Berkeley. He even completed an accepted thesis on the microfauna from a drill core on

Midway Island. He finally gave up after failing the (ridiculous, in my opinion) language requirement for German—after failing the test, as I recall, twelve times. Holden, his talent lost to science, remains today a sociable, but now virtually incommunicado, hermit somewhere in the outback of Washington state. Before moving even further into the boondocks (and beyond telephone range) Holden was known to many geologists as the self-appointed President of the tongue-in-cheek *Stop Continental Drift Society*, complete with newsletter.

ARIZONA STATE UNIVERSITY (1976–present)

Although I thoroughly enjoyed my governmental career, there were many drawbacks and constraints. One never has free reign, the level of stimulus is low, and there are too many mundane assignments. As in the military, one gets paid just as much for marching as for fighting. The rules of Civil Service employment are not conducive to high quality selection of scientists. In 1992, I was one of the three alumni given a Special Achievement Award by the College of Liberal Arts and Sciences of the University of Illinois. It was of more than passing interest to me to observe that all three of us were from scientific academia—apparently the place where worthwhile contributions can be most readily made.

Throughout my federal career, tempting opportunities arose to join academia. I turned these down until 1975 when I had completed my 30 years of duty which provided immediate retirement benefits. On leave with generous rights to re-employment, I accepted temporary assignments at the University of Illinois, Washington University in St. Louis, Washington State University, and Arizona State University. After a brief return to NOAA, I accepted an appointment with Arizona State University in 1977. In 1985 I became active emeritus. I never gave any thought to retiring to the quiet life. Being a professor in a research-oriented university strikes me as the best of all possible worlds even as an emeritus without compensation. I find that people my age are too old for me. Working with college students helps one to remain young, at least in mind. The university has been sufficiently generous in providing me office space to continue pursuing my research interests. An end to age discrimination is one of the new rights that has enhanced the quality of life. A principal interest for coming to ASU was the presence of a strong planetary group and its Center for Meteorite Studies which houses one of the world's largest collections of meteorites. This association has been stimulating. One interesting example has been the recognition with John McHone of the 20-km-across El Gygytgyn crater in Siberia, not as a volcanic caldera, but as the world's largest Neogene meteorite impact crater.

As one ages there is a normal tendency to indulge in criticism and I am no exception. An example is the geologic concept made especially popular by Kenneth Hsu that the Mediterranean Sea dried up to the point of complete desiccation in the late Miocene. It is apparently now a consensus model. It seems to be both intuitively and scientifically unlikely that this could happen. In my view the proper explanation for the undersea salt deposits is by precipitation from a saturation brine basin creating "precipitites" rather than evaporites. Remarkable ideas require extraordinary proof and must be challenged. More importantly, on the fringe of science I have locked horns with the resurgent creationist movement and especially to the claim of "scientific" creationism. The integrity of science must be defended from the onslaught of pseudoscience. I find it particularly galling for religion to attempt to trade on the prestige of science. As one of the sponsors for the National Center for Science Education, I have been active in the creation versus evolution controversy. In the mid-1980s, Arizona was the scene of a strong effort to introduce creationism into the science curriculum of public schools. So far only a few scientists have played a role in preventing this from happening.

A project that brought considerable satisfaction to me (along with Troy Pewe and Mitchell Woodhouse) was getting Arizona petrified wood (*Araucariaxylon arizonicum*) designated as the official state fossil in 1988. This spectacular rainbow-hued Triassic silicified wood displays exquisite beauty and is a significant link in the evolution of vascular plants. Now, more than a score of states have state fossils. The effort was not easy as it took four years of background work. The then Governor of Arizona, Evan Mecham, a Mormon creationist, vetoed the bill for reasons that seemed to be dissembling. In reality, I assume that a 225 million-year-old fossil interfered with his belief in a 6,000-year-old Earth. His veto has also been termed a retaliation against state senator Doug Todd, sponsor of the state fossil bill, who was apparently critical of Mecham. Mecham also attempted to get equal time for creationism in the public schools. Fortunately, Mecham was involuntarily removed from office and a new governor signed the resubmitted Arizona fossil bill in 1988.

MOON, METEORITE CRATERS, AND ASTROBLEMES (1945–present)

I will treat this aspect of my scientific life separately as astrogeology has been a continuing interest largely as a hobby until I retired from federal employment to join Arizona State University in 1977. My interest in astronomy and especially in the Moon commenced in high school with the dream of one day being an astronomer, a desire entirely self-generated and of course never fulfilled. At the University of Illinois I took the only two courses available. As a graduate student I proposed writing my PhD thesis about the surface features of the Moon, but it was turned down as not a subject suitable for scientific contemplation. The idea was chided as totally bizarre and besides "there was no one to check my field work." I was told that speculation does not become a young student and it would be better to do some real geology like mapping a quadrangle in Vermont. However, during my WWII years as a pilot in the Army Air Corps I wrote a paper entitled "Meteoritic Origin of the Moon's Surface Features." Delays were inevitable, especially because R. T. Chamberlin, editor of the *Journal of Geology*, had recently written a paper on the tectonic origin of the lunar geomorphology; but eventually in 1946 he did publish my paper. It was the first such paper in a geological journal to suggest neither a volcanic nor a tectonic origin since that of G. K. Gilbert in 1895. Although volcanic and endogenic explanations remained the consensus view until about 1970, the cosmic bombardment interpretation for lunar craters is now unquestioned. I also recognized the dark maria as giant impact basins but erred in the belief that their basaltic infill was created at the time of impact and not much later.

Flying long navigation training missions caused me to reflect about the Earth below in a generalized way. It is a form of remote sensing not unlike contemplating the Moon. One day in 1943, it occurred to me that the disrupted nest of lower Paleozoic strata in the Kentland quarry in Indiana might not be cryptovolcanic as commonly supposed but an astroidal impact scar. Could not the orientation of shatter cones exposed there resolve this uncertainty? To test this idea I later stopped over at the Air Force Base in Rantoul, Illinois, and hitchhiked to Kentland. There, indeed, was a preferred orientation such that, when the vertically-dipping strata were rotated to an assumed original horizontal position (a counterclockwise rotation in this case), the cones pointed upward suggesting by Hartman's Law that the fracturing impulse came from above and hence was cosmic rather than volcanic. A year later, when I was based in Nashville, I searched the Flynn Creek and Wells Creek Basin cryptoexplosion structures in Tennessee with a mine detector for a possible remnant meteoritic debris. This was a direct but naive effort resulting only in my finding some nails and horseshoes. I already was inclined to believe that the eight cryptovolcanic structures described by Walter Bucher in 1935 as abortive volcanic explosion features were most likely impact structures.

With the launching of *Sputnik* in 1958 there was a sudden surge of interest in space. I was invited along by Gerard Kuiper, along with Eugene Shoemaker and two other geologists, for four nights of moon observation with the McDonald Observatory telescope in west Texas. Once there, I suggested we spend one day examining the Sierra Madera, a deranged mountainous structure, as a possible terrestrial analog of a lunar crater. Claude Albritton, although he had never visited the site, suggested it to be a candidate astrobleme based upon its damped wave structural style as mapped by Philip King. We found it to be nicely shatter-coned, which already for me was a definitive criterion for astroblemes. Years earlier I had studied them at the Steinheim Basin, Wells Creek Basin, Flynn Creek, and Kentland. Two publications resulted. This started me on a worldwide search for shatter-coned structures. About 70 are now known out of about 140 putative impact structures worldwide. Two U. S. Geological Survey geologists subsequently were assigned to map Sierra Madera. They were initially avowed skeptics of its impact origin but came away convinced of its astrobleme status.

In a *Scientific American* article in 1960 entitled "Astroblemes," I coined this term from Greek roots meaning "star" and a "wound by a thrown object." The term has been widely accepted and I still regard it as appropriate for the Earth but not for the Moon or Venus where impact structures retain their pristine form. Certainly, most ancient terrestrial impact sites are not crater-form and many like Sierra Madera are actually mountains. It is useful to have a prototype or a type-locality. Accordingly, I have suggested Meteor Crater as the prototype meteorite crater and the twin and simultaneous created, but remarkably different structurally, Steinheim and Ries basins of Miocene age in Germany as jointly the prototype astrobleme. The German sites are actually transitional between a crater-form and an eroded scar but together they provided the criteria needed for identifying ancient impact scars. I first visited these sites twice in the mid-1950s. The second visit was with Preston Cloud who at the time could not accept my interpretation but later would become an ardent supporter. Then in 1978, as an Alexander von Humboldt awardee at Tübingen, I had a chance to examine these astroblemes in some detail.

While searching for the terrestrial equivalent of the lunar crater Copernicus on Earth, I was drawn to a giant bull's-eye in South Africa, the Vredefort Ring, a central Archean granite body 40 km across surrounded by a thick upturned collar of Precambrian sediments. It seemed a prime suspect as an astrobleme created by recoil uplift following a hypervolocity asteroidal impact. I wrote letters to South African geologists seeking to know if this structure was shatter-coned. Answers were at first negative but eventually Robert Hargraves replied "Yes!" and sent definitive photographs to prove it. With that information in hand I committed a cardinal sin: writing a paper before visiting the site entitled "Vredefort Ring Structure: Meteorite Impact Scar" espousing an impact scenario. This model has stood the test of time pretty much intact although the local geologists immediately rejected it. A few years later (1964) I was able to visit the Vredefort Ring and test my ideas. My hour-long invited talk was immediately followed by an even longer critique offered by Louis Nicolaysen who totally rejected my interpretation. It was an interesting exchange of views and a good example, I believe, of South African geologists being ultra-conservative and bound by traditional views.

Following my Vredefort paper I turned my interest to searching for a terrestrial analog of a lunar mare—a "wet" impact with a central melt sheet or triggered volcanism. The large (35 × 60 km) kidney-shaped Sudbury Igneous Complex (S.I.C.) in Canada seemed a potential tectonized or squashed example. Of course, as an astrobleme one needs to scrap the classical model of an intrusive lopolith and consider the S.I.C. as an open pool of chilled magma crusted over with a fall back suevite (impact microbreccia) rather than a volcanic tuff. In the spring of 1962, to investigate my hunch I took leave from my position at the Navy Electronics Laboratory to visit Sudbury. A brief field study convinced me of the reasonable reality of my model based largely on geological relationships. It was not until the end of my visit that I discovered definitive shatter coning, as this fracturing is degraded in Precambrian rocks, unlike its development in limestone terranes. The resulting paper entitled "Sudbury Structure as an Astrobleme" eventually appeared in 1964. Publication had been delayed by a "pocket veto" by the reviewer—who finally returned the manuscript only long after being prodded to do so. Even then the only comment was "nonsense." Fortunately the editor of the *Journal of Geology* decided to override this negative appraisal. In 1972, I amplified my model to argue that the Sudbury nickel ores found in the sublayer were cosmogenic. The view has yet to be accepted although geologic relationships, especially the emplacement timing, support this view. In 1992, I participated in a NASA symposium at Sudbury on large terrestrial impacts with Sudbury being the type example. It was good to hear one's ideas become mainstream.

My inborn yen to see the world has been amply fulfilled by sailing the seven seas (North and South Atlantic, North and South Pacific, Indian, Arctic and Antarctic Oceans) as an oceanographer. I have spent a total of about four years at sea. This has also permitted me to visit and do geological research on all seven continents. Several of my visits to impact sites were made possible only enroute to join or leave various oceanographic expeditions. Thus, I was able to study the Serra da Cangalha and Araguainha astroblemes in Brazil enroute to work in the South Atlantic. Similarly, out of Dakar I organized an expedition into the Richat Dome in central Mauritania. This most wondrous giant dome, a giant bull's-eye from space, proved not to be meteoritic. A French report of coesite (a silica polymorph created by shock) proved to be barite and a central breccia is not an impact breccia. The centers of impact structures are upturned, disrupted, and eviscerated but Richat Dome has horizontal beds in the central eye. Oceanographic travels also permitted a study of both the Lonar Crater in India and the Vredefort Ring in South Africa enroute to work on the Indian Ocean. And I studied Gosses Bluff astrobleme in Australia enroute to leading a research cruise out of Perth and across the Great Bight of Australia.

Much of 1978 was spent at the University of Tübingen in Germany as a Humboldt Prize awardee. It was a fruitful and broadening experience offering especially a chance to study the Steinheim Basin and Ries Basin astroblemes in some detail. Remarkably, although the product of a twin impact event about 15 Ma, they are wholly different in structural style and shock effects. More importantly, I organized a month-long expedition with John McHone, Philippe Lambert, and my son Rex to reconnoiter possible impact sites in Algeria. It was exceedingly frustrating to set up this fieldwork but once in Algiers, almost on speculation, we had full cooperation and a joint program with Algerian geologists. It was a good example of comraderie achieved among scientists at the working level once the political difficulties are ignored even though these are never solved. We visited many sites deep in the Sahara; Talemzane and Amguid geomorphically appear to be meteorite craters, while Tin Bider and Ouarkzis are probable astroblemes.

Mention must be made here of the original significant chance encounter with John McHone (then a recent MS from Old Dominion University and later a PhD from the University of Illinois) at NOAA in Miami. I had arranged a trip to study Laguna Guatavita in Colombia, a sacred high Andean crater lake famous as the supposed site of the El Dorado golden trove. My initial companion canceled out at the last moment, so on the spur of the moment I invited McHone. This lake remains a site for treasure hunters as the pre-Columbian Indians were said to have littered the bottom with sacrificial gold icons. We assessed this feature as not meteoritic but, in all probability, the collapsed summit of a salt dome following ground water solution. Incidentally, we discovered that the lake had been secretly and successfully drained about 1905. Apparently no significant treasure was discovered—or at least none to speak of. Collaboration with McHone has now extended over two decades with other successful studies in South America, Europe, Africa, Canada, the United States, and from space imagery.

One of our more exotic exploits was gaining entry in June 1991 into politically-closed Cuba under the guise of attending a marine science meeting. Our real purpose was to check out the circular and domal Isle of

Pines (location of a maximum security prison) as the possible ground-zero site for the Cretaceous/Tertiary cosmic impact believed to have caused the great extinction of life including the last dinosaurs. Our interest had been aroused by the report of such things as cone beds. These, however, turned out to be cannon-barrel-shaped concretions and not shock-created shatter cones. There is a saying that one cannot prove a negative but we actually did. We found no evidence of shock in any rocks from the region. Co-operation by Cuban geologists proved once again that comraderie among scientists transcends political differences. Subsequently, the correct ground zero was located in Yucatan, the site of the buried Chicxulub crater.

As with poets, the Moon was a source of wonder to me early in life. In the early 1940s when I first studied selenography, no one in their wildest dreams even thought that twenty-five years later Man would walk on our satellite. Especially startling to me was the pre-*Apollo Ranger* 7 missile impact on the Moon which transmitted live video images of the crater Alphonsus back to Earth as it crashed. With the drift of the continents over the aeons, Mother Earth has slowly changed her expression while the Man-in-the-Moon has looked down with a fixed stare since the terminal bombardment. Plate tectonics has been the game plan for Earth but cosmic impacts have been the wild cards. Their significant role, even in perturbing evolution, ranks as a new geological paradigm.

SOME RECOGNITIONS AND HONORS

What drives people to do what they do? In my case, wealth, power, and fame have been unimportant. I have regarded gathering material goods as an encumbrance and preferred the simple life or, as Picasso once advised "Be a poor man with money in your pocket." As a journeyman scientist my stimulus has been delving into the nature of Nature. Along this pathway, it has been satisfying to be overly recognized by my peers and colleagues with various honors. I will list some of them here for the record, and perhaps this will also provide some credentials for the subsequent venting of my sure-to-be unpopular recipe for curing Man's basic problems: Phi Beta Kappa, University of Illinois, 1937; World War II, multi-engine pilot, five medals; Brevet Captain, Illinois State National Guard; Honorary Pilot, Chilean Air Force; Lt. Col., U.S. Air Force Reserve (Ret.); Antarctica Service Medal, Navy-Byrd Expedition 1946–47; U.S. Navy Civilian Service Award (1960) for bathyscaph *Trieste* ultra-deep diving program; Walter Bucher Medal of American Geophysical Union for geo-tectonics, 1971; Gold Medal of U.S. Department of Commerce for Exceptional Service, 1972; Alexander von Humboldt Prize (West German) 1978; Francis P. Shepard Medal of Society of Economic Paleontologists and

Mineralogists for Marine Geology, 1979; Barringer Medal for terrestrial meteorite crater research, 1985; Founder of Plate Tectonics Award, Texas A & M University, 1987; Doctor of Science (*honoris causa*), Arizona State University, 1988; Penrose Medal, Geological Society of America, 1988; Distinguished Achievement Award, Arizona State University, 1990; Distinguished Alumni Award, University of Illinois, 1992; Honorary Fellow Geological Society of London, Geological Society of Brazil, and Canadian Society of Petroleum Geologists; Fellow, Geological Society of America, Meteoritical Society, American Geophysical Union, Mineralogical Society of America.

PERSONAL PHILOSOPHY AND GRATUITOUS ADVICE

I will end this account of my life and times with my worldview—my philosophy of life. This, of course, has been molded by my lifetime of experience and my scientific (and hopefully objective) outlook. I define myself as a naturalistic materialist, a no-nonsense scientist. The natural to me is sufficiently awesome with its three "infinities": the infinitely large celestial universe, the infinitely small world of the atom and quantum mechanics, and the infinitely complex realm of life. I reject the supernatural realm of miracles and faith. I accept the material—the reality of matter, but I reject mind and spirit. The mind is what the brain does. I am certainly a skeptic but I do not qualify as an agnostic—this, in my book, is a gutless atheist. Nor am I an atheist as this term carries a lot of negative connotations; I regard the terms nontheist or ethical culturist, as better. Besides an overt admission of atheism would disqualify one from holding public office in many states. I am a secular humanist, a term of derision coined by the TV evangelists, but it describes me well. Humankind must accept the here-and-now and the natural scheme of things in the real world by resolutely solving our own problems through reason and evidence. And since I have recently written a book entitled *Creation/Evolution Satiricon: Creationism Bashed (Did the Devil Make Darwin Do It?)* with illustrator John Holden, perhaps this qualifies me also as a secular humorist. I believe my views are typical of most scientists with a philosophical bent, although most prefer to remain silent as being outspoken is not the way to win friends and influence people.

All of the "infinites" mentioned above have philosophical import. This applies also to the "finite"—the nominal or meter-sized world. The cosmological paradigms from Copernicus to Newton have revealed that the Earth is not special (the principle of mediocrity). Subatomic quantum mechanics with its uncertainty principle has torpedoed determinism. But evolution as the paradigm of life most profoundly affects one's worldview.

I hold no brief for the "good old days" as the human condition has improved each year of my lifetime. Science and technology are responsible as the prime movers advancing our improved circumstances. Scientists and engineers, not wars and generals, have fashioned our present society. For example, I believe the world would be much the way it is even if Hitler had won World War II. Nevertheless, I remain pessimistic about the future. I envision that our civilization is even now descending on a downward spiral as man forsakes his proper stewardship of the Earth and our environment. Remarkably, solutions to the world basic problems, although stark, are simple. They can be defined by the acronym ZEES—Zero population growth, Evolution, Eugenics, and Secularization. Unfortunately, these are currently all politically incorrect concepts. I also hold no brief for the wisdom of the ages, or of the aged. People of my age are too old (in their rigidity of mind) for me. I have lectured as a naturalist on about a score of deluxe cruises all over the world. These mostly retired people seem not to respond to any intellectual stimulation above the level of playing bingo.

In 1981, during the early days of the resurgent creation/evolution controversy, the Council of the National Academy of Science stated in a resolution, "Religion and science are separate and mutually exclusive realms of human thought whose presentation in the same context leads to misunderstanding of both scientific theory and religious belief." This official statement, apparently a peace offering in response to creationism parading as science, is off the mark. Science and religions are not separate and mutually exclusive realms of thought. They are overlapping and irreconcilable modes of thought with science being uniquely the way of knowing. The tenets of religion can and should be subject to exacting scrutiny and testing. Miracles, the power of prayer, life after death, heaven versus hell, and the god concept should not be placed off-limits for investigation. Religion should not remain exempt from criticism and science should not avoid confrontation. After a millenium, men of the cloth still have not resolved, "How many angels can dance on the head of a pin?" Perhaps the answer is: All of them. Let the theologians refute each other!

In the past century an attempt was made to measure the weight of the soul by weighing a dying person just before and after death—a crude but reasonable first approach. The answer, as I recall, was three ounces. Modern scientists and theologians both laugh at the result as they believe, for different reasons, that the answer is zero. To scientists the soul is nonexistent; to the theologian the soul exists but is weightless—although even light has mass, and one photon is measurable nowadays, so a soul cannot even wear a halo. This question is not trivial because it bears upon when a zygote becomes a person and this is central to the pro-life/pro-choice debate. As President Clinton remarked at a press conference in early 1993, "theologians disagree about when the soul enters the fertilized cell and personhood is achieved." For workable public policy purposes, we should accept "never," as the burden of proof logically falls on those who propose.

The greatest immediate threat to this world are the religious fundamentalists, especially the Muslims and the Christians. There are countless zealots who stand ready to terrorize the world and "do what God would have done, if only He had facts of the case." Witness the 1993 Branch Davidian holocaust in Waco, Texas, or the bombing of the World Trade Center in New York. Over the millennia, religious conflicts have been and continue to be the major cause of war. We even have a word for it—holy terror. While the religious right remains the principal problem, the mainstream churches as well need to shift away from the supernatural and become centers of ethical culture. Our ethics in turn must be based on the realities of the human condition and these are Darwinian. We must not appeal to some Santa-Claus-in-the-Sky. This would be like throwing a drowning man both ends of a piece of rope.

In bizarre contrast to conventional geology, the creationist young-Earth scenario compresses all world class geologic events into Noah's Flood year about 2400 B.C. by Archbishop Ussher's calendar, some 1600 years after Earth's creation. Accordingly, in 375 days all Phaneorozoic strata were laid down by the raging Deluge, then Ararat was erupted, and this was followed by the docking of the ark. Whew! Elsewhere in the world, the ocean basins were catastrophically opening up, mountains were folded, volcanoes erupted, continents drifted apart at 45 miles per hour, and in the final days the Grand Canyon was cut. It was a good year for geology! The Ice Ages then waxed and waned in the century or so following the Flood. Creationists like to emphasize that some classical geologists believed in the reality of Flood deposits or diluvium. True, but this interpretation only applied to the regolithic Pleistocene till plains of Great Britain. Modern creationists accept the modern geological interpretation of their deposition by continental glaciation. Instead, however, they lay claim to the entire lithified Phanerozoic sedimentary sequence, with a composite thickness of at least 150 km, down to the Basement complex as Noachian. This would be high humor except that many Americans accept this nonsense—and this is sad.

So what should scientists do other than be amused? In view of the evident fragility of the human condition, I submit that we should enter the campaign against creationism for the human mind which must be won one issue at a time. Although holdouts remain, science has defeated flatearthism and the geocentric Solar System. Can we not also return the Noah's ark story to being an ancient legend in the heroic tradition and convince the public that Earth is more than 6,000 to 10,000 years old, a remarkable uncertainly for so recent an event? The geologists' age (4.54 ± 1% Ga) is nearly one million times longer, hardly a trivial disagreement. These issues, on a par with the flat Earth, are the Achilles' heel of the creationist movement which pits pseudoscience against the real thing. Somewhere out in this great land of ours there must be geologists ready to stand tall and defend our science. As the Bible tells us: "When I was a child, I spake as a child; I understood as a child, I thought as a child: but when I became a man, I put away childish things" (Cor. 12: 11). Young-Earth and Flood creationism are glaring nonsense and nonscience. We must sink the ark and the young Earth as well. Inaction is also taking action.

The larger question is: Can the findings and realities of science change our cultural ethic as it already has our technology? Most futurists agree that our civilization is on a downward spiral owing to the population explosion. The ultimate solutions such as zero population growth, eugenics, and secularization are so stark that even their mere mention is a media taboo. Because such ideas are not politically correct, no advocate could ever win political power—the ultimate *Catch* 22. Even these could never occur anyway until an ethic based on Darwinian evolution is accepted. It is the most important of all paradigms and is the mirror we must use to see ourselves. We must play the hand we were dealt even if evolution has a dark side. As top primate, Man is super ape, not a fallen angel; the world belongs not to Man but we belong to the world; and the good of the species takes precedence over the rights of the individual. Evolution is a random walk through time with no ultimate goal in mind. Opportunism is the password and, if a new structure works, build on it through natural selection. We do not live in the greatest of all possible worlds but only the real world. The Greek gods had it better. Truth may be stranger than fiction but it is not as popular. In the long run, the reality of evolution must prevail over make-believe and feel-good creationism.

Not far from my home in Tempe, Arizona is Biosphere II, a giant glasshouse enclosure which is a one-trillionth scale model of the Earth (Biosphere I). It is a fascinating environmental research project concerned with Man's survival on Earth and perhaps eventually on Mars. But, in a sense, there have been many earlier "experiments" where in all cases society has crashed. A good example is Easter Island where the early Polynesian settlers died out apparently soon after they cut down the last tree. Already heavily damaged by a plague of people, the Earth is a fragile oasis in space. I have enjoyed a minor association with the Biosphere II project.

Environmental problems now head the list of human concerns with the

Malthusian specter of mass starvation standing offstage. In my worldwide travels I have been increasingly impressed that most countries are already "basket cases," which certainly applies to much of Africa, India, Indonesia, and South America. It has been a decade since a Caribbean cruise ship has dared to visit Haiti. The rampant population explosion is, of course, the world's major problem; it is aided and abetted by religion, but any direct mention by the media is taboo. The recently held Rio de Janeiro conference on the environment could not even be convened until all parties agreed that demographics would not be on the agenda. The threat of nuclear war has ebbed but our social order remains archaic relative to the advances of science and technology.

Eugenics was a popular idea at the turn of the century; today most persons do not even know the meaning of the word. Among those that do there is usually a knee-jerk reaction concerning Adolf Hitler and racism; the word has been redefined with entirely negative connotations. We all take pride in being distinctly different from anyone else but when it comes to larger groups, like races, we must accept them all as clones. I believe that any individual's persona is dominantly controlled by genes rather than environment. This view must remain intuitive until information is collected but science must not be prohibited from collecting this data base. The human gene pool must be protected and improved. This does not mean homogenized but instead diversified. Bringing a child into this world should become not a right but a special privilege. Would it not be wonderful to raise a child who is free of all the 4,000 plus genetic diseases, as bright as a Nobel Laureate, capable of winning a decathalon, and as handsome as, say Harry Belafonte?

Recently media mogol Ted Turner funded a competition for a book offering a workable plan for a sustainable peaceful world. Ten thousand manuscripts were submitted, but not one passed muster. This is understandable as no one is willing to bite the bullet because the obvious solutions are stark. Scientists must lead the way first in getting rid of backward religions and cults (the world of make-believe) and replacing them with an ethical culture (reality based on Darwinian evolution). Experimental societies are needed free of gurus and books of revelation to show the way. Such a plan could never be implemented within the USA because of its limited executive power. The most socially advanced society I have ever visited is the city-state of Singapore, a model to be emulated, including its severe restrictions on free-wheeling democracy.

A United Nations study group on demographics reported that the present population growth curve produces a one hundred-fold increase in the world population to 697 billion by the year 2150. This is clearly an extrapolation to the impossible and foretells of catastrophic perturbations by such events as pestilence, starvation, ethnic cleansings, or nuclear wars. Even today 99.9 percent of our overall population would starve were it not for artificial selection by plant physiologists of the Green Revolution, but such scientific fixes cannot cope with exponential peopling. We must thank Charles Darwin for discovering the malleability of species, the Franciscan monk Gregor Mendel for studying peas rather than piety, and God for not meddling while evolution produces its wondrous works. There is no limit to what science can do if left unfattened.

Again, let me emphasize that the solution to the apparent degradation of the human condition is ZEES. This is Zero population growth, accepting Evolution as the ruling paradigm of life, Eugenics, and Secularization. Spaceship Earth is hurtling through space and no one, or any collective conscience, is at the helm. Time is of the essence and it is already late. Planets with biospheres where conditions are just right (The Goldilocks's paradox) for higher life are exceedingly rare in the cosmos. Planet Earth may even be unique. Nature is amoral and does not care. Regarding secularization, let me emphasize that I am not opposed to churches per se, as they serve many useful social needs. But churches should eschew infallibility, faith, magic, the supernatural, liturgy, dogmas, vestments (especially funny hats), books of revelation, special diets, heaven and hell, life after death, mind-stomping chants, miracles, and petitionary prayers to an old-man-in-the-sky—and that is just for starters. But there is still a role for ethical culture and humanism.

A Japanese billionaire recently assembled a book on the meaning of life by polling famous people for their opinions. The answers were facetious, mystic, religious, irrelevant, or psychobabble: Life has no meaning; intelligent life is an accident of evolution; living well and ethically is the best revenge. No one got it right. The true Darwinian "purpose" (nonteleological) of all life is to reproduce—and in excess numbers, fueling natural selection. It is one of the dark sides of evolution that the drive to reproduce is greater than the will to live. So what about the eternal questions: Where did we come from? Why are we here? And *quo vadis*? Darwinian evolution, the greatest of all paradigms, explains it all (even though the real world is not the best of all possible worlds) and leaves God unemployed. Since the first spark of life (biopoeisis) four aeons ago by mutations and natural selection, man has evolved by ascent with modification. We are here at this point in time because one has to be somewhere. As for *quo vadis*, sentient man is no longer subject to natural selection. We can control our future but probably do not have the collective will to do so. Exit *Homo sapiens*?

CONCLUDING REMARKS

Now in my so-called Golden Years, I am thankful that one's mind ages better than one's body. My hero in this respect is Frank Lloyd Wright (Wrong?) who did his best work in his cantakerous late years. I must confess that I am a bit seduced by the words of the late William Saroyan who, when he was told his days were numbered, said "I expect that I know we all have to go sometime but I sort of hope that they might make an exception in my case."

My eventual wish is to be struck by a meteorite and then fossilized. This would be a privilege. I would be the first human to meet such a heroic demise although this apparently happened to an Ordovician cephalopod in Sweden. Hopefully the meteorite might even be a fragment of the Phoceaid family asteroid #4666, so kindly named Asteroid Dietz by its discoverers Caroline and Eugene Shoemaker. As a memorial it is certainly better than any headstone. They assure me that the number 666 is fortuitous but I sometimes wonder. In May 1993, it was a profound experience for me to be able to see an image of this cosmic cannonball acquired by a telescope in Colorado by my astronomer nephew Richard Dietz. It is commonly supposed that old scientists like old soldiers, should just fade away. I, for one, refuse to do so. Although my mountain climbing days are past, I have yet to meet a rock I did not like. I remain anxious to contribute and especially to defend the integrity of the scientific method as the only road even to proximal, small-truth. As the circle of light expands, the perimeter of darkness grows ever larger.

Annu. Rev. Earth Planet. Sci. 1994. 22:33–61

THE FATE OF DESCENDING SLABS

Thorne Lay

Earth Science Department and Institute of Tectonics, University of California, Santa Cruz, California 95064

KEY WORDS: subduction, mantle convection, mantle discontinuities, mantle phase transitions, mantle evolution

INTRODUCTION

Oceanic lithosphere, the stiff plate formed by the cooling of chemically differentiated material that rises and melts under mid-ocean ridges and spreads laterally in the process called sea-floor spreading, has an ephemeral residence at the surface of the Earth. Typically, within 100–200 Ma, densification of the old oceanic lithosphere leads to a dynamic instability causing it to sink into the interior, descending as a layered slab with highly anomalous thermal and chemical characteristics relative to the surrounding ambient mantle. The slab heats as it sinks, but the warming is sluggish, slow to overcome the effect of more than 100 Ma of cooling at the surface, and the negative buoyancy of the slab continues to drive its descent.

Increasing pressure and temperature with depth drive various chemical reactions in the slab, including expulsion of water and other volatiles accumulated at the surface, and destabilization of associated hydrous phases. Suites of mineralogical phase transformations occur within the slab's crustal and depleted-mantle components. Because the slab has been processed through the melting zone at the ridge, partially hydrated, and cooled significantly, the phase equilibria within the slab differ from those of the surrounding mantle. Near 400 km depth an exothermic phase transformation occurs, with $(Mg,Fe)_2SiO_4$ olivine in the slab converting to the high-pressure modified (β) spinel structure. The spinel transition helps to drive the slab onward, although a central core of metastable olivine may persist to greater depths, with eventual transformational faulting

0084–6597/94/0515–0033$05.00

producing deep earthquakes. Other phase transitions occur within the transition zone (400–720 km deep). The slab continues to sink rather readily, with the leading edge reaching a depth near 660 km about 10 Ma after leaving the surface. But now the slab encounters resistance to further penetration, sometimes resulting in distortion of the slab, and there is a build up of internal compressional stresses oriented in the down-dip direction of the flow. The slab at 660 km depth is still colder than the surrounding mantle by many hundreds of degrees, and appears to have sufficiently high viscosity to retain its integrity as a stress guide. An abrupt cessation of all earthquake activity occurs by 700 km depth. Any slab material that penetrates deeper will undergo an endothermic dissociative transformation of spinel-structured $(Mg,Fe)_2SiO_4$ into perovskite-structured $(Mg,Fe)SiO_3$ and (Mg,Fe)O. Pyroxenes and garnets will also transform to the perovskite structure, which is the predominant mineral form in the lower mantle.

There is no major disagreement on any aspect of this scenario up to this point, but what happens next? Does the slab continue to descend, penetrating as deep as the core-mantle boundary and thus defining the downwelling in a whole-mantle convection system? Is it deflected and retained within the upper mantle of a strongly stratified convective system, isolated from the deeper mantle? Or is there some intermediate convective configuration, involving either variable penetration of the lower mantle or piling up and catastrophic overturn of the accumulated mass? These questions are followed by the issue of how slab material—which eventually gets thermally assimilated but retains its anomalous chemical signature—ultimately cycles through the mantle?

This issue of the fate of subducting slabs has aroused great interest in the earth sciences, as it is intimately linked to the chemical cycling, and thermal and dynamical evolution of the planet (Silver et al 1988, Jordan et al 1989, Olson et al 1990). Fundamental questions about the causes and persistence of geochemical heterogeneity in the mantle, the possible existence of a thermal boundary layer at the transition zone, and episodicity of tectonic motions at the surface are all linked to the issue of whether slabs do or do not sink into the lower mantle. Observations from numerous disciplines have been brought to bear on this question, and periodically, seemingly compelling arguments have been articulated both for and against deep mantle penetration by the subducting downwellings, only to be softened by subsequent recognition of overlooked complexities. Slab penetration debates have often ended in a frustrating draw, and it is fair to say that this remains one of the foremost unresolved questions in the earth sciences. Research activity continues at an intense pace, with new perspectives on the problem emerging from unexpected directions almost annually. The recent developments in this arena will be highlighted here, and the current weight of evidence will be assessed, although any conclusions found here may prove as ephemeral as the slabs themselves.

SEISMOLOGICAL CONSTRAINTS ON DEEP SLAB STRUCTURE

While there are many indirect lines of evidence invoked in favor of layered or whole-mantle convection scenarios based on observed geochemical heterogeneities or notions of early mantle melting and chemical fractionation (e.g. Silver et al 1988, Anderson 1987b), direct detection of subducted oceanic lithosphere is still the most attractive way to assess the fate of slabs. This places a primacy on seismological observations, which are the most precise means available to determine the deep three-dimensional structure within the Earth. For the deep slab issue, seismological information comes in three forms: 1. earthquakes occur within the anomalous thermal/chemical environment of the subducted slab, illuminating the flow by their very presence, at least to their maximum depth extent; 2. the strain release in deep earthquakes reflects the deep slab deformational environment; and most importantly, 3. the seismic waves radiated from the deep events propagate through the surrounding medium, conveying information about the deep structure to surface sensors. Recent seismological investigations of each of these aspects are shedding light on the fate of deep slabs.

Seismicity Tracers of Deep Slabs

The recognition that intermediate and deep focus earthquakes occur in regions where cold oceanic lithosphere has sunk into the mantle played a major role in the development of the theory of plate tectonics in the 1960s (e.g. Isacks et al 1968). The maximum depth to which seismic events occur is about 700 km (Stark & Frohlich 1985), although many Benioff-Wadati seismicity zones do not extend to even this deep. Earthquakes below 100 km are only found in regions of current or recent subduction; anomalously low temperatures are required for any form of abrupt strain energy release. While pore-fluid assisted brittle failure or frictional sliding are probably responsible for most earthquake activity down to depths of 300 km, earthquake activity at greater depths appears to require a different mechanism. Recent demonstrations that, in the presence of deviatoric stresses, some important upper mantle and slab component mineral phase transformations involve shear faulting-type instabilities (e.g. Green & Burnley 1989, Green et al 1990, Kirby et al 1991, Burnley et al 1991, Meade & Jeanloz

1991) may provide a mechanism for the occurrence of earthquakes at transition zone depths (400–660 km).

Particularly important phase transitions exhibiting transformational faulting involve olivine, as it transforms to high-pressure modified spinel or spinel structures (Green et al 1992a,b). Olivine is expected to be a significant component of the oceanic lithosphere below the basaltic crust. Experimental work has shown that kinetic effects may inhibit the olivine-spinel phase transitions, causing them to occur below their equilibrium depths for a normal mantle geotherm (near 400 km for olivine-modified spinel and 520 km for modified spinel-spinel) (e.g. Sung 1979, Sung & Burns 1976). A viable explanation for deep earthquake occurrence is that transformational faulting takes place in a kinetically-overdriven tongue of metastable olivine within the colder central portions of the downwelling slab. Travel time patterns from earthquakes in the Japan slab provide some evidence for the postulated low-velocity wedge in the core of the slab at transition zone depths (Iidaka & Suetsugu 1992), although studies in other regions have not revealed such a structure. From this perspective, deep earthquake activity can be viewed as tracking the coldest regions of the slab, and is expected to terminate as the slab heats up sufficiently to overcome the kinetic barriers to transformation or as it goes through higher pressure phase transitions that do not involve faulting instabilities, such as the spinel-perovskite transition in olivine expected near a depth of 660 km (e.g. Ito & Yamada 1982, Ito & Takahashi 1989).

While seismicity distributions intrinsically only illuminate the slab where physical conditions allow earthquakes to occur, analysis of the three-dimensional spatial configuration of the slab in the seismogenic region (see reviews by Burbach & Frohlich 1986, Yamaoka et al 1986, Fukao et al 1987, Chiu et al 1991) can constrain deformations above the 660 km mantle discontinuity. While some slabs, such as Tonga, appear to display buckling and imbrication (e.g. Giardini & Woodhouse 1984, Fischer & Jordan 1991), others, such as the Japan slab show surprisingly straight orientations all the way to 660 km depth. Of particular interest are regions of isolated deep earthquake activity, displaced laterally from the well-defined deep seismic zones in several arcs. Figure 1 (Okino et al 1989) shows an isolated event offset from the deep seismicity in the Izu slab, similar to events found under the northern Kurile arc (Glennon & Chen 1993), Spain, Peru, and the New Hebrides (Ekström et al 1990, Lundgren & Giardini 1992b). Given the special environment required for deep earthquake occurrence, these rare events are usually taken as evidence for horizontal deflection of the deep slab, although the possibility that independent detached slab remnants are responsible for the anomalous events complicates such arguments (Hamburger & Isacks 1987). While the spatial distribution of seismicity intrinsically cannot resolve the maximum depth extent of subduction, the shape of the downwelling revealed by the seismicity does provide a boundary condition for modeling slab kinematics for the aseismic extension of the slab (e.g. Creager & Boyd 1991).

Strain Orientations in Deep Slabs

Seismic waves from deep earthquakes not only reveal the occurrence of the strain release, but also indicate the strain orientation at the source (e.g. Isacks and Molnar 1971, Vassiliou 1984, Apperson & Frohlich 1987) and the total energy release (Richter 1979). A robust feature of deep earthquake focal mechanisms is their strong tendency to have roughly down-dip compressional strain axes, usually interpreted as indicating resistance to penetration into the lower mantle. With the accumulation of increasing numbers of earthquake focal mechanisms, recent efforts have been directed at assessing the overall slab deformation associated with the deep seismicity. The underlying assumption is that the deep earthquake strain tensors are indicative of the overall slab deformation and can be interpreted in the context of viscous loading of coherent slab structures (e.g. Vassiliou et al 1984, Vassiliou & Hager 1988).

Giardini & Woodhouse (1986) related focal mechanisms in the Tonga subduction zone to the contorted seismicity distribution in the region to infer systematic along-strike shear deformation of the entire slab induced by horizontal flow in the surrounding mantle. The structure of shear zones within the slab that accommodate such distortion can be revealed by seeking alignments of seismicity with focal mechanism nodal planes (Giardini & Woodhouse 1984, Lundgren & Giardini 1992a). Zhou (1990b) conducted a systematic study of focal mechanisms for earthquakes in the northwest Pacific and Tonga-Kermadec, and found evidence for flattening of the compressional strain axes near 660 km depth beneath the Izu-Bonin and Tonga arc, but no clear evidence for this in the Kurile, Japan, or Marianas slabs.

The strain rate of the seismogenic portion of a deep slab can be estimated by summing the contributions from all earthquake mechanisms in the slab. Fischer & Jordan (1991) find that along-arc strain rate decreases relative to the cross-strike strain rate from north to south in the Tonga slab—a result generally compatible with the slab shearing model of Giardini & Woodhouse (1986). A 50% thickening of the seismogenic core of the slab can be accounted for by the seismic moment release. While this indicates that slab thickening can occur due to the resistance to lower mantle penetration, it appears that the main volume of subducted material reaching 660 km depth is aseismic, thus seismic strain release cannot quantify the associated deformation. Another complexity is that thermal and phase

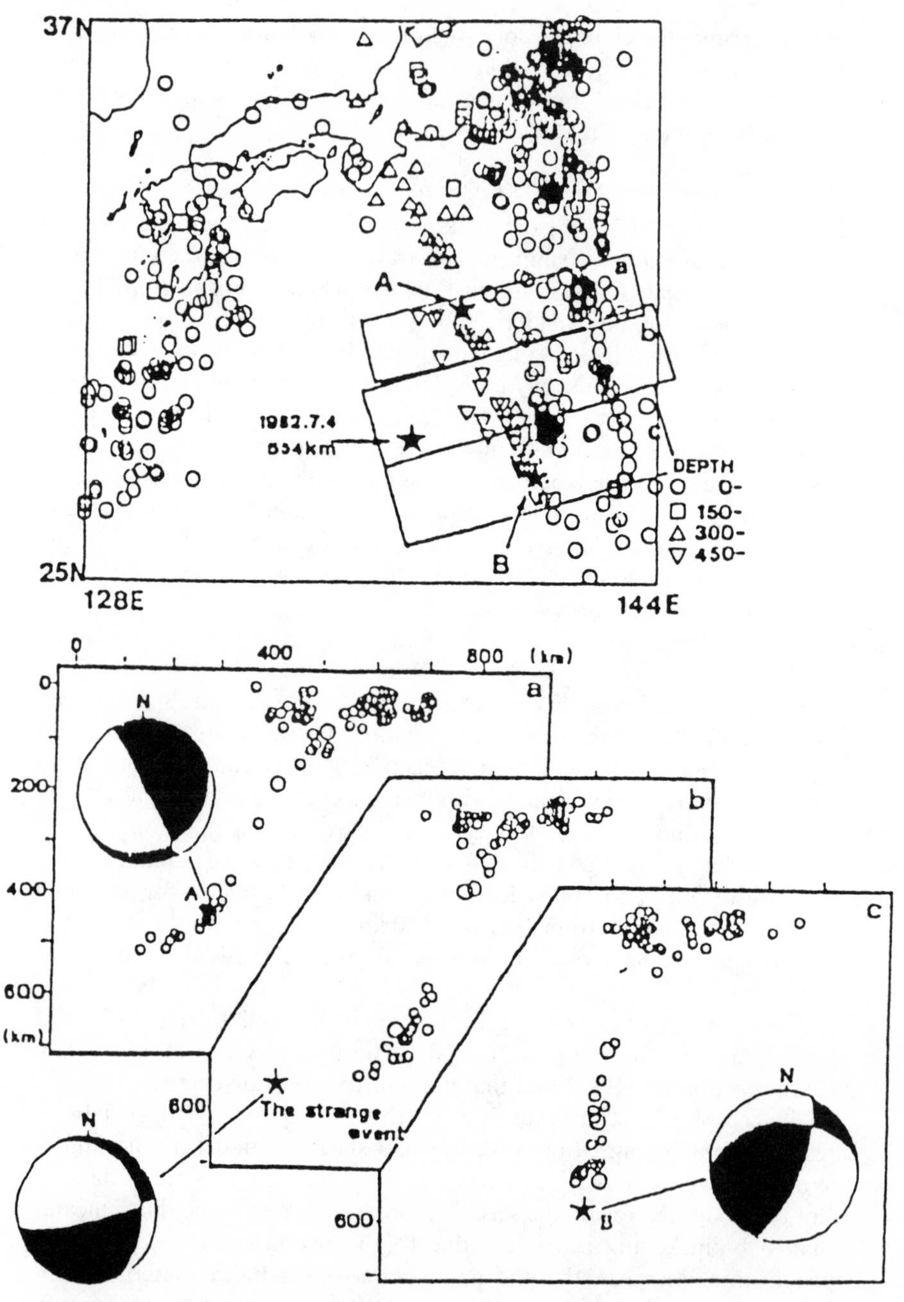

Figure 1 Example of a rare, isolated deep focus earthquake. The July 4, 1982 event occurred at a depth of 534 km, offset about 200 km west from the primary zone of deep activity associated with subduction of the Izu (Pacific) slab (from Okino et al 1989).

transformation induced stresses may overprint any viscous deformational stresses caused by resistance to subduction (e.g. Goto et al 1983, 1985, 1987; Kirby et al 1991). In this case, stresses generated within the slab may account for the down-dip alignment of compressional strain axes.

Seismic Velocity Heterogeneity

Huge volumes of slab have descended in the current subduction zones, greatly exceeding the volumes that are presently seismically active (e.g. Richards & Engebretson 1992, Scrivner & Anderson 1992); thus seismological techniques for imaging the aseismic slab material must be used to assess where the slab material has gone. Since slabs are thermally and chemically distinct from the surrounding mantle, seismic waves transmitted through the slab incur amplitude and travel-time anomalies. For 25 years seismologists have been analyzing wave propagation characteristics of descending slabs, addressing both the seismogenic regions and the deep extension of the slab (Lay 1993). Current approaches to the problem of resolving the fate of slabs include: 1. modeling of travel-time and amplitude anomalies from individual events; 2. tomographic imaging of the mantle around deep earthquakes; 3. waveform modeling of distortions produced by gradients in seismic velocity structure near the deep slab; and 4. tomographic imaging of lower mantle structures and mantle discontinuity topography on both large and small scales.

The basic idea underlying travel-time modeling of deep slabs is that the cold slab should have higher elastic velocities than the ambient mantle, with the tabular geometry of the slab (at least above 660 km) giving rise to simple symmetries in the patterns of relatively early and late arrivals (the former have longer path lengths in the slab material). Numerous studies indicate that in the transition zone slabs do have about 3–5% faster seismic velocities than the surrounding mantle. Thermal models, combined with low-pressure values for temperature derivatives of seismic velocity, may account for this level of heterogeneity (e.g. Creager & Jordan 1986); stronger velocity gradients are expected on the crustal side of the slab. There is, unfortunately, still substantial uncertainty in seismic velocity dependence on temperature at high pressures, complicating predictions of deep slab velocity heterogeneity (Anderson 1987a, 1988). The chemically layered nature of the slab adds detailed velocity structure, such as a high-velocity eclogitic crustal layer, the possible presence of low temperature phases such as $MgSiO_3$-ilmenite, as well as significant ($\pm 5\%$) variations associated with either elevated or depressed equilibrium phase boundaries and/or kinetically suppressed boundaries. Anisotropic characteristics of the slab and the surrounding flow may also contribute to the seismic velocity signature of a deep slab (Anderson 1987b, Kendall & Thomson

1993). Finally, there is the major limitation of seismology: Seismic wave heterogeneity may have multiple causes. For example, it may be very difficult to distinguish aseismic slab extensions into the lower mantle from induced downwellings in a thermally-coupled convection system. Consideration of dynamical models when interpreting seismic images is always crucial.

Most studies of travel-time patterns associated with deep slabs utilize deep earthquakes. The greatest challenge in isolating the slab signature in the seismic wave arrival times is that the location of the seismic source is not known independently. Standard earthquake location procedures project as much as possible of the observed travel-time deviations from the reference Earth model into shifting the source location and origin time. Routine earthquake locations are performed with no slab structure in the Earth model, or at best a very simplified slab model, leading to biased locations of deep events. Patterns of anomalies with respect to the biased locations are expected to be quite different from patterns relative to the actual location (e.g. Creager & Boyd 1992), thus one must account for this effect in any modeling effort.

Numerous studies have used the residual sphere method introduced by Davies & McKenzie (1969) to explore systematic patterns in deep earthquake travel times. Residual spheres are projections of travel-time anomalies onto a source focal sphere, indicating the azimuth and take-off angle of each ray path corresponding to observed or calculated anomalies. Jordan (1977), Creager & Jordan (1984, 1986), Fischer et al (1988, 1991) and Boyd & Creager (1991) have conducted a series of residual-sphere modeling efforts, all of which favor tabular high-velocity (3–4%) extensions of slabs four to five hundred kilometers below the deepest seismicity in the Kurile, Japan, Marianas, Tonga, and Aleutian arcs (Figure 2). The smoothed residual patterns for relocated sources, corrected for station statics and low-degree aspherical velocity models of the lower mantle (e.g. Dziewonski 1984), were compared with three-dimensional ray-tracing calculations for thermal slab models embedded in a homogeneous mantle, similarly relocated and filtered. Excellent fits to the data are achieved in most cases, using very simple models. In all of these studies the preferred slab models steepen in dip in the lower mantle, but are otherwise not strongly distorted by passing through the 660 km boundary (Figure 2). Sensitivity tests (Fischer et al 1988) indicate that broadening of the slab by up to a factor of three is allowable, without seriously degrading the fit to the data, but the preferred models do not broaden.

Various concerns have been raised about the surprisingly clean results obtained in this group of residual sphere studies, involving details of the processing (e.g. the heavy filtering used to enhance the smooth slowly-varying patterns in the data, which are assumed to be induced by near-source structure), the data selection (e.g. the use of "noisy" catalog travel-time measurements from nonstandard stations for only down-going P arrivals which span a very limited portion of the focal sphere), the model parameterization (e.g. neglect of possible slab anisotropy), and the heterogeneous Earth correction procedures (e.g. use of low resolution Earth models which probably underestimate actual deep heterogeneity to correct for distant path effects). Several recent studies have attempted to overcome some of the possible difficulties.

The possibility of systematic bias in catalog data sets associated with variations in instrument magnification was raised by Grand (1990), who found that low gain stations tend to report late arrivals. A greater concern about the catalog travel times is the intrinsically large scatter in the data. Gudmundsson et al (1990) analyzed ISC travel-time statistics, separating spatially coherent and incoherent variance, and found low signal to noise ratios for P waves at teleseismic distances. Efforts to manually measure arrival times (e.g. Takei & Suetsugu 1989, Ding & Grand 1992) tend to substantially reduce the range of anomalies in residual spheres for deep events relative to the catalog values.

To isolate the near-source signature of deep slab heterogeneity on travel-times, it is important to suppress contributions from mantle heterogeneity far removed from the source region. Given the predominance of large-scale heterogeneity in both the upper and lower mantles (e.g. Su & Dziewonski 1991, 1992), one cannot safely assume that any slowly varying pattern in a residual sphere must be caused by near-source structure. Zhou & Anderson (1989b) showed that residual spheres for shallow earthquakes in eastern China actually resemble those for deep events in the Kurile slab, suggesting a common component of lower mantle or near-receiver heterogeneity on paths to teleseismic stations. Lay (1983), Schwartz et al (1991a,b), and Gaherty et al (1991) have presented clear evidence for lower mantle and near-receiver contributions to deep earthquake travel-time residuals being underpredicted by standard station corrections or integration through low resolution aspherical Earth models. These distant effects can produce smooth patterns in residual spheres that can bias deep slab modeling. Models with substantially broadened slabs, involving relatively weak lower-mantle heterogeneity, are indicated by these studies.

Confronted with the difficulty of isolating near-source travel-time anomalies, there are two different procedures. As they become available, one can use higher resolution aspherical Earth models (e.g. Inoue et al 1990) to make deep path corrections, or one can adopt an empirical calibration strategy, using data from nearby events to establish improved path corrections, free from the damping and smoothing effects of tomographic

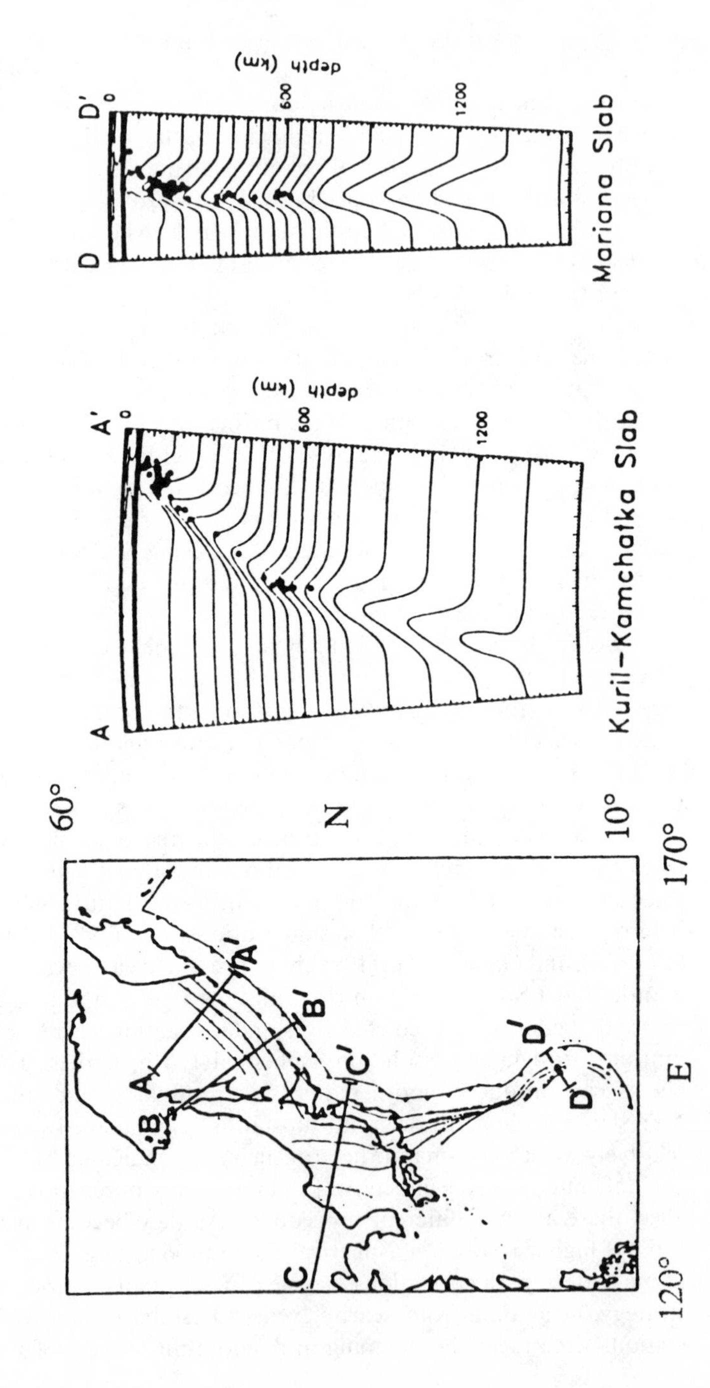

60°
N
10°
170°
E
120°
A
A'
B
B'
C
C'
D
D'
Kuril-Kamchatka Slab
Mariana Slab
depth (km)
0
600
1200

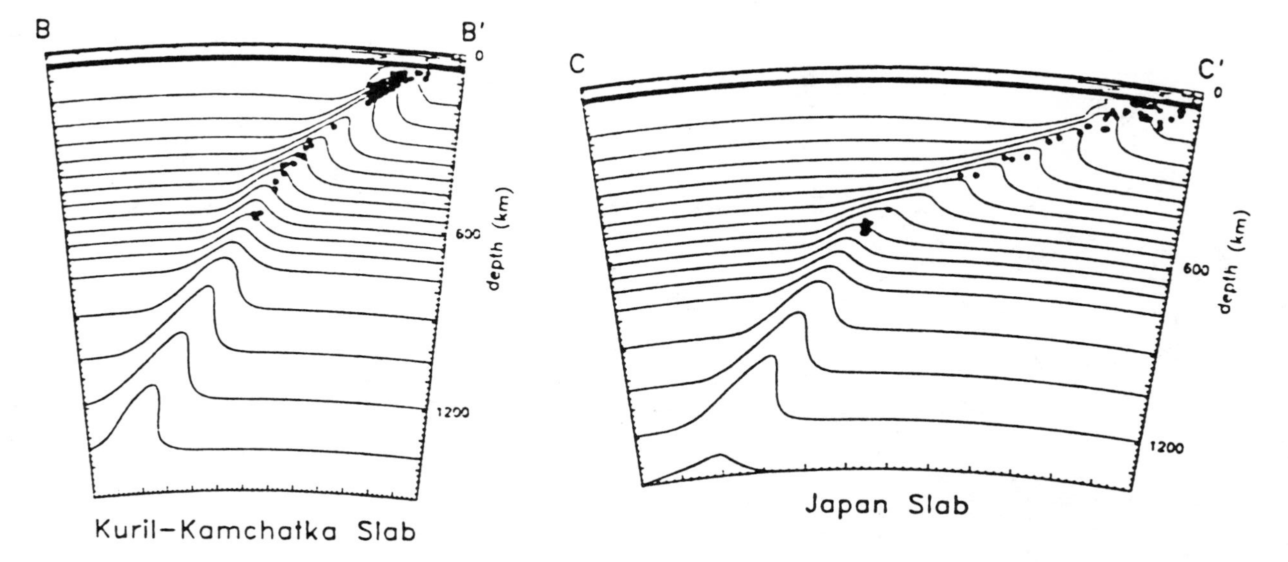

Figure 2 Cross sections through three-dimensional slab velocity models derived by residual sphere analysis of several intermediate and deep focus earthquakes in the western Pacific subduction zones. The seismic zone in the map is contoured in 100-km intervals. In all cases, slab models extend to 1350 km depth, although penetration depths below 1000–1200 km cannot be resolved by the data. Contour interval is 0.25 km/s (from Creager & Jordan 1986).

inversions. Schwartz et al (1991b) applied corrections for a high-resolution lower-mantle shear velocity model beneath North America, finding that it strongly reduced relative S wave travel-time patterns for Kurile slab sources. Zhou & Anderson (1989a) and Zhou et al (1990) used whole-mantle and regional northwest Pacific tomographic models to correct residual spheres for many events in the western Pacific. The latter authors extended the ray parameter coverage of the focal sphere substantially by including close-in observations as well. Consistently, many of the coherent features in residual spheres that give rise to the simple slab structures shown in Figure 2 are substantially reduced, weakening the evidence for simple slab penetration models. There is also evidence for horizontal flattening of some slabs, revealed only in the extended ray parameter coverage.

Since all tomographic models are potentially flawed in terms of providing detailed corrections for specific paths, empirical corrections have also been attempted. Zhou & Anderson (1989a) found that removing average path anomalies computed for many events in the western Pacific weakened slab-like signals in individual event residual spheres, especially for very deep events. Gaherty et al (1991) and Schwartz et al (1991a,b) found that using surface reflected phases to make distant path corrections also greatly reduced the patterns for deep events.

Differential residual spheres, first introduced by Toksöz et al (1971), can be computed for events at different positions in the same slab to cancel distant path effects. While the travel-time anomalies that result are substantially reduced relative to those found by the early residual sphere analyses, these studies have tended to favor the presence of high-velocity material in the lower mantle along portions of the down-dip extension of the Japan and Kurile slabs (Takei & Suetsugu 1989, Okano & Suetsugu 1992, Ding & Grand 1992). For the Marianas, the differential residual sphere method appears to eliminate the high-velocity region below the deepest events (Okano & Suetsugu 1992). The most complete analysis of this type has been for the Kurile slab, where Ding & Grand (1992) favor a model with substantial advective thickening of the plate and lower mantle penetration in the northern part of the arc, and flattening of the plate to subhorizontal in the southern arc. This is consistent with most of the tomographic results discussed later.

While most of these procedures for improving corrections for distant propagation effects lack the comprehensive three-dimensional ray tracing and careful handling of relocation artefacts that characterize the early work on residual spheres (Ding & Grand 1992 is an exception), it is clear that reliance on low resolution models for corrections is inadequate. Interpretation of differential patterns is somewhat more complicated than for individual events, but the confidence gained in the isolation of near-source contributions is substantial. Of course, there is a danger in differential residual sphere and other empirical path calibration procedures of eliminating the desired near-source signal if the path separation is not sufficient. Given the complex three-dimensional geometry of all subduction zones, this is a serious consideration.

If aseismic extensions of slabs have simple geometries, the residual-sphere modeling approach is quite attractive. However, if the slabs are strongly distorted so that no simple slab parameterization can be justified, it is preferable to image the slab heterogeneity by seismic tomography. Numerous tomographic studies of subducted slabs have been built on basic strategies introduced by Aki & Lee (1976) and Aki et al (1977), expanded to handle vastly increasing data sets and model sizes (e.g. Spakman & Nolet 1988). While early models (e.g. Hirahara 1977) addressed the upper mantle structure alone, many studies include the lower mantle below the seismogenic regions. Accounting for source mislocations and deep mantle and near-receiver effects are still major problems, as is the vertical smearing in tomographic images associated with limited crossing ray coverage below the deepest sources (Spakman et al 1989, Zhou 1988). While limited efforts have been made to solve the simultaneous location and velocity determination problem (e.g. Engdahl & Gubbins 1987), most tomographic studies utilize an iterative procedure: The initial catalog locations are relocated in a homogeneous model; then, at least in some studies, they are relocated in a first-step tomographic model using simplified ray tracing. A major problem is that much of the slab signal may be lost in the initial location process, unless the ray path coverage is extensive enough to eliminate the trade-off between slowly varying slab anomalies and slowly varying relocation effects. In the future, inversions starting with an initial slab structure may overcome much of this difficulty.

Tomographic images have been produced recently for the northwest Pacific subduction zones (Kamiya et al 1988, 1989; Suetsugu 1989; Spakman et al 1989; Zhou & Clayton 1990; Nenbai 1990; van der Hilst et al 1991, 1993; Fukao et al 1992; Yamanaka et al 1992), Java (Fukao et al 1992, Puspito et al 1993), Tonga and the New Hebrides (Zhou 1990a), Central America and the Caribbean (van der Hilst & Spakman 1987, 1989), and the Mediterranean (e.g. Spakman et al 1988, 1993; Ligdas et al 1991; Spakman 1991). Strategies for eliminating distant effects have included the omission of teleseismic ray paths altogether (Zhou & Clayton 1990), correction for mean station residuals at teleseismic distances (e.g. Kamiya et al 1988, van der Hilst et al 1991), and simultaneous inversion for the complete global structure using a variable block size for the near-source and deep mantle portions of the model (Nenbai 1990, Fukao et al

1992). Improvements in the source location problem have been achieved by including both regional and teleseismic data, as well as surface reflections (pP, PP), which enhance the focal sphere coverage and reduce errors in source depth (van der Hilst & Engdahl 1991, 1992).

The resulting tomographic images of deep slabs are relatively consistent in detecting fast velocity slab structures encompassing the deep earthquake locations. However, the general characteristics of deep slabs are significantly more complex than the models shown in Figure 2. The most recent models (Figure 3), which utilize the most reasonable strategies for controlling mislocation effects and distant contributions, are suggestive of a vertical extension of the central Kurile slab to depths of about 1100 km under the Sea of Okhotsk (van der Hilst et al 1991, 1993), flattening to subhorizontal for the southern Kurile slab (van der Hilst et al 1991, 1993; Fukao et al 1992), and flattening and thickening of the Japan slab (van der Hilst et al 1991, 1993; Fukao et al 1992). Much of the flattening in the southern Kuriles and Japan appears to occur above 660 km depth, but there is evidence for some high-velocity material extending below this depth in the model of Fukao et al (1992), which could be consistent with earlier studies by Kamiya et al (1989), Creager & Jordan (1986), and Takei & Suetsugu (1989), but this proves to be a minor feature relative to the horizontally deflected material. The central Izu-Bonin slab appears to deflect horizontally above the 660 km discontinuity, extending across the entire width of the Philippine plate to the Ryukyu arc (van der Hilst et al 1991, 1993; Fukao et al 1992), while a broadened high-velocity anomaly extends vertically below the Marianas subduction zone to depths of about 1200 km. The Middle American slab appears to extend well below the cut-off of seismicity (van der Hilst et al 1993), possibly connecting up to a broad fast velocity region more than 1000 km deep beneath the Caribbean imaged in earlier work (Jordan & Lynn 1974, Lay 1983, Grand 1987). Large-scale tomographic images of the entire mantle, while lacking resolution of small structures such as slabs, do suggest generally fast lower-mantle velocities below the circum-Pacific (e.g. Dziewonski & Woodhouse 1987, Su & Dziewonski 1991), but a direct connection of these very large-scale features with slab anomalies is not yet established. Distinguishing continuous downwellings from thermally-coupled layered systems is particularly difficult for large-scale tomographic models.

In addition to seismic wave travel times, amplitude and waveform anomalies have been analyzed to detect any effects of deep slab velocity heterogeneity. Ray path deflections caused by slab structure have been analyzed for more than 20 years (Sleep 1973), but new broadband data offer improved resolution of wavefield effects. Theoretical calculations predict that the high-velocity heterogeneity of slabs should cause complex diffrac-

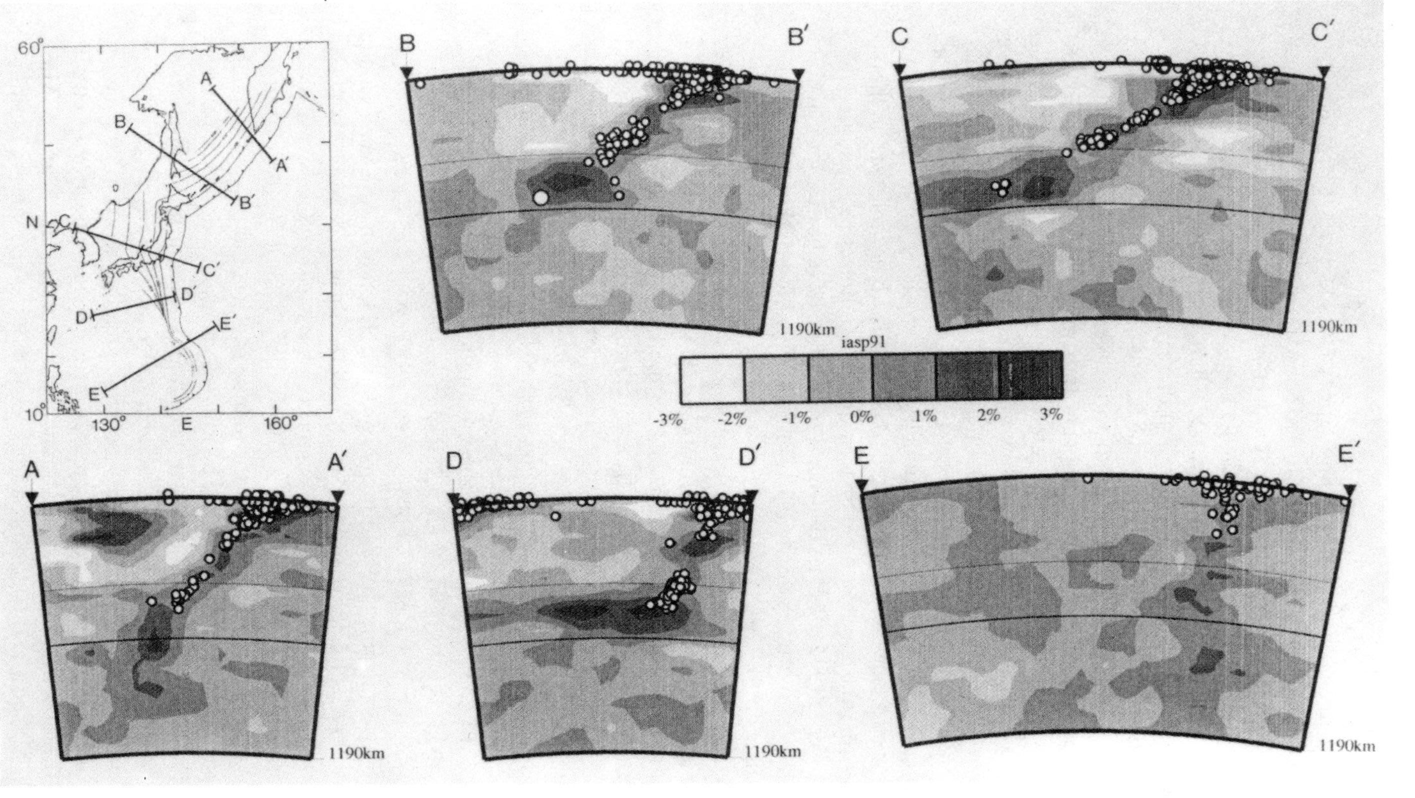

Figure 3 Cross sections through three-dimensional tomographic P wave velocity models for western Pacific subduction zones. The seismic zone in the map is contoured in 100-km intervals. Faster velocity material (relative to the reference iasp91 model) is darker, with shading in 1% velocity variations. Dots are earthquake hypocenters along each of the cross sections. Mantle discontinuities at depths of 400 and 660 km are shown. Note that high velocity material correlates with the deep seismic zones, and shows evidence for both penetration (AA′, EE′) and flattening (CC′, DD′) at the top of the lower mantle (from van der Hilst et al 1991, 1993).

tion, defocusing, and multipathing effects (Vidale 1987, Cormier 1989, Witte 1989, Weber 1990, Sekiguchi 1992, Vidale et al 1991). Some observations of waveform distortions for deep earthquakes do indicate the presence of complex velocity gradients along the ray paths (Beck & Lay 1986, Choy & Cormier 1986, Silver & Chan 1986, Lay & Young 1989, Schwartz et al 1991a, Fischer 1990), but there is still no convincing demonstration that these effects constrain the fate of deep slabs. The deep mantle has many heterogeneities that can distort waveforms (Lay 1983, Vidale & Garcia-Gonzalez 1988), and it appears that many waveform effects are superimposed to produce the scattered behavior that is observed. While there is strong evidence that shallow slab structures do affect teleseismic waveforms (Engdahl & Kind 1986, Engdahl et al 1988, Gubbins & Snieder 1991), it appears that deep slab velocity gradients are not as strong as indicated by the early thermal models for deep slabs. In particular, defocusing effects, which are predicted to be readily observable for the simple tabular thermal models, have proven difficult to detect (e.g. Vidale 1987, Gaherty et al 1991, Fischer 1990), although there have been some attempts to include amplitude information in the construction of deep slab models (Suetsugu 1989).

One additional area in which seismology is contributing to our understanding of deep slabs is in the mapping of topography on the 660 km discontinuity. While this seismic discontinuity is most likely the result of the spinel-perovskite phase transformation, rather than a chemical boundary, its undulations do provide constraints on the lateral temperature variations near the boundary. Several studies have attempted to detect perturbations of the boundary in the vicinity of an impinging slab, using converted and reflected phases. Depressions of the phase transition by tens of kilometers have been detected under Tonga (Bock & Ha 1984, Richards & Wicks 1990, Wicks & Richards 1991, Vidale & Benz 1992), the Izu-Bonin/Marianas (Vidale & Benz 1992, Wicks & Richards 1993), the southwest Pacific (Revenaugh & Jordan 1991), South America (Vidale & Benz 1992) and the Kuriles (Shearer 1991, Shearer & Masters 1992). While these studies support the evidence for a negative Clapeyron slope for this phase transition (Ito et al 1990), they cannot be used to argue against chemical stratification of the mantle which could prevent the slabs from penetrating. This is because postulated chemical differences between the upper and lower mantle that are needed to account for differences in the average densities (Jeanloz & Knittle 1989, Stixrude et al 1992) may not have a strong seismic velocity contrast (Jeanloz 1991). The primary application of boundary topography is instead as an indicator of cumulative thermal anomaly near the boundary. For example, Shearer & Masters (1992) show a 1500 km broad depression of the 660 km discontinuity landward from the southern Kurile arc. This is consistent with the cooling effect of a horizontally deflected slab, as imaged in the tomographic and differential residual sphere analyses mentioned above. If all recently subducted slabs are piled up in the transition zone, one would expect broad depressions of the 660 km discontinuity around all margins of the Pacific ocean, which is not generally observed (Shearer & Masters 1992), although current topographic models are very limited in resolution.

DYNAMICAL FACTORS INFLUENCING SLAB FATE

While seismology provides tantalizing images of heterogeneous elastic properties of the interior, quantification of the actual subduction process requires insights from numerical modeling. There have been great advances in our ability to simulate viscous flow in the mantle for increasingly realistic parameter ranges, and a wide variety of possible influences on the mantle system have now been explored. One of the surprising perspectives to emerge is that there are many factors that should act to inhibit deep mantle penetration by the downwelling slab, even if the mantle is not chemically layered. Some of the recent developments will be discussed below.

Effects of Chemical and Viscous Stratification

Given sufficient intrinsic chemical density contrast between the upper and lower mantle, it is not difficult to develop rigorously stratified mantle convection. Christensen & Yuen (1984) estimate that a 5% contrast is sufficient to deflect the slab, preventing mixing between the upper and lower mantle. This value is at the upper end of estimates of a bulk chemical density contrast between the upper and lower mantle based on mineral physics experiments on silicate perovskite and magnesiowüstite (e.g. Knittle et al 1986, Jeanloz & Knittle 1989, Stixrude et al 1992). The critical physical parameter underlying these arguments is the thermal expansion coefficient for silicate perovskite at high pressure, critical to estimating the density predicted for various lower mantle compositions. While there is controversy over the experimental determination of perovskite thermal expansion (Chopelas & Boeler 1989), most measurements have been made either at atmospheric pressure (Knittle et al 1986, Ross & Hazen 1989, Parise et al 1990), or at pressures below the stability field of perovskite (Mao et al 1991, Wang et al 1991). New measurements at simultaneous high pressure and high temperature that are just beginning to emerge (Funamori & Yagi 1993) indicate less decrease in thermal expansion with pressure than previously thought. Confirmation of these results will be critical for resolving the argument for a component of chemical contrast between the upper and lower mantles.

Density contrasts lower than 5% allow partial penetration of the downwelling slab, and the slab can sink all the way through the lower mantle for density contrasts less than 2%, which is at the lower range of the estimated chemical density contrast. The relative viscosity of the subducting slab influences its behavior, with higher viscosity allowing it to retain a concentrated negative buoyancy that abets penetration. Experimental work (Kincaid & Olson 1987) indicates that slab geometry also influences slab penetration, with steeper dipping slabs penetrating more effectively. Silver et al (1988) argue that penetrative convection, in which the compositionally-induced density increase across the transition zone is 2% or less, is a viable explanation for a variety of deep earth constraints. Continued efforts to quantify any bulk chemistry difference, between the upper and lower mantle are critical for assessing the role of any chemical layering, as are efforts to detect any seismic boundary that can be attributed to a compositional contrast. Qualitatively, the evidence for thickened and flattened slab structures presented above is consistent with the range of behavior expected for moderate chemical stratification, but it is now clear that other effects can also produce such slab distortions.

Viscous stratification of the mantle provides an effective mechanism for deforming subducting slabs, and can readily lead to broadened downwellings in the deep mantle. An increase in viscosity across the 660 km boundary by a factor of 10 to 100 has been invoked in several efforts to model the geoid (Hager et al 1985, Hager & Richards 1989, Hager & Clayton 1989), although less abrupt viscosity increases have also been proposed (e.g. Forte et al 1992). An abrupt increase in viscosity can strongly deform the subducting slab, causing it to fold over on itself in the uppermost part of the lower mantle (Gurnis & Hager 1988). The viscous flow models can produce many features similar to those seen in tomographic images. Thus, viscous stratification is a viable mechanism for distorting the deep slab and advectively thickening it; but alone it can not prevent eventual sinking of the subducted mass unless extreme viscosity contrasts are invoked. A combination of a vicosity increase and a chemical contrast can be reconciled with the geoid, but requires a substantial accumulation of subducted material at the top of the upper mantle, and a deeply (several hundred kilometer) depressed compositional discontinuity. This could be reconciled with the broad high-velocity regions seen tomographically beneath the Marianas, Java, and Japan.

Effects of Chemical Buoyancy of the Slab

The distinct phase equilibria of slabs must also be considered when assessing their dynamics. Ringwood & Irifune (1988) and Anderson (1987b) evaluate the intrinsic density of the enriched, eclogitic crustal and depleted, harzburgitic mantle portions of subducted slabs as a function of depth. If thermally equilibrated with the surrounding mantle, these components could be dynamically stabilized in the transition zone. The intrinsic chemical buoyancy can affect subduction, and perhaps lead to separation of the slab components. However, numerical calculations of viscous slabs including chemical buoyancy effects indicate that thermal buoyancy predominates, and even if an increase in viscosity with depth leads to strong folding and deformation of the slab, the chemical buoyancy effects are secondary in the overall dynamics of the slab (Richards & Davies 1989, Gaherty & Hager 1993). These calculations indicate that unless the slab material can be held stagnant in the transition zone for sufficient time so that thermal equilibration is achieved, chemical buoyancy effects will not play a major role in determining the fate of the slab. It is important to note that if the slab material does penetrate below the transition zone, the basaltic component may again become denser than the surrounding mantle, adding negative buoyancy to the slab (Ringwood & Irifune 1988). Continued work on the high-pressure density of slab materials is needed to address this issue.

Effects of Phase Transitions in the Slab

There has been extensive work recently on the dynamic effects of the endothermic and exothermic phase transformations affecting upwellings and downwellings in the transition zone. Early work on the 660 km endothermic transition (e.g. Christensen 1982, Christensen & Yuen 1984, 1985) indicated that a large negative Clapeyron slope (-4 to -8 MPa/K) would be required to induce stratified convection, which was viewed as unlikely. Recent estimates of the Clapeyron slope vary from -2.8 to -4.0 MPa/K (Ito & Takahashi 1989, Ito et al 1990), and updated computations at higher Rayleigh number (*Ra*) have incorporated these values. Calculations with $Ra = 10^6$–5×10^7 have been conducted in a two-dimensional Cartesian geometry to assess the effects of the 660 and 400 km phase transitions (Liu et al 1991, Zhao et al 1992, Steinbach & Yuen 1992). Increasing Rayleigh number tends to promote layering of the system, as does the coexistence of two phase transitions (Figure 4). Spherical axisymmetric calculations indicate similar phenomena (Machetel & Weber 1991, Peltier & Solheim 1992), as well as a tendency for time-dependent behavior involving catastrophic flushing out of the accumulated downwellings in the upper mantle. Catastrophic overturn is observed in spherical axisymmetric (Machetel & Weber 1991, Solheim & Peltier 1993), two-dimensional Cartesian (Weinstein 1993), and three-dimensional Cartesian (Honda et al 1993) geometries, but the degree of episodicity of these instabilities is reduced in fully three-dimensional spherical flow regimes

(Tackley et al 1993), where flushing events are more localized phenomena. The latter calculations are for a realistic mix of bottom and internal heating, with a volume-averaged Rayleigh number from internal heating and superadiabaticity of 1.9×10^7. The occurrence of intermittent over-turning events is an efficient mechanism for producing a heterogeneity spectrum dominated by long-wavelength features, as observed for the Earth. However, it is not clear that upper and lower mantle heterogeneities are as decorrelated as expected for the quasi-stratified flow regimes that have been computed (Jordan et al 1993). At this time, the effects of high Rayleigh number flow with strongly temperature-dependent viscosity and phase transitions have yet to be established. It is likely that the stiffness of the slab will strongly influence the ability of the phase transitions to impede their downward motions, so it is premature to base too many conclusions on these calculations.

Historical Slab Accumulations

The history of subduction places important constraints on the interpretation of seismic models for deep slabs. Estimates of the cumulative subducted material in circum-Pacific regions have been found to correlate with long-wavelength seismic velocity heterogeneity in both the transition zone (Scrivner & Anderson 1992) and the radially-averaged lower mantle (Richards & Engebretson 1992). Given the long time required for thermal equilibration of the slab material (e.g. Shiono & Sugi 1985), one expects large volumes of fast-velocity material near or below subduction zones, unless the slabs readily sink deep in the lower mantle. Degree 2 harmonic terms of aspherical velocity heterogeneity and accumulated slab volume do correlate rather well in the transition zone and uppermost part of the lower mantle (Scrivner & Anderson 1992), which can be used to argue either for or against slab penetration into the lower mantle, as long as broadening and thickening of the slab is allowed. Of course, degree 2 is extremely large scale, even relative to the volumes of accumulated slab material, and higher degree components (admittedly, less well resolved in the current seismic models) show far less correlation. Efforts to relate specific deep slab structures to the history of subduction are just beginning (Engebretson & Kirby 1992, Grand & Engebretson 1992, van der Hilst et al 1992), but this promises to be a fruitful area of research.

SUMMARY AND SYNTHESIS

There is as yet no unambiguous resolution of the fate of deep slabs. The collective seismological, geodynamical, and mineralogical evidence summarized above can perhaps best be reconciled with a model of the

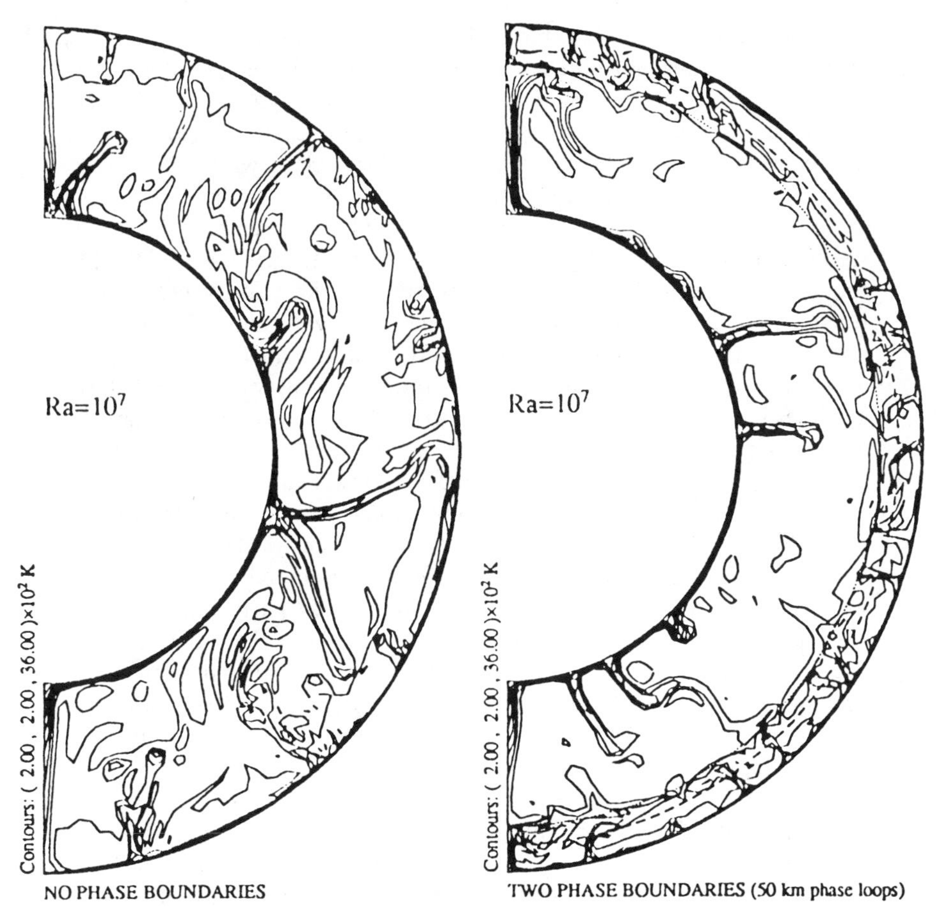

Figure 4 Instantaneous temperature fields from two axisymmetric simulations of mantle convection at a Rayleigh number of 10^7. Heating is entirely from below. The calculations on the left do not include any phase transitions, while those on the right include both a model olivine-β spinel transition and a model spinel-perovskite transition, with 50 km thick phase loops. Note that neither hot ascending plumes nor cold descending plumes appear to penetrate the 660 km discontinuity on the right (from Peltier & Solheim 1992).

Earth in which descending slabs encounter resistance to lower mantle penetration, as a result of increasing viscosity with depth, intrinsic resistance from the endothermic perovskite phase transitions, and possibly a few percent chemical density contrast between the upper and lower mantles. Apparently, the termination of seismicity near 660 km depth can be attributed to combined effects of thermal assimilation and completion of phase transformations in the slab (recall that seismicity in many slabs does not even penetrate this far), and, for seismogenic slabs that exceed this depth, silicate material transforming into perovskite, extinguishing the mechanisms causing deep earthquakes. The aseismic portion of the slab broadens and deflects at and below the phase boundary. Slabs with rapid trench motions, such as the Izu slab, may flatten above the boundary.

Because the thermal inertia of the slab is still high at 660 km depth, the intrinsic chemical buoyancy of the slab components, which would tend to retain the slab in the transition zone if thermal equilibration were achieved, is inadequate to keep the slab from sinking. The lower mantle downwelling appears to be slowed relative to the upper mantle rates, with large slab accumulations and entrained material producing long-wavelength heterogeneities in the lower mantle below active subduction zones. This heterogeneity is imaged in residual sphere and tomographic models. Broadening, and perhaps buckling of the slab, reduces the lower-mantle velocity gradients involved, making the deep slab anomalies inefficient diffractors of seismic waves. The variability of tabular extensions of deep slabs, some lying horizontal and others penetrating vertically, can be attributed to the variable tectonic history and specific geometry of particular zones, as material piles up and deforms near the top of the lower mantle. While the slab material may tend to flush out catastrophically, the images of accumulations above the 660 km discontinuity do not show the massive volumes expected for the total history of subduction. Instead, continuous piling up of material just above, and within the uppermost portion of the lower mantle is suggested by the collective seismic images. Of course, some flushing events may be ongoing, while in other places accumulations are just beginning.

Other interpretations can of course be made, but it seems that most of them have serious flaws. Undistorted slab penetration into the lower mantle is implied by some residual sphere models, but this is hard to reconcile with the abundant evidence for significantly broadened anomalies below 660 km depth, as well as the clear evidence for horizontal deflection of some slabs. A penetrative convection scenario, with relatively little deformation of the slab can be invoked to accommodate the duality of slab behavior (e.g. Silver et al 1988), but this requires that the Earth system resides within a fairly restrictive range of conditions. If the evidence for a bulk difference in chemistry between the upper and lower mantle holds up, a modified version of the penetrative convection scenario—in which a significant broadening of deep slab material is included—may still be the best option. Horizontal deflections of all slabs in a rigorously stratified model is difficult to reconcile with the evidence for tabular lower-mantle velocity anomalies down-dip of several slab trajectories. While several-hundred-kilometer deflections of a chemical discontinuity are expected, there is evidence for much deeper slab-like structures which cannot be readily explained, unless one invokes downwellings in a thermally coupled convection system. Horizontally deflected slabs are intrinsically more difficult to detect than steeply dipping extensions, but given the large cumulative volumes of subducted material, it is surprising that more dramatic transition zone anomalies near subduction zones are not apparent if this scenario is correct. Improved tomography of the transition zone and mapping of mantle discontinuity topography can contribute significantly to this problem.

Clearly, improved seismic images of deep slab structure, additional experimental work on chemical contrasts between the upper and lower mantles, and three-dimensional spherical convection calculations with phase transitions and temperature-dependent viscosity (to better incorporate slab properties) are all needed before there will be any consensus on the fate of descending slabs.

ACKNOWLEDGMENTS

I thank Rob van der Hilst for providing Figure 3. Quentin Williams, Charles Ammon, and Kristine Eckhardt provided helpful comments on the manuscript. My research on subducting slabs is supported by NSF Grant EAR9104764, with facilities support provided by the W. M. Keck Foundation. This is contribution number 215 of the Institute of Tectonics and C. F. Richter Seismological Laboratory.

Literature Cited

Aki K, Christoffersson A, Husebye ES. 1977. Determination of the three-dimensional seismic structure of the lithosphere. *J. Geophys. Res.* 82: 277–96

Aki K, Lee W. 1976. Determination of three-dimensional velocity anomalies under a seismic array using first P arrival times from local earthquakes. *J. Geophys. Res.* 81: 4381–99

Anderson DL. 1987a. A seismic equation of state. II. Shear properties and thermodynamics of the lower mantle. *Phys. Earth Planet. Inter.* 45: 307–23

Anderson DL. 1987b. Thermally induced phase changes, lateral heterogeneity of the mantle, continental roots and deep slab anomalies. *J. Geophys. Res.* 92: 13,968–80

Anderson DL. 1988. Temperature and pressure derivatives of elastic constants with application to the mantle. *J. Geophys. Res.* 93: 4688–700

Apperson KD, Frohlich C. 1987. The relationship between Wadati-Benioff zone geometry and P, T and B axes of intermediate and deep focus earthquakes. *J. Geophys. Res.* 92: 13,821–31

Beck SL, Lay T. 1986. Test of the lower mantle slab penetration hypothesis using broadband S waves. *Geophys. Res. Lett.* 13: 1007–10

Bock G, Ha J. 1984. Short period S-P conversion in the mantle at a depth near 700 km. *Geophys. J. R. Astron. Soc.* 77: 593–615

Boyd TM, Creager KC. 1991. The geometry

of Aleutian subduction: three-dimensional seismic imaging. *J. Geophys. Res.* 96: 2267–91

Burbach GV, Frohlich C. 1986. Intermediate and deep seismicity and lateral structure of subducted lithosphere in the circum-Pacific region. *Rev. Geophys.* 24: 833–74

Burnley PC, Green HW II, Prior DJ. 1991. Faulting associated with the olivine to spinel transformation in Mg_2GeO_4 and its implications for deep-focus earthquakes. *J. Geophys. Res.* 96: 425–43

Chiu J-M, Isacks BL, Carwell RK. 1991. 3-D configuration of subducted lithosphere in the western Pacific. *Geophys. J. Int.* 106: 99–111

Chopelas A, Boehler R. 1989. Thermal expansion measurements at very high pressure, systematics and a case for a chemically homogeneous mantle. *Geophys. Res. Lett.* 16: 1347–50

Choy GL, Cormier VF. 1986. Direct measurement of the mantle attenuation operator from broadband P and S waveforms. *J. Geophys. Res.* 91: 7326–42

Christensen U. 1982. Phase boundaries in finite amplitude mantle convection. *Geophys. J. R. Astron. Soc.* 68: 487–97

Christensen UR, Yuen DA. 1984. The interaction of a subducting lithospheric slab with a chemical or phase boundary. *J. Geophy. Res.* 89: 4389–402

Christensen UR, Yuen DA. 1985. Layered convection induced by phase transitions. *J. Geophys. Res.* 90: 10,291–300

Cormier VF. 1989. Slab diffraction of S waves. *J. Geophys. Res.* 94: 3006–24

Creager KC, Boyd TM. 1991. The geometry of Aleutian subduction: three-dimensional kinematic flow model. *J. Geophys. Res.* 96: 2293–307

Creager KC, Boyd TM. 1992. Effects of earthquake mislocation on estimates of velocity structure. *Phys. Earth Planet. Sci.* 75: 63–76

Creager KC, Jordan TH. 1984. Slab penetration into the lower mantle. *J. Geophys. Res.* 89: 3031–49

Creager KC, Jordan TH. 1986. Slab penetration into the lower mantle beneath the Mariana and other island arcs of the northwest Pacific. *J. Geophys. Res.* 91: 3573–89

Davies D, McKenzie DP. 1969. Seismic travel time residuals and plates. *Geophys. J. R. Astron. Soc.* 18: 51–63

Ding X, Grand S. 1992. Slab related velocity anomaly below Kurile subduction zone. *Eos, Trans. Am. Geophys. Union* 73: 385 (Abstr.)

Dziewonski AM. 1984. Mapping the lower mantle: determination of lateral heterogeneity in P velocity up to degree and order 6. *J. Geophys. Res.* 89: 5929–52

Dziewonski AM, Woodhouse JH. 1987. Global images of the Earth's interior. *Science* 236: 37–48

Ekström, G., Dziewonski AM, Ibanez J, 1990. Deep earthquakes outside slabs. *Eos, Trans. Am. Geophys. Union* 71: 1462 (Abstr.)

Engdahl ER, Gubbins D. 1987. Simultaneous travel time inversion for earthquake location and subduction zone structure in the central Aleutian islands. *J. Geophys. Res.* 92: 13,855–62

Engdahl ER, Kind R. 1986. Interpretation of broad-band seismograms from central Aleutian earthquakes. *Ann. Geophys.* 4: 233–40

Engdahl ER, Vidale JE, Cormier VF. 1988. Wave propagation in subducted lithospheric slabs. *Proc. 6th course: Digital Seismology and Fine Modeling of the Lithosphere, Int. School of Appl. Geophys., Majorana Center, Erice, Sicily*, pp. 139–55. New York: Plenum

Engebretson D, Kirby S. 1992. Deep Nazca slab seismicity: why is it so anomalous? *Eos, Trans. Am. Geophys. Union* 73: 379 (Abstr.)

Fischer KM. 1990. Waveforms from slab earthquakes: deep slab structure in the Kurils and Java. *Eos, Trans. Am. Geophys. Union* 71: 1460 (Abstr.)

Fischer KM, Creager KC, Jordan TH. 1991. Mapping the Tonga slab. *J. Geophys. Res.* 96: 14,403–27

Fischer KM, Jordan TH. 1991. Seismic strain rate and deep slab deformation in Tonga. *J. Geophys. Res.* 96: 14,429–44

Fischer KM, Jordan TH, Creager KC. 1988. Seismic constraints on the morphology of deep slabs. *J. Geophys. Res.* 93: 4773–83

Forte AM, Dziewonski AM, Woodward RL. 1992. Aspherical structure of the mantle, tectonic plate motions, nonhydrostatic geoid, and topography of the core-mantle boundary. In *Dynamics of the Earth's Deep Interior and Earth Rotation*, ed. JL LeMouël, AGU Geodyn. Ser. In press

Fukao Y, Obayashi. M., Inoue H, Nenbai M. 1992. Subducting slabs stagnant in the mantle transition zone. *J. Geophys. Res.* 97: 4809–22

Fukao Y, Yamaoka K, Sakurai T. 1987. Spherical shell tectonics: buckling of the subducted lithosphere. *Phys. Earth Planet. Inter.* 45: 59–67

Funamori N, Yagi T. 1993. High pressure and high temperature in situ x-ray observation of $MgSiO_3$ perovskite under lower mantle conditions. *Geophys. Res. Lett.* 20: 387–90

Gaherty JB, Hager BH. 1993. Compositional vs. thermal buoyancy and the evolution of subducted lithosphere. *Geophys. Res. Lett.* Submitted

Gaherty J, Lay T, Vidale JE. 1991. Investigation of deep slab structure using long period S waves. *J. Geophys. Res.* 96: 16,349–67

Giardini D. 1992. Space-time distribution of deep seismic deformation in Tonga. *Phys. Earth Planet. Inter.* 74: 75–88

Giardini D, Woodhouse JH. 1984. Deep seismicity and modes of deformation in Tonga subduction zone. *Nature* 307: 505–9

Giardini D, Woodhouse JH. 1986. Horizontal shear flow in the mantle beneath the Tonga arc. *Nature* 319: 551–55

Glennon MA, Chen W-P. 1993. Systematics of deep-focus earthquakes along the Kuril-Kamchatka arc and their implications on mantle dynamics. *J. Geophys. Res.* 98: 735–69

Goto K, Hamaguchi H, Suzuki Z. 1983. Distribution of stress in descending plate in special reference to intermediate and deep focus earthquakes, I. Characteristics of thermal stress distribution. *Tohoku Geophys. J.* 29: 81–105

Goto K, Hamaguchi H, Suzuki Z. 1985. Earthquake generating stresses in a descending slab. *Tectonophysics* 112: 111–28

Goto K, Suzuki Z, Hamaguchi H. 1987. Stress distribution due to olivine-spinel phase transition in descending plate and deep focus earthquakes. *J. Geophys. Res.* 92: 13,811–20

Grand SP. 1987. Tomographic inversion for shear velocity beneath the North American plate. *J. Geophys. Res.* 92: 14,065–90

Grand SP. 1990. A possible station bias in travel time measurements reported to ISC. *Geophys. Res. Lett.* 17: 17–20

Grand SP, Engebretson DC. 1992. The relation between subduction history and lower mantle heterogeneity. *Eos, Trans. Am. Geophys. Union* 73: 385 (Abstr.)

Green HW II, Burnley PC. 1989. A new, self-organizing mechanism for deep-focus earthquakes. *Nature* 341: 733–37

Green HW II, Scholz CH, Tingle TN, Young TE, Koczynski TA. 1992a. Acoustic emissions produced by anticrack faulting during the olivine $\rightarrow$ spinel transformation. *Geophys. Res. Lett.* 19: 789–92

Green HW II, Young TE, Walker D, Scholz CH. 1990. Anticrack-associated faulting at very high pressure in natural olivine. *Nature* 348: 720–22

Green HW II, Young TE, Walker D, Scholz CH. 1992b. The effect of nonhydrostatic stress on the $\alpha \rightarrow \beta$ and $\alpha \rightarrow \gamma$ olivine phase transformations. *Geophys. Monogr.* In press

Gubbins D, Snieder R. 1991. Dispersion of P waves in subducted lithosphere: evidence for an eclogite layer. *J. Geophys. Res.* 96: 6321–33

Gudmundsson O, Davies JH, Clayton RW. 1990. Stochastic analysis of global traveltime data: Mantle heterogeneity and random errors in the ISC data. *Geophys. J. Int.* 102: 25–43

Gurnis M, Hager BH. 1988. Controls on the structure of subducted slabs. *Nature* 335: 317–21

Hager BH, Clayton R. 1989. Constraints on the structure of mantle convection using seismic observations, flow models, and the geoid. In *Mantle Convection: Plate Tectonics and Global Dynamics*, ed. WR Peltier, pp. 657–763. New York: Gordon and Breach

Hager BH, Clayton RW, Richards MA, Comer RP, Dziewonski AM. 1985. Lower mantle heterogeneity, dynamic topography and the geoid. *Nature* 313: 541–45

Hager BH, Richards MA. 1989. Long-wavelength variations in Earth's geoid: physical models and dynamical implications. *Phil. Trans. R. Soc. London Ser. A* 328: 309–27

Hamburger MH, Isacks BL. 1987. Deep earthquakes in the southwest Pacific: a tectonic interpretation. *J. Geophys. Res.* 92: 13,841–54

Hirahara K. 1977. A large-scale three-dimensional seismic structure under the Japan islands and the Sea of Japan. *J. Phys. Earth* 25: 393–417

Honda S, Yuen DA, Balachandar S, Reuteler D. 1993. Three-dimensional instabilities of mantle convection with multiple phase transitions. *Science* 259: 1308–11

Iidaka T, Suetsugu D. 1992. Seismological evidence for metastable olivine inside a subducting slab. *Nature* 356: 593–95

Inoue H, Fukao Y, Tanabe K, Ogata Y. 1990. Whole mantle P wave travel time tomography. *Phys. Earth Planet. Inter.* 59: 294–328

Isacks B, Molnar P. 1971. Distribution of stresses in the descending lithosphere from a global survey of focal mechanism solutions of mantle earthquakes. *Rev. Geophys. Space Phys.* 9: 103–74

Isacks B, Oliver J, Sykes LR. 1968. Seismology and the new global tectonics. *J. Geophys. Res.* 73: 5855–99

Ito E, Akaogi M, Topor L, Navrotsky A.

1990. Negative pressure-temperature slopes for reactions forming $MgSiO_3$ perovskite from calorimetry. *Science* 249: 1275–78

Ito E, Takahashi E. 1989. Postspinel transforms in the system Mg_2SiO_4-Fe_2SiO_4 and some geophysical implications. *J. Geophys. Res.* 94: 10,637–46

Ito E, Yamada H. 1982. Stability relations of silicate spinels, ilmenites, and perovskites. In *High Pressure Research in Geophysics*, ed. S Akimoto, MH Manghani, pp. 405–19. Tokyo: Center for Acad. Publ.

Jeanloz R. 1991. Effects of phase transitions and possible compositional changes on the seismological structure near 650 km depth. *Geophys. Res. Lett.* 18: 1743–46

Jeanloz R, Knittle E. 1989. Density and composition of the lower mantle. *Phil. Trans. R. Soc. London Ser. A* 328: 377–89

Jordan TH. 1977. Lithospheric slab penetration into the lower mantle beneath the Sea of Okhotsk. *J. Geophys.* 43: 473–96

Jordan TH, Lerner-Lam AL, Creager KC. 1989. Seismic imaging of mantle convection: the evidence for deep circulation. In *Mantle Convection: Plate Tectonics and Global Dynamics*, ed. WR Peltier, pp. 87–201. New York: Gordon and Breach

Jordan TH, Lynn WS. 1974. A velocity anomaly in the lower mantle. *J. Geophys. Res.* 79: 2679–85

Jordan TH, Puster P, Glatzmaier GA, Tackley PJ. 1993. Comparisons between seismic earth structures and mantle flow models based on radial correlation functions. *Science* 261: 1427–31

Kamiya S, Miyatake T, Hirahara K. 1988. How deep can we see the high velocity anomalies beneath the Japan Islands? *Geophys. Res. Lett.* 15: 828–31

Kamiya S, Miyatake T, Hirahara K. 1989. Three dimensional P wave velocity structure beneath the Japanese Islands. *Bull. Earthq. Res. Inst., Univ. Tokyo* 64: 457–85

Kendall J-M, Thomson CJ. 1993. Seismic modeling of subduction zones with inhomogeneity and anisotropy—I. Teleseismic P-wavefront tracking. *Geophys. J. Int.* 112: 39–66

Kincaid C, Olson P. 1987. An experimental study of subduction and slab migration. *J. Geophys. Res.* 92: 13,832–40

Kirby SH, Durhan WB, Stern LA. 1991. Mantle phase changes and deep earthquake faulting in subducting lithosphere. *Science* 252: 216–25

Knittle E, Jeanloz R, Smith GL. 1986. The thermal expansion of silicate perovskite and stratification of the Earth's mantle. *Nature* 319: 214–16

Lay T. 1983. Localized velocity anomalies in the lower mantle. *Geophys. J. R. Astron. Soc.* 72: 483–516

Lay T. 1993. Seismological constraints on the velocity structure and fate of subducting lithospheric slabs: 25 years of progress. *Adv. Geophys.* In press

Lay T, Young CJ. 1989. Waveform complexity in teleseismic broadband SH displacements: slab diffractions or deep mantle reflections? *Geophys. Res. Lett.* 16: 605–8

Ligdas CN, Main IG, Adams RD. 1990. 3-D structure of the lithosphere in the Aegean region. *Geophys. J. Int.* 102: 219–29

Liu M, Yuen DA, Zhao W, Honda S. 1991. Development of diapiric structures in the upper mantle due to phase transitions. *Science* 252: 1836–39

Lundgren PR, Giardini D. 1992a. Seismicity, shear failure and modes of deformation in deep subduction zones. *Phys. Earth Planet. Inter.* 74: 63–74

Lundgren PR, Giardini D. 1992b. Isolated deep earthquakes and the fate of subduction in the mantle. *Eos, Trans. Am. Geophys. Union* 73: 386 (Abstr.)

Machetel P, Weber P. 1991. Intermittent layered convection in a model mantle with an endothermic phase change at 670 km. *Nature* 350: 55–57

Mao HK, Hemley RJ, Fei Y, Shu JF, Chen LC, et al. 1991. Effect of pressure, temperature and composition on lattice parameters and density of (Mg,Fe)SiO_3-perovskite to 30 GPa. *J. Geophys. Res.* 96: 8069–79

Meade C, Jeanloz R. 1991. Deep focus earthquakes and recycling of water into the Earth's mantle. *Science* 252: 68–72

Nenbai M. 1990. *How deep does a lithospheric slab descend into the mantle?* MS thesis. Nagoya Univ.

Okano K, Suetsugu D. 1992. Search for lower mantle high velocity zones beneath the deepest Kurile and Mariana earthquakes. *Geophys. Res. Lett.* 19: 745–48

Okino K, Ando M, Kaneshima S, Hirahara K. 1989. The horizontally lying slab. *Geophys. Res. Lett.* 16: 1059–62

Olson P, Silver PG, Carlson RW. 1990. The large-scale structure of convection in the Earth's mantle. *Nature* 344: 209–15

Parise JB, Wang Y, Yeganeh-Haeri A, Cox DE, Fei Y. 1990. Crystal structure and thermal expansion of (Mg,Fe)SiO_3 perovskite. *Geophys. Res. Lett.* 17: 2089–92

Peltier WR, Solheim LP. 1992. Mantle phase transitions and layered chaotic convection. *Geophys. Res. Lett.* 19: 321–24

Puspito NT, Yamanaka Y, Miyatake T, Shimazuki K, Hirahara K. 1993. Three-dimensional P wave velocity structure beneath the Indonesian regin. *Tectonophysics* 220: 175–92

Revenaugh J, Jordan TH. 1991. Mantle layering from ScS reverberations 2. The transition zone. *J. Geophys. Res.* 96: 19,763–80

Richards MA, Davies GE. 1989. On the separation of relatively buoyant components from subducted lithosphere. *Geophys. Res. Lett.* 16: 831–43

Richards MA, Engebretson DC. 1992. Large-scale mantle convection and the history of subduction. *Science* 355: 437–40

Richards MA, Wicks CW. 1990. S-P conversion from the transition zone beneath Tonga and the nature of the 670 km discontinuity. *Geophys. J. Int.* 101: 1–35

Richter FM. 1979. Focal mechanisms and seismic energy release of deep and intermediate earthquakes in the Tonga-Kermadec region and their bearing on the depth extent of mantle flow. *J. Geophys. Res.* 84: 6783–95

Ringwood AE, Irifune T. 1988. Nature of the 650 km seismic discontinuity: implications for mantle dynamics and differentiation. *Nature* 331: 131–36

Ross NL, Hazen RM. 1989. Single-crystal x-ray diffraction study of $MgSiO_3$ perovskite from 77 to 400 K. *Phys. Chem. Miner.* 4: 299–305

Schwartz SY, Lay T, Beck SL. 1991a. Shear wave travel time, amplitude, and waveform analysis for earthquakes in the Kurile slab: constraints on deep slab structure and mantle heterogeneity. *J. Geophys. Res.* 96: 14,445–60

Schwartz SY, Lay T, Grand S. 1991b. Seismic imaging of subducted slabs: trade-offs with deep path and near-receiver effects. *Geophys. Res. Lett.* 18: 1265–68

Scrivner C, Anderson DL. 1992. The effect of post Pangea subduction on global mantle tomography and convection. *Geophys. Res. Lett.* 19: 1053–56

Sekiguchi S. 1992. Amplitude distribution of seismic waves for laterally heterogeneous structures including a subducting slab. *Geophys. J. Int.* 111: 448–64

Shearer PM. 1991. Constraints on upper mantle discontinuities from observations of long-period reflected and converted phases. *J. Geophys. Res.* 96: 18,147–82

Shearer PM, Masters TG. 1992. Global mapping of topography on the 660-km discontinuity. *Nature* 355: 791–96

Shiono K, Sugi N. 1985. Life of an oceanic plate: Cooling time and assimilation time. *Tectonophysics* 112: 35–50

Silver P, Carlson RW, Olson P. 1988. Deep slabs, geochemical heterogeneity and the large-scale structure of mantle convection: investigation of an enduring paradox. *Annu. Rev. Earth Planet. Sci.* 16: 477–541

Silver PG, Chan WW. 1986. Observations of body wave multipathing from broadband seismograms: evidence for lower mantle slab penetration beneath the Sea of Okhotsk. *J. Geophys. Res.* 91: 13,787–802

Sleep NH. 1973. Teleseismic P-wave transmission through slabs. *Bull. Seismol. Soc. Am.* 63: 1349–73

Solheim LP, Peltier WR. 1993. Avalanche effects in phase transition modulated thermal convection: a model of the Earth's mantle. *J. Geophys. Res.* In press

Spakman W. 1991. Delay time tomography of the upper mantle below Europe, the Mediterranean, and Asia Minor. *Geophys. J. Int.* 107: 309–32

Spakman W, Nolet G. 1988. Imaging algorithms, accuracy and resolution in delay time tomography. In *Mathematical Geophysics: A Survey of Recent Developments in Seismology and Geodynamics*, ed. NJ Vlaar, pp. 155–88. Dordrecht: Reidel

Spakman W, Stein S, van der Hilst R, Wortel R. 1989. Resolution experiments for NW Pacific subduction zone tomography. *Geophys. Res. Lett.* 16: 1097–100

Spakman W, van der Lee S, van der Hilst R. 1993. Travel-time tomography of the European-Mediterranean mantle down to 1400 km. *Phys. Earth Planet. Inter.* 79: 3–74

Spakman W, Wortel MJR, Vlaar NJ. 1988. The Hellenic subduction zone: a tomographic image and its geodynamic implications. *Geophys. Res. Lett.* 15: 60–63

Stark PB, Frohlich C. 1985. The depth of the deepest deep earthquakes. *J. Geophys. Res.* 90: 1859–69

Steinbach V, Yuen DA. 1992. The effects of multiple phase transitions on Venusian mantle convection. *Geophys. Res. Lett.* 19: 2243–46

Stixrude L, Hemley RJ, Fei Y, Mao HK. 1992. Thermoelasticity of silicate perovskite and magnesiowüstite and stratification of the earth's mantle. *Science* 257: 1099–101

Su W-J, Dziewonski AM. 1991. Predominance of long-wavelength heterogeneity in the mantle. *Nature* 352: 121–26

Su W-J, Dziewonski AM. 1992. On the scale of mantle heterogeneity. *Phys. Earth Planet. Inter.* 74: 29–54

Suetsugu D. 1989. Lower mantle high velocity zone beneath the Kurils as inferred from P wave travel time and amplitude data. *J. Phys. Earth* 37: 265–95

Sung CM. 1979. Kinetics of the olivine-spinel transition under high pressure and temperature: experimental results and geophysical implications. In *High-Pressure Science and Technology*, ed. KD Timmerhaus, MS Barber, pp. 31–41. New York: Plenum

Sung CM, Burns RG. 1976. Kinetics of high pressure phase transformations: implications to the evolution of the olivine-spinel transition in the downgoing lithosphere and its consequences on the dynamics of the mantle. *Tectonophysics* 31: 1–32

Tackley PJ, Stevenson DJ, Glatzmaier G, Schubert G. 1993. Effects of an endothermic phase transition at 670 km depth on spherical mantle convection. *Nature* 361: 699–704

Takei Y, Suetsugu D. 1989. A high-velocity zone in the lower mantle under the Japan subduction zone inferred from precise measurements of P-wave arrival times. *J. Phys. Earth* 37: 225–31

Toksöz, N., Minear JW, Julian BR. 1971. Temperature field and geophysical effects of a downgoing slab. *J. Geophys. Res.* 76: 1113–38

van der Hilst R, Engdahl ER. 1991. On the use of pP and PP in delay time tomography. *Geophys. J. Int.* 106: 169–88

van der Hilst R, Engdahl ER. 1992. Stepwise relocation of ISC earthquake hypocenters for linearized tomographic imaging of slab structure. *Phys. Earth Planet. Inter.* 75: 39–54

van der Hilst RD, Engdahl ER, Spakman W. 1992. What can we learn about deep slab structure and mantle dynamics from tomographic images? *Eos, Trans. Am. Geophys. Union* 73: 385 (Abstr.)

van der Hilst R, Engdahl ER, Spakman W. 1993. Tomographic inversion of P and pP data for aspherical mantle structure below the northwest Pacific region. *Geophys. J. Int.* 115: 264–302

van der Hilst R, Engdahl ER, Spakman W, Nolet G. 1991. Tomographic imaging of subducted lithosphere below northwest Pacific island arcs. *Nature* 353: 37–43

van der Hilst R, Spakman W. 1987. A tomographic image of the Lesser Antilles subduction zone. *Eos, Trans. Am. Geophys. Union* 68: 1376 (Abstr.)

van der Hilst RD, Spakman W. 1989. Importance of the reference model in linearized tomography and images of subduction below the Caribbean plate. *Geophys. Res. Lett.* 16: 1093–96

Vassiliou MS. 1984. The state of stress in subducting slabs as revealed by earthquakes analyzed by moment tensor inversion. *Earth Planet. Sci. Lett.* 69: 195–202

Vassiliou MS, Hager BH. 1988. Subduction zone earthquakes and stress in slabs. *Pageoph.* 128: 574–624

Vassiliou MS, Hager BH, Raefsky A. 1984. The distribution of earthquakes with depth and stress in subducting slabs. *J. Geodyn.* 1: 11–28

Vidale JE. 1987. Waveform effects of a high velocity, subducted slab. *Geophys. Res. Lett.* 14: 542–45

Vidale JE, Benz HM. 1992. Upper-mantle seismic discontinuities and the thermal structure of subduction zones. *Nature* 356: 678–83

Vidale JE, Garcia-Gonzalez D. 1988. Seismic observation of a high velocity slab 1200–1600 km in depth. *Geophys. Res. Lett.* 15: 369–72

Vidale JE, Williams Q, Houston H. 1991. Waveform effects of a metastable olivine tongue in subducting slabs. *Geophys. Res. Lett.* 18: 2201–04

Wang Y, Weidner DJ, Liebermann RC, Liu X, Ko J, et al. 1991. Phase transition and thermal expansion of $MgSiO_3$ perovskite. *Science* 251: 410–13

Weber M. 1990. Subduction zones—their influence on traveltimes and amplitudes of P waves. *Geophys. J. Int.* 101: 529–44

Weinstein SA. 1993. Catastrophic overturn of the earth's mantle driven by multiple phase changes and internal heat generation. *Geophys. Res. Lett.* 20: 101–4

Wicks CW Jr, Richards MA. 1991. Effects of source radiation patterns on the phase $S_{670}P$ beneath the Tonga subduction zone. *Geophys. J. Int.* 107: 279–90

Wicks CW Jr, Richards MA. 1993. A detailed map of the 660 km discontinuity beneath the Izu-Bonin subduction zone. *Science* 261: 1424–27

Witte D. 1989. *The pseudospectral method for simulating wave propagation.* PhD dissertation. Columbia Univ., New York

Yamanaka Y, Miyatake T, Hirahara K. 1992. Fingering and lower mantle penetration of the Kurile slab. Preprint

Yamaoka K, Fukao Y, Kumazawa M. 1986. Spherical shell tectonics: effects of sphericity and inextensibility on the geometry of the descending lithosphere. *Rev. Geophys.* 24: 27–55

Zhao W, Yuen DA, Honda S. 1992. Multiple phase transitions and the style of mantle convection. *Phys. Earth Planet. Inter.* 72: 185–210

Zhou H-W. 1988. How well can we resolve the deep seismic slab with seismic tomography? *Geophys. Res. Lett.* 15: 1425–28

Zhou H-W. 1990a. Mapping of P wave slab anomalies beneath the Tonga, Kermadec and New Hebrides arcs. *Phys. Earth Planet. Inter.* 61: 199–229

Zhou H-W. 1990b. Observations on earthquake stress axes and seismic morphology of deep slabs. *Geophys. J. Int.* 103: 377–401

Zhou H-W, Anderson DL. 1989a. Search for deep slabs in the northwest Pacific mantle. *Proc. Natl. Acad. Sci., USA* 86: 8602–6

Zhou H-W, Anderson DL. 1989b. Teleseismic contributions to focal residual spheres and Tangshen earthquake sequence. *Eos, Trans. Am. Geophys. Union* 70: 1322 (Abstr.)

Zhou H-W, Anderson DL, Clayton RW. 1990. Modeling of residual spheres for subduction zone earthquakes, 1. Apparent slab penetration signatures in the NW Pacific caused by deep diffuse mantle anomalies. *J. Geophys Res.* 95: 6799–827

Zhou H-W, Clayton RW. 1990. P and S wave travel time inversions for subducting slab under the island arcs of the northwest Pacific. *J. Geophys. Res.* 95: 6829–51

Annu. Rev. Earth Planet. Sci. 1994. 22:63–91

CLADISTICS AND THE FOSSIL RECORD: The Uses of History

Kevin Padian, David R. Lindberg, and Paul David Polly

Department of Integrative Biology and Museum of Paleontology, University of California, Berkeley, California 94720

KEY WORDS: systematics, paleontology, diversity, biogeography

INTRODUCTION

Cladistics, or phylogenetic systematics, was developed by the German entomologist Willi Hennig in the 1940s and 1950s and culminated in his book *Grundzuge einer Theorie der Systematik* (1950) and a later manuscript translated by D. Dwight Davis and Rainer Zangerl in 1966 as *Phylogenetic Systematics*. Though at first uninfluential outside a group of largely European entomologists, its rediscovery and promulgation in the 1970s by American ichthyologist Gareth Nelson and his colleagues eventually swept the major centers of biological taxonomy and systematics in North America and Europe [Hull (1988) reviews the history and sociology of this movement]. Its initial success was in the fields of fossil and recent fishes, notably in museums in New York, London, and Paris; but it soon spread to the other vertebrates, the invertebrates, plants, and unicellular forms, and has now become the dominant paradigm for systematic methodology.

The basic premises of cladistics are that only strictly monophyletic groups (i.e. those comprising *all* the descendants of a common ancestor) should be recognized and used, and that these groups (called *clades*) must be diagnosed by new evolutionary features (synapomorphies) that evolved in the common ancestor and were passed to its descendants. The nested patterns in which these synapomorphies are shared by different taxa are used to group them in an order of evolutionary descent and relationship. Classification becomes an objective exercise determined by the order of branching.

This apparently simple procedure has profound implications for the

0084-6597/94/0515-0063$05.00

studies of phylogeny (evolutionary relationships), taxonomy (the basis on which organisms are recognized and sorted), and classification (the ranking of taxa, traditionally in hierarchical but not necessarily parallel Linnean groups). For evolutionary biologists, cladistics also forces a specific phylogenetic structure upon typical questions posed about the evolution of function, physiology, adaptation, and behavior, as well as ecology (Brooks & McLennan 1990, Harvey & Pagel 1991). For paleontologists and geologists, a complementary set of questions and implications has arisen, and these form the focus of our review.

We begin with a brief summary of the principles and methods of cladistics, and discuss how these differ from traditional ones. We then show some implications of new cladistic analyses for paleontology and geology. First, phylogenetic relationships of many fossil groups have been elucidated or altered, and accordingly classifications have changed. For example, it is no longer possible (and it never was accurate) to speak of "fishes giving rise to amphibians, which gave rise to reptiles, which gave rise to mammals and birds." Second, we provide some examples of paleobiological questions for which cladistics has provided new insights, a different perspective, or a more powerful test or explanatory framework. These include questions about diversity through time, rates of taxonomic change and turnover, problems of origination and extinction of groups, calibration of molecular evolutionary trees, the evolution of adaptations, and trends in ecological and biogeographic change. Third, and most importantly, we develop some ideas for questions and fields in which cladistics can alter traditional paradigms and provide new ways of thinking about old problems in paleobiology and macroevolution.

PRINCIPLES AND METHODS OF CLADISTICS

Several texts and standard references (Eldredge & Cracraft 1980, Wiley 1981, Wiley et al 1991) explain in detail the basics of cladistics, and more advanced methods, computer programs, and philosophical questions are discussed in leading journals such as *Cladistics*, *Systematic Biology* (formerly *Systematic Zoology*), *Systematic Botany*, and the *Annual Review of Ecology and Systematics*. Here we give only the barest of outlines using the simplest vocabulary possible.

A basic principle of evolution is that organisms are descended from their parents with heritable modifications that are then further passed on and altered in future generations. This implies that a particular modification seen in a given organism and its descendants, but not in its ancestors and relatives, is a character that has been introduced into the evolutionary pattern at a given point. Such characters, shared by organisms that derived them from a common ancestor, are called *synapomorphies* (shared derived characters). The common possession of synapomorphies is the criterion used to unite taxa into *monophyletic* groups—the Holy Grail of cladistics. To be monophyletic, a group must include an ancestor and *all* its descendants. If it includes only some descendants, it is not monophyletic, but paraphyletic (Figure 1). For example, the earliest known bird, *Archaeopteryx*, shares dozens of synapomorphies with small carnivorous dinosaurs (Ostrom 1976; Padian 1982; Gauthier 1984, 1986). Because birds evolved from dinosaurs, which are reptiles, birds are cladistically considered members of the Reptilia, and "reptiles" are not monophyletic unless they include birds. Because monophyletic groups contain all the descendants of an ancestor, they cannot be said to "give rise" to any descendant groups. When monophyletic groups are found to be linked by synapomorphies, they are called *sister groups*. Among living reptiles, birds are the sister group of crocodiles, united with them in the Archosauria (Gauthier 1984, 1986).

In cladistics, taxa are *diagnosed* by synapomorphies, and *defined* by ancestry (Rowe 1987). Because any two organisms must share a common ancestor at some point, however remote in evolutionary history, they and all descendants of that common ancestor are defined cladistically to form a monophyletic taxon. We use synapomorphies to diagnose monophyletic groups, which we hypothesize share these features because they have inherited them from a common ancestor. It follows that all characters must be synapomorphies at some taxonomic level. Feathers are unique to birds, and so help to diagnose them as a group; but the possession of feathers cannot tell us the difference between a scrub jay and a Steller's

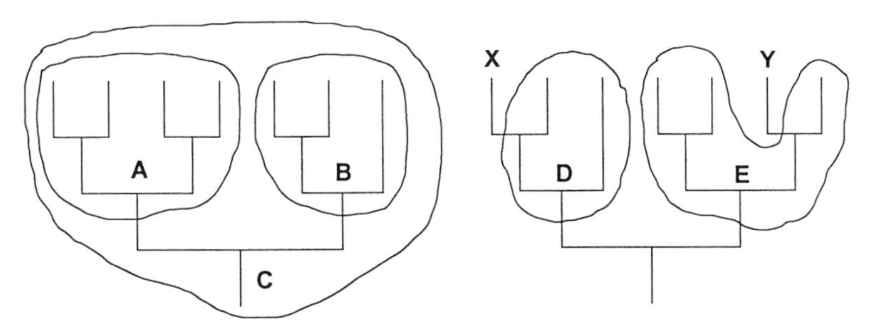

Figure 1 In the diagram at left, groups A, B, and C are monophyletic, because they include *all* the descendants of a common ancestor. In the diagram at right, groups D and E are paraphyletic (not monophyletic), because they exclude some descendants (X and Y) of their common ancestor.

blue jay: The character is too general. However, specific *features* of the feathers (character states) may indeed differentiate the two species. Hence every character is derived at some level, but general (original or "primitive") at another.

A cladistic analysis of a group of taxa is generally begun by selecting characters with varying states that can be analyzed to determine which state is ancestral or original for the group and which is (or are) derived. This is called *polarization* of characters, and it is usually done with reference to the characters of one or more outgroups (Maddison et al 1984). These data are arranged in a matrix in which the rows and columns represent taxa and characters. Cells are coded with numbers or symbols representing the character states of each character for each taxon. The codes can be used to signify general and derived states. Computer programs such as HENNIG86 (Farris 1988), PAUP (Swofford 1991), PHYLIP (Felsenstein 1991), and MacCLADE (Maddison & Maddison 1992) group taxa successively according to the distribution of the synapomorphies that they share, and link succeedingly remote groups to these until all are linked. Many programs help to deal with conflicting and unknown character information—subjects that are beyond our scope here [but see Wiley et al (1991) for further details and an introduction to the primary literature].

According to its proponents, the principal advantage of the cladistic system over others is that it is unrelentingly evolutionary; that is, it is fundamentally concerned with the branching order of evolution. Its insistence on monophyletic groups means that many familiar and long-hallowed taxa must be completely reorganized or eliminated entirely. To its detractors, cladistics unnecessarily eliminates traditionally useful and often ecologically or morphologically cohesive groups in the name of monophyly, but the benefits are not worth the losses. Rather than discuss that question here, we choose to show some examples of how cladistic methods have changed our understanding of the evolution of life and of important events in its history. We do not pretend that cladistics is a panacea for all paleontological questions, but we think that the test is as much in the pragmatics as in the ideology.

SOME NEW CLADISTIC PHYLOGENIES AND THEIR IMPLICATIONS

We noted above that one major implication of cladistics is that its insistence on monophyly tends to invalidate many long-familiar groups, or restructure them radically. Some major examples follow. These are based on de Queiroz & Gauthier's (1990, 1992) terminology from their phylogenetic taxonomy of craniates.

Cambrian Arthropods

Significant evolutionary changes in the fossil record traditionally have been reflected by expansions of taxonomic rank. The appearance of new phyla or other higher taxa in the fossil record indisputably focuses attention on events in the history of life. Sudden radiations such as the "Cambrian explosion," or the presence of taxa that are not readily included in living groups (e.g. the Ediacaran and some components of the Burgess Shale), are typically the subject of tremendous speculation and scenarios. These fossil problematica are often assessed in terms of either their morphological or taxonomic differences from living taxa, but important phylogenetic questions are seldom based on an explicit framework of relationships.

A recent exception to this has been the work of Briggs & Fortey (1989) and Briggs et al (1992), who have addressed the question of Cambrian arthropod phylogeny and whether or not disparity was greater in the Cambrian than now (see Gould 1991, Foote & Gould 1992). Although Gould (1989) clearly distinguished between taxonomic diversity and morphological disparity, his conclusion that morphological disparity was greater in the Burgess Shale arthropods than in living groups was not tested phylogenetically (see also Gould 1991). Subsequent phylogenetic analyses by Briggs & Fortey (1989) and Briggs et al (1992) failed to support greater disparity. Their analyses suggested that extinct Cambrian taxa dominate one arthropod subclade while the remaining taxa are scattered among living groups (Figure 2). Traditional overemphasis of unique character combinations (*autapomorphies*), combined with the use of paraphyletic grades such as the "Trilobitomorphs," have obscured detailed relationships and quantifications of disparity.

Monoplacophorans

Monoplacophorans are another Cambrian group whose evolutionary history needs to be reinterpreted by cladistic analysis. These cap-shaped Early Paleozoic molluscs were well-studied by Knight (1952), and his reconstructions have proven remarkably accurate. Formerly known only from Cambrian-Silurian deposits, monoplacophorans were recovered alive and well in the deep sea in 1952. Because of the antiquity of the group and its unique molluscan morphology, it was immediately christened "a living fossil" and incorporated into molluscan phylogeny as the putative ancestor of the living molluscan classes. However, without a rigorous cladistic analysis it has not been possible to sort out primitive from derived characters; nor has there been sufficient consideration of which autapomorphies might be misinterpreted as ancestral characters.

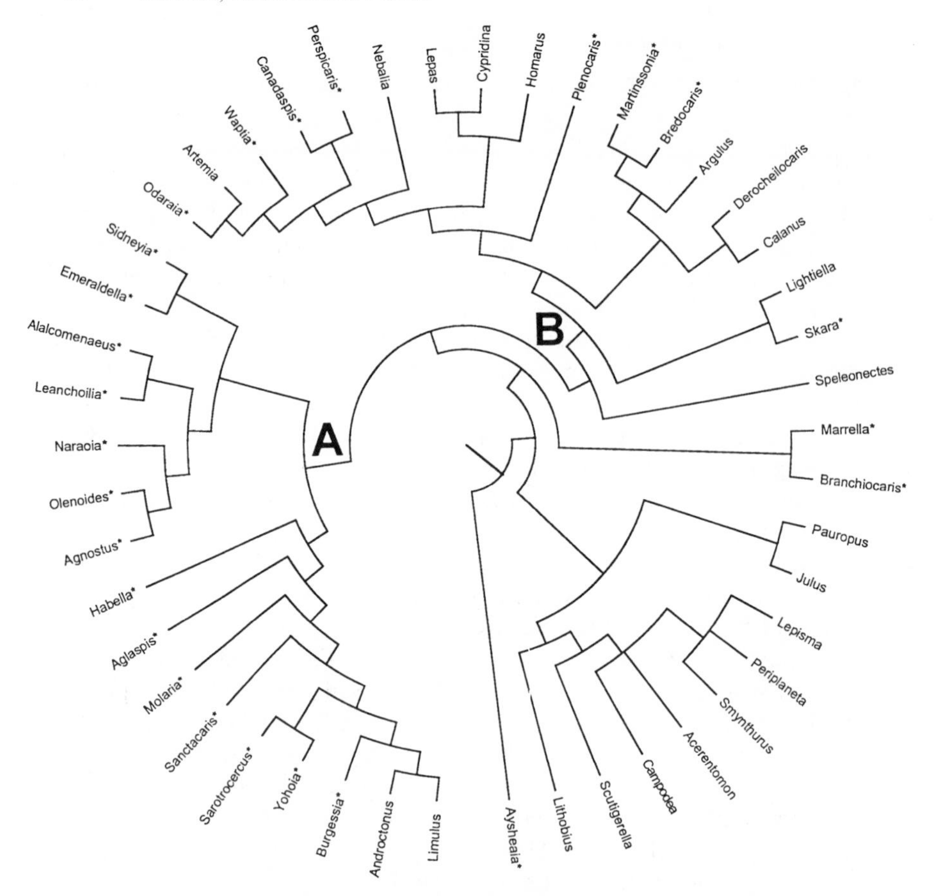

Figure 2 Cladogram of Cambrian and living arthropods. Cambrian arthropods are marked with an asterisk. In subclade A 87.5% of the taxa are extinct Cambrian taxa; in subclade B only 44.4% of the taxa are extinct. Differential extinction rates in different subclades exacerbate patterns of disparity. After Briggs et al (1992).

Although paleontologists, geologists and zoologists do not dispute that monoplacophoran habitats have changed considerably through geologic time, there has been little discussion of the potential for correlated anatomical changes. Cladistic analysis provides a framework against which to evaluate character distributions and transformations in the Monoplacophora. These analyses (Stuber & Lindberg 1989) advise that components of the radulae, shell structure, nervous system, excretory system, buccal mass, and other features all display the autapomorphic condition:

Their character states are unique to the monoplacophorans and not primitive for the Mollusca. It appears that monoplacophorans did not experience descent without modification for the last 350 million years as previously thought. Based on a cladistic analysis they are highly modified animals with numerous autapomorphies, and possible character reversals. Thus, the use of monoplacophorans as "primitive" taxa to polarize characters in the other molluscan classes is a dangerous practice.

Fishes

It has long been held that tetrapods evolved from some aquatic form that would normally be called a "fish." Hence in the phylogenetic system of cladistics, "fishes" must be paraphyletic, unless all vertebrates are called "fishes." But we already have a word for this: vertebrates. The closest living forms to tetrapods are the lungfishes, and they and the tetrapods together form a clade called Choanates (because they all have internal choanae, or nostrils that open inward). The coelacanth joins the choanates as the Sarcopterygia ("fleshy-fins"). These join the rayfinned fish, which includes nearly all the forms usually considered "fishes," in a group called Osteichthyes (bony fish). The sharks and their allies (Chondrichthyes, or cartilage fish) join the Osteichthyes in a group called Gnathostomi (jawed mouths), and they are thereby separated from other vertebrates that lack jaws, the lampreys. All these groups together form a clade called Vertebrata, because they have vertebrae; the hagfishes do not, but they do have at least a rudimentary cranium, so they and the vertebrates together form the Craniata.

Figure 3 demonstrates this branching sequence, together with the new names of groups in the phylogenetic system and some of the synapomorphies that diagnose them. We suggest that the name "Pisces" (colloquially, "fishes"), if it be used at all, be restricted to Actinopterygii (the living rayfins) and all the extinct bony fishes (Osteichthyes) closer to them than to Sarcopterygii. This would simplify and clarify the use of the term "fishes" in both technical and common senses, inasmuch as the vast majority of non-tetrapod craniate species ("fishes") are actinopterygians, and their biology differs in fundamental ways from those of the other non-tetrapods. It can be seen readily that, whether or not this usage is adopted, "fishes" cannot have given rise to tetrapods. Rather, tetrapods evolved from within a group of choanate sarcopterygians—fleshy-finned forms with internal choanae, neither of which features are shared by rayfins. This is less confusing because it obviates trying to figure out how a trout or perch could have given rise to a tetrapod, what sorts of features it would have had to lose or gain, how the ecological transition was made, and so on. Instead, the rayfins are seen for what they are, a sidebranch of vertebrate evolution quite distinct from the sidebranch including tetrapods.

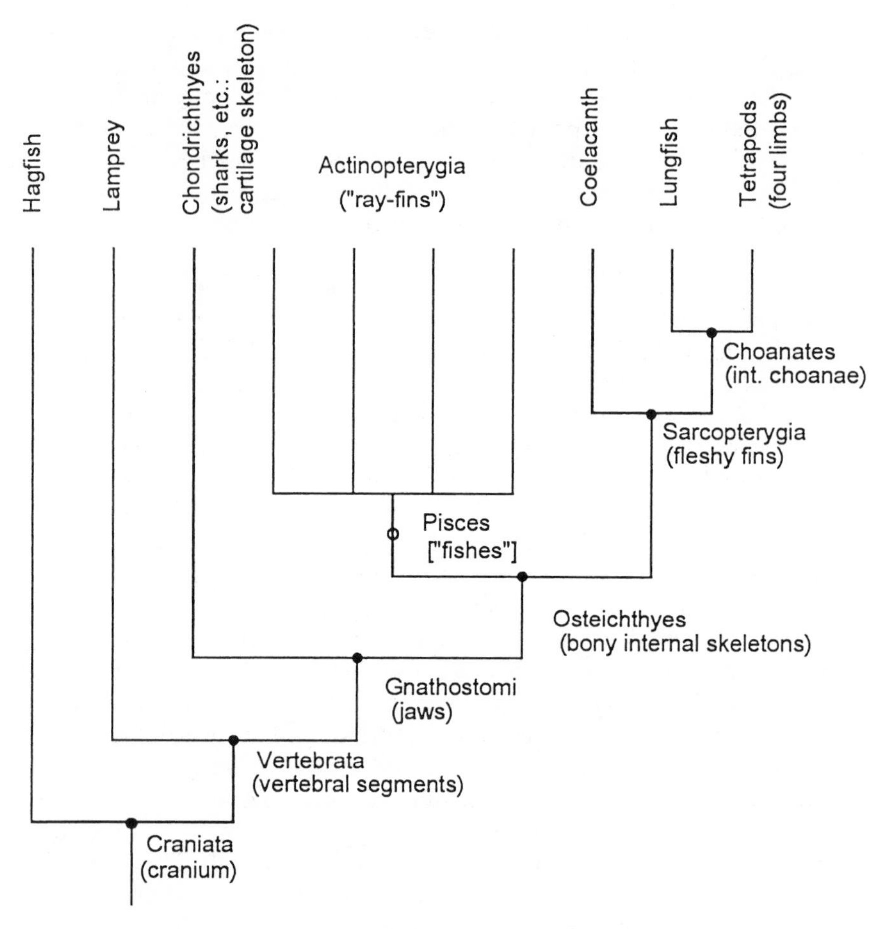

Figure 3 Tree of the basal vertebrate groups, including some major features of their evolution (synapomorphies, in parentheses), based on cladistic analyses. After de Queiroz & Gauthier (1992) and other references.

Tetrapods

Similarly, the tetrapods have been restructured and renamed according to cladistic analysis (Gaffney 1980; de Queiroz & Gauthier 1992; Benton 1988; Novacek & Wyss 1986; Estes & Pregill 1988; Prothero & Schoch 1989; Gauthier 1984, 1986; Gauthier et al 1988a,b, 1989). A recent summary (after de Queiroz & Gauthier 1992) is diagrammed in Figure 4.

Amphibians are rightly seen as a sidebranch of tetrapod evolution, not as an amorphous basal mass of tetrapods out of which reptiles evolved. Amphibians are restricted to the living groups and all extinct forms closer to them than to living reptiles. Amphibians, in this sense, share features such as a broad flat skull, a double occipital condyle, the loss of a finger, and vertebrae in which the intercentra dominate the pleurocentra. All these features would have to be lost or reversed for amphibians to be the "ancestors" of reptiles, as traditionally taught.

In fact paleontologists have recognized for some years now (Panchen 1980, Heaton 1980, Milner 1988, Panchen & Smithson 1988, Gauthier et al 1989) that a collection of Carboniferous tetrapods called "anthracosaurs" (from their discovery in coal deposits) are less closely related to proper amphibians than they are to *Seymouria*, *Diadectes*, and other forms traditionally considered "reptiles" or near-reptiles. Placing them all in a clade

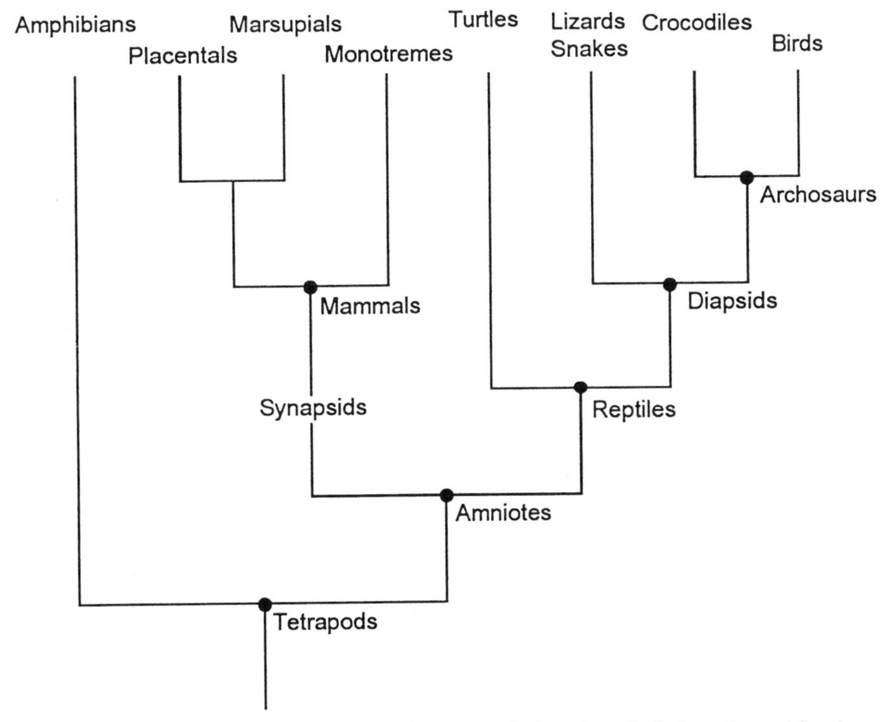

Figure 4 Tree of the major groups of living tetrapods, based on cladistic analyses. After de Queiroz & Gauthier (1992) and other references.

called Anthracosauria emphasizes this shared ancestry. However, we no longer consider that this group gave rise to "reptiles." Reptiles are one branch of a more inclusive clade called Amniota, named for the amniote egg that improves nourishment, resists desiccation, and frees the life cycle from an aquatic or humid stage necessary to keep the egg from drying out. Amniotes evolved from anthracosaurs that were largely terrestrial, with improved limbs and vertebrae for locomotion and a lengthened, more mobile neck for feeding.

The first amniotes diversified quickly during the later Carboniferous into several groups that appeared almost simultaneously in the fossil record (Carroll 1988, Gauthier et al 1988b). One group is distinguished from the others by a single opening in the temporal region (the side of the skull behind the eye); animals to this point had no such openings. This "one-holed" group is called the Synapsida, and its members include the familiar fin-backed *Dimetrodon*, the two-tusked dicynodonts, and the mammals. Another group evolved enlarged suborbital openings in the bottom of the skull, and a suite of other features; these are the Reptilia proper, which include turtles, lizards, crocodiles, and so on. Within this group a substantial Permo-Carboniferous diversification (Carroll 1988) was succeeded by differential extinctions from which two major groups now survive. One group retained the no-holed temporal region, lost teeth, and evolved a shell; these are the turtles (Chelonia). Another group evolved not just one, but two temporal openings; these are called diapsids, and they contain the remaining reptiles. Their relationships were cladistically sorted by Gauthier and co-workers (Gauthier 1984, 1986; Gauthier & Padian 1985; Gauthier et al 1988a,b,c) and substantial progress continues in this group (Sereno 1986, 1991; Benton & Clark 1988; Evans 1988; Rieppel 1988; Cracraft 1988; Frost & Etheridge 1989; de Queiroz 1987). Perhaps of greatest interest is that the conclusion that birds evolved from small theropod (carnivorous) dinosaurs in the mid- or Late Jurassic has been firmly cemented by over 200 synapomorphies, thanks to subsequent cladistic analyses (Gauthier 1984, 1986; Gauthier & Padian 1985; Cracraft 1988; Sereno & Rao 1992; Perle et al 1993). This phylogenetic grounding provides a strong basis for approaching more process-oriented questions, such as how feathers and flight evolved, as we discuss below.

With new cladistic analyses such as these, the long-familiar and perpetually confusing term "mammal-like reptile" has now gone by the boards. This misnomer arose from the conception that the earliest amniotes were reptiles. As cladistic analysis shows, living reptile groups and their extinct relatives did not appear until well after their split with the group leading to mammals. Hence the ancestors of mammals never were reptiles, but were amniotes, like the ancestors of reptiles. To call these forms "mammal-like reptiles" is as nonsensical as calling the ancestors of turtles and diapsids "reptile-like mammals." One implication of this revision is that no living reptile can be considered a model for "primitive" amniote structure, function, physiology, or ecology. By studying characteristics commonly shared by living reptiles and mammals, and tracing the evolution of these characters in the extinct relatives of both branches, we can arrive at a more reasonable picture of what the first amniotes were like (Heaton 1980, Gauthier et al 1988a,b).

Mammals themselves have been subjected to extensive phylogenetic analysis using morphological evidence from living and fossil forms, as well as molecular evidence (Novacek 1992; papers in Benton 1988), though the congruence between morphology and molecules is not always ideal (Patterson 1987, McKenna 1987, Novacek 1992). Among fossil mammals, much work has been done within the major orders of mammals (e.g. Prothero & Schoch 1989, Benton 1988), as well as with determining the relationships of mammalian orders to each other. Much of this work is inconclusive because phylogenetic analyses fail to produce a clear branching pattern: Some taxa are too transformed to relate easily to a nearest known neighbor; there is rampant convergence in dental, cranial, or post-cranial features; crucial fossil taxa are missing parts or missing entirely; molecules have been found to be as plastic and potentially misleading as anatomy when it comes to indicating phylogenetic pathways. Nonetheless, some clear signals are emerging. Some mammalian groups, such as Condylarthra and Insectivora, have been broken up or greatly restricted in membership, because some of their members have been found to be closer to other groups, or because they were held together as a group only by primitive characters, not shared derived ones. Insectivora is generally abandoned in favor of the use of the several component taxa, such as Lipotyphla and Menotyphla, that can be constituted as monophyletic groups (Novacek & Wyss 1986, Novacek 1992). The fossil Creodonta, long thought to contain the ancestors of Carnivora, now turn out to be an entirely different group with no ancestral role (P. D. Polly, in preparation). Both Carnivora and Ungulata (hoofed mammals) were traditionally thought to have evolved from early Tertiary Condylarthra, but this latter group as traditionally constituted contains many forms lumped together only by primitive features. The hoofed mammals and some other taxa (sirenians, tubulidentate anteaters, and whales) appear to form a natural group, and the most primitive known fossil whales have limbs with a mesaxonic (even-toed) pattern seen in artiodactyl ungulates, as well as some early Tertiary forms usually included in the Condylarthra. Carnivorans and creodonts, traditionally grouped together, now appear more closely related to different mammalian radiations.

IMPLICATIONS OF CLADISTICS FOR PALEONTOLOGY AND PALEOBIOLOGY

Cladistics and Stratigraphy

As any geologist can readily appreciate, fine-scaled biostratigraphy depends on fine-scaled taxonomic identifications. Where fossils are abundant, and the record of index fossils is good, as in the records of most planktonic marine organisms used in correlation, identifications can normally be made to the species level. However, for deposits in which the fossil record is poorer, the specimens rarer, and the representation largely incomplete, as in many or most terrestrial environments (Padian & Clemens 1985), correlation at the species level may not be possible. It is therefore all the more important that biostratigraphic correlations that are forced to use supraspecific taxonomic levels be restricted to closely analyzed monophyletic groups (Padian 1989a). Paraphyletic groups may suggest both unnaturally short and long ranges, which can suggest spurious correlations. Revising paraphyletic groups and forming monophyletic groups from them frequently changes substantially our picture of taxonomic diversity across major boundaries, for example those associated with mass extinctions (Padian & Clemens 1985; Benton 1988, 1989a,b).

For example, consider a paraphyletic group "A" (Figure 5), which comprises the taxa P, Q, R, and S, but not T. The stratigraphic ranges of the individual taxa in "A" are not congruent: P, Q, and R span various ranges in time period 1, while S and T are found in both periods 1 and 2. In traditional usage, the range of taxon "A" (P+Q+R+S) is hardly distinguishable from that of T, because both "groups" are found in both time periods. But now consider the cladogram in Figure 5. S is the sister taxon of T, which suggests that S has been traditionally included in "A" on the basis of primitive features, not synapomorphies (shared derived features). In fact, no features characterize group "A" that are not shared by T. "A" therefore has an artificially extended range. Using stratigraphy, we see that no members of "A" except S survive into time period 2. Cladistics tells us that this is logical, because S and T are sister groups. If S is included in the paraphyletic group "A," it is less useful for stratigraphic purposes. The survival of S into time period 2 does not imply anything more; but if S is considered a member of the spurious group "A," the stratigraphic picture is muddied.

Other direct inferences about stratigraphic and temporal ranges can be made on the basis of cladistic analysis. For example, if X and Y are sister groups, and a member of group X is found in the Oligocene, then group Y must also have been present by the Oligocene, even if we have no fossils of its members (Figure 6). Norell (1992) has discussed these "ghost

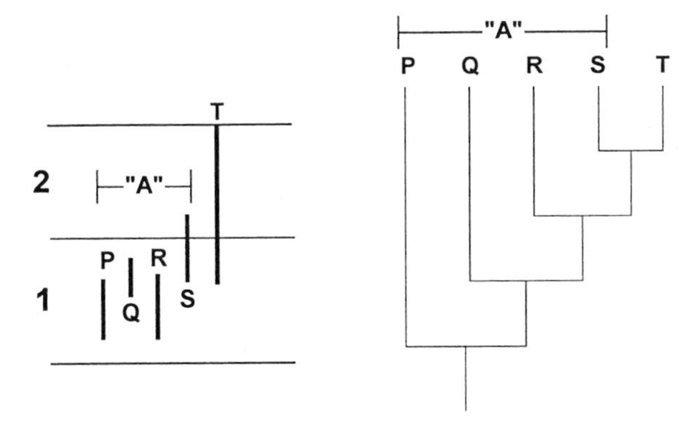

Figure 5 How paraphyletic groups confound biostratigraphic refinement. (*Left*) The stratigraphic ranges of the individual taxa in A are not congruent: P, Q, and R span various ranges in time period 1, while S and T are found in both periods 1 and 2. Thus, the range of taxon "A" (P+Q+R+S) is hardly distingushable from that of T, because both "groups" are found in both time periods. (*Right*) However, when a cladistic approach is taken, S is the sister taxon of T, which suggests that T has been traditionally included in "A" on the basis of primitive features, not synapomorphies (shared derived features). In fact, no features characterize group "A" that are not shared by T. "A" therefore has an artificially extended range.

lineages," and suggested further interpretations of the completeness of the fossil record of any group as implied by the relative degree of congruence of its preserved fossil record with its phylogeny. One implication of Norell's analysis is that we should expect to find the earliest records of many groups pushed back in time, as we determine that their sister groups were already extant. Norell (1992) notes that although the earliest records of most lizard

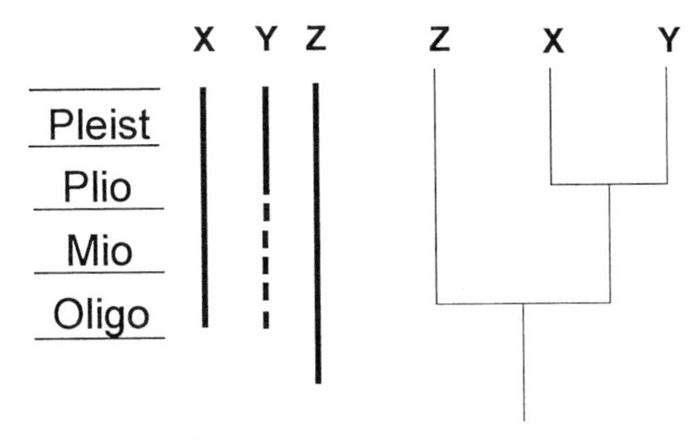

Figure 6 X and Y are sister taxa, and Z is their outgroup. The fossil record of X is older than that of Y, but the early presence of members of group X implies that members of group Y were also present then.

families do not predate the latest Cretaceous (Estes 1983), the lineages were separate by no later than the Early Cretaceous, according to cladistic analysis, and this has implications for the biogeographic spread of lizards (both before and after the breakup of Pangaea). Ghost lineages can be assumed to have existed, and in some cases can be counted as real, if unknown, taxa when dealing with some kinds of macroevolutionary problems. Much depends on the completeness of the stratigraphic record. Norell & Novacek (1992a,b) and Norell (1993) surveyed dozens of vertebrate groups, comparing known stratigraphic ranges with divergence patterns expected on the basis of cladistic phylogenies. They found highly variable patterns, depending as much on superpositional as on cladistic resolution (Figure 7). For example, if members of a group did not separate well stratigraphically, the phylogeny could not be expected to be recovered in this sequence. As Norell & Novacek (1992b) note, "a good fossil record is one that can be empirically demonstrated to contain evolutionary pattern, as opposed to information that only reflects preservational bias." Testing the fossil record against cladistic phylogenies determines how complete the record probably is, and therefore how useful the taxon might be for stratigraphic purposes.

Beyond these considerations is the understanding of how the use of paraphyletic versus monophyletic groups affects our perception of biotic

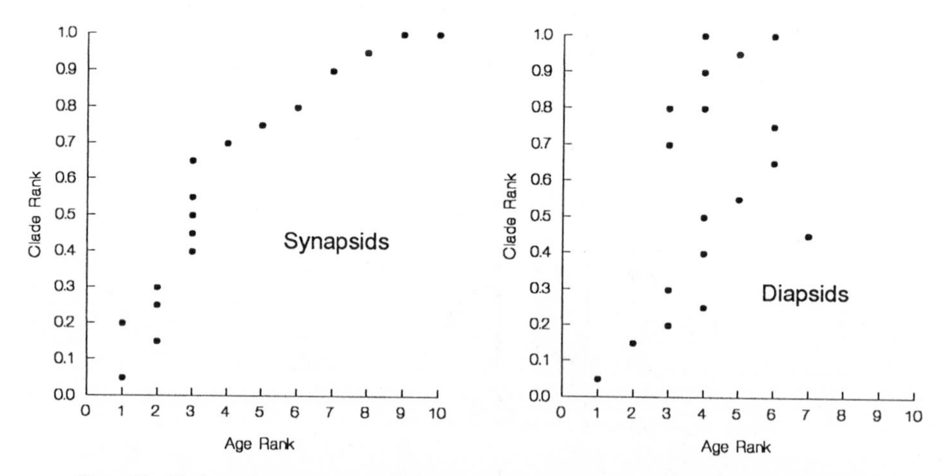

Figure 7 Clade rank versus age rank for the two major branches of amniotes—Synapsida and Diapsida. In this example the high degree of scatter in the diapsids indicates that the correlation between their phylogeny and the appearance of diapsid taxa in the fossil record is poor relative to the synapsids which show a higher correlation. After Norell & Novacek (1992b).

turnover. There seems to be general agreement among workers in all paleontological fields that traditional taxonomy does not reflect actual phylogenetic relationships very well. (If there are four "families" in an order, this says nothing about the relationship of these "families" to each other, or whether any or all are paraphyletic.) For some time, phylogeneticists have wondered whether the substantial portion of paraphyletic taxa in the fossil record could be giving misleading signals about the pace of diversity through time. Sepkoski & Kendrick (1991) modeled phylogenetic diversity with interesting results: They found that few changes were detectable in the major events of life through time when paraphyletic and monophyletic model clades were compared. But as one might expect, the fidelity of correspondence was higher at lower taxonomic levels. The signals for origination and extinction patterns were somewhat damped, compared to a paraphyletic pattern. Overall, however, Sepkoski & Kendrick found that traditional paraphyletic taxa can yield a meaningful signal, if sampling is sufficient. But these conclusions about model taxa have not been sustained by other workers studying actual data sets (Donoghue 1991, Smith & Patterson 1988, Archibald 1993, Edgecombe 1992).

An example of the importance of monophyletic groups to the assessment of patterns of biotic turnover through time can be readily provided by the history of tetrapods (Padian & Clemens 1985, Benton 1989a, Padian 1989b). As noted above in our first section, traditional classification holds that both mammals and birds evolved from (separate) "reptile" ancestors. Under this view, "reptiles" constitute a paraphyletic group. Cladistic revision of tetrapod groups, discussed above, instead focuses on the earliest amniotes, and restricts "reptiles" to living reptiles (including birds) and all taxa closer to them than to mammals. Under this view, the amniotes have two great branches, one leading to the living reptiles and birds (Reptilia or Sauropsida), and another leading to mammals and their relatives (Synapsida). This requires us to stop thinking about the extinct relatives of mammals as "reptiles." If we do so, and plot the respective histories of these clades through time (Figure 8), we find that there has been something of a relay race between the two, with the synapsids predominant in the Late Paleozoic, the reptiles coming into prominence in the Late Triassic and dominating until the Late Cretaceous, and the synapsids again (in the form of mammals) surging in the Tertiary, as do the reptiles once more (in the form of birds, plus the lizards, snakes, crocodiles, and turtles). This is substantially different from the old division between the "Age of Reptiles" (the Mesozoic) and the "Age of Mammals" (the Cenozoic), which is more a reflection of which land animals were among the largest or most conspicuous at a particular time.

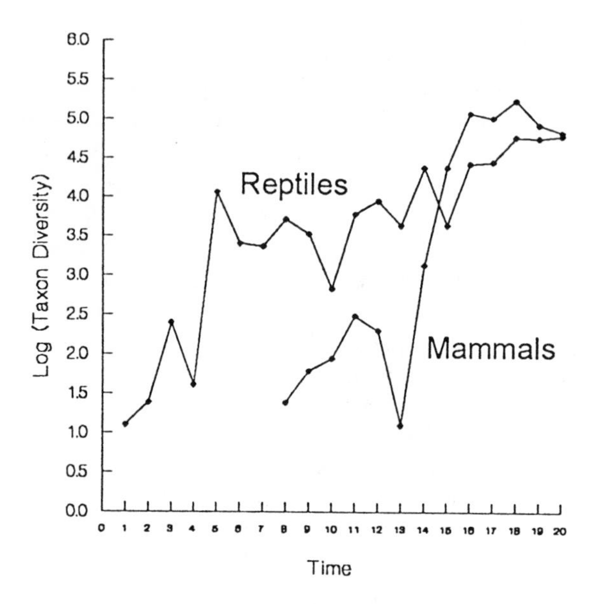
Reptiles
Mammals
Log (Taxon Diversity)
6.0
5.5
5.0
4.5
4.0
3.5
3.0
2.5
2.0
1.5
1.0
0.5
0.0
0 1 2 3 4 5 6 7 8 9 10 11 12 13 14 15 16 17 18 19 20
Time

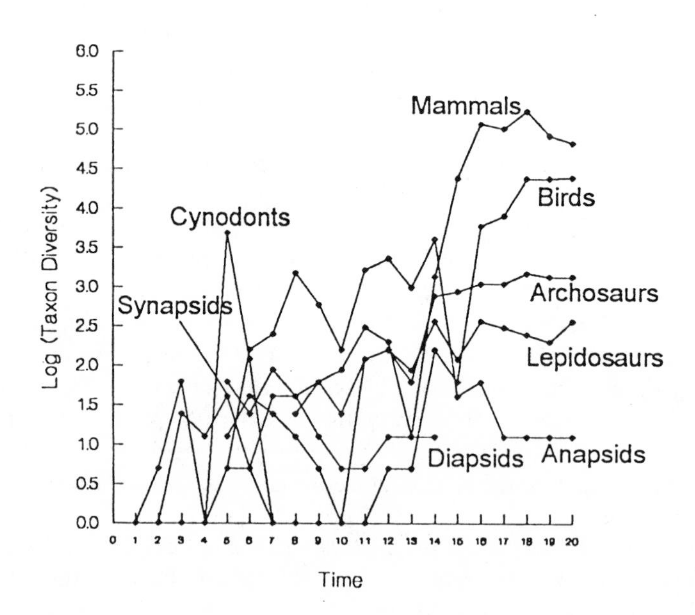
Mammals
Birds
Cynodonts
Archosaurs
Synapsids
Lepidosaurs
Diapsids
Anapsids
Log (Taxon Diversity)
6.0
5.5
5.0
4.5
4.0
3.5
3.0
2.5
2.0
1.5
1.0
0.5
0.0
0 1 2 3 4 5 6 7 8 9 10 11 12 13 14 15 16 17 18 19 20
Time

In protistan paleontology, stratigraphy and morphology provide the framework on which most systematic and evolutionary studies are done. This is not surprising given the relatively fine resolution that is attainable within samples. However, a strictly stratigraphic approach can lead to circular arguments regarding relationships. For example, the appearance of distinct morphological sequences through time, combined with the presence of possible morphological intergrades, is often interpreted as an evolutionary series connecting ancestral and derived taxa. How characters change is also determined from these sequences, and independent, non-stratigraphic evidence is wanting. It is neither sufficient nor robust to: (*a*) recognize a relationship between two morphologically similar taxa based on their sequence in the geological column, (*b*) describe the character transformations based on this sequence, and (*c*) then construct an evolutionary scenario to explain the change. While the protist fossil record is the best available, its resolution does not abrogate the need for independent assessment of the relationships among protist taxa.

Cladistic analyses of protists are beginning to appear in the literature (D'Hondt 1991, MacLeod 1988, MacLeod & Kitchell 1988), and recently these analyses have been used as hypotheses against which to test adaptational scenarios. MacLeod (1993) examined speciation events in the microfossil record using the Maastrichtian-Danian radiation of triserial and biserial planktic foraminifera. Although this event has been traditionally considered one of the best examples of adaptive radiation, MacLeod found that while there was congruence between patterns of morphological change and habitat shifts, speciation rates were higher than expected given the rates of habitat diversification. MacLeod therefore rejected the "vacant niche" model and considered the "taxon-pulse" model to be more plausible. Rigorous testing of these alternative models of

Figure 8 The top graph treats diversity through time (arbitrary units) by breaking amniotes into mammals and "reptiles," in which "reptiles" include all amniotes that aren't mammals (except birds, not pictured here). The bottom graph divides amniotes cladistically: Many former "reptiles" now group with the mammals in Synapsida, and birds are included in Reptilia. A "relay race" now becomes more apparent: Synapsids dominated the Permian and Early Triassic, but diapsids (mostly archosaurs) took over in the mid-Triassic and dominated until the Late Cretaceous extinctions. Turtles took up a small role in the Late Triassic and continue to the Recent; lizards and their relatives (lepidosaurs) maintained a low profile through the Mesozoic but lately their record has improved. Marine reptiles are mostly responsible for the Mesozoic diversity of diapsids apart from archosaurs and lepidosaurs. The synapsids resurged again in the Tertiary, but it cannot strictly be called the "Age of Mammals" because surviving reptiles, especially birds, constitute a considerable portion of Tertiary diversity.

speciation and adaptation were only possible because of the use of cladistic analysis.

Calibration of the Molecular Clock

One of the most topical applications of cladistics in paleontology is to the question of rates of evolution, particularly molecular evolution. After molecular sequence data became available, first for proteins and later for DNA, it was noticed that rates of sequence divergence were so regular (Zuckerkandl & Pauling 1962, 1965; Margoliash 1963) that they could be used to estimate the time of divergence between two lineages. The relative constancy of molecular sequence evolution contradicted some previous assessments of evolutionary rates based on morphology preserved in the fossil record. Morphology is known to evolve at varying rates, in part with varying selection pressures (Simpson 1944, 1953). Sarich & Wilson (1967a,b) used fossils to put an absolute time scale on serum albumin evolution in Primates. They found that the tree of Primates based on immunological distances among the albumins corresponded to the accepted morphological tree of the primate groups they were studying. They calibrated the time of divergence between anthropoids and Old World monkeys at about 30 million years ago, based on fossil evidence, and they proportionally estimated the probable time of divergence between humans and other apes at about 5 million years ago. This age was much younger than that estimated by paleontologists at the time, but fossil evidence later corroborated a more recent divergence between hominids and other apes, although not as late as 5 million years ago (Pilbeam 1978, Greenfield 1980).

Current debates on molecular clocks question the constancy and group-specificity of the evolution of particular genes. Calibrations of some molecular clocks have been based on a few poorly controlled divergence dates from the fossil record. The timing of the Old World Monkey/anthropoid split is still controversial (Simons 1990), the timings of eubacteria divergences used to calibrate 16S rRNA have wide margins of error (Wilson et al 1987), and divergence times of rodent groups are also based on sketchy phylogenetic positions of fossils (Catzeflis et al 1992). Some early molecular phylogenies set divergence points at the first appearance of a group member in the fossil record; so, for example, the crocodile-bird split was calibrated by the first known crocodile (Late Triassic, about 210 my ago). But cladistic analysis shows that the crocodile-bird split occurred at least 235 my ago (Middle Triassic)—a 10% difference in calibration. Similarly, the first turtles appear in the Late Triassic (about 210 million years ago), but cladistic analysis shows that their lineage must have split from the other reptiles (their sister group, the Diapsida) at least by the Middle Pennsylvanian (about 350 my ago)—a 50% difference in calibration. Fossil representatives of turtle sister groups are now beginning to fill this temporal gap (Gaffney 1980, Reisz & Laurin 1991).

To improve the utility of molecular clocks, comprehensive studies of extant taxonomic groups with excellent fossil records are necessary. The groups should have a variety of living representatives that have both shallow and deep branch-points in the group phylogeny, and there should be well dated fossil taxa associated with the nodes to provide minimum estimates of divergence times. Such data would provide a strict time control for comparing divergence rates amongst the lineages containing extant forms. Good examples of such groups include marine molluscs and mammalian carnivores, whose respectively long (about 550 my) and recent (about 50 my) fossil records could test slowly and rapidly evolving molecules, respectively.

The Evolution of Major Adaptations

Historically, the evolution of major adaptations in the history of life has been treated by looking for "key innovations" in functional complexes that frequently characterize a diverse group, such as gastropods or mammals. Frequently, this approach has centered on identifying such key features and then devising an adaptive scenario to explain their evolution in terms of orthodox Darwinian natural selection (e.g. Bock 1986). However, this logic is circular and so is inherently incapable of testing any hypotheses about evolution (Padian 1994). Cladistic approaches to phylogeny provide an independent test of such hypotheses, because the methods of cladistics naturally separate the clusters of adaptive characters into a hierarchical sequence of acquisition. So, for example, as noted above, cladistic analysis has strongly supported the hypothesis that birds evolved from small carnivorous dinosaurs (Ostrom 1976; Padian 1982; Gauthier 1984, 1986). What does this tell us about the origins of flight, a major adaptation?

The theropod dinosaurs from which birds evolved already had a hand reduced to the first three fingers, a foot reduced to the middle three toes (with the first offset to the side), a long, S-shaped neck, and gracile hindlimbs with a long shank and reduced fibula. In the lineage leading to birds, the forelimbs increased in length, the tail shortened and stiffened slightly, and eventually the bones of the pelvis began to shift backwards while the hands grew longer and began to flex sideways (Padian 1982, Gauthier 1984, Gauthier & Padian 1985, Padian 1987). This sequence of evolution of features is available as an independent test of how flight evolved. For example, the view that birds were initially arboreal climbers that glided and eventually took active flight finds no support in the record of phylogeny (cf

Bock 1986). On the contrary, the inferred habits of bird relatives are terrestrial, cursorial, and predatory, which suggests an origin of flight from the ground up, by incremental running leaps (Padian 1987, 1994). Hypotheses of functional origin should be kept separate from phylogenetic hypotheses, so that they may be tested independently (Padian 1982, 1987, 1994). In this way the sequence of assembly of major adaptive complexes may be dissected, and the pattern of evolution is made available to test hypotheses of evolutionary process.

Gastropod molluscs are one of the most diverse groups of invertebrate animals, both in form and habitat. They have figured prominently in paleontological and biological studies, and have served as study organisms in numerous evolutionary, biomechanical, ecological, physiological, and behavioral studies. The higher classification of gastropods has received scant attention for much of this century but recently there has been a comprehensive reconsideration of gastropod phylogeny (e.g. Graham 1985, Haszprunar 1988). An increasing awareness of the importance of phylogeny in the construction of evolutionary scenarios and explanations requires a rigorous analysis of the data, so that alternative theories and interpretations of molluscan evolution can be evaluated.

For more than 70 years, gastropods have been divided into a tripartite classification scheme that correlates the abundance of various members of these groups with the divisions of the Phanerozoic. Thus, in the Paleozoic the Archaeogastropoda are the predominant group, and the Mesozoic is the age of the Mesogastropoda; during the Cenozoic the Neogastropoda become dominant. While these taxonomic divisions are often treated as clades in terms of adaptive prowess, ecological diversity, and so on, they are clearly grades of evolution and not clades that share exclusive ancestors. Moreover, members of these groups are all present in the fossil record by the end of the Paleozoic, and it is only differential turnover in taxonomic rates that produce the stratigraphically-stepped patterns of the groups. Like the patterns seen in the vertebrates (Figure 8), the use of gastropod clades rather than grades produces a more complex and interesting history for the group.

This is not to assert that questions asked about "grades" are always noninformational. For example, there may be questions about herbivory in gastropods that make little difference if asked cladistically; but *within* "herbivores" the kind of herbivory used, or the multiple evolutions of radiations and specializations, can only be revealed by phylogenetic analysis. In other words, a phylogenetic hypothesis may not be needed to measure or describe a taxon, its features, performance, interactions, etc. However, when one begins talking about how it got that way, a phylogenetic hypothesis is implicitly necessary (see also O'Hara 1988).

Biogeography

In addition to temporal information, the fossil record also provides spatial patterns of taxa. Paleobiogeographic studies seek to document the changing distributions of animals and plants through time and correlate these changes with physical and biotic events and perturbations. Evolutionary scenarios to explain these disjunct distributions are moot if the "species pair" does not share a common ancestor. And what is meant by a region having "low generic diversity"? Are all genera created equal and therefore a comparable form of organismal currency? Clearly, these studies require independent assessment and demonstration of the phyletic relationships of the organisms.

Historically, changes in the distribution of organisms have been interpreted as the results of dispersal events (Forbes 1846, Darwin 1859, Depéret 1908, Matthew 1915). In the marine realm these changes were easily explained by the dispersal of pelagic larvae or rafted adults. Terrestrial taxa were more problematic: Fragmented distributions of taxa often required complex networks of land bridges between continents. The discovery and acceptance of plate tectonics in the late 1960s also sparked a revolution in biogeographic studies. Plants and animals now rode plates into new configurations rather than sauntered between them on mythical bridges. Plate tectonics provided a mechanism by which distributions could be both divided or connected. And again it was Gareth Nelson, and his colleagues, Donn Rosen and Norman Platnick at the American Museum of Natural History in New York, who championed the approach that has become known as vicariance biogeography (see Nelson & Platnick 1981).

Inherent in any biogeographic study are assumptions of relationship (e.g. gene flow among individuals of a population or common ancestry of taxa that compose a clade) (see Wiley 1981 for an overview). Whether a taxon has been divided or rafted into a new region via crustal plate segments or the water column, the "smoking gun" of such an event is the recognition of a relationship between the disjunct taxa. However, most taxa lack a modern phylogenetic treatment. Thus, sister taxa status (i.e. two taxa that share a common ancestor) has not been demonstrated for many of the classical examples cited in the literature. Instead of a phylogenetic hypotheses, most workers use the Linnean classification as a proxy for relatedness. However, there remains the danger of confusing common ancestry with convergence.

Although dispersal and vicariance are often cast as diametrically opposed processes, Hallam's (1981, p.340) point that vicariance and dispersal are "two sides of the same coin" appears to apply even within a single biotope. For example, while the closing of the Panamic portal produced a vicariance event that separated Caribbean and tropical eastern

Pacific marine organisms, the perturbations to tropical current patterns caused by the emerging isthmus appears to have facilitated the interchange of temperate marine organisms between the northeastern and southeastern Pacific Oceans (Lindberg 1991) as well as the great American vertebrate interchange (Stehli & Webb 1985).

Biogeographical patterns may be elucidated through the use of area cladograms. Like taxon cladograms, area cladograms also reflect relationships, albeit relationships between areas or regions. Figure 9 illustrates the relationships and the patterns of divergence between the major crustal plates. Taxa with relationships that are congruent with this area cladogram would not falsify a vicariance model of their biogeographic history. Moreover, the addition of fossils often provides critical resolution of area cladograms, especially in those situations where the taxon has become extinct in one or more areas.

Taxon and area cladograms, used in conjunction with fossil occurrences, can also provide support for the biogeographic history of dispersing marine invertebrates. Many marine gastropod and bivalve taxa of the northeastern Pacific Ocean have sister taxa with earlier geological occurrences in the northwestern Pacific (MacNeil 1965, Marincovich 1983, Vermeij 1991, and references therein). Some of these taxa are also present in the southeastern Pacific (Lindberg 1991). For several of these groups, taxon and area cladograms and the fossil record all support a dispersal history from the northwestern Pacific to the northeastern Pacific and subsequently into the southeastern Pacific (Figure 10).

Fossil horses are perhaps the most commonly used textbook story of evolution. They attracted the attention of early evolutionists because their history combined aspects of adaptation, morphological change, and biogeography in a single, familiar example demonstrating gradual evolu-

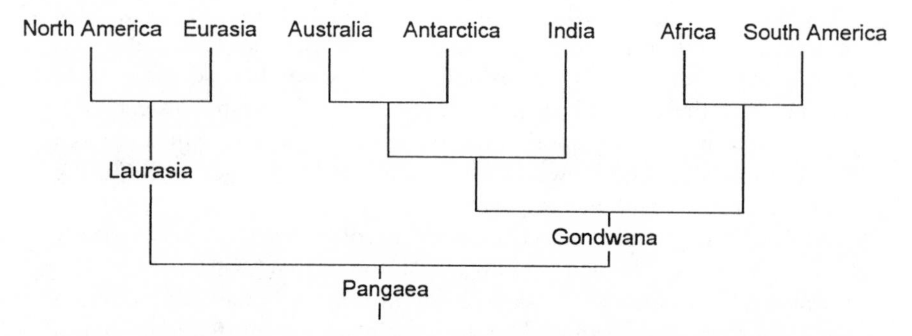

Figure 9 Area cladogram for the breaking up of Pangaea during the last 220 million years.

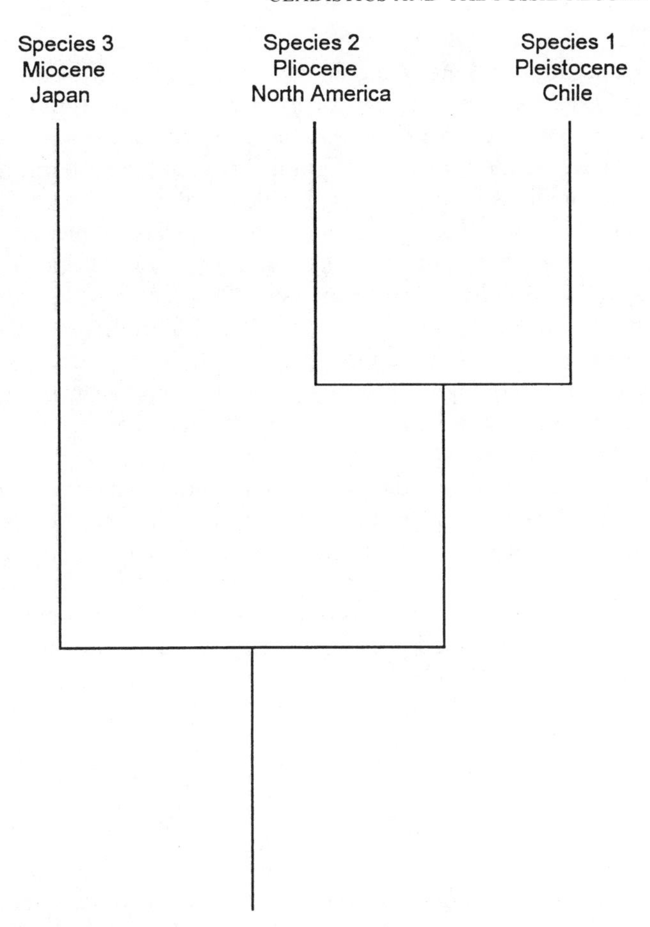

Figure 10 Congruence of area and taxon cladograms for the dispersal of a nearshore marine gastropod taxon from the northwestern Pacific → northeastern Pacific → southeastern Pacific during the Tertiary. The first fossil occurrence of the taxon in each region is also congruent with these relationships. See Lindberg (1991) for examples.

tionary change through time. They had a small, short-faced, five-toed ancestor with a low-crowned dentition suited for browsing, but as climates changed and grasslands appeared during the Miocene, the number of toes was gradually reduced and the teeth became higher crowned and more useful for grazing and grinding grass. Pre-cladistic analyses of horse phylogeny and biogeography indicated that horses originated in North America, with *Hyracotherium* ("Eohippus"), and that they dispersed on four separate occasions to the Old World via the Bering Bridge during the late Cenozoic and once to South America via Panama during the Pleistocene (Simpson 1951). Cladistic analyses of horse relationships (Hooker 1989; Hulbert 1989; MacFadden 1976, 1992) have shown that, while essentially correct, horse evolution and biogeography are considerably more complicated than previously thought. MacFadden (1992) argues that while the horse family (Equidae) may have originated in North America if their closest relatives were the Eocene Phenacodontidae, it is also possible that they were related to *Radinskya* from Asia. If the latter is the case, then equids must have immigrated to North America from Asia. Cladistic analyses also show that equids underwent a vicariant event (i.e. their group naturally divided geographically) in the mid-Eocene, when the DeGeer land bridge between North America and Scandanavia broke apart (MacFadden 1992). As a result, the true "equidae" were isolated in North America and the Palaeotheriidae were isolated in Eurasia. Following the rupture of the DeGeer bridge, the Bering bridge became the dominant route of dispersal for horses. By plotting the geographic and stratigraphic occurrences of equids on a cladogram of their relationships it is possible to reconstruct four major dispersals of equids out of North America. The early hyracotheres emigrated in the Eocene, the Anchitheres emigrated in the early Miocene, the Hipparions in the late Miocene, and the true horses in the Plio-Pleistocene (MacFadden 1992). Only by having a detailed knowledge of horse relationships is it possible to reconstruct the history of their migrations and evolution in all of its complexity.

CONCLUSIONS

Cladistics has revolutionized the way we look at historical biology, or what has traditionally been called "evolution." Cladistics elucidates the patterns of evolution by forcing scientists to state explicitly and hierarchically the evidence that appears to link biological taxa to other taxa. This method is far more exposed to direct evaluation than are the methods of traditional evolutionary biology. Scientists cannot simply invoke Darwin and hide behind assertions of adaptive value or necessary selective pressures; instead, they must first demonstrate the patterns shown by the characteristics of the organisms. Hypotheses about evolutionary processes can then be evaluated against the evolutionary patterns revealed by the organisms themselves.

We have tried to show how cladistic analysis of the pattern of biological history works as a starting point for the analysis of other patterns, such as biogeographic history, and processes, such as the evolution of major adaptive features. Cladistics is an endpoint in historical biology only when investigators are interested in nothing further than the most parsimonious evaluation of relationships among groups. While important, this function leaves unexplored the power of cladistic analysis to approach broader evolutionary problems.

In this equation, the role of fossils is complex and has sometimes been underestimated. Because fossils are necessarily less complete than living organisms, they preserve only a portion of their original information. Nonetheless, to ignore fossils, or to treat them simply as taxa that should be "added later" to historical analyses based on living forms, is a mistake, as several studies have shown (Donoghue et al 1989). Gauthier et al (1988a) showed that if certain early extinct taxa related to mammals are omitted from phylogenetic analysis, mammals and birds show up as sister groups (to the exclusion of crocodiles and other reptiles), or birds and crocodiles are the sister group of mammals (to the exclusion of turtles and lizards). Doyle & Donoghue (1987) showed that without including the extinct taxa *Caytonia* and the Bennettitales, some crucial features in angiosperm plant evolution could not be recovered accurately and the polarity of some characters would be reversed at certain hierarchical levels. From these examples it can easily be seen that cladistic analysis constrains certain kinds of scenarios in historical biology—not just of the evolution of characters and groups, but of their adaptive features, their biogeography, and the vicissitudes of their diversity through time.

ACKNOWLEDGMENTS

We thank Drs. J. W. Valentine and J. A. Gauthier for comments on the manuscript. This is contribution No. 1613 from the University of California Museum of Paleontology.

Literature Cited

Archibald JD. 1993. The importance of phylogenetic analysis for the assessment of species turnover: a case history of Paleocene mammals in North America. *Paleobiology* 19: 1–27

Benton MJ, ed. 1988. *Phylogeny and Classification of the Tetrapods*, Vols. 1, 2. Oxford: Clarendon. 377 pp. 329 pp.

Benton MJ. 1989a. Patterns of evolution and extinction in vertebrates. In *Evolution and the Fossil Record*, ed. KC Allen, DEG Briggs, pp. 218–41. London: Belhaven

Benton MJ. 1989b. Mass extinctions among tetrapods and the quality of the fossil record. *Phil. Trans. R. Soc. London Ser. B* 325: 369–86

Benton MJ, Clark JM. 1988. Archosaur phylogeny and the relationships of the Crocodylia. See Benton 1988, Vol. 1, pp. 289–332

Bock WJ. 1986. The arboreal origin of avian flight. In *The Origin of Birds and the Evolution of Flight*, ed. K Padian, pp. 57–82. *Mem. Calif. Acad. Sci.* 8

Briggs DEG, Fortey RA. 1989. The early radiation and relationships of the major Arthropod groups. *Science* 246: 241–43

Briggs DEG, Fortey RA, Wills MA. 1992. Morphological disparity in the Cambrian. *Science* 256: 1670–73

Brooks DJ, McLennan DA. 1990. *Phylogeny, Ecology, and Behavior: A Research Program in Comparative Biology*. Chicago: Univ. Chicago Press. 434 pp.

Carroll RL. 1988. *Vertebrate Paleontology and Evolution*. New York: Freeman. 698 pp.

Catzeflis FM, Aguilar J, Jaeger J-J. 1992. Muroid rodents, phylogeny and evolution. *Trends Ecol. Evol.* 7(4): 122–26

Cracraft J. 1988. The major clades of birds. See Benton 1988, Vol. 1, pp. 339–61

Darwin C. 1859. *On the Origin of Species by Natural Selection, or the Preservation of Favoured Races in the Struggle for Life*. London: John Murray

Depéret C. 1908. The evolution of Tertiary mammals, and the importance of their migrations. *Amer. N.* 42(494): 109–14, (495): 166–70, (497): 303–7

de Queiroz K. 1987. Phylogenetic systematics of iguanine lizards: a comparative study. *Univ. Calif. Publ. Zool.* 118: 1–203

de Queiroz K, Gauthier J. 1990. Phylogeny as a central principle in taxonomy: phylogenetic definitions of taxon names. *Syst. Zool.* 39: 307–22

de Queiroz K, Gauthier J. 1992. Phylogenetic taxonomy. *Annu. Rev. Ecol. Syst.* 23: 449–80

D'Hondt S. 1991. Phylogenetic and stratigraphic analysis of earliest Paleoeocene biserial and triserial planktonic foraminifera. *J. Foraminiferal Res.* 21: 168–81

Donoghue MJ. 1991. The use of phylogenies in studying diversification, with examples from plants. *Geol. Soc. Am. Abstr. Prog.* 23: A281 (Abstr.)

Donoghue MJ, Doyle JA, Gauthier J, Kluge AG, Rowe T. 1989. The importance of fossils in phylogeny reconstruction. *Annu. Rev. Ecol. Syst.* 20: 431–60

Doyle JA, Donoghue MJ. 1987. The importance of fossils in elucidating seed plant phylogeny and macroevolution. *Rev. Palaeobot. Palynol.* 53: 321–431

Edgecombe GD. 1992. Trilobite phylogeny and the Cambrian-Ordovician "Event": Cladistic reappraisal. In *Extinction and Phylogeny*, ed. MJ Novacek, QD Wheeler, pp. 144–177. New York: Columbia Univ. Press

Eldredge N, Cracraft J. 1980. *Phylogenetic Patterns and the Evolutionary Process*. New York: Columbia Univ. Press. 349 pp.

Estes R. 1983. Sauria terrestria, Amphisbaenia. *Handbuch der Palaeoherpetologie*, Vol. 10A. Stuttgart: Gustav Fischer

Estes R, Pregill G, eds. 1988. *Phylogenetic Relationships of Lizard Families*. Palo Alto, CA: Stanford Univ. Press. 631 pp.

Evans SE. 1988. The early history and relationships of the Diapsida. See Benton 1988, Vol. 1, pp. 221–60

Farris JS. 1988. *HENNIG86*. Version 1.5. Computer. program distributed by author: 41 Admiral St., Port Jefferson Station, NY

Felsenstein J. 1991. *PHYLIP*. Phylogeny Inference Package Ver. 3.4. Seattle: Univ. Washington

Foote M, Gould SJ. 1992. Cambrian and Recent morphological disparity. *Science* 258: 1816

Forbes E 1846. On the connexion between the distribution of the existing fauna and flora of the British Isles, and the geological changes which have affected their area especially during the epoch of the great northern drift. *Mem. Geol. Surv. Great Britain* 1: 336–432

Frost DR, Etheridge R. 1989. A phylogenetic analysis and taxonomy of iguanian lizards (Reptilia: Squamata). *Misc. Publ. Univ. Kansas Mus. Nat. Hist.* 81: 1–65

Gaffney ES. 1980. Tetrapod monophyly: a phylogenetic analysis. *Bull. Carnegie Mus. Nat. Hist.* 13: 92–105

Gauthier JA. 1984. *A cladistic analysis of the higher systematic categories of the Diapsida*. PhD Thesis. Univ. Calif., Berkeley

Gauthier JA. 1986. Saurischian monophyly and the origin of birds. In *The Origin of Birds and the Evolution of Flight*, ed. K Padian, pp. 1–55. *Mem. Calif. Acad. Sci.* 8

Gauthier JA, Cannatella D, de Queiroz K, Kluge AG, Rowe T. 1989. Tetrapod phylogeny. In The *Hierarchy of Life*, ed. B Fernholm, K Bremer, H Jornvall, 25: 337–53. Amsterdam: Elsevier. 499 pp.

Gauthier J, Estes R, de Queiroz K. 1988. A phylogenetic analysis of Lepidosauromorpha. See Estes & Pregill 1988, pp. 15–99

Gauthier JA, Kluge AG, Rowe T. 1988a. Amniote Phylogeny and the importance of fossils. *Cladistics* 4: 105–209

Gauthier JA, Kluge AG, Rowe T. 1988b. The early evolution of the Amniota. See Benton 1988, Vol. 1, pp. 103–56

Gauthier J, Padian K. 1985. Phylogenetic, functional, and aerodynamic hypotheses of the origin of birds and their flight. In *The Beginnings of Birds*, ed. MK Hecht, JH Ostrom, G Viohl, P Wellnhofer, pp. 185–97. Eichstatt, Germany: Freunde des JuraMuseums. 382 pp.

Gould SJ. 1989. *Wonderful Life: The Burgess Shale and the Nature of History*. New York: Norton. 347 pp.

Gould SJ. 1991. The disparity of the Burgess Shale arthropod fauna and the limits of cladistic analysis: Why we must strive to quantify morphospace. *Paleobiology* 17(4): 411–23

Graham A. 1985. Evolution within the Gastropoda: Prosobranchia In *The Mollusca Evolution*, ed. ER Trueman, MR Clark, 10: 151–86. New York: Academic

Greenfield LO. 1980. A late divergence hypothesis. *Am. J. Phys. Anthropol.* 52: 351–65

Hallam A. 1981. Response. In *Vicariance Biogeography: A Critique*, ed. G Nelson, D Rosen, p. 340. New York: Columbia Univ. Press

Harvey PH, Pagel MD. 1991. *The Comparative Method in Evolutionary Biology*. Oxford: Oxford Univ. Press. 239 pp.

Haszprunar G. 1988. On the orgin and evolution of major gastropod groups, with special reference to the Streptoneura (Mollusca). *J. Moll. Stud.* 54: 367–441

Heaton MJ. 1980. The Cotylosauria: a reconsideration of a group of archaic tetrapods. See Panchen 1980, pp. 497–552

Hennig W. 1950. *Grundzuge einer Theorie der Phylogenetischen Systematik*. Berlin: Deutscher Zentralverlag

Hennig W. 1966. *Phylogenetic Systematics*. Urbana: Univ. Ill. Press. 263 pp.

Hooker JJ. 1989. Character polarities in early perissodactyls and their significance for *Hyracotherium* and infraordinal relationships. See Prothero & Schoch 1989, pp. 79–101

Hulbert RC Jr. 1989. Phylogenetic interrelationships and evolution of North American late Neogene Equidae. See Prothero & Schoch 1989, pp. 176–96

Hull DL. 1988. *Science as a Process*. Chicago: Univ. Chicago Press

Knight JB. 1952. Primitive fossil gastropods and their bearing on gastropod classification. *Smithsonian Misc. Coll.* 117(13): 1–56

Lindberg DR. 1991. Marine biotic interchange between the northern and southern hemispheres. *Paleobiology* 17(3): 308–24

MacFadden BJ. 1976. Cladistic analysis of primitive equids, with notes on other perissodactyls. *Syst. Zool.* 25: 1–14

MacFadden BJ. 1992. *Fossil Horses: Systematics, Paleobiology, and Evolution of the Family Equidae*. Cambridge: Cambridge Univ. Press. 369 pp.

MacLeod N. 1988. Lower and middle Jurassic *Perispyridium* (Radiolaria) from the Snowshoe Formation, east-central Oregon. *Micropaleontology* 34: 289–315

MacLeod N. 1993. The Maastrichtian-Danian radiation of triserial and biserial planktic foraminifera: testing phylogenetic and adaptational hypostheses in the (micro) fossil record. *Mar. Micropaleontol.* 21: 47–100

MacLeod N, Kitchell, JA. 1988. The origin of *Hantkenina*: a phylogenetic analyusis of alternative hypotheses. *Geol. Soc. Am. Abstr. Progr.* 20(6): A228 (Abstr.)

MacNeil FS. 1965. Evolution and distribution of the genus *Mya*, and Tertiary migrations of Mollusca. *U.S. Geol. Surv. Prof. Pap.* 483-G: 1–51

Maddison WP, Donoghue MJ, Maddison DR. 1984. Outgroup analysis and parsimony. *Syst. Zool.* 33: 83–103

Maddison WP, Maddison DR. 1992. *MacClade: Analysis of Phylogeny and Character Evolution. Version* 3.0. Sunderland, England: Sinauer Assoc.

Margoliash E. 1963. Primary structure and evolution of cytochrome c. *Proc. Natl. Acad. Sci. USA* 50: 672–79

Marincovich L Jr. 1983. Molluscan paleontology, paleoecology, and North Pacific correlations of the Miocene Tacxhilni Formation, Alaska Peninsula, Alaska. *Bull. Am. Paleontol.* 84(317): 1–155

Matthew WD. 1915. Climate and evolution. *Ann. New York Acad. Sci.* 24: 171–318

McKenna MC. 1987. Molecular and morphological analysis of high-level mammalian interrelationships. See Patterson 1987, pp. 55–93

Milner AR. 1988. The relationships and origin of living amphibians. See Benton 1988, pp. 59–102

Nelson GJ, Platnick NI. 1981. *Cladistics and Vicariance: Patterns in Comparative Biology*. New York: Columbia Univ. Press

Norell MA. 1992. Taxic origin and temporal diversity: the effect of phylogeny. In *Phylogeny and Extinction*, ed. MJ Novacek, QD Wheeler, pp. 89–118. New York: Columbia Univ. Press

Norell MA. 1993. Tree-based approaches to understanding history: comments on ranks, rules, and the quality of the fossil record. *Am. J. Sci.* 293-A: 407–17

Norell MA, Novacek MJ. 1992a. The fossil record: comparing cladistic and paleontologic evidence for vertebrate history. *Science* 255: 1690–93

Norell MA, Novacek MJ. 1992b. Congruence between superpositional and phylogenetic patterns: comparing cladistic patterns with fossil records. *Cladistics* 8: 319–38

Novacek MJ. 1992. Mammalian phylogeny: shaking the tree. *Nature* 356: 121–25

Novacek MJ, Wyss AR. 1986. Higher-level relationships of the recent Eutherian orders: morphological evidence. *Cladistics* 2: 257–87

O'Hara RJ. 1988. Homage to Clio, or, toward an historical philosophy for evolutionary biology. *Syst. Zool.* 37(2): 142–55

Ostrom JH. 1976. *Archaeopteryx* and the origin of birds. *Biol. J. Linn. Soc.* 8: 91–182

Padian K. 1982. Macroevolution and the origin of major adaptations: vertebrate flight as a paradigm for the analysis of patterns. *Proc. Third. N. Am. Paleontol. Conv.* 2: 387–92

Padian K. 1987. A comparative phylogenetic and functional approach to the origin of vertebrate flight. In *Recent Advances in the Study of Bats*, ed. B Fenton, PA Pacey, JMV Rayner, pp. 3–22. Cambridge: Cambridge Univ. Press

Padian K. 1989a. Did "thecodontians" survive the Triassic? In *Dawn of the Age of Dinosaurs in the American Southwest*, ed. SG Lucas, AP Hunt, pp. 401–14. Albuquerque: New Mex. Mus. Nat. Hist.

Padian K. 1989b. Rebounds and relays in vertebrate evolution. *Geol. Soc. Am. Abstr. Prog.* 21: A31 (Abstr.)

Padian K. 1994. Form vs. Function: the evolution of a dialectic. In *Functional Morphology and Vertebrate Paleontology*, ed. JJ Thomason. Cambridge: Cambridge Univ. Press

Padian, K., Clemens, W.A. 1985. Terrestrial vertebrate diversity: episodes and insights. In *Phanerozoic Diversity Patterns*, ed. J.W. Valentine, pp. 41–96. Princeton: Princeton Univ. Press. 441 pp.

Panchen AL. 1980. *The Terrestrial Environment and the Origin of Land Vertebrates*. London: Academic. 633 pp.

Panchen AL, Smithson TR. 1988. The relationships of the earliest tetrapods. See Benton 1988, Vol. 1, pp. 1–32

Patterson C. 1987. *Molecules and Morphology in Evolution: Conflict or Compromise?* Cambridge: Cambridge Univ. Press. 229 pp.

Perle A, Norell MA, Chiappe LM, Clark JM. 1993. Flightless bird from the Cretaceous of Mongolia. *Nature* 362: 623–28

Pilbeam DR. 1978. Rethinking human origins. *Discovery* 13(1): 2–9

Prothero DR, Schoch RM. 1989. *The Evolution of Perissodactyls*. Oxford: Clarendon. 537 pp.

Reisz R, Laurin M. 1991. *Owenetta* and the origin of turtles. *Nature* 349: 324–26

Rieppel O. 1988. The classification of the Squamata. See Benton 1988, Vol. 1, pp. 261–94

Rowe T. 1987. Definition and diagnosis in the phylogenetic system. *Syst. Zool.* 36: 208–11

Sarich VM, Wilson AC. 1967a. Rates of albumin evolution in Primates. *Proc. Natl. Acad. Sci. USA* 58: 142–47

Sarich VM, Wilson AC. 1967b. Immunological time scale for hominid evolution. *Science* 158: 1200–4

Sepkoski JJ, Kendrick DC. 1991. Numerical experiments with model paraphyletic taxa. *Geol. Soc. Am. Abstr. Prog.* 23: A281 (Abstr.)

Sereno PC. 1986. Phylogeny of the bird-hipped dinosaurs (Order Ornithischia). *Natl. Geogr. Soc. Res.* 2: 234–56

Sereno PC. 1991. Basal archosaurs: phylogenetic relationships and functional implications. *Mem. Soc. Vert. Paleontol.* 2: 1–53

Sereno PC, Rao C. 1992. Early evolution of avian flight and perching: new evidence from the Lower Cretaceous of China. *Science* 255: 845–48

Simons E.L. 1990. Discovery of the oldest known anthropoidean skull from the Paleogene of Egypt. *Science* 237: 1567–69

Simpson GG. 1944. *Tempo and Mode in Evolution*. New York: Columbia Univ. Press. 237 pp.

Simpson GG. 1951. *Horses: The Story of the Horse Family in the Modern World and through Sixty Million Years of History*. New York: Oxford Univ. Press. 247 pp.

Simpson GG. 1953. *The Major Features of Evolution*. New York: Columbia Univ. Press. 434 pp.

Smith AB, Patterson C. 1988. The influence of taxonomic method on the perception of patterns of evolution. *Evol. Biol.* 23: 127–216

Stehli FG, Webb SD. 1985. *The Great American Biotic Interchange*. New York: Plenum

Stuber, RA, Lindberg DR. 1989. Is the radula of living monoplacophorans primitive? *Geol. Soc. Am. Abstr. Prog.* 21(7): A289 (Abstr.)

Swofford DL. 1991. *PAUP: Phylogenetic Analysis Using Parsimony*. Version 3.1. Computer program distributed by Ill. Nat. Hist. Surv., Champaign, Ill

Vermeij GJ. 1991. When biotas meet—understanding biotic interchange. *Science* 253: 1099–104

Wiley EO. 1981. *Phylogenetics: The Theory and Practice of Phylogenetic Systematics*. New York: Wiley-Interscience. 439 pp.

Wiley EO, Siegel-Causey D, Brooks DR, Funk VA. 1991. The compleat cladist: A primer of phylogenetic procedures. *Univ. Kans. Spec. Publ.* 19: 1–158

Wilson AC, Ochman H, Prager EM. 1987. Molecular time scale for evolution. *Trends Genet*. 3: 241–47

Zuckerkandl E, Pauling L. 1962. Molecular disease, evolution, and genetic diversity. In *Horizons in Biochemistry*, ed. M Kasha, B Pullman, pp. 189–225. New York: Academic

Zuckerkandl E, Pauling L. 1965. Evolutionary divergence and convergence in proteins. In *Evolving Genes and Proteins*, ed. V Bryson, HJ Vogel, pp. 97–166. New York: Academic

Annu. Rev. Earth Planet. Sci. 1994. 22:93–117

STRUCTURAL DYNAMICS OF SALT SYSTEMS

Martin P. A. Jackson, Bruno C. Vendeville, and Daniel D. Schultz-Ela

Bureau of Economic Geology, The University of Texas at Austin, Austin, Texas 78713

KEY WORDS: extension, modeling, salt diapir, structural geology, tectonics

INTRODUCTION

The structural dynamics of salt systems—salt tectonics—encompasses any deformation involving salt or other evaporites. It includes halokinesis, in which deformation is driven primarily by salt upwelling and withdrawal, not by regional tectonics. A salt-tectonic system is composed of a source layer of rock salt or other evaporites (collectively referred to as "salt"), overlain by sedimentary overburden, and overlying a basement or subsalt strata. The salt acts as a lubricant, aiding decoupling of the overburden, and accommodates the potential gaps and overlaps between shifting fault blocks in the deforming overburden. This review focuses on the structural dynamics of such salt systems, with an emphasis on diapir formation in both extensional and contractional settings. We also consider subsiding diapirs, salt welds, and allochthonous salt sheets.

BASIC INTERACTIONS

Because the weight of overlying sediments tends to expel salt upward, salt tectonics is largely confined to shallow crust above the ductile-brittle transition zone some 8–15 km deep. In this brittle domain, the overburden deforms not by creep flow—as has been widely assumed by modelers of salt diapirism—but by frictional slip along faults or penetrative slip surfaces (e.g. see compilations in Brace & Kohlstedt 1980 and Weijermars et

0085–6597/94/0515–0093$05.00

al 1993). In marked contrast, salt typically deforms as a viscous or power-law fluid with viscosity typically ranging between 10^{17} and 10^{19} Pa s, depending mainly on grain size and water content (van Keken et al 1993). Salt contains shear zones, but faults within it form only during rapid strain rates accompanying seismic or certain igneous intrusive events.

Except in salt glaciers, dry salt is constrained to flow slowly. However, unconfined, damp salt can flow unusually fast (Table 1, Talbot & Rogers 1980, Talbot & Jarvis 1984). Because of the low viscosity of damp salt, processes controlling salt mechanics mostly depend on (*a*) pressure (pressure in salt, friction in overburden rocks) rather than time, or (*b*) parameters independent of salt rheology (e.g. rates of regional shortening, extension, and sediment progradation or aggradation). These multiple parameters can be simply combined into three ratios: V/B, P/B, and P/V, where V, P, and B are stresses due to salt viscosity, salt pressure, and overburden brittle strength, respectively (Vendeville & Jackson 1993).

V/B reflects the relative strengths of salt and overburden. A system containing a very thin overburden or that is deformed rapidly has a high V/B. The salt layer thins uniformly and stretches the pervasively faulted overburden skin. Viscous forces also increase markedly if the source layer is greatly thinned by extension or salt withdrawal. In contrast, a system comprising a thick overburden above moderately thick, slowly deformed salt has a low V/B because of salt's low viscosity. Here, salt tends not to stretch its overburden, which is decoupled from its basement. The overburden layer faults and tilts above the lubricating salt.

P/B determines whether a diapir can autonomously pierce its overburden. Salt flow driven by pressure gradients can overcome the resisting brittle strength of the overburden (high P/B) if the diapir roof does not exceed a critical thickness (see *Active Piercement*).

Table 1 Strain rates and velocities in salt compared with other tectonic systems[a]

Type of flow	Strain rate s^{-1}	Velocity mm a^{-1}	Velocity
Lava flow	10^{-5} to 10^{-4}	5×10^{11} to 3×10^{13}	1 to 60 km hr^{-1}
Ice glacier	10^{-10} to 5×10^{-8}	3×10^{5} to 2×10^{7}	1 to 60 m day^{-1}
Salt glacier	1×10^{-11} to 2×10^{-9}	2×10^{3} to 2×10^{6}	10 to 100 km Ma^{-1}
Mantle currents	10^{-14} to 10^{-15}	10 to 1×10^{3}	2 m a^{-1} to 5 m day^{-1}
Salt tongue spreading (<30 km wide)	8×10^{-15} to 1×10^{-11}	2 to 20	2 to 20 km Ma^{-1}
Salt tongue spreading (>30 km wide)	3×10^{-16} to 1×10^{-15}	0.5 to 3	0.5 to 3 km Ma^{-1}
Salt diapir rise	2×10^{-16} to 8×10^{-11}	1×10^{-2} to 2	10 m to 2 km Ma^{-1}

[a] From Jackson & Talbot 1991.

P/*V* controls the rise rate of passive diapirs emergent at the sea floor or land surface (see *Passive Piercement*). Driving pressure is proportional to diapir height and to the salt-overburden density ratio. Viscous forces are proportional to salt viscosity and inversely proportional to diapir width and source-layer thickness. A high *P*/*V* (high density contrast, low salt viscosity, wide diapir, or thick source layer and overburden) favors rapid salt upwelling. Conversely, low *P*/*V* increases viscous drag and retards diapir rise.

Regional extension can initiate reactive diapirs (see *Reactive Diapirism*). Their rise rate is controlled by the regional extension rate. Early in the reactive stage the brittle strength of the graben floor still exceeds the underlying diapiric pressure (low, but gradually increasing *P*/*B*) and prevents diapiric breakout. Further extensional thinning of the graben floor increases *P*/*B*, eventually allowing the diapir to actively pierce and perhaps emerge at the surface.

Diapir subsidence rate depends inversely on *P*/*V* (see SUBSIDING DIAPIRS). Thinning of the source layer by extension and withdrawal lowers *P*/*V*. Rapid widening of diapirs during extension also lowers *P*/*V*.

Most of the concepts summarized here originated or were tested by modeling. Weijermars et al (1993) reviewed the history, methodology, and dynamic scaling laws of modeling salt tectonics.

MODES OF DIAPIRIC PIERCEMENT

Salt diapirs have discordant contacts with overlying strata, as opposed to concordant structures such as salt pillows and salt-cored anticlines. To become discordantly encased in the overburden, a diapir faces a "room problem." Three modes of piercement solve the room problem: active, passive, and reactive diapirism (Figure 1).

Early in this century, different researchers argued whether diapirs rose because of density inversion, crystallization forces, or lateral contraction. Regardless of the elusive driving force, most agreed that the diapir was forcefully intruded by the process now called *active piercement* (Nelson 1991). Such a diapir forces its roof upward and sideways, thereby solving the room problem (Figure 1, *middle panel*).

This dogma was shattered by Barton's (1933) concept of a downbuilding or *passive diapir* (Nelson 1991, Jackson & Talbot 1991) that remains at or near the surface while sediments accumulate discordantly around it. The diapir grows taller because its base sinks relative to the surface. No overburden is displaced, so there is no room problem (Figure 1, *bottom panel*).

A third mechanism of piercement is *reactive diapirism* (Vendeville & Jackson 1992a). Unlike the other two modes, reactive piercement requires

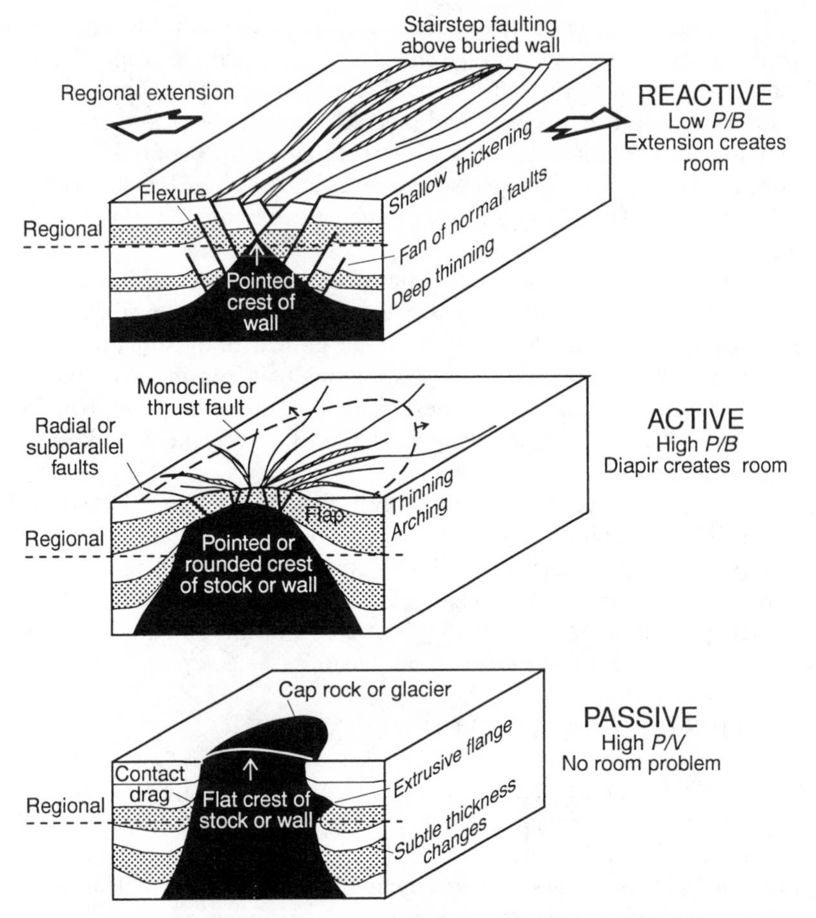

Figure 1 Three piercement modes for salt diapirs (*black*) and their characteristic structures. Regional datum (*dashed line*) is base of upper stippled layer. *P*, *V*, and *B* refer to stresses due to salt pressure, salt viscosity, and overburden brittle strength, respectively.

regional extension caused by rifting or by thin-skinned gravity spreading or gliding. Regional extensional faulting creates space for this type of diapir (Figure 1, *top panel*).

Initiation of Diapirism

Initiation of diapirs would pose no problem if their overburden were a fluid lacking yield strength. Diapirs would rise from any subtle bulge of salt, such as that created by facies changes. The only possible impediment would be a geologically insignificant rise rate (Jackson & Talbot 1986). But the finite strength of brittle sedimentary rocks impedes penetration driven by the weak differential pressures in a subtle bulge of salt. Without regional extension, a realistic brittle overburden can only be pierced if it is thin and flanked by much thicker sediments. This sedimentary differential loading could create enough pressure at the crest of a bulging pillow for active breakthrough and emergence.

Until the 1990s the role of thin-skinned extension in initiating and promoting diapiric rise was generally ignored. Grabens associated with diapirs were attributed to intrusion, withdrawal, or dissolution of salt diapirs. Regional faults lacing around the diapirs were regarded as resulting from, or superposed on, an independent mechanism of diapiric rise. This view became suspect when the first computerized structural restorations from the Gulf of Mexico were published (Worrall & Snelson 1989).

Physical modeling by Vendeville & Jackson (1992a) showed that regional extension can initiate and promote diapirism regardless of the thickness, lithology, and density of the overburden. The overburden necks to form grabens or half grabens, and pressurized salt rises as diapiric walls into these structurally-thinned zones. Faulting and loading of the overburden drive salt flow rather than vice versa. Tectonic differential loading by extension initiates salt diapirism more effectively than does sedimentary differential loading because extension (*a*) pervades rifts and divergent margins where salt typically accumulates, (*b*) weakens the overburden by fracturing and thinning, and (*c*) by creating open-cast structures, differentially loads salt much more effectively than do sedimentary facies changes, which are more gradual and of lower density contrast. Regional extension provides the only means for tabular salt to pierce thick or thin overburdens of initially uniform thickness. A survey of 18 of the world's salt basins documents a close, consistent temporal link between the onset of diapirism and regional extension (Jackson & Vendeville 1994). Extending salt basins typically develop salt structures, whereas nonextending basins typically do not. In some basins containing thick salt (e.g. SW Iran), diapirism was delayed as long as 400 Ma until the basin extended regionally. In other salt provinces (e.g. French Maritime Alps), episodic growth of salt diapirs correlates with episodic regional extension. Once initiated, salt diapirism can continue if contraction or quiescence follows regional extension. Thus, even in salt basins overprinted by inversion or orogenic contraction (e.g. Morocco, Lusitania, Basque-Cantabrian, North Sea), diapirs were initiated during extension on divergent continental margins or in intracontinental rifts.

Reactive Diapirism

A reactive diapir progressively pierces an initial half graben or graben. As extension proceeds, the fault block slides into the source layer until supported by the pressure of salt and by flexural resistance of the block. Continued extension generates new faults from older fracture zones that activate successively inward above the diapir crest. Each new fault slices smaller pieces off the diminishing graben floor. Faulted strata above the diapir crest diverge over time down the flanks of the rising diapir. These supercrestal strata lie below regional datum (Figure 1), which is the reference line connecting undeformed (apart from compaction) points on a horizon (see Figure 6). Strata drop below regional datum during extension, reactive diapirism, salt withdrawl, or salt dissolution; strata rise above regional datum during contraction or active diapirism.

Extension thins the overburden and promotes underlying diapirism, whereas sediment accumulation in the graben thickens the overburden and retards diapirism. The ratio between the rates of sediment aggradation, $\dot{A}$, and regional extension, $\dot{\varepsilon}$, controls structural style. Rapid extension and slow deposition (low $\dot{A}/\dot{\varepsilon}$) promote reactive diapir rise by progressively thinning the diapir roof, thereby increasing P/B. Conversely, slow extension and rapid deposition opposes diapirism by thickening the overburden and decreasing P/V, leading to normal growth faulting and salt rollers—sharp-crested low diapirs (Vendeville & Jackson 1993). The rate of regional extension controls the rate of reactive diapirism. Extension is typically much slower than unconstrained salt flow, so deformation is virtually independent of salt viscosity. Whenever regional extension ceases, reactive diapirs stop growing.

Active Piercement

If a diapir becomes sufficiently tall and its roof sufficiently thin, pressure at the diapiric crest impels it to the active stage. The diapir lifts its roof above regional datum, then rotates, breaks through, shoulders aside, and disperses its remains while dragging up adjoining strata (Figure 1). Driven by diapir pressure, an active diapir forcefully intrudes its overburden (Schultz-Ela et al 1993). During reactive piercement, in contrast, the strength of the overburden is overcome by the tangential forces driving regional extension.

Active diapirism is enhanced by: (*a*) high density contrast between overburden and salt, (*b*) weak overburden, (*c*) large diapiric height and width relative to regional overburden thickness, (*d*) elongated rather than circular planform, and (*e*) topographic relief rather than salt relief. For a density ratio of 1.1, and a height/width diapiric ratio of 1, a salt wall must be encased by roughly three-quarters of the overburden thickness before it can intrude actively. Active diapirism becomes progressively more difficult for diapiric crests whose cross-sectional shape range from a rectangle, a round-cornered rectangle, a semicircle, to an upright triangle.

Modeling suggests the following evolution for active piercement (Vendeville & Jackson 1992a, Schultz-Ela et al 1993): The overburden roof begins to arch upward as the diapir transmutes from reactive to active. Forceful intrusion extends the roof's outer arc to produce a central graben flanked by upward and outward rotating flaps. At their outer ends the flaps hinge by (*a*) bending, which is enhanced by layer-parallel slip, or (*b*) either inward-dipping reverse or outward-dipping normal faults. Reverse faults curve outward as they propagate upward from an initially near-vertical initiation. Meanwhile, normal faults in the central graben propagate downward with new faults created outward (in contrast, reactive faults propagate inward). Antithetic fault slip maintains a flat crest on the graben. Thin, wide roofs develop paired grabens separating an undeformed flat roof from adjoining wedge-shaped flaps. Active diapirs are surrounded by radial faults if regional extension has ceased (Figure 1) or by subparallel normal faults if extension accompanies piercement (Withjack & Scheiner 1982).

As the diapir nears emergence, its pointed crest rounds and widens. Erosional thinning and weakening of the roof stimulate further active rise, while adjoining redeposition increases the pressure for salt breakout. The displaced flaps oversteepen and slump onto the flanking overburden as chaotically resedimented, recumbently folded, imbricately thrusted strata. Finally, the flaps overturn or are entrained by outward salt flow or are removed by erosion. The diapir then emerges as a narrow crest then with its full width. The diapir can also propagate along strike of the graben that initiated it reactively. A density inversion promotes emergence but is not essential if the graben floor is below the pressure head of the source layer.

The overburden load on the source layer can pump salt from the diapiric vent to form an exposed bulge of salt ranging from a low mound (Jackson et al 1990) to a mountain 1 km high (Talbot & Jarvis 1984). The cross-sectional shape of the salt bulge and whether it extrudes glacially depend on the relative rates of upward salt flow and dissolution of its crest.

Passive Piercement

A diapir becomes passive when it emerges (Figures 1, 2). Strata accumulate around the diapir and thin over the crest. A thin roof has little mechanical influence on a passive diapir and is periodically destroyed by active breakout. This lack of vertical constraint has two important implications. First, there is no room problem because flanking strata are not displaced. Faulting and folding are thus negligible around passive diapirs, apart from a narrow drag zone along their contacts (Figure 1). Thickness changes in synkinematic strata (those deposited during the flow of underlying salt) may not be apparent if the overburden resists flexure, in which case salt is withdrawn from wide surroundings. Peripheral sinks are deeper if the overburden can flex and the diapir taps mainly nearby salt. Second, accumulating sediments directly contact salt. Changes in sedimentation rate can profoundly influence diapiric shape (Vendeville & Jackson 1993). The balance between rates of (*a*) aggradation and (*b*) net diapiric rise (gross rise diminished by dissolution and erosion) determines the cross-sectional shape of a passive diapir. If the sediment aggradation rate equals the diapiric rise rate, the diapiric margins grow vertically. However, if the diapir rises faster than the surrounding sediment aggrades, the laterally unsupported salt repeatedly overflows its margins, thus widening upward with time. Conversely, if aggradation is relatively fast, onlapping strata encroach the diapir, which thus narrows upward with time.

Sedimentation rate also controls the planform of a passive diapir. A diapir's flux increases with increasing width. At the diapir's widest part, salt flow rates are high relative to aggradation rates (Figure 2). Conversely, contact drag affects the narrow ends of an elliptical diapir more than wider parts, creating *relatively* high aggradation rates. Thus, salt spreads at the widest part of the diapir and retreats from its elliptical ends. Accordingly, the originally-elongated salt wall evolves toward a string of pluglike passive stocks rising from the deeply buried reactive ridge. The number of exposed plugs declines as they exhaust their supplies and become buried (Lin 1992).

The gross flow rate of passive diapirs reflects the balance between driving

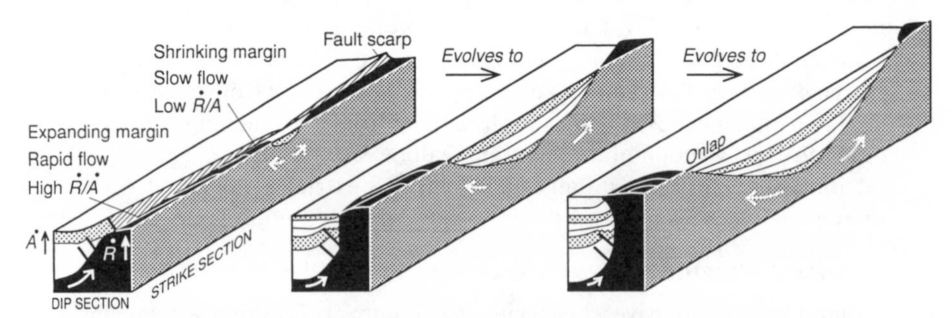

Figure 2 Newly emergent salt wall (*left*, transforming from active to passive) evolves to separate passive stocks (*right*) by along-strike flow of salt. Stocks become more rounded as they spread at their wider, more-vigorously flowing zones and shrink at their narrower, more sluggish zones as strata onlap across them. $\dot{A}$ is local aggradation rate, and $\dot{R}$ is net rise rate of diapiric salt (qualitatively contoured in white on diapir crest).

forces, retarding forces, and salt supply (Vendeville et al 1993b). The flow rate of salt increases with diapir width and also changes over time (Figure 3). As the passive diapir initially grows taller and the overburden thickens, increased pressurization of the source layer increases the flow rate proportionally to overburden thickness. Later, the flow rate is controlled by source-layer thickness because as this layer is depleted, resistance to flow climbs steeply in inverse proportion to the cube of its thickness. Rise rates in nature are much lower than the scaled peak experimental rates, suggesting that either (*a*) much salt must dissolve from passive diapirs or (*b*) most passive diapirs would rapidly extrude to form salt sheets unless dissolved. The long phase of flow deceleration results in low flow rates like those in nature (Figure 3).

Superposing constant sedimentation rate on this change in salt flow rate creates a characteristic diapiric shape (Figure 3, *bottom panel*). Its base narrows upward as salt initially flows slowly and sediments encroach; the diapir then widens upward as salt flow accelerates and diapiric overhangs spread over the sediments. Finally, deceleration of salt flow from the depleting source layer tapers, then completely buries the diapir's shoulders. This evolution produces a diapiric stem without necking or otherwise deforming the overburden. Salt flows along the diapiric wall from below depressions into intervening culminations that evolve into increasingly pluglike, rounded stocks (Figure 2).

Evolutionary Paths of Diapirs

During constant moderate aggradation, a salt diapir would evolve from reactive to active to passive. This progression can be changed by either depletion of source layer or variations in regional extension or aggradation rates. Various evolutionary paths are possible:

- A diapir could appear to be initiated passively. If the overburden is thin and unevenly deposited, the reactive and active stages of growth are brief. Strata faulted during these early stages could be too thin to be seismically visible.
- A diapir could appear to have been continuously passive, although it actually passed through countless cycles. Each cycle could comprise an episode of passive growth during slow aggradation, followed by brief burial during a pulse of high aggradation, and terminated by brief active breakout. Each active stage destroys the domed sedimentary veneer by local extension, entrainment, slumping, erosion, and dissolution collapse.
- An active diapir can partially break through its roof but never emerge, for any of three reasons. First, the source layer may become depleted (Figure 3, *right*). Second, piracy of salt by an emergent segment of a wall

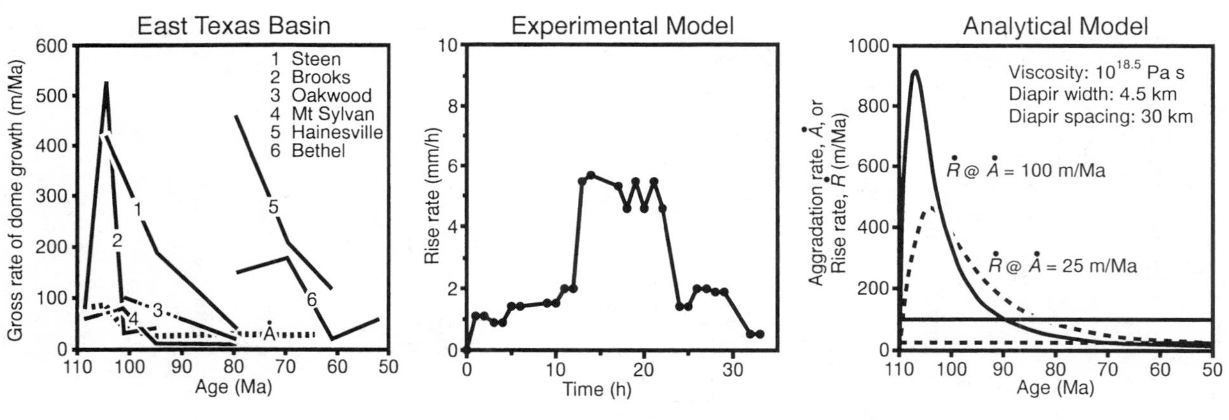

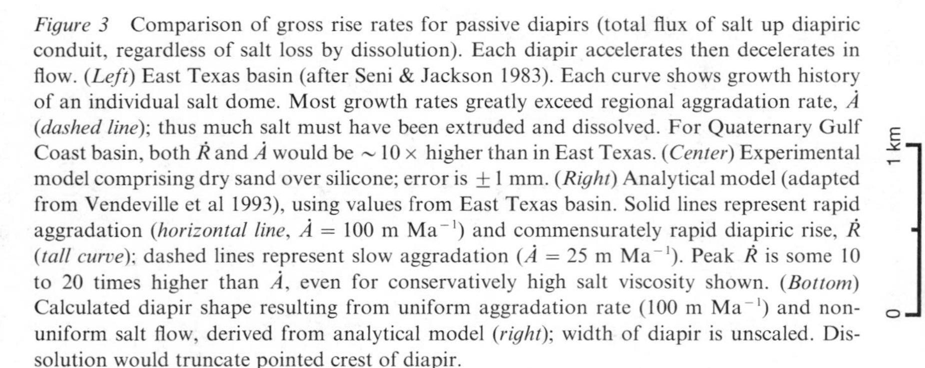

Figure 3 Comparison of gross rise rates for passive diapirs (total flux of salt up diapiric conduit, regardless of salt loss by dissolution). Each diapir accelerates then decelerates in flow. (*Left*) East Texas basin (after Seni & Jackson 1983). Each curve shows growth history of an individual salt dome. Most growth rates greatly exceed regional aggradation rate, $\dot{A}$ (*dashed line*); thus much salt must have been extruded and dissolved. For Quaternary Gulf Coast basin, both $\dot{R}$ and $\dot{A}$ would be $\sim 10\times$ higher than in East Texas. (*Center*) Experimental model comprising dry sand over silicone; error is ± 1 mm. (*Right*) Analytical model (adapted from Vendeville et al 1993), using values from East Texas basin. Solid lines represent rapid aggradation (*horizontal line*, $\dot{A} = 100$ m Ma^{-1}) and commensurately rapid diapiric rise, $\dot{R}$ (*tall curve*); dashed lines represent slow aggradation ($\dot{A} = 25$ m Ma^{-1}). Peak $\dot{R}$ is some 10 to 20 times higher than $\dot{A}$, even for conservatively high salt viscosity shown. (*Bottom*) Calculated diapir shape resulting from uniform aggradation rate (100 m Ma^{-1}) and non-uniform salt flow, derived from analytical model (*right*); width of diapir is unscaled. Dissolution would truncate pointed crest of diapir.

can halt growth of other adjoining sections (Figure 2). Third, numerical modeling suggests that without erosion or deposition, an active diapir only partially penetrates all but the thinnest roofs before reaching equilibrium (Schultz-Ela et al 1993).

THIN-SKINNED EXTENSIONAL SALT TECTONICS

Evaporites accumulate almost exclusively in extensional regimes such as divergent continental margins and rifts. Thin-skinned extension involves stretching of the cover decoupled by a basal layer of salt or other evaporites from an essentially undeformed basement.

An unambiguously extensional structure is a salt roller—a sharp-crested low diapir. One flank is concordant with the overburden, whereas the other discordant flank is the basal segment of a listric growth fault. Structural relief and width of a roller gradually increase over time, as shown by physical modeling (Vendeville & Cobbold 1987, Cobbold et al 1989, Vendeville & Jackson 1992a,b). Palinspastic reconstructions (Worrall & Snelson 1989, Wu et al 1989, Duval et al 1992, Schultz-Ela 1992) and this modeling favor regional extension triggering salt upwelling rather than the upwelling causing the faulting.

Aggradation rates vary in time and space. For example, Paleogene accumulation rates on the stable shelf of the Gulf of Mexico varied by a factor of ten, largely due to a pulsating supply of sediment (Galloway & Williams 1991). Accumulation rates could vary as much as 100-fold on the adjoining slope, where most salt structures are initiated, because structures create accommodation space and concentrate sediment pathways.

The influence of aggradation rate on the structural style of extensional salt tectonics was first noted by Vendeville & Cobbold (1987) and systematically modeled by Lin (1992), who simulated a tenfold variation in aggradation rate during regional extension. Markedly different structural styles resulted (Figure 4). High $\dot{A}/\dot{\varepsilon}$ (rapid aggradation, slow extension) favors major listric growth faults rather than diapiric growth. Initially, grabens pierced by symmetric, buried diapiric walls are formed. As the overburden thickens, only the downslope-dipping boundary fault of each graben remains active, evolving into a listric growth fault with a deeply buried, low diapiric roller in its footwall. In contrast, low $\dot{A}/\dot{\varepsilon}$ promotes salt diapirism. Under slow aggradation, triangular salt walls penetrate the initial grabens and evolve into tall walls having steep or overhanging, discordant contacts. Their crests are typically shallowly buried or emergent, in which case they become flat-topped and pluglike. In the extensional spectrum, these tall stocks are polar to salt rollers (Figure 4). Intermediate aggradation rates resulted in intermediate diapiric structures (moderately buried, large triangular salt walls).

Where salt walls penetrate only the deep overburden, the deep cover extends mainly by widening of salt walls, whereas the shallow cover extends by faulting in the overburden; extension is partly cryptic (Figure 5, *left*). Where passive salt walls penetrate the entire overburden, the section can extend almost entirely by widening of walls between separating glide blocks of overburden. Lack of visible faults makes even large extensions subtle or cryptic (Figure 5, *right*) (Vendeville & Jackson 1992a).

Walls and faults are typically parallel to each other and perpendicular to the dip of the basin margin, as in the Kwanza basin of Angola (Duval et al 1992). On an irregular margin, downslope extension can be convergent

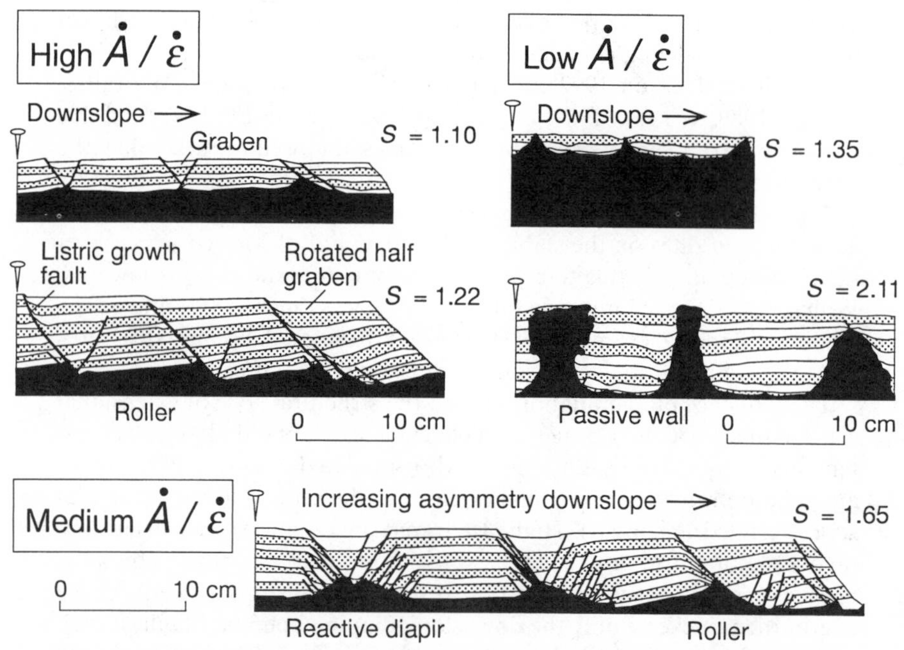

Figure 4 Ratio of aggradation rate to extension rate ($\dot{A}/\dot{\varepsilon}$) controls structural style of experimental diapirs during extension. (*Upper left*) High $\dot{A}/\dot{\varepsilon}$ (34 cm) results in deeply buried, low salt rollers bounded by lower parts of listric growth faults and overlain by rotated half grabens. (*Upper right*) Low $\dot{A}/\dot{\varepsilon}$ (4.5 cm) results in shallowly buried, tall stocks and walls and negligible faults. (*Bottom*) Medium $\dot{A}/\dot{\varepsilon}$ (6.4 cm) results in moderately tall buried diapiric walls whose triangular shape becomes increasingly asymmetric toward the right, where the overburden starts wedging out. S = finite stretch. (After Lin 1992, Vendeville & Jackson 1992a.)

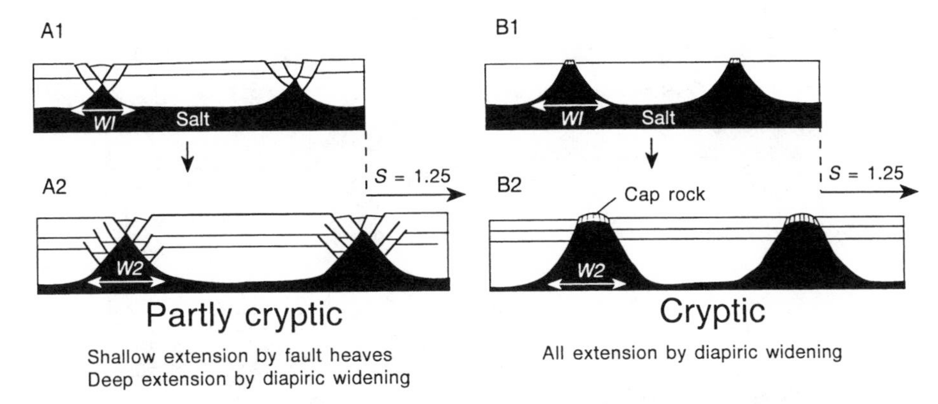

Figure 5 Partly cryptic extension (*left*) and cryptic extension (*right*) for same finite stretch S. W = width of diapiric base.

over indentations or divergent over salients, as on the Campos basin of Brazil (Cobbold & Szatmari 1991, Rouby et al 1993).

THICK-SKINNED EXTENSIONAL SALT TECTONICS

Thick-skinned extension stretches both basement and cover. Even trivial normal faults offsetting the base of salt have long been thought to initiate diapirs and thus to influence their location and shape. However, modeling indicates that the effect of basement faults can propagate upward through the salt into the overburden only if lateral salt flow is hindered by unusually high salt viscosity, very rapid regional extension, or extreme thinness of the salt (Figure 6*a*) (Vendeville et al 1993a).

Where the basement fault slips at moderate rates or the salt is moderately thin (Figure 6*b*), the cover over the downthrown block sags, forming a monocline above the basement fault. Local stretching of the upper monoclinal hinge is accentuated by regional extension to form a graben above the upthrown basement block next to (not above) the basement fault. This thinning triggers reactive diapiric piercement into the faulted graben.

During slow extension, a thick (say >500 m) salt layer effectively decouples its brittle overburden from a faulting basement (Figure 6*c*). Salt flows from the upthrown block toward the downthrown block, allowing the overburden to subside uniformly across the basement fault. Extension widens the salt basin and stretches the cover. Faults and walls in the overburden form subparallel to the direction of regional extension, which may or may not be parallel to the basement faults. But as long as the salt

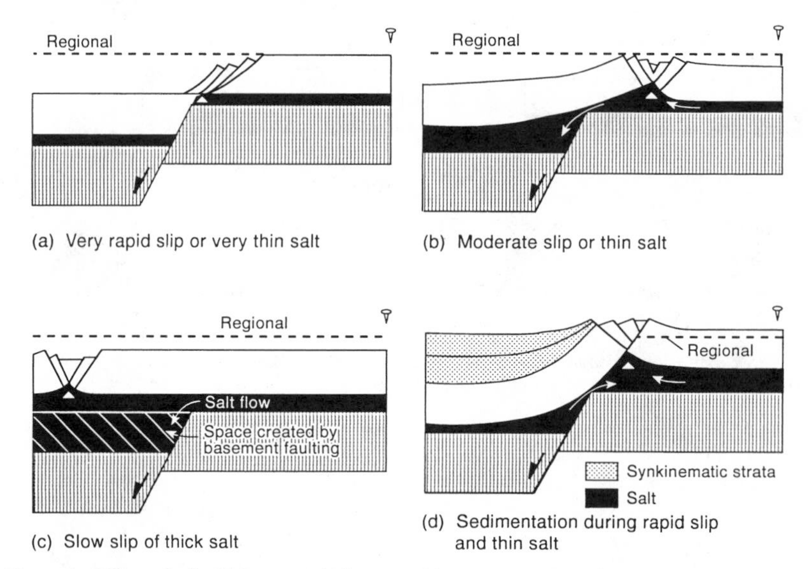

Figure 6 Effect of salt thickness and slip rate of basement fault on structural style of cover. Regional datum delineates original top of overburden before deformation began. All rock units are prekinematic (originally uniform thickness before deformation began), except where labeled otherwise. Diapirs are initiated above white triangles. (After Vendeville et al 1993.)

layer is thick, the location, spacing, and throw of the overburden structures are effectively independent of all but the larger basement faults. Only after the source layer greatly thins does the basement footwall indent the overburden, forming new faults and folds directly above the basement fault.

Rather than being filled by salt, the space created by basement downthrow can be balanced by locally thickened synkinematic sediments situated above draped overburden (Figure 6*d*). The load of accumulating sediments can reverse flow in the salt, squeezing salt from above the downthrown block to above the upthrown block (Vendeville et al 1993a).

In summary, extension of the overburden directly causes diapirism, regardless of basement faulting, unless salt is thin or the slip rate is high. A basement fault affects diapirism only indirectly by causing or allowing the overburden to extend.

SUBSIDING DIAPIRS, SALT WELDS, AND FAULT WELDS

Regional extension eventually retards the rise of diapirs and even causes them to subside. Salt walls must widen between diverging blocks of overburden, thereby increasing demand on salt supply. The supply rate diminishes as the source layer thins due to extension and withdrawal into the diapir. Eventually salt import is too slow to supply salt for expanding walls. The diapir begins to sag. Sedimentary thicks fill either (*a*) half-graben rollovers above the diapir's subsiding flanks, leaving the crest unaffected (Figure 7*A*) or (*b*) a linear, or even circular, crestal graben above the sagging crest of the diapir, which inverts from a topographic bulge to a subsiding graben (Figure 7*B*). Adjoining horns of salt project into each fault bounding the graben. Salt dikes can intrude faults (Hale et al 1992), but these extensional horns are not injections; they are residual structures whose apices record the original diapiric crest.

Other products of source-layer depletion during regional extension include *turtle-structure anticlines* (Figure 7*A*), which form because each end of an overburden slab subsides as the diapir supporting it sags. This anticlinal bending of the slab forms a crestal fan of faults. Turtle-structure anticlines have been attributed solely to salt withdrawal from the periphery of a pillow as it evolves into a *rising* diapir (Trusheim 1960), but turtle structures also characterize a *subsiding* diapir during regional extension (Duval et al 1992, Vendeville & Jackson 1992b). Extensional turtle structures are subparallel and elongated, whereas classic withdrawal turtles are equant.

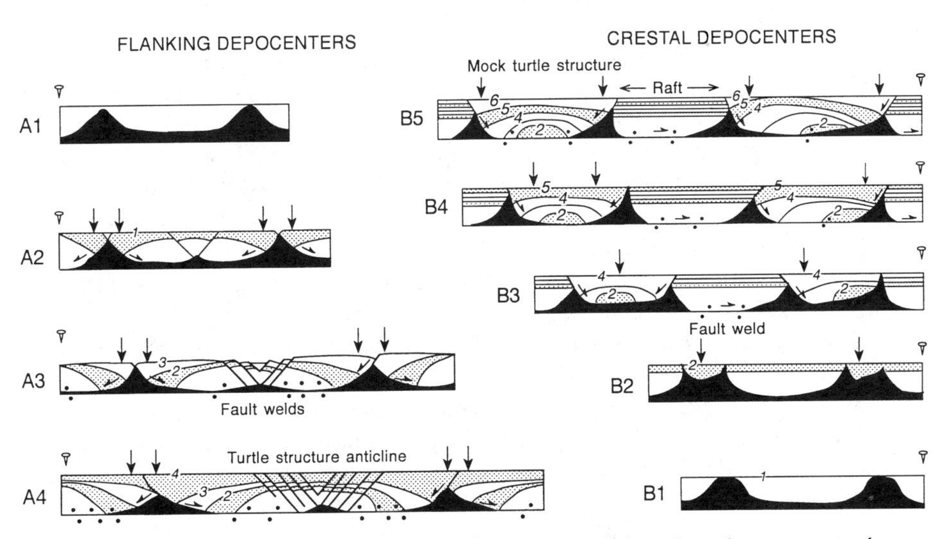

Figure 7 Formation of turtle structure anticline (A1–A4) and mock turtle structures and rafts (B1–B5) during extensional salt tectonics. Vertical arrows show depocenters extensionally created by subsiding diapirs.

During extreme thin-skinned extension, another type of turtle structure can form (Figure 7*B*). A crestal depocenter indents a subsiding diapir so much that the graben grounds onto the basement, segmenting the original diapir into two triangular relics separated by a deep, synclinal depocenter. If extension continues, the flanks of these diapiric relics subside, and the synclinal depocenter inverts into a crestally-faulted arch called a *mock-turtle anticline* (Vendeville & Jackson 1992b). A mock-turtle anticline can be distinguished from a turtle anticline because it initiates *above* a diapir rather than *between* diapirs; also, below a mock-turtle anticline a large stratigraphic section—representing the duration that the precursor diapir was at the surface—is missing.

The formation of mock-turtle anticlines is part of *raft tectonics*. Normal-fault blocks separate entirely into glide blocks called rafts (Figure 7*B*). The Kwanza basin of Angola is the archetypical area for raft tectonics (Duval et al 1992). There, transgressive carbonate overburden began to extend when only a few hundred meters thick, forming many small, tilted, Phase 1 rafts over relatively thick salt. These rafts were yoked together by sedimentation before rupturing into much larger nontilted Phase 2 rafts 10 to 20 km long, which slid concordantly over thinned salt. Younger sediments accumulated asymmetrically in strike-parallel depocenters created by widening grabens separating Phase 2 rafts. Apart from sea-floor spreading, lateral space for extension was created locally by downdip fold-and-thrust-belts and by transfer of salt to higher stratigraphic levels in the form of allochthonous sheets.

Some of the structures of extensionally-induced subsidence have been interpreted as arising purely by dissolution. Subsurface dissolution cannot form large crestal grabens or flanking half grabens wherever the salt diapir is encased in impermeable shale, deeply buried, lacking cap rock, in an environment of rapid sedimentation, or where palinspastic restoration indicates abrupt, deep subsidence of a diapir crest after a much longer time at the surface (Duval et al 1992, Vendeville & Jackson 1992b).

Salt welds are surfaces or thin zones joining strata originally separated by thicker salt that is mostly or completely removed by lateral creep or dissolution (Jackson & Cramez 1989). Because salt can seal petroleum systems, the thickness of salt remaining in the weld affects the vertical migration of hydrocarbons. Welds commonly separate discordant or dis-conformable strata. Primary, secondary, and tertiary welds, respectively, join strata originally separated by gently dipping bedded salt, steep-sided salt diapirs, or gently dipping allochthonous salt sheets.

Salt welds are difficult to differentiate from fault welds, which have significant fault slip or shear along the smeared layer of salt. Both types of welding can create discordances and disconformities, and both create accommodation space (Figure 8). Most cross sections of salt structures can be ambiguously restored by assuming either salt welds and salt withdrawal or fault welds and extension (Hossack & McGuinness 1990, Diegel et al 1993). However, the type of welding can be distinguished by local structural criteria (Figure 8) or by knowing the total extension of the section or the original salt thickness.

CONTRACTIONAL SALT TECTONICS

Contractional salt tectonics is produced by convergent (including transpression) plate tectonics—often superposed on a divergent rift or continental margin—or is restricted to the narrow seaward margin of a divergent continental margin. Contractional salt tectonics conveniently divides into two types, cases in which: (*a*) thin salt merely lubricates decoupling of the overburden from its basement and fills anticlinal cores and (*b*) thicker salt forms diapirs before contraction.

Examples of thin-skinned contraction over thin salt are in Arctic Canada; the Franklin Mountains of northwest Canada; the Appalachians; and the deep-water fringes of largely divergent salt basins such as the Perdido and Mississippi Fan fold belts of the Gulf of Mexico, the Kwanza basin of Angola, the Campos basin of Brazil, and the southern Red Sea.

These contractional belts over thin salt are exemplified by the eastern Parry Islands fold belt of Arctic Canada (Harrison & Bally 1988, Harrison et al 1991), which contains two décollements: a lower evaporite and an upper shale. The sinuously arcuate fold belt comprises upright folds cored by thrust complexes. Anticlines were initiated by buckling of the competent carbonate-dominated strata between the weak salt and shale (Figure 9*a*). The salt then migrated laterally into the cores of the anticlines, forming pinched ridges called *salt welts*. The less lithified sequence above the strong carbonates contracted by compaction and uniform thickening. The carbonates then contracted by thrusting as the overlying strata buckled (Figure 9*b*). Shale was squeezed into the anticlinal cores to form sharp-crested shale welts stacked over the salt welts (Figure 9*c*). A second generation of thrusts subsequently ramped from the shale into the competent layers above (Figure 9*d*). Anticlinal cores also contain pop-up structures, thrust splays, and faults that zigzag up-section, reversing sense of slip. The folds and thrusts verge equally forward (to south) and backward (to north), a characteristic of décollement over salt (Davis & Engelder 1987).

The amount of shortening in diapiric provinces ranges from mild inversion of an extensional system (North Sea, Germany) to full fold-and-thrust belts in various Alpine chains of Europe, North Africa, southwest Asia, and elsewhere. Inversion in the North Sea and Germany is the foreland

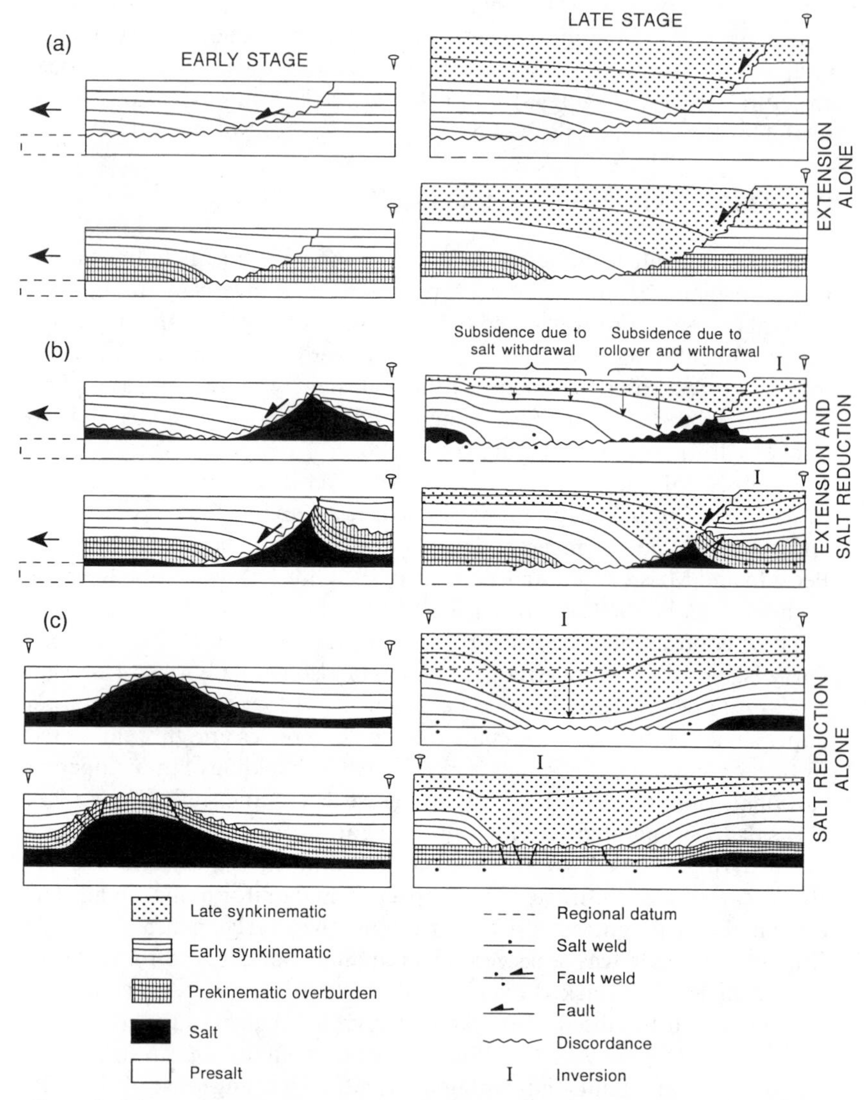

Figure 8 Diagnostic features of (*a*) growth faults, (*b*) fault welds, and (*c*) salt welds. Upper section of each pair shows only synkinematic strata, whereas lower section shows basal prekinematic unit of overburden. Vertical arrows show subsidence below regional datum (*dashed line*).

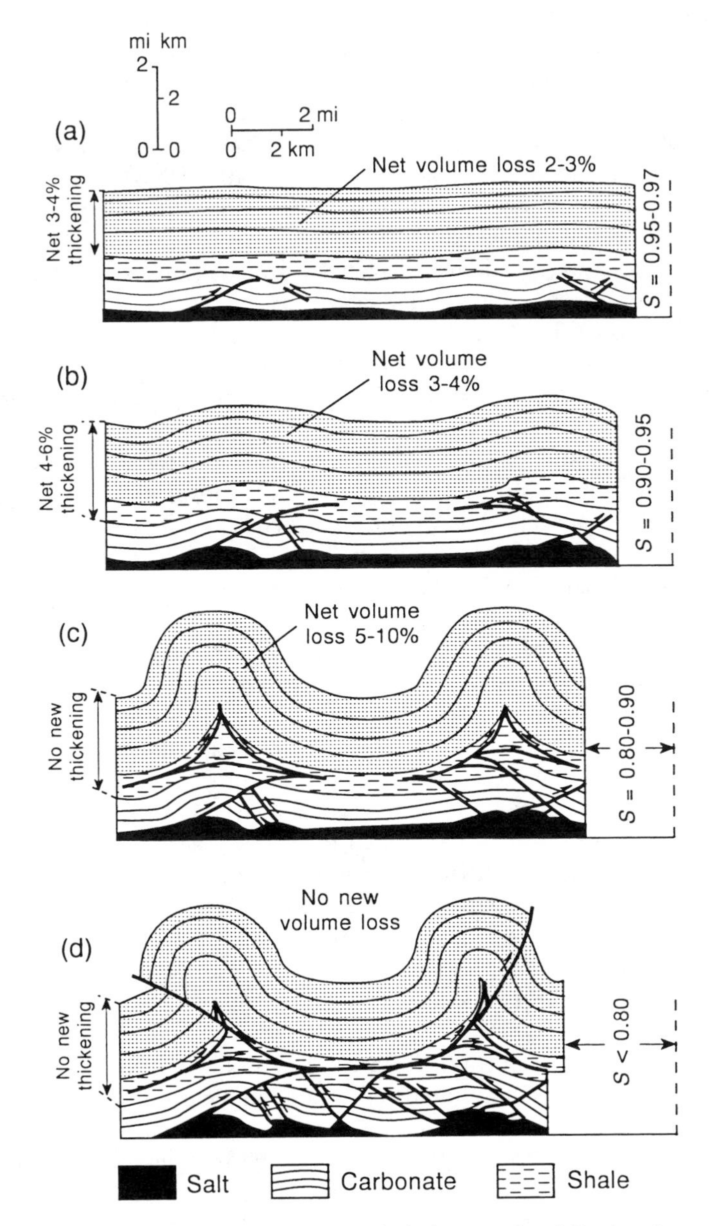

Figure 9 Evolution of salt and shale welts (pinched cores of anticlines) and associated thrusts during contractional salt tectonics in the eastern Parry Islands fold belt, Arctic Canada (after Harrison & Bally 1988, Harrison et al 1991).

expression of severe Alpine shortening to the south. Many concordant salt structures in this region (e.g. Silver Pit region in the UK North Sea) are still widely attributed merely to buoyant upwelling of Zechstein salt by gravity-driven halokinesis. Yet these swells are overlain by domes or anticlines of overburden, whose imposing thickness and uniformity exclude a purely halokinetic origin by active doming. Rather, these folds must have formed by regional contraction during inversion. Salt merely collected in the lower-pressure anticlinal cores.

Salt lubrication promotes the lateral propagation of fold-and-thrust belts. Kastens (1991) attributed accelerated advance (<10 km Ma^{-1} increasing to 12–22) of the Mediterranean Ridge accretionary prism to the Neogene deposition of Messinian salt across the ridge.

Most salt diapirs in fold-and-thrust belts were apparently initiated in extensional systems long before contraction began (Jackson & Vendeville 1994). For example, both the Spanish Prebetic and the French Maritime Alps were rifted several times. Palinspastic removal of the overprint of Alpine folding reveals these rift-phase structures (de Ruig 1992, Dardeau & Graciansky 1990). Salt diapirism began during rifting soon after the salt accumulated. The diapirs rose reactively into tears of the less-dense overburden. During ensuing crustal shortening, the preexisting diapirs were distorted, truncated, and further injected into the contracting overburden, and new salt-cored folds formed. Diapirs in these Alpine contractional belts show no consistent relationship to anticlines or synclines.

Physical modeling shows that whether the salt is thin or thick, merely buckling the overburden cannot produce diapirs (Vendeville 1991). Salt could diapirically break out from anticlinal cores if (*a*) the anticlinal crest was thinned by erosion or local extension and (*b*) the anticlinal core of salt was confined and pressurized as adjoining synclines grounded onto the basement and blocked lateral escape of salt. Thrusting is generally more effective than buckling in forming diapirs.

ALLOCHTHONOUS SALT SHEETS

One of the most important discoveries in salt tectonics has been that irregular masses of salt hundreds of square kilometers in area are actually allochthonous sheets rather than rooted, autochthonous massifs. Salt allochthons characterize divergent continental margins such as the Gulf of Mexico and are also known from offshore Brazil, West Africa, and the Red Sea. Instead of extending down several kilometers to the source layer, these salt bodies overlie younger strata. This realization opened the possibility of prodigious subsalt hydrocarbon reservoirs, particularly in the Gulf of Mexico, where a vast complex of salt allochthons increases in size and degree of coalescence westward (Liro 1992, Wu 1993). Most salt sheets there have partly or completely coalesced into composite diapirs called canopies (Jackson et al 1990). The leading edge of this complex climbed 8 km vertically and up to 80 km laterally to form the Sigsbee Scarp at the base of the continental slope.

Thin allochthonous sheets gravitationally spread over or through weak, unconsolidated, mud-rich, low-density sediments within a few 100 meters of the sediment surface (Nelson 1991, Fletcher et al 1993, McGuinness & Hossack 1993). The observed shallow emplacement levels indicate that the mechanical control is overburden strength rather than overburden density; were density the controlling factor, sheets would spread at a level of neutral buoyancy 1 to 1.5 km deep.

Stratigraphic markers record the timing and direction of allochthonous salt emplacement (Figure 10). Sedimentary or structural truncation of underlying strata against the lower contact of a salt tongue forms a basal cutoff. Each cutoff marks the former leading edge of the advancing salt tongue. *Salt ramps* and *flats* are the steeply inclined and gently inclined segments of the stairstep base of a salt tongue, which cut up the stratigraphic section in the direction of emplacement. Ramps form where the ratio of aggradation to salt spreading is high. Salt spreads and flats form when sedimentation is very slow. The basal cutoffs of salt sheets suggest that their injection rate is up to 100 times faster than the coeval aggradation rate (Jackson & Talbot 1991). In the Gulf of Mexico, salt sheets appear to spread after each episode of downdip contraction associated with updip extension (Peel et al 1993).

An allochthonous sheet can act as a source layer for a second cycle of salt tectonics. During progradation, the rear (landward) margin of the sheet gradually segments. Intrasalt minibasins evolve from slope basins above tabular salt allochthons to shelf basins, which are bounded by arcuate growth faults formed largely by salt withdrawal downslope rather than by extension (Diegel et al 1993). The sheet is segmented into discrete salt structures separated by salt welds and fault welds, together constituting a *roho system* (Schuster 1993). Initially the allochthon acts as a deflating cushion that accommodates sediment deposition in intrasalt basins. After much time, loading, or extension, the original sheet transforms into thin but broad welds, which act as major subhorizontal detachments for deep growth faults whose slip creates space for further sedimentation (Diegel et al 1993, Peel et al 1993). Second-cycle salt structures rise from these deformed sheets (Worrall & Snelson 1989, Hossack & McGuinness 1990, Seni 1992).

As the sheet segments, the weight of overlying sediments squeezes salt toward the shallow leading edge of the allochthon. The salt sheet must

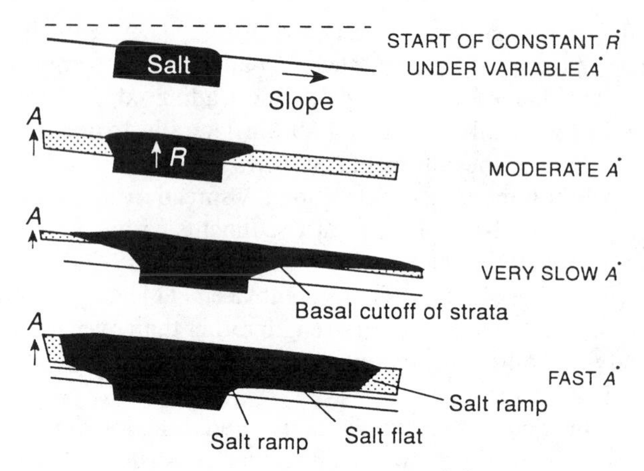

Figure 10 Asymmetric spreading of a passive salt sheet over or just below the sedimentary surface. Highly variable aggradation rate produces salt ramps, salt flats, and basal cutoffs along the base of the salt allochthon. A = aggradation increment, $\dot{A}$ = aggradation rate, R = salt rise increment, $\dot{R}$ = salt rise rate.

continually climb stratigraphic section and break through the continually accumulating sedimentary veneer. Physical modeling (Jackson & Vendeville 1993) suggests that salt is expelled seaward from beneath a prograding wedge into the core of a box fold fringing the wedge. Fore- and back-grabens form along the hinges of the box fold. Pressurized by the prograding overburden, a diapiric wall emerges through the structurally-thinned back-graben. Extrusions from the fore-graben flow down and invert the forelimb then spread radially to coalesce as a lobate canopy. The roof of the box fold detaches by arcuate slumps and ruptures into rafts, which are overthrust on the back of the spreading salt over the inverted limb of the isocline, locally stacking and repeating stratigraphy. Well intersections of repeated stratigraphy and structural reconstructions prove that the sedimentary roof of a spreading salt sheet is carried as an overthrust over equivalent strata below the salt sheet (McGuinness & Hossack 1993). The stretching carapace protects the underlying salt from dissolution, as does the residual crust of insoluble gypsum originally disseminated in salt.

The allochthons are fed by landward-dipping stalks, which flatten into salt welds by contraction and by withdrawal into the salt sheet. Once welded, these huge landward-dipping discontinuities, known as stepped counter-regional systems (Schuster 1993), resemble listric normal growth faults; but little slip has occurred along them, and the adjoining rollover anticlines are produced by salt withdrawal rather than by extension (Jackson & Vendeville 1993, Schuster 1993).

The cycle of sheet emplacement, inflation, segmentation, and climb can be repeated several times, creating several tiers of allochthonous salt. Three or four such tiers have been recognized in the northern Gulf of Mexico (Diegel & Schuster 1990, Peel et al 1993, Wu 1993).

CONCLUSION

Salt tectonics represents the interplay between salt, acting as a weak but highly pressurized fluid and its overburden, acting as a strong, brittle roof, which faults in response to the tangential forces of plate tectonics, to its own weight, and to salt pressure. Varied responses allow salt to smear as a lubricant, to rise as massive diapirs, or to spread as vast glaciers. Although some astonishing concepts have been revealed in the past few years, research on salt tectonics proceeds apace, and new concepts are rapidly overhauled by even newer ones. Several topics are still murky, but some are likely to clarify soon: 1. establishing whether extension above and below salt layers can be cryptically coupled, 2. quantifying the roles of extension versus salt withdrawl in creating accommodation space, 3. distinguishing between synclines produced by salt withdrawal (bending folds) and by regional contraction (buckling folds), 4. determining the role of contraction in the emplacement of salt sheets, 5. establishing the degree of exposure and protection from dissolution of submarine salt glaciers, and 6. examining suture processes where diapirs coalesce into canopies.

ACKNOWLEDGMENTS

Our research on this topic at the Applied Geodynamics Laboratory was funded, inspired, and animated by continual interaction with the following companies: Agip, Amoco, Arco, British Petroleum, Chevron, Conoco/DuPont, Elf Aquitaine, Eurosim, Exxon, Marathon, Mobil, Petrobras, Phillips, Shell, Texaco, and Total. Jake Hossack inculcated the importance of regional datum upon us. Figures were prepared by Michele LaHaye and Susan Krepps, directed by Richard Dillon, and by the authors and Hongxing Ge. The paper was edited by Tucker F. Hentz, Amanda R. Masterson, and Kitty Challstrom and is published by permission of the Director, Bureau of Economic Geology, The University of Texas at Austin.

Literature Cited

Barton DC. 1933. Mechanics of formation of salt domes with special reference to Gulf Coast salt domes of Texas and Louisiana. *Am. Assoc. Petrol. Geol. Bull.* 17: 1025–83

Brace WF, Kohlstedt DL. 1980. Limits on lithospheric stress imposed by laboratory experiments. *J. Geophys. Res.* 85: 6248–62

Cobbold P, Rossello E, Vendeville BC. 1989. Some experiments on interacting sedimentation and deformation above salt horizons. *Bull. Soc. Geol. France* 8(3): 453–60

Cobbold PR, Szatmari P. 1991. Radial gravitational gliding on passive margins. *Tectonophysics* 188: 249–89

Dardeau G, Graciansky PC. 1990. Halokinesis and Tethyan rifting in the Alpes-Maritimes (France). *Bull. Centres Rech. Explor.-Prod. Elf-Aquitaine* 14(2): 443–64

Davis DM, Engelder T. 1987. Thin-skinned deformation over salt. In *Dynamical Geology of Salt and Related Structures*, ed. I. Lerche, J. J. O'Brien, pp. 301–37. Orlando, Florida: Academic

de Ruig MJ. 1992. *Tectono-sedimentary evolution of the Prebetic fold belt of Alicante (SE Spain): a study of stress fluctuations and foreland basin deformation.* PhD thesis. Vrije Univ., Amsterdam. 207 pp.

Diegel FA, Karlo JF, Schuster DC, Shoup RC, Tauvers PR. 1993. Cenozoic structural evolution and tectonostratigraphic framework of the northern Gulf Coast continental margin. *Am. Assoc. Petrol. Geol. Annu. Conv. Off. Progr., New Orleans*, p. 91 (Abstr.)

Diegel FA, Schuster DC. 1990. Regional cross sections and palinspastic reconstructions, northern Gulf of Mexico. *Geol. Soc. Am. Abstr. with Programs* 22(7): A66 (Abstr.)

Duval B, Cramez C, Jackson MPA. 1992. Raft tectonics in the Kwanza basin, Angola. *Mar. Petrol. Geol.* 9: 389–404

Fletcher RC, Hudec MR, Watson IA. 1993. Salt glacier model for the emplacement of an allochthonous salt sheet. *Am. Assoc. Petrol. Geol. Int. Hedberg Res. Conf., Salt Tectonics, Bath, England* (Abstr.)

Galloway WE, Williams TA. 1991. Sediment accumulation rates in time and space: Paleogene genetic stratigraphic sequences of the northwestern Gulf of Mexico basin. *Geology* 19: 986–89

Hale D, Hill NR, Stephani, J. 1992. Imaging salt with turning seismic waves. *Geophysics* 57(11): 1453–62

Harrison JC, Bally AW. 1988. Cross-section of the Parry Islands fold belt on Melville Island, Canadian Arctic Islands: implications for the timing and kinematic history of some thin-skinned décollement systems. *Can. Petrol. Geol. Bull.* 36(3): 311–32

Harrison JC, Fox FG, Okulitch AV. 1991. Late Devonian–Early Carboniferous deformation of the Parry Islands and Canrobert Hills fold belts, Bathurst and Melville Islands. In *Geology of the Innuitian Orogen and Arctic Platform of Canada and Greenland*, ed. H. P. Trettin, pp. 321–41. Ottawa: Geol. Surv. Can.

Hossack JR, McGuinness DB. 1990. Balanced sections and the development of fault and salt structures in the Gulf of Mexico (GOM). *Geol. Soc. Am. Abstr. with Programs* 22(7): A48 (Abstr.)

Jackson MPA, Cornelius RR, Craig CH, Gansser A, Stöcklin J, Talbot CJ. 1990. Salt diapirs of the Great Kavir, Central Iran. *Geol. Soc. Am. Mem.* 177, 139 pp.

Jackson MPA, Cramez C. 1989. Seismic recognition of salt welds in salt tectonics regimes. *Proc. Gulf Coast Sec. Soc. Econ. Paleontol. Mineral. Found. Res. Conf., 10th, Houston, Texas*, pp. 66–71

Jackson MPA, Talbot CJ. 1986. External shapes, strain rates, and dynamics of salt structures. *Geol. Soc. Am. Bull.* 97(3): 305–23

Jackson MPA, Talbot CJ. 1991. A glossary of salt tectonics. *Univ. Texas at Austin, Bur. Econ. Geol., Geol. Circ.* 91-4, 44 pp.

Jackson MPA, Vendeville BC. 1993. Extreme overthrusting and extension above allochthonous salt sheets emplaced during experimental progradation. *Am. Assoc. Petrol. Geol. Annu. Conv. Off. Progr., New Orleans*, pp. 122–23 (Abstr.)

Jackson MPA, Vendeville BC. 1994. Regional extension as a geologic trigger for diapirism. *Geol. Soc. Am. Bull.* 94(1) In press

Kastens KA. 1991. Rate of outward growth of the Mediterranean Ridge accretionary complex. *Tectonophysics* 199: 25–50

Lin S-T. 1992. *Experimental study of syndepositional and postdepositional gravity spreading of a brittle overburden and viscous substratum.* Master's thesis. Univ. Tex., Austin. 196 pp.

Liro LM. 1992. Distribution of shallow salt structures, lower slope of the northern Gulf of Mexico, USA. *Mar. Petrol. Geol.* 9(4): 433–51

McGuinness DB, Hossack JR. 1993. The development of allochthonous salt sheets as controlled by the rates of extension, sedimentation, and slat supply. *Am. Assoc. Petrol. Geol. Int. Hedberg Res. Conf., Salt Tectonics, Bath, England* (Abstr.)

Nelson TH. 1991. Salt tectonics and listric-normal faulting. In *The Gulf of Mexico Basin*, ed. A. Salvador, pp. 73–89. Boulder: Geol. Soc. Am.

Peel FJ, Travis CJ, Hossack JR, McGuinness DB. 1993. Structural provinces in the cover sediments of the US Gulf of Mexico basin: linked systems of extension, compression and salt movement. *Am. Assoc. Petrol. Geol. Annu. Conv. Off. Progr., New Orleans*, 164 pp. (Abstr.)

Rouby D, Cobbold PR, Szatmari P, Demercian S, Coelho D, Rici JA. 1993. Least-squares palinspastic restoration of regions of normal faulting—application to the Campos basin (Brazil). *Tectonophysics*. 221: 439–52

Schultz-Ela DD. 1992. Restoration of cross sections to constrain deformation processes of extensional terranes. *Mar. Petrol. Geol.* 9(4): 372–88

Schultz-Ela DD, Jackson MPA, Vendeville BC. 1993. Mechanics of active salt diapirism. *Tectonophysics*. In press

Schuster DC. 1993. Deformation of allochthonous salt and evolution of related structural systems, eastern Louisiana Gulf Coast. *Am. Assoc. Petrol. Geol. Annu. Conv. Off. Progr., New Orleans*, p. 179 (Abstr.)

Seni SJ. 1992. Evolution of salt structures during burial of salt sheets on the slope, northern Gulf of Mexico. *Mar. Petrol. Geol.* 9: 452–68

Seni SJ, Jackson MPA. 1983. Evolution of salt structures, East Texas diapir province, Part 2: Patterns and rates of halokinesis. *Am. Assoc. Petrol. Geol. Bull.* 67: 1245–74

Talbot CJ, Jarvis RJ. 1984. Age, budget and dynamics of an active salt extrusion in Iran. *J. Struct. Geol.* 6: 521–33

Talbot CJ, Rogers EA. 1980. Seasonal movements in a salt glacier in Iran. *Science* 208: 395–97

Trusheim F. 1960. Mechanism of salt migration in northern Germany. *Am. Assoc. Petrol. Geol. Bull.* 44(9): 1519–40

van Keken PE, Spiers CJ, van den Berg AP, Muyzert EJ. 1993. The effective viscosity of rocksalt: implementation of steady state creep laws to numerical models of salt diapirism. *Tectonophysics* 225: 457–76

Vendeville BC. 1991. Thin-skinned compressional structures above frictional-plastic and viscous décollement layers. *Geol. Soc. Am. Abstr. with Programs* 23: A423 (Abstr.)

Vendeville BC, Cobbold PR. 1987. Synsedimentary gravitational sliding and listric normal growth faults: insights from scaled physical models. *C. R. Acad. Sci. Paris* 305: 1313–19

Vendeville BC, Jackson MPA. 1991. Deposition, extension, and the shape of downbuilding diapirs. *Am. Assoc. Petrol. Geol. Bull.* 75: 687–88 (Abstr.)

Vendeville BC, Jackson MPA. 1992a. The rise of diapirs during thin-skinned extension. *Mar. Petrol. Geol.* 9: 331–53

Vendeville BC, Jackson MPA. 1992b. The fall of diapirs during thin-skinned extension. *Mar. Petrol. Geol.* 9: 354–71

Vendeville BC, Jackson MPA. 1993. Rates of extension and deposition determine whether growth faults or salt diapirs form. *Proc. Gulf Coast Sec. Soc. Econ. Paleontol. Mineral. Found. Res. Conf., 14th, Houston, Texas*. In press

Vendeville BC, Jackson MPA, Ge H. 1993a. Detached salt tectonics during basement-involved extension. *Am. Assoc. Petrol. Geol. Annu. Conv. Off. Progr., New Orleans*, p. 195 (Abstr.)

Vendeville BC, Jackson MPA, Weijermars, R. 1993b. Rates of salt flow in passive diapirs and their source layers. *Proc. Gulf Coast Sec. Soc. Econ. Paleontol. Mineral. Found. Res. Conf., 14th, Houston, Texas*. In press

Weijermars R, Jackson MPA, Vendeville BC. 1993. Rheological and tectonic modelling of salt provinces. *Tectonophysics* 217: 143–74

Withjack MO, Scheiner C. 1982. Fault patterns associated with domes—an experimental and analytical study. *Am. Assoc. Petrol. Geol. Bull.* 66: 302–16

Worrall DM, Snelson S. 1989. Evolution of the northern Gulf of Mexico, with emphasis on Cenozoic growth faulting and the role of salt. In *The Geology of North America—An Overview*, ed. A. W. Bally, A. R. Palmer, pp. 97–138. Boulder: Geol. Soc. Am.

Wu S. 1993. *Salt and slope tectonics offshore Louisiana*. PhD thesis. Rice Univ., Houston. 251 pp.

Wu S, Bally AW, Cramez C. 1989. Allochthonous salt, structure and stratigraphy of the northeastern Gulf of Mexico. Part II: Structure. *Mar. Petrol. Geol.* 7: 334–71

Annu. Rev. Earth Planet. Sci. 1994. 22:119–44

GIANT HAWAIIAN LANDSLIDES[1]

James G. Moore and William R. Normark

US Geological Survey, Menlo Park, California 94025

Robin T. Holcomb

US Geological Survey, University of Washington, Seattle, Washington 98195

KEY WORDS: Hawaiian Ridge, slump, debris avalanche

INTRODUCTION

Dozens of major landslides have recently been discovered on the submarine flanks of the Hawaiian Ridge (Figures 1 and 2). These landslides are among the largest on Earth, attaining lengths greater than 200 km and volumes of several thousand cubic kilometers. Two general types of giant landslides, slumps and debris avalanches, are identified, but many intermediate forms occur. These were revealed during a 1986–1991 swath sonar mapping program of the United States Hawaiian Exclusive Economic Zone, a cooperative venture by the U.S. Geological Survey and the British Institute of Oceanographic Sciences. Employed in that work was the long-range side-looking sonar system GLORIA, which records acoustic back-scatter from the seafloor in an effective swath 25–30 km wide centered on the survey ship track (Somers et al 1978). The area mapped along the Hawaiian Ridge—2200 km long by 600 km wide—includes about 1.3 million km^2.

This article reviews general features of the landslides, including criteria for identification, types, relation to volcano structure, mechanism of formation, and degradation with age. Detailed information is presented for a few examples. Discovery of the remarkably common occurrence of giant

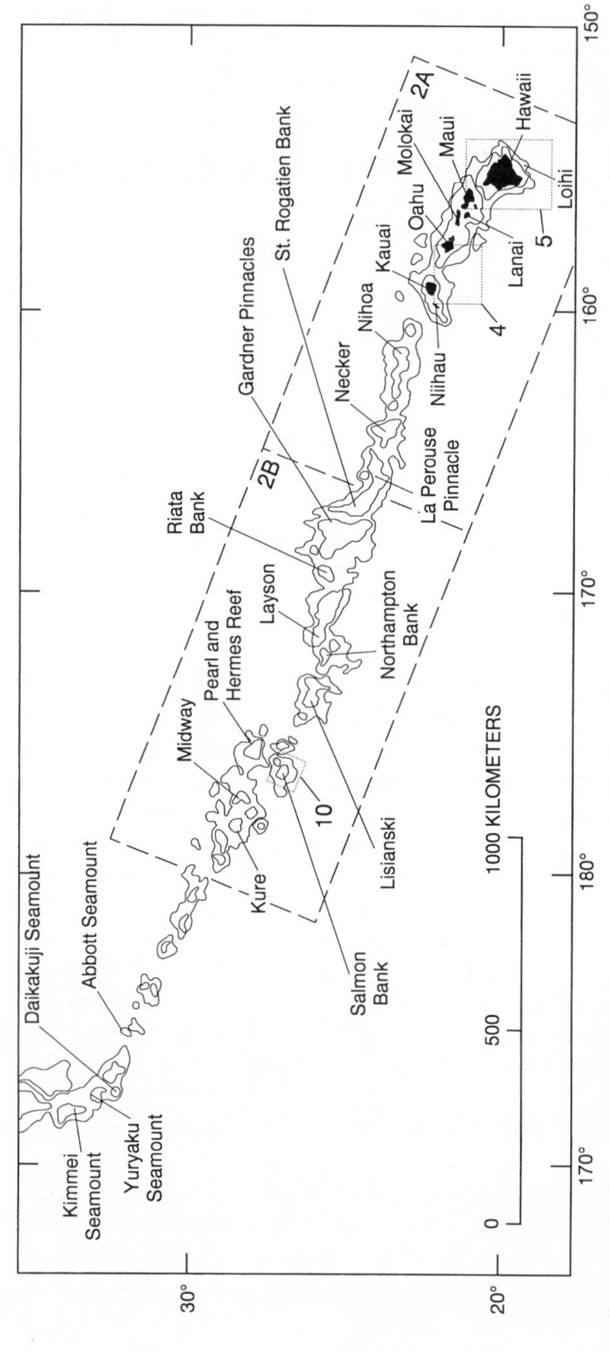

Figure 1 Bathymetry of the Hawaiian volcanic chain with contours at 1 and 2 km depths shown in area of the chain only (after Clague & Dalrymple 1987). Age of volcanoes at the bend, near Kimmei seamount, is about 40 Ma; volcanoes on the island of Hawaii are still active. Dashed boxes indicate areas of Figures 2A, 2B, 4, 5, and 10.

landslides on the Hawaiian Ridge has led to the realization that such forms of mass wasting occur on the submarine flanks of other large marine volcanoes the world over.

GEOLOGIC SETTING OF HAWAIIAN RIDGE

The Hawaiian Ridge was produced by ascent of heat and fluids from a mantle hotspot centered beneath the present island of Hawaii at the southeast end of the ridge. Eruptions of basaltic lava above the hot spot built a succession of giant marine volcanoes. As each volcano grew, movement of the Pacific lithospheric plate toward the west-northwest carried the volcanic edifice away from the hotspot, eventually causing it to become extinct when the feeding conduit was cut. As older volcanoes approached extinction, newer ones grew adjacent to them on oceanic crust newly brought over the hotspot. Long continuation of this process built the present Hawaiian Ridge into a volcanic chain thousands of kilometers long with a few still active volcanoes at its southeast end.

The load of the volcanoes has depressed the underlying lithosphere forming a moat-like depression flanking the ridge—the Hawaiian Trough, and an outer bulge seaward of the moat—the Hawaiian Arch. The axis of the Trough, with sea-floor depths commonly exceeding 5 km, is about 140 km from the ridge axis and is partially filled with volcanoclastic material derived from mass wasting of the adjacent volcanoes. The axis of the broad Arch is about 250 km from the ridge axis and is generally about 4.5 km deep. The morphologic expression of both the Trough and Arch are lost with age primarily as a result of sedimentary infilling of the Trough.

During growth of the Hawaiian Ridge the Pacific plate moved about 9 cm/yr west-northwest. This rate is based on radiometric ages of lava flows erupted near the end of each volcano's shield-building period (Clague & Dalrymple 1987). Hence the volcanoes of the ridge form a continuous age sequence. The volcano of Midway Island grew about 28 million years ago and the volcanoes of Hualalai, Mauna Loa, and Kilauea, of the island of Hawaii 2000 km southeast, grew only during the last few hundred thousand years and are still active. The subaerial form of the volcanoes clearly reflects this age difference. The older northwest volcanoes are deeply incised or eroded to sea level, capped by coral reefs, and deeply submerged. In contrast, the submarine flanks of the volcanoes age slowly and are relatively well-preserved. Age-related differences in volcano morphology of the older volcanoes as compared to the younger and active ones are much less below sea level than above.

Although submarine landsliding occurs throughout the growth of the

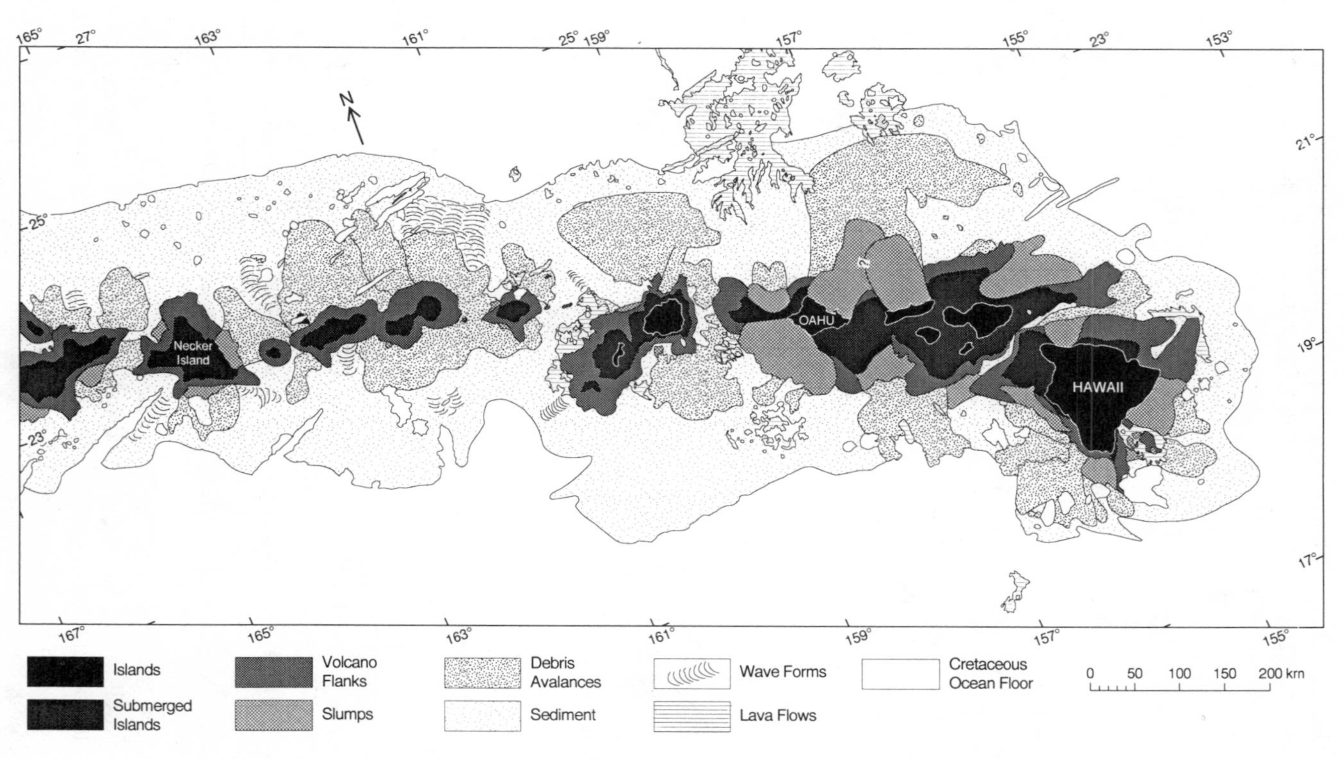

Figure 2A Generalized geologic map of the Hawaiian Ridge. See Figure 1 for location.

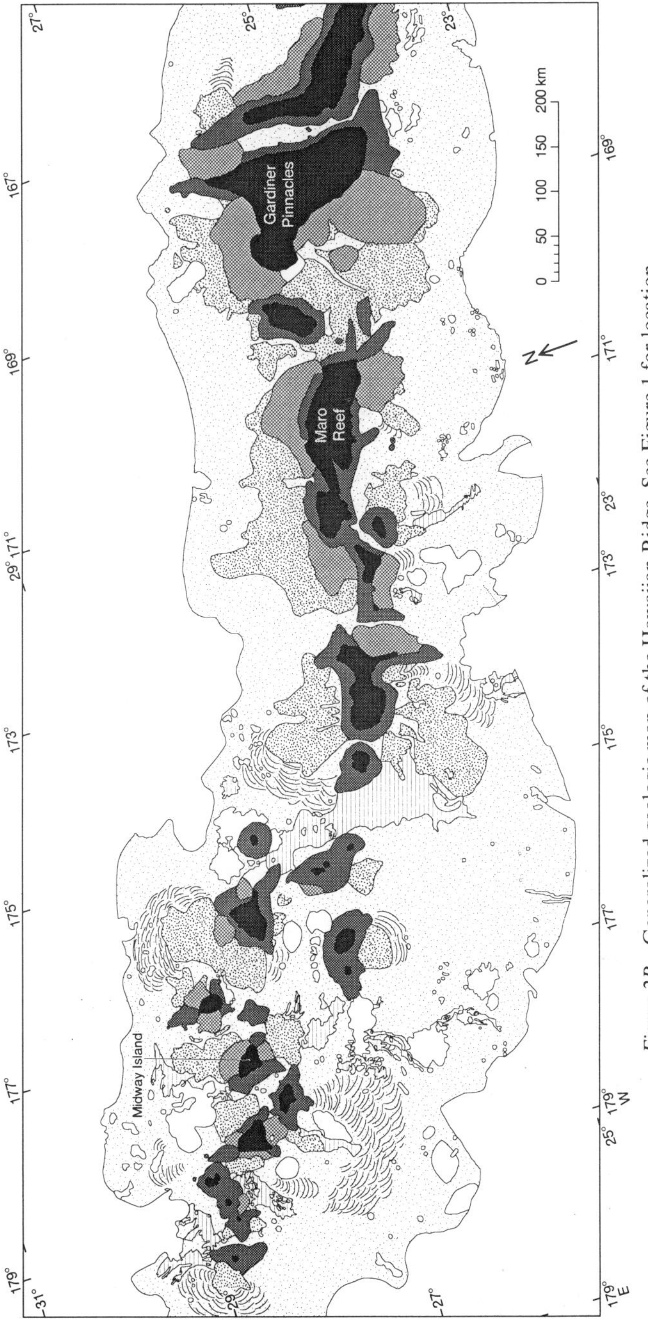

Figure 2B Generalized geologic map of the Hawaiian Ridge. See Figure 1 for location.

volcanoes (and is present on the embryonic Loihi volcano), the largest landslides apparently occur late in the period of active shield growth when the volcanoes are close to their maximum size, are young and unstable, and when seismic activity is high. Hence, the ages of volcanoes along the ridge are mirrored by the ages of their associated landslides.

DISTRIBUTION OF LANDSLIDES

At least 68 major landslides more than 20 km long occur along a 2200 km stretch of the Hawaiian Ridge from 200 km northwest of Midway (180° W longitude) to the island of Hawaii (154° W longitude; Figure 2). Hence the giant landslides average one every 32 km or one every 350 ka. Landslides cover about one half of the flanks of the ridge. Surprisingly the western landslides, which are nearly 30 million years older than the younger ones on the eastern end of the ridge, appear well preserved on sonar side-scan images. The maximum length of landslides apparently increases from 50–100 km at the older west end of the mapped area to 150–300 km at the younger east end (Figure 3*a*). This trend may result in part from thicker post-sliding sediment covering the lower-relief distal parts of older debris fields. A similar relation, however, is apparent for slumps (Figure 3*a*), which have higher relief that is less likely to be obscured by sediment cover; this circumstance suggests that indeed the landslides tend to be longer and larger toward the younger end of the ridge.

Although landslides are equally numerous on the northeast and southwest flanks of the ridge, the larger ones tend to be directed more toward the older (west-northwest) end of the ridge, rather than normal to the ridge (Figure 3*b*). This tendency may result from the greater chance of preservation of landslides toward the trailing, rather than leading, end of the propagating volcanic ridge. All landslides that move approximately east-southeast, for instance, off the actively-growing end of the ridge, become covered by continued growth of the ridge.

In addition to the giant landslides described here that are readily detectable in deep water on small-scale GLORIA images, medium-sized landslides having volumes of tens of km^3 are common in shallower water. However, these potentially active landslides of intermediate size have not been adequately mapped because of the difficulties of conducting side-scan sonar surveys in shallow water near the coast. As a result, the hazard of submarine landsliding cannot be estimated well from available data; a better assessment must await detailed studies of smaller but more frequent landslides (Normark et al 1993).

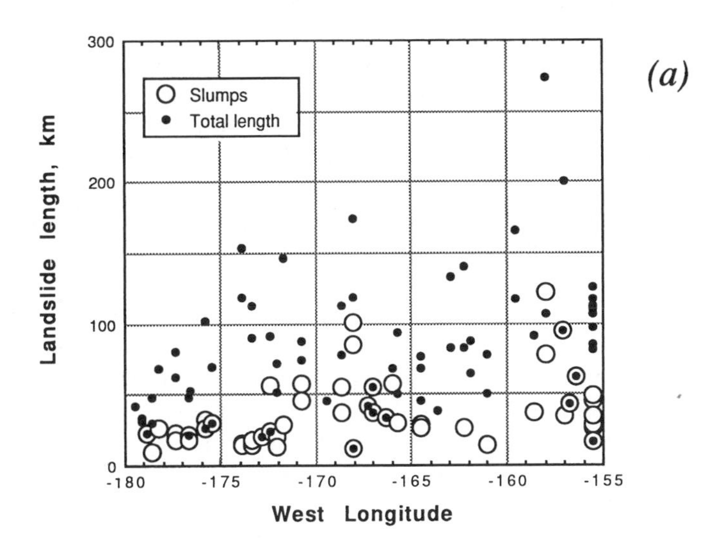

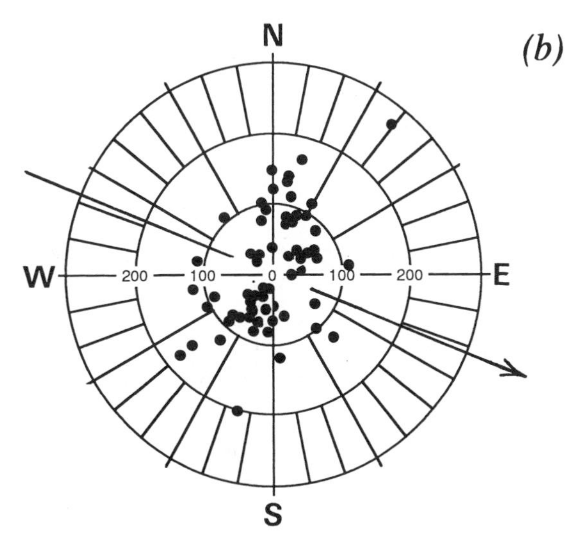

Figure 3 Features of Hawaiian submarine landslides longer than 20 km. (*a*) Length of slumps and total length of landslides relative to longitude. (*b*) Direction of slide movement relative to total length of landslide in km. Arrow shows trend of Hawaiian Ridge.

CHARACTER OF LANDSLIDES

The mass wasting features on the Hawaiian Ridge comprise a continuum ranging in size over several orders of magnitude. The general term "landslide" is used herein for all forms of mass movement from incipient slumping that has only slightly disrupted the structural coherence of the volcano flanks, through various degrees of disaggregation, to long-runout avalanches in which fragmentation has totally disrupted and dispersed original volcanic structures. The large landslides described here (more than 20 km long) have been grouped by morphology and degree of dislocation into two major types: slumps and debris avalanches. Because oversteepened parts of slumps may fail as debris avalanches, facies of both types commonly occur in the same landslide. This terminology (after Moore et al 1989) is based on criteria derived from GLORIA sonar images and will no doubt be modified as we learn more about these giant landslides.

Landslide morphology is better preserved below sea level than above because of the lower likelihood of erosion or cover by younger lava. However, many subaerial topographic features owe their origin to landslide processes. All of the major fault systems on the island of Hawaii, as well as subtle changes of slope that probably reflect buried fault scarps, can be related to landslides. The rift zones that radiate from many volcano summits may reflect the primary pull-apart regions at the heads of giant slumps (Moore & Krivoy 1964) or zones of gravitational spreading. Large erosional canyons on several of the islands seem to form within the oversteepened amphitheaters at the heads of giant debris avalanches (Moore et al 1989). The three largest historic earthquakes on the island—those of 1868, 1951, and 1975, all of magnitude 7 or greater—were apparently related to incremental movement of giant slumps (Moore & Mark 1992).

Slumps

Large slumps are deeply rooted in the volcanic edifice and may extend back to the volcanic rift zone and down to the base of the volcanic pile. Slumps may creep over an extended period as they keep pace with the load of volcanic material erupted on their upper part. The slumps are as much as 110 km wide and 10 km thick, and have an overall gradient greater than 3°. In the upper tensional part of the slumps, transverse normal faults marked by scarps commonly bound a few large tilted blocks that may be tens of km in length and several km in width (Figure 4). The compressional regime in the lower part of the slumps is marked by broad bulges, closed depressions, and steep toes. Despite faulting and dislocation, the slumps generally maintain an overall coherent lobate shape by which they are identified in GLORIA images (Figure 4). Parts of the slumps may collapse and produce a debris avalanche that extends downslope beyond the lower end of the slump (Compare Figures 4 and 2). Such compound landslides are common.

The active Hilina slump on the south flank of Kilauea (Figures 5 and 6) may serve as a model of giant slump behavior. However, we know only that the upper few kilometers are the site of repeated dike injection; processes and structures below are still a matter of uncertainty. During a magnitude 7.2 earthquake in 1975, a 60-km length of Kilauea's south coast subsided as much as 3.5 m and moved seaward as much as 8 m. The Hilina fault system was reactivated over a length of 30 km with normal faults downthrown toward the sea (Tilling et al 1976, Lipman et al 1985). In addition to such episodes of rapid movement, the south flank of the volcano also is creeping seaward continuously as indicated by frequently

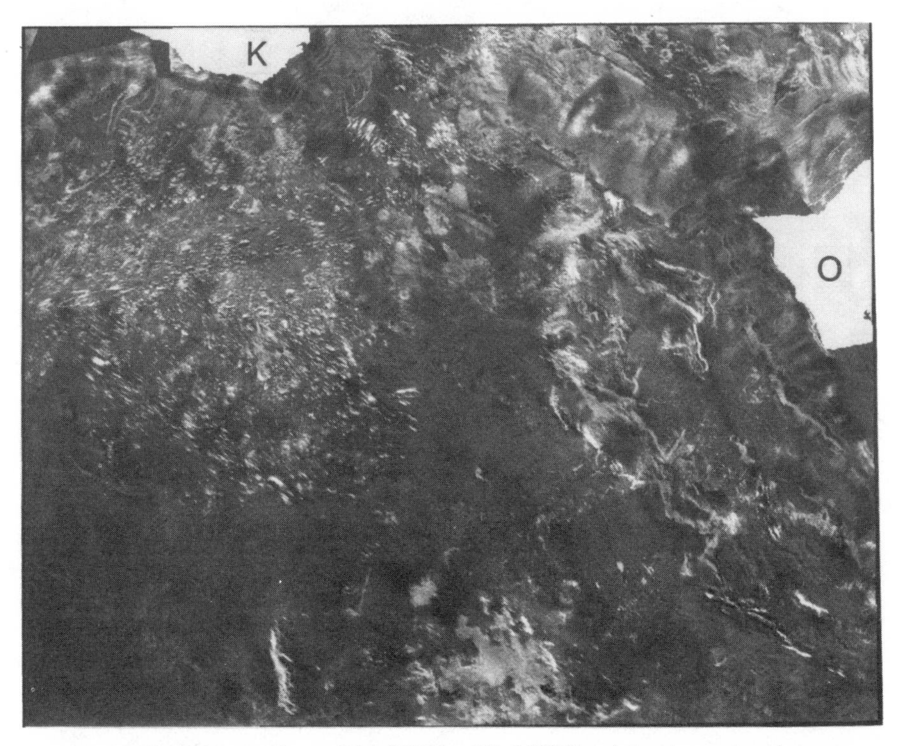

Figure 4 GLORIA sonar image (20.5–22°N lat; 158–160°W long) showing contrast between debris avalanche and slump. The south Kauai debris avalanche (with speckled reflectors) extends 100 km south of the island of Kauai (K) and the Waianai slump (with northwest-trending scarps) extends 65 km southwest of Oahu (O). Vertical dimension (north-south) of image is 167 km. See Figure 1 for location.

repeated geodetic measurements. The east rift zone itself seems to define the pull-apart zone; the region north is stable whereas the region south, extending to the coast and beyond, is moving south at about 6 cm/year (Segall et al 1992).

Drill holes near the axis of the rift zone have penetrated coral reef

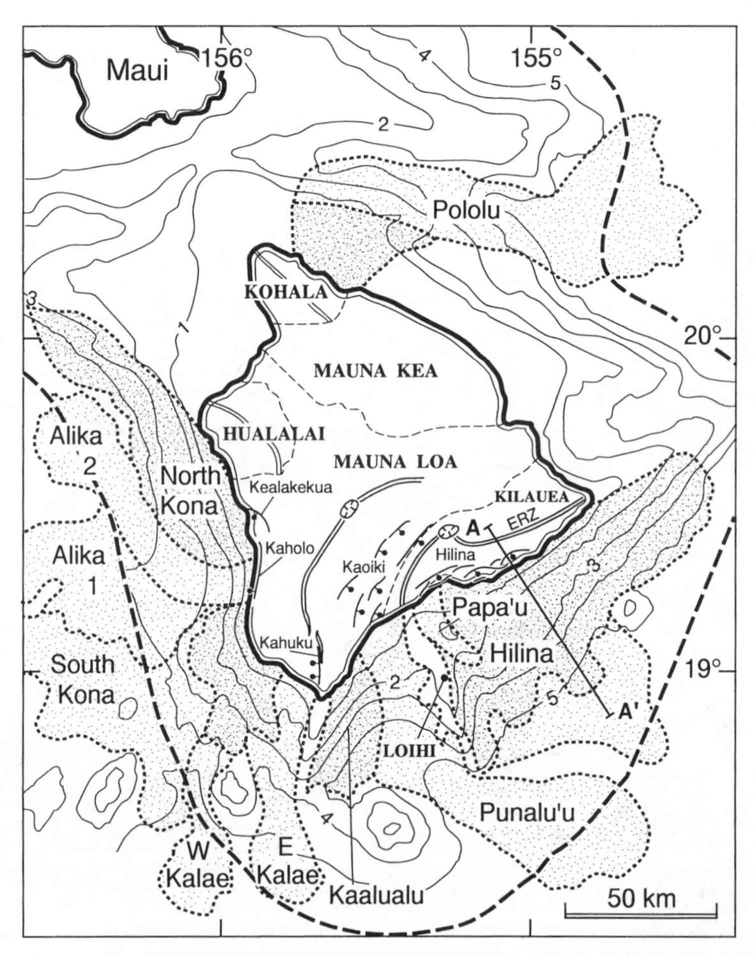

Figure 5 Island of Hawaii and offshore bathymetry; depth contours in kilometers. Dashed lines outline the five volcanoes that compose the island. Dark pattern—major slumps; light pattern—debris avalanches; fine lines—major fault systems (ball on downthrown side); heavy dashed line—axis of Hawaiian Trough; double lines—volcanic rift zones (ERZ, Kilauea's east rift zone). Profile located along line A-A′ shown in Figure 6.

deposits more than 1300 m below sea level—deeper than can be attributed to regional subsidence of Kilauea (Thomas 1992). These near-shore deposits mark a period of stability and lack of neighboring volcanism, and indicate that dramatic changes have recently occurred in this region.

Most earthquake hypocenters near the east rift zone occur in two depth-group bands that parallel the rift zone at depths of 0–5 and 5–13 km. The shallower band coincides with the surface trace of vents and cones of the rift zone, whereas the deeper band is centered about 4 km seaward of the rift zone (Klein et al 1987). The seismic gap 1–3 km wide between the bands (as viewed in map plan) is believed to reflect the presence of an elongate mass or master dike of hot magma (or zone of plasticity) along the core of the rift zone (Figure 6). Such a fluid core to the rift zone has been postulated earlier because of the need for secondary magma storage zones to account for the greater volumes of erupted lava as compared to the volume of summit subsidence (Swanson et al 1976). The master dike serves both as the conduit conducting magma from the sub-summit magma chamber out the rift zone, and as the primary pull-apart zone at the head of the Hilina slump permitting spreading of the mobile south flank of the volcano (Delaney et al 1990, Borgia & Treves 1992).

The shallow band of earthquakes above this master dike results from brittle fracturing as cracks propagate to the surface to accommodate spreading of the rift zone. These cracks conduct magma upward from the master dike in the core of the rift zone through thin feeder dikes to eruptive vents along the rift zone. Major aseismic decoupling takes place in the liquid or plastic core of the master dike. The deeper earthquakes on the

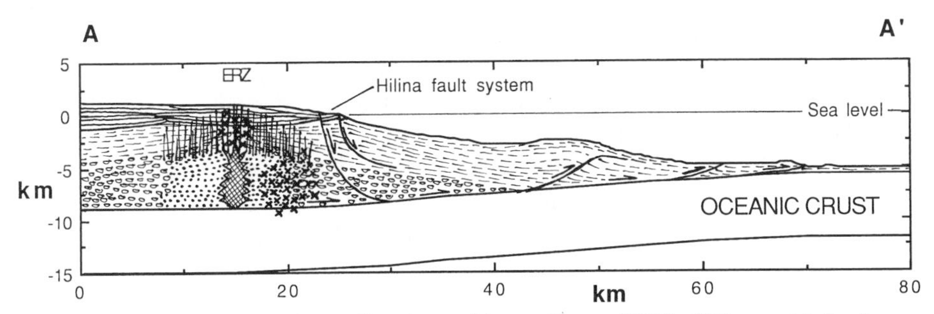

Figure 6 Cross section of the Hilina slump with east rift zone (ERZ) of Kilauea at its head; no vertical exaggeration. Horizontal layering—subaerial lava; dashed layering—fragmental lava debris (hyaloclastite); ellipses—pillow lava; vertical lines—sheeted dikes; dotted pattern—gabbro; cross-hatched pattern—magma; crosses—generalized location of earthquake hypocenters (Klein et al 1987). Modified from Lipman et al 1985, Delaney et al 1990, Borgia & Treves 1992. See Figure 5 for location.

seaward side of the master dike result from fracturing and faulting in the south flank as it decouples from the unmoving lower regions and adjusts in zones of stress concentration where the rock is cool enough to fracture.

The slump is bounded below by a major décollement that probably slopes upward from near the base of the volcanic pile at about 10 km below sea level to the ocean floor 5 km deep 50 km seaward. This sliding surface probably coincides in part with the base of the Hawaiian volcanic edifice where it rests on Cretaceous oceanic crust. At its shallower southern part the slump movement is probably facilitated by poorly consolidated, volcanogenic and pelagic sediment on the oceanic crust, and on its deeper, northern part by a hot plastic zone related to magma within the rift zone (Figure 6). The depth extent of the Hilina fault system and whether it connects with the base of the slump is debatable because offshore movement proceeds almost aseismically. These faults may reach to the décollement or curve to join major lithologic discontinuities within the slump (Figure 6).

A steady state movement of the slump by spreading away from the molten core of the rift-zone master dike and solidification of the margins of the master dike will cause the rift zone to migrate seaward at one half the rate of rift-zone opening or slump movement (Hill & Zucca 1987). This spreading will develop a three-fold layering in both the stable or nonsliding part of the volcano and the upper part of the slump, consisting of: a shallow layer of lava flows and hyaloclastites fed from eruptive vents on the rift zone, an intermediate sheeted dike complex of volcanic feeder dikes, and a deep gabbroic layer of solidified walls of the master rift-zone dike (Borgia & Treves 1992).

A major submarine bench about 3000 m deep on the Hilina slump is 35 km long and as wide as 15 km and parallels the subaerial Hilina fault system. This bench is widest downslope from the region where south-flank earthquakes are most concentrated. Elevated ridges near the outer (southern) edge of the bench bound several closed depressions, some of which are more than 100 m deep. The presence of these unfilled, closed depressions below an area where lava flows frequently enter the sea and feed fragmental debris downslope, demonstrates that the bench is tectonically active. The depressions are forming faster than they are being filled.

Growth of the bench is apparently related to a major thrust fault in the lower, compressional part of the Hilina slump (Figure 6). Thrusting elevates the seaward part of the bench and creates a steep scarp 2 km high along the southeast side of the bench. Borgia et al (1990) suggest that such thrusts are probably blind thrusts surmounted by folds.

Three arcuate ridges that are convex southeast occur at the lowest submarine reaches of the Hilina Slump immediately adjacent to the undisturbed deep sea floor near the axis of the Hawaiian Trough. These ridges are 8 to 13 km long and 500 to 700 m high along their southern flanks. They are apparently uplifted by additional thrust faults branching from the primary décollement at the base of the Hilina slump, and probably mark the farthest extension of the slump onto the apron of debris around the foot of the volcano (Figure 6). Hummocks from related debris avalanches extend somewhat farther seaward.

The similarity of the east rift zone of Kilauea, and Hawaiian rift zones in general, to oceanic spreading ridges has been discussed by Holcomb & Clague (1983), Hill & Zucca (1987), and Lonsdale (1989). The general model of formation of the massive slumps by spreading from an elongate rift-zone magma chamber at their upper part, and overriding undisturbed seafloor at their lower part, has led Borgia & Treves (1992) to view the Hawaiian volcanoes as a link between volcanic processes and oceanic tectonic plates. The spreading volcanic rift zone, where extrusive and intrusive layers are generated, corresponds to the oceanic spreading ridges, and the lower part that overrides the ocean floor is "analogous to the overthrusting of a subduction zone accretionary prism" (Thurber & Gripp 1988).

Debris Avalanches

The debris avalanches are more surficial features as compared with the giant slumps. They are commonly longer, thinner, and less steep, and have a well-defined amphitheater at their head and hummocky terrain in their lower part. The debris avalanches are 0.05–2 km thick, as much as 230 km long, and possess an overall gradient of less than 3°. Some have apparently formed by catastrophic failure of oversteepened slumps. Rapid movement of avalanches in single events is indicated by their thinness and great length, by movement up the slope of the Hawaiian Arch of their distal reaches in some places (Moore et al 1989), and by their hummocky, fragmented distal regions that closely resemble those of known catastrophic subaerial volcanic landslides, such as those at Mount St. Helens (Voight et al 1981) and Mt. Shasta (Crandell et al 1984).

Amphitheaters commonly are better developed at the heads of the debris avalanches, especially below sea level, than on the slumps. The slow, intermittent nature of slump movement allows the amphitheaters to be filled more completely by the volcanic products erupted during and following movement. The upper part of the East Ka Lae landslide is fed from a large amphitheater (Moore & Clague 1992), the east wall of which is the subaerial Kahuku fault scarp and its offshore extension (Figure 5). Dikes parallel to the rift zone are exposed along the submarine segment of the scarp indicating that the East Ka Lae debris avalanche cut and steepened the west flank of the submarine rift zone ridge after the southwest rift was established (Moore & Clague 1992).

The upper reaches of the Alika debris avalanches are smooth talus slopes extending from the shoreline down to 2 km depth (Figures 5 and 7). The landslide amphitheaters are not obvious above sea level because they are covered by extensive young lava flows from Mauna Loa. Anomalously steep slopes, however, in a zone 15 km wide extending 8 km inland upslope from the Alika debris avalanches may reflect a landslide amphitheater that is now largely filled and covered by younger lava flows (Moore & Mark 1992). The entire west flank of Mauna Loa is steeper than normal shield slopes and probably results from lava flows filling amphitheaters from the earlier North and South Kona and Alika landslides (Figure 7).

Large submarine canyons offshore from the Hawaiian islands commonly incise the amphitheaters of major debris avalanches, and can be traced into some of the major subaerial canyons of Hawaii. Submarine canyons occur on the upper parts of the north Kauai, northeast Oahu, north Molokai, and northeast Hawaii debris avalanches and extend down 800 to 1300 m below sea level. Subaerial valleys between the Ninole Hills, on the southeast flank of Mauna Loa volcano, have been interpreted as partly buried canyons in the amphitheater of a major southeast-directed landslide (Lipman et al 1990). The submarine canyons were apparently carved subaerially after landsliding and were later submerged by regional subsidence. Subaerial canyon cutting in the basaltic bedrock of the amphitheaters was promoted by the oversteepened, and recently stripped slopes, especially in the amphitheaters formed on the windward (northeast) volcano slopes that receive high rainfall.

The Alika-2 debris avalanche, among the youngest recognized, moved west in its upper reaches, but turned north 60° in its middle course and followed the base of the west slope of the volcano (Figures 5, 7). This change in direction of movement could have resulted from deflection by a debris ridge left by the Alika-1 debris avalanche as well as by the small seamount that separates the Alika-1 and -2 landslide tongues. The middle course of the avalanche has a 40 km long flat-floored channel that is 10 km wide flanked by natural levees. Individual levee segments can be traced as far as 12 km, are more than 1 km wide, and rise from 20 to 100 m above the channel floor (Moore et al 1992). Deep-tow side-scan images show that the levees are discontinuous and are commonly composed of piles of individual blocks and hummocks. Other large levees on GLORIA images

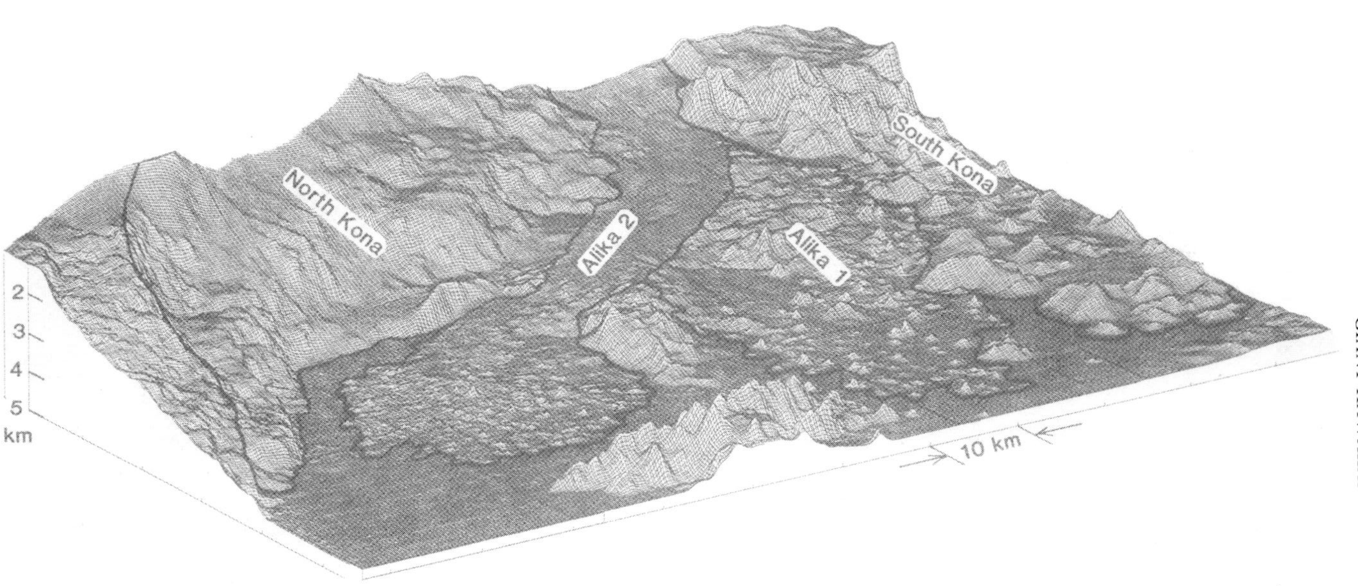

Figure 7 Physiographic view toward the southeast, based on NOAA multibeam bathymetry, of the west submarine flank of Mauna Loa volcano showing four major landslides with 4-fold vertical exaggeration. See area of diagram in Figure 5.

occur in the upper parts of debris avalanches west of Gardner Pinnacles, west of Lisianski, and north of Kauai (Figure 1).

Fields of hummocks or blocky hills are common on the distal parts of debris avalanches. These fields produce a unique speckled pattern on sonar images that is the most defining characteristic of debris avalanches (Figure 4). Multibeam sonar mapping has defined about 190 hummocks from 0.5 to 2 km in size on the 34–km diameter distal lobe of the Alika-2 debris avalanche.

Available data from the Alika landslides suggest that the size distributions of hummocks differ from one avalanche to another. On the Alika-2 avalanche many of the larger hummocks are about 1 km in diameter (Figures 7, 8), whereas on the South Kona avalanche the hummocks are 10 times this size (Figure 7). A full documentation of the range of hummock sizes requires a combination of several scale-dependant techniques (Figure 8). Multibeam echosounding data, gridded at 200 m spacing for production of bathymetric maps, can resolve only those hummocks larger than several hundred meters. Deep-towed side-looking sonar can effectively define blocks from about 5 m to 1 km wide. Photographs by deep-towed cameras can depict blocks and fragments from a few meters to a few centimeters in size. Hundreds of photographs from within the lateral margin of the Alika-2 debris avalanche indicate that smaller fragments occupy the space between the larger blocks and hummocks, and that considerably more than one half the area is littered with fragmental rock material. The volcanic rock fragments are commonly intricately fractured, sharply angular, and show sub-vertical layering that is probably rotated from an original sub-horizontal attitude.

The combined measurements from photographs, deep-towed vehicles, and multibeam bathymetry indicate that a continuum of hummock-block sizes occur from <1 to 1500 m in size (Figure 8). Widths of blocks or hummocks are several times their height and this factor increases with increasing hummock size so that hummocks exceeding about 1 km in diameter show a diameter:height ratio of about 10:1.

Except for a cover of post-sliding sediment <0.5 m thick on the Alika-2 debris avalanche, the spacing of the hummocks and blocks is remarkably similar to that of the 1980 Mount St. Helens subaerial volcanic landslide immediately after emplacement. In addition, the diameter-height relations of the hummocks are similar to those of subaerial volcanic debris avalanches such as those of Mount St. Helens (Voight et al 1981) and Mount Shasta (Crandell et al 1984). This similarity further suggests that submarine debris avalanches can move rapidly like the Mount St. Helens avalanche, and therefore pose the secondary hazard of tsunami production (Moore & Moore 1988).

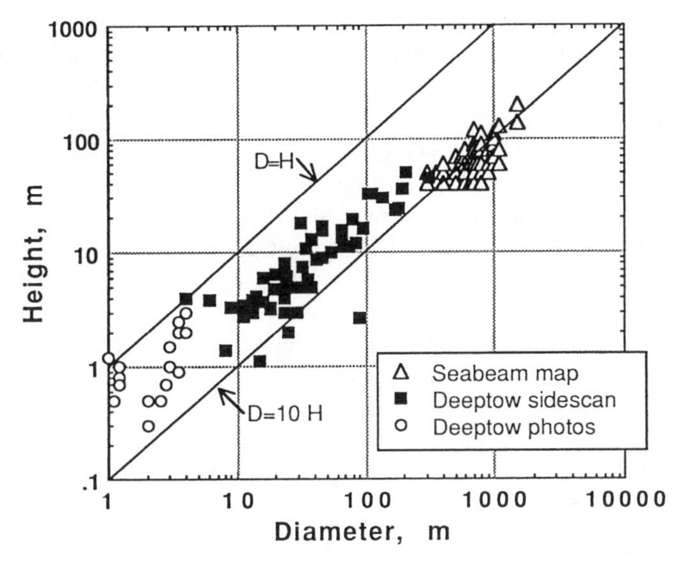

Figure 8 Diameter relative to height of hummocks from distal lobe of Alika-2 slide (see Figure 7 for location). Hummocks range through 3 orders of magnitude in size requiring 3 different methods to image and measure them.

SEDIMENTATION OF LANDSLIDE DEPOSITS

The large debris avalanches provide a major part of the material that fills the Hawaiian Trough. Smaller, and much more frequent, landslides that involve mass wasting of volcanic, coral-reef, and pelagic sediment that accumulates on the flanks of volcanoes provide much of the rest of the fill of the Trough. Along the older part of the Hawaiian Ridge, these mass failures of accumulated sediment are commonly triggered by storm surges and internal waves rather than by earthquakes as is common with the younger volcanoes. The transformation of these smaller landslides to debris flows and turbidity currents effectively distributes the sediment throughout the Trough. The relation of these finer-grained sediment types to the larger debris avalanches and slumps can provide information on the nature and age of landsliding; in turn, the effect of large landslides on the shape of the flanks of the volcanoes can provide longterm controls on the source and transport pathways for the finer-grained sediment.

A seismic reflection survey from Kauai to Hawaii indicates that the northern Trough is filled with as much as 2 km of material deposited in four sequential stages (Rees et al 1993). From the bottom up these stages are represented by 1. a basal pelagic layer, 2. a layer of volcanoclastic sediments, 3. a layer of landslide debris, and 4. a final layer of turbidite and pelagic sediment. The basal unit of pelagic sediment 50–100 m thick was deposited on the 80-Ma oceanic crust prior to flexural depression. This slowly-deposited sediment comes from a variety of sources including wind blown material, slowly settling fine-grained sediment, and biogenic debris. Layer 2 is bedded sediment that fills the Trough as it begins to rapidly subside due to loading by adjacent downstream volcanoes; much of this transport is by turbidity currents flowing along and across the axis of the moat transporting sediment from erosion and mass wasting of older islands in the chain. The third layer, which locally constitutes the bulk of the fill, includes massive volumes of fragmental volcanic rock that represent an average of four major debris avalanche units, each up to 700 m thick. The top layer of ponded sediment in the deepest part of the trough was apparently largely deposited by turbidity currents after volcanism and subsidence effectively ceased.

Pelagic Sedimentation

The Cretaceous crust as well as the volcanic ridge built upon it in the Hawaiian region is subject to slow and continual pelagic sedimentation. The rate of accumulation of this sediment layer varies from place to place and through geologic time partly because of changes in the amount of fine-grained sediment derived from the growing volcanic ridge. Airborne volcanic ash and wind-blown dust, fine-grained material stirred up by submarine landslides, and material eroded from volcanoes and fringing coral reefs all contribute. An understanding of the general rate of accumulation of pelagic sediment can provide information on the age of the giant landslides.

The sedimentation rate on the Hawaiian Trough and Arch can be determined from sediment thicknesses on the large North Arch lava flow (North of Oahu, Figure 2), which has been dated by magnetostratigraphy of sediment cores and the thickness of the alteration layer (palagonite) formed on the surface of basalt glass exposed to seawater (Clague et al 1990). The rate of sediment accumulation on the lava varies both with age and with distance from the islands. At a distance of about 140 km from the islands, somewhat beyond the axis of the Trough, the rate is 5.8 m/my for the period 1.7 to 1 Ma, and 1.8 m/my for the period 1 to 0.7 Ma. At a distance of about 180 km from the islands, near the axis of the Hawaiian Arch, the rates are less than half those determined near the Trough.

These rates can be compared to those at the Ocean Drilling Program Site 842, 320 km west of the island of Hawaii on the Arch. Since 3.5 Ma an average rate of 3.8 m/my, with variations between 2.5 and 11.6 m/my, was determined by relating remnant magnetization of the core to the magnetic polarity time scale (Shipboard Scientific Party 1992). Some of the volcanic silt beds at this site, however, have been interpreted as turbidity current deposits generated by the large debris avalanches from Mauna Loa (Garcia et al 1992); turbidite deposition could account for the short periods of higher sedimentation rate.

A third method of estimating pelagic sedimentation rates is by measuring sediment thickness on debris avalanches assuming that the landslide ages approximate the ages of the end of shield-building of their host volcano as determined by K-Ar dating of lava flows (Clague & Dalrymple 1987). The thickness of sediment on the tops of hummocks, where turbidity flow sedimentation is minimized, were measured on 3.5 kHz echo-sounding records for a number of debris avalanches along the ridge. Sediment thickness systematically increases toward the older end of the ridge and the rate approximates 2.5 m/my (Figure 9), which is in general agreement with other estimates described above.

The approximate mean sedimentation rate of 2.5m/my can be employed to estimate the relative and absolute ages of some of the landslides where sediment thickness can be measured on the tops of hummocks provided that other forms of sediment transport, such as turbidity flows, are not important. The Alika-2 debris avalanche (Figure 5) is one of the youngest

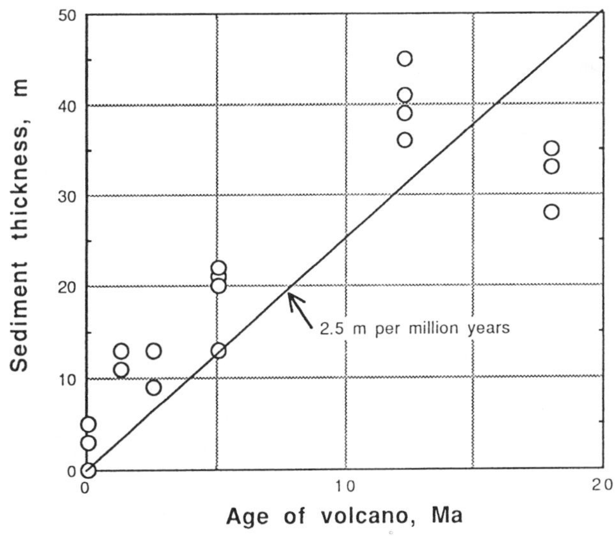

Figure 9 Sediment thickness on top of hummocks of selected landslides as determined from 3.5 kHz echo-sounding records.

for which sediment thicknesses are available. Hundreds of seafloor photographs in the distal hummocky debris field show a sediment cover of only a few tens of centimeters (Moore et al 1992), indicating a maximum age of several 100 ka. Both the Clark and Nuuanu debris avalanches, which are southwest of Lanai and northeast of Oahu respectively (Figures 1 and 2), carry about 9 m of sediment atop distal hummocks suggesting that both are about the same age, roughly 3.6 Ma.

Turbidites

Turbidity currents, which generally flow downslope on the flanks of the ridge, are responsible for transporting large volumes of fine sediment locally over great distances. Geophysical evidence indicates that a large proportion of sediment in the Hawaiian Trough 100 km from the ridge axis consists of turbidites (Normark & Shor 1968). Core from an ODP hole on the Hawaiian Arch 320 km west of Hawaii contains turbidite layers containing glass fragments of a composition similar to that of Mauna Loa volcano on Hawaii (Garcia et al 1992); this suggests long distance transport and upslope flow.

Partly covering and extending downslope from several of the landslides are lobate areas that include a few to dozens of large lobate and crescent-shaped wave forms that resemble sediment waves or dunes in GLORIA images (Figure 10). We prefer the term *mud waves*, following the usage of Flood & Shor (1988), because the reflector character on 3.5-kHz reflection profiles is typical of fine-grained sediment. Flood & Shor (1988) also noted that the term "sediment waves" can be correctly used in describing most sinusoidal sediment topography which can include areas of submarine slumps or even basement-controlled topographic relief (e.g. see Dadisman et al 1992). More than two dozen areas of mud waves have been mapped that are 5–75 km in length, and as much as 100 km in width (Figure 2).

The long axes of the mud waves are generally parallel with, and on the downslope extension of, the toe of landslides (Figure 2). The wave fields occur as a distal areole on the lower part of debris avalanches so that the axes of the waves are perpendicular to the movement direction of the adjacent avalanche or are parallel to regional depth contours. The mud waves are common in 4.5 to 5.2 km depth. Wavelengths range from 1.5 to 3.5 km with amplitudes (as measured on 3.5-kHz echosounding records) of 10–30 m. These dimensions are typical of mud waves previously described from a variety of deep-sea depositional environments (Normark et al 1980, Flood & Shor 1988). Although individual wave forms can commonly be traced for 4–10 km, smaller ones that are not well depicted on GLORIA images probably occur. The wavelength-to-amplitude ratio of the wave forms is about 100 to 1 suggesting that their distinctive

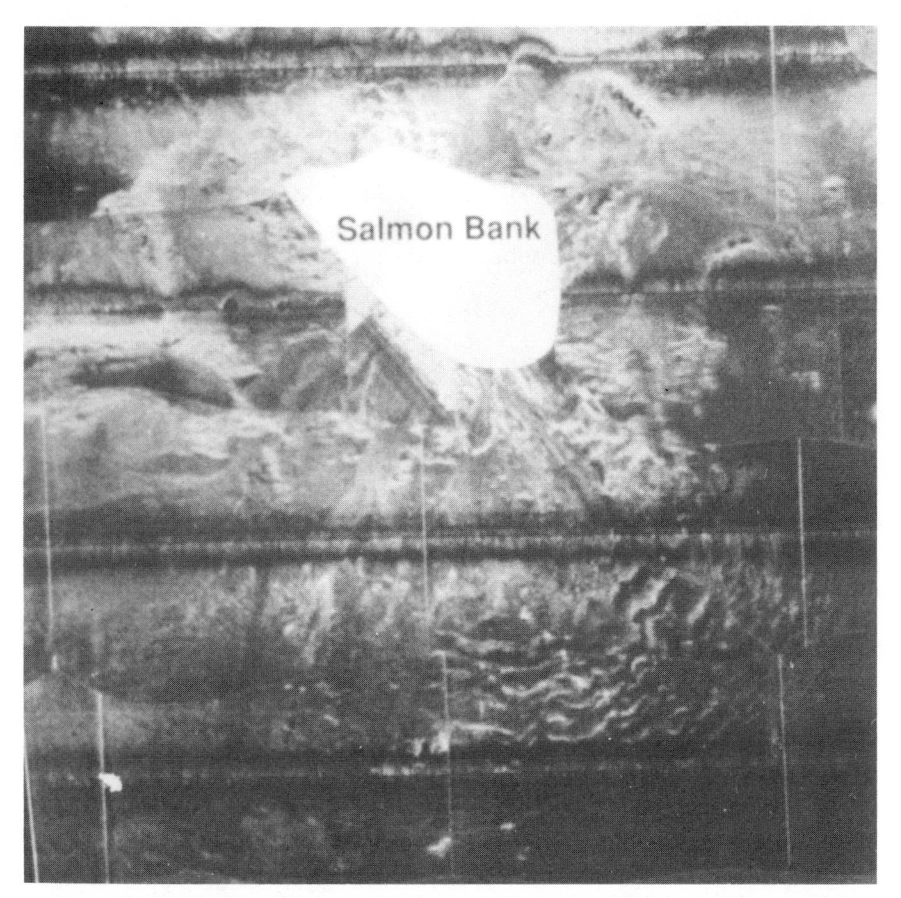

Figure 10 GLORIA sonar images of mud waves with wavelength of about 2.5 km on debris avalanche on south flank of Salmon Bank. East-west lines are ship tracks that are about 35 km apart. See Figure 1 for location.

appearance in the GLORIA imagery results from variation in acoustic backscatter of sediment types from the troughs to the crests of the dunes and not to bathymetric relief.

Prominent chutes, which appear to conduct sediment downslope from now submerged islands and reefs, occur upslope from the wave forms and may be the principal feeding channels for the sediment of the mud waves. Mud waves are more abundant along the western segment of the ridge (Figure 2) and do not appear east of the island of Kauai (160° W long). Restriction of the mud waves to the older part of the ridge suggests that they form slowly by ongoing processes (Flood 1988) after sufficient sediment has accumulated to cover earlier local relief. Because most of the mud waves occur downslope from now submerged islands that were deeply eroded and thickly capped by coral reefs, it is likely that nearshore processes and pelagic sediment deposition on the upper flanks of the volcanoes contributed the abundant sediment necessary to construct the wave forms. This debris was perhaps added incrementally by downward moving turbidity currents that must have swept over broad areas in order to have deposited and shaped the sediment producing the large coherent fields of mud waves.

Coral Reefs

Waning of volcanic activity at the end of shield building causes relatively passive shoreline conditions that favor the growth and preservation of coral reefs. The rapid subsidence of the volcanoes, resulting from the load they impose on the lithosphere, drown the reefs as much as several km below sea level. Recognition of coral reefs in GLORIA sidescan images has permitted the mapping of the former shorelines of now partly or totally submerged islands (Figure 2).

Commonly, reefs post-date landslides because they grow and are preserved only after the end of vigorous volcanic activity, which appears also to be the primary period of landslide movement. However, continued investigations will no doubt reveal places where landslides have disrupted coral reefs. In the cases examined, the reefs lie on top of the upper part of landslides and therefore the age of the reef will provide a minimum age for landslide movement. For example, an unbroken −150 m submerged reef at the head of the northern Alika landslide complex and the North Kona landslide has been dated at 18 ka (Moore & Clague 1992) and hence indicates that the Kealakekua fault at the head of these landslides (Figure 5) and subordinate breakaway scarps have been inactive since that time despite minor historic ground cracking (Lipman et al 1988).

LANDSLIDES ON OTHER MARINE VOLCANOES

Large-scale landsliding of marine volcanoes is not restricted to the Hawaiian Ridge. Many examples have been recently discovered at other locations, stimulated in part by the Hawaiian discoveries. Multibeam bathymetric surveys off the island of La Reunion have supported the previously proposed notion that the east side of the active Piton de la Fournaise volcano has failed by massive landsliding (Lenat et al 1989). Many features of Mount Etna, both above and below sea level, have recently been explained by gravitational spreading (Borgia et al 1992). Multibeam bathymetric maps from the northwestern Pacific reveal evidence for numerous large-scale landslides on the Emperor and Michelson Ridges as well as on seamounts of the Mapmakers and the Marcus-Wake groups (Smoot & King 1992). Limited multibeam surveys of the southern Galapagos platform reveal morphologic similarities with major Hawaiian slumps (Chadwick et al 1992).

Re-examination of early GLORIA images has revealed large landslides on the submarine slopes of the Canary Islands and Tristan de Chuna that had previously gone unrecognized (Holcomb & Searle 1991). Peculiar morphology of the subaerial Marquesas volcanoes coupled with recent bathymetric surveys has led Barsczus et al (1992) and Filmer et al (1992) to propose major collapse of eight of the volcanoes.

In addition to these recent discoveries, giant landslides on many other island volcanoes have long been suspected from topographic anomalies both above and below sea level (Fairbridge 1950). Island asymmetries including broad embayments and high coastal cliffs, commonly on the leeward side, hint at major submarine landslides. A general survey of oceanic volcanoes led Holcomb & Searle (1991) to conclude that many single landslides have removed 10–20% of their source volcanoes, and that larger fractions have been removed from a single edifice by multiple landslides. Large landslides have affected oceanic volcanoes ranging widely in size, geologic setting, and climatic environment.

SUMMARY

Sixty-eight landslides more than 20 km long are present along a 2200 km segment of the Hawaiian Ridge from near Midway to Hawaii. Some of the landslides exceed 200 km in length and 5000 km^3 in volume, ranking them among the largest on Earth. Most of these giant landslides were discovered during a mapping program of the U. S. Hawaiian Exclusive Economic Zone from 1986 to 1991 utilizing the GLORIA side-looking sonar mapping system.

Two general types of landslides are present: slumps and debris avalanches. Many intermediate forms occur and some debris avalanches form from oversteepened slumps. The slumps are deeply rooted in the volcanoes and may extend back to volcanic rift zones and down to the base of the volcanic pile at about 10 km depth. Tension at the upper part is accommodated by major normal faults as well as by pull-apart structures of the volcanic rift zones, and the lower compressional regime is marked by broad bulges, closed depressions, and steep toes. Incremental movement of a few meters on historic slumps has produced major earthquakes (magnitude >7). Magma or associated hot zones of plasticity probably enhance basal movement of the upper part, and unconsolidated sediment promotes movement of the lower part. Debris avalanches are thinner, longer, and move on lower gradients than slumps. Their rapid movement is indicated by the fact that some have moved uphill for tens of kilometers, and are believed to have produced major tsunamis. The debris avalanches left large amphitheaters at their heads and produce broad hummocky distal lobes at their toes. Commonly, major canyons have incised the amphitheaters.

Giant landslides have recently been discovered on many other marine volcanoes where they also can be related to volcanic structure and eruptive activity. It is now clear that large-scale collapses of the flanks of oceanic volcanoes are as important as the basic volcanic processes in determining the growth history and final form of the volcanoes. Future studies need to obtain details of the morphology of the landslide deposits and the nature of the transported material to learn about the landslide mechanisms and the interaction of landsliding to the volcanic processes of the host volcano. Detailed multibeam bathymetry and extensive sampling is needed not only for the large debris avalanches but also for the smaller, nearshore landslides that probably occur with more frequency and, therefore, pose a more immediate hazard than the larger failures. Such detailed studies should focus on young and currently active landslides on volcanic islands, such as the Hilina slump on the south flank of Kilauea volcano of Hawaii, where subaerial monitoring of landslide movement, volcanic activity, and seismicity can provide critical clues to the processes involved that are difficult to obtain in deep-marine settings.

ACKNOWLEDGMENTS

This work is largely based on the 1986–1991 swath sonar mapping program of the United States Hawaiian Exclusive Economic Zone, a cooperative venture of the U. S. Geological Survey and the British Institute of Oceanographic Sciences. We are indebted to the crews and technical staffs of the M/V *FARNELLA* for their labors during this work and have benefited from discussions with our colleagues associated with the program. We thank Ellen Lougee who drafted Figures 1, 2, and 5, Carolyn Degnan who prepared Figure 7 from NOAA digital data, and Paul Delaney and David Scholl who critically reviewed the manuscript and provided helpful comments.

Literature Cited

Barsczus HG, Filmer PE, Desonie D. 1992. Cataclysmic collapses and mass wasting processes in the Marquesas. *Eos, Trans. Am. Geophys. Union* 73: 313 (Abstr.)

Borgia A, Burr J, Moniero W, Morales LD, Alvarado GE. 1990. Fault propagation folds induced by gravitational failure and slumping of the Central Costa Rica Volcanic Range: implications for large terrestrial and Martian volcanic edifices. *J. Geophys. Res.* 95: 14,357–82

Borgia A, Ferrari L, Pasquare G. 1992. Importance of gravitational spreading in the tectonic and volcanic evolution of Mount Etna. *Nature* 357: 231–35

Borgia A, Treves B. 1992. Volcanic plates overriding the oceanic crust: structure and dynamics of Hawaiian volcanoes. *Geol. Soc. London Spec. Publ.* 60: 277–99

Chadwick WW Jr, Moore JG, Fox CG, Christie DM. 1992. Morphologic similarities of submarine slope failures: south flank of Kilauea, Hawaii, and the southern Galapagos platform. *Eos, Trans. Am. Geophys. Union* 73: 507 (Abstr.)

Clague DA, Dalrymple GB. 1987. The Hawaiian–Emperor volcanic chain, Part I. *US Geol. Surv. Prof. Pap.* 1350: 5–54

Clague DA, Holcomb RT, Sinton JM, Detrick RS, Torresan ME. 1990. Pliocene and Pleistocene alkalic flood basalts on the seafloor north of the Hawaiian islands. *Earth Planet. Sci. Lett.* 98: 175–91

Crandell DR, Miller CD, Glicken HX, Christiansen RL, Newhall CG. 1984. Catastrophic debris avalanche from ancestral Mount Shasta volcano, California. *Geology* 12: 143–46

Dadisman SV, Marlow MS, Rothwell RG, Harris M. 1992. Cruise report: GLORIA survey of the Hawaiian Island Chain, F2–90-HW. *US Geol. Surv. Open-File Rep.* 92-206, 65 pp.

Delaney PT. 1992. Motion of Kilauea volcano during sustained eruption from the Puu Oo and Kupaianaha vents 1983–1991. *Eos, Trans. Am. Geophys. Union* 73: 506 (Abstr.)

Delaney PT, Fiske RS, Miklius A, Okamura AT, Sako MK. 1990. Deep magma body beneath the summit and rift zones of Kilauea volcano, Hawaii. *Science* 247: 1311–16

Fairbridge RW. 1950. Landslide patterns on oceanic volcanoes and atolls. *Geophys. J.* 115: 84–88

Filmer PE, McNutt MK, Webb H, Dixon DJ. 1992. Volcanism and archipelagic aprons: a comparison of the Marquesan and Hawaiian islands. *Eos, Trans. Am. Geophys. Union* 73: 313 (Abstr.)

Flood RD. 1988. A lee wave model for deep-sea mudwave activity. *Deep-Sea Res.* 35: 973–83

Flood RD, Shor AN. 1988. Mud waves in the Argentine Basin and their relationship to regional bottom circulation patterns. *Deep-Sea Res.* 35: 943–71

Garcia MO, Scientific Party ODP Leg 136. 1992. Volcanic sands from ODP site 842, 300 km west of Hawaii: turbidites from giant debris flows. *Eos, Trans. Am. Geophys. Union* 73: 513–14 (Abstr.)

Hill DP, Zucca JJ. 1987. Geophysical constraints on the structure of Kilauea and Mauna Loa volcanoes and some implications for seismomagmatic processes. *US Geol. Surv. Prof. Pap.* 1350: 903–17

Holcomb RT, Clague DA. 1983. Volcanic eruption patterns along submarine rift zones. *Proc. Oceans '83 Conf.* 2: 787–90. San Francisco: IEEE/MTS

Holcomb RT, Searle RC. 1991. Large landslides from oceanic volcanoes. *Mar. Geotechnol.* 10: 19–32

Klein FW, Koyanagi RY, Nakata JS, Tanigawa WR. 1987. The seismicity of Kilauea's magma system. Part II. *US Geol. Surv. Prof. Pap.* 1350: 1019–185

Lenat J-F, Vincent P, Bachelery P. 1989. The off-shore continuation of an active basaltic volcano: Piton de la Fournaise (Reunion Island, Indian Ocean). *J. Volcanol. Geotherm. Res.* 36: 1–36

Lipman PW, Lockwood JP, Okamura RT, Swanson DA, Yamashita KM. 1985. Ground deformation associated with the 1975 magnitude-7.2 earthquake and resulting changes in activity of Kilauea volcano, Hawaii. *US Geol. Surv. Prof. Pap.* 1276. 45 pp.

Lipman PW, Normark WR, Moore JG, Wilson JB, Gutmacher CE. 1988. The giant submarine Alika debris slide, Mauna Loa, Hawaii. *J. Geophys. Res.* 93: 4279–99

Lipman PW, Rhodes JM, Dalrymple GB. 1990. The Ninole Basalt—implications for the structural evolution of Mauna Loa volcano, Hawaii. *Bull. Volcanol.* 53: 1–19

Lonsdale P. 1989. A geomorphological reconnaissance of the submarine part of the east rift zone of Kilauea volcano, Hawaii. *Bull. Volcanol.* 51: 123–44

Moore GW, Moore JG. 1988. Large-scale bedforms in boulder gravel produced by giant waves in Hawaii. *Geol. Soc. Am. Spec. Pap.* 229: 101–10

Moore JG, Clague DA. 1992. Volcano growth and evolution of the island of Hawaii. *Bull. Geol. Soc. Am.* 104: 1471–84

Moore JG, Clague DA, Holcomb RT, Lipman PW, Normark WR, Torresan ME. 1989. Prodigious submarine landslides on the Hawaiian Ridge. *J. Geophys. Res.* 94: 17,465–84

Moore JG, Krivoy H. 1964. The 1962 flank eruption of Kilauea Volcano and structure of the east rift zone. *J. Geophys. Res.* 69: 2033–45

Moore JG, Mark RK. 1992. Morphology of the island of Hawaii. *GSA Today* 2: 257–59

Moore JG, Normark WR, Gutmacher CE. 1992. Major landslides on the submarine flanks of Mauna Loa volcano, Hawaii. *Landslide News* 6: 13–15

Normark WR, Hess GR, Stow DAV, Bowen AJ. 1980. Sediment waves on the Monterey Fan levees: a preliminary physical interpretation. *Mar. Geol.* 37: 1–18

Normark WR, Moore JG, Torresan ME. 1993. Giant volcano-related landslides and the development of the Hawaiian Islands. In *Submarine Landslides: Selected Studies in the US Exclusive Economic Zone*, ed. WC Schwab, HJ Lee, DC Twichell. *US Geol. Surv. Bull.* 2002: 184–96

Normark WR, Shor GG Jr. 1968. Seismic reflection study of the shallow structure of the Hawaiian Arch. *J. Geophys. Res.* 73: 6991–98

Rees BA, Detrick RS, Coakley BJ. 1993. Seismic stratigraphy of the Hawaiian flexural moat. *Geol. Soc. Am. Bull.* 105: 189–205

Segall P, Delaney P, Arnadottir T, Freymueller J, Owen S. 1992. Deformation of the south flank of Kilauea Volcano. *Eos, Trans. Am. Geophys. Union* 73: 505–6 (Abstr.)

Shipboard Scientific Party. Site 842. 1992. In *Proc. Ocean Drilling Prog. Initial Rep.*, ed. A Dziewonski, R Wilkens, J Firth, et al, 136: 37–63. College Station, TX: Ocean Drilling Prog.

Somers ML, Carson RM, Revie JA, Edge RH, Barrow VJ, Andrews, AG. 1978. GLORIA II—An improved long range sidescan sonar. *Proc. IEE/IERE Subconf. on Ocean Instr. and Commun., Oceanol. Int.*, pp. 16–24. London: BPS

Smoot NC, King RE. 1992. Three-dimensional secondary surface geomorphology of submarine landslides on northwest Pacific plate guyots. *Geomorphology* 6: 151–74

Stearns HT, Macdonald GA. 1946. Geology and ground-water resources of the island of Hawaii. *Hawaii Div. Hydrogr. Bull.* 9. 363 pp.

Swanson DA, Jackson DB, Koyanagi RY, Wright TL. 1976. The February 1969 east rift eruption of Kilauea Volcano, Hawaii. *US Geol. Surv. Prof. Pap.* 891. 30 pp.

Thomas DM. 1992. Thermal and structural conditions within the Kilauea east rift zone as indicated by deep drilling data. *Eos, Trans. Am. Geophys. Union* 73: 513 (Abstr.)

Thurber CH, Gripp AE. 1988. Flexure and seismicity beneath the south flank of Kilauea volcano and tectonic implications. *J. Geophys. Res.* 93: 4271–78

Tilling RI, Koyanagi RY, Lipman PW, Lockwood JP, Moore JG, Swanson DA. 1976. Earthquakes and related catastrophic events, Island of Hawaii, November 29, 1975. A preliminary report. *US Geol. Surv. Circ.* 740. 33 pp.

Voight B, Glicken H, Janda RJ, Douglass PM. 1981. Catastrophic rockslide avalanche of May 18. In *The* 1980 *eruptions of Mount St. Helens, Washington*, ed. PW Lipman, DR Mullineaux. *US Geol. Surv. Prof. Pap.* 1250: 347–400

Annu. Rev. Earth Planet. Sci. 1994. 22:145–65

THE LATE EOCENE-OLIGOCENE EXTINCTIONS

Donald R. Prothero

Department of Geology, Occidental College, Los Angeles, California 90041

KEY WORDS: Antarctic, glaciation, fossil mammals, foraminifera, climate change

INTRODUCTION

The transition from the Eocene to the Oligocene Epochs, from about 40 to 30 Ma (million years ago), was the most significant interval in Earth history since the dinosaurs died out 65 Ma. From the warm, equable "greenhouse" climate of the early Eocene (a relict of the age of dinosaurs), the Earth experienced major climatic changes. Global temperature plummeted, and the first Antarctic ice sheets appeared. These climatic stresses triggered extinctions in plants and animals, both on land and in the oceans. By the early Oligocene (33 Ma), the Earth had a much cooler, more temperate climate, with a much lower diversity of organisms. Indeed, the Eocene-Oligocene transition marked the change from the global "greenhouse" world of the Cretaceous and early Cenozoic to the glaciated "ice house" world of today.

Despite the intense research interest in mass extinctions over the past two decades, the Eocene-Oligocene extinctions have been relatively understudied and misunderstood. While hundreds of papers have been published on the terminal Cretaceous extinction of dinosaurs and ammonites since the discovery of the iridium anomaly in 1980, only a few dozen articles have been published on the Eocene-Oligocene extinctions. Much of this work has now been invalidated by new data.

In the enthusiasm to force the Eocene-Oligocene extinctions into the mold of the Cretaceous-Tertiary impact hypothesis and the periodic extinction hypothesis (Raup & Sepkoski 1984), a lot of misinformation has appeared. Typically, impact proponents treat the Eocene-Oligocene tran-

0084–6597/94/0515–0145$05.00

sition as a single catastrophic "Terminal Eocene Event" that happened precisely 26 million years after the terminal Cretaceous event. The discovery of iridium and other extraterrestrial impact debris near the end of the Eocene was also much publicized (Alvarez et al 1982, Asaro et al 1982, Ganapathy 1982, Glass et al 1982), as if this evidence was sufficient by itself to explain late Eocene events.

In the 1980s, however, scientists began to conduct detailed studies which allowed sober reassessment of the various explanations of Eocene-Oligocene events. The 1985 International Geological Correlation Project 174 Symposium which focused only on the "Terminal Eocene Event" (Pomerol & Premoli-Silva 1986) was followed by a 1989 Penrose Conference which brought together much new data from both the terrestrial and marine realms over the entire 10 million years of the Eocene-Oligocene transition (Prothero & Berggren 1992). New information from the deep sea, especially from Ocean Drilling Project (ODP) cruises in the Southern Ocean, greatly improved our understanding of Antarctic climatic changes. Detailed studies of the changes in terrestrial soils, plants, snails, and mammals were finally undertaken.

The most important breakthrough, however, did not come from new records of the transition, but from improved dating and correlation of the existing data base. By means of $^{40}Ar/^{39}Ar$ dating and magnetic stratigraphy, major errors in the correlation and calibration of the time scale were discovered and corrected (Swisher & Prothero 1990, Prothero & Swisher 1992, Berggren et al 1992). As a result, all correlations and numerical age estimates (and the conclusions based upon them) published before 1990 are now out of date. This review updates the current state of our understanding of this fascinating period in Earth history, based on the data summarized previously (Prothero & Berggren 1992, Prothero 1994) and new information that has emerged since then.

THE TIME SCALE

Before discussing the data on the Eocene-Oligocene transition, let us review the latest developments in dating this episode. The correlation and calibration of Eocene and Oligocene rocks are not as simple and straightforward as one might expect. Radiometrically datable materials are only available from scattered volcanic ash layers in a few restricted localities. Correlation and dating must be done by detailed biostratigraphic studies in marine sections and cores which only occasionally contain volcanic materials for dating (Hardenbol & Berggren 1978, Montanari 1990). Most of the dates in marine rocks have come from the greenish K-rich clay mineral glauconite, which has been unreliable in some cases in the past (Obradovich 1989, Aubry et al 1989). Consequently, several different time scales have been published for the marine Eocene and Oligocene, with significant discrepancies. For example, the Eocene-Oligocene boundary has been placed as young as 32 Ma (Odin 1982) and as old as 38 Ma (Lowrie et al 1982, Harland et al 1982), with the most commonly cited value at 36.5 Ma (Berggren et al 1985). This is an enormous range of age estimates for a stratigraphic boundary that is so recent in the geologic past.

The situation becomes even more difficult when it comes to correlating terrestrial rocks to the marine standard. Eocene-Oligocene marine and non-marine rocks rarely interfinger to allow direct correlation. The best, most fossiliferous terrestrial sequences occur in continental interior basins, with no chance of direct correlation to sequences bearing marine fossils. Thus, the terrestrial record has its own provincial time scale, such as the North American land mammal "ages" (Wood et al 1941, Woodburne 1987) and their Eurasian and South American counterparts (Savage & Russell 1983). A similar time scale was established for fossil plants in the Pacific Northwest of North America (Wolfe 1981). Both were calibrated with potassium-argon (K-Ar) dates from volcanic ash layers (Evernden et al 1964, Evernden & James 1964, Krishtalka et al 1987, Emry et al 1987). Fresh, datable ash layers are much more abundant on land than they are in the ocean, since they are closer to their volcanic source and do not experience submarine weathering.

Nevertheless, the sparse K-Ar dates on terrestrial sequences gave only a rough equivalence to the marine record (Berggren et al 1978, 1985; Krishtalka et al 1987; Emry et al 1987). In North America, the Eocene and Oligocene land mammal "ages" were not based on biostratigraphic stages, nor were they subdivided into increments of less than 2–3 million years in duration. These crude correlations gave relatively low resolution, although the overall pattern of change was sufficient for the "big picture" to emerge (Lillegraven 1972, Webb 1977).

The scarcity of reliable radiometric dates in marine sequences forced biostratigraphers to tie in terrestrial volcanic ash dates wherever possible (Berggren et al 1978, 1985). This was greatly facilitated when magnetic stratigraphy emerged as a method that could correlate terrestrial and marine records (Prothero et al 1982, 1983; Flynn 1986). But such magnetic correlations were only as good as the radioisotopic dates on which they were based. As long as potassium-argon was the only dating technique available, there was no way to check its reliability.

The development of $^{40}Ar/^{39}Ar$ dating represented a major breakthrough in Cenozoic geochronology (McDougall & Harrison 1988). Not only could the same volcanic ashes be double-checked with a different isotopic system, but there were many other advantages. $^{40}Ar/^{39}Ar$ dating allows much greater precision (error estimates of only ±100,000–200,000 years, rather than the ±700,000–1,000,000 years typical of K-Ar), and dates can be determined from much smaller amounts of material. This made it possible to analyze many previously undatable layers for the first time. Most importantly, $^{40}Ar/^{39}Ar$ dating can detect problems of contamination and alteration much more successfully than the old K-Ar methods. In the stepwise heating method, the contamination appears as a spectrum of apparent ages, before the true age (from the interior of the crystal) plots as a plateau. In the single-crystal laser-fusion method, each crystal is individually picked by hand, evaluated for freshness, and then vaporized by laser to release all of its argon. With the new automated equipment in several laboratories, it is possible to measure dozens of crystals, determine whether the dates cluster with a reasonable statistical average, and eliminate dates that are widely divergent.

When $^{40}Ar/^{39}Ar$ dating was applied to many of the terrestrial volcanic ashes that were dated 30 years ago with K-Ar (Evernden et al 1964), significant errors became apparent (Swisher & Prothero 1990). For example, K-Ar dating of bulk samples of Flagstaff Rim Ash J produced a date of about 32.4 Ma (Evernden et al 1964, Emry 1973). When individual crystals of the same ash were redated by $^{40}Ar/^{39}Ar$ methods, the ash yielded dates of 34.5±0.087 Ma on biotite and 34.7±0.036 Ma on anorthoclase (Swisher & Prothero 1990, Prothero & Swisher 1992)—more than 2 million years older than previous estimates! These new dates radically shortened the apparent age span of the critical Flagstaff Rim section in central Wyoming, which in turn forced recalibration of the magnetic stratigraphy, and shifted the correlation of the land sequences by two whole magnetic polarity chrons (Prothero et al 1982, 1983; Prothero & Swisher 1992).

The consequence of these new dates, recalibrations, and revised correlations has been a radical rethinking of the entire late Paleogene time scale. The calibration of the marine time scale was once very controversial, but now most scientists agree that the best age estimate of Eocene-Oligocene boundary is about 33.5 Ma, rather than the estimates of 38 or 36.6 or 32 Ma published previously (Berggren et al 1992). The revisions of the marine time scale have forced recalibration of the magnetic polarity time scale, which had previously been calibrated with erroneous K-Ar dates (Cande & Kent 1993). The most radical change occurred in terrestrial mammal-bearing sequences in North America and Asia. The classic "late Eocene" Uintan land mammal "ages" in North America became middle Eocene, and the "early Oligocene" Chadronian land mammal "age" in North America (and its equivalent in Asia) became late Eocene. The "middle" and "late" Oligocene Orellan and Whitneyan land mammal "ages" in North America are now early Oligocene, and even the "early Miocene" Arikareean land mammal "age" in North America has become mostly late Oligocene. For years, mammalian biostratigraphers had identified the Eocene-Oligocene boundary with the Duchesnean-Chadronian boundary, and were trained to think that "Chadronian equals early Oligocene." With the new correlations, the Eocene-Oligocene boundary shifts up one whole land mammal "age" (to the Chadronian-Orellan boundary), and paleontologists had to rethink their lifelong assumptions about Eocene and Oligocene (Swisher & Prothero 1990, Prothero & Swisher 1992).

The current time scale is shown in Figure 1, and further detail about its development can be found in Berggren et al (1992) and Cande & Kent (1993).

THE CLIMATIC AND BIOTIC RECORD

Before we discuss the various hypotheses for the Eocene-Oligocene extinctions, we should examine the data that these hypotheses attempt to explain. New studies on marine and terrestrial sequences have produced much more detailed evidence about the nature of the Eocene-Oligocene transition than previously known. Some of this information was summarized previously (Prothero 1985, 1989), but the new time scale makes those papers out of date in some respects. Most of the data that follows come from papers in the volume edited by Prothero & Berggren (1992), especially the summary chapter by Berggren & Prothero (1992).

The Marine Record

The most detailed and informative record of climatic changes and extinctions occurs in the marine realm. There the record is relatively complete, especially in deep-sea sections, which were deposited far below base level and therefore rarely eroded. In addition, marine rocks typically reflect global oceanic conditions, so the record of their oxygen and carbon isotopes, and the presence of unusual sediment types (such as ice-rafted detritus) are as informative as the changes that occur in the biota.

From the marine record, it is clear that there were several pulses of climatic change and extinction during the Eocene-Oligocene transition. Contrary to common misconceptions that the extinctions occurred during the "Terminal Eocene Event" (about 34 Ma), most of the extinctions in the marine organisms (especially among the warm-water taxa) were concentrated at the end of the middle Eocene (about 37 Ma). Particularly severe extinctions occurred in the calcareous nannoplankton (Aubry 1992) and the planktonic foraminifera (Keller 1983, Boersma et al 1987, Keller & MacLeod 1992). Many tropical taxa disappeared as mid-latitude faunas

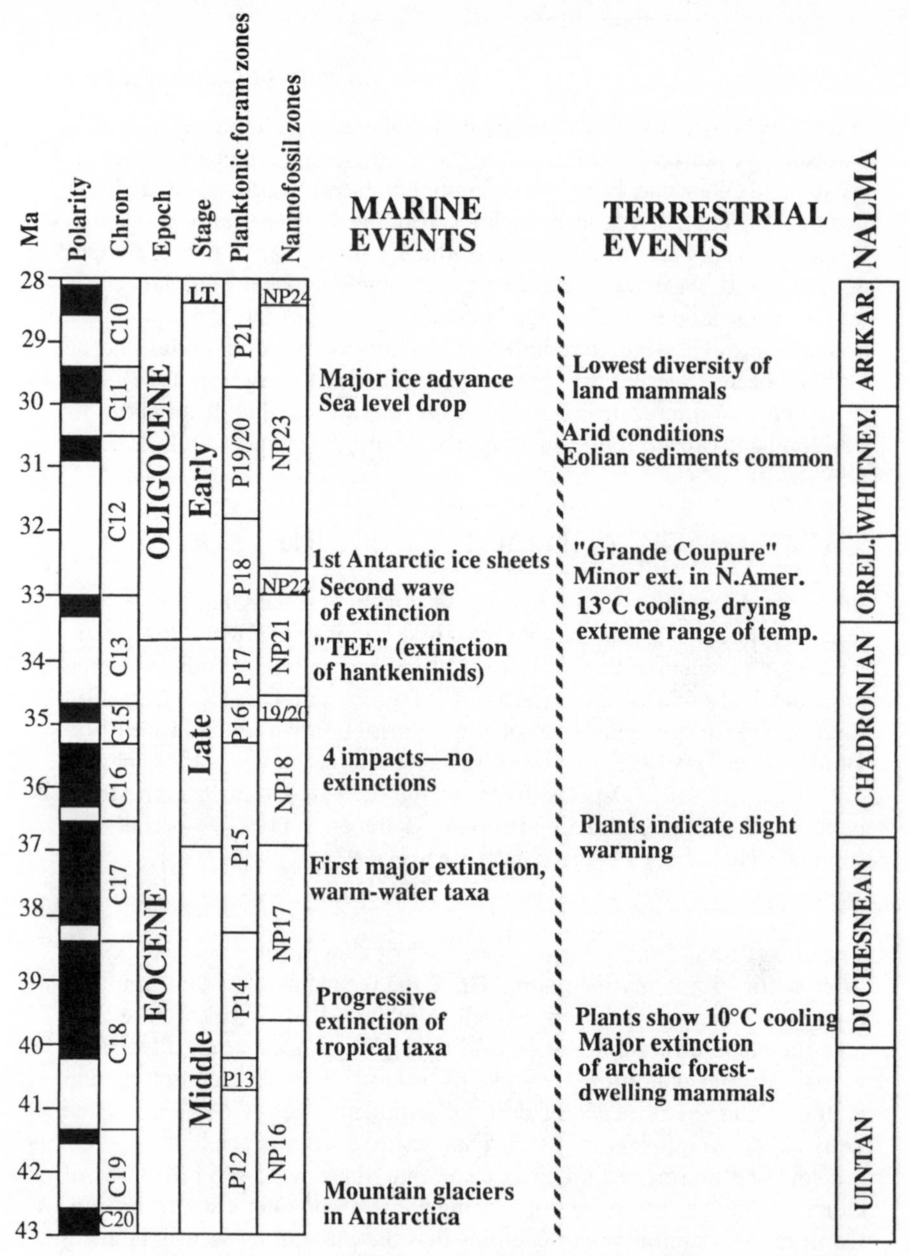

Figure 1 Chronology of events during the Eocene-Oligocene transition. Magnetic polarity time scale after Cande & Kent (1993); North American land mammal "ages" (NALMA) after Prothero & Swisher (1992); marine chronology updated from Berggren et al (1985). The major extinction events occurred at the end of the middle Eocene (about 37–38 Ma) and in the earliest Oligocene (about 33 Ma). The Eocene-Oligocene boundary ("Terminal Eocene Event," or "TEE") occurred at 33.5 Ma and was not so important. Evidence of impacts is dated between 35.5 and 36 Ma. Abbreviations are as follows: Arikar. = Arikareean; Orel. = Orellan; Whitney. = Whitneyan.

migrated toward the equator to escape cooler high-latitude water masses. Bottom-dwelling organisms also suffered severe extinctions during the middle Eocene as well. This can be seen in groups as disparate as benthic foraminiferans (Adams et al 1990, Gaskell 1991, Thomas 1992) and molluscs, which lose 89% of the gastropod species and 84% of the bivalve species at the middle-late Eocene transition (Hansen 1987, 1992). Some groups (such as echinoids and ostracodes) apparently do not show a striking middle Eocene extinction, but among taxa with a sufficiently detailed record, this middle-late Eocene extinction is the most severe of all mass extinctions in the Cenozoic. The most fundamental biotic division of the Cenozoic is not between the Tertiary and Quaternary, or Paleogene and Neogene, but between the middle and late Eocene.

A few minor extinctions in the planktonic foraminiferans (Keller 1983) and the radiolarians (Riedel & Sanfilippo 1986) occurred during the middle of the late Eocene (middle Priabonian, about 35 Ma). Minor extinctions are also reported for the molluscs (Hansen 1987, 1992). No other marine group shows any significant extinction at this time.

In contrast to the the end of the middle Eocene, the Eocene-Oligocene boundary (about 34 Ma, or late magnetic Chron C13R as currently defined) is a relatively minor event. Only a few planktonic foraminifera became extinct at this boundary (particularly the index genus, the spiny foraminiferan *Hantkenina*). Despite the publicity given to the "Terminal Eocene Event" in the past, it is an embarrassingly small extinction. This is partly a problem of resolution and also one of definition. In the past, low-resolution studies tended to lump all events near the Eocene-Oligocene boundary as one "event." But recent research has shown that the major extinctions and climatic changes that occurred near the boundary are actually found a million years later in the earliest Oligocene (about 33 Ma, or mid Chron C13N). Ironically, this discrepancy became apparent in 1989, just after the International Subcommission of Paleogene Stratigraphy voted to establish the Eocene-Oligocene boundary at the last occurrence of *Hantkenina* in the Massignano section near Ancona, Italy. Subsequent research by Brinkhuis (1992) showed that the stratotype section of the upper Eocene Priabonian Stage was partly lower Oligocene (as currently defined), so that the current definition of the boundary may have to be revised. Many scientists feel that the dramatic global events at 33 Ma make a more natural Eocene-Oligocene boundary, but until the vote of the Subcommission is reconsidered, the current boundary stands.

The earliest Oligocene event is reflected in the extinction of many surviving species of calcareous nannofossils (Aubry 1992), planktonic foraminiferans (Keller 1983), benthic foraminiferans (Gaskell 1991, Thomas 1992), diatoms (Baldauf 1992), ostracodes (Benson 1975), molluscs (Hansen 1987, 1992) and echinoids (McKinney et al 1992). Like the middle

Eocene extinctions, the attrition was heaviest in warm-water taxa, leaving a relatively low diversity of organisms to survive into the early and late Oligocene.

Some extinctions also occurred during the middle Oligocene, particularly among planktonic foraminiferans that were less tolerant of cold conditions (Keller 1983) and the calcareous nannoplankton (Haq et al 1977). However, the middle Oligocene event was not nearly as severe as previous extinctions, perhaps because the surviving fauna was already cold-adapted from the earlier attrition of warm-climate taxa.

The Terrestrial Record

Although the new calibrations and time scales change the marine story somewhat, their effect on our understanding of terrestrial events is truly profound. For example, earlier papers (Prothero 1985, 1989) which treated the Duchesnean-Chadronian boundary as the Eocene-Oligocene boundary, and labeled the Chadronian-Orellan boundary as the "mid-Oligocene event" are now completely out of date, and conclusions based on these correlations are invalid.

Once the new correlations and calibrations are taken into account, the terrestrial record shows striking similarities to the marine record of extinctions (Figure 1). The major changes take place in the late middle Eocene (about 40 Ma), and again in the earliest Oligocene (33 Ma). Relatively few changes take place in the middle of the late Eocene, or in the mid-Oligocene.

NORTH AMERICA The best documentation and dating of the terrestrial Eocene-Oligocene transition comes from North America. This is primarily because it is the home of most paleontologists, but also because there is an excellent record of this transition preserved and exposed in many places in the Rocky Mountains and High Plains. The clearest signal comes from the land plants, as described by Wolfe (1971, 1978, 1992). Based on his method of estimating temperature from the percentage of entire leaf margins, Wolfe documents a severe cooling event of about 10°C at the end of the middle Eocene, followed by a slight warming in the late Eocene, and then an even more extreme cooling of about 13°C in the earliest Oligocene. Wolfe (1971) originally called this chill the "Oligocene deterioration," but in 1978 he coined the phrase "Terminal Eocene Event" (it is now correlated with the early Oligocene). In addition to the cooling of mean annual temperatures, there was also a great increase in the mean annual range of temperatures, from about 3–5°C during the warm, tropical middle Eocene to almost 25°C in the Oligocene. These cooling events result in the replacement of subtropical early and middle Eocene floras (typical of modern Central America) with plants that are mostly typical of broad-leaved deciduous forests (such as those of the present-day New England and eastern Canada).

Ancient soil horizons in the Big Badlands of South Dakota and elsewhere show a similar trend (Retallack 1983, 1992). Late Eocene paleosols were formed under a dense forest canopy with about 1000 mm of annual precipitation. By the Orellan (early Oligocene), there was a wooded grassland in the Badlands with about 500–900 mm of rainfall, and in the early Arikareean (late Oligocene) the region was covered with an open grassland receiving only 350–450 mm of rainfall per year.

Living amidst this changing vegetation were a variety of climatically-sensitive animals. Land snails from the Chadronian of Douglas, Wyoming, were large-shelled taxa similar to those found in subtropical climates with seasonal precipitation, such as that of the present-day central Mexican plateau (Evanoff et al 1992). They indicate a mean annual temperature of about 16°C and a mean annual precipitation of about 450 mm. By the Orellan (early Oligocene), these snails had been replaced by drought-tolerant small-shelled forms typical of a warm-temperate open woodland habitat with a long dry season, such as that of Baja California today. The deposits that produce these snail fossils also show a change from floodplains to sand dune deposition. Amphibians and reptiles exhibit a similar trend toward cooling and drying (Hutchison 1982, 1992). The aquatic salamanders, crocodilians, and turtles, which were so common in the middle Eocene, gradually became scarcer through the late Eocene and by the early Oligocene were replaced by land tortoises.

Although the climatic implications of the changes in land mammals are not so obvious, there are striking differences (Prothero 1985, 1989; Stucky 1990, 1992). As we saw in the marine record, the biggest wave of extinctions took place near the end of the middle Eocene (end of the Uintan, about 40 Ma), when about 25% of the genera of land mammals disappeared. Most were members of archaic groups of mammals typical of the Paleocene and early Eocene, adapted to forest browsing or an arboreal life. When the drying climate began to break up the forest canopy, both the leaf-eaters and the tree-dwellers were most severely affected.

In the late Eocene (Chadronian), there were relatively few extinctions, but there was much immigration of new groups of mammals, apparently from Asia. These included the earliest dogs, camels, rhinos, pocket gophers, beavers, squirrels, rabbits, and shrews. Along with a number of native groups (such as horses and oreodonts) these mammals came to dominate faunas during the Oligocene and Miocene and formed a stable entity known as the "White River Chronofauna" that persisted for about 20 million years (Emry 1981, Emry et al 1987).

The earliest Oligocene climatic crash was accompanied by surprisingly minor extinctions in North American land mammals. A few archaic groups, such as the huge browsing brontotheres, the camel-like oromerycids, and a few archaic rodents did die out, but most mammals typical of the Chadronian persisted into the Orellan with only minor changes in species or relative abundance. However, two groups (cricetid rodents and leptauchenine oreodonts), with relatively high-crowned teeth for eating tough vegetation, did appear in the Orellan and flourished in the Whitneyan.

EUROPE The European record has been relatively well studied, although the chronology there is not based on radiometric dates or magnetostratigraphy. Instead, the mammal-bearing beds frequently interfinger with shallow marine beds (which sometimes contain fossil mammals that were washed out to sea), so that there is direct correlation with the European marine standard.

Floral evidence (Collinson & Hooker 1987, Collison 1992) shows that the middle Eocene forests of Europe were tropical, but in the late Eocene subtropical evergreens (taxodiaceous swamps and reed marshes) were dominant. By the early Oligocene, these plants were replaced by mixed deciduous/evergreen plants indicating a warm-temperate seasonal climate. European floras never reached the extremes of cooling or drying seen in North America, possibly because Europe was an archipelago on the fringe of the tropical Tethys Sea, with the moderating effects of coastal climates and warm waters nearby.

European land mammals underwent several changes during this time (Collinson & Hooker 1987, Hooker 1992, Legendre & Hartenberger 1992). At the end of the middle Eocene (Bartonian), many of the large soft-browsing perissodactyls (odd-toed hoofed mammals) became extinct and were replaced by coarse-browsing palaeotheres (distant relatives of horses). There were also major extinctions in arboreal primates and apatemyids as well as small mammals and insectivores. This pattern closely parallels the extinctions at the end of the middle Eocene (Uintan) in North America.

The most significant change, however, is known as the Grande Coupure, or "great cutoff." First recognized in 1909 by Stehlin, the Grande Coupure marks the end of a large number of archaic European mammalian groups, including many of the same types of tree-dwellers and leaf-eaters seen in North America in the late Eocene, as well as many European endemics. These archaic mammals were abruptly replaced by Asian immigrants which included rabbits, rhinos, advanced rodents, artiodactyls, and carnivorans. Arboreal mammals disappeared completely, and large ground-dwelling mammals and grain-eating rodents dominated the fauna.

For years, the Grande Coupure was correlated with the Eocene-Oligocene boundary (Savage & Russell 1983). But Hooker (1992) has shown that the Grande Coupure occurs slightly above the European marine Eocene, and probably correlates with the dramatic climatic changes that occurred in the earliest Oligocene. The abruptness of the Grande Coupure is not primarily due to the climatic changes in the early Oligocene. Rather, it was caused by the immigration of so many Asian groups, possibly because lower sea level in the early Oligocene opened up corridors from Asia and eastern Europe.

ASIA Correlation of the Asian record with the rest of the world has been hampered by several factors. Stratigraphic analysis is still in its early stages, with almost no magnetostratigraphy or radioisotopic dating to tie Asian terrestrial sequences to the global standard. Most of the age assignments of the stages of the Asian Eocene and Oligocene have been based on comparisons to North American mammals. Now that the North American calibration has changed, the entire correlation of Asian terrestrial sequences needs to be reconsidered (Berggren & Prothero 1992). Consequently, the age assignments of faunas in many papers on Chinese or Mongolian mammals (e.g. Li & Ting 1983, Wang 1992, Dashzeveg & Devyatkin 1986) will probably be revised when the full effect of the changes in the time scale are assimilated.

Adjusting for the changes in time scale, the Asian mammalian record closely parallels the changes seen in North America. The largest extinction occurred at the end of the middle Eocene (Sharamurunian), when 45 genera of archaic Eocene mammals died out, including tillodonts, arctocyonids, helohyids, and eurymylids. The earliest Oligocene (end of the Houldjinian) saw the disappearance of brontotheres, mesonychids, and archaic tapiroids, along with some Asian endemics. This pattern closely parallels the extinction of brontotheres and many tapiroids in North America and Europe at the same time. Following this extinction, cricetid rodents with high-crowned teeth diversified in the early Oligocene (Kekeamuan), as occurred during the Orellan and Whitneyan in North America.

The floral record has been less studied and published, but some features are beginning to emerge (Leopold et al 1992). At the end of the middle Eocene in China, there was a striking change of pollen types, which indicated the development of mesic forests in the coastal part of southeastern China while arid types retreated to Mongolia. Eventually a subtropical woody savanna developed in northern China in the late Eocene. In the Oligocene, floral diversity decreased, and the increase in temperate deciduous trees and conifers suggest further cooling.

OTHER CONTINENTS The other continents have a relatively poor record of the Eocene-Oligocene transition. Rasmussen et al (1992) reviewed the record for Africa (mainly the Fayum region of Egypt) and found little evidence of major changes in the middle or late Eocene, or Oligocene. They pointed out that these land faunas bordered the warm tropical Tethys seaway, which buffered them from the climatic extremes seen in higher latitudes.

The South American mammal record is much better developed than that of Africa, but unfortunately it contains a gap between 41 and 21 Ma which is just beginning to be filled (Wyss et al 1990). Nevertheless, there are notable differences between middle Eocene (Mustersan) and late Oligocene (Divisaderan-Deseadan) faunas that suggest major climatic changes. The most obvious differences occur in South America's endemic ungulate faunas, which change from archaic browsers with low-crowned teeth to much more hypsodont grazing mammals; rodent-like marsupials and primitive edentates also disappeared. Marshall & Cifelli (1989) suggest that this change was due to a change from subtropical woodlands to seasonally-arid savanna woodlands.

Australia's Cenozoic mammal record is virtually nonexistent before the early Miocene. The recently discovered early Eocene Tinga Marra local fauna (Godthelp et al 1992) only begins to fill the gap. However, the paleobotanical record is excellent (Kemp 1978). In the middle Eocene, tropical rainforests covered Australia, but these began to decline in diversity in the late Eocene. In the earliest Oligocene, there was a dramatic cooling in Australia as cold water began to circulate between Australia and Antarctica for the first time (see below). Australian Oligocene floras were dominated by cool-temperate plants tolerant of high seasonality, with increased rainfall in the coal swamp regions of southern Australia, and increased aridity with open forests and a more herbaceous understory elsewhere on the continent (Truswell & Harris 1982).

Middle Eocene floras and pollen are known from Seymour Island on the Antarctic Peninsula (Case 1988). They suggest a cool-temperate rain forest with large trees, similar to those found today on Tasmania, New Zealand, and southern South America. Although there is evidence that glaciers may have appeared in some parts of Antarctica by the late Eocene, they did not yet cover the continent.

THE SEARCH FOR CAUSES

Now that we have summarized the patterns of diversity change and extinctions, we can evaluate different models that attempt to explain them. Recall that we have described a complex transition which spans almost ten million years. Most of the extinctions occurred in tropical taxa at the end of the middle Eocene (about 37–38 Ma), and in the earliest Oligocene (33 Ma), with very minor extinctions in the middle of the late Eocene (about 35–36 Ma), and in the late Oligocene (about 30–31 Ma). The suggested causes typically fall into three categories: extraterrestrial, volcanic, and tectonic-climatic. We shall examine each in order of increasing plausibility.

Asteroids, Comets, and Volcanoes

Although there is strong evidence that asteroid impacts and/or volcanic eruptions may have caused the Cretaceous-Tertiary extinctions, such is not the case for the Eocene-Oligocene transition. The initial 1982 reports (Alvarez et al 1982, Asaro et al 1982, Ganapathy 1982, Glass et al 1982) of iridium, microtektites, and glassy impact spherules at the end of the Eocene were met with great fanfare, but since then the impact advocates have all but abandoned the Eocene-Oligocene extinctions. There are several reasons for this. First, the iridium and microtektites occur in the *middle* of the late Eocene, associated with no extinctions of significance. Although there was once much controversy over the dating of the microtektite layers (and just how many there were), recent investigations (Miller et al 1991b) found just four discrete horizons ranging in age from 36 to 35.4 Ma. This is too late for the middle Eocene crash, and too early for the early Oligocene crisis.

Secondly, the protracted pattern of Eocene-Oligocene extinctions cannot be explained by any sudden impact, especially when that transition is spaced out over about ten million years. Hut et al (1987) avoided this difficulty by postulating a series of comet showers over the interval, but this hypothesis falters on the same problems that the asteroid impact does. In addition, many planetary scientists doubt that comets could produce the impact droplets or iridium anomalies seen in the middle late Eocene.

Finally, the obvious climatic changes (discussed below) were clearly not caused by extraterrestrial impacts, yet they must have had a profound effect on Eocene biotas. If the late Eocene impacts had any effect whatsoever, they were very minor compared to the profound effect of climatic change.

Volcanism has also been blamed for the Eocene-Oligocene extinctions. If gigantic volcanic eruptions were spaced over a long enough period of time, and spewed enough gases into the atmosphere to change global climate, then such an explanation might be plausible. Kennett et al (1985) found many volcanic ash layers in deep-sea cores in the southwest Pacific, mostly from volcanoes in New Zealand and along the boundary of the Australian and Indo-Pacific plates. However, most of these ash layers occur in the uppermost Eocene and lower Oligocene—too late for the middle Eocene extinctions. Nor was there a single great pulse of volcanism that might explain the early Oligocene refrigeration event (see below).

Rampino & Stothers (1988) blamed the Eocene-Oligocene extinctions on massive flood basalt eruptions which covered 750,000 square kilometers of Ethiopia, supposedly during the late Eocene. However, the dates on these lavas are too late to explain the middle Eocene extinctions, and spaced out over too large a time interval to match the patterns of extinction in the early Oligocene. It is conceivable that they may have had some effect on the atmosphere and climate, but their chronology does not match the known pattern of extinctions closely enough to establish a direct link.

Tectonic and Climatic Changes

Throughout the discussion of the biotic evidence above, it is clear that there was a strong signal of protracted climatic change during the Eocene-Oligocene transition. Clearly, the search for proximate causes must examine reasons for this climatic change, and extraterrestrial and/or volcanic causes are insufficient to explain the pattern we see. Fortunately, the past decade of oceanic drilling has given us an excellent record of global ocean isotopic chemistry and sedimentation on which to base our paleoclimatic interpretations.

The most important data are the abundant evidence of changes in oceanic circulation from the oxygen isotopes (Miller et al 1987, Miller 1992). For a long time, it was dogma that Antarctica had no significant ice sheets before the middle Miocene. Matthews & Poore (1980) challenged this assumption, and began to reinterpret the oxygen-isotope record with some Paleogene Antarctic ice included in their calculations. Since then, most scientists (e.g. Miller & Fairbanks 1983, 1985; Shackleton 1986; Miller et al 1987; Miller 1992) have found that this assumption makes sense. Depending upon the corrections for the ice volume effect, the oxygen isotopes indicate a global temperature drop of about 5–6°C in the earliest Oligocene, which is as large as the changes between glacial and interglacial worlds during the Pleistocene ice ages (Miller et al 1987). A dramatic change in the carbon isotopes at the same time is thought to indicate the influx of cold bottom waters from both the northern and southern high latitudes (Miller 1992). From isotopic evidence such as this, a number of interesting models of oceanic circulation have been advanced (Boersma et al 1987, Kennett & Stott 1990, Miller 1992, Aubry 1992).

Once the "ice-free" assumption was abandoned, reports of Eocene and Oligocene Antarctic glaciers became more acceptable. In 1987 Birkenmajer reported evidence of middle Eocene glacial deposits from King George Island on the Antarctic Peninsula. Wei (1989) argued that ice-rafted sediments from the Pacific side of Antarctica are also middle Eocene in age, although others question this age assignment. Even more striking is the evidence from several deep-sea cores around the Antarctic (Miller et al 1991a; Zachos et al 1992, 1993) and the CIROS-1 drill hole in the Ross Sea (Barrett et al 1989) that there was a major early Oligocene ice advance on that continent. However, this pulse of glaciation may have only lasted a million years, and apparently was concentrated on the Indian and Pacific Ocean sectors of Antarctica (Miller et al 1991a), with a limited effect on the South Atlantic sector or East Antarctica (Kennett & Barker 1990).

The largest episode of pre-Miocene Antarctic glaciation occurred in the mid-Oligocene (about 30 Ma), and may have lasted about 4 million years (Miller et al 1991a). Thick mid- and late Oligocene glacial deposits are reported from all over Antarctica (Barrett et al 1989, LeMasurier & Rex 1982, Bartek et al 1992). This major ice advance is responsible for the largest sea level drop on the entire Vail onlap-offlap curve (Haq et al 1987), and major mid-Oligocene unconformities all over the world (Poag & Ward 1987, Miller et al 1987).

What triggered the growth of these southern ice sheets, and the changes in global oceanic circulation and climate that resulted? Several authors (e.g. Frakes & Kemp 1972, Kennett et al 1975, Kennett 1977) have suggested that the development of the circum-Antarctic current may have been a critical factor. Today, the circum-Antarctic current is the largest of oceanic currents, with a volume 1000 times that of the Amazon (Callahan 1971). It causes upwelling of deep bottom nutrients, which leads to enormous oceanic productivity. Once these circumpolar waters have welled up, they are chilled and sink to form the deep, cold bottom waters (the psychrosphere) that flow north along the bottom of the Atlantic and Pacific. The circum-Antarctic current also locks in the cold of the South Pole as it circles clockwise around Antarctica, and prevents these waters from mixing with more equatorial waters, which might warm up the polar waters.

All of the climatic and isotopic evidence suggest that this current did not exist in the early Eocene, which was the warmest period of the entire Cenozoic. Instead, tropical and polar waters routinely mixed, ameliorating the extremes between pole and equator (Frakes & Kemp 1972, Kennett et al 1975, Kennett 1977, Kennett & Stott 1990). Both Australia and South America were still attached to Antarctica, preventing any circum-Antarctic circulation. Australia began to move away from Antarctica during the Cretaceous, and a shallow marine gulf developed between the two continents by the middle Eocene (Weissel et al 1977, Mutter et al 1985, McGowran 1973, Kennett et al 1975). By the late Eocene, the connection between Tasmania and Antarctica had separated enough to allow shallow marine circulation into the Pacific (Murphy & Kennett 1986, Kamp et al

1990). Full-scale deep-water circulation through this gap did not occur until the mid-Oligocene (Murphy & Kennett 1986, Kamp et al 1990), and may have been responsible for the massive glaciation and global sea level drop. The circum-Antarctic current was completed when South America pulled away from Antarctica in the latest Oligocene ($\sim$25 Ma), opening up the Drake Passage to deep polar waters (Barker & Burell 1982, Sclater et al 1986).

Another source of cold bottom waters was the Arctic Ocean, which was isolated from the rest of the world's oceans in the Eocene. In the early Oligocene, a deep-water passage in the Norwegian-Greenland Sea apparently opened up (Talwani & Eldholm 1977, Berggren 1982), allowing the North Atlantic Deep Water to flow down from the Arctic through the Atlantic (Miller 1992). Clearly the combination of cold waters from both polar regions was critical in the global refrigeration and climatic changes that occurred beginning in the middle Eocene.

Although these tectonic causes are much more plausible than impacts or volcanoes, they do not solve every problem. For example, if there was no circum-Antarctic circulation before the late Eocene, then why was there a cooling event at the end of the middle Eocene, and glaciers on the Antarctic Peninsula? Using computerized climatic modeling, Bartek et al (1992) have suggested that the Antarctic was already cold enough in the middle Eocene for glaciation. All that was lacking was a source of moisture, and their models show that the opening of the gulf between Antarctic and Australia during the middle Eocene would have provided it.

Others are bothered by the apparent abruptness of the change, especially when the causes are long-term tectonic changes. Kennett & Stott (1990) couple the late Eocene cooling with the earliest Eocene "greenhouse" warming in a complex oceanic circulation model which postulates radical changes in oceanic chemisty and flow patterns. Zachos et al (1993) suggest that the gradual changes in oceanic circulation and greenhouse gases may have reached climatic thresholds, where abrupt changes would result. Clearly, there are many more interesting ways of interpreting the late Paleogene climatic data, even as new data emerge from further exploration in and around the Antarctic, and from the oceans.

CONCLUSIONS

The Eocene-Oligocene extinctions took place over about 10 million years, starting with a major extinction in tropical organisms at the end of the middle Eocene (about 37–38 Ma), and followed by a significant global cooling event and a lesser extinction event in the earliest Oligocene (about 33 Ma). Very minor extinctions occurred in the late Eocene (about 35 Ma) and in the mid-Oligocene (about 30 Ma). In spite of the evidence of four impacts around 35–36 Ma, no short-term extraterrestrial events or volcanic eruption is sufficient to explain this pattern of extinction. The overwhelming evidence for global cooling and oceanic circulation changes argue that these must have been the proximal cause of extinction. The refrigeration of the Antarctic and development of cold bottom waters from both poles were critical to this change. The likely triggers of this cooling were the development of the circum-Antarctic current and the opening of the Norwegian-Greenland Sea. Changes in greenhouse gases may also have been critical as certain climatic thresholds were exceeded.

ACKNOWLEDGMENTS

I thank Bill Berggren for teaching me so much about the Eocene-Oligocene transition, for reviewing this manuscript, and for all his help in convening several symposia and editing our volume. I learned much about the marine record from Marie-Pierre Aubry, Gerta Keller, Jim Kennett, Brian McGowran, Ken Miller, and Jim Zachos. This research was sponsored by NSF grant EAR91-17819.

Literature Cited

Adams CG, Lee DE, Rosen BR. 1990. Conflicting isotopic and biotic evidence for tropical sea-surface temperatures during the Tertiary. *Palaeogeogr. Palaeoclimatol. Palaeoecol.* 77: 289–313

Alvarez W, Asaro F, Michel HV, Alvarez LW. 1982. Iridium anomaly approximately synchronous with terminal Eocene extinctions. *Science* 216: 886–88

Asaro F, Alvarez LW, Alvarez W, Michel HV. 1982. Geochemical anomalies near the Eocene/Oligocene and Permian/Triassic boundaries. *Geol. Soc. Am. Spec. Pap.* 190: 517–28

Aubry M-P. 1992. Late Paleogene calcareous nannoplankton evolution: a tale of climatic deterioration. See Prothero & Berggren 1992, pp. 272–309

Aubry M-P, Berggren WA, Kent DV, Flynn JJ, Klitgord KD, et al. 1989. Paleogene geochronology: an integrated approach. *Paleoceanography* 3(6): 707–42

Baldauf JG. 1992. Middle Eocene through early Miocene diatom floral turnover. See Prothero & Berggren 1992, pp. 310–26

Barker PF, Burrell J. 1982. The influence upon Southern Ocean circulation sedimentation and climate of the opening of Drake Passage. In *Antarctic Geoscience*, ed. C. Craddock, pp. 377–85. Madison: Univ. Wis. Press

Barrett PJ, Hambrey MJ, Harwood DM, Pyne AR, Webb PN. 1989. Synthesis. In *Antarctic Cenozoic History from the CIROS-1 Drillhole, McMurdo Sound*, ed. PJ Barrett, *Sci. Inf. Publ. Cent. Dept. Sci. Ind. Res. Bull.* 245: 241–51. Wellington, New Zealand

Bartek LR, Sloan LC, Anderson M, Ross MI. 1992. Evidence from the Antarctic continental margin of late Paleogene ice sheets: a manifestation of plate reorganization and synchronous changes in atmospheric circulation over the emerging Southern Ocean? See Prothero & Berggren 1992, pp. 131–159

Benson RH. 1975. The origin of the psychrosphere as recorded in changes of deep sea ostracode assemblages. *Lethaia* 8: 69–83

Berggren WA. 1982. Role of ocean gateways in climatic change. In *Climate in Earth History*, ed. W Berger, JC Crowell, pp. 118–285. Washington DC: Natl. Acad. Sci

Berggren WA, Kent DV, Flynn JJ. 1985. Paleogene geochronology and chronostratigraphy. *Mem. Geol. Soc. London* 10: 141–95

Berggren WA, Kent DV, Obradovich JD, Swisher CC III. 1992. Toward a revised Paleogene geochronology. See Prothero & Berggren 1992, pp. 29–45

Berggren WA, McKenna MC, Hardenbol J, Obradovich JD. 1978. Revised Paleogene polarity time-scale. *J. Geol.* 86: 67–81

Berggren WA, Prothero DR. 1992. Eocene-Oligocene climatic and biotic evolution: an overview. See Prothero & Berggren 1992, pp. 1–28

Birkenmajer K. 1987. Tertiary glacial and interglacial deposits, South Shetland Islands, Antarctica: Geochronology versus biostratigraphy (a progress report). *Bull. Pol. Acad. Sci. Earth Sci.* 36: 133–45

Boersma A, Premoli-Silva I, Shackleton NJ. 1987. Atlantic Eocene planktonic foraminiferal paleohydrographic indicators and stable isotope paleoceanography. *Paleoceanography* 2(3): 287–331

Brinkhuis H. 1992. Late Paleogene dinoflagellate cysts with special reference to the Eocene/Oligocene boundary. See Prothero & Berggren 1992, pp. 327–40

Callahan JE. 1971. Velocity structure and flux of the Antarctic Circumpolar Current of South Australia. *J. Geophys. Res.* 76: 5859–70

Cande SC, Kent DV. 1993. A new geomagnetic polarity timescale for the late Cretaceous and Cenozoic. *J. Geophys. Res.* 97(B10): 13,917–53

Case JA. 1988. Paleogene floras from Seymour Island, Antarctic Peninsula. *Geol. Soc. Am. Mem.* 169: 523–30

Collinson ME. 1992. Vegetational and floristic changes around the Eocene/Oligocene boundary in western and central Europe. See Prothero & Berggren 1992, pp. 437–50

Collinson ME, Hooker JJ. 1987. Vegetational and mammalian faunal changes in the early Tertiary of southern England. In *The Origins of Angiosperms and their Biological Consequences*, ed. EM Friis, WG Chaloner, PR Crane, pp. 259–304. Cambridge: Cambridge Univ. Press

Dashvezeg D, Devyatkin EV. 1986. Eocene-Oligocene boundary in Mongolia. See Pomerol & Premoli-Silva 1986, pp. 153–57

Emry RJ. 1973. Stratigraphy and preliminary biostratigraphy of the Flagstaff Rim area, Natrona County, Wyoming. *Smithsonian Contrib. Paleobiol.* 18

Emry RJ. 1981. Additions to the mammalian fauna of the type Duchesnean, with comments on the status of the Duchesnean. *J. Paleontol.* 55: 563–70

Emry RJ, Bjork PR, Russell LS. 1987. The Chadronian, Orellan, and Whitneyan land mammal ages. See Woodburne 1987, pp. 118–52

Evanoff E, Prothero DR, Lander RH. 1992. Eocene-Oligocene climatic change in North America: the White River Formation near Douglas, east-central Wyoming. See Prothero & Berggren 1992, pp. 116–30

Evernden JF, James GT. 1964. Potassium-argon dates and the Tertiary floras of North America. *Am. J. Sci.* 262: 945–74

Evernden JF, Savage DE, Curtis GH, James GT. 1964. Potassium-argon dates and the Cenozoic mammalian chronology of North America. *Am. J. Sci.* 262: 145–98

Flynn J.J. 1986. Correlation and geochronology of middle Eocene strata from the western United States. *Palaeogeogr. Palaeoclimatol. Palaeoecol.* 55(1986): 335–406

Frakes LA, Kemp EM. 1972. Influence of continental positions on early Tertiary climates. *Nature* 240: 97–100

Ganapathy R. 1982. Evidence for a major meteorite impact on the Earth 34 million years ago: implication for Eocene extinctions. *Science* 216: 885–86

Gaskell PA 1991. Extinction patterns in Paleogene benthic foraminiferal faunas: relationship to climate and sea level. *Palaios* 6: 2–16

Glass BP, DuBois DL, Ganapathy R. 1982. Relationship between an iridium anomaly and the North American microtektite layer in core RC9-58 from the Caribbean Sea. *J. Geophys. Res.* 87: 425–28

Godthelp H, Archer M, Cifelli RL, Hand SJ, Gilkerson CF. 1992. Earliest known Australian Tertiary mammal fauna. *Nature* 356: 514–16

Hansen TA. 1987. Extinction of late Eocene to Oligocene molluscs: relationship to shelf area, temperature changes, and impact events. *Palaios* 2: 69–75

Hansen TA. 1992. The patterns and causes of molluscan extinction across the Eocene/Oligocene boundary. See Prothero & Berggren 1992, pp. 341–48

Haq BU, Hardenbol J, Vail PR. 1987. The chronology of fluctuating sea level since the Triassic. *Science* 235: 1156–67

Haq BU, Premoli-Silva I, Lohmann GP. 1977. Calcareous plankton paleobiogeographic evidence for major climatic fluctuations in the early Cenozoic Atlantic Ocean. *J. Geophys. Res.* 82: 3861–76

Hardenbol J, Berggren WA. 1978. A new Paleogene numerical time scale. *Am. Assoc. Petrol. Geol. Stud. Geol.* 6: 213–34

Harland WB, Cox AV, Llewellyn PG, Pickton CAG, Smith AG, Walters R. 1982. *A Geologic Time Scale*. Cambridge: Cambridge Univ. Press. 131 pp.

Hooker JJ. 1992. British mammalian paleocommunities across the Eocene-Oligocene transition and their environmental implications. See Prothero & Berggren 1992, pp. 494–515

Hut P, Alvarez W, Elder WP, Hansen T, Kauffman EG, et al. 1987. Comet showers as a cause of mass extinctions. *Nature* 329: 118–26

Hutchison JH. 1982. Turtle, crocodilian and champsosaur diversity changes in the Cenozoic of the north-central region of the western United States. *Palaeogeogr. Palaeoclimatol. Palaeoecol.* 37: 149–64

Hutchison JH. 1992. Western North American reptile and amphibian record across the Eocene/Oligocene boundary and its climatic implications. See Prothero & Berggren 1992, pp. 451–63

Kamp PJJ, Waghorn DB, Nelson CS. 1990. Late Eocene-early Oligocene integrated isotope stratigraphy and biostratigraphy for paleoshelf sequences in southern Australia: paleoceanographic implications. *Palaeogeogr. Palaeoclimatol. Palaeoecol.* 80: 311–23

Keller G. 1983. Paleoclimatic analyses of middle Eocene through Oligocene planktic foraminiferal faunas. *Palaeogeogr. Palaeoclimatol. Palaeoecol.* 43: 73–94

Keller G, MacLeod N, 1992. Eocene-Oligocene faunal turnover in planktic foraminifera, and Antarctic glaciation. See Prothero & Berggren 1992, pp. 218–44

Kemp EM. 1978. Tertiary climatic evolution and vegetation history in the southeast Indian Ocean region. *Palaeogeogr. Palaeoclimatol. Palaeoecol.* 24: 169–208

Kennett JP. 1977. Cenozoic evolution of Antarctic glaciation, the Circum-Antarctic Ocean, and their impact on global paleoceanography. *J. Geophys. Res.* 82: 3843–60

Kennett JP, Barker PF. 1990. Latest Cretaceous to Cenozoic climate and oceanographic developments in the Weddell Sea, Antarctica: An ocean drilling perspective. *Proc. Ocean Drill. Prog.* 113, Part B: 937–60

Kennett JP, Houtz RE, Andrews PB, Edwards AR, Gostin VR, et al. 1975. Cenozoic paleoceanography in the southwest Pacific Ocean, Antarctic glaciation and the development of the circum-Antarctic current. *Init. Rep. Deep Sea Drill. Proj.* 29: 1155–69

Kennett J.P., Stott L.D. 1990. Proteus and Proto-Oceanus: ancestral Paleogene oceans as revealed from Antarctic stable isotopic results; ODP Leg 113. *Proc. Ocean Drill. Prog. Sci. Res.* 113: 865–80

Kennett JP, von der Borch C, Baker PA, Barton CE, Boersma A, et al. 1985. Paleotectonic implications of increased late Eocene-early Oligocene volcanism from South Pacific DSDP sites. *Nature* 316: 507–11

Krishtalka L, Stucky RK, West RM, McKenna MC, Black CC, et al. 1987. Eocene (Wasatchian through Duchesnean) biochronology of North America. See Woodburne 1987, pp. 77–117

Legendre S, Hartenberger J-L. 1992. The evolution of mammalian faunas in Europe during the Eocene and Oligocene. See Prothero & Berggren 1992, pp. 516–28

LeMasurier WE, Rex DC. 1982. Volcanic record of Cenozoic glacial history in Marie Byrd Land and western Ellsworth Land: revised chronology and evaluation of tectonic factors. In *Antarctic Geoscience*, ed. C Craddock, pp. 725–34. Madison: Univ. Wis. Press

Leopold EB, Liu G, Clay-Poole S. 1992. Low-biomass vegetation in the Oligocene? See Prothero & Berggren 1992, pp. 399–420

Li C-K, Ting S-Y. 1983. The Paleogene mammals of China. *Bull. Carnegie Mus. Nat. Hist.* 21: 1–93

Lillegraven JA. 1972. Ordinal and familial diversity of Cenozoic mammals. *Taxon* 21: 261–74

Lowrie W, Napoleone G, Perch-Nielsen K, Premoli-Silva I, Toumarkine M. 1982. Paleogene magnetic stratigraphy in Umbrian pelagic carbonate rocks: the Contessa sections, Gubbio. *Geol. Soc. Am. Bull.* 92: 414–32

Marshall LG, Cifelli RL. 1989. Analysis of changing diversity patterns in Cenozoic land mammal age faunas, South America. *Palaeovertebrata* 19: 169–210

Matthews RK, Poore RZ. 1980. Tertiary $\delta^{18}O$ record and glacio-eustatic sea-level fluctuations. *Geology* 8: 501–4

McDougall I, Harrison CGA. 1988. *Geochronology and Thermochronology by the* $^{40}Ar/^{39}Ar$ *Method*. New York: Oxford Univ. Press

McGowran B. 1973. Observation Borehole No. 2, Gambier Embayment of the Otway Basin: Tertiary micropaleontology and stratigraphy. *S. Aust. Dept. Mines Min. Res. Rev.* 135: 43–55

McKinney ML, Carter BD, McNamara KJ, Donovan SK. 1992. Evolution of Paleogene echinoids: a global and regional view. See Prothero & Berggren 1992, pp. 348–67

Miller KG. 1992. Middle Eocene to Oligocene stable isotopes, climate and deep-water history: the Terminal Eocene Event? See Prothero & Berggren 1992, pp. 160–77

Miller KG, Berggren WA, Zhang J, Palmer-Julson AA. 1991b. Biostratigraphy and isotope stratigraphy of upper Eocene microtektites at Site 612: How many impacts? *Palaios* 6: 17–38

Miller KG, Fairbanks RG. 1983. Evidence for Oligocene-middle Miocene abyssal circulation changes in the western North Atlantic. *Nature* 306: 250–53

Miller KG Fairbanks R.G. 1985. Oligocene to Miocene global carbon isotope cycles and abyssal circulation changes. *Am. Geophys. Union Geophys. Monogr.* 32: 469–86

Miller KG, Fairbanks RG, Mountain GS. 1987. Tertiary oxygen isotope synthesis, sea level history, and continental margin erosion. *Paleoceanography* 2: 1–19

Miller KG, Wright JD, Fairbanks RG. 1991a. Unlocking the Ice House: Oligocene-Miocene oxygen isotopes, eustasy, and margin erosion. *J. Geophys. Res.* 96: 6829–48

Montanari A. 1990. Geochronology of the terminal Eocene impacts; an update. *Geol. Soc. Am. Spec. Pap.* 247: 607–16

Murphy MG, Kennett JP. 1986. Development of latitudinal thermal gradients during the Oligocene: oxygen-isotope evidence from the southwest Pacific. *Init. Rep. Deep Sea Drill. Proj.* 90: 1347–60

Mutter JC, Hegarty KA, Cande SC, Weissel JK. 1985. Breakup between Australia and Antarctica: a brief review in light of new data. *Tectonophysics* 114: 255–79

Obradovich JD. 1989. A different perspective on glauconite as a chronometer for geologic time scale studies. *Paleoceanography* 3: 757–70

Odin GS, ed. 1982. *Numerical Dating in Stratigraphy*. New York: Wiley

Poag CW, Ward LW. 1987. Cenozoic unconformities and depositional supersequences of North Atlantic continental margins: testing the Vail model. *Geology* 15: 159–62

Pomerol C, Premoli-Silva I, eds. 1986. *Terminal Eocene Events*. Amsterdam: Elsevier. 414 pp.

Prothero DR. 1985. North American mammalian diversity and Eocene-Oligocene extinctions. *Paleobiology* 11(4): 389–405

Prothero DR. 1989. Stepwise extinctions and climatic decline during the later Eocene and Oligocene. In *Mass Extinctions: Processes and Evidence*, ed. SK Donovan, pp. 211–34. New York: Columbia Univ. Press

Prothero DR. 1994. *The Eocene-Oligocene Transition: Paradise Lost*. New York: Columbia Univ. Press

Prothero DR, Berggren WA. 1992. *Eocene-Oligocene Climatic and Biotic Evolution*. Princeton: Princeton Univ. Press. 568 pp.

Prothero DR, Denham CR, Farmer HG. 1982. Oligocene calibration of the magnetic polarity time scale. *Geology* 10: 650–53

Prothero DR, Denham CR, Farmer HG. 1983. Magnetostratigraphy of the White River Group and its implications for Oligocene geochronology. *Palaeogeogr. Palaeoclimatol. Palaeoecol.* 42: 151–66

Prothero DR, Swisher CC III. 1992. Magnetostratigraphy and geochronology of the terrestrial Eocene-Oligocene transition in North America. See Prothero & Berggren 1992, pp. 46–73

Rampino MR, Stothers RB. 1988. Flood basalt volcanism during the past 250 million years. *Science* 241: 663–68

Rasmussen DT, Bown TM, Simons EL. The Eocene-Oligocene transition in continental Africa. See Prothero & Berggren 1992, pp. 548–66

Raup DM, Sepkoski JJ Jr. 1984. Periodicity of extinctions in the geologic past. *Proc. Nat. Acad. Sci.* 81: 801–5

Retallack GJ. 1983. A paleopedological approach to the interpretation of terrestrial sedimentary rocks: the mid-Tertiary fossil soils of Badlands National Park, South Dakota. *Geol. Soc. Am. Bull.* 94: 823–40

Retallack GJ. 1992. Paleosols and changes in climate and vegetation across the Eocene/Oligocene boundary. See Prothero & Berggren 1992, pp. 382–98

Riedel WR, Sanfilippo A, 1986. Radiolarian events and the Eocene-Oligocene boundary. See Pomerol & Premoli-Silva 1986, pp. 253–57

Savage DE, Russell DE. 1983. *Mammalian Paleofaunas of the World*. Reading, Mass.: Addison-Wesley. 432 pp.

Sclater JG, Meinke L, Bennett A, Murphy C. 1986. The depth of the ocean through the Neogene. *Geol. Soc. Am. Mem.* 163: 1–19

Shackleton NJ. 1986. Paleogene stable isotope events. *Palaeogeogr. Palaeoclimatol. Palaeoecol.* 57: 91–102

Stehlin HG. 1909. Remarques sur les faunules de mammifères des couches éocènes et oligocènes du Bassin de Paris. *Bull. Soc. Geol. France* 9: 488–520

Stucky RK. 1990. Evolution of land mammal diversity in North America during the Cenozoic. *Curr. Mammal.* 2: 375–432

Stucky RK. 1992. Mammalian faunas in

North America of Bridgerian to early Arikareean "ages" (Eocene and Oligocene). See Prothero & Berggren 1992, pp. 464–93

Swisher CC III, Prothero DR. 1990. Single-crystal $^{40}Ar/^{39}Ar$ dating of the Eocene-Oligocene transition in North America. *Science* 249: 760–62

Talwani M, Eldholm O. 1977. Evolution of the Norwegian-Greenland Sea. *Geol. Soc. Am. Bull.* 88: 969–99

Thomas E. 1992. Middle Eocene-late Oligocene bathyal benthic foraminifera (Weddell Sea): faunal changes and implications for oceanic circulation. See Prothero & Berggren 1992, pp. 245–71

Truswell EM, Harris WK. 1982. The Cainozoic palaeobotanical record in arid Australia: fossil evidence for the origins of arid-adapted flora. In *Evolution of the Flora and Fauna of Arid Australia*, ed. WR Barker, PJM Greenslade, pp. 67–76. Adelaide: Peacock

Wang B. 1992. The Chinese Oligocene: a preliminary review of mammalian localities and local faunas. See Prothero & Berggren 1992, pp. 529–47

Webb SD. 1977. A history of savanna vertebrates in the New World. Part I: North America. *Annu. Rev. Ecol. Syst.* 8: 355–80

Wei W. 1989. Reevaluation of the Eocene ice-rafting record from subantarctic cores. *Antarctic J. U. S.* 1989: 108–9

Weissel JK, Hayes DE, Herron EM. 1977. Plate tectonics synthesis: the displacements between Australia, New Zealand, and Antarctica since the late Cretaceous. *Mar. Geol.* 25: 231–77

Wolfe JA. 1971. Tertiary climatic fluctuations and methods of analysis of Tertiary floras. *Palaeogeogr. Palaeoclimatol. Palaeoecol.* 9: 27–57

Wolfe JA. 1978. A paleobotanical interpretation of Tertiary climates in the Northern Hemisphere. *Am. Sci.* 66: 694–703

Wolfe JA. 1981. A chronologic framework for Cenozoic megafossil floras of northwestern North America and its relation to marine geochronology. *Geol. Soc. Am. Spec. Pap.* 184: 39–47

Wolfe JA. 1992. Climatic, floristic, and vegetational changes near the Eocene/Oligocene boundary in North America. See Prothero & Berggren 1992, pp. 421–36

Wood HE II, Chaney RW Jr, Clark J, Colberst EH, Jepsen GL, et al. 1941. Nomenclature and correlation of the North American continental Tertiary. *Geol. Soc. Am. Bull.* 52: 1–48

Woodburne MO, ed. 1987. *Cenozoic Mammals of North America: Geochronology and Biostratigraphy*. Berkeley: Univ. Calif. Press. 336 pp.

Wyss AR, Norell MR, Flynn JJ, Novacek MJ, Charrier R, et al. 1990. A new early Tertiary mammal fauna from central Chile: implications for Andean stratigraphy and tectonics. *J. Vert. Paleontol.* 10(4): 518–22

Zachos JC, Breza JR, Wise SW. 1992. Early Oligocene ice-sheet expansion on Antarctica: stable isotope and sedimentological evidence from Kerguelen Plateau, southern Indian Ocean. *Geology* 20: 569–73

Zachos JC, Lohmann KG, Walker JCG, Wise SW. 1993. Abrupt climate change and transient climates during the Paleogene: a marine perspective. *J. Geol.* 101: 191–213

Annu. Rev. Earth Planet. Sci. 1994. 22:167–205

QUANTUM GEOPHYSICS

M. S. T. Bukowinski

Department of Geology and Geophysics, University of California, Berkeley, California 94720

KEY WORDS: Earth composition, equations of state, high pressure, mineral physics, theoretical geochemistry

1. INTRODUCTION

The microscopic world of quantum mechanics appears at first to have little relevance in the domain of planetary physics. Earthquakes, volcanism, the geodynamo, mantle convection, tides, and many other Earth processes are manifestations of classical laws of mechanics, electromagnetism, and thermodynamics. But these laws operate on and within the Earth, pieces of which mineralogists have been studying for centuries. Minerals, and hence their aggregates like rocks and planets, owe their stability and properties to the strange laws of quantum mechanics. Obeying these laws, atoms bind together and form planets, inside of which they are forced closer together to form a host of materials. Many of these materials may remain forever hidden within planetary depths, but send a variety of clues that mineralogists, in concert with planetary scientists, hope to decipher.

Brief History

Mineralogy as a science can be said to have started in 1556 with the publication of De Re Metallica by Georgius Agricola. By the nineteenth century the theory of lattices was essentially completed and simple atomistic models were devised in an effort to account for crystal forms. Crystal chemistry was born when W. L. Bragg (1920) demonstrated that interatomic distances in crystals could be estimated from sums of well-defined atomic radii. This deceptively simple observation helped to rationalize numerous crystal structures and, in the hands of Goldschmidt (1926, 1937), provided an early link between atomic models and geochemical processes.

Concurrently, during the first three decades of this century, the evolution

0084–6597/94/0515–0167$05.00

of quantum mechanics brought about a conceptual revolution. Rutherford's discovery of the nucleus in 1911, and Bohr's quantized orbital model of the hydrogen atom in 1913, showed that matter's building blocks have an inner structure of their own. With Schrödinger's development of his famous wave equation in 1926, the evolution of the quantum theory of atoms, molecules, and solids proceeded rapidly (see Slater 1965 for an insightful review). In the earth sciences, however, the need for microscopic concepts that go beyond Goldschmidt's ion did not arise till much later.

Geology was in the grip of Hutton and Lyell's uniformitarianism until about the end of the nineteenth century. Whether periodic or uniform, the processes that shaped the surface of the Earth were believed to be unchanging. Hence, geologists confined themselves to the study of the accessible surface and felt no need to look back beyond the last few hundred million years. However, the twentieth century would prove as revolutionary in geology as it was in physics.

Several developments forever changed the way geologists think about the Earth. Seismologists discovered that the Earth has a radial structure, and that the interior is much denser than crustal rocks. By 1936, when Lehmann discovered the liquid outer core, the majore features of the Earth's internal structure were established. The geologists' suspicion that the Earth is much older than Kelvin's 100 million years was proven true after the discovery of radioactivity and radiogenic dating (Patterson 1956). Geochemical and isotopic studies of meteorites, the exploration of the Moon and planets by spacecraft, and rapid advances in theoretical understanding of planet formation following the pioneering work of Safronov (1969) resulted in the unification of geology and planetary astronomy. Lastly, the theory of plate tectonics, derived from Wegener's long rejected model of continental drift, was sanctioned as the surficial manifestation of convective motions in the Earth's interior. The Earth's magnetic field, through its imprint on rocks in the continental and oceanic crusts helped to establish the reality of plate tectonics. Furthermore, the magnetic field's genesis was found in motions within the outer core (Elsasser 1950, Bullard & Gellman 1954). Thus, by the 1950s, the object of the geologist's study was a dynamic planet whose radial structure, and its evolution from a nebula to its present state, became leading topics of inquiry.

Central in these inquiries are the atoms whose interactions caused the initial condensation. They dictate which minerals exist within planets, how these minerals behave individually and in their various aggregates, and how elements are partitioned among these minerals. Standing on Goldschmidt's impressive early achievements, geologists developed more elaborate geochemical models based on well-established chemical principles like acidity and basicity, and simple quantum mechanical concepts like covalency and polarity of bonds (e.g. Ramberg 1952). But it took the full weight of Pauling's rules to permanently inject quantum mechanics into the mainstream of geochemistry and mineralogy. Distilled by Pauling (1939) from insightful observations and the then newfangled quantum mechanics, these rules provided practical guidelines on the relationship between bonding and the structure of molecules and minerals. They dominated theoretical mineralogy and geochemistry until the 1960s, when computers became powerful enough to carry out quantum mechanical calculations that would eventually rival experiments in their credibility (Richards 1979, Goddard 1985). Mineralogists were quick to avail themselves of these new tools in an attempt to rationalize structural and spectroscopic properties of minerals in terms of bonding models (see Gibbs 1982, Tossell & Vaughan 1992, Hess & McMillan 1993 for reviews).

While most geologists dealt with minerals formed near the Earth's surface, others sought to interpret the Earth's deep structure, as revealed by seismology. Until the pioneering high-pressure experiments of Bridgman, there were those who thought that only gases are compressible; nothing was known about the behavior of materials subjected to planetary pressures. Bridgman showed that all materials compress, and that pressure-induced phase transformations, often accompanied by large density increases, were common. This first hint of a rich connection between Earth structure and the physics of minerals was successfully exploited by Birch, a disciple of Bridgman. In his classic 1952 paper, Birch presented persuasive arguments that much of the Earth's interior structure could be accounted for in terms of compressibility and pressure-induced transformations in minerals known to exist near the surface of the Earth. Thus was born a new extension of mineralogy known today as mineral physics.

Early mineral physics research relied heavily on the conceptual tools introduced by Birch. Geophysicists used his finite-strain theory of equations of state (Birch 1938, 1952) and his velocity-density systematics to interpolate and extrapolate equation-of-state data to pressures of geophysical interest. But soon questions were asked for which classical physics and phenomenological quantum models provided no answers. Not surprisingly, these questions addressed the least accessible portion of the Earth: the core. Before the latter was convincingly shown to be made primarily of iron (Birch 1952), Ramsay (1949) speculated that the core-mantle boundary might be the site of a pressure-induced metallization of the silicate minerals in the mantle. At about the same time, Elsasser & Isenberg (1949) asked whether the inner-outer core discontinuity might not be caused by an electronic collapse of Fe from a $3d^6 4s^2$ to a $3d^8$ electronic configuration. Although similar transformations had been observed in other elements, this question remained unanswered until 1976, when a detailed quantum mechanical calculation showed that the electronic structure of Fe changes very little within the pressure range found in the Earth (Bukowinski & Knopoff 1976, Bukowinski 1976a). These calculations also provided the first realistic estimates of the electronic specific heat needed to interpret shock-wave data (Bukowinski 1976b). More modern calculations have since confirmed and extended these first quantum mechanical examinations of the Earth's core (Boness et al 1986, Boness & Brown 1990, Stixrude & Cohen 1993).

Mineral Physics Today

Modern mineral physics plays a key role in the interpretation of geophysically derived properties of the Earth's interior. The gross features of the seismic compressional and shear wave velocities and the density are well accounted for in terms of plausible chemical composition models. Major structural features, such as the 400 and 670 km discontinuities, correspond closely to phase transformations found in the laboratory. However, the finer details of chemical and mineral composition remain a matter of dispute, largely because of a lack of data at simultaneous high pressure and temperature, and the absence of data for shear wave velocity at lower-mantle pressures. For example, accurate high-pressure/high-temperature values of the elastic constants of silicate spinels, pyroxenes, and garnets, as well as accurate phase diagrams, are needed to test the hypothesis that changes in chemical composition are implied by the radial structure in the transition zone (Ita & Stixrude 1992). Mantle tomography yields tantalizing glimpses into the lateral structure of the Earth's interior, much of which may be due to temperature gradients associated with mantle convection, but may also be due to partial melting and/or chemical heterogeneity. A quantitative model that accounts for all these features requires a strong understanding of the thermodynamic and elastic properties of minerals and melts, their relationship to mineral structure, and the effect that pressure and temperature has on these properties and relationships.

Such a conceptual framework cannot be derived from experiment alone. A theoretical foundation is needed to bring unity to the great variety of measurements that can be made. Model Hamiltonians that approximate the total energy of materials can yield estimates of thermoelastic properties, and can elucidate the effects of pressure and temperature. At a fundamental level, this energy must be known as a function of atomic configuration, and must perforce involve the physics of electronic bonding. Hence, theoretical mineral physics aims to provide a conceptual framework, based on quantum mechanics, within which such apparently diverse properties as structure, compressibility, phase transformations, vibrational and optical spectra. heat capacity, elastic wave velocities, defect properties, rheology, and others can be related to each other and to fundamental concepts of interatomic bonding.

It is impossible to give a meaningful complete review of all of quantum mineral physics in a short article. A great number of reports have been written on quantum mechanical investigations of bonding, structure, and spectroscopic properties of atomic clusters that are representative of local environments in various minerals. Fortunately, several excellent reviews are available in the literature (Gibbs 1982, Geisinger et al 1984, Navrotsky et al 1985, Hess & McMillan 1993). Tossell & Vaughan (1992) list over 1000 references in an up-to-date and detailed review of quantum mineralogy and geochemistry. The quadrennial *Reviews of Geophysics* supplements include extensive overviews of the mineral physics literature and brief discussions of recent theoretical accomplishments (Jeanloz 1983, Bass 1987, Wolf et al 1991). In this article I present an overview of the fundamental concepts—along with selected recent applications—that guide theoretical investigations of the effects of pressure on bonding, and hence structure and properties of materials. The consideration of pressure effects biases the discussion towards condensed materials, either crystalline or amorphous, at the expense of molecules or crystal fragments. Apart from their role in providing insights into the nature of local bonding effects, and hence potential approximations to the total energy, molecular-type calculations provide little direct information on pressure effects.

2. INNER EARTH ENVIRONMENT

In this section I briefly summarize the type of minerals that mineral physicists must try to understand, and the pressure and temperature conditions to which these minerals are subjected inside the Earth. Although the discussion is limited to the Earth, many of the results are equally applicable to the other terrestrial planets. I do not discuss the giant planets for their composition is dominated by H, He, and gaseous compounds like methane, ammonia, etc. While they probably have cores of compositions that are similar to the terrestrial planets, the pressures are much higher and greatly exceed current experimental capabilities.

Chemical and Mineral Composition

Cosmochemical arguments, plus petrological studies of igneous rocks and mantle xenoliths, place significant constraints on the relative abundances of nonvolatile elements in the Earth. O, Fe, Si, Mg, Al, Ca, and Na make up more than 99% of the bulk of the Earth. Except for K, U, and Th, whose radioactive isotopes generate about 3/4 of the observed surface heat flow (Jeanloz & Morris 1986), and elements like C and H whose oxides form volatile compounds that affect melt behavior, the remaining elements have no significant impact on the Earth's structure and evolution.

The chemical composition of various layers of the Earth grows less certain as their depth increases beyond that sampled by volcanism. Minerals known to exist in the upper mantle have been subjected to high pressures and found to transform to more closely packed structures. Some of these transformations are associated with seismic discontinuities. Thus, the transformation of olivine to wadsleyite occurs at a pressure and temperature that correspond to the 400 km seismic discontinuity. The 670 km discontinuity can be explained in terms of the γ-spinel to perovskite plus magnesiowüstite transformation (Jeanloz & Richter 1979). Whether these transformations are sufficient to account for the detailed properties of the seismic discontinuities is a subject of debate (Jackson 1983, Jeanloz & Knittle 1986, Bukowinski & Wolf 1990, Stixrude et al 1992). However, except for the crust-mantle and core-mantle boundaries, no major changes in chemical composition are presently resolved by the combined seismic and mineral data.

Figure 1 shows the seismic density and velocity profiles of the Earth along with the principal minerals believed to exist in each of the seismic layers. The actual composition may be far more complex. For example, recent high-pressure petrology experiments suggest that there may be as many as 25 distinct phases present in the transition zone (Ita & Stixrude 1992). Detailed studies of seismograms give hints of minor discontinuities in the lower mantle that may implicate additional minerals and/or changes in chemical composition (Petersen et al 1993). As experimental technology advances, there is little doubt that new phases will be found. There are already experimental hints, supported by theoretical calculations, that there may be several perovskite-type silicates in the lower mantle, instead of the single phase indicated in Figure 1. Minor amounts of water may also stabilize some new phases (Finger et al 1991).

From a theoretical mineral physics perspective, the Earth consists of well oxidized mantle and crust, and a metallic alloy core. The ratio of oxygen atoms to all other atoms combined is between 4/3 and 3/2 in the mantle, and significantly higher in the crust. Since O is also the largest of the abundant ions, it tends to play a dominant role in determining the molar volume of minerals, and their relative thermodynamic stability. As I discuss later, this dominant role of O presents some nontrivial obstacles on the road towards quantitatively accurate theoretical models of minerals.

Pressure and Temperature

The pressure at the center of the Earth is approximately 3.6 Mbar, and at the bottom of the silicate mantle it is about 1.4 Mbar (Figure 1). We may

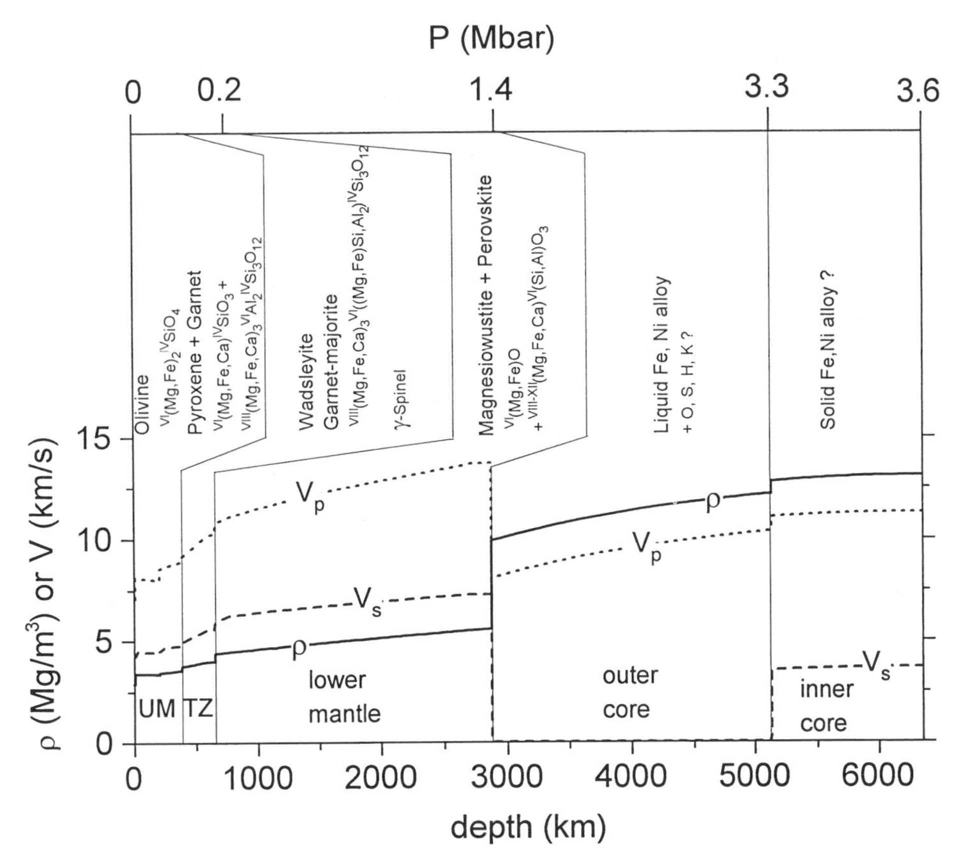

Figure 1 Density (ρ) and velocity (V_P: longitudinal, V_S: shear) profiles in the Earth, and approximate mineral assemblages believed to make up the indicated layers.

therefore take 1 Mbar as a suitable unit of pressure in the Earth. To put these numbers in an electronic context, consider an O^{2-} ion. At surface conditions, the radius of an O^{2-} ion is about $2.5a_0$, where a_0 ($=0.529$ Å) is the Bohr radius. Seismology tells us that the density of silicates increases by about 50% over this pressure, which translates into a change of the O ion volume of about $20a_0^3$. Multiplying this by 1 Mbar yields an energy of just under 2 eV. This is a rather interesting number, since it is comparable to typical bond strengths, and valence electron energies. Hence pressure within the Earth, and the other terrestrial planets, may cause significant changes in valence electron orbitals, which may in turn affect the nature of bonding, structure, and physical properties of minerals.

If pressures within planets may disrupt the valence structure of atoms, can we continue to use the familiar language of atoms and ions to describe the properties of minerals believed to exist there? Or are we compelled to always treat these materials as aggregates of electrons and nuclei? To what extent is the intuition we build on observing chemical and physical phenomena at the surface a reliable guide to deep planetary mineralogy? These are the kind of questions that Ramsay raised when he suggested that the core might consist of a metallic phase of mantle silicates.

A simple dimensional analysis indicates that only valence electrons are likely to be significantly modified, and that most electrons continue to be bound in discrete atomic-like shells. The analysis is simplified if we introduce atomic units. The unit of length is the Bohr radius defined above; the corresponding unit of atomic volume is a_0^3. The unit of mass is taken as the mass of the electron m_e, the unit of angular momentum is $\hbar$ (Planck's constant divided by 2π), and the unit of charge is the electronic charge e. With these definitions, the unit of energy is the Hartree, twice the ground state energy of the hydrogen atom, given by $E_0 \equiv e^2/a_0 = 27.2$ eV. The corresponding unit of pressure is then $P_0 \equiv e^2/a_0^4 = 294$ Mbar. We interpret this as the magnitude of the pressure needed to seriously disrupt the shell structure of atoms. For example, calculations of the electronic structure of Al show that the $2s$ and $2p$ core electrons start being ionized at about 50 Mbar, and do not become free-electron-like until 750 Mbar (McMahan & Ross 1979). Since P_0 is about three times the pressure at the center of Jupiter, the largest planet, and almost 100 times larger than the pressure at the Earth's center, we may safely assume that the "damage" done by planetary pressures to atomic structure is limited to the valence-shell modifications discussed earlier.

Thermal energy makes its presence known by exciting various "ons," i.e. electrons, excitons, phonons, magnons, etc. All such excitations are quantized with energy separations that range from essentially continuous electron states in metals to the several electron volts separating valence from conduction bands in insulators. Excitonic states affect the transport properties of minerals and may thus contribute to finite electrical conductivity and opacity to radiative heat in mantle minerals. However, our primary concern here is with the dominant role of electrons and phonons.

A significant electronic temperature is of the order of $E_0/k_B \cong 3.2 \times 10^5$ K, where k_B is Boltzmann's constant. Since temperatures throughout the mantle are unlikely to exceed 4.5×10^3 K (Jeanloz & Morris 1986), electrons generally remain in their ground states. In the core, the width of the conduction band of iron is of the order of $E_0/2$ (Bukowinski 1976a) and the temperature is well below 10^4 K. Hence the fraction of thermally excited electrons around the Fermi energy is about $2kT/E_0 < 0.06$. These can contribute significantly to the specific heat of transition metals like Fe (Bukowinski 1976b, Boness et al 1986), but cause essentially negligible perturbations to ground state bonding properties (McMahan & Ross 1977). For practical purposes, temperature has virtually no effect on the electronic structure, and hence bonding, of minerals within the Earth.

Phonons are readily excited in the Earth and other planets. Although discrete, phonon energies are separated by energies of the order of $(1/M)^{1/2}/N$, where M is the atomic mass and N^3 is the number of atoms in the crystal. For minerals with a mean atomic weight of 20, i.e. those in the mantle of the Earth, this difference is about 10^{10} times smaller than electronic level separations. Thus, at planetary temperatures, the phonon population is enormous and plays an important role in the thermodynamics of planets. However, since m_e/M is typically much less than 10^{-4}, the Born-Oppenheimer approximation holds: The electronic eigenstates at any time are in equilibrium with the instantaneous positions of the nuclei. It follows that the density, elastic properties, and most other thermodynamic and chemical properties of minerals within the Earth can be analyzed by computing their static lattice energetics.

3. FUNDAMENTAL CONCEPTS

Under the pressures and temperatures of interest, the materials that make up the Earth are assemblages of electrons and nuclei. Based on arguments such as those in Section 2, we expect that most of the electrons condense around the nuclei to form atomic-like structures that differ little from normal atoms or ions. Quantum mechanics provides the qualitative explanation for the familiar material form that such assemblages take, and the quantitative principles according to which the properties of the materials can be computed.

In all many-electron systems the motion of each electron is correlated with that of every other electron. To minimize the Coulomb energy each electron must try to stay as close as possible to the nucleus, while simultaneously avoiding other electrons. Quantum mechanics, however, imposes severe restrictions on the extent and manner in which this energy minimization can be achieved.

In a classical world, the electrons in an atom would simply collapse onto the nucleus, and matter as we know it would not exist. What prevents this collapse from happening is the kinetic energy of the electrons, which increases rapidly when their positions are increasingly localized in space. As an example consider a simple model of the H atom. Assuming that the electron is at some average distance r from the nucleus, the total energy is $E = p^2/2 - 1/r$ Hartrees, where p is the electron's momentum. According to Heisenberg's uncertainty principle, $\Delta r \Delta p \geq 1$, where Δ represents uncertainty. Making the reasonable approximation that $\Delta r \approx r$, and $\Delta p \approx p$, the energy becomes $1/2r^2 - 1/r$, and has a minimum of $-1/2$ Hartree at $r = 1$ Bohr radius. These happen to be (rather coincidentally) the correct energy and size of the H atom. The important thing to note is that the kinetic energy, which scales as $1/r^2$, balances the negative potential energy at a stable equilibrium configuration, and rapidly overwhelms it as the atomic size diminishes. The same quantum mechanical principle stabilizes many electron atoms and condensed materials.[1]

But the atom thus obtained is mostly empty space. Virtually all of its mass is concentrated in a nucleus that has a radius of about $10^{-5}a_0$. What prevents the atoms from freely interpenetrating each other when a crystal or liquid is subjected to pressure? Put another way, what makes condensed matter so hard to compress? Whatever it is that keeps atoms apart is ultimately responsible for the densities of planetary interiors, the speed of seismic waves, and the radial structure of planets which is punctuated by pressure-induced phase transformations that, to a large extent, occur to decrease atomic overlap.

The answer to these questions lies in Pauli's exclusion principle which prevents more than one electron from occupying the same quantum state. It is highly instructive to consider these effects in the context of the microscopic contributions to the energy of a system of nuclei and electrons.

Total Energy Calculations

Most mineral properties of interest to geophysicists can be derived from total energy calculations. Under hydrostatic pressure P and temperature T, materials adopt a configuration that minimizes the Gibbs free energy $G(P, T)$. Because of the Born-Oppenheimer theorem (e.g. Ziman 1972), the electronic and nuclear motions can be decoupled. Then, in the quasi-harmonic approximation, the free energy of a material with structure described by parameter $\mathbf{x}$ is (Wallace 1972)

$$G(P, T) = E(\mathbf{x}) + F_{\text{th}}(\mathbf{x}, T) + PV(\mathbf{x}), \tag{1}$$

where $E(\mathbf{x})$ is the static energy, V is the volume, and

$$F_{\text{th}}(\mathbf{x}, T) = \sum_{\mathbf{k},\lambda} [(1/2)\omega_{\mathbf{k},\lambda} + k_B T \ln(1 - e^{-(\omega_{\mathbf{k},\lambda}/k_B T)})] \tag{2}$$

is the vibrational free energy; $\omega_{\mathbf{k},\lambda}$ is a phonon frequency, and the summation is over all phonon branches λ and all wavevectors $\mathbf{k}$ in the first

[1] The general proof of the stability of materials is subtle and requires more than the uncertainty principle. See Lieb, E., Rev. Mod. Phys. 48, 553 (1976) for a detailed discussion.

Brillouin zone. The frequencies are derived from the classical equations of motion of a set of nuclei responding to changes in $E(\mathbf{x})$ induced by their motion. Hence an accurate calculation of $E(\mathbf{x})$ is the most important component in the derivation of equilibrium bulk properties of materials. In a similar manner, it is possible to use equilibrium energy changes to study surface properties, point defects, phase transformations, elastic constants, certain phonon frequencies, etc. (e.g. Pickett 1989).

THE MANY-ELECTRON PROBLEM Consider a system with N nuclei of charge Z_α at positions $\mathbf{R}_\alpha$, and n electrons with coordinates $\mathbf{r}_i$, where $\mathbf{r}_i$ stands for the spatial and spin coordinates. The spin s can have values of $\pm 1/2$. Let $\mathbf{r}$ and $\mathbf{R}$ stand for the set of all the electron and nuclear coordinates, respectively. Then, according to quantum mechanics:

$$E(\mathbf{x}) \rightarrow E(\mathbf{R}) = \frac{\displaystyle\int \Psi^*(\mathbf{r};\mathbf{R})\hat{\mathbf{H}}\Psi(\mathbf{r};\mathbf{R})d^n\mathbf{r}}{\displaystyle\int \Psi^*(\mathbf{r};\mathbf{R})\Psi(\mathbf{r};\mathbf{R})d^n\mathbf{r}}, \tag{3}$$

where the exact nonrelativistic Hamiltonian is (in atomic units)

$$\hat{\mathbf{H}} = -\frac{1}{2}\sum_{i=1}^{n}\nabla_i^2 - \sum_{i=1}^{n}\sum_{\alpha=1}^{N}\frac{Z_\alpha}{|\mathbf{r}_i - \mathbf{R}_\alpha|} + \sum_{i<j}\frac{1}{|\mathbf{r}_i - \mathbf{r}_j|} + \frac{1}{2}\sum_{\alpha=1}^{N}\sum_{\beta=1}^{N}\frac{Z_\alpha Z_\beta}{|\mathbf{R}_\alpha - \mathbf{R}_\beta|}, \tag{4}$$

and E is the total energy of the electrons. $\Psi(\mathbf{r};\mathbf{R})$, the electronic wavefunction, is a solution of Schrödinger's equation:

$$\hat{\mathbf{H}}\Psi = E\Psi. \tag{5}$$

The first term in Equation (4) gives the total kinetic energy of the electrons, while the second and third terms represent the Coulomb interaction of the electrons with the nuclei and other electrons, respectively. The last term, representing the Coulomb interaction among the nuclei, is constant for any given atomic configuration. Therefore, $\Psi(\mathbf{r};\mathbf{R})$ depends on $\mathbf{R}$ only parametrically. $\Psi(\mathbf{r};\mathbf{R})$ is interpreted to be a probability amplitude, that is $\Psi^*(\mathbf{r};\mathbf{R})\Psi(\mathbf{r};\mathbf{R})d^n\mathbf{r}$ is the probability that the electrons will be found within the n-dimensional volume $d^n\mathbf{r}$ when the nuclei are at positions $\mathbf{R}$.

Exact solutions of Schrödinger's equation are rarely obtainable. The electron-electron interaction disallows separation of the electrons' coordinates. Thus, for a typical crystal containing about 10^{23} atoms per cm^3, we face the impossible task of solving an equation with that many coordinates. Even if a solution were possible, the staggering amount of numerical information thus generated would surely defy comprehension! It is thus essential that a number of approximations be made to allow analytical or numerical solutions.

Fortunately this process is greatly facilitated by the Rayleigh-Ritz variational theorem, which states that if $\Psi(\mathbf{r};\mathbf{R})$ is an approximate wavefunction, then an extremum of $E(\mathbf{R})$ with respect to variations of $\Psi(\mathbf{r};\mathbf{R})$ is an upper bound to the true ground state energy, and that the error in $E(\mathbf{R})$ is of second order in the error in $\Psi(\mathbf{r};\mathbf{R})$ (Szabo & Ostlund 1982). Carrying out such a variation, subject to the condition that $\int \Psi^*(\mathbf{r};\mathbf{R})\Psi(\mathbf{r};\mathbf{R})d^n\mathbf{r}$ remain constant (conservation of electrons) recovers Equation (5). However, a far more interesting and fruitful approach is to start with *approximate* Hamiltonians that yield, through the variational principle, a Schrödinger's equation that can be solved and whose energies and wavefunctions are sufficiently accurate for the purpose at hand. In practice this means reducing the *many-electron* problem to an *effective single-electron* problem. In what follows, I review two distinct approaches to this goal: the Hartree-Fock method and the density functional method.

HARTREE-FOCK APPROXIMATION The Hartree-Fock method was the first truly successful reduction of the many-electron problem to an effective one-electron model. It is worth reviewing in some detail because it laid the foundation for further one-electron type theories, and because it offers an exact treatment of exchange correlations among electrons.

In systems with many electrons we expect that each electron "sees" a smeared average of the remaining electrons. The electrons' positions are then uncorrelated, and the total wavefunction is a product of single electron states Φ_i:

$$\Psi(\mathbf{r}) = \prod_{i=1}^{n} \Phi_i(\mathbf{r}_i).$$

The corresponding total charge density is

$$\rho(\mathbf{r}) = \sum_{i=1}^{n} |\Phi_i(\mathbf{r})|^2. \tag{6}$$

This formalism was used with considerable success by Hartree, who found it to work reasonably well even for atoms with few electrons. There is, however, a fundamental principle of quantum mechanics that is ignored by a simple product wavefunction. Wavefunctions of particles with half-integer spin, i.e. fermions, must be antisymmetric with respect to the interchange of two particles. The simplest such wavefunction that can be constructed from single electron states is a Slater determinant (Slater 1930):

$$\Psi(\mathbf{r}_1, \mathbf{r}_2, \ldots, \mathbf{r}_n) = \frac{1}{\sqrt{n!}} \begin{vmatrix} \Phi_1(\mathbf{r}_1) & \Phi_1(\mathbf{r}_2) & \ldots & \Phi_1(\mathbf{r}_n) \\ \Phi_2(\mathbf{r}_1) & \Phi_2(\mathbf{r}_2) & \ldots & \Phi_2(\mathbf{r}_n) \\ \vdots & \vdots & \ddots & \vdots \\ \Phi_n(\mathbf{r}_1) & \Phi_n(\mathbf{r}_2) & \ldots & \Phi_n(\mathbf{r}_n) \end{vmatrix} \tag{7}$$

where the prefactor normalizes the function to 1.

Substitution of Equation (7) into Equation (3) yields the total energy expression:

$$\begin{aligned} E = \sum_{i=1}^{n} \int d\mathbf{r} \Phi_i^*(\mathbf{r}) \left(-\frac{1}{2}\nabla^2 - \sum_{\alpha=1}^{N} \frac{Z_\alpha}{|\mathbf{r}-\mathbf{R}_\alpha|} \right) \Phi_i(\mathbf{r}) + \frac{1}{2} \sum_{\alpha=1}^{N} \sum_{\beta=1}^{N} \frac{Z_\alpha Z_\beta}{|\mathbf{R}_\alpha - \mathbf{R}_\beta|} \\ + \frac{1}{2} \sum_{i=1}^{n} \sum_{j \neq i}^{n} \int d\mathbf{r}_1 d\mathbf{r}_2 \frac{1}{|\mathbf{r}_1 - \mathbf{r}_2|} |\Phi_i(\mathbf{r}_1)|^2 |\Phi_j(\mathbf{r}_2)|^2 \\ - \frac{1}{2} \sum_{i=1}^{n} \sum_{j \neq i}^{n} \delta_{s_i s_j} \int d\mathbf{r}_1 d\mathbf{r}_2 \frac{1}{|\mathbf{r}_1 - \mathbf{r}_2|} \Phi_i^*(\mathbf{r}_1) \Phi_j(\mathbf{r}_1) \Phi_j^*(\mathbf{r}_2) \Phi_i(\mathbf{r}_2), \end{aligned} \tag{8}$$

where s_i and s_j are the electronic spins. Application of the variational principle then yields the Hartree-Fock equations for the one-electron wavefunctions Φ_i (see Szabo & Ostlund 1982 for a detailed treatment):

$$\begin{aligned} \varepsilon_i \Phi_i(\mathbf{r}) = \left[-\frac{1}{2}\nabla^2 - \sum_{\alpha=1}^{N} \frac{Z_\alpha}{|\mathbf{r}-\mathbf{R}_\alpha|} + \sum_{j \neq i}^{n} \int d\mathbf{r}_2 \frac{\Phi_j^*(\mathbf{r}_2)\Phi_j(\mathbf{r}_2)}{|\mathbf{r}_2 - \mathbf{r}|} \right] \Phi_i(\mathbf{r}) \\ - \sum_{j \neq i}^{n} \delta_{s_i s_j} \int d\mathbf{r}_2 \frac{\Phi_j^*(\mathbf{r}_2)\Phi_i(\mathbf{r}_2)}{|\mathbf{r}_2 - \mathbf{r}|} \Phi_j(\mathbf{r}), \end{aligned} \tag{9}$$

where ε_i is the energy of the ith single electron state, and equals its ionization energy, provided we assume that the other electrons' states do not change in response to the new electronic configuration (Koopmans 1933). The first three terms on the right-hand side of both equations represent the electronic kinetic energy, the Coulomb potential due to the nuclei, and the Coulomb potential due to all other electrons, respectively. The fourth term in the total energy has no classical counterpart. Known as the exchange energy, it is a correction to the Coulomb interaction brought about by the antisymmetry requirement on the wavefunction. Note that the sum includes only states with spins equal to that of the ith electron. The corresponding term in the Hartree-Fock equation takes the form of a nonlocal potential operator. A simple transformation shows it to cause electrons of like spin to avoid each other: Each electron carries with it a "hole" from which electrons with the same spin are excluded (Slater 1951). The exact form of the hole depends on the electronic states but always integrates to exactly one electron. By keeping like-spin electrons apart, the holes lower the Coulomb repulsion energy of the system.

Effect of Pauli's exclusion principle In spite of the negative contribution of the exchange interaction to the total energy, the net effect of the antisymmetry of the total wavefunction is to greatly increase the total energy of materials. Antisymmetry gives us Pauli's exclusion principle: No two electrons can be in the same quantum state. If they were, then the corresponding columns in Equation (7) are equal and $\Psi(\mathbf{r}_1, \mathbf{r}_2, \ldots, \mathbf{r}_n)$ vanishes identically.

The effect of Pauli's principle is best seen by expressing the total energy in terms of the single electron energies. Combining Equations (8) and (9) we obtain

$$E = \sum_{i=1}^{n} \varepsilon_i - \frac{1}{2}\sum_{i,j}^{n} \int d\mathbf{r}_1 d\mathbf{r}_2 \frac{[|\Phi_i(\mathbf{r}_1)|^2 |\Phi_j(\mathbf{r}_2)|^2 - \delta_{s_i s_j} \Phi_i^*(\mathbf{r}_1)\Phi_i(\mathbf{r}_2)\Phi_j^*(\mathbf{r}_2)\Phi_j(\mathbf{r}_1)]}{|\mathbf{r}_1 - \mathbf{r}_2|}. \qquad (10)$$

The $i \neq j$ restriction was removed since the corresponding Coulomb and exchange terms are identical and cancel each other out. The second term corrects for the fact that the sum of energies counts all electron interactions twice.

Without Pauli's exclusion principle, the energy is minimized by placing all electrons into the lowest single electron state with energy ε_1. Because of Pauli's principle, however, only one electron can go into the first state. Hence, as many states must be occupied as there are electrons. The effect can be huge: For n noninteracting fermions confined to a box, in the absence of Pauli's principle the energy per particle is just ε_1, but is proportional to $n^{2/3}\varepsilon_1$ with Pauli's principle (e.g. Ashcroft & Mermin 1976). For one cm^3 of material the ratio is of the order of 10^{15}! The ratio of the corresponding pressures is of the same order. In normal materials much of this "Pauli pressure" is canceled by a net electrostatic attraction, but what is left accounts for the finite density of all common matter.

Correlation energy There are some obvious improvements that can be made to the Hartree-Fock method. The antisymmetry of the wavefunction caused electrons of like spin to be correlated. However, the use of a single Slater determinant wavefunction was motivated by the assumption that electrons moved in some average distribution of all other electrons. In fact, the Coulomb repulsion correlates the motion of individual electrons, regardless of their spin. Such correlations further lower the Coulomb repulsive energy of the system.

It is possible to account for all electron correlations, and in principle to approach the exact nonrelativistic quantum energy of a system, by a technique known as Configuration Interaction (CI). In the Hartree-Fock approach, the wavefunction is built up from just those single particle orbitals that are occupied by the electrons. In a variational solution, the energy can always be lowered towards the true energy by increasing the degrees of freedom of the wavefunction. This can be done by using a set of $S > n$ single electron orbitals and expanding $\Psi(\mathbf{r}; \mathbf{R})$ in terms of all possible nth order determinants that can be constructed from them. The exact solution is obtained in the limit $S \rightarrow \infty$. The difference between this exact energy and the Hartree-Fock energy is defined as the *correlation energy*. In practice, even for systems with few electrons, the correlation energy can only be calculated approximately, since the number of single particle wavefunctions must be limited to a finite set (see Dykstra 1983, Lawley 1987, McWeeney 1978, Szabo & Ostlund 1989, Schaefer 1977, Wilson 1984 for detailed accounts). For more complex systems like crystals, the correlation energy remains virtually intractable, although some progress has been made through quantum Monte Carlo simulations of electron gases (Ceperley & Alder 1980, Fahy et al 1988, Li et al 1991).

To obtain the best possible energies with a finite amount of computational labor requires considerable skill and insight. Although it is possible to obtain very accurate equilibrium geometries and energies for small molecules, truly quantitative agreement with experimental potential energy surfaces proves elusive even for something as simple as the H_2 molecule (Szabo & Ostlund 1982). For solid state calculations the method becomes extremely cumbersome even at the Hartree-Fock level. Approximate methods that can be successfully applied to solids were developed only recently (Pisani et al 1988).

DENSITY-FUNCTIONAL THEORY The difficulties with Hartree-Fock theory arise from the nonlocality of the exchange interaction term and the complexity of the correlation energy correction. For the sake of simplicity, it is highly desirable to replace them with some effective local potential $V_{xc}(\mathbf{r})$ so that Schrödinger's equation acquires the simpler form:

$$\varepsilon_i \Phi_i(\mathbf{r}) = \left[-\frac{1}{2}\nabla^2 + V_N(\mathbf{r}, \mathbf{R}) + \int d\mathbf{r}_2 \frac{\rho(\mathbf{r}_2)}{|\mathbf{r}_2 - r|} + V_{xc}(\mathbf{r}) \right] \Phi_i(\mathbf{r}), \tag{11}$$

where we used Equation (6) to simplify the electron-electron Coulomb term, and $V_N(\mathbf{r}, \mathbf{R})$ stands for the nucleon-electron Coulomb potential of Equation (9). It happens that a simple application of the Hartree-Fock equation provides the basic idea on which such a theory may be developed.[2]

[2] Attempts to express the energy of atoms in terms of the charge density date back to work by Thomas and Fermi. See Jones & Gunnarson (1989) for a review.

A many-electron problem for which the Hartree-Fock equation can be solved exactly is the uniform electron gas of density ρ neutralized by a uniform positive charge density. The direct Coulomb interactions of the electrons with each other and the positive background cancel out, and we are left with just the kinetic energy and exchange interactions in Equation (8). The exchange energy per electron turns out to be proportional to $-\rho^{1/3}$, suggesting an effective exchange potential of the same form (Ashcroft & Mermin 1976). Slater (1951) suggested that, if we ignore for the moment the fact that electrons in atoms or condensed materials are far from uniformly distributed, the Hartree-Fock equations can be reduced to Equation (11) by replacing the complicated exchange term with a free-electron-like formula in which ρ is replaced by the local electron density $\rho(\mathbf{r})$.

That the quantum state of a system with highly correlated electrons can be determined from the single electron charge density may appear unreasonable. And yet Hohenberg & Kohn (1964) provided rigorous support for the concept by proving that the total many-electron energy, subjected to external potentials, is a unique functional of the electron density. They showed that such a functional exists (though its exact form remains unknown), and that it is minimized by the exact ground-state single-particle density. Kohn & Sham (1965) then made the method practical by showing that the many-electron problem is exactly equivalent to a system of single electrons whose states are controlled by a set of self-consistent single-particle equations.

The Kohn-Sham total energy functional is

$$E[\rho] = -\frac{1}{2}\sum_{i=1}^{n} \int d\mathbf{r}\Phi_i^*\nabla^2\Phi_i + \int d\mathbf{r}\, V_N(\mathbf{r})\rho(\mathbf{r}) + \frac{1}{2}\int d\mathbf{r}d\mathbf{r}' \frac{\rho(\mathbf{r})\rho(\mathbf{r}')}{|\mathbf{r}-\mathbf{r}'|} + E_{xc}[\rho(\mathbf{r})] + E_{NN}(\mathbf{R}), \quad (12)$$

where $E_{NN}(\mathbf{R})$ is the Coulomb interaction energy of the nuclei, $E_{xc}[\rho(\mathbf{r})]$ is the exchange-correlation functional, and $\rho(\mathbf{r})$ is given by Equation (6). As in the Hartree-Fock case, the Φ_is that minimize the energy are found by applying the variational theorem to Equation (12), yielding Equation (11) with $V_{xc}(\mathbf{r})$ given by the functional derivative:

$$V_{xc}(\mathbf{r}) = \frac{\delta E_{xc}[\rho(\mathbf{r})]}{\delta\rho(\mathbf{r})}. \quad (13)$$

If $E_{xc}[\rho(\mathbf{r})]$ were known exactly then Equation (13) would give a potential that represents the effects of exchange and correlation exactly. However, this exact form is not known for arbitrary electron densities. A workable approach is based on the local density approximation (LDA) proposed by Kohn & Sham (1965). In the LDA, the exchange correlation energy per electron at point $\mathbf{r}$, $\varepsilon_{xc}(\mathbf{r})$, is assumed to be the same as the exchange correlation energy of a uniform electron gas with a density equal to that of the actual gas at point $\mathbf{r}$. Then

$$E_{xc}[\rho(\mathbf{r})] = \int d\mathbf{r}\varepsilon_{xc}[\rho(\mathbf{r})]\rho(\mathbf{r}) \tag{14}$$

and

$$V_{xc}(\mathbf{r}) = \frac{\partial\{\rho(\mathbf{r})\varepsilon_{xc}[\rho(\mathbf{r})]\}}{\partial\rho(\mathbf{r})}. \tag{15}$$

Many attempts have been made to find suitable approximations to $\varepsilon_{xc}[\rho]$. Most are based on exact results for the limiting cases of very low and very high uniform electron densities (see Jones & Gunnarson 1989 for review). von Barth & Hedin (1972) introduced a local spin density approximation (LSDA), in which independent densities are assigned to the two possible spin polarizations. They thus made possible the study of magnetic materials and general open-shell systems. Perdew & Zunger (1981) proposed an exchange correlation interaction based on the Monte Carlo simulations of Ceperley & Alder (1980). The latter are believed to provide the closest approximation to the true energy of a uniform electron gas. The various forms give very similar energies, suggesting that little improvement can be made without going beyond the LDA. However, a very large number of calculations attests to the fact that the LDA gives very accurate ground-state properties of crystals, surfaces, and amorphous materials.

PRACTICAL MATTERS The fundamental concepts outlined above can, at least in principle, be used to study any material within the Earth. They can account for the formation of molecules and crystals from isolated atoms in the solar nebula, and for their behavior upon being subjected to pressure and high temperature following burial within planets. However, even with the reduction of electron-electron interactions to an effective single body exchange-correlation potential, there remain significant practical problems that can only be surmounted by additional simplifications.

The equations that determine the electronic states are nonlinear. Since the potentials that constitute the effective Hamiltonians depend on the wavefunctions, it is necessary to guess the initial potentials (or, equivalently, wavefunctions) and then iterate the solutions until the potentials and charge density become self consistent. Known as Self Consistent Field

(SCF) methods, they are further referred to as "first principles" or "ab initio" if they start with Schrödinger's equation and the identity of the atoms, but no empirical information.

If $\hat{\mathbf{H}}_{\text{eff}}\Phi_i = \varepsilon_i\Phi_i$ stands for either the Hartree-Fock or density functional integro-differential equations, then ε_i is given by Equation (3) with $\hat{\mathbf{H}}_{\text{eff}}$ and Φ_i in place of $\hat{\mathbf{H}}$ and Ψ. The iterative solution is made tractable by expanding the wavefunctions in terms of some convenient basis set that is complete and, usually, orthogonal. Substituting

$$\Phi_i = \sum_l c_{il}\chi_l$$

and minimizing ε_i with respect to the c_{il}^* yields a set of algebraic equations for the c_{il}:

$$\sum_l (H_{kl} - \varepsilon_i S_{kl})c_{il} = 0,$$

where

$$H_{kl} \equiv \int d\mathbf{r}\chi_k^* \mathbf{H}_{\text{eff}}\chi_l$$

and

$$S_{kl} \equiv \int d\mathbf{r}\chi_k^*\chi_l.$$

The solutions are given by the *secular equation*:

$$|H_{kl} - \varepsilon_i S_{kl}| = 0 \tag{16}$$

and the normalization condition

$$\sum_k \sum_l c_{ik}^* S_{kl} c_{li} = 1. \tag{17}$$

The detailed form of these equations depends on the specific problem being studied. In molecular studies it is customary to express the molecular orbitals as a Linear Combination of Atomic Orbitals (LCAO) belonging to the various atoms in the molecule, or in terms of local atomic orbital combinations that concentrate electronic charge along bond directions (e.g. Coulson 1961, Parr 1963, McWeeney 1978, Szabo & Ostlund 1989). The molecular energy spectra are discrete, but come in groups of closely spaced levels separated by energy gaps. As the number of atoms in the molecule increases, so does the number and density of levels. In applications to crystals, which are essentially infinite molecules, translational symmetry requires that the wavefunctions be Bloch states:

$$\Phi_k(\mathbf{r}) = u_k(\mathbf{r})e^{i\mathbf{k}\cdot\mathbf{r}},$$

where **k** is a vector in reciprocal space, and $u_k(\mathbf{r})$ has the translational symmetry of the lattice. Energy levels now form essentially continuous bands, separated by gaps, but often overlapping. This sets up what is known as a band structure problem. $u_k(\mathbf{r})$ can be expanded in various basis sets, leading to a number of techniques. In the LDA, these include LCAO (Bullett 1980, 1987), Augmented Plane Waves (APW; Mattheiss et al 1968, Moruzzi 1978), Linearized Augmented Plane Waves (LAPW; Wimmer et al 1981, Jansen & Freeman 1984), Linear Muffin Tin Orbitals (LMTO; Andersen 1975, Skriver 1984), pseudopotentials (Yin & Cohen 1982), etc. Recently, a periodic Hartree-Fock code was developed that overcomes many of the computational difficulties (Pisani 1987, Pisani et al 1988). Readers who are interested in side by side discussions of several band structure methods and their applications may consult Ziman (1972), Callaway (1974), Godwal et al (1983) and Srivastava & Weaire (1987), among others.

There are many occasions when it is convenient or necessary to resort to further simplifications in total energy calculations. Atomistic models often provide the only practical way to study structural and thermodynamic properties of complex minerals. First principles and ab initio methods can be used to constrain model interatomic potentials. Because of their intuitive appeal and computational efficiency, such potentials allow considerable numerical experimentation. I discuss some of these methods in the next section, after I review a few recent examples of the uses of electronic structure calculations in geophysics.

4. SOME GEOPHYSICAL APPLICATIONS

Starting with the first-principles examination of the electronic structure and equation of state of Fe at high pressures (Bukowinski & Knopoff 1976; Bukowinski 1976a,b), there has been a steady increase in the complexity of the minerals studied. This has been made possible by the great advances in computer technology, and the consequent evolution of electronic structure methods whose accuracy is limited almost entirely by theoretical, rather than numerical, approximations.

Nevertheless, the complexity of crystals whose behavior under pressure can be studied from first principles remains limited. Accordingly, theoretical studies have generally focused on relatively simple minerals that can be studied accurately enough to make meaningful comparisons with data. Fortunately, most of the complex silicate minerals are found in the crust and upper mantle, where the pressure and temperature are low enough to be accessible to a variety of experimental techniques. As we penetrate deeper into the Earth, the high pressure that starts causing experimental difficulties also replaces the ubiquitous SiO_4 tetrahedra with SiO_6 octahedra, and mineral structures become accessible to accurate first principles and ab initio techniques. These minerals are believed to include oxides like MgO, CaO, SiO_2, Al_2O_3, FeO, the perovskite-type silicates $MgSiO_3$, $CaSiO_3$, and possibly various solid solutions of all of the above. Deeper yet, we come to the core and its Fe alloy. Recent experimental evidence on the properties of Fe has instigated new theoretical studies of its high-pressure structure.

Band Structure Applications

First-principles band structure calculations have yielded insights into the effects of pressure on the electronic structure of simple oxide minerals. Equations of state computed from the electronic structures are in very good to excellent agreement with data (Bukowinski 1985, Aidun et al 1984, Chang & Cohen 1984, Froyen & Cohen 1984, Bukowinski & Aidun 1985, Mehl et al 1988, Cohen 1991). Figure 2 shows selected calculated and experimental equations of state of cubic alkali halides and alkaline earth oxides. The curves labeled MPIB are based on ab initio pair potentials, discussed later. Although the alkali halides are of no geophysical interest, their compressibility allows large experimental compression, and hence strong tests of the theoretical equations of state. The reliability of the theoretical results is well illustrated by the CaO (B2) APW equation of state, which was calculated several years before the measurements were made by Richet et al (1988). Theoretically derived values of the compressibility and its pressure derivative are also accurate. The APW calculations predict that CaO transforms from the B1 (NaCl) to the B2 (CsCl) structure at about 300 kbar, which is only about half the experimental transformation pressure (Bukowinski 1985). In MgO, this transformation is predicted to occur at about 2.05 Mbar, but values of 10 Mbar and about 5 Mbar were predicted from pseudopotential (Chang & Cohen 1984) and LAPW (Mehl et al 1988) calculations, respectively. There is no experimental observation with which to compare, but the LAPW and pseudopotential predictions are more likely to be accurate since they make no shape approximations to the electronic potential and charge density. This example illustrates the extreme accuracy that is needed to correctly predict solid state phase transformations. Free energy curves for neighboring phases often differ from each other by tiny fractions of their absolute value.

Calculated ground state electronic structures and densities agree well with available X-ray diffraction and spectroscopic data, and provide an

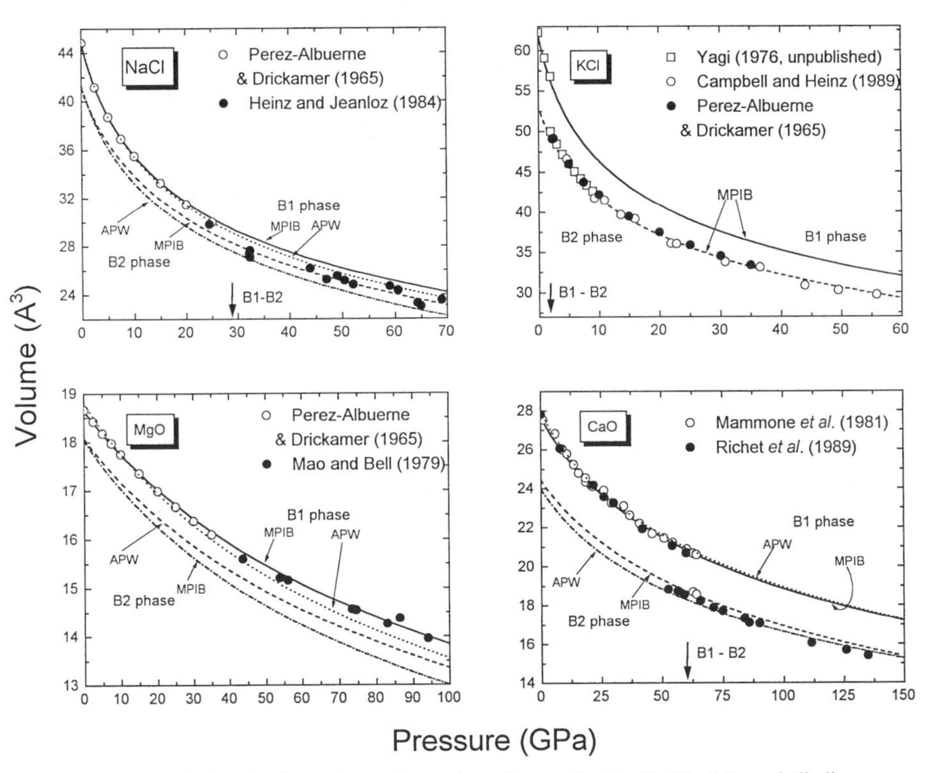

Figure 2 Calculated and experimental equations of state of cubic alkali halides and alkaline-earth oxides. Vertical arrows indicate MPIB calculated B1 to B2 transformation pressures. MPIB calculations are from Zhang & Bukowinski (1991); APW calculations for NaCl from Bukowinski & Aidun (1985); others from Bukowinski (1985).

interesting illustration of the strong influence of cations on the behavior of oxygen-dominated valence bands (Bukowinski 1980, 1982). Consider, for example, MgO and CaO. CaO is somewhat more ionic than MgO since Ca is the less electronegative cation. The valence band of CaO is about 2.5 eV wide, while that of MgO is about twice that. This is consistent with expected trends for ionic salts: The large radius of Ca^{2+} results in a weaker crystal field at the O^{2-} site and smaller O-O overlap—both effects contribute to less splitting of the O $2p$ states, and hence a narrower band.

The subtle differences between MgO and CaO become more pronounced under applied pressure. The band gap of MgO increase with pressure (Liberman 1978, Bukowinski 1980) while that of CaO decreases (Bukowinski 1982). The former is consistent with the experimental observation that, in contrast to the great majority of materials, the index of refraction of

MgO decreases with compression (Vedam & Schmidt 1966). The decrease of the gap in CaO is due to the presence of *d*-electron states near the bottom of the conduction band (recall that Ca is next to Sc, the first transition metal in the periodic chart). It turns out that the energy of the *d*-electron states does not increase as rapidly with compression as does the energy of the O $2p$ states, leading to a narrowing gap. This in turn leads to increased virtual interactions between valence and conduction states and an enhancement of the polarizability of the valence states and the compressibility of the crystal.

We learn from these examples that the properties of oxides are controlled by complicated electronic interactions. The O valence states are greatly affected by the nature of the electronic state of the cations, and by the proximity of the oxygens to each other. In effect, the nature of O is strongly affected by its environment, as are the interactions between oxygen and other ions in the crystal.

All ions are to some extent affected by their environment. The effects are stronger in many oxides because the O^{2-} ion is not stable in the gas state: The tenth electron is not bound. And yet, calculations and experimental binding energies suggest that, to a very good approximation, crystals like MgO are made up of overlapping Mg^{2+} and O^{2-} ions (Mehl et al 1988). It must be that the O^{2-} ion is stabilized by the crystal field around it, which therefore influences the detailed form of the oxygen charge density. This leads to many-body interactions, which are exhibited, among other properties, by the large violation of the Cauchy relations. The latter require that the elastic constants C_{12} and C_{44} be equal in cubic crystals with atoms/ions that interact via potentials that depend only on their separation. For example, $C_{12}/C_{44} = 95/156, 59/81, 47/56, 49/34$ in MgO, CaO, SrO, and BaO, respectively (Sumino & Anderson 1984). The largest violation occurs in MgO, which has the smallest cation and hence the largest O-O overlap and valence band width.

The valence band widths also correlate with the ability of simple oxides to transform to more closely packed structures. Since in most geophysically important minerals O^{2-} is the largest ion, phases that have significant O-O overlap can only form under high pressure. The valence band is generally derived from O valence electrons, so that the width of this band gives an indication of the strength of the O-O overlap interactions. The calculated valence band widths for the B2 phases of CaO and MgO, and the fluorite phase of SiO_2, are 2.8 eV, 6.7 eV, and 11 eV, respectively. CaO transforms to the B2 phase at about 600 kbar; a similar transformation has not been observed in MgO but is predicted to occur at pressures in excess of 2 Mbar. Based on oxide and fluoride mineral systematics, the fluorite structure was suggested for the next high-pressure phase of SiO_2 after stishovite

(Al'tshuler et al 1973, Simakov et al 1973, Liu 1982, Jamieson 1977). The large valence bandwidth suggests that this is very unlikely to occur within Earth pressures. Indeed, band-structure calculations by Carlson et al (1984) and Bukowinski & Wolf (1986) showed that the fluorite phase is thermodynamically unstable relative to stishovite. Lattice-dynamics calculations indicated that it is also dynamically unstable until pressures that significantly exceed those in the lower mantle (Bukowinski & Wolf 1986). High pressure is required to cause the Si and O ions to overlap and thus stabilize what is at lower pressure an inherently unstable cubic lattice of oxygen ions, with the small Si ions sitting ineffectively in the middle of large oxygen cubes.

Recent LAPW calculations of the properties of stishovite and possible post-stishovite phases of silica offer excellent examples of the power of modern band-structure techniques. Calculations by Park et al (1988) suggested that stishovite transforms to a pyrite-like phase with symmetry $Pa\bar{3}$ at about 600 kbar. Cohen (1991, 1992) also used this method to study stishovite, the $Pa\bar{3}$ phase, and a new phase with the $CaCl_2$ structure observed experimentally by Tsuchida & Yagi (1989). Based on static energy calculations, he found that stishovite transforms into the $CaCl_2$ at 450 kbar, which in turn transforms into the $Pa\bar{3}$ structure at about 1.56 Mbar. The experimental transformation pressure to the $CaCl_2$ phase is about twice the calculated one. However, the experiments involved large uniaxial stresses, and the changes in the structure are very small, which suggests that the transformation pressure is not well constrained. Cohen also argues that temperature would significantly increase the theoretical transformation pressure. The different pressures of the predicted transformation to the $Pa\bar{3}$ structure are probably due to incomplete convergence of the Park et al calculations.

Cohen was also able to reproduce the experimental pressure dependence of the A_{1g} and B_{1g} Raman modes of stishovite. The latter is soft and causes the related shear elastic modulus $c_{11}-c_{12}$ to become negative in the stishovite phase at about 450 kbar, which leads to the stishovite to $CaCl_2$ transformation. Hemley (1987) had used his observed softness of the B_{1g} mode to predict that stishovite would become unstable at a higher pressure, but was unable to get an accurate estimate of the pressure because of the limited pressure range of the Raman measurements. However, the agreement of theory with experiment argues strongly for the accuracy of the calculated transition pressure.

To end the discussion of band-structure methods we come back to the core of the Earth, and the high-pressure phase diagram of its principal component Fe. There are four solid phases of Fe below 120 kbar: δ (bcc), α (bcc), γ (fcc), and ε (hcp). The last three coexist at the triple point at 120 kbar. Beyond this pressure, the only known phases are the γ, ε, and liquid phases. The slope of the γ-ε phase boundary is uncertain at high pressure, and there are contradictory experimental measurements of the γ-liquid boundary (Brown & McQueen 1980, 1986; Bass 1987; Boehler et al 1990; Williams et al 1987; Mao et al 1987). From what is a rather complicated story, two principal problems emerge: The position of the γ-ε-liquid triple point is highly uncertain, and it is difficult to reconcile the shock-wave data of Brown & McQueen without introducing yet another triple point at which the γ, liquid, and a new phase coexist. Ross et al (1990) used an empirically constrained model (Young & Grover 1983) to propose that the new phase could be a nonmagnetic bcc phase that is stabilized at high temperature by its high entropy.

Previous first principles studies of high-pressure Fe did not include the bcc phase, since it was believed to be relevant only at low pressure. Stixrude & Cohen (1993) undertook to examine the proposal of Ross et al (1990). They used the LAPW LDA approximation for the γ and ε phases, and the spin-polarized LSDA approach for the bcc phase. Figure 3 summarizes some of the results. The stable low-pressure phase is correctly found to be bcc, and the hcp phase is found to be stable at high pressure and low temperature, again correctly. Note, however, the bcc to hcp transformation is predicted to occur at negative pressures. The energy of the bcc phase, whose calculated magnetic moment approaches zero at inner core pressures, is found to remain rather high relative to the hcp phase. This suggests that the bcc phase is not present in the core. However, the calculated equations of state overestimate the densities of all phases of iron: that of the bcc phase is very close to the experimental equation of state for the hcp phase (Mao et al 1990). Better agreement might be obtained by considering improvements in the exchange-correlation functional, and by including thermal excitations of phonons and electrons. It is possible that the local density approximation is not accurate enough to treat transition metals. Bukowinski (1976a,b) needed to use an unusually low exchange potential to obtain agreement with the zero-pressure density of the fcc phase.[3]

FeO offers another example of the limitations of LDA theory and band-structure theory. FeO is of geological interest since it may be incorporated into the core and is an end member of the magnesiowüstite $Mg_xFe_{1-x}O$ solid solution series, a candidate lower-mantle component (Kato & Ring-

[3] Indeed, use of generalized gradient corrections to the exchange-correlation functional gives total energies and equations of state that are in much better agreement with data for Fe. However, the conclusion regarding the presence of the bcc phase in the core remains unchanged (L. Stixrude, personal communication).

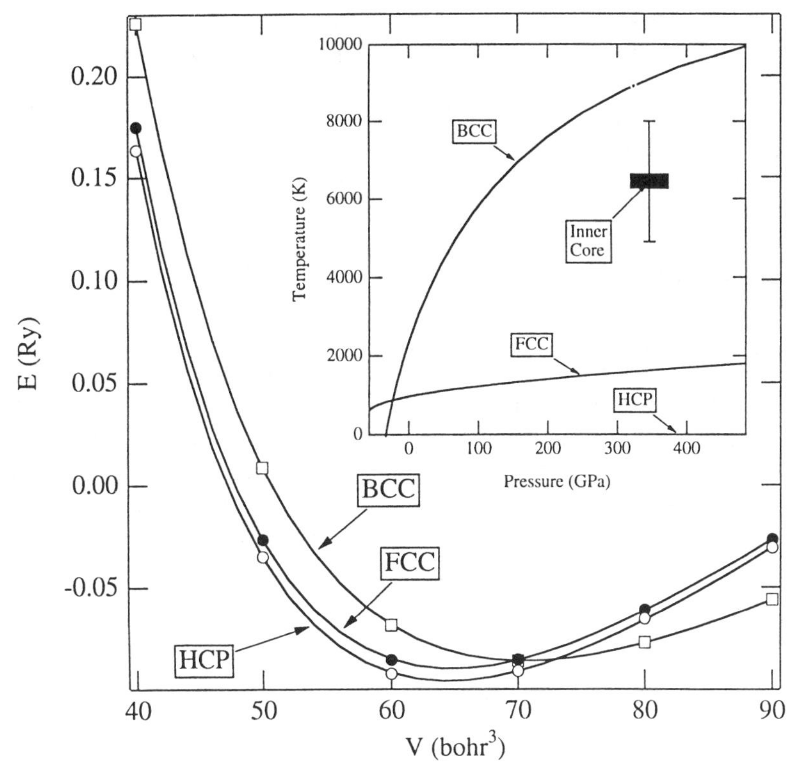

Figure 3 Calculated total energies for iron. Inset shows the energy of the fcc and bcc phases relative to that of the hcp phase. Energy differences are expressed as temperature per atom. Also shown is the inner core pressure, and the range of proposed temperatures. The bcc-hcp energy difference is much larger than the available thermal energy, suggesting that the inner core cannot be made of bcc iron. 1 Rydberg is 13.6 eV. (Figure courtesy of L. Stixrude).

wood 1989, Knittle & Jeanloz 1991). Apart from its nonstoichiometry, the presence of microstructural domains, and its complex mechanical properties (Hazen & Jeanloz 1984), FeO has complicated electronic properties which it shares with other third period transition metal monoxides. While standard band-structure calculations predict MnO, FeO, CoO, and NiO to be metallic, they are in fact insulators. One explanation of this holds that they are Mott insulators, in which the tendency of electrons to lower their kinetic energy by delocalization, and hence hopping from atom to atom, is frustrated by the large repulsion between the hopping electron and those on neighboring atoms. This is clearly an electron correlation effect that is akin to that which causes a homonuclear diatomic molecule to break up into neutral atoms as the bond length is increased, rather than into positive and negative ions.

Such localized electrons are not well described by LDA band theory. However, LSDA band calculations that take into account the magnetic structure have shown that a correct theory of the insulating behavior of these transition metal monoxides lies somewhere between the Mott insulator model and LSDA band theory (see Pickett 1989, Cox 1992, Isaak et al 1993 for reviews).

In spite of the limitations of LSDA band theory, Isaak et al (1993) were able to use it to elucidate some properties of stoichiometric FeO. They find that the local magnetic moments greatly diminish with pressure which, since local magnetic order appears to be intimately associated with the insulating behavior, suggests that band theory is adequate at high pressure. Their calculations show that the B1 phase is unstable to rhombohedral strains at low temperatures and high pressures. The magnitude of the strain appears to be consistent with experimental observations. The strain increases bonding among nearest Fe ions at the expense of Fe-O bonding, an effect that lowers the energy of the crystal sufficiently to stabilize it against a transformation to the B2 structure. The latter is predicted to occur at about 5 Mbar, which is consistent with the lack of evidence for any first-order transformations in FeO up to at least 2.2 Mbar. Isaak et al argue that temperature may remove the strain energy that stabilizes the rhombohedrally distorted B1 phase, and thus allows a transformation to the B2 phase at lower pressures. This offers an appealing explanation of the phase transformation to a metallic phase observed in shock-wave experiments (Jeanloz & Ahrens 1980, Knittle & Jeanloz 1991).

Atomistic Models

Finding equilibrium structures often requires that the energy be optimized with respect to tens of structural parameters. The use of first-principles approaches under these conditions can be prohibitively time consuming and expensive. Atomistic models allow much more efficient calculations of total energies, and can also be readily used to calculate dynamic properties of materials. By atomistic I mean any model in which the material is assumed to be made up of well-defined units, be they atoms or ions, which then interact with each other. The total energy can then be written quite generally as (e.g. Schatz 1989)

$$E = \sum_{i<j} \phi_2(r_{ij}) + \sum_{i<j<k} \phi_3(r_{ij}, r_{ik}, r_{jk}) + \dots \tag{18}$$

where r_{ij} is the distance between atoms i and j, $\phi_2(r_{ij})$ represents the interaction between these atoms in the absence of other atoms, and ϕ_3 is the three-body interaction associated with the ijk cluster—i.e. the change in the interactions between two atoms associated with the perturbation exerted on them by a third atom. Equation (18) is useful if the summation can be truncated at the two- or three-body term. It should be apparent that such truncations use *effective* potentials, i.e. a two-body model uses potentials that are not likely to give accurate interactions for isolated pairs.

Simple parametric potentials have been in use for many decades. The disadvantage of such models is that they require accurate data to constrain the parameters, and their validity is usually limited to the conditions at which the data were obtained. For a review of empirical potentials see Catlow & Mackrodt (1985). In keeping with the title of this review, the following discussion is limited to *ab initio* pair potentials.

ELECTRON GAS POTENTIALS Approximate atomistic quantum mechanical models are more easily extended to complex minerals and have shown great promise in the *ab initio* modeling of their structure and thermoelastic properties. The electron gas theory of pair potentials, introduced by Gordon & Kim (1972), has proven particularly fruitful. In the electron gas model, the total charge density is taken to be a superposition of spherical atomic/ionic charge densities. The total energy of the system is assumed to be given by the density functional as defined by Equations (12) and (14), with the exception that the kinetic energy is taken to be that of a free electron gas: $T[\rho] = \int \rho(\mathbf{r})\varepsilon_k[\rho(\mathbf{r})]d\mathbf{r}$, where $\varepsilon_k[\rho] = (3/10)(3\pi^2)^{2/3}\rho^{2/3}$ is the average kinetic energy per electron. The interaction potential between atoms i and j is then given by the total energy of the overlapping atoms less the sum of the energies of the isolated atoms:

$$\phi_{ij} = E[\rho_i+\rho_j]-E[\rho_i]-E[\rho_j]. \tag{19}$$

The total energy E_T of the material then is

$$E_T = \sum_{i<j} \phi_{ij}+\sum_i \phi_i, \tag{20}$$

where ϕ_i is the self energy of the ith atom/ion, defined as the change in the ionic energy associated with spherical deformation of its charge density caused by interactions with the rest of the system.

In Gordon & Kim's original model, the ionic charges, which were assumed to be rigid and spherically symmetric, were obtained from atomic Hartree-Fock calculations. The rigidity of the charge densities eliminates the need for the self-energy terms. The exchange-correlation energy was taken as a logarithmic interpolation between the known low- and high-density expressions. This method produced fairly accurate equations of state and thermal properties of simple alkali halides, alkaline-earth oxides, and some fluoride perovskites (Kim & Gordon 1974; Cohen & Gordon 1975, 1976; Boyer 1981, 1983, 1984). Waldman & Gordon (1979) introduced scaling factors for the kinetic and exchange-correlation energies in an attempt to make the atomic energies calculated with the electron gas method as close as possible to the more accurate Hartree-Fock energies. Thus corrected, the method became known as the Modified Electron Gas (MEG) theory.

The MEG method was used successfully to study the structure of $MgSiO_3$ and $CaSiO_3$ perovskites as functions of temperature and pressure (Wolf & Bukowinski 1985, 1987). Equilibrium structures for a given temperature and pressure were found by minimizing the Gibbs free energy, as given by Equations (1) and (2). The phonon spectrum was calculated with quasi-harmonic lattice dynamics. These calculations gave accurate values of the compressibility and thermal expansivity, and nicely accounted for the orthorhombic and cubic structures of $MgSiO_3$ and $CaSiO_3$, respectively. The calculations also indicated that $MgSiO_3$ could undergo temperature-induced phase transformations from orthorhombic to tetragonal and then cubic. The effect of pressure on the transformation temperatures places them near the top of the lower mantle. There is some indirect evidence that such transformations do occur in $MgSiO_3$ perovskite (Wang et al 1990), and they are known to occur in the isoelectronic analog compound $NaMgF_3$ (Chao et al 1961). The calculated densities, however, were significantly underestimated, which was consistent with other MEG calculations (e.g. Cohen & Gordon 1975, 1976).

The key to improving the early electron gas models already existed within the MEG model. We saw in the section on band-structure methods that the free O^{2-} ion is unstable, but appears to be stabilized by the crystal field. In the MEG model, the effect of the crystal was approximated by placing the ion in a Watson sphere (Watson 1958). The Watson sphere is a charged shell of radius R_W and total charge Q_W which produces the electrostatic potential:

$$V_W(r) = \frac{Q_W}{r} \quad \text{for} \quad r > R_W$$
$$= \frac{Q_W}{R_W} \quad \text{for} \quad r \leq R_W. \tag{21}$$

The oxygen charge distribution was obtained with this potential added to the Hartree-Fock Hamiltonian. Watson's fixed values of Q_W and R_W were used. This gave rigid ions and resulted in central potentials that were independent of the crystal configuration, were too repulsive, and could not yield elastic constants that violated the Cauchy relations. The solution is to allow the radius of the Watson sphere to vary in response to changes in crystal configuration. Muhlhausen & Gordon (1981) proposed that the charge of the sphere be chosen to neutralize the ionic charge, and the radius be such that the potential inside the sphere is equal to the site Coulomb potential due to the rest of the crystal. The new method was dubbed the "Shell-Stabilized MEG" model, or SSMEG. Since the ion responds to the changes in the Watson sphere, the self energy must be included in the total energy. Notable applications of geophysical interest include another study of $MgSiO_3$ and $CaSiO_3$ perovskites by Hemley et al (1987), and a study of MgO (Hemley et al 1985). The new perovskite calculations also yielded the correct structures, gave much better absolute densities, but significantly underestimated the compressibilities.

Several variants of the SSMEG-type model have since been proposed. The Potential Induced Breathing model (PIB, Boyer et al 1985) differs from the SSMEG model in that the ionic charge densities are computed with a density functional implementation of the relativistic Dirac Hamiltonian, instead of the Hartree-Fock equation. One advantage of this approach is that the same density functional is used to calculate the charge density and interionic interactions, thus eliminating the need for scaling factors. The Hamiltonian also included a self-energy correction, which attempts to correct for the fact that in density functional theory the self terms in the Coulomb and exchange interactions do not cancel exactly, as they do in Hartree-Fock theory (see discussion of Pauli's principle in Section 3).

In all the models thus far discussed, the ionic charge densities are either rigid, or respond only to the point-ion Coulomb potential, as approximated by a Watson sphere. In real materials, electrons relax to minimize the total energy, with contributions from kinetic and exchange-correlation corrected electrostatic energies. Given an ionic model, this relaxation could be accomplished by minimizing the crystal energy with respect to variations of all the ionic wavefunctions. However, such a procedure would vitiate much of the simplicity of the electron gas models. As a practical approximation one could minimize the energy with respect to changes in the Watson sphere radius. This process allows the charge density to relax in response to two competing effects: contraction of the charge density lowers the repulsive overlap energy of adjacent ions, but raises the self energy. Equilibrium is obtained when the corresponding forces cancel. This idea was originally suggested by Muhlhausen & Gordon (1981) and implemented in the Variationally Stabilized MEG (VSMEG) model (Wolf & Bukowinski 1988). This model retained the Hartree-Fock equation for the ionic wavefunctions. The MVIB (Modified Variationally Induced Breathing) model (Zhang 1993) is essentially identical to the VSMEG, but uses identical nonrelativistic density functional Hamiltonians to compute the ionic self energies and ionic overlap energies.

More accurate electron gas potentials were obtained from the Modified Potential-Induced Breathing (MPIB) model (Zhang & Bukowinski 1991), which attempted to obtain approximate self consistency between the total charge density and the crystal potential. The Watson sphere potential was replaced by the spherically averaged total crystal potential, including Coulomb and exchange-correlation contributions, due to the superposed ionic charge densities. The calculation was iterated until the potential and charge densities were self consistent. Although the spherical averaging prevents full self consistency, this method approaches full variational relaxation of the charge density more closely than the others. More rigorous self-consistent calculations were made by LeSar (1988) and Edwardson (1989), but these were never used to study geologically interesting materials.

The PIB, VSMEG, MVIB, and MPIB models proved quite successful in geophysical applications. Sample MPIB equations of state are shown in Figure 2; their accuracy rivals that of accurate band structure–derived results, and exceeds the latter in total energy differences, as exemplified by the accurate calculated phase transformation pressures and binding energies (Zhang & Bukowinski 1991). Figure 4 compares calculated and experimental equations of state for some geophysically important minerals. The $MgSiO_3$ example illustrates the importance of charge relaxation: The rigid ion densities are in error by more than 20%, while those that allow relaxation differ from data and each other by at most a few percent. Models that allow some degree of optimization of the energy with respect to the wavefunctions (MPIB and MVIB) are the most accurate. Although elastic constants calculated with the non-rigid-ion models can exhibit errors of the order of 10 to 20%, they violate the Cauchy relationships in the same manner as the experimental constants (Mehl et al 1986, Wolf & Bukowinski 1988), and yield temperature derivatives of elastic constants that are in good agreement with available data (Isaak et al 1990).

Minerals with tens of atoms per unit cell can be modeled efficiently with electron gas potentials. A study of numerous possible high-pressure phases of SiO_2 confirms that the $CaCl_2$ structure is the first post-stishovite phase, and suggests that a monoclinically distorted $Pa\bar{3}$ structure becomes stable at about 1.7 Mbar (Figure 5). Other simulations show that, in accord with available data, there is little solubility among $MgSiO_3$, $CaSiO_3$, and Al_2O_3 at high pressures, thus indicating the possible presence of several perovskite-like phases in the lower mantle (Zhang 1993). Earlier examples suggest that the results of these calculations are accurate enough to warrant

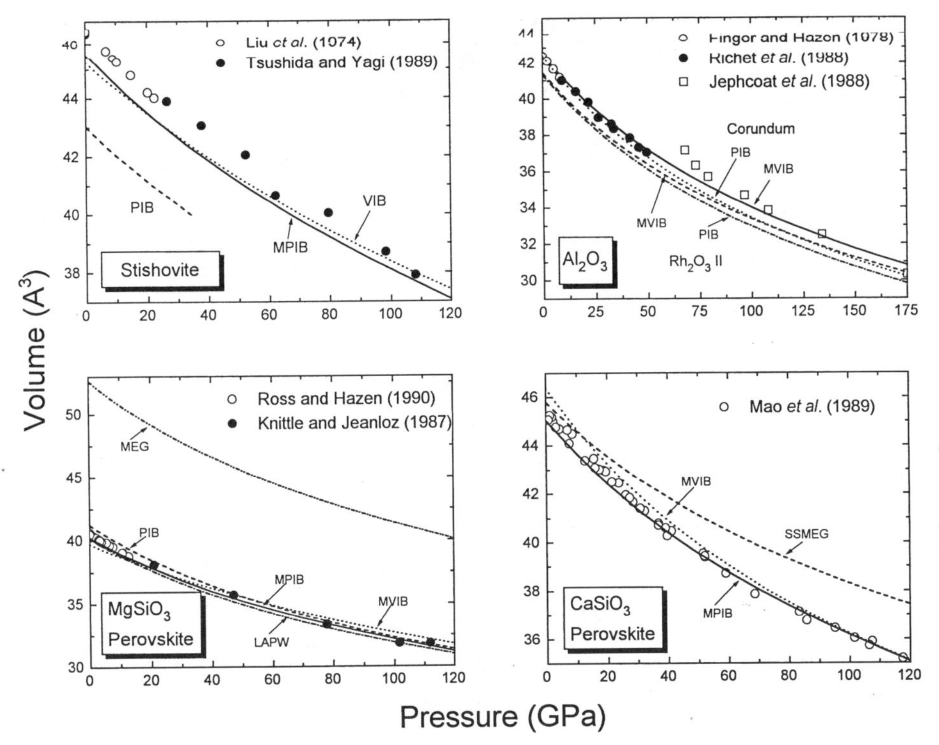

Figure 4 Calculated and experimental equations of state for selected lower mantle phases. MVIB and MPIB calculations are from Zhang (1993). Stishovite: PIB from Cohen (1987a); MPIB from Zhang & Bukowinski (1991). Al_2O_3: PIB from Cynn et al (1990). $MgSiO_3$: PIB from Cohen (1987b); LAPW from Stixrude & Cohen (1993); MEG from Wolf & Bukowinski (1987). $CaSiO_3$: SSMEG from Hemley et al (1987).

serious experimental searches for minerals uncovered in such numerical "experiments."

5. CONCLUDING REMARKS

Applications of quantum mechanics to geologically motivated problems have grown rapidly in recent years. Most of the significant contributions to geophysics are based on first principles or ab initio methods like those described above. A longer review would have to discuss analytic potential models with parameters constrained by ab initio calculations. These have contributed some useful insights into the structure of silica crystal phases (Lasaga & Gibbs 1988; Stixrude & Bukowinski 1988, 1990a; Tsuneyuki et

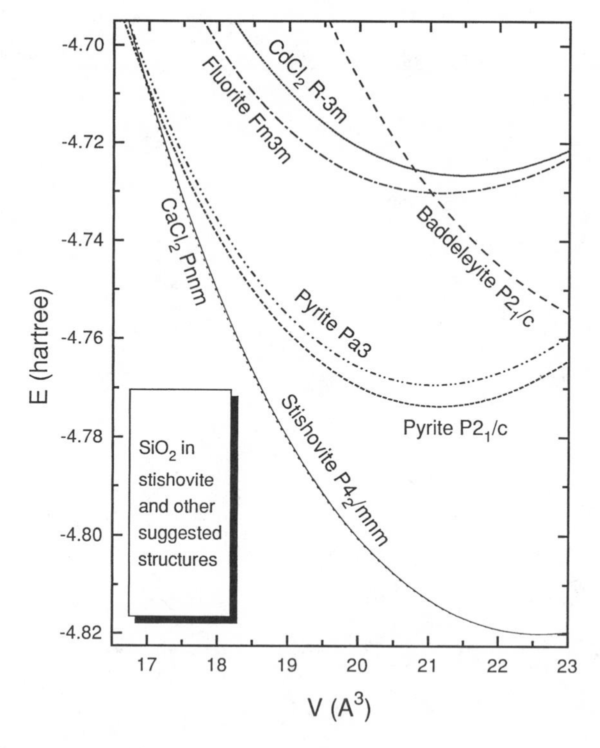

Figure 5 Lowest energy high-pressure phases found in a search of 17 potential structures. Total energies were computed with the MVIB method (Zhang 1993). The total energies of stishovite and Pa3 phases are very similar to those obtained with first principles band structure calculations (Cohen 1991).

al 1988, 1989) as well as liquids and glasses (Rustad et al 1990, Stixrude & Bukowinski 1990b, Stixrude et al 1991). A parametric many-body potential for alkaline-earth oxides could account for their zero-pressure elasticity and predicted accurate pressure derivatives of elastic constants, phase transformation pressures, and high-pressure thermal properties, and provide a possible explanation for the high seismic value of $(\partial \ln V_S/\partial \ln V_P)_P$ (Agnon & Bukowinski 1988, 1990a,b). Studies of the effect of pressure on silicate liquid structure will doubtlessly motivate more serious attempts to construct potentials that reflect the effects of local structure on bonding.

Much of the theoretical effort has endeavored to establish that quantum methods can yield quantitatively accurate material properties. There is no doubt that very accurate equations of state can be calculated by anyone willing to make the effort. Structures of crystals can also be predicted quite accurately, although detailed comparisons with data often require that these structures be obtained from dynamical, rather than static, calculations. Elastic constants of relatively simple materials can be obtained accurately with band-structure methods. More complex crystals still require very significant computational efforts. There is a need for atomistic models that can yield accurate elastic constants either ab initio or by extrapolating available low-pressure data. High-pressure silicate liquid simulations currently rely on parametric potentials. Progress is needed to improve the accuracy of such potentials. Practical methods for dealing with transition metal cations are needed to analyze their partitioning among various phases at high pressure, and their site preferences. Efficient means for optimizing structures with few symmetry constraints will help computer searches for potential high-pressure multicomponent phases. The more covalent structures found in the shallower parts of the Earth deserve much more attention from a first-principles crystal physics perspective than they have seen thus far (Hess & McMillan 1993).

New techniques are being developed in computational physics that will allow mineral physicists to address most of the above questions in a rigorous fashion. A combination of molecular dynamics with density-functional theory provides a means to study static and dynamic properties of large quantum mechanical systems (Car & Parrinello 1985). The method optimizes the total energy with respect to ionic coordinates and electronic wavefunctions simultaneously. Coupled with pseudopotential techniques, it has proven to be extremely powerful and far more efficient than standard algebraic approaches like band-structure theory (Payne et al 1992). It can, and will, be applied to a great variety of problems, including those mentioned above. Periodic Hartree-Fock (Pisani et al 1988, Hess & McMillan 1993) methods yield very accurate mineral structures, and will provide a benchmark for accurate exchange interaction treatments and, by extension, the correlation energy. Density-functional methods are likely to benefit from these developments.

The principal theme of this article is the demonstrated potential of modern quantum methods to accurately determine properties of materials subjected to conditions that are either difficult or impossible to achieve with existing laboratory techniques. It is an indisputable fact that whenever the quantum mechanical equations can be solved accurately, the results always agree with observation to as many digits as the data provide. The power of modern computers, including some desktop models, provides unprecedented opportunities to solve these equations for systems of ever increasing complexity. The day is near when most material processes within planets will be modeled on computers with an accuracy that will render many measurement techniques obsolete. Some of the materials will be discovered on the computer! All the while, insights gained from all the numerical experiments are accumulating and, one hopes, leading to intuitive models such as those that were so common before the power of computers took over.

ACKNOWLEDGMENTS

I am grateful to my wife Halina for her loving help when this paper was already late and I was temporarily incapacitated. I thank Huoyi Zhang for his help with references and figures, and Lars Stixrude for sending results of his iron calculations prior to publication. This work was supported by grants from the National Science Foundation and the Institute of Geophysics and Planetary Physics of the University of California.

Literature Cited

Agnon A, Bukowinski MST. 1988. High pressure shear moduli—a many body model for oxides. *Geophys. Res. Lett.* 15: 209–12

Agnon A, Bukowinski MST. 1990a. Thermodynamic and elastic properties of many-body model for simple oxides. *Phys. Rev.* B1: 41,7755–66

Agnon A, Bukowinski MST. 1990b. δ_s at high pressure and $d \ln V_s / d \ln V_p$ in the lower mantle. *Geophys. Res. Lett.* 17: 1149–52

Aidun J, Bukowinski MST, Ross M. 1984. Equation of state and metallization of CsI. *Phys. Rev.* B29: 2611–21

Al'tshuler LV, Podurets MA, Simakov GV, Trunin RF. 1973. High density forms of fluorite and rutile. *Sov. Phys. Solid State* 15: 969–71 (From Russian)

Andersen OK. 1975. Linear methods in band theory. *Phys. Rev.* B12: 3060–83

Ashcroft NW, Mermin ND. 1976. *Solid State Physics*. New York: Holt, Rinehart and Winston

Bass JD. 1987. Mineral and melt physics. *Rev. Geophys.* 25: 1265–76

Birch F. 1938. The effect of pressure upon the elastic properties of isotropic solids, according to Murnaghan's theory of finite strain. *J. Appl. Phys.* 9: 279–88

Birch F. 1952. Elasticity and constitution of the earth's interior. *J. Geophys. Res.* 57: 227–86

Boehler R, von Bargen N, Chopelas A. 1990. Melting, thermal expansion, and phase transitions of iron at high pressures. *J. Geophys. Res.* 95: 21,731–36

Boness DA, Brown JM. 1990. The electronic band structures of iron, sulfur, and oxygen at high pressures and the earth's core. *J. Geophys. Res.* 95: 21,721–30

Boness DA, Brown JM, McMahan AK. 1986. The electronic thermodynamics of iron under Earth core conditions. *Phys. Earth Planet. Int.* 42: 227–40

Boyer LL. 1981. Origin of superionicity in the alkaline earth halides. *Solid State Ionics* 5: 581–84

Boyer LL. 1983. Bonding and equation of state of MgO. *Phys. Rev.* B: 27,1271–75

Boyer LL. 1984. Parameter-free equation-of-state calculations for $CsCaF_3$. *J. Phys C: Solid State* 17: 1825–32

Boyer LL, Mehl MJ, Feldman JL, Hardy JR, Flocken JW, Fong CY. 1985. Beyond the rigid-ion approximation with spherically symmetric ions. *Phys. Rev. Lett.* 54: 1940–43

Bragg WL. 1920. The arrangement of atoms in crystals. *Philos. Mag.* 40: 169

Brown JM, McQueen RG. 1980. Melting of iron under core conditions. *Geophys. Res. Lett.* 7: 533–36

Brown JM, McQueen RG. 1986. Phase transitions, Gruneisen parameter, and elasticity for shocked iron between 77 GPa and 400 GPa. *J. Geophys. Res.* 91: 7485–94

Bukowinski MST. 1976a. On the electronic structure of iron at core pressures. *Phys. Earth Planet. Int.* 13: 57–66

Bukowinski MST. 1976b. A theoretical equation of state for the inner core. *Phys. Earth Planet. Int.* 14: 333–44

Bukowinski MST. 1980. Effect of pressure on bonding in MgO. *J. Geophys. Res.* 85: 285–92

Bukowinski MST. 1982. Pressure effects on bonding in CaO: comparison with MgO. *J. Geophys. Res.* 87: 303–10

Bukowinski MST. 1985. First principles equation of state of MgO and CaO. *Geophys. Res. Lett.* 12: 536–39

Bukowinski MST, Aidun J. 1985. First principles versus spherical ion models of the B1 and B2 phases of NaCl. *J. Geophys. Res.* 90: 1794–800

Bukowinski MST, Knopoff L. 1976. Electronic structure of iron and models of the earth's core. *Geophys. Res. Lett.* 3: 45–48

Bukowinski MST, Wolf GH. 1986. Equation of state and stability of fluorite-structured SiO_2. *J. Geophys. Res.* 91: 4704–10

Bukowinski MST, Wolf GH. 1990. Thermodynamically consistent decompression: implications for lower mantle composition. *J. Geophys. Res.* 95: 12,583–93

Bullard EC, Gellman H. 1954. Homogeneous dynamo and terrestrial magnetism. *Phil. Trans. R. Soc. London Ser. A* 247: 213

Bullett DW. 1980. The renaissance and quantitative development of of the tight binding method. In *Solid State Physics, Advances in Research and Applications*, Vol. 35, ed. H Seitz, D Turnbull, pp. 129–214. New York: Academic

Bullett DW. 1987. Applications of atomic orbital methods to the structure and properties of complex transition-metal compounds. *Phys. Chem. Miner.* 14: 485–91

Callaway J. 1974. *Quantum Theory of the Solid State*. New York: Academic

Campbell AJ, Heinz DL. 1989. Equation of state of the B2 phase of potassium chloride. *Eos, Trans. Am. Geophys. Union* 43: 1356

Car R, Parrinello M. 1985. Unified approach for molecular dynamics and density-functional theory. *Phys. Rev. Lett.* 55: 2471–74

Carlsson AE, Ashcroft NW, Williams AR. 1984. Properties of SiO_2 in a high-pressure fluorite structure phase. *Geophys. Res. Lett.* 11: 617–19

Catlow CRA, Mackrodt WC, eds. 1985. *Computer Simulations of Condensed Matter*. Amsterdam: North-Holland

Ceperley DM, Alder BJ. 1980. Ground state of the electron gas by a stochastic method. *Phys. Rev. Lett.* 45: 566–69

Chang KJ, Cohen ML. 1984. High-pressure behavior of MgO: structural and electronic properties. *Phys. Rev.* B30: 4774–81

Chao ECT, Evans H, Skinner B, Milton C. 1961. Neighborite, $NaMgF_3$, a new mineral from the Green River formation, South Ouray, Utah. *Am. Mineral.* 46: 379–93

Cohen AJ, Gordon RG. 1975. Theory of the lattice energy, equilibrium structure, elastic constants, and pressure-induced phase transformations in alkali-halide crystals. *Phys. Rev.* B12: 3228–41

Cohen AJ, Gordon RG. 1976. Modified electron-gas study of the stability, elastic properties, and high-pressure behavior of MgO and CaO crystals. *Phys. Rev.* B14: 4593–605

Cohen RE. 1987a. Calculation of elasticity and high pressure instabilities in corundum and stishovite with the potential induced breathing model. *Geophys. Res. Lett.* 14: 37–40

Cohen RE. 1987b. Elasticity and equation of state of $MgSiO_3$ perovskite. *Geophys. Res. Lett.* 14: 1053–56

Cohen RE. 1991. Bonding and elasticity of stishovite SiO_2 at high pressures: linearized augmented Plane Wave calculations. *Am. Mineral.* 76: 733–42

Cohen RE. 1992. First-principles predictions of elasticity and phase transitions in high pressure SiO_2 and geophysical implications. In *High-Pressure Research: Applications to Earth and Planetary Sciences*, ed. Y Syono, MH Manghnani, pp. 425–31. Tokyo: TERRAPUB/Washington DC: Am. Geophys. Union

Coulson CA. 1961. *Valence*. London: Oxford Univ. Press

Cox PA. 1992. *Transition Metal Oxides—An Introduction to Their Electronic Structure and Properties*. Oxford: Oxford Univ. Press

Cynn H, Isaak DG, Cohen RE, Nicol MF, Anderson OL. 1990. A high-pressure phase transition of corundum predicted by the potential induced breathing model. *Am. Mineral.* 75: 439–42

Dykstra CE, ed. 1983. *Advanced Theories and Computational Approaches to the Electronic Structure of Molecules*. Dordrecht: Reidel

Edwardson PJ. 1989. Corridors-between-adjacent-sites model of the four phases of $KNbO_3$. *Phys. Rev. Lett.* 63: 55–58

Elsasser WM. 1950. The earth's interior and geomagnetism. *Rev. Mod. Phys.* 22: 1

Elsasser WM, Isenberg I. 1949. Electronic phase transition in iron at extreme pressures. *Phys. Rev.* 76: 469 (Abstr.)

Fahy S, Wang XW, Louie SG. 1988. Variational quantum Monte Carlo nonlocal pseudopotential approach to solids: cohesive and structural properties of diamond. *Phys. Rev. Lett.* 61: 1631–34

Finger LW, Hazen RM. 1978. Crystal structure and compressibility of ruby to 80 kbar. *Yearbook, Geophys. Lab., Carnegie Inst.*, pp. 525–27

Finger LW, Hazen RM, Prewitt CT. 1991. Crystal structure of $Mg_{12}Si_4O_{19}(OH)_2$ (phase B) and $Mg_{14}Si_5O_{24}$ (phase AnhB). *Am. Mineral.* 76: 1–7

Froyen S, Cohen ML. 1984. Structural properties of NaCl. *Phys. Rev.* B29: 3770–72

Geisinger KL, Gibbs GV, Navrotsky A. 1984. A molecular orbital study of bond length and angle variations in framework silicates. *J. Non-cryst. Solids* 68: 401–2

Gibbs GV. 1982. Molecules as models for bonding in silicates. *Am. Mineral.* 67: 421–50

Goddard WA. 1985. Theoretical chemistry comes alive: full partner with experiment. *Science* 227: 917–23

Godwal BK, Sikka SK, Chidambaram R. 1983. Equation of state theories of condensed matter up to about 10 TPa. *Phys. Rep.* 102: 121–97

Goldschmidt VM. 1926. *Geochemische Verteilungsgesetze der Elemente*. Oslo: Skrifter Norske Videnskaps Akad Oslo I. Mat-Naturv. Kl

Goldschmidt VM. 1937. The principles of distribution of chemical elements in minerals and rocks. *J. Chem. Soc. London* 655–73

Gordon RG, Kim YS. 1972. A theory for the forces between closed shell atoms and molecules. *J. Chem. Phys.* 56: 3122–33

Hazen R, Jeanloz R. 1984. Wüstite ($Fe_{1-x}O$): a review of its defect structure and physical properties. *Rev. Geophys. Space Phys.* 122: 37–46

Heinz DL, Jeanloz R. 1984. Compression of the B2 high-pressure phase of NaCl. *Phys. Rev. B* 30: 6045–50

Hemley RJ. 1987. Pressure dependence of Raman spectra of SiO_2 polymorphs: α-quartz, coesite, and stishovite. See Manghnani & Syono 1987, pp. 347–59

Hemley RJ, Jackson MD, Gordon RG. 1985. First-principles theory for the equations of state of minerals at high pressures and temperatures: application to MgO. *Geophys. Res. Lett.* 12: 247–50

Hemley RJ, Jackson MD, Gordon RG. 1987. Theoretical study of the structure, lattice dynamics, and equations of state of perovskite-type $MgSiO_3$ and $CaSiO_3$. *Phys. Chem. Miner.* 14: 2–12

Hess AC, McMillan PF. 1993. Ab initio methods in geochemistry and mineralogy. In *Advances in Electronic Structure Theory* 2, ed. TH Dunning Jr. New York: Jai. In press

Hohenberg P, Kohn W. 1964. Inhomogeneous electron gas. *Phys. Rev.* B136: 864–71

Isaak DG, Cohen RE, Mehl MJ. 1990. Calculated elastic and thermal properties of MgO at high pressures and temperatures. *J. Geophys. Res.* 95: 7055–67

Isaak DG, Cohen RE, Mehl MJ, Singh DJ. 1993. Phase stability of wustite at high pressure from first-principles linearized augmented plane-wave calculations. *Phys. Rev. B* 47: 7720–31

Ita J, Stixrude L. 1992. Petrology, elasticity, and composition of the mantle transition zone. *J. Geophys. Res.* 97: 6849–66

Jackson I. 1983. Some geophysical constraints on the chemical composition of the earth's lower mantle. *Earth Planet. Sci. Lett.* 62: 91–103

Jamieson JC. 1977. Phase transitions in rutile type structures. In *High Pressure Research in Geophysics*, ed. S Akimoto, MH Manghnani, pp. 209–18. Tokyo: Cent. Acad. Publ.

Jansen HJF, Freeman HJ. 1984. Total energy full-potential linearized-augmented-plane-wave method for bulk solids: electronic and structural properties of tungsten. *Phys. Rev.* B30: 561–69

Jeanloz R. 1983. Mineral and melt physics. *Rev. Geophys. Space Phys.* 21: 1487–503

Jeanloz R, Ahrens TJ. 1980. Equations of state of FeO and CaO. *Geophys. J. R. Astron. Soc.* 62: 529–49

Jeanloz R, Knittle E. 1986. Reduction of mantle and core properties to a standard state by adiabatic decompression. In *Advances in Physical Geochemistry: Chemistry and Physics of Terrestrial Planets*, ed. SK Saxena, 6: 275–309. New York: Springer-Verlag

Jeanloz R, Morris S. 1986. Temperature distribution in the mantle and crust. *Annu. Rev. Earth Planet. Sci.* 14: 377–415

Jeanloz R, Richter FM. 1979. Convection, composition, and the thermal state of the lower mantle. *J. Geophys. Res.* 84: 5497–504

Jephcoat AP, Hemley RJ, Mao H-K. 1988. X-ray diffraction of ruby (Al_2O_3: Cr^{3+}) to 175 GPa. *Physica* B150: 115–21

Jones RO, Gunnarson O. 1989. The density functional formalism. *Rev. Mod. Phys.* 61: 689–746

Kato T, Ringwood AE. 1989. Melting relationships in the system Fe-FeO at high pressures: implications for the composition and formation of the earth's core. *Phys. Chem. Miner.* 16: 524–38

Kim YS, Gordon RG. 1974. Theory of binding of ionic crystals: applications to alkali-halide and alkaline-earth-dihalide crystals. *Phys. Rev.* B9: 3548–54

Knittle E, Jeanloz R. 1987. Synthesis and equation of state of (Mg,Fe)SiO_3 perovskite to over 100 GPa. *Science* 235: 669–70

Knittle E, Jeanloz R. 1991. The high-pressure phase diagram of $Fe_{0.94}O$: a possible constituent of the earht's core. *J. Geophys. Res.* 96: 16,169–80

Kohn W, Sham LJ. 1965. Self-consistent equations including exchange and correlation effects. *Phys. Rev.* A140: 1133–38

Koopmans T. 1933. Über die Zuordnungvon wellenfunktinen und eigenwerten zu den einzelnen elektronen eines atoms. *Physica* 1: 104–13

Lasaga AC, Gibbs GV. 1988. Quantum mechanical potential surfaces and calculations on minerals and molecular clusters. *Phys. Chem. Miner.* 16: 29–41

Lawley KP, ed. 1987. *Ab Initio Methods in Quantum Chemistry*. New York: Wiley

LeSar R. 1988. Equations of state of dense helium. *Phys. Rev. Lett.* 61: 2121–24

Li XP, Ceperley DM, Martin RM. 1991. Cohesive energy of silicon by the Green's function Monte Carlo method. *Phys. Rev.* B44: 10,929–32

Liberman DA. 1978. Self-consistent-field electronic structure calculations for compressed magnesium oxide. *J. Phys. Chem. Solids* 39: 255–57

Lieb, E. 1976. The stability of matter. *Rev. Mod. Phys.* 48: 553

Liu LG. 1982. High pressure phase transformations of the dioxides: implications for structures of SiO_2 at high pressure. In *High Pressure Research in Geophysics*, ed. S Akimoto, MH Manghnani, pp. 349–60. Tokyo: Cent. Acad. Publ.

Liu LG, Bassett WA, Takahashi T. 1974. Effect of pressure on the lattice parameters of stishovite. *J. Geophys. Res.* 79: 1160–64

Mammone JF, Mao H-K, Bell PM. 1981. Equation of state of CaO under static pressure conditions. *Geophys. Res. Lett.* 79: 140–43

Manghnani MH, Syono Y, eds. 1987. *High-Pressure Research in Mineral Physics*. Tokyo: Terra Sci./Washington, DC: Am. Geophys. Union

Mao H-K, Bell PM. 1979. Equations of state of MgO and ε-Fe under static pressure conditions. *J. Geophys. Res.* 84: 4533–36

Mao H-K, Chen LC, Hemley RJ, Jephcoat AP, Wu Y, Bassett WA. 1989. Stability and equation of state of $CaSiO_3$-perovskite to 134 GPa. *J. Geophys. Res.* 94: 17,889–94

Mao H-K, Bell PM, Hadidiacos C. 1987. Experimental phase relations of iron to 360 kbar, 1400 C, determined in an internally heated diamond-anvil apparatus. See Manghnani & Syono 1987, pp. 135–38

Mao H-K, Wu Y, Chen LC, Shu JF, Jephcoat AP. 1990. Static compression of iron to 300 GPa and $Fe_{0.8}Ni_{0.2}$ alloy to 260 GPa: implications for composition of the core. *J. Geophys. Res.* 95: 21,737–42

McMahan AK, Ross M. 1977. High-temperature electron-band calculations. *Phys. Rev.* B15: 718–25

McMahan AK, Ross M. 1979. Shell structure effects in compressed aluminum. In *High-Pressure Science and Technology*, Vol. 6, ed. KD Timmerhaus, MS Barber, pp. 920–26. New York: Plenum

McWeeney R. 1978. *Methods of Molecular Quantum Mechanics*. New York: Academic

Mehl MJ, Cohen RE, Krakauer H. 1988. Linearized Augmented Plane Wave electronic structure calculations for MgO and CaO. *J. Geophys. Res.* 93: 8009–22

Mehl MJ, Hemley RJ, Boyer LL. 1986. Potential-induced breathing model for the elastic moduli and high-pressure behavior of the cubic alkaline-earth oxides. *Phys. Rev.* B33: 8685–95

Moruzzi VL, Janak JF, Williams AR. 1978. *Calculated Electronic Properties of Metals*. New York: Pergamon

Muhlhausen C, and Gordon RG. 1981. Density-functional theory for the energy of crystals: test of the ionic model. *Phys. Rev.* B24: 2147–60

Navrotsky A, Geisinger KL, McMillan P, Gibbs GV. 1985. The tetrahedral framework in glasses and melts-Inferences from molecular orbital calculations and implications for structure, thermodynamics, and physical properties. *Phys. Chem. Miner.* 11: 284–98

Park KT, Terakura K, Matsui Y. 1988. Theoretical evidence for a new ultra-high-pressure phase of SiO_2. *Nature* 336: 670–72

Parr RG. 1963. *The Quantum Theory of Molecular Electronic Structure*. New York: Benjamin

Patterson C. 1956. Age of meteorites and the earth. *Geochim. Cosmochim. Acta* 10: 230–37

Pauling L. 1939. *The Nature of the Chemical Bond*. New York: Cornell Univ. Press

Payne MC, Teter MP, Allan DC, Arias TA, Joannopoulos JD. 1992. Iterative minimization techniques for ab initio total-energy calculations: molecular dynamics and conjugate gradients. *Rev. Mod. Phys.* 64: 1045–97

Perdew, JP, Zunger A. 1981. Self-interaction correction to density-functional approximations for many-electron systems. *Phys. Rev.* B23: 5048–79

Perez-Albuerne EA, Drickamer HG. 1965. Effect of high pressure on the compressibility of seven crystals having the NaCl or CsCl structure. *J. Chem. Phys.* 43: 1381–87

Petersen N, Gossler J, Kind R, Stammler K, Vinnik L. 1993. Precursors to SS and structure of transition zone of the northwest Pacific. *Geophys. Res. Lett.* 20: 281–84

Pickett WE. 1989. Electronic structure of the high-temperature oxide superconductors. *Rev. Mod. Phys.* 61: 433–512

Pisani C. 1987. Hartree-Fock ab initio approaches to the solution of some solid-state problems. *Int. Rev. Phys. Chem.* 6: 367–84

Pisani C, Dovesi R, Roetti C. 1988. *Hartree-Fock Ab Initio Treatments of Crystalline Systems. Lect. Notes in Chem.*, Vol. 48, ed. G Berthier, MJS Dewar, H Fisher, K Fukai, GG Hall, et al. Berlin: Springer-Verlag

Ramberg H. 1952. Chemical bonds and distribution of cations in silicates. *J. Geol.* 61: 318–52

Ramsey WH. 1949. On the nature of the earth's core. *Mon. Not. R. Astron. Soc. Geophys. Suppl.* 5: 409

Richards G. 1979. Third age of quantum chemistry. *Nature* 78: 507

Richet P, Mao H-K, Bell PM. 1988. Static compression and equation of state of CaO to 1.35 Mbar. *J. Geophys. Res.* 93: 15,279–88

Richet P, Xu J-A, Mao H-K. 1988. Quasi-hydrostatic compression of ruby to 500 Kbar. *Phys. Chem. Miner.* 16: 207–11

Ros M, Young DA, Grover R. 1990. Theory of the iron phase diagram at earth core conditions. *J. Geophys. Res.* 95: 21,713–16

Ross NL, Hazen RM. 1990. High-pressure crystal chemistry of $MgSiO_3$ perovskite. *Phys. Chem. Miner.* 17: 228–37

Rustad JR, Yuen DA, Spera FJ. 1990. Molecular dynamics of liquid SiO_2 under high pressure. *Phys. Rev.* A42: 2081–89

Safronov VS. 1969. *Evolution of the Protoplanetary Cloud and Formation of the Earth and Planets.* Moscow: Nauka. Transl. 1972, for NASA and NSF by Isr. Prog. Sci. Transl. as *NASA TT F-677*

Schaefer HF. 1977. *Methods of Electronic Structure Theory*. New York: Plenum

Schatz GC. 1989. The analytical representation of electronic potential-energy surfaces. *Rev. Mod. Phys.* 61: 669–88

Simakov GV, Podurets MS, Trunin RF. 1973. New data on the compressibility of oxides and fluorides and the theory of hamogeneous composition of the earth. *Dokl. Akad. Nauk SSSR* 211: 1330–32

Skriver H. 1984. *The LMTO Method.* Berlin: Springer-Verlag

Slater JC. 1930. Note on Hartee's method. *Phys. Rev.* 35: 210

Slater JC. 1951. A simplification of the Hartree-Fock method. *Phys. Rev.* 81: 385–90

Slater JC. 1965. *Quantum Theory of Molecules and Solids*, Vol. 2. New York: McGraw-Hill

Srivastava GP, Weaire D. 1987. The theory of the cohesive energies of solids. *Adv. Phys.* 36: 463–517

Stixrude L, Bukowinski MST. 1988. Simple covalent potential models of tetrahedral SiO_2: applications to α-quartz and coesite at pressure. *Phys. Chem. Miner.* 16: 199–206

Stixrude L, Bukowinski MST. 1990a. Rings, topology, and the density of tectosilicates. *Am. Mineral.* 75: 1159–69

Stixrude L, Bukowinski MST. 1990b. A novel topological compression mechanism in a covalent liquid. *Science* 250: 541–43

Stixrude L, Cohen RE. 1993. First principles investigation of the electronic structure and physical properties of solid iron at core pressures. *Eos, Trans. Am. Geophys. Union* 74: 305 (Abstr.)

Stixrude L, Hemley RJ, Fei Y, Mao H-K. 1992. Thermoelasticity of silicate perovskite and magnesiowstite and stratification of the Earth's mantle. *Science* 257: 1099–101

Stixrude L, Oshagan A, Bukowinski MST. 1991. Coordination changes and the vibrational spectrum of SiO_2 glass at high pressure. *Am. Mineral.* 76: 1761–64

Sumino Y, Anderson OL. 1984. Elastic constants of minerals. In *CRC Handbook of Physical Properties of Rocks*, Vol. III, ed. RS Carmichael, pp. 39–138. Boca Raton, Fla.: CRC

Szabo A, Ostlund NS. 1982. *Quantum Chemistry: Introduction to Advanced Electronic Structure Theory*. New York: Macmillan

Tossell JA, Vaughan DJ. 1992. *Theoretical Geochemistry: Applications of Quantum Mechanics in the Earth and Mineral Sciences.* New York: Oxford Univ. Press

Tsuchida Y, Yagi T. 1989. A new, post-stishovite high-pressure polymorph of silica. *Nature* 340: 217–20

Tsuneyuki S, Matsui Y, Aoki H, Tsukada M. 1989. New pressure-induced structural transformations in silica obtained by computer simulation. *Nature* 339: 209–11

Tsuneyuki S, Tsukada M, Aoki H, Matsui Y. 1988. First-principles interatomic potential of silica applied to molecular dynamics. *Phys. Rev. Lett.* 61: 869–72

Urey HC. 1952. *The Planets, Their Origin*

and Development. New Haven: Yale Univ. Press

Vedam K, Schmidt EDD. 1966. Variation of refractive index of MgO with pressure to 7 kbar. *Phys. Rev*. 146: 548–54

von Barth U, Hedin L. 1972. A local exchange-correlation potential for the spin-polarized case: I. *J. Phys. C Solid State Phys*. 5: 1629–42

Waldman M, Gordon RG. 1979. Scaled electron gas approximation for intermolecular forces. *J. Chem. Phys*. 71: 1325–39

Wallace DC. 1972. *Thermodynamics of Crystals*. New York: Wiley

Wang Y, Guyot F, Yeganeh-Haeri A, Liebermann RC. 1990. Twinning in $MgSiO_3$ perovskite. *Science* 248: 468–71

Watson RE. 1958. Analytic Hartree-Fock solutions for O^{-2}. *Phys. Rev*. 111: 1108–10

Williams Q, Jeanloz R, Bass J, Svendsen R, Ahrens TJ. 1987. The melting curve of iron to 2.5 Mbar: first experimental constraint on the temperature at the earth's center. *Science* 236: 181–82

Wilson EB Jr. 1984. *Electron Correlation Methods in Molecules*. Oxford: Clarendon

Wimmer E, Krakauer H, Weinert M, Freeman AJ. 1981. Full-potential self-consistent linearized augmented-plane-wave method for calculating the electronic structure of molecules and surfaces: O_2 molecule. *Phys. Rev*. B24: 864–75

Wolf GH, Bukowinski MST. 1985. Ab initio structural and thermoelastic properties of orthorhombic $MgSiO_3$ perovskite. *Geophys. Res. Lett*. 12: 809–12

Wolf GH., Bukowinski MST. 1987. Theoretical study of the structural properties and equations of state of $MgSiO_3$ and $CaSiO_3$ perovskites: implications for lower mantle composition. See Manghnani & Syono 1987, pp. 313–31

Wolf GH, Bukowinski MST. 1988. Variational stabilization of the ionic charge densities in the electron-gas theory of crystals: applications to MgO and CaO. *Phys. Chem. Miner*. 15: 209–20

Wolf GH, Mackwell SJ, Bassett WA. 1991. Mineral and melt physics. *Rev. Geophys*. Suppl.: 844–63

Yin MT, Cohen ML. 1982. Theory of ab initio pseudopotential calculations. *Phys. Rev*. B25: 7403–11

Young DA, Grover R. 1983. Theory of the iron equation of state and melting curve to very high pressures. In *Shock Waves in Condensed Matter*, ed. JR Asay, RA Graham, GK Straub, pp. 65–67. New York: Elsevier

Zhang H. 1993. *A study of lower mantle mineralogy by ab initio potential models*. PhD thesis. Univ. Calif., Berkeley. 239 pp.

Zhang H, Bukowinski MST. 1991. Modified potential-induced breathing model of potentials between closed-shell ions. *Phys. Rev*. B44: 2495–503

Ziman JM. 1972. *Principles of the Theory of Solids*. Cambridge: Cambridge Univ. Press

Annu. Rev. Earth Planet. Sci. 1994. 22:207–37

MECHANICS OF EARTHQUAKES

H. Kanamori

Seismological Laboratory, California Institute of Technology, Pasadena, California 91125

KEY WORDS: stress drop, static stress drop, dynamic stress drop, kinetic friction

INTRODUCTION

An earthquake is a sudden rupture process in the Earth's crust or mantle caused by tectonic stress. To understand the physics of earthquakes it is important to determine the state of stress before, during, and after an earthquake. There have been significant advances in seismology during the past few decades, and some details on the state of stress near earthquake fault zones are becoming clearer. However, the state of stress is generally inferred indirectly from seismic waves which have propagated through complex structures. The stress parameters thus determined depend on the specific seismological data, methods, and assumptions used in the analysis, and must be interpreted carefully.

This paper reviews recent seismological data pertinent to this subject, and presents simple mechanical models for shallow earthquakes. Scholz (1989), Brune (1991), Gibowicz (1986), and Udias (1991) recently reviewed this subject from a different perspective, and we will try to avoid duplication with these papers as much as possible. Because of the limited space available, this review is not intended to be an exhaustive summary of the literature, but reflects the author's own view on the subject.

Throughout this paper we use the following notation unless indicated otherwise: α = P-wave velocity, β = S-wave velocity, V_r = rupture velocity, $\dot{U}$ = fault particle-motion velocity, σ_0 = tectonic shear stress on the fault plane before an earthquake, σ_1 = tectonic shear stress on the fault plane after an earthquake, $\Delta\sigma = \sigma_0 - \sigma_1$ = static stress drop, σ_f = kinetic frictional stress during faulting, $\Delta\sigma_d = \sigma_0 - \sigma_f$ = dynamic (kinetic) stress drop, S = fault area, D = fault offset, $\dot{D} = 2\dot{U}$ = fault offset particle vel-

0084–6597/94/0515–0207$05.00

ocity, M = earthquake magnitude, μ = rigidity, $M_0 = \mu DS$ = seismic moment.

TECTONIC STRESS

The rupture zones of earthquakes are usually planar (fault plane), but occasionally exhibit a complex geometry. The stress distribution near a fault zone varies as a function of time and space in a complex manner. Before and after an earthquake (interseismic period), the stress varies gradually over a time scale of decades and centuries, and during an earthquake (coseismic period) it varies on a time scale of a few seconds to a few minutes.

The stress variation during an interseismic period can be considered quasi-static. It varies spatially with stress concentration near locations with complex fault geometry. We often simplify the situation by considering static stress averaged over a scale length of kilometers. We call this stress field the "macroscopic static stress field." In contrast, we call the stress field with a scale length of local fault complexity the "microscopic static stress field."

During an earthquake, stress changes very rapidly. It decreases in most places on the fault plane, but it may increase at some places, especially near the edge of a fault where stress concentration occurs. We call the stress field averaged over a time scale of faulting the "macroscopic dynamic stress field," and that with a time scale of rupture initiation, the "microscopic dynamic stress field."

Macroscopic Static Stress Field

Figure 1 shows a schematic time history of macroscopic static stress over three earthquake cycles. After an earthquake the shear stress on the fault

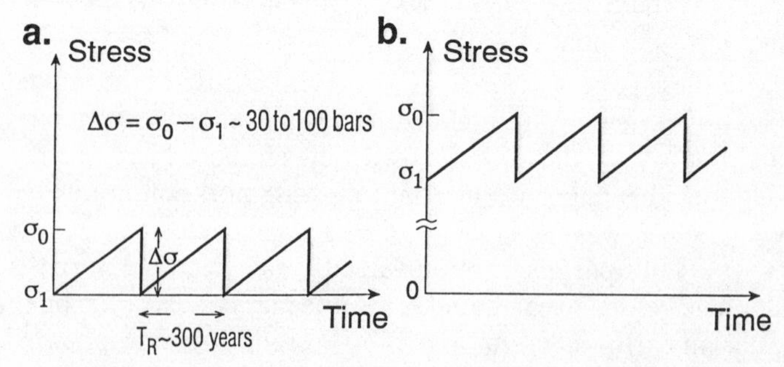

Figure 1 Schematic figure showing temporal variations of macroscopic (quasi-)static stress on a fault plane. (*a*) Weak fault model. (*b*) Strong fault model.

monotonically increases from σ_1 to σ_0 during an interseismic period. When it approaches σ_0, the fault fails causing an earthquake, and the stress drops to σ_1, and a new cycle begins. The stress difference $\Delta\sigma = \sigma_0 - \sigma_1$ is the static stress drop, and T_R is the repeat time. For a typical sequence along active plate boundaries, $\Delta\sigma \approx 30$ to 100 bars, and $T_R \approx 300$ years. (The numerical values given in the text are representative values for illustration purposes only; more details will be given in the section for each parameter.)

The absolute value of σ_0 and σ_1 cannot be determined directly with seismological methods; only the difference, $\Delta\sigma = \sigma_0 - \sigma_1$, can be determined. If fault motion occurs againist kinetic (dynamic) friction, σ_f, repeated occurrence of earthquakes should result in a local heat flow anomaly along the fault zone. From the lack of a local heat flow anomaly along the San Andreas fault, a relatively low value, 200 bars or less, has been suggested for σ_f (Brune et al 1969; Henyey & Wasserburg 1971; Lachenbruch & Sass 1973, 1980). More recent studies on the stress on the San Andreas fault zone also suggest a low stress—less than a few hundred bars (Mount & Suppe 1987, Zoback et al 1987). However, the strength of rocks (frictional strength) measured in the laboratory suggests that shear stress on faults is high, probably higher than 1 kbar (Byerlee 1970, Brace & Byerlee 1966). Figures 1*a* and 1*b* show the two end-member models, the weak fault model ($\sigma_0 \approx 200$ bars), and the strong fault model ($\sigma_0 \approx 2$ kbars). In these simple models "strength of fault" refers to σ_0. Actually, σ_0 and σ_1 may vary significantly from place to place and from event to event; the loading rate may also change as a function of time so that the time history is not expected to be as regular as indicated in Figure 1.

Microscopic Static Stress Field

An earthquake fault is often modeled with a crack in an elastic medium. Figure 2 shows the distribution of shear stress near a crack tip (e.g. Knopoff

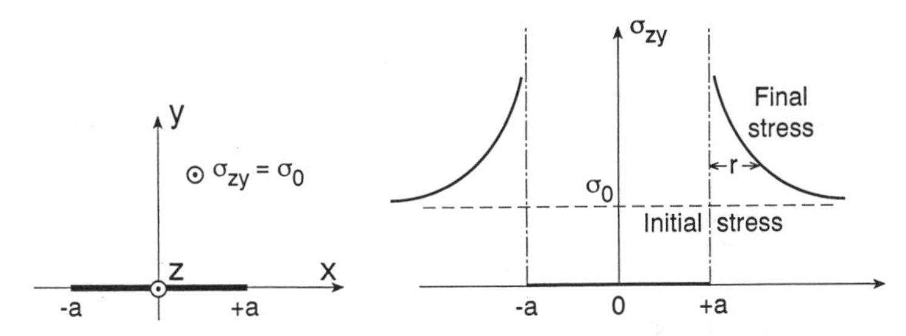

Figure 2 Static stress field near a crack tip. (*Left*) Geometry. A 2-dimensional crack with a width of $2a$ extending from $z = -\infty$ to $+\infty$ is formed under shear stress $\sigma_{zy} = \sigma_0$. (*Right*) Shear stress σ_{zy} before (*dashed line*) and after (*solid curves*) crack formation.

1958). For an infinitely thin crack in a purely elastic medium, the stress (σ_{zy} in Figure 2) is unbounded at the crack tip, and decreases as $1/\sqrt{r}$ as the distance r from the crack tip increases (i.e. inverse square-root singularity). In a real medium, the material near the crack tip yields at a certain stress level (yield stress) causing the stress near the crack tip to be finite. Nevertheless, the behavior shown in Figure 2 is considered to be a good qualitative representation of static stress field near a fault tip.

In actual fault zones, the strength is probably highly nonuniform, and many local weak zones ("micro-faults") and geometrical irregularities are distributed as shown in Figure 3. As the fault system is loaded by tectonic stress σ, stress concentration occurs at the tip of many micro-faults as shown in Figure 3. Near the areas of stress concentration, the stress can be much higher than the loading stress σ. As the stress near the fault tip reaches a threshold value determined by some rupture criteria (e.g. Griffith 1920), and if the friction characteristic is favorable for unstable sliding (e.g. Dieterich 1979, Rice 1983, Scholz 1989), the fault ruptures. As mentioned earlier, the strength of the fault refers to the tectonic stress σ at the time of rupture initiation ($=\sigma_0$), but not to the stress near the fault tip where the stress is much higher than σ_0.

The microscopic stress distribution is mainly controlled by the distribution of micro-faults and is very complex, but the average over a scale length of kilometers is probably close to the loading stress.

Macroscopic Dynamic Stress Field

Although the dynamic stress change during faulting can be very complex, its macroscopic behavior can be described as follows (Brune 1970). If, at $t = 0$, the fault ruptures instantaneously under tectonic stress σ_0, then the displacement of a point just next to the fault will be as shown in Figure

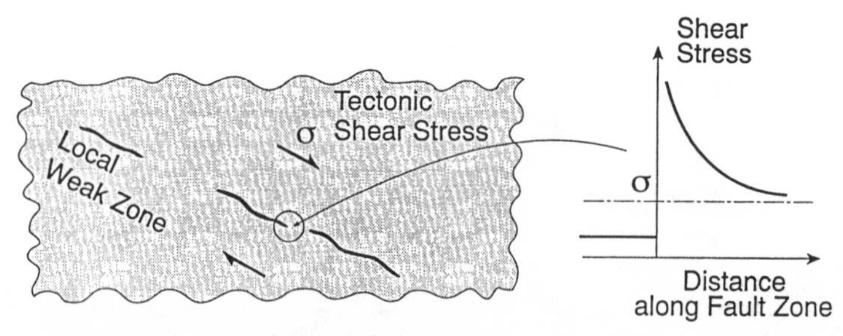

Figure 3 Schematic figure showing a fault zone in the Earth's crust. Heavy solid curves indicate local weak zones (micro-faults) under tectonic shear stress σ. The figure on the right shows stress concentration near the tip of micro-faults. The shear stress on the micro-fault is not necessarily 0, but is significantly smaller than the loading stress σ.

4*a*. Fault motion is resisted by kinetic friction σ_f during slippage so that the difference $\Delta\sigma_d = \sigma_0 - \sigma_f$ is the effective stress that drives fault motion, and is called the dynamic stress drop. In general $\Delta\sigma_d$ varies with time. Following Brune (1970), $\Delta\sigma_d$ can be related to the particle velocity $\dot{U}$ of one side of the fault. After rupture initiation, the shear disturbance propagates in the direction perpendicular to the fault (Figure 4*a*). At time t it reaches the distance βt, beyond which the disturbance has not arrived. Denoting the displacement on the fault at this time by $u(t)$, the instantaneous strain is $u(t)/\beta t$. Since this is caused by $\Delta\sigma_d$,

$$\Delta\sigma_d = \mu u(t)/\beta t$$

from which we obtain

$$u(t) = \Delta\sigma_d \beta t/\mu \quad \text{and} \quad \dot{u}(t) = (\Delta\sigma_d/\mu)\beta = \dot{U} = \text{constant}. \tag{1}$$

Curve (1) in Figure 4*b* shows $u(t)$ for this case.

As the fault rupture encounters some obstacle or the end of the fault, the fault motion slows down and eventually stops as shown by curve (2) in Figure 4*b*. This result is in good agreement with the numerical result by Burridge (1969).

Since the fault rupture is not instantaneous, but propagates with a finite rupture velocity V_r, which is usually about 70 to 80% of β, the actual macroscopic particle motion is slower than that given by (1). Also, the beginning of rupture can no longer be given by a linear function of time (e.g. Ida & Aki 1972). Nevertheless, the macroscopic behavior can be described by (1) with deceleration of about a factor of 2, as shown by

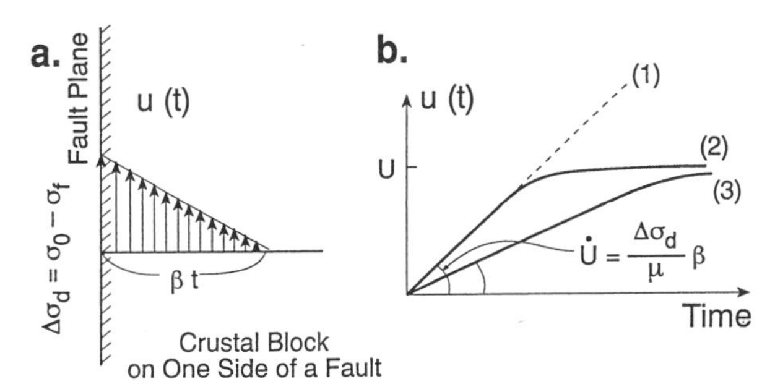

Figure 4 (*a*) Displacement at time t as a function of fault-normal distance from the fault. The stress on the fault is the effective stress (dynamic stress drop) $\Delta\sigma_d = \sigma_0 - \sigma_f$. The disturbance has propagated to a distance of βt. (*b*) Particle motion of one side of the fault as a function of time. Curve (1) is for an infinitely long fault when the $\Delta\sigma_d$ is applied instantaneously. Curve (2) is for a finite fault. Curve (3) is for a finite fault when $\Delta\sigma_d$ is applied as a propagating stress.

curve (3) in Figure 4*b*. Numerical results by Hanson et al (1971), Madariaga (1976), and Richards (1976) support this conclusion. If we denote the offset and duration of slip by D $(=2U)$ and T_r, then the average macroscopic particle velocity is $\langle \dot{U} \rangle = D/(2T_r)$. Thus, Equation (1) suggests

$$\langle \dot{U} \rangle = \frac{1}{c_1}\left(\frac{\Delta\sigma_d}{\mu}\right)\beta, \quad \text{and} \quad \Delta\sigma_d = c_1\left(\frac{\mu}{\beta}\right)\langle \dot{U} \rangle, \tag{2}$$

where c_1 is a constant which is of the order of 2.

Microscopic Dynamic Stress Field

Since the theory of cracks in an elastic medium is well established, it is most convenient to use the results from crack theory to understand the dynamic stress field during faulting. In a real fault zone, the fault geometry is complex, and the strength and material properties are heterogeneous so that the stress field is very different from that computed for a simple crack. Nevertheless, the results for a simple crack provide a useful insight regarding the general property of stress and particle motion during faulting. Many theoretical studies have been made on this subject (Kostrov 1966, Burridge 1969, Takeuchi & Kikuchi 1971, Kikuchi & Takeuchi 1971, Ida 1972, Richards 1976, Madariaga 1976). Here we briefly describe the results for steady-state crack propagation described by Freund (1979). The geometry of the crack is shown in Figure 5*a*. (This is called the antiplane shear mode III problem.) The crack expands in the $+x$ direction under uniform far-field stress $\sigma_{yz} = \sigma_0$. The crack propagates at a constant velocity $V_r < \beta$ (steady propagation). The crack surface extends from $-a$ to a over which the stress is equal to the kinetic friction σ_f. Then Equations (21) and (22) of Freund (1979) yield:

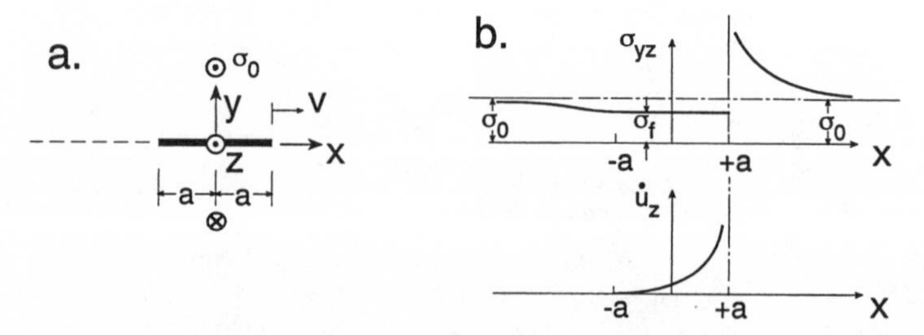

Figure 5 Steady crack propagation. (*a*) Geometry (same as Figure 2*a*). A crack propagating in the x direction at a velocity of V_r. (*b*) The shear stress σ_{zy} and particle velocity $\dot{u}_z$ as a function of distance.

$$\dot{u}_z = 0, \quad \sigma_{yz} = (\sigma_0 - \sigma_f)\left(\sqrt{\frac{x+a}{x-a}} - 1\right) + \sigma_0 \quad \text{for} \quad x > a$$

$$\dot{u}_z = \frac{V_r}{\mu}(\sigma_0 - \sigma_f)\sqrt{\frac{a+x}{a-x}}, \quad \sigma_{yz} = \sigma_f \qquad \text{for} \quad -a < x < a$$

$$\dot{u}_z = 0, \quad \sigma_{yz} = (\sigma_0 - \sigma_f)\left(\sqrt{\frac{x+a}{x-a}} - 1\right) + \sigma_0 \quad \text{for} \quad x < -a. \tag{3}$$

From the condition that there be no stress singularity at the trailing edge at $x = -a$,

$$(\sigma_0 - \sigma_f) = \frac{\mu U}{2\pi a}\sqrt{1 - \left(\frac{V_r}{\beta}\right)^2}, \tag{4}$$

where U is the total displacement of one side of the crack. The stress difference $\sigma_0 - \sigma_f$ is the dynamic stress drop $\Delta\sigma_d$ defined earlier.

These results are graphically shown in Freund (1979) and are reproduced in Figure 5*b*. The typical inverse square-root singularity $(1/\sqrt{x-a})$ is seen for σ_{yz} ahead of the leading edge, and for $\dot{u}_z$ just behind it.

The degree of singularity depends on the physical condition near the crack tip, e.g. the dependence of the cohesive force on velocity and displacement. In a real fault zone, because of the finite strength of the material, the velocity and stress must be finite.

In principle, we should be able to determine the time history of particle velocity from seismological observations and compare it to Equation (3), but in practice it is difficult to determine this uniquely. Seismologically, one observes convolution of the local slip function shown in Figure 4*b* and the rupture propagation effect, and it is difficult to separate these two factors. Most commonly, seismologists can determine only the average particle velocity. Using (3) we obtain the average particle velocity

$$\langle \dot{U} \rangle = \frac{1}{2a}\int_{-a}^{+a} \dot{u}_z \, dx = \frac{\pi V_r}{2\mu}(\sigma_0 - \sigma_f),$$

from which

$$(\sigma_0 - \sigma_f) = \Delta\sigma_d = \frac{2\mu}{\pi V_r}\langle \dot{U} \rangle \approx \left(\frac{\mu}{\beta}\right)\langle \dot{U} \rangle, \tag{5}$$

where $V_r = 0.7\beta$ is assumed. Equation (5) agrees with Equation (2) except for the factor c_1, which is of the order of 2. Considering all the uncertainties in the determination of $\langle \dot{U} \rangle$ and the model, this much of uncertainty is inevitable. For simplicity's sake we will use Equation (5) in the following discussion, but this uncertainty (a factor of 2) must be borne in mind in interpreting the particle velocity in terms of $\Delta\sigma_d$. Husseini (1977) obtained a similar expression.

The large particle velocity near the leading edge contributes to excitation of high-frequency accelerations, but the actual mechanism is complex. For example, Yamashita (1983) explained the observed accelerations in terms of abrupt changes in rupture propagation, but exactly how high-frequency accelerations are excited is still an unresolved problem. High-frequency accelerations can be excited by irregular rupture velocity, sudden changes in material strength or sudden changes in frictional characteristics (Aki 1979). Chen et al (1987) showed that heterogeneities of both stress drop and cohesion are the main factors that control the growth, cessation, and healing of the crack, and that the complexities in seismic radiation are caused by the complex healing process as well as complex rupture propagation.

In the discussion above, V_r is assumed constant. In a more realistic case of spontaneous crack propagation, however, V_r is determined by the property of the material (cohesive energy, surface energy) and the geometry of the crack, and the degree of stress concentration near the crack tip changes drastically (Kostrov 1966, Kikuchi & Takeuchi 1971, Burridge 1969, Richards 1976). However, in most seismological applications, $V_r/\beta \approx 0.7$ to 0.8 and the model of steady subsonic crack propagation is considered reasonable.

SEISMOLOGICAL OBSERVATIONS

Static Stress Drop

Static stress drop $\Delta\sigma$ can be determined by the ratio of displacement u to an appropriate scale length $\tilde{L}$ of the area over which the displacement occurred:

$$\Delta\sigma = c\mu(u/\tilde{L}). \tag{6}$$

This scale length could be the fault length L, the fault width W, or the square root of fault area S, depending on the fault geometry.

Since the stress and strength distributions near a fault are nonuniform, the slip and stress drop are in general a complex function of space. In most applications, we use the stress drop averaged over a certain area, e.g. the entire fault plane. Locally, the stress drop can be much higher than the average (Madariaga 1979). To be exact, the average stress drop is the spatial average of the stress drop. However, the limited resolution of seismological methods often allows determinations of only the average displacement over the fault plane, which in turn is used to compute the average stress drop.

Stress drops have been estimated using the following methods:

1. From D and $\tilde{L}$ estimated from geodetic data.
2. From D estimated from surface break, and S estimated from the aftershock area.
3. From seismic moment M_0, and S estimated from either the aftershock area, surface break, or geodetic data.
4. From M_0, and $\tilde{L}$, estimated from the source pulse width τ, or the characteristic frequency (often called the corner frequency) f_0 of the source spectrum.
5. From the slip distribution on the fault plane determined from high-resolution seismic data.
6. From a combination of the above.

Method 1 was used by Tsuboi (1933) for the 1927 Tango, Japan, earthquake. Tsuboi concluded that the strain associated with the earthquake is of the order of 10^{-4} which translates to $\Delta\sigma$ of 30 bars ($\mu = 3 \times 10^{11}$ dyne/cm^2 is assumed). Chinnery (1964) also used this method to conclude that the stress drops of earthquakes are about 10 to 100 bars.

Method 2 is used when geodetic data are not available. Unfortunately the surface break does not necessarily represent the slip at depth (some earthquakes do not produce a surface break, e.g. the 1989 Loma Prieta, California, earthquake). In general, the overall extent of the aftershock area can be taken as the extent of the rupture zone. Although this interpretation is not correct in detail (for many earthquakes, the aftershocks do not occur in the area of large slip, but in the surrounding areas), the overall distribution of the aftershocks appears to coincide with the extent of the rupture zone. However, the aftershock area usually expands as a function of time, and there is always some ambiguity regarding identification of aftershocks and the aftershock area. Most frequently, the aftershock area defined at about one day after the main shock is used for this purpose (Mogi 1968), but this definition is somewhat arbitrary.

Method 3 is most commonly used for large earthquakes. The seismic moment M_0 can be reliably determined from long-period surface waves and body waves for most large earthquakes in the world using the data from seismic stations distributed worldwide. When the fault geometry is fixed, the seismic moment is a scalar quantity given by $M_0 = \mu DS$. From M_0 and S, D can be determined. If we define the scale length of the fault by $\tilde{L} = S^{1/2}$, the average strain change is $\varepsilon = c_1 D/S^{1/2}$ where c_1 is a constant determined by the geometry of the fault, and is usually of the order of 1. Then the stress drop is

$$\Delta\sigma = \mu\varepsilon = c_1\mu D/S^{1/2} \tag{7}$$

Method 4 is frequently used for relatively small ($M < 5$) earthquakes. For these events, the seismic moment can be usually determined from body waves. For these small earthquakes, the shape of the fault plane is not known so that a simple circular fault model with radius r is often used. If the source is simple, the pulse width τ is approximately equal to r/V_r. Since $V_r \approx 0.7\beta$, $\tau = c_2 r/\beta$, where c_2 is a constant of the order of 1 (Geller 1976, Cohn et al 1982). Another way of determining the source dimension is to use the frequency spectrum of seismic waves. Brune (1970) related the corner frequency f_0 of the S wave spectrum to r. Theoretically, if the source is simple, the pulse width τ can be translated to a corner frequency f_0, but if the source is complex, the interpretation of f_0 is not straightforward. Because of its simplicity, this method is widely used. However, many assumptions were built into this method (e.g. circular fault etc), so that the values determined for individual events are subject to large uncertainties, but the average of many determinations is considered significant.

Method 5 is most straightforward in concept, but is difficult to use unless high-quality data are available, preferably in both near- and far-field. With the increased availability of strong-motion records, this method is now widely used (e.g. Hartzell & Helmberger 1982). The slip function on the fault plane is determined directly, which can be used to estimate not only the average stress drop but also local stress drop.

Many determinations of stress drops have been made by combining these methods. Figure 6*a* shows the relation between M_0 and S for large and great earthquakes (Kanamori & Anderson 1975). In general, log S is proportional to (2/3) log M_0. Since $M_0 = \mu SD = c\Delta\sigma S^{3/2}$, Figure 6*a* indicates that $\Delta\sigma$ is constant over a large range of M_0. The straight lines in Figure 6*a* show the trends for circular fault models with $\Delta\sigma = 1$, 10, and 100 bars. The actual value of the stress drop depends on the fault geometry and other details, but the overall trend appears well established. Stress drop $\Delta\sigma$ varies from 10 to 100 bars for large and great earthquakes.

For smaller earthquakes, it is necessary to use higher frequency waves to determine source dimensions, but the strong attenuation and scattering of high-frequency waves make the determination of source dimensions more difficult. Because of this difficulty, whether the trend shown in Figure 6*a* continues to very small source dimensions or not has been debated. Several studies indicate that it breaks down at $r = 100$ m, but a recent result obtained by Abercrombie & Leary (1993) from down-hole (2.5 km) observations near Cajon Pass, California, suggests that the trend continues to at least $r = 10$ m, as shown in Figure 6*b*.

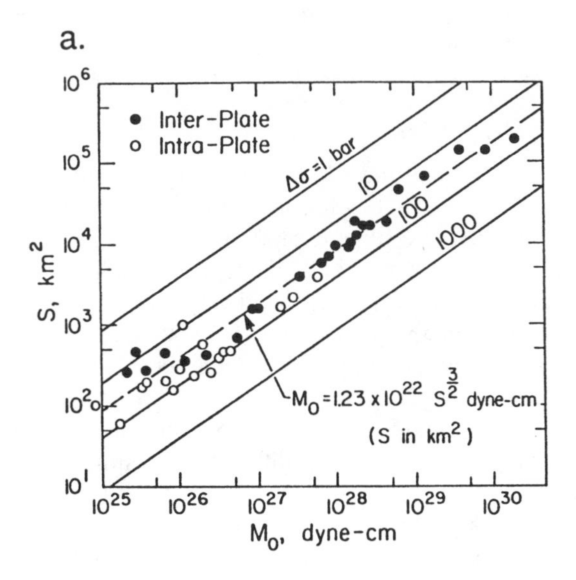

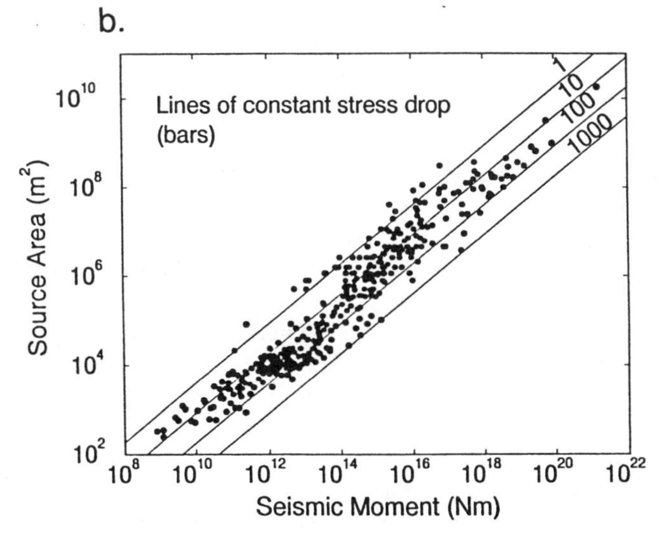

Figure 6 (*a*) Relation between fault area S and seismic moment M_0, for large and great earthquakes (Kanamori & Anderson 1975). (*b*) Relation between seismic moment M_0 and source area for small and large earthquakes (Abercrombie & Leary 1993).

These results suggest that $\Delta\sigma$ averaged over a distance of 100 m or longer appears to be within a range of 1 to 1000 bars, for a range of M_0 from 10^{16} to 10^{30} dyne-cm. The implications of the constant stress drop have been discussed by many investigators (e.g. Aki 1971, Hanks 1979).

High Stress-Drop Events

As shown in Figures 6*a* and 6*b*, the static stress drops of most large ($M > 5.5$ or $M_0 > 2.2 \times 10^{24}$ dyne-cm) earthquakes are in the range of 10 to 100 bars, but there are some exceptions. Moderate earthquakes with very high stress drops (300 bars to 2 kbars) are occasionally observed. For example, Munguia & Brune (1984) found very high stress-drop (up to 2.5 kbars) earthquakes in the area of the 1978 Victoria, Baja California, earthquake swarm. These earthquakes are characterized by high-frequency source spectra. To identify these high stress-drop earthquakes, near-field observations are necessary. As the distance increases, the attenuation of high-frequency energy makes it difficult to identify high stress-drop earthquakes.

Recently several high stress-drop earthquakes were observed with close-in wide-dynamic range seismographs. For example, Kanamori et al (1990, 1993) estimated $\Delta\sigma$ of the 1988 Pasadena, California, earthquake ($M = 4.9$) to be 300 bars to 2 kbars over a source dimension of about 0.5 km. Another example is the 1991 Sierra Madre, California, earthquake ($M = 5.8$). Wald (1992) and Kanamori et al (1993) estimated $\Delta\sigma$ to be 150 to 300 bars over a source dimension of about 4 km. The large range given to these estimates is due to the uncertainty in the source dimension and rupture geometry. Nevertheless, there is little doubt that these earthquakes have a significantly larger stress drop than most earthquakes. These results indicate that $\Delta\sigma$ can be very large over a scale length of a few km. In most large earthquakes, regions of high and low stress drops are averaged out resulting in $\Delta\sigma$ of 10 to 100 bars.

Although these high stress-drop earthquakes may occur only in special tectonic environments, we consider that they represent an end-member of earthquake fault models, as we discuss later.

Dynamic Stress Drop

As discussed earlier, the dynamic stress drop $\Delta\sigma_d$ is the stress that drives fault motion, and controls the particle velocity of fault motion. The particle velocity $\dot{U}$ of fault motion is thus an important seismic source parameter that provides estimates of $\Delta\sigma_d$, through relations like (2) or (5).

Maximum ground motion velocities recorded by strong motion instruments provide crude estimates of the particle velocity of fault motion. Brune (1970) suggested, using the data from six earthquakes, an upper limit of the particle velocity of 1 m/sec. A compilation of strong-motion data by Heaton et al (1986, figure 20) also indicates that the upper limit of the observed ground motion velocity is about 1 m/sec.

Unfortunately, direct determination of $\dot{U}$ is not very easy because the observed waveform is the convolution of the local dislocation function and the fault rupture function. Kanamori (1972) estimated $\dot{U}$ for the 1943 Tottori, Japan, earthquake to be about 42 cm/sec using a very simple fault model. A similar method was used to determine the particle velocities for several Japanese earthquakes: 1 m/sec for the 1948 Fukui earthquake (Kanamori 1973), 50 cm/sec for the 1931 Saitama earthquake (Abe 1974a), 30 cm/sec for the 1963 Wakasa Bay earthquake (Abe 1974b), and 92 cm/sec for the 1968 Saitama earthquake (Abe 1975). These results indicate a range of $\Delta\sigma_d$ from 40 to 200 bars (using 2) and 20 to 100 bars (using 5). Boatwright (1980) developed a method to determine dynamic stress drops from seismic body waves.

Some eyewitness reports suggest somewhat larger particle velocities, but even if we allow for the uncertainties in the measurements, $\dot{U}$ appears to be bounded at about 2 m/sec.

For more recent earthquakes, the distribution of slip and particle velocity is determined by seismic inversion. Heaton (1990) estimated $\Delta\sigma_d$ for several earthquakes from the particle motion velocities thus determined. His estimate ranges from 12 to 40 bars for the average $\Delta\sigma_d$, and from 22 to 84 bars for the local $\Delta\sigma_d$.

Quin (1990) and Miyatake (1992a,b) attempted to determine $\Delta\sigma_d$ from the slip time history estimated by seismic inversion. Quin (1990) modeled the dynamic stress release pattern of the 1979 Imperial Valley, California, earthquake using the slip distribution determined by Archuleta (1984). Miyatake (1992a,b) used the slip models for the Imperial Valley earthquake and several Japanese earthquakes determined by Takeo (1988) and Takeo & Mikami (1987), and estimated the static stress drop $\Delta\sigma$ on the fault plane from the slip distribution. Assuming that $\Delta\sigma_d = \Delta\sigma$, he computed the local slip function using the method developed by Mikumo et al (1987). A good agreement between the computed slip function and that determined by seismic inversion led him to conclude that $\Delta\sigma_d \approx \Delta\sigma$ (within a factor of 2), which is in good agreement with Quin's (1990) result. Since the rise time of local slip function determined by seismic inversion is usually considered to be the upper limit (a very short rise time cannot be resolved with the available seismic data), the conclusion by Quin and Miyatake indicates that $\Delta\sigma_d$ is comparable to $\Delta\sigma$, or possibly larger. However, $\Delta\sigma_d$ is unlikely to be much higher than 200 bars, because the observed particle velocity seems to be bounded at about 2 m/sec.

McGuire & Hanks (1980) and Hanks & McGuire (1981) related the root-mean-square (RMS) acceleration to stress drop. Using this relation, Hanks & McGuire (1981) estimated stress drops for many California earthquakes to be about 100 bars. This estimate, obtained from the radiated wave field rather than the static field, can be regarded as a measure of the average $\Delta\sigma_d$.

In summary, $\Delta\sigma_d$ varies over a range of 20 to 200 bars, which is approximately the same as that for $\Delta\sigma$.

Asperities and Barriers

Many studies have shown that slip distribution on a fault is very complex, i.e. most large earthquakes are multiple events at least on a time scale of a few seconds to a few minutes (e.g. Imamura 1937, Miyamura et al 1964, Wyss & Brune 1967, Kanamori & Stewart 1978). Recent seismic inversion studies have shown this complexity in great detail for earthquakes in both subduction zones (e.g. Ruff & Kanamori 1983; Lay et al 1982; Beck & Ruff 1987, 1989; Schwartz & Ruff 1987; Kikuchi & Fukao 1987) and in continental crusts (see Heaton 1990 for a summary). Two examples are shown in Figure 7 (Wald et al 1993, Mendoza & Hartzell 1989).

These models have been often interpreted in terms of *barriers*—areas where no slip occurs during a main shock (Das & Aki 1977), and *asperities*—areas where large slip occurs during a main shock (e.g. Kanamori 1981). The mechanical properties of the areas between asperities are poorly understood. One possibility is that the slip there occurs gradually in the form of creep and small earthquakes during the interseismic periods. If this is the case, the same asperities break in every earthquake cycle, producing a "characteristic" earthquake sequence. Another possibility is that the areas between asperities remain locked (i.e. barriers) until the next major sequence when they fail as asperities for that sequence. In this case, the rupture pattern would be very different from sequence to sequence resulting in a "noncharacteristic" earthquake sequence. It is also possible that asperities and barriers are not permanent features, but are controlled by nonlinear frictional characteristics so that the distribution of asperities and barriers can vary in a chaotic fashion (Rice 1991). The distributions of barriers and asperities could also change due to redistribution of water and pore pressure before and during earthquakes.

Since the physical nature of asperities and barriers is not well understood, here we use the terms simply to describe complexity of fault rupture patterns. Regardless of their physical nature, it is important to recognize that the mechanical properties (strength and frictional characteristics) of fault zones are spatially very heterogeneous and the degree of heterogeneity varies significantly for different fault zones.

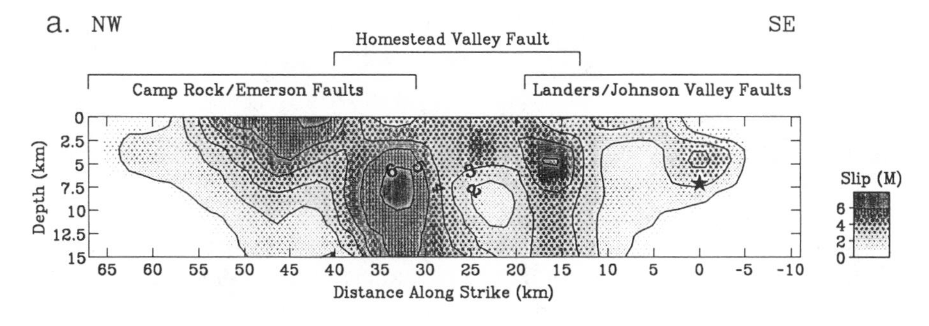

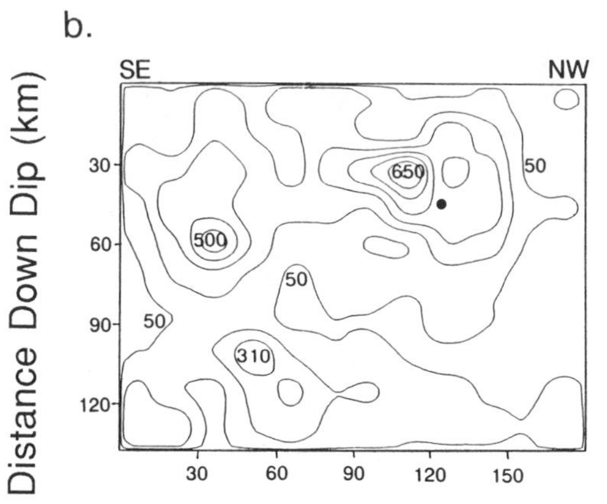

Figure 7 (*a*) Rupture pattern of the 1992 Landers, California, earthquake determined from strong-motion, teleseismic, and geodetic data (after Wald & Heaton 1993). (*b*) Rupture pattern of the 1985 Michoacan, Mexico, earthquake determined from strong-motion and teleseismic data (after Mendoza & Hartzell 1989). In both (*a*) and (*b*), the contour lines show the total amount of displacement. In these studies, in addition to displacements, slip velocities are also approximately determined.

Energy Release

Since energy release in an earthquake is caused by fault motion driven by dynamic stress, the energy budget of an earthquake must provide a clue to the stress change during an earthquake. The simplest way to investigate this problem is to consider a crack in an elastic medium on which the stress drops from σ_0 to σ_1. During slippage frictional stress acts against motion. This type of intuitive model was first used by Orowan (1960), and has been subsequently used by many investigators (Savage & Wood 1971, Wyss & Molnar 1972).

The total energy change is then given by

$$W = \tfrac{1}{2}S(\sigma_0+\sigma_1)\bar{D} = S\bar{\sigma}\bar{D}, \tag{8}$$

where S is the surface area of the crack, $\bar{\sigma}$ is the average stress, and $\bar{D}$ is the displacement averaged over the crack surface (Knopoff 1958, Kostrov 1974, Dahlen 1977, Savage & Walsh 1978).

Since slip occurs against frictional stress σ_f, the energy $H = \sigma_f S\bar{D}$ will be lost to heat. If we ignore the energy necessary to create new surfaces at the crack tip (surface energy), we can assume that the difference will be radiated by elastic waves. Thus

$$E_S = W - H. \tag{9}$$

The importance of surface energy has been discussed by Husseini (1977) and Kikuchi & Fukao (1988). In general, if $V_r/\beta = 0.7$ to 0.8, the surface energy is about 1/4 of E_S (Husseini 1977), so that the radiated energy is about 3/4 of the E_S given by (9). Kikuchi & Fukao (1988) showed that this ratio also depends on the aspect ratio of the fault, and in an extreme case, the radiated energy can be only 10% of the E_S given by (9). Considering the limited accuracy of the energy estimate, we will ignore the surface energy in the following discussion. However, the surface energy could be important under certain circumstances.

The relations (8) and (9) are most conveniently illustrated in Figure 8 which was used by Kikuchi & Fukao (1988) and Kikuchi (1992). The vertical axis is the stress on the fault plane (crack surface) and the horizontal axis is the displacement measured in $S\bar{D}$. The total energy release W is given by the trapezoid OABC (Equation 8). In the simplest case (Case I) we assume that $\sigma_f =$ const and $\sigma_1 = \sigma_f$, i.e. the shear stress on the fault after an earthquake is equal to σ_f. In this case heat loss $H = \sigma_f S\bar{D}$ is given

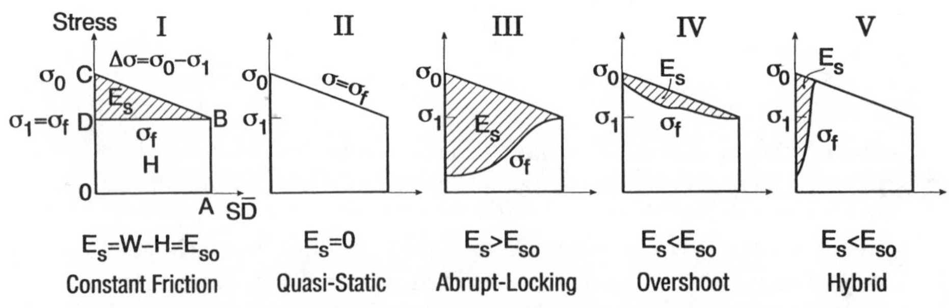

Figure 8 Schematic representation of energy budget for a stress relaxation model (modified from Kikuchi & Fukao 1988). Case I: Constant Friction model; Case II: Quasi-Static model; Case III: Abrupt-Locking model; Case IV: Overshoot model; Case V: Hybrid model.

by the rectangular area OABD, and E_S is given by the triangular area DBC. Thus

$$E_S = W - H = \tfrac{1}{2}S\bar{D}[(\sigma_0+\sigma_1)-2\sigma_f] = \tfrac{1}{2}S\bar{D}[(\sigma_0-\sigma_1)-2(\sigma_f-\sigma_1)]. \quad (10)$$

If we assume $\sigma_1 = \sigma_f$, then the second term in the brackets vanishes, and (10) is reduced to

$$E_S = \tfrac{1}{2}S\bar{D}(\sigma_0-\sigma_1) = \frac{\Delta\sigma}{2\mu}M_0. \quad (11)$$

Since both M_0 and $\Delta\sigma$ can be determined with seismological methods, the radiated energy E_S can be estimated using (11). Kanamori (1977) used this relationship to estimate the amount of energy released by large earthquakes. Although no direct evidence is available for the validity of the assumption $\sigma_1 = \sigma_f$, the relation (11) seems to hold for many large earthquakes for which M_0 and E_S have been independently estimated. However, since many assumptions and simplifications have been made in obtaining (11), E_S thus estimated should be considered only approximate.

It is useful to consider a few alternatives using the diagrams shown in Figure 8. The most extreme is a quasi-static case (Case II in Figure 8) in which frictional stress is adjusted so that it is always equal to the stress on the fault plane. In this case the frictional stress is given by the straight line CB, and no energy is radiated (i.e. $E_S = 0$). The entire strain energy is expended to generate heat and to create new crack surfaces. The other extreme case (Case III in Figure 8) involves a sudden drop in friction, possibly at the time slippage begins. In this case a larger stress is available for driving the fault motion, and more seismic energy will be radiated than in Case I. Case IV in Figure 8, which is intermediate between Case I and Case II, represents less wave energy radiation than Case I. Kikuchi & Fukao (1988) and Kikuchi (1992), using the data on E_S, M_0, and $\Delta\sigma$, favored this case. In this case the contribution of surface energy is important. The dynamic stress drop during faulting, $\sigma - \sigma_f$, is smaller than the static stress drop $\Delta\sigma$. It is also possible that dynamic stress can be very large during a short period of time, but then drops quickly to a low level so that E_S is smaller than that for Case I. This is shown as Case V in Figure 8. In Figure 8 the large $\Delta\sigma_d$ occurs at the beginning, but it can happen at any time during faulting.

These models are useful for understanding the basic behavior of complex earthquake faulting. However, because actual fault zones may be very different, both between different tectonic provinces (e.g. subduction zones, transform faults, intra-plate faults, etc) and between faults with different characteristics in the same province (e.g. faults with slow slip rate vs fast slip rate, etc), it is likely that more than one of the mechanisms discussed above are involved in real faulting.

As shown by Figure 8 and Equation (11), the ratio $2\mu E_S/M_0$ gives a measure of the average dynamic stress drop $\Delta\sigma_d$ during slippage. If σ_f = const and $\sigma_1 = \sigma_f$ (Case I), then $\Delta\sigma_d = \Delta\sigma$. However, for Cases III, IV and V,

$$2\mu \frac{E_S}{M_0} = \Delta\sigma_d = \eta_S \Delta\sigma, \tag{12}$$

when $\eta_S > 1$ for Case III and $\eta_S < 1$ for Cases IV and V.

Unfortunately, estimating energy E_S is not easy. Traditionally, E_S has been estimated from the earthquake magnitude M. The most commonly used relation is the Gutenberg-Richter relation (Gutenberg & Richter 1956):

$$\log E_S = 1.5M_S + 11.8 \quad (E_S \text{ in ergs}),$$

where M_S is the surface-wave magnitude. However, this is an average empirical relation, and is not meant to provide an accurate estimate of E_S. The total energy must be estimated by an integral of the entire wave train, rather than from M_S which is determined by the amplitude at a single period of 20 sec.

Currently, radiated energy is estimated directly from seismograms. Two methods are being used. In the first method (Thatcher & Hanks 1973, Boatwright 1980, Boatwright & Choy 1985, Bolt 1986, Houston 1990a,b), the ground-motion velocity of radiated waves, either body or surface waves, is squared and integrated to estimate E_S. Sometimes equivalent computation is done on the frequency domain. In this method, the major difficulties are obtaining complete coverage of the focal sphere and in the correction of the propagation effects, i.e. geometrical spreading, attenuation, waveguide effects, and scattering. If a large amount of data is available, one can estimate E_S fairly accurately with several empirical corrections and assumptions.

The second method involves determination of the source function by inversion of seismograms (Vassiliou & Kanamori 1982, Kikuchi & Fukao 1988). In this case, the propagation effects are removed through the process of inversion, but the solution is usually band-limited in frequency. Nevertheless, with the advent of sophisticated inversion algorithms, this method has been used with considerable success (Kikuchi & Fukao 1988).

Kanamori et al (1993) estimated E_S using the high-quality broadband data that has recently become available at short distances from earthquakes in southern California. Figure 9 shows the relation between E_S and M_0 thus obtained for recent earthquakes in southern California. The dynamic stress drops shown in Figure 9 are computed using (12) with $\mu = 3 \times 10^{11}$ dynes/cm^2. The earthquakes shown in Figure 9 [the 1989 Montebello earthquake ($M = 4.6$), the 1988 Pasadena earthquake ($M = 4.9$), the 1991 Sierra Madre earthquake ($M = 5.8$), the 1992 Joshua Tree earthquake ($M = 6.1$), the 1992 Big Bear earthquake ($M = 6.4$), and the 1992 Landers earthquake ($M = 7.3$)] have stress drops in a range of 50 to 300 bars—significantly higher than those for many large earthquakes elsewhere computed by Kikuchi & Fukao (1988) from the E_S/M_0 ratios. As will be discussed later, this difference can be interpreted as due to the long repeat times of the earthquakes shown in Figure 9.

The values of $\Delta\sigma_d$ shown in Figure 9 are smaller than $\Delta\sigma$ for the same earthquakes (not shown here; see Kanamori et al 1993) by a factor of about 3. Kikuchi & Fukao (1988) found an even larger difference for the earthquakes they examined. They attributed this difference to surface energy, and favored the Case IV stress release model shown in Figure 8. However, estimates of $\Delta\sigma_d$ and $\Delta\sigma$ are subject to large uncertainties so that whether this difference is significant or not is presently unresolved.

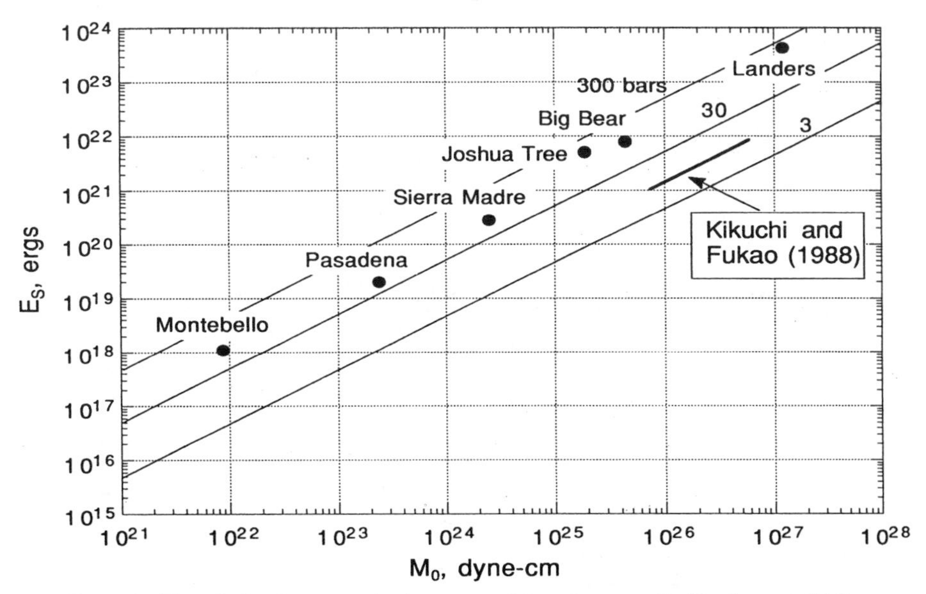

Figure 9 The relation between seismic moment M_0 and energy E_S. The heavy solid line indicates the range of the E_S/M_0 ratios of large earthquakes determined by Kikuchi & Fukao (1988).

Relationship between Stress Drop and Slip Rate

The slip rate of faults varies significantly from less than 1 mm/year to several cm/year. Faults with slow and fast slip rates usually have long and short repeat times, respectively. Kanamori & Allen (1986) and Scholz et al (1986) independently found that earthquakes on faults with long repeat times radiate more energy per unit fault length than those with short repeat times. Houston (1990b) also found evidence for this. Figure 10 shows the results obtained by Kanamori & Allen (1986). A typical earthquake on a fault with fast slip rate and short repeat time is the 1966 Parkfield, California, earthquake ($M = 6$, slip rate = 3.5 cm/year, repeat time = 22 years). In contrast, a typical earthquake on a fault with slow slip rate is the 1927 Tango, Japan, earthquake ($M = 7.6$, repeat times > 2000 years). Even if the Tango earthquake has about the same fault length as the Parkfield earthquake, its magnitude is more than 1.5 units larger. A more recent example is the 1992 Landers, California, earthquake. Despite the relatively large magnitude, its fault length is only 70 km. The repeat time

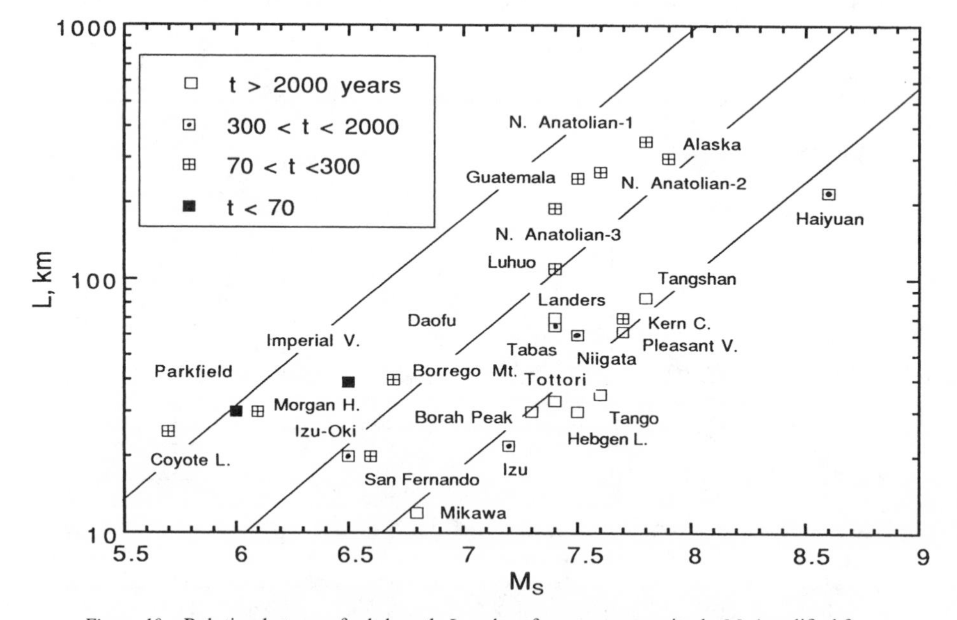

Figure 10 Relation between fault length L and surface-wave magnitude M_S (modified from Kanamori & Allen 1986). The fault length is shorter for earthquakes with long repeat times than for those with short repeat times with the same M_S. Since the energy released E_S is proportional to $10^{1.5M_S}$, the above relation means that earthquakes on faults with long repeat times radiate more energy per unit fault length than those with short repeat times.

of the Landers earthquake is believed to be very long (see below). As shown in the previous section, a larger amount of energy release per unit fault length suggests a larger dynamic stress drop. Thus, this observation suggests that dynamic stress drop increases as fault slip rate decreases.

The implication is that the strength of a fault increases with the time during which the two sides of the fault have been locked. This may be viewed as manifestation of an evolutionary process of a fault (Scholz 1989).

The results for the E_S/M_0 ratio shown in Figure 9 can be interpreted in the same way. The repeat time of major earthquakes on the frontal fault system where the Pasadena and the Sierra Madre earthquakes occurred is believed to be very long—a few thousand years (e.g. Crook et al 1987). Also, the repeat time of the faults in the eastern Mojave desert where the Joshua Tree, the Big Bear, and the Landers earthquakes occurred is thought to be very long (e.g. Sieh et al 1993). Since no measurements of E_S/M_0 have been made yet for earthquakes on faults with short repeat times in southern California, we cannot directly compare $\Delta\sigma_d$ for earthquakes with long and short repeat times in southern California. Nevertheless, Figure 9 provides evidence that earthquakes with long repeat times have larger dynamic stress drops.

STRESS RELEASE MODELS FOR EARTHQUAKES

Housner (1955) and Brune (1970) presented a model in which only part of the total available stress is released during an earthquake. Partial stress drop could be caused by some obstacles (e.g. interlocking asperities) prematurely stopping fault motion. Brune (1976) later called this the abrupt-locking model. An important consequence of this model is a significantly larger dynamic stress drop than static stress drop.

Many abrupt-locking mechanisms can be considered. Fault rupture may encounter a strong spot on the fault which prevents the fault from rupturing further. Another possibility is that rapid healing of the rupture surface during slippage, which slows down the particle motion of the fault, eventually brings it to a halt. Rapid healing results in a change of kinetic friction, which plays a key role in controlling slip behavior (Dieterich 1979, Scholz 1989). For most materials, kinetic friction is considerably smaller than static friction (Bowden & Taber 1964, Rabinowicz 1965); a power law of the form, $\sigma_f = av^{-b}$ (v = sliding velocity, a and b = constants), is often used in material science.

Many mechanisms for the reduction of kinetic friction have been suggested, for example: 1. melting on the fault plane, 2. acoustic fluidization (Melosh 1979), 3. infinitesimal motion normal to the slip plane (Schallamach 1971, Brune et al 1993). Since most experimental data on dynamic friction have been obtained for sliding velocities much lower than those present during seismic faulting, about 1 m/sec, the exact behavior of kinetic friction in real fault zones can only be assumed.

Heaton (1990), recognizing that the duration of slip at a given point on a fault is much shorter than the duration of rupture over the entire fault, attempted to explain this observation using a velocity-dependent kinetic friction law ($\sigma_f = \sigma_{f_0} - c\dot{D}$, c = constant). As shown in Figure 5*b*, a crack tip causes the square-root stress singularity ahead of the rupture front. Just behind the rupture front the fault particle velocity becomes very high, thus decreasing kinetic friction. As the particle velocity decreases away from the crack tip, kinetic friction increases again, and the fault motion eventually stops. Thus, at a given time during faulting, slip is occurring over a short distance (slip pulse). This slip pulse propagates on a fault at the rupture velocity. Heaton (1990) called this model the "slip-pulse model." In the slip-pulse model, it is the velocity-dependent kinetic friction rather than the static stress near the fault that controls dynamic fault motion.

Brune et al (1993) found evidence for fault-normal motion during slippage in their foam rubber models of rupture. The fault-normal motion can effectively reduce the normal stress which in turn reduces kinetic friction during slippage.

The energy release pattern for the abrupt-locking model or slip-pulse model can be represented by Case III or Case V in Figure 8.

The details of the frictional characteristics are still unknown, and a better understanding of kinetic friction during slippage under the conditions that prevail during seismic faulting [high normal stress, high (1 m/sec) particle velocity, etc] is critically important for understanding the physics and mechanics of earthquake faulting.

MODELS FOR HETEROGENEOUS FAULTS

As summarized in the previous sections, any model for earthquake process must take into account the following:

1. The static stress drop $\Delta\sigma$ is, on the average, 1 to 100 bars. Some events, however, have very high stress drops.
2. The dynamic stress drop $\Delta\sigma_d$ is, on the average, about the same order as the static stress drop $\Delta\sigma$. However, it may vary significantly as a function of time and space, and could exceed the static stress drop for a short period of time during faulting. In general, both dynamic and static stress drops appear to be higher for faults with slow slip rates (long repeat times) than those with fast slip rates (short repeat times).
3. The slip distribution on a fault plane is generally very complex, sug-

gesting heterogeneity in the strength or frictional characteristics of the material in the fault zone.

End-Member Models

To understand the complexity of fault rupture patterns, we first consider simple end-member models characterized by different magnitudes of $\Delta\sigma_d$ and $\Delta\sigma$.

Table 1 summarizes four end-member models thus introduced. In this table "high" and "low" typically mean 300 and 30 bars, respectively, but they are meant to be representative values. Type 1 has high $\Delta\sigma_d$ and high $\Delta\sigma$, and is a model for small to moderate size earthquakes on a fault with slow slip rate, such as the 1988 Pasadena, the 1991 Sierra Madre, and the 1992 Landers earthquakes. Type 2 has high $\Delta\sigma_d$ and low $\Delta\sigma$. This is essentially the abrupt-locking partial stress-drop model (Brune 1970, 1976) and the slip-pulse model (Heaton 1990). Fault slip motion occurs very rapidly, but it locks up prematurely by either encountering an obstacle or sudden healing. Type 3 has low $\Delta\sigma_d$ and high $\Delta\sigma$. This corresponds to the overshoot model described by Madariaga (1976). The model suggested by Kikuchi & Fukao (1988) and Kikuchi (1992) belongs to this type in which the effective driving stress is smaller than the static stress drop. The most extreme case of this type would be creep. If the sliding condition is such that kinetic friction is always equal to shear traction on the fault plane, the fault motion becomes quasi-static without seismic radiation. Type 4 has both low $\Delta\sigma_d$ and low $\Delta\sigma$. Variations of pore pressures (Sibson 1973, Rice 1992), rock types (Allen 1968), fault geometry (Sibson 1986), and chemical process may be responsible for these different types of faults.

Composite Models

Evidence for abrupt-locking and slip-pulse models has been discussed by Heaton (1990) and Brune (1991). Brune et al (1986) presented many examples of ω^{-1} roll-off of the source spectrum which suggests partial stress drop. Heaton (1990) presented examples of slip-pulse ruptures. For

Table 1 Four end-member models

	$\Delta\sigma_d$	$\Delta\sigma$	$\dot{U}$	U^a
Type 1	high	high	large	large
Type 2	high	low	large	small
Type 3	low	high	small	large
Type 4	low	low	small	small

[a] A uniform scale length is assumed.

these models, $\Delta\sigma_d$ must be significantly larger than $\Delta\sigma$. The results of Quin (1990) and Miyatake (1992a,b), however, show $\Delta\sigma_d \approx \Delta\sigma$. It is possible that $\Delta\sigma_d$ determined in these inversion studies is a lower bound because of the limited resolution of the method. Kikuchi & Fukao (1988) and Kikuchi (1992) favor the Case IV energy release pattern (Figure 8) which is not consistent with the partial stress-drop model or slip-pulse model. However, Case V which is a modification of Case IV can probably satisfy the data presented by Kikuchi & Fukao (1988) and Heaton (1990).

As discussed earlier, actual earthquake sequences are extremely complex, and are likely to involve more than one mechanism. In view of this complexity, we now try to interpret earthquake sequences using combinations of the end-member models described above.

Figure 11*a* shows a combination of Type 1 and Type 4 behavior. Since $\Delta\sigma_d$ and $\Delta\sigma$ are directly proportional to particle velocity $\dot{U}$ and displacement U, respectively, the variations of $\dot{U}$ and U along the fault will be as schematically shown in Figure 11*a*. This combination explains the rupture patterns of the 1979 Imperial Valley earthquake (Quin 1990, Miyatake 1992a) for which $\Delta\sigma_d \approx \Delta\sigma$. Other earthquakes such as the 1984 Morgan Hills, California, earthquake (Hartzell & Heaton 1986, Beroza & Spudich 1988), the 1987 Superstition Hills, California, earthquake (Wald et al 1990), and the 1989 Loma Prieta, California, earthquake (e.g. Hartzell et al 1991, Wald et al 1991, Steidl et al 1991, Beroza 1991) probably belong to this category.

If we have a combination of Type 1 and Type 2 behavior, $\dot{U}$ and U may appear as shown in Figure 11*b*. In this case, an abrupt-locking or slip-

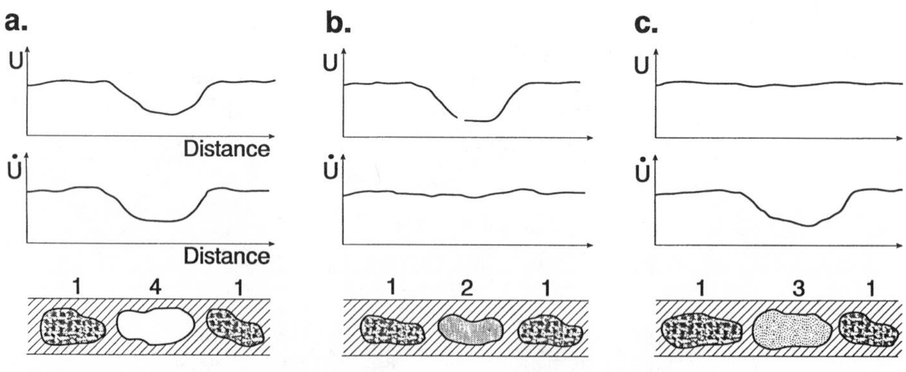

Figure 11 Composite fault models. (*a*) Combination of Type 1 (high $\Delta\sigma_d$ and high $\Delta\sigma$) and Type 4 (low $\Delta\sigma_d$ and low $\Delta\sigma$). Schematic distributions of slip and slip velocity are shown. (*b*) Combination of Type 1 and Type 2 (high $\Delta\sigma_d$ and low $\Delta\sigma$). (*c*) Combination of Type 1 and Type 3 (low $\Delta\sigma_d$ and high $\Delta\sigma$).

pulse mechanism must prevail over a significant section of the fault. The results obtained by Quin (1990) and Miyatake (1992a,b) do not exhibit this pattern. However, as T. Heaton (personal communication, 1993) suggests, if the rise time determined by inversion is the upper bound, then this combination represents the rupture patterns of earthquakes mentioned above. In any event, for these combinations the low and high static-stress drops are averaged out to yield the moderate static-stress drop typical of crustal earthquakes occurring on active plate boundaries.

The combination of Type 1 and Type 3 models would yield a pattern as shown in Figure 11*c*. The slip would be relatively large and uniform over the entire fault, but the slip velocity is high only at limited places. This could be a model for the 1906 San Francisco, California, earthquake. Wald et al (1993) concluded that relatively short-period (20 sec or less) seismic waves were excited from a fault segment about 100 km long, while the geodetic data and the surface break indicate fairly uniform slip over a distance of 350 km (Thatcher 1975). Although no direct seismic data are available, the 1857 Fort Tejon, California, earthquake had a relatively uniform and large slip over 350 km (Sieh 1978), and could be of this type. The creeping section of the San Andreas fault north of the 1857 rupture zone is the extreme Type 3 case.

Three other combinations can be made, but the two end-member models (Type 1 for high stress-drop earthquakes and Type 3 for creep events) and the three combinations described above seem to explain the important features of most earthquake sequences.

CONCLUSION

Fault Strength

The particle velocity of fault motion appears to be bounded at about 1 to 2 m/sec which corresponds to an upper bound of $\Delta\sigma_d$ of about 100 to 200 bars (Equation 5). The values of $\Delta\sigma_d$ determined from the E_S/M_0 ratios (Figure 9) support this. Thus, if σ_f is bounded at 200 bars, as suggested by the heat flow arguments, then, considering the observation $\Delta\sigma_d \approx \Delta\sigma$, the strength of fault σ_0, defined in Figure 1, cannot be much higher than 400 bars. A corollary of this is that the strength of the crust varies significantly. As Jeffreys (1959) showed, the existence of high mountains such as the Himalayas indicates a large stress difference of at least 1.5 kbars within the outermost 50 km of the crust. Thus, although the strength of active fault zones such as the San Andreas is low, high stresses must be prevailing in the crust away from it. In other words, earthquakes occur where the crust is weak (Kanamori 1980, Zoback et al 1987). As mentioned earlier, some earthquakes on faults with low slip rates tend to have high stress drops. Since the slip rates are low, these earthquakes would not produce a significant heat flow anomaly even if the stress drop is high.

The above argument hinges on the premise that the lack of heat flow anomaly indicates low kinetic friction during faulting. If the heat flow problem can be interpreted differently (e.g. Scholz 1992), however, the above conclusion must be modified significantly.

Static Stress Drop vs Dynamic Stress Drop

The values of $\Delta\sigma_d$ obtained from particle velocities of fault motion are, on the average, of the same order as $\Delta\sigma$. The available data of the E_S/M_0 ratio, however, suggest that the radiated energy E_S appears to be significantly less than expected for a simple model in which $\Delta\sigma_d \approx \Delta\sigma$. This appears to be inconsistent with the results obtained from slip and slip-velocity data. Since large uncertainties still exist in the estimates of any of these parameters, this difference may not be significant. However, if the difference is real, one way to explain it is to introduce a drastic time dependence of $\Delta\sigma_d$: $\Delta\sigma_d$ is very large only within a short time interval during faulting but decreases rapidly. This behavior is illustrated by the Case V energy release pattern shown in Figure 8, and is expected for the abrupt-locking or slip-pulse model. Iio (1992) found evidence for initial slow slip before the major fault motion, which suggests a rapid change in kinetic friction. Wald et al (1991) found evidence of a slow rupture about 2 sec before the main rupture of the 1989 Loma Prieta earthquake. These slow precursory events may be a manifestation of time-dependent kinetic friction during faulting, and merit further studies.

To resolve this problem completely, it would be necessary to determine the slip function on the fault plane and the energy release more accurately using broadband high-resolution data.

State of Stress on a Fault Plane

As many studies have demonstrated, the heterogeneity of mechanical properties along a fault inferred from the complexity of rupture patterns is probably one of the most important elements of earthquake faults. Since different segments may have drastically different strengths and frictional characteristics, it is necessary to consider different rupture mechanisms for different segments. When a rupture occurs over a long fault, different segments interact with each other in a complex fashion, thereby causing complex seismic radiation as observed.

Mechanical properties of faults may be controlled by many factors: lithology in the fault zone, temperature, pore pressure, fault geometry, fault orientation with respect to tectonic stress, and slip rates. Rupture initiation, propagation, and cessation are all controlled by the mechanical properties of each segment. The overall rupture patterns in each seismic cycle (characteristic versus noncharacteristic) depend on how different fault segments interact with each other. A better understanding of these would require detailed studies of seismic radiation from earthquakes that occur in different tectonic environments. Because the quality of seismic data has dramatically improved recently, we can now determine the distribution of slip and particle velocity on a fault and the energy release during an earthquake much better than before. These improved data and the variations of these source parameters between earthquakes from different tectonic environments and from faults with different slip rates will provide an important clue to the mechanics of earthquakes.

ACKNOWLEDGMENTS

I benefited from discussions with Tom Heaton, Don Anderson, Masayuki Kikuchi, Takeshi Mikumo, Larry Ruff, and Dave Wald. This paper was built upon many hours of discussions during the Coffee Break of the Seismological Laboratory. I thank all the participants. This research was partially supported by the U.S. Geological Survey Grant 14-08-0001-G1774. This paper is Contribution No. 5214, Division of Geological and Planetary Sciences, California Institute of Technology.

Literature Cited

Abe K. 1974a. Seismic displacement and ground motion near a fault: the Saitama earthquake of September 21, 1931. *J. Geophys. Res.* 79: 4393–99
Abe K. 1974b. Fault parameters determined by near- and far-field data: The Wakasa Bay earthquake of March 26, 1963. *Bull. Seismol. Soc. Am.* 64: 1369–82
Abe K. 1975. Static and dynamic parameters of the Saitama earthquake of July 1, 1968. *Tectonophysics* 27: 223–38
Abercrombie R, Leary P. 1993. Source parameters of small earthquakes recorded at 2.5 km depth, Cajon Pass, Southern California: implications for earthquake scaling. *Geophys. Res. Lett.* 20: 1511–14
Aki K. 1971. Earthquake mechanism. *Tectonophysics* 13: 423–46
Aki K. 1979. Characterization of barriers on an earthquake fault. *J. Geophys. Res.* 84: 6140–48
Allen CR. 1968. The tectonic environment of seismically active and inactive areas along the San Andreas fault system. In *Proc. Conf. Geol. Probl. San Andreas Fault Syst.*, pp. 70–82. Palo Alto: Stanford Univ. Publ. Geol. Sci.
Archuleta RJ. 1984. A faulting model for the 1979 Imperial Valley earthquake. *J. Geophys. Res.* 89: 4559–85
Beck S, Ruff L. 1987. Rupture process of the great 1963 Kurile Islands earthquake sequence: asperity interaction and multiple event rupture. *J. Geophys. Res.* 92: 14,123–38
Beck S, Ruff L. 1989. Great earthquakes and subduction along the Peru trench. *Phys. Earth Planet. Int.* 57: 199–224
Beroza GC. 1991. Near-source modeling of the Loma Prieta earthquake: evidence for heterogeneous slip and implications for earthquake hazard. *Bull. Seismol. Soc. Am.* 81: 1603–21
Beroza GC, Spudich P. 1988. Linearized

inversion for fault rupture behavior: application to the 1984 Morgan Hill, California, earthquake. *Bull. Seismol. Soc. Am.* 78: 6275–96

Boatwright J. 1980. A spectral theory for circular seismic sources; simple estimates of source dimension, dynamic stress drop, and radiated seismic energy. *Bull. Seismol. Soc. Am.* 70: 1–27

Boatwright J, Choy GL. 1985. Teleseismic estimates of the energy radiated by shallow earthquakes. In *US Geol. Surv. Open File Rep.* 85-0290-*A*, *Workshop XXVIII on the Borah Peak, Idaho Earthquake*, ed. RS Stein, RC Bucknam, ML Jacobson, pp. 409–48. Menlo Park: US Geol. Surv.

Bolt BA. 1986. Seismic energy release over a broad frequency band. *Pageoph* 124: 919–30

Bowden FP, Tabor D. 1964. *The Friction and Lubrication of Solids*. Oxford: Clarendon. 374 pp.

Brace WF, Byerlee JD. 1966. Stick-slip as a mechanism for earthquakes. *Science* 153: 990–92

Brune JN. 1970. Tectonic stress and spectra of seismic shear waves from earthquakes. *J. Geophys. Res.* 75: 4997–5009

Brune JN. 1976. The physics of earthquake strong motion. In *Seismic Risk and Engineering Decisions*, ed. C Lomnitz, E Rosenblueth, pp. 141–71. New York: Elsevier

Brune JN. 1991. Seismic source dynamics, radiation and stress. *Rev. Geophys.* 29: 688–99 (Suppl.)

Brune JN, Brown S, Johnson PA. 1993. Rupture mechanism and interface separation in foam rubber models of earthquakes: A possible solution to the heat flow paradox and the paradox of large overthrusts. *Tectonophysics* 218: 59–67

Brune JN, Fletcher J, Vernon F, Haar L, Hanks T, Berger J. 1986. Low stress-drop earthquakes in the light of new data from the Anza, California telemetered digital array. In *Earthquake Source Mechanics*, ed. S Das, J Boatright, CH Scholz, pp. 237–45. Washington, DC: Am. Geophys. Union

Brune JN, Henyey TL, Roy RF. 1969. Heat flow, stress, and the rate of slip along the San Andreas fault,California. *J. Geophys. Res.* 74: 3821–27

Burridge R. 1969. The numerical solution of certain integral equations with non-integrable kernels arising in the theory of crack prpagation and elastic wave diffraction. *Phil. Trans. R. Soc. London Ser. A* 265: 353–81

Byerlee JD. 1970. Static and kinetic friction of granite at high normal stress. *Inst. J. Rock Mech. Min. Soc.* 7: 577–82

Chen YT, Chen XF, Knopoff L. 1987. Spontaneous growth and autonomous contraction of a two-dimensional earthquake fault. *Tectonophysics* 144: 5–17

Chinnery MA. 1964. The strength of the Earth's crust under horizontal shear stress. *J. Geophys. Res.* 69: 2085–89

Cohn SN, Hong TL, Helmberger DV. 1982. The Oroville earthquakes: a study of source characteristics and site effects. *J. Geophys. Res.* 87: 4585–94

Crook RJ, Allen CR, Kamb B, Payne CM, Proctor RJ. 1987. Quaternary geology and seismic hazard of the Sierra Madre and associated faults, western San Gabriel Mountains. In *Recent Reverse Faulting in the Transverse Ranges, California, US Geol. Surv. Prof. Pap.* 1339, ed. DM Morton, RF Yerkes, pp. 27–64

Dahlen FA. 1977. The balance of energy in earthquake faulting. *Geophys. J. R. Astron. Soc.* 48: 239–61

Das S, Aki K. 1977. Fault planes with barriers: a versatile earthquake model. *J. Geophys. Res.* 82: 5658–70

Dieterich JH. 1979. Modeling of rock friction 2. Simulation of preseismic slip. *J. Geophys. Res.* 84: 2169–75

Freund L. 1979. The mechanics of dynamic shear crack propagation. *J. Geophys. Res.* 84: 2199–209

Geller RJ. 1976. Scaling relations for earthquake source parameters and magnitudes. *Bull. Seismol. Soc. Am.* 66: 1501–23

Gibowicz SJ. 1986. Physics of fracturing and seismic energy release: a review. *Pageoph* 124: 611–58

Griffith AA. 1920. The phenomena of rupture and flow in solids. *Phil. Trans. R. Soc. London Ser. A* 221: 169–98

Gutenberg B, Richter CF. 1956. Magnitude and energy of earthquakes. *Ann. Geofis. Rome* 9: 1–15

Hanks TC. 1979. b-value and w-g seismic source models: implications for tectonic stress variations along active crustal fault zones and the estimation of high-frequency strong ground motions. *J. Geophys. Res.* 84: 2235–42

Hanks TC, McGuire RK. 1981. The character of high-frequency strong ground motion. *Bull. Seismol. Soc. Am.* 71: 2075–95

Hanson ME, Sanford AR, Shaffer R. 1971. A source function for a dynamic bilateral brittle shear fracture. *J. Geophys. Res.* 76: 3375–83

Hartzell SH, Heaton TH. 1986. Rupture history of the 1984 Morgan Hill, California, earthquake from the inversion of strong motion records. *Bull. Seismol. Soc. Am.* 76: 649–74

Hartzell S, Helmberger DV. 1982. Strong-motion modeling of the Imperial Valley earthquake of 1979. *Bull. Seismol. Soc. Am.* 72: 571–96

Hartzell S, Stewart GS, Mendoza C. 1991. Comparison of L1 and L2 norms in a teleseismic waveform inversion for the slip history of the Loma Prieta, California, earthquake. *Bull. Seismol. Soc. Am.* 81: 1518–39

Heaton T. 1990. Evidence for and implications of self-healing pulses of slip in earthquake rupture. *Phys. Earth Planet. Int.* 64: 1–20

Heaton T, Tajima F, Mori AW. 1986. Estimating ground motions using recorded accelerograms. *Surv. Geophys.* 8: 25–83

Henyey TL, Wasserburg GJ. 1971. Heat flow near major strike-slip faults in California. *J. Geophys. Res.* 76: 7924–46

Housner GW. 1955. Properties of strong ground motion earthquakes. *Bull. Seismol. Soc. Am.* 45: 197–218

Houston H. 1990a. Broadband source spectra, seismic energy, and stress drop of the 1989 Macquarie Ridge earthquake. *Geophys. Res. Lett.* 17: 1021–24

Houston H. 1990b. A comparison of broadband source spectra, seismic energies, and stress drops of the 1989 Loma Prieta and 1988 Armenian earthquakes. *Geophys. Res. Lett.* 17: 1413–16

Husseini MI. 1977. Energy balance for formation along a fault. *Geophys. J. R. Astron. Soc.* 49: 699–714

Ida Y. 1972. Cohesive force across the tip of a longitudinal-shear crack and Griggith's specific surface energy. *J. Geophys. Res.* 77: 3796–805

Ida Y, Aki K. 1972. seismic source time function of propagating longitudinal-shear cracks. *J. Geophys. Res.* 77: 2034–44

Iio Y. 1992. Slow initial phase of the P-wave velocity pulse generated by microearthquakes. *Geophys. Res. Lett.* 19: 477–80

Imamura A. 1937. *Theoretical and Applied Seismology*. Tokyo: Maruzen. 358 pp.

Jeffreys H. 1959. *The Earth*. Cambridge: Cambridge Univ. Press. 420 pp.

Kanamori H. 1972. Determination of effective tectonic stress associated with earthquake faulting. The Tottori earthquake of 1943. *Phys. Earth Planet. Int.* 5: 426–34

Kanamori H. 1973. Mode of strain release associated with major earthquakes in Japan. A. *Rev. Earth Planet. Sci.* 1: 213–39

Kanamori H. 1977. The energy release in great earthquakes. *J. Geophys. Res.* 82: 2981–87

Kanamori H. 1980. The state of stress in the Earth's lithosphere. In *Phys. Earth's Int., Course LXXVIII*, ed. AM Dziewonski, E Boschi, pp. 531–54. Amsterdam: North-Holland

Kanamori H. 1981. The nature of seismicity patterns before large earthquakes. In *Earthquake Prediction*, ed. DW Simpson, PG Richards, Maurice Ewing Ser. 4: 1–19. Washington, DC: Am. Geophys. Union

Kanamori H, Allen CR. 1986. Earthquake repeat time and average stress drop. In *Earthquake Source Mechanics, Geophys. Monogr.*, ed. S Das, J Boatwright, CH Scholz, pp. 227–35. Washington, DC: Am. Geophys. Union

Kanamori H, Anderson DL. 1975. Theoretical basis of some empirical relations in seismology. *Bull. Seismol. Soc. Am.* 65: 1073–95

Kanamori H, Hauksson E, Hutton LK, Jones LM. 1993. Determination of Earthquake Energy Release M_L Using TERRAscope. *Bull. Seismol. Soc. Am.* 83: 330–46

Kanamori H, Mori J, Heaton TH. 1990. The 3 December 1988, Pasadena earthquake ($M_L = 4.9$) recorded with the very broadband system in Pasadena. *Bull. Seismol. Soc. Am.* 80: 483–87

Kanamori H, Stewart G. 1978. Seismological aspects of the Guatemala earthquake of February 4, 1976. *J. Geophys. Res.* 83: 3427–34

Kikuchi M. 1992. Strain drop and apparent strain for large earthquakes. *Tectonophysics* 211: 107–13

Kikuchi M, Fukao Y. 1987. Inversion of long-period P-waves from great earthquakes along subduction zones. *Tectonophysics* 144: 231–47

Kikuchi M, Fukao Y. 1988. Seismic wave energy inferred from long-period body wave inversion. *Bull. Seismol. Soc. Am.* 78: 1707–24

Kikuchi M, Takeuchi H. 1971. Unsteady propagation of longitudinal-shear crack. *Zisin* (*J. Seismol. Soc. Jpn.*) 23: 304–12

Knopoff L. 1958. Energy release in earthquakes. *Geophys. J.* 1: 44–52

Kostrov BV. 1966. Unsteady propagation of longitudinal shear cracks. *Appl. Math. Mech.* 30: 1241–48

Kostrov BV. 1974. Seismic moment and energy of earthquakes, and seismic flow of rock. *Izv. Earth Phys.* 1: 23–40 (From Russian)

Lachenbruch AH, Sass JH.1973. Thermomechanical aspects of the San Andreas fault system. In *Proc. Conf. Tectonic Problems of the San Andreas Fault System*, ed. A. Nur, pp. 192–205. Stanford, CA: Stanford Univ. Press

Lachenbruch AH, Sass JH. 1980. Heat flow and energetics of the San Andreas fault zone. *J. Geophys. Res.* 85: 6185–222

Lay T, Kanamori H, Ruff L. 1982. The asperity model and the nature of large subduction zone earthquakes. *Earthquake Predict. Res.* 1: 3–71

Madariaga R. 1976. Dynamics of an expanding circular fault. *Bull. Seismol. Soc. Am.* 66: 639–66

Madariaga R. 1979. On the relation between seismic moment and stress drop in the presence of stress and strength heterogeneity. *J. Geophys. Res.* 84: 2243–50

McGuire RK, Hanks TC. 1980. RMS acceleration and spectral amplitudes of strong ground motion during the San Fernando, California earthquake. *Bull. Seismol. Soc. Am.* 70: 1907–19

Melosh J. 1979. Acoustic fluidization: a new geologic process? *J. Geophys. Res.* 84: 7513–20

Mendoza C, Hartzell SH. 1989. Slip distribution of the 19 September 1985 Michoacan, Mexico, earthquake: near-source and teleseismic constraints. *Bull. Seismol. Soc. Am.* 79: 655–69

Mikumo T, Hirahara K, Miyatake T. 1987. Dynamical fault rupture processes in heterogeneous media. *Tectonophysics* 144: 19–36

Miyamura S, Omote S, Teisseyre R, Vesanen E. 1964. Multiple shocks and earthquake series pattern. *Int. Inst. Seismol. Earthquake Eng. Bull.* 2: 71–92

Miyatake T. 1992a. Dynamic rupture process of inland earthquakes in Japan: weak and strong asperities. *Geophys. Res. Lett.* 19: 1041–44

Miyatake T. 1992b. Reconstruction of dynamic rupture process of an earthquake with constraints of kinematic parameters. *Geophys. Res. Lett.* 19: 349–52

Mogi K. 1968. Development of aftershock areas of great earthquakes. *Bull. Earthquake Res. Inst. Tokyo Univ.* 46: 175–203

Mount VS, Suppe J. 1987. State of stress near the San Andreas fault: implications for wrench tectonics. *Geology* 15: 1143–46

Munguia L, Brune JN. 1984. High stress drop events in the Victoria, Baja California earthquake swarm of 1978 March. *Geophys. J. R. Astron. Soc.* 76: 725–52

Orowan E. 1960. Mechanism of seismic faulting in rock deformation. *Geol. Soc. Am. Mem.* 79: 323–45

Quin H. 1990. Dynamic stress drop and rupture dynamics of the October 15, 1979 Imperial Valley, California, earthquake. *Tectonophysics* 175: 93–117

Rabinowicz E. 1965. *Friction and Wear of Materials*. New York: Wiley. 244 pp.

Rice JR. 1983. Constitutive relations for fault slip and earthquake instabilities. *Pageoph* 121: 443–75

Rice JR. 1991. Spatio-temporally complex fault slip: 3D simulations with rate- and state-dependent friction on a fault surface between elastically deformable continua. *Eos Trans. Am. Geophys. Union* 72: 278 (Abstr.)

Rice JR. 1992. Fault stress states, pore pressure distributions, and the weakness of the San Andreas fault. In *Fault Mechanics and Transport Properties of Rocks: A Festschrift in Honor of WF Brace*, ed. B Evans, W Teng-Fong, pp. 475–503. New York: Academic

Richards PG. 1976. Dynamic motions near an earthquake fault: a three-dimensional solution. *Bull. Seismol. Soc. Am.* 66: 1–32

Ruff L, Kanamori H. 1983. The rupture process and asperity distribution of three great earthquakes from long-period diffracted P-waves. *Phys. Earth Planet. Int.* 31: 202–30

Savage JC, Walsh JB. 1978. Gravitational energy and faulting. *Bull. Seismol. Soc. Am.* 68: 1613–22

Savage JC, Wood MD. 1971. The relation between apparent stress and stress drop. *Bull. Seismol. Soc. Am.* 61: 1381–88

Schallamach A. 1971. How does rubber slide? *Wear* 17: 301–12

Scholz CH. 1989. Mechanics of faulting. *Annu. Rev. Earth Planet. Sci.* 17: 309–34

Scholz CH. 1992. Paradigms or small change in earthquake mechanics. In *Fault Mechanics and Transport Properties of Rocks: A Festschrift in Honor of WF Brace*, ed. B Evans, W Teng-Fong, pp. 505–17. New York: Academic

Scholz CH, Aviles CA, Wesnousky SG. 1986. Scaling differences between large interplate and intraplate earthquakes. *Bull. Seismol. Soc. Am.* 76: 65–70

Schwartz S, Ruff L. 1987. Asperity distribution and earthquake occurrences in the southern Kurile Islands arc. *Phys. Earth Planet Int.* 49: 54–77

Sibson RH. 1973. Interactions between temperature and pore fluid pressure during faulting and a mechanism for partial or total stress relief. *Nature* 243: 66–68

Sibson RH. 1986. Rupture interaction with fault jogs. In *Earthquake Source Mechanics*, ed. S Das, J Boatright, CH Scholz, pp. 157–67. Washington, DC: Am. Geophys. Union

Sieh K, Jones L, Hauksson E, Hudnut K, Eberhart-Phillips D, et al. 1993. Near-field investigations of the Landers earthquake sequence, April to July 1992. *Science* 260: 171–76

Sieh KE. 1978. Slip along the San Andreas fault associated with the great 1857 earthquake. *Bull. Seismol. Soc. Am.* 68: 1421–57

Steidl JH, Archuleta RJ, Hartzell S. 1991. Rupture history of the 1989 Loma Prieta, California, earthquake. *Bull. Seismol. Soc. Am.* 81: 1573–602

Takeo M. 1988. Rupture process of the 1980 Izu-Hanto-Toho-Oki earthquake deduced from strong motion seismograms. *Bull. Seism. Soc. Am.* 78: 1074–91

Takeo M, Mikami N. 1987. Inversion of strong motion seismograms for the source process of the Naganoken-Seibu Earthquake of 1984. *Tectonophysics* 144: 271–85

Takeuchi H, Kikuchi M. 1971. On Kostrov's theory on crack propagation. *Geol. Eng.* 7: 13–19

Thatcher W. 1975. Strain accumulation and release mechanism of the 1906 San Francisco earthquake. *J. Geophys. Res.* 80: 4862–72

Thatcher W, Hanks TC. 1973. Source parameters of southern California earthquakes. *J. Geophys. Res.* 78: 8547–76

Tsuboi C. 1933. Investigation of deformation of the crust found by precise geodetic means. *Jpn. J. Astron. Geophys.* 10: 93–248

Udias A. 1991. Source mechainsm of earthquakes. *Adv. Geophys.* 33: 81–140

Vassiliou MS, Kanamori H. 1982. The energy release in earthquakes. *Seismol. Soc. Am. Bull.* 72: 371–87

Wald D. 1992. Strong motion and broadband teleseismic analysis of the 1991 Sierra Madre, California, earthquake. *J. Geophys. Res.* 97: 11,033–46

Wald DJ, Heaton TH. 1993. Spatial and temporal distribution of slip for the 1992 Landers, California, earthquake. *Bull. Seismol. Soc. Am.* Submitted

Wald DJ, Helmberger DV, Hartzell SH. 1990. Rupture process of the 1987 Superstition Hills earthquake from the inversion of strong-motion data. *Bull. Seismol. Soc. Am.* 80: 1079–98

Wald DJ, Helmberger DV, Heaton TH. 1991. Rupture model of the 1989 Loma Prieta earthquake from the inversion of strong-motion and broadband teleseismic data. *Bull. Seismol. Soc. Am.* 81: 1540–72

Wald DJ, Kanamori H, Helmberger DV, Heaton TH. 1993. Source study of the 1906 San Francisco earthquake. *Bull. Seismol. Soc. Am.* 83: 981–1019

Wyss M, Brune JN. 1967. The Alaska earthquake of 28 March 1964: a complex multiple rupture. *Bull. Seismol. Soc. Am.* 57: 1017–23

Wyss M, Molnar P. 1972. Efficiency, stress drop, apparent stress, effective stress, and frictional stress of Denver, Colorado, earthquakes. *J. Geophys. Res.* 77: 1433–38

Yamashita T. 1983. Peak and root-mean-square accelerations radiated from circular cracks and stress-drop associated with seismic high-frequency radiation. *J. Phys. Earth* 31: 225–49

Zoback MD, Zoback ML, Mount VS, Suppe J, Eaton J, et al. 1987. New evidence on the state of stress of the San Andreas fault system. *Science* 238: 1105–11

Annu. Rev. Earth Planet. Sci. 1994. 22:239–71

ACTIVE TECTONICS OF THE AEGEAN REGION

James Jackson

Department of Earth Sciences, Bullard Laboratories, University of Cambridge, Madingley Road, Cambridge CB3 0EZ, United Kingdom

KEY WORDS: continental tectonics, active deformation, Mediterranean, earthquakes, Greece

1. INTRODUCTION

The Aegean Sea and its surrounding regions comprise one of the most rapidly deforming parts of the Alpine-Himalayan mountain belt. Though the deformation of the belt as a whole is related to the northward movement of Africa, Arabia, and India relative to Eurasia, the tectonics of the Aegean region itself is dominated by strike-slip and extensional motions. The actively deforming part of the Aegean region is not large (about 700×700 km^2) compared with that in Asia, yet it has had an influence on our views of continental tectonics disproportionate to its size, and comparable perhaps only to Tibet. Fundamental contributions from the Aegean have been made to many topics, including:

1. *The three-dimensional nature of continental tectonics.* McKenzie (1970, 1972) demonstrated that present-day motions in the Aegean could be understood as a consequence of those 1500 km farther east in the Caucasus and eastern Turkey, and that large-scale movement of material along the strike of mountain belts could be an important feature of continental tectonics.
2. *Sedimentary basin formation by lithosphere stretching.* This concept (McKenzie 1978b), which provides the quantitative framework for many studies of rifted continental basins and margins, originated from the Aegean, where McKenzie (1978a) noticed the association of thinned crust with high heat flow, normal faulting, and subsidence.

0084–6597/94/0515–0239$05.00

3. *Obtaining average stress tensors from the slip directions on distributed faults.* This technique, much publicized by French geologists, was first used and developed in the Aegean (Carey & Brunier 1974, Angelier & Goguel 1979).
4. *The role of normal faulting in accommodating crustal extension.* An important debate of the 1980s concerned the relative importance of high- vs low-angle and planar vs listric normal faulting in sedimentary basins. It arose from the inability to reconcile the stretching required by crustal thinning and subsidence observations in basins such as the North Sea with that estimated from normal faulting (which was usually much lower). In the Aegean, it could be demonstrated from a combination of surface observations and earthquake seismology that some large active normal faults were approximately planar throughout the seismogenic upper crust and dipped at $\sim 45°$: within the dip range of ~ 30–$60°$ that is now recognized globally (e.g. Jackson 1985, Jackson & White 1989).
5. *How distributed deformation is accommodated by faulting.* Although the Aegean was not the first deforming region in which block rotations about a vertical axis were demonstrated by paleomagnetic measurements, it was one of the first in which the relation between such rotations, the slip directions on the faults, and the overall deformation could be addressed (McKenzie & Jackson 1983, 1986). This highlighted the question of the interplay between the seismogenic upper crust and the rest of the lithosphere, which is the main topic of this review.
6. *Vertical motions accompanying normal faulting.* Since many of the active normal faults in the Aegean are at sea level, this provides a reference against which one can observe vertical motions, which dominate the drainage and sedimentation and hence ultimately the distribution and preservation of potential syn-rift hydrocarbon reservoirs (e.g. Collier 1990, Roberts & Jackson 1991, Leeder & Jackson 1993). The Aegean is much used by the oil industry as an analog of sedimentation in older extensional basins.
7. *The driving forces of continental tectonics.* In the eastern Mediterranean some of the driving forces that are potentially responsible for the deformation, such as the buoyancy forces arising from crustal thickness contrasts and the sinking of a subducting slab, are readily identified, and led to early speculation on their importance (McKenzie 1972, Le Pichon 1982).

Certainly other regions were also influential in debates of these topics. But the Aegean has contributed profoundly to all of them for numerous reasons: It is deforming rapidly, the overall motions are well understood, there are many earthquakes with a long record of historical seismicity, it is at sea level, there is an abundance of clear geological and geomorphological data on land, and there are abundant geophysical data from offshore. It is a special place—a natural laboratory for continental tectonics, and has a significance beyond its purely geographical context.

The Aegean is poised to contribute once again to a fundamental question in continental tectonics; that of the relation between the deformation of the upper part of the lithosphere which is visible at the Earth's surface, and the deformation of the lower lithosphere. The scale on which large topographic features, such as mountain belts, plateaus, and basins, occur as well as their thermal evolution, suggests that the whole lithosphere is involved in their formation (e.g. McKenzie 1978a,b, England & Jackson 1989). The lithosphere is defined by its thermal structure (Parsons & McKenzie 1978) and is about 125 km thick, but the nature of its deformation is unlikely to be uniform with depth. Earthquakes are generally restricted to the upper 10–20 km of the continental crust and indicate the importance of discontinuous, fault-dominated deformation within this upper seismogenic layer. The absence of earthquakes within the lower 80–100 km of the continental lithosphere, together with the likely rheology at these depths (e.g. Brace & Kohlstedt 1980), suggest that the lower lithosphere deforms in a more distributed, continuous fashion, which can be represented by a flow field. Until recently, our estimates of the rates and styles of deformation in the eastern Mediterranean came from arguments based on seismicity, geology, and geomorphology. We are now entering a time when these motions can be measured directly by space-based geodetic techniques. This review summarizes what we expect these new techniques to be able to see, the extent to which early geodetic results are compatible with the nongeodetic estimates, and the insight these studies give into the question of what controls the overall deformation of the continental lithosphere: the behavior of the seismogenic upper crust or the flow in the mantle lithosphere beneath?

2. GEOLOGICAL AND KINEMATIC BACKGROUND

Although this review is concerned with the active tectonics of the region, the earlier geological history is thought to influence the present-day deformation. In essence, the region was shortened by a series of collisional events in the late Mesozoic and early Tertiary, which imparted a strong structural fabric in the form of folds, thrust faults, and sutures that trend NW-SE in mainland Greece, and then change to a more E-W or ENE-WNW orientation across the central Aegean and into western Turkey (see e.g. Smith & Moores 1974, Robertson & Dixon 1984, Şengör et al 1984 for summaries). The region then began to extend, as early as 25 Ma ago in parts of the southern Aegean (Angelier et al 1982, Jolivet et al 1993), and extension in some form, combined with strike-slip motions, has been the dominant feature of the deformation ever since. There is much uncertainty about how the extension has varied in time, space, and magnitude throughout the Aegean region, and whether the earliest manifestations of extension in the southern Aegean are part of the same kinematic processes (i.e. in the same orientation and driven by the same forces) that are operating today. We know much more about what is occurring today than what occurred in the past, so we begin by discussing present-day motions and then return to the question of how long they have been operating.

The active deformation in the Aegean region is dominated by two effects (Figure 1): (*a*) the westward motion of Turkey and (*b*) the southwestward motion of the southern Aegean, both relative to Eurasia (McKenzie 1972, 1978a; Le Pichon & Angelier 1979, 1981). McKenzie (1972) was the first to estimate the likely rates at which these motions occur. He noticed that the central plateau of Turkey is a relatively flat and aseismic region, bounded by the North Anatolian fault zone on its northern side and by the East Anatolian fault zone to the southeast. The localized right-lateral

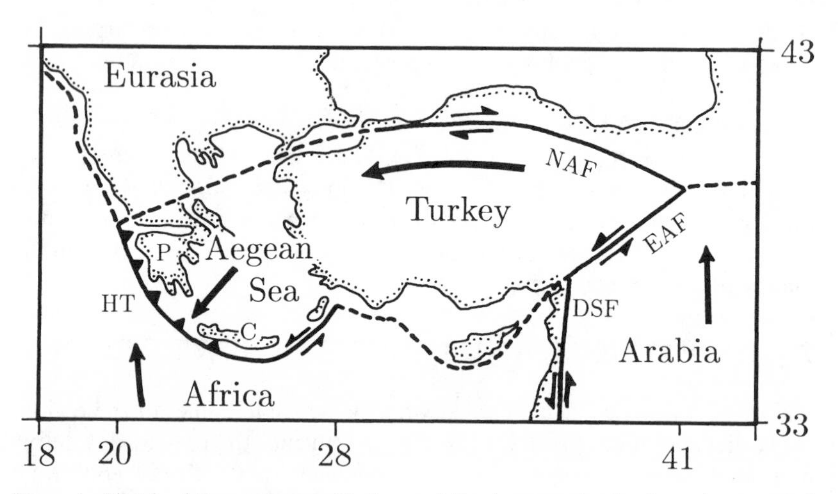

Figure 1 Sketch of the motions in Turkey and the Aegean, showing how the westward motion of Turkey relative to Eurasia is accommodated by strike-slip faulting on the North (NAF) and East (EAF) Anatolian fault zones and by shortening in the Hellenic Trench (HT). Black arrows are the approximate directions of motion of Arabia, Turkey, the southern Aegean, and Africa relative to Eurasia. DSF is the Dead Sea Fault zone, C is Crete, and P is the Peloponnese.

character of the North Anatolian fault zone is beyond doubt, since it has ruptured virtually its entire length in a series of large earthquakes this century (Ketin 1948, Ambraseys 1970). The East Anatolian fault zone is more diffuse, containing a variety of active faults and other structures (Lyberis et al 1992). The historical record reveals several large earthquakes in this fault zone, even though the seismicity of this century has been low (Ambraseys 1989; Figure 2). The few moderate-sized recent earthquakes show a variety of focal mechanisms, but all of them have slip vectors directed roughly 063°, parallel to the strike of the zone (Taymaz et al 1991a). There seems little doubt that the East Anatolian fault zone accommodates principally left-lateral strike-slip motion, though it may also involve some shortening.

McKenzie (1972) estimated the motions on the North and East Anatolian fault zones by constructing a velocity triangle for the motions between Arabia, Eurasia, and Turkey at 41°E. This is redrawn in Figure 3*a* using

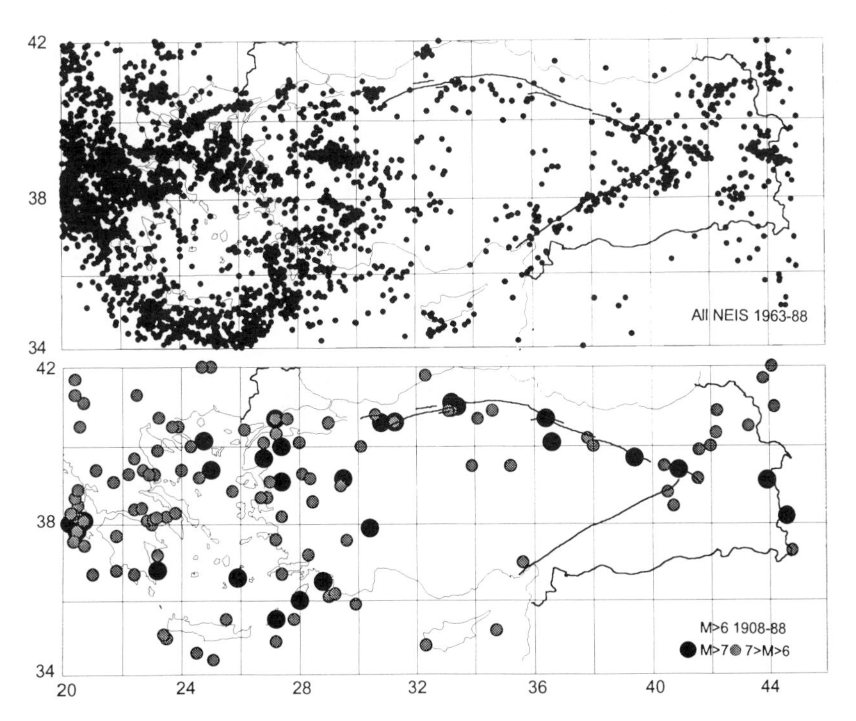

Figure 2 Seismicity of Turkey and the Aegean Sea. The top map shows all earthquakes shallower than 50 km reported by the National Earthquake Information Service (NEIS) in the period 1963–1988. The bottom map shows the earthquakes of $M_s \geqslant 6$ in the period 1908–1988. The North and East Anatolian fault zones are marked with thick lines.

the revised motion for Eurasia–Arabia from Jestin et al (1993), who point out the deficiency of the NUVEL-1 model (de Mets et al 1990) for these two plates. The velocity triangle assumes that the shortening between Eurasia and Arabia west of 41°E is accommodated purely by strike-slip motion on the North and East Anatolian fault zones, and hence by the westward motion of Turkey relative to Eurasia (Figure 1). This argument yields slip rates across the North and East Anatolian fault zones of ~36 mm/yr and ~33 mm/yr, which are likely to be upper bounds, as some shortening may occur between the two strike-slip faults. This analysis is simplistic but has been the basis for all predicted estimates of deformation rates farther west in the Aegean. As yet, there is little substantial evidence to test these rates. The seismicity in the period 1909–1981 accounts for ~39 mm/yr on the North Anatolian fault zone (Jackson & McKenzie 1988a), but may not be representative of longer periods. Preliminary results from GPS (Global Positioning System) surveys are consistent with the predicted senses of motion on these fault zones, but the rates are not yet well resolved (Oral et al 1991, 1992).

McKenzie (1972) realized that the westward motion of Turkey relative

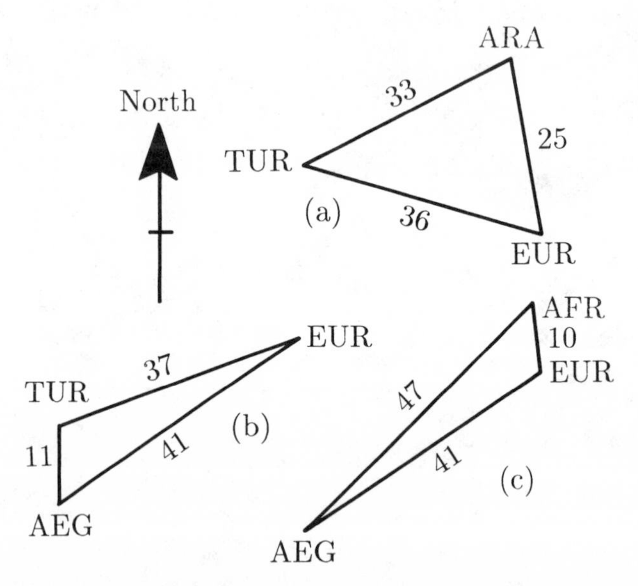

Figure 3 Velocity triangles used to estimate the velocities (*a*) in eastern Turkey at the junction of the North and East Anatolian Faults, (*b*) in the southern Aegean Sea, (*c*) in the Hellenic trench. ARA: Arabia; EUR: Eurasia; TUR: Turkey; AEG: southern Aegean; AFR: Africa. Velocities are in mm/yr.

to Eurasia was ultimately accommodated in the subduction zone of the Hellenic Trench, though the North Anatolian fault zone does not cross the Aegean Sea in the simple manner indicated by Figure 1. He observed that the southern Aegean Sea was relatively aseismic (Figure 2), and thought that this block was moving SW relative to Eurasia, because of the right-lateral strike-slip faults with a NE-SW strike in the northern Aegean (Figures 4 and 5). He also observed that the deformation in SW Turkey, which separates the central Turkish plateau from the southern Aegean block, occurs mostly on normal faults with an E-W strike. He therefore reasoned that the motion of the southern Aegean relative to Turkey can not have an easterly component, or some E-W shortening would occur in SW Turkey. These arguments led to the construction of a velocity triangle, which is produced in Figure 3*b*, updated with the revised Turkey-Eurasia motion from Figure 3*a* and the revised average slip vector in the northern Aegean (056°) from Taymaz et al (1991b). Assuming a N-S motion between the southern Aegean and Turkey gives a minimum predicted velocity of 41 mm/yr for the southern Aegean block relative to Eurasia in the direction 236°. As we will see later (Figure 11), this predicted rate is comparable to the 33–36 mm/yr velocities relative to Europe of SLR sites on the SW border of the Aegean. Le Pichon & Angelier (1979) also estimated an extension rate in the Aegean using different arguments based on the length and probable age of the subducted slab beneath the Hellenic Trench. Their rate of $\sim$35 mm/yr is averaged over $\sim$13 Ma.

Much of the predicted motion between the southern Aegean and Eurasia appears to be accommodated seismically. Jackson & McKenzie (1988a,b) and Ekström & England (1989) used the method of Kostrov (1974) to obtain the average extension rate across the whole extending region by summing the moment tensors of earthquakes of $M_s \geqslant 6.0$ in the period 1909–1981. Their results depend on an empirical relation between seismic moment (M_0) and surface wave magnitude (M_s) for older earthquakes whose moments have not been calculated directly; the precise nature of this relation is the main source of uncertainty. The $M_s - M_0$ relation in the Aegean appears to be different from the globally-averaged relation. Using the continental $M_s - M_0$ relation of Ekström & Dziewonski (1988), which yields the smallest seismic moment for a given M_s, gives an extension rate of about 30 mm/yr across the whole Aegean (Table 1), but this average value obscures important spatial variations in the strain field (Eyidoğan 1988, Jackson et al 1992), which are discussed later.

Along the western coast of Albania and Greece, and seaward of an arc from SW Greece through Crete to SW Turkey, is a relatively narrow ($\sim$100 km wide) belt of active shortening, with folding and earthquakes that have thrust and reverse fault mechanisms (Figures 4 and 5). Between

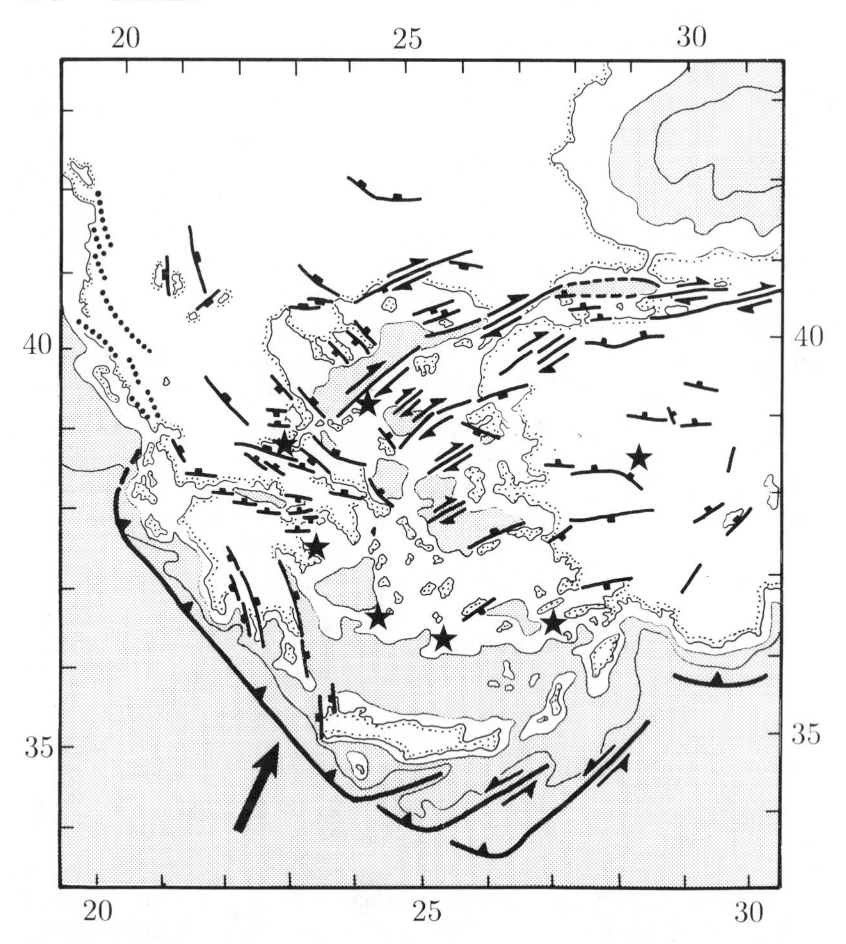

Figure 4 Summary tectonic map showing the orientations of the principal active structures in the Aegean region. The large black arrow shows the direction of convergence in the Hellenic Trench (*thick solid line*) between the Mediterranean sea floor and the over-riding Aegean Sea. The dotted lines in coastal Albania and NW Greece are anticlinal axes. Other thick lines are major active faults: Those with a NE-ENE strike are mostly right-lateral strike-slip faults with some normal faulting component; those with a WNW-NW strike are predominantly normal faults, with slip vectors trending SSE-SSW (see Figure 6). Stars are Quaternary volcanoes. Bathymetric contours are at 1000 m and 2000 m, shaded below 1000 m.

the longitudes of 20°E and 28°E, oceanic lithosphere to the SW is subducted in the Hellenic Trench beneath the continental lithosphere of the Aegean to the NE, leading to the formation of an inclined seismic zone dipping NE to a depth of about 150–200 km (Hatzfeld & Martin 1992) with active volcanoes above it. Northwest of about 39°N 20°E, along the

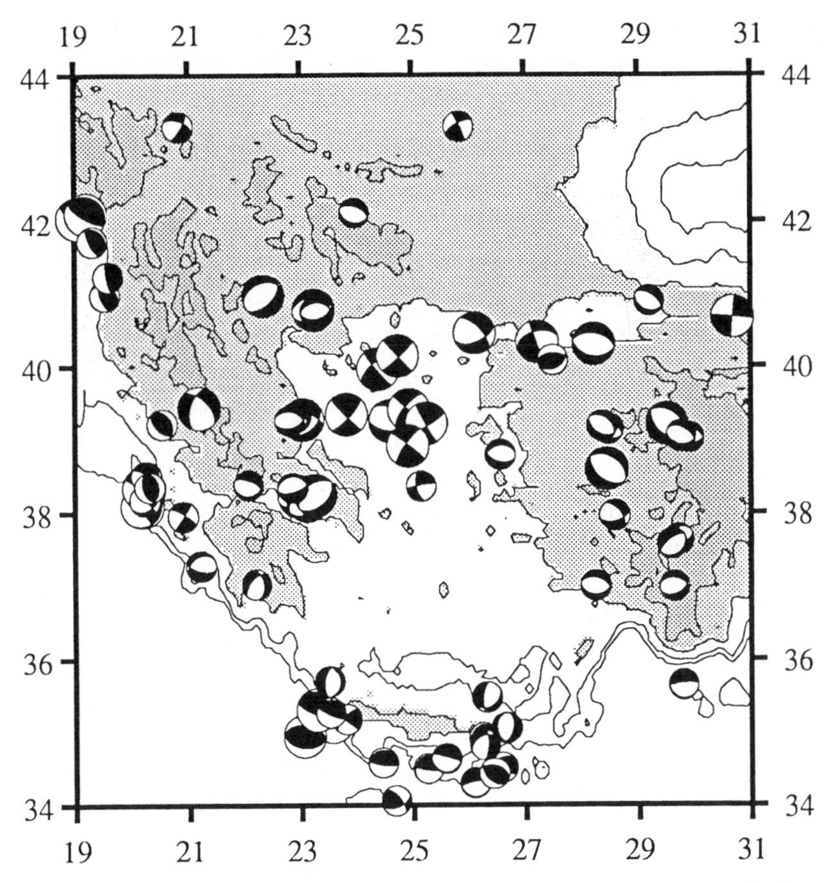

Figure 5 Summary map of earthquake focal mechanisms in the Aegean region. Mechanisms are mainly from Taymaz et al (1990, 1991b), Taymaz & Price (1992), Braunmiller (1991), and Centroid Moment Tensor solutions published by Harvard. All have been constrained by modeling long-period P and SH body waves and are of $M_w \geqslant 5.5$. The larger symbols are earthquakes of $M_w \geqslant 6.0$. Contours are at 1000 m intervals.

coast of NW Greece and Albania, the shortening is between the two continental regions of the Adriatic-Apulia platform to the SW and Greece-Albania to the NE (see Figure 12 for location), and there is no evidence for subcrustal earthquakes. The extension in the Aegean Sea must be accommodated by shortening in the Hellenic Trench system, since there is no doubt that Africa and Eurasia are converging at this longitude. The shortening rate in the trench system is easily estimated (Figure 3*c*), but depends almost entirely on the estimated extension rate in the Aegean, since this is so much greater than the 10 mm/yr convergence rate between

Table 1 Estimated velocity of the SW Aegean relative to Eurasia

Method	Velocity (mm/yr)		Reference
Kinematic	~41	(upper bound)	Figure 3, McKenzie (1972)
Seismicity 1909–1981	29^{+18}_{-9}	(lower bound)	Jackson & McKenzie (1988b) Ekström & England (1989)
Velocity field from earthquakes 1909–83	30	(lower bound)	Figure 9, Jackson et al (1992)
Broken Slat Model	~38		Figure 8, Taymaz et al (1988b)
Geodetic (SLR)	33–36		Figure 11, Robbins et al (1993) Smith et al (1993) Noomen et al (1993) Sahin et al (1993)

Africa and Eurasia. These arguments yield an expected present-day shortening rate in the trench of ~50 mm/yr, though much of this deformation must occur aseismically, as the earthquakes account for only a small fraction of this rate (Jackson & McKenzie 1988a,b).

It is clear from the seismicity (Figure 2) and focal mechanisms (Figure 5) that the motion of the southern Aegean relative to Eurasia is not accommodated by a single fault, as indicated in Figure 1, but is distributed widely over many faults in the central and northern Aegean Sea. It is a general feature of active continental tectonics that the earthquakes are spread over regions several times wider than the lithosphere thickness, yet are restricted vertically to the upper 10–20 km of the continental crust. At large length scales, i.e. at dimensions comparable or larger than the lithosphere thickness, it is reasonable to approximate the deformation of the continental lithosphere as that of a continuum, which can then be described by a continuous velocity field (e.g. England & McKenzie 1982, McKenzie & Jackson 1983, England & Jackson 1989). An adequate description of the kinematics requires us to address two questions: 1. What is the continuous velocity field that describes the deformation of the lithosphere as a whole?, and 2. How is this velocity field accommodated by discontinuous motion on faults in the upper crust? As we will see, the answers to these questions in the Aegean provide insights into how the motions in the upper crust and the rest of the lithosphere interact.

3. FAULTING

Since the early studies of Mercier et al (1976, 1979), McKenzie (1978a), Dewey & Şengör (1979), and Le Pichon & Angelier (1979), a great deal more has become known about the faulting in the northern and central Aegean region. A simplified summary of the faulting is shown in Figure 4, and extensive reviews are given by Mercier et al (1989), Mascle & Martin (1990), Roberts & Jackson (1991), and Taymaz et al (1991b). Figure 4 should be viewed together with the focal mechanisms in Figure 5. West of about 31°E the North Anatolian Fault splays into a series of distributed sub-parallel right-lateral strike-slip faults that cross NW Turkey and the northern Aegean with a NE to ENE strike (Barka & Kadinsky-Cade 1988, Roussos & Lyssimachou 1991). These faults are associated with deep basins offshore (Lybéris 1984, and Figure 4). The strike-slip faults do not cross central Greece to meet the Hellenic Trench, but end abruptly in the western part of the northern Aegean Sea, where they meet the NW- to WNW-striking normal fault systems that dominate the structure of central Greece. These normal faults have slip vectors that are directed SSE to SSW (Figure 6). The change in direction of the earthquake slip vectors that accompanies the abrupt change in strike of the active faults near the eastern coast of mainland Greece suggests that the distributed right-lateral shear crossing the northern Aegean Sea is accommodated by clockwise rotation of the fault-bounded blocks in central Greece relative to Eurasia (McKenzie & Jackson 1983, 1986). This inference has been confirmed by measurements of paleomagnetic declinations (see reviews by Kissel & Laj 1988 and Kissel et al 1989).

The way in which these strike-slip and normal fault systems are thought to move is illustrated schematically by a simple model proposed by Taymaz et al (1991b) and shown in Figure 7. The model consists of two sets of slats, which represent fault-bounded blocks, that are in relative motion. The slats are joined to the eastern and western margins of the model by pivots whose separation cannot change. Where the two sets of slats meet, they are joined by pivots that can move to change their separation. The eastern margin of the model rotates clockwise, representing the distributed right-lateral shear at the western end of the North Anatolian Fault (see Taymaz et al 1992). The relative motion between the eastern set of slats is dominantly right-lateral strike-slip, and they all rotate slowly counter-clockwise relative to the top of the model (Eurasia). The western set of slats rotate faster and clockwise, and the motion between them is mostly extensional.

Taymaz et al (1991b) calculated the instantaneous velocities of all points on these slats and the slip vectors between the slats (fault blocks) for a configuration that best resembles the north Aegean (Figure 8). This configuration differs from the simple cartoon in Figure 7 in that the western margin of the model is also allowed to rotate clockwise, though at a slower rate than the eastern margin. This is in agreement with paleomagnetic

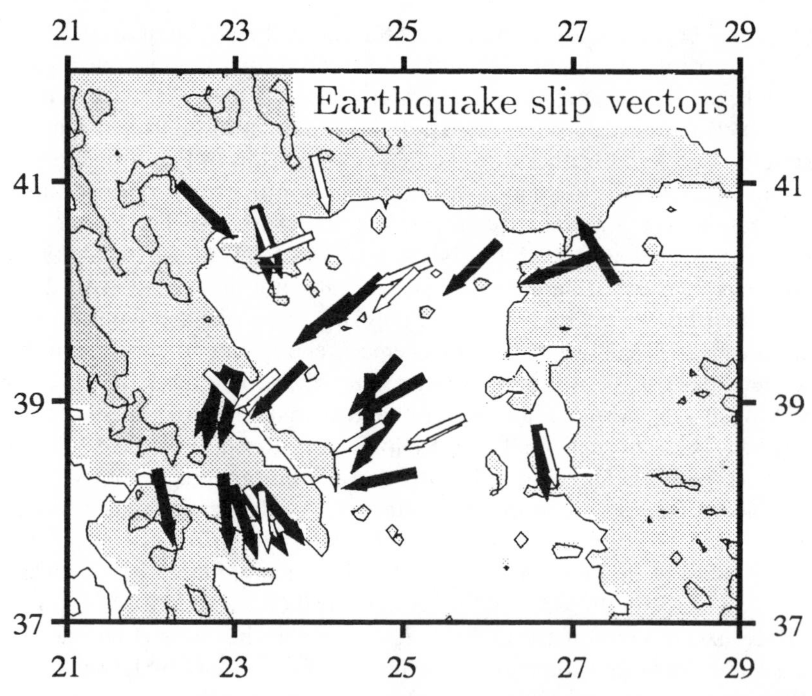

Figure 6 Earthquake slip vectors in the northern Aegean region. Directions are shown for the relative motion of the south side of the fault relative to the north side. Only earthquakes whose fault orientations are constrained by long-period P and SH body wave modeling are included. Those with black arrows are for earthquakes of $M_w \geqslant 5.5$. The white arrows are for events of $5.2 \leqslant M_w < 5.5$. The pattern shows the SW-directed slip vectors on the strike-slip faults in the central-north Aegean Sea and the SSE- to SSW-directed slip vectors on the normal fault system of mainland Greece. Note the anomalous NW-directed slip vector in NW Turkey (40.0°N, 27.5°E), which is from a reverse faulting earthquake that may have occurred on a restraining bend in a large strike-slip fault that moved in an earthquake of $M_s = 7.2$ in 1953 (see Figure 5 and Taymaz et al 1991b).

declinations in Pliocene rocks, which show that the western seaboard of Greece has rotated clockwise relative to Europe at a rate of about 5°/Ma for the last 5 Ma (Kissel & Laj 1988). Taymaz et al (1991b) discuss how well this simple model describes the faulting, paleomagnetic, geodetic, and geological observations in the northern Aegean. A limitation of the model is that, because the pivots on its eastern margin cannot separate, it takes no account of the N-S extension of western Turkey, which is manifest by the major E-W graben bounded by normal faults that dominate the

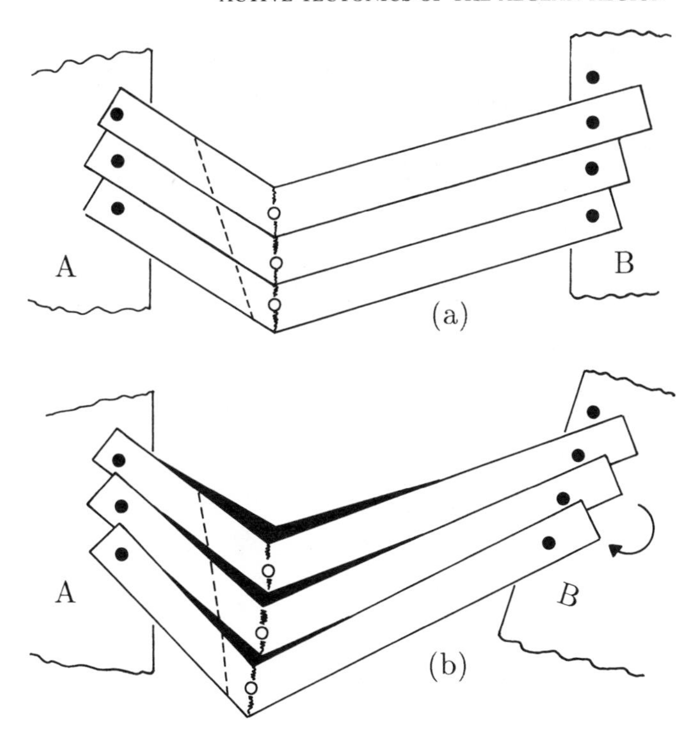

Figure 7 Sketch of a simple model involving two sets of slats that are in relative motion. The model can easily be constructed from wood or card. Solid circles are pivots (screws) attached to plate A on the left or to plate B on the right. Open circles are screws joining two slats, but which are otherwise free to move. The configuration in (*a*) moves to that in (*b*) following a clockwise rotation of the right-hand margin. In this illustration the left-hand pivots, attached to A, do not rotate. In the analogy with the Aegean they rotate clockwise, but more slowly, than those on B (see Figure 8). The black regions in (*b*) represent new surface area created by extension. The dashed line on the left-hand slats shows that the *relative* rotation between segments of an offset passive marker (such as an early Tertiary fold axis) is much less than the bulk rotation of the slats.

geological structure there (Figures 4 and 5). With this proviso, the model adequately describes how distributed strike-slip and normal faulting, combined with rotations about a vertical axis, take up the westward motion of Turkey and the N-S extension of the Aegean. From a comparison between the seismic moment tensor elements calculated for each set of slats in the model and those observed for earthquakes over the period 1909–1983, Taymaz et al (1991b) suggest that most of the deformation predicted by their model occurs by seismic slip on faults.

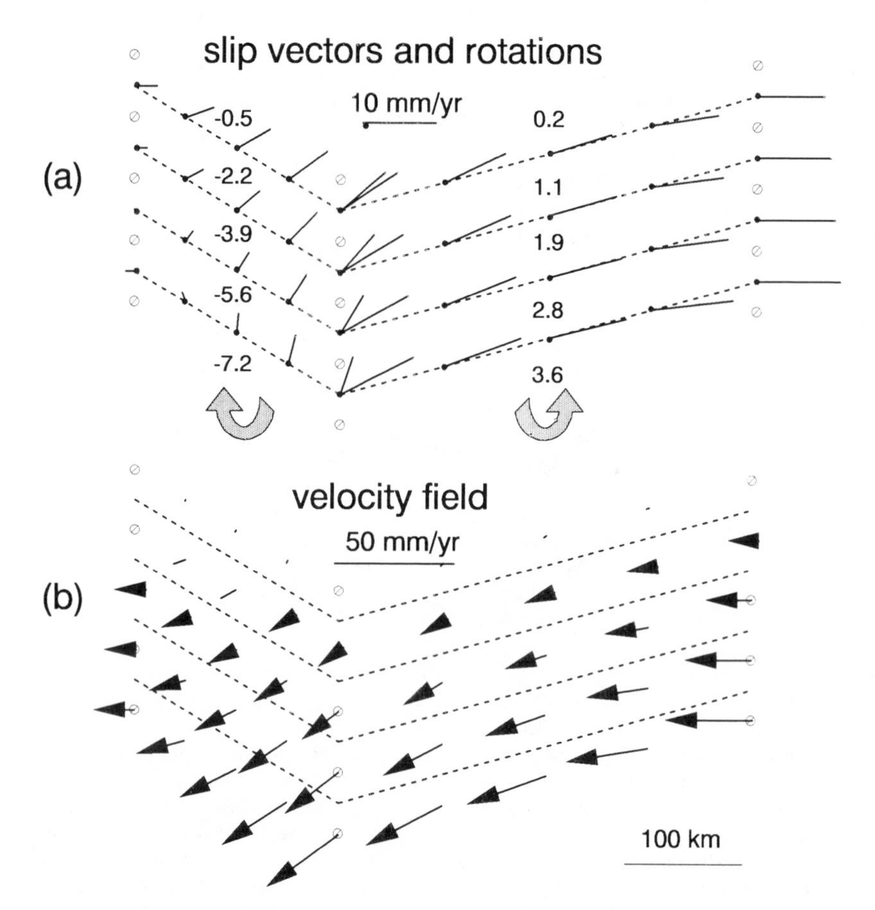

Figure 8 The broken slat model of Taymaz et al (1991b), illustrating the kinematics of fault motion in the central and northern Aegean Sea (roughly the area between 38–42°N and 22–28°E). (*a*) Block boundaries (faults) are dashed. Slip vectors between blocks are shown by solid lines attached to points on the boundaries: Their lengths are proportional to their velocities. Pivots (screws) are shown by open circles with a diagonal line. Rotation rates relative to the northern boundary are given in degrees per million years (counterclockwise positive). (*b*) Arrows show the velocities of points on the slats (the feet of the arrows) relative to the top left-hand screw, with lengths proportional to the magnitude of the velocity. Note that, in the west, the velocity of a point relative to the top of (*b*) is not the same as the direction of slip on an adjacent fault in (*a*).

4. THE VELOCITY FIELD

Whereas Figure 8*a* shows the slip vectors between the slats (i.e. the rates and directions of slip on the faults that separate them), Figure 8*b* shows the velocities of points on the slats, relative to the top pivot on the left-hand margin (i.e. conceptually relative to Eurasia). Figure 8*b* is the velocity field accommodated by the faulting in the model. Note that the westerly component of velocity increases southwards (accommodating the E-W right-lateral shear between Turkey and Eurasia), and that the N-S component of velocity also increases southwards (accommodating the N-S extension of the Aegean). With the rather few points shown in Figure 8*b* the velocity field looks continuous, in that the length and direction of the velocity vectors vary smoothly over the model. In fact the velocity field is discontinuous across the faults (slat boundaries). The model in Figure 8 illustrates two different but complementary ways of looking at the deforming lithosphere in the north Aegean. On the one hand, we can imagine that a smoothed version of the velocity field in Figure 8*b* is a reasonable representation of the flow in the mantle part of the lithosphere that deforms by distributed creep. On the other hand, the slip (faulting) between the slats in Figure 8*a* shows how discontinuous deformation in the seismogenic upper crust is able to achieve the same overall motion as the flow beneath it.

Is it possible to obtain a velocity field that represents the long wavelength deformation of the lithosphere directly from observations? One of the difficulties we face here is that we can only make quantitative measurements of deformation rates at the Earth's surface and within the seismogenic upper crust that deforms discontinuously. Some progress can be made if we make the assumption that, provided we average over length scales comparable with the lithosphere thickness, the deformation of the upper crust approximates the distributed flow in the rest of the lithosphere beneath it. This is the simplest assumption we can make, and there is some evidence that it is reasonable: The Moho is often elevated under regions of extension in the upper crust and depressed beneath mountain ranges, and some of the thermal effects observed in sedimentary basins indicate that the deformation is not restricted to the crust, but also involves the mantle directly beneath it (McKenzie 1978b). With this assumption, we can, in principle, use deformation rates observed at the surface to obtain a velocity field that represents the average deformation of the whole lithosphere. The most useful measurements come from geodesy, the slip rates on faults, rates of vertical motion (uplift and subsidence), and the seismic moment release in earthquakes.

A problem with the use of these data types is the choice of a reference frame. Most observations measure the symmetric part of the velocity field (i.e. the strain rate tensor) and are unaffected by a rigid body rotation (see Jackson & McKenzie 1988a). However, abundant paleomagnetic data in deforming regions attest to the importance of rotations about a vertical axis. This problem may potentially be overcome in two ways: 1. by extending the area of geodetic observations out of the deforming region into a rigid plate, which is now possible with space-based techniques; or 2. by splitting the deforming region into small enough sub-regions that the spatial variations in the components of the strain rate tensor may be described. These spatial variations contain information about relative rotations, which can be used to reconstruct a velocity field relative to some given frame (see Haines 1982, Walcott 1984, Holt et al 1991, Jackson et al 1992, Holt & Haines 1993). The question is whether the data are of sufficient quality and density to justify such treatment.

Figure 9 shows a velocity field in the Aegean obtained by Jackson et al (1992) from the distribution of earthquake moment tensors over the time interval 1909–1983. Velocities are shown relative to the top of the picture (Eurasia). Figure 9 was obtained by fitting low-order polynomial surfaces to a strain rate field obtained by summing the seismic moment tensors of earthquakes within a grid of box size 1.5°. This operation involves considerable smoothing as the earthquakes between 1909 and 1983 are not uniformly distributed: Most of the seismic moment release comes from NW Turkey, the central-north Aegean, and central Greece. It is important to consider whether these spatial variations in strain rate are a short-term phenomenon or likely to be true also of longer periods. From our knowledge of the historical earthquake record, the distribution of active faults, and the bathymetry, it is probable that the pattern of seismic strain release between 1909 and 1983 is broadly representative of longer periods at the length scale of ~ 150 km which was chosen to derive the velocity field in Figure 9. Given the strain rate field itself, the velocity field is robust as it is related to the strain rates by integration (Jackson et al 1992). On the other hand, because Jackson et al (1992) used a $M_s - M_0$ relation that gave the lowest likely seismic moments for a given M_s, and only included earthquakes of $M_s \geqslant 6.0$, and because there may be an aseismic contribution to the overall motion, the magnitudes of the velocities are likely to be underestimated.

The velocities in Figure 9 are comparable in orientation and magnitude with those predicted by the simple model of Taymaz et al (1991b) in Figure 8*b*, with the exception of those in western Turkey, which are directed SW (in Figure 9) rather than W (in Figure 8*b*). This discrepancy arises because the velocity field in Figure 9 includes the N-S extension in western Turkey which is ignored in the slat model. In the southern part of the Aegean Sea,

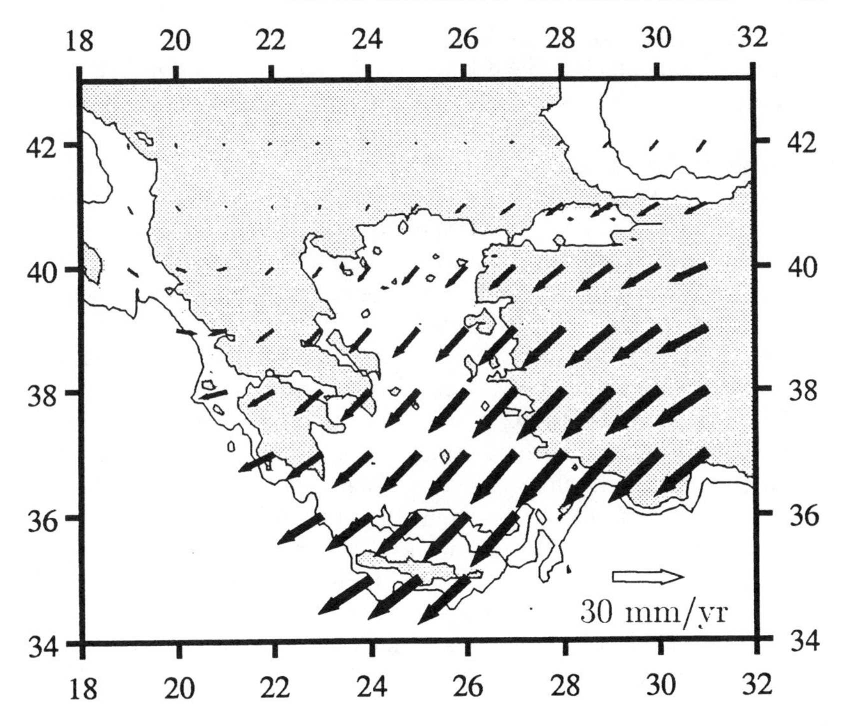

Figure 9 The velocity field in the Aegean Sea determined from the spatial distribution of seismic moment tensors during the period 1909–1983 (adapted from Figure 16*d* of Jackson et al 1992). Velocities are shown relative to the top of the figure (stable Eurasia). The lengths of the vectors are proportional to their magnitudes (a scale is shown by the white arrow at the bottom of the figure).

both show velocities toward the SW at ~30 mm/yr (Figure 9) and ~38 mm/yr (Figure 8*b*). These velocities swing westward in the western part of the region, and both the westward and the southward components of velocity increase toward the south. The main features of the velocity fields in these figures are an E-W shortening in the northern Aegean, combined with an E-W right-lateral shear, a N-S extension, and a clockwise rotation of the western seaboard of Greece.

5. PRELIMINARY GEODETIC RESULTS

Space-based geodesy in the Aegean region is still in its infancy, but is starting to produce measurements that can be compared with nongeodetic estimates of the deformation. Billiris et al (1991) used GPS receivers to reoccupy a triangulation network installed in central Greece in 1890–1900. They estimated an extension rate of 11–14 mm/yr across the main graben systems of central Greece—comparable with that calculated in this position from the slat model in Figure 8*b* (Taymaz et al 1991b). The actual velocities of points in central Greece relative to Eurasia are not well resolved because the original triangulation survey measured angles only and had no reference frame with which to establish orientation.

Satellite Laser Ranging (SLR) measurements in the Aegean began in 1986, and of the six mobile stations in Greece (Figure 10), all have been occupied four times between 1986 and 1992 except for Karitsa in the NW, which has been occupied only twice. Various groups have analyzed these SLR data (Robbins et al 1992, 1993; Smith et al 1993; Noomen et al 1993; Sahin et al 1993), and all produce similar results. The actual measurements are of baseline length changes between stations, and an example (from Smith et al 1993) is shown in Figure 10. Since the SLR observations in Greece are made simultaneously with other observations elsewhere, it is possible to convert these length changes into velocities relative to some other reference frame. The velocities in Figure 11 are calculated relative to a frame defined by SLR stations in stable Europe. The general features of the SLR baseline length changes and velocities in Figures 10 and 11 are similar those in the velocity field of Figure 9, though the SLR velocities are somewhat higher, which is expected from the way in which Figure 9 was calculated from the seismicity (see above). In particular, the SLR velocities in the SW Aegean are ~33–36 mm/yr towards the SW, relative to Eurasia. Figure 10 shows one feature that is not contained in the velocity field in Figure 9: an E-W extension of ~20 mm/yr across parts of the relatively aseismic southern Aegean Sea. This extension is manifest by N-S striking normal faults along the overriding margin of the Hellenic Trench (Angelier et al 1982, Armijo et al 1992) and in the earthquake focal mechanisms (Figure 5). It is also expected if the divergence of slip vectors on thrust faults in the trench, which change from an azimuth of ~205° near Crete (Taymaz et al 1990) to ~230° near 38°N 21°E, is accommodated in the overriding Aegean lithosphere (Hatzfeld et al 1993), though the baseline between Crete and the Peloponnese shows anomalously little change (Figure 10).

Figure 11 also shows velocities calculated from GPS measurements in western Turkey by Oral et al (1993). These are relative to the SLR site at Yigilca in Turkey (41.0°N, 31.5°E), which is not quite in the same frame as the other SLR velocities in Figure 11, as all analyses of SLR data show that Yigilca has a small (~4 mm/yr) velocity relative to Europe. The GPS velocities are based on only two sets of measurements over a two-year period and are certainly preliminary. They show similar orientations to those derived from the seismicity in Figure 9 but are directed WSW rather than SW. Their magnitudes are larger than expected, given the SLR velocities in SW Greece.

Various geodetic projects are currently underway in the Aegean, and will surely modify details of our present understanding of the deformation. However, when we compare the preliminary geodetic estimates of the velocity of the southern Aegean relative to Europe with earlier nongeodetic estimates (Table 1), it is clear that these earlier estimates were not fundamentally wrong. They differ from the SLR velocities in ways that are expected. The velocities determined from the seismicity are too small

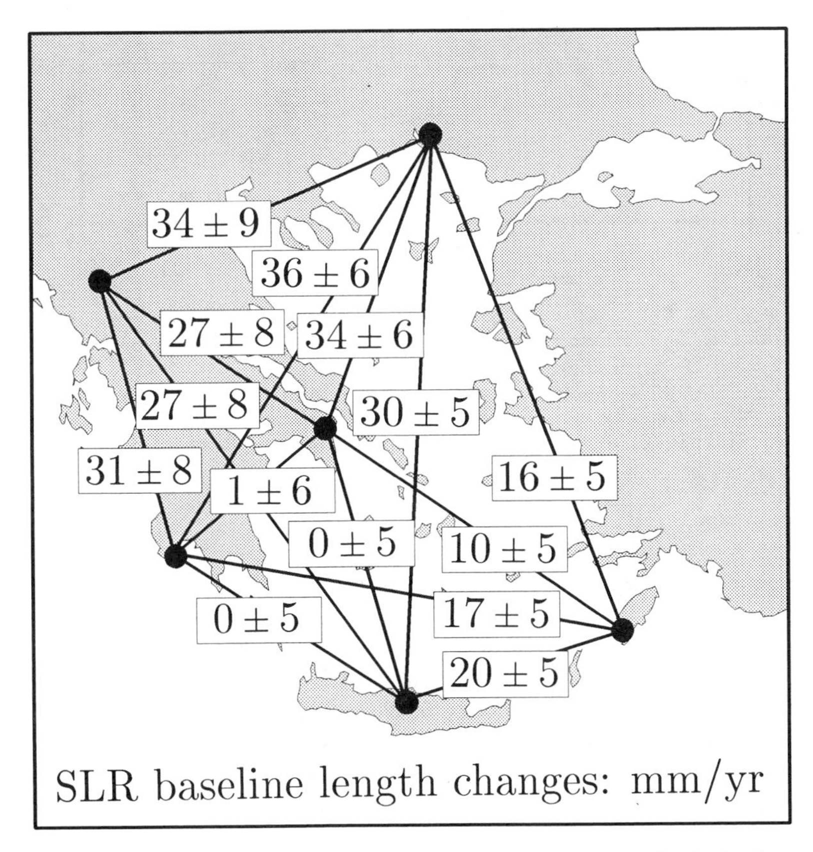

Figure 10 Measured rates of change in the length of lines joining SLR sites in the Aegean (from Smith et al 1993). All sites have been occupied at least four times between 1986 and 1992, except for the station in the NW (Karitsa), which has been occupied only twice.

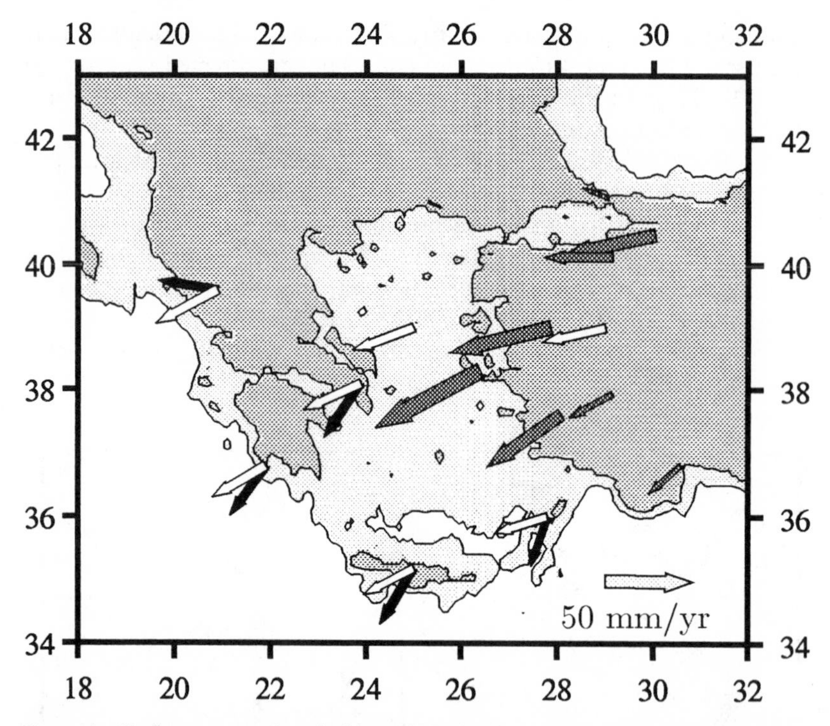

Figure 11 Black arrows are the velocities of SLR sites in Greece and NW Turkey, calculated relative to a frame in northern Europe (Smith et al 1993). Shaded arrows are velocities of GPS sites in Turkey, relative to the SLR site at Yigilca in Turkey (41.0°N, 31.5°E), from Oral et al (1993). The white arrows are the velocities predicted at various points if they were on a Turkish "plate" rotating anticlockwise relative to Europe about the pole given by Jackson & McKenzie (1984). The scale is given by the arrow pointing east in the lower right corner.

because of the way in which moments were estimated, because only earthquakes of $M_s \geqslant 6.0$ were used, and because there may be some contribution from aseismic creep. The E-W extension on the southern margin of the Aegean has not been expressed in earthquakes of $M_s \geqslant 6.0$ this century, so is not manifest in the velocity field in Figure 9, but it is nonetheless clear from the smaller earthquakes and observed faulting (Armijo et al 1992). The early agreement between the geodetically and nongeodetically estimated motions supports the following assertions:

1. That the bulk of the motion in the north and central Aegean regions is expressed by slip on faults in earthquakes of $M_s \geqslant 6.0$ over the past

century (Jackson & McKenzie 1988a,b; Taymaz et al 1991b), and that any aseismic motion has roughly the same orientation as the seismic motion.

2. That McKenzie's (1972) analysis of the motions in the eastern Mediterranean—though simplistic, based on limited data, and on an edifice of assumptions—was basically correct.
3. That most of the shortening in the Hellenic Trench occurs aseismically (Jackson & McKenzie 1988a).

6. RELATIONS BETWEEN FAULTING AND FLOW IN THE LITHOSPHERE

The most active part of the region is the north and central Aegean Sea and its surrounding land areas. Figure 9 shows an estimate of the velocity field that represents the overall deformation of the lithosphere, and Figure 8 shows schematically how that deformation is accommodated discontinuously by faulting in the upper seismogenic crust. We have thus provided preliminary answers to the two questions asked at the end of Section 2. What can we do with this information?

To begin with, we can ask why the NE-SW right-lateral shear that crosses the north Aegean is accommodated in central Greece in so complicated a fashion (by normal faulting that rotates clockwise), and not by a continuation of the strike-slip faulting through central Greece to join up with the Hellenic Trench, as envisaged by McKenzie (1972) and indicated in Figure 1. The answer to this probably lies in the strong structural fabric imparted to Greece by the Mesozoic and early Tertiary shortening, which occurred mostly on thrusts that now have a NW strike and dip to the NE. In eastern mainland Greece, where most of the active normal faults dip NE, it is conceivable that this structural grain was easier to reactivate than to truncate by new faults (McKenzie & Jackson 1983), though there is no clear evidence for reuse of old thrust planes by normal faults, at least at shallow depths. But this suggestion of reactivation cannot be the whole story. Kinematically, it would be possible for the normal faults to move with the same slip vector azimuth as the strike-slip faults in the northern Aegean, in the manner of ridge-transform fault systems in the oceans. In fact, the slip vector azimuths on the normal faults are different from those on the strike-slip faults (Figure 6), so that the normal faults can only achieve the overall motion if they and the blocks they bound rotate clockwise about a vertical axis. Why should they do this?

One conceptual framework for viewing the rotation of blocks in the upper crust is to imagine they are rigid floating inclusions responding to forces on their bases that result from the flow in the lithosphere beneath

(McKenzie & Jackson 1983). The instantaneous rotation rate of such bodies about a vertical axis depends on their orientation and shape as well as on both the symmetric and antisymmetric components of the velocity field (Lamb 1987). At present, and until densely-spaced geodetic measurements become available, measurements of paleomagnetic declinations in rocks provide the only estimates of the rotation rates of crustal blocks in the Aegean. These measurements are not ideal: Suitable rocks are not widespread, and a measureable rotation is usually obtained only for rocks several million years old, so that the rotation rates are averaged over time, and thus are not instantaneous rates. The paleomagnetic data are reviewed by Kissel & Laj (1988) and Kissel et al (1989). Within the error of these measurements, the rotation rates about a vertical axis in the north and central Aegean region are matched in sense and magnitude by the calculated rotation rates of elongated rigid blocks (resembling sticks) aligned parallel to the major faults and floating in the velocity field of Figure 9 (Jackson et al 1992). Jackson & Molnar (1990) show that the same conceptual framework adequately predicts the observed rotation rates in the western Transverse Ranges of southern California. Such a point of view is obviously simplistic: The crustal blocks are not isolated inclusions in a fluid, nor, to geologists in the field, do they appear to be perfectly rigid. The assumption of rigidity is probably not a serious error: Although some minor faulting occurs within the blocks bounded by the major faults, it is probable (and has been argued above) that most of the deformation occurs by slip on the large faults, which are organized into coherent patterns. The assumption that the blocks are isolated implies that the interaction between adjacent blocks is less important than their interaction with the flow field beneath them, and that the faults themselves are relatively weak.

We have here the makings of a coherent picture. The distributed WSW-ENE right-lateral shear between Turkey and Eurasia does not continue through central Greece on strike-slip faults of this strike because the pre-existing structural grain of Greece is so strong that it is easier to move faults with a NW to WNW strike. These faults are themselves weak, but bound relatively strong elongated blocks that rotate clockwise in the underlying continuous velocity field. In order to achieve the overall WSW motion, the slip vectors between the blocks in central Greece must have an azimuth counterclockwise of this; in fact, they vary between SSW and SSE (Figure 6). In this picture, the upper crust responds passively to the flow field beneath, which controls the rotation rate of the fault-bounded blocks, but the structural anisotropy of the upper crust influences the deformation to the extent that it controls the orientation of the faults that become activated (or reactivated). But this, too, may not be the whole story. A hint that some other effect may play a part is contained in the observation that the velocity fields in Figures 8*b* and 9 are rather similar, particularly in their central and western parts.

The slat model used to calculate Figure 8*b* contains the obviously artificial constraint that the slats cannot change in length. Given a strain rate field in which the shear part of the horizontal components exceeds the dilatational part, as it does throughout the deforming area in Figure 9, there will be two horizontal directions in which the deformation produces no change in length. It is possible to reproduce the horizontal components of the strain rate tensor by faulting with strikes parallel to either of these directions (Holt & Haines 1993). Jackson et al (1992) show that in the velocity field of Figure 9 one or other of these directions of zero length change, which vary with position, closely matches the observed strikes of the active faults in the north-central Aegean. Although the slips on faults in earthquakes were used to obtain the velocity field, the argument is not entirely circular: The velocity field is obtained from averaged and smoothed strain rates, and the requirement that the velocity field is continuous constrains the way in which the strain rates can vary spatially. Using the same technique, Holt & Haines (1993) obtained a similar result from seismic strain rates in central Asia, where they were able to reproduce the curvature of major strike-slip faults in eastern Tibet. However, we expect the fault blocks, which are elongated parallel to these directions of zero length change, to be rotating about a vertical axis, and more importantly *to have already rotated* by a finite amount. If the blocks have rotated and the directions of zero length change have also rotated to follow them, this is a first indication that structural anisotropy in the upper crust, which probably controls the orientation of the faults that form, may also influence the nature of the average velocity field in the rest of the lithosphere beneath (Jackson et al 1992, Lamb 1993). This conclusion has implications for the dynamics of the deformation in the Aegean.

7. DYNAMICS OF THE DEFORMATION

McKenzie (1972) realized that the present westward motion of Turkey relative to Eurasia could be maintained by the potential energy difference between the thick crust in eastern Turkey and the Caucasus and the low elevations of the Aegean and western Mediterranean. The presence of thrust surfaces in the Hellenic Trench allows the Aegean to be pushed easily over the Mediterranean sea floor, which is lower and denser as it is oceanic in character (de Voogd et al 1992). Turkey is presumably not pushed over the equally deep (and probably oceanic) Black Sea, because there is no equivalent surface of weakness between the two. As well as the buoyancy force arising from crustal thickness contrasts, there is the possible effect of the downgoing slab in the Hellenic Trench itself, which, if it sinks with a component of velocity out of its plane, can provide a south-directed force that contributes to the extension of the Aegean (Le Pichon 1982, McKenzie & Yılmaz 1991). The relative importance of the push from the east and the pull from the south is uncertain. Although Le Pichon & Angelier (1979) considered the trench effects to be dominant, that view has been modified by the geodetic results (e.g. Figure 11) which suggest that the major contribution to the motion of the southern Aegean relative to Eurasia comes from the Turkey-Eurasia motion, rather than from the southward migration of the subduction zone towards Africa (Le Pichon et al 1993). Nonetheless, SLR-determined velocities (Figure 11) and earthquake slip vectors in both the north Aegean (Taymaz et al 1991b) and the Hellenic Trench near Crete (Taymaz et al 1990) are directed more to the south than would be expected from the Turkey-Eurasia motion alone. Significant N-S extension manifestly occurs in the Aegean. Clearly, something is happening in addition to the westward motion of Turkey.

Taymaz et al (1991b) point out that north of about 39°N the shortening along the western seaboard of Greece and Albania changes from a continent-ocean subduction zone to a continent-continent collision between Greece and the Apulian platform to the west (Figure 12). The thickened crust in this collision belt will act to resist the westward motion of Turkey and prevent the NW seaboard of Greece from rotating clockwise sufficiently quickly to accommodate the distributed right-lateral shear between Turkey and Europe, leading to E-W shortening in the northern Aegean. Taymaz et al (1991b) suggest that the geometry of the faulting in the northern Aegean resembles that of a single gigantic kink fold with a vertical hinge axis, which forms as the E-W shortening is partially accommodated by N-S extension (analogous to the slats in Figure 7), but without preserving surface area as crustal thinning also occurs. In this scheme, the continental collision in NW Greece and Albania plays an essential role, forcing material towards the south and over the Mediterranean sea floor. Le Pichon & Angelier (1979) and Angelier et al (1982) produced a reconstruction of the Aegean before the onset of the present-day extensional regime. This was based mainly on the length of the seismically active subducted slab in the Hellenic Trench (which provided an estimate of the total extension) and bathymetry variations in the Aegean (which provided an estimate of the spatial variations of crustal thickness and extension). Their reconstruction restores the convex bulge of the Aegean arc to an essentially linear feature. Perhaps this feature, which presumably contained a strong structural grain from the Mesozoic and early Tertiary shortening, has indeed "buckled" to form the present kink-like geometry seen in Figures 4, 8, and 12, modified by arc-parallel extension along the southern margin.

This possibility of a gigantic kink fold returns us to the discussion in the previous section. If the velocity field has changed with time to maintain the directions of zero length change parallel to the limbs of the kink fold, then clearly the upper crust can exert an important influence on the deformation of the lithosphere as a whole, and the view that the upper

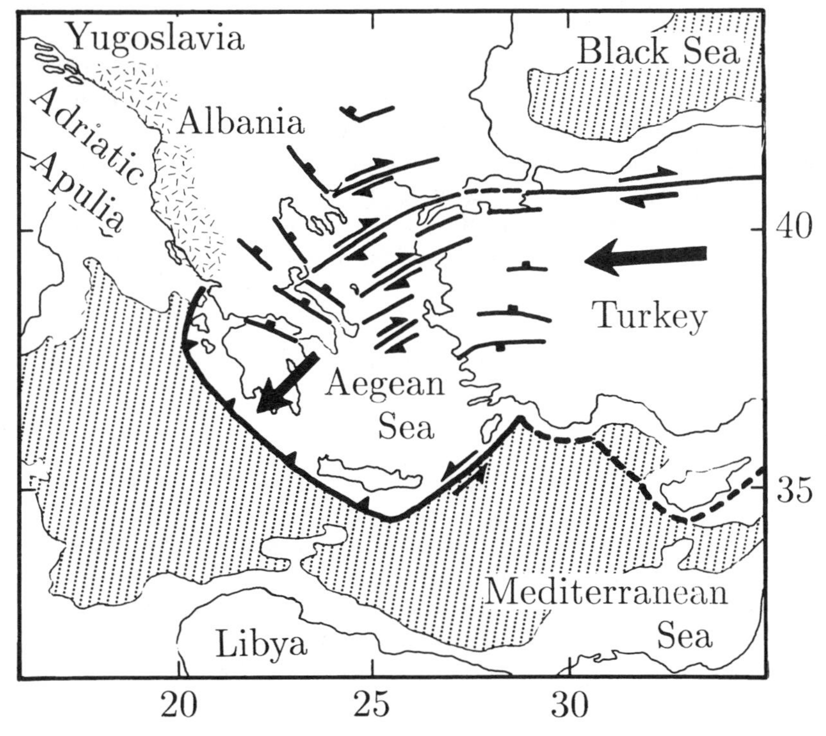

Figure 12 Summary sketch of the active tectonics in the Aegean region. Faults are shown simplified. The shaded regions of the Mediterranean and Black Seas have water depths greater than 1000 fathoms (1800 m) and are approximately coincident with those regions underlain by oceanic crust. Continental crust is white. Large arrows show the westward motion of Turkey and SW motion of the southern Aegean, both relative to Eurasia. The stippled region of coastal Albania and Yugoslavia is where active continental shortening and thickening are occurring, caused by the collision of this region with the Adriatic and Apulia platforms. It is this collision that resists the westward motion of Turkey and forces the Aegean Sea to the SW, where it can easily override the Mediterranean oceanic crust in the Hellenic Trench.

crust simply responds in a passive way to the flow beneath it is too naive. Is the location of the abrupt change from strike-slip to normal faulting in the western Aegean too far away from the collision in NW Greece and Albania to be an effect of the viscous stresses induced by this resisting margin (see e.g. England et al 1985), or must the stresses be transmitted in the upper crust along the limbs of the kink fold? This question is not easily answered: The viscous stresses may be modified by topography, and the abrupt change in the faulting, although it occurs in the western north Aegean, is not perfectly aligned between all limbs of the possible kink fold. There may also have been pre-existing transverse weaknesses in the originally linear structural belt that predetermined where the kink would initiate.

It is probable that the westward motion of Turkey, the pull from the sinking slab in the trench, the resistance to shortening in the NW, and the strength of the pre-existing anisotropy of the crust all play some part in the dynamics. The clues to their relative importance may lie in the geological history and the sequence of deformational events (see e.g. Şengör et al 1985). A review of that history is not appropriate here, but the question of how long the present motions have been operating for, is one that is worth asking.

8. SOME ASPECTS OF THE NEOGENE GEOLOGICAL HISTORY

Geological observations permit only an imperfect reconstruction of tectonic history and in some cases allow ambiguous interpretations: For example, does the onset of shortening in the western Hellenic arc necessarily imply the onset of subduction beneath it? Much of the evidence for the onset of Neogene extension or strike-slip motion in Greece and Turkey relies on the biostratigraphic dating of terrestrial or lacustrine sediments and is not well constrained. The Aegean region contains widespread Neogene volcanic rocks, some of which are related to subduction (Fytikas et al 1984, Paton 1993), but not necessarily beneath the currently active arc. Nevertheless, most authors believe that the present-day association of strike-slip motion in Turkey, extension in the Aegean, and subduction in the Hellenic Trench was established, perhaps in a slightly different form, by about 10–16 Ma ago. The North Anatolian Fault appears to have begun to move about 10 Ma ago in the late Serravallian to Tortonian (Şengör 1979), and the earliest sediments in the graben systems of western Turkey are at least this old (Seyitoğlu & Scott 1991, 1992; Seyitoğlu et al 1992). Le Pichon & Angelier (1979) estimated that the present subduction system began about 13 Ma ago, based on the age of the oldest volcanic rocks in the present arc and an assumption about the shape of the underlying slab. It is immediately apparent that there are problems reconciling these dates with the present day rates of motion.

The length of the seismically active part of the slab in the Hellenic subduction zone is about 300 km. All of this material could have been subducted in the last 5–6 Ma at the present convergence rate of 50–60 mm/yr in the Hellenic Trench. If the length of this slab corresponds roughly to the amount of extension in the overriding Aegean Sea, as envisaged by Le Pichon & Angelier (1979), then all of this extension could have occurred in that time. The probable present-day slip rate of 30–40 mm/yr on the North Anatolian Fault, if operative over 5–6 Ma, would produce an offset of 150–240 km, which is greater than the total known offset of about 85 km (Şengör 1979), and far greater than the post–early Pliocene offset of 25 km estimated by Barka & Hancock (1984). It is appealing to suggest that the rapid present-day rates of motion were established within in the last 5 Ma, as suggested by Mercier et al (1989) on geological grounds, and by Kissel & Laj's (1988) interpretation of paleomagnetic declination data, though this does not diminish the problem of the offset on the North Anatolian Fault. Why should the rates increase in this way? Assuming that the extension in the Aegean is related to the motion of Turkey, we once again return to the geometry of the deformation in eastern Turkey. In Figure 3*a* the magnitude of Turkey's westward motion relative to Europe depends on the azimuth of the strike-slip motion on the East Anatolian fault zone. McKenzie & Yılmaz (1991) liken this geometry to the expulsion of a pip between the two closing blades of a pair of scissors. If the angle between the North and East Anatolian fault zones has decreased with time, the expulsion velocity of the pip (Turkey) would have increased. The angle between these strike-slip faults can only decrease if there is some deformation of the material between them. This may have occurred by earlier motion on several NE-SW left-lateral strike-slip faults that are sub-parallel and NW of the East Anatolian Fault (Şengör et al 1985, Dewey et al 1986, Perinçek et al 1987). Motion on a parallel set of these faults may have allowed them to rotate clockwise, thereby decreasing the angle between the North and East Anatolian fault zones (Figure 13), but there is no direct evidence that this occurred.

The possibility that the present geometry of the deforming Aegean region may have been established relatively recently does not explain the significance of much earlier extension $\sim$25 Ma ago in the southern Aegean (Jolivet et al 1993), and of earlier subduction that might be revealed in tomographic studies (Spakman et al 1988). Nor does it explain why the southernmost Aegean Sea north of Crete, which is the deepest and most

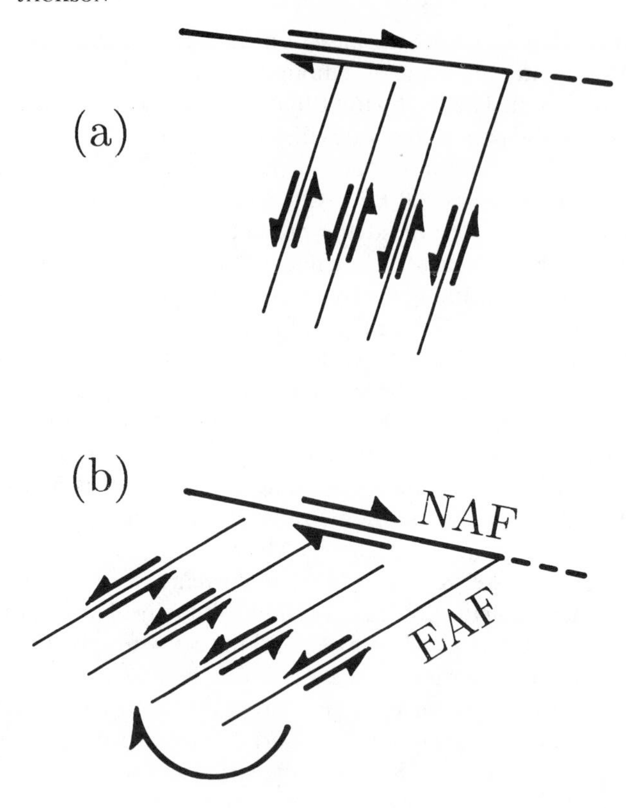

Figure 13 Cartoon to illustrate how the angle between the North (NAF) and East (EAF) Anatolian faults might have decreased through time if some of the faults that are sub-parallel to the East Anatolian Fault were also active, allowing the left-lateral set of faults to rotate clockwise.

extended part of the region (Makris 1978), is virtually aseismic and possibly inactive today (Sonder & England 1989, Wdowinski et al 1989). Jolivet et al (1993) emphasize that Miocene stretching lineations within ductilely deformed rocks in the southern Aegean are similar in azimuth to the earthquake slip vectors farther north in the central Aegean. The Miocene lineations may well have rotated and changed orientation since they formed. Is this another indication that the presently active deformation is following an older structural fabric?

9. CONCLUSIONS

Two fundamental questions in continental tectonics concern the velocity field that describes the deformation of the lithosphere at large length scales and how this velocity field is accommodated by faulting in the upper crust. The rapid rates of deformation and the abundance of observational data make the Aegean Sea a particularly suitable region in which to address these questions. Much of this review has been concerned with the relation between the velocity field and the faulting, and with what that tells us about the interaction between the upper crust and the creeping lithosphere beneath it. This interaction is perhaps more complex than was originally thought. It is probable that the structural anisotropy of the upper crust influences the direction in which faults form or become reactivated, and that the overall velocity field may adjust with time in order to keep these faults active as they and the blocks they bound rotate. The mechanical anisotropy of the lithosphere may not be restricted to the upper crust, where it is most visible and obvious, but may extend into the mantle lithosphere as well. The discussion of this topic is necessarily tentative, as it is based on velocity fields that are themselves preliminary and may well improve (or change) with more abundant geodetic data. Nonetheless, one conclusion is firm: that space-based geodesy will not make redundant the geological and seismological observations on which most of our understanding has so far been based. The velocity field is only part of the picture: Knowledge of how faulting achieves the overall motions is essential to our understanding of continental tectonics.

ACKNOWLEDGMENTS

I thank P. Cross, D. Hatzfeld, L. Jolivet, X. Le Pichon, B. Oral, and R. Reilinger for sending me preprints of their work, and P. England and D. McKenzie for helping me with some of the figures. I am particularly grateful to J. Robbins and D. E. Smith for making available their SLR results, which are used in Figures 10 and 11. My views about the tectonics of the Aegean have been greatly influenced by innumerable discussions with N. Ambraseys, P. England, A. J. Haines, D. Hatzfeld, W. Holt, C. Kissel, C. Laj, B. Parsons, X. Le Pichon, D. McKenzie, J. Mercier, P. Molnar, A. M. C. Şengör, T. Taymaz, and A. D. Smith. I thank them all for their generosity, but emphasize that the views expressed here are my own, and that I alone am responsible for any errors or misconceptions. This review is Cambridge Earth Sciences contribution 3314.

Literature Cited

Ambraseys NN. 1970. Some characteristic features of the North Anatolian Fault zone. *Tectonophysics* 9: 143–65

Ambraseys NN. 1989. Temporary seismic quiescence: SE Turkey. *Geophys. J. Int.* 96: 311–31

Angelier J, Goguel J. 1979. Sur une méthode simple de détermination des axes principaux des contraintes pour une population de failles. *C. R. Acad. Aci. Paris* 288: 307–10

Angelier J, Lyberis N, Le Pichon X, Huchon P. 1982. The tectonic development of the Hellenic Trench and Sea of Crete: a synthesis. *Tectonophysics* 86: 159–96

Armijo R, Lyon-Caen H, Papanastassiou D. 1992. East-west extension and Holocene normal-fault scarps in the Hellenic arc. *Geology* 20: 491–94

Barka A, Hancock P. 1984. Neotectonic deformation patterns in the convex-northwards arc of the North Anatolian fault zone. See Dixon & Robertson 1984, pp. 763–74

Barka A, Kadinsky-Cade K. 1988. Strike-slip fault geometry in Turkey and its influence on earthquake activity. *Tectonics* 7: 663–84

Billiris H, Paradissis D, Veis G, England P, Featherstone W, et al. 1991. Geodetic determination of the strain of Greece in the interval 1900 to 1988. *Nature* 350: 124–29

Brace WF, Kohlstedt DL. 1980. Limits on lithosphere stress imposed by laboratory experiments. *J. Geophys. Res.* 85: 6248–52

Braunmiller J. 1991. *Down-dip geometry and depth extent of normal faults in the Aegean—evidence from earthquakes*. MS thesis. Oregon State Univ. 225 pp.

Carey E, Brunier B. 1974. Analyse théorique et numerique d'un modèle mécanique élémentaire appliqué a l'etude d'une population de failles. *C. R. Acad. Aci. Paris* 279: 891–94

Collier RE. 1990. Eustatic and tectonic controls upon Quaternary coastal sedimentation in the Corinth Basin, Greece. *J. Geol. Soc. London* 147: 301–14

de Mets C, Gordon RG, Argus DF, Stein S. 1990. Current plate motions. *Geophys. J. Int.* 101: 425–78

de Voogd B, Truffert C, Chamot-Rooke N, Huchon P, Lallemant S, Le Pichon X. 1992. Two-ship deep seismic soundings in the basins of the Eastern Mediterranean Sea (Pasiphae cruise). *Geophys. J. Int.* 109: 536–52

Dewey JF, Hempton MR, Kidd WSF, Şaroğlu F, Şengör AMC. 1986. Shortening of continental lithosphere: the neotectonics of Eastern Anatolia—a young collision zone. In *Collision Tectonics*, ed. MP Coward, AC Ries, *Geol. Soc. London Spec. Publ.* 19: 3–36

Dewey JF, Şengör AMC. 1979. Aegean and surrounding regions: complex multiplate and continuum tectonics in a convergent zone. *Bull. Geol. Soc. Am.* 90: 84–92

Dixon JE, Robertson AHF, eds. 1984. *The Geological Evolution of the Eastern Mediterranean, Geol. Soc. London Spec. Publ.* 17

Ekström GA, Dziewonski A. 1988. Evidence of bias in the estimation of earthquake size. *Nature* 332: 319–23

Ekström GA, England PC. 1989. Seismic strain rates in regions of distributed continental deformation. *J. Geophys. Res.* 94: 10,231–57

England PC, Houseman GA, Sonder LJ. 1985. Length scales for continental deformation in convergent, divergent and strike-slip environments: analytical and approximate solutions for a thin viscous sheet model. *J. Geophys. Res.* 90: 3351–57

England PC, Jackson JA. 1989. Active deformation of the continents. *Annu. Rev. Earth Planet. Sci.* 17: 197–226

England PC, McKenzie D. 1982. A thin viscous sheet model for continental deformation. *Geophys. J. R. Astron. Soc.* 70: 295–321; also correction to the above. 1983. *Geophys. J. R. Astron. Soc.* 73: 523–32

Eyidoğan H. 1988. Rates of crustal deformation in western Turkey as deduced from major earthquakes. *Tectonophysics* 148: 83–92

Fytikas M, Innocenti F, Manetti P, Mazzuoli R, Peccerillo A, Villari L. 1984. Tertiary to Quaternary evolution of volcanism in the Aegean region. See Dixon & Robertson 1984, pp. 687–99

Haines AJ. 1982. Calculating velocity fields across plate boundaries from observed shear rates. *Geophys. J. R. Astron. Soc.* 68: 203–9

Hatzfeld D, Besnard M, Makropoulos K, Hatzidimitriou P. 1993. Microearthquake seismicity and fault plane solutions in the southern Aegean and its tectonic implications. *Geophys. J. Int.* In press

Hatzfeld D, Martin C. 1992. Intermediate depth seismicity in the Aegean defined by teleseismic data. *Earth Planet. Sci. Lett.* 113: 267–75

Holt WE, Haines AJ. 1993. Velocity field in deforming Asia from inversion of earthquake released strains. *Tectonics* 12: 1–20

Holt WE, Ni JF, Wallace TC, Haines AJ. 1991. The active tectonics of the eastern Himalayan syntaxis and surrounding regions. *J. Geophys. Res.* 96: 14,595–632

Jackson JA. 1985. Active normal faulting and crustal extension. In *Continental Extensional Tectonics*, ed. MP Coward, JF Dewey, PL Hancock, *Geol. Soc. London Spec. Publ.* 28: 3–17

Jackson JA, Haines AJ, Holt WE. 1992. The horizontal velocity field in the deforming Aegean Sea region determined from the moment tensors of earthquakes. *J. Geophys. Res.* 97: 17,657–84

Jackson JA, McKenzie D. 1984. Active tectonics of the Alpine-Himalayan Belt between western Turkey and Pakistan. *Geophys. J. R. Astron. Soc.* 77: 185–264

Jackson JA, McKenzie D. 1988a. The relationship between plate motions and seismic moment tensors, and the rates of active deformation in the Mediterranean and the Middle East. *Geophys. J.* 93: 45–73

Jackson JA, McKenzie D. 1988b. Rates of active deformation in the Aegean Sea and surrounding areas. *Basin Res.* 1: 121–28

Jackson JA, Molnar P. 1990. Active faulting and block rotations in the western Transverse ranges, California. *J. Geophys. Res.* 95: 22,073–87

Jackson JA, White NJ. 1989. Normal faulting in the upper continental crust: observations from regions of active extension. *J. Struct. Geol.* 11: 15–36

Jestin F, Huchon P, Gaulier JM. 1993. The Somalia plate and the East African rift system: present kinematics. *Geophys. J. Int.* In press

Jolivet L, Brun JP, Gautier S, Lallemand S, Patriat M. 1993. 3-D kinematics of extension in the Aegean from the early Miocene to the present: insight from the ductile crust. *Earth Planet. Sci. Lett.* In press

Ketin I. 1948. Uber die tektonisch-mechanischen Folgerungen aus den grossen anatolischen Erdbeden des letzten Dezenniums. *Geol. Rundsch.* 36: 77–83

Kissel C, Laj C. 1988. The Tertiary geodynamical evolution of the Aegean arc: a paleomagnetic reconstruction. *Tectonophysics* 146: 183–201

Kissel, C, Laj C, Poisson, A, Simeakis K. 1989. A pattern of block rotations in central Aegea. In *Paleomagnetic Rotations and Continental Deformation*, ed. C Kissel, C Laj, pp. 115–29. London: Kluwer

Kostrov VV. 1974. Seismic moment and energy of earthquakes, and seismic flow of rock. *Izv. Acad. Sci. USSR Phys. Solid Earth* (English transl.) 1: 23–44

Lamb SH. 1987. A model for tectonic rotations about a vertical axis. *Earth Planet. Sci. Lett.* 84: 75–86

Lamb SH. 1993. Going with the flow. *Nature* 362: 294–95

Leeder MR, Jackson JA. 1993. The interaction between normal faulting and drainage in active extensional basins, with examples from the western United States and Greece. *Basin Res.* 5: 79–102

Le Pichon X. 1982. Land-locked oceanic basins and continental collision: the eastern Mediterranean as a case example. In *Mountain Building Processes*, ed. K Hsu, pp. 201–11. London: Academic

Le Pichon X, Angelier J. 1979. The Hellenic arc and trench system: a key to the evolution of the Eastern Mediterranean area. *Tectonophysics* 60: 1–42

Le Pichon X, Angelier J. 1981. The Aegean Sea. *Phil. Trans. R. Soc. London Ser. A* 300: 357–72

Le Pichon X, Chamot-Rooke N, Huchon P, Luxey P. 1993. Implications des nouvelles mesures de géodésie spatiale en Grèce et en Turquie sur l'extrusion latérale de l'Anatolie et de l'Égée. *C. R. Acad. Aci. Paris* 316: 983–90

Lybéris, N. 1984. Tectonic evolution of the North Aegean Trough. See Dixon & Robertson 1984, pp. 708–25

Lybéris N, Yürür T, Chorowicz J, Kasapoğlu E, Gündoğdu N. 1992. The East Anatolian Fault: an oblique collisional belt. *Tectonophysics* 204: 1–15

Makris J. 1978. The crust and upper mantle of the Aegean region from deep seismic soundings. *Tectonophysics* 46: 269–84

Mascle J, Martin L. 1990. Shallow structure and recent evolution of the Aegean Sea: a synthesis based on continuous reflection profiles. *Mar. Geol.* 94: 271–99

McKenzie D. 1970. The plate tectonics of the Mediterranean region. *Nature* 226: 239–43

McKenzie D. 1972. Active tectonics of the Mediterranean region. *Geophys. J. R. Astron. Soc.* 30: 109–85

McKenzie D. 1978a. Active tectonics of the Alpine-Himalayan belt: the Aegean Sea and surrounding regions. *Geophys. J. R. Astron. Soc.* 55: 217–54

McKenzie D. 1978b. Some remarks on the development of sedimentary basins. *Earth Planet. Sci. Lett.* 40: 25–32

McKenzie D, Jackson JA. 1983. The relationship between strain rates, crustal thickening, paleomagnetism, finite strain and fault movements within a deforming zone. *Earth Planet. Sci. Lett.* 65: 182–202, also correction to the above. 1984. *Earth Planet. Sci. Lett.* 70: 444

McKenzie D, Jackson JA. 1986. A block model of distributed deformation by faulting. *J. Geol. Soc. London* 143: 249–53

McKenzie D, Yılmaz Y. 1991. Deformation and volcanism in western Turkey and the Aegean. *Bull. Tech. Univ. Istanbul* 44: 345–73

Mercier JL, Carey E, Philip HR, Sorel D.

1976. La néotectonique plio-quaternaire de l'arc egéen externe et de la Mer égéen et ses relations avec séismicité. *Bull. Soc. Geol. Fr.* 18: 159–76
Mercier JL, Delibassis N, Gautier A, Jarrige JJ, Lemille F, et al. 1979. La néotectonique de l'arc égéen. *Rev. Geol. Dyn. Geogr. Phys.* 21: 67–92
Mercier JL, Sorel D, Vergely P, Simeakis K. 1989. Extensional tectonic regimes in the Aegean basins during the Cenozoic. *Basin Res.* 2: 49–71
Noomen R, Ambrosius B, Wakker K. 1993. Crustal motions in the Mediterranean region determined from laser ranging to Lageos. In *Crustal Dynamics Project, Geophys. Monogr. Am. Geophys. Union.* In press
Oral M, Reilinger R, Toksöz, M. 1992. Deformation of the Anatolian block as deduced from GPS measurements. *Eos Trans. Am. Geophys. Union, Fall AGU Meet. Suppl.* 73: 120 (Abstr.)
Oral M, Reilinger R, Toksöz M, Barka. 1993. Preliminary results of 1988 and 1990 GPS measurements in western Turkey and their tectonic implications. In *Crustal Dynamics Project, Geophys. Monogr. Am. Geophys. Union* In press
Oral M, Reilinger R, Toksöz M, Barka A, Kinık I. 1991. Preliminary results of 1988 and 1990 GPS measurements in western Turkey. *Eos Trans. Am. Geophys. Union, Fall AGU Meet. Suppl.* 72: 115 (Abstr.)
Parsons B, McKenzie D. 1978. Mantle convection and the thermal structure of the plates. *J. Geophys. Res.* 83: 4485–96
Paton SM. 1993. *The relationship between extension and volcanism in western Turkey, the Aegean Sea, and central Greece.* PhD thesis. Cambridge Univ. 300 pp.
Perinçek D, Günay Y, Kozlu H. 1987. New observations on strike-slip faults in east and southeast Anatolia. *7th Biannual Petroleum Congr. of Turkey*, pp. 89–103. Ankara: Turkish Assoc. Petrol. Geol.
Robbins J, Torrence M, Dunn P, Smith D, Kolenkiewicz R. 1993. Tectonic deformation at European and Mediterranean SLR stations. *Ann. Geophys.* 11: 114 (Abstr.)
Robbins J, Torrence M, Dunn P, Williamson R, Smith D, Kolenkiewicz R. 1992. SLR dtermined velocities in the Mediterranean region: implications for regional tectonic models. *Eos Trans. Am. Geophys. Union, Fall AGU Meet. Suppl.* 73: 122 (Abstr.)
Roberts S, Jackson JA. 1991. Active normal faulting in central Greece: An overview. In *The Geometry of Normal Faults*, ed. AM Roberts, G Yielding, B Freeman, *Geol. Soc. London Spec. Publ.* 56: 125–42
Robertson A, Dixon J. 1984. Introduction: aspects of the geological evolution of the Eastern Mediterranean. See Dixon & Robertson 1984, pp. 1–74
Roussos N, Lyssimachou T. 1991. Structure of the central North Aegean Trough: an active strike-slip deformation zone. *Basin Res.* 3: 37–46
Sahin M, Rands P, Cross P. 1993. Crustal dynamics in Türkiye from the WEGNER/MEDLAS satellite laser ranging data. *Geol. J.* In press
Şengör AMC. 1979. The North Anatolian transform fault: its age, offset and tectonic significance. *J. Geol. Soc. London* 136: 269–82
Şengör AMC, Görür N, Şaroğlu F. 1985. Strike slip faulting and related basin formation in zones of tectonic escape: Turkey as a case study. In *Strike-slip Deformation, Basin Formation, and Sedimentation. Soc. Econ. Paleontol. Mineral. Spec. Publ.* 37: 227–64
Şengör AMC, Yılmaz Y, Sungurlu O. 1984. Tectonics of the Mediterranean Cimmerides: nature and evolution of the western termination of Palaeo-Tethys. See Dixon & Robertson 1984, pp. 77–112
Seyitoğlu G, Scott B. 1991. Late Cenozoic crustal extension and basin formation in west Turkey. *Geol. Mag.* 128: 155–66
Seyitoğlu G, Scott B. 1992. The age of the Büyük Menderes graben (west Turkey) and its tectonic implications. *Geol. Mag.* 129: 239–42
Seyitoğlu G, Scott B, Rundle C. 1992. Timing of Cenozoic extensional tectonics in west Turkey. *J. Geol. Soc. London* 149: 533–38
Smith AG, Moores EM. 1974. The Hellenides. In *Mesozoic and Cenozoic Orogenic Belts*, ed. AM Spencer. *Geol. Soc. London Spec. Publ.* 4: 159–86
Smith DE, Kolenkiewicz R, Robbins JW, Dunn PJ, Torrence MH. 1993. Horizontal crustal motion in the central and eastern Mediterranean inferred from Satellite Laser Ranging measurements. *Geophys. Res. Lett.* In review
Sonder LJ, England PC. 1989. Effects of temperature-dependent rheology on large-scale continental extension. *J. Geophys. Res.* 94: 7603–19
Spakman W, Wortel M, Vlaar N. 1988. The Hellenic subduction zone: a tomographic image and its geodynamical implications. *Geophys. Res. Lett.* 15: 60–63
Taymaz T, Eyidoğan H, Jackson J. 1991a. Source parameters of large earthquakes in the East Anatolian fault zone (Turkey). *Geophys. J. Int.* 106: 537–50
Taymaz T, Jackson JA, McKenzie D. 1991b. Active tectonics of the north and central Aegean Sea. *Geophys. J. Int.* 106: 433–90

Taymaz T, Jackson JA, McKenzie D. 1992. Reply to the comment by R. Westaway on "Active tectonics of the north and central Aegean Sea" by T. Taymaz, J. Jackson and D. McKenzie. *Geophys. J. Int.* 110: 623

Taymaz T, Jackson JA, Westaway R, 1990. Earthquake mechanisms in the Hellenic Trench near Crete. *Geophys. J. Int.* 102: 695–731

Taymaz T, Price S. 1992. The 1971 May 12 Burdur earthquake sequence, SW Turkey: a synthesis of seismological and geological observations. *Geophys. J. Int.* 108: 589–603

Walcott RI. 1984. The kinematics of the plate boundary through New Zealand: a comparison of the short and long term deformations. *Geophys. J. R. Astron. Soc.* 79: 613–33

Wdowinski S, O'Connell RJ, England P. 1989. A continuum model of continental deformation above subduction zones: application to the Andes and the Aegean. *J. Geophys. Res.* 94: 10,331–46

Annu. Rev. Earth Planet. Sci. 1994. 22:273–317

GEOMORPHOLOGY AND IN-SITU COSMOGENIC ISOTOPES

T. E. Cerling

Department of Geology and Geophysics, University of Utah,
Salt Lake City, Utah 84112

H. Craig

Isotope Laboratory, Scripps Institution of Oceanography,
University of California at San Diego, La Jolla, California 92093

KEY WORDS: dating, geochronology, isotopes, landforms, surface processes

> How many years must a mountain exist
> Before it is washed to the sea?
>
> Robert Zimmerman

INTRODUCTION

The answer, of course, is literally "blowing in the wind" for any geomorphologist with a dust-collecting network and infinite patience. But those results are not yet in, and if we wish to make quantitative studies of exposure ages and erosion rates along geomorphic surfaces in various climatic, tectonic, and geographic regimes, it is necessary to have a methodology something like the use of nuclear geochronology for dating the rocks of the Earth's crust. Thus we require a protocol based on nuclear geophysics: Current studies of terrestrial in-situ cosmogenic isotopes are developing precisely such a protocol. Like the early stages of the development of modern geochronology, however, there are sampling and analytical problems, interpretive difficulties, and some distance yet to travel up the mountain before a well-defined and clear science emerges from the forest.

Given our current reasonably good understanding of the present-day

0084–6597/94/0515–0273$05.00

cosmic-ray flux in which the Earth is bathed, the formulation of the fundamental problem is clear. A geomorphic surface, fixed in geomagnetic coordinates on the Earth's surface and with a defined dip and strike, is bombarded by the incident cosmic radiation, creating new (in-situ) isotopes, both stable and radioactive, by an exponentially attenuated flux in the substrate of the surface. This is a "double-moving-boundary process": The surface is in general undergoing erosion or ablation, thus decreasing in altitude, or is increasing in altitude due to sedimentation, etc. Additionally, the entire surface is subject to tectonic motion so that it increases or decreases in altitude due to large-scale motions. Because the incident cosmic-ray flux in the atmosphere is itself attenuated by the atmospheric depth through which it passes en route to the surface, changes in altitude of the surface alter the production rate throughout the range of nucleon production in the matrix.

The initial problem was formulated for stable cosmogenic species by Craig & Poreda (1986a): We give the result here in order to fix our ideas for the above physical features without the complications of radioactive decay. The general equation for, e.g. ^{3}He or ^{21}Ne concentration in the surface layer of the exposed matrix can be written, simply, as

$$C_0(t) = J_0 T^*[1-\exp(-t/T^*)]. \tag{1}$$

Here J_0 is the production rate of the cosmogenic species at the altitude of the surface of the rock ($z = 0$), and the concentration of the species ($C = C_0$ at $z = 0$) is zero at $t = 0$. The "time constant" T^* in (1) is defined (for stable isotopes) as

$$T^* = 1/(U^*/h^* + E/z^*), \tag{2}$$

in which $U^* = (U-E)$, where U is a rate of regional uplift (positive or negative) in an absolute reference frame, and E is the rate of erosion (or accumulation) in the same units (m/My in the original). The remaining terms, h^* and z^*, are attenuation path lengths for the incident cosmic radiation in the atmosphere and in the rock matrix, where we assume for simplicity an exponential behavior for the attenuation in each substance. For the path length for rocks, z^*, this is not a source of error, and the value of 0.55 m for z^* is derived, for basalts, from the value of the atmospheric attenuation per g/cm^2 and the mean density of basalt. In the atmosphere the assumption of exponential attenuation in the lower troposphere is not so readily justified, but because it allows the solution of Equation (1) in closed form, it is very useful. In fact, although neither the atmospheric pressure nor the cosmic-ray production rates in air are exactly exponential, to first order $h^* = 1800$ m, and the approximation is justified up to altitudes of ~4500 m, with ~20% error at 5000 m (Craig & Poreda 1986a).

At $t = 0$, the rock-surface elevation is

$$h(0) = h(t) - U^*t, \tag{3}$$

where "t" is the time since the initial exposure of a pristine surface (e.g. by lava eruption, faulting, etc). The original cosmogenic isotope production rate is then

$$J_0(0) = J_0(t)\exp(-U^*t/h^*), \tag{4}$$

and these equations then suffice to fix the surface-layer ($z = 0$) concentration $C_0(t)$ as a function of time as given by the general parameters U and E. Below the surface, the concentration at any depth z below the surface is readily seen to be simply $C_0 \exp(-z/z^*)$ because of the exponential nature of the attenuation of the production rate.

In Figure 1, the concentration of a cosmogenic stable isotope as a

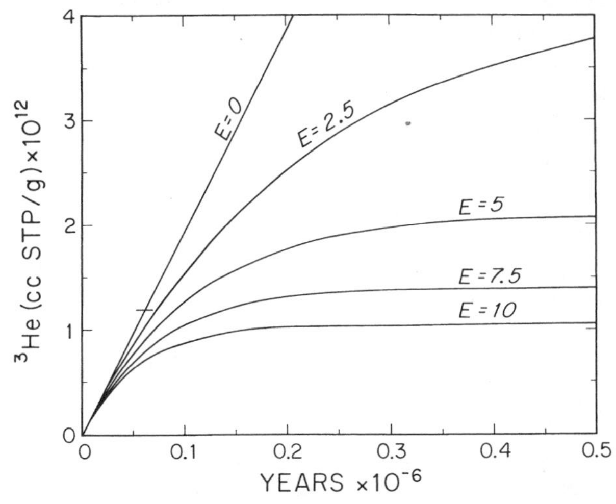

Figure 1 Cosmogenic ^{3}He as a function of time and erosion rate (Craig & Poreda 1986). The curves are calculated for various values of the erosion rate (E: m/My) for the Maui Haleakala lavas, and extend over the ~500,000 y age of these flows. [A regional "uplift" rate $U = -45$ m/My (subsidence) has been used, corresponding to a subsidence of ~40 m during the interval since Maui moved off hotspot, but the value of U has only a minor effect on the E values.] The cosmogenic ^{3}He in the olivines (1.22 $\mu\mu$cc/g $= 35.0 \times 10^6$ atoms/g, shown as a tick mark crossing the $E = 0$ line) corresponds to an "exposure age" (defined for $E = 0$) estimated as 64,000 years in the original paper: using the new calibrated value of the ^{3}He production rate (Cerling & Craig 1993), which is 46% greater than the original assumption, the exposure age is now 44,000 years, and the erosion-rate values (E, m/My) on the curves are now 46% greater. Thus the calculated maximum erosion rate at Haleakala Crater becomes 12.5 m/My (vs the original value of 8.6 m/My).

function of time is plotted for several values of erosion rate (E). The graph is the original display of the cosmogenic ^{3}He data for lavas on the crest of Haleakala Volcano, Maui (Craig & Poreda 1986a), but the figure is general for all similar species. The defining limit of the envelope of concentrations is the straight line for $E = 0$, i.e. no erosion, and t is small vs T^*. In this particular case the measured cosmogenic ^{3}He concentration in the surface lavas intersected locus $E = 0$ at $t = 64{,}000$ years. This is the "exposure age," corresponding to zero erosion rate and small rates of uplift or subsidence: It is the time required to produce the observed concentration of the cosmogenic species (^{3}He in this case) for zero erosion.

The actual exposure age of these lavas, however, is known to be of the order of 500,000 years—almost an order of magnitude greater than the "cosmogenic exposure age" given by the concentration intersect on $E = 0$. The curves in Figure 1, drawn for erosion rates from 2.5 to 10 m/My, show that for finite erosion rates, longer times are required to reach an observed concentration because of the continual removal of the surface (Equation 1): In this case, for long actual exposure ages (since eruption of the lavas), the required erosion rate is 8.6 m/My. The uncertainties are seen to be asymmetrical, but they can hardly be more than 25%, so that this first determination of a "cosmogenic" erosion rate has a reasonable claim to a useful precision. We are currently continuing similar studies on the olivine basalts on Mauna Kea.

In a continuing study of the Maui Haleakala lavas, Craig & Poreda (1986b) analyzed helium in ~50 analyses of both olivine and cpx phenocrysts, from nine separate flows in four different Kula-age formations and ten other flows with Tertiary to Recent eruption ages. ^{3}He/^{4}He ratios as high as 99 R_A (where R_A is the ^{3}He/^{4}He ratio in air, 1.40×10^{-6}) in the total He in olivines, and 70 R_A clinopyroxene total He, have been measured, and ratios up to 1200 R_A (olivines) and 334 R_A (cpx) have been obtained by fusion of the vacuum-crushed grains, in which essentially pure cosmogenic ^{3}He has been extracted. Most significantly for the methodology used above, the cosmogenic ^{3}He concentration in olivines and clinopyroxenes in the same flows is the same (olivine/cpx ratio = 0.987), and the erosion rate, based on all the data, was found to be ~9 m/My. Studies of the associated cosmogenic ^{21}Ne (Marti & Craig 1987) and ^{10}Be and ^{26}Al (Nishiizumi et al 1990) have now been made, so that a complete set of data for the Maui "type locality" is beginning to become established.

When the radioactive cosmogenic isotopes ^{10}Be and ^{26}Al are considered, the equations are somewhat more complicated, as detailed in a following section, but the essential result is the same. Erosion ($E = +$) and Uplift ($U = +$) of the surface act in the same way to require longer times to produce a given concentration of an isotope (surficial removal, or raising the surface to levels of higher values of J_0). For subsidence ($U = -$, as in the present case) the effect has the same sign as for net accumulation on the surface ($E = -$). For details of the analytical work and treatment of subsidence rates, erosion, etc., refer to Craig & Poreda (1986b): Here we are concerned primarily with demonstrating the effects of these processes.

HISTORICAL BACKGROUND

The study of terrestrial cosmogenic isotopes (in the general sense) goes back almost fifty years, to the earliest measurements on radiocarbon and tritium. During the succeeding decades several schools carried out intensive research programs on other atmospheric cosmogenic isotopes, but of all these studies only the isotopes ^{10}Be, ^{39}Ar, ^{85}Kr, and ^{81}Kr, in addition to radiocarbon, have remained as important contributors to geophysics, and the Ar and Kr isotopes are still awaiting the development of large-scale and routine analytical techniques to realize their potential. On the other hand, the development of analytical methods capable of detecting and analyzing the cosmogenic isotopes produced in rock surfaces (the in-situ cosmogenic isotopes) and the applications of these methods over the past two decades (Table 1) have resulted in what is clearly destined to become a permanently entrenched new field of geophysics. This review is restricted to the developments of this field since the beginnings that date back to the late 1970s, and in one case, even as far back as 1955 when the focus of cosmogenic isotope studies was almost entirely on radiocarbon in the atmosphere.

R. Davis and O. A. Schaeffer were the first to study an in-situ cosmogenic isotope, when they measured the ^{36}Cl activity in a phonolite from 3260 m altitude in Cripple Creek, Colorado, selected because of its high Cl content

Table 1 Cosmogenic isotopes for in-situ exposure age studies of terrestrial rocks

Isotope	Half-life (years)	Principal targets (lithosphere)
^{3}He	stable	O, Si, Al, Mg, etc
^{10}Be	1.5×10^6	O, Si, Al
^{14}C	5730	C, O
^{21}Ne	stable	Mg, Na, Si, Al
^{26}Al	0.71×10^6	Si, Al
^{36}Cl	0.30×10^6	Cl, K, Ca
^{53}Mn	3.7×10^6	Fe
^{131}Xe	stable	Ba

of 0.35% (Davis & Schaeffer 1955, Schaeffer & Davis 1956). ^{36}Cl (half-life = 308,000 years) activity was determined to be 0.12 dpm/g within an uncertainty of 16%, relative to zero activities assumed for measurements of oceanic and Great Salt Lake Cl. In this original work the concepts we use today were fully explored in terms of exposure ages, and the understanding was that such ages could only be evaluated for surfaces that had been subject to insignificant erosion over the exposure time. The ^{36}Cl production rate was estimated for the Cripple Creek site, and an exposure age of $24{,}000 \pm 4000$ years was calculated. The authors discuss the processes by which fresh rock surfaces usable for dating are formed, and evaluate the possible effects of erosion. From the standpoint of both the development of the necessary analytical techniques and the understanding and evaluation of the results, this work is both the discovery paper for in-situ cosmogenic geophysics, and a classic contribution to the annals of nuclear geophysics.

The work of Davis and Schaeffer was not only a classic; it was, typically, far ahead of its time. It was not until twenty years later that the second major paper in this field was published: a study of in-situ cosmogenic xenon isotopes in sedimentary barites by Srinivasan (1976). Cosmogenic Xe isotopes, produced by spallation reactions on Ba in a Swaziland barite from the famous Fig Tree Shale (3000 My), were readily measured by mass spectrometry for the isotopes ^{124}Xe and ^{128}Xe, and even larger enrichments were found for ^{131}Xe which is also formed by muon-induced reactions. Srinivasan then evaluated the production rate and showed that the isotopic Xe excesses corresponded to "surface residence times" (=exposure ages) in the range of 120,000 to 270,000 years. This paper also contains a study of Kr isotopes in the samples and detailed discussions of the production mechanisms and rates of all the Kr and Xe isotopes: like the work of Davis and Schaeffer on ^{36}Cl, it is a geophysical classic that bears reading and study today.

It is curious, looking back, that so little attention was paid to these papers, especially the more recent work by Srinivasan (1976) which was based on what had by then become well-established high-sensitivity analytical techniques for rare-gas isotopes. Most major reviews in the field, even of recent vintage, pay little if any attention to Srinivasan's pioneering work. In the following year, development of the field was continued in a paper by Yokoyama et al (1977) in which the in-situ production of ^{22}Na and ^{24}Na was measured on Mont Blanc and two other Alps, produced by cosmic-ray interactions with Al from a piece of a TV tower, a pot-cover, and from the 20-year-old wreck of an airplane. In this third major study, the production rates were calculated not only for the Na isotopes, but also for ^{3}H, and Be, C, Al, Cl, Ar, Mn, and Fe isotopes as well, and as a function of latitude and altitude over the Earth's surface. With the publication of this paper, the in-situ cosmogenic field had come of age, despite the fact that almost no geophysicists were aware of it.

Two years later Craig et al (1979) studied the native metal josephinite together with the Greenland native irons, and showed that claims of incredibly high isotopic excesses were due to analytical artifacts during gas extraction. However, ^{3}He and ^{21}Ne were observed, albeit much scaled down from the supposed superlatives, and were tentatively attributed to in-situ cosmogenic production. At that time the papers of Srinivasan and Yokoyama were either unknown or forgotten: ^{3}He and ^{21}Ne production rate calculations for Fe-Ni metals were made by two of the authors (K. Marti, S. Regnier) and by request were recalculated by D. Lal as a check. For josephenite (1 km altitude), a ^{3}He exposure age of 1.8 My was obtained from the data—a time very short compared to the actual age of ~ 150 My. However, it was realized that the argument for cosmogenic ^{3}He was rendered nonviable by the lack of documented "shielded" samples in which the inherited ^{3}He content could be measured (this is similar to problems of present-day diamond studies): Thus it is possible that all, or even none, of the josephenite He is cosmogenic. The entire subject of the expected cosmogenic ^{3}He and ^{21}Ne contents of these native metals has now been covered in exhaustive detail (Poreda et al 1990), and this is clearly a subject worth investigating in the future for cosmogenic exposure ages over the entire josephenite formation.

Finally, in 1986, the entire range of in-situ cosmogenic studies was packaged and delivered to the journals in a single year. Cosmogenic ^{3}He in the lavas of Maui was measured at La Jolla, in continuation of the trail from josephenite through oceanic basalts (Craig & Poreda 1986), and at WHOI by Kurz following Japanese reports of high $^{3}He/^{4}He$ ratios obtained by fusion of the Maui samples (Kurz 1986a,b): Both these studies drew heavily on the innovation of crushing techniques for He extraction introduced by Kurz at WHOI. These papers established the definite existence of cosmogenic ^{3}He, and shortly afterwards Marti & Craig (1987) showed the existence of cosmogenic Ne isotopes in the same lavas. Klein et al (1986) and Nishiizumi et al (1986) demonstrated the existence of in-situ cosmogenic ^{10}Be and ^{26}Al in Libyan Desert glass and terrestrial quartz, respectively. Shortly afterwards, Nishiizumi et al (1990) measured ^{10}Be, ^{26}Al, and ^{3}He on the Maui Haleakala olivines, establishing the relative production rates of these three isotopes in olivine basalts. Finally, Phillips et al (1986) measured ^{36}Cl in a number of volcanic lavas and tuffs, and discussed the results in terms of applications to exposure age determinations (although the fundamental process—erosion—was neglected in their considerations).

"Cosmogenic in-situ nuclear geophysics" began in 1955 like most of the rest of the subject, and lay fallow until rejuvenated in 1976 by Srinivasan, and by Yokoyama and colleagues in 1977. Josephenite nudged it a little in 1979, and the whole field came together in 1986 with six papers in one year, covering stable rare-gas isotopes, ^{10}Be and ^{26}Al, and reaching back and extending forward the original impetus from ^{36}Cl. Cosmogenic nuclear geophysics had come of age and was ready to go to work.

PRODUCTION OF COSMOGENIC ISOTOPES

Cosmic ray–derived nucleons decrease exponentially with depth through the atmosphere, hydrosphere, and lithosphere. The principal nucleons of the cosmic-ray flux are protons, alpha particles, secondary neutrons, and muons. In addition, other subsurface reactions (e.g. radiogenic) may produce the isotopes of interest. The cosmic-ray energy spectrum in the atmosphere is invariant below 12 km (atmospheric depth = 200 g/cm^2) (Lal & Peters 1967), although the cosmic-ray flux decreases by approximately 3X per 1.5 km altitude to sea level (atmospheric depth = 1033 g/cm^2). Atmospheric shielding is an important difference between terrestrial and extraterrestrial exposures studies because low angle incident cosmic rays are shielded by the Earth's atmosphere whereas no such shielding takes place for extraterrestrial samples, and production of secondary neutrons from interaction with matter takes place in both the atmosphere and the sample in the terrestrial environment but takes place principally in the sample for extraterrestrial samples. In the following section, we discuss cosmic-ray sources, principles of production of in-situ cosmogenic isotopes, calculated and measured production rates, attenuation coefficients of cosmogenic isotope production, and effects of both spatial and temporal variations of the Earth's geomagnetic field.

Cosmic-Ray Spectrum

Cosmogenic isotopes are produced by spallation reactions induced by high energy nucleons, secondary thermal neutron capture reactions, and by muon-induced reactions (Lal & Peters 1967, Lal 1988a). Nucleons in both the solar cosmic-ray and the Galactic cosmic-ray spectra are primarily protons and alpha particles (Lal & Peters 1967, Pomerantz & Duggel 1974, Reedy 1987). Upon entry of the Earth's atmosphere they strongly interact with other nuclei; this attenuates the cosmic-ray flux and produces a shower or cascade of secondary protons, neutrons, and other particles, many of which have very high energies. Thus the original cosmic "ray" gives rise to spallation products and/or secondary neutrons which can further interact with atomic nuclei; the cosmic-ray flux is significantly diminished when it reaches the Earth's surface.

Solar cosmic rays are produced by the Sun and have energies from about 1 to 50 MeV, with some particles having energies as high as 100 MeV. The Solar cosmic-ray flux is primarily from protons (98%) and some helium and larger nuclei. The flux is variable over several orders of magnitude, reaching 10^6 protons cm^{-2} s^{-1} during large events, but averaging about 100 protons cm^{-2} s^{-1} (Reedy 1987). The flux is lowest during periods of low solar activity and highest during periods of high solar activity. The flux must be integrated over the 11-year sunspot cycle and over the long-term solar sunspot activity. Most of the primary cosmic rays interact with the Earth's upper atmosphere producing a cascade or shower of secondary neutrons, μ-mesons (muons), and other secondary particles (i.e. cosmogenic isotopes such as ^3H, ^3He, ^{10}Be, etc). The production of in-situ cosmogenic isotopes due to Solar cosmic radiation is minimal on Earth compared to that derived from higher energy Galactic cosmic rays.

Galactic cosmic rays originate outside of our solar system, have much higher energies than solar cosmic rays (up to 100 GeV) and have time-averaged particle fluxes of about 3 protons cm^{-2} s^{-1} (Reedy 1987). The shape of the energy spectrum at high energies (> 10 GeV) is generally approximated by $E^{-\gamma}$ where γ is between 2.4 and 2.7 (Reedy et al 1983, Simpson 1983). The flux is modulated by the magnetic field of the Sun which varies with solar activity, and shows a negative correlation with sunspot number (Pomerantz & Duggel 1974). Galactic cosmic rays penetrate more deeply into the atmosphere where they also produce a shower of secondary particles. Secondary particles derived from Galactic cosmic rays produce most of the in-situ cosmogenic isotopes found in terrestrial rocks. The average Galactic cosmic ray produces about seven neutrons in its cascade of particles; whereas charged particles are slowed by ionization, neutrons travel until they interact with matter or lose energy by inelastic scattering (Reedy & Arnold 1972, Reedy 1987).

Production of In-Situ Terrestrial Cosmogenic Isotopes

High energy nucleons are required for spallation reactions because the binding energy of atomic nuclei is between 7.4 and 8.8 MeV for all but the lightest isotopes (A < 11; Friedlander et al 1981). The Solar cosmic-ray spectrum has energies from 1 to 100 MeV and the Galactic cosmic-ray spectrum has energies from 1 to 100 GeV (Reedy 1987), providing nucleons with ample energy for spallation reactions. Spallation reactions produce daughter isotopes with smaller A (sum of protons plus neutrons) than the target isotope. Particularly important spallation reaction products for in-situ geomorphic studies include ^{3}He, ^{10}Be, ^{14}C, ^{21}Ne, ^{26}Al, and ^{36}Cl. These reactions are target dependent: Oxygen, which is present in all silicate rocks and many other minerals, produces ^{3}He, ^{10}Be, and ^{14}C by spallation; spallation off silicon and aluminum produces significant amounts of ^{3}He and ^{26}Al, as well as ^{10}Be and ^{21}Ne, but significantly less ^{10}Be than oxygen in silicates; magnesium and sodium are very important in producing spallation $^{20,21,22}Ne$ by reactions such as $^{24}Mg(n,\alpha)^{21}Ne$ or $^{23}Na(n,p2n)^{21}Ne$; ^{36}Cl is produced in significant amounts by spallation off potassium and calcium. Some spallation products (e.g. ^{4}He) have high absolute production rates but are not easy to measure because the background values are so high, while others (e.g. ^{7}Be) have short half-lives so that they are of little use for in-situ studies although they are of use for other surficial problems.

The total neutron flux includes thermal neutrons (energy <0.5 eV) which have been slowed by elastic scattering in the atmosphere or the rock matrix. While fast neutrons are important in spallation reactions, thermal neutrons are very important for neutron capture reactions. The thermal neutron flux is also attenuated as it passes through matter, but in a different manner than most cosmic ray–associated nucleons. Although fast neutrons are attenuated in an exponentially decreasing fashion, some thermal neutrons escape from the lithosphere into the atmosphere. Thus thermal neutrons have a maximum flux below the surface of the lithosphere (calculated to be at a depth of 45 g/cm^2; Reedy & Arnold 1972, Fabryka-Martin 1988) which was corroborated by measurements of the neutron flux distribution (Yamashita et al 1966) and ^{36}Cl in lunar rocks (Nishiizumi et al 1984a). Thermal neutron capture reactions are very important for the production of certain isotopes from targets with high capture cross sections. Important thermal neutron capture reactions for in-situ studies include $^{35}Cl(n,\gamma)^{36}Cl$, $^{39}K(n,\alpha)^{36}Cl$, and $^{6}Li(n,\alpha)^{3}He$. For rocks with certain compositions, thermal neutron capture reactions can make up a significant proportion of the total cosmogenic isotope production.

Secondary cosmic-ray muons (μ^+, μ^-) are subatomic particles (mass of 207 m_e, where m_e is the mass of an electron) produced by cosmic-ray interaction with matter in the atmosphere or lithosphere. The negative muon can be captured by a nucleus to form a μ-mesic atom which then decays to a long-lived or stable isotope. Muons have a much longer attenuation length than neutrons so that at shielding depths greater than about 1000 g/cm^2 (sum of atmospheric shielding and vertical rock depth) fast muon interactions and slow muon capture must be evaluated for the total production rate of cosmogenic isotopes (Lal 1987a,b). Muon production at depth could be significant for stable cosmogenic isotopes (^{3}He, ^{21}Ne) in samples with long burial ages, or in samples from areas with high erosion rates.

Other mechanisms also can produce the same isotopes as are produced by the cosmic-ray flux. In particular, in certain cases radiogenic production can be important because of the neutron flux associated with natural fission, or from (α,n) reactions [e.g. $^{18}O(\alpha,n)^{21}Ne$]. For stable isotopes (^{3}He, ^{21}Ne) or those that exhibit a significant fraction of isotope production by thermal neutrons (^{36}Cl) the amount of radiogenic production may be non-negligible. For ^{36}Cl this can be evaluated because the capture cross section is well known (Phillips et al 1986, Zreda et al 1991). However, for the stable isotopes, assumptions must be made about the age of the sample and the distribution of the neutron and α flux throughout the lifetime of the mineral.

Attenuation Coefficients

The mass attenuation coefficient for neutron flux or for in-situ cosmogenic isotope production describes the change in the flux as cosmic rays pass through matter:

$$J = J_0 e^{-M/\Lambda}, \tag{5}$$

where J and J_0 are the production or disintegration rates at the depth of interest and the surface, respectively, Λ is the mass attenuation coefficient in g/cm^2, and M is the accumulated mass in g/cm^2. It is also convenient to consider the characteristic attenuation length for material of constant density:

$$z^* = \Lambda/\rho, \tag{6}$$

where ρ is the density so that

$$J = J_0 e^{-z/z^*}, \tag{7}$$

where z is depth in the constant density medium. For a mass attenuation coefficient of 170 g/cm^2, water ($\rho = 1.0$ g/cm^3), granite ($\rho = 2.7$ g/cm^3), and basalt ($\rho = 3.2$ g/cm^3) have characteristic attenuation lengths of 170, 63, and 53 cm, respectively.

The value of the attenuation coefficient depends primarily on the energy spectrum of the incident cosmic rays and secondaries (Lingenfelter 1963, Lal & Peters 1967). Most of the in-situ cosmogenic isotopes produced in the lithosphere are derived from Galactic cosmic rays which are of much higher energy than solar cosmic rays (Lingenfelter 1963, Lal & Peters 1967, Reedy et al 1987). Although the Galactic cosmic-ray flux to the Solar System is essentially constant over the long time periods considered for in-situ cosmogenic exposure age studies, the flux to the Earth's surface changes as a result of the variation of the Earth's magnetic field and because of solar activity. Two important observations have been made concerning the mass attenuation length of the cosmic-ray flux: 1. The mass attenuation length does not significantly change with solar activity, although the cosmic-ray flux is higher at high latitudes during periods of low solar activity (Lingenfelter 1963); and 2. the mass attenuation length changes with latitude—while the cosmogenic flux is higher at high latitudes, the attenuation length is greater at low latitudes because the incident cosmic-ray flux is harder (Lingenfelter 1963). Mass attenuation coefficients for neutron flux in air or for in-situ isotope production in various media are given in Table 2, which shows that the measured attenuation coefficients vary from about 145 to 190 g/cm^2 for the lithosphere with values being distributed about a value of approximately 170 g/cm^2. Atmospheric values range from 164 to 212 g/cm^2. Middleton & Klein (1987) estimate the μ-attenuation coefficient to be about 1500 g/cm^2.

Samples collected at significant depths below the surface should be corrected to the surface value. However, it is important to note that the

Table 2 Mass attenuation coefficients for different materials

Nuclide measured	Material (latitude)	Attenuation coefficient Λ (g/cm^2)	Reference
n	air (>10°)	212	Lingenfelter, 1963
n	air (20°)	205	Lingenfelter, 1963
n	air (40°)	180	Lingenfelter, 1963
n	air (>60°)	164	Lingenfelter, 1963
^{3}He	basalt (19°)	170	Kurz, 1986b
^{7}Be	air (25°)	165	Nakamura et al 1972
^{7}Be	air (50°)	158	Nakamura et al 1972
^{10}Be	rhyolite (36°)	172	Olinger et al 1992
^{10}Be	laterite (12°)	190	Bourlès et al 1992
^{10}Be	quartz (78°)	145	Brown et al 1992
^{10}Be	air (78°)	142	Brown et al 1991
^{10}Be	lunar basalt	163	Nishiizumi et al 1984a
^{21}Ne	rhyolite (36°)	178	Olinger et al 1992
^{26}Al	rhyolite (36°)	173	Olinger et al 1992
^{26}Al	lunar basalt	165	Nishiizumi et al 1984b
^{26}Al	laterite (12°)	190	Bourlès et al 1992
^{26}Al	quartz (78°)	156	Brown et al 1992
^{32}P	air (33°)	144	Rama & Honda, 1961
^{32}P	air (25°)	160	Mabuchi et al 1971
^{36}Cl	air (20°)	152	Zreda et al 1991
^{36}Cl	lunar basalt	180	Nishiizumi et al 1984a
^{53}Mn	lunar basalt	166	Nishiizumi et al 1983

mass attenuation coefficient assumes that the production rate is modified only by mass absorption. For the special cases where a significant fraction of the nuclide is produced by thermal neutron capture (e.g. ^{36}Cl) or where significant attenuation has not taken place in the atmosphere (e.g. the Moon), the production in the near subsurface increases relative to the surface value to an accumulated mass depth of about 45 g/cm^2 (45, 16, and 14 cm for water, granite, and basalt, respectively).

Figure 2 shows the depth dependence in the lithosphere (depth less than 200 g/cm^2) of cosmogenic isotope production from two localities, one at high latitude and one at low latitude. The mass attenuation coefficients calculated from these data, 145 and 179 g/cm^2, respectively, show the expected trends: that high-latitude sites have a lower mass attenuation coefficient because the cosmic-ray energy spectrum is harder at low latitudes than at high latitudes. For deep lithosphere samples (greater than 200 g/cm^2) significant μ-induced isotopes may be present (Kurz 1986b). However, for depths less than 1 meter in the lithosphere, the contribution of radiogenic and μ-induced isotopes is small (but sometimes significant) compared to the total cosmogenic flux (Kurz 1986b, Fabryka-Martin 1988, Nishiizumi et al 1989).

Variations due to Changes in Solar Activity

The cosmic-ray flux to the Earth is inversely related to solar activity (Forbush 1954). The solar cosmic-ray flux is higher during periods of

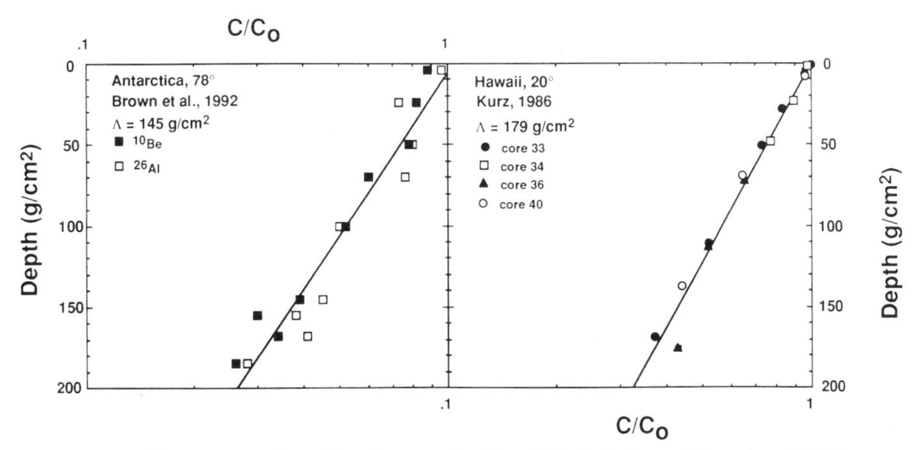

Figure 2 Mass attenuation effects for samples from high latitude and from low latitude. Data from Brown et al (1992) and Kurz (1986b) for rock sample depths less than 200 g/cm^2. Concentrations are normalized to the Earth's surface.

sunspot activity; most solar cosmic rays in an 11-year sunspot cycle are produced during a few solar flare events (Lal & Peters 1967, Pomerantz & Duggel 1974, Reedy et al 1983). However, the solar cosmic-ray flux is of much lower energy than the Galactic cosmic-ray flux, so that the solar cosmic-ray flux makes little contribution to the total cosmic-ray flux at the Earth's surface, even at high latitudes. While the solar cosmic-ray component may be important in some extraterrestrial studies, it contributes only a few percent of the total atmospheric ^{14}C production (Lingenfelter 1963) and is not significant for in-situ surface studies of the Earth's surface.

The Galactic cosmic-ray flux to the Earth's surface, on the other hand, is greatly affected by the solar cycle. Solar plasma clouds derived from periods of solar activity interact with the Earth's magnetosphere to produce geomagnetic storms that modulate the Galactic cosmic-ray flux; in addition the plasma clouds sweep cosmogenic particles from their path (Lal & Peters 1967). The aggregate result of these periods of increased solar activity is to significantly decrease the Galactic cosmic-ray flux to the Earth and its atmosphere. The low-energy part of the cosmic-ray spectrum is most strongly affected so that the most pronounced decrease during periods of high solar activity occurs at high latitudes and altitudes.

Geomagnetic Field Variations: Spatial and Temporal Effects

The Earth's geomagnetic field deflects incoming cosmic rays and hence affects the in-situ production rate. The deflection is related to the incident angle and the rigidity of the cosmic ray, which is defined as $r = pc/q$, where p is momentum, q is the charge of the particle, and c is the velocity of light (O'Brien 1979). The cutoff rigidity for any particular angle of incidence is the lowest rigidity for which a cosmic ray can enter the Earth's atmosphere. The present vertical cutoff rigidity varies between 0 and 17 GV; vertical cutoff rigidities are lowest at the magnetic pole and highest at the equator (Shea et al 1987). The higher cutoff rigidity near the equator means that the cosmic-ray spectrum is harder at the equator and hence the attenuation coefficient of the atmosphere is higher at low latitudes than at high latitudes.

The Earth's geomagnetic pole is essentially the geographic pole for periods greater than about 2000 years (Champion 1980, Ohno & Hamano 1992), so that in the following discussion it is assumed that the geographic pole represents the average geomagnetic pole position. Figure 3 shows production rates as a function of latitude and elevation based on the models from Lal & Peters (1967) and Lal (1988c, 1991); preliminary data on absolute production rates from different altitudes and latitudes are consistent with these results (Cerling & Craig 1994, Zreda et al 1991). The

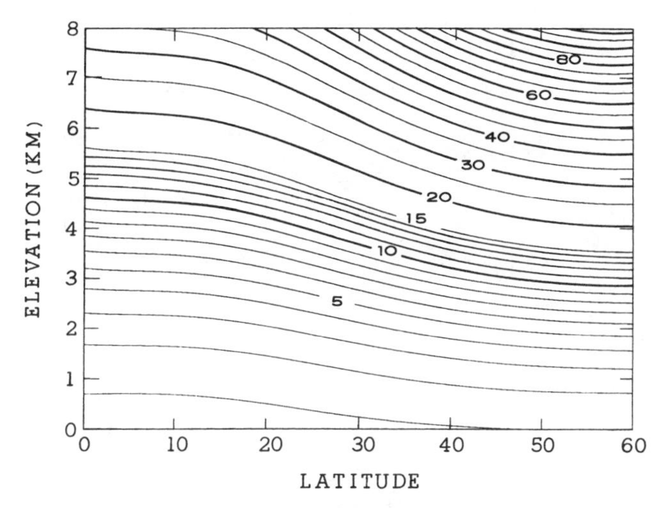

Figure 3 Production of cosmogenic isotopes for the present cosmic ray flux as a function of geomagnetic latitude and altitude (from Lal & Peters 1967 and Lal 1991). Values are normalized to 1.0 for sea level and high latitudes. The geomagnetic pole averages the geographic pole for times longer than a few thousand years (Champion 1980, Ohno & Hamano 1992).

relationship of Lal & Peters (1967) is similar to that of Lingenfelter (1963) and Yokoyama et al (1979), but differs in detail.

Temporal changes in the Earth's geomagnetic field are very important for the total production of cosmogenic isotopes. The ^{14}C radiocarbon production rate is known to vary inversely with the strength of the magnetic field (Damon et al 1978, Suess 1986). For reasons described above, a weaker magnetic field allows greater penetration of cosmic rays into the atmosphere resulting in a increase in the production of cosmogenic nuclides. Because the cutoff rigidity is low at high latitudes and high at low latitudes, the change in the cosmic-ray flux at high latitudes is significantly less than the change at low latitudes for a given change in the Earth's magnetic moment. Likewise the change in the production rate at high altitudes will be greater than at low altitudes for a given change in the magnetic moment. Therefore, the greatest change in production rates due to changes in the dipole moment will be at high altitudes and low latitudes, and will be much less at low altitudes and high latitudes (Kurz et al 1990). As the production rate for cosmogenic isotopes is integrated over the exposure period, changes in the dipole moment are attenuated with time. Figure 4 shows the integrated production rate of ^{3}He for the last 80,000 years calculated for changes in the Earth's dipole moment.

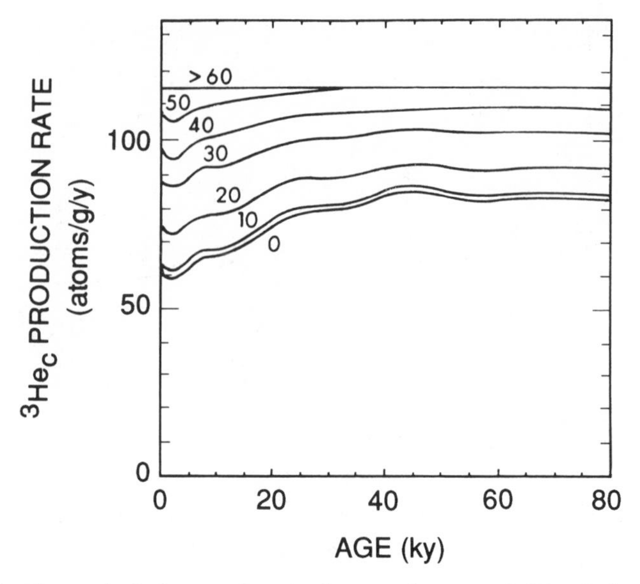

Figure 4 Changes in the integrated production rate of cosmogenic ^{3}He calculated from Lal (1991) and Mazaud et al (1991) and normalized to the production rate measured at Tabernacle Hill, Utah (Cerling & Craig 1994). The date of Tabernacle Hill was corrected for changes in the radiocarbon production rate using the relationship observed by Bard et al (1990).

Non-Cosmogenic Isotopes and Atmospherically-Derived Cosmogenic Isotopes

For an accurate determination of the concentration of cosmogenic isotopes within a sample the non-cosmogenic component of the isotope of interest must be assessed; this must be subtracted from the total amount measured to give the cosmogenic component. For different isotopes different origins are possible, but in any case the possibility of a non-cosmogenic component must be considered. Three types of non-cosmogenic isotope corrections need to be considered: inherited, in-situ nucleogenic, and contamination by atmospherically-derived (common) cosmogenic nuclides.

For the stable cosmogenic isotopes such as ^{3}He and ^{21}Ne, an original amount of trapped non-cosmogenic component may be present, often in vesicles, which can be released by crushing (Craig & Poreda 1986a; Kurz 1986a,b; Cerling 1990; Kurz et al 1990). Correction can be made for the non-cosmogenic component by measuring the $^{3}He/^{4}He$ ratio and the ^{4}He concentration (Craig & Poreda 1986a; Kurz 1986a,b). $^{21}Ne/^{20}Ne$ corrections are made assuming that the non-cosmogenic $^{21}Ne/^{20}Ne$ ratio is atmospheric (Marti & Craig 1987, Staudacher & Allègre 1991, Poreda & Cerling 1992). Brook & Kurz (1993) suggest that the assumption that the gas released during crushing is a trapped non-cosmogenic component may be inappropriate for quartz because they observed high and variable $^{3}He/^{4}He$ ratios in crushing Antarctic quartz samples. For the cosmogenic isotopes that are radioactive (^{10}Be, ^{26}Al, ^{36}Cl), the inherited component in minerals crystallized from magmas is nil, except in a few rare circumstances. For example, Tera et al (1986) and Monaghan et al (1988) report primary ^{10}Be in some young island arc volcanic rocks from the Aleutians that was incorporated into the magma prior to phenocryst formation.

STABLE COSMOGENIC ISOTOPES: RESOLUTION OF THE COSMOGENIC COMPONENT ^{3}He is the cosmogenic stable isotope that has been studied in some detail. Olivine and clinopyroxene phenocrysts in basaltic lavas generally have fluid inclusions containing helium and other rare gases from the mantle with characteristic isotopic signatures. On exposure after eruption, cosmogenic ^{3}He and ^{21}Ne are created essentially uniformly throughout the matrix of a crystal. Craig & Poreda (1986a) resolved the mantle and cosmogenic components by first crushing the phenocrysts in vacuum to release the total helium for mass spectrometric analysis. The crushed powders were then heated in vacuum to 1800°C, to release the cosmogenic helium, plus some remaining inclusion helium, for analysis. Plotting ^{3}He vs ^{4}He for the extracted helium fractions results in two graphs: 1. the mantle-component linear array, with a $^{3}He = 0$ intercept, and 2. a parallel linear array with a nonzero intercept equal to the cosmogenic ^{3}He component. The cosmogenic component in the Haleakala lavas was thus found to range from 0.80 to 1.22×10^{-12} cm^{3}/g ($\mu\mu cc/g$) at standard temperature and pressure and to amount to $\sim 75\%$ of the total ^{3}He in the phenocrysts.

Alternatively, helium may have been extracted from phenocrysts entirely by high-temperature fusion, so that the inherited mantle helium stored in the phenocryst inclusions is not separately determined. This is the case for the suites of ultramafic xenoliths analyzed by Porcelli et al (1986), just prior to the discovery of cosmogenic helium. The original data include samples of ultramafic nodules erupted from the Lashaine tuff cone in northern Tanzania, in which the $^{3}He/^{4}He$ ratios and total ^{4}He were measured on the fusion-extracted gases. In this case, a different procedure may be applied. Assuming that the individual nodules have the same cosmogenic ^{3}He concentration, but variable inherited helium contents, a plot of the $^{3}He/^{4}He$ ratio vs the reciprocal ^{4}He concentration will give a linear array in which the intercept is the inherited helium isotope ratio, and the slope is the cosmogenic ^{3}He content of the olivine and cpx crystals.

Figure 5 shows such a plot for the Lashaine nodule samples of Porcelli et al (1986). The $^3He/^4He$ ratios are plotted in the conventional R/R_A format, where $R = {^3He}/{^4He}$, and R_A is the atmospheric helium ratio (1.40×10^{-6}). The straight line is fit to the three points for sample BD-738 (two olivines and one cpx), a garnet lherzolite, but the fit is reasonable for all five data points. The intercept defines the inherited mantle helium component with $R/R_A = 5.9$, while the slope (57.8) gives the cosmogenic ^3He concentration as 0.081 $\mu\mu$cc/g, some 15 times lower than the Haleakala value from 2 km greater altitude. These Lashaine values are similar to those determined in a second paper by Porcelli et al (1987), wherein the data points were treated separately and a mean inherited $^3He/^4He$ ratio was assumed: The graphical method is used here to test consistency of the data and derived mean values for the parameters.

The appropriate ^3He production rate for the Lashaine nodules is based on the base-line rate $J = 115$ atoms g^{-1} y^{-1} (Cerling & Craig 1994) in Table 3, modified for an altitude of ~1200 m, and adjusted for the mean production rate over the past 20 Ky at latitude 3°S (Figure 4): This gives a rate for Lashaine of $J = 180$ atoms g^{-1} y^{-1} = 6.70 $\mu\mu$cc(STP) g^{-1} My^{-1}. [Porcelli et al use a value of $J = 10$ $\mu\mu$cc(STP) g^{-1} My^{-1}, based on the

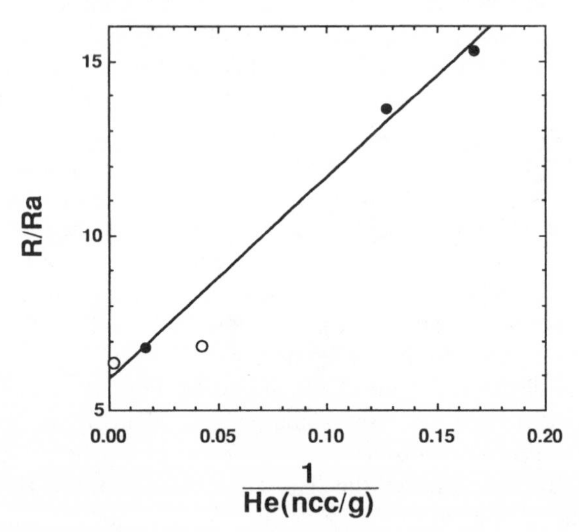

Figure 5 Measured $^3He/^4He$ ratios (as R/R_A) for the Lashaine, Tanzania nodules plotted against the reciprocal ^4He concentrations. The data were measured by Porcelli et al (1986). The intercept in this plot defines the "inherited" mantle He ratio (5.9 R_A), while the slope gives the cosmogenic ^3He concentration (0.081 $\mu\mu$cc/g, corresponding to an exposure age of 12,100 years). The straight line is fitted to three points (*solid circles*) from the same nodule. See Craig & Poreda (1994) for discussion of this figure.

Table 3 Measured production rates of in-situ terrestrial cosmogenic isotopes†

	Mineral/rock	Composition	Production rate atoms/g/year	Reference
$^{3}He^{a}$	olivine	Fo_{81}	115^{b}	Cerling & Craig, 1993
	olivine	Fo_{81}	109^{b}	Poreda & Cerling, 1992
	olivine		94^{c}	Kurz et al 1990
	clinopyroxene		115^{b}	Cerling, 1990
	quartz		>85	Brown et al 1991
^{10}Be	quartz		6.0	Nishiizumi et al 1989
	quartz		6.4	Brown et al 1991
	olivine		5.3	Nishiizumi et al 1990
^{14}C	quartz		21	Donahue et al 1990
$^{21}Ne^{a}$	olivine	Fo_{81}	45^{d}	Poreda & Cerling, 1992
	clinopyroxene	$En_{44}Fs_{13}Wo_{42}$	23^{d}	Poreda & Cerling, 1992
	plagioclase	Ab_{36}	17^{d}	Poreda & Cerling, 1992
^{26}Al	quartz		37	Nishiizumi et al 1989
	quartz		42	Brown et al 1991
	olivine		15.4	Nishiizumi et al 1990
^{36}Cl	basalt		7^{e}	Zreda et al 1991
	granite		9^{e}	Zreda et al 1991

† All values given for sea level ($M = 1033$ g/cm^2) at high latitudes (>60°), and have been corrected for the difference between ^{14}C scale and the absolute ages (Pearson & Stuiver 1986, Stuiver & Pearson 1986, Bard et al 1990). Values given are averaged over long periods of time (>5000 years). Some of the differences in rates may be due to their being averaged over different time periods because the integrated production rate varies with time (Figure 4).

[a] 100 atoms/g/y = 3.72 $\mu\mu$cc/g/My.

[b] Integrated over the last 17,500 years.

[c] Integrated over the last 5000 years (see discussion in Cerling & Craig 1994).

[d] Estimated using the production rate at Tabernacle Hill (115 atoms/g/year) normalized to high latitudes and altitudes, and using ^{3}He concentration as a cosmogenic flux monitor.

[e] Compositionally dependent (K, Ca, Cl).

original calculated production rates of Yokoyama et al (1977) which were remarkably accurate considering the complete lack of actual data at that time.] Because the nodules are not significantly eroded, the Lashaine exposure age for the eruption of the tuff cone is thus found to be 12,100 years. We estimate that this age is probably determined to ~1500 years either way—an excellent return for so few data.

IN-SITU NUCLEOGENIC ISOTOPE PRODUCTION Natural radioactivity producing neutrons or alpha particles results in additional isotope production. Alpha production in the U- and Th-decay series results in (α,n) reactions, and spontaneous fission of uranium (the basis for fission track dating; Fleischer et al 1975) provides an additional source of neutrons. Thus, many rocks have a significant flux of α particles or neutrons which may produce significant quantities of isotopes over sufficient time. Very few of the neutrons produced have energies high enough to cause spallation but they are very important for (n,α), (n,T), or (n,γ) reactions. Some significant reactions include ^{6}Li(n,α)t $\rightarrow$ ^{3}He, ^{10}B(n,2α)t $\rightarrow$ ^{3}He, ^{17}O(n,α)^{14}C, ^{18}O(α,n)^{21}Ne, ^{23}Na(n,t $\rightarrow$ ^{3}He)^{21}Ne, ^{24}Mg(n,α)^{21}Ne, ^{35}Cl(n,γ)^{36}Cl, and ^{39}K(n,α)^{36}Cl (Kurz 1986a,b; Lal 1987b; Fabryka-Martin 1988). Production of these radiogenically-derived isotopes is usually small compared to those derived from cosmic-ray neutrons or α particles over the surface exposure period, and hence is usually small compared to the cosmogenic component during the surface exposure period. However, for the stable cosmogenic isotopes, such as ^{3}He or ^{21}Ne, the long term accumulation (10^{7} years, or more) could be a significant contribution to the total ^{3}He or ^{21}Ne in the sample. Poreda & Cerling (1992) report high concentrations of ^{21}Ne from Mesozoic plutonic quartz found in young glacial moraines that were attributed to nucleogenic production over ca. 10^{8} years. Kohl & Nishiizumi (1992) suggest that nucleogenic ^{21}Ne contamination by the ^{18}O(α,n)^{21}Ne reaction can be removed by leaching because of the relatively short path length of α particles (10 s of microns).

ATMOSPHERICALLY-DERIVED COSMOGENIC ISOTOPES These are distinguished from in-situ cosmogenic isotopes because the former are produced in the atmosphere and swept out by rainwater whereas the latter are produced in-situ within the rock (the former are also called "garden-variety" by Lal 1986). For some cosmogenic isotopes the atmospherically-derived flux is high enough and the absorption by rock is high enough that a significant amount of atmospherically-derived cosmogenic isotope is present, leading to their use in exposure age studies of soils, soil formation rates, and soil erosion rates (Monaghan et al 1983, 1992; Brown et al 1988; McKean et al 1993). The flux of atmospherically-derived cosmogenic isotopes is high enough for ^{10}Be and ^{36}Cl so that samples must be leached to separate it from the in-situ component which may be only 1% or less of the total. This is usually done by successive leaching in acid (Brown et al 1991, Kohl & Nishiizumi 1992, Zreda et al 1991). Atmospherically-derived ^{3}He and ^{21}Ne is insignificant in minerals because they are not transferred from the air to the soil, and the atmospheric production of ^{26}Al is very low because of a low abundance of suitable target isotopes. ^{14}C contamination by atmospheric ^{14}C is a problem for samples contaminated with organic carbon or with pedogenic carbonate. It can be removed by oxidation or by acid treatment.

This discussion points out that for most samples the total abundance of an isotope must be corrected to get the in-situ cosmogenic component. The total isotopic abundance can include inherited, nucleogenic, and atmospherically-derived components that need to be evaluated for each sample.

Production Rates

The production of cosmogenic isotopes is a function of the cosmic-ray flux including secondaries and the composition of the rock. The cosmic-ray flux integrated over the exposure history of the sample is related to latitude, altitude (including depth below the surface), and the length of time the sample has been exposed to cosmic rays. Until a few years ago, there were no direct measurements of the production rates of terrestrial in-situ cosmogenic isotopes. However, in the past few years various research groups have measured the terrestrial in-situ production rates of different cosmogenic isotopes by measuring the concentrations of cosmogenic isotopes in pure materials such as laboratory chemicals (Lal et al 1960, Mabuchi et al 1971) or wrecked airplanes (Yokoyama et al 1977), dated lava flows (e.g. Phillips et al 1986, Kurz et al 1990, Cerling 1990, Zreda et al 1991, Cerling & Craig 1993), in moraines deposited during the last glaciation (Nishiizumi et al 1989, Zreda et al 1991), by measuring the concentrations of several radioactive cosmogenic isotopes (Brown et al 1991), and by measuring concentrations of cosmogenic isotopes in co-existing minerals with different compositions (Poreda & Cerling 1992). Terrestrial production rates are not directly comparable to extraterrestrial production rates (e.g. Nishiizumi et al 1980) because of attenuation of both solar and Galactic cosmic rays by the Earth's atmosphere and magnetosphere, or to theoretical production rates based upon nuclear cross section measurements (e.g. Raisbeck & Yiou 1974, Reedy & Arnold 1972, Reedy et al 1979). However, measured production rates are similar to calculated production rates when these differences are taken into account (compare production rates below to those given by Yokoyama et al 1977).

Table 3 shows terrestrial production rates of some of the important cosmogenic isotopes for certain mineral or rock compositions and Table 4 shows production rates for quartz. All of the given production rates are calculated for high latitudes ($>60^\circ$) and sea level ($M = 1033$ g/cm^2) and integrated over a specified time interval. Most cosmogenic isotopes show a compositional dependence of the production rate as is evident in Table 3 and Figure 6, which shows the compositional dependence of ^{21}Ne on the concentration of Mg and Na. It is clear that considerably more work needs to be done on establishing the production rate of in-situ cosmogenic isotopes. Efforts have been hampered by the lack of samples of well established age (see discussion by Cerling & Craig 1993), because most lava flows have insufficient ^{14}C dates necessary to establish their age and K-Ar is not sufficiently accurate for samples younger than about 100,000 years because of problems of precision and inherited argon. Figure 7 shows cosmogenic ^{3}He from North American sites with well established ages. In

Table 4 Isotope production rates in quartz (sea level at high latitude)

Isotope	Target	Production rate atoms/g/year	Reference
^{3}He	O, Si	115.0	[a]
^{10}Be	O, Si	6.0	Nishiizumi et al 1989
^{10}Be	O, Si	6.4	Brown et al 1991
^{14}C	O, Si	21.0	Donahue et al 1990
^{21}Ne	Si	14.5	[b]
^{21}Ne	Si	14.7	[c]
^{26}Al	Si	36.8	Nishiizumi et al 1989
^{26}Al	Si	42.0	Brown et al 1991
^{36}Cl	^{35}Cl	4.0	[d] Zreda et al 1991

[a] Estimated to be same as Tabernacle Hill (Cerling & Craig 1994).
[b] Calculated from Poreda & Cerling (1992).
[c] Calculated from Graf et al (1991) and Nishiizumi et al (1989).
[d] Compositionally dependent (Cl).

some cases it is possible to use the presence of two isotopes to calculate production rates (Brown et al 1991) because the ratio of isotopes with different half-lives changes with time.

MEASUREMENT OF COSMOGENIC ISOTOPES

Cosmogenic isotopes are measured either by conventional noble-gas mass spectrometry or by accelerator mass spectrometry (AMS). For any of the isotopes discussed here, the realistic detection limit is on the order of 10^4 to 10^5 atoms per sample. Sampling strategies and isotope extraction methods are determined by these limitations, and by the local production rate and time the sample has been exposed to cosmic radiation.

For the noble gases ^{3}He and ^{21}Ne, which are analyzed by conventional mass spectrometry, the inherited component must be determined separately from the cosmogenic component (Craig & Poreda 1986a; Kurz 1986a,b; Cerling 1990; Brook & Kurz 1993) which is released by heating to high temperatures from a pure mineral separate. Mass spectrometric corrections including isotope discrimination or resolution of similar peaks such as HD (for ^{3}He) and $^{40}Ar^{++}$, $H_2{}^{18}O$, $C_3H_6^{++}$, $^{20}NeH^{+}$ and CO_2^{++} (for Ne) must be made (Marti & Craig 1987, Graf et al 1991, Poreda & Cerling 1992, Niedermann et al 1993).

The radioactive cosmogenic isotopes are measured by AMS which allows for reduction in sample size by several orders of magnitude compared to conventional counting methods. AMS requires that the isotope be ionized with minimal production of interfering isobars; the ions are

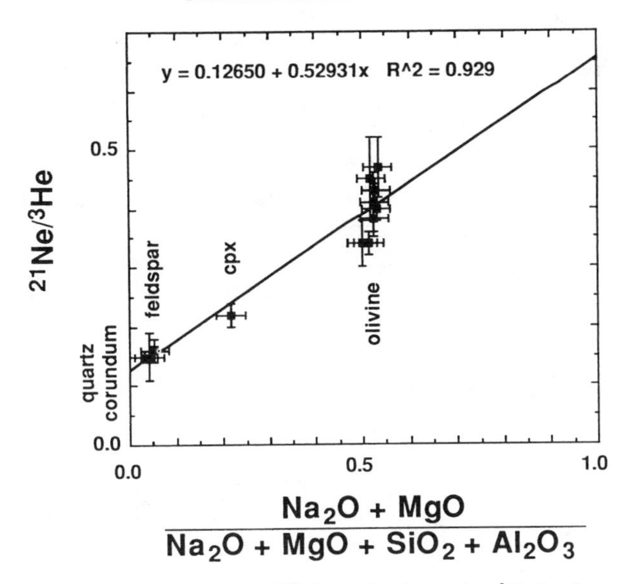

Figure 6 Compositional dependence of ^{21}Ne production, using ^{3}He in phenocryst olivine as a flux monitor. From data of Poreda & Cerling (1992).

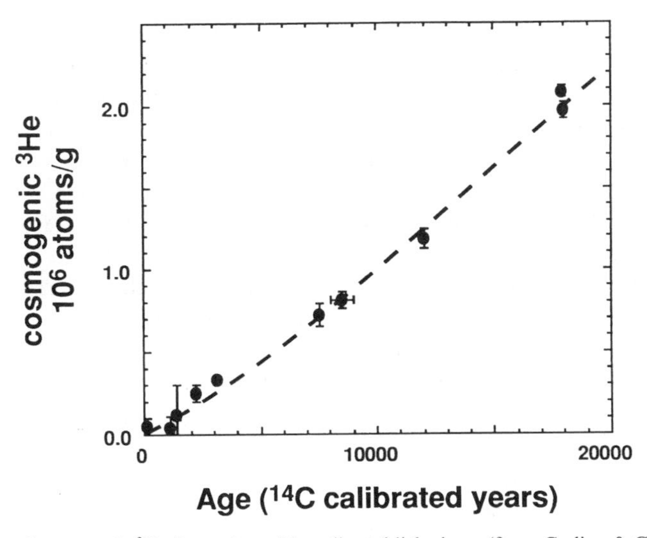

Figure 7 Cosmogenic ^{3}He from sites with well established age (from Cerling & Craig 1994, and unpublished data).

then accelerated through a potential up to several MeV, the isotopes are separated by the mass to charge ratio in several magnetic fields along the flight path, and finally the interfering isobars are detected and resolved (Elmore & Phillips 1987). Samples must be prepared to minimize isobaric interference (e.g. ^{10}B with ^{10}Be). Mineral separates often have an advantage over bulk samples because the production rates are easier to characterize for simple chemistries. Sample preparation is discussed by Srinivasan (1976), Lal et al (1987a, 1990), Nishiizumi et al (1989), Jull et al (1989, 1992), Brown et al (1991), Zreda et al (1991), and Kohl & Nishiizumi (1992).

IN-SITU COSMOGENIC ISOTOPES: EXPOSURE AGES AND EROSIONAL HISTORY

The potential for terrestrial in-situ cosmogenic isotopes to be used for exposure age dating or for modeling erosion has been covered many times previously (e.g. Davis & Schaeffer 1955; Srinivasan 1976; Lal 1986, 1987a,b, 1988a, 1991; Craig & Poreda 1986a; Klein et al 1986; Kurz 1986b; Nishiizumi et al 1986, 1991a; Lal et al 1987a; Brown et al 1991; Zreda & Phillips 1993). Several different equations are necessary to describe the concentration of in-situ cosmogenic isotopes within a rock. For a horizontal surface with a constant production rate over the time of interest, the production rate within rock varies as

$$J(z) = J(0)\, e^{-\rho z/\Lambda}, \tag{8}$$

where J is the production rate (atoms g^{-1} s^{-1}) at depth z (cm) in the rock; ρ and Λ are as defined above. The mass attenuation coefficient Λ is treated as constant although Kurz (1986b) treats the case where a significant fraction of the cosmogenic isotopes are produced by μ^- particles, which have a different attenuation coefficient, and Zreda & Phillips (1993) discuss the special case for ^{36}Cl where the total production reaches a maximum at a depth of about 45 g/cm^2 because of neutron activation of ^{35}Cl by thermal neutrons. The value for Λ/ρ represents the characteristic length z^* for cosmogenic production, sometimes called the e-folding depth, and is generally about 50 to 70 cm for most rocks. The concentration of in-situ cosmogenic isotopes is given by the equation:

$$dC(z,t)/dt = -C(z,t)\lambda + J(z,t), \tag{9}$$

where

$$J(z,t) = J(0)\, e^{-(z/z^*)t}, \tag{10}$$

t is time (s), and λ is the decay constant (s^{-1}) for radionuclides. For significant exposure periods, it is important to consider the effect of erosion which brings previously shielded rocks to the surface. In the following discussion a constant erosion rate E (cm/s) is assumed so that

$$\frac{\partial C}{\partial t} = J_0 e^{-z/z^*} + E\frac{\partial C}{\partial z} - \lambda C, \tag{11}$$

which has the general solution (Lal 1991, Brown et al 1991):

$$C(t, z) = \frac{J_0}{E/z^* + \lambda}[e^{-(z/z^*)}][1 - e^{-t(E/z^* + \lambda)}] + C_0 e^{-\lambda t} + \Sigma[J(\text{other})], \tag{12}$$

where C_0 is the initial concentration of the isotope and $\Sigma[J(\text{other})]$ represents other production mechanisms as discussed above. Setting $\lambda = 0$ for the case of nonradioactive species, Equation 12 reduces to Equation 1 when $U = E$ in the latter, i.e. when uplift is balanced by erosion (or subsidence by accumulation), so that $U^* = 0$. [Note, however, this is not the case for setting $U = 0$ in Equation 1, because to be completely rigorous Equations 9 to 12 should include the term E/h^* even when one takes $U = 0$, because erosion (accumulation) changes the altitude of the surface, and thus the production rate of the cosmogenic isotope.] For a sample collected at the surface this equation becomes:

$$C(t) = \frac{J_0}{E/z^* + \lambda}[1 - e^{-t(E/z^* + \lambda)}] + C_0 e^{-\lambda t} + \Sigma[J(\text{other})]. \tag{13}$$

For the case where the initial concentration C_0 and the production by other mechanisms are evaluated, the cosmogenic component $C_c(t)$ is

$$C_c(t) = \frac{J_0}{E/z^* + \lambda}[1 - e^{-t(E/z^* + \lambda)}]. \tag{14}$$

When erosion is negligible ($E \approx 0$), two further special cases are of interest—stable cosmogenic isotopes and radioactive cosmogenic isotopes:

$$C_c(t) = J_0 t \tag{15}$$

for stable isotopes, and

$$C_c(t) = \frac{J_0}{\lambda}(1 - e^{-\lambda t}) \tag{16}$$

for radioactive cosmogenic isotopes.

These equations illustrate some important aspects of surface exposure dating using cosmogenic isotopes. Stable cosmogenic isotopes increase with time unless chemical or diffusional loss occurs (e.g. Cerling 1990, Trull et al 1991, Brook & Kurz 1993, Brook et al 1993) so that previous exposure or long-term production by radiogenic or μ-induced radiation may be significant for samples whose crystallization age is much greater than the exposure age. Radiogenic isotopes, on the other hand, have an inherently low background for previous exposure events or long-term production because radioactive decay causes the C_0 and C(other) terms to decrease according to the half-life of the isotope. Likewise, it is unlikely that a significant fraction of the isotope of interest crystallized when the rock formed, except in unusual circumstances such as for island arc volcanic rocks incorporating ^{10}Be (e.g. Tera et al 1986, Monaghan et al 1988) or carbonates accumulating ^{14}C near the surface such as in soils.

Figure 8 shows the concentrations of in-situ cosmogenic isotopes for the case where there is no erosion. Radioactive isotopes approach a steady-state concentration after 4 to 5 half-lives, limiting the application of those isotopes to cases where the exposure age is within 3 to 4 half-lives. For cases where there is significant erosion steady-state values are reached earlier. Figure 9 shows that for constant erosion rates even the stable isotopes can reach steady-state values on the time scale of 100s of thousands of years for low erosion rates (ca. 10 m/My). For regions with erosion rates exceeding 100 m/My it is unlikely that the assumption of constant erosion rates is valid.

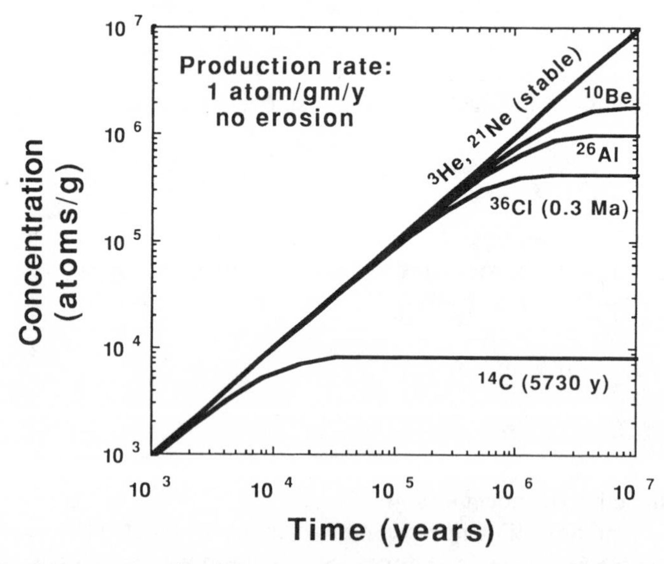

Figure 8 Concentration of in-situ cosmogenic isotopes for a sample with a constant production rate for the case with no erosion. Radioactive cosmogenic isotopes approach a steady-state concentration after about 5 half-lives.

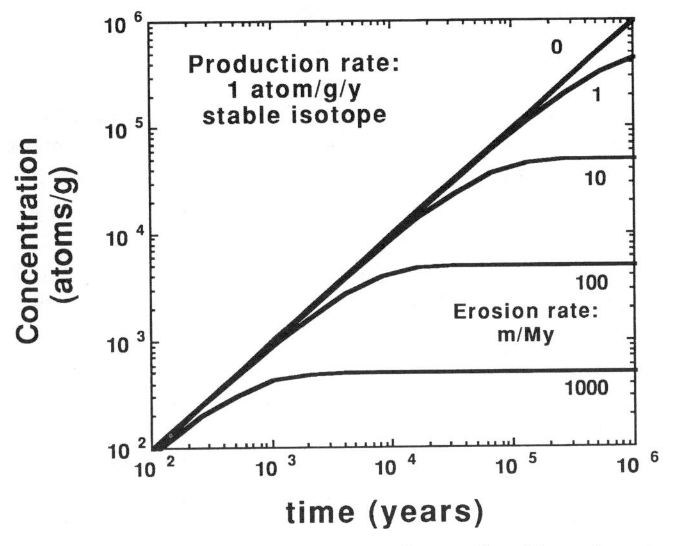

Figure 9 Concentrations of in-situ stable isotopes for samples with erosion rates ranging from 0 to 1000 m/My.

Klein et al (1986), Nishiizumi et al (1991a,b), and Lal (1991) discuss the advantages of using multiple isotopes, especially the ^{10}Be–^{26}Al pair, both of which are commonly measured in quartz samples. ^{10}Be and ^{26}Al appear to be produced in the same ratio at depth (Brown et al 1992) although such may not be the case for all isotope pairs because of production by thermal neutrons (e.g. ^{36}Cl) or by μ^- at depth. These papers show that the $^{26}Al/^{10}Be$ ratio is constant for samples with no burial history except for extremely old surfaces that have low erosion rates. Therefore, significant deviation of the $^{26}Al/^{10}Be$ ratio below its production ratio of about 6.1 implies that the sample is very old (>ca. 250,000 years) *and* has had a very low erosion rate (<1 m/My), that the sample has been buried and shielded from cosmic rays for a time sufficient for ^{26}Al depletion compared to ^{10}Be, or both. If one assumes a simple history for the latter case (a single cycle of exposure followed by burial) then the original exposure period as well as the burial time can be computed (see Klein et al 1986, Nishiizumi et al 1991a, Lal 1991).

Models have also been developed for: ice accumulation or ablation (Lal et al 1987b, Lal & Jull 1990), significant changes in altitude (Craig & Poreda 1986a, Lal 1991), alternating periods of erosion and stability (Klein et al 1986, Lal 1991), and differential erosion of moraines producing a population of boulders at the surface which have different initial burial depths (Zreda & Phillips 1993).

REALITY OF SAMPLING CONDITIONS

Condition of Sample Surface

While the above discussion indicates that the in-situ cosmogenic surface exposure method has great potential for solving some problems basic to geomorphology, archeology, erosion, sediment transport, fault recurrence intervals, etc, the reality of the field situation tempers enthusiasm when it is time to collect samples. A clear problem is the identification of "original" surfaces. Glacially striated surfaces (Figure 10A) are a welcome but rare case of preserved surfaces that could have had only minimal erosion (<1 cm) as the delicate scratches are only a few mm deep. Volcanic features that form at the Earth's surface include pahoehoe structures, but excavation of lava flows often shows that some flows are literally layer upon layer of onlapping pahoehoe features. However, new volcanic flows have numerous closed depressions that eventually get filled with volcanic rubble which provide evidence for significant erosion. Samples collected from rubble-free closed depressions are likely to have experienced little erosion. Likewise, lava flows are generally more vesicular at the top of the flow and become more dense at depth, so that dense lavas are unlikely to be primary surface features of lava flows. Large boulders are often a tempting target to collect, but it should be remembered that trees can turn over even large boulders (Figure 10B), and that exfoliation by heat (such as range fires) can cause degradation of the original surface of the boulder (Bierman & Gillespie 1991). In some cases the surface of the boulder is sporadically exfoliated giving rise to "hat" boulders. Such disintegration probably explains why Bull Lake moraines in the Yellowstone region are virtually devoid of surface boulders whereas the younger Pinedale moraines have numerous well-exposed boulders. Surface heating without exfoliation

---

Figure 10 Exposure age surfaces and their problems.

(*A*) Glacially striated surface on basalt from Mauna Kea, Hawaii. Striations indicate that minimal surface modification has occurred.

(*B*) Large volcanic boulder turned over by a tree.

(*C*) Desert pavement surface (Nevada) made up of volcanic clasts underlain by loess.

(*D*) Large boulder deposited by the Bonneville Flood (Idaho) showing water-smoothed surface, indicating that the surface has not exfoliated since transport and deposition. Boulder is at least 7 km downstream from the outcrop from which it was derived.

(*E*) Provo shoreline showing a wave-cut platform and "sea stacks" cut into basalt at Dunderburg Butte in Utah.

(*F*) Tabernacle Hill flow surface near Fillmore, Utah. Pahoehoe surface is in the foreground and pressure ridges are in the background. The Tabernacle Hill flow erupted during the short-lived Provo stillstand and is an important calibration point for in-situ production of cosmogenic isotopes.

A

B

C

D

Figure 10.—(continued)

E

F

Figure 10.—(continued)

could cause selective loss of some in-situ cosmogenic isotopes (e.g. ^{3}He or ^{21}Ne). Clearly such problems are far more likely in some vegetational settings and climates than others, but such considerations must always be taken into account. A previous exposure history is a potential problem for many samples, especially any redeposited samples (e.g. cataclysmic flood deposits, alluvial deposits, glacial scours or moraines). The problems are different for different isotopes (e.g. stable versus radioactive) and for different problems and should be evaluated for each case.

Spatial or Temporal Shielding of Cosmic Rays

Spatial shielding of cosmic rays results in a lower cosmic ray flux and hence lower production rate, except in the case where there is significant production by thermal neutrons in which there can be enhanced production. All of the above discussion has assumed that the target surface was horizontal and that there was no shielding of incident cosmic rays. However, in certain circumstances cosmic rays are shielded by rocks (or trees), usually at relatively low angles above the horizon. Because incident cosmic rays are attenuated by the atmosphere and the atmosphere is thickest for low angle incident radiation, the correction for production due to primary and secondary cosmic rays is relatively small if the shielding angles are low (Figure 11). Effects of shielding can be obtained by inte-

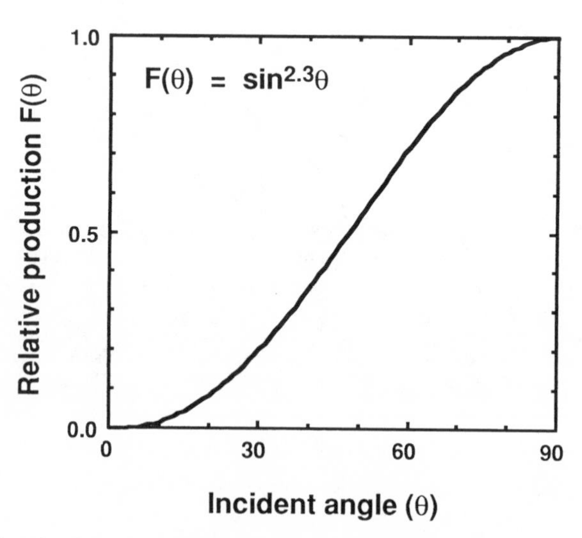

Figure 11 Angular dependence of cosmic-ray production: $F(\theta) = \sin^{2.3}\theta$ (Nishiizumi et al 1989).

grating the free access of cosmic radiation for the angular dependence of incident cosmic rays (Nishiizumi et al 1989):

$$F(\theta) = \sin^{2.3} \theta \tag{17}$$

compared to the case for no shielding (2π). For many cases shielding of incident cosmic radiation is minimal; however, for regions of high relief it may be very important.

Cosmogenic isotopes with a significant component due to thermal neutrons warrant discussion because the production of isotopes by thermal neutrons does not have the simple depth dependence due to the air-lithosphere boundary effect (Yamashita et al 1966, O'Brien et al 1978, Fabryka-Martin 1988, Zreda et al 1991). For certain geometries this may require a significant correction.

Temporally, several corrections should be considered. First, burial by wind blown ash, dust, or sand could cause a significant shielding for uncertain periods of time if it occurred in the past history during different dust flux conditions. Likewise, burial by ice or snow could have an important effect for total shielding because of attenuation by slow neutrons which is enhanced in the presence of hydrogen (Zreda & Phillips 1993). Tree cover should also be considered in some cases (Cerling & Craig 1993).

While this discussion has focused on realizing the effect of shielding for cosmogenic applications, it is important to note that samples collected to evaluate the mantle He or Ne component must avoid any possible cosmogenic contributions. In such cases it is preferable to sample several meters below the surface. For either application careful field notes and photographs are important.

Gain or Loss of Isotopes from Minerals

An important consideration is the possible gain or loss of cosmogenic isotopes from the sample. Gain by contamination from atmospherically-derived isotopes has already been discussed, as has the inherent background of pre-exposure isotopes including those produced by deeply penetrating muons or radiogenic contamination. Loss of cosmogenic isotopes is not significant for most isotopes except ^{3}He, which can be lost by diffusion from certain minerals such as quartz or feldspar on time scales relevant to exposure age studies (Cerling 1990, Trull et al 1991, Brook & Kurz 1993, Brook et al 1993). Olivine and pyroxene are retentive with respect to both ^{3}He and ^{21}Ne, and plagioclase is retentive for ^{21}Ne (Craig & Poreda 1986a, Cerling 1990, Poreda & Cerling 1993). Cerling (1990), Brook & Kurz (1993), and Brook et al (1993) show that significant diffusional loss of ^{3}He in quartz takes place even at low temperatures in Antarctica—up to 100% for fine-grained quartz, although in large grains the loss may be minimal. It is unlikely to be retained in quartz or feldspars in warmer climates. Quartz grains in plutonic rocks have been problematical for ^{21}Ne studies, possibly because of $^{18}O(\alpha,n)^{21}Ne$ reactions over long periods of time (Poreda & Cerling 1992).

APPLICATIONS TO SURFACE EXPOSURE PROBLEMS

Dating of geomorphic surfaces is still a major problem in geomorphology. In the past few years, the in-situ cosmogenic exposure age method has provided some dates on a variety of geomorphic surfaces. A number of these are discussed below to give a flavor of the sorts of surfaces that can be dated using this method. There is still some inconsistency in production rates used by various authors, which results in minor inconsistencies between different workers (or the same workers at different times). Much of this is because the production rates are not well enough known, in part because the ^{14}C age (as opposed to the calendar age) was used to calibrate production rates in some studies but not in others (see discussions in Bard et al 1990, Kurz et al 1990, and Cerling & Craig 1994). In the following discussion we use the dates given in the original work unless specified. "^{14}C BP" refers to ages based on the radiocarbon time scale without correction (Stuiver & Pearson 1986, Pearson & Stuiver 1986, Bard et al 1990).

Glaciation

The direct dating of glacial events is virtually impossible except by the in-situ exposure method. Previously, most attempts at dating glacial events used bracketing ^{14}C dates on organic material (bogs, trees, etc) and tephrochronology. Several groups have dated morainal material using in-situ cosmogenic isotopes. Table 5 gives data measured at SIO on glacially-scoured surfaces or glacial erratics from the Makanakan glaciation on Mauna Kea, Hawaii (Figure 10A). Three different samples give exposure ages of 13,200 $\pm$ 200 years. Dorn et al (1991) estimate deglaciation of the younger Makanakan glaciation of Mauna Kea, Hawaii to be complete by ca. 15,000 years based on ^{36}Cl dates. Work by Phillips et al (1990) in Bloody Canyon (California) gives ages of 21,400 and 24,300 years for the Tioga and Tenaya moraines, respectively, and a significantly older age for the Tahoe moraine. The older moraines (such as Tahoe) were complicated by the natural excavation of glacial erratic boulders which results in a distribution of ages, with the oldest ages being closest to the age of the glacial event (Phillips et al 1990, Zreda & Phillips 1993). Cerling (1990) found a volcanic flow in the Owens Valley overlain by Tioga age glacial

Table 5 Our cosmogenic ^{3}He data from the Makanakan glaciation on Mauna Kea, Hawaii

Sample (elevation)[a]	^3He_c[b] 10^6 atoms/g	% ^3He_c[c]	Age[d] (years)
8119 (3752 m)	11.3±0.4	39.1	13,360±480
8119 (3752 m)	11.0±0.4	35.9	13,040±520
8123 (3720 m)	10.8±0.2	95.8	12,960±240
8123 (3720 m)	11.1±0.2	93.1	13,310±220
8125 (3720 m)	11.0±0.2	86.5	13,290±210

[a] Measured at Scripps Institution of Oceanography with the elevation of each sample given.
[b] ^3He_c is the cosmogenic ^{3}He after correction for inherited ^{3}He.
[c] % ^3He_c is the percent of the total ^{3}He in the sample.
[d] Calculated using the production rate from Figure 4.

outwash; the flow gave three exposure age ^{3}He dates averaging 12,600 years ^{14}C BP (ca. 15,700 years calibrated ^{14}C BP). Very old glacial deposits for Antarctica are indicated by the work of Cerling (1990) and Nishiizumi et al (1991a) in the Allan Hills region, and by Brown et al (1991), Nishiizumi et al (1991a), Brook & Kurz (1993), and Brook et al (1993) for the Dry Valleys region. Because of the very old exposure ages, erosion rates can be estimated using the ^{10}Be–^{26}Al pair. Glacial episode Taylor IVb is estimated to be about 2 Ma, with erosion rates of 0.1 m/My or lower (Brown et al 1991, Nishiizumi et al 1991a, Brook et al 1993).

The in-situ exposure age method has great promise for determining the glacial history of the late Pleistocene and Holocene. Rates of glacial retreat, the global extent of the younger Dryas, and the synchroneity of glacial stages are tractable problems that can be solved using cosmogenic isotope measurements.

Erosion Rates

Craig & Poreda (1986a) and Kurz (1986b) discuss erosion rates calculated from stable noble gases and conclude that the summit of Haleakala Volcano (Hawaii) is eroding at a rate on the order of 10 m/My (Figure 1). Cerling (1990) estimates erosion rates on the order of 2 m/My for stable basalt surfaces in the Coso volcanics of California. Using both ^{10}Be and ^{26}Al, Albrecht et al (1993) estimate erosion rates for stable surfaces of the Bandelier Tuff (New Mexico) to be on the order of 1 to 10 m/My. Although higher than the rates reported by Brook et al (1993) such erosion rates are still considerably lower than regional erosion rates, as would be expected for relatively stable surfaces in an arid landscape.

Inselbergs, Kopjes, and Ancient Erosion Surfaces

Anyone viewing the numerous bustling kopjes (residual hills) that punctuate the Serengeti savannah at convenient intervals must immediately think of the application of cosmogenic isotopes to erosion rates and exposure ages over this broad region. These small kopjes are, however, simply the end-member of a range of majestic features such as the classic inselberg of Ayers Rock in Central Australia: almost 3 km of exposed Precambrian sandstone rising some 300 meters or more above the almost flat surrounding landscape. Inselbergs and kopjes, the spectacular granite tors of Dartmoor, and the half-drowned Brazilian Sugar Loaf, all represent obvious targets for quantitative studies of erosion rates and exposure ages in climatic regimes ranging from temperate and dry to tropical and humid, with erosional processes varying from exfoliation to chemical weathering.

Kopjes and inselbergs dominate the spectacular erosional surfaces of the southern African continent that will someday be the subject of detailed isotopic studies. No better introduction to the geology and geography of these inselberg-dominated landscapes can be found than the classic test *Principles of Physical Geology* (Holmes 1965). The ancient erosion surfaces dating back almost 140 million years comprise five well-characterized systems ranging from Madagascar to Angola and the Congo, and from Capetown to Uganda, shaped by the classic Penckian processes of pedimentation and scarp retreat encroaching on the flattened upland surfaces. The photographs and diagrams of Holmes (1965, pp. 600–15) must be consulted for a proper appreciation of these majestic escarpments, rising to more than 3 km altitude on the Gondwana surface. These great inselberg landscapes are in fact an entire subject for a major program of cosmogenic isotope studies of variations in erosion rates with climatic, tectonic, and rock compositional features on a grand scale.

Volcanic Events

On obvious application for the in-situ cosmogenic exposure method is the dating of young basaltic or andesitic landforms, which are difficult to date by K-Ar because of their low potassium content and inherited argon problems. Rhyolitic landforms are not so amenable to in-situ methods as it is often difficult to recognize original surfaces. Some basaltic landforms that have recently been dated include the Owens Valley flow at Aberdeen, California ($12{,}600 \pm 1000$ ^{14}C BP: ca. $15{,}700 \pm 1100$ years using the production rate of Figure 4) and Red Hill in the Coso volcanic field, California [$57{,}000 \pm 2000$ ^{14}C BP (Cerling 1990): $= 69{,}000 \pm 2000$ years using Figure 4]; Lashaine volcano in East Africa (see Figure 5 and earlier discussion);

the Potrillo volcanic field in New Mexico (Anthony & Poths 1992); and the Lathrop Wells cone in Nevada (Zreda et al 1993). The age of the Lathrop Wells basaltic cinder cone and its associated flows is controversial because of its proximity to the proposed radioactive waste repository at Yucca Mountain. The age of $81{,}000 \pm 8000$ years reported by Zreda et al (1993) using in-situ ^{36}Cl, which has been corroborated by ^{3}He measurements by T. E. Cerling & R. J. Poreda (unpublished) and by Poths & Crowe (1992), is older than estimated using landform degradation rates (Wells et al 1990) but younger than K-Ar results (Turrin et al 1991) which have a very large age distribution and may suffer from inherited argon (Poths & Crowe 1992). So far, the in-situ cosmogenic isotope method gives the most consistent results although the age estimate may change as better production rates are established (see Table 3 and Figure 4).

Alluvial Deposits (Of Time and the River)

Oh give me a land
Where the bright diamond sand
Flows leisurely down the stream . . .

"Home on the Range" (Anon)

The anonymous author of the song beautifully describes the transport of heavy minerals down the courses of rivers fortunate enough to flow from diamond-rich sources. For a long time it was known that a small fraction of analyzed diamonds had very high $^{3}He/^{4}He$ ratios relative to mantle basalts, but because almost all analyzed diamonds are of unknown provenance, conventional wisdom could only suppose that such high ratios were truly representative of Earth's primordial mantle material. One of the present authors, however, supposed otherwise, and suggested that these diamonds contained cosmogenic ^{3}He. Thus in a cooperative study, a "High-^{3}He" diamond was found to contain significant amounts of ^{10}Be proving that both it and the high ^{3}He content were due to cosmic-ray production during transport and surface exposure in the rivers of Zaire (Lal et al 1987b). Diamonds are the classic heavy mineral tracers for fluvial transport because their morphology changes progressively downstream from the source to the seacoast: We now have the possibility of a direct measure of the "leisurely" times involved.

Measurements of cosmogenic ^{3}He will ultimately supply a method for measuring relative and perhaps absolute residence times of alluvial diamonds in river courses, although the "lingering times" in local gravel accumulations at slope and course changes will introduce some interesting problems to the analysis. It may be noted that ^{3}He and ^{10}Be are the only cosmogenic isotope that can be measured in alluvial diamonds, but the complete spectrum will be available in such heavy minerals as cassiterites, gold, and other placer gravels. Other more common minerals, such as quartz (Lal & Arnold 1985), can be used to estimate regional transport and erosion rates although in the end such studies will require numerous samples to overcome the stochastic nature of sediment transport and surface residence history.

Ice Ablation Rates

Ice ablation rates have been determined in the Allan Hills region, Antarctica. Fireman & Norris (1982) measured anomalously high ^{14}C in ablating ice from the Allan Hills and suggested that it may result from in-situ spallation of oxygen in the ice. Ice from this region in 1987 had ^{14}C concentrations of several 1000 atoms/g which correspond to ablation rates of ca. 6 and 8 cm/year for two different sites (Lal et al 1990). Those values agree with earlier estimates of ice ablation rates based on the stake method.

Meteorite Impact

Meteor Crater in Arizona is a late Pleistocene impact crater that provides an excellent example of some of the aspects of cosmogenic dating approaches. The meteor impact exposed large blocks of the Kaibab dolomite which had been previously shielded by about 10 m of the overlying Moenkopi sandstone. From sampling the tops of ejecta blocks, Nishiizumi et al (1991b) find an impact age of $49{,}200 \pm 1700$ years using ^{10}Be and ^{26}Al, and Phillips et al (1991) find an impact age of $49{,}700 \pm 850$ years using ^{36}Cl. These dates agree with an earlier thermoluminescence age of $49{,}000 \pm 3000$ years (Sutton 1985) and is considerably older than previous estimates based on soil development (Shoemaker 1983). Nishiizumi et al (1991b) also collected samples from the base of a large ejecta block and used the age derived from it to calculate erosion rates around the base of the block; they also measured ages of talus within the crater and found them to be significantly younger than the age of the crater itself. These studies illustrate the importance of studying multiple samples to better assure that the age of the event is determined.

Desert Pavement

One of the ubiquitous features of desert landscapes, and one with a widely misunderstood origin, is desert pavement (Figure 10C). Desert pavement is the interlocking pebble- to gravel-size surface material, often with a very flat surface, that can withstand even heavy vehicular traffic until the surface is disrupted. Loess is usually found below the pavement, sometimes to a depth of several meters. Most introductory textbooks give the explanation that the fine material has been blown away, giving a deflationary lag deposit. Never is an explanation given as to why the peculiar mixture of loess with gravel was deposited in the first place! When wetted, smectites in the loess expand giving the landscape a "puffy" appearance. McFadden et al (1987) suggest that desert pavements are actually inflation surfaces, starting with a coarse pebble or gravel surface layer that is detached from the surface as aeolian silt accumulates in cracks and crevices adjacent to pebble or gravel surfaces. Thus, desert pavement actually stays at the atmosphere-lithosphere interface as inflationary dust lifts it above the original surface. Wells et al (1991) showed that the concentrations of cosmogenic ^{3}He and ^{21}Ne in desert pavement clasts were the same as in outcropping basalt flows underlying the pavement, which supports the accretionary model for desert pavement formation. This leads to the speculation that the oldest surfaces on the continents (except Antarctica) may well be inflationary desert pavements.

Cataclysmic Floods

Cataclysmic floods represent another important category of potential geomorphic events that can be dated by the in-situ cosmogenic exposure method. Extensive flood events resulting from overflow of a nonresistant sill (e.g. Bonneville Flood, Owens Valley Flood) or glacial outburst (e.g. Channeled Scablands of the Columbia River, Big Lost River) have been recognized in the western U.S. and elsewhere. Dating of these deposits cannot be done directly using conventional dating techniques, but have had to rely on extensive stratigraphic studies. Large boulders deposited by these floods (Figure 10D) and water-scoured surfaces are amenable to dating by the in-situ cosmogenic method. Ages of 15,500 ^{14}C BP ($=20{,}500 \pm 800$ years using the production rates of Figure 4) for the Owens River flood (Cerling 1990) and 20,500 years for the Big Lost River flood (Cerling et al 1994) are compatible with other geological arguments. The well-dated Bonneville Flood (Oviatt et al 1992) has been used as a calibration point for determining cosmogenic production rates (Cerling 1990, Cerling & Craig 1994; see also Figure 10E). However, the concentration of cosmogenic ^{3}He in the Bonneville Flood deposits is the same as from the basalt of Tabernacle Hill (Figure 10F), which has a very similar age, when corrected for altitude and latitude (Cerling & Craig 1994).

The Final Riddle of the Sphinx (and other Archaeological Studies)

The Sphinx is currently the subject of controversy with respect to ages of the individual components. The proposed ages for this monumental artifact are, on the one hand, $\sim$4500 ^{14}C BP (conventional wisdom: Pharaoh Khafre in charge), and on the other, 7000–9000 ^{14}C BP (with restoration by Khafre at the conventional date; Schock et al 1991). The latter estimate is based on subsurface sonic patterns attributed to weathering of the limestone to depths as great as two meters or more at the front of the big cat, but much less elsewhere on the feline corpus. Thus the question at hand is whether the head and forequarters of the Sphinx, as well as the original quarry floor (possibly 12,000 ^{14}C BP?) predate the hindquarters by 2500 to as much as 7500 years. The question is precisely the type best suited in principle for study by cosmogenic isotopes. The problem is simply whether the solution lies within the scope of the present state of the art.

First, the problem is not simply the exposure age, because erosion has proceeded fairly rapidly in terms of the desire for preservation. Holmes (1965, p. 15) shows the state of this feline artifact, elegantly linked with Khafre's pyramid, and labeled as "two celebrated products of the Biosphere." Both are nummulitic limestone: The Sphinx was carved from the outcropping living rock and the Pyramids were made of massive blocks of the same material. (The application of cosmogenic techniques to the Pyramids is difficult because the surfaces were once faced with granite, but the Sphinx has been effulgently nude since birth.) Secondly, the only possible candidate isotope is cosmogenic ^{14}C (Jull et al 1993) or ^{36}Cl. Donahue et al (1990) estimate that the sea-level production rate of ^{14}C is $\sim$21 atoms g^{-1} y^{-1}, so that at, e.g. 8000 ^{14}C BP and 4500 ^{14}C BP respectively, the expected cosmogenic activities are $\sim$170,000 and 95,000 atoms/g, with a difference, therefore, of the order of 25,000 atoms/g for 3500 years in age. One wants to determine (*a*) the actual exposure age (with due attention to erosion), and (*b*) the difference in age between component sections. One has to dissolve the limestone matrix to extract cosmogenic ^{14}C: A. J. T. Jull (personal communication to HC) estimates that the blank contribution in this process will be about the same as the expected activities per gram cited above, so that some 50 grams of material per sample will be required. This is by no means a simple undertaking, considering that the above predicted activities will be lower because of the finite erosion rates. It is only feasible because about half of the cosmogenic ^{14}C is present as ^{14}CO (Lal et al 1990), making it possible to extract the cosmogenic component selectively from limestone.

Egyptologists, relax: You have nothing to fear for the moment. (But let your scientific progeny beware! The time will come for this study.) This is only one example of the many archaeological problems that lend themselves to investigation by the in-situ cosmogenic method.

CONCLUSIONS

In-situ cosmogenic isotopes can be used to study the exposure history of the Earth's surface. Production rates of different cosmogenic isotopes are known and are being refined. The production rate of an individual isotope is determined by the rock composition and the integrated cosmic-ray flux at its locality. At high latitudes the variation with time is small and is expected to be much higher at low latitudes; the signal is attenuated with time. The production rate of all isotopes increases significantly with elevation. The production rate for ^{3}He at sea level and high latitudes is about 115 atoms g^{-1} y^{-1} over the last 17,500 years; other isotopes have lower production rates for sea level and high latitudes, and are more compositionally dependent. For many studies, a thousand years of exposure history produces enough in-situ cosmogenic isotopes to be measured.

In-situ cosmogenic isotopes have great potential to establish rates and dates of geomorphic processes and events. The exposure age of many geomorphic surfaces can be determined, including glacial or flood deposited boulders or scoured surfaces, lava flows, quarried building stone, and alluvial fans or terraces. Rates of erosion, glacial retreat, ice ablation, cliff retreat, or inflation of desert pavement can also be studied. Multiple cosmogenic isotopes can be used to unravel the complexity of the exposure histories of many geomorphic features, both constructive and destructive in origin. The answers are indeed "blowing in the cosmogenic wind," and the coming decade will reap a harvest of new understanding and ideas over scales representing many orders of magnitude of the processes that shape and define the Earth's surface.

Acknowledgments

This work was supported by National Science Foundation grants to TEC and HC. We thank R. J. Poreda for many discussions over many years.

Literature Cited

Albrecht A, Herzog GF, Klein J, Dezfouly-Arjomandy B, Goff F. 1993. Quaternary erosion and cosmic-ray-exposure history derived from ^{10}Be and ^{26}Al produced in situ—an example from Pajarito plateau, Valles caldera region. *Geology* 21: 551–54

Anthony EY, Poths J. 1992. ^{3}He surface exposure dating and its implications for magma evolution in the Potrillo volcanic field, Rio Grande Rift, New Mexico, USA. *Geochim. Cosmochim. Acta* 56: 4105–8

Bard E, Hamelin B, Fairbanks RG, Zindler A. 1990. Calibration of the ^{14}C time scale over the past 30,000 years using mass spectrometric U-Th ages from Barbados corals. *Nature* 345: 405–10

Bierman P, Gillespie A. 1991. Range fires: a significant factor in exposure-age determination and geomorphic surface evolution. *Geology* 19: 641–44

Bourlès DL, Brown ET, Raisbeck GM, Yiou F, Grandin G, et al. 1992. Constraining the history of iron crust laterite systems using in situ produced cosmogenic ^{10}Be and ^{26}Al. *Eos, Trans. Am. Geophys. Union* 73: 173 (Abstr.)

Brook EJ, Kurz MD. 1993. Surface exposure chronology using in situ cosmogenic ^{3}He in Antarctic quartz sandstone boulders. *Quat. Res.* 39: 1–10

Brook EJ, Kurz MD, Ackert RP, Denton GH, Brown ET, et al. 1993. Chronology of Taylor Glacier advances in Arena Valley, Antarctica, using in situ cosmogenic ^{3}He and ^{10}Be. *Quat. Res.* 39: 11–23

Brown ET, Brook EJ, Raisbeck GM, Yiou F, Kurz MD. 1992. Effective attenuation lengths of cosmic rays producing ^{10}Be and ^{26}Al in quartz: implications for exposure age dating. *Geophys. Res. Lett.* 19: 369–72

Brown ET, Edmond JM, Raisbeck GM, Yiou F, Kurz MD, et al. 1991. Examination of surface exposure ages of Antarctic moraines using in situ ^{10}Be and ^{26}Al. *Geochim. Cosmochim. Acta* 55: 2269–83

Brown L, Pavich MJ, Hickman RE, Klein J, Middleton R. 1988. Erosion of the eastern United States observed with ^{10}Be. *Earth Surf. Proc. Landforms* 13: 441–57

Cerling TE. 1990. Dating geomorphic surfaces using cosmogenic ^{3}He. *Quat. Res.* 33: 148–56

Cerling TE, Craig H. 1994. Cosmogenic production rates of ^{3}He from 39 to 46°N latitude, western USA and France. *Geochim. Cosmochim. Acta* 58: 249–55

Cerling TE, Poreda RJ, Rathburn SL. 1994. Cosmogenic ^{3}He and ^{21}Ne age of the Big Lost River Flood, Snake River Plain, Idaho. *Geology*. In press

Champion DE. 1980. Holocene geomagnetic secular variation in the western United States: implications for the global geomagnetic field. *U. S. Geol. Surv. Open-File Rep.* 80-824: 1–314

Craig H, Poreda RJ. 1986a. Cosmogenic ^{3}He in terrestrial rocks: the summit lavas of Maui. *Proc. Natl. Acad. Sci.* 83: 1970–74

Craig H, Poreda RJ. 1986b. Cosmogenic ^{3}He and model erosion rates in terrestrial rocks. *Eos, Trans. Am. Geophys. Union* 67: 414 (Abstr.)

Craig H, Poreda RJ. 1994. Cosmogenic ^{3}He in olivines and clinopyroxenes from Maui basalts. *Geochim. Cosmochim. Acta* In press

Craig H, Poreda R, Lupton JE, Marti K, Regnier S. 1979. Rare gases and hydrogen in josephinite. *Eos, Trans. Am. Geophys. Union* 60: 970 (Abstr.)

Damon PE, Lerman JC, Long A. 1978. Temporal fluctuations of atmospheric ^{14}C: causal factors and implications. *Annu. Rev. Earth Planet. Sci.* 6: 457–94

Davis R, Schaeffer OA. 1955. Chlorine-36 in nature. *Ann. NY Acad. Sci.* 62: 105–22

Donahue DJ, Jull AJT, Toolin LJ 1990. Radiocarbon measurements at the University of Arizona AMS facility. *Nucl. Instr. Meth. Phys. Res.* B52: 224–28

Dorn RI, Phillips RM, Zreda MG, Wolfe EW, Jull AJT, et al. 1991. Glacial chronology. *Natl. Geogr. Res. Expl.* 7: 456–71

Elmore D, Phillips FM. 1987. Accelerator mass spectrometry for measurement of long-lived radioisotopes. *Science* 236: 543–50

Fabryka-Martin JT. 1988. *Production of radionuclides in the earth and their hydrologic significance, with emphasis on chlorine*-36 *and iodine*-129. PhD dissertation. Univ. Ariz. 400 pp.

Fireman EL, Norris TL. 1982. Ages and composition of gas trapped in Allan Hills and Byrd core ice. *Earth Planet. Sci. Lett.* 60: 339–50

Fleischer RL, Price PB, Walker RM. 1975. *Nuclear Tracks in Solids: Principles and Applications*. Berkeley: Univ. Calif. Press. 605 pp.

Forbush SE. 1954. World-wide cosmic-ray variations. *J. Geophys. Res.* 59: 525–42

Friedlander G, Kennedy JW. Macias ES, Miller JM. 1981. *Nuclear and Radiochemistry*. New York: Wiley. 684 pp.

Graf T, Kohl CP, Marti K, Nishiizumi K. 1991. Cosmic-ray produced neon in Antarctic rocks. *Geophys. Res. Lett.* 18: 203–6

Holmes A. 1965. *Principles of Physical Geology*. New York: Ronald. 2nd ed.

Jha R, Lal D. 1982. On cosmic ray production of isotopes in surface rocks. In *National Radiation Environment*, ed. KG Vohra, KC Pillai, UC Mishra, S Sadasivan, pp. 629–35. New Dehli: Eastern Wiley

Jull AJT, Donahue, DJ, Linick TW, Wilson GC. 1989. Spallogenic ^{14}C in high-altitude rocks and in Antarctic meteorites. *Radiocarbon* 31: 719–24

Jull AJT, Wilson AE, Burr GS, Toolin LJ, Donahue DJ. 1992. Measurements of cosmogenic ^{14}C produced by spallation in high-altitude rocks. *Radiocarbon* 34: 737–44

Klein J, Giegengack R, Middleton R, Sharma P, Underwood JR, et al. 1986. Revealing histories of exposure using in situ produced ^{26}Al and ^{10}Be in Libyan desert glass. *Radiocarbon* 28: 547–55

Kohl CP, Nishiizumi K. 1992. Chemical isolation of quartz for measurement of in-situ-produced cosmogenic nuclides. *Geochim. Cosmochim. Acta* 56: 3583–87

Kurz MD. 1986a. Cosmogenic helium in a terrestrial igneous rock. *Nature* 320: 435–39

Kurz MD. 1986b. In situ production of terrestrial cosmogenic helium and some applications to geochronology. *Geochim. Cosmochim. Acta* 50: 2855–62

Kurz MD, Colodner D, Trull TW, Moore RB, O'Brien K. 1990. Cosmic-ray exposure dating with in situ produced cosmogenic ^{3}He: results from young Hawaiian flows. *Earth Planet. Sci. Lett.* 97: 177–89

Lal D. 1986. On the study of continental erosion rates and cycles using cosmogenic ^{10}Be and ^{26}Al and other isotopes. In *Dating Young Sediments*, ed. AJ Hurford, E Jäger, JAM Ten Cate, pp. 285–97. Bangkok: CCOP Tech. Secretariat

Lal D. 1987a. Cosmogenic nuclides produced in situ in terrestrial solids. *Nucl. Instr. Meth. Phys. Res.* B29: 238–45

Lal D. 1987b. Production of ^{3}He in terrestrial rocks. *Chem. Geol.* (*Isotope Geosci. Sect.*) 66: 89–98

Lal D. 1988a. In situ produced cosmogenic isotopes in terrestrial rocks. *Annu. Rev. Earth Planet. Sci.* 16: 355–88

Lal D. 1988b. Interesting geophysical applications of terrestrial long-lived cosmogenic stable and radioactive isotopes. *Proc. 45th Solar-Terrestrial Relationships and the Earth Environment in the Last Millennia*, pp. 234–45. Bologna, Italy: Soc. Ital. Fis.

Lal D. 1988c. Theoretically expected variations in the terrestrial cosmic-ray production of isotopes. *Proc. 45th Solar-Terrestrial Relationships and the Earth Environment in the Last Millennia*, pp. 216–33. Bologna, Italy: Soc. Ital. Fis.

Lal D. 1991. Cosmic ray labeling of erosion surfaces: in situ nuclide production rates and erosion models. *Earth Planet. Sci. Lett.* 104: 424–39

Lal D, Arnold JR. 1985. Tracing quartz through the environment. *Proc. Ind. Acad. Sci.: Earth Planet. Sci.* 94: 1–5

Lal D, Arnold JR, Honda M. 1960. Cosmic-ray production rates of Be^{7} in oxygen, and P^{32}, P^{33}, S^{35} in argon at mountain altitudes. *Phys. Rev.* 118: 1626–32

Lal D, Jull AJT. 1990. On determining ice accumulation rates in the past 40,000 years using in situ cosmogenic ^{14}C. *Geophys. Res. J.* 17: 1303–6

Lal D, Jull, AJT, Donahue DJ, Burtner D, Nishiizumi K. 1990. Polar ice ablation rates measured using in situ cosmogenic ^{14}C. *Nature* 346: 350–52

Lal D, Nishiizumi K, Arnold JR. 1987a. In situ cosmogenic ^{3}H, ^{14}C, and ^{10}Be for determining the net accumulation and ablation rates of ice sheets. *J. Geophys. Res.* 92: 4947–52

Lal D, Nishiizumi K, Klein J, Middleton R, Craig H. 1987b. Cosmogenic ^{10}Be in Zaire alluvial diamonds: implications for ^{3}He contents of diamonds. *Nature* 328: 139–41

Lal D, Peters B. 1967. Cosmic-ray produced radioactivity on the earth. *Handb. Phys.* 46: 551–612

Lingenfelter RE. 1963. Production of 14-C by cosmic ray neutrons. *Rev. Geophys.* 1: 35–55

Mabuchi H, Gensho Y, Wada Y, Hamaguchi H. 1971. Phosphorus-32 induced by cosmic rays in laboratory chemicals. *Geochem. J.* 4: 105–10

Marti K, Craig H. 1987. Cosmic-ray produced neon and helium in the summit lavas of Maui. *Nature* 325: 335–37

Mazaud A, Laj C, Bard E, Arnold M, Tric E. 1991. Geomagnetic field control of ^{14}C production over the last 80 Kyr: implications for the radiocarbon time-scale. *Geophys. Res. Lett.* 18: 1885–88

McFadden LD, Wells SG, Jercinovich MJ. 1987. Influences of eolian and pedogenic processes on the origin and evolution of desert pavements. *Geology* 15: 504–8

McKean J, Dietrich WE, Finkel RC, Southon JR, Caffee MC. 1993. Quantification of soil production and downslope creep rates from cosmogenic ^{10}Be accumulations on a hillslope profile. *Geology* 21: 343–46

Middleton R, Klein, J. 1987. ^{26}Al: measurement and applications. *Phil. Trans. R. Soc. London Ser. A* 323: 121–43

Monaghan MC, Klein J, Measures CI. 1988. The origin of ^{10}Be in island-arc volcanic rocks. *Earth Planet. Sci. Lett.* 89: 288–98

Monaghan MC, Krishnaswami S, Thomas, JH. 1983. ^{10}Be concentrations and the long-term fate of particle-reactive nuclides in five soil profiles from California. *Earth Planet. Sci. Lett.* 65: 51–60

Monaghan MC, McKean J, Dietrich W, Klein J. 1992. ^{10}Be chronometry of bedrock-to-soil conversion rates. *Earth Planet. Sci. Lett.* 111: 483–92

Nakamura Y, Mabuchi H, Hamaguchi H. 1972. ^{7}Be production from oxygen by atmospheric cosmic rays. *Geochem. J.* 6: 43–47

Niedermann S, Graf Th, Marti K. 1993. Mass spectrometric identification of cosmic-ray-produced neon in terrestrial rocks with multiple neon components. *Earth Planet. Sci. Lett.* 118: 65–73

Nishiizumi K, Elmore D, Ma XZ, Arnold JR. 1984a. ^{10}Be and ^{36}Cl depth profiles in an Apollo drill core. *Earth Planet. Sci. Lett.* 70: 157–63

Nishiizumi K, Klein J, Middleton R, Arnold JR. 1984b. ^{26}Al depth profile in Apollo 15 drill core. *Earth Planet. Sci. Lett.* 70: 164–68

Nishiizumi K, Klein J, Middleton R, Craig H. 1990. Cosmogenic ^{10}Be, ^{26}Al, and ^{3}He in olivine from Maui lavas. *Earth Planet. Sci. Lett.* 98: 263–66

Nishiizumi K, Kohl CP, Arnold JR, Klein J, Fink D, et al. 1991a. Cosmic ray produced ^{10}Be and ^{26}Al in Antarctic rocks: exposure and erosion history. *Earth Planet. Sci. Lett.* 104: 440–54

Nishiizumi K, Kohl CP, Shoemaker EM, Arnold JR, Klein J, et al. 1991b. In situ ^{10}Be–^{26}Al exposure ages at Meteor Crater, Arizona. *Geochim. Cosmochim. Acta* 55: 2699–703

Nishiizumi K, Lal D, Klein J, Middleton R, Arnold JR. 1986. Production of ^{10}Be and ^{26}Al by cosmic rays in terrestrial quartz in situ and implications for erosion rates. *Nature* 319: 134–36

Nishiizumi K, Lal D, Klein J, Middleton R, Arnold JR. 1989. Cosmic ray production rates of ^{10}Be and ^{26}Al in quartz from glacially polished rocks. *J. Geophys. Res.* 94: 17,907–15

Nishiizumi K, Murrell MT, Arnold JR. 1983. ^{53}Mn profiles in four Apollo surface cores. *J. Geophys. Res.* 88: 211–19

Nishiizumi K, Regnier S, Marti K. 1980. Cosmic ray exposure ages of chondrites, pre-irradiation and constancy of cosmic ray flux in the past. *Earth Planet. Sci. Lett.* 50: 156–70

O'Brien K, Sandmeier HA, Hansen GE, Campbell JE. 1978. Cosmic ray induced neutron background sources and fluxes for geometries over water, ground, iron, and aluminum. *J. Geophys. Res.* 83: 114–20

O'Brien KA. 1979. Secular variations in the production of cosmogenic isotopes in the earth's atmosphere. *J. Geophys. Res.* 84: 423–31

Ohno M, Hamano Y. 1992. Geomagnetic poles over the past 10,000 years. *Geophys. Res. Lett.* 19: 1715–18

Olinger CT, Poths J, Nishiizumi K, Kohl CP, Finkel RC, et al. 1992. Attenuation lengths of cosmogenic production of ^{26}Al, ^{10}Be, and ^{21}Ne in Bandelier Tuff. *Eos, Trans. Am. Geophys. Union* 73: 185 (Abstr.)

Oviatt CG, Currey DR, Sack D. 1992. Radiocarbon chronology of Lake Bonneville, eastern Great Basin, USA. *Palaeogeogr. Palaeoclim. Palaeoecol.* 99: 225–41

Pearson GW, Stuiver M. 1986. High-precision calibration of the radiocarbon time scale, 500–2500 BC. *Radiocarbon* 28: 839–62

Phillips FM, Leavy BD, Jannik NO, Elmore D, Kubik PW. 1986. The accumulation of cosmogenic chlorine-36 in rock: a method for surface exposure dating. *Science* 231: 41–43

Phillips FM, Zreda MG, Smith SM., Elmore D, Kubik PW. et al. 1991. Age and geomorphic history of Meteor Crater, Arizona, from cosmogenic ^{36}Cl and ^{14}C in rock varnish. *Geochim. Cosmochim. Acta* 55: 2695–98

Phillips FM, Zreda MG, Smith SS, Elmore D, Kubik PW, et al. 1990. Cosmogenic chlorine-36 chronology for glacial deposits at Bloody Canyon, eastern Sierra Nevada. *Science* 248: 1529–32

Pomerantz MA, Duggel SP. 1974. The sun and cosmic rays. *Rev. Geophys. Space. Phys.* 12: 343–61

Porcelli DR, O'Nions RK, O'Reilly SY. 1986. Helium and strontium isotopes in ultramafic xenoliths. *Chem. Geol.* 54: 237–49

Porcelli DR, Stone JOH, O'Nions RK. 1987. Enhanced ^{3}He/^{4}He ratios and cosmogenic helium in ultramafic xenoliths. *Chem. Geol.* 64: 25–36

Poreda RJ, Cerling TE 1992. Cosmogenic neon in recent lavas from the western United States. *Geophys. Res. Lett.* 19: 1863–66

Poreda RJ, Marti K, Craig H. 1990. Rare gases and hydrogen in native metals. In *From Mantle to Meteorites*, ed. K Gopalan, VK Gaur, BLK Somayajulu, JD MacDougall, pp. 153–72. Bangalore: Ind. Acad. Sci.

Poths J, Crowe BM. 1992. Surface exposure ages and noble gas components of volcanic units at the Lathrop Wells volcanic center, Nevada. *Eos, Trans. Am. Geophys. Union* 73(43): 610

Raisbeck GM, Yiou F. 1974. Cross sections for the spallation production of ^{10}Be in targets of N, Mg, and Si and their astrophysical applications. *Phys. Rev.* C9: 1385

Rama, Honda, M. 1967. Cosmic-ray induced radioactivity in terrestrial materials. *J. Geophys. Res.* 66: 3533–39

Reedy RC. 1987. Predicting the production rates of cosmogenic nuclides in extraterrestrial matter. *Nucl. Instr. Meth. Phys. Res.* B29: 251–61

Reedy RC, Arnold JR. 1972. Interaction of solar and galactic cosmic-ray particles with the moon. *J. Geophys. Res.* 77: 537–55

Reedy RC, Arnold JR, Lal D. 1983. Cosmic-ray record in solar system matter. *Annu. Rev. Earth Planet. Sci.* 33: 505–37

Reedy RC, Herzog GF, Jessberger EK. 1979. The reaction Mg(n,α)Ne at 14.1 and 14.7 MeV: cross sections and implications for meteorites. *Earth Planet. Sci. Lett.* 44: 341–48

Schock RM, West JA. 1991. Redating the great Sphinx of Giza, Egypt. *Geol. Soc. Am. Abstr. with Programs* 23: 253

Schaeffer OA, Davis R. 1956. Chlorine-36 in

Nature. In *Nuclear Processes in Geologic Settings, Proc. 2nd Conf.*, NAS-NRC Publ. 400, pp. 172–80. Washington, DC: Natl. Acad. Sci.

Shea MA, Smart DF, Gentile LC. 1987. Estimating cosmic ray vertical cutoff rigidities as a function of the McIlwain *L*-parameter for different epochs of the geomagnetic field. *Earth Planet. Sci. Lett.* 48: 200–5

Shoemaker EM. 1983. Asteroid and comet bombardment of the earth. *Annu. Rev. Earth Planet. Sci.* 11: 461–94

Simpson JA. 1983. Elemental and isotopic composition of the galactic cosmic rays. *Annu. Rev. Nucl. Part. Sci.* 33: 323–81

Srinivasan B. 1976. Barites: anomalous xenon from spallation and neutron-induced reactions. *Earth Planet. Sci. Lett.* 31: 129–41

Staudacher T, Allègre CJ. 1991. Cosmogenic neon in ultramafic nodules from Asia and in quartzite from Antarctica. *Earth Planet. Sci. Lett.* 106: 87–102

Stuiver M, Pearson GW. 1986. High-precision calibration of the radiocarbon time scale, AD 1950–500 BC. *Radiocarbon* 18: 805–38

Suess HE. 1986. Secular variations of cosmogenic ^{14}C on earth: their discovery and interpretation. *Radiocarbon* 28: 259–65

Sutton SR. 1985. Thermoluminescence measurements on shock-metamorphosed sandstone and dolomite from Meteor Crater, Arizona. 2. Thermoluminescence age of Meteor Crater. *J. Geophys. Res.* 90: 3690–700

Tera F, Brown L, Morris J, Sacks IS, Klein J, et al. 1986. Sediment incorporation in island-arc magmas: inferences from ^{10}Be. *Geochim. Cosmochim. Acta* 50: 535–50

Trull TW, Kurz MD, Jenkins WJ. 1991. Diffusion of cosmogenic ^{3}He in olivine and quartz: implications for surface exposure dating. *Earth Planet. Sci. Lett.* 103: 241–56

Turrin BD, Champion D, Fleck RJ. 1991. ^{40}Ar/^{39}Ar age of the Lathrop Wells volcanic center, Yucca Mountain, Nevada. *Science* 253: 654–57

Wells SG, McFadden LD, Olinger CT. 1991. Use of cosmogenic ^{3}He and ^{21}Ne to understand desert pavement formation. *Geol. Soc. Am. Abstr. with Programs* 23: 206

Wells SG, McFadden LD, Renault CE, Crowe BM. 1990. Geomorphic assessment of late Quaternary volcanism in the Yucca Mountain area, southern Nevada: implications for the proposed high-level radioactive waste repository. *Geology* 18: 549–53

Yamashita M, Stephens LD, Patterson, HW. 1966. Cosmic ray-produced neutrons at ground level: neutron production rate and flux distribution. *J. Geophys. Res.* 71: 3817–34

Yokoyama Y, Reyss J, Guichard F. 1977. Production of radionuclides by cosmic rays at mountain altitudes. *Earth Planet. Sci. Lett.* 36: 44–56

Zreda MG, Phillips FM. 1993. Surface exposure dating by cosmogenic chlorine-36 accumulation. In *Dating in Surface Context*, ed. C Beck. Albuquerque: Univ. New Mex. Press. In press

Zreda MG, Phillips FM, Elmore D, Kubik PW, Sharma P, et al. 1991. Cosmogenic chlorine-36 production rates in terrestrial rocks. *Earth Planet. Sci. Lett.* 105: 94–109

Zreda MG, Phillips FM, Kubik PW, Sharma P, Elmore D. 1993. Cosmogenic ^{36}Cl dating of a young basaltic eruption complex, Lathrop Wells, Nevada. *Geology* 21: 57–60

Annu. Rev. Earth Planet. Sci. 1994. 22:319–51

ARC ASSEMBLY AND CONTINENTAL COLLISION IN THE NEOPROTEROZOIC EAST AFRICAN OROGEN: Implications for the Consolidation of Gondwanaland

Robert J. Stern

Center for Lithospheric Studies, University of Texas at Dallas, Richardson, Texas 75083-0688

KEY WORDS: tectonics, crustal evolution

INTRODUCTION

Some of the most important, rapid, and enigmatic changes in our Earth's environment and biota occurred during the Neoproterozoic Era (1000–540 million years ago; Ma). Paramount among these changes are the rapid evolution of eukaryotes and appearance of metazoa (Knoll 1992, Conway Morris 1993), major episodes of continental glaciation that may have extended to low latitudes (Hambrey & Harland 1985), marked increases in the oxygen concentration of the atmosphere and hydrosphere (Derry et al 1992), the reappearance of sedimentary banded iron formations (BIF; James 1983), and striking temporal variations in the isotopic composition of C and Sr (Asmerom et al 1991, Derry et al 1992). Understanding the causes of and relationships between these changes is a challenging focus of interdisciplinary research, and there are compelling indications that the most important causes were tectonic (Des Marais et al 1992, Veevers 1990). For example, development of ocean basins may have been accompanied by the development of seafloor hydrothermal systems, which lowered the $^{87}Sr/^{86}Sr$ of seawater, led to the development of BIF, and formed anoxic basins where organic carbon could be buried, thus leading to an increase in O_2. Continental collision and formation of a supercontinent may have led to continental glaciation and an increase in the $^{87}Sr/^{86}Sr$ of seawater,

0084–6597/94/0515–0319$05.00

and could also have formed sites (high sedimentation rate submarine fans) where organic carbon was buried. Understanding the style and sequence of Neoproterozoic tectonic events is fundamental if we are to establish the cause-and-effect relationship between tectonics and Neoproterozoic global change.

The Neoproterozoic Era encompasses a protracted orogenic cycle referred to here as the "Pan-African" orogenic cycle. Kennedy (1964) coined the term "Pan-African Thermo-Tectonic Episode" to characterize the structural differentiation of Africa into cratons and mobile belts during the latest Precambrian and earliest Paleozoic, but the Pan-African has been redefined (Kröner 1984) as involving a protracted orogenic cycle from 950 to 450 Ma. This tectonism was not limited to Africa—diastrophic events of similar age and style are common throughout Gondwanaland (Figure 1) and in many parts of Laurasia. The time span of 500 million years is much longer than any Phanerozoic orogeny, so reference to a Pan-African "orogenic cycle" must suffice until the timing and regional extent of discrete tectonic events is better constrained.

Not only will questions about the evolving Neoproterozoic environment be answered when we better understand Pan-African tectonics, but the Pan-African orogenic cycle is interesting in its own right. The nature of the episode supports arguments that modern plate tectonic systems had begun by about 1000 Ma (Davies 1992). In contrast to crust produced between 1.0 and 1.8 Ga, Pan-African juvenile (i.e. mantle-derived) crust preserves most of the hallmarks of modern plate tectonic regimes, including abundant ophiolites, calc-alkaline batholiths and volcanic sequences, and immature clastic sediments. Horizontal tectonic displacements are proved by ophiolitic nappes which traveled at least several tens and perhaps a few hundreds of kilometers. In many instances, Pan-African metamorphism resulted in granulite formation, suggesting crustal overthickening due to continent-continent collision (Burke & Dewey 1972).

McWilliams (1981) argued that Gondwanaland was assembled during the Neoproterozoic from two fragments—East and West Gondwanaland—along the Pan-African Mozambique Belt of East Africa (Figure 1). This suggestion is supported by the lithologic associations characteristic of the Mozambique Belt and its extension to the north, the Arabian-Nubian Shield (ANS). Following the arguments of Berhe (1990) that ophiolite-decorated meridional sutures can be traced in E. Africa, I accept these entities as along-strike correlatives, and refer to the larger structure as the East African Orogen (EAO; Figure 2). The ophiolites, granulites, and structures of the EAO are fossil fragments of a Neoproterozoic Wilson cycle, representing the opening and closing of an ocean basin that lay between the older crustal blocks of East and West Gondwanaland. The

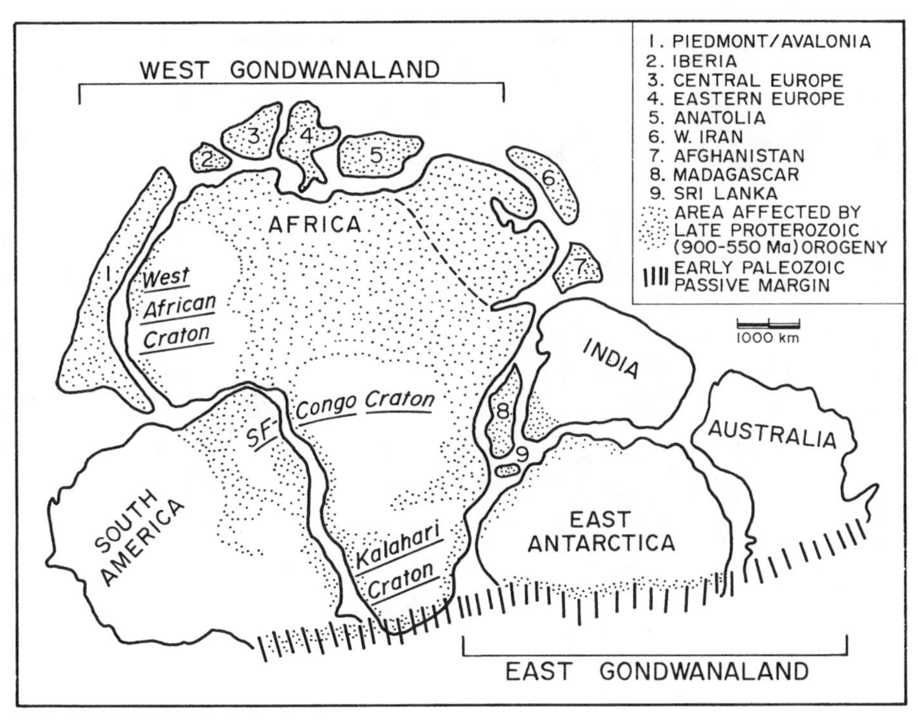

Figure 1 Map of a part of Gondwanaland, showing the position of the continents and smaller continental fragments in the early Mesozoic (De Wit et al 1988). This is the surviving nucleus of the supercontinent of Greater Gondwanaland which at the end of the Precambrian was about 25% larger in area, mostly on the Australian, Indian, Arabian, and NE African margins. Stippled area shows the region affected by the 950–450 Ma Pan-African event. SF = Sao Francisco

length and breadth of the EAO are similar to those of more recent orogenic belts, such as the North America cordillera (Oldow et al 1989) and of the mountain belt that stretches from western Europe to eastern Asia—which Şengör (1987) calls the "Tethysides orogenic collage" (Figure 3). The time span involved in all these orogens is similar—several hundreds of millions of years. Other Wilson cycles of similar or younger age affected West Gondwanaland, but the abundance of ophiolites and juvenile arc assemblages indicates that the EAO formed during a protracted period of juvenile crust formation in intra-oceanic arcs culminating in continental collision. The Neoproterozoic assembly of Gondwanaland was more complicated than simply bringing two halves together, and West Gondwanaland in particular was assembled from several pieces (see Figure 2 of

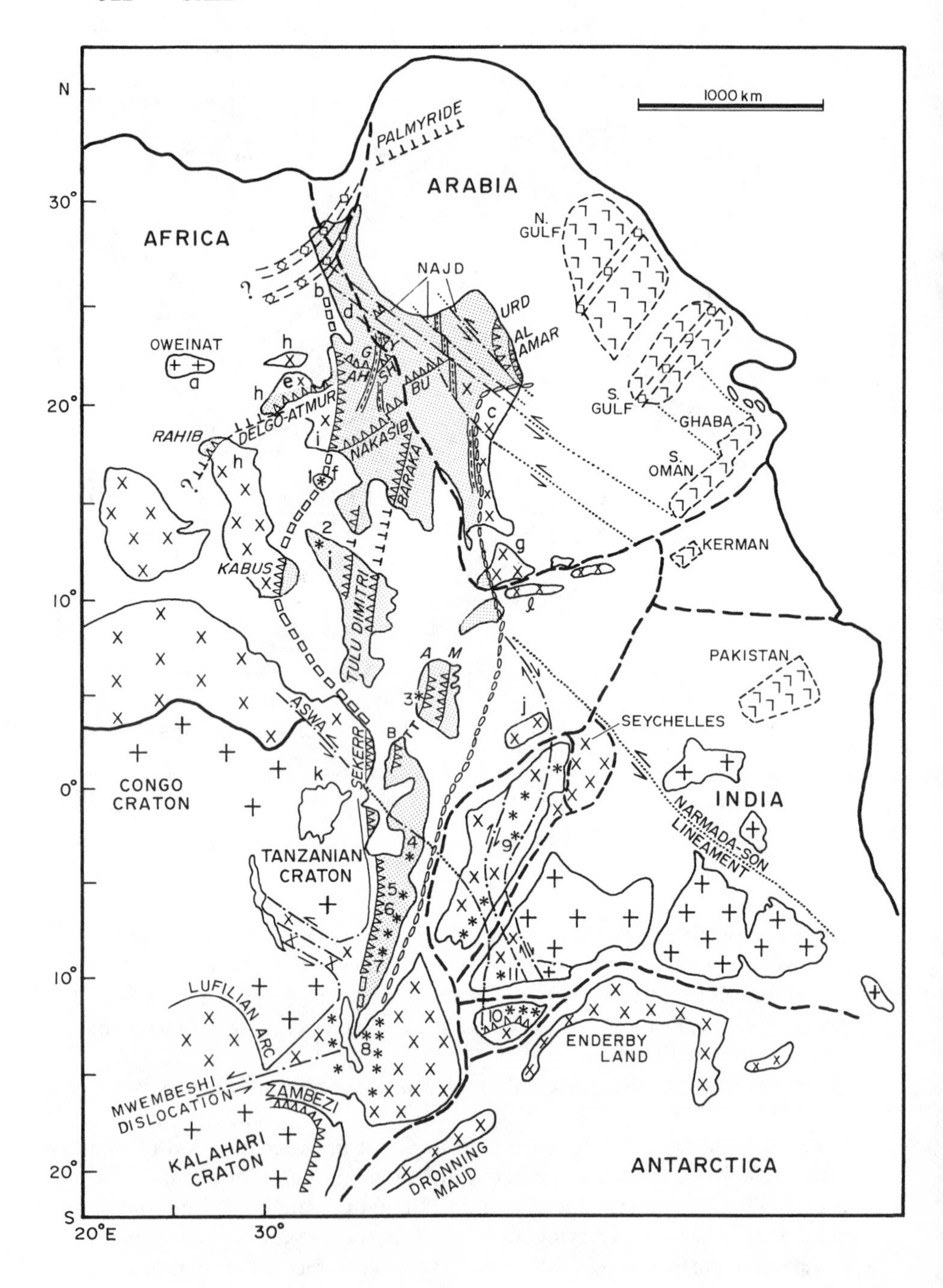

N
1000 km
PALMYRIDE
ARABIA
AFRICA
N. GULF
NAJD
URD
AL AMAR
OWEINAT
DELGO-ATMUR
RAHIB
NAKASIB
BARAKA
S. GULF
GHABA
S. OMAN
KERMAN
KABUS
TULU DIMITRI
PAKISTAN
SEYCHELLES
ASWA
SEKERR
CONGO CRATON
INDIA
NARMADA-SON LINEAMENT
TANZANIAN CRATON
LUFILIAN ARC
ENDERBY LAND
MWEMBESHI DISLOCATION
ZAMBEZI
KALAHARI CRATON
DRONNING MAUD
ANTARCTICA
30°
20°
10°
0°
10°
20°
S
20°E
30°

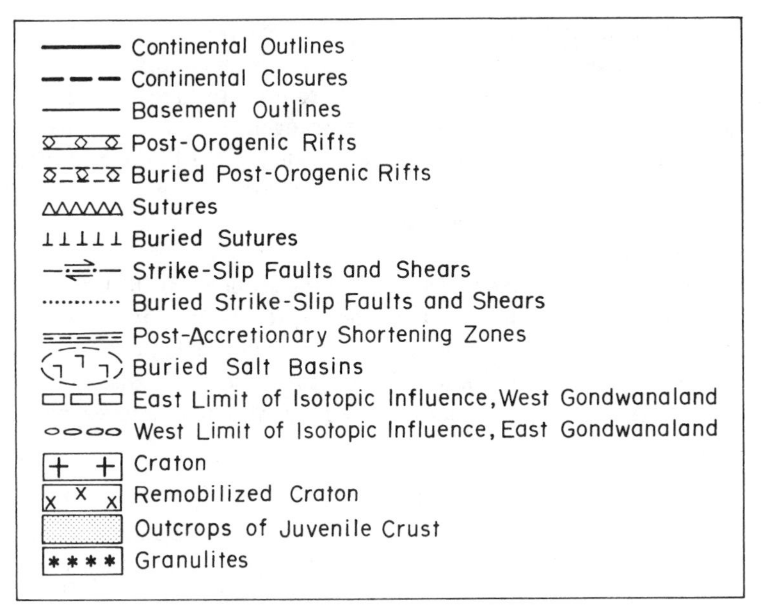

Figure 2 Tectonic map of the East African Orogen. Continental fragments are shown as configured at the end of the Precambrian.

Small letters (a–l) are places where pre–Pan African crustal components can be documented: a = Oweinat (2.5 Ga basement; Klerkx & Deutsch 1977); b = Nakhil (1.6 Ga xenocrystic zircons in 580 Ma granite; Sultan et al 1990); c = Afif terrane (1.6–1.8 Ga crust; Stacey & Hedge 1984, Stoeser et al 1991); d = E. Desert of Egypt (1.1–2.3 Ga cobbles in Atud conglomerate, Dixon 1981; 1.5–2.7 Ga detrital zircons in wackes, Wust et al 1987); e = Wadi Halfa (1.2–2.5 Ga basement; Stern et al 1993); f = Sabaloka (1.0–2.6 Ga detrital zircons in metasedimentary granulites; Kröner et al 1987); g = Eastern Yemen (<3.0 Ga T_{DM} Nd model ages; Stoeser et al 1991, Whitehouse et al 1993); h = W. Egypt and Sudan (1.5–2.7 Ga T_{DM} Nd model ages, Harms et al 1990; 1.9–2.6 Ga zircon ages, Sultan et al 1992b); i = central Sudan (1.5 Ga T_{DM} Nd model age; Harris et al 1984); j = S. Somalia (1.8–2.5 Ga inherited zircons in 540 Ma granites; Küster et al 1990); k = Uganda (2.6–2.9 Ga basement; Leggo 1974); l = N. Somalia (1.8 Ga xenocrystic zircons in 715–820 Ma granites; Kröner et al 1989).

Numbers 1–11 denote Neoproterozoic granulites: 1 = Sabaloka (710 Ma zircon age; Kröner et al 1987); 2 = Jebel Moya (740 Ma zircon age; Stern & Dawoud 1991); 3 = Bergudda Complex (545 Ma zircon age; Ayalew & Gichile 1990); 4 = Pare Mountains (645 ± 10 Ma U-Pb zircon age; Muhongo & Lenoir 1993) and Wami River (715 Ma zircon age; Maboko et al 1985); 5 = Uluguru (695 ± 4 Ma U-Pb zircon age, Muhongo & Lenoir 1993; 630 Ma hornblende Ar closure age, Maboko et al 1989); 6 = Furua (650 Ma zircon age, Coolen et al 1982; 615–665 hornblende Ar closure age, Andriessen et al 1985); 7 = Songea; 8 = Malawi and Mozambique (800–1100 Ma Rb-Sr whole rock; Andreoli 1991); 9 = Madagascar (860 Ma; Cahen et al 1984); 10 = Sri Lanka (520–550 Ma; Burton & O'Nions 1990, Kröner 1991); 11 = S. India (Choudhary et al 1992).

Capital letters refer to ophiolites and/or sutures: G = Gerf ophiolite; AH = Allaqi-Heiani suture; SH = Sol Hamed suture; BU = Bir Umq suture; A = Adola ophiolite; M = Moyale ophiolite; B = Baragoi ophiolite.

Dalziel 1992), but the timing and sequence of events preserved in the EAO are representative of the consolidation of Gondwanaland.

The Pan-African orogenic cycle figures prominently in recent hypotheses regarding the breakup and collisional rearrangement of a Neoproterozoic supercontinent (Moores 1991, Hoffman 1991, Dalziel 1991). Specifically, East and West Gondwanaland must have collided after supercontinent breakup, (Storey 1993). so the evolution of the EAO provides important constraints on these hypotheses.

The purpose of this paper is to review the sequence of tectonic events responsible for forming the EAO and thereby to demonstrate when Gondwanaland was born. Our knowledge of the EAO lags behind most other orogens of similar size and importance, and this is partly due to the fact that the orogen cuts across many developing African countries, where it

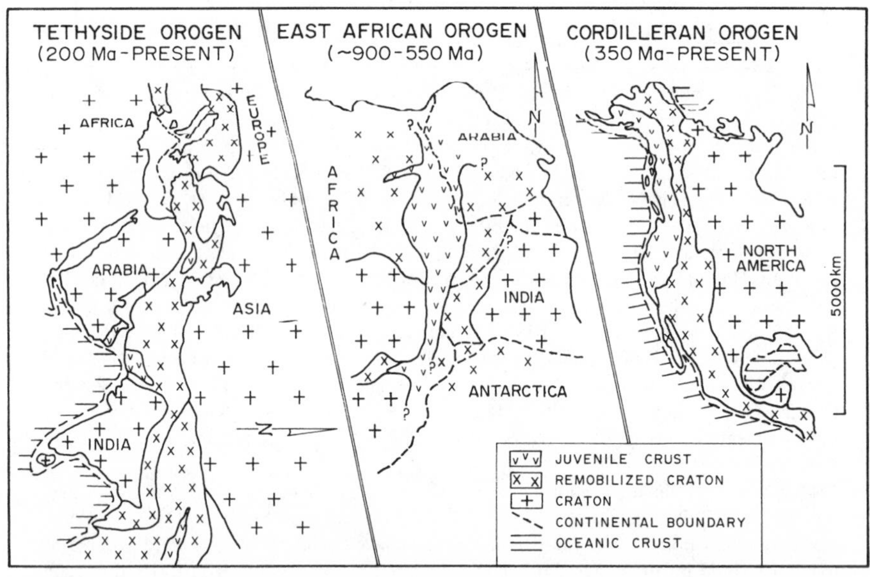

Figure 3 Comparison at the same scale of the size and style of the Neoproterozoic East African Orogen with the Phanerozoic Cordilleran (Oldow et al 1989) and Tethyside (Şengör 1987) orogens. For the East African Orogen, subdivision of juvenile, cratonic, and reworked cratonic crust is based on geochronologic and radiogenic isotopic data. For Phanerozoic orogens, distinction of craton from reworked craton is based on topography, while subdivision of reworked craton from juvenile crust is based on isotopic data ("0.706 line") in the Cordilleran and interpretation of regional geologic maps (Şengör 1987). Note that the East African Orogen is like the Cordilleran Orogen in the abundance of juvenile crust and like the Tethyside Orogen in manifesting the effects of continent-continent collsion.

may be difficult to work, or where only limited resources can be spent by national geologic surveys on academic pursuits, no matter how important. This situation is unlikely to improve in the near future. Nonetheless, how and when Gondwanaland was assembled has important implications for the evolution of the Neoproterozoic environment and Precambrian tectonics, and a summary at this time serves to emphasize problems that could reward further work with fundamental insights. To enliven the tale, this review is heavily spiced with interpretation, but my intent is to present the EAO so that both the constraints that we have and the problems that persist are apparent. Because so many of the fundamental questions begin with "When?," the story is told from beginning to end of the EAO orogenic cycle: the beginning, arc evolution and consolidation, continental collision, and escape tectonics. Towards the end, two of the more pressing unresolved problems are outlined: Where does the EAO continue to the south (or does it)?; and what happened to the EAO forelands in the north? Finally, the implications of EAO evolution for Neoproterozoic continental reconstructions are explored.

THE BEGINNING

The main question is, "Does the EAO manifest a Wilson cycle orogeny?"; that is, did it begin with rifting of an earlier continental assembly? Direct evidence for rifting may be preserved in sedimentary successions in the EAO that have been interpreted as passive margin deposits in Kenya (Vearncombe 1983, Key et al 1989, Mosley 1993) and in Sudan (Kröner et al 1987). These sequences are difficult to date. Metasedimentary rocks from Sudan were metamorphosed to granulite facies at about 720 Ma, so deposition must be earlier. A thick succession of multiply-deformed slope- and basinal-facies carbonates is found in northern Sudan along the line of the Keraf zone, and these are most readily interpreted as remnants of a passive margin sequence (Schandelmeier et al 1993). These rocks are also undated, but regional considerations indicate deposition prior to 750 Ma. Arguments for the existence of a passive margin are clearest for Kenya. Key et al (1989) argued that the Kenyan sediments were deposited during a period of crustal extension sometime between 840 and 770 Ma. These sediments have been isoclinally folded, thrust, and metamorphosed to amphibolite facies, so that their stratigraphic relationships and original dimensions are poorly known. Nevertheless, the metasediments appear to become finer as one moves eastward, and characteristic lithologies include quartzites, semipelitic gneiss, and carbonates. These rocks often preserve primary sedimentary structures, including ripple marks and cross-bedding (Mosley 1993). Along strike to the south in Tanzania, Prochaska & Pohl

(1983) interpreted Neoproterozoic amphibolite-facies mafic and ultramafic rocks as having formed during early rifting, an interpretation that also applies to ca. 790 Ma mafic-ultramafic complexes with negative initial $\varepsilon Nd(t)$ in Madagascar (Guerrot et al 1993). [Note: $\varepsilon Nd(t)$ refers to the difference in the $^{143}Nd/^{144}Nd$ of a rock from that of the bulk earth, in parts per 10,000, at the time of formation t. For further discussion, see DePaolo (1988).] Vearncombe (1983) recognized facies in western Kenya indicating eastward transitions from continental margin to shelf and finally deep sea conditions. Complications due to metamorphism and orogeny and incomplete exposure along strike make it difficult to decide whether all of the metasediments preserved in the Kenyan sector of the EAO were once laterally equivalent or whether they have been juxtaposed along an intervening suture. Shackleton (1986) argued that the Kenyan ophiolites are rooted on the east side of the EAO. If this is true, then the telescoped Kenyan metasedimentary basin is about 400 km wide, consistent with it being interpreted as a passive margin.

Two possible aulacogen-like structures (rifts extending inland from the continental margins) are found along and west of the EAO. In south-central Africa, the E-W trending Zambezi belt (Figure 2) has been interpreted as an intensely deformed aulacogen exposed at mid-crustal levels (Hanson et al 1989). The Zambezi belt contains extensive tracts of remobilized sialic basement structurally overlain by thick supracrustal rocks. Bimodal volcanics (U-Pb zircon age of 879 ± 19 Ma; Wilson et al 1993, Hansen et al 1993) inferred to record initial extension-related magmatism are structurally overlain by clastic sediments and then shallow marine carbonate rocks. This sequence is interpreted by Hanson et al (1989, 1993) to represent extension and thermal subsidence. It was metamorphosed to amphibolite facies and penetratively deformed at about 820 Ma (Hanson et al 1988b). Crust and structural trends of the Mesoproterozoic Irumide belt are correlated across the Zambezi belt, apparently precluding a large Pan-African ocean basin (Hanson et al 1988a), and supporting the interpretation of an aulacogen [although many workers disagree with this conclusion, such as Burke & Dewey (1972)]. If so, the age of 880 Ma for rift-related volcanism implies a similar age of rifting in the EAO. In Sudan, the recently discovered Atmur-Delgo suture (Figure 2; Schandelmeier et al 1993) appears to represent the third arm of a RRR triple junction, floored by oceanic crust. The Sm-Nd age of about 750 Ma obtained from the ophiolite (Denkler et al 1993) is a minimum age for the rifting event in northeast Africa.

Finally, consideration of the oldest Pan-African igneous rocks in the Arabian-Nubian Shield suggests that rift-related igneous activity began about 880 Ma in the northern part of the EAO. There, the oldest Pan-

African igneous rocks are a bimodal sequence of moderately high-titanium tholeiitic basalts and subordinate rhyodacites of the Baish Group, interpreted by Reischmann et al (1984) to have been erupted in an intra-oceanic arc setting. These, and similar volcanics in Sudan, yield single zircon evaporation ages of 840–870 Ma (Kröner et al 1991, 1992a) and can be interpreted as manifesting early stages of rifting in the EAO, for the following reasons: 1. They are compositionally similar to other rift-related volcanic sequences in the Arabian-Nubian Shield, such as the Arbaat Group (Abdelsalam & Stern 1993a); 2. volcanic sequences that form during continental breakup show many of the chemical characteristics of arc lavas (e.g. depletion in Nb) and typically plot in the field of arc lavas on discriminant diagrams (Wang & Glover 1992); and 3. these volcanics occur in the Haya terrane in Sudan and in the Asir terrane in Arabia (Figure 4), adjacent to older crust of the Afif terrane and interbedded with pelitic and quartzitic rocks. The T_{DM} age (crust formation Nd model ages; Nelson & DePaolo 1985) for one rhyolite from a similar sequence in Sudan is 1.3 Ga, indicating a significant contribution from older crust, plausibly incorporated as a result of contamination during the initial stages of rifting. If this interpretation is correct, then rifting in the northern part of the EAO occurred about 840–870 Ma, and the Asir-Haya terranes traveled as the conjugate margin of the eastern continental fragment.

In summary, an argument can be made that the EAO began by rifting of a continent, beginning about 870 Ma. Rifting led to the development of a passive margin and perhaps two aulacogen-like structures, one of which evolved into a narrow ocean basin. The passive margin is best preserved in the west, where it nonetheless has been multiply deformed and metamorphosed, and is poorly developed in the east. This may be due to poor exposure and more intense deformation and metamorphism, or it may be due to the fact that the eastern margin of the EAO developed later as an Andean-type convergent margin.

The hypothesis that the EAO began by rifting, with the implication that a larger continental block was broken up into West Gondwanaland and, perhaps, East Gondwanaland about 870 Ma, is consistent with global considerations. Hoffman (1989, 1991) marshaled evidence that a supercontinent (Rodinia) formed by about 1.1 Ga. Application of the model of supercontinent cyclicity (Gurnis 1988, Veevers 1990) suggests that Rodinia should begin to break up after about 150–200 Ma. This inference is supported from the subsidence histories of some Neoproterozoic basins in Australia, which preserve evidence for a regional rifting event at 900 Ma (Lindsay et al 1987). Evidence that this was followed by a major episode of sea-floor spreading is contained in the EAO ophiolite record (discussed later) and supported by the global record of $^{87}Sr/^{86}Sr$ in carbonates, which

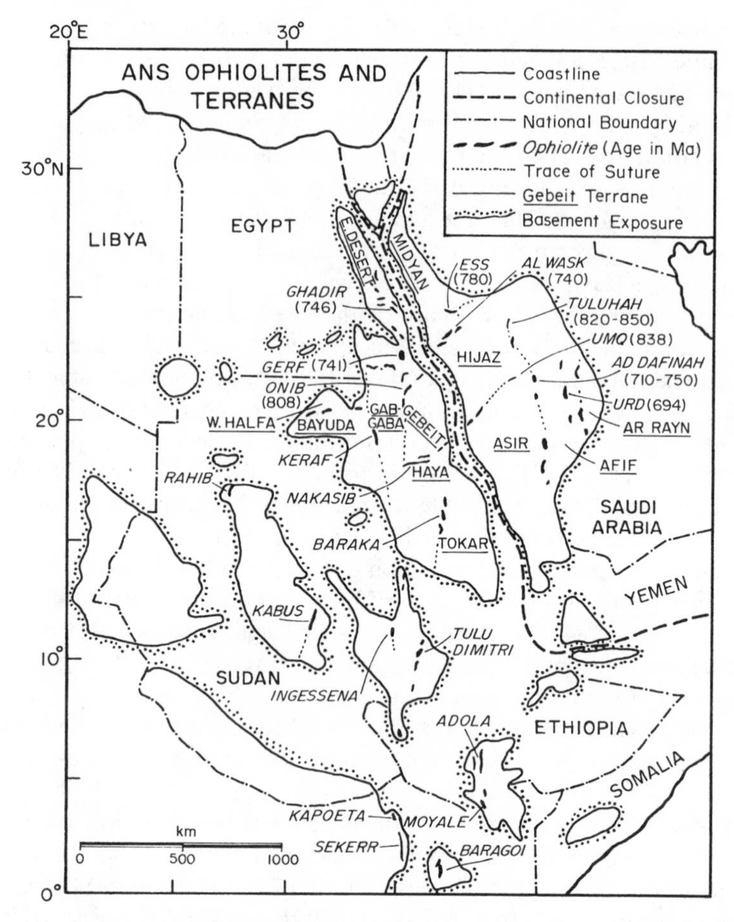

Figure 4 Distribution of ophiolites and terranes in the Arabian-Nubian Shield, northern EAO. Not all of the ophiolites are shown, but the ones that are shown are drawn as close to scale as possible. Ages (in Ma) are given in parentheses. Dotted lines show the inferred extensions of sutures. Terrane names are underlined.

reaches a minimum—indicating that the seafloor hydrothermal flux dominated over continental runoff—sometime around 800 and 900 Ma (Figure 5).

OPHIOLITES, ARCS, AND THE SIZE OF THE MOZAMBIQUE OCEAN

While older ophiolites are known from elsewhere on Earth, the earliest record of abundant ophiolites is preserved in the Arabian-Nubian Shield

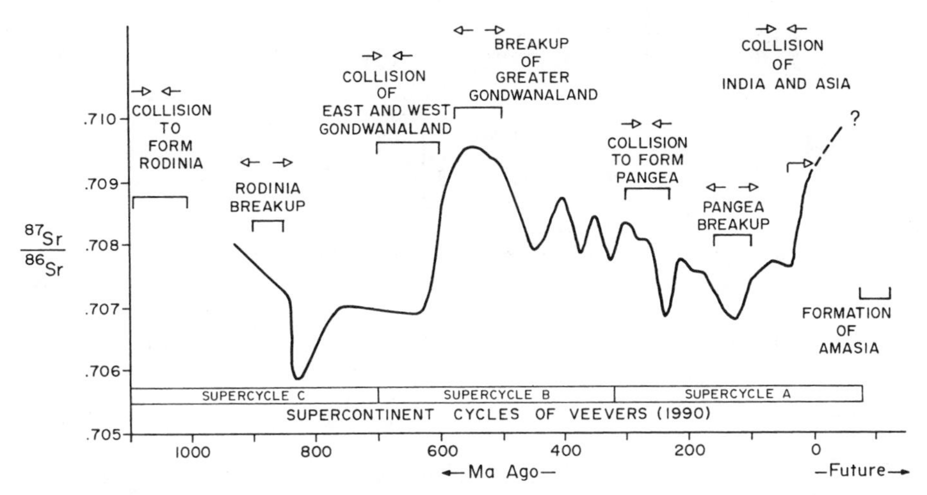

Figure 5 Generalized variation in the isotopic composition of Sr in the oceans through time, based on studies of $^{87}Sr/^{86}Sr$ in carbonate rocks (Burke et al 1982, Veizer et al 1983, Asmerom et al 1991). Temporal and compositional resolution decrease with increasing age. Nonradiogenic Sr reflects high inputs of seafloor hydrothermal Sr, whereas radiogenic Sr reflects high riverine runoff from old continental crust. Supercontinent cycles of Veevers (1990) and approximate times of formation and dispersal of supercontinents are shown; note that the breakup of "Greater Gondwanaland" continued throughout the Paleozoic. "Amasia" refers to the supercontinent (America plus Asia) expected to form from closure of the Pacific. A high flux from seafloor hydrothermal activity is indicated at about 800 Ma. The rise in seawater $^{87}Sr/^{86}Sr$ did not begin until about 600 Ma.

(Figure 4). These occur as variably dismembered nappes associated with suturing either between arc terranes or between the juvenile crust of the ANS and the palimpsest older crust to the west. Where dated, these range in age from about 880 to 690 Ma (Figure 6). Unequivocal ophiolites have not been reported from south of Kenya (Price 1984, Ries et al 1992). Shanti & Roobol (1979) first recognized ophiolitic nappes in the ANS, and since that time nearly all significant ophiolitic occurences have been interpreted as nappes (Church 1988, Abdelsalam & Stern 1993b). ANS ophiolites demonstrate two important aspects of EAO evolution: (*a*) that oceanic crust formed and was partly preserved; and (*b*) that ophiolitic and other nappes traveled as much as 150 km or more (Stern et al 1990), indicating that horizontal transport was an important aspect of crustal shortening.

ANS ophiolites are similar to many Phanerozoic ophiolites in having a strong "supra-subduction zone" (SSZ; Pearce et al 1984) chemical signature. These characteristics—along with the abundance of immature and volcaniclastic sediments deposited on the ophiolites—indicate ANS ophi-

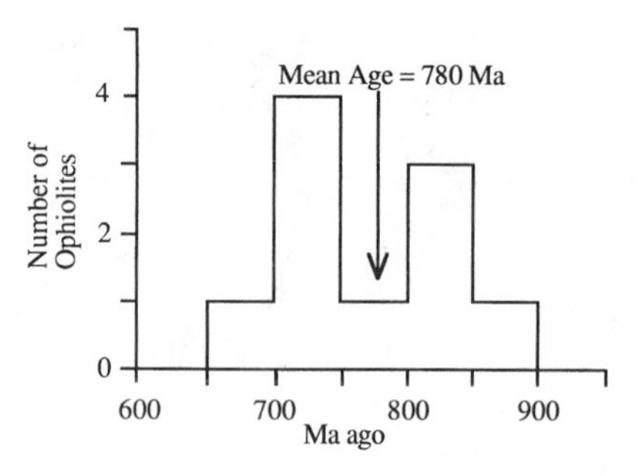

Figure 6 Histogram of robust ages for 10 ophiolites from Egypt and Sudan (Kröner et al 1992b) and Saudi Arabia (Claesson et al 1984, Pallister et al 1988).

olites for the most part formed in arc, back-arc, or fore-arc settings (Price 1984, Pallister et al 1988, Berhe 1990, Abdel Rahman 1993). Judging from the chemistry of the ophiolites and the associated sediments, we have no ophiolites that preserve the seafloor of the "Mozambique Ocean" (Dalziel 1991)—the greater ocean that may have existed between East and West Gondwanaland, perhaps 700–850 Ma. This is not surprising, because Phanerozoic ophiolites usually come from fringing SSZ environments and only rarely from the main ocean basin (Pearce et al 1984). This means that we have no *direct* information about the size, age, structure, or even the existence of the Mozambique Ocean. However, two lines of evidence indicate that the Mozambique Ocean must have been a large basin: The first comes from consideration of the implications of extensional stress regimes at convergent margins; the second concerns the significance of crustal growth rates.

Studies of stress regimes associated with modern convergent margins indicate that arcs under extension—those most likely to spawn back-arc basins—are typically those that are subducting old seafloor. This is because an important driving force for plate motion is the excess density of subducted lithosphere (Davies & Richards 1992). Lithospheric age and density correlate strongly; consequently the stresses in the over-riding plate vary from strong compression to strong extension with age of the subducting plate. Molnar & Atwater (1978) found that all modern Pacific arcs that rifted to form back-arc basins subduct seafloor older than 100 Ma. This result can be used to infer from the abundance of back-arc basin ophiolites in the ANS that the Mozambique Ocean was likely to have contained seafloor that was on the order of 100 My or older when it was subducted. Alternatively, some ANS SSZ ophiolites may have formed in fore-arcs, an interpretation which is supported by the local occurrence of boninites (Al-Shanti & Gass 1983, Reischmann 1986, Berhe 1990, Abdel Rahman 1993), in which case these ophiolites are best interpreted as forming during subduction zone initiation (Stern & Bloomer 1992). This also requires dense, old lithosphere within the plate to be digested. In either case, the inference from abundant SSZ ophiolites indicates strong extension in the associated convergent margin, suggesting that old seafloor was subducted. Old seafloor requires ocean basins that are at least as old. If seafloor spreading is continuous, age implies size. For example, given a moderate full spreading rate of 10 cm/yr, 100 Ma seafloor implies an ocean 10,000 km wide. That the Mozambique Ocean was a large ocean basin is the logical conclusion of this line of reasoning.

Calc-alkaline and tholeiitic plutons, lavas, pyroclastics, volcaniclastics, and associated immature sediments make up a major portion of the upper crust of the ANS. These are interpreted to comprise arc terranes, separated one from the other by ophiolite-decorated suture zones which mark the positions of fossil subduction zones (Figure 4; Vail 1985, Stoeser & Camp 1985). It has proven difficult to identify all of the components of individual arcs such as their accretionary prisms, fore-arc basins, or frontal arcs. The only well-documented accretionary prism is in the Al Amar suture of the eastern Arabian shield (Figure 2; Al-Shanti & Gass 1983). Blueschist occurences have not been documented, nor have any paired metamorphic belts been convincingly demonstrated. Nevertheless, the chemical and isotopic compositions, largely submarine nature of the igneous and immature sedimentary rocks of the ANS, and their occurrence in ophiolite-bounded terranes are most consistent with their origin in immature arcs [but see alternative explanations of Stern et al (1991) for one volcano-sedimentary belt].

Because the rock record indicates that processes of Neoproterozoic crustal growth were similar to modern plate tectonics, the rate of growth of Neoproterozoic continental crust may have been similar as well. Phanerozoic continental crust formation occured at convergent margins for the most part and is estimated to be about 1.1 km^3/yr (Reymer & Schubert 1984). Several efforts have been made to estimate growth rates over an estimated 300 Ma history for the ANS. These estimates range from about 20% to nearly 80% of the global Phanerozoic crustal growth rates (Figure 7). This means that, if the assumption of similar Neoproterozoic and Phanerozoic growth rates is correct, between 20% and 80% of all Neoproterozoic crustal growth occurred in the ANS. These estimates apply only to the Arabian-Nubian Shield, and the range largely reflects the uncertainties about how much of this crust is juvenile. A similar estimate can be made for the area mapped as juvenile crust in Figure 2, i.e. the area between the limits of isotopic influence of East and West

Gondwanaland. It is not generally recognized that much juvenile crust exists south of Sudan and Ethiopia, but initial $^{87}Sr/^{86}Sr$ ratios for rocks from the EAO in Kenya and Tanzania are moderately low, with a mean of about 0.7041 (Key et al 1989, Maboko et al 1985). The proportion of juvenile contributions is certainly higher in the north than in the southern EAO, but a mean of 80% juvenile contribution over the part of the EAO shown in the stippled pattern in Figure 2 is probably conservative. These considerations lead to an estimate of about 30% of the Phanerozoic growth rate (*I* in Figure 7).

The regions mapped as remobilized craton in Figure 2 contain a significant juvenile component. This is demonstrated by the discovery of an ophiolite belt striking WSW into the interior of Africa in northern Sudan (Schandelmeier et al 1993); by the presence of ca. 560 Ma batholiths with moderately low initial $^{87}Sr/^{86}Sr$ (0.705–0.706; Fullagar 1978) in the Tibesti massif, Libya, to the west; and by the presence of 700–850 Ma igneous rocks to the east, with low initial $^{87}Sr/^{86}Sr$ (0.7025–0.7041; Gass et al 1990). In many places as much as two thirds of the crust mapped as remobilized craton may be Neoproterozoic juvenile contributions; nevertheless, the crustal growth rate used here is based on a conservative 30% juvenile addition. This estimate, plus that previously calculated, suggests that about

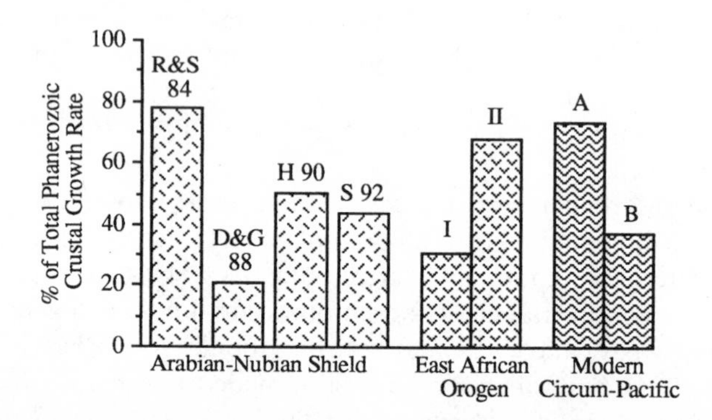

Figure 7 Estimates for crustal growth rates in the Arabian-Nubian Shield indicating that a large proportion of Neoproterozoic crutal growth occured in the northern East African Orogen [I = 80% of volume beneath "Juvenile Crust" in Figure 2 (35 km thick crust − 6 km thick oceanic crust + 6 km erosion over 300 My); II = I plus 30% of volume beneath area marked "Remobilized Craton" in Figure 2], and Modern Circum-Pacific [A = fraction of world's present convergent margins (Forsyth & Uyeda 1975); B = one half of A, estimated mean for an opening and then closing Pacific-type ocean]. The present growth rate of the continental crust is estimated to be 1.1 km^3/yr (Reymer & Schubert 1984). R&S 84 = Reymer & Schubert 1984, D&G 88 = Dixon & Golombek 1988, H 90 = Harris et al 1990, S 92 = Sultan et al 1992a.

65% of the Phanerozoic growth rate was responsible for the formation of the region affected by the Pan African in Figure 3 (*II* in Figure 7). This estimate is similar to the estimated growth rate of the modern circum-Pacific (*A* in Figure 7), and is considerably greater than the mean crustal growth expected for opening (no crustal growth) and closing of a Pacific-sized ocean basin (*B* in Figure 7). Insofar as total crustal growth is a function of the volume of seafloor subducted, the implication is that the Mozambique Ocean was large, perhaps comparable in size to that of the modern Pacific. These results also indicate that a disproportionately large fraction of Neoproterozoic juvenile crust formation occurred in the EAO, similar to the present situation in the western Pacific.

Recent crustal refraction studies indicate that ANS crust is composed of mafic lower and felsic upper crust, separated by a mid-crustal Conrad discontinuity (Gettings et al 1986, El-Isa et al 1987, Rihm et al 1991). Geochemical studies of xenoliths confirm that the lower crust is mafic (Table 1) and records equilibration pressures up to 12 kbar and temperatures of 830° to 1200°C (McGuire & Stern 1993), but have not resolved whether the mafic crust is residual or cumulate. McGuire & Stern (1993) interpret the isotopic and chemical data to indicate that the mafic lower crust formed during the Neoproterozoic, an argument that is supported by isotopic studies of spinel peridotites indicating that the mantle lithosphere beneath Arabia formed about 700 Ma (Henjes-Kunst et al 1990). The ANS lower crust preserves little evidence for the involvement of pre-existing crustal materials, and the absence of metasedimentary protoliths is noteworthy; it must be juvenile. It appears that mafic lower crust is a characteristic of the ANS and contrasts with the deeper crust recorded to the south and west, where pre-Pan African crust was involved and metasedimentary protoliths are found. In at least two of these localities [Sabaloka ("f" in Figure 2; Kröner et al 1987), and northeast Tanzania (Shackleton 1993)], however, there is evidence of mafic intrusion into the lower crust during the Neoproterozoic, and similar inferences are made from refraction studies in Kenya (KRISP Working Party 1991).

Estimates of the composition of the ANS upper, lower, and bulk continental crust are listed in Table 1. The ANS upper crust has a composition similar to that of the upper continental crust estimated by Taylor & McClennan (1985). The "Bulk ANS Crust" approximates a Mg-enriched andesite and is very similar to the estimate of the "Bulk Continental Crust" of Taylor & McClennan (1985).

Formation and coalescence of ANS arc terranes must have predated continental collision. Terrane accretion was completed by about 700 Ma (Ayalew et al 1990, Pallister et al 1988, Kröner et al 1992b, Stern & Kröner 1993), except along the Urd and Al Amar sutures in the eastern part of

Table 1 Composition of the ANS crust

Oxide	Mean ANS UC[a]	WAALC[b]	Bulk crust ANS[c]	Bulk continental crust[d]
SiO_2	66.4	50.6	58.2	57.3
TiO_2	0.63	0.90	0.77	0.90
Al_2O_3	14.9	16.4	15.6	15.9
FeO*	4.39	9.87	7.23	9.1
MgO	3.31	8.04	5.76	5.3
CaO	4.20	10.77	7.60	7.4
Na_2O	4.21	3.03	3.60	3.1
K_2O	1.86	0.32	1.06	1.1
P_2O_5	0.15	0.13	0.14	—

[a] Mean Upper Crust of the Arabian-Nubian Shield, estimated using 32,634 points of mapped basement lithologies in the Eastern Desert of Egypt (Stern 1979) and mean compositions for various lithologies reported in Stern (1979, 1981), Abdel-Rahman (1993), Greenberg (1981), and Neary et al (1976).

[b] Weighted Average Arabian Lower Crust (McGuire & Stern 1993).

[c] 48.2% mean ANS UC plus 51.8% WAALC, relative masses estimated from crustal structure and densities of Gettings et al (1986).

[d] Bulk continental crust composition (Taylor & McClennan 1985).

the Arabian shield (Figure 2) where suturing occurred between 640 and 680 Ma (Stacey et al 1984). Also, there is field evidence that the Keraf structure in northern Sudan (Figure 4) is younger than the NE-SW sutures of the ANS (Schandelmeier et al 1993). Thus, suturing of the terranes within the ANS from 700 to 760 Ma was followed by suturing of the ANS composite terrane between East and West Gondwanaland, about 640 to 680 Ma.

THE COLLISION OF EAST AND WEST GONDWANALAND

The EAO in Kenya and Tanzania has long been recognized as manifesting an episode of Tibetan-style continental collision and crustal thickening (Burke & Dewey 1972). The most direct evidence for the timing of collision between the continental blocks of East and West Gondwanaland comes from the age of granulites exposed in the EAO. Insofar as their exposure reveals 15–45 km of uplift and erosion, EAO granulites show where the greatest thickening occurred and lead directly to inferences about where continent-continent collision was most intense. Granulites are not found north of central Sudan and southern Ethiopia but are common in southern Kenya, Tanzania, Malawi, and Mozambique.

The lower crust of the southern EAO contrasts with that to the north in two fundamental ways. First, whereas that to the north is largely intact and lies preserved 20 or more km deep in the present-day crust, that of the southern EAO crops out as tectonic slices. We do not know what composes the present lower crust of the southern EAO, although in Kenya the crust east of the Tanzanian craton and the East African rift is about 35 km thick, has a well-defined Conrad discontinuity, and contains seismic suggestions of mafic underplating (KRISP Working Party 1991). Second, whereas the lower crust of the northern EAO is mafic igneous rock, southern EAO granulites include a variety of igneous and metasedimentary rocks, including marbles and quartzites. These differences indicate the different histories of the northern EAO, dominated by igneous processes and relatively mild terrane accretion events, and that of the southern EAO, the focus of continental collision between East and West Gondwanaland.

Two granulite occurrences in Sudan have been dated with U-Pb zircon techniques. At Sabaloka (locality 1, Figure 2) meta-igneous and meta-sedimentary protoliths were metamorphosed to granulite facies at about 710 Ma (Kröner et al 1987). The metasedimentary granulites retain isotopic evidence of an Archean to mid-Proterozoic provenance, while the metamafic granulites may manifest igneous underplating similar to that of the ANS. Igneous bodies of charnockite, enderbite, and granite were intruded at 740 Ma at Jebel Moya (Stern & Dawoud 1991; locality 2, Figure 2). The "Samburuan" granulite-facies thrusting and metamorphism of Kenya is interpreted as manifesting plate collision across the N-S orogenic strike and is given an age of ~820 Ma based on an Rb-Sr whole-rock errorchron, but this is probably partially reset (Key et al 1989). (Note: An errorchron is an alignment of data points on an isochron diagram where the correspondence of the data to a least-squares fit is worse than that expected from analytical uncertainty alone.) The best known EAO granulites occur in the Eastern Granulite Complexes of Tanzania, a 900 km long discontinuous belt of metasedimentary and meta-igneous granulites, eclogites, metapyroxenites, meta-anorthosites, and ultramafics. These complexes display a complicated structural history, but the youngest foliation is subhorizontal and is interpreted as resulting from the development of huge, deep crustal nappes accompanying crustal thickening due to collision (Malisa & Muhongo 1990, Shackleton 1993). U-Pb zircon ages from the eastern granulites of Tanzania are interpreted to approximate the time of granulite-facies metamorphism and range from 650 to 715 Ma (Maboko et al 1985, Coolen et al 1982). This metamorphism occurred at 8 to 13 kbar and 700 to 900°C (Coolen 1982, Maboko et al 1989, Appel et al 1993). It is possible that metamorphism was related to the intrusion of basalts deep in the crust and that some of the older granulites formed prior to collision (Appel et al 1993). Uplift of the granulite terranes may better bracket the time of collision. This was accompanied by hydration and retrogression under amphibolite-facies conditions (ca. 475°–500°C), dated using hornblende K/Ar and $^{40}Ar/^{39}Ar$ techniques at about 630 Ma (Andriessen et al 1985, Maboko et al 1989). The only thermochronometric study of the EAO, for the Uluguru complex, Tanzania, has been presented by Maboko et al (1989). Their study indicates uplift (ca. 0.16 mm/a) from granulite formation until closure of hornblende to Ar diffusion (ca. 475°C) at 630 ± 10 Ma. Uplift then slowed to 0.05 to 0.08 mm/yr until muscovite closure (ca. 300°C) at 490 ± 10 Ma, with continued slow cooling and uplift until K-feldspar closure (ca. 170–200°C), at 420–450 Ma. These results indicate that, after a tectonic crustal thickening event and granulite-facies metamorphism, cooling followed an exponentially decaying path that is consistent with 20 to 40 km of isostatically-driven unroofing. There are two other granulite belts in Tanzania: the central and western granulite complexes. These have not been dated using U-Pb zircon techniques, and interpretation of Rb-Sr whole-rock ages for these bodies is problematic.

Younger granulites are found in the eastern part of the EAO. In southern Ethiopia (locality 3 in Figure 2), Gichile (1992) describes mafic granulites which indicate temperatures of 670–900°C and pressures of 9 ± 1.5 kbar. Similar charnockites from the nearby Bergudda complex yield a U-Pb zircon age of 545 Ma (Ayalew & Gichile 1990). The granulites of southernmost India and Sri Lanka (localities 10 and 11, Figure 2) were once thought to be Archean and Paleoproterozoic, but have since been shown to be Neoproterozoic, forming at 660 to 550 Ma (Burton and O'Nions 1990, Baur et al 1991, Choudhary et al 1992). Zircon evaporation ages for granulites from southeast Madagascar indicate granulite-facies metamorphism at 570–580 Ma (Paquette et al 1993). These younger ages may either indicate that a second, younger continent-continent collision occurred in the eastern part of the EAO or that crustal thickening progressed from west to east across the collision zone.

Arguments for a younger age of collision between East and West Gondwanaland have been advanced. For example, Kröner (1993) argued that the aforementioned granulites in southernmost India, Sri Lanka, and Madagascar indicated collision at the end of the Neoproterozoic. Powell et al (1993) concluded from paleomagnetic data that Gondwanaland formed no earlier than the end of the Neoproterozoic and perhaps as late as the mid-Cambrian. These arguments are also attractive insofoar as they are consistent with a rapid rise in seawater $^{87}Sr/^{86}Sr$ at the end of the Precambrian (Figure 5), but they are not consistent with the sequence of tectonic events observed along the EAO. Specifically, the consolidation of juvenile arcs was completed by about 700 Ma, and this should predate terminal collision by at most a few tens of millions of years. Also, it will be shown below that tectonic escape was underway by about 640 Ma, and this should not have commenced until terminal collision was underway.

WHERE DOES THE EAO CONTINUE TO THE SOUTH—OR DOES IT?

To the south, there are abundant granulitic rocks in Mozambique, Malawi, and Madagascar (Andreoli 1984). Rb-Sr geochronologic data indicate that metamorphism and deformation in the Malawi-Mozambique region occurred about 850 to 1100 Ma (Andreoli 1981, Costa et al 1992, Pinna et al 1993). Andreoli (1981) also reported U-Pb zircon ages of 840 and 920 Ma, but the common occurrence of cores indicates that inheritance is a likely problem and these ages may provide only a maximum age constraint. K-Ar results indicate that this region experienced thermal resetting ~450 to 650 Ma (Cahen et al 1984).

If the basement of Mozambique and Malawi last experienced major deformation and metamorphism about 1 Ga, and the interpretation that the EAO manifests closure of a major ocean basin is correct, then the EAO must continue either to the southwest through the Zambezi belt or southeast through Madagascar and India. Some (e.g. Hoffman 1991, Unrug 1993) argue that the Zambezi and Damaran structures comprise a suture between East and West Gondwanaland, but I question this hypothesis for 3 reasons: 1. The Irumide orogen can be traced through and across the Zambezi belt, indicating that continental separation either did not occur or was minor (Hanson et al 1988b); 2. isotopic data for the Damaran belt indicate that little juvenile crust was produced, suggesting that it represents a collapsed intracontinental basin (Hawkesworth et al 1986); and 3. tectonism was diachronous, with rifting beginning at 870 Ma in the Zambezi belt (Wilson et al 1993, Hanson et al 1993) and at about 750 Ma in the Damaran (Miller & Burger 1983), and thrusting beginning at about 820 Ma in the Zambezi belt (Hanson et al 1988b), continuing until about 550 Ma in the Damaran belt (Hawkesworth et al 1986).

The EAO could extend to the southeast, south of the multiply-deformed rocks of Madagascar, but no likely candidate for a suture has been identified. Similarities in basement geology between South Africa and Antarctica led Groenewald et al (1991) to conclude that these were a single entity by 1 Ga. Hoffman (1991) made an intriguing suggestion: The Mozambique Ocean closed about a pivot in what is now South Africa. This would result in the creation and subduction of large areas of oceanic crust in the north but vanishingly little in the south. This model is supported by the apparent absence of ophiolites south of the equator, where the preserved mafic-ultramafic complexes manifest rifts which did not open to the dimensions of large ocean basins (Prochaska & Pohl 1983, Guerrot et al 1993).

The problem of the southern extension of the EAO is a first-order unresolved question. For now, I prefer to interpret the EAO as continuing south from Tanzania into Mozambique and then south along the present Antarctica-Africa margin. The suture in Mozambique thus may be cryptic, with its trace plausibly defined by the N-S trending Cobué-Geci-Nkalapa-Luatize thrusts and shear zones (Pinna et al 1993). The following arguments support this view:

1. Granulites from Tanzania indicate that a major suture exists, and the strike of this suture appears to continue either southwest into the Zambezi belt or to the south into Mozambique. The continuation into the Zambezi belt was rejected above, leaving only the latter hypothesis.
2. Structural trends can be traced from the southern part of the Zambezi belt southeast and then south into the western deformation front of Mozambique (Vail 1965), suggesting contemporaneity of these two deformations. Deformation in the Zambezi belt is dated at 820 Ma, thus that in the western deformation front should be similar in age.
3. Structures at the deformation front, between the Kalahari Craton and the Mozambique belt south of the Zambezi belt, are remarkably similar to those described along the deformation front in Tanzania and Kenya (compare Johnson 1968 with Hepworth & Kennerly 1970 and Sanders 1965).

WHAT HAPPENED TO THE FORELANDS OF THE ARABIAN-NUBIAN SHIELD?

Another important unresolved set of questions concerns the fate of the forelands to the ANS, both to the east in Arabia and to the west in North Africa. In both instances we have a poor understanding of the tectonic setting in which extensive crustal remobilization occurred, and thick Phanerozoic cover complicates the problem for Arabia. In North Africa, there is geochronologic and isotopic evidence that extensive tracts of continental crust existed in Mesoproterozoic and earlier times (Klerkx & Deutsch 1977, Harms et al 1990, Sultan et al 1992b). This tract has been subjected to intense remobilization, including widespread intrusion and anatexis, deformation, metamorphism, uplift, and erosion, to the extent that Black & Liegois (1993) call this the "Central Saharan Ghost Craton." It now extends from the eastern Hoggar to the Nile and from the Mediterranean to the Congo craton, covering nearly half of Africa; it is simultaneously an excellent example of cratonic reactivation and one of the most poorly understood among all tracts of continental crust. The situation for Arabia is grossly similar; a large tract of remobilized Mesoproterozoic and older crust exists in the eastern Arabian shield and Yemen (Figure 2), but juvenile Neoproterozoic crust occurs in several localities to the east in Oman (Gass et al 1990).

The timing and cause of cratonic remobilization are presently unresolved. Geologic studies in the region of interest in North Africa are politically and logistically difficult, and this is compounded by the fact that large amounts of inheritance make it difficult to interpret what geochronologic data do exist. For example, we do not know if this ghost craton collided with (Pin & Poidevin 1987), or is a remobilized part of (Sultan et al 1992b), the Congo craton. There are three general classes of explanations for the remobilization: 1. convergent margin or collisional processes; 2. extensional processes; and 3. lithospheric delamination. According to the first class of explanations, westward-directed thrusting from the ANS led to crustal thickening and cratonic reworking (Schandelmeier et al 1988). A variant of this argues that at least one ocean basin and related arc batholiths formed in the interior of North Africa (Fullagar 1978). The second class of models has not been generally considered, but crustal extension can remobilize very large crustal tracts, such as the Basin-and-Range province of the western United States. Finally, regional Neoproterozoic lithospheric delamination has been advocated (Ashwal & Burke 1989, Black & Liegeois 1993). Inasmuch as crustal extension and lithospheric delamination happen most readily where the lithosphere and/or crust are warm and thickened, all three models are consistent with early thickening of the foreland. The presence of ophiolite belts that trend westward into the interior of North Africa (Stern et al 1990, Schandelmeier et al 1993) requires early extension and sea-floor spreading followed by compressional emplacement of ophiolitic nappes. This demonstrates that both crustal extension and collision occured in the foreland, although the timing remains contentious. These models need to be assessed; our understanding of EAO evolution will remain incomplete until we have a better understanding of the cause and timing of this cratonic remobilization.

RIGID INDENTORS AND ESCAPE TECTONICS

Collision along the EAO led to the development of strike-slip shear zones and faults and related extensional basins, similar to those of Cenozoic Asia. For the collision of India with Asia, Harrison et al (1992) estimate that one third of the convergence was accomodated by underthrusting and a slightly smaller amount was taken up by the lateral extrusion of Southeast Asia. If similar processes accompanied formation of the EAO (and an oceanic "free-face" existed), then the 20 to 40 km of uplift recorded in

EAO granulites must have been accompanied by a significant episode of "escape tectonics." Several variants of this have been proposed for the EAO. The first was proposed by Schmidt et al (1979) to explain the Najd fault system of Arabia (Figure 2) as the result of a continental "rigid indentor" to the east being subducted beneath the ANS to the west. The specific predictions of this model fail, as pointed out by Stern (1985). Burke & Şengör (1986) developed the concept of tectonic escape and applied it to the entire EAO, implying that much of the along-strike dichotomy resulted from the extrusion of trapped juvenile terranes from in front of the "hard" collision between East and West Gondwanaland towards an oceanic free-face in the north. Berhe (1990) noted that the entire EAO was affected by NW-SE trending strike-slip faults and suggested that these resulted from an oblique collision. Bonavia & Chorowicz (1992) argued that the Tanzania craton acted as the rigid indentor, consistent with an interpretation of an east-dipping subduction zone beneath East Gondwanaland (Shackleton 1986).

The timing of EAO terminal collision is constrained by when tectonic escape began and how long it continued. These ages also yield insights into the kinematics of collision (for example, how long convergence continued between East and West Gondwanaland). Faulting and related deformation along the Najd fault system of Saudi Arabia began about 630 Ma (Stacey & Agar 1985) and continued until about 530 Ma (Fleck et al 1976), indicating that collision was well underway prior to 630 Ma, although it is not clear how much earlier than this collision began. This uncertainty is partly due to our ignorance about whether or not (and if so, how long) there is a lag time between collision and escape tectonics. We are also uncertain when, in the escape history of the EAO, the Najd and other NW-trending sinistral shear systems formed. For example, in Kenya and Tanzania there are N-S trending "straightening zones" (Hepworth 1967) which fold the older nappes along isoclinal, vertical axes. This fabric constitutes the most obvious N-S fabric of the EAO south of Sudan. The straightening zones clearly post-date emplacement of the ophiolitic nappes and preceed development of the NW-SE shear zones, but it is not clear whether straightening zones demonstrate continued shortening of the nappe stack or reveal the first stages in tectonic escape. In Ethiopia and Sudan, these zones are interpreted to manifest E-W shortening with little strike-slip shearing (Beraki et al 1989, Miller & Dixon 1992). In Sudan, Abdelsalam (1994) documents the evolution of a N-S zone of pure strain into a NW-SE oriented strike-slip fault and interprets this as the result of progressive E-W shortening. In Kenya and in Arabia, similar structures are interpreted as sinistral shear zones (Key et al 1989, Duncan et al 1990, Mosley 1993).

Regardless of whether the straightening zones indicate pure or simple shear, their ages provide an independant younger limit to the time of collision. In Sudan, deformation along the Hamisana shear zone has been bracketed between 610 and 660 Ma (Stern & Kröner 1993). In Kenya, two phases of post-accretionary deformation are found that predate development of the NW-trending strike-slip faults: the ca. 620 Ma Baragoian and the ca. 580 Ma Barsaloian events. Both Baragoian and Barsaloian structures are characterized by upright tight folds, but the former trend NW-SE to NNW-SSE and record sinistral displacements whereas the latter trend N-S and record dextral displacements. Both are interpreted as post-collisional ductile deformation parallel to the orogen (Key et al 1989).

Tectonic escape resulted in the formation of a large area of ca. 600 Ma old rifts (Figure 2), in NE Egypt (Stern 1985), in Oman (Wright et al 1990), and in the Persian Gulf (Husseini & Husseini 1990). This episode of collision-related extension led in turn to the rifting away of some fragments of greater Gondwanaland and the formation of a passive margin in North Africa and Arabia during latest Neoproterozoic time (Bond et al 1984). The supercontinent that resulted from collision between East and West Gondwanaland must have been significantly larger than the Gondwanaland that persisted throughout the Paleozoic, although we do not know how much larger or where the rifted fragments are now. Thus the discussion by Burke & Şengör (1986) regarding whether the Persian Gulf salt basins are related to oceanic closing or opening is moot: They are related to both.

A better understanding of the timing and style of EAO escape tectonics should be a goal of future research in the East African Orogen, but for the present synthesis the following points are accepted:

1. Collision between East and West Gondwanaland involved the Tanzanian craton as the rigid indentor and the western flank of East Gondwanaland as the region of crustal thickening and plastic deformation.
2. Escape was predominantly to the north.
3. Escape began sometime before 610 Ma and after 660 Ma and continued until about 530 Ma, implying that convergence between East and West Gondwanaland continued for 120 to 170 My after initial collision.
4. Tectonic escape led to the formation of rifted basins in northeast Africa and Arabia which, in turn, led to continental separation and the formation of a passive margin on the north flank of Gondwanaland at the very end of the Precambrian (Husseini 1989, Brookfield 1993). The size and location of the rifted fragments are unknown but may be found in central Europe, Turkey, and environs.

These relationships are shown in Figure 8, which compares the inferred tectonic relationships around the EAO with that of a mirror image of the modern India-Asia collision.

SIGNIFICANCE OF THE EAO FOR NEOPROTEROZOIC PALEOGEOGRAPHIC RECONSTRUCTIONS

Several models for the evolution and disposition of the continents during the Neoproterozoic have been advanced in the past few years (see review by Storey 1993). The above review of the EAO provides the following constraints for these models:

1. The formation of the Mozambique Ocean about 800–850 Ma reflects breakup of the Mesoproterozoic supercontinent Rodinia.
2. The abundance of ophiolites and volume of juvenile continental crust preserved in the northern EAO indicate that this rift evolved into a Pacific-sized "Mozambique Ocean" basin. The formation of this and other Neoproterozoic ocean basins was responsible for the excursion of the seawater Sr isotope curve to nonradiogenic values about 800 Ma.
3. The concentration of juvenile crust and ophiolites in the Arabian-Nubian Shield are consistent with Hoffman's (1991) model that the Mozambique Ocean closed like a fan, with a hinge in South Africa.
4. The deformation sequence—arc accretion → terminal collision → tectonic escape—indicates that the Mozambique Ocean closed to form Greater Gondwanaland around 640–700 Ma. The significance of the younger (ca. 550 Ma) granulite-facies metamorphic event of regional extent is not understood, but the absence of associated antecedent juvenile crust and ophiolites or subsequent escape tectonics implies that it does not date terminal collision along the EAO between East and West Gondwanaland.
5. Rifting at the Neoproterozoic-Cambrian boundary followed collision along the EAO and is related to initial disruption of Greater Gondwanaland. As Murphy & Nance (1991) emphasize, there are two episodes of supercontinent breakup during the Neoproterozoic, but both are protracted. The first is related to disaggregation of Rodinia; the formation of the Mozambique Ocean is related to this. Rifting of Laurentia away from Antarctica about 750 Ma (Storey 1993) may be the last stage of this episode. The second stage of rifting begins a few tens of millions of years after the formation of Greater Gondwanaland, beginning at the very end of the Neoproterozoic and continuing throughout the Paleozoic. This is the rifting episode identified by Bond et al (1984).

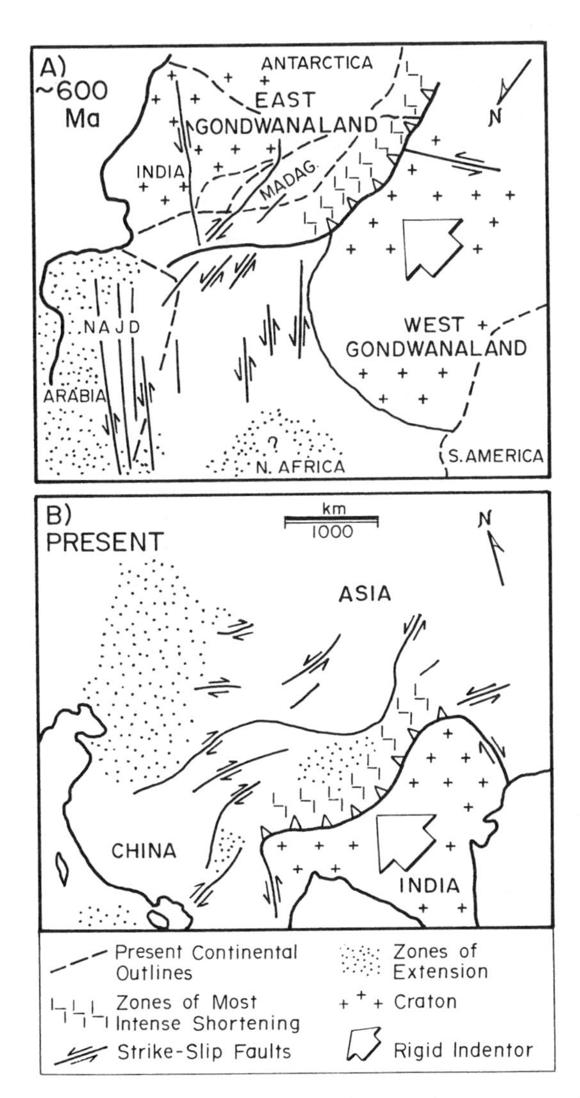

Figure 8 Comparison of continental collisions at the same scale. Both are oriented so that the rigid indentor is moving towards the upper left corner: (*A*) East African Orogen, ca 600 Ma ago. Areas without ornamentation are juvenile or were remobilized during the Neoproterozoic. (*B*) Modern India-Asia collision, shown as mirror-image so that the free face and principal zone of tectonic escape are on the same side of the rigid indentor as is the case for the EAO.

SUMMARY

Our understanding of the style and timing of collision between East and West Gondwanaland to form the East African Orogen is incomplete, but the general outline of this important event in Earth history is slowly emerging. A crude model (Figure 9) embodies present understanding of this evolution. Present reconstructions of the earliest stages in the EAO orogenic cycle may be more speculation than understanding, but initiation by rifting of a continent is simple, logical, and consistent with the data at hand. It is attractive to identify this rifting as part of the breakup of Rodinia, and we can place it in time about 850–900 Ma. In contrast, the sequence of events that begins with sea-floor spreading and formation of arcs and back-arc basins and continues with the accretion of these tectonic cells into juvenile crust is coming into sharper focus, at least for the

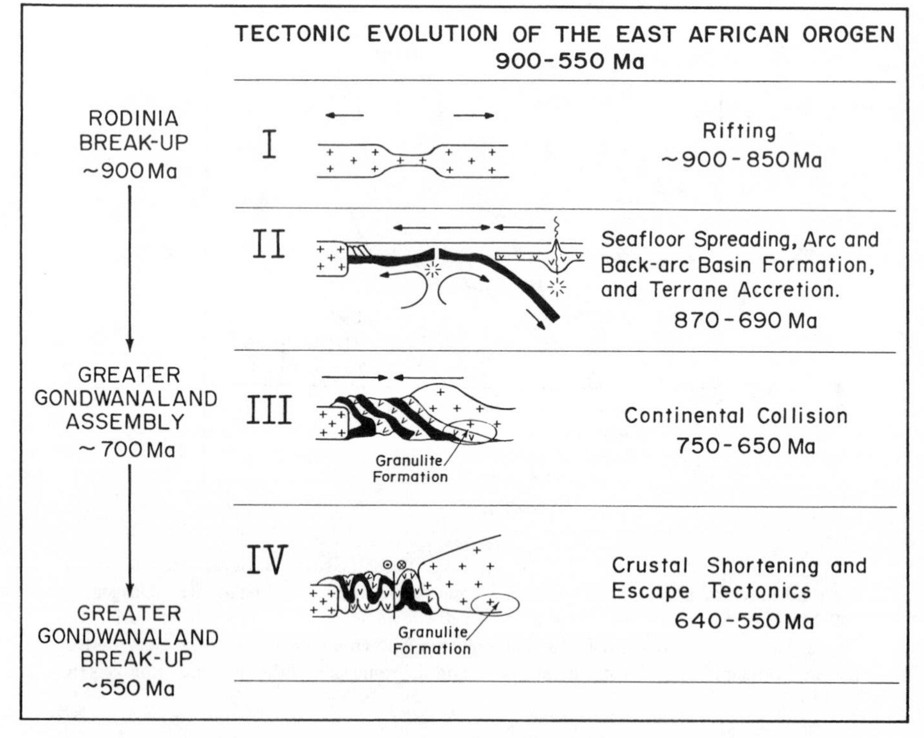

Figure 9 Two-dimensional summary of the tectonic evolution of the East African Orogen, as described in the text.

Arabian-Nubian Shield. There is good evidence that these processes were underway as early as 870 Ma and continued until at least 690 Ma. Assuming that modern crustal production rates are approximately valid for the Neoproterozoic as well, the volume of juvenile crust that formed during this interval suggests that a Pacific-sized ocean was opened and closed. Accretion of numerous juvenile arcs and a few older continental fragments continued during this interval. Continental collision first led to crustal thickening and uplift, beginning perhaps as early as 750 Ma but certainly by 700 Ma, and continued with orogenic collapse and escape tectonics until the end of the Precambrian. At least as far as the EAO is concerned, this dates the consolidation of East and West Gondwanaland, although several oceanic basins of unknown dimensions persisted within West Gondwanaland. Crustal thickening and uplift propagated with time eastward, now marked by zones of ~550 Ma granulites in southernmost India, Sri Lanka, and Madagascar. Tectonic escape led to the development of major rift basins in the northern EAO and environs which led directly to sea-floor spreading and formation of a passive margin on the remnants of Gondwanaland and the formation of an ocean basin to the north, at about 550 Ma ago.

ACKNOWLEDGMENTS

This manuscript benefits from the constructive criticisms of Kevin Burke, Tim Dixon, Richard Hanson, Alfred Kröner, Sospeter Muhongo, John Rogers, Heinz Schandelmeier, Robert Shackleton, Mohamed Sultan, Raphael Unrug, and John Vail. I especially appreciate Al Engel's early encouragement and the opportunity that he provided to begin studying the basement of northeast Africa. My work in northeast Africa is supported by NASA. This is UTD Programs in Geosciences contribution #751.

Literature Cited

Abdel Rahman ESM. 1993. Geochemical and geotectonic controls of the metallogenic evolution of selected ophiolite complexes from the Sudan. *Berliner Geowiss. Abh.* A: 145. 175 pp.

Abdelsalam MG. 1994. Structural evolution of the Late Proterozoic (Pan-African) Oko Shear Zone, Red Sea Hills, Sudan: implications for the evolution of post-accretionary structures in the Arabian-Nubian Shield. *J. Geol. Soc. London* Submitted

Abdelsalam MG, Stern RJ. 1993a. Tectonic evolution of the Nakasib suture, Red Sea Hills, Sudan: evidence for a Wilson Cycle in the Late Precambrian of the Arabian-Nubian Shield. *J. Geol. Soc. London* 150: 393–404

Abdelsalam MG, Stern RJ. 1993b. Structural styles of late Proterozoic sutures in the Arabian-Nubian Shield: the Nakasib suture, Red Sea Hills, Sudan. *J. Geol. Soc. London* In press

Al-Shanti AM, Gass IG. 1983. The Upper Proterozoic ophiolite mélange zones of the easternmost Arabian shield. *J. Geol. Soc. London* 140: 867–76

Andreoli MA. 1981. *The amphibolite and the granulite facies rocks of southern Malawi.* PhD thesis. Univ. Witwatersrand, S. Afr. 351 pp.

Andreoli MA. 1984. Petrochemistry, tectonic evolution and metasomatic mineralisations of Mozambique belt granulites from S. Malawi and Tete (Mozambique). *Precambrian Res.* 25: 161–86

Andreoli MA. 1991. Petrological markers in terrane analysis: massif anorthosites, high pressure granulites and scapolitized rocks from the Mozambique belt, southern Malawi. *UNESCO, Geol. Econ. Dev., Newslett.* 8: 43–55

Andriessen PAM, Coolen JJMMM, Hebeda EH. 1985. K-Ar hornblende dating of late Pan-African metamorphism in the Furua granulite complex of southern Tanzania. *Precambrian Res.* 30: 351–60

Appel P, Möller A, Schenk V, Muhongo S. 1993. Granulite facies metamorphism and P-T evolution in the Mozambique Belt of Tanzania. See Thorweihe & Schandelmeier 1993, pp. 171–75

Ashwal LD, Burke K. 1989. African lithospheric structure, volcanism, and topography. *Earth Planet. Sci. Lett.* 96: 8–14

Asmerom Y, Jacobsen SB, Knoll AH, Butterfield NJ, Swett K. 1991. Strontium isotopic variations of Neoproterozoic seawater: implications for crustal evolution. *Geochim. Cosmochim. Acta* 55: 2883–94

Ayalew T, Bell K, Moore JM, Parrish RR. 1990. U-Pb and Rb-Sr geochronology of the Western Ethiopian Shield. *Geol. Soc. Am. Bull.* 102: 1309–16

Ayalew T, Gichile S. 1990. Preliminary U-Pb ages from southern Ethiopia. In *Recent Data in African Earth Sciences*, ed. G Rocci, M Deschamps, pp. 127–30. *CIFEG Occ. Publ.* 22

Baur N, Kröner A, Liew TC, Todt W. Williams IS, Hofmann AW. 1991. U-Pb isotopic systematics of zircons from prograde and retrograde transition zones in high-grade orthogneisses, Sri Lanka. *J. Geol.* 99: 527–45

Beraki WH, Bonavia FF, Getachew T, Schmerold R, Tarekegn T. 1989. The Adola Fold and Thrust Belt, southern Ethiopia: a re-examination with implications for Pan-African evolution. *Geol. Mag.* 126: 647–57

Berhe SM. 1990. Ophiolites in Northeast and East Africa: implications for Proterozoic crustal growth. *J. Geol. Soc. London* 147: 41–57

Black B, Liegeois J-P. 1993. Cratons, mobile belts, alkaline rocks and continental lithospheric mantle: the Pan-African testimony. *J. Geol. Soc. London* 150: 89–98

Bonavia FF, Chorowicz J. 1992. Northward expulsion of the Pan-African of northeast Africa guided by a reentrant zone of the Tanzania craton. *Geology* 20: 1023–26

Bond GC, Nickeson PA, Kominz MA. 1984. Breakup of a supercontinent between 625 Ma and 555 Ma: new evidence and implications for continental histories. *Earth Planet. Sci. Lett.* 70: 325–45

Brookfield ME. 1993. The Himalayan passive margin from Precambrian to Cretaceous times. *Sediment. Geol.* 84: 1–35

Burke K, Dewey JF 1972. Orogeny in Africa. In *Orogeny in Africa*, ed. TFJ Dessauvagie, AJ Whiteman, pp. 583–608. Ibadan, Nigeria: Geol. Dept., Univ. Ibadan

Burke K, Sengör C. 1986. Tectonic escape in the evolution of the continental crust. In *Reflection Seismology: The Continental Crust*, ed. M Barazangi, L Brown, pp. 41–53. *Am. Geophys. Union, Geodyn. Ser.* 14 Washington, DC: Am. Geophys. Union

Burke WH, Denison RE, Hetherington EA, Koepnick RB, Nelson HF, Otto JB. 1982. Variation of seawater $^{87}Sr/^{86}Sr$ throughout Phanerozoic time. *Geology* 10: 516–19

Burton KW, O'Nions RK. 1990. Fe-Ti oxide chronometry: with application to granulite formation. *Geochim. Cosmochim. Acta* 54: 2593–602

Cahen L, Snelling NJ, Delhal J, Vail JR, Bonhomme M, Ledent D. 1984. *The Geochronology and Evolution of Africa.* Oxford: Clarendon. 512 pp.

Choudhary AK, Harris NBW, Van Calstern P, Hawkesworth CJ. 1992. Pan-African charnockite formation in Kerala, south India. *Geol. Mag.* 129: 257–67

Church WR. 1988. Ophiolites, sutures, and micro-plates of the Arabian-Nubian Shield. In *The Pan-African Belt of Northeast Africa and Adjacent Areas*, ed. S El-Gaby, RO Greiling, pp. 289–316. Braunschweig: Vieweg

Claesson S, Pallister JS, Tatsumoto M. 1984. Samarium-neodymium data on two late Proterozoic ophiolites of Saudi Arabia and implications for crustal and mantle evolution. *Contrib. Mineral. Petrol.* 85: 244–52

Conway Morris S. 1993. The fossil record and the early evolution of the Metazoa. *Nature* 361: 219–25

Coolen JJMMM. 1982. Carbonic fluid inclusions in granulites from Tanzania—a comparison of geobarometric methods

based on fluid density and mineral chemistry. *Chem. Geol.* 37: 59–77
Coolen JJMMM, Priem HNA, Verdurmen EAT, Verschure RH. 1982. Possible zircon U-Pb evidence of Pan-African granulite facies metamorphism in the Mozambique belt of southern Tanzania. *Precambrian Res.* 17: 31–40
Costa M, Ferrara G, Sacchi R, Tonarini S. 1992. Rb/Sr dating of the Upper Proterozoic basement of Zambesia, Mozambique. *Geol. Rundsch.* 81: 487–500
Dalziel IWD. 1991. Pacific margins of Laurentia and East Antarctica–Australia as a conjugate rift pair: evidence and implication for an Eocambrian supercontinent. *Geology* 19: 598–601
Dalziel IWD. 1992. Antarctica: a tale of two supercontinents? *Annu. Rev. Earth Planet. Sci.* 20: 501–26
Davies GF. 1992. On the emergence of plate tectonics. *Geology* 20: 963–66
Davies GF, Richards MA. 1992. Mantle convection. *J. Geol.* 100: 151–206
Denkler T, Harms U, Franz G, Darbyshire DPF, Pilot J, Schandelmeier H. 1993. Evolution of the south-western part of the Late Proterozoic Atmur-Delgo suture zone (Northern Province, Sudan). See Thorweihe & Schandelmeier 1993, pp. 87–89
DePaolo DJ. 1988. *Neodymium Isotope Geochemistry*. Berlin: Springer-Verlag. 187 pp.
Derry LA, Kaufman AJ, Jacobsen, SB. 1992. Sedimentary cycling and environmental change in the Late Proterozoic: evidence from stable and radiogenic isotopes. *Geochim. Cosmochim. Acta* 56: 1317–29
Des Marias DJ, Strauss H, Summons RE, Hayes JM.1992. Carbon isotope evidence for the stepwise oxidation of the Proterozoic environment. *Nature* 359: 605–9
De Wit M, Jeffery M, Bergh H, Nicolaysen L 1988. *Geological map of sectors of Gondwana reconstructed to their disposition* ~150 *Ma* (1: 10,000,000).Tulsa: Am. Assoc. Petrol. Geol.
Dixon TH. 1981. Age and chemical characteristics of some pre-Pan African rocks in the Egyptian shield. *Precambrian Res.* 14: 119–33
Dixon TH, Golombek MP 1988. Late Precambrian crustal accretion rates in northeast Africa and Arabia. *Geology* 16: 991–94
Duncan IJ, Rivard B, Arvidson RE, Sultan M. 1990. Structural interpretation and tectonic evolution of a part of the Najd Shear Zone (Saudi Arabia) using Landsat thematic mapper data. *Tectonophysics* 178: 309–35
El-Isa Z, Mechie J, Prodehl C, Makris J, Rihm R. 1987. A crustal structure study of Jordan derived from seismic refraction data. *Tectonophysics* 138: 235–53
Fleck RJ, Coleman RG, Cornwall HR, Greenwood WR, Hadley DG, et al. 1976. Geochronology of the Arabian Shield, western Saudi Arabia: K-Ar results. *Geol. Soc. Am. Bull.* 87: 9–21
Forsyth D, Uyeda S. 1975. On the relative importance of the driving forces of plate tectonics. *Geophys. J. R. Astron. Soc.* 43: 163–200
Fullagar, P.D. 1978. Pan-African age granites of Northeastern Africa: new or reworked sialic materials? In *Geology of Libya*, Vol. 3, ed. MJ Salem, MT Busrewil, pp. 1051–58. New York: Academic
Gass IG, Ries AC, Shackleton RM, Smewing JD. 1990. Tectonics, geochronology and geochemistry of the Precambrian rocks of Oman. In *The Geology and Tectonics of the Oman Region*, ed. AHF Robertson, MP Searle, AC Ries, pp. 585–99. *Geol. Soc. London Spec. Publ.* 49
Gettings ME, Blank HR, Mooney WD, Healey JH. 1986. Crustal structure of southwestern Saudi Arabia. *J. Geophys. Res.* 91: 6491–512
Gichile S. 1992. Granulites in the Precambrian basement of southern Ethiopia: geochemistry, P-T conditions of metamorphism and tectonic setting. *J. Afr. Earth Sci.* 15: 251–63
Greenberg JK. 1981. Characteristics and origin of Egyptian Younger Granites: summary. *Geol. Soc. Am. Bull.* 92: 224–32
Groenewald PB, Grantham GH, Watkeys MK. 1991. Geological evidence for a Proterozoic to Mesozoic link between southeastern Africa and Dronning Maud Land, Antarctica. *J. Geol. Soc. London* 148: 1115–23
Guerrot C, Cocherie A, Ohnenstetter M. 1993. Origin and evolution of the West Andriamena Pan African mafic-ultramafic complexes in Madagascar as shown by U-Pb, Nd isotopes and trace element constraints. *Terra Abstr.* 5: 387
Gurnis M. 1988. Large-scale mantle convection and the aggregation and dispersal of supercontinents. *Nature* 332: 695–99
Hambrey MJ, Harland WB. 1985. The Late Proterozoic glacial era. *Palaeogeogr. Palaeoclimatol. Palaeoecol.* 51: 255–72
Hanson RE, Donovan RN, Wilson TJ. 1989. The late Proterozoic Zambezi belt in southern Africa: a model for the deeper levels of the southern Oklahoma aulacogen. *Geol. Soc. Am., Abstr. with Programs* 21: 13
Hanson RE, Wilson TJ, Brueckner HK, Onstott TC, Wardlaw MS, et al. 1988a. Reconnaissance geochronology, tectono-

thermal evolution, and regional significance of the middle Proterozoic Choma-Kalomo block, southern Zambia. *Precambrian Res.* 42: 39–61

Hanson RE, Wilson TJ, Munyanyiwa H. 1993. Geologic evolution of the Late Proterozoic Zambezi Belt in Zambia. *J. Afr. Earth Sci.* In press

Hanson RE, Wilson TJ, Wardlaw MS. 1988b. Deformed batholiths in the Pan-African Zambezi belt, Zambia: age and implications for regional Proterozoic tectonics. *Geology* 16: 1134–37

Harms U, Schandelmeier H, Darbyshire DPF. 1990. Pan-African reworked early/middle Proterozoic crust in NE Africa west of the Nile: Sr and Nd isotope evidence. *J. Geol. Soc. London* 147: 859–72

Harris NBW, Gass IG, Hawkesworth CJ. 1990. A geochemical approach to allochthonous terranes: a Pan-African case study. *Phil. Trans. R. Soc. London Ser. A* 331: 533–48

Harris NBW, Hawkesworth CJ, Ries AC. 1984. Crustal evolution in northeast and east Africa from model Nd ages. *Nature* 309: 773–76

Harrison TM, Copeland P, Kidd WSF, Yin A. 1992. Raising Tibet. *Science* 255: 1663–70

Hawkesworth CJ, Menzies MA, van Calsteren P. 1986. Geochemical and tectonic evolution of the Damara belt, Namibia. In *Collision Tectonics*, ed. MP Coward, AC Ries, pp. 305–19. London: Geol. Soc.

Henjes-Kunst F, Altherr R, Baumann A. 1990. Evolution and composition of the lithospheric mantle underneath the western Arabian peninsula: constraints from Sr-Nd isotope systematics of mantle xenoliths. *Contrib. Mineral. Petrol.* 195: 460–72

Hepworth JV. 1967. The photogeological recognition of ancient orogenic belts in Africa. *J. Geol. Soc. London* 123: 253–92

Hepworth JV, Kennerly JB. 1970. Photogeology and structure of the Mozambique orogenic front near Kolo, north-east Tanzania. *J. Geol. Soc. London* 125: 447–79

Hoffman PF. 1989. Speculations on Laurentia's first gigayear (2.0 to 1.0 Ga). *Geology* 17: 135–38

Hoffman PF. 1991. Did the Breakout of Laurentia turn Gondwanaland inside-out? *Science* 252: 1409–12

Husseini MI. 1989. Tectonic and deposition model of late Precambrian–Cambrian Arabian and adjoining plates. *Am. Assoc. Petrol. Geol. Bull.* 73: 1117–31

Husseini MI, Husseini SI. 1990. Origin of the Infracambrian Salt basins of the Middle East. In *Classic Petroleum Provinces*, ed. J Brooks, pp. 279–92. *Geol. Soc. London Spec. Publ.* 50

James HL. 1983. Distribution of banded iron formation in space and time. In *Iron Formations: Facts and Problems*, ed. AF Trendall, RC Morris, pp. 471–90. Amsterdam: Elsevier

Johnson RL. 1968. Structural history of the western front of the Mozambique Belt in northeast Southern Rhodesia. *Geol. Soc. Am. Bull.* 79: 513–26

Kennedy WQ. 1964. The structural differentiation of Africa in the Pan-African (±500 m.y.) tectonic episode. *Leeds Univ. Res. Inst. Afr. Geol. Annu. Rep.* 8: 48–49

Key RM, Charsley TJ, Hackman BD, Wilkinson AF, Rundle CC. 1989. Superimposed Upper Proterozoic collision-controlled orogenies in the Mozambique orogenic belt of Kenya. *Precambrian Res.* 44: 197–225

Klerkx J, Deutsch S. 1977. Resultats preliminaires obtenus par la methode Rb/Sr sur l'age des formations Precambriennes de la region d'Uweinat (Libye). *Mus. R. Afr. Centrale, Tervuren (Belg.) Dept. Geol. Min. Rapp. Annu.* 1976, pp. 83–94

Knoll A. 1992. The early evolution of Eukaryotes: a geological perspective. *Science* 256: 622–27

KRISP Working Party. 1991. Large-scale variation in lithospheric structure along and across the Kenya rift. *Nature* 354: 223–27

Kröner A. 1984. Late Precambrian plate tectonics and orogeny: a need to redefine the term Pan-African. In *African Geology*, ed. J Klerkx, J Michot, pp. 23–28. Tervuren: Musée. R. l'Afrique Centrale

Kröner A. 1991. African linkage of Precambrian Sri Lanka. *Geol. Rundsch.* 80: 429–40

Kröner A. 1993. The Pan African belt of northeastern and Eastern Africa, Madagascar, southern India, Sri Lanka and East Antarctica: terrane amalgamation during the formation of the Gondwana supercontinent. See Thorweihe & Schandelmeier 1993, pp. 3–9

Kröner A, Linnebacher P, Stern RJ, Reischmann T, Manton W, Hussein IM. 1991. Evolution of Pan-African island arc assemblages in the southern Red Sea Hills, Sudan, and in southwestern Arabia as exemplified by geochemistry and geochronology. *Precambrian Res.* 53: 99–118

Kröner A, Pallister JS, Fleck RJ. 1992a. Age of initial oceanic magmatism in the Late Proterozoic Arabian Shield. *Geology* 20: 803–6

Kröner A, Reischmann T, Todt W, Zimmer M, Stern RJ, et al. 1989. Timing, mechanism, and geochemical patterns of arc accretion in Arabian-Nubian Shield and its extensions into Israel and Somalia. *Abstr. 28th Int. Geol. Congr.* 2: 231

Kröner A, Stern RJ, Dawoud AS, Compston W, Reischmann T. 1987. The Pan-African continental margin in northeastern Africa: evidence from a geochronological study of granulites at Sabaloka, Sudan. *Earth Planet. Sci. Lett.* 85: 91–104

Kröner A, Todt W, Hussein IM, Mansour M, Rashwan AA. 1992b. Dating of late Proterozoic ophiolites in Egypt and the Sudan using the single grain zircon evaporation technique. *Precambrian Res.* 59: 15–32

Küster D, Utke A, Leupolt L, Lenoir JL, Haider A. 1990. Pan-African granitoid magmatism in northeastern and southern Somalia. *Berliner Geowiss. Abh.* 120: 519–36

Leggo PJ. 1974. A geochronological study of the basement complex of Uganda. *J. Geol. Soc. London* 130: 263–77

Lindsay JF, Korsch RJ, Wilford JR. 1987. Timing the breakup of a Proterozoic supercontinent: evidence from Australian intracratonic basins. *Geology* 15: 1061–64

Maboko MAH, Boelrijk NAIM, Priem HNA, Verdurmen EAT. 1985. Zircon U-Pb and biotite Rb-Sr dating of the Wami River granulites, Eastern Granulites, Tanzania: evidence for approximately 715 Ma old granulite-facies metamorphism and final Pan-African cooling approximately 475 Ma ago. *Precambrian Res.* 30: 361–78

Maboko MAH, McDougall I, Zeitler PK. 1989. Dating late Pan-African cooling in the Uluguru granulite complex of Eastern Tanzania using the ^{40}Ar-^{39}Ar technique. *J. Afr. Earth Sci.* 9: 159–67

Malisa E, Muhongo S. 1990. Tectonic setting of gemstone mineralization in the Proterozoic metamorphic terrane of the Mozambique belt in Tanzania. *Precambrian Res.* 46: 167–76

McGuire AV, Stern RJ. 1993. Granulite xenoliths from western Saudi Arabia: the Lower Crust of the Late Precambrian Arabian-Nubian Shield. *Contrib. Mineral. Petrol.* 114: 395–408

McWilliams MO. 1981. Palaeomagnetism and Precambrian tectonic evolution of Gondwana. In *Precambrian Plate Tectonics*, ed. A Kröner, pp. 649–87. Amsterdam: Elsevier

Miller MM, Dixon TH. 1992. Late Proterozoic evolution of the northern part of the Hamisana zone, northeast Sudan: constraints on Pan-African accretionary tectonics. *J. Geol. Soc. London* 149: 743–50

Miller RMcG, Burger AJ. 1983. U-Pb zircon age of the early Damaran Naauwpoort formation. In *Evolution of the Damara Orogen of Southwest Africa/Namibia*, ed. RMcG Miller, pp. 267–72. *Spec. Publ. Geol. Soc. S. Afr.* 11

Molnar P, Atwater T. 1978. Interarc spreading and Cordilleran tectonics as alternatives related to the age of subducted oceanic lithosphere. *Earth Planet. Sci. Lett.* 41: 330–40

Moores EM. 1991. The southwest U.S.—East Antarctica (SWEAT) connection: a hypothesis. *Geology* 19: 425–28

Mosley PN. 1993. Geological evolution of the Late Proterozoic "Mozambique Belt" of Kenya. *Tectonophysics* 221: 223–50

Muhongo S, Lenoir J-L. 1993. Pan-African granulite-facies metamorphism in the Mozambique Belt of Tanzania: evidence from U-Pb on zircon geochronology. *J. Geol. Soc. London* In press

Murphy JB, Nance RD. 1991. Supercontinent model for the contrasting character of Late Proterozoic orogenic belts. *Geology* 19: 469–72

Neary CR, Gass IG, Cavanagh BJ. 1976. Granitic association of northeastern Sudan. *Geol. Soc. Am. Bull.* 87: 1501–12

Nelson BK, DePaolo DJ. 1985. Rapid production of continental crust 1.7 to 1.9 b.y. ago: Nd isotopic evidence from the basement of the North American mid-continent. *Geol. Soc. Am. Bull.* 96: 746–54

Oldow JS, Bally AW, Avé Lallemant HG, Leeman WP. 1989. Phanerozoic evolution of the North Ameican Cordillera; United States and Canada. In *The Geology of North America—An overview*, ed. AW Bally, AR Palmer, pp. 139–232. Boulder: Geol. Soc. Am.

Pallister JS, Stacey JS, Fischer LB, Premo WR. 1988. Precambrian ophiolites of Arabia: geologic settings, U-Pb geochronology, Pb-isotope characteristics, and implications for continental accretion. *Precambrian Res.* 38: 1–54

Paquette J-L, Nédélec A, Moine B, Rakotondrazafy M. 1993. U-Pb, single zircon Pb-evaporation and Sm-Nd dating of the S.E. Madagascar domain: a prevailing Panafrican event. *Terra Abstr.* 5: 393

Pearce JA, Lippard SJ, Roberts S. 1984. Characteristics and tectonic significance of supra-subduction zone ophiolites. In *Marginal Basin Geology*, ed. BP Kokelaar, MF Howell, pp. 77–94. *Geol. Soc. London Spec. Publ.* 16

Pin C, Poidevin JL. 1987. U-Pb zircon evidence for a Pan-African granulite facies metamorphism in the Central African Republic. A new interpretation of the high-grade series of the northern border of the Congo Craton. *Precambrian Res.* 36: 303–12

Pinna P, Jourde G, Calvez JY, Mroz JP,

Marques JM. 1993. The Mozambique Belt in northern Mozambique: Neoproterozoic (1100–850 Ma) crustal growth and tectogenesis, and superimposed Pan-African (800–550 Ma) tectonism. *Precambrian Res.* 62: 1–59

Powell C, Li ZX, McElhinny M, Meert J, Park J. 1993. Paleomagnetic constraints on timing of the Neoproterozoic breakup of Rodinia and the Cambrian formation of Gondwanaland. *Geology* 21: 889–92

Price RC. 1984. *Late Precambrian mafic-ultramafic complexes in northeast Africa.* PhD dissertation. Open Univ., Milton Keynes, UK. 326 pp.

Prochaska W, Pohl W. 1983. Petrochemistry of some mafic and ultramafic rocks from the Mozambique Belt, northern Tanzania. *J. Afr. Earth Sci.* 1: 183–93

Reischmann T. 1986. *Geologie und Genese spätproterozoischer Vulkanite der Red Sea Hills, Sudan.* PhD dissertation. Johannes Gutenberg Univ., Mainz

Reischmann T, Kröner A, Basahel A.1984. Petrography, geochemistry, and tectonic setting of metavolcanic sequences from the Al Lith area, southwestern Arabian Shield. In *IGCP Proj.* 164, *Proc., Vol.* 6, *Pan-African crustal evolution in the Arabian-Nubian Shield*, pp. 365–79. Jeddah: Faculty of Sci., King Abdulaziz Univ.

Reymer A, Schubert G. 1984. Phanerozoic addition rates to the continental crust and crustal growth. *Tectonics* 3: 63–77

Ries AC, Vearncombe JR, Price RC, Shackleton RM. 1992. Geochronology and geochemistry of the rocks associated with a Late Proterozoic ophiolite in West Pokot, NW Kenya. *J. Afr. Earth Sci.* 14: 25–36

Rihm R, Makris J, Möller L. 1991. Seismic surveys in the northern Red Sea: asymmetric crustal structure. *Tectonophysics* 198: 279–95

Sacchi R, Marques J, Costa M, Casati C. 1984. Kibaran events in the southernmost Mozambique Belt. *Precambrian Res.* 25: 141–59

Sanders LD. 1965. Geology of the Contact between the Nyanza Shield and the Mozambique Belt in western Kenya. *Geol. Surv. Kenya Bull.* 7 45 pp.

Schandelmeier H, Darbyshire DPF, Harms U, Richter A. 1988. The East Saharan Craton: evidence for pre-Pan African crust in NE Africa west of the Nile. In *The Pan-African Belt of Northeast Africa and Adjacent Areas*, ed. S El-Gaby, RO Greiling, pp.69–94. Braunschweig: Vieweg

Schandelmeier H, Küster D, Wipfler E, Abdel Rahman EM, Stern RJ, et al 1993. Evidence for a new Late Proterozoic suture in Northern Sudan. See Thorweihe & Schandelmeier 1993, pp. 83–86

Schmidt DL, Hadley DG, Stoeser DB. 1979. Late Proterozoic crustal history of the Arabian Shield, southern Najd Province, Kingdom of Saudi Arabia. In *Evolution and Mineralization of the Arabian-Nubian Shield*, Vol. 2, ed. SA Tahoun, pp. 41–58. New York: Pergamon

Şengör CA. 1987. Tectonics of the Tethysides: orogenic collage development in a collisional setting. *Annu. Rev. Earth Planet. Sci.* 15: 213–44

Shackleton RM. 1986. Precambrian collision tectonics in Africa. In *Collision Tectonics*, ed. MP Coward, AC Ries, pp. 329–49. London: Geol. Soc.

Shackleton RM. 1993. Tectonics of the lower crust: a view from the Usambara Mountains, NE Tanzania. *J. Structural Geol.* 15: 663–71

Shanti M, Roobol MJ. 1979. A late Proterozoic ophiolite complex at Jabal Ess in northern Saudi Arabia. *Nature* 279: 488–91

Stacey JS, Agar RA. 1985. U-Pb isotopic evidence for the accretion of a continental microplate in the Zalm region of the Saudi Arabian Shield. *J. Geol. Soc. London* 142: 1189–204

Stacey JS, Hedge CE. 1984. Geochronologic and isotopic evidence for Early Proterozoic crust in the Eastern Arabian Shield. *Geology* 12: 310–13

Stacey JS, Stoeser DB, Greenwood WR, Fischer LB. 1984. U-Pb geochronology and geological evolution of the Halaban-Al Amar region of the Eastern Arabian Shield, Kingdom of Saudi Arabia. *J. Geol. Soc. London* 141: 1043–55

Stern RJ. 1979. *Late Precambrian ensimatic volcanism in the Central Eastern Desert of Egypt.* PhD dissertation. Univ. Calif., San Diego, 210 pp.

Stern RJ. 1981. Petrogenesis and tectonic setting of Late Precambrian ensimatic volcanic rocks, Central Eastern Desert of Egypt. *Precambrian Res.* 16: 195–230

Stern RJ. 1985. The Najd Fault System, Saudi Arabia and Egypt: a Late Precambrian rift-related transform system? *Tectonics* 4: 497–511

Stern RJ, Bloomer SH. 1992. Subduction zone infancy: examples from the Eocene Izu-Bonin-Mariana and Jurassic California arcs. *Geol. Soc. Am. Bull.* 104: 1621–36

Stern RJ, Dawoud AS. 1991. Late Precambrian (740 Ma) charnockite, enderbite, and granite from Jebel Moya, Sudan: a link between the Mozambique Belt and the Arabian-Nubian Shield? *J. Geol.* 99: 648–59

Stern RJ, Kröner A. 1993. Geochronologic and isotopic constraints on late Precambrian crustal evolution in NE Sudan. *J. Geol.* 101: 555–74

Stern RJ, Kröner A, Bender R, Dawoud AS. 1993. Juvenile Pan-African and remobilized ancient crustal components preserved in the basement around Wadi Halfa, N. Sudan: insights into the evolution of the foreland to the Arabian-Nubian Shield. See Thorweihe & Schandelmeier 1993, pp. 79–81.

Stern RJ, Kröner A, Rashwan AA. 1991. A late Precambrian (~710 Ma) high volcanicity rift in the southern Eastern Desert of Egypt. *Geol. Rundsch.* 80: 155–70

Stern RJ, Nielsen KC, Best E, Sultan M, Arvidson RE, Kröner A. 1990. Orientation of late Precambrian sutures in the Arabian-Nubian shield. *Geology* 18: 1103–6

Stoeser DB, Camp VE. 1985. Pan African microplate accretion of the Arabian Shield. *Geol. Soc. Am. Bull.* 96: 817–26

Stoeser DB, Whitehouse MJ, Agar RA, Stacey JS. 1991. Pan-African accretion and continental terranes of the Arabian Shield, Saudi Arabia and Yemen. *Eos, Trans. Am. Geophys. Union* 72: 299 (Abstr.)

Storey BC. 1993. The changing face of late Precambrian and early Palaeozoic reconstructions. *J. Geol. Soc. London* 150: 665–68

Sultan M, Bickford ME, El Kaliouby B, Arvidson RE. 1992a. Common Pb systematics of Precambrian granitic rocks of the Nubian Shield (Egypt) and tectonic implications. *Geol. Soc. Am. Bull.* 104: 456–70

Sultan M, Chamberlin KR, Bowring SA, Arvidson RE, Abuzied H, El Kaliouby B. 1990 Geochronologic and isotopic evidence for involvement of pre-Pan-African crust in the Nubian shield, Egypt. *Geology* 18: 761–64

Sultan M, Tucker RD, Gharbawi RI, Ragab AI, El Alfy Z. 1992b. On the extension of the Congo craton into the Western Desert of Egypt. *Geol. Soc. Am., Abstr. with Programs* 24: 52 (Abstr.)

Taylor SR, McClennan SM. 1985. *The Continental Crust: Its Composition and Evolution.* Oxford: Blackwell Sci. 312 pp.

Thorweihe U, Schandelmeier H, eds. 1993. *Geoscientific Research in Northeast Africa (abstracts)* Rotterdam: Balkema

Unrug R. 1993. Neoproterozoic assembly of the Gondwana supercontinent: definition of component cratons and the keystone position of the Salvador-Congo craton. *Precambrian Res.* In press

Vail JR. 1965. Zones of progressive regional metamorphism across the western margin of the Mozambique Belt in Rhodesia and Mozambique. *Geol. Mag.* 103: 231–39

Vail JR. 1985. Pan-African (Late Precambrian) tectonic terrains and the reconstruction of the Arabian-Nubian Shield. *Geology* 13: 839–42

Vearncombe JR. 1983. A proposed continental margin in the Precambrian of western Kenya. *Geol. Rundsch.* 72: 663–70

Veevers JJ. 1990. Tectonic-climatic supercycles in the billion-year plate-tectonic eon: Permian Pangean icehouse alternates with Cretaceous dispersed-continents greenhouse. *Sediment. Geol.* 68: 1–16

Veizer J, Compston W, Clauer N, Schidlowski M. 1983. $^{87}Sr/^{86}Sr$ in Late Proterozoic carbonates: evidence for a "mantle" event at ~900 Ma ago. *Geochim. Cosmochim. Acta* 47: 295–302

Wang P, Glover L III. 1992. A tectonics test of the most commonly used geochemical discriminant diagrams and patterns. *Earth Sci. Rev.* 33: 111–31

Whitehouse MJ, Windley BF, Ba-Bitat MA. 1993. Tectonic framework and geochronology of the southern Arabian Shield, Yemen. *Eos, Trans. Am. Geophys. Union* 74: 300–1 (Abstr.)

Wilson TJ, Hanson RE, Wardlaw MS. 1993. Late Proterozoic evolution of the Zambezi belt, Zambia: implications for regional Pan-African tectonics and shear displacements in Gondwana. In *Gondwana 8—Assembly, Evolution, and Dispersal,* ed. RH Findlay, R Unrug, MR Banks, JJ Veevers, pp. 69–82. Rotterdam: Balkema

Wright VP, Ries AC, Munn SG. 1990. Intraplatformal basin-fill deposits from the Infracambrian Huqf Group, east Central Oman. In *The Geology and Tectonics of the Oman Region,* ed. AHF Robertson, MP Searle, AC Ries, pp. 601–16. *Geol. Soc. London Spec. Publ.* 49

Wust HJ, Todt W, Kröner A. 1987. Conventional and single grain zircon ages for metasediments and granite clasts from the Eastern Desert of Egypt: evidence for active continental margin evolution in Pan-African times. *Terra Cognita* 7: 165 (Abstr.)

Annu. Rev. Earth Planet. Sci. 1994. 22:353–83

THE INITIATION OF NORTHERN HEMISPHERE GLACIATION

M. E. Raymo

Department of Earth, Atmospheric, and Planetary Sciences, Massachusetts Institute of Technology, Cambridge, Massachusetts 02139

KEY WORDS: Pliocene, ice ages, climate change, Neogene

INTRODUCTION

For well over a century, scientists have speculated that variations in the atmospheric concentration of radiatively important trace gases, such as carbon dioxide, could control the Earth's climate (e.g. Arrhenius 1896). Recent observations of significant glacial-interglacial variation in atmospheric CO_2 in ice cores put this idea on a firm observational as well as theoretical footing (e.g. Barnola et al 1987). In this century, as atmospheric CO_2 levels creep steadily upwards in response to the burning of fossil fuels and forests, the specter of manmade global climate change looms before us (IPCC 1990)—a possibility that has heightened interest in the natural variability of climate in the past. In particular, the $\sim 3°C$ warming predicted for the next century has stimulated interest in past warm climates, such as the Middle Pliocene around 3.0 million years (Ma) ago, an interval that has often been invoked as a "greenhouse" world (Budyko et al 1985).

In this paper, the climate transition from the warm mid-Pliocene (around 3.2 Ma) to the onset of northern hemisphere ice ages around 2.4 Ma is examined. Evidence for the initiation of significant northern hemisphere glaciation is examined as well as how this event affected climate around the globe. While the cause of individual glacial-interglacial oscillations is tied to Milankovitch variations in the Earth's orbit around the sun (e.g. Imbrie et al 1992), these insolation changes cannot account for the long-term cooling trend which culminated in northern hemisphere glaciation. In the final section of this paper, mechanisms of long-term climate change

0084–6597/94/0515–0353$05.00

are examined with special emphasis on those that have been proposed to explain the late Neogene cooling of the northern hemisphere.

Throughout this paper, data and figures are presented and discussed in accordance with the magnetic time scale of Berggren et al (1985). Although a revised magnetic time scale (Shackleton et al 1990, Cande & Kent 1992) is now available to the geologic community, almost all of the studies discussed in this review were published using the Berggren et al time scale, hence its adoption here. For the interval between 2 and 3 Ma examined below, one can approximate the more recent magnetic time scale by adding 6% to all ages. Isotopic stage designations, some of which are indicated on Figures 1 and 3, are obviously not affected by the choice of time scales. They refer to specific features of the oxygen isotope record, features whose age depends on the time scale being used. In this paper, the isotopic stage designations of Raymo et al (1989) are used.

EVIDENCE FOR NORTHERN HEMISPHERE GLACIATION

The general outline of Pliocene-Pleistocene climate history has been known since the 1970s. This knowledge derives primarily from the measurement of oxygen isotope ratios ($^{18}O/^{16}O$) in the shells of calcitic foraminifera. When ice sheets grow on land, ^{16}O, the light isotope of oxygen, is preferentially extracted from the oceans and concentrated in continental ice sheets. This causes the ocean isotopic ratio $\delta^{18}O$:

$$\delta^{18}O = \frac{(^{18}O/^{16}O)_{sample} - (^{18}O/^{16}O)_{standard}}{(^{18}O/^{16}O)_{standard}} \times 1000$$

to get correspondingly heavier, or more positive. In addition, a temperature-dependent fractionation between water and calcite also causes the $\delta^{18}O$ value of calcite to increase as ocean temperatures cool. Thus, glacial climates are associated with more positive $\delta^{18}O$ values while more negative values are associated with warmer interglacial climates (Figure 1). For a detailed review of this methodology see Mix (1987).

Shackleton & Opdyke (1977) applied this oxygen isotope technique to the question of when the Ice Ages started. Using an extremely low sedimentation rate piston core (<1 cm/Kyr) from the equatorial Pacific Ocean, these authors were able to generate a $\delta^{18}O$ record that extended back to the Gauss magnetic chron (2.47–3.40 Ma). Although the V28-179 record was of poor quality and resolution due to the low sedimentation rate, intense bioturbation, and low carbonate content of the core, many of Shackleton & Opdyke's conclusions are still valid today. They proposed

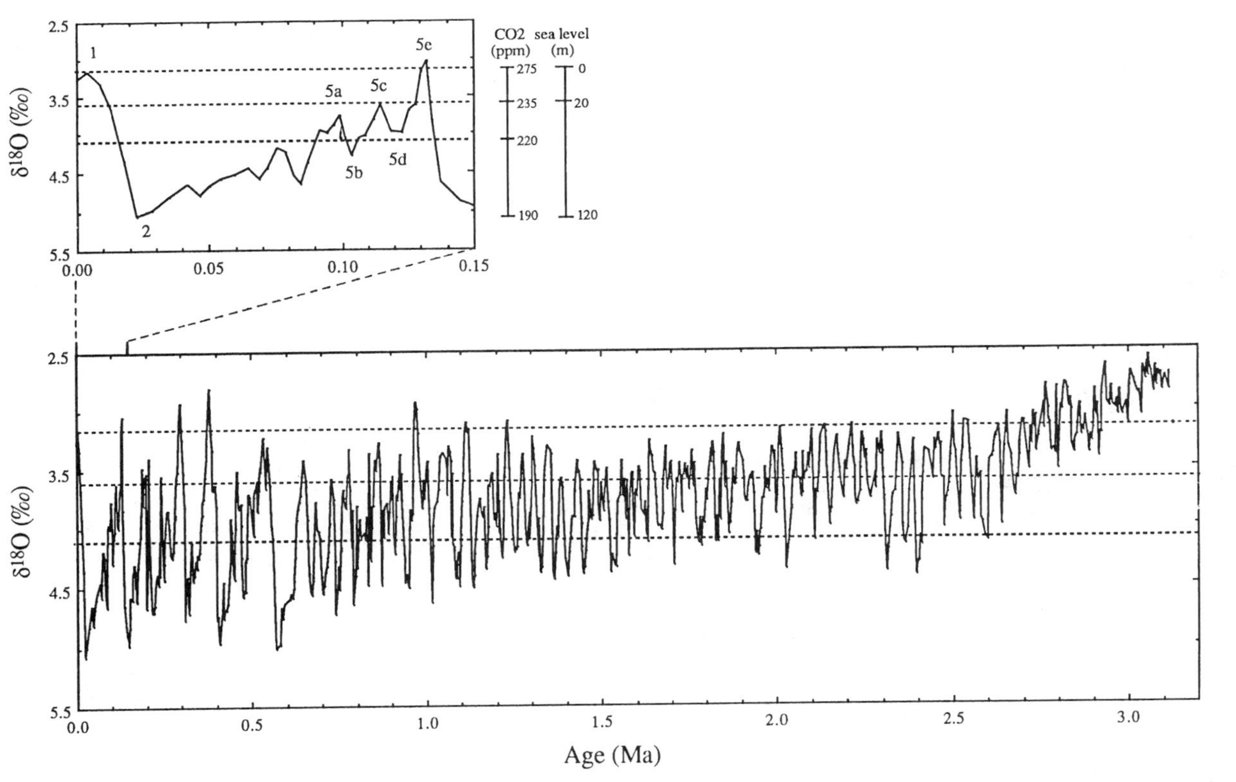

Figure 1 Benthic δ^{18}O record from DSDP Site 607 (41°N, 33°W) plotted to paleomagnetic timescale of Berggren et al (1985). Dashed horizontal lines (from top to bottom) represent Holocene, stage 5c, and the stage 2/1 boundary at this site. The last climate cycle is expanded to show estimated sea level change and atmospheric CO_2 change. For additional discussion of these data see Raymo (1992).

that prior to 3.1 Ma little evidence for glaciation was observed; that between 2.4 and 3.1 Ma, evidence for small ice sheets equivalent to approximately 40 m of sea level change was observed; and that subsequent to 2.4 Ma, isotopic excursions approximately two thirds of late Pleistocene values were observed, marking "a major change in the character of glaciations." While an apparent age correlation to ice-rafted detritus (IRD) deposits in the North Alantic (Berggren 1972) suggested that the $\delta^{18}O$ variations in V28-179 reflected northern hemisphere climate history, Shackleton & Opdyke were unable to rule out the possibility that the $\delta^{18}O$ record was reflecting ice volume changes in Antarctica.

This opportunity came with the recovery of Site 552, a DSDP (Deep Sea Drilling Project) hydraulic piston core from the North Atlantic, which had relatively higher sedimentation rates and also extended back to the Gauss magnetic chron. At this site, Shackleton et al (1984) were able to show that negative excursions in $\delta^{18}O$ correlated with influx of IRD—unequivocal evidence for nearby continental ice sheets. With this new record, Shackleton et al (1984) reaffirmed their age estimate of $\sim$2.4 Ma for the initiation of major northern hemisphere glaciation. However, the lack of any significant IRD prior to 2.5 Ma led them to conclude that northern glaciation was minimal prior to this time although the $\delta^{18}O$ record suggested that "there was considerable climatic variability, somewhere on the globe, even before 2.5 Myr."

The most recent step forward in our understanding of the evolution of the northern hemisphere ice ages, with respect to the structure of the $\delta^{18}O$ record, was again associated with an improvement in core recovery techniques. Ruddiman et al (1986a) demonstrated that significant amounts of core material could be lost during the hydraulic piston coring process due to ship heaving and other factors. To circumvent this problem they double-cored each site, correlated between holes, and then used material from the offset hole to patch in missing sections from the main hole. Using offset cores, Raymo et al (1989) generated a late Pliocene composite section from Site 607 in the North Atlantic. The more complete $\delta^{18}O$ (Figure 1) and %carbonate records from this site suggested that hiatuses were indeed present at each of the core breaks in the Site 552 record, including a 130 Kyr break from about 2.52 to 2.65 Ma. The presence of this hiatus at Site 552 made the initiation of northern hemisphere glaciation appear more abrupt than it actually was.

The Site 607 record clearly shows "Milankovitch" climate oscillations which get progressively "colder" (i.e. more positive in $\delta^{18}O$) during the late Pliocene (Figures 1 and 2). Between 3.1 and 2.95 Ma, $\delta^{18}O$ values oscillated between minimum values around 2.6‰ and maximum values around 3.1‰. Thus even the cold extremes during this interval were more negative in $\delta^{18}O$ (warmer) than Holocene values ($\sim$3.2‰). Within this interval, data indicate that bottom waters were up to 3.5°C warmer than present or that there was significantly less ice on Antarctica (or some combination of these two effects). Complete deglaciation of the modern Antarctic ice cap would decrease the mean ocean $\delta^{18}O$ value by $\sim$0.9‰ relative to today (Shackleton & Kennett 1975). At 2.95 Ma, slight but obvious steps in both the mean and amplitude of the $\delta^{18}O$ signal occurred at Site 607, and between 2.95 and $\sim$2.7 Ma, minimum $\delta^{18}O$ values were slightly more positive ($\sim$2.8‰) while maximum values hovered around 3.4‰. Thus, it is not until after 2.95 Ma that "cold" episodes were colder than today. Warm intervals remained significantly warmer than the Holocene.

Between 2.7 and 2.4 Ma, small amounts of IRD are observed coincident with positive $\delta^{18}O$ excursions in North Atlantic cores (e.g. 607, 609, and 610) providing direct evidence for continental ice sheets in the northern hemisphere and melting icebergs in the North Atlantic Ocean (Raymo et al 1989, Jansen et al 1990). Large-scale IRD fluxes also appeared in the Norwegian and Barents Seas at this time documenting expansion of northern European ice sheets (Vorren et al 1988, Jansen et al 1988, Jansen & Sjøholm 1991). Interglacial $\delta^{18}O$ values averaged about 3.1‰ at Site 607 while, at the cold extreme, values were $\sim$4.0‰; the overall amplitude was approximately half of the late Pleistocene $\delta^{18}O$ signal. During glaciations between 2.7 and 2.4 Ma, high-latitude climate is inferred to have been significantly colder than at present. However, $\delta^{18}O$ values still fell within the range of oxygen isotopic stage 5 indicating that although continental ice sheets expanded on land, these cold events were more analogous to substages 5b and 5d than to glacial stages 2–4 (Figure 1).

Subsequent to 2.4 Ma, three $\delta^{18}O$ events (stages 96, 98, and 100) with a typical amplitude of $\sim$1.2‰ are correlated with the first major influxes of IRD into the open North Atlantic Ocean. These events, traditionally thought to mark the "onset" of significant northern hemisphere glaciation, reach approximately two thirds of late Pleistocene glacial values (as originally proposed by Shackleton & Opdyke 1977). Minimum values around 3.2‰ indicate that warm periods at this time were very similar to present (with Antarctic ice volumes similar to today). Lastly, the interval between 2.3 and 2.0 Ma was characterized by lower $\delta^{18}O$ amplitudes ($\sim$0.6‰) suggesting that glacial episodes were not as pronounced as between 2.3 and 2.4 Ma (Raymo et al 1989).

All published high resolution $\delta^{18}O$ records (Figures 2 and 3; Shackleton & Hall 1989, Sikes et al 1991, Raymo et al 1992) confirm the basic ice volume history outlined above. The gradual increase in $\delta^{18}O$ values between 3.0 and 2.4 Ma is associated with a gradual increase in IRD in

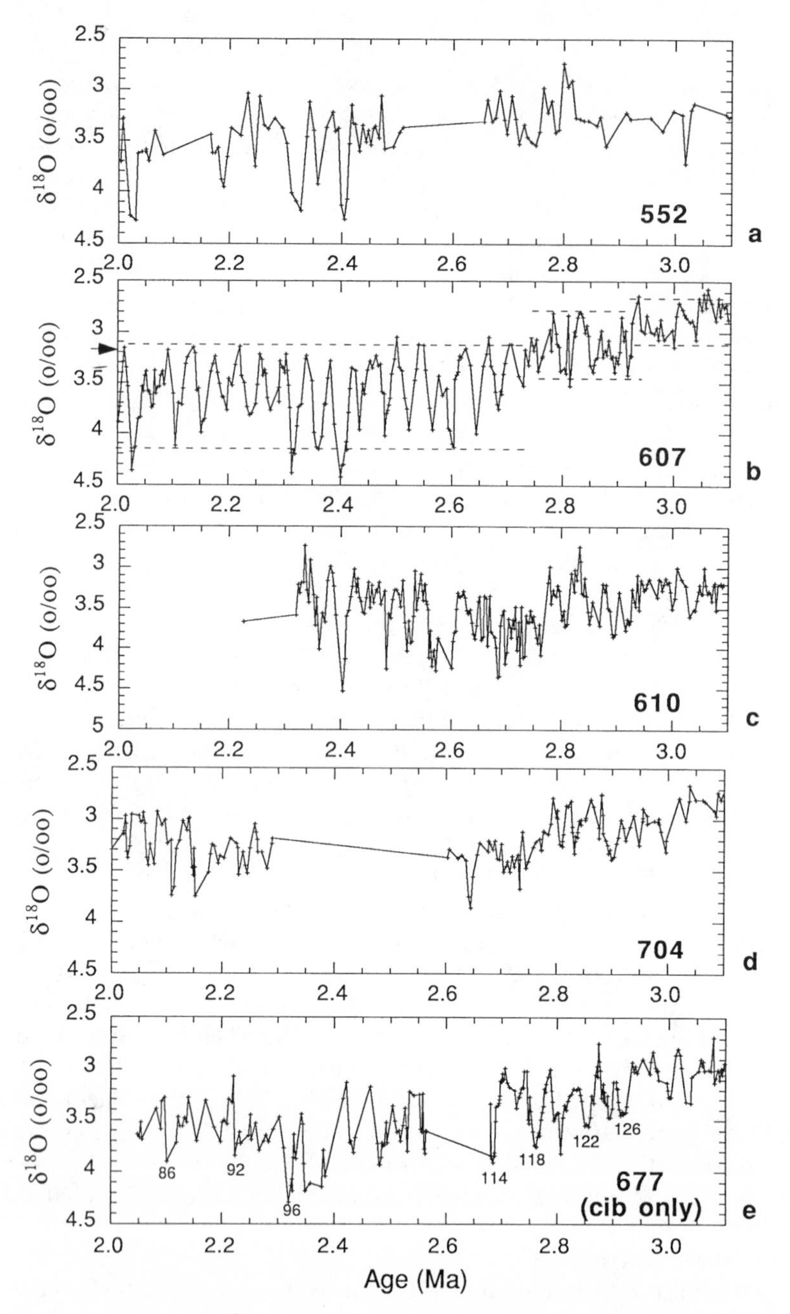

Figure 2 Late Pliocene $\delta^{18}O$ data plotted vs age (see Raymo et al 1992). Gaps represent material inferred to be missing at core breaks. Note that only data from *Cibicidoides* are used in this figure. Arrow indicates mean Holocene $\delta^{18}O$ values.

the Norwegian Sea and the open North Atlantic. While minor amounts of IRD are observed in the Norwegian Sea as early as 5.45 Ma (Jansen et al 1990), it is not till 2.57 Ma (2.44 Ma in open North Atlantic) that the first large fluxes, reflecting widespread continental glaciation, are observed in the Norwegian Sea. Likewise, in the northern Pacific Ocean, the onset of significant ice-rafting episodes is dated at ~2.48 Ma (Rea & Schrader 1985). The de facto onset of northern hemisphere glaciation can be considered to be stage 100, or 2.44 Ma (2.57 Ma by the time scale of Shackleton et al 1990).

The terrestrial record of continental glaciation in the northern hemisphere is obviously less complete and more poorly dated. Early efforts in dating till in Iceland suggested that glacial activity started as early as 3.2 Ma (McDougall & Wensink 1966; see also discussion in Raymo et al 1986). A number of younger tills, dating between 1.7 and 2.5 Ma have been found in the Pacific Northwest and Yellowstone regions (Thompson 1991). In a comprehensive review of vegetation records from the American west, Thompson (1991) documents a trend from generally warmer and wetter conditions between 4.8 and 2.4 Ma to colder, more arid conditions after 2.4 Ma. He also discusses evidence for the transient development of steppe-like environments in Idaho as early as 3.0 Ma. Freshwater ostracode data from the American west also suggest wetter environments between 3.5 and 2.5 Ma (Forester 1991). In north central China, the first loess deposits, formed under extremely cold, dry conditions, are dated at ~2.35 Ma (Kukla 1987, Kukla & An 1989). The Chinese loess is slightly younger than the estimated age of 2.6 Ma for increased dust fluxes to the Pacific Ocean from this region (Rea & Schrader 1985)—a disagreement that may indicate uncertainty in the land-ocean age correlation.

Eolian loess deposits, typically associated with glacial conditions, have also been identified in sediments as old as 3.0 Ma in central Alaska (Westgate et al 1990). These deposits may have been derived from mountain glaciers in the Alaska Range. However, in agreement with oxygen isotope evidence, fossil vegetation from the circum-Arctic shows that the climate was, at least intermittently, significantly warmer than at present in the late Pliocene. A pollen-based climate reconstruction from the Kap Kobenhavn formation of northern Greenland (Funder et al 1985) suggests that conditions similar to those in modern Labrador existed at this time and that the Greenland ice sheet was either absent or very reduced in size. Likewise, sedimentologic structures along the coast suggest at least seasonally ice-free conditions at this time. Although the dating of this section is uncertain and could range fron 3.1 to 1.6 m.y., an age of about 2.0 Ma BP has been proposed for this site (Funder et al 1985, Brigham-Grette & Carter 1992). At Ocean Point in northern Alaska, pollen recon-

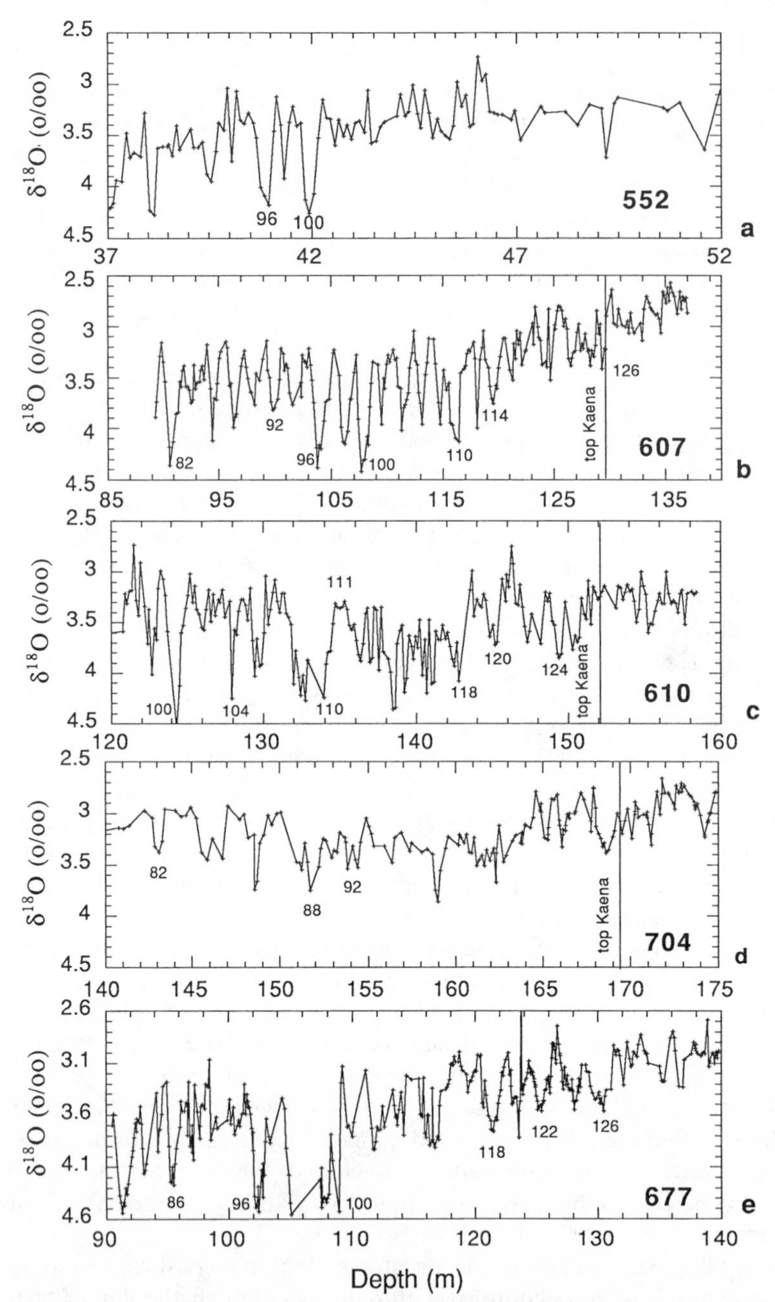

Figure 3 Late Pliocene $\delta^{18}O$ records from Sites 552, 610, 607, 704, and 677 plotted against actual sub-bottom depth (552, 610) or to composite sub-bottom depth (607, 704, 677). The top of the Kaena magnetic subchron (2.92 Ma) is indicated in those cores with a magnetic record. Selected isotopic stages are identified according to Raymo et al (1989, 1992).

structions from the Gubik Formation also suggest a warmer climate, represented by an open boreal forest, in the late Pliocene (Nelson & Carter 1985, Brigham-Grette & Carter 1992). The absolute age of this section is again not well-constrained and is placed sometime between 2.7 Ma and 2.1 Ma (Brigham-Grette & Carter 1992). During warm intervals it is unlikely that seasonal Arctic sea ice existed at this coastal location.

In the Fish Creek section, also in northern Alaska, invertebrate (mollusc) and vertebrate (sea otter) evidence, shown to be younger than 2.5 Ma by magnetostratigraphy, indicate that the Arctic margin could not have been frozen for more than one month a year at this location. Similarly, geologic evidence from a number of other sites, summarized by Repenning et al (1987) and Brigham-Grette & Carter (1992) reflects a generally warmer circum-Arctic climate in the late Pliocene, with some indications by pollen of alternating warmer and colder intervals. Unfortunately, the ages of most of these sites are poorly constrained but are believed to have been deposited prior to the Pliocene-Pleistocene boundary (1.6 Ma). Because no land sections have been firmly dated in the critical late Pliocene/early Pleistocene interval after 2.5 Ma, it remains uncertain when the present-day perennial sea ice cover developed in the Arctic. The deep sea record from this ocean is also ambiguous, primarily due to low sedimentation rates and difficulties in accurately dating these sediments. Herman & Hopkins (1980), Herman et al (1989), and Gilbert & Clark (1982/1983), all proposed that modern Arctic sea ice cover formed after 0.9 Ma ago. However, permanent perennial sea ice may have developed as early as the Pliocene/Pleistocene boundary (1.6 Ma; Scott et al 1989). It appears that the early period of northern hemisphere glaciation may have been associated with only seasonal Arctic sea ice cover, at least during Milankovitch interglacial extremes.

RESPONSE OF GLOBAL CLIMATE TO GLACIATION

Sea Surface Temperatures

A number of studies have examined the response of sea surface temperatures (SSTs) to northern hemisphere glaciation. In addition, climate modelers have recently expressed an interest in the response of ocean temperatures to the warm climates of 3.0 Ma ago, warmth possibly caused by enhanced atmospheric CO_2 levels (e.g. Crowley 1991). Studies of foraminiferal assemblages from the late Pliocene (3.4–1.6 Ma; Berggren 1972, Poore & Berggren 1975, Loubere & Moss 1986, Raymo et al 1986) suggest that North Atlantic sea surface temperatures were warmer prior to 2.5 Ma. This inference was based primarily on the fact that larger, more ornate species (e.g. *N. atlantica*) were replaced at high latitudes by smaller, denser, more compact species (e.g. *N. pachyderma* sinistral) as ice sheets developed in the late Pliocene. Unfortunately, because many of the most abundant late Pliocene species (*N. atlantica*, *G. puncticulata*, *N. acostoensis*, etc) became extinct by the Pleistocene (1.6 Ma), a direct calibration of the temperature and salinity preferences of these species is impossible. With this in mind, I discuss below recent studies that reconstruct sea surface temperatures for the mid-Pliocene North Atlantic. In particular, as climate modelers are now using these reconstructions as input into general circulation model experiments, an assessment of the assumptions these studies are based on seems warranted.

Many readers are probably familiar with the CLIMAP (1981) global SST reconstruction for the last glacial maximum (18,000 years ago). The CLIMAP group used transfer functions developed by Imbrie & Kipp (1971) and Kipp (1976) to infer past sea surface temperatures by looking at the geographic distribution in the past of modern species whose environmental tolerances are well-known. Specifically, in Kipp (1976), the percent abundance of 29 foraminiferal species was estimated in 191 core-top (modern) samples from the North Atlantic. These data were factor-analyzed into varimax assemblages and a least-squares regression technique was used to relate the varimax assemblages to observed sea surface temperatures and salinities at each core site. From these relationships, transfer functions were produced. Downcore species abundance data are then described in terms of the core-top assemblages, which are then used with the paleoecological equations to estimate past environmental conditions at a site. The standard errors of the transfer function were estimated by Kipp (1976) to be $\pm 1.16°C$ for the cold season and $\pm 1.38°C$ for the warm. On rare occasions, one can find assemblages in late Pleistocene sediments that have no modern analog (due to dissolution, stratigraphic mixing, taxonomic misidentification, local ecological variation, or other factors). These samples are easily identified by their low communalities—a statistical measure of how well the assemblage fits the varimax model. Imbrie et al (1973) suggest that communalities greater than 0.8 are necessary for accurate environmental estimates.

Of course, the key assumption in this and other transfer functions that have been developed to study Pleistocene climates is that species do not change their environmental preferences with time. If we know that left-coiling *N. pachyderma* has a maximum abundance in water at less than 5°C today, then we can assume that the same holds true for *N. pachyderma* sinistral which lived one million years ago. Difficulties arise when components of the faunal assemblage are extinct since we cannot be assured that they have the same environmental preferences as their extant relatives.

Dowsett & Poore (1990) and Dowsett (1991) developed and tested a foraminiferal transfer function that would be applicable to the Pliocene. The standard errors for their GSF18 transfer function were estimated to be ±1.47°C for the cold season and ±1.36°C for the warm season, only slightly different from those of the Kipp (1976) transfer function. To develop GSF18, Dowsett made two important changes from the CLIMAP transfer function. First, he simplified the taxonomy slightly by lumping certain abundant species, such as right and left-coiling *N. pachyderma*, thus reducing the number of taxonomic catagories that needed to be counted. [Ruddiman & Esmay (1986) carried out a similar exercise developing a five "species" transfer function that could be applied much more rapidly to the long Pleistocene sections recovered by DSDP and ODP (Ocean Drilling Program) drilling.] Dowsett (1991) correctly tested the effects of this change in the structure of the transfer function by taking existing faunal data from the 18 Kyr and 122 Kyr time-slice data bases and regrouping them according to his taxonomic scheme. He found, as expected, some loss of resolution in the SST estimates. In particular, lumping cold *Neogloboquadrina* species resulted in an overestimate of high-latitude SSTs by 1–2°C and an underestimate of SSTs at mid-latitudes (30–40°N) by 1–3°C over CLIMAP estimates.

While the above lumping of certain species inevitably leads to a loss of information, this is not the major source of error in GSF18. To generate a Pliocene transfer function, one other assumption is made by Dowsett & Poore (1990), namely that extinct species occupy the same environmental niche as their presumed modern descendents (lineages are inferred from similar gross morphology and biogeographic ranges). This assumption obviously begs the question of why species evolved. Two rationales are offered in support of this assumption: similar morphology and similar geographic ranges. However, while *G. ruber* and the extinct *G. obliquus* have a similar morphology and are assumed to have the same environmental tolerances, a study of their evolutionary lineages (Kennett & Srinivasan 1983) show them to be completely separate over the Neogene (the last 24 Ma). How likely is it that they occupied the same enviromental niche in the Pliocene (*G. obliquus*) and Pleistocene (*G. ruber*)? In most cases, Pliocene species are larger, more delicate, or more ornate than their presumed Pleistocene counterparts (e.g. *G. puncticulata*—*G. inflata*, *N. atlantica*—*N. pachyderma*, *G. miocenica*—*G. menardii*, *G. fistulosis*—*G. sacculifer*) suggesting, by analogy with the modern equator to pole changes in morphology, that the extinct species lived in warmer water. The second rationale used for assuming similar temperature preferences—similar biogeographic ranges—is clearly circular reasoning. How would one recognize a regional water mass change if the de facto assumption is made that extinct species which lived in a region had the same temperature and salinity preferences as the modern species that live there?

How pervasive are now-extinct species in mid-Pliocene planktonic foraminiferal assemblages? At Site 552, one of the sites for which faunal census data were published (Dowsett & Poore 1990), extinct species comprise over 50% of the total fauna in the interval between 2.85 and 3.15 Ma. However, because the transfer function artificially treats these species as their modern-day counterparts, the samples all have communalities above 0.7. At many other sites used in the Dowsett et al (1992) reconstruction (e.g. 502, 548, 606) communalities below 0.7 are quite common (38–56% of samples; Dowsett & Poore 1991), due primarily to "no-analog" percent abundances of the assumed "modern" species. These data, obviously a significant fraction of the Pliocene faunal record, are excluded from the SST reconstruction, possibly imparting a systematic bias to the results.

Taking GFS18 results at face value, and not assigning any error to the substitution of extinct for modern species, Dowsett et al (1992) presented a mid-Pliocene sea surface temperature reconstruction that showed SSTs at 50°N approximately 4°C warmer and SSTs at low latitudes approximately 1°C warmer than at present (Figure 2 in Dowsett et al 1992). Considering the above assumptions, the true estimation error on the GFS18 transfer function is unclear. For instance, the species-lumping discussed above may impart a 1–2°C overestimatation of SSTs at the higher latitudes; likewise, many of the extinct tropical species of the late Pliocene (*G. obliquus*, *G. altispira*, *G. trilobus*, *N. humerosa*, *G. pseudopima*, *G. decoraperta*, *G. miocenica*, *S. seminulina*, *G. limbata*, etc) may actually have preferred warmer water than their Pleistocene replacements. Given the importance of accurate SST reconstructions for the evaluation of climate change mechanisms (e.g. Rind & Chandler 1991, Crowley 1991; see also below), it is critical that vigorous investigation into the preferred habitats of extinct species begin. In particular, estimates of past CO_2 levels using the $\delta^{13}C$ of marine organic matter (e.g. Rau et al 1991, Freeman & Hayes 1992) require accurate estimates of SST to calculate CO_2 solubility in past oceans.

North Atlantic sea surface temperatures have also been estimated using marine ostracod fauna (Cronin & Dowsett 1990, Cronin 1991), which live in shallow water above the seasonal thermocline. In other words, ostracod live at the same depths as planktonic foraminifera, only in a near-shore environment. Using a transfer function based on 100 modern sediment samples divided into 59 ostracode taxa, Cronin (1991) examined Pliocene sections from the tropics (9°N) to polar regions (66°N) along the western Atlantic margin. Based on the distribution of genera at eight well-dated sites, Cronin (1991) concluded that, during the Middle Pliocene (3.5–3.0 Ma), tropical and subtropical shelf-water temperatures were slightly cooler than today, while at temperate to subpolar latitudes temperatures were up to 10°C warmer than today. Temperatures off northern Iceland were up to 6°C warmer than today in summer (4°C in winter) suggesting very little influence by the East Greenland Current or seasonal sea ice which dominates this region today.

Unlike foraminiferal transfer functions, which are based primarily on species, the ostracod transfer function analyzes the fauna at a generic level. By breaking the fauna down into 59 primarily generic groupings, this approach may be more robust with respect to extinct species than the foraminiferal transfer function. Unlike some foraminiferal genera (e.g. the Neogloboquadrinids), ostracode genera are generally restricted to one or only a few climatic zones, possibly resulting in less of a systematic error due to species-level evolution. Cronin (1991) gives an error on the transfer function temperature estimates of ± 2°C.

A study of dynocyst abundances in the northern Atlantic also suggests warmer SSTs prior to 2.4 Ma, although Edwards et al (1991) discuss the pitfalls of inferring environmental conditions from "no-analog" fauna. Based on the presence of subtropical taxa at Site 552 (56°N, 23°W), they suggest a greater northward extension of the Gulf Stream/North Atlantic Current system around 3.0 Ma. Edwards and coworkers also propose that outflow of low-salinity polar water from the Greenland Sea was greatly reduced at this time.

The various methods used to estimate SSTs all generally conclude that the North Atlantic was warmer in the mid-Pliocene around 3.0 Ma ago. This warmth is attributed to enhanced heat transport in the Gulf Stream/North Atlantic Drift system and a weakening of the cold East Greenland Current (Cronin 1991, Edwards et al 1991, Dowsett et al 1992). A weakening, or warming, of the East Greenland Current is also supported by evidence for Arctic coastal warmth discussed in the previous section. Although warmer SSTs could also be ascribed to decreased heat removal by the atmosphere (especially in winter), further evidence suggesting stronger ocean heat transport to the North Atlantic is discussed in the following section on thermohaline circulation. In detail, the ostracod transfer function suggests colder low-latitude regions (although cooler termperatures could be due to stronger summer upwelling along the coast) and much warmer mid- to high-latitude SSTs than the foraminiferal-based estimates. At present, it is impossible to say which estimates are more accurate.

In the North Pacific, the SST response to glacial inception in the late Pliocene has been investigated using siliceous flora (diatoms) and fauna (radiolaria) (Morley & Dworetzky 1991). Based on assemblage abundances, these authors suggest a pattern of climate change very similar to that inferred from North Atlantic fauna as well as oxygen isotopes: a period of relatively warmer, mild conditions between 3.0 and 2.7 Ma, followed by a major cooling at 2.46 Ma, roughly coincident with the major enrichment in benthic oxygen isotope records discussed above as well as with large inputs of IRD into both the North Atlantic (Shackleton et al 1984, Raymo et al 1989) and Pacific Oceans (Rea & Schrader 1985). As in the North Atlantic $\delta^{18}O$ and IRD records, these authors note a brief return to milder, warmer conditions between $\sim$2.1 and 2.3 Ma.

Thermohaline Circulation

Ocean thermohaline circulation plays an important role in controlling the global distribution of heat and in modulating exchange of CO_2 between the deep ocean and the atmosphere (e.g. Broecker & Denton 1989, Rind & Chandler 1991, Boyle 1988, Broecker & Peng 1989, Charles & Fairbanks 1992). Today, most deep water forms in two locations: the North Atlantic Ocean and the Southern Ocean around Antarctica. The sinking of deep water in the Norwegian-Greenland Seas and Labrador Sea sets up a "conveyor belt" which then draws additional warm, salty thermocline water northward. As this water cools and sinks, heat released to the atmosphere is advected eastward over Europe and Scandanavia warming these regions. During the last glacial maximum, North Atlantic Deep Water (NADW) formation was greatly suppressed (Curry & Lohmann 1983; Boyle & Keigwin 1982, 1987; Oppo & Fairbanks 1987) and sea surface temperatures dropped as the polar front, or sea ice limit, migrated southward to the latitude of Spain (CLIMAP Project Members 1981). The decline in NADW formation may also have led to the reduction of atmospheric CO_2 observed in ice cores, thereby causing further cooling (Boyle 1988, Keir 1988, Broecker & Peng 1989).

Because ocean thermohaline circulation affects both the global surface heat distribution as well as atmospheric CO_2 concentrations, the response of NADW formation to global "greenhouse" warming is an important question. Modern oceanographic studies by Brewer et al (1983), Aagaard (1988), and Schlosser et al (1991) suggest that the salinity of water in the North Atlantic plays a vital role in controlling thermohaline overturn. For instance, a low-salinity anomaly in the North Atlantic, resulting from increased freshwater flow from the Arctic Ocean, was associated with a reduction in deep-water formation in the 1970s. Likewise, an ocean general circulation model suggested that thermohaline overturn would decrease in the North Atlantic in a doubled CO_2 world (Mikolajewicz et al 1990). This change was attributed to a decrease in surface water salinity caused by increased regional precipitation. Because weaker NADW production could decrease the transport of heat to high latitudes as well as lower atmospheric CO_2 (as observed during last glacial maximum), the simplest interpretation of these model results is that weaker thermohaline circulation could act as a negative feedback as the climate warmed. However, the geologic record suggests the opposite, warmer climates are associated with stronger NADW formation, at least for the last 3.2 Ma.

The history of glacial-interglacial change in deep ocean circulation is reconstructed with the use of carbon isotopes recorded in the calcite tests of foraminifera that live on the ocean floor (see Curry et al 1988 for an extensive review). Because NADW forms with high initial $\delta^{13}C$ values ($>1.0‰$), its presence is clearly seen in vertical cross sections of $\delta^{13}C$ in the modern ocean (Kroopnick 1985). Upper Circumpolar Deep Water (UCDW) and Antarctic Bottom Water (AABW), the other two major deep-water masses in the Atlantic, are more negative in $\delta^{13}C$. By reconstructing deep-water $\delta^{13}C$ gradients in the past, paleoceanographers can determine how the path of deep-water flow and the relative contribution of different source waters to the deep ocean changed through time.

By examining the evolution of $\delta^{13}C$ gradients between three cores from the North Atlantic and Pacific oceans, Raymo et al (1990a) reconstructed the history of NADW formation back to 2.5 Ma. They showed that NADW formation typically decreased during glaciations and that the suppression of NADW became particularly pronounced over the last million years with the development of larger ice sheets. Raymo et al (1992) studied the response of NADW to global cooling and the intensification of northern hemisphere glaciation between 3.1 and 2.0 Ma. They showed that global cooling led to gradually stronger suppression of NADW production and that the North Atlantic, prior to major northern hemisphere glaciation, was probably characterized by vigorous thermohaline circulation. (For a discussion of compatible and conflicting sedimentological studies see that paper.) This interpretation, also reached by Hodell & Venz (1992), rests heavily on the carbon isotopic record of ODP Site 704 from the South Atlantic (Figure 4). Unfortunately, this record has a significant hiatus around the late Gauss/early Matuyama and, as a result, stratigraphic control is poor. In addition, Mix et al (1994) suggest that Site 704 may be sensitive to regional variations in Antarctic circumpolar circulation which mask the true flux of NADW from the North Atlantic. Work in progress on ODP Leg 108 cores at the equator should provide additional constraints on the response of NADW formation to northern hemisphere glaciation.

What caused the inferred decrease in NADW formation over the late Pliocene? By analogy to the late Pleistocene study of Boyle & Keigwin (1987), Raymo et al (1992) proposed that lowered sea surface temperatures

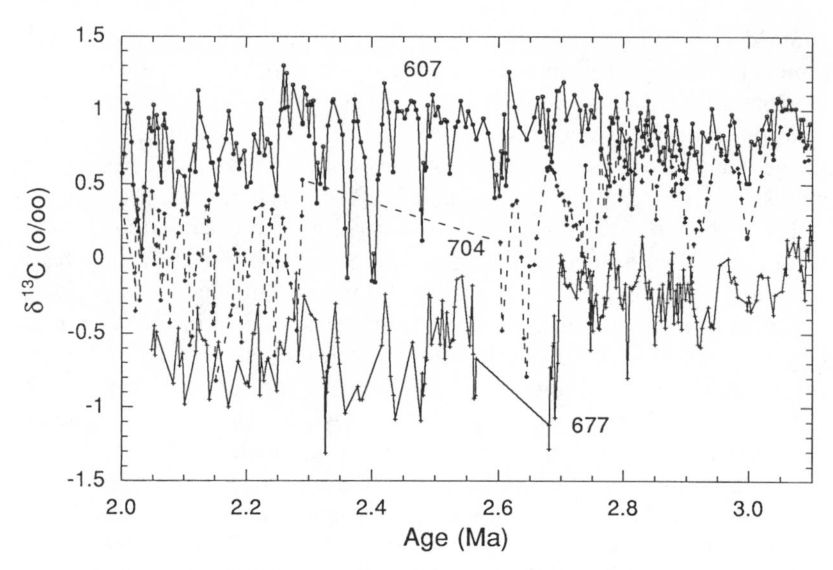

Figure 4 *Cibicidoides* δ^{13}C data from Sites 607 (North Atlantic), 704 (South Atlantic), and 677 (Equatorial Pacific). Site 704 δ^{13}C values become increasingly Pacific-like after 2.75 Ma suggesting decreased production of NADW.

reduced evaporation and, hence, decreased the surface salinity and potential density of surface waters. These waters were therefore less likely to sink and form a deep-water mass. As discussed above, prior to 2.6 Ma, planktonic fauna and ostracod data indicate warmer SSTs in the North Atlantic (Raymo et al 1986; Loubere & Moss 1986; Dowsett & Poore 1990, 1991; Cronin & Dowsett 1990; Cronin 1991), while many other high-latitude sites suggest that the Arctic was seasonally (or completely) ice free at this time (Funder et al 1985, Brouwers et al 1991, Matthews & Ovenden 1990, Carter et al 1986). Subsequent cooling is observed at many of these locations (North Atlantic and Arctic) in conjunction with the intensification of northern hemisphere glaciation around 2.4 Ma.

Two mechanisms that could decrease SST and deep-water production rates have been suggested by general circulation model (GCM) results. Manabe & Broccoli (1985) showed that orographic diversion of winds by a continental ice sheet on Canada would result in a significant cooling of the surface North Atlantic. Likewise, a GCM experiment in which Arctic sea ice extent was reduced (Raymo et al 1990b) (reflecting climate conditions in the pre-glacial Pliocene) showed three changes that would promote greater thermohaline overturn in the Norwegian-Greenland and Labrador Seas and, thus, more vigorous formation of NADW: (*a*) an enhancement of surface water salinities in the North Atlantic region resulting from an increase in evaporation relative to precipitation; (*b*) a localized strengthening of the Icelandic low over the Norwegian-Greenland Sea which would enhance northward advection of salty water into the area of deep-water formation; and (*c*) an increase in the salinity of water leaving the Arctic, driven by increased evaporative fluxes in this region when sea ice limits are reduced. Further studies, both of the paleoceanographic record and with coupled ocean-atmosphere climate models, are needed to evaluate the above possibilities.

Although here we consider changes in thermohaline circulation to be a response to glaciation, they can just as easily be considered as a potential cause for glaciation (see later in this paper). A decrease in the production of NADW would decrease heat transport to high latitudes as well as possibly result in a decrease in atmospheric CO_2—both factors that would enhance cooling. Thus, at the least, NADW production can act as a strong positive feedback and possibly as an independent climate forcing mechanism (e.g. Rind & Chandler 1991).

Antarctic Glacial History

As discussed earlier, benthic $\delta^{18}O$ records from the mid-Pliocene are consistent with a major deglaciation of Antarctica around 3.0 Ma (although more negative $\delta^{18}O$ values at this time could also be ascribed to warmer deep ocean temperatures). However, estimates that eustatic sea level was up to 40 m higher in the Middle Pliocene also point to significant deglaciation of Antarctica at this time (Haq et al 1987, Dowsett & Cronin 1990, Krantz 1991). A pronounced transgression, which deposited the Rushmere and Morgarts Beach members of the Yorktown Formation, occurred between 4.0 and 3.2 Ma. Krantz (1991) estimates sea level up to 35 meters higher than present during this interval. Studies of the Orangeburg Scarp in North and South Carolina also suggest sea levels up to 35 meters higher than present between 3.5 and 3.0 Ma (Dowsett & Cronin 1990).

More direct evidence for Antarctic deglaciation comes from terrestrial plant remains which were deposited in the Transantarctic Mountains after 3.0 Ma ago (Webb et al 1987). These deposits contain *Nothofagus*, or southern beech, a plant which suggests a climate more analogous to that of southern Patagonia. Biostratigraphic diatom markers (Webb et al 1984) and radiometric dating of coexisting ash layers in Antarctica (Barrett et al 1992) both confirm that the warmer conditions existed later than three million years ago. Such a major deglaciation of East Antarctica (possibly more than half the current ice sheet volume), occurring when global climate was only slightly warmer than at present, implies a cryosphere much less stable than previously thought (Kennett 1977). In particular, Barrett et al (1992) caution that ice sheet stability needs to be evaluated more seriously in face of the greenhouse warming predicted for the next century. They propose that the Antarctic ice sheet reached its present size simultaneously with the development of large ice sheets in the northern hemisphere ($\sim$2.4 Ma).

The evidence for a major Pliocene deglaciation of Antarctica is not universally accepted. Geomorphologic features and glacial landforms show evidence for continuous cold environments for at least the last 10 Ma with no suggestion of meltwater activity or temperate climate conditions (Denton et al 1984, Clapperton & Sugden 1990). Likewise, a number of paleoclimatologists argue that since Antarctica is currently the driest continent on Earth with mean annual temperatures well below zero, any warming would actually increase ice volume by increasing snowfall (e.g. Prentice & Matthews 1991, Oglesby 1989). This line of reasoning would predict that Antarctic ice volume decreased over the late Pliocene interval of polar cooling ($\sim$3.0–2.4 Ma).

In contrast, studies of subantarctic deep-sea sediments (Hodell & Ciesielski 1990, 1991; Hodell & Venz,1992; Warnke & Allen 1991) indicate increased delivery of ice-rafted sediment after 2.46 Ma, suggesting ice sheet expansion on Antarctica. This followed an interval of progressive cooling from 3.25 Ma to 2.4 Ma over which the Polar Front Zone migrated northward in the Atlantic sector of the Southern Ocean and silica productivity increased (Hodell & Warnke 1991). Froelich et al (1991) inferred from high rates of carbonate deposition that Site 704, located just north of the polar front today, was also north of the polar front between 2.9 and 2.7 Ma (e.g. conditions similar to today). Perennial sea ice cover was also established in the Weddell Sea at this time (Abelmann et al 1990). Carbon isotopic data indicate decreased ventilation of deep waters in the circum-Antarctic after 2.75 Ma, undoubtedly due to a reduction in North Atlantic Deep Water flux during glaciations (Hodell & Ciesielski 1990, 1991; Hodell & Venz 1992; Raymo et al 1992). Hodell & Ceisielski (1990) propose that the climate of the southern polar regions is linked to that of the northern hemisphere through a series of positive feedbacks. In their model, increased suppression of NADW during northern hemisphere glaciations decreases the flux of salt and heat to the Southern Ocean allowing the northward expansion of sea ice. In addition, they point out that a decrease in the flux of NADW could result in lowered atmospheric CO_2 [via the polar alkalinity model of Broecker & Peng (1989)] further strengthening the climate coupling between the northern and southern hemispheres.

Low Latitude Climate

One of the most pronounced features of low to mid-latitude climate is the seasonal change in winds and precipitation associated with the Asian monsoon. The monsoon, caused by the differential heating of land (primarily the Tibetan Plateau) and sea (the Indian Ocean), is characterized by rising air masses over land in summer and moisture-laden surface winds that flow off the Indian Ocean into southern Asia. These southwesterly winds bring the summer monsoon rains. In winter, as land surfaces cool, the seasonal wind pattern reverses. It has been shown that throughout Africa, the Mediterranean, and Asia, regional precipitation and aridity variations are tied to the strength of the Asian monsoon (Street-Perrott & Harrison 1984, Pokras & Mix 1985, Prell & Van Campo 1986, Clemens & Prell 1990). In addition, the intensity of oceanic upwelling in the Arabian Sea also reflects the strength of the summer monsoon winds (Prell 1984, Prell & Kutzbach 1987).

The strength of the monsoon, in turn, has been shown to vary as a function of three primary factors: the strength of seasonal insolation over the Tibetan plateau, plateau elevation, and the presence of glacial or interglacial surface boundary conditions (see review by Prell & Kutzbach 1992). Using GCM experiments, Kutzbach (1981) showed that stronger insolation heating at the latitude of the plateau, induced by Milankovitch variations in precession, resulted in a stronger monsoon—a prediction matched by observations. Likewise, a series of GCM experiments that varied Tibetan Plateau elevation (Hahn & Manabe 1975, Ruddiman & Kutzbach 1989, Prell & Kutzbach 1992) suggest that monsoon intensity increases with the elevation of the plateau. Lastly, the presence of continental ice sheets in the northern hemisphere acts to suppress the monsoon as indicated by evidence for a weaker monsoon during the Last Glacial Maximum, a time of similar low-latitude insolation (Prell 1984, Street-Perrott & Harrison 1984, Prell & Van Campo 1986). This suppression may be due to glacial changes in SST, atmospheric CO_2, land albedo, and/or the extent of seasonal snow in Eurasia and Tibet rather than the extent and height of the Laurentide or Scandanavian ice sheets per se (Prell & Kutzbach 1992).

Studies of the long-term evolution of the Asian monsoon indicate that the onset of northern hemisphere glaciation resulted in increased aridity in northeast Africa after 2.4 Ma, possibly due to glacial suppression of monsoon intensity. This trend is seen in palynological studies (Bonnefille 1983), terrestrial isotopic studies (Cerling et al 1977, Abell 1982), and macrofaunal studies (Vrba 1985, Wesselman 1985, Grine 1986). In the eastern equatorial Atlantic, a pronounced increase in the input of terrigenous dust is observed at 2.4 Ma—an increase associated with strengthened northern hemisphere winter trade winds (Ruddiman & Janecek 1989). Enhanced opal fluxes suggest that upwelling intensity also increased after 2.5 Ma, again probably due to strengthened trade winds (Ruddiman & Janecek 1989). After 2.4 Ma, the modulation of dust fluxes to the Arabian Sea at the 41,000 year obliquity period of ice sheet variablility also points to high-latitude influence on African climate and monsoonal variability after this time (de Menocal et al 1991). de Menocal et al (1993) propose that low-latitude African climate was dominated by monsoon forcing, at precessional frequencies, prior to the initiation of major northern hemisphere glaciation and that, after that time, regional precipitation and aridity patterns at low latitudes were more strongly influenced by remote forcing from high latitudes at the obliquity frequency. Additional high-resolution records of Pliocene low-latitude climate are needed to test this hypothesis.

CAUSE OF NORTHERN HEMISPHERE GLACIATION

As outlined above, northern hemisphere glaciation began gradually, rather than abruptly, between 2.9 and 2.4 Ma. Mechanisms to explain the onset of the ice ages fall within two broad catagories: terrestrial and extraterrestrial. Proposed extraterrestrial causes of climate change include variations in solar output, collisions with asteroids, passage through interstellar dust clouds, as well as supernova explosions, to name a few (see Pollack 1982 for an excellent summary). However, not one of these factors has provided a convincing explanation for the onset of northern hemisphere glaciation. While Milankovitch variations in insolation caused by changes in the Earth's obliquity, precession, and eccentricity, obviously play a critical role in pacing the sequence of glacial/interglacial oscillations, they can not explain the long-term cooling trend observed in the Pliocene-Pleistocene. Below, explanations for northern hemisphere glaciation that invoke terrestrial changes in boundary conditions are presented and discussed. In evaluating these mechanisms, keep in mind two possible models of the climate system: one in which climate responds linearly to changes in forcing; and one in which climate "thresholds" are crossed. In the second case some kind of climate instability results in a nonlinear response to external forcing (e.g. North 1984).

Plate Movements

Many of the terrestrial mechanisms proposed as explanations for long-term climate variation are related to plate tectonics and the dynamic forces within the Earth that are continually modifying the Earth's surface.

Probably one the simplest ideas put forth to explain the timing of glaciations is polar positioning of continents. Very few studies of Paleozoic climate history fail to mention this popular idea (e.g. Crowell & Frakes 1970, Caputo & Crowell 1985, Crowley et al 1987), however, as we move toward the present, it becomes less certain that polar continental position has any strong effect on climate. Barron (1981) and Barron & Washington (1985) conclude that while long-term plate motions could be responsible for the general climate cooling observed over the last 150 Ma, such movements would be too slight to account for the relatively faster cooling observed since the Eocene. The only way plate positions can effect a relatively rapid change in climate is by invoking a "critical point" in the climate system. North (1984) and North & Crowley (1985) discuss how small icecap instabilities could result in a nonlinear response to subtle variations in seasonal insolation. They speculate that such an instability could lead to the rapid formation of the Greenland ice sheet. If one then invoked a series of positive feedbacks, such as the development of perennial Arctic sea ice cover or a decrease in NADW production, such a change could conceivably then result in widespread northern hemisphere glaciation.

Sills and Gateways

For many years researchers have speculated that the onset of northern hemisphere glaciation was due to the uplift of the Panamanian Isthmus or the subsidence of the Bering Straits. Age estimates for the final closure of the Panamanian Isthmus range from 2.5 to 3.7 Ma. Based on the appearance of North American large mammal assemblages in South America (and vice-versa), faunal exchange between the two continents happened between 2.5 and 2.8 Ma (Lundelius 1987, Marshall 1988, MacFadden et al 1993). By contrast, marine evidence suggests that uplift of the Isthmus had a significant impact on ocean circulation and biotic exchange much earlier (Duque-Caro 1990, Brunner 1984, Emiliani et al 1972). Based on carbon and oxygen isotopic data and coiling ratios of planktonic foramanifera, Keigwin (1978 1982) estimated that surface waters of the Caribbean and Pacific were isolated by 3 Ma, while a study of shallow-water fauna on both sides of the Isthmus concluded that it was effectively closed by 3.5 Ma (Coates et al 1992). It is unclear how the marine and terrestrial estimates can be reconciled although one can speculate that mammal migrations could not occur until the bridge was emergent everywhere.

The relationship between the development of the Isthmus of Panama and the initiation of northern hemisphere glaciation is also unclear. The formation of the Isthmus could predate major northern hemisphere glaciation by as much as a million years or, if the mammal data are taken at face value, it could have occurred almost simultaneously with global cooling. Early researchers suggested that a closed isthmus would deflect warm Gulf Stream water to the northwest Atlantic providing a ready source of moisture for ice growth (e.g. Stokes 1955). However, an ocean circulation model (Maier-Reimer & Mikolajewicz 1990) suggests the opposite: The North Atlantic region warms significantly when the Isthmus is closed due to stronger ocean heat transport. The absence of a strong meridional current system results in regional cooling and the growth of sea ice in the Norwegian Sea. To explain higher polar temperatures in the early Pliocene, Maier-Reimer & Mikolajewicz conclude that some other factor, such as higher atmospheric CO_2, is needed to provide warmth when the Isthmus was open.

In the northern Pacific Ocean, the opening of the Bering Straits appears to have occurred earlier than the climate cooling which began around 2.9 Ma and culminated in major continental ice growth by 2.4 Ma. Atlantic and Arctic marine species in Middle Pliocene deposits from eastern Kamchatka indicate that faunal migrations were occuring through the Bering Straits by 4.1 Ma (Gladenkov et al 1991). Additional migrations are documented between 3.6 and 3.2 Ma when many Pacific species invade the Arctic and North Atlantic Oceans (Hopkins 1967, Gladenkov 1981). Thus, changes in moisture and heat transport through the Bering Straits are not a likely explanation for late Pliocene cooling of the northern hemisphere.

Topographic Changes

Another important aspect of the Earth's geography is its vertical dimension, e.g. mountains and plateaus. Many early ice age theories focused on mountains, undoubtedly inspired by the presence of majestic alpine glaciers in many of the mid- to high-latitude regions of the world. One such mechanism invokes epierogenic uplift of northern Canada causing nucleation of continental ice sheets and, ultimately, glaciation (Flint 1957, Birchfield et al 1982), although relatively little data exist to support this hypothesis. Ruddiman et al (1986b) and Ruddiman & Raymo (1988) proposed that Pliocene uplift of the Tibetan and Colorado Plateaus could have initiated northern hemisphere glaciation via the effects of uplift on planetary wave structure. By enhancing southward meanders of the upper westerlies and the outbreak of cold polar air masses over North America and Europe, plateau uplift could have led to glacial inception. However, a problem with this hypothesis is that the evidence is weak for recent rapid plateau uplift in Tibet and the American west. Molnar & England (1990) argued that much of the evidence for Pliocene-Pleistocene uplift could be accounted for by enhanced erosion and generation of relief rather than by a dramatic increase in mean elevations. In particular, the Tibetan Plateau probably attained much of its present elevation by the late Miocene (Sorkabi & Stump 1993). More recent work by Ruddiman and colleagues have emphasized the role of plateau uplift in explaining the evolution of regional precipitation and temperature patterns and global climate over the last 40 Ma, rather than the last 3 Ma (Ruddiman & Kutzbach 1989, Ruddiman et al 1989, Prell & Kutzbach 1992).

Volcanism

From the above discussion it would appear that in addition to geography and topography, other factors must play a role in explaining the onset of northern hemisphere glaciation. One possibility is that global cooling resulted from enhanced volcanism and the corresponding increase in ash and aerosol concentrations in the atmosphere which would reflect sunlight. Based on the distribution of ash layers in ocean sediments, Kennett & Thunell (1977) noted a general increase in global volcanism in the Quaternary. However, this and other early surveys (Rea & Scheidegger 1979) suggest that the increase in volcanism occurred after the Pliocene-Pleistocene boundary at 1.6 Ma. Likewise, it was argued that conversion of ash to bentonites and the movement of sea floor crust towards island arc sources of ash would bias the ash records toward a late Neogene increase in volcanism (Ninkovich & Donn 1976, Hein et al 1978). It was also questioned whether volcanism could provide the persistent climate forcing needed to explain millions of years of cold glacial climates or whether it acted as a trigger, tipping the system into a new regime.

The above questions have recently come to the forefront with the presentation of initial drilling results from North Pacific ODP Leg 145. Rea et al (1993) show evidence for a several-fold increase in regional volcanism essentially coincident with the initiation of northern hemisphere glaciation. Undoubtedly the volcanism-climate link will be examined anew as these results are published over the next few years.

Atmospheric Composition

It has long been recognized that variations in radiatively important trace gases such as carbon dioxide or water could have significant effects on the thermal radiation balance of the Earth's atmosphere and, hence, global temperatures (e.g. Chamberlin 1899). This link is supported by the strong covariance of atmospheric CO_2 and temperature observed in the Vostok ice core record of the last 140,000 years (Barnola et al 1987). A number of hypotheses have been proposed that link the longer-term evolution of global climate to changes in the composition of the Earth's atmosphere;

in all cases, these mechanisms have plate tectonic motions as their root cause (e.g. Walker et al 1981, Raymo et al 1988, Berner 1990).

Of particular relevance is the study of Raymo et al (1988) which summarized evidence that global chemical weathering rates had increased significantly over the last 5 Ma. Because chemical weathering is the major sink of atmospheric CO_2 on geologic time scales, they proposed that this increase in weathering led to a "reverse greenhouse" and global cooling since the late Miocene. This idea found support in the study of Crowley (1991) who compared evidence for Middle Pliocene warmth with the results of general circulation model experiments and concluded that the climate of three million years ago was consistent with doubled CO_2 levels (e.g. approximately 550 ppm CO_2 by volume). Recent attempts to test this hypothesis using a proxy method based on the $\delta^{13}C$ of marine organic carbon [see Rau et al (1991) for a description] suggest that mid-Pliocene atmospheric CO_2 levels were closer to $\sim$360 ppm, approximately 25% higher than preindustrial values (Raymo & Rau 1992). These preliminary results await confirmation.

Ocean Heat Transport

An alternative hypothesis to "reverse greenhouse" cooling was proposed by Rind & Chandler (1991). They suggested that pre-Pleistocene warmth at high latitudes was due to stronger ocean thermohaline circulation which increased surface ocean heat transport to the poles. This hypothesis is supported by evidence presented earlier suggesting stronger NADW formation prior to the intensification of northern hemisphere glaciation (Raymo et al 1992). Cronin (1991) also found evidence for a stronger Gulf Stream/North Atlantic Drift system at this time. Likewise, Dowsett et al (1992) proposed that the relative constancy of low-latitude SSTs in the mid-Pliocene implied stronger ocean heat transports (although the error of their method is potentially greater than the increase in SST that they were trying to exclude).

Of course, with both decreased CO_2 or decreased ocean heat transports, one must ask why these changes took place? Why would chemical weathering rates increase or why would thermohaline circulation weaken over the past few million years? These processes themselves are sensitive to climate. Chemical weathering rates can change as a function of precipitation, vegetation, or temperature. Likewise, ocean circulation patterns are controlled by evaporation-precipitation pattern as well as regional wind fields. One possibility is that increased uplift and erosional activity in mountain ranges such as the Himalaya could have resulted in a significant increase in chemical weathering rates over the Neogene (Raymo et al 1988). Sorkhabi & Stump (1993) summarize tectonic evidence for a Pliocene-Pleistocene phase of uplift and denudation in the Himalayas, and an inflection in the ocean strontium isotope record at ~2.5 Ma (Capo & DePaolo 1990) also suggests an increase in chemical weathering occurred at this time. However, the possibility that these changes are due to the erosional action of mountain glaciers, which expanded in response to global cooling, can not be ruled out (Molnar & England 1990).

THRESHOLDS AND FEEDBACKS

Ultimately, the intensification of northern hemisphere glaciation which took place between 2.9 and 2.4 Ma was either a linear response to a specific change in boundary conditions taking place within this interval—such as the closing of the Panamanian Isthmus or an episode of pronounced exhumation and weathering which drew down atmospheric CO_2, or the global cooling reflected a threshold response to a longer, more gradual forcing with positive feedbacks playing an important role. For instance, a CO_2 decline occurring over a million plus years reaches a point at which perennial Arctic sea ice forms; this causes NADW formation to decrease which weakens ocean heat transport to high latitudes as well as causes a further down draw of CO_2 [e.g. via mechanisms described by Boyle (1988), Keir (1988), and/or Broecker & Peng (1989)]. Continental glaciation ultimately results.

With available data, these two scenarios are extremely difficult to distinguish. In addition, given the number of important climate feedbacks, the behavior of the system could reflect some combination of linear and nonlinear responses. Two important climate change contenders—decreases in atmospheric CO_2 and decreases in ocean heat transport to the poles—need further investigation. Application of proxy ρCO_2 methods to the Pliocene needs to be refined, as do studies of Pliocene thermohaline circulation. Lastly, the history of Arctic sea ice and its influence on deep-water formation, regional surface temperatures, and atmospheric pressure patterns requires investigation. The recent Ocean Drilling Program leg in the Arctic and Nordic Seas should provide a wealth of information on these and other topics relevant to the cause and mechanisms of northern hemisphere glaciation in the Pliocene.

ACKNOWLEDGMENTS

I thank B. Ruddiman, D. Oppo, and D. Norris for their thoughtful reviews and suggestions. NSF grant OCE9257191 provided support for this work.

Literature Cited

Aagaard K. 1988. The Arctic thermohaline circulation. *Eos, Trans. Am. Geophys. Union* 69: 1043

Abell PI. 1982. Paleoclimates at Lake Turkana, Kenya, from oxygen isotope ratios of gastropod shells. *Nature* 297: 321–23

Abelmann AR, Gersonde R, Speiss V. 1990. Pliocene-Pleistocene paleoceanography of the Weddell Sea—siliceous microfossil evidence. In *Geologic History of the Polar Oceans: Arctic versus the Antarctic*, ed. U Bleil, J Theide, pp. 729–59. Amsterdam: Kluwer

Arrhenius S. 1896. The influence of the carbonic acid in the air upon the temperature of the ground. *Philos. Mag.* 41: 237–76

Barnola JM, Raynaud D, Korotevich YS, Lorius C. 1987. Vostok ice core provides 160,000 year record of Atmospheric CO_2. *Nature* 329: 408–14

Barrett PJ, Adams CJ, McIntosh WC, Swisher CC, Wilson GS. 1992. Geochronological evidence supporting Antarctic deglaciation three million years ago. *Nature* 359: 816–18

Barron EJ. 1981. Paleogeography as a climatic forcing factor. *Geol. Rundsch.* 70: 737–47

Barron EJ, Washington WM. 1985. Warm Cretaceous Climates: high atmospheric CO_2 as a plausible mechanism. In *The Carbon Cycle and Atmospheric CO_2: Natural Variations Archean to Present*, ed. ET Sundquist, WS Broecker, pp. 546–53, *Geophys. Monogr.* 32. Washington, DC: Am. Geophys. Union

Berggren WA. 1972. Late Pliocene-Pleistocene glaciation. *Init. Rep. Deep Sea Drill. Proj.* 12: 953–63

Berggren WA, Kent DV, Flynn JJ, Van Couvering JA. 1985. Cenozoic geochronology. *Geol. Soc. Am. Bull.* 96: 1407–18

Berner RA. 1990. Atmospheric carbon dioxide levels over Phanerozoic time. *Science* 249: 1382–86

Birchfield GE, Weertmann J, Lunde AT. 1982. A model study of high-latitude topography in the climatic response to orbital insolation anomalies. *J. Atmos. Sci.* 39: 71–87

Bonnefille R. 1983. Evidence for a cooler and drier climate in the Ethiopian uplands, 2.5 Ma ago. *Nature* 303: 487–91

Boyle EA. 1988. Vertical oceanic nutrient fractionation and glacial/interglacial CO_2 cycles. *Nature* 331: 55–56

Boyle EA, Keigwin LD. 1982. Deep circulation of the North Atlantic over the last 200,000 years: geochemical evidence. *Science* 218: 784–87

Boyle EA, Keigwin LD. 1987. North Atlantic thermohaline circulation during the past 20,000 years linked to high-latitude surface temperature. *Nature* 330: 35–40

Brewer PG, Broecker WS, Rooth CG, Swift JH, Takahashi T, Williams RT. 1983. A climatic freshening of the Deep Atlantic north of 50°N over the past 20 years. *Science* 222: 1237–39

Brigham-Grette J, Carter LD. 1992. Pliocene marine transgression of northern Alaska: circumarctic correlations and paleoclimatic interpretations. *Arctic* 45: 74–89

Broecker WS, Denton GH. 1989. The role of ocean-atmosphere reorganizations in glacial cycles. *Geochim. Cosmochim. Acta* 53: 2465–501

Broecker WS, Peng T-H. 1989. The cause of the glacial to interglacial atmospheric CO_2 change: a polar alkalinity hypothesis. *Global Biogeochem. Cycles* 3: 215–39

Brouwers EM, Jorgensen MO, Cronin TM. 1991. Climatic significance of the ostracode fauna from the Pliocene Kap Kobenhavn formation, North Greenland. *Micropaleontology* 37: 245–67

Brunner CA. 1984. Evidence for increased volume transport of the Florida Current in the Pliocene and Pleistocene. *Mar. Geol.* 54: 223–35

Budyko MI, Ronov AB, Yanshin AL. 1985. *The History of the Earth's Atmosphere.* Leningrad: Gidrometeoizdat. 209 pp; Engl. transl. 1987. Berlin: Springer-Verlag. 139 pp.

Cande SC, Kent DV. 1992. A new geomagnetic polarity time scale for the late Cretaceous and Cenozoic. *J. Geophys. Res.* 97: 13,917–51

Capo RS, DePaolo DJ. 1990. Seawater strontium isotopic variations from 2.5 million years ago to the present. *Science* 249: 51–55

Caputo MV, Crowell JC. 1985. Migration of glacial centers across Gondwana during the Paleozoic Era. *Geol. Soc. Am. Bull.* 96: 1020–36

Carter LD, Brigham-Grette J, Marincovich L, Pease VL, Hillhouse JW. 1986. Late Cenozoic Arctic Ocean sea ice and terrestrial paleoclimate. *Geology* 14: 675–78

Cerling TE, Hay RL, O'Neil JR. 1977. Isotopic evidence for dramatic climate changes in east Africa during the Pleistocene. *Nature* 267: 137–38

Chamberlin, TC. 1899. An attempt to frame a working hypothesis of the cause of glacial periods on an atmospheric basis. *J. Geol.* 7: 545–84, 667–85, 751–87

Charles CD, Fairbanks RG. 1992. Evidence

from Southern Ocean sediments for the effect of North Atlantic deep-water flux on climate. *Nature* 355: 416–19
Clapperton CM, Sugden DE. 1990. Late Cenozoic glacial history of the Ross Embayment, Antarctica. *Quat. Sci. Rev.* 9: 253–72
Clemens SC, Prell WL. 1990. Late Pleistocene variability of Arabian Sea summer monsoon winds and dust source aridity: an eolian record from the lithogenic component of deep-sea sediments. *Paleoceanography* 5: 109–45
CLIMAP Project Members. 1981. Seasonal reconstructions of the earth's surface at the last glacial maximum. *Geol. Soc. Am. Map Chart Ser.* MC-36: 1–18
Coates AG, Jackson JBC, Collins LS, Cronin TM, Dowsett HJ, et al. 1992. Closure of the Isthmus of Panama: the near-shore marine record of Costa Rica and western Panama. *Geol. Soc. Am. Bull.* 104: 814–28
Cronin TM. 1991. Pliocene shallow water paleoceanography of the North Atlantic Ocean based on marine ostracods. *Quat. Sci. Rev.* 10: 175–88
Cronin TM, Dowsett HJ. 1990. A quantitative micropaleontologic method for shallow marine paleoclimatology: application to Pliocene deposits of the Western North Atlantic Ocean. *Mar. Micropal.* 16: 117–47
Crowell JC, Frakes LA. 1970. Phanerozoic ice ages and the cause of ice ages. *Am. J. Sci.* 268: 193–224
Crowley TC. 1991. Modeling Pliocene warmth. *Quat. Sci. Rev.* 10: 275–82
Crowley TJ, Mengel JG, Short DA. 1987. Gondwanaland's seasonal cycle. *Nature* 329: 803–7
Curry WB, Duplessy JC, Labeyrie LD, Oppo D, Kallel N. 1988. Quaternary deep-water circulation changes in the distribution of $\delta^{13}C$ of deep water SCO_2 between the last glaciation and the Holocene. *Paleoceanography* 3: 317–42
Curry WB, Lohmann, GP. 1983. Reduced advection into the Atlantic Ocean eastern basins during the last glacial maximum. *Nature* 306: 577–80
de Menocal P, Bloemendal J, King J. 1991. A rock magnetic record of monsoonal dust deposition to the Arabian Sea: evidence for a shift in the mode of deposition at 2.4 Ma. *Proc. Ocean Drill. Prog., Sci. Results* 117: 389–407
de Menocal PB, Ruddiman WF, Pokras EM. 1993. Influences of high- and low-latitude processes on African terrestrial climate: Pleistocene eolian records from equatorial Atlnatic Ocean Drilling Program Sites 663. *Paleoceanography* 8: 209–41
Denton GH, Prentice ML, Kellogg DE, Kellogg TB. 1984. Late Tertiary history of the Antarctic ice sheet: evidence from the Dry Valleys. *Geology* 12: 263–67
Dowsett HJ. 1991. The development of a long-range foraminifer transfer function and application to late Pleistocene North Atlantic climatic extremes. *Paleoceanography* 6: 259–73
Dowsett HJ, Cronin TM. 1990. High eustatic sea level during the middle Pliocene: evidence from the southeastern U.S. Atlantic Coastal Plain. *Geology* 18: 435–38
Dowsett HJ, Cronin TM, Poore RZ, Thompson RS, Whatley RC, Wood AM. 1992. Micropaleontological evidence for increased meridional heat transport in the North Atlantic Ocean during the Pliocene. *Science* 258: 1133–35
Dowsett HJ, Poore RZ. 1990. A new planktic foraminifer transfer function for estimating Pliocene through Holocene sea surface temperatures. *Mar. Micropal.* 16: 1–23
Dowsett HJ, Poore RZ. 1991. Pliocene sea surface temperatures of the North Atlantic Ocean at 3.0 Ma. *Quat. Sci. Rev.* 10: 189–203
Duque-Caro H. 1990. Neogene stratigraphy, paleoceanography and paleobiogeography in northwest South America and the evolution of the Panama Seaway. *Palaeogeogr. Palaeoclimatol. Palaeoecol.* 77: 203–34
Edwards LE, Mudie PJ, de Vernal A. 1991. Pliocene paleoclimatic reconstruction using dinoflagellate cysts: comparison of methods. *Quat. Sci. Rev.* 10: 259–73
Emiliani C, Gartner S, Lidz B. 1972. Neogene sedimentation on the Blake Plateau and the emergence of the Central American Isthmus. *Palaeogeogr. Palaeoclimatol. Palaeoecol.* 11: 1–10
Flint RF. 1957. *Glacial and Pleistocene Geology*. New York: Wiley
Forester RM. 1991. Pliocene-climate history of the western United States derived from lacustrine ostracodes. *Quat. Sci. Rev.* 10: 133–45
Freeman KH, Hayes JM. 1992. Fractionation of carbon isotopes by phytoplankton and estimates of ancient CO_2 levels. *Global Biogeochem. Cycles* 4: 185–98
Froelich PN, Malone PN, Hodell DA. 1991. Biogenic opal and carbonate accumulation rates in the subantarctic South Atlantic: the late Neogene of Meteor Rise Site 704. *Proc. Ocean Drill. Prog., Sci. Results* 114: 97–122
Funder S, Abrahamsen N, Bennike O, Feyling-Hanssen RW. 1985. Forested Arctic: evidence from North Greenland. *Geology* 13: 542–46

Gilbert MW, Clark DL. 1982/83. Central Arctic Ocean paleoceano-graphic interpretations based on late Cenozoic calcareous dinoflagellates. *Mar. Micropal.* 7: 385–401

Gladenkov YB. 1981. Marine Plio-Pleistocene of Iceland and problems of its correlation. *Quat. Res.* 15: 18–23

Gladenkov YB, Barinov KB, Basilian AE, Cronin TM. 1991. Stratigraphy and paleoceanography of Pliocene deposits of Karaginsky Island, Eastern Kamchatka, U.S.S.R. *Quat. Sci. Rev.* 10: 239–45

Grine FE. 1986. Ecological causality and the pattern of Plio-Pleistocene hominid evolution in Africa. *S. Afr. J. Sci.* 82: 87–89

Hahn DG, Manabe S. 1975. The role of mountains in the south Asian monsoon circulation. *J. Atmos Sci.* 32: 1515–41

Haq BU, Hardenbol J, Vail PR. 1987. Chronology of fluctuating sea levels since the Triassic. *Science* 235: 1156–67

Hein JR, Scholl DW, Miller J. 1978. Episodes of Aleutian Ridge explosive volcanism. *Science* 199: 137–41

Herman Y, Hopkins DM. 1980. Arctic oceanic climate in late Cenozoic time. *Science* 209: 557–62

Herman Y, Osmond JK, Sonayajuly BLK. 1989. Late Neogene Arctic paleoceanography: micropaleontology, stable isotopes, and chronology. In *The Arctic Seas: Climatology, Oceanography, Geology, and Biology*, ed. Y Herman, pp. 581–656. New York: Van Nostrand Reinhold

Hodell DA, Ciesielski PF. 1990. Southern Ocean response to the intensification of northern hemisphere glaciation at 2.4 Ma. In *Geological History of the Polar Oceans: Arctic versus Antarctic*, ed. U Bleil, J Thiede, pp. 707–28. Amsterdam: Kluwer

Hodell DA, Ciesielski PF. 1991. Stable isotopic and carbonate stratigraphy of the late Pliocene and Pleistocene of Hole 704A: eastern subantarctic South Atlantic. *Proc. Ocean Drill. Prog., Sci. Results* 114: 409–35

Hodell DA, Venz K. 1992. Toward a high-resolution stable istopic record of the southern ocean during the Plio-Pleistocene (4.8–0.8 Ma). In *Antarctic Research Series*, ed. JP Kennett, 56: 265–310. Washington, DC: Am. Geophys. Union

Hodell DA, Warnke DA. 1991. Climate evolution of the Southern Ocean during the Pliocene Epoch from 4.8 to 2.6 million years ago. *Quat. Sci. Rev.* 10: 205–14

Hopkins DM, ed. 1967. *The Bering Land Bridge*. Stanford, CA: Stanford Univ. Press.

Imbrie J, Boyle EA, Clemens SC, Duffy A, Howard WR, et al. 1992. On the structure and origin of major glaciation cycles. 1. Linear responses to Milankovich forcing. *Paleoceanography* 7: 701–38

Imbrie J, Kipp NG. 1971. A new micropaleontological method for quantitative paleoclimatology: application to a late Pleistocene Caribbean core. In *The Late Cenozoic Glacial Ages*, ed. KK Turekian, pp. 71–131. New Haven: Yale Univ. Press

Imbrie J, van Donk J, Kipp NG. 1973. Paleoclimatic investigation of late Pleistocene Caribbean deep-sea cores: comparison of isotopic and faunal methods. *Quat. Res.* 3: 10–38

IPCC, Intergovermental Panel on Climate Change. 1990. *Climate Change, the IPCC Scientific Assessment*. ed. JT Houghton, GJ Jenkins, JJ Ephraums. Cambridge: Cambridge Univ. Press. 365 pp.

Jansen E, Bleil U, Henrich R, Kringstad L, Slettermark B. 1988. Paleoenvironmental changes in the Norwegian Sea and the Northeast Atlantic during the last 2.8 Ma: DSDP/ODP sites 610, 643, and 644. *Paleoceanography* 3: 563–81

Jansen E, Sjøholm J. 1991. Reconstruction of glaciation over the past 6 Myr from ice-borne deposits in the Norwegian Sea. *Nature* 349: 600–3

Jansen E, Sjøholm J, Bleil U, Erichsen JA. 1990. Neogene and Pleistocene glaciations in the northern hemisphere and late Miocene-Pliocene global ice volume fluctuations: evidence from the Norwegian Sea. In *Geological History of the Polar Oceans: Arctic versus Antarctic*, eds U Bleil, J Thiede, pp. 677–705. Amsterdam: Kluwer

Jansen E, Veum T. 1990. Evidence for two-step deglaciation and its impact on North Atlantic deep-water circulation. *Nature* 343: 612–16

Keigwin LD. 1978. Pliocene closing of the Isthmus of Panama based on biostratigraphic evidence from nearby Pacific Ocean and Caribbean Sea cores. *Geology* 6: 630–34

Keigwin LD. 1982. Isotope paleoceanography of the Caribbean and east Pacific: role of Panama uplift in late Neogene time. *Science* 217: 350–53

Keir RS. 1988. On the late Pleistocene ocean geochemistry and circulation. *Paleoceanography* 3: 413–45

Kennett JP. 1977. Cenozoic evolution of Antarctic glaciation, the circum-Antarctic Ocean, and their impact on global paleoceanography. *J. Geophys. Res.* 82: 3843–60

Kennett JP, Srinivasan S. 1983. *Neogene Planktonic Formaninifera*. 265 pp. Stroudsburg, Pa: Hutchinson Ross

Kennett JP, Thunell RC. 1977. Global increase in Quaternary explosive volcanism. *Science* 187: 497

Kipp NG. 1976. New transfer function for estimating past sea-surface conditions from sea-bed distribution of planktonic foraminiferal assemblages in the North Atlantic. In *Investigation of Late Quaternary Paleoceanography and Paleoclimatology*, ed. RM Cline, JD Hays, *Geol. Soc. Am. Mem.* 145: 3–42

Krantz DE. 1991. A chronology of Pliocene sea-level fluctuations: the U.S. middle Atlantic coastal plain record. *Quat. Sci. Rev.* 10: 163–73

Kroopnick P. 1985. The distribution of carbon-13 in the world oceans. *Deep-Sea Res.* 32: 57–84

Kukla G. 1987. Loess stratigraphy in central China. *Quat. Sci. Rev.* 6: 191–219

Kukla G, An Z. 1989. Loess stratigraphy in central China. *Palaeogeogr. Palaeoclimatol. Palaeoecol.* 72: 203–25

Kutzbach JE. 1981. Monsoon climate of the early Holocene: climate experiment with the earth's orbital parameters for 9000 years ago. *Science* 214: 59–61

Loubere P, Moss K. 1986. Late Pliocene climatic change and the onset of northern hemisphere glaciation as recorded in the northeast Atlantic Ocean. *Geol. Soc. Am. Bull.* 97: 818–28

Lundelius EL Jr. 1987. The North American quaternary sequence. In *Cenozoic Mammals of North America*, ed. MO Woodburne, pp. 211–35. Berkeley: Univ. Calif. Press

MacFadden BJ, Anaya F, Argollo J. 1993. Magnetic polarity stratigraphy of Inchasi: a Pliocene mammal-bearing locality from the Bolivian Andes deposited just before the Great American Interchange. *Earth Planet. Sci. Lett.* 114: 229–41

Maier-Reimer E, Mikolajewicz U. 1990. Ocean general circulation model sensitivity experiments with an open central American isthmus. *Paleoceanography* 5: 349–66

Manabe S, Broccoli AJ. 1985. The influence of continental ice sheets on the climate of an ice age. *J. Geophys. Res.* 90: 2167–90

Marshall LG. 1988. Land mammals and the Great American interchange. *Am. Sci.* 76: 380–88

Matthews JV, Ovenden LE. 1990. Late Tertiary plant macrofossils from localities in northern North America (Alaska, Yukon, and Northwest Territories). *Arctic* 43: 364–92

McDougall I, Wensink H. 1966. Paleomagnetism and geochronology of the Piocene-Pleistocene lavas in Iceland. *Earth Planet. Sci. Lett.* 1: 232–36

Mikolajewwicz U, Santer BD, Maier-Reimer E. 1990. Ocean response to greenhouse warming. *Nature* 345: 589–93

Mix A. 1987. The oxygen-isotope record of glaciation. In *North America and Adjacent Oceans During the Last Deglaciation*, ed. WF Ruddiman, HE Wright, pp. 111–36. Denver: Geol. Soc. Am.

Mix AC, Pisias NG, Rugh W, Wilson J, Morey A, Hagelberg T. 1994. Benthic foraminiferal stable isotope record from Site 849, 0–3.2 Ma: local and global climate changes. *Proc. Ocean Drill. Prog., Sci. Results* In press

Molnar P, England P. 1990. Late Cenozoic uplift of mountain ranges and global climate change: chicken or egg? *Nature* 346: 29–34

Morley JJ, Dworetzky BA. 1991. Evolving Pliocene-Pleistocene climate: a North Pacific perspective. *Quat. Sci. Rev.* 10: 225–38

Nelson RE, Carter LD. 1985. Pollen analysis of a late Pliocene and early Pleistocene section from the Gubik Formation of Arctic Alaska. *Quat. Res.* 24: 195–306

Ninkovitch D, Donn WL. 1976. *Science* 194: 899–906

North GR. 1984. The samll ice cap instability in diffusive climate models. *J. Atmos. Sci.* 41: 3390–95

North GR, Crowley TC. 1985. Application of a seasonal climate model to Cenozoic glaciation. *J. Geol. Soc.* 142: 475–82

Oglesby RJ. 1989. A GCM study of Antarctic glaciation. *Clim. Dyn.* 3: 135–56

Oppo DW, Fairbanks RG. 1987. Variability in the deep and intermediate water circulation of the Atlantic Ocean: Northern Hemisphere modulation of the Southern Ocean. *Earth Planet. Sci. Lett.* 86: 1–15

Pokras EM, Mix AC. 1985. Eolian evidence for spatial variability of late Quaternary climates in tropical Africa. *Quat. Res.* 24: 137–49

Pollack JB. 1982. Solar, astronomical, and atmospheric effects on climate. In *Climate in Earth History*, pp. 68–76. Washington, DC: US Natl. Res. Council

Poore RZ, Berggren WA. 1975. The morphology and classification of Neogloboquadrina atlantica (Berggren). *J. Foram. Res.* 5: 76–84

Prell WL. 1984. Monsoonal climate of the Arabiaan Sea during the late Quaternary: a response to dchanging solar radiation. In *Milankovitch and Climate*, ed. AL Berger, J Imbrie, J Hayes, G Kukla, B Saltzman. pp. 349–66, Dordrecht: Reidel

Prell WL, Kutzbach JE. 1987. Monsoon variability over the past 150,000 years. *J. Geophys. Res.* 92: 8411–525

Prell WL, Kutzbach JE. 1992. Sensitivity of

the Indian monsoon to forcing parameters and implications for its evolution. *Nature* 360: 647–52

Prell WL, Van Campo W. 1986. Coherent response of Arabian Sea upwelling and pollen transport to late Quaternary monsoonal winds. *Nature* 323: 526–28

Prentice ML, Matthews RK. 1991. Tertiary ice sheet dynamics: the snow gun hypothesis. *J. Geophys. Res.* 96: 6811–27

Rau GH, Froelich PN, Takahashi T, DesMarais DJ. 1991. Does sedimentary organic $\delta^{13}C$ record variations in Quaternary ocean [CO_2(aq)]? *Paleoceanography* 6: 335–47

Raymo ME. 1992. Global climate change: a three million year perspective. In *Start of a Glacial*, ed. G Kukla, E Went, *Proc. Mallorca NATO ARW, NATO ASI Ser. I*, 3: 207–23. Heidelberg: Springer-Verlag

Raymo ME, Hodell D, Jansen E. 1992. Response of deep ocean circulation to initiation of northern hemisphere glaciation (3–2 Ma). *Paleoceanography* 7: 645–72

Raymo ME, Rau G. 1992. Plio-Pleistocene Atmospheric CO_2 levels inferred from POM $\delta^{13}C$ at DSDP Site 607. *Eos, Trans. Am. Geophys. Union* 73: 95

Raymo ME, Rind D, Ruddiman WF. 1990b. Climatic effects of reduced Arctic sea ice limits in the GISS II general circulation model. *Paleoceanography* 5: 367–82

Raymo ME, Ruddiman WF. 1992. Tectonic forcing of late Cenozoic climate. *Nature* 359: 117–22

Raymo ME, Ruddiman WF, Backman J, Clement BM, Martinson DG. 1989. Late Pliocene variation in northern hemisphere ice sheets and North Atlantic deepwater circulation. *Paleoceanography* 4: 413–46

Raymo ME, Ruddiman WF, Clement BM. 1986. Pliocene-Pleistocene paleoceanography of the North Atlantic at Deep Sea Drilling Project Site 609. *Init. Rep. Deep Sea Drill. Proj.* 94: 895–901

Raymo ME, Ruddiman WF, Froelich PN. 1988. Influence of late Cenozoic mountain building on ocean geochemical cycles. *Geology* 16: 649–53

Raymo ME, Ruddiman WF, Shackleton NJ, Oppo DW. 1990a. Evolution of Atlantic-Pacific $\delta^{13}C$ gradients over the last 2.5 m.y. *Earth Planet. Sci. Lett.* 93: 353–68

Rea DK, Basov IA, Janecek TR, Shipboard Scientific Party. 1993. Cenozoic paleoceanography of the North Pacific Ocean: results of ODP leg 145, the North Pacific Transect. *Eos, Trans. Am. Geophys. Union* 74: 173

Rea DK, Scheidegger KF. 1979. Eastern Pacific spreading rate fluctuation and its relation to Pacific area volcanic episodes. *J. Volcanol. Geotherm. Res.* 5: 135–48

Rea DK, Schrader H. 1985. Late Pliocene onset of glaciation: ice-rafting and diatom stratigraphy of North Pacific DSDP cores. *Palaeogeogr. Palaeoclimatol. Palaeoecol.* 49: 313–25

Repenning CA, Brouwers EM, Carter LD, Marincovich L, Ager TA. 1987. The Beringian ancestry of *Phenacomys* (Rodentia: Cricetidae) and the beginning of the modern Arctic Ocean borderland biota. *U.S. Geol. Surv. Bull.* 1687 29 pp.

Rind D, Chandler M. 1991. Increased ocean heat transports and warmer climates. *J. Geophys. Res.* 96: 7437–61

Ruddiman WF, Cameron D, Clement BM. 1986a. Sediment disturbance and correlation of offset holes drilled with the hydraulic piston corer: Leg 94. *Init. Rep. Deep Sea Drill. Proj.* 94: 615–34

Ruddiman WF, Esmay A. 1986. A streamlined foraminiferal transfer function for the subpolar north Atlantic. *Init. Rep. Deep Sea Drill. Proj.* 94: 1045–59

Ruddiman WF, Janecek T. 1989. Pliocene-Pleistocene biogenic and terrigenous fluxes at equatorial Atlantic sites 662, 663, and 664. *Proc. Ocean Drill. Prog., Sci. Results* 108: 211–40

Ruddiman WF, Kutzbach JE. 1989. Forcing of late Cenozoic northern hemisphere climate by plateau uplift in southeast Asia and the American Southwest. *J. Geophys. Res.*94(D15): 18,409–27

Ruddiman WF, Prell WL, Raymo ME. 1989. History of late Cenozoic uplift on Southeast Asia and the American Southwest: rationale for general circulation modeling experiments. *J. Geophys. Res.* 94: 18,379–91

Ruddiman WF, Raymo ME. 1988. Northern hemisphere climate regimes during the past 3 Ma: possible tectonic connections. In *The Past Three Million Years: Evolution of Climatic Variability in the North Atlantic Region*, ed. NJ Shackleton, RG West, DQ Bowen, pp. 227–34. Cambridge: Cambridge Univ. Press

Ruddiman WF, Raymo M, McIntyre A. 1986b. Matuyama 41,000-year cycles: North Atlantic Ocean and northern hemisphere ice sheets. *Earth Planet. Sci. Lett.* 80: 117–29

Schlosser P, Bonisch G, Rhein M, Bayer, R. 1991. Reduction of deepwater formation in the Greenland Sea during the 1980s: evidence from tracer data. *Science* 251: 1054–56

Scott DB, Mudie PJ, Baki V, MacKinnon KD, Cole FE. 1989. Biostratigraphy and the late Cenozoic paleoceanography of the Arctic Ocean: foraminiferal, lithostratigraphic, and isotopic evidence. *Geol. Soc. Am. Bull.* 101: 260–77

Shackleton NJ, Backman J, Zimmerman H, Kent DV, Hall MA, Roberts DG, et al. 1984. Oxygen isotope calibration of the onset of ice-rafting and history of glaciation in the North Atlantic region. *Nature* 307: 620–23

Shackleton NJ, Hall MA. 1989. Stable istope history of the Pleistocene at ODP Site 677. *Proc. Ocean Drill. Prog., Sci. Results* 111: 295–316

Shackleton NJ, Kennett JP. 1975. Paleotemperature history of the Cenozoic and the initiation of Antarctic glaciation: oxygen and carbon isotope analyses in DSDP sites 277, 279, and 281. *Init. Rep. Deep Sea Drill. Proj.* 29: 743–55

Shackleton NJ, Opdyke ND. 1977. Oxygen isotope and paleomagnetic evidence for early northern hemisphere glaciation. *Nature* 270: 216–9

Shackleton NJ, Berger A, Peltier WR. 1990. An alternative astronomical calibration of the lower Pleistocene timescale based on ODP site 677. *Trans. R. Soc. Edinburgh: Earth Sci.* 81: 251–61

Sikes EL, Keigwin LD, Curry WB. 1991. Pliocene paleoceanography: circulation and oceanographic changes associated with the 2.4 Ma glacial event. *Paleoceanography* 6: 245–57

Sorkhabi RB, Stump E. 1993. Rise of the Himalaya: a geochronologic approach. *GSA Today* 3: 86–92

Stokes WL. 1955. Another look at the ice age. *Science* 122: 815–21

Street-Perrott FA, Harrison, SP. 1984. Temporal variations in lake levels since 30,000 years B.P.: an index of the global hydrological cycle. In *Climate Processes and Climate Sensitivity*, ed JE Hansen, T Takahashi, *AGU Geophys. Monogr.* 29: 118–29. Washington, DC: Am. Geophys. Union

Thompson RS. 1991. Pliocene environments and climates in the western United States. *Quat. Sci. Rev.* 10: 115: 31

Vorren TO, Hald M, Lebesbye E. 1988. Late Cenozoic environments in the Barents Sea. *Paleoceanography* 3: 601–12

Vrba ES. 1985. African Bovidae: evolutionary events since the Miocene. *S. Afr. J. Sci.* 81: 263–66

Walker JCG, Hays PB, Kasting JF. 1981. A negative feedback mechanism for the long-term stabilization of Earth's surface temperature. *J. Geophys. Res.* 86: 9976–82

Warnke DA, Allen CP. 1991. Ice rafting, glacial-marine sediments, and siliceous oozes: South Atlantic/subantarctic Ocean. *Proc. Ocean Drill. Prog., Sci. Results* 114: 589–98

Webb PN, Harwood DM, McKelvey BC, Mercer JH, Stott LD. 1984. Cenozoic marine sedimentation and ice-volume variation on the East Antarctic craton. *Geology* 12: 287–91

Webb PN, McKelvey BC, Harwood DM, Mabin MCG, Mercer JH. 1987. Sirius Formation of the Beardmore glacier region. *Antarctic J. US* 22: 8–13

Wesselman HB. 1985. Fossil micromammals as indicators of climatic change about 2.4 m.y. ago in the Omo Valley, Ethiopia. *S. Afr. J. Sci.* 81: 260–61

Westgate JA, Stemper BA, Pewe TL. 1990. A 3 m.y. record of Pliocene-Pleistocene loess in interior Alaska. *Geology* 18: 858–61

Annu. Rev. Earth Planet. Sci. 1994. 22:385–417

GLOBAL VARIATIONS IN CARBON ISOTOPE COMPOSITION DURING THE LATEST NEOPROTEROZOIC AND EARLIEST CAMBRIAN

R. L. Ripperdan

Department of Geological Sciences, University of California, Santa Barbara, California 93106

KEY WORDS: carbon isotope stratigraphy, Precambrian-Cambrian boundary

INTRODUCTION

The rapid rise of skeletonized organisms at the beginning of the Cambrian Period traditionally has separated the geologic time scale into two fundamental units—the Precambrian and the Phanerozoic. Intercontinental correlations using this event as a global chronostratigraphic marker have suffered, however, from a lack of continuously fossiliferous sequences outside the Siberian Platform and China; from concerns that the first occurrences of biomineralized fossils are not temporally equivalent in all geographic localities; and from the paucity of biostratigraphic data recovered from the generally barren underlying strata. This, in turn, has frustrated attempts to evaluate the numerous theories proposed to explain the rise of Cambrian mineralized faunas (e.g. Brasier 1982, 1985; Sepkoski 1983; McMenamin 1987).

The recognition of variations in the carbon isotopic composition of marine carbonates during the Phanerozoic (e.g. Keith & Weber 1964) raises hopes of overcoming these difficulties through the use of a stratigraphic component that is extremely abundant within the geologic record. Carbon isotope profiles from a number of Precambrian-Cambrian boundary sections show similarities that suggest each locality preserves a contemporaneous record of secular variations in the carbon isotope com-

0084–6597/94/0515–0385$05.00

position of the global ocean system. Unifying these results into a global reference curve suitable for chronostratigraphic correlation, and using this reference curve to broadly evaluate evolutionary and lithostratigraphic relationships during the boundary interval, are the main objectives of this paper.

FUNDAMENTALS

Variations in the isotopic composition of carbon ($\delta^{13}C$) in the marine environment are a function of: the fraction of available CO_2 undergoing photosynthesis reduction or synthesis to organic carbon, the amount of fixed organic carbon buried in sediments, and the fractionation between the inorganic and organic carbon reservoirs, which is ultimately regulated by pCO_2 and temperature (Figure 1). The $\delta^{13}C$ value of marine bicarbonate ion, HCO_3^-, is sampled by $CaCO_3$ deposition of marine limestones and carbonaceous skeletons with a minor fractionation of 1–2‰ that is only slightly dependent on temperature (Anderson & Arthur 1983). Since the burial flux of organic compounds with low $\delta^{13}C$ values ($\delta^{13}C_{org} \sim -25‰$) ultimately controls the isotopic composition of HCO_3^-, the $\delta^{13}C$ of shallow-water marine carbonates ($\delta^{13}C_{carb}$) serves as a measure of net worldwide carbon deposition during long-term steady-state conditions. The value of $\delta^{13}C_{carb}$ changes through geologic time due to factors influencing carbon burial, including levels of anoxia, rates of sedimentation, and levels of primary productivity (e.g. Woodruff & Savin 1985, Schlanger et al 1987, Magaritz & Stemmerik 1989). Isotopic shifts in the global ocean-atmosphere system are superimposed on regional and local variability, so that carbon isotope profiles from widely separated localities are often strikingly similar despite differences in the absolute value of $\delta^{13}C_{carb}$.

A number of competing effects can bias the preserved $\delta^{13}C_{carb}$ signature away from the original isotopic composition of HCO_3^-. Isotopic shifts can be generated through disequilibrium deposition of skeletal material by some shelled organisms. These "vital effects" may lead to shifts of 1‰ or more (Grant 1992). Unstable carbonate phases, such as aragonite or magnesian calcite, may experience alteration of the original $\delta^{13}C$ signature during stabilization as low-Mg calcite. Some of the myriad post-depositional processes that can contribute secondary components to primary isotopic signatures include: bacterial mediation of pore-water chemistries, early diagenetic replacement of primary carbonate mineralogy, occlusion of original porosity by late-stage carbonate cements, crystallization of late-phase carbonate as vein or cavity infillings, and fluid migrations at elevated temperatures and salinities. Post-depositional effects are particularly acute for oxygen; however, carbon isotope ratios are often virtually unaffected

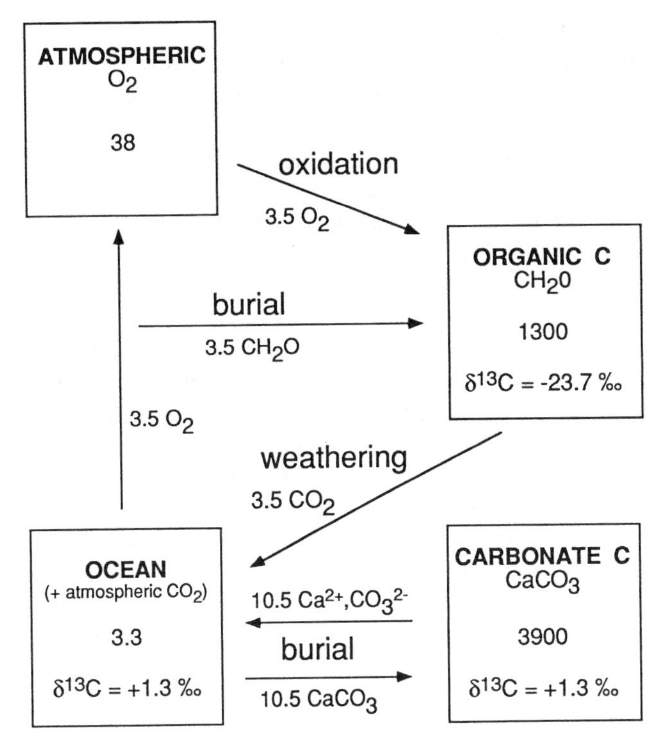

Figure 1 The marine carbon cycle. Reservoirs are in 10^{18} moles; fluxes are in 10^{18} moles/Ma. Modified from Kump & Garrels (1986).

because of the low water/rock ratios for carbon prevalent in carbonate strata (e.g. Magaritz 1983, Banner & Hanson 1990).

The influence of post-depositional processes can be evaluated using several different methodologies. Trace element compositions of elements such as strontium and manganese provide useful information, since Sr is preferentally removed during recrystallization of original metastable carbonate phases, while Mn becomes enriched during the formation of late-stage ferroan calcite cements (Al-Aasm & Veizer 1986; see also Gao & Land 1991, Ripperdan et al 1992). Higher Mn/Sr values are therefore often indicative of complex or severe diagenetic histories. Petrographic relationships can be particularly informative in the identification of various mineral phases developed during diagenesis. Other methods for evaluating the effects of diagenesis include comparisons of carbonate mineral pairs and determinations of covariance between $\delta^{13}C$ and $\delta^{18}O$; and comparisons of $\delta^{13}C_{carb}$ and $\delta^{13}C_{org}$ profiles and determinations of their relative difference.

In principle, microsampling selected rock components is the best method for minimizing the inclusion of some types of post-depositional components as part of the isotopic signature. However, studies utilizing whole rocks have often proven to be highly effective (e.g. Magaritz et al 1986, Fairchild & Spiro 1987; see also Kaufman et al 1991), especially when the relative proportions of components remain approximately constant and finer-grained, monofacial units are preferentially sampled.

PREVIOUS RESULTS

Carbon isotope profiles are now available from a number of areas with important Upper Neoproterozoic and Lower Cambrian sections, including the Siberian Platform (Magaritz et al 1986, 1991; Kirschvink et al 1991), Morocco (Tucker 1986, Kirschvink et al 1991, Magaritz et al 1991), Svalbard and East Greenland (Knoll et al 1986), the Lesser Himalayas (Aharon et al 1987), southern China (Hsü et al 1985, Lambert et al 1987, Brasier et al 1990), Iran (Brasier et al 1990), Namibia (Kaufman et al 1991, Derby et al 1992), South Australia (Tucker 1991), and Newfoundland (Brasier et al 1992) (Figure 2). Many of these are summarized below.

To compare these results and develop a global reference curve, a major

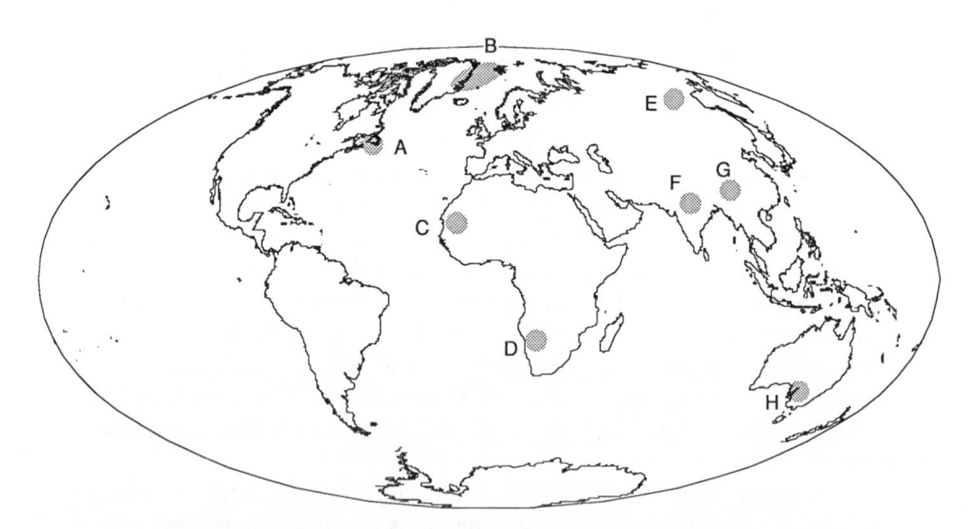

Figure 2 Locations of Upper Neoproterozoic-Lower Cambrian localities used in this compilation. A = southeastern Newfoundland/Avalonia; B = Greenland/Spitsbergen; C = Anti-Atlas Mtns., Morocco; D = Namibia; E = Lena and Aldan Rivers, Siberia; F = Lesser Himalayas, India; G = Meishucun, China; H = Flinders Ranges, Australia.

nomenclatural difficulty must be overcome. The base of the Tommotian Stage in Siberia traditionally has been placed at the first appearance of mineralized molluscan and archaeocyathid fossils, and underlying trace and skeletal fossils, including the Nemakit-Daldyn fauna, have been considered Vendian in age. The definition of the base of the *Phycodes pedum* ichnofossil zone as the Precambrian-Cambrian boundary level has elevated much of the known Siberian Platform Vendian sequence into the Cambrian. The retention of "Vendian" as a Neoproterozoic chronostratigraphic unit being assured, various names such as "Nemakit-Daldynian" and "Rovnian" have sought to fill the nomenclatural vacuum, but these have yet to appear widely in the literature and because of uncertainty in their application, currently provide little assistance to intercontinental correlation. Therefore, strata demonstrably below the first archaeocyathids and molluscs but above Ediacaran-grade faunas will, for the purposes of this paper, generally be referred to simply as "pre-Tommotian" (although it is noted that this definition may prove to be untenable; see Dorzhnamzhaa & Gibsher 1990). Age assignments in sequences where biostratigraphic control is lacking or inconclusive will generally retain those of the most relevant earlier references.

Siberian Platform

A highly fossiliferous succession of flat-lying platform carbonates of Late Proterozoic through Cambrian age rests unconformably on Archean basement along the Lena and Aldan Rivers of eastern Siberia. The first mineralized archaeocyathid faunas are found near the top of the dolomite-rich Yudoma Formation, and are generally considered to mark the base of the Tommotian Stage on the Siberian Platform (Rozanov 1984; see also Khomentovsky & Karlova 1993). The overlying Pestrotsvet and Tumuldur Formations reflect the subsequent transgression that occurred through the Tommotian and into the Atdabanian. The Tommotian-Atdabanian boundary is marked by the appearance of mineralized arthropods (trilobites) at the base of the Tumuldur Formation.

The first carbon isotope profile recovered from the southeastern Siberia Platform outlined cyclical changes in $\delta^{13}C$ (whole rock) within the Dvortsy section along the Aldan River, Yakutia, Russia (Magaritz et al 1986). The profile is characterized by a steady 7‰ rise in $\delta^{13}C$ through dolomites of the pre-Tommotian Yudoma Formation, followed by a rapid 4‰ drop through the uppermost Yudoma Formation and lowermost Pestrotsvet Formation (Figure 3). The base of the Tommotian, drawn at the base of the *A. sunnaginicus* archaeocyathid zone within the upper 1 m of the Yudoma Formation, is found just below the culmination of this drop. Younger strata show a sharp rise of approximately 3‰ followed by a

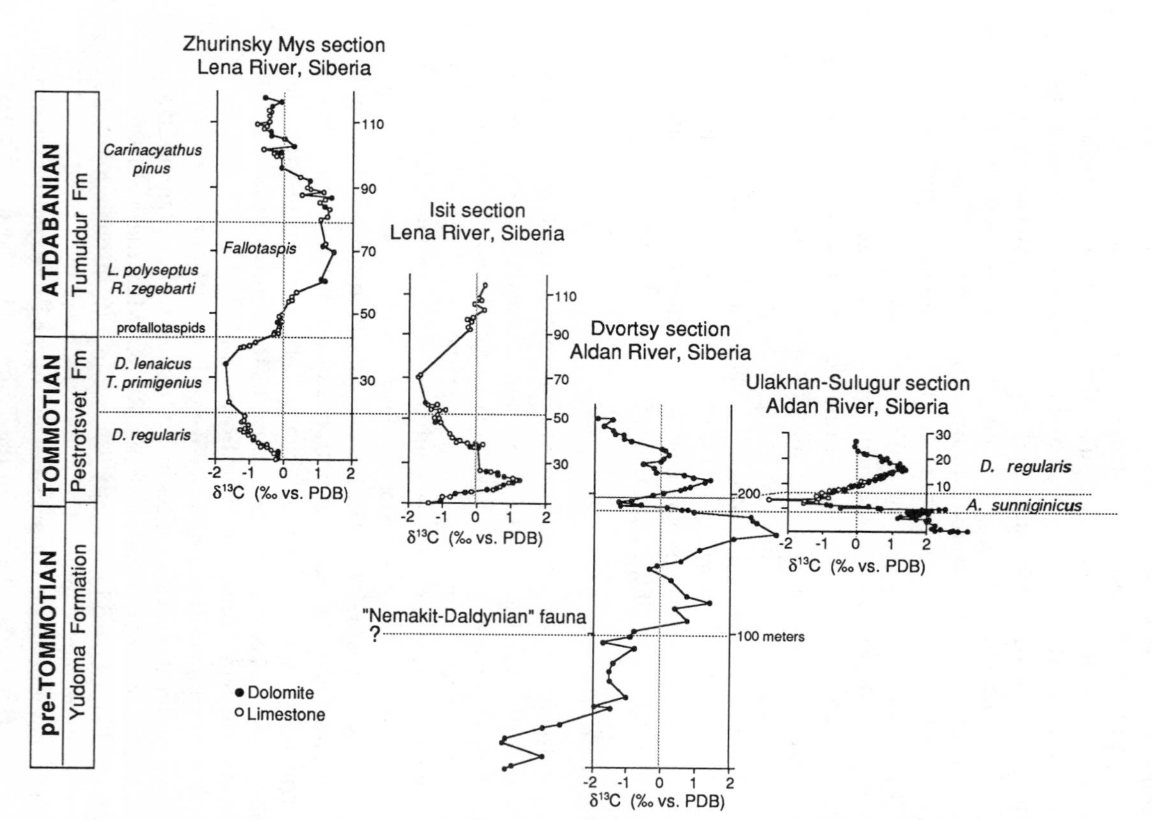

Figure 3 Carbon isotope profiles from sections along the Lena and Aldan Rivers, Siberian Platform, Russia. From Magaritz et al (1986), Kirschvink et al (1991), and Magaritz et al (1991). Filled symbols denote dolomite mol% > 40; open symbols represent limestones.

broad decline of about 3‰ through the top of the sampled section.

Subsequent papers by Magaritz (1989), Magaritz et al (1991), and Kirschvink et al (1991) substantially improved the Early Cambrian portion of the original $\delta^{13}C$ profile with results from additional sections along the Aldan and Lena Rivers. The sharp drop immediately above the base of the Tommotian was confirmed by results from the section at Ulakhan-Sulugur, 50 km downstream from the Dvortsy section. The sharpness of the drop led Magaritz (1989) to conclude that a small stratigraphic break exists at this level, as was previously suggested by Rozanov (1984). The results from Ulakhan-Sulugur also indicated that $\delta^{13}C$ values briefly remained constant at $\sim +2$‰ immediately before the drop—an observation that was missed at Dvortsy because of larger sampling intervals. The subsequent rise to values of ~ 1‰ (dubbed $\delta^{13}C$ Cycle II) was identified in extremely detailed profiles from Ulakhan-Sulugur and Isit (Lena River), as was the subsequent drop in $\delta^{13}C$ marking the end of the Tommotian ($\delta^{13}C$ Cycle III). Data from the Zhurinsky Mys section (Lena River) demonstrated an additional rise in $\delta^{13}C$ ($\delta^{13}C$ Cycle IV) beginning nearly coincident with the base of the Atdabanian Stage.

Magaritz et al (1991) and Kirschvink et al (1991) argued that the preserved $\delta^{13}C$ values represent secular variation in seawater because widely-spaced contemporaneous sections yielded parallel isotope profiles within a good biostratigraphic framework. $\delta^{18}O$ values provide additional support for this conclusion. $\delta^{18}O$ and $\delta^{13}C$ show only a weak negative correlation (-0.36) at Dvortsy (Magaritz et al 1986), whereas a positive correlation would be expected if both values had been distorted by freshwater diagenesis. Also, $\delta^{18}O$ values are relatively heavy for ancient carbonates, ranging between -7‰ to -4‰ (vs PDB). Similarly heavy $\delta^{18}O$ values are typical of Cambrian-Ordovician carbonates with low conodont alteration indices, which indicate cool thermal histories (Epstein et al 1977), while contemporaneous sections with high index values often have $\delta^{18}O < -10$‰ (R. Ripperdan, unpublished data).

Anti-Atlas Mountains, Morocco

Thick Neoproterozoic and Cambrian units are widespread near the village of Tiout in the western Anti-Atlas Mountains of southern Morocco. The sequence consists of the basal Serie de Base, followed by the Adoudon Formation (Dolomié Inferieur), the Lie de Vin Formation, the Igoudine Formation (Calcaire Supérieur), and the Serie Schistose-Calcaire. The succession is apparently complete and represents primarily shallow-water shelf sedimentation, although parts of the lower Lie de Vin Formation may contain terrigenous units (Latham 1990).

Attempts to determine a precise location for the Precambrian-Cambrian boundary at Tiout and in surrounding sections have proven problematic. Unfortunately, small shelly fossils heralding the pre-Tommotian and Tommotian have yet to be identified from the sequence, despite many attempts (Schmitt & Monninger 1977). Initial studies based on acritarchs and stromatolites seemed to indicate a Tommotian and/or Vendian age for the Lie de Vin Formation (Schmitt & Monninger 1977, Bertrand-Sarfati 1981). Algal and trace fossil discoveries suggest that the Tommotian-Atdabanian boundary lies within the upper quarter of the Lie de Vin, and that the base of the Tommotian is near the top of the Adoudan Formation (Latham & Riding 1990). These interpretations are supported by correlations between $\delta^{13}C$ profiles from the Siberian Platform and Morocco (Magaritz et al 1991), and by a U-Pb age of 522 $\pm$ 4 Ma determined using zircons extracted from an ash-fall tuff in the upper Lie de Vin (Compston et al 1992). Archaeocyathid biomeres near the top of the Igoudine Formation and the trilobite *Fallotaspis* in basal strata of the Serie Schisto-Calcaire suggest a mid-Atdabanian age for upper units of the Igoudine (Debrenne & Debrenne 1978, Sdzuy 1978). The Adoudan Formation (Dolomié Inferieur) has been variously attributed to the late Riphean (Monninger & Schmitt 1977) and the latest Vendian (Latham & Riding 1990, Magaritz et al 1991). An Ediacaran-type soft-bodied fauna is found within the Serie de Base (Houzay 1979), and a tilloid occurs in Precambrian strata lying unconformably beneath (Leblanc 1981).

The first carbon isotope profile from the area utilized the classic Precambrian-Cambrian boundary section near Tiout (Tucker 1986). The profile was further refined within the upper Lie de Vin Formation by Kirschvink et al (1991) using the Tiout section, and was extended downward throughout the Adoudon Formation by additional results from an adjacent section near Oued Sdas (Magaritz et al 1991). The composite $\delta^{13}C$ profile is marked by a more than 10‰ rise through almost 900 meters of the Adoudon, to values of nearly +7‰ (Figure 4). A drop of more than 9‰ occurs in the upper 200 m of the Adoudon, with values nearing −3‰ at the base of the overlying Lie de Vin Formation. Spotty recoveries from the 1000 meters of the Lie de Vin suggest generally negative $\delta^{13}C$ values through this interval, with a shift to positive values in the lower Igoudine Formation. Values stabilize at about 0‰ immediately below the the first appearance of trilobites in the upper Igoudine, and then vary within a narrow range throughout the overlying 300 m at Tiout.

Tucker (1986) argued that $\delta^{13}C$ values determined from the Tiout section were probably primary because adjacent limestones and dolomites did not preserve markedly different $\delta^{13}C$ signatures, suggesting that little of the long-range variation was due to diagenesis. Magaritz et al (1991) noted that the relatively continuous pattern of $\delta^{13}C$ shifts and the apparently

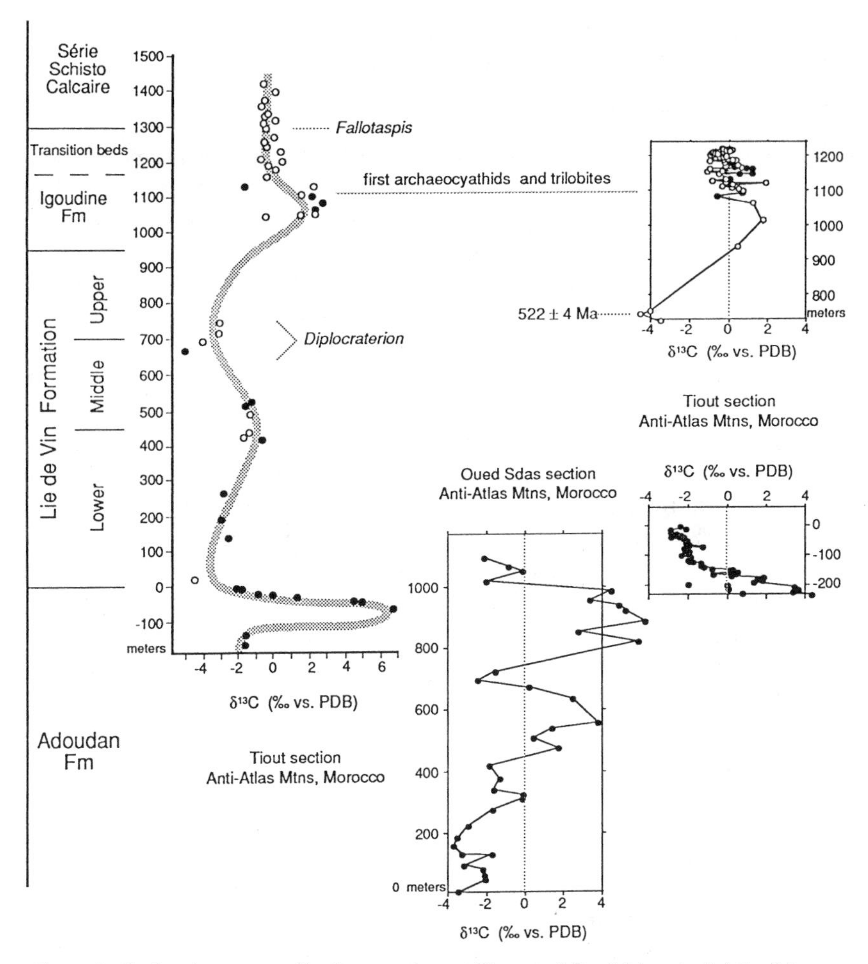

Figure 4 Carbon isotope profiles from sections at Tiout and Oued Sdas, Anti-Atlas Mountains, Morocco. From Tucker (1986) (*left*), Magaritz et al (1991) (*middle, lower right*) and Kirschvink et al (1991). Biostratigraphic data from Latham & Riding (1990); geochronology from Compston et al (1992). [Note: Magaritz et al (1991) ascribe the difference in stratigraphic positions of the positive carbon isotope event near the base of the Lie de Vin Formation to a problematic offset in the earlier reference.]

good correlation of $\delta^{13}C$ profiles from widely separated localities made a diagenetic origin implausible. The postulated occurrence of continental or lacustrine sediments within the lower Lie de Vin Formation (Latham 1990) indicate that $\delta^{13}C$ signals from this part of the sequence should be interpreted with great caution, since such units often have substantial components of meteroic or lacustrine origin.

Svalbard/East Greenland

Strong regional similarities between Upper Proterozoic/lowermost Cambrian sedimentary successions in East Greenland, northeastern Spitsbergen, and Svalbard prompted comparisons of their carbon isotope profiles by Knoll et al (1986). The extremely thick (>7 km) sequences can be broadly characterized as a pair of quartz arenite-carbonate successions separated by a Vendian-aged glaciogenic interval (e.g. Swett 1981), with hiatuses bounding the latter. Detailed lithostratigraphic and micropaleontological studies identified tiepoints that permitted intersectional correlations.

Most of the $\delta^{13}C$ profile determined by Knoll et al (1986) was from units significantly older than latest Neoproterozoic. A limited number of samples from the upper portion of their $\delta^{13}C$ profile (Figure 5) indicate a rapid drop from the stable, relatively high ($\sim +5$‰) values characterizing the late Riphean, to values approaching -7‰. The lowest values occur in conjunction with major tillite units that were interpreted to be Vendian (Varanger) in age. A steady rise ensued, with two samples approaching $\delta^{13}C \sim +10$‰ immediately below a hiatus at the top of the Vendian. Cambrian values, which were obtained only from the Spitsbergen section, range between -1.8 to $+1.1$‰, and are associated with *Salterella* and late *Holmia*-age acritarchs (Knoll & Swett 1987).

Unfortunately, the major unconformities occurring within the youngest part of the sequence make it difficult to compare the Arctic profile with contemporaneous profiles from other localities. The general pattern of depleted to enriched $\delta^{13}C$ following a glacial episode, seen at the top of the sequence, is broadly similar to other localities, especially Morocco. The younger unconformity is potentially correlative to the suggested unconformity in the uppermost pre-Tommotian of the Siberian Platform. However, the discontinuous nature of the Arctic sequence makes it impossible to argue that $\delta^{13}C$ values are primary through comparisons with other profiles.

Since secondary processes are assumed to be able to shift the isotopic compositions of organic or carbonate carbon, but not simultaneously both, systematic changes in the isotopic fractionation between them can help assess the influence of diagenesis on $\delta^{13}C$ signatures. The difference, ε ($=\delta^{13}C_{carb}-\delta^{13}C_{org}$), was used as a reliability criterion, with $\varepsilon = 28.5$ taken as the midpoint of a 4 unit range. Using ε as a guide, the $\delta^{13}C_{org}$ and $\delta^{13}C_{carb}$ profiles appear to be reasonably parallel, showing similarly rapid drops and ensuing rises in $\delta^{13}C$ through the Vendian-Cambrian interval.

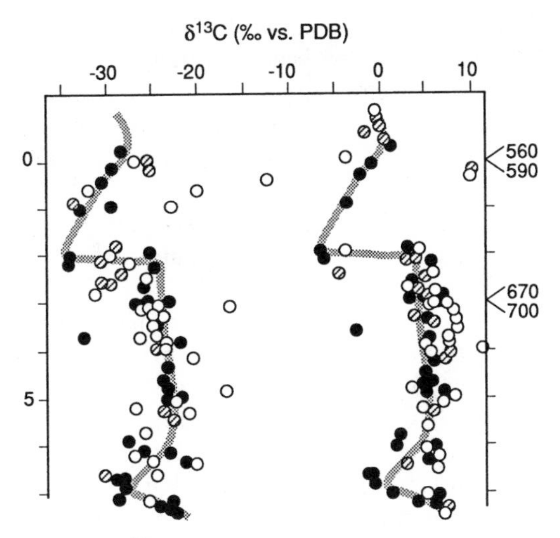

Figure 5 $\delta^{13}C_{org}$ (*left*) and $\delta^{13}C_{carb}$ isotope (*right*) profiles from East Greenland Svalbard. Dashed lines represent $\varepsilon = 28.5$. Filled circles represent carbonate-organic pairs within ± 2‰ of $\varepsilon = 28.5$‰; other pairs are represented by open and crosshatched circles—the latter used when evidence indicates that the companion point is aberrant. Tiepoints on left vertical axis are based on intrabasinal lithologic correlations; hiatuses have been identified at 0 and 3. The Vendian-Cambrian boundary is presumed to be at 0. Absolute ages on right vertical axis are estimates. From Knoll et al (1986).

This suggests that many of the samples are recording primary variations in the marine environment. Importantly, it also appears to rule out secondary mechanisms such as methane fermentation (e.g. Irwin et al 1977) for the high values of $\delta^{13}C_{carb}$ seen in much of the profile.

Lesser Himalayas

The age of the Krol-Tal succession, preserved along the southern margin of the Himalaya Mountains, has been a subject of long-standing controversy. Gansser (1974) reported the calcareous algae *Renalcis* and *Oleckmia* in the Krol B Member, which implies a Nemakit-Daldyn or younger age (Riding & Voronova 1984). Singh & Rai (1983) reported archaeocyathids from near the top of the Krol E Member, with Vendian-aged stromatolites in the underlying Krol D Member and a shelly phosphatic microfauna in the overlying lower Chert Member of the Tal Formation. Brasier & Singh (1987) disputed the occurrence of archaeocyathids in the Krol E Member and questioned the reported fauna in the Krol B Member. They suggested that the Krol D Member is latest pre-Tommotian, while the Chert Member is either earliest Tommotian or very slightly older.

The Upper Krol Formation consists of planar stromatolites with interbedded tidal flat deposits, and has been regionally dolomitized. The uppermost units of the Krol E Member grade conformably into the Tal Formation, which contains black shales, cherts, and phosphorites within its basal units. The lowest unit of the Tal is highly fossiliferous and contains the first small shelly fossils found in the succession. Units higher in the Tal grade upwards into terrigenous clastic deposits. The first polymeroid trilobites are found within the Tal Formation, approximately 250 m above the Chert Member.

Aharon et al (1987) presented a $\delta^{13}C$ profile from the Krol-Tal succession (Figure 6). The most noteworthy feature of their profile was high in the Krol D Member, where $\delta^{13}C$ values approached +6‰. Subsequent values dropped steadily to ~0‰ at the first appearance datum (FAD) of biomineralized fossils near the top of the Krol E Member. $\delta^{13}C$ values reported from the overlying Tal Member plummet to as low as −18‰, and probably reflect post-depositional modification.

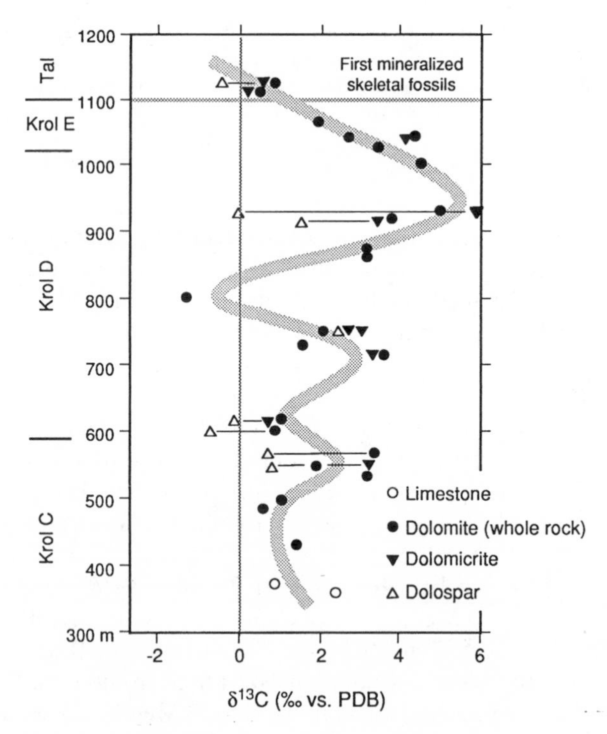

Figure 6 Carbon isotope profile from the Upper Krol and Tal Formations, Lesser Himalaya, India. From Aharon et al (1987).

Their study utilized a number of different lines of evidence to argue that $\delta^{13}C$ values from the Krol Formation and the Chert Member had not been significantly altered by diagenesis, and therefore preserved secular variations in $\delta^{13}C$. Fabric analysis indicated that dolomitization was an early event. Isotopic heterogeneities of coexisting dolomite phases argued against significant isotopic resetting after their original partitioning. $^{87}Sr/^{86}Sr$ ratios from the section were in agreement with accepted values for the composition of seawater (Veizer et al 1983). On the other hand, values from the organic-rich Tal Member probably reflected bacterially mediated pore-water chemistries during post-depositional modification (e.g. Irwin et al 1977), and as such, most likely contain little or no primary $\delta^{13}C$ signal.

Meishucun, South China

The section at Meishucun, near Kunming, Yunnan Province, China has been the object of intense study since its identification as a candidate stratotype for the Precambrian-Cambrian boundary by an International Commission on Stratigraphy working group (Cowie 1985). It was subsequently abandoned as a candidate because of questions concerning the stratigraphic and faunal continuity of the sequence near important markers suggested as boundary horizons (Cowie 1985, Brasier 1989, Brasier et al 1990, Kirschvink et al 1991). It remains, however, a key study area because it preserves one of only a very few relatively continuous faunal successions spanning the appearance of the earliest skeletal fossils to the first trilobites.

The sequence at Meishucun consists of fine-grained dolomites, phosphatic dolomites, phosphorites, and siliciclastics, and is richly fossiliferous (Luo et al 1984, Li 1986, Qian & Bengston 1989). The lower 40 meters are predominantly dolomite, with numerous intercalated phosphatic units. Sandy siltstones and argilllites dominate the remainder of the section. The succession is punctuated by a series of marker horizons (Markers A–D) denoting important biological and chemical developments in the section, and by five biozones (see Brasier et al 1990).

Early carbon isotope studies of the Meishucun section and nearby contemporaneous sections, though tantalizing, were not well-constrained by biostratigraphy (Hsü et al 1985, Lambert et al 1987). A subsequent study by Brasier et al (1990) determined a $\delta^{13}C$ profile from the Meishucun section that was characterized by generally negative values except between Marker B and Marker C, where values rose by nearly 5‰ to $\sim +2$‰ (Figure 7). $\delta^{13}C$ values of samples above Marker C demonstrated a marked discordance with the underlying profile, suggesting a major stratigraphic break at the top of the Dahai dolomites, and are relatively depleted, with $\delta^{13}C$ values ranging between -3‰ and -7‰.

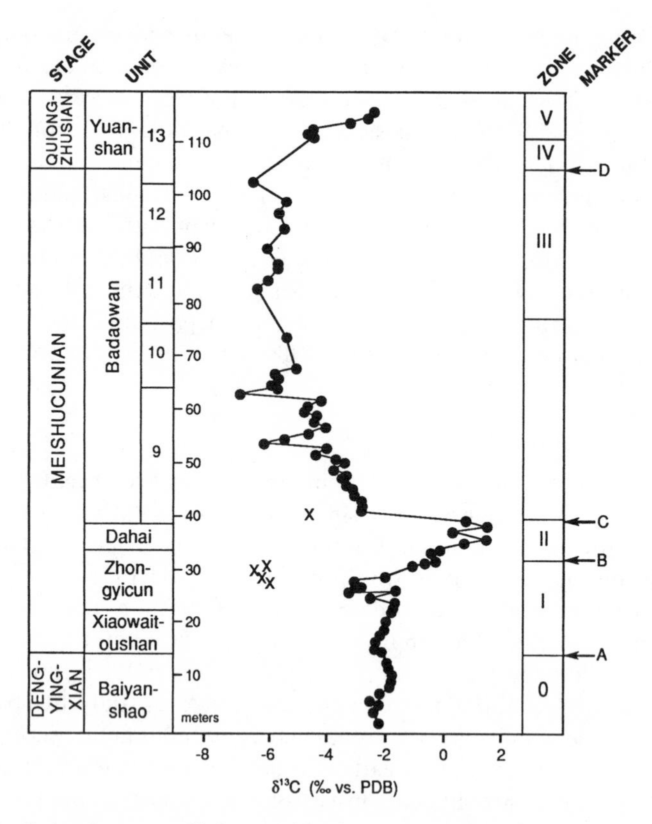

Figure 7 Carbon isotope profile from Meishuchun, southern China. The locations of marker horizons and biozones are given in the right column; Marker D corresponds to the first trilobites. Results from phosphatic samples between Marker B and Marker C are denoted by "x"s. Modified from Brasier et al (1990).

A broadly positive correlation exists between $\delta^{13}C$ and $\delta^{18}O$ at Meishucun, elevating suspicions that diagenetic components may be present. However, trends in the Meishucun $\delta^{13}C$ profile were not distinct from sections where weak negative correlations between $\delta^{18}O$ and $\delta^{13}C$ were found (Magaritz et al 1986), suggesting the Meishucun profile may at least partially represent secular variations during deposition, with some important exceptions. Phosphatic samples displayed a strong positive correlation between $\delta^{13}C$ and $\delta^{18}O$, and $\delta^{18}O$ results were $< -13‰$, while $\delta^{18}O$ values from dolomitic samples generally ranged between about $-10‰$ to $-8‰$. Brasier et al (1990) noted that $\delta^{13}C$ values from phosphate units were likely the result of diagenesis, and a subsequent paper by Kirschvink et al (1991) using this data set removed them from the profile.

Interpretation of post-Marker C (Badaowan Member) $\delta^{13}C$ values is complicated by the organic-rich nature of the strata, raising the possibility that prominent post-depositional components are present. The shift towards markedly lighter $\delta^{13}C$ values is also directly correlated with a major change from predominantly carbonate to predominantly siliclastic units at Marker C, so lithology-dependent diagenetic components may also be present. However, provocative correlations with $\delta^{13}C$ profiles from the Flinders Ranges, Australia, and Siberia raise the strong possibility that $\delta^{13}C$ variations in the Badaowan Member preserve some component of secular variation in the composition of seawater. The depleted $\delta^{13}C$ values in the Badaowan appear to correlate to the severe $\delta^{13}C$ depletion above the Krol-Tal boundary in the Lesser Himalayas. This fosters the speculation that the very low $\delta^{13}C$ values in these sections may represent a strong regional $\delta^{13}C$ signature superimposed upon global variation.

Flinders Range, South Australia

Well-exposed Neoproterozoic sequences in the Flinders Ranges of South Australia are overlain unconformably by the Lower Cambrian Hawker Group, beginning with the Parachilna Formation and followed by the Woodendina Dolomite and Wilkawillina Limestone (Dalgarno 1964). Shelf sandstones of the Parachilna Formation contain the trace fossil *Diplocraterion*, apparently indicating a Tommotian age (Riding & Voronova 1984). The overlying Woodendina Dolomite and Wilkawillina Limestones are suggested to be Upper Tommotian to Botomian (Daily 1972, Brasier 1976). The Parachilna and Woodendina have been interpreted to represent a transgressive system, with the Wilkawillina representing an open-marine highstand system (Haslett 1975). Under the Hawker Group are the Wilpena Group, which contains the famous Ediacara soft-body fauna of the Pound Quartzite, and lower down, the Umberatana Group, which contains the tillite-bearing Elatina Formation.

Tucker (1991) presented a $\delta^{13}C$ profile using overlapping sections from the Hawker Group (Figure 8). The major features of the profile are a 5‰ drop in $\delta^{13}C$ through the Parachilna Formation, followed by a steady rise in $\delta^{13}C$ values from about -5‰ to values near 0‰ throughout the Woodendina Dolomite. The rise is interrupted by four points in the upper Woodendina that define a more sharply positive rise. Four data points from the Wilkawillina Limestone have values between -1‰ and $+1$‰.

It was suggested that the $\delta^{13}C$ profile closely represented the composition of seawater because (*a*) dolomite fabrics indicated that pervasive dolomitization was of one generation and probably the result of early diagenesis; and (*b*) analysis of 4 ooids gave $\delta^{13}C$ results within 0.1‰ of the whole rock or matrix signature (Tucker 1991). The validity of the latter

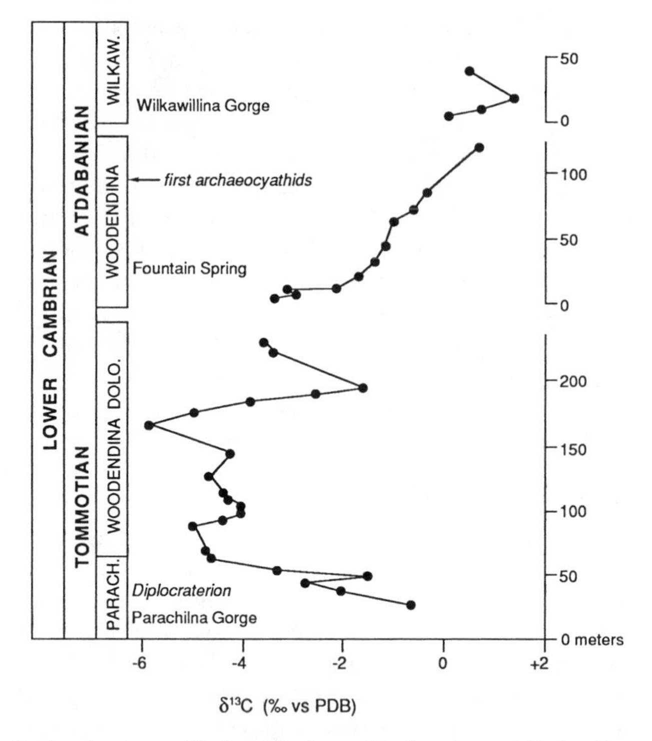

Figure 8 Carbon isotope profile from the Lower Hawker Group, Flinders Ranges, South Australia. From Tucker (1991).

argument is bolstered by the poor correlation of δ^{18}O values with δ^{13}C; δ^{13}C values of matrix and ooids can be homogenized by pervasive diagenesis, but this would be expected to have a similar (and strong) effect on δ^{18}O. The results from the Wilkawillina Limestone show sharply lowered δ^{18}O. However, dolomites commonly have higher δ^{18}O values than contemporaneous limestones, with little effect on δ^{13}C (e.g. Gao & Land 1991), so the heavier δ^{18}O values of the Woodendina Dolomite do not necessarily imply stronger diagenetic δ^{13}C components within the Wilkawillina Limestone.

Namibia

The identification of chronostratigraphic markers within the Damara Supergroup, found on both the Congo and Kalahari Cratons of northern and southern Namibia, are key to understanding the complex history of the two cratons during their initial separation and later amalgamation, and provide critical constraints on Pan-African events throughout West

Gondwana. Numerous attempts to correlate within the Damara Supergroup have been based largely on tillite units, with varying results. A primary focus has been the identification of the Kalahari Craton–correlate for the Otavi Group Chuos tillite. McMillan (1968) and Hoffman (1989) considered the Chuos to be correlative with the Kaigas tillite at the base of the Hilda Formation in the Gariep Group (Figure 9). Kröner (1971) correlated the Chuos to a putative tillite (the Vingerbreek) in the lower part of the Schwarzrand Subgroup of the Nama Group. This suggestion later found some support from paleomagnetic results (Kröner et al 1980), but the evidence is permissive. Germs (1974) considered a tillite within the Numees Formation (Gariep Group) to be equivalent to the Chuos tillite.

Paleontological evidence permits a reasonable estimate of the age of the lower Nama Group as pre-Tommotian (Germs 1972 a,b,c; Crimes & Germs 1982; Germs et al 1986; Crimes 1987). The Nomtsas Formation (uppermost Schwarzrand Subgroup) contains the trace fossil *Phycodes pedum*, indicating a Cambrian age (Germs 1972c). The Nomtsas is apparently separated from underlying rocks by a locally profound unconformity (Germs 1983). Ediacaran-grade faunas are found in the Kuibis and Schwarzrand Subgroups (e.g. Gürich 1930, Richter 1955, Glaessner 1963, Pflug 1966, Germs 1973a,b). The age of the Nomtsas and overlying Fish River Subgroup has been bracketed between 570 and 530–500 Ma based on K-Ar determinations of detrital white mica (Horstmann et al 1990).

The age of the Otavi Group is less well-constrained. Paleontological evidence from the Otavi Group is limited to stromatolites, which are

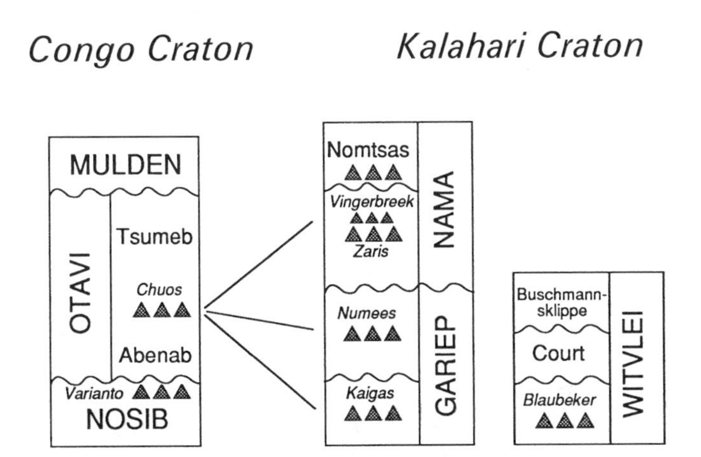

Figure 9 Regional geologic relationships within the Damara Supergroup, Namibia. Wavy lines represent unconformities; triangles represent glacial deposits.

notoriously difficult to use for dating. Existing U-Pb radiometric dates from the top of the Nosib Group immediately below the Otavi indicate a maximum age of about 750 Ma (Miller & Burger 1983) for onset of Otavi deposition. Although Smith (1961) and Jacob (1974) argued that the Nosib sequence is transitional upwards into the Otavi within the central Damara Belt, the pronounced unconformity typically separating the groups in other areas limits the efficacy of this date as a constraint for the Otavi Group on the Congo Craton.

Kaufman et al (1991) used analyses of nonluminescent microspar and dolomicrospar to outline $\delta^{13}C$ profiles from Damara Supergroup successions on the Kalahari and Congo Cratons. A number of different methods were used to evaluate the effects of diagenesis, including petrographic relationships and cathodoluminescence, comparisons of $\delta^{13}C$ and $\delta^{18}O$ compositions of subsamples of microspar and dolomicrospar to total carbonate, and determinations of $\delta^{13}C$ contents of total organic carbon. The results, coupled with the use of nonluminescent carbonate phases, suggested that the $\delta^{13}C$ signatures from most of the samples are primary. Their results from the Nama-Witvlei and Otavi Groups, summarized in Figure 10, are surprisingly similar. Both profiles are characterized by rising $\delta^{13}C$, from values of about $-6‰$ to greater than $+5‰$, and, in the case of the Otavi, greater than $+9‰$. The Nama-Witvlei Group profile begins immediately above an unconformity between the Court and Buschmannsklippe Subgroups. A trend of steadily rising values is interrupted by the Vingerbreek tillite and siliciclastic units of the Nasep Formation. Upon resumption in the middle Schwarzrand Subgroup, $\delta^{13}C$ values oscillate between $+1‰$ and $+2‰$ up through the Early Cambrian Nomtsas Formation.

Following a period of relatively high $\delta^{13}C$ during the Abenab Subgroup, the Otavi Group profile begins a 16‰ rise immediately above the Chuos Tillite at the base of the Tsumeb Subgroup. The rise ends in the lower Huttenberg Formation, and is followed by a marked 9‰ decline to values near 0‰ at the top of the unit, where the stratigraphic succession is punctuated by a major unconformity.

Previous attempts to correlate glaciogenic units within the Damara Supergroup can be evaluated using the data from Kaufman et al (1991). The suggested correlation of the Numees and Chuos tillites (Germs 1974) is highly favorable if the unconformity between the Buschmannsklippe and Court Subgroups represents the nearshore equivalent of the lower Numees Formation of the Gariep Group, as has been suggested by Hoffman (1989). On the other hand, the match between the Vingerbreek and Chuos tillites proposed by Kröner (1971) is highly unlikely, since the pattern of $\delta^{13}C$ variations above each of these levels is entirely different.

Proposed correlations between the Kaigas and Chuos tillites are impossible to test rigorously because of a paucity of data from the Gariep Group, and because of insecure lithostratigraphic correlations between the Gariep and Witvlei Groups. Kaufman et al (1991) tentatively equated them using the argument that a highly positive $\delta^{13}C_{carb}$ value from a single sample with an anomalous $\delta^{13}C_{org}$ signature in the the Hilda Formation was the only known Kalahari Craton-equivalent to the very high $\delta^{13}C$ values found in the Huttenberg Formation, and that highly negative values found in the overlying Ovamboland Formation of the Mulden Group (Congo Craton) were therefore equivalent to the low $\delta^{13}C$ samples of the lower Nama Group. This suggestion requires that at least two extended periods of strong $\delta^{13}C$ enrichment occurred during the late Proterozoic. It is worth noting, however, that $\delta^{13}C$ values obtained from near the base of the Otavi Group were also highly enriched, and that if these correlate to the Hilda Formation, then the correlation of the Kaigas and Chuos tillites based on $\delta^{13}C$ loses its support, as does the requirement for two separate $\delta^{13}C$ depletion events. The addition of radiometric constraints to the Otavi and Nama Group profiles will undoubtedly do much to clarify their relationships to each other and to other Neoproterozoic sequences.

Southeastern Newfoundland/Avalonia

The faunal and lithological successions on the Burin Peninsula, southeastern Newfoundland, have been extensively described as part of the effort to identify a Precambrian-Cambrian boundary stratotype (Cowie 1985, Crimes & Anderson 1985, Narbonne et al 1987, Landing et al 1989, Myrow 1987, Myrow et al 1988). Unlike most of the other classic areas, the boundary interval here lies within predominantly siliciclastic sediments. The recent establishment of the boundary at the base of the *Phycodes pedum* ichnofossil zone, within member 2 of the Chapel Island Formation, dashed hopes that $\delta^{13}C_{carb}$ correlations of the boundary from its stratotype locality could be made, since carbonate facies are completely absent from the Burin Peninsula near the boundary horizon. Fortunately, overlying Cambrian strata contain material suitable for sampling, and a relatively detailed $\delta^{13}C$ curve has been assembled from a number of sections within the Avalon Zone (Brasier et al 1992) (Figure 11).

The upper members of the Chapel Island Formation show a general pattern towards heavier $\delta^{13}C$ in younger units, although variability is quite high. A barren interval corresponding to the Random Formation separates this trend from the rest of the profile. A single sample from near the base of the Petley Formation suggests a second period of light values, with $\delta^{13}C < -5‰$, but the high potential for diagenetic overprinting hampers its usefulness for stratigraphic purposes. Variability is markedly reduced

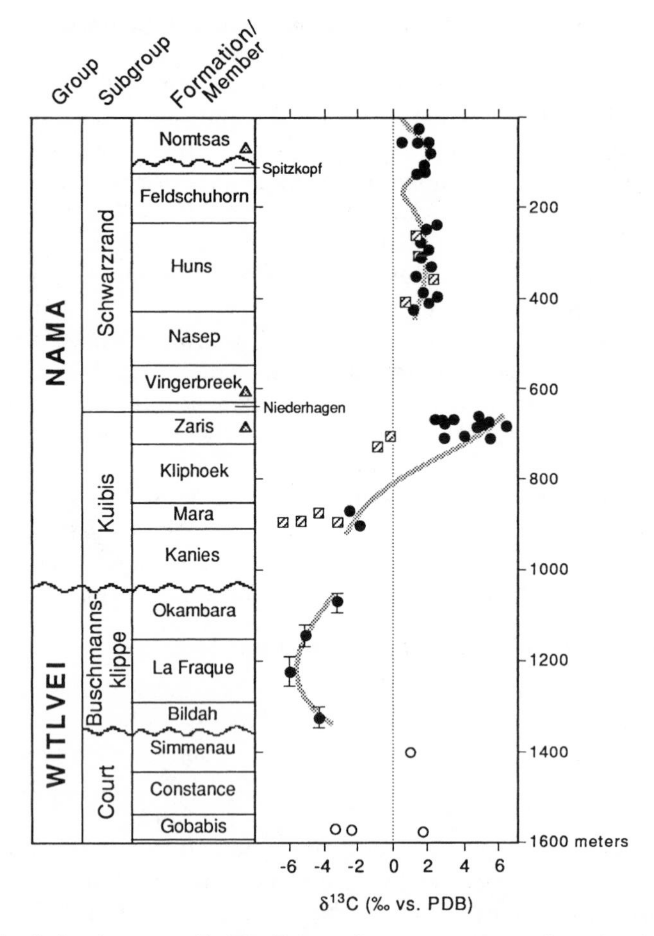

Figure 10 Carbon isotope profile from Damara Supergroup strata on the Kalahari (*left*) and Congo (*right*) Cratons. Filled and crosshatched circles represent nonluminescent microspar and dolomicrospar, respectively. Filled symbols represent samples from marine environments; open symbols are apparently lacustrine. Cross-hatched squares represent samples from restricted environments. Modified from Kaufman et al (1991).

in portions of the profile from the overlying Bonavista Group, and a clear negative cycle is apparent. The profile trends markedly positive in the upper Bonavista, and then reverses itself again to lighter values in the Brigus Formation.

Unfortunately, the record assembled from the Burin Peninsula is plagued by a number of potential problems. Extremely low $\delta^{18}O$ values

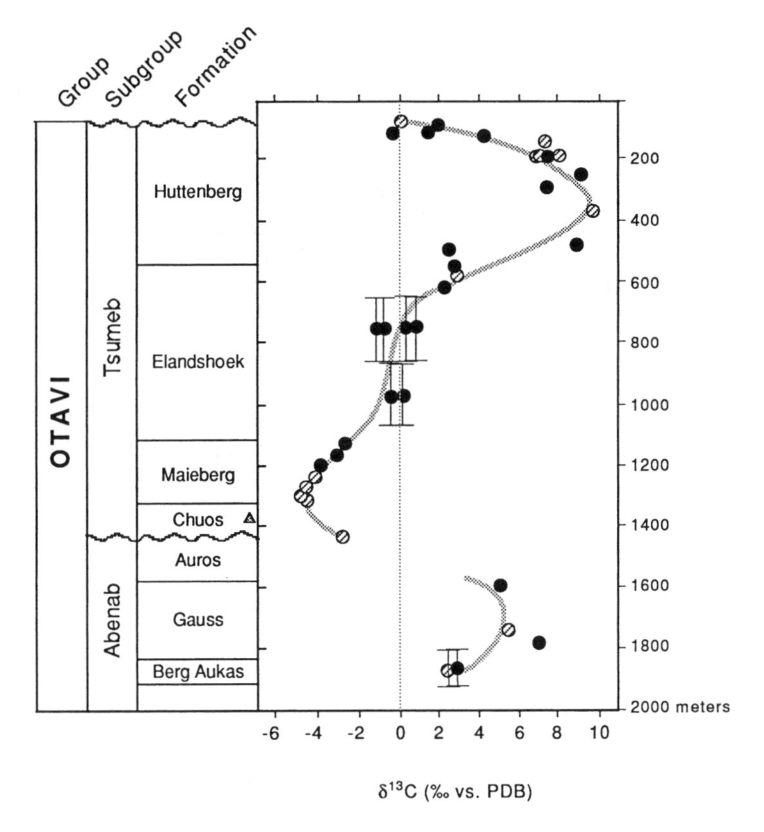

Figure 10—(continued)

are ubiquitous and probably reflect the relatively deep burial and higher temperatures experienced by the region. Ferroan concretions (indicative of diagenesis) are abundant, and often have light $\delta^{13}C$ values compared to those from bedded calcites. The possible introduction of carbon from neighboring siliciclastics during early diagenesis cannot be discounted, and may be partially responsible for the low $\delta^{13}C$ values seen in the Chapel Island Formation. Although great care was taken in the selection of sampling material, and simultaneous measurement of fossils and matrix material were performed as a check, there is little direct evidence that the measured $\delta^{13}C$ values are primary (Brasier et al 1992).

Despite these problems, the $\delta^{13}C$ curve from the Burin Peninsula compares very favorably with the record from the Siberian Platform, especially in the Tommotian-aged Bonavista Group. Their similarity is most readily

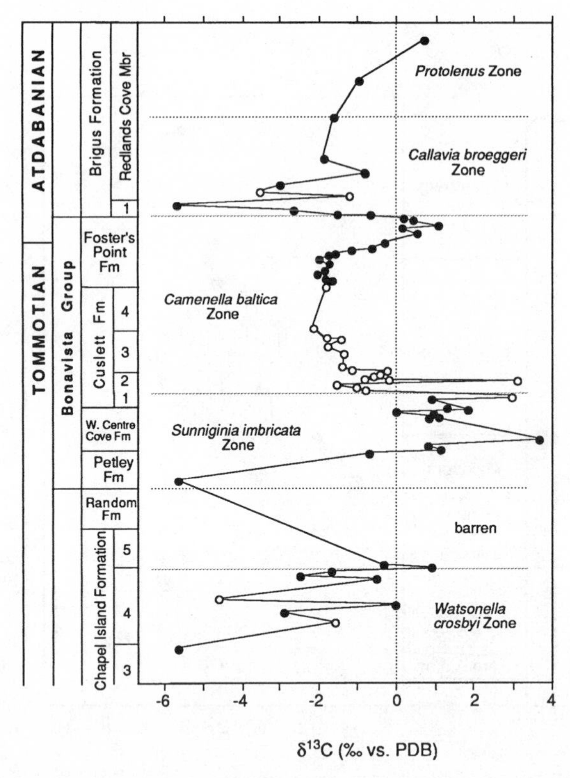

Figure 11 Reconstructed carbon isotope profile from Avalonia (southeastern Newfoundland and England). Filled circles represent samples from non-ferroan bedded micrites; open symbols represent non-ferroan micritic nodules. Profile begins immediately above the first observed *Phycodes pedum* marking the Proterozoic-Phanerozoic boundary. From Brasier (1992).

explained by assuming that both sections preserve a record of secular variation in δ^{13}C during deposition. The addition of $\delta^{13}C_{org}$ data from the Burin Peninsula should greatly contribute to understanding the nature of δ^{13}C variations within this critical area.

CORRELATIONS

Similar variations in δ^{13}C occur repeatedly throughout the geologic record, and on this basis alone, one could with little effort find Mesozoic and Cenozoic records that correlate with high confidence into Vendian, pre-Tommotian, and Tommotian strata. This fundamental danger is particularly conspicuous when attempting to compare carbon isotope profiles from localities where supporting fossil data are conflicting or sparse, as is the case for many Upper Neoproterozoic and Lower Cambrian sections. Although the problem is partially relieved by the lithostratigraphic correlations available when working within single marine basins (e.g. Knoll et al 1986), intercontinental correlations using $\delta^{13}C$ profiles currently must be packaged with a healthy measure of skepticism.

Nonetheless, it can be informative to advance correlations based heavily on chemostratigraphic constraints. A suggested $\delta^{13}C$ reference curve and correlation of the reviewed sections is given in Figure 12. It has already

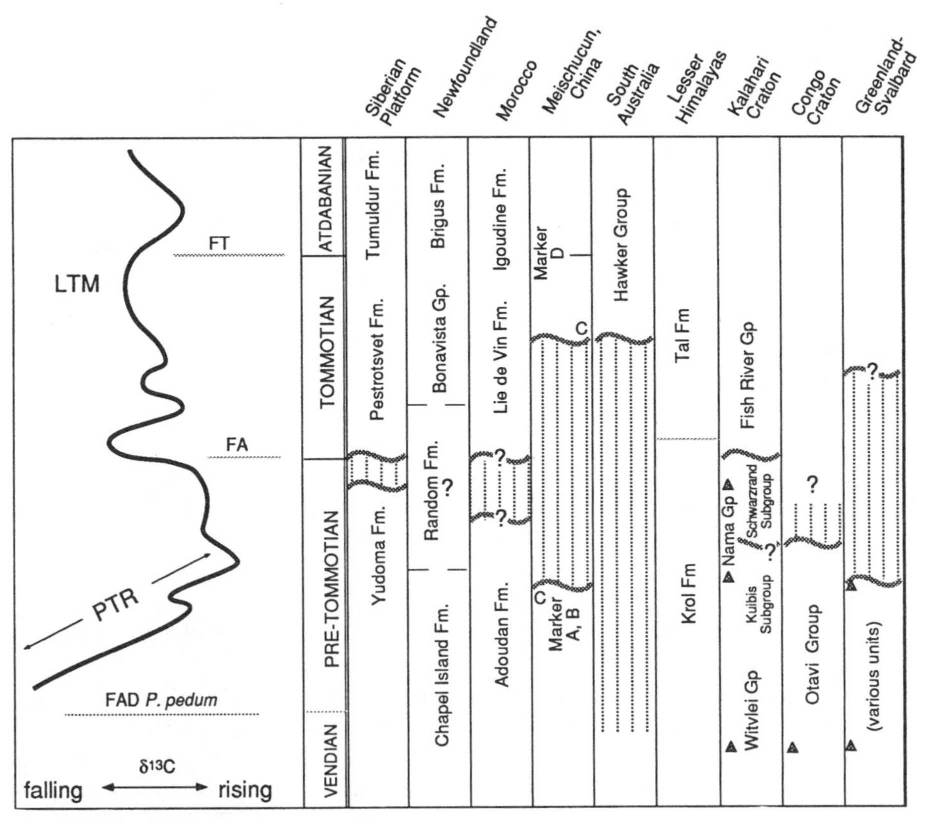

Figure 12 $\delta^{13}C$ reference curve and interpretive correlations through the Precambrian-Cambrian boundary interval. PTR = Pre-Tommotian $\delta^{13}C$ Rise; LTM = Late Tommotian $\delta^{13}C$ Minimum. Wavy lines indicate unconformities; triangles indicate glacial units. FA = first archaeocyathids; FT = first trilobites. Level of speculation approximately increasing from left to right.

been noted that the $\delta^{13}C$ profiles from the Siberian Platform and southeastern Newfoundland have strong similarities that are correlatable within the good biostratigraphic controls available for the two sections. Their $\delta^{13}C$ profiles together serve as a de facto reference curve. The principle features of the two profiles are a steady rise in $\delta^{13}C$ from relatively depleted values of ~ -6‰ to positive values in excess of $+3$‰, culminating just below the base of the Tommotian Stage in Siberia; and a broad negative cycle of lighter $\delta^{13}C$ values immediately below the first trilobites marking the base of the Atdabanian. Because the older feature has been best described from the Yudoma Formation on the Siberian Platform, where it occurs immediately below the base of the Tommotian, it is herein termed the "Pre-Tommotian $\delta^{13}C$ Rise," or PTR. The younger feature is herein termed the "Late Tommotian $\delta^{13}C$ Minimum," or LTM, because of its culmination (on the Siberian Platform) immediately below the Tommotian-Atdabanian boundary. Based on the profile from Newfoundland, the first appearance datum of *Phycodes pedum* either predates or is very early in the development of the PTR, indicating that the PTR must be almost entirely Cambrian. The end of the PTR occurs immediately before the first archaeocyathids in the Siberian Platform succession, so that the PTR is exclusively pre-Tommotian in this area. Its ending in the Newfoundland succession is apparently obscured by the barren strata of the Random Formation, suggesting the base of the Tommotian Stage here is probably very near the base of the Bonavista Group.

The record of the PTR in Morocco appears to be nearly complete, spanning the entire Adoudan Formation, and the straightforward correlation proposed by Magaritz et al (1991) is retained. The correlation of the LTM between the upper Lie de Vin Formation and the Siberian Platform offered by Kirschvink et al (1991) is plausible and fits the available biostratigraphic data (see Latham & Riding 1990), although it is imprecise because of low sample density in the Moroccan section. These correlations indicate that the lower 700 meters of the Lie de Vin are Tommotian, which is compatible with recent suggestions based on biostratigraphic evidence (Latham & Riding 1990).

Kirschvink et al (1991) correlated the LTM to the base of the Meishucun section, basing their suggestion heavily on the putative presence of Atdabanian faunal elements near Marker B (Cowie 1985) and on magnetic polarity correlation. Although intriguing, this alternative contradicts a substantial biostratigraphic data base, including the first and last appearances of key Nemakit-Daldynian taxa below Marker B (Brasier et al 1990). They further suggested that Marker C correlates above mid-Atdabanian strata because counterparts to the light values of the "Badaowan $\delta^{13}C$ Minimum" at Meishucun have yet to be found in post-Tommotian Cambrian strata on the Siberian Platform. This ignores the possibility that post-Marker C $\delta^{13}C$ values at Meishucun may reflect significant local depletion of seawater $\delta^{13}C$, or may contain strong diagenetic components. Brasier et al (1990) correlated Badaowan strata with the $\delta^{13}C$ minimum in the *A. sunniginicus* zone immediately overlying the PTR in Siberia. Although supported by the available biostratigraphic data, this correlation suffers because it fails to explain the absence of correlatives to the positive $\delta^{13}C$ values found between the PTR and LTM in Siberia at Dvortsy and Isit.

The correlations for Meishucun preferred here equate the LTM to the low $\delta^{13}C$ values within the Badaowan Member. Rising values within the overlying Yuanshan Member suggest a further correlation to the basal Atdbanian $\delta^{13}C$ profile in Siberia. One of the more appealing aspects of this alternative is that it indicates the first Chinese trilobites found at Marker D are temporally equivalent with the first profallotaspid trilobites in Siberia, and therefore, that the base of the Qiongzhusian Stage is equivalent to the base of the Atdabanian. It also implies that much of the Tommotian profile is absent from the Meishucun section—a hypothesis compatible with observations that an unconformity exists at Marker C (Brasier et al 1990)—and that the unconformity is temporally equivalent to a pre-Tommotian/Tommotian unconformity on the Siberian Platform.

Brasier et al (1990) equated the base of the Parachilna Formation, Flinders Ranges, Australia, with the end of the PTR, and the lowest $\delta^{13}C$ values in the overlying Woodendina Dolomite with the LTM. Although their correlation of the LTM is retained, correlation of the PTR was made without any supporting evidence of a $\delta^{13}C$ rise in underlying strata and may be untenable. The presence of *Diplocraterion* in the Parachilna suggests a Tommotian or younger age, since *Diplocraterion* has not yet been found in pre-Tommotian or older strata (Riding & Voronova 1984, Latham & Riding 1990). A more parsimonious explanation is that deposition of the Hawker Group initiated during the Middle Tommotian, after the interval of rapid variations in $\delta^{13}C$ preserved on the Siberian Platform. This suggestion is supported by the recognition that depositional system tracts in the Hawker Group are transgressive; most localities record regressive cycles and shoaling lower in the Tommotian and in the pre-Tommotian, with a shift to transgressive units higher in their sequences. This also implies that the onset of Tommotian deposition in the Flinders Range and at Meishucun may be temporally equivalent (see Figure 12). In what may be merely a striking coincidence, correlation of the LTM between South Australia and Meishucun suggests local values of $\delta^{13}C$ were virtually identical.

Correlation of the PTR to the $\delta^{13}C$ maximum in the Krol D Member

(Aharon al 1987) relies heavily on the proximity of the first appearance of shelly faunas to an interval of declining $\delta^{13}C$ values at the top of the Krol E member. The virtual absence of negative values below the $\delta^{13}C$ maximum in the Krol D Member strongly suggests that the sequence here contains an abbreviated record of the PTR. The nature of that abbreviation is unknown; sampling intervals were too large to assess the stratigraphic completelness of the Krol. The absence of an unconformity would suggest that the lower and middle PTR must be below the Krol C Member, which would lend some credence to the suggestion that the Krol B Member contains a Nemakit-Daldyn or younger fauna (Gansser 1974). However, the $\delta^{13}C$ and biostratigraphic results from this area sponsor many alternative hypotheses, and conclusions based on them must be considered highly speculative.

Kaufman et al (1991) suggested the steady rise in $\delta^{13}C$ seen on the Kalahari Craton is a nearly complete record of the PTR, and this has been retained. More detailed comparisons with the Siberian Platform sequence are complicated by the lack of an obvious Siberian correlate to the extended period of stable $\delta^{13}C$ values in the upper Nama Group, although a short interval immediately below the stratigraphic break at Ulakhan-Sulugur is a plausible candidate. The correlation equates the occurrence of *Phycodes pedum* in the Nomtsas Formation to the base of the Tommotian in Siberia and the Ediacaran-grade faunas of the Nama Group with at least part of the Nemakit-Daldyn fauna, which (if correct) would imply that the chronostratigraphic utility of these biostratigraphic elements may be restricted. It is noteworthy that recent reports from the Siberian Platform indicate the presence of Ediacaran faunas above the first skeletal fossils (Dorzhnamzhaa & Gibsher 1990).

The proposed correlations of the PTR into the Otavi Group and East Greenland-Svalbard $\delta^{13}C$ profiles are the most speculative. Correlation to the Otavi profile above the Chuos tillite is permitted by allowing the Chuos to be correlative to the Numees tillite, as suggested by Germs (1974). This equates an erosional surface at the top of the Otavi Group with the Vingerbreek Tillite of the Nama Group (see Figure 9). Kaufman et al (1991) and Knoll & Walter (1992) favored an older age for the Otavi and postulated the existence of at least two periods of $\delta^{13}C$ depletion at the end of the Neoproterozoic. However, their interpretations are heavily based on lithostratigraphic correlations within the Damara Supergroup and on stromatolite occurrences within the Otavi, and both lines of evidence are open to criticism. The $\delta^{13}C$ profile from East Greenland-Svalbard has the stratigraphic range necessary to test their interpretation, and it appears to have only one plausible PTR-equivalent, but major unconformities within the Vendian to Cambrian portion of the East Greenland-Svalbard profile severely limit the utility of this profile as a constraint. At the moment, the stratigraphic uniqueness of the PTR is an open question, and therefore, correlation of the Otavi Group to Neoproterozoic sequences in other areas remains largely enigmatic.

A MASS EXTINCTION AT THE BEGINNING OF THE TOMMOTIAN?

A strong relationship between carbon isotope shifts, evolutionary events, and sea level has been noted from other major boundary intervals, including the Paleozoic-Mesozoic and Mesozoic-Cenozoic (e.g. Raup 1986; Magaritz 1989, 1991; Holser & Magaritz 1987), the Cambrian-Ordovician (Ripperdan et al 1992), and the Ordovician-Silurian (Orth et al 1986). Although no direct link has been conclusively demonstrated between these phenomena, their common association at important global event horizons raises suspicions that durable mechanistic links may exist between them.

The interpretive $\delta^{13}C$ reference curve and correlations shown in Figure 12 indicate that the unconformity found at the base of the Tommotian in Siberia is correlative to unconformities in China, Australia, and possibly Namibia; to the occurrence of continental deposits in the lower Lie de Vin Formation, Morocco; and to the switch to siliciclastic deposition in China and India. This supports the notion that global sea level was lowered during the PTR, and that a switch to rising sea level began sometime during the Tommotian (e.g. Brasier 1992). If the proposed correlations are correct, evidence from the Nama Group (the Vingerbreek tillite) suggests that a glacial episode occurred at the end of the PTR.

Seilacher (1984) argued that Vendian-Ediacaran faunas were lost abruptly in a major extinction event at the end of the Vendian (pre-Tommotian of the Siberian Platform), rather than earlier during the Neoproterozoic. If this hypothesis is accepted, then the combined changes in sea level (possibly glacially-driven) and $\delta^{13}C$, and the loss of a major fauna at the base of the Tommotian, are identical to the sequence of events seen across the Permian-Triassic boundary (Figure 13). The similarity of the relationship between falling $\delta^{13}C$ and the appearance of replacement faunas is particularly compelling. The nearly simultaneous disappearance of pre-Tommotian dolomites in Siberia, Morocco, India, the Otavi Group(?), and China indicated by the proposed correlations allows speculation that a major change in marine chemistries accompanied the termination of the PTR. This may have played a role both in the loss of the existing faunas, and in the radiation of the new, biomineralized faunas marking the beginning of the Tommotian Stage.

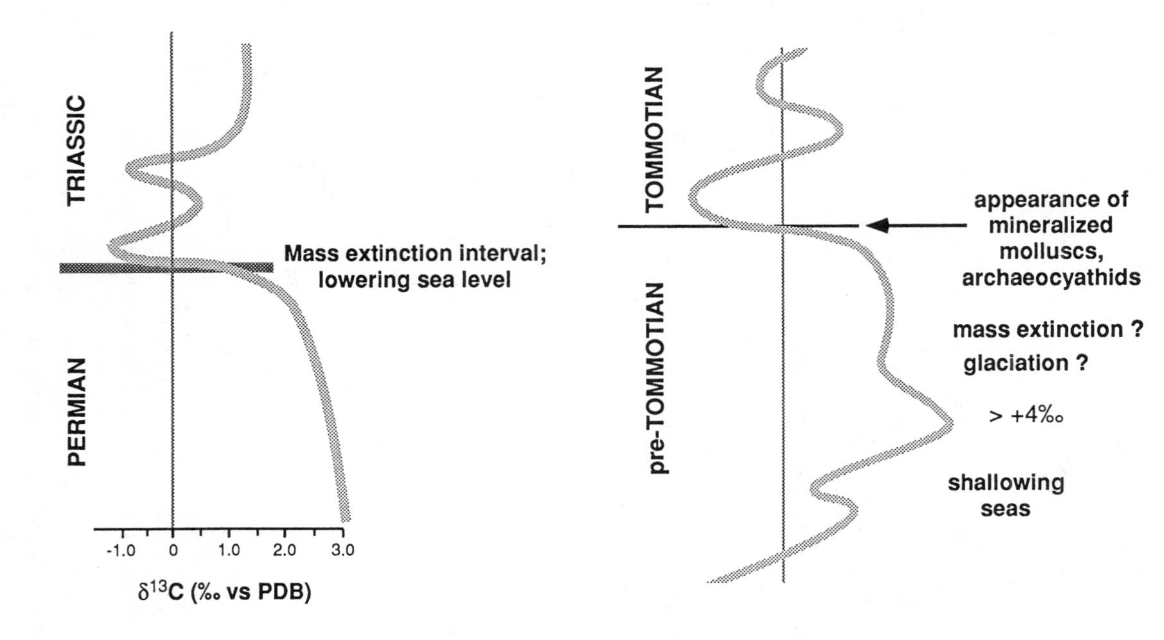

Figure 13 Comparison of events during the Permian-Triassic (*left*) and Precambrian-Cambrian (*right*) boundary intervals. Permian-Triassic data from Holser & Magaritz (1992).

CONCLUSIONS

Carbon isotope-based correlations between a number of sections spanning the Precambrian-Cambrian boundary interval indicate that an extended period of rising $\delta^{13}C$ values occurred between the base of the Cambrian and the base of the Tommotian Stage, followed by predominantly low $\delta^{13}C$ values immediately below the base of the Atdabanian Stage. Hiatuses near the base of the Tommotian are suggested by many of the profiles, and probably reflect a lowering of global sea level. A putative glacial tillite in the lower Schwarzrand Subgroup, northern Namibia, may be correlative with this event. The first biomineralized archaeocyathids and molluscans appear immediately after the end of the pre-Tommotian $\delta^{13}C$ Rise, perhaps in response to changing marine chemistries and the loss of existing faunas. The widespread and coincident rapid changes in $\delta^{13}C$ values, sea level, evolutionary innovation, and lithology at the base of the Tommotian facilitate speculation that a global event similar to that at the Permian-Triassic boundary may have occurred.

If correlations between the Kalahari Craton, Namibia, and the Siberian Platform are correct, the reported first appearance in Namibia of *Phycodes pedum*, the marker taxon for the Precambrian-Cambrian, is demonstrably younger than in the stratotype locality in Newfoundland. This would place a limitation on the utility of *Phycodes pedum* as a chronostratigraphic tiepoint. The correlation also suggests that Ediacaran-grade faunas in Namibia coexisted with the Nemakit-Daldyn fauna in Siberia. Correlations in younger units suggest that the first appearance of mineralized arthropods in Siberia and China may be nearly isochronous, thus equating the bases of the Tommotian and Quiongzhusian stages. Older units allow speculative correlations between possibly Varangian tillite units and unconformities at the base of sections in Siberia and Morocco. The evidence, however, is permissive, and is based on the assumption that the $\delta^{13}C$ rise during the pre-Tommotian is unique within the Neoproterozoic. This view could change dramatically with additional carbon isotope and radiometric data from Neoproterozoic sequences.

ACKNOWLEDGMENTS

This paper is dedicated to the memories of Professors Mordeckai Magaritz and Heinz Lowenstam. The author wishes to thank F. Corsetti, J.L. Kirschvink, and P. Myrow for their constructive comments.

Literature Cited

Aharon P, Schidlowski M, Singh IB. 1987. Chronostratigraphic carbon isotope record of the Lesser Himalaya. *Nature* 327: 699–702

Al-Aasm IS, Veizer J. 1986. Diagenetic stabilization of aragonite and low-Mg calcite, I. Trace elements in rudists. *J. Sediment. Petrol.* 56: 138–52

Anderson TF, Arthur MA. 1983. Stable isotopes of oxygen and carbon and their application to sedimentological and paleoenvironmental problems. In *SEPM Short Course: Stable Isotopes in Sedimentary Petrology*, ed. MA Arthur, TF Anderson, IR Kaplan, J Veizer, LS Land, 10: 1–151

Banner JL, Hanson GN. 1990. Calculation of simultaneous isotopic and trace element variations during water-rock interaction with applications to carbonate diagenesis. *Geochim. Cosmochim. Acta* 54: 3123–37

Barnes CR, Petryk AA, Bolton TE. 1981. Anticosti Island, Quebec. In *Volume I: Guidebook*, ed. PJ Lesperance, pp. 1–25. IUGS Subcomm. Silurian Stratigr., Ord.-Silurian Boundary Work. Group Field Meet., Anticosti-Gaspé, Quebec, 1981

Bertrand-Sarfati J. 1981. Probleme de la limite Precambrien-Cambrien: la section de Tiout (Maroc); les stromatolites et leur biostratigraphie (Schmidt 1979): critiques et observations. *Newslett. Stratigr.* 10: 20–26

Brasier MD. 1976. Early Cambrian intergrowths of archaeocyathids, *Renalcis* and pseudostromatolites from South Australia. *Palaeontology* 19: 223–45

Brasier MD. 1982. Sea-level changes, facies changes and the Late Precambrian-Early Cambrian evolutionary explosion. *Precambrian Res.* 17: 105–23

Brasier MD. 1985. Evolutionary and geological events across the Precambrian-Cambrian boundary. *Geol. Today* 1: 141–46

Brasier MD. 1989. On mass extinctions and faunal turnover near the end of the Precambrian. In *Mass Extinctions: Processes and Evidence*, ed. SK Donovan, pp. 73–88. New York: Columbia Univ. Press

Brasier MD. 1990. Nutrients in the early Cambrian. *Nature* 347: 521–2

Brasier MD. 1992. Global ocean-atmosphere change across the Precambrian-Cambrian transition. *Geol. Mag.* 129: 161–68

Brasier MD, Anderson MM, Corfield RM. 1992. Oxygen and carbon isotope stratigraphy of early Cambrian carbonates in southeastern Newfoundland and England. *Geol. Mag.* 129: 265–79

Brasier MD, Magaritz M, Corfield R, Luo H-L, Wu X, et al. 1990. The carbon- and oxygen-isotope record of the Precambrian-Cambrian boundary interval in China and Iran and their correlation. *Geol. Mag.* 127: 319–32

Brasier MD, Singh P. 1987. Microfossils and Precambrian-Cambrian boundary stratigraphy at Maldeota, Lesser Himalaya. *Geol. Mag.* 124: 323–45

Compston W, Williams I, Kirschvink JL, Zhang Z-C, Ma G-G. 1992. Zircon U-Pb ages for the Early Cambrian time scale. *J. Geol. Soc. London* 149: 171–84

Cowie JC. 1985. Continuing work on the Precambrian-Cambrian boundary. *Episodes* 8: 93–97

Crimes TP. 1987. Trace fossils and correlation of late Precambrian and early Cambrian strata. *Geol. Mag.* 124: 97–119

Crimes TP, Anderson MM. 1985. Trace fossils from late Precambrian-Early Cambrian strata of southeastern Newfoundland (Canada): temporal and environmental implications. *J. Paleontol.* 59: 548–60

Crimes TP, Germs GJB. 1982. Trace fossils from the Nama Group (Precambrian-Cambrian) of South West Africa/Namibia. *J. Paleontol.* 65: 890–907

Daily B. 1972. The base of the Cambrian and the first Cambrian faunas. *Centre Prec. Res., Univ. Adelaide Spec. Pap.* 1: 13–41

Dalgarno CR. 1964. Lower Cambrian stratigraphy of the Flinders Ranges. *Trans. R. Soc. S. Austr.* 88: 129–44

Debrenne F, Debrenne M. 1978. Archaeocyathid fauna of the lowest fossiliferous levels of Tiout (Lower Cambrian, Southern Morocco). *Geol. Mag.* 115: 101–19

Derby LA, Kaufman AJ, Jacobsen SB. 1992. Sedimentary cycling and environmental change in the Late Proterozoic: evidence from stable and radiogenic isotopes. *Geochim. Cosmochim. Acta* 56: 1317–29

Dorzhnamzhaa D, Gibsher AS. 1990. New data on stratigraphic allocation of vendediakarian fauna of Mongolia. *Third Int. Symp. Cambrian Sys., Novosibirsk*, pp. 28 (Abstr.)

Epstein AG, Epstein JB, Harris LD. 1977. Conodont color alteration- an index of organic metamorphism. *USGS Prof. Pap.* 995: 1–27

Fairchild IJ, Spiro B. 1987. Petrological and isotopic implications of some contrasting late Precambrian carbonates, N.E. Spitsbergen. *Sedimentology* 34: 973–89

Gansser A. 1974. The Himalayan Tethys. *Riv. Ital. Paleontol. Stratigr. Mem.* XIV: 393–414

Gao G, Land LS. 1991. Geochemistry of Cambrian-Ordovician Arbuckle Limestone, Oklahoma: implications for diagenetic $\delta^{18}O$ alteration and secular $\delta^{13}C$ and $^{87}Sr/^{86}Sr$ variation. *Geochim. Cosmochim. Acta* 55: 2911–20

Germs GJB. 1972a. The stratigraphy and paleontology of the lower Nama Group, South West Africa. *Bull. Precambrian Res. Unit, Univ. Cape Town* 12. 250 pp.

Germs GJB. 1972b. New shelly fossils from the Nama Group, South West Africa. *Am. J. Sci.* 272: 752–61

Germs GJB. 1972c. New trace fossils from the Nama Group, South West Africa. *J. Paleontol.* 46: 864–76

Germs GJB. 1973a. A reinterpretation of *Rangea schneiderhoehni* and the discovery of a related new fossil from the Nama Group, South West Africa. *Lethaia* 6: 1–10

Germs GJB. 1973b. Possible sprigginid worms and a new trace fossil from the Nama Group, South West Africa. *Geology* 1: 69–70

Germs GJB. 1974. The Nama Group in South West Africa and its relationship to the Pan-African geosyncline. *J. Geol.* 82: 301–17

Germs GJB. 1983. Implications of a sedimentary facies and depositional environmental analysis of the Nama Group in South West Africa/Namibia. *Geol. Soc. S. Afr. Spec. Publ.* 11: 89–114

Germs GJB, Knoll AH, Vidal G. 1986. Latest Proterozoic microfossils from the Nama Group, Namibia (South West Africa). *Precambrian Res.* 32: 45–62

Glaessner MF. 1963. Zur Kenntris der Nama-fossilien Südwest-Afrikas. *Ann. Natushist. Mus. Wein* 66: 113–20

Grant SWF. 1992. Carbon isotopic vital effects and organic diagenesis, Lower Cambrian Forteau Formation, northwest Newfoundland: implications for $\delta^{13}C$ chemostratigraphy. *Geology* 20: 243–46

Gürich G. 1930. Die bislang ältesten spuren von organismen in Süd-Afrika. *C. R.* 15*th Int. Geol. Cong., S. Afr.* 1929: 670–80

Hallam A. 1984. Pre-Quaternary sea-level changes. *Annu. Rev. Earth Planet. Sci.* 12: 205–43

Haslett PG. 1975. The Woodendina Dolomite and Wirrapowie Limestone—two new Lower Cambrian formations, Flinders Ranges, South Australia. *Trans. R. Soc. S. Austr.* 99: 21–219

Hoffman KH. 1989. New aspects of lithostratigraphic subdivision and correlation of late Proterozoic to early Cambrian rocks of the southern Damara Belt and their correlation with the central and northern Damara Belt and the Gariep Belt. *Commun. Geol. Serv. Namibia* 5: 59–67

Holser WT, Magaritz M. 1987. Events near the Permian-Triassic boundary. *Mod. Geol.* 11: 155–180

Holser WT, Magaritz M. 1992. Cretaceous/Tertiary and Permian/Triassic boundary events compared. *Geochim. Cosmochim. Acta* 56: 3297–309

Horstmann UE, Ahrendt H, Clauer N, Porada H. 1990. The metamorphic history of the Damara Orogen based on K/Ar data of detrital white micas from the Nama Group, Namibia. *Precambrian Res.* 48: 41–61

Houzay JP. 1979. Imprints related to jellyfishes in the series at the base of the Adoudounian; uppermost Precambrian of the Anti-Atlas, Morocco. *Geol. Mediterr.* 6: 379–84

Hsü KJ, Oberhansli H, Gao J-Y, Sun S, Chen H-H, Krahenbuhl U. 1985. "Strangelove ocean" before the Cambrian explosion. *Nature* 316: 809–11

Irwin H, Curtis C, Coleman N. 1977. Isotopic evidence for source of diagenetic carbonates formed during burial of organic-rich sediments. *Nature* 269: 209–13

Jacob RE. 1974. Geology and metamorphic petrology of the southern margin of the Damara Orogen around Omitara, southwest Africa. *Bull. Precambrian Res. Unit, Univ. Cape Town* 17. 185 pp.

Kaufman AJ, Hayes JM, Knoll AH, Germs GJB. 1991. Isotopic compositions of carbonates and organic carbon from upper Proterozoic successions in Namibia: stratigraphic variation and the effects of diagenesis and metamorphism. *Precambrian Res.* 49: 301–27

Keith ML, Weber, J.N. 1964. Carbon and oxygen isotopic compositions of selected limestones and fossils. *Geochim. Cosmochim. Acta* 28: 1787–816

Khomentovsky VV, Karlova GA. 1993. Biostratigraphy of the Vendian-Cambrian beds and the lower Cambrian boundary in Siberia. *Geol. Mag.* 130: 29–45

Kirschvink JL, Magaritz M, Ripperdan RL, Zhuravlev AYu, Rozanov AYu. 1991. The Precambrian/Cambrian boundary: magnetostratigraphy and carbon isotopes resolve correlation problems between Siberia, Morocco, and South China. *GSA Today* 1: 69–71,87,91

Knoll AH, Hayes JM, Kaufman AJ, Swett K, Lambert IB. 1986. Secular variations in carbon isotope ratios from Upper Proterozoic successions of Svalbard and East Greenland. *Nature* 321: 832–38

Knoll AH, Swett K. 1987. Micropaleontology across the Precambrian-Cambrian boundary in Spitsbergen. *J. Paleontol.* 61: 898–926

Knoll AH, Walter MR. 1992. Latest Proterozoic stratigraphy and Earth history. *Nature* 356: 673–77

Kröner A. 1971. Late Precambrian correlation and the relationship between the Damara and Nama Systems of South West Africa. *Geol. Rundsch.* 67: 688–705

Kröner A, McWilliams MO, Germs GJB, Reid AB, Schalk KEL. 1980. Paleomagnetism of late Precambrian to early Paleozoic mixtite-bearing formations in Namibia (South West Africa): the Nama Group and Blaubeker Formation. *Am. J. Sci.* 280: 942–68

Kump LR, Garrels RM. 1986. Modeling atmospheric O_2 in the global sedimentary redox cycle. *Am. J. Sci.* 286: 337–60

Lambert IB, Walter MR, Zang W-L, Lu S-N, Ma G-G. 1987. Paleoenvironment and carbon isotope study of the Yangtze Platform. *Nature* 325: 140–42

Landing E, Myrow PM, Benus A, Narbonne GM. 1989. The Placentian Series: appearance of the oldest skeletalized faunas in southeasern Newfoundland. *J. Paleontol.* 63: 739–69

Latham AJ. 1990. *The Precambrian/Cambrian transition in Morocco.* PhD thesis. Univ. Wales, Cardiff. 240 pp.

Latham AJ, Riding R. 1990 Fossil evidence for the location of the Precambrian/Cambrian boundary in Morocco. *Nature* 344: 752–54

Leblanc M. 1981. The late Precambrian Tiddiline Tilloid of the Anti-Atlas, Morocco. In *Earth's Pre-Pleistocene Glacial Record,* ed. MJ Hambrey, WB Harland, pp. 120–22. Cambridge: Cambridge Univ. Press

Li Y-Y. 1986. Proterozoic and Cambrian phosphorites—regional review: China. In *Proterozoic and Cambrian Phosphorites*, ed. PJ Cook, JH Shergold, pp. 42–62. Cambridge: Cambridge Univ. Press

Luo H-L, Jiang Z-W, Wu X-C, Song X-L, Lin O-Y, et al. 1984. *Sinian-Cambrian Boundary Stratotype Section at Meishucun, Hinning, Yunnan, China.* Yunnan: People's Publ. House. 154 pp. (In Chinese)

Magaritz M. 1983. Carbon and oxygen isotope composition of recent and ancient coated grains. In *Coated Grains*, ed. TM Peryt, pp. 27–37. Berlin: Springer-Verlag

Magaritz M. 1989. $\delta^{13}C$ follow extinction events: a clue to faunal radiation. *Geology* 17: 337–40

Magaritz M. 1991. Carbon isotopes, time boundaries and evolution. *Terra Nova* 3: 251–56

Magaritz M., Holser WT, Kirschvink JL. 1986. Carbon-isotope events across the Precambrian/Cambrian boundary on the Siberian Platform. *Nature* 320: 258–59

Magaritz M, Kirschvink JL, Latham AJ, Zhuravlev AYu, Rozanov AYu. 1991. Precambrian/ Cambrian boundary problem: carbon isotope correlations for Vendian and Tommotian time beween Siberia and Morocco. *Geology* 19: 847–50

Magaritz M, Stemmerik L. 1989. Oscillations of carbon and oxygen isotope compositions of carbonate rocks between evaporative and open marine environments, upper Permian of East Greenland. *Earth. Planet. Sci. Lett.* 93: 233–40

McMenamin MAS. 1987. The emergence of animals. *Sci. Am.* 255: 94–102

McMillan MD. 1968. Geology of the Witputs-Sendelingsdrif area. *Bull. Precambrian Res. Unit, Univ. Cape Town* 4. 177 pp.

Miller RM, Burger AJ. 1983. U-Pb zircon age of the early Damaran Naauwpoort Formation. *Geol. Soc. S. Afr. Spec. Publ.* 11: 431–515

Myrow PM. 1987. *Sedimentology and depositional history of the Chapel Island Formation (Late Precambrian to Early Cambrian), southeast Newfoundland.* PhD thesis. Memorial Univ. Newfoundland, St. John's. 512 pp.

Myrow PM, Narbonne GM, Hiscott RN. 1988. Storm-shelf and tidal deposits of the Chapel Island and Random Formations, Burin Peninsula: facies and trace fossils. *Geol. Assoc. Can. Annu. Meet. Field Trip Guidebook B6.* 108 pp.

Narbonne GM, Myrow PM, Landing E, Anderson MM. 1987. A candidate stratotype for the Precambrian-Cambrian boundary, Fortune Head, Burin Peninsula, southeastern Newfoundland. *Can. J. Earth Sci.* 24(7): 1277–93

Orth CJ, Gilmore JS, Quintana LR, Sheehan PM. 1986. Terminal Ordovician extinction: geochemical analyses of the Ordovician/Silurian boundary, Anticosti Island, Quebec. *Geology* 44: 433–36

Pflug HD. 1966. Neue fossilreste aus den Nama-Schichten in Südwest-Afrika. *Palaontol. Zool.* 40: 14–25

Qian Y, Bengston S. 1989. Paleontology and biostratigraphy of the Early Cambrian Meishucunian State in Yunnan Province, South China. *Fossils and Strata* 24, 156 pp.

Raup DM. 1986. Biological extinction in earth history. *Science* 231: 833–36

Richter R. 1955. Die ältesten fossilien Süd-Afrikas. *Sencken. Lethaea* 36: 243–89

Riding R, Voronova V. 1984. Assemblages of calcareous algae near the Precambrian/Cambrian boundary in Siberia and Mongolia. *Geol. Mag.* 121: 205–10

Ripperdan RL, Magaritz M, Nicoll RS, Shergold JH. 1992. Simultaneous changes in carbon isotopes, sea level, and conodont biozones within the Cambrian-Ordovician boundary interval at Black Mountain, Australia. *Geology* 20(11): 1039–42

Rozanov AYu. 1984. The Precambrian-Cambrian boundary in Siberia. *Episodes* 7: 20–24

Schlanger SO, Arthur MA, Jenkyns HC, Scholle PT. 1987. The Cenomanian-Turonian oceanic anoxic event, 1. Stratigraphy and distribution of organic-rich beds and the marine $\delta^{13}C$ excursion. In *Marine Petroleum Source Rocks*, ed. J Brooks, AJ Fleet, pp. 371–99. Geol. Soc. London Spec. Publ. 26

Schmitt MO, Monninger WEW. 1977. Stromatolites and thrombolites in Precambrian/Cambrian boundary beds of the Anti-Atlas, Morocco; preliminary results. In *Fossil Algae*, ed. E Fluegel, pp. 80–85. Berlin: Springer-Verlag

Sdzuy K. 1978. The Precambrian-Cambrian boundary beds in Morocco (preliminary report). *Geol. Mag.* 115: 83–94

Seilacher A. 1984. Late Precambrian and Early Cambrian metazoa: preservational or real extinctions. In *Patterns of Change in Earth Evolution*, ed. HD Holland, AF Trendall, pp. 159–68. Berlin: Springer-Verlag

Sepkoski JJ. 1983. Precambrian-Cambrian boundary: the spike is driven and the monolith crumbles. *Paleobiology* 9: 199–206

Singh IB, Rai V. 1983. Fauna and biogenic structures in Krol-Tal succession (Vendian-Early Cambrian), Lesser Himalaya: their biostratigraphic and palaeoecological significance. *J. Palaeontol. Soc. India* 28: 67–90

Smith DAM. 1961. *The geology of the area around the Khan and Swakop Rivers in South West Africa*. Thesis. Univ. Witswatersrand, S. Afr. 77 pp.

Swett K. 1981. Cambro-Ordovician strata in Ny Friesland, Spitsbergen and their palaeotectonic significance. *Geol. Mag.* 118: 225–50

Tucker ME. 1986. Carbon isotope excursions in Precambrian/Cambrian boundary beds, Morocco. *Nature* 319: 48–50

Tucker ME. 1991. Carbon isotopes and Precambrian-Cambrian boundary geology, South Australia: ocean basin formation, seawater chemistry and organic evolution. *Terra Nova* 1: 573–82

Veizer J, Compston W, Clauer N, Schidlowski M. 1983. $^{87}Sr/^{86}Sr$ in late Proterozoic carbonates; evidence for a "mantle" event at 900 Ma ago. *Geochem. Cosmochim. Acta* 47: 295–302

Woodruff F, Savin SM. 1985. $\delta^{13}C$ values of Miocene Pacific benthic foraminifera: correlation with sea level and biological productivity. *Geology* 13: 119–22

Annu. Rev. Earth Planet. Sci. 1994. 22:419–55

THE EVOLUTIONARY HISTORY OF WHALES AND DOLPHINS

R. Ewan Fordyce

Department of Geology, University of Otago, Dunedin, New Zealand

Lawrence G. Barnes

Natural History Museum of Los Angeles County, 900 Exposition Boulevard, Los Angeles, California 90007

KEY WORDS: marine mammal, Cetacea, fossil, systematics

INTRODUCTION

Cetaceans—the whales, dolphins, and porpoises—are the taxonomically most diverse clade of aquatic mammals, with a fossil record going back at least to Middle Eocene time (52 Ma—millions of years before present) (Figure 1). There are 75 to 77 living species in 13 or 14 families and two suborders: Mysticeti (baleen whales) and Odontoceti (toothed whales, dolphins, and porpoises) (Rice 1984, Evans 1987). The extinct Archaeoceti are a third suborder. Extant cetaceans are ecologically diverse; sizes range from under 2 m to over 25 m, and habitats range from shelf and surface water to abyssal settings in tropical to polar oceans; some species live in fresh water. Living species attract marked public and scientific interest, but fossils also have an important role. The excellent fossil record helps us to understand morphological transition series and homologies of structures in living taxa. A resurgence of interest in anatomy has led to pioneering, albeit preliminary, cladistic analyses of fossil and extant Cetacea, with pivotal input from paleontologists. [For a discussion of cladistics, see Padian et al (1993) in this volume.] Although volatile and clearly needing more study, the resulting classifications provide an adequate foundation for broader studies. Cetacean constructional morphology and related aspects of paleobiology are still in their infancy. Fossil taxa are potentially

0084–6597/94/0515–0419$05.00

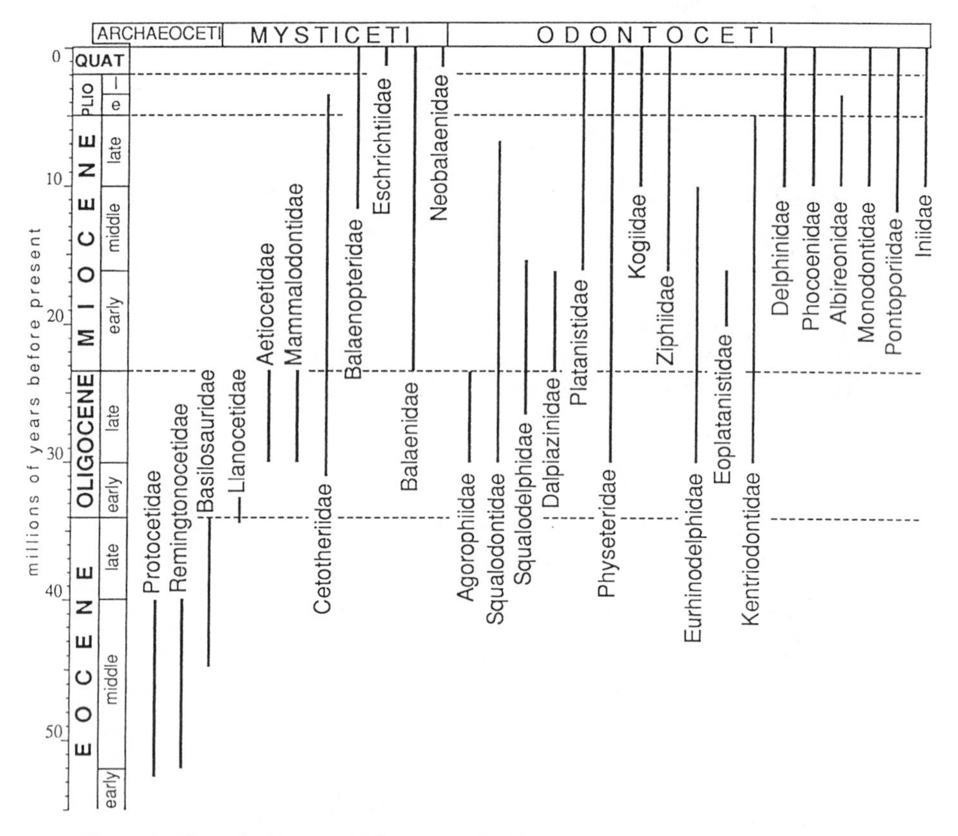

Figure 1 Chronologic ranges of cetacean families worldwide, based on most recently published accurate data and modified slightly based on unpublished data available to the authors. Solid bars show reported or likely maximum ranges. Where ranges fall within a sub-epoch but are not known precisely, range bars are extended to sub-epoch boundaries. Formally named families of uncertain status are not shown.

valuable in stratigraphy. Changes in abundance with changing facies help in the interpretation of ancient depositional settings, while broader changes in diversity through time help us interpret broader changes in currents and climates in the global oceans. We explore these topics below.

Other general reviews of cetacean evolution include Miller (1923), and Kellogg (1928, 1936). Fordyce (1980, 1989b, 1992), Barnes (1984b, 1990, 1992), Barnes & McLeod (1984), and Barnes et al (1985) provided recent synopses. Works on extant Cetacea include those by Evans (1987), Slijper (1979), Rice (1977, 1984), and contributions edited by Ridgway & Harrison

(1985, 1989). Hershkovitz (1966) cataloged extant species; Simpson (1945) listed fossils. We stress references that are accessible, not necessarily the earliest authority.

ANATOMY AND BIOLOGY OF CETACEANS

Fossil Cetacea probably had body forms similar to those of modern species (e.g. Figure 2). Anatomy and biology are documented widely (Rice 1984, Evans 1987). Living Cetacea are streamlined, fusiform, and the smooth body is hairless with few vibrissae (facial hairs). Color patterns vary widely between species; a dark dorsal surface and lighter ventral surface are common. The body is enveloped in blubber which largely obliterates expressions from the facial muscles. Teeth occur in most adult Odontoceti and in embryonic Mysticeti, but toothless adult Mysticeti have baleen plates used in filter-feeding. Relatively small eyes lie laterally. There is no external ear pinna, and the auditory canal is closed. One or two nostrils (blowholes) open dorsally on the head, in contrast to a more anterior position in most mammals. The laterally flattened forelimbs are usually short and rigid, without a flexible elbow. In most species the neck is indistinct and rather inflexible. A dorsal fin supported by connective tissue is normally present. The body is subcylindrical posteriorly to about level with the ventral genital slit and anus, but beyond this the tail peduncle is laterally compressed. Bilateral, rigid tail flukes are supported by connective tissue, with vertebrae only in the midline.

Skeletal and particularly cranial anatomy is a prime basis for cetacean taxonomy. General features of cetacean skulls (Kellogg 1928, Fraser & Purves 1960, Barnes 1990, McLeod et al 1989) include the following, many of which are synapomorphies (see Padian et al 1993) for Cetacea: a distinct and usually long rostrum (upper jaw) formed by elongated prenarial portions of premaxillae and maxillae; paired bony nares displaced back on the skull; discrete origins for muscles associated with the blowholes; bony orbits placed widely apart under long, flat and transversely broad supraorbital processes; postglenoid processes and glenoid fossae (cavities for jaw articulation) shaped for simple hinge movement; large origins for temporalis muscles; robust basioccipital crests with each bulla having a thin outer lip, a sigmoid process with fused malleus, a large tympanic cavity, and a dense involucrum; periotics (earbones) widely separated and loosely attached to the skull, each with distinct anterior and posterior (mastoid) processes, within the ear, a large stapedius muscle fossa, smooth dorsal face on the cochlear portion of the periotic, and small origin for tensor tympani muscle; and reduced foramina for the internal carotid arteries.

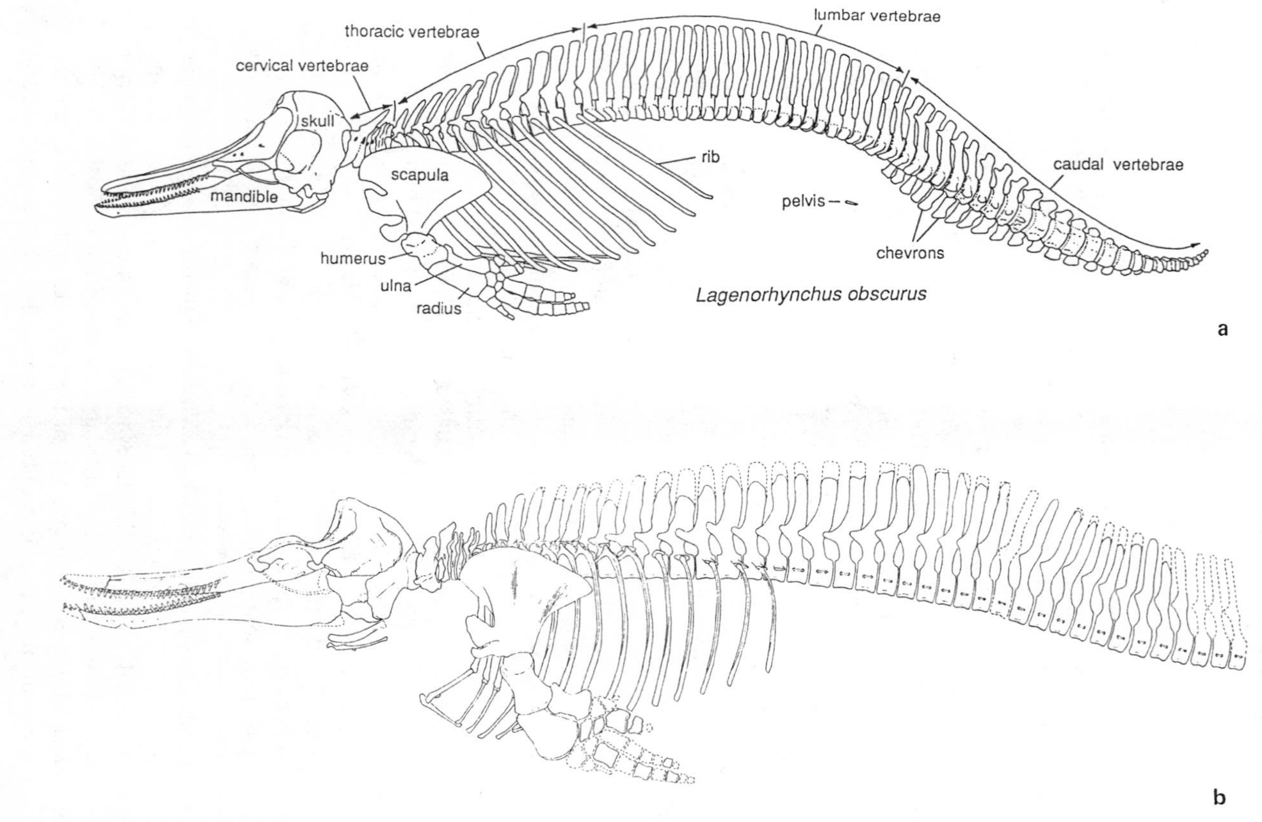

Figure 2 The cetacean skeleton: (*a*) left lateral view, with major parts labeled, of *Lagenorhynchus obscurus*, the living dusky dolphin of the south Atlantic and South Pacific; (*b*) left lateral view of *Albireo whistleri*, an extinct latest Miocene dolphin from the eastern North Pacific, partial skeletal reconstruction based on the holotype fossil skeleton from Isla de Cedros, Baja California, Mexico.

Fossil and Recent cetacean taxa are diagnosed cranially by proportions and size of the skull components including the rostrum, teeth, facial and temporal muscle origins, the basicranial (pterygoid) sinuses that form as outgrowths of the eustachian tubes, and tympano-periotics. In all Odontoceti and Mysticeti, rostral bones and the supraoccipital are "telescoped" (Miller 1923) toward each other, although such telescoping is not demonstrably homologous in these two groups.

The seven cervical (neck) vertebrae are usually compressed anteroposteriorly, and are sometimes fused. There are usually more post-cervical vertebrae than in other mammals. The ribs usually attach loosely to the thoracic vertebrae and sternum, and in extant Cetacea, lumbar vertebrae grade back to caudal (tail) vertebrae without obvious sacral vertebrae. Well-developed vertebral spines and transverse processes on the lumbar vertebrae reflect massive tail muscles. A subspherical "ball" vertebra marks the point of maximum bending at the peduncle-fluke junction (Watson 1991). Caudal vertebrae in the flukes have distinct rectangular profiles. In the forelimb, the scapula has a reduced supraspinatus fossa and subparallel anteriorly directed acromion and coracoid processes, without a clavicle. Cetaceans normally show hyperphalangy (marked increase in number of phalanges). The pelvis in the modern animals is usually present as two separate bilaterally paired simple elements, sometimes with vestigial hind limb bones, although early whales had more complete hind limbs (Gingerich et al 1990).

OPPORTUNITIES AND LIMITS TO THE STUDY OF FOSSIL CETACEA

Fossil Cetacea are conspicuous in outcrop, but rare. The large size and the intrinsic fragility of porous bones make some fossils hard to collect and preserve. Many fossil species are represented by only one published specimen, and stratophenetic approaches, best suited to fossils from successive closely spaced horizons, have dealt poorly with a patchy record that reveals only a few good ancestor-to-descendant sequences. Conversely, information provided by the multiple elements of one skeleton is valuable in cladistic and functional studies. Most skeletal parts have been used at some time as type specimens, with species based on part or whole skulls, teeth, mandibles, vertebrae, or limb bones. Unfortunately, many species based on isolated bones are poorly defined and diagnosed by modern standards, and many cannot be compared directly with one another. Ideally, type specimens should include skulls and associated earbones. Most higher taxa need to be reviewed in light of changing concepts of taxonomy, and many groups are acknowledged to be paraphyletic. For these reasons, and because some epochs or geographic regions are represented poorly in the fossil record, published lists of taxa must be interpreted carefully.

DISTRIBUTION

Chronologic Distribution

The chronologic record of cetacean families is shown in Figure 1. Such charts are limited by taxonomy (i.e. are the groups real?) and geology (Barnes 1977, Fordyce 1992). Bars mark known limits of stratigraphic ranges, but do not necessarily indicate a continuous record. For most taxa, published information does not allow age resolution below the level of the stage or subepoch. Many age determinations lack lists of age-diagnostic taxa, and few articles cite ages based on the planktic microfossils used in long-distance correlations. Nevertheless, trends are apparent, and are reviewed below. Fossil taxa are indicated by a dagger † prefix.

Geographic Distribution—Regional Studies

Early studies were based in Europe and North America, with few developments occurring elsewhere until early in the twentieth century. Important marine sequences around the Mediterranean (and former Paratethys-Tethys) include the Neogene in Italy and France (Dal Piaz 1977, de Muizon 1988a), Oligocene in Austria (Rothausen 1968, 1971), and Caucasus (Mchedlidze 1984, 1989; Barnes 1985c) and Eocene in Egypt (Gingerich 1992, Kellogg 1936). Eastern North Atlantic faunas are known mainly from the North Sea margins (Abel 1905, Rothausen 1986), but there are a few from Britain (Hooker & Insole 1980). Eocene to Pliocene Cetacea from the western North Atlantic include those documented by Kellogg (listed by Whitmore 1975), Whitmore & Sanders (1977), Hulbert & Petkewich (1991), and Sanders & Barnes (1993). Southwest Atlantic faunas from Argentina include middle Cenozoic taxa (Cabrera 1926, de Muizon 1987) currently under study by M. A. Cozzuol. Eocene presumed Tethyan species from Pakistan and India represent some of the oldest Cetacea (Gingerich et al 1983, Kumar & Sahni 1986, Gingerich & Russell 1990, Thewissen & Hussain 1993), but there are no other significant described faunas from around the Indian Ocean. The Pacific, the largest ocean during cetacean history, deserves more attention. Japanese Neogene fossils are well documented (Oishi 1985, Kimura et al 1992), and studies of Oligocene species are under way (Okazaki 1988, Kimura et al 1992). Eastern North Pacific Neogene assemblages are also well known (Barnes 1977, 1984a), but Oligocene species are mainly undescribed (Whitmore & Sanders 1977).

Peru has yielded Neogene odontocetes (de Muizon 1984, 1988c) and mysticetes (Pilleri 1989). New Zealand assemblages, from the Southern Ocean margin, span from Eocene to Quaternary (Fordyce 1991). A scattered Oligocene to Neogene record from Australia (Fordyce 1984) also hints at the composition of Southern Ocean faunas. Only one Paleogene and one Pliocene site are known from Antarctica (Fordyce 1989b). Extant fluvio-lacustrine odontocetes include species of Platanistidae, Pontoporiidae, Iniidae, and Lipotidae. Fossil odontocetes from fluvio-lacustrine sediments include Miocene †Eurhinodelphidae from Australia (Fordyce 1983), a Miocene ziphiid from Kenya (Mead 1975b), Miocene pontoporiids and iniids from Argentina (Cozzuol 1985), and a possible Miocene lipotid from China (Zhou et al 1984).

Depositional Settings and Taphonomy

Although fossil Cetacea are found mainly in proximal marine sedimentary rocks now exposed on land, distributions are best stated in terms of oceans. Perhaps only early archaeocetes, with amphibious seal-like habits, were linked strongly to land. Most living species occupy coastal waters only occasionally, and strand rarely. Sporadic worn or broken bones from shallow sublittoral inner shelf facies suggest that strandings are at most a minor potential source of fossils. Fossils from more distal shallow, mid- and outer-shelf settings are common, while specimens from bathyal settings are rarely collected. Bone-bearing sediments include muddy conglomerate, quartzose and calcareous sandstones, siltstone and mudstone, limestone, greensand, diatomite, and concretionary mudstone (Barnes 1977, Myrick 1979, de Muizon 1984, Fordyce 1991, Gingerich 1992). Reworked bone-bearing clasts are known from debris flows. Lag accumulations mark unconformities, and remanié elements (Boreske et al 1972) may occur in nodule beds. A few specimens are known from ocean dredgings (Whitmore et al 1986).

Most fossils occur as isolated bones or small clusters of bones, presumably dropped from floating carcasses (Schäfer 1972). Articulated or semi-articulated specimens are found in distal settings, represented by massive mudstones (Squires et al 1991); indeed, cetacean carcasses, oases for obligate bone-dwelling invertebrates, may be an important abyssal energy source both in the past and at the present time (Squires et al 1991, Allison et al 1991). Fossil skeletons are often partly articulated, surrounded by bones scattered by scavenging or currents. A common ventral-up orientation of skeletons reflects the influence on burial position of a gas-filled and thus buoyant abdomen in a decomposing carcass. Semi-articulated fossils are sometimes abundant in bone beds (e.g. Sharktooth Hill, Middle

Miocene, California; Barnes 1977). Fossils in burial position are published widely, but explicit taphonomic studies are few (Myrick 1979, Lancaster 1986).

CLASSIFICATION

Post-Darwinian approaches to cetacean classification have long involved evolutionary systematics, championed by Simpson (1945), and an eclectic approach is still employed (e.g. Mchedlidze 1984, Mitchell 1989). Cladistics is becoming more common (e.g. Barnes 1985b, 1990; de Muizon 1988a, 1991), and has already proliferated names for higher-level taxa (families, superfamilies, and infraorders; Table 1, Figure 3). Only Heyning (1989) has published a computer-aided analysis of cetacean taxonomy (Figure 3). Manual cladistic analyses identify problems, but computer-based analyses make better sense of complex data, thus leading to more stable clades and clade ranks. Phenetic and biochemical techniques are currently unimportant.

Monophyly, Diphyly, and Relationships with Other Eutherian Mammals

Living cetaceans are so well adapted to an obligate aquatic lifestyle that there are few structures that initially reveal close relatives to other Eutheria. Kellogg's (1936) review of relationships was inconclusive, while Simpson (1945) placed the Order Cetacea in an isolated Cohort, Mutica. Such uncertainty arose partly through perceptions that there are few structural intermediates between †Archaeoceti, Odontoceti, and Mysticeti, and thus that Cetacea are diphyletic (Miller 1923, Yablokov 1965). Van Valen's (1968) succinct review led to rapid acceptance of monophyly (McLeod et al 1989), a view further reinforced by an expanding fossil record. Odontoceti and/or Mysticeti have been viewed as originating from †Protocetidae (Van Valen 1968), †Remingtonocetidae (Kumar & Sahni 1986), or, most likely, †Basilosauridae (Barnes & Mitchell 1978). Synapomorphies for Cetacea and for Odontoceti + Mysticeti have been discussed or listed widely (Rice 1984; Barnes 1984b, Barnes 1990; Barnes & McLeod 1984; Heyning 1989; Heyning & Mead 1990; McLeod et al 1989, Milinkovitch et al 1993; Novacek 1993). From the 1950s on, biochemical, karyological, cytological, and other techniques have repeatedly clustered Cetacea close to Artiodactyla (literature cited by Barnes & Mitchell 1978, Heyning 1989, Novacek 1993, Milinkovitch et al 1993). The structure of teeth, skulls and vestigial hind limbs in fossils (Van Valen 1966, Gingerich & Russell 1990, Gingerich et al 1990, Novacek 1993) further supports relationships with ungulates. Significantlly, Flower (1883) long ago proposed a relationship of cetaceans with ungulates on anatomical grounds.

Osteology vs Molecules in Taxonomy

Milinkovitch et al (1993) suggested, on the basis of DNA analysis, that sperm whales, Physeteridae + Kogiidae, are more closely related to rorquals, Balaenopteridae (Mysticeti) than to other Odontoceti (Figure 3), and that physeterids and balaenopterids had a common ancestor about 10–15 Ma. This would imply that the Odontoceti are paraphyletic, and that the ability to echolocate was probably lost secondarily in rorquals or evolved independently in different odontocete groups. Conversely, osteological and other anatomical studies (Fraser & Purves 1960, Kasuya 1973, Heyning 1989, Heyning & Mead 1990, Barnes 1990, de Muizon 1991) indicate that the traditional Mysticeti and Odontoceti are clades with ancient origins well demonstrated by the fossil record. Many synapomorphies unite physeterids with other odontocetes, including features of the face, basicranium, and tympano-periotic (Barnes 1990, de Muizon 1991). Undisputed Late Oligocene physeterids are known. Balaenopterids have a shorter (Late Miocene to Recent) record. They are universally accepted as Mysticeti, and thus are members of an early Oligocene to Recent clade. Balaenopterids probably originated among the Oligocene to Pliocene †Cetotheriidae. If, as suspected, physeterids are a basal clade of odontocetes, they could be phenetically close to some mysticetes, so that some techniques may not resolve cladistic affinities clearly. Myoglobin DNA, used by some analyses as evidence for relationships, is notoriously unreliable because of numerous cases of convergences. DNA cladistic analyses, like other taxonomic procedures, are probably sensitive to choice and number of data and the interpretation of outgroups. The concept of a sister-group relationship between sperm and baleen whales needs more study; meanwhile, the fossil record provides a valuable check on rates of evolution in cetacean mitochondrial DNA.

REVIEW OF CETACEAN TAXA

†Archaeoceti

Archaeocetes are a paraphyletic group of archaic toothed Cetacea that lack cranial features of Odontoceti and Mysticeti (Kellogg 1936). The included grade families †Protocetidae and †Basilosauridae and clade †Remingtonocetidae are Eocene only, while younger supposed archaeocetes are either misidentified or are too incomplete to place conclusively. Mitchell (1989) raised all three families to the rank of superfamily (†Protocetoidea, †Remingtonocetoidea, and †Basilosauroidea) to accommodate

Table 1 Classification of Cetacea[a]

Order Cetacea Brisson 1762.
  Suborder †Archaeoceti Flower 1883.
    Family †Protocetidae Stromer 1908. Early-Middle Eoc., Tet.–Caribbean.
      Subfamily †Pakicetinae Gingerich and Russell 1990.
    Family †Remingtonocetidae Kumar and Sahni 1986. Middle Eoc., Tet.
    Family †Basilosauridae Cope 1868. Middle-Late Eoc. Tet., N. Atl., SW. Pac.
      Subfamily †Dorudontinae (Miller 1923) Slijper 1936.
      Subfamily †Basilosaurinae (Cope 1868) Barnes & Mitchell 1978.
  Suborder Mysticeti Flower 1864.
    Family †Llanocetidae Mitchell 1989. Late Eoc. or Early Olig., SE. Pac.–SW. Atl.
    Family †Aetiocetidae Emlong 1966. Late Olig., N. Pac.
    Family †Mammalodontidae Mitchell 1989. Late Olig., SW. Pac.
    Family †Kekenodontidae (Mitchell 1989). Fordyce 1992.
        Late Olig., SW. Pac.
    Family †Cetotheriidae (Brandt 1872) Miller 1923.
        Early?, Late Olig.-Early or Late? Plio. Pac., Atl., Med.–Par.
    Family Balaenopteridae Gray 1964. Middle?, Late.-Rec.
        Fossil: Pac., Atl., Med. Rec.: cos.
      Subfamily Megapterinae (Gray 1866) Gray 1868.
      Subfamily Balaenopterinae (Gray 1864) Brandt 1872.
    Family Eschrichtiidae Ellerman and Morrison-Scott 1951.
        Quat.–Rec. Fossil: N. Pac., Rec.: N. Pac.;
        recently extinct, N. Atl.
    Family Neobalaenidae Gray 1873. Rec.: SOc.
    Family Balaenidae Gray 1825. Early Mio.–Rec.
        Fossil: Atl., Pac., SOc. Rec: temperate–polar oceans.
  Suborder Odontoceti Flower 1864.
  Superfamily unresolved.
    Family †Agorophiidae Abel 1913. Late Olig., N. Atl.
  Superfamily Physeteroidea (Gray 1821) Gill 1872.
    Family Physeteridae Gray 1821. Sperm whales.
        Late Olig.–Rec. Fossil: Atl., Med.–Par., Pac., SOc., Rec.: cos.
    Subfamily †Hoplocetinae Cabrera 1926.
    Subfamily Physeterinae (Gray 1821) Flower 1867.
  Family Kogiidae (Gill 1871) Miller 1923. Pygmy sperm whales.
        Late Mio–Rec. Fossil: E. Pac., SW. Pac., NW. Atl. Rec., temperate–tropical oceans.
  Superfamily Ziphioidea (Gray 1865) Gray 1868.
    Family Ziphiidae Gray 1865. Beaked whales: Middle Mio.–Rec.
        Fossil: cos., FW. Africa. Rec.: cos.
      Subfamily Ziphiinae (Gray 1865) Fraser & Purves 1960.
      Subfamily Hyperoodontinae (Gray 1846) Muizon 1991.

[a] Infraorders, discussed in the text, are not cited. Only common synonyms are cited. For emended names, original authors are cited in parentheses, followed by revisers. Taxonomic references are not cited in the bibliography unless the articles are also cited in the text.

Abbreviations: N—north; S—south; SW—southwest; SE—southeast; E—east, Eoc—Eocene; Olig—Oligocene; Mio—Miocene; Plio—Pliocene; Quat—Quaternary; Rec—Recent; e—early; m—middle; l—late; FW—fresh water; Pac—Pacific; Tet—Tethys; Med—Mediterranean; Atl—Atlantic; cos—cosmopolitan; Par—Paratethys; SOc—Southern Ocean.

Table 1—(*continued*)

Superfamily Platanistoidea (Gray 1863) Simpson 1945.
  Family †Squalodontidae Brandt 1872. Late Olig.–Middle Mio.,
      Atl., Med.–Par., Pac.
    Subfamily †Patriocetinae (Abel 1913) Rothausen 1968.
    Subfamily †Squalodontinae (Brandt 1872) Rothausen 1968.
  Family †Squalodelphidae Dal Piaz 1916. Early Mio.,
      Med., SW. Atl., SW. Pac.
  Family †Dalpiazinidae de Muizon 1988a. Early Mio.,
      Med., Sw. Atl., SW. Pac.
  Family Platanistidae (Gray 1863).
      Middle Mio.–Rec., Fossil: NE. Atl. Rec.: FW. India.
Superfamily †Eurhinodelphoidea (Abel 1901) de Muizon 1988.
  Family †Eoplatanistidae de Muizon 1988a. Early Mio. Med.
  Family †Eurhinodelphidae Abel 1901. (=Rhabdosteidae Gill 1871).
      Late Olig.?, Early–Middle Mio. Atl., Med., ?Par., Pac., FW. Australia.
Superfamily Delphinoidea (Gray 1821) Flower 1864.
      (including Monodontoidea Fraser & Purves 1960)
  Family †Kentriodontidae (Slijper 1936) Barnes 1978.
      Late Olig.–Late Mio. Atl., Med.–Par., Pac., SOc.
    Subfamily †Kampholophinae Barnes 1978.
    Subfamily †Kentriodontinae (Slijper 1936).
    Subfamily †Lophocetinae Barnes 1978.
  Family †Albireonidae Barnes 1984a. Late Mio.–Early Plio. NE. Pac.
  Family Monodontidae Gray 1821.
      Fossil: NE. Pac., N. Atl. Rec.: Arctic–N. Pac.–N. Atl.
    ?Subfamily Orcaellinae (Nishiwaki 1963) Kasuya 1973.
    Subfamily Delphinapterinae Gill 1871.
    Subfamily Monodontinae (Gray 1821) Miller & Kellogg 1955.
  Family Delphinidae Gray 1821.
      (including Holodontidae Brandt 1873, Hemisyntrachelidae Slijper 1936).
      Late Mio.-Rec. Fossil: Atl., Med.–Par., Pac., SOc. Rec.: cos.
    Subfamily Steninae (Fraser & Purves 1960) Mead 1975.
    Subfamily Delphininae (Gray 1821) Flower 1867.
    Subfamily Globicephalinae (Gray 1866) Gill 1872.
  Family Phocoenidae (Gray 1825) Bravard 1885. Late Mio.-Rec.
      Fossil: Pacific, Rec.: Cos.
    Subfamily Phocoenoidinae Barnes 1985a.
    Subfamily Phocoeninae (Gray 1825) Barnes 1985a.
Superfamily unresolved.
  Family Iniidae Flower 1867. Late Mio.–Rec. Fossils:
      FW. Argentina. Rec.: Brazil–Venezuela.
  Family Pontoporiidae (Gill 1871) Kasuya 1973. Late Mio.–Rec.
      Fossils: SW. Atl., N. Pac. Rec.: SW. Atl., FW. China.
    Subfamily Lipotinae (Zhou et al 1979) Barnes 1985b.
    Subfamily †Parapontoporiinae Barnes 1985b.
    Subfamily Pontoporiinae Gill 1871.

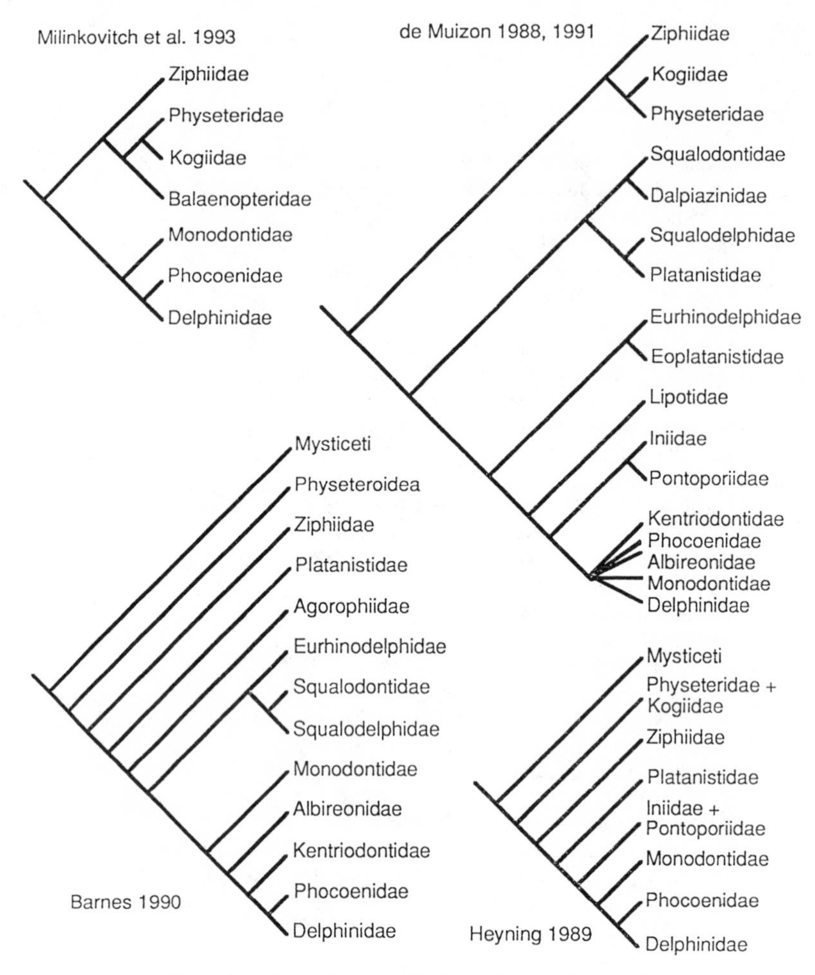

Figure 3 Alternative classifications of the Odontoceti.

"a wide diversity of species greatly different in morphology," but this move needs to be supported by careful cladistic analysis.

†PROTOCETIDAE The ancient grade †Protocetidae, which Kellogg (1936) regarded as "an unnatural assemblage," derives its identity from the small long-beaked †*Protocetus atavus* (Middle Eocene, Egypt-Tethys) which has sacral vertebrae that might have articulated with pelvic bones. The reputed oldest and perhaps amphibious cetacean, †*Pakicetus inachus* (Early or

Middle Eocene, Pakistan), has ears adapted for underwater hearing (Gingerich et al 1983, Gingerich & Russell 1990, Thewissen & Hussain 1993). The great variety of protocetid teeth suggests diverse feeding modes not yet substantiated by described skulls or skeletons. Protocetids include Middle Eocene taxa reported only from the Tethys–equatorial Atlantic–Caribbean (Kellogg 1936, Barnes & Mitchell 1978, Kumar & Sahni 1986, Hulbert & Petkewich 1991). These records support the idea of a Tethyan, presumed warm-water origin for cetaceans.

†REMINGTONOCETIDAE Species of †*Remingtonocetus* and †*Andrewsiphius* formed the bizarre remingtonocetids (Middle Eocene, India), a short-lived archaeocete clade characterized by a long, narrow skull and jaws, with cheek-teeth placed relatively far forward of the eyes. Kumar & Sahni (1986) suggested that remingtonocetids have pterygoid sinuses in the orbits, like those of odontocetes, and thus that the group is ancestral to odontocetes. Illustrations of the skull base of remingtonocetids suggest that the sinuses are not expanded, and relationships with odontocetes are unproven.

†BASILOSAURIDAE The paraphyletic Middle to Late Eocene †Basilosauridae are highly evolved taxa because they have cheek-teeth with multiple accessory denticles (small cusps) and expanded pterygoid sinus fossae in the skull base. More sophisticated feeding and hearing capabilities are indicated. These features suggest a sister group relationship with Odontoceti+Mysticeti. Barnes & Mitchell (1978), Gingerich (1992), and Gingerich et al (1990) suggested revisions to the taxonomy proposed by Kellogg (1936). The subfamily †Basilosaurinae appears to form a clade of large species with elongate vertebral bodies, of which †*Basilosaurus cetoides* (latest Eocene, North American Gulf Coast) is the largest (to >15 m). Large size and elongate vertebral bodies are perhaps synapomorphies which preclude basilosaurines from being the ancestors of Odontoceti or Mysticeti. The diverse paraphyletic †Dorudontinae includes species without elongate vertebral bodies, perhaps even the immediate ancestors of odontocetes and mysticetes (Barnes & Mitchell 1978). At least some, and probably all basilosaurids had hind legs (Gingerich et al 1990). Genera and species are diagnosed mainly on proportions of the skull and teeth.

Mysticeti

All living Mysticeti are large filter-feeders (Slijper 1979, Rice 1984), and filter-feeding can be inferred for most, perhaps all, fossil Mysticeti. Major evolutionary trends within the group include the loss of teeth, development of large body size and large heads, shortening of the intertemporal region as rostral bones and the supraoccipital approach each other, and shortening of the neck. Mysticetes have a uniquely "telescoped" skull in which the maxilla uniquely extends posteriorly under the orbit to form a plate-like infraorbital process (Miller 1923). In living Mysticeti the expanded maxilla forms the origin for the epithelially-derived baleen plates used in filter-feeding—a unique and characteristic behavior of the group (Pivorunas 1979), although archaic toothed mysticetes perhaps lacked baleen. Other skull features, many related functionally to filter-feeding, help diagnose Mysticeti (e.g. Barnes 1990, Barnes & McLeod 1984). The evolution of filter-feeding was a key factor in the origin of mysticetes (Fordyce 1980, 1989b).

HIGHER CLASSIFICATION Family-level subdivisions of the Mysticeti (see Table 1) are currently not as contentious as those of odontocetes. This generally reflects a lack of recent detailed work; the mysticetes await cladistic reappraisal. Mitchell (1989) proposed a non-cladistic subdivision of Mysticeti into two infraorders. A grade Infraorder †Crenaticeti, which was not strictly diagnosed, includes only †*Llanocetus denticrenatus*. †Aetiocetidae and †Mammalodontidae are of uncertain infraordinal position. Infraorder Chaeomysticeti, interpreted here as a clade, encompasses all baleen-bearing taxa, including superfamilies Eschrichtioidea, Balaenopteroidea and Balaenoidea. Mitchell's proposed subdivisions and ranks are not used in Table 1; they await justification by careful cladistic study.

TOOTHED ARCHAIC MYSTICETES: †AETIOCETIDAE, †LLANOCETIDAE, †MAMMALODONTIDAE, AND †KEKENODONTIDAE †*Aetiocetus cotylalveus* Emlong 1966 (Late Oligocene, northeast Pacific) gives its name to the †Aetiocetidae, the first toothed mysticete family established. Emlong identified †*A. cotylalveus* as an archaeocete because it has teeth, but Van Valen (1968) placed it more appropriately in the Mysticeti. The mix of archaeocete-like and mysticete skull features seen in †*Aetiocetus* influenced cetologists into accepting that mysticetes arose from archaeocetes (Barnes 1987, 1989; Kimura et al 1992). Barnes (1987) referred the enigmatic late Oligocene †*Chonecetus sookensis* (also northeast Pacific; Russell 1968) to the †Aetiocetidae. Mchedlidze (1976) identified †*Mirocetus* and †*Ferecetotherium* (late Oligocene, Paratethys) as aetiocetids, but †*Mirocetus* is not clearly so and †*Ferecetotherium* is probably a physeterid (Barnes 1984b).

The oldest described mysticete is the toothed †*Llanocetus denticrenatus*, the only species in the family †Llanocetidae. It was based on a fragment of large inflated mandible (Mitchell 1989) of Late Eocene or probable Early Oligocene age (Seymour Island, Antarctica; southeast Pacific–southwest Atlantic). The relatively complete but undescribed holotype skull (Fordyce 1989b) is still under preparation. The †Llanocetidae has yet to be strictly diagnosed, and its cladistic relationships are uncertain.

The small †*Mammalodon colliveri* (Late Oligocene or Early Miocene, southwest Pacific), for which Mitchell (1989) proposed the monotypic family †Mammalodontidae, has a relatively very short rostrum, flat palate, and heterodont (differentiated) teeth. Only the holotype is described formally (Pritchard 1939, Fordyce 1984), but other Late Oligocene specimens occur in the southwest Pacific (Fordyce 1992). Mitchell (1989) offered a synoptic description of the †Mammalodontidae, but did not diagnose it on synapomorphies. Its possible relationships with the trans-Pacific †Aetiocetidae are uncertain.

†*Kekenodon onamata* is an enigmatic and long-debated cetacean (Late Oligocene, southwest Pacific) known only from large teeth, earbones, and a few other fragments. Mitchell (1989) provided a synoptic description for a new subfamily †Kekenodontinae, which he placed in the †Archaeoceti. Fordyce (1992, Figure 18.2) placed †*K. onamata* in Mysticeti: †Kekenodontidae. Cladistic relationships are uncertain. Other enigmatic probable mysticetes are known mainly from teeth (Fordyce 1992). Such fossils include two unnamed Early Oligocene species (Fordyce 1989a), several named Oligocene species wrongly placed in the odontocete genus †*Squalodon*, and probably †*Phococetus vasconum* (early Oligocene?, northeast Atlantic) which Kellogg (1936) and Mitchell (1989) regarded as an archaeocete.

In summary, primitive toothed mysticetes are more widespread and diverse than formerly understood. Their teeth may have been used, in the mysticete fashion, for bulk feeding rather than for selecting individual prey, but functional studies are needed to confirm this. Fossils include Early Oligocene basal Mysticeti which may prove critical in interpreting their archaeocete ancestry and the later Cenozoic diversification of mysticetes. Most of the fossils described so far are too fragmentary to be placed cladistically.

†CETOTHERIDAE Cetotheres are a paraphyletic group of archaic, baleen-bearing, toothless mysticetes whose name derives from the Late Miocene †*Cetotherium rathkii* (Paratethys; Kellogg 1928). Cetotheres have flat rostra with prominent ventral nutrient foramina for baleen, and a frontal that slopes gently (Miller 1923). The earliest cetotheres, †*Mauicetus* and †*Cetotheriopsis*, appeared by Late Oligocene time (Rothausen 1971, Fordyce 1992, Sanders & Barnes 1993). More than 30 genera and about 60 named Late Oligocene to Late Pliocene species are known (Kellogg 1931, Barnes & McLeod 1984, Fordyce 1992). Studies are plagued by taxa based on noncomparable and, often undiagnostic elements (Kellogg (1968). Cabrera (1926) and others identified some potential but as yet unformalized subfamilial groups. Cetotheres include taxa apparently close to

the ancestry of balaenopterids, but not putative ancestors for balaenids or eschrichtiids (McLeod et al 1993).

BALAENOPTERIDAE Rorquals include six extant species placed in two subfamilies, Balaenopterinae and Megapterinae (Rice 1984, Ridgway & Harrison 1985). Balaenopterids differ from cetotheres in their more complex interdigitation of rostral and cranial bones, and particularly in the supraorbital process, which is depressed abruptly from the cranial vertex to form a deeper origin for the temporalis muscle. There are firm Late Miocene records (Barnes 1977), and Bearlin (1988) reported an apparent Middle Miocene species from the western South Pacific. Some of the many named fossil genera and species have been placed occasionally with cetotheres, whence rorquals probably arose.

ESCHRICHTIIDAE Gray whales are represented by only one extant species, *Eschrichtius robustus*, which has only a fossil Quaternary record in the North Pacific and a prehistoric record in the North Atlantic (Barnes & McLeod 1984). It is not clear whether relationships are closer to balaenopterids or balaenids (McLeod et al 1993). Barnes & McLeod (1984) discounted other fossil records of *Eschrichtius*, noting that these represent balaenopterids or undiagnosable taxa. There is no close relationship between *Eschrichtius* and cetotheres.

NEOBALAENIDAE The small living pygmy right whale, *Caperea marginata*, found only in the Southern Hemisphere, is variously placed with the balaenids or in its own family (Miller 1923, Kellogg 1928, Barnes & McLeod 1984). The one reported fossil record of a neobalaenid, the Chilean †*Balaena simpsoni*, is dubious. They seem to be a primitive sister taxon of the Balaenidae (McLeod et al 1993).

BALAENIDAE Living right whales and bowheads are large slow swimming mysticetes with a narrow, highly arched rostrum and long baleen (Rice 1984). These are adaptations to "skimming" filter-feeding (Pivorunas 1979). There are many later Neogene records, mostly fragmentary specimens from around the North Atlantic and North Pacific (Barnes & McLeod 1984, McLeod et al 1993). The early Miocene †*Morenocetus parvus* (southwest Atlantic; Cabrera 1926) is the oldest known balaenid; it reveals no obvious clues to the origins of the right whales.

Odontoceti

All species in the clade Odontoceti (toothed whales, dolphins, and porpoises) have skulls in which the maxilla uniquely "telescopes" or extends posteriorly over the orbit to form an expanded bony supraorbital process (Miller 1923). In living odontocetes this supraorbital process forms an origin for facial (maxillonasolabialis) muscle (Mead 1975a), which inserts around the single blowhole and associated complex nasal diverticula. The facial muscle complex and nasal apparatus probably generate the high-frequency sounds used by living odontocetes (Wood & Evans 1980) to echolocate in navigation and hunting. Fordyce (1980) suggested that the evolution of echolocation was critical in the origin of odontocetes. Other cranial features help diagnose Odontoceti (e.g. Barnes 1990), but teeth, long used in odontocete taxonomy, seem unreliable (Barnes 1977). The oldest certain odontocetes that are accurately dated and described are from the Late Oligocene. Supposed Early Oligocene species (Fordyce 1992) are recorrelated as Late Oligocene, while other reported Early Oligocene odontocetes (Squires et al 1991) have yet to be described. A diverse Late Oligocene record (Fordyce 1992) indicates a significant earlier Oligocene radiation.

Evolutionary trends within Odontoceti include expansion and increase in size of the face, shortening of the intertemporal region, elevation of the cranial vertex posterior to the nasals, increased facial asymmetry, enlargement of basicranial pterygoid sinus fossae, and isolation of the earbones from the skull. The jaws may become extremely long, narrow, and polydont, or short and blunt, or toothless.

HIGHER CLASSIFICATION The higher classification of odontocetes (Table 1) is currently volatile, with little chance of a consensus view that would allow a detailed correlation between evolution and geological processes. Volatility has arisen through advances in alpha taxonomy of fossils and neontological anatomy. Traditional subdivisions of the Odontoceti have been questioned mostly by paleontologists, creating disagreement about a basic framework for the extant species. Alternative cladograms are shown in Figure 3. Broader issues in odontocete classification include the ranks of taxa, the role and definition of paraphyletic taxa, the typological status of some fossil families, and the value of traditional superfamily subdivisions. Contentious issues include the placement of Physeteridae, Ziphiidae, †Eurhinodelphidae, and species of river dolphins (traditionally joined in the Platanistoidea), and the definition and diagnosis of the †Agorophiidae and †Squalodontidae. Some of these issues were covered by Barnes (1984a, 1985b, 1990), Heyning (1989), de Muizon (1987, 1988a, 1988b, 1991), Milinkovitch et al (1993), and Novacek (1993).

†AGOROPHIIDAE Traditionally, the grade †Agorophiidae Abel 1914 encompases heterodont odontocetes which primitively retain parietals exposed across the intertemporal region. Rothausen (1968) viewed †Agorophiidae as a structural and temporal grade between †Archaeoceti and the odontocete family †Squalodontidae, whence most later odontocetes

arose. Supposed agorophiids include †*Xenorophus*, †*Archaeodelphis*, †*Microzeuglodon*, †*Atropatenocetus*, and †*Mirocetus*, and undescribed fossils from the Atlantic and Pacific margins of North America (Whitmore & Sanders 1977); all are apparently Late Oligocene. Fordyce (1981) rejected the notion of a grade Agorophiidae, suggesting that the only certain member of the family is †*Agorophius pygmaeus* (Late Oligocene, northwest Atlantic) of uncertain cladistic relationships. Supposed "agorophiids" probably represent a plethora of low-diversity and perhaps high rank archaic taxa, such as would be expected during early phases of evolutionary radiations.

PHYSETEROIDEA, PHYSETERIDAE, AND KOGIIDAE Sperm whales—Physeteridae—have an ancient and diverse record, although only one species, *Physeter catodon*, survives. Primitive sperm whales had both upper and lower functional teeth; more highly evolved sperm whales, like the living species, have reduced or vestigial upper teeth. The Early Miocene †*Diaphorocetus poucheti* (western South Atlantic; Kellogg 1925b), one of the oldest described physeterids, has a skull with a distinctive supracranial basin which presumably held a large fatty melon and spermaceti organ. The older †*Ferecetotherium* (Late Oligocene, Paratethys; Mchedlidze 1984) is probably also a physeterid (Barnes 1984b, 1985c). It is likely that †*Diaphorocetus*, †*Idiorophus*, and other early physeterids were, like *Physeter*, deep diving squid-eaters. Sperm whales are widespread in Pliocene and Miocene sediments (Kellogg 1925b), whence come many dubious genera and species based on isolated teeth.

The extant pygmy sperm whales (see Caldwell & Caldwell 1989)—Kogiidae—are closely related to Physeteridae. As in physeterids, there is a supracranial basin, but kogiids differ markedly in their small size, short rostrum, and skull details. The oldest clearly identified kogiids are †*Praekogia* (Barnes 1973, Early Pliocene) and †*Scaphokogia* (de Muizon 1988c, Late Miocene), both from the subtropical eastern Pacific. Isolated fossil teeth, reportedly those of kogiids, do not clearly belong to the family.

ZIPHIIDAE Extant beaked whales are medium to large, semi-solitary pelagic cetaceans which are mostly near-toothless squid-eaters. *Mesoplodon* is one of the most diverse extant genera of odontocetes (Mead 1989). Ziphiids range back to the Middle Miocene (Mead 1975b), with many specimens represented by robust, dense rostra that are sometimes recovered from the sea floor (Whitmore et al 1986). †*Squaloziphius emlongi* de Muizon 1991 (Early Miocene) was placed in a new subfamily †Squaloziphiinae, although its skull lacks convincing ziphiid features and appears more reminiscent of †Eurhinodelphidae. Ziphiids have been classified either with sperm whales in a superfamily Physeteroidea, or as a sister group to extant odontocetes other than physeterids (Figure 3).

PLATANISTOIDEA The extant Asiatic river dolphins, *Platanista* spp., are the basis for the family Platanistidae and superfamily Platanistoidea—taxa with a long and confusing history of use (Kellogg 1928; Simpson 1945; Fraser & Purves 1960; Barnes 1984b, 1985b; Zhou et al 1979; Heyning 1989). de Muizon (1987, 1988a, 1991) suggested that †Squalodontidae, †Squalodelphidae, ‡Dalpiazinidae, and Platanistidae form a clade unified by features of the scapula. Key features cannot be seen in some squalodontids and squalodelphid-like species, and more study is needed to justify this concept of Platanistoidea. Platanistoids sensu de Muizon have a longer fossil record than suspected previously, with moderate species diversity from the late Oligocene to about Middle Miocene, but they declined as delphinoids radiated dramatically late in the Miocene. Among platanistoids, only the extant *Platanista* spp. inhabit fresh waters.

†SQUALODONTIDAE The taxonomic limits of the Late Oligocene to Late Miocene family †Squalodontidae are not clear. In key reviews, Kellogg (1923) and Rothausen (1968) used the †Squalodontidae as a grade. Rothausen identified a need for cladistic review, but despite recent study (de Muizon 1991), clear synapomorphies have not been published. Squalodontids are still identified mainly by their close topographic match with well-documented skulls of †*Squalodon*. Probably only the long-beaked †*Eosqualodon*, †*Squalodon*, †*Kelloggia*, and †*Phoberodon* (and possibly †*Patriocetus*) are actually members of the †Squalodontidae, while many nominal squalodontids, including some named species of †*Squalodon*, probably belong in other families. The robust-snouted †*Prosqualodon australis* (Early Miocene, Southern Ocean) is enigmatic; Cozzuol & Humbert-Lan (1989) suggested affinities with dolphins (Delphinida), but relationships need more (cladistic) study.

†SQUALODELPHIDAE Three Early Miocene taxa are known: †*Notocetus* (southwest Atlantic), †*Medocinia* (northeast Atlantic), and †*Squalodelphis* (Mediterranean) (de Muizon 1988a). These have small slightly asymmetrical skulls with moderately long rostra, near-homodont teeth, and pterygoid lateral laminae (de Muizon 1987). Late Oligocene to Early Miocene species of †*Microcetus* and †*Prosqualodon* (southwest Pacific) and †*Sulakocetus* (Late Oligocene, Paratethys; Mchedlidze 1984) formerly referred to the †Squalodontidae probably belong within or close to the †Squalodelphidae.

†DALPIAZINIDAE de Muizon (1988a) established a new family, †Dalpiazinidae, and new genus, †*Dalpiazina*, for "†*Acrodelphis*" *ombonii* (Early

Miocene, Mediterranean). †Dalpiazina ombonii has a small symmetrical skull, reduced median dorsal exposure of frontals, and a long rostrum with many near-homodont teeth. Fordyce & Samson (1992) reported an undescribed earliest Miocene species from the southwest Pacific. †*Dalpiazina* is not firmly placed within the Platanistoidea (de Muizon 1991, Figure 15), and more study is needed.

PLATANISTIDAE The blind endangered Ganges and Indus River dolphins, *Platanista* spp., have no fossil record, and the time of invasion of fresh waters is unknown. Middle to Late Miocene marine species of †*Zarhachis* and †*Pomatodelphis* (both eastern North Atlantic) are closely related to *Platanista*, although they differ in rostral profiles and cranial symmetry, and in their development of pneumatized bony facial crests. These taxa have sometimes been placed in the †Acrodelphidae, a family that de Muizon (1988a) regarded as too poorly defined to use.

†EURHINODELPHOIDEA, †EURHINODELPHIDAE, AND †EOPLATANISTIDAE Dramatically long-beaked polydont dolphins in the extinct family †Eurhinodelphidae were widespread and moderately diverse during the Early to Middle Miocene (Kellogg 1925a, Barnes 1977, Myrick 1979). Late Oligocene provisional records (Fordyce 1992) include †*Iniopsis* (Paratethys), and an apparently Miocene species occurs in lacustrine sediments in central Australia (Fordyce 1983). Eurhinodelphid relationships are contentious (Figure 3; Barnes 1984b, 1990; de Muizon 1988a, 1991). The †Eoplatanistidae (Early Miocene, Mediterranean), which is unrelated to Platanistidae (Barnes 1984), includes only two species of †*Eoplatanista* Dal Piaz (1917; de Muizon 1988a). Like eurhinodelphids, these have small and virtually symmetrical skulls and long, polydont upper and lower jaws.

DOLPHINS—DELPHINIDA AND DELPHINOIDEA Dolphins are taxonomically diverse Odontoceti, mostly species of small to medium size. Formal names (Table 1) derive from the living common dolphin, *Delphinus delphis*. Most of the fossils are Late Miocene or younger, although the delphinoidean record extends back to the Late Oligocene. Simpson (1945) used a superfamily Delphinoidea to include Delphinidae, Monodontidae, and Phocoenidae, to which Barnes (1978, 1984a) added †Kentriodontidae and †Albireonidae. de Muizon (1988b) placed Pontoporiidae and Iniidae in a new superfamily Inioidea, the Lipotidae in the new Lipotoidea, and joined these superfamilies together with Delphinoidea in a new infraorder Delphinida. Diagnostic features of the skull include pterygoid sinus fossae and earbones (de Muizon 1988b, Barnes 1990).

†KENTRIODONTIDAE Most early dolphins, including many previously supposed Delphinidae, are placed in the grade family †Kentriodontidae. Ken-

triodontids are archaic dolphins with polydont teeth, elaborate basicranial sinuses, and symmetrical cranial vertices (Barnes 1978); most were probably 2–4 m long (see Figure 4). The oldest records, of Late Oligocene age [†*Oligodelphis* (Barnes 1985c), †*Kentriodon*? (Fordyce 1992)], provide few clues as to their sister group; an origin among †Squalodontidae (Barnes et al 1985) seems unlikely. Early to Late Miocene taxa, such as †*Kentriodon* and †*Pithanodelphis*, are generically diverse and widespread (Barnes & Mitchell 1984, Barnes 1985d). Kentriodontids seem related to the living Delphinidae, and perhaps were similarly pelagic.

†ALBIREONIDAE This Late Miocene–Pliocene family is known from the superficially porpoise-like †*Albireo whistleri* and an undescribed species from the temperate eastern North Pacific (Barnes 1984a). Barnes discounted close relationships with Phocoenidae because of differences in skull sutures, basicranial sinuses, and periotics, and noted that †*Albireo* is too specialized to have given rise to any extant phocoenid, delphinid, or monodontid. Barnes suggested that †*Albireo* was derived from kentriodontids, and de Muizon (1988b) placed †*Albireo* as a sister group to phocoenids.

MONODONTIDAE The living narwhal (*Monodon*) and beluga (*Delphinapterus*) live in Arctic waters (Rice 1984), and are known from Atlantic Subarctic Quaternary records, but Late Miocene to Late Pliocene monodontids, including †*Denebola*, occur in temperate to subtropical settings in the East Pacific (Barnes 1977, 1984a; de Muizon 1988c). Recent habitat shifts are indicated. Kasuya (1973), Barnes (1984b), and others regarded the Australian-Indonesian *Orcaella* as a monodontid, with intriguing paleozoogeographic and evolutionary implications, but de Muizon (1988b) and Heyning (1989) placed *Orcaella* firmly as a delphinid.

DELPHINIDAE Living delphinids are ecologically diverse. Their habits may be neritic (the small *Cephalorhynchus*) or oceanic (*Lissodelphis*, *Stenella*); some dolphins are near-cosmopolitan (the large *Orcinus*). A key diagnostic feature is cranial asymmetry, particularly involving the premaxillae (Barnes 1977, 1978). The oldest firm records of delphinids thus defined are from the later Middle or early Late Miocene (Barnes 1977). Abundant published older records (Kellogg 1928, Simpson 1945), often based on isolated elements, are mostly misidentifications or are based on undiagnostic specimens (see Barnes 1978). Barnes (1990) provided a general overview of delphinids as part of a review of *Tursiops*.

PHOCOENIDAE Porpoises, like Delphinidae, have a record that extends back to the Late Miocene, although with only six small extant species, the porpoises are less diverse than living delphinids (Barnes 1985a). †*Pisco-*

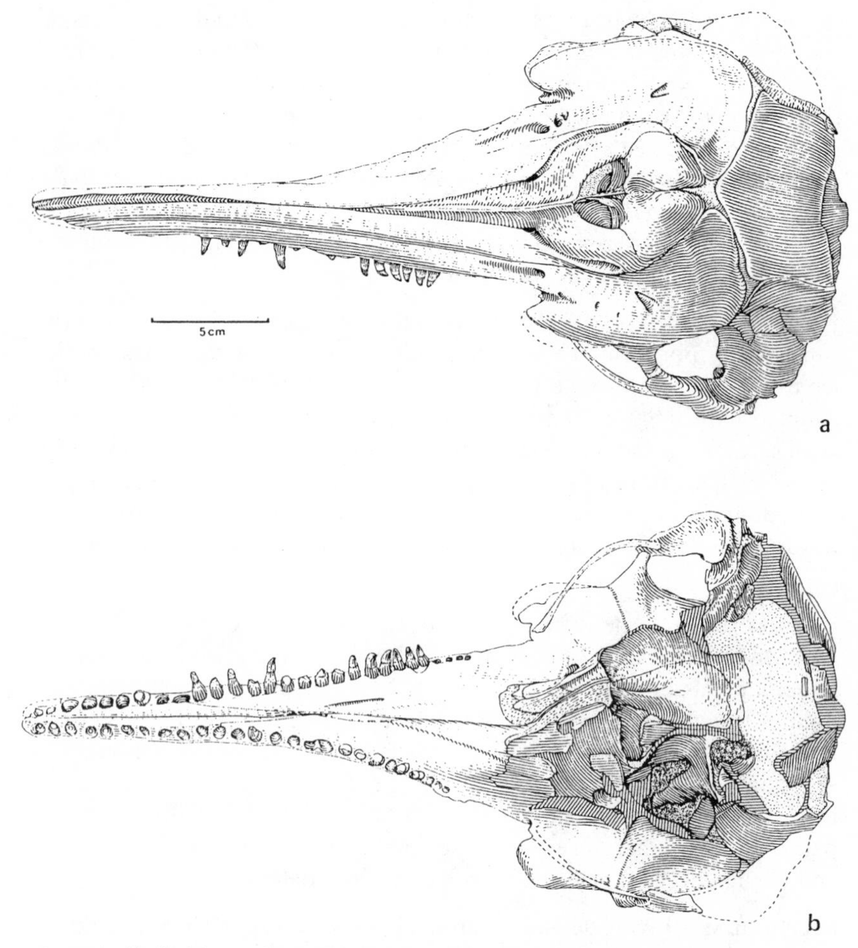

Figure 4 Skull of the small Late Miocene kentriodontid dolphin, *Atocetus nasalis* (originally named as *Pithanodelphis nasalis*), an extinct delphinoid from the eastern North Pacific Ocean. Such relatively complete fossil specimens are responsible for recent advances in the interpretation of cetacean morphology, phylogeny, and systematics. Views of the holotype cranium from Orange County, California: (*a*) dorsal view; (*b*) ventral view; modified from Barnes 1985d.

lithax tedfordi Barnes 1984a seems to be the most primitive fossil phocoenid and is much like generalized delphinids (Barnes 1993a, 1993b). The oldest members are East Pacific taxa such as †*Salumiphocaena* Barnes 1985a and †*Australithax* de Muizon 1988c; these have rather symmetrical skulls, premaxillary eminences, and basicranial sinus fossae like extant phocoenids. Earlier records of supposed phocoenids are misidentifications (Barnes 1978, Fordyce 1981).

PONTOPORIIDAE, INIIDAE, AND LIPOTIDAE The genus- and species-level status of living "non-platanistoid river dolphins" is fairly clear. The small, long-beaked species *Pontoporia blainvillei* (Pontoporiidae), which lives nearshore in the western South Atlantic, is the only extant pontoporiid. *Inia geoffrensis* (Iniidae) is a fresh-water species found in Amazon drainages, and *Lipotes vexillifer* (Lipotidae of Zhou et al 1979; Lipotinae of Barnes 1985b) lives in the Yangtze River, China. Relationships above the genus level are complicated; different arrangements affect the placement of fossils and interpretations of evolution and paleozoogeography.

Fossil *Pontoporia*-like taxa include species of †*Pliopontos* and †*Parapontoporia* from temperate to subtropical marine settings in the east Pacific (Barnes 1977, 1984a; de Muizon 1983, 1988c). Late Miocene or Pliocene †*Pontistes* and *Pontoporia* species come from marine sediments in Argentina (Cozzuol 1985). All pontoporiidas except for †*Parapontoporia* have virtually symmetrical cranial vertices, and most have long rostra and many tiny teeth. Barnes (1984a, 1985b) recognized †*Parapontoporia* as morphologically intermediate between *Pontoporia* and *Lipotes*, and thus subdivided Pontoporiidae into Pontoporiinae, †Parapontoporiinae, and Lipotinae. Conversely, de Muizon (1988b) placed †*Parapontoporia* with *Lipotes* in his Lipotoidea, uniting Pontoporiidae and Iniidae on features including markedly triangular transverse processes of the lumbar vertebrae. The only fossil placed close to *Lipotes* is †*Prolipotes*, based on an unrevealing fragment of mandible from freshwater Neogene sediments in China (Zhou et al 1984).

Inia geoffrensis (Amazon and Orinoco rivers) is the only extant iniid. Despite the lack of a significant fossil record, Grabert (1983) linked the evolution of *I. geoffrensis* to tectonic changes associated with Andean uplift. Like the Delphinidae, the Iniidae has been a repository for many fossils (Simpson 1945, Barnes 1978, Pilleri & Gihr 1979), but few species are placed there now. Cozzuol (1985) reviewed the Late Miocene or Pliocene fluvial †*Ischyrorhynchus* and †*Saurodelphis*, listing most other named iniid genera and species as synonyms.

INTERPRETING THE RECORD

Broader Patterns

The 10 or 11 species of early Middle Eocene Cetacea indicate a modest early radiation (Fordyce 1992), but cetacean family-level diversity was low during the Eocene. Major times of change occurred at the family level (Figure 1) during the Oligocene, when Odontoceti and Mysticeti appeared and radiated rather explosively, and during the Middle to Late Miocene, when extant delphinoid and mysticete groups appeared. There are no certain large-scale extinctions in cetacean fossil history. Important turnover at the genus level, for example, later in the Pliocene, is not revealed by family ranges. Clades below the suborder but above the family level are defined too poorly to be indicators of sure trends.

Clade durations vary (Figure 1), with 25 million years for Odontoceti and Mysticeti and, for families, about 10 million years (for example, Delphinidae, Phocoenidae, and Monodontidae) to over 20 million years (Physeteridae, Balaenidae). Short geologic histories for families such as Eschrichtiidae, †Mammalodontidae, and †Albireonidae, are probably artificial; a longer but unrecorded time span is more likely (Fordyce 1992), while others may ultimately be reduced to subfamily rank. Genera need care because some, such as †*Cetotherium* and †*Squalodon*, have been used as "scrap basket" grades. The extant *Balaenoptera* and *Megaptera* range to the late Middle Miocene (Barnes 1977), about 11 million years, and the extant dolphin *Tursiops* has middle Pliocene records (Barnes 1990). The reported Early to Late Miocene range (Rothausen 1968) for †*Squalodon* seems anomalously long and the family probably died out in Middle Miocene time. At the species level, Fordyce (1992) noted that no Eocene or Oligocene cetacean clearly ranges through more than one stage. Fossil records for living species of *Eschrichtius* and *Tursiops* are Pleistocene (Barnes & McLeod 1984, Barnes 1990). Overall, species durations of 1–2 million years seem likely, so that geographically separate occurrences of the same species may have stratigraphic value.

Diversity Trends

Discussion of cetacean diversity usually assumes that global diversity reflects broad oceanic heterogeneity, with local diversity patterns reflecting restricted ecological opportunities. Geographic diversity patterns for extant Cetacea include: the tropics—about 48 species; temperate regions—about 55 species; poles—about 28 species; and the eastern North Pacific—about 30 species (Barnes 1977). Faunal diversity for fossil assemblages along the northeast Pacific margin may exceed 20 species, comparable to the Recent, and with comparable ecological partitioning in terms of inferred habit (Barnes 1977).

Changes in diversity at genus and species level potentially reveal links between cetacean evolution and geological events. The record is too poor and too poorly calibrated to reveal global taxonomic diversity over short intervals (on the order of 1–2 million years), which would best indicate ecological and habitat opportunities, but more crude assessments are possible. For the Paleogene, Orr & Faulhaber (1975) plotted diversity changes at the genus level to show a marked drop in diversity from Late Eocene (ten genera) to Early Oligocene (two genera), followed by three genera in the Late Oligocene; diversity increased in the Miocene. Changes were attributed to changing paleotemperatures and plankton diversity (Barnes 1977). Previous characterizations of Oligocene time as having low cetacean diversity have now been dispelled. Fordyce (1992; Figure 1) emphasized species level rather than generic diversity. He reported an increase in species diversity for most of the Eocene, a drop in the latest Eocene, and a marked increase from Early to Late Oligocene. The latter was attributed to increasing oceanic heterogeneity, particularly circulation changes associated with the breakup of Gondwana and the creation of the Southern Ocean. A high diversity for Late Oligocene time (35–50+ species) is not fully comparable with the Recent, since the former reflects an over 5 my long sample. Data have not been published for Neogene species.

Structure and Function

Cetacean structures can be viewed as constrained by interacting components of constructional morphology (Fordyce 1989b). Historical factors include the plesiomorphies of cladists. These conservative features, many of which have long been fixed for functional reasons, reveal distant ancestry (for example, with artiodactyls; above), but little about the immediate adaptations of a species. Functional factors are reflected in profound adaptations for aquatic life, such as locomotory, feeding, and acoustic complexes. Fossils may reveal the minimum age of such complexes, in which innovations may reflect adaptation to new physical environments and geological change (Fordyce 1980, 1989b, 1992). Ecological interaction may be indicated; mirror-image changes in diversity, for example, the decline of platanistoids concomitant with the radiation of delphinoids (especially delphinids), could reflect ecological displacement through functional superiority. Structural or fabricational changes include those constrained by geometry, such as surface area: volume ratios, and area of the feeding apparatus relative to body mass.

Locomotion

Modern (e.g. post Oligocene) cetaceans are typically streamlined, have no external hind limbs, and have nonrotational elbow joints, paddle-like forelimbs for steering, elongate tail stocks, and horizontal propulsive tail flukes that are somewhat stiff and rigid. Terminal caudal vertebrae within the tail fluke are nearly square and the vertebra at the point of up-and-down rotation at the front of the fluke has a rounded intervertebral face.

These features exist in fossil skeletons, including those of archaeocetes, which indicates the presence of typical horizontal tail flukes early in cetacean evolution. Dorsal fins vary in size, shape, and presence among living species, and animals with missing or damaged fins can still navigate. Dorsal fins are not indicated by any skeletal structure, and are therefore not determinable from the fossil record.

The pectoral flipper in modern Cetacea, composed of the mammalian forelimb elements, has a rigid elbow joint and the flipper is used as a rudder in locomotion. In Archaeoceti, the more primitive elbow joint is flexible (rotatable), the humerus long, and the flipper not as streamlined as in extant Cetacea. Some (possibly all) archaeocetes also had external hind limbs [the pelvic facets on sacral vertebrae of †*Protocetus* suggest a well developed pelvis; hind limbs have been discovered on †*Prozeuglodon*; an undescribed protocetid from the Eocene of Georgia also has a pelvis (Hulbert & Petkewich 1991)]. Early archaeocetes might have been able to haul out on beaches as do pinnipeds. Requirements of amphibious land-breeding inferred from this probably compromised adaptation for long distance locomotion; if amphibiousness persisted in later archaeocetes, it was doubtless lost in early odontoceti and mysticeti whose key adaptations (echolocation, filter-feeding) allowed exploitation of food farther offshore (Fordyce 1980). The small functional hind limbs of †*Prozeuglodon isis* (Middle Eocene, Egypt) perhaps aided copulation (Gingerich et al 1990) or locomotion in shallow waters. Dense pachyostotic ribs in †*Basilosaurus cetoides* and other extinct cetacea possibly helped buoyancy control (de Buffrénil et al 1990) as in sirenians.

Hearing and Echolocation

All cetaceans have a modified ear structure that allows them to hear directionally in water. Cetaceans can also hear out of water. The auditory structure in archaeocetes and mysticetes is more primitive and less modified from the structures in terrestrial mammals than the anatomy of odontocetes, which is highly modified. No fossil or living mysticete shows bone features that might be construed as echolocation adaptations, undermining the idea (Milinkovitch et al 1993) of secondary loss of echolocation in mysticetes. Underwater acoustic communication is important in Odontoceti and Mysticeti (Wood & Evans 1980, Evans 1987) and, given the development of the tympanic bulla and auditory ossicles, was probably also important in the Archaeoceti.

All modern cetaceans lack an external ear, and only a tiny hole remains. Therefore, sound is received principally through other parts of the head. Underwater sound travels into the head and reaches the ears differentially, thus allowing directional hearing. The ear bones are isolated, to varying degrees among different groups, within a fat capsule, and this isolates the ear for hearing by impedence mismatching.

Active echolocation—the ability to detect the distance and size of objects underwater using reflected sound produced by the animal—has been evolved in the odontocetes. It involves the highly evolved hearing apparatus coupled with a unique sound making system (Norris 1968). An odontocete produces high-frequency sound ("clicks") by moving recycled air in a network of sacs and valves of the nasal passages, focuses it and projects it into the environment through the fatty melon (which acts as an acoustic lens) on the face (Norris 1968, Norris & Evans 1967). Asymmetrical structures of the face may be involved in the sophisticated production of both high- and low-frequency sounds.

Elaborate air sinuses have invaded the basicranium and orbit in several odontocete lineages (Fraser & Purves 1960), and may prevent sound generated in the nasal passages from directly impacting the ears and brain. Sound that reflects off objects in the environment and returns to the animal is transmitted through the side of the lower jaw via a thin area called the "pan bone," to the ear (Norris 1968).

This development may be a key feature of the successful Later Neogene oceanic delphinoid radiation (Fraser & Purves 1960). Perhaps the better echolocation of delphinoids, through more elaborate pterygoid sinuses and better-isolated earbones, allowed them to replace the more primitive groups of the Miocene. Most fossil odontocetes do have the same basic skull structure as modern odontocetes, and they undoubtedly could echolocate. However, most of the primitive Oligocene and Miocene odontocetes do not have asymmetrical crania, and this suggests that they had not acquired a sophisiticated level of echolocation. Squalodontids have a symmetrical facial region, and its depression indicates the presence of moderately developed facial muscles that could have been used to control the melon and nasal sacs. Periotics of Late Oligocene odontocetes show adaptations for receiving high-frequency sound (Fleischer 1976).

Body Size

Cetaceans include the largest living animals; even the smallest Cetacea are rather large for mammals. Minimum body size, which governs surface area to volume ratio, is constrained by rates of heat loss in water, and there is a predictable lower limit to body size for aquatic endotherms (Downhower & Blumer 1988). Large size, such as seen in Physeter and the migratory Mysticeti, could minimize heat loss in cold waters, could be an anti-predator strategy, and could minimize drag per unit mass during swimming or deep diving. The upper limit to size is probably constrained by the need to lose heat in proportion to mass, by the feeding apparatus (which must be scaled in proportion to body volume unless feeding method or type of food changes), and by the surface area of flukes (which must provide propulsion capable of moving the body mass).

Some theories of vertebrate evolutionary strategies view large species as K-selected indicators of highly stable environments, with small generalized species more characteristic of the early phases of radiations. This appears to be true of Cetacea. Among Paleogene Cetacea (Fordyce 1992), early protocetids at the base of the cetacean radiation were small, although probably larger than the smallest extant odontocetes. Contrary to published suggestions (reviewed by Fordyce 1992), Early Oligocene Cetacea include large species, discounting the notion that an episode of Late Eocene gigantism (†Basilosaurinae) was followed by times of smaller Oligocene species. Known Oligocene Odontoceti are small but some contemporaneous Mysticeti were large. There is little published evidence of large species early in the Miocene, but one late Miocene balaenopterid (Barnes et al 1987) is comparable in size with the living blue whale, *Balaenoptera musculus*.

In most odontocetes the males are larger than the famales; exceptions include only about six species among the beaked whales (Ziphiidae), porpoises (Phocoenidae), and Platanistidae, Pontoporiidae, and Iniidae. In all living mysticetes the females are larger than the males (Ralls 1976). While the reasons for this patterns are not known, it might be related to the unusually large size of cetacean newborn and the need to provide huge amounts of milk for rapid growth. Some cetacean species mature in less than five years.

Paedomorphism

Various species of living cetaceans in different families are paedomorphic. Paedomorphosis is the persistence of fetal or juvenile characters in reproductive age adults, and it appears evolutionarily as a derived character among cetaceans; it is less prevalent or nonexistent in earlier fossil members of the order. Paedomorphism appears in skulls of various derived mysticetes and odontocetes, and is extreme in the living species of phocoenids (Barnes 1985a), in which it is universally present and pronounced.

Primitive Cetacea lack paedomorphism, and most Miocene and earlier cetaceans have heavily built skulls with relatively small braincases, large zygomatic arches, large occipital and lambdoidal crests, prominent tuberosities, long rostra, large deeply socketed teeth, and an absence of cranial vacuities. Most of these structures are enhanced during maturation. Kentriodontids, the Late Oligocene to Late Miocene family of probable basal delphinoid dolphins, show little evidence of paedomorphism. Adults of

Recent phocoenids have relatively short rostra, large braincases, small zygomatic arches, and small occipital and lambdoidal crests. These cranial characters and proportions of adult phocoenid skulls are characteristic of newborn skulls of some other odontoceti, and their appearance in some groups is evidence of paedomorphism.

Feeding

Protocetids (Early Middle Eocene) have simple heterodont teeth with prominent anterior diastemata. †*Pakicetus* could probably shear and grind as well as snap (Gingerich & Russell 1990), in contrast to later Cetacea. †*Protocetus* has a long narrow rostrum and simple jaw articulation, presumably for forceps-like quick grasping of single prey, as also seen in many later Cetacea. Longer jaws in remingtonocetids perhaps indicate pursuit of fast prey. Basilosaurids include small to large species with complex denticulate diphyodont teeth, but are not polydont. Their cheek teeth commonly show apical wear, but shearing and crushing were probably minor aspects of their feeding. Basilosaurids show a greater range of tooth and jaw form, and body size, as expected for a geographically more widespread group living in increasingly heterogeneous, cooling oceans during later Eocene. Basilosaurids probably exploited resources not cropped by protocetids, for example, offshore or deep-dwelling prey.

Extant Odontoceti and Mysticeti are polydont with a single set of teeth (monophyodont). Mysticetes resorb their multiple simple tooth buds while in utero. Because many early fossil Odontoceti and Mysticeti are not polydont, it is possible that the marked polydonty in extant forms is not synapomorphous. Tooth and rostrum structure in the earliest mysticetes (Early Oligocene) is consistent with filter-feeding on a mass of prey. The large †*Llanocetus denticrenatus* has large diastemata between superficially basilosaurid-like teeth which carry palmate denticles (Mitchell 1989) and it probably filter-fed as do living crab eater seals; other toothed mysticetes are interpreted as filter-feeders (Barnes & McLeod 1984, Fordyce 1989a). Fordyce (1980, 1989b, 1992) suggested that filter-feeding in Mysticeti evolved in response to availability of new food and, in turn, to new oceanic circulation patterns associated with the creation of the Southern Ocean and the final breakup of Gondwana. Migration perhaps evolved at the same time, to allow seasonal high-latitude feeding alternating with breeding in thermally less-stressful temperate-tropical latitudes.

Baleen-bearing filter-feeding mysticetes evolved by the late Oligocene, 4–5 million years after toothed mysticetes appeared. Rostra of such animals as the early cetotheres are more similar to those of rorquals than to right whales or gray whales, which suggests that gulp-feeding (Pivorunas

1979, Lambertson 1983) was used. Different sizes and proportions of rostra among early cetotheres suggest that ecological (feeding) partitioning was comparable to that of living balaenopterids. Living balaenids skim-feed (Pivorunas 1979), using long baleen in a narrow arched rostrum. Early Miocene balaenids reveal that this is an ancient behavior. It is not known when the "bottom-ploughing" feeding method of gray whales (Eschrichtiidae) evolved because no primitive members of this group are recognized. Balaenopterids appeared by the Late Miocene, dramatically separated from cetotheres in having an abruptly depressed frontal which suggests a marked functional shift in origin and action of the temporalis muscle and perhaps in "gulp-feeding."

The earliest Odontoceti probably used echolocation to help hunt single prey. Fordyce (1980, 1990) suggested that, as with mysticete feeding, echolocation evolved about the Early Oligocene in response to changing food resources (especially below the photic zone), changing oceans, and continental rearrangement. Most odontocetes have a moderately attenuated rostrum like that seen in basal Cetacea, but a short, robust, broad rostrum has evolved in some groups (†*Prosqualodon* and Globicephalinae) sometimes with marked loss of teeth. Tooth number and tooth form vary widely in the long-beaked Odontoceti. There is a general trend toward homodonty and increased polydonty. Squalodontidae have long, deep, robust rostra and stout denticulate cheek-teeth; these features suggest more than just simple grasp-and-swallow-whole, and perhaps prey such as seabirds were taken. Extremely long rostra, often with, extreme polydonty, have evolved repeatedly and convergently (†Eurhinodelphidae, †Dalpiazinidae, Platanistidae, Pontoporiidae, Lipotidae, Iniidae; see ranges in Figure 1), although details vary in different taxa. The robust, wide pterygoid lateral lamina found in many of these taxa may be a homoplasy and related functionally to the muscles for the long beak, rather than a synapomorphy. Platanistid rostra may be laterally compressed (*Platanista*) or dorsoventrally compressed (†*Zarhachis*), while †*Eurhinodelphis* has a long subcylindrical rostrum in which the bizarrely toothless tip far overhangs the mandibles. Early Miocene physeterids and Middle Miocene ziphiids perhaps fed at depth on oceanic squid, as do their extant descendants. A reduced role for teeth in processing food is shown by convergent tooth loss in diverse groups, including some Physeteridae, Ziphiidae, Kogiidae, Monodontidae, and Globicephalinae. No odontocetes are reportedly filter-feeders, bottom-ploughers, or durophagous (crushers of hard-shelled prey) feeders. There is no clear ecological (feeding) overlap with other marine mammal groups.

Diverse feeding methods were attained quickly in odontocete history. By the end of the Oligocene, temperate-latitude odontocetes showed a range of feeding strategies comparable with those of living temperate-latitude faunas. Major changes in taxonomic and thus ecological patterns that occurred late in the Neogene may have been driven by feeding strategies. Squalodontids and eurhinodelphids became extinct, the formerly marine iniids, pontoporiids (and lipotids?), and platanistids gradually disappeared from the marine record, and monodontids disappeared from low to mid-latitudes. At about this time, the delphinids radiated rapidly, and it is possible that they ecologically displaced other groups. Delphinids are now the most diverse and widespread of the oceanic temperate-tropical smaller odontocetes, and indeed of all Cetacea.

ENVIRONMENTAL CHANGE AND PALEOZOOGEOGRAPHY

Major geological events that probably influenced cetacean evolution and distribution include (Figure 5): late stages in the closure of the Tethys seaway, with Africa and India suturing to Eurasia about the time the earliest protocetids appeared; the Paleogene opening of the Southern Ocean, culminating in the Oligocene with circum-Antarctic flow south of Australia and South America; closure of the Indo-Pacific seaway in the Neogene when Australia collided with the Indonesian arc; and closure of the Panamanian seaway in the Pliocene (references in Fordyce 1989b, 1992). Some events in cetacean history broadly correlate with these changes (Fordyce 1989b; see Figure 5).

Among living Cetacea, large species (e.g. most Mysticeti) are generally more widely distributed than small species (e.g. dolphins) (Barnes 1977, Evans 1987). In terms of evolution, large species were probably influenced only by large-scale geological changes, while small species were probably susceptible to, for example, the isolation of small basins through regression.

Many marine organisms have antitropical (or bitemperate or bipolar) disjunct distributions (Berg 1933, Hubbs 1952). North-south population pairs among large whales or taxon pairs among small cetaceans are typical (Davies 1963, Barnes 1985a). Cetacean diversity is highest in temperate latitudes and low in tropical and polar latitudes (Barnes 1977). Warm, tropical waters appear to separate populations, and Pleistocene glacial-interglacial oscillations may have separated populations (Davies 1963) for varying durations of time and brought about genetic isolation, and in some groups speciation. Even though few fossil cetaceans are known from more than one locality, some, such as the porpoise †*Piscolithax*, seem to have also had north-south taxon pairs. Further collection of specimans might show the phenomenon to be more common in the fossil record.

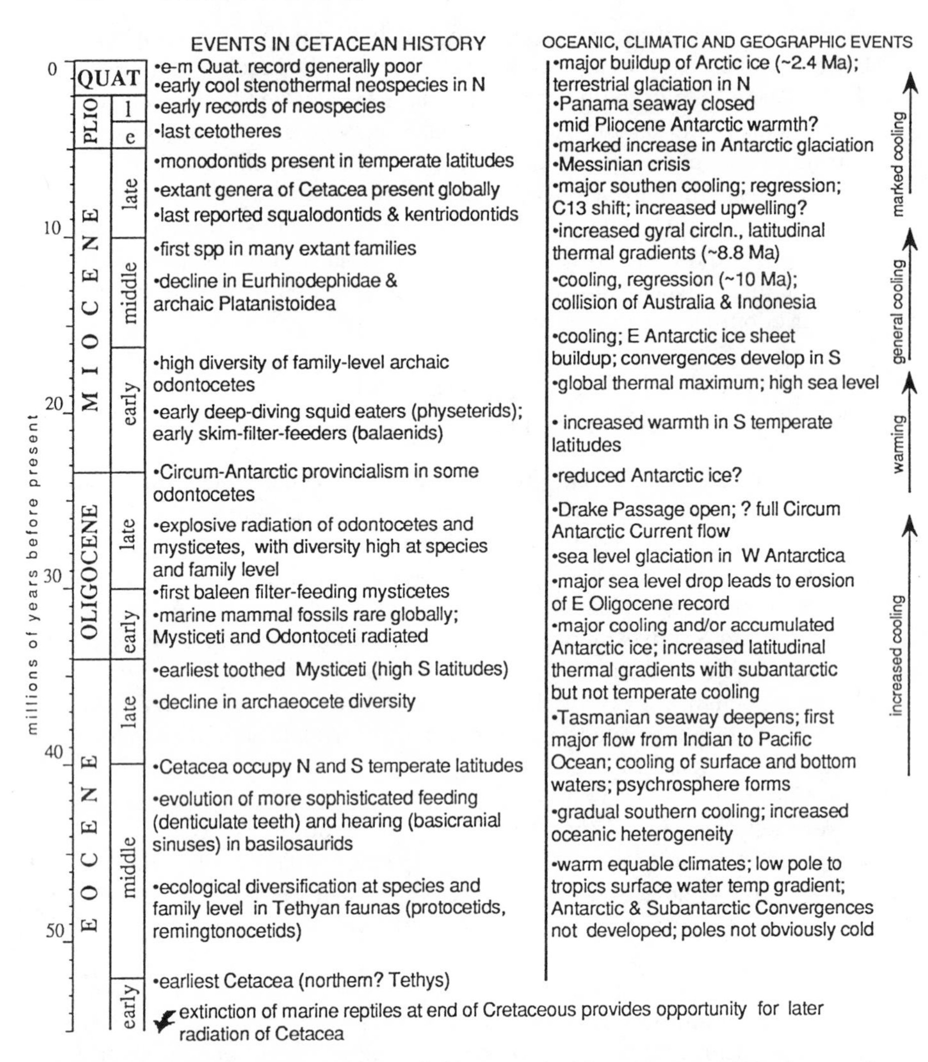

Figure 5 Events in cetacean and oceanic history, revised from Fordyce (1989). Sources of information listed by Fordyce (1989).

Acknowledgments

We thank Alastair G. Watson for critical comments, and Bob Connell and Phillip A. Maxwell for their help. Wendy Smith-Griswold, former Earth Sciences Division Scientific Illustrator at the Natural History Musuem of

Los Angeles County, prepared the illustration of the fossil dolphin skeleton in Figure 2*b* from a composite of photos of various bones. Linda Staniec helped with typing and editing the manuscript.

Literature Cited

Abel O. 1905. Les Odontocètes du Boldérien (Miocène supérieur) d'Anvers. *Mém. Mus. R d'Hist. Nat. Bel.* 3: 1–155

Allison PA, Smith CR, Kukert H, Deming JW, Bennett BA. 1991. Deep-water taphonomy of vertebrate carcasses: a whale skeleton in the bathyal Santa Catalina Basin. *Paleobiology* 17(1): 78–89

Barnes LG. 1973. *Praekogia cedrosensis*, a new genus and species of fossil pygmy sperm whale from Isla Cedros, Baja California, Mexico. *Contrib. Sci. Nat. Hist. Mus. Los Angeles Cty.* 247: 1–20

Barnes LG. 1977. Outline of eastern North Pacific fossil cetacean assemblages. *Syst. Zool.* 25: 321–43

Barnes LG. 1978. A review of *Lophocetus* and *Liolithax* and their relationships to the delphinoid family Kentriodontidae (Cetacea: Odontoceti). *Nat. Hist. Mus. Los Angeles Cty. Sci. Bull.* 28: 1–35

Barnes LG. 1984a. Fossil odontocetes (Mammalia: Cetacea) from the Almejas Formation, Isla Cedros, Mexico. *Paleobios* 42: 1–46

Barnes LG. 1984b. Whales, dolphins and porpoises: origin and evolution of the Cetacea. In *Mammals. Notes for a Short Course Organizied by P. D. Gingerich and C. E. Badgley*, ed. TW Broadhead, pp. 139–53. Knoxville: Univ. Tenn. Dept. Geol. Sci. 234 pp.

Barnes LG. 1985a. Evolution, taxonomy and antitropical distributions of the porpoises (Phocoenidae, Mammalia). *Mar. Mammal Sci.* 1(2): 149–65

Barnes LG. 1985b. Fossil pontoporiid dolphins (Mammalia: Cetacea) from the Pacific coast of North America. *Contrib. Sci. Nat. Hist. Mus. Los Angeles Cty.* 363: 1–34

Barnes LG. 1985c. Review: G. A. Mchedlidze, General features of the paleobiological evolution of Cetacea, 1984. English translation. *Mar. Mamm. Sci.* 1(1): 90–93

Barnes LG. 1985d. The Late Miocene dolphin *Pithanodelphis* Abel, 1905 (Cetacea: Kentriodontidae) from California. *Contrib. Sci. Nat. Hist. Mus. Los Angeles Cty.* 367: 1–27

Barnes LG. 1987. *Aetiocetus* and *Chonecetus*, primitive Oligocene toothed mysticetes and the origin of baleen whales. *J. Vertebr. Paleontol. Suppl.* 7(3): 10A

Barnes LG. 1989. *Aetiocetus* and *Chonecetus* (Mammalia: Cetacea); primitive Oligocene toothed mysticetes and the origin of baleen whales. *Fifth Int. Theriol. Congr., Abstr. Pap. and Posters, Rome, Italy* 1: 479. Rome: Int. Theriol. Congr.

Barnes LG. 1990. The fossil record and evolutionary relationships of the genus *Tursiops*. In *The Bottlenose Dolphin*, ed. S Leatherwood, RR Reeves, pp. 3–26. San Diego: Academic. 653 pp.

Barnes, LG. 1992. Whale. In *McGraw-Hill Yearbook of Science and Technology* 1993, pp. 482–84. New York: McGraw-Hill

Barnes LG. 1993a. Nuevo información sobre el marsopa fósil primitiva, *Piscolithax tedfordi* (Cetacea; Phocoenidae), de Isla de Cedros, México. [New information on the primitive fossil porpoise, *Piscolithax tedfordi* (Cetacea; Phocoenidae), from Isla de Cedros, México.] *Mem. Segunda Réunion Int. sobre Geol. de la Peninsula de Baja Calif., Univ. Autónoma de Baja Calif., Ensenada*

Barnes LG. 1993b. Nueva información sobre el esqueleto de la marsopa fósil primitiva, *Piscolithax tedfordi* (Cetacea; Odontoceti; Phocoenidae), de Isla de Cedros, México. [New information on the skeleton of the primitive fossil porpoise, *Piscolithax tedfordi* (Cetacea; Odontoceti; Phocoenidae), from Isla de Cedros, Mexico.] p. 2. *Resum.—Abstr., XVIII Réunion Int. Estud. Mamíferos Marinos*, Soc. Mex. Estud. Mamíferos Marinos, Univ. Autónoma de Baja Calif. Sur, La Paz, Baja California Sur, México

Barnes LG, Domning DP, Ray CE. 1985. Status of studies on fossil marine mammals. *Mar. Mamm. Sci.* 1(1): 15–53

Barnes LG, McLeod SA. 1984. The fossil

record and phyletic relationships of gray whales. In *The Gray Whale: Eschrichtius Robustus*, ed. ML Jones, SL Swartz, S Leatherwood, pp. 3–32. Orlando: Academic. 600 pp.

Barnes LG, Mitchell ED. 1978. Cetacea. Chapter 29, p. 582–602. In *Evolution of African Mammals*, ed. VJ Maglio, HBS Cooke, chap. 29, pp. 582–602. London/Cambridge. Mass.: Harvard Univ. Press

Barnes LG, Mitchell ED. 1984. *Kentriodon obscurus* (Kellogg, 1931), a fossil dolphin (Mammalia: Kentriodontidae) from the Miocene Sharktooth Hill Bonebed in California. *Contrib. Sci. Nat. Hist. Mus. Los Angeles Cty.* 353: 1–23

Barnes LG, Raschke RE, Brown JE. 1987. A fossil baleen whale. *Whalewatcher* 21(4): 7–10

Bearlin RK. 1988. The morphology and systematics of Neogene Mysticeti from Australia and New Zealand. *N. Z. J. Geol. Geophys.* 31: 257 (Abstr.)

Boreske JR, Goldberg L, Cameron B. 1972. A reworked cetacean with clam borings: Miocene of North Carolina. *J. Paleontol.* 46(1): 130–39

Cabrera A. 1926. Cetáceos fosiles del Museo de la Plata. *Rev. Mus. La Plata* 29: 363–411

Caldwell DK, Caldwell MC. 1989. Pygmy sperm whale—*Kogia breviceps* (de Blainville, 1838); dwarf sperm whale—*Kogia simus* (Owen, 1866). See Ridgway & Harrison 1989, pp. 235–60

Cozzuol MA. 1985. The Odontoceti of the "Mesopotamiense" of the Parana River ravines. Systematic review. *Invest. Cetacea* 7: 39–53

Cozzuol MA, Humbert-Lan G. 1989. On the systematic position of the genus *Prosqualodon* Lydekker, 1893 and some comments on the odontocete family Squalodontidae. *Fifth Int. Theriol. Congr., Abstr. Pap. and Posters, Rome, Italy.* 1: 483–84

Dal Piaz G. 1917. Gli Odontoceti del Miocene Bellunese. Parte quarta. *Eoplatanista italica. Mem. Inst. Geol. Univ. Padova* 5(2): 1–23

Dal Piaz G. 1977. Gli Odontoceti del Miocene Bellunese. Parte Quinta-Decima. *Cyrtodelphis, Acrodelphis, Protodelphinus, Ziphiodelphis, Scaldicetus. Mem. Inst. Geol. Univ.* Padova: 1–128

Davies JL. 1963. The antitropical factor in cetacean speciation. *Evolution* 17: 107–16

de Buffrénil V, de Ricqlés A, Ray CE, Domning DP. 1990. Bone histology of the ribs of the archaeocetes (Mammalia: Cetacea). *J. Vertebr. Paleontol.* 10(4): 455–66

de Muizon C. 1983. †*Pliopontos littoralis* un nouveau Platanistidae Cetacea du Pliocéne de la côte péruvienne. *C. R. Acad. Sci. Paris, Ser. II* 296: 1101–4

de Muizon C. 1984. Les vertébrés fossiles de la Formation Pisco (Pérou). Deuxième partie: Les odontocètes (Cetacea, Mammalia) du Pliocène inférieur de Sud-Sacaco. *Inst. Fr. Etudes Andines Mem.* 50: 1–188

de Muizon C. 1987. The affinities of *Notocetus vanbenedeni*, an Early Miocene platanistoid (Cetacea, Mammalia) from Patagonia, southern Argentina. *Am. Mus. Novitates* 2904: 1–27

de Muizon C. 1988a. Le polyphylétisme des Acrodelphidae, odontocètes longirostres du Miocène européen. *Bull. Mus. Natl. Hist. Nat. Paris* 4 Série 10 C(1): 31–88

de Muizon C. 1988b. Les relations phylogénétiques des Delphinida (Cetacea, Mammalia). *Ann. Paléontol.* 74(4): 159–227

de Muizon C. 1988c. Les vertébrés fossiles de la Formation Pisco (Pérou). Troisième partie: les odontocètes (Cetacea, Mammalia) du Miocène. *Inst. Fr. Etudes Andines Mem.* 78: 1–244

de Muizon C. 1991. A new Ziphiidae (Cetacea) from the Early Miocene of Washington State (USA) and phylogenetic analysis of the major groups of odontocetes. *Bull. Mus. Natl. Hist. Nat. Paris* 4 Série 12 C(3–4): 279–326

Downhower JF, Blumer LS. 1988. Calculating just how small a whale can be. *Nature* 335: 675

Evans PGH, 1987. *The Natural History of Whales and Dolphins*. London: Helm. 343 pp.

Fleischer G. 1976. Hearing in extinct cetaceans as determined by cochlear structure. *J. Paleontol.* 50: 133–52

Flower WH. 1883. On whales, past and present, and their probable origin. *Not. Proc. R. Inst. Great Brit.* 10: 360–376

Fordyce RE. 1980. Whale evolution and Oligocene Southern Ocean environments. *Palaeogeogr. Palaeoclimatol. Palaeoecol.* 31: 319–36

Fordyce RE. 1981. Systematics of the odontocete *Agorophius pygmaeus* and the family Agorophiidae (Mammalia: Cetacea). *J. Paleontol.* 55: 1028–45

Fordyce RE. 1983. Rhabdosteid dolphins (Mammalia: Cetacea) from the Middle Miocene, Lake Frome area, South Australia. *Alcheringa* 7: 27–40

Fordyce RE. 1984. Evolution and zoogeography of cetaceans in Australia. In *Vertebrate Zoogeography and Evolution in Australasia*, ed. M Archer, G Clayton, pp. 929–48. Carlisle: Hesperian. 1203 pp.

Fordyce RE. 1989a. Problematic Early Oligocene toothed whale (Cetacea,

?Mysticeti) from Waikari, North Canterbury, New Zealand. *N. Z. J. Geol. Geophys.* 32(3): 385–90

Fordyce RE. 1989b. Origins and evolution of Antarctic marine mammals. *Spec. Publ. Geol. Soc. London* 47: 269–81

Fordyce RE. 1991. A new look at the fossil vertebrate record of New Zealand. In *Vertebrate Palaeontology of Australasia*, ed. PV Rich, JM Monaghan, RF Baird, TH Rich, pp. 1191–316. Melbourne: Pioneer Design Studio/Monash Univ. 1437 pp.

Fordyce RE. 1992. Cetacean evolution and Eocene/Oligocene environments. In *Eocene-Oligocene Climatic and Biotic Evolution*, ed. D Prothero, W Berggren, pp. 368–81. Princeton, NJ: Princeton Univ. Press. 568 pp.

Fordyce RE, Samson CR. 1992. Late Oligocene platanistoid and delphinoid dolphins from the Kokoamu Greensand–Otekaike Limestone, Waitaki Valley region, New Zealand: an expanding record. *Geol. Soc. N. Z. Misc. Publ.* 63a: 66 (Abstr.)

Fraser FC, Purves PE. 1960. Hearing in cetaceans: evolution of the accessory air sacs and the structure of the outer and middle ear in recent cetaceans. *Bull. Brit. Mus. (Nat. Hist.), Zool.* 7: 1–140

Gingerich PD. 1992. Marine mammals (Cetacea and Sirenia) from the Eocene of Gebel Mokattam and Fayum, Egypt: stratigraphy, age and paleoenvironments. *Univ. Mich. Pap. Paleontol.* 30: 1–84

Gingerich PD, Russell DE. 1990. Dentition of early Eocene *Pakicetus* (Mammalia, Cetacea). *Contrib. Mus. Paleontol. Univ. Mich.* 28(1): 1–20

Gingerich PD, Smith BH, Simons EL. 1990. Hind limbs of Eocene *Basilosaurus*: evidence of feet in whales. *Science* 249: 154–57

Gingerich PD, Wells NA, Russell DE, Shah SM. 1983. Origin of whales in epicontinental remnant seas: new evidence from the Early Eocene of Pakistan. *Science* 220: 403–6

Grabert H. 1983. Migration and speciation of the South American Iniidae (Cetacea, Mammalia). *Z. Säugetierkd.* 49(6): 334–41

Hershkovitz P. 1966. Catalog of living whales. *Smithsonian Inst. Bull.* 246: 1–259

Heyning JE. 1989. Comparative facial anatomy of beaked whales (Ziphiidae) and a systematic revision among the families of extant Odontoceti. *Contrib. Sci. Nat. Hist. Mus. Los Angeles Cty.* 405: 1–64

Heyning JE, Mead JG. 1990. Evolution of the nasal anatomy of cetaceans. In *Sensory Abilities of Cetaceans*, ed. J Thomas, R Kastelein, pp. 67–79. New York: Plenum

Hooker JJ, Insole AN. 1980. The distribution of mammals in the English Palaeogene. *Tert. Res.* 3: 31–45

Hulbert RC, Petkewich RM. 1991. Innominate of a Middle Eocene (Lutetian) protocetid whale from Georgia. *J. Vertebr. Paleontol. Suppl.* 11(3): 36A

Kasuya T. 1973. Systematic consideration of Recent toothed whales based on the morphology of tympanoperiotic bone. *Sci. Rep. Whales Res. Inst.* 25: 1–103

Kellogg AR. 1923. Description of two squalodonts recently discovered in the Calvert Cliffs, Maryland; and notes on the shark-toothed dolphins. *Proc. US Natl. Mus.* 62(6): 1–69

Kellogg AR. 1925a. On the occurrence of remains of fossil porpoises of the genus *Eurhinodelphis* in North America. *Proc. US Natl. Mus.* 66(26): 1–40

Kellogg AR. 1925b. Two fossil physeteroid whales from California. *Carnegie Inst. Washington Publ.* 348(1): 1–34

Kellogg AR. 1928. History of whales—their adaptation to life in the water. *Q. Rev. Biol.* 3: 29–76, 174–208

Kellogg AR. 1931. Pelagic mammals from the Temblor Formation of the Kern River region, California. *Proc. Calif. Acad. Sci. Ser.* 4 19: 217–389

Kellogg AR. 1936. A review of the Archaeoceti. *Carnegie Inst. Washington Publ.* 482: 1–366

Kellogg AR. 1968. Fossil marine mammals from the Miocene Calvert formation of Maryland and Virginia. Part 5. Miocene Calvert mysticetes described by Cope. *US Natl. Mus. Bull.* 247: 103–32

Kimura M, Barnes LG, Furusawa H. 1992. Classification and distribution of Oligocene Aetiocetidae (Mammalia, Mysticeti) from western North America and Japan. *29th Int. Geol. Congr., Kyoto, Japan*, 2: 348 (Abstr.)

Kumar K, Sahni A. 1986. *Remingtonocetus harudiensis*, new combination, a middle Eocene archaeocete (Mammalia, Cetacea) from western Kutch, India. *J. Vertebr. Paleontol.* 6(4): 326–49

Lambertson RH. 1983. Internal mechanism of rorqual feeding. *J. Mamm.* 64(1): 76–88

Lancaster WC. 1986. The taphonomy of an archaeocete skeleton and its associated fauna. *Gulf Coast Assoc. Geol. Soc. Publ.* 119: 119–31

Mchedlidze GA. 1976. Osnovnyye cherty paleobiologicheskoy istorii Kitoobraznykh. (Basic features of the paleobiological history of the Cetacea.) *Akad. Nauk Gruz. SSR*, Inst. Paleobiologii, "Metsniereba" Press. 136 pp. (In Russian); English Summary. Transl. under the title "General features of the paleobiological evolution of Cetacea," 1984. Smithsonian Inst. Lib. and Natl. Sci. Found., TT 78-52026. New Delhi: Amerind. 139 pp.

Mchedlidze GA, 1984. *General Features of the Paleobiological Evolution of Cetacea*. New Delhi: Amerind. 136 pp.

Mchedlidze GA, 1989. *Fossil Cetacea of the Caucasus*. New Delhi: Amerind. 150 pp.

McLeod SA, Brownell RL Jr, Barnes LG. 1989. The classification of cetaceans: are all whales related? *Fifth Int. Theriol. Congr., Abst. Pap. and Posters, Rome, Italy* 1: 490

McLeod SA, Whitmore FC Jr, Barnes LG 1993. Evolutionary relationships and classification. In *The Bowhead Whale*, ed. JJ Burns, JJ Montague, CJ Cowles, pp. 45–70. Soc. Mar. Mamm. Spec. Publ. No. 2

Mead JG. 1975a. Anatomy of the external nasal passages and facial complex in the Delphinidae (Mammalia: Cetacea). *Smithsonian Contrib. Zool.* 207: 1–72

Mead JG. 1975b. A fossil beaked whale (Cetacea: Ziphiidae) from the Miocene of Kenya. *J. Paleontol.* 49(4): 745–51

Mead JG. 1989. Beaked whales of the genus *Mesoplodon*. See Ridgway & Harrison 1989, pp. 349–430

Milinkovitch MC, Ortí G, Meyer A. 1993. Revised phylogeny of whales suggested by mitochondrial ribosomal DNA sequences. *Nature* 361: 346–48

Miller GS. 1923. The telescoping of the cetacean skull. *Smithsonian Misc. Coll.* 76: 1–71

Mitchell ED. 1989. A new cetacean from the Late Eocene La Meseta Formation, Seymour Island, Antarctic Peninsula. *Can. J. Fish. Aquat. Sci.* 46: 2219–35

Myrick AC Jr. 1979. *Variation, taphonomy, and adaptation of the Rhabdosteidae (=Eurhinodelphidae) (Odontoceti, Mammalia) from the Calvert Formation of Maryland and Virginia*. PhD thesis. Univ. Calif., Los Angeles. 411 pp. (Available through Univ. Microfilms, Ann Arbor, Mich.)

Norris, KS. 1968. The evolution of acoustic mechanisms in odontocete cetaceans. In *Evolution and Environment: A Symposium Presented on the Occasion of the 100th Anniversary of the Foundation of the Peabody Museum of Natural History at Yale University*, ed. ET Drake, pp. 297–324, New Haven, Conn.: Yale Univ. Press. 470 pp.

Norris, KS, Evans, WE. 1967. Directionality of echolocation clicks in the rough-tooth porpoise, *Steno bredanensis* (Lesson). In *Marine Bio-Acoustics, Proc. Second Symp. on Marine Bio-Acoustics, New York*, ed. W.N. Tavolga, pp. 305–16, Oxford: Pergamon

Novacek MJ. 1993. Genes tell a new whale tale. *Nature* 361: 298–99

Oishi M. 1985. Outline of studies on the fossil cetaceans in Japan. *Monogr. Assoc. Geol. Collab. Jpn.* 30: 127–35

Okazaki Y. 1988. Oligocene squalodont (Cetacea/Mammalia) from the Ashiya Group, Japan. *Kitakyushu Shiritsu Shizenshi Hakubutsukan Kenkyu Hokoku* 8: 75–80

Orr WN, Faulhaber J. 1975. A middle Tertiary cetacean from Oregon. *Northw. Sci.* 49: 174–81

Padian K, Lindberg DR, Polly PD. 1993. Cladistics and the fossil record: the uses of history. *Annu. Rev. Earth Planet. Sci.* 22: 63–91

Pilleri G, 1989. *Beiträge sur Palontologie der Cetaceen Perus*. Ostermundigen: Hirnanatomisches Inst. 233 pp.

Pilleri G, Gihr M. 1979. Skull, sonar field and swimming behavior of *Ischyrorhynchus vanbenedeni* (Ameghino 1891) and taxonomical position of the genera *Ischyrorhynchus, Saurodelphis, Anisodelphis* and *Pontoplanodes*. *Invest. Cetacea* 10: 17–70

Pivorunas A. 1979. The feeding mechanisms of baleen whales. *Am. Sci.* 67(4): 432–40

Pritchard BG. 1939. On the discovery of a fossil whale in the older Tertiaries of Torquay, Victoria. *Victorian Nat.* 55: 151–59

Ralls K. 1976. Mammals in which females are larger than males. *Q. Rev. Biol.* 51: 245–76

Rice DW. 1977. A list of the marine mammals of the world. *Natl. Mar. Fish. Serv. Spec. Sci. Rep.* 711: 1–15

Rice DW. 1984. Cetaceans. In *Orders and Families of Recent Mammals of the World*, ed. S Anderson, J Knox-Jones, pp. 447–90. New York: Wiley

Ridgway SH, Harrison RJ, eds. 1985. *Handbook of Marine Mammals. Vol. 3. The Sirenians and Baleen Whales*. London: Academic. 362 pp.

Ridgway SH, Harrison RJ, eds. 1989. *Handbook of Marine Mammals. Vol 4. River Dolphins and the Larger Toothed Whales*. London: Academic. 442 pp.

Rothausen K. 1968. Die systematische Stellung der europäischen Squalodontidae (Odontoceti: Mamm.). *Paläontol. Z.* 421/2: 83–104

Rothausen K. 1971. *Cetotheriopsis tobieni* n. sp., der erste paläogene Bartenwale (Cetotheriidae, Mysticeti, Mamm.) nördlich des Tethysraumes. *Abh. Hess. Landesmtes Bodenforsch.* 60: 131–48

Rothausen K. 1986. Marine Tetrapoden im

tertiären Nordsee-Becken. 1. Nord- und mitteldeutscher Raum ausschliesslich Niederrheinische Bucht. In *Nordwestdeutschland im Tertiär*, 1 (Beitr. Reg. Geol. Erde 18): 510–57. Berlin-Stuttgart: Gebrüder Borntraeger

Russell LS. 1968. A new cetacean from the Oligocene Sooke Formation of Vancouver Island, British Columbia. *Can. J. Earth Sci.* 5: 929–33

Sanders AE, Barnes LG. 1993. Late Oligocene *Cetotheriopsis*-like mysticetes (Mammalia, Cetacea) from near Charleston, South Carolina. Proc. 1991 Marine Mammal Symp., Soc. Vertebr. Paleontol. Annu. Meet., Proc. San Diego Soc. Nat. Hist. In press.

Schäfer W, 1972. *Ecology and Paleoecology of Marine Environments*. Edinburgh: Oliver and Boyd. 568 pp.

Simpson GG. 1945. The principles of classification and a classification of mammals. *Bull. Am. Mus. Nat. Hist.* 85: i–xvi, 1–350

Slijper EJ, 1979. *Whales*. London: Hutchinson. 511 pp.

Squires RL, Goedert JL, Barnes LG. 1991. Whale carcasses. *Nature* 349: 574

Thewissen JGM, Hussain ST. 1993. Origin of underwater hearing in whales. *Nature* 361: 444–45

Van Gelder RG. 1977. Mammalian hybrids and generic limits. *Am. Mus. Novi.* 2635: 1–25

Van Valen L. 1966. Deltatheridia, a new order of mammals. *Bull. Am. Mus. Nat. Hist.* 132(1): 1–126

Van Valen L. 1968. Monophyly or diphyly in the origin of whales. *Evolution* 22: 37–41

Watson AG. 1991. Where does the dolphin bend its tail? The hinge vertebra in *Tursiops truncatus*. *Abstr. Biennial Conf., Biol. of Marine Mammals, Chicago*, 1991, p.71

Whitmore FC. 1975. Remington Kellogg, 1892–1969. *Biogr. Mem. Natl. Acad. Sci.* 46: 159–89

Whitmore FC, Morejohn GV, Mullins HT. 1986. Fossil beaked whales—*Mesoplodon longirostris* dredged from the ocean bottom. *Natl. Geogr. Res.* 2(1): 47–56

Whitmore FC, Sanders AE. 1977. Review of the Oligocene Cetacea. *Syst. Zool.* 25: 304–20

Wood FG, Evans WE. 1980. Adaptiveness and ecology of echolocation in toothed whales. In *Animal Sonar Systems*, ed. RG Busnel, JF Fish, pp. 381–425. New York: Plenum

Yablokov AV. 1965. Convergence or parallelism in the evolution of cetaceans. *Int. Geol. Rev.* 7: 1461–68

Zhou K, Qian W, Li Y. 1979. The osteology and the systematic position of the baiji, *Lipotes vexillifer*. *Acta Zool. Sin.* 25: 58–74

Zhou K, Zhou M, Zhao Z. 1984. First discovery of a Tertiary platanistoid fossil from Asia. *Sci. Rep. Whales Res. Inst. Tokyo* 35: 173–81

Annu. Rev. Earth Planet. Sci. 1994. 22:457–97

METEORITE AND ASTEROID REFLECTANCE SPECTROSCOPY: Clues to Early Solar System Processes

Carlé M. Pieters

Department of Geological Sciences, Brown University, Providence, Rhode Island 02912

Lucy A. McFadden

California Space Institute, University of California at San Diego, La Jolla, California 92093 and Department of Astronomy, University of Maryland, College Park, Maryland 20742

KEY WORDS: absorption bands, asteroid composition, mineralogy, remote compositional analyses

INTRODUCTION

Although Earth has a magnetic shield that protects it from energetic ions and a cushioning atmosphere that catches the tons of small particles (< 1 mm) that bombard it daily, solid material from extraterrestrial sources nevertheless reaches the surface of Earth many times a year (Dodd 1981). When the largest of these bodies impact the surface with great velocity, they release immense energy, usually form a crater, and often disrupt the terrestrial biosphere. Fortunately, such events occur relatively infrequently, although individual impacts are currently unpredictable beyond a statistical accounting (Shoemaker et al 1979, Grieve 1982). Cosmic material between the size of small dust particles and the larger crater-forming bodies is slowed by the terrestrial atmosphere upon entry. The outermost portion of these objects is melted and ablated during passage forming a fusion crust which protects the interior, and they land

0084–6597/94/0515–0457$05.00

as solid objects on the surface. These are the meteorites, our free samples of other parts of the solar system.

A handful of well-documented meteorites are now known to have originated from the Moon and probably Mars. The vast majority of these extraterrestrial samples, however, have been shown to have originated on the smaller bodies of the solar system—i.e. asteroids. These meteorites all have ancient radiometric ages $\sim$4.5 Aeons indicating that their principal formation age is that of the solar system as a whole. Study of the major, minor, and isotope element chemistry of meteorites provides a wealth of information related to their starting composition and subsequent processes involved in their chemical evolution. The petrographic texture documents the thermal and physical processes through which the samples have evolved. These clues form the basis of our understanding of the pre-planet forming stages of the solar system and the subsequent evolution of small bodies. While the meteorites provide invaluable compositional information, they do not provide a full inventory of small bodies in the solar system, nor do they provide information about their heritage and the regions of the asteroid belt, or the solar system in general, which they represent. Gaffey et al (1993a) describe the meteorites as solar system "float," aptly employing geologic terminology for scattered rock fragments not found in direct association with their source regions or outcrops.

When one takes an astronomical perspective, there are thousands of small bodies located predominantly between Mars and Jupiter which comprise the main asteroid belt. The instantaneous distribution of these small bodies is shown in Figure 1. Also easily seen in the plane view of the solar system are the inner planet-crossing asteroids and the Trojan asteroids at the leading (L4) and trailing (L5) lagrangian points of Jupiter's orbit. In addition, a growing number of small bodies have been recently discovered beyond the orbit of Jupiter and even Pluto (e.g. Steel et al 1992, Jewitt & Luu 1993). The other primary structural components of the asteroid belt are its resonances, regions where perturbations have resulted in zones of depletion of planetary materials. Commensurate resonances with Jupiter (e.g. 3:1 and 2:1) form some of the most prominent low-density zones in the main asteroid belt, called the Kirkwood gaps (e.g. Froeschle & Greenberg 1989). Secular resonances that form from the perturbations of all planets (e.g. ν_6, ν_{16}) are more complex but are also clearly evident in the structure of the main belt. The theory and simulations of the dynamics of secular resonances were reviewed by Scholl et al (1989) and later by Milani & Knežević (1992). Both types of resonances are notable features of the asteroid belt and are zones that are dynamically unstable for planetary materials.

In the past few decades the combined studies of meteorite composition

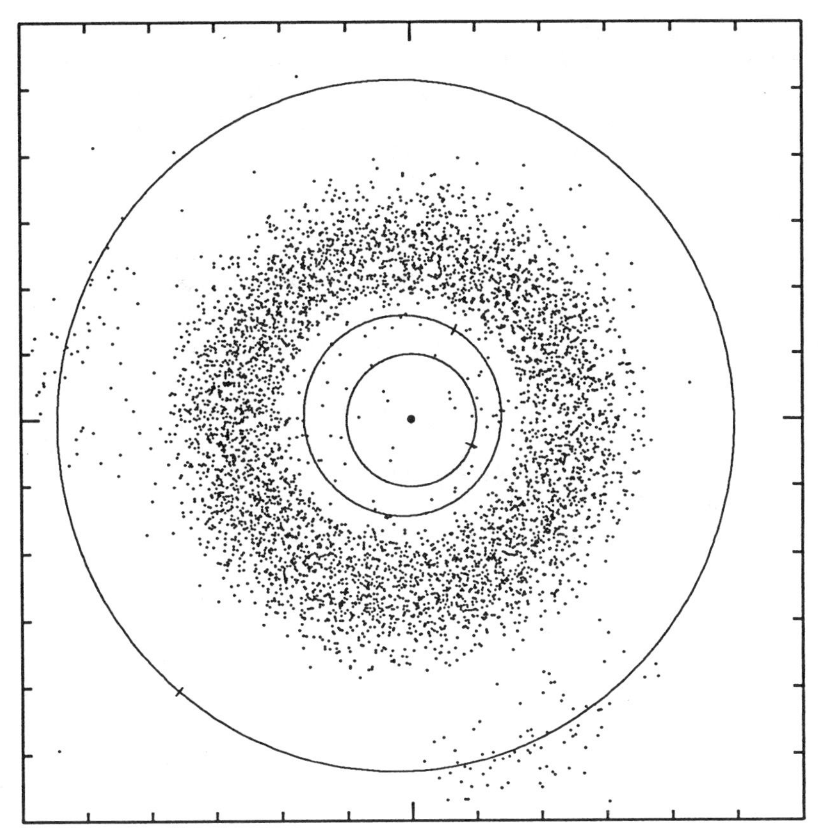

Figure 1 Plane view of the solar system on August 31, 1994, the scheduled date of asteroid 1620 Geographos flyby by the *Clementine* spacecraft. The Sun is the larger dot at one focus of the solar system. Concentric circles represent the orbits of Earth, Mars, and Jupiter from the focus outward. Each dot represents the location of one of 5531 numbered asteroids on the above date.

and petrology, the physical properties of asteroids, and the dynamical evolution of the asteroids, have documented several fundamental links between specific meteorite and asteroid classes. A more detailed understanding of the relation of meteorites to asteroids and derived information on the composition, structure, and evolution of the solar system focuses on two basic questions. First, which asteroids are the parent bodies of the different types of meteorites studied in terrestrial laboratories and what is the spatial distribution and abundance of such asteroids? A related fundamental question is: To what extent does the meteorite collection represent extraterrestrial material in the asteroid belt and the solar system? Second, what are the composition and basic physical properties of small bodies in the solar system which are not represented in the meteorite collection and how abundant are such bodies? As reviewed here, significant progress has been made addressing these issues, and the physical processes which produced the asteroid belt as it exists today are better understood. The same fundamental questions remain for many of the meteorite and asteroid classes, and it is the task of the next several decades to accurately survey and identify the composition and physical characteristics of materials formed at the very beginning of solar system evolution.

Spacecraft encounters with several selected asteroids will clearly play a central role in this endeavor, but because of the great diversity of asteroids and meteorites, remote sensing approaches from the Earth or Earth orbit must necessarily provide the baseline information. This review summarizes the current capabilities of, advances in, and issues surrounding reflectance spectroscopy and its use as a tool to evaluate the mineralogy of meteorites and asteroids.

Meteoriticists and asteroid scientists have developed independent (and largely unrelated) classification schemes for meteorites and asteroids. These classification schemes provide an overview of the diversity of both groups of materials. The primary classes of meteorites are summarized in Table 1. This classification is based principally on compositional distinctions between meteorites. Petrographic texture of the silicates is often used for subclasses, and both aqueous and metamorphic alteration is recognized (McSween 1979, Scott et al 1989). The abundance of specific meteorite types is normally evaluated through the statistics of observed meteorite falls (and recovery). Since stony meteorites weather more readily than iron meteorites in the terrestrial environment, the use of fall statistics eliminates the bias that weathering imparts on the statistics of meteorites that are simply found in a field long after their arrival.

The most commonly used classification scheme for asteroid taxonomy is summarized in Table 2. The basic unit of classification is a spectral class identified by a letter designation (D,P,C, . . .). Members of a class are often referred to as having a characteristic spectral type. Determination of a class designation is well defined in fields such as botanical taxonomy and requires that specific boundaries between classes can be defined and that variations between classes are discontinuous. The analogue to an asteroid class in botanical taxonomy, for example, is the species (Jeffrey 1982). The most frequently used class designation adopted for asteroids is defined by a statistical minimal tree algorithm using 8-color extended visible data and albedo information developed by Tholen (1984) and discussed in Tholen

Table 1 Summary of meteorite classes, their mineralogy, and fall statistics

Class*	Mineralogy[a]	Falls[b]	Antarctic[c]
Chondrites (containing chondrules)			
Carbonaceous—roughly solar composition; abundant Carbon			
CI	hydrous; sheet silicates, no chondrules, no Fe^0	5	0
CM	hydrous; sheet silicates, no Fe^0	18	14
CO	mafic minerals; small chondrules, minor Fe^0	5	4
CV	mafic minerals; Ca-Al-inclusions, minor Fe^0	7	5
Ordinary—olivine + low-Ca Pyroxene + feldspar + metal			
H	high in total Fe; > 50% Fe^0	276	904
L	low in total Fe; ~ 1/3 Fe^0	319	490
LL	low in total Fe, most oxidized	66	83
Enstatite—Enstatite mineralogy + metal			
E	iron highly reduced, mostly Fe^0	13	9
Achondrites (melted; differentiated)			
Howardites	breccias of Eucrite-, Diogenite-like material	18	4
Eucrites	Low-Ca pyroxene + plagioclase	25	18
Diogenites	Mg-pyroxene	9	9
Urelites	C-rich; olivine + pyroxene + volatiles	4	11
Aubrites	Enstatite	9	2
SNC[d]	basaltic, olivine	4	3
Lunar[e]	anorthositic breccia	0	(9)
Primitive[f]	olivine + pyroxene + metal + troilite ± plagioclase	(4)	(14)
Stony-irons (metal + silicates)			
Mesosiderites	Fe-Ni + HED achondrites	6	3
Pallasites	network of Fe-Ni + olivine	3	0
Irons	Fe-Ni (kamacite and taenite) + triolite	42	21

* There are also several unclassified or unusual meteorites.
[a] After Sears & Dodd (1988).
[b] After Sears & Dodd (1988), Graham et al (1985).
[c] After Lipschutz et al (1989).
[d] Shergottites, Nakhlites, Chassignites; probable Mars parent body (McSween 1985).
[e] Lindstrom et al (1991).
[f] McCoy et al (1993). Acapulcoites and Lodranites: achondritic texture, chondritic composition. (Numbers in parentheses were tabulated later and are derived from larger statistics.)

& Barucci (1989). The minimal tree analysis of these data does a good job of defining class boundaries, and as such it fulfills the principal requirement of a class. Although other asteroid classification systems have been devised [see review in Tholen (1984) for historical perspective, and Tholen & Barucci (1989) for a review of recent systems], Tholen's classification has gained the widest acceptance because it encompasses the largest asteroid data set and has been resilient as new data become available. It should be noted that this asteroid taxonomy is based on purely statistical relationships within a few sets of asteroid data (albedo and 8-color measurements).

Table 2 Taxonomy of main belt asteroids with associated albedo and spectral features

Class	Measured #	IRAS albedo	8-color photometry description	Additional features often observed
P	46	<0.04	featureless spectrum increasing with increasing wavelength	
C	227	0.04–0.06	weak UV band, flat-reddish reflectance vis-NIR	0.7, 0.6 and 0.43 μm, some have 3-μm bands
B	21	0.04–0.08	weak UV band, decreasing reflectance toward NIR	NIR variation, some have 3-μm bands
F	50	0.04–0.08	weak UV band, decreasing reflectance toward NIR	possibly broad, weak NIR feature, some have 3-μm bands
D	59	0.04–0.07	strongly increasing with increasing wavelength	possibly a band at 2.2 μm
G	16	0.09	strong UV band <0.4 mm, flat vis-NIR	0.6, 0.67, 0.7, 3.0 μm bands
T	13	<0.10	weak, broad UV-vis absorption flat NIR	significant variation in NIR reflectance
M	42	0.12–0.22	flat-slightly reddish vis-NIR flat vis-NIR	significant NIR variation, high radar albedo broad band >1.6 μm
S	347	0.14–0.17	UV band <0.7 μm, 1.0 μm band, red slope vis-NIR	2.0 μm band, significant variations
A	7	0.12–0.39	strong UV and 1 μm band	composite olivine band, no 2.0 μm band
Q	1	0.16–0.21*	strong UV, 1.0 μm band no red slope	2.0 μm band no IR red slope
R	1	0.34	strong UV and 1.0 μm band red slope	strong 2.0 μm band IR red slope
V	1	0.38	strong UV and 1 μm band	strong 2.0 μm, weak 1.25 μm inflection
E	13	0.38	slightly increasing reflectance with increasing wavelength	weak features NIR; variability

From PDS data base of 844 asteroids contributed by Tholen, with no observational bias correction. [PDS Small Bodies Node, 8/93.]

* Upper range is from ground-based observation of 1862 Apollo, a near-Earth asteroid.

The relevance of the classification scheme to the scientific endeavor—in this case clues to early solar system processes—is dependent on the degree to which the taxa represent discrete compositional groups. Although composition affects the 8-color data, it is not a unique relation. As will be seen in subsequent sections, detailed spectroscopic measurements that are more sensitive to compositional properties identify asteroid compositions that do not necessarily correspond directly to the taxonomy.

Excellent and relatively up-to-date broad discussions of meteorite research can be found in *Meteorites and the Early Solar System* (Kerridge & Matthews 1988). Meteorite surveys include Wasson (1985), Dodd (1981), and Lipschutz (1993). Broad overviews of asteroid research can be found in *Asteroids* (Gehrels 1979) and *Asteroids II* (Binzel et al 1989). A thoughtful interpretation of asteroid spectra and the structure of the solar system can be found in Bell et al (1989). The most recent summary of asteroid spectral measurements and integration of compositional interpretations has been prepared by Gaffey et al (1993a).

FOUNDATIONS OF REMOTE MINERALOGICAL ANALYSIS

Although asteroids and meteorites have been studied for centuries, it is only during the past several decades that analytical tools have become available with increasing sophistication to link them compositionally. Broad-band color measurements, such as the classic UBV system used in astronomy, and the later 8-color survey, allowed the asteroids to be distinguished and classified in groups. The more extensive spectroscopic measurements that encompass the visible through near-infrared, on the other hand, provide the principal diagnostic information for compositional analyses.

Historic Overview of Early Work

In the late 1960s astronomers began to use sensitive photoelectric detectors that allowed high precision measurements of solar system bodies using a large number of filters across the extended visible (0.32–1.1 μm) spectrum . Tom McCord pioneered this field with spectroscopic observations of the Moon (McCord 1968), then turned the instrument to look at the asteroid Vesta with great success (McCord et al 1970). At the same time, crystal field theory was being perfected to explain the physical principles accounting for absorptions due to transition metal ions in mineral structures (Burns & Fyfe 1967, Burns 1970a), and reflectance measurements were being made of geologic materials in the laboratory demonstrating the existence of diagnostic absorption bands (Adams & Felice 1967; Adams 1974, 1975; Hunt & Salisbury 1970; Hunt et al 1971, 1973a,b). The success of these ventures led to several major pieces of research that laid the foundation for much subsequent meteorite and asteroid spectroscopic analyses.

In the early 1970s a photoelectric assessment of the spectral properties of asteroids was driven by scientific interest and available technology. C. R. Chapman led a major asteroid survey (Chapman 1972) that resulted in a series of papers describing the spectra of asteroids based on extended visible (0.3–1.1 μm) spectra (Chapman et al 1973 a,b; Chapman 1976;

Chapman & Gaffey 1979). Chapman & Salisbury (1973) made an initial comparison between available data on asteroid and meteorite properties and showed this to be a very productive area of investigation. In the early 1970s Johnson & Fanale (1973) performed a laboratory survey of the spectral reflectance properties of low-albedo meteorites. In the mid-1970s M. J. Gaffey undertook a detailed laboratory investigation of the spectral properties of carefully selected meteorites (Gaffey 1974). The resulting manuscript (Gaffey 1976) remains one of the most complete overviews of meteorite spectral properties from the visible to the near-infrared (0.35–2.5 μm). These two fields—asteroid and meteorite spectral studies—converged to allow more detailed interpretations of the mineral character of asteroids, and also resulted in the conclusion that mineral assemblages generally similar to most meteorite types can be found in the asteroid belt, with the exception of the abundant ordinary chondrites (Gaffey & McCord 1978, 1979).

There were two additional important milestones that laid the foundation for much subsequent research linking asteroids and meteorites. The first was a survey conducted by McFadden (1983) of near-Earth asteroids, a prime source for meteorites currently falling to Earth. These asteroids have orbits that cross or are near to that of Earth. They are thus good potential sources for Earth-destined meteorites, but are dynamically short lived and cannot remain in near-Earth orbit for longer than 10^7–10^8 years (Wetherill 1974). Near-Earth asteroids need to be replenished regularly. McFadden et al (1984, 1985) showed that almost all the types of asteroids observed in the main asteroid belt are seen in the near-Earth asteroid population, thus strengthening the link between the two populations. A notable exception was 1862 Apollo, a single member of the Q class of objects, which provided a close analogue to the most abundant type of meteorites, the ordinary chondrites. 1862 Apollo provided a possible source for ordinary chondrites, but its apparent uniqueness provided little information about the significance to the structure of the solar system of this abundant meteorite type, thus underlining what remains one of the largest mysteries in asteroid/meteorite science.

A second milestone in addressing the character and structure of the main belt of asteroids was reached when a significant amount of information on asteroid colors had been compiled. This cumulative asteroid data allowed Gradie & Tedesco (1982, using data subsequently published in Zellner et al 1985) to recognize that the main belt of asteroids was not well mixed. As shown in Figure 2, each asteroid class appears to exhibit a distinct distribution as a function of distance from the Sun. This suggests that in spite of the likely reworking of asteroidal surfaces by other asteroidal material, some semblance of order of these primitive materials has been maintained throughout the age of the solar system.

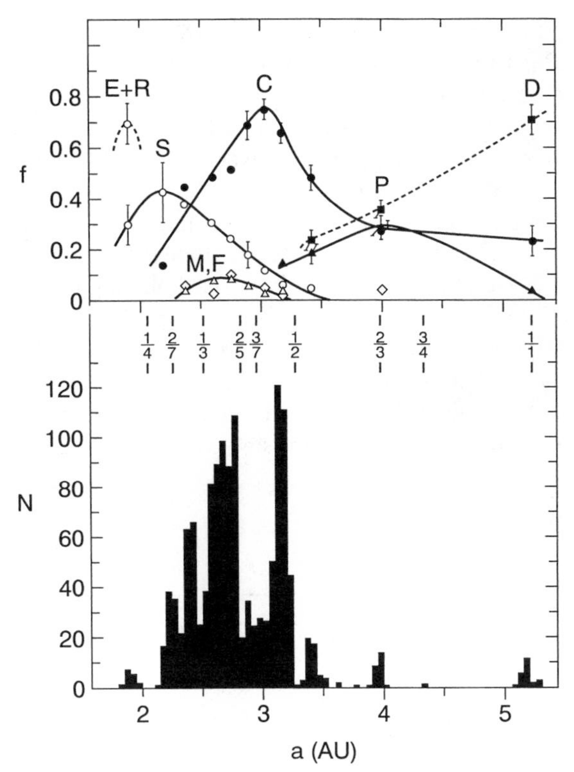

Figure 2 The distribution of major taxonomic classes with respect to heliocentric distance is shown in the top figure. Bias corrections for observation circumstances and family membership have been included. The peak occurrence of Ss in the inner belt, Cs in the mid-belt, and Ps and Ds in the outer belt have been interpreted as evidence of compositional variation across the asteroid belt. The bottom histogram shows the number distribution of 1373 asteroids and the Kirkwood gaps of the main belt. (After Gradie & Tedesco 1982).

Principles of Reflectance Spectroscopy and Mineral Interpretation

The compositional interpretation of reflectance spectra for meteorites and asteroids is based upon the principles of molecular and crystal field theories (e.g. Cotton 1971, Burns 1970a). The success of remotely sensing the composition of asteroid surfaces rests on the fact that there are well-characterized absorption bands in the visible and near-infrared regions of the electromagnetic spectrum. These absorptions are diagnostic of the presence of rock-forming minerals from which cosmochemically significant inferences can be made about the evolution of materials during and after solar system formation.

The interpretation of asteroid spectra has been made from the premise that the surface material of asteroids is composed of cosmochemically abundant minerals. Based primarily on meteorite studies, the origin of asteroid mineral assemblages were believed to be either derived from the solar nebula and survived intact, or were thermally and/or aqueously processed after formation during the first few hundred million years of solar system evolution. The most common components of asteroid surface materials that can be identified remotely include (listed in the approximate order of most to least degree of certainty based on diagnostic parameters): pyroxene, olivine, phyllosilicates, organic material, and opaques (which include metallic iron, graphite, troilite, and magnetite). Visible to near-infrared spectra of these components are shown in Figure 3. The combinations of these minerals on any particular asteroid surface reflect the formation and post-formation processing to which the asteroid has been subjected.

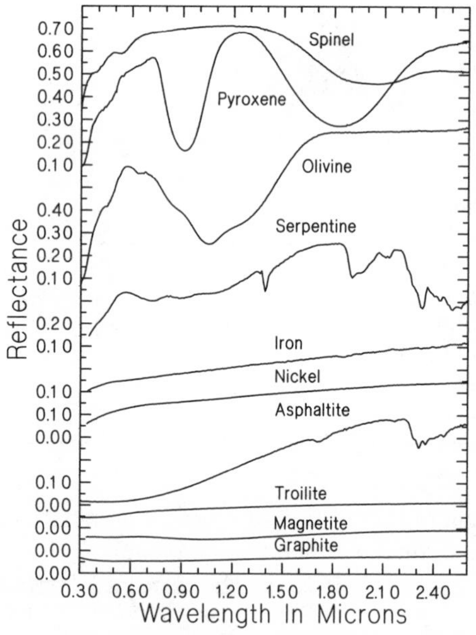

Figure 3 Representative bidirectional reflectance spectra of mineral components of asteroids and meteorites measured in the RELAB ($i = 30°$, $e = 0°$). All samples are particulate (< 500 μm). Each spectrum has been offset vertically for clarity. These data were obtained for E. Cloutis (spinel, iron, nickel, troilite, magnetite, graphite), L. Moroz (asphaltite, an organic compound), J. Mustard (serpentine, a phyllosilicate), and general RELAB collection (pyroxene, olivine).

Significant limiting factors affecting the mineralogical interpretation of meteorite and asteroid spectra include the signal to noise of the system (S/N), spectral range of the measurement, and spectral resolution. Since diagnostic mineral absorption features occur from the visible through the near-infrared (Figure 3) and mineral assemblages include several superimposed features that require deconvolution (see the next section), it is essential to acquire high spectral resolution data ($\Delta\lambda/\lambda = 1\%$) over the broadest spectral range (0.3–3.5 μm) available for reflectance measurements to maximize their interpretive value.

In the laboratory environment, the signal to noise ratio of a reflectance measurement is a combination of the S/N measured for the standard (typically 3000/1) and for the sample. For assemblages that are intrinsically bright, that is with reflectance of 10% or brighter at 0.56 μm, S/N in current laboratory measurements is nominally 1000 or better between 0.3–2.6 μm. For very dark material (reflectance $<5\%$), which is abundant in the asteroid belt, reflectance S/N ratios of 100 can be achieved between 0.3–2.6 μm. Shown in Figure 4 are examples of laboratory reflectance measurements of particulate meteorite samples. Although the spectra are shown scaled to 0.56 μm to allow comparison with asteroid spectra, their albedos fall into two groups: (*a*) moderate to bright meteorites ($R_{0.56} = 21$ to 34%) which include basaltic achondrites and ordinary chondrites, and (*b*) dark meteorites, ($R_{0.56} = 5$–9%) including the various types of carbonaceous chondrites.

At the telescope, the observer usually has more noise to contend with than in the laboratory and the quality of the data are not as high. Available spectra of asteroids are limited by the sensitivity of detectors and the faintness of the signal from the asteroid combined with the addition of atmospheric and instrumental noise contributions. Overall, the quality of asteroid spectra is poorer than laboratory spectra and the reliability of the interpretations is less than for laboratory samples. It is always wise to obtain independent data to confirm any unusual or important spectral feature. In the past decade technological advances have resulted in the availability of instruments that provide better asteroid spectra over a wider spectral range, and in some cases have higher spectral resolution and better S/N than was available previously. Shown in Figure 5 are representative spectra for each of the major asteroid classes. These spectra were derived from composites of data acquired with different visible and near-infrared instruments (each data set should soon be available in the Planetary Data System, Spectroscopy subnode of the Geoscience node).

Major Technical Advances

Most asteroid and meteorite research has been confined to Earth-based telescopes and laboratories. There are several significant technical

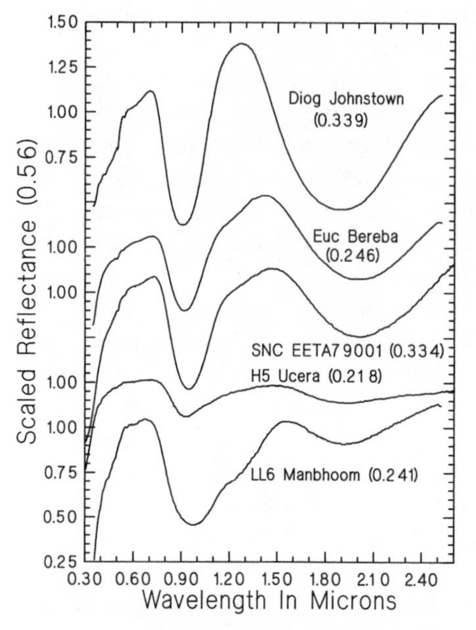

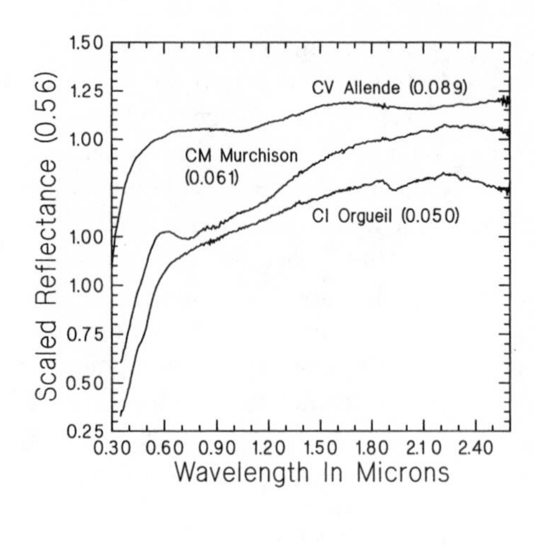

Figure 4 Representative reflectance spectra of meteorite classes. The spectra are scaled to unity at 0.56 μm and the number in parentheses is the laboratory reflectance at that wavelength. The spectra for Johnstown, Bereba, and Manbhoom are from the Gaffey collection of directional hemispheric spectra (Gaffey 1976). The remaining spectra were measured in the RELAB ($i = 30°$, $e = 0°$) for D. Britt (Ucera), L. McFadden (EETA79001), L. Moroz (Murchison, Orgueil), and T. Hiroi (Allende).

advances that have broadened the character of this research and continue to bring new discoveries and excitement to the field. These advances are summarized below and specific results are discussed in the following two sections.

SENSOR TECHNOLOGY Observations of asteroids have dramatically improved with the development of instruments with higher sensitivity, higher spectral resolution, and broader spectral range. The sensitivity allows smaller asteroids to be observed with sufficient precision for classification, thus both increasing the number of asteroids measured and expanding the data base to include small, faint asteroids. As discussed above, the higher spectral resolution and broader spectral range that includes the near-infrared is essential to progress beyond classification of asteroids based on color and albedo in order to allow mineral identification and abundance estimates of surface material based on diagnostic absorption features.

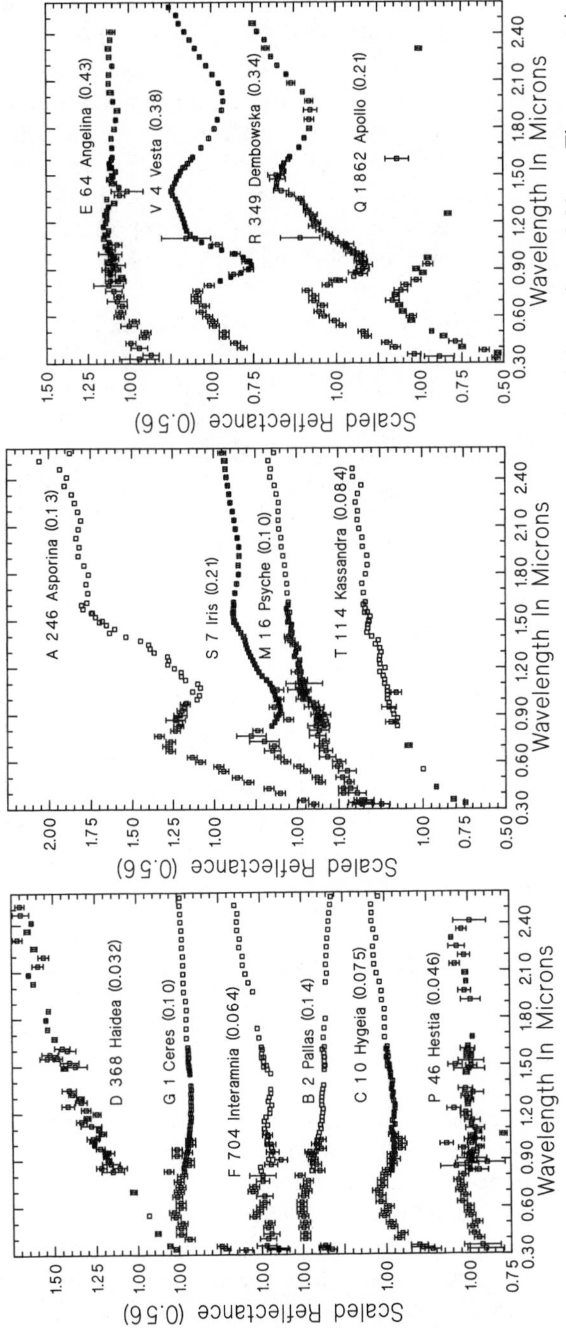

Figure 5 Representative reflectance spectra of asteroid taxonomic classes from 0.3 to 2.6 μm and scaled to unity at 0.56 μm. These composite spectra are derived from collections by Chapman & Gaffey (1979), Zellner et al (1985), and Bell et al (1988). Taxonomy is derived from Tholen (1989). Albedos (indicated in parentheses) are from *IRAS* measurements (Tedesco 1989), except for that of 1862 Apollo, which is from Lefofsky et al (1981b) ground-based radiometry.

ANTARCTIC METEORITES A multinational effort has recovered thousands of meteorites from the ice sheets of Antarctica, almost tripling the number of well-preserved meteorites available for study (see review by Dennison et al 1986). Since there has been very little opportunity for significant macroscopic degradation of the Antarctic meteorites, they are believed to be statistically representative of the falls in that part of the world over a several million year period. As can be seen in Table 1, the ordinary chondrites still dominate the population, but the proportion of specific groups is somewhat different, possibly suggesting an evolving population of meteorites that fall to Earth. When the first lunar meteorite, ALHA81005, was recognized in the Antarctic collection, it was a bit of a surprise. The geophysicists had long argued the lack of meteorites from the Moon to be evidence that the SNC meteorites (shergottites, nakhlites, and Chassigny) could not have originated on Mars since it would be harder to launch material from Mars than from the Moon. The existence of lunar meteorites in the Antarctic collection stimulated further review of ejection mechanisms during an impact event (e.g. Melosh 1984, 1985) and helped soften a principal argument against meteorites from Mars.

ANALYTICAL APPROACHES The interpretive base for evaluating the composition of asteroidal surfaces and linking them to meteorite assemblages has been improved through both empirical and quantitative approaches. A larger number and wider representation of high quality spectra of meteorites, their analogues, and possible primitive material measured in the laboratory has set bounds on the range of possible mineral assemblages present on asteroids. Perhaps equally important, the development and testing of models that accurately characterize mixtures of materials as well as models that quantitatively assess individual absorption bands has been re-evaluated and significantly expanded.

ADVANCES IN INTERPRETIVE CAPABILITIES

There are several approaches to compositional interpretations of asteroid reflectance spectra, all of which have merit, but some of which have progressed more extensively than others. Initially, the objective was simply to identify the type of minerals present or possibly present on an asteroid surface and to compare the results to the known mineral assemblages of meteorites. As mentioned above, this approach was successfully used to identify the type of pyroxene for Vesta (McCord et al 1970, McFadden et al 1977) and continues to be very productive if diagnostic absorption bands are observed in a remotely measured spectrum (e.g. identification of organics on Pholus by Cruikshank et al 1993, discussed in a later section).

Recognizing that laboratory samples of meteorites may not fully reproduce all the characteristics of unsampled asteroid surfaces, much recent effort has focused on limiting the range of possible interpretations through experimental, theoretical, and analytical approaches. The combined results of the improved interpretive capabilities discussed below coupled with better and more abundant asteroid observations discussed in the next section are summarized in Table 3.

Table 3a Weighted mineralogical interpretations of asteroid classes*

Class	Presence certain	Probable	Consistent/possible
P	—	anhydrous silicates	organics
C#	hydrated silicates, phyllosilicates	—	?
B	hydrated silicates	—	?
F	hydrated silicates	—	?
D	—	anhydrous silicates, organics	?
G	hydrated silicates, phyllosilicates	—	?
T	—	—	troilite, metal
M#	metal	—	enstatite
S#	low-Ca pyroxene + olivine	metal	(spinel)
A	olivine	—	metal
Q	low-Ca pyroxene + olivine	—	?
R	low-Ca pyroxene	olivine	?
V	low-Ca pyroxene	plagioclase	?
E	—	enstatite	metal

Table 3b Weighted meteorite analogue interpretations of asteroid classes*

Class	Certain	Probable	Consistent/possible
P	—	—	
C#	CM	CI	shock darkened OC, metamorphosed
B	—	—	metamorphosed CC
F	—	—	metamorphosed CC
D	—	—	—
G	—	—	metamorphosed CC
T	—	—	—
M#	Fe-Ni metal	enstatite	—
S#	—	primitive achondrites	pallasites, OC regolish, urelites
A	—	olivine achondrites, pallasites	—
Q	ordinary chondrites	—	—
R	—	—	—
V	HED	—	—
E	—	enstatite C, aubrites	—

HED = Howardites, Eucrites, Diogenites; OC = ordinary chondrites; CC = carbonaceous chondrites.
* For at least some members.
# Groups quite diverse; M class may not all contain metal.

Empirical methods have been used extensively and rely on spectroscopic surveys of materials measured in the laboratory and systematic manipulation of spectral parameters. Compositional inferences are made by comparing observed systematic relations within the laboratory data (e.g. chemistry vs spectral parameters) with measured properties of asteroids. This approach, of course, is limited to the range of materials analyzed in the laboratory and assumes singularity of diagnostic criteria. With the computer capabilities now available, several quantitative approaches have also evolved that model artificial mixtures of meteorite and analogue materials to reproduce observed characteristics of asteroids. In addition, quantitative deconvolution of individual absorption features is now possible and this analytical capability allows features in a remotely acquired spectrum to be characterized without actually having a sample of the same assemblage. A certain degree of humility, however, must be superimposed on all interpretive approaches since our knowledge of the space environment is limited and Nature may be able to alter a normally well-characterized material beyond recognition or our starting assumptions (such as laws that govern the evolution of the solar system) may be inaccurate.

Empirical Mineralogy

In addition to an understanding of the physical causes of diagnostic absorptions, all mineralogical interpretations of reflectance spectra are based in part on a library of laboratory spectroscopic data of relevant materials. After the initial survey of meteorite spectra by Johnson & Fanale (1973) and by Gaffey (1976), several researchers began to document the spectral properties of a number of meteorites and analogue materials that comprise meteorite assemblages. The most commonly used or referenced laboratory studies relevant to asteroids and meteorites are summarized below. A recent review by S. J. Gaffey et al (1993) provides a broader summary of laboratory analyses.

MAFIC SILICATES The landmark investigation of the reflectance properties of pyroxenes by Adams in 1974 demonstrated that not only do (anhydrous) pyroxenes exhibit two very diagnostic ferrous absorptions (near 1 and 2 μm; see Figure 3), but that the wavelength minimum of both of these absorptions varies systematically with composition (relative amounts of Mg, Fe, Ca). This study was later refined by Cloutis & Gaffey (1991) using a more diverse set of pyroxene compositions. A similar analysis of the reflectance properties of a suite of chemically well-characterized olivines of different compositions was undertaken by King & Ridley (1987), after initial work by Burns (1970b) using transmission data. King & Ridley showed that the minimum of the broad ferrous composite olivine band (see Figure 3) varies with Fe-Mg composition between 1.05 and 1.08 μm. A more detailed characterization of olivine compositional properties is discussed below in the section entitled Quantitative Diagnostic Absorptions.

A number of laboratory studies were also performed to investigate systematic variations of absorption features when known proportions of mafic minerals were mixed with other minerals (e.g. Nash & Conel 1974, Singer 1981, Crown & Pieters 1987). One of the studies most relevant to asteroid/meteorite spectral analysis was the measurement of mixtures of a low-Ca pyroxene (orthopyroxene) and an olivine (Singer 1981, Cloutis et al 1986). Olivine exhibits a composite band near 1.05 μm but no features near 2 μm, and orthopyroxenes exhibit two diagnostic absorptions near 0.9 and 1.80 μm (see Figure 3). Singer showed that for a material containing both these minerals, the relative strength (band area) of absorptions near 1 and 2 μm varied systematically, and this relation should allow the relative abundance of the two minerals to be predicted. Cloutis et al (1986) later derived a laboratory calibration of the relationship between band area ratios and relative proportion of olivine and orthopyroxene. Although particles <63 μm in size were not included in the Cloutis et al analysis, the relationship appears to hold consistently for a range of large particle sizes. This empirical calibration has been used, for example, to estimate the olivine/orthopyroxene ratio for S-class asteroids (Gaffey 1984; Gaffey et al 1989, 1993b). It should be noted that this relationship is only strictly applicable for the binary system involving olivine and orthopyroxene. Since clinopyroxenes exhibit a much broader range of compositions and site preferences, the Cloutis et al (1986) empirical relations cannot be used to predict olivine/clinopyroxene mixtures. In addition, the presence of opaques such as metal appears to add a nonlinearity to the band ratio method (see subsection on iron metal).

While much research has focused on the characteristics of olivines and pyroxenes which have been readily identified in asteroid and meteorite spectra, the diagnostic properties of less optically active minerals received less attention. Fortunately, detailed studies of several sheet silicates (King & Clark 1989, Clark et al 1990) were concluded about the same time telescopic instruments were available to detect more subtle features in asteroid spectra (see next section). Of particular importance to asteroid analyses were the measurements of King & Clark who documented the ~0.7 μm iron absorption in serpentines and chlorites (Figure 3) and showed this feature to be increasingly prominent in finer size fractions. As shown in Figure 4, this phyllosilicate feature is clearly observed in most CM meteorites (Johnson & Fanale 1973, Gaffey & McCord 1979, Moroz & Pieters 1991, Hiroi et al 1993c).

IRON METAL The original survey of meteorite spectral properties by Gaffey (1976) showed Fe-Ni metal to exhibit a featureless red-sloped spectrum (increasing reflectance with wavelength). These diffuse reflectance spectra were acquired using an integrating sphere. Subsequent analysis of the bidirectional properties of metal and surface roughness (Britt & Pieters 1988) showed that it is the strong specular component of reflectance that provides the red slope in iron reflectance. Nonspecular geometries of measurements from iron surfaces are characterized by less intense and relatively flat spectra. If a surface consists of randomly oriented metallic particles, the specular component still dominates the reflectance spectrum and a red continuum is evident. Particulate samples of Fe-Ni metal were prepared and measured by Cloutis et al (1990a) and the spectral properties of mixtures of 45–90 μm particulate iron with olivine and pyroxenes were documented (Cloutis et al 1990b). In these studies the continuum for Fe-metal bearing samples was consistently red sloped. The presence of particulate metal was also shown to affect band area ratios of metal/pyroxene mixtures, requiring independent criteria to be developed to distinguish between the presence of metal and the relative abundance of pyroxene in olivine/pyroxene mixtures.

One of the more perplexing laboratory results is that of Gaffey (1986) who measured the spectral properties of separates of the metal component in ordinary chondrites. Gaffey showed that, instead of the expected red-sloped continuum, the chondritic metal separates exhibit a flat featureless spectrum, and a laboratory spectrum of particulate ordinary chondritic material containing metal shows no red slope. Gaffey hypothesizes that the metal grains in ordinary chondrites are coated with an optically thick layer that masks the normal features of Fe-Ni. Since metal is a significant component in ordinary chondrites, a principal issue in identifying a potential parent body for these abundant meteorites is therefore whether such iron coatings exist and are maintained during regolith formation. As discussed below, the effect of Ni-Fe and reduced Fe in asteroidal regoliths is currently ambiguous, but regolith processes is an area of active research.

OPAQUES The early work of Nash & Conel (1974) and Pieters (1974) showed the nonlinear, disproportionately large effect opaques have on a reflectance spectrum of an intimate mixture. This is of particular interest to asteroid analyses since a large fraction of asteroids are exceptionally dark by any terrestrial standard (Tedesco et al 1989a. A series of detailed studies by R. N. Clark (Clark 1981a,b, 1983; Clark & Lucey 1984) evaluated the dramatic effect even a fraction of a percent opaques could have on suppressing spectral features of semitransparent silicates or ices. Much depends on the size of the opaque (smaller, micrometer sized, are very efficient absorbers) and whether it is dispersed (making a more effective absorber) or concentrated (making a less effective absorber) in the matrix material. The general effects of typical meteoritic opaques, carbon and magnetite, were documented in laboratory experiments with mafic minerals by Cloutis et al (1990b) and Miyamoto et al (1982).

Mixture Systematics

Conceptually, one should be able to mix together the diagnostic spectra for various minerals shown in Figure 3 to form the spectrum of an assemblage of minerals. A mixture of two or more materials should exhibit weighted properties of each, thus allowing not only the identification of individual components but also an estimation of their abundance. There are, however, both linear and nonlinear mixing approaches for combining spectra of materials to form a reflectance spectrum. Linear mixing occurs when the components are widely separated and each material contributes to the bulk spectrum in proportion to its areal extent (e.g. Singer & McCord 1979, Adams et al 1993). An analogous situation for an asteroid is when there are compositional and/or physical variations across the surface and different regions contribute different spectra to the integrated whole. The size of these regions could be anywhere from centimeters to several kilometers in scale.

Nonlinear mixing is more difficult to accurately model (Hapke 1981; Clark & Roush 1984, Mustard & Pieters 1987, 1989; Johnson et al 1992; Hiroi & Pieters 1992) and occurs when grains are intimately mixed so that radiation can interact with multiple components before being scattered from the surface. These models have been shown to accurately predict mineral abundances in a mixture to about 5%, provided photometrically accurate measurements of individual components (endmembers) are available. Although appropriate for many environments, nonlinear intimate mixture models are often difficult to use in asteroid studies because detailed photometric information for all required endmembers is not available. It has been shown (e.g. J. F. Mustard. in preparation) that using the simpler linear mixture approach for materials that are intimately mixed nevertheless produces very useful results that are accurate for *relative* comparisons of similar materials (but not accurate for absolute abundances). In other words, even if it is not known whether materials are spatially separate or intimately mixed, comparisons of relative abundances of similar materials derived with a linear mixing approach is valid, but the derived abundance values may not be accurate.

Linear mixing models have been frequently used when modeling asteroid spectra with mixtures of meteorite spectra. In a typical approach, a series of meteorite spectra is selected for mixing and some limiting goodness of fit criteria are defined. Different proportions of the meteorite spectra (endmembers) are combined together to form a composite spectrum which is then compared to the asteroid spectrum. Computer algorithms then iterate until a minimum error is produced between the asteroid spectrum and the composite meteorite spectrum. The results typically provide information about which meteorite assemblages might contribute to the observable properties of an asteroid and in what relative abundance. Examples of this approach can be found in Hiroi & Takeda (1991) and Hiroi et al (1993a,b). These studies have shown that some (but not all) olivine-rich asteroids (A class) can be modeled with spectra from the metal and the silicate portion of pallasites and that some (but not all) S-class asteroids can be modeled with various proportions of primitive achondrites and metal.

A number of applications using intimate mixture modeling have also recently been evaluated. B. E. Clark et al (1993) used a Hapke mixing model to try to reproduce near-infrared spectra for 36 S-class asteroids from the Bell et al (1988) survey with endmembers such as low-Ca orthopyroxene, clinopyroxene, olivine, plagioclase, iron metal, and a generic flat absorbing material. Preliminary results showed all but two of the asteroid spectra could be reasonably modeled with different proportions of the endmembers, but that mineral assemblage solutions were nonunique. These results underscore a primary limitation of any form of mixture modeling, namely that an independent criterion for uniqueness often remains unidentified.

A different intimate mixture model was used by Hiroi et al (1993d) to compare visible to near-infrared spectra of C, G, B, and F asteroids with spectra of a variety of carbonaceous chondrites and thermally metamorphosed samples of Murchison (CM). It has been recognized since the early work by Johnson & Fanale (1973) that the location of the ultraviolet absorption edge in spectra of these dark asteroids is at a much shorter wavelength than that observed for carbonaceous chondrites in the laboratory (compare the ultraviolet properties of the C, G, and F asteroids in Figure 5 with the CV, CM, and CI meteorites of Figure 4). Until spectra of three unusual thermally metamorphosed meteorites from the Antarctic collection were measured, no carbonaceous chondrites exhibited the same ultraviolet spectral characteristics as that observed in the asteroids (Hiroi et al 1993c). The results of the Hiroi et al mixture modeling indicate that the surfaces of these dark asteroids are best modeled with a mixture that contains a significant amount ($> 50\%$) of materials similar to the thermally metamorphosed carbonaceous chondrites. The same unusual spectral properties of these thermally metamorphosed meteorites is closely mimicked by portions of the Murchison CM meteorite that were thermally processed in the laboratory (Hiroi et al 1993c,d). This strengthens the case for thermal metamorphism on many of the dark asteroids.

Quantitative Diagnostic Absorptions

A complementary approach to characterizing components in an intimate mixture involves quantitatively modeling individual mineral absorptions as combinations of modified Gaussian functions (e.g. Sunshine et al 1990), including those that form composite absorption bands (e.g. the three bands of the broad composite band in olivine, Figure 3). This approach is very powerful in that it provides quantitative band information (band center, width, strength) that is directly associated with the physical process of absorption. The procedure, the modified Gaussian model (MGM), requires high spectral resolution and high-precision data, and simultaneously fits multiple absorption features (see Sunshine et al 1990 for algorithm details).

An example of an MGM application is demonstrated in Figure 6 (from Sunshine et al 1993) for the EETA79001 shergottite spectrum shown in Figure 4. EETA79001 exhibits two non-brecciated basaltic lithologies (McSween & Jarosewich 1983). The predominance of pyroxene is evident in the EETA79001 spectrum by the two diagnostic absorption bands near 0.95 and 2.05 μm. Petrographic analyses reveal the presence of significant proportions of both a high-Ca pyroxene and a low-Ca pyroxene in this sample (McSween & Jarosewich 1983). Without more detailed analysis of the spectrum, however, it would appear that the two pyroxene bands observed near 0.95 and 2.05 μm represent the presence of an intermediate composition pyroxene. Shown in Figure 6 (*top*) is an MGM analysis of the absorptions in lithology A that assumes a single pyroxene composition in the sample. Both the magnitude and the asymmetry of the residual error and the systematic deviations of the modeled spectrum to the actual spectrum indicate this meteorite spectrum is not well modeled with a single pyroxene (see discussion in Sunshine & Pieters 1993a). On the other hand, when MGM is allowed to fit bands appropriate for two pyroxenes to the EETA 79001 spectrum (Figure 6, *bottom*) the fit is accurate to within the noise of the system. Furthermore, the relative proportions of the low-Ca (large arrows) and the high-Ca (small arrows) pyroxenes derived from band strength relationships agree well with what is observed petrographically (Sunshine et al 1993).

The MGM approach has also been used in analysis of olivine reflectance spectra to quantify the systematic variations of the absorptions with Mg-Fe composition (Sunshine & Pieters 1990, and in preparation). The band centers and strength of the three olivine bands that form the composite feature near 1.05 μm are shown to vary with composition, but the band

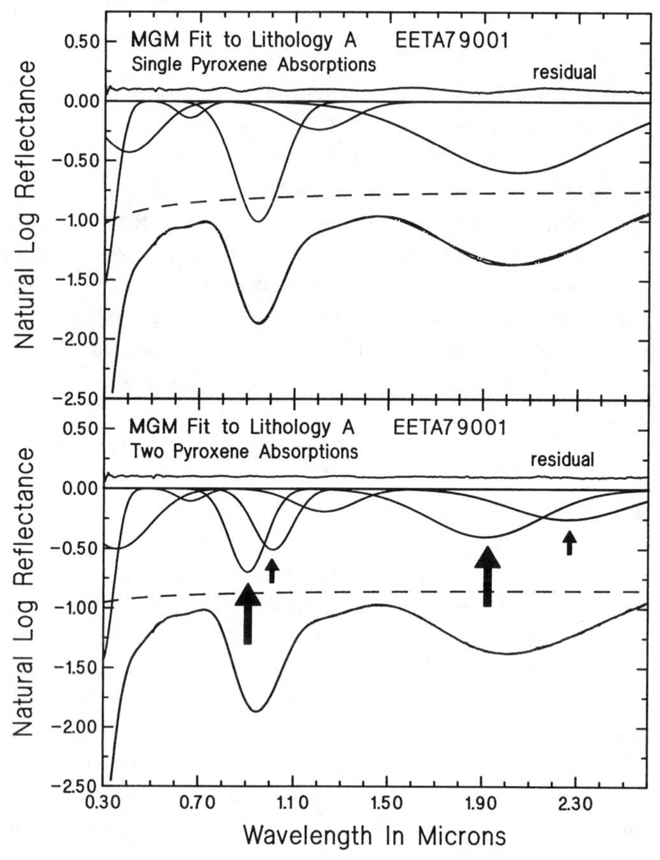

Figure 6 MGM analysis of shergottite EETA79001 lithology A (after Sunshine et al 1993). The four components of the analysis include (from top to bottom in each figure): the residual error between the log of the modeled spectrum and the log of the actual spectrum (offset 10% for clarity), the individual modified Gaussian distributions representing absorption bands, the continuum or baseline onto which these distributions are added (*dashed line*), and the modeled spectrum superimposed onto the actual spectrum. The upper figure contains MGM fits to EETA79001 using absorption bands in the 1 and 2 μm region representing a single pyroxene composition. Note the systematic pattern of the residual errors and that the peaks of the residual errors are uncorrelated with characteristics of the derived absorption bands. This is diagnostic of a fit that requires additional absorption bands (see Sunshine & Pieters 1993a). The lower figure contains MGM fits to the same spectrum using absorption bands in the 1 and 2 μm region representing two pyroxenes of different composition. Large arrows indicate absorption bands from the low-calcium component, and small arrows indicate the absorption bands from the high-calcium (clinopyroxene) component.

widths remain constant. The observed systematic variation of band centers allows a determination of olivine composition from an accurate deconvolution of a reflectance spectrum of a material with an unknown composition. This approach has been used to show that the olivine composition of A-class asteroid 246 Asporina is fosteritic, or Mg-rich (Sunshine & Pieters 1993b).

Space Weathering

Although there are obvious spectral similarities between meteorites measured in the laboratory and asteroids observed using remote sensors (spectra of Figures 4 and 5), the optical properties of the two types of materials do not exactly correspond (with the notable exception of 4 Vesta and basaltic achondrites). Principal component analyses of the two populations sampled to the 8-color standard show that the statistical measurement of spectral variability for prominent asteroid types do not readily overlay those for meteorite classes (Britt et al 1992). Individual or composite features can be identified in spectra of each that are readily interpreted (olivine, pyroxene, phyllosilicate), but there are unaccounted differences between asteroid spectra and meteorite spectra in such characteristics as strength and shape of absorption features, near-infrared continuum, nature of the spectrum below 0.5 μm, and overall albedo.

These differences could, and probably do, have several causes. Physical preparation of a meteorite for measurement in the laboratory may not accurately reproduce the physical form of material on an asteroid surface. In addition, a meteorite several centimeters in size may not represent its parent body due to the 10^6–10^7 difference in scale between meteorites and asteroids. Furthermore, there are most certainly asteroids that have not contributed to our meteorite collection and may thus not correspond to any meteorite class. There may be a selection process such that the solid material removed from an asteroid to become a meteorite does not represent surface material. And most worrisome, the only documented samples from an extraterrestrial body exposed to the space environment (the Moon) have shown that solid rocks and soil are quite different optically. Significant optical alteration occurs as a fine-grained regolith develops from local lithologies. Below, we summarize various aspects of the research concerning space weathering of surface materials.

Space weathering processes active in the lunar environment result in a lower albedo, weaker absorptions, and a red-sloped continuum. It is important to note, however, that although diagnostic mineral absorptions may be altered by such effects, they are not eliminated (e.g. Pieters 1993). Although the optical effects of space weathering were thought to be due to glass and/or agglutinate formation, more detailed evaluation of glass

(Bell et al 1976, Wells & Hapke 1977) and agglutinates (Pieters et al 1993) showed that neither exhibited the required optical effects. Recent analyses of lunar soil separates have shown that it is only the finest fraction of naturally developed lunar soils (which constitute <25‰) that accounts for the red-sloped continuum characteristic of lunar alteration, and that the processes producing optical alteration in the space environment appear to be surface correlated (Pieters et al 1993). A leading hypothesis of the physical cause of lunar optical alteration is linked to small particles of reduced iron such as those recently produced experimentally by Allen et al (1993) and the submicroscopic single-domain iron measured by ferromagnetic resonance (Morris 1977, 1980) that preferentially accumulates in the finest fraction. Understanding the properties of this reduced iron, the conditions of formation, and the optical effects on soil products are areas of active research. Since the finest fraction appears to be most strongly affected by space weathering, the degree to which small particles accumulate in asteroid regolith must be evaluated. In particular, the relative effects of gravitational, magnetic, and molecular forces on small particles adhering to surfaces may produce quite different results depending on the size of the asteroid and the physical characteristics of its surface (composition, texture, impact history).

The effects of irradiation is an additional important topic that requires controlled experiments. The solar wind is likely to play a pivotal role in space weathering, but its effect on geologic materials is poorly known. Hapke et al (1975) showed that there is a differential retention of elements sputtered using 2 keV hydrogen ions. Enrichment ratios of sputtered deposits appeared to follow atomic weight with Fe > Ca > Al > O. Preliminary experiments have been performed by Benedix (1993) in which silicates such as olivine were irradiated with 10 keV hydrogen ions in a vacuum environment. Minor optical effects were observed for the irradiated samples, but the active physical processes require further investigation.

Two recent laboratory experiments were performed to independently evaluate the optical effects of glass formation during regolith development. In the simulation by Clark et al (1992) chondritic meteorites were ground, a portion melted, and a pulverized fraction of the quenched material remixed with the original chondritic powder. In the experiments of Cloutis & Gaffey (1993) synthetic glasses of varying compositions were prepared and mixed with a variety of mafic silicates. In both cases the glass lowered the albedo and weakened the original absorption features. Both experimenters found it difficult to produce glass free of microcrystals. The results are useful in a qualitative sense, but since the effects of the low oxygen fugacity of the space environment on glasses (Bell et al 1976) was not reproduced, they cannot be directly compared to space weathering.

Some meteorites may provide clues to the processes active on the surface of meteorite parent bodies. Bell & Keil (1988) attempted to evaluate the spectral properties of gas-rich meteorite breccias, which are believed to contain components that have accumulated exposure to the space environment. They observed minor spectral variations between the gas-rich matrix and clasts across a cut slab of the meteorite. Because spectra of non-particulate samples (slabs) are strongly dominated by the specular reflection component (Fresnel reflection) and less light is able to interact with and scatter from individual grains, direct comparison of these experiments with particulate samples and the effects of space weathering on regoliths can not be made.

Shock-darkened ordinary chondrites ("black chondrites") are another type of meteorite that have clearly undergone alteration effects that occurred either on the surface of the meteorite parent body or as the meteorite was ejected from the parent body. Britt & Pieters (1991) showed that these altered meteorites constitute 14–17% of the ordinary chondrite falls. Britt later documented that it is the finely dispersed metal and opaques (~ 2 μm in diameter) in the matrix of these shocked meteorites that causes the notable darkening (Britt & Pieters 1993). These shock-darkened meteorites exhibit a lower albedo and weaker absorption bands than their unshocked counterparts of the same composition, but, unlike the lunar case, they exhibit a flat continuum. The physical conditions necessary to transform an ordinary chondrite into a black chondrite are currently not well constrained.

Although the details of both the causes and effects of space weathering remain unclear, the multispectral observations of the S-class asteroid Gaspra by the *Galileo* spacecraft clearly show spectral differences across the surface (Belton et al 1992). Small bright craters are less red than their surroundings and exhibit stronger 1 μm absorptions—the very characteristics expected for space weathering from the lunar experience. It is thus likely that some form of space weathering has been observed on an asteroid, but the magnitude of these effects is clearly less on an asteroid like Gaspra than it is on the Moon.

ANTARCTIC METEORITES

The collection of Antarctic meteorites has dramatically increased the number of well-preserved meteorites available for analysis. Spectral reflectance studies of the Antarctic meteorites were initially carried out to search for spectral types that might be rare among the meteorites but at the same time similar to some asteroid spectral types. McFadden et al (1980) examined a suite of eight Yamato meteorites that were texturally different from existing members of their meteorite classes, basaltic achondrites, ureilites, and carbonaceous chondrites. While the variations in the spectral features of these Antarctic meteorites were mineralogically and petrologically significant, this first search did not yield any major new meteorite spectral types. More recently, successful examples of identifying and using unique properties of the Antarctic collection include the studies by Hiroi and associates mentioned in the previous section. Most primitive achondrites (Lodranites, Acapulcoites) come principally from the Antarctic collection. The thermally metamorphosed carbonaceous chondrites, which best mimic the ultraviolet properties of C-class asteroids, are unique to the Antarctic collection.

Among the significant discoveries from Antarctica are meteorites originating from the Moon (Lindstrom et al 1991). Spectral reflectance studies of the first recognized lunar meteorite ALHA81005 (Pieters et al 1983) was carried out in an effort to constrain the source region on the Moon from where the meteorite was ejected. The assemblage brought to Earth and named ALHA81005 exhibited features diagnostic of low-Ca pyroxene and olivine, characteristics which at that time had not been observed for craters on the Moon which had been studied from ground-based reflectance spectroscopy and which could serve as possible source craters. Based on the known compositional properties of nearside lunar units, the meteorite most likely was ejected from the nearside limb or the far side of the Moon. A small crater on the limb near Mare Crisium was in fact later found to have the required spectral characteristics (Pieters 1986, 1993). Spectral studies of a second lunar meteorite, Y791197 (McFadden et al 1986), showed it also to have a highland composition, but to probably originate from a separate ejection event. Spectroscopic studies have been initiated for a third lunar meteorite, A881757 Asuka 31, an unusual gabbroic sample. Preliminary studies of mineral separates from this sample have provided the first spectroscopic analysis of uncontaminated maskelynite, shocked plagioclase (Pieters & Rode 1993). Maskelynite was shown to exhibit the spectral properties of glass, but with an exceptionally strong ultraviolet absorption below 0.4 μm.

The suite of meteorites called the Shergottites-Nakhlites-Chassignites (SNCs), which are believed to have been ejected from Mars (McSween 1985), was significantly enlarged as a result of the Antarctic meteorite expeditions. Spectral reflectance measurements of these meteorites may provide constraints on their source regions when there are spectral reflectance data of Mars with sufficiently high spatial resolution to provide meaningful constraints. Singer & McSween (1991) and Mustard et al (1993) have discussed the constraints on the composition of the martian crust based on remote measurements of the surface and on reflectance studies of SNC meteorites by McFadden (1987) and McFadden & Pratt (1989). While there is evidence for pyroxene-bearing material (like shergottite) on the martian surface from telescopic and spacecraft investigations, no surface exposures of the cumulate phases of the SNCs, the nakhlites or chassignites, have been observed remotely.

ADVANCES IN ASTEROID OBSERVATIONS

Photoelectric detectors were used for the 8-color asteroid survey which forms the basis for asteroid taxonomy. Since the initial productive age of photoelectric photometry using a series of filters, most advances in asteroid observations have resulted from the use of more sensitive and efficient sensors which cover a much larger spectral range.

Extended Spectral Range of Near-IR Spectroscopy

As the 25-filter two-beam photoelectric photometer was being retired, circular variable filters (CVFs) with 1.5% wavelength resolution and InSb detectors were coming on-line. Near-infrared measurements covering 0.8–2.6 μm (52-color) were made, first of the larger asteroids (McFadden et al 1981, Gaffey 1984) and then as a survey of all asteroid types with $V_{mag} < 13$ (Bell et al 1985). Inner belt asteroids near the 3:1 Kirkwood gap were targeted in a study by McFadden & Chamberlin (1992) using a cooled grating InSb array at the Infrared Telescope Facility (IRTF). Data from the expanding Bell et al survey provide a mother lode of information that continues to form the basis for new ideas related to asteroid surface composition.

MAFIC SILICATE MINERALOGY AND CHEMISTRY One of the first major discoveries enabled by the technical advances of near-infrared spectrometers and the IRTF was the discovery of olivine-rich asteroids in the main asteroid belt (Cruikshank & Hartmann 1984), predicted from cosmochemical principles. These asteroids (A class) exhibit well-defined olivine absorption features (e.g. 246 Asporina in Figure 5) and are believed to represent the subcrustal, mantle region of bodies which have undergone chemical differentiation.

Using the same instrumentation augmented by visible and infrared measurements, Cruikshank et al (1991a) discovered three additional basaltic achondrite meteorite analogues among the near-Earth asteroids. The orbits of these near-Earth V-class asteroids are remarkably similar, indicating that if they have a similar origin, the disruption event must have occurred recently. This discovery brought to four the number of possible basaltic achondrite parent bodies among the dynamically short-lived near-Earth asteroids.

Other near-Earth asteroids have been the subject of further study as well. Gaffey et al (1992) present visible and near-infrared evidence confirming that the Apollo asteroid 3103 1982 BB is an E-class asteroid. Unlike the V-class near-Earth asteroids studied by Cruikshank et al, 1982BB is in a relatively long-lived orbit with an orbital resonance (3:5) with the Earth. Gaffey et al (1992) suggest that this asteroid is the principal source of the enstatite achondrite meteorites (Aubrites).

Asteroids with previously unidentified mineralogies have also been discovered as a result of the 52-color IR survey. A few S-class main belt asteroids which do not have a 1.0 micron band were observed to exhibit a prominent broad feature near 2 μm. The lack of a prominent feature near 1.0 μm in an asteroid with moderate to high albedo implies that ferrous silicates are low in abundance or absent. Very few materials, except for various forms of spinel, exhibit a singular 2 μm absorption. The observed unusual spectral character of these asteroids led to an interpretation that they contain spinel-bearing assemblages, perhaps similar to the Ca-Al inclusions in CV meteorites (Burbine et al 1992).

The near-infrared properties of S-class asteroids has been under study for over a decade and two very different, but equally important, hypotheses on their origin remain open for discussion. From the character of the absorption features near 1 and 2 μm (see Figure 5) the surfaces of S-class asteroids are known to contain a mineral assemblage consisting of orthopyroxene and olivine. S-class asteroids also exhibit a red-sloped near-infrared continuum which normally indicates the presence of a metallic component (although there are exceptions such as the red continuum of lunar soils which may or may not be linked to fine-grained metal and, conversely, laboratory spectra of metal-bearing ordinary chondrites which do not exhibit a red continuum). This basic information is consistent with at least two types of material: a stony iron assemblage and an ordinary chondrite assemblage. The literature is full of arguments for and against each possibility. Most of the unresolved issues concern 1. the character and effects of alteration processes and regolith development active on an asteroid surface (e.g. Pieters et al 1976, 1993) and 2. the degree to which compositional components derived from laboratory spectra of meteorites and meteorite analogues (see previous section) can be used to interpret spectral properties measured for an asteroid surface (e.g. Gaffey & McCord 1978, Gaffey 1984, Gaffey et al 1993b). This is obviously an important dilemma to resolve since the S-class asteroids are abundant in the inner part of the main belt (see Figure 2 and Table 2) and the stony iron interpretation implies that pervasive differentiation and igneous processes affected the inner belt, while the chondrite interpretation suggests the opposite.

The most complete survey of visible and near-infrared spectral properties of 39 S-class asteroids has been compiled by Gaffey et al (1993b). In this investigation Gaffey et al divide the S-class asteroids into eight different groups based on wavelength of the composite 1 μm band and a band area ratio parameter (2 μm band to that of the 1 μm band). Using empirically derived interpretive algorithms Gaffey et al characterize the compositions of these groups as ranging from undifferentiated to fully differentiated bodies. They also present comparisons of spectral parameters with asteroid size. Assuming smaller asteroids would be less affected by optical alteration, they argue that the size information refutes most regolith alteration processes. One of the most important results unveiled from Gaffey et al's discussion is that the S class of asteroids are not one type of material, but are quite compositionally diverse. [It is interesting to note a bit of convergence of the two hypotheses for S-class asteroids in this discussion. Gaffey originally proposed the stony iron hypothesis (Gaffey 1984), but with more detailed and expanded data and analyses Gaffey et al (1993b) recognize that one of their eight groups of S-class asteroids is "not inconsistent with" ordinary chondrites.]

ORGANICS An excellent overview of organic compounds in carbonaceous chondrites can be found in Cronin et al (1988). Several investigators (e g. Gradie & Veverka 1980; Vilas & Smith 1985; Cruikshank et al 1991b; Moroz et al 1991, 1992) have suggested that organics could be the cause of the low albedo and red coloration of the P and D outer solar system asteroids, but until recently no conclusive spectroscopic evidence for organics had been observed. The high sensitivity of several current spectroscopic instruments has resulted in the confirming evidence. Cruikshank et al (1993) integrated new telescopic and laboratory data identifying an absorption feature at 1.7 μm and a stronger absorption at 2.27 μm for the small planetesimal 5145 Pholus (which has a perihelion inside the orbit of Saturn and a aphelion outside the orbit of Neptune). Because Pholus is also one of the reddest objects in the solar system (Mueller et al 1992, Fink et al 1992, Binzel 1992) Cruikshank et al were able to convincingly show that the surface of Pholus contains aliphatic-rich and high H/C solid asphaltite-like organics. They also suggested that Pholus may represent an endmember body containing unaltered primitive materials, and that comparable materials on asteroids closer to the Sun have been transformed to more closely resemble organics seen in meteorites.

Observations in the 3 μm Region

In the mid-1980s a cooled grating array infrared spectrometer became available permitting measurements of asteroids in the 3.0 μm spectral region with greater sensitivity and spectral resolution than in the previous decade. This spectral region is particularly difficult observationally due to an extremely strong atmospheric water absorption. For comparison, the spectral characteristics of a suite of meteorites in this spectral region can be found in Salisbury et al (1991). Lebofsky et al (1981a) first interpreted the 3.0 μm absorption band as a water of hydration feature observed in large C-class asteroids. Jones (1988) measured 19 low-albedo asteroids using the cooled-grating infrared array spectrometer at the IRTF at Mauna Kea, Hawaii to search for and characterize the 3.0 μm water of hydration band. Examples of such spectra are shown in Figure 7. Of the 32 asteroids available for analysis (Jones et al 1990), 66% of the C class (and their subsets) contain 3-μm absorption bands, diagnostic of the presence of hydrated silicates. Of the P and D class, which populate the Trojan and Cybele regions of the asteroid belt, only 1 (324 Bamberga) has a water of hydration band.

A significant result from the work of Jones et al (1990) is the cosmochemical implication that the distribution of asteroids with water of hydration bands peaks in the middle of the main asteroid belt, and decreases both inward and outward from there. The P- and D-class asteroids which dominate the outer belt apparently do not contain any form of water. This result is somewhat counter-intuitive as there is ample water ice among the satellites of Jupiter and Saturn. The explanation proposed by Jones et al (1990) is that the temperatures were too low for aqueous alteration, and over the age of the solar system volatiles have not been retained on the surface for bodies the size of outer belt asteroids (100s of km in diameter). Additional measurements obtained in the 3 μm region will provide important information on the spatial distribution of hydrated materials and will help constrain the processes that have created and/or destroyed them.

A recent intriguing result is the interpretation of King et al (1992) that a 3.07 μm absorption band in the largest main belt asteroid, 1 Ceres, is due to an ammoniated form of phyllosilicate rather than water frost as interpreted by Lebofsky et al (1981a). Their interpretation uses new, high resolution spectra in the 3–5 μm region as well as laboratory studies of the ammoniated form of a saponite clay mineral. Their argument is primarily based on the difference in band widths between the observed band in the Ceres spectrum and laboratory spectra of water frost. Mixing analyses combining an opaque material with phyllosilicate produces a spectrum

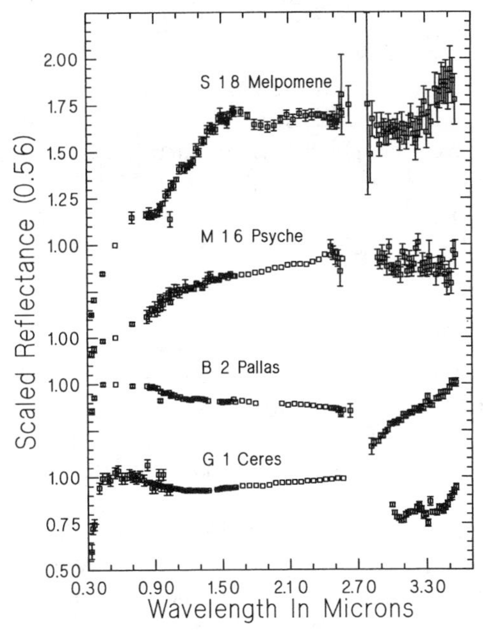

Figure 7 Examples of composite asteroid reflectance spectra from the visible through 3.6 μm. The long wavelength data (2.6–3.6 μm) are from Jones et al (1990). The 3 μm region is an important part of the spectrum for identifying fundamental vibrational modes of such volatile species as OH, H_2O and H_2O frost. Main belt asteroids exhibit diverse characteristics and the presence of these bands in asteroid spectra (Pallas and Ceres) demonstrates the existence of hydrated species which could indicate either a primary accretion composition or a secondary alteration product.

reasonably similar to Ceres. If this interpretation is correct and other ammoniated assemblages are found in the main belt, it implies that secondary heating processes have not exceeded 400 K in the asteroid belt since this ammoniated assemblage was formed. This result, however, must be reconciled with independent evidence for the presence of water from ultraviolet spectra showing OH emission, presumably formed from dissociation of water vapor (A'Hearn & Feldman 1992).

High S/N CCD

In the past decade CCD spectrometers replaced most of the visible photoelectric systems bringing significant increases in sensitivity and increased spectral resolution. Several applications of this technology have brought a diversity of new results.

DARK ASTEROIDS/PHYLLOSILICATES Vilas first used CCD spectrometers to measure asteroid spectra and took advantage of the improved sensitivity to study the faint asteroids of the outer asteroid belt (Vilas & Smith 1985). In this study it was noted that there is a general spectral reddening trend as a function of heliocentric distance. This observational program was expanded by Sawyer (1991) who also studied low-albedo asteroids of the main belt. These two programs resulted in the discovery of weak absorption features in low-albedo asteroids. The most common and prominent feature is a broad weak absorption that occurs near 0.7 μm and is believed to arise from Fe^{2+}–Fe^{3+} charge transfer absorptions in phyllosilicate minerals (Vilas & Gaffey 1989, Vilas et al 1993b). Examples of this absorption are shown in Figure 8*a*. Possible additional weak features have been observed at 0.43 and 0.6 μm (Vilas et al 1993a,b). Prior to the availability of CCD spectrometers, no diagnostic absorption bands of the dark C-class asteroids were observed in the spectral region between 0.4 and 1.0 μm. The discovery of these bands increases the specificity of the mineralogical interpretation of dark asteroids.

The identification of features in C-class asteroid spectra, such as the pervasive 0.7 μm phyllosilicate band, is noteworthy not only because of the specificity of the mineralogical identification which they afford, but because these minerals are secondary products of aqueous alteration. During the next decade it is anticipated that weak features of dark asteroids in the visible will be more thoroughly documented. This will allow valuable comparisons between the visible features and the water of hydration band near 3 μm, adding an important dimension to our understanding of the compositional makeup of the asteroid belt and outer solar system.

SMALL BASALTIC ASTEROIDS Whereas CCDs were first used to target faint asteroids in the outer solar system, a recent spectroscopic survey targeting faint (and thus small) asteroids in the inner part of the asteroid belt has also been scientifically productive (Binzel & Xu 1993, Binzel et al 1993).

Until recently, only 4 Vesta, the second largest asteroid in the main belt ($\sim$500 km in diameter), and four small near-Earth asteroids were known to have surface compositions similar to the howardite, eucrite, and diogenite (HED) meteorites. The V class of asteroids in the main belt contained a single member, Vesta, until 1993 when Binzel & Xu (1993) reported their discovery of 14 small asteroids ($\leqslant$10 km in diameter) of basaltic achondrite composition forming a trail from Vesta to the 3:1 Kirkwood gap. The visible spectra of two of these small V-class main-belt asteroids are shown in Figure 8*b*. The observations of Binzel & Xu have provided a possibly significant breakthrough in solving the puzzle of how Vesta could be the parent body of the HED meteorites—a preference among geochemists (e.g. Drake 1979). Large fragments apparently ejected from Vesta during

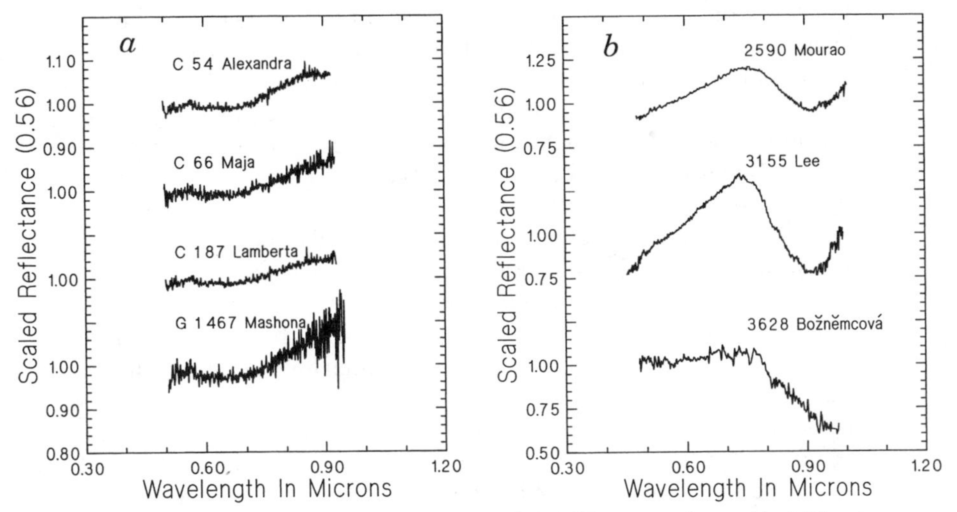

Figure 8 Examples of high spectral resolution visible CCD spectra of asteroids. (*a*) Spectra of several low albedo asteroids (Vilas et al 1993b). These dark asteroids exhibit a weak but definite 0.7 μm feature that can be attributable to the presence of phyllosilicates. (*b*) Spectra of selected small main belt asteroids. Mourao and Lee are two of the suite of small asteroids that appear to be associated with Vesta and the HED meteorites (Binzel & Xu 1993). Bŏzněmcov́ is currently the only small asteroid near the 3:1 Kirkwood gap that exhibits a prominent Fe^{2+} absorption centered beyond 0.9 μm and a relatively flat spectrum between 0.5 and 0.7 μm, consistent with spectral properties of ordinary chondrites (Binzel et al 1993).

a major impact event have been placed in orbits approaching the 3:1 Kirkwood Gap, where the probability of gravitational instabilities are high and from which Earth-crossing orbits can readily evolve (Wisdom 1985). From near the 3:1 Kirkwood gap, small asteroids are readily perturbed into Earth-crossing orbits becoming Apollo asteroids. Associated fragments evolve dynamically and, if they collide with Earth, become meteorites.

Conventional wisdom has previously described 4 Vesta as a compositionally unique asteroid of the main belt with a surface that exhibits little, if any space weathering (e.g. Matson et al 1977). If significant space weathering does occur on asteroidal surfaces, however, then the apparently unaltered optical properties of Vesta's current regolith indicate that material across the body has been freshly exposed by a recent event, such as a large impact. In support of this scenario are three pieces of evidence: 1. the unweathered appearance of the regolith on Vesta, 2. the existence of a family of small V-class asteroids (also with spectra indicating freshly exposed material) that string from Vesta to the 3:1 resonance, and 3. the similar (basaltic achondrite) V-class near-Earth asteroids in poorly evolved orbits (suggesting recent arrival to their current positions).

MAIN BELT ORDINARY CHONDRITE CANDIDATE In the same ongoing survey of small main belt asteroids Binzel et al (1993) discovered a small main belt asteroid with characteristics that are directly comparable to those observed for the ordinary chondrite meteorites in the laboratory. The spectrum of this asteroid, 3628 Božněmcová, is also shown in Figure 8*b*. 3628 Božněmcová is a 7 km diameter object that is located near the 3:1 Kirkwood gap. Its CCD reflectance spectrum has a strong 1 μm absorption band and, according to the available broadband near-infrared measurements, probably has a 2.0 μm band as well. What makes this an unusual spectrum are the strength of the ferrous bands and the apparent lack of optical effects of space weathering. Unlike the V-class basaltic achondrite asteroids, however, a *large* parent body has not been found for the ordinary chondrites.

To account for the abundance of ordinary chondrites in the meteorite collection, either (*a*) large parent bodies must be found to supply material to the appropriate resonance and/or near-Earth orbit, (*b*) a large number of small bodies must be located, or (*c*) cosmochemists and dynamists must accept a paucity of ordinary chondrite material in the main belt. Marti & Graf (1992) discuss evidence from cosmic rays indicating that the collisions that produced the ordinary chondrite classes were stochastic events. This suggests that there is more than one ordinary chondrite parent body. Because most large main belt asteroids have been included in the various asteroid surveys, but no ordinary chondrite-like material has been directly identified, if such large bodies exist they must exhibit a surface not directly comparable to those of ordinary chondrites. If this is the case, it is most likely a result of space weathering. On the other hand, if very small asteroids are the principal supplier of ordinary chondrites, then more must be discovered. It is significant that Binzel et al have found only one ordinary chondrite in the main belt out of the 80 thus far measured. The next phase of observations may provide the determinative answer.

CURRENT PERSPECTIVES ON THE ASTEROID-METEORITE LINK

The above discussion highlights many of the recent advances in spectroscopic observations of meteorites and asteroids and our ability to infer compositional implications from these data. Due to the limited nature of the data, the questions posed in the Introduction of this review are not fully resolved. Nevertheless, the compositional complexities of small bodies in the solar system are becoming more readily apparent and additional patterns may emerge as new investigations are pursued and more complete data are analyzed in an integrated manner. There are several points, some more obvious than others, that are important to consider at this stage of our understanding of the asteroid-meteorite link:

1. The asteroids are a more diverse group of bodies than is suggested from the suite of materials represented by the meteorites. This is simply another way of saying that the meteorites have sampled some, but not all types of asteroids. As the data for Pholus exemplifies, *we clearly have an incomplete sample of primitive solar system materials.*
2. The relative abundances of meteorites that fall to Earth are determined by processes that affect a small number of asteroids in any given time period. *Our current meteorite collection does not necessarily reflect the bulk composition of the asteroid belt, but it reflects the interplay of random collisions and dynamical forces that control delivery of material from the resonances.*
3. The general order originally observed throughout the main belt of asteroids using 8-color data appears to be an inherent part of the structure even as more complete spectroscopic data have been acquired. However, asteroids with compositional properties important to understanding solar system processes (hydrated components, organics, phyllosilicates, etc) do not necessarily correspond to the current classification of asteroids. As a more diverse range of compositional information for a larger number of asteroids accumulates, *it is clear that the original asteroid taxonomy will need to be revised to reflect the true complexity seen from compositionally related parameters.*

ACKNOWLEDGMENTS

NASA and NSF support for this research is gratefully acknowledged (CMP: NAGW-28 and NAG9-184; LAM: NAGW-2215 and NSF-HRD-9253136). Several spectra were acquired using RELAB, a multiuser facility supported under NASA grant NAGW-748. The assistance of Steve Pratt, Trudy Johnson, and Jeff Bytof in preparation of this manuscript is much appreciated.

Literature Cited

Adams JB. 1974. Visible and near-infrared diffuse reflectance spectra of pyroxenes as applied to remote sensing of solid objects in the solar system. *J. Geophys. Res.* 79: 4829–36

Adams JB. 1975. Interpretation of visible and near-infrared diffuse reflectance spectra of pyroxenes and other rock forming minerals. In *Infrared and Raman Spectroscopy of Lunar and Terrestrial Materials*,. pp. 91–116. New York: Academic

Adams JB, Filice AL. 1967. Spectral reflectance 0.4 to 0.2 microns of silicate rock powders. *J. Geophys. Res.* 72: 5705–15

Adams JB, Smith MO, Gillespie AR. 1993. Imaging spectroscopy: interpretation based on spectral mixture analysis. In *Remote Geochemical Analysis: Elemental and Mineralogical Composition*, ed. CM Pieters, PAJ Englert, pp. 145–66. New York: Cambridge Univ. Press

Allen CC, Morris RV, Lauer HV Jr, McKay DS. 1993. Effects of microscopic iron metal on the reflectance spectra of glass and minerals. *Lunar Planet. Sci. XXIV* 17–18 (Abstr.)

A'Hearn MF, Feldman PD. 1992. Water vaporization on Ceres. *Icarus* 98: 54–60

Bell JF, Davis DR, Hartmann WK, Gaffey MJ. 1989. Asteroids: The big picture. See Binzel et al 1989, pp. 921–45

Bell JF, Hawke PD, Owensby PD, Gaffey MJ. 1985. The 52-color asteroid survey: results and interpretation. *Bull. Am. Astron. Soc.* 17: 729

Bell JF, Keil K. 1988. Spectral alteration effects in chondritic gas-rich breccias: implications for S-class and Q-class asteroids. *Proc. Lunar & Planet. Sci. Conf.*, 18*th* Houston, pp. 573–80. Houston: Lunar Planet. Inst.

Bell JF, Owensby PD, Hawke BR, Gaffey MJ. 1988. The 52-color asteroid survey: final results and interpretation. *Lunar Planet. Sci. XIX* 57–58 (Abstr.)

Bell PM, Mao HK, Weeks RA. 1976. Optical spectra and electron paramagnetic resonance of lunar and synthetic glasses: a study of the effects of controlled atmosphere, composition, and temperature. *Proc. Lunar Sci. Conf.*, 7*th* pp. 2543–59. Houston: Lunar Planet. Inst.

Belton MJS, Veverka J, Thomas P, Helfenstein P, Simonelli D, et al. 1992. Galileo encounter with 951 Gaspra: first pictures of an asteroid. *Science* 257: 1647–52

Benedix GK. 1993. *Effects of* 10 *keV hydrogen ion irradiation on silicate minerals: implications for asteroid surface alteration.* PhD thesis. Univ. Calif., San Diego

Binzel RP. 1992. The optical spectrum of 5145 Pholus. *Icarus* 99: 238–40

Binzel RP, Gehrels T, Matthews MS, eds. 1989. *Asteroids II.* Tucson: Univ. Ariz. Press. 1258 pp.

Binzel RP, Xu S. 1993. Chips off of asteroid 4 Vesta: evidence for the parent body of basaltic achondrite meteorites. *Science* 260: 186–91

Binzel RP, Xu S, Bus SJ, Skrutskie MF, Meyer MR, et al. 1993. Discovery of a main belt asteroid resembling ordinary chondrite meteorites. *Science* 262: 1541–43

Britt DT, Pieters CM. 1988. Bidirectional reflectance properties of iron-nickel meteorites. *Proc. Lunar Planet. Sci.*, 18*th*, pp. 503–12. Houston: Lunar Planet. Inst.

Britt DT, Pieters CM. 1991. Black ordinary chondrites: an analysis of abundance and fall frequency. *Meteoritics* 26: 279–85

Britt DT, Pieters CM. 1993. Darkening in black and gas-rich ordinary chondrites: the spectral effects of opaque morphology and distribution. *Geochim. Cosmochim. Acta.* In press

Britt DT, Tholen DJ, Bell JF, Pieters CM. 1992. Comparison of asteroid and meteorite spectra: classification by principle component analysis. *Icarus* 99: 153–66

Burbine TH, Gaffey MJ, Bell JF. 1992. S-asteroids 387 Aquitania and 908 Anacostiaa: possible fragments of the breakup of a spinel-rich C3V/0 parent body. *Meteoritics* 27: 424–34

Burns RG. 1970a. *Mineralogical Applications of Crystal Field Theory*. Cambridge: Cambridge Univ. Press. 224 pp.

Burns RG. 1970b. Crystal field spectra and evidence of cation ordering in olivine minerals. *Am. Mineral.* 55: 1608–32

Burns RG, Fyfe WS. 1967. Crystal field theory and the geochemistry of transition elements. In *Researches in Geochemistry*, ed. PH Abelson, 2: 259–85. New York: Wiley

Chapman CR. 1972. *Surface properties of asteroids.* PhD thesis. Mass. Inst. Technol., Cambridge

Chapman CR. 1976. Asteroids as meteorite parent bodies: the astronomical perspective. *Geochim. Cosmochim. Acta.* 40: 701–19

Chapman CR, Gaffey MJ. 1979. Reflectance spectra for 277 asteroids. See Gehrels 1979, pp. 655–87

Chapman CR, McCord TB, Johnson TV. 1973a. Asteroid spectral reflectivities. *Astron. J.* 78: 126–40

Chapman CR, McCord TB, Pieters C. 1973b. Minor planets and related objects. X Spectrophometric study of the composition of (1685) Toro. *Astron. J.* 78: 502–5

Chapman CR, Salisbury JW. 1973. Comparison of meteorite and asteroid spectral reflectivities. *Icarus* 19: 507–22

Clark BE, Fanale FP, Salisbury JW. 1992. Meteorite-asteroid spectral comparison: the effects of comminution, melting, and recrystallization. *Icarus* 97: 288–97

Clark BE, Lucey PG, Bell JF, Fanale FP. 1993. Spectral mixing models of S-type asteroids. *Lunar Planet. Sci. XXIV* 303–4 (Abstr.)

Clark RN. 1981a. The spectral reflectance of water-mineral mixtures at low temperatures. *J. Geophys. Res.* 86: 3074–86

Clark RN. 1981b. Water frost and ice: the near-infrared spectral reflectance 0.625–2.5 μm. *J. Geophys. Res.* 86: 3087–96
Clark RN. 1983. Spectral properties of mixtures of montmorillonite and dark carbon grains: implications for remote sensing minerals containing chemically and physically adsorbed water. *J. Geophys. Res.* 88: 10,635–44
Clark RN, King TVV, Klejwa M, Swayze GA. 1990. High spectral resolution relectance spectroscopy of minerals. *J. Geophys. Res.* 95: 12,653–80
Clark RN, Lucey P. 1984. Spectral properties of ice-particulate mixtures and implications for remote sensing: 1. Intimate mixtures. *J. Geophys. Res.* 89: 6341–48
Clark RN, Roush. 1984. Reflectance spectroscopy: quantitative analysis techniques for remote sensing applications. *J. Geophys. Res.* 89: 6329–40
Cloutis EA, Gaffey MJ. 1991. Pyroxene spectroscopy revisited: spectral-compositional correlations and relationship to geothermometry. *J. Geophys. Res.* 96: 22,809–26
Cloutis EA, Gaffey MJ. 1993. Lunar regolith analogues: spectral reflectance properties of compositional variations. *Icarus* 102: 203–24
Cloutis EA, Gaffey MJ, Jackowski TL, Reed JL. 1986. Calibrations of phase abundance, composition, and particle size distribution for olivine-orthopyroxene mixtures from reflectance spectra. *J. Geophys. Res.* 91: 11,641–53
Cloutis EA, Gaffey MJ, Smith DGW, Lambert RSJ. 1990a. Metal silicate mixtures: spectral properties and applications to asteroid taxonomy. *J. Geophys. Res.* 95: 8323–38
Cloutis EA, Gaffey MJ, Smith DGW, Lambert RSJ. 1990b. Reflectance spectra of mafic silicate-opaque assemblages with applications to meteorite spectra. *Icarus* 84: 315–33
Cotton FA. 1971. *Chemical Applications of Group Theory*. New York: Wiley-Interscience. 386 pp.
Cronin JR, Pizzarello S, Cruikshank DP. 1988. Organic matter in carbonaceous chondrites, planetary satellites, asteroids and comets. In *Meteorites and the Early Solar System*, ed. JF Kerridge, MS Matthews, pp. 819–57. Tucson: Univ. Ariz. Press
Crown DA, Pieters CM. 1987. Spectral properties of plagioclase and pyroxene mixtures and the interpretation of lunar soil spectra. *Icarus* 72: 492–506
Cruikshank DP, Allamandola LJ, Hartmann WK, Tholen DJ, Brown RH, et al. 1991b. Solid C≡N bearing material on outer solar system bodies. *Icarus* 94: 345–53
Cruikshank DP, Hartmann WK. 1984. The meteorite-asteroid connection: two olivine-rich asteroids. *Science* 223: 281–83
Cruikshank DP, Moroz LV, Geballe TR, Pieters CM, Bell JF. 1993. Asphaltite-like organic solids on planetesimal 5145 Pholus. *Bull. Am. Astron. Soc.* 25: 1125–26
Cruikshank DP, Tholen DJ, Hartmann WK, Bell JF, Brown RH. 1991a. Three basaltic Earth-approaching asteroids and the source of the basaltic meteorites. *Icarus* 89: 1–13
Dennison JE, Lingner DW, Lipschutz ME. 1986. Antarctic and non-Antarctic meteorites form different populations. *Nature* 319: 390–93
Dodd RT. 1981. *Meteorites: A Petrologic-Chemical Synthesis*. New York : Cambridge Univ. Press. 368 pp.
Drake MJ. 1979. Geochemical evolution of the eucrite parent body: possible nature and evolution of asteroid 4 Vesta? See Gehrels 1979, pp. 765–82
Fink U, Hoffman M, Grundy W, Hicks M, Sears W. 1992. The steep red spectrum of 1992 AD: an asteroid covered with organic material? *Icarus* 97: 145–49
Froeschle Cl, Greenberg R. 1989. Mean motion resonances. See Binzel et al 1989, pp. 827–44
Gaffey MJ. 1974. *A systematic study of the spectral reflectivity characteristics of the meteorite classes with applications to the interpretation of asteroid spectra for mineralogical and petrological information*. PhD thesis. Mass. Inst. Tech., Cambridge
Gaffey MJ. 1976. Spectral reflectance characteristics of the meteorite classes. *J. Geophys. Res.* 81: 905–20
Gaffey MJ. 1984. Rotational spectral variations of asteriod (8) Flora: implications for the nature of the S-type asteroids and for the parent bodies of the ordinary chondrites. *Icarus* 60: 83–114
Gaffey MJ. 1986. The spectral and physical properties of metal in meteorite assemblages: implications for asteroid surface materials. *Icarus* 66: 468–86
Gaffey MJ, Bell JF, Cruikshank DP. 1989. Reflectance spectroscopy and asteroid surface mineralogy. See Binzel et al 1989, pp. 98–127
Gaffey MJ, Burbine TH, Binzel RP. 1993a. Asteroid spectroscopy: progress and perspectives. *Meteoritics* 28: 161–87
Gaffey MJ, Bell JF, Brown RH, Burbine TH, Piatek JL, et al. 1993b. Mineralogical variations within the S-type asteroid class. *Icarus* 106: In press

Gaffey MJ, McCord TB. 1978. Asteroid surface materials: mineralogical characterizations from reflectance spectra. *Space Sci. Rev.* 21: 555–628

Gaffey MJ, McCord TB. 1979. Mineralogical and petrological characterizations of asteroid surface materials. See Gehrels 1979, pp. 687–723

Gaffey MJ, Reed KL, Kelley MS. 1992. The relationship of E-type Apollo asteroid (3103) 1982 BB to the enstatite achondrite meteorites and the Hungaria asteroids. *Icarus* 100: 95–109

Gaffey SJ, McFadden LA, Nash D, Pieters CM. 1993. Ultraviolet, visible, and near-infrared reflectance spectroscopy: laboratory spectra of geologic materials. In *Remote Geochemical Analysis: Elemental and Mineralogical Composition*, ed. CM Pieters, PAJ Englert, pp. 43–77. New York: Cambridge Univ. Press

Gehrels T, ed. 1979. *Asteroids*. Tucson: Univ. Ariz. Press

Gradie J, Tedesco E. 1982. Compositional structure of the asteroid belt. *Science* 216: 1405–7

Gradie JC, Veverka J. 1980. The composition of the Trojan asteroids. *Nature* 283: 840–42

Graham AL, Bevan AWR, Hutchison R. 1985. *Catalogue of Meteorites*. Tucson: Univ. Ariz. Press. 4th ed.

Grieve RAF. 1982. The record of impact on Earth: implications for a major Cretaceous/Tertiary impact event. *Spec. Pap. Geol. Soc. Am.* 190: 25–37

Hapke B. 1981. Bidirectional reflectance spectroscopy, 1. Theory. *J. Geophys. Res.* 86: 3039–54

Hapke B, Cassidy W, Wells E. 1975. Effects of vapor-phase deposition processes on the optical, chemical, and magnetic properties of the lunar regolith. *The Moon* 13: 339–53

Hiroi T, Bell JF, Takeda H, Pieters CM. 1993a. Spectral comparison between olivine-rich asteroids and pallasites. *NIPR Symp. on Antarctic Meteorites, Tokyo*, 6: 234–45. Tokyo: Natl. Inst. Polar Res.

Hiroi T, Bell JF, Takeda H, Pieters CM. 1993b. Modeling of S-type asteroid spectra using primitive achondrites and iron meteorites. *Icarus* 102: 107–16

Hiroi T, Pieters CM. 1992. Effects of grain size and shape in modeling reflectance spectra of mineral mixtures. *Proc. Lunar Planet. Sci.* 22: 313–25

Hiroi T, Pieters CM, Zolensky ME, Lipschutz ME. 1993c. Evidence of thermal metamorphism on the C, G, B, and F asteroids. *Science* 261: 1016–18

Hiroi T, Pieters CM, Zolensky ME, Lipschutz ME. 1993d. Possible thermal metamorphism on the C, G, B, F asteroids detected from their reflectance spectra in comparison with carbonaceous chondrites. *NIPR Symp. on Antarctic Meteorites, Tokyo*. Submitted. Tokyo: Natl. Inst. Polar Res.

Hiroi T, Takeda H. 1991. Reflectance spectroscopy and mineralogy of primitive achondrites-lodranites. *NIPR Symp. on Antarctic Meteorites, Tokyo*, 4: 163–77. Tokyo: Natl. Inst. Polar Res.

Hunt GR, Salisbury JW. 1970. Visible and near-infrared spectra of minerals and rocks: I. Silicate minerals. *Mod. Geol.* 1: 283–300

Hunt GR, Salisbury JW, Lenhoff CJ. 1971. Visible and near-infrared spectra of minerals and rocks: III. Oxides and hydroxides. *Mod. Geol.* 2: 195–205

Hunt GR, Salisbury JW, Lenhoff CJ. 1973a. Visible and near-infrared spectra of minerals and rocks: VI. Additional silicates. *Mod. Geol.* 4: 85–106

Hunt GR, Salisbury JW, Lenhoff CJ. 1973b. Visible and near-infrared spectra of minerals and rocks: VIII. Intermediate igneous rocks. *Mod. Geol.* 4: 237–44

Jeffrey C. 1982. *An Introduction to Plant Taxonomy*. Cambridge : Cambridge Univ. Press. 154 pp.

Jewitt D, Luu J. 1993. Discovery of the candidate Kuiper belt object 1992 QB1. *Nature* 362: 730–32

Johnson PE, Smith MO, Adams JB. 1992. Simple algorithms for remote determination of mineral abundances and particle sizes from reflectance spectra. *J. Geophys. Res.* 97: 2649–58

Johnson TV, Fanale FP. 1973. Optical properties of carbonaceous chondrites and their relationship to asteroids. *J. Geophys. Res.* 78: 8507–18

Jones T. 1988. *An infrared reflectance study of water in outer belt asteroids: clues to composition and origin*. PhD thesis. Univ. Ariz., Tucson

Jones TD, Lebofsky LA, Lewis JS, Marley MS. 1990. The composition and origin of the C, P, and D asteroids: water as a tracer of thermal evolution in the outer belt. *Icarus* 88: 172–92

Kerridge JF , MS Matthews, eds. 1988. *Meteorites and the Early Solar System*. Tucson: Univ. Ariz. Press. 1269 pp.

King TVV, Clark RN. 1989. Spectral characterization of chlorites and Mg-serpentines using high-resolution reflectance spectroscopy. *J. Geophys. Res.* 94: 13,997–4008

King TVV, Clark RN, Calvin WM, Sherman DM, Brown RH. 1992. Evidence for ammonium-bearing minerals on Ceres. *Science* 255: 1551–53

King TVV, Ridley WI. 1987. Relation of the spectroscopic reflectance of olivine to mineral chemistry and some remote sensing applications. *J. Geophys. Res.* 92: 11,457–69

Lebofsky LA, Feierberg MA, Tokunaga AT, Larson HP, Johnson JR. 1981a. The 1.7 to 4.2 μm spectrum of asteroid 1 Ceres: evidence for structural water in clay minerals. *Icarus* 48: 453–59

Lebofsky LA, Veeder GJ, Rieke GH, Lebofsky MJ, Matson DL, et al. 1981b. The albedo and diameter of 1862 Apollo. *Icarus* 48: 335–38

Lindstrom MM, Schwarz C, Score R, Mason B. 1991. MacAlpine Hill 88104 and 88105 lunar highland meteorites: general description and consortium overview. *Geochim. Cosmochim. Acta.* 55: 2999–3007

Lipschutz ME. 1993. Meteorites. In *The Encyclopedia of the Solar System*, ed. TV Johnson, P Weissman. New York: Academic In press

Lipschutz ME, Gaffey MJ, Pellas P. 1989. Meteoritic parent bodies: nature, number, size and relation to present-day asteroids. See Binzel et al 1989, pp. 740–77

Marti K, Graf T. 1992. Cosmic-ray exposure history of ordinary chondrites. *Annu. Rev. Earth Planet. Sci.* 20: 221–43

Matson DL, Johnson TV, Veeder GJ. 1977. Soil maturity and planetary regoliths: the Moon, Mercury, and the asteroids. *Proc. Lunar Sci. Conf., 8th* 1001–11

McCord TB. 1968. A double beam astronomical photometer. *Appl. Opt.* 7: 475–78

McCord TB, Adams JB, Johnson TV. 1970. Asteroid Vesta: spectral reflectivity and compositional implications. *Science* 168: 1445–47

McCoy TJ, Keil K, Clayton RN, Mayeda TK. 1993. Classificational parameters for Acapulcoites and Lodranites: the case of FRO 90011, EET 84302 and ALH A81187/84190. *Lunar Planet. Sci. XXIV* 945–46

McFadden LA. 1983. *Spectral reflectance of near-Earth asteroids: implications for composition, origin, and evolution.* PhD thesis. Univ. Hawaii, Honolulu

McFadden LA. 1987. Spectral Reflectance of SNC Meteorites: relationships to Martian surface composition. *MEVTV Workshop on Nature and Composition of Surface Units of Mars*, ed. JR Zimbelman, SC Solomon, VL Sharpton, p. 144. LPI Tech. Rep. 88-05. Houston: Lunar Planet. Inst.

McFadden LA, Chamberlin AB. 1992. Near-infrared reflectance spectra: applications to problems in asteroid-meteorite relationships. In *Asteroids, Comets, Meteors* 1991, N. Ariz. Univ., Flagstaff, Ariz. pp. 413–16. Houston: Lunar Planet. Inst.

McFadden LA, Gaffey MJ, McCord TB. 1984. Mineralogical-petrological characterization of near-Earth asteroids. *Icarus* 59: 25–40

McFadden LA, Gaffey MJ, McCord TB. 1985. Near-Earth asteroids: possible sources from reflectance spectroscopy. *Science* 229: 160–63

McFadden LA, Gaffey MJ, Takeda H. 1980. Reflectance spectra of some newly found, unusual meteorites and their bearing on surface mineralogoy of asteroids. *Lunar Planet. Symp., 13th*, Tokyo, pp. 173–280

McFadden LA, Gaffey MJ, Takeda H. 1981. The layered crust model and the surface of Vesta. *Lunar Planet. Sci. XII* 685–87

McFadden LA, McCord TB, Pieters C. 1977. Vesta: the first pyroxene band from new spectroscopic measurements. *Icarus* 31: 439–46

McFadden LA, Pieters CM, Huguenin RL, Hawke BR, King TVV, Gaffey MJ. 1986. Reflectance spectroscopy of lunar meteorite Y791197: relation to remote sensing data bases of the moon. *Mem. Natl. Inst. Polar Res.* No. 41: 140–51

McFadden LA, Pratt S. 1989. Remote sensing and the Shergottite-Nakhlite-Chassignite meteorite parent body. *Bull. Am. Astron. Soc.* 21: 967 (Abstr.)

McSween HY. 1979. Are carbonaceous chondrites primitive or processed? A review. *Rev. Geophys. Space Phys.* 17: 1059–77

McSween HY. 1985. SNC meteorites: clues to martian petrologic evolution? *Rev. Geophys. Res* 23: 319–416

McSween HY, Jarosewich E. 1983. Petrogenesis of the Elephant Moraine A79001 meteorite: multiple magma pulses on the shergottite parent body. *Geochim. Cosmochim Acta* 47: 1501–13

Melosh HJ. 1984. Impact ejection, spallation, and the origin of meteorites. *Icarus* 59: 234–60

Melosh HJ. 1985. Ejection of rock fragments from planetary bodies. *Geology* 13: 144–48

Milani A, Knežević Z. 1992. Asteroid proper elements and secular resonances. *Icarus* 98: 211–32

Miyamoto M, Mito A, Takano Y. 1982. An attempt to reduce the effects of black material from the spectral reflectance of meteorites or asteroids. *Seventh Symp. on Antarctic Meteorites, Tokyo*, pp. 291–307. Tokyo: Natl. Inst. Polar Res.

Moroz LV, Pieters CM. 1991. Reflectance spectra of some fractions of Migei and Murchinson CM chondrites in the range of 0.3–2.6 μm. *Lunar Planet. Sci. XXII* 923–24 (Abstr.)

Moroz LV, Pieters CM, Akhmanova MV. 1991. Spectroscopy of solid carbonaceous materials: implications for dark surfaces of outer belt asteroids. *Lunar Planet. Sci. XXII* 925–26 (Abstr.)

Moroz LV, Pieters CM, Akhmanova MV. 1992. Why the surfaces of outer belt asteroids are dark and red? *Lunar Planet. Sci. XXIII* 931–32 (Abstr.)

Morris RV. 1977. Origin and evolution of the grain-size dependence of the concentration of fine-grained metal in lunar soils: the maturation of lunar soils to a steady-state stage. *Proc. Lunar Sci. Conf., 8th*, pp. 3719–47

Morris RV. 1980. Origins and size distribution of metallic iron particles in the lunar regolith. *Lunar Planet. Sci. Conf., 11th*, pp. 1697–712

Mueller BEA, Tholen DJ, Hartmann WK, Cruikshank DP. 1992. Extraordinary colors of asteroidal object (5145) 1992 AD. *Icarus* 97: 150–54

Mustard JF, Erard S, Bibring J-P, Head JW, Hurtrez S, et al. 1993. The surface of Syrtis Major: composition of the volcanic substrate and mixing with altered dust and soil. *J. Geophys. Res.* 98(E2): 3387–400

Mustard JF, Pieters CM. 1987. Quantitative abundance estimates from bidirectional reflectance measurements. *Proc. Lunar Planet. Sci. Conf., 17th*. In *J. Geophys. Res.* 92: E617–26

Mustard JF, Pieters CM. 1989. Photometric phase functions of common geologic minerals and application to quantitative analysis of mineral mixture reflectance spectra. *J. Geophys. Res.* 94: 13,619–34

Nash DB, Conel JE. 1974. Spectral reflectance systematics for mixtures of powdered hypersthene, labradorite, and ilmenite. *J. Geophys. Res.* 79: 1615–21

Pieters CM. 1974. Polarization in a mineral absorption band. In *Planets, Stars, and Nebulae Studied with Photopolarimetry*, ed. T Gehrels, pp. 405–18. Tucson: Univ. Ariz. Press

Pieters CM 1986. Composition of the lunar highland crust from near-infrared spectroscopy. *Rev. Geophys.* 24(3): 557–78

Pieters CM. 1993. Compositional diversity and stratigraphy of the lunar crust derived from reflectance spectroscopy. In *Remote Geochemical Analysis: Elemental and Mineralogical Composition*, ed. CM Pieters, PAJ Englert, pp. 309–42. New York: Cambridge Univ. Press

Pieters CM, Fischer EM, Rode O, Basu A. 1993. Optical effects of space weathering: the role of the finest fraction. *J. Geophys. Res.* 98: 20,817–24

Pieters CM, Gaffey MJ, Chapman CR, McCord TB. 1976. Spectrophotometry (0.33 to 1.07 μm) of 433 Eros and compositional implications. *Icarus* 28: 105–15

Pieters CM, Hawke BR, Gaffey MJ, McFadden LA. 1983. Possible lunar source areas of meteorite ALHA 81005: geochemical remote sensing information. *Geophys. Res. Lett.* 10: 813–16

Pieters CM, Rode O. 1993. Spectral properties of maskelenite from the gabbroic meteorite A881757. *Bull. Am. Astron. Soc.* 25: 1134–35

Salisbury JW, D'Aria DM, Jarosewich EJ. 1991. Midinfrared (2.5–13.5 μm) reflectance spectra of powdered stony meteorites. *Icarus* 92: 280–97

Sawyer SR. 1991. *A high resolution CCD spectroscopic survey of low albedo main belt asteroids*. PhD thesis. Univ. Texas, Austin

Scholl H, Froeschle C, Kinoshita H, Yoshikawa M, Williams JG. 1989. Secular resonances. See Binzel et al 1989, pp. 845–61

Scott ERD, Taylor GJ, Newsom HE, Herbert F, Zolensky M, Kerridge JF. 1989. Chemical, thermal, and impact processing of asteroids. See Binzel et al 1989, pp. 701–39

Sears DWG, Dodd RT. 1988. Overview and classification of meteorites. In *Meteorites and the Early Solar System*, ed. JF Kerridge, MS Matthews, pp. 3–31. Tucson: Univ. Ariz. Press

Shoemaker EM, Williams JG, Helin EF, Wolfe RF. 1979. Earth-crossing asteroids: orbital classes, collision rates with Earth and origin. See Gehrels 1979, pp. 253–82

Singer RB. 1981. Near-infrared spectral reflectance of mineral mixtures: systematic combinations of pyroxenes, olivine, and iron oxides. *J. Geophys. Res.* 87: 7967–82

Singer RB, McCord TB. 1979. Mars: large scale mixing of bright and dark surface materials and implications for analysis of spectral reflectance. *Proc. Lunar Planet. Sci. Conf., 10th*. pp. 1835–48

Singer RB, McSween HY Jr. 1991. The Composition of the Martian crust: evidence from remote sensing and SNC meteorites. In *Resources of Near-Earth Space*. Tucson: Univ. Ariz. Press

Steel D, RH McNaught, Asher D. 1992. 1991 DA: an asteroid in a bizarre orbit. In *Asteroids, Comets, Meteors* 1991, N. Ariz. Univ., Flagstaff, Ariz. pp. 573–576. Houston: Lunar Planet. Inst.

Sunshine JM, McFadden LA, Pieters CM. 1993. Reflectance spectra of the Elephant Moraine A79001 meteorite: implications for remote sensing of planetary bodies. *Icarus* 105: 79–91

Sunshine JM, Pieters CM. 1990. Extraction of compositional information from olivine reflectance spectra: a new capability for lunar exploration. *Lunar Planet. Sci. XXI* 1223–24 (Abstr.)

Sunshine JM, Pieters CM, Pratt SF. 1990.

Deconvolution of mineral absorption bands: an improved approach. *J. Geophys. Res.* 95: 6955–66
Sunshine JM, Pieters CM. 1993a. Estimating modal abundances from the spectra of natural and laboratory pyroxene mixtures using the modified Gaussian model. *J. Geophys. Res.* 98(E5): 9075–87
Sunshine JM, Pieters CM. 1993b. Determining the composition of olivine on asteroidal surfaces. *Lunar Planet. Sci. XXIV* 1379–80 (Abstr.)
Tedesco EF, Matson DL, Veeder GJ. 1989a. Classification of IRAS asteroids. See Binzel et al 1989, pp. 290–97
Tedesco EF, Williams JG, Matson DL, Veeder GJ, Gradie JC, Lebofsky LA . 1989b. Three-parameter asteroid taxonomy classifications. See Binzel et al 1989, pp. 1151–61
Tholen DJ. 1984. *Asteroid taxonomy from cluster analysis of photometry*. PhD thesis. Univ. Ariz., Tucson
Tholen, DJ. 1989. Asteroid taxonomic classifications. See Binzel et al 1989, pp. 1139–53
Tholen DJ, Barucci MA. 1989. Asteroid taxonomy. See Binzel et al 1989, pp. 298–315
Vilas F, Gaffey MJ. 1989. Identification of phyllosilicate absorption features in main-belt and outer-belt asteroid reflectance spectra. *Science*. 246: 790–92
Vilas F, Hatch EC, Larson SM, Sawyer SR, Gaffey MJ. 1993a. Ferric iron in primitive asteroids: a 0.43-μm absorption feature. *Icarus* 102: 225–31
Vilas F, Larson SM, Hatch EC, Jarvis KS. 1993b. CCD reflectance spectra of selected asteroids. II. Low-albedo asteroid spectra and data extraction techniques. *Icarus* 105: 67–78
Vilas F, Smith BA. 1985. Reflectance spectroscopy (-0.5–1.0 μm) of outer belt asteroids: implications for primitive, organic Solar System material. *Icarus* 64: 503–16
Wasson JT. 1985. *Meteorites: Their Record of Early Solar-System History*. New York: Freeman. 316 pp.
Wells EB, Hapke B. 1977. Lunar soil: iron and titanium bands in the glass fraction. *Science* 195: 977–79
Wetherill GW. 1974. Solar system sources of meteorites and large meteoroids. *Annu. Rev. Earth and Planet. Sci.* 2: 303–31
Wisdom J. 1985. A perturbative treatment of motion near the 3/1 commensurability. *Icarus* 63: 272–89
Zellner B, Tholen DJ, Tedesco EF. 1985. The eight-color asteroid survey: results for 589 minor planets. *Icarus* 61: 355–416

Annu. Rev. Earth Planet. Sci. 1994. 22:499–551

MARINE BLACK SHALES: Depositional Mechanisms and Environments of Ancient Deposits

Michael A. Arthur

Department of Geosciences and Earth System Science Center, Pennsylvania State University, University Park, Pennsylvania 16802

Bradley B. Sageman

Department of Geological Sciences, Northwestern University, Evanston, Illinois 60201

KEY WORDS: anoxia, organic carbon burial, biofacies, geochemical tools, oceanic anoxic events (OAE)

INTRODUCTION

> The well-mixed oxygenated ocean of today seems not to be the model for the past 600 m.y.
>
> (Degens & Stoffers 1976)

Organic-carbon-rich strata or "black shales," including dark gray to black, laminated, carbonaceous mudrocks characterized by impoverished benthonic faunas, or devoid of metazoan life, have long intrigued geologists, both because of their widespread distribution at certain times in the past and their early recognition as potential hydrocarbon source rocks. One of the major obsessions of many early workers, to the mid-1900s, was the application of uniformitarian principles to depositional models for black shales. Thus, although it was recognized that some black shale units were unusually widespread, many early workers sought less extensive, modern analogues for environments of formation of dark-colored, rela-

0084–6597/94/0515–0499$05.00

tively organic-carbon (OC)-rich muds. In fact, the main debate focused on shallow- vs deep-water origins; most workers assumed that "black shale" deposition required a supply of organic matter and conditions conducive for preservation of that organic material, including depletion of dissolved oxygen in waters overlying the sediment/water interface. However, the "barred (silled) basin" depositional model dominated most interpretations of ancient black shales.

With the advent of the Deep Sea Drilling Project in 1968, a wealth of new data became available from the previously unsampled ocean basins. This led to the realization that the deep-sea record of the past 150 million years (m.y.) also contains several widespread "black shale" sequences and, more important, that OC-rich shales of Cretaceous age appear to be confined to particular stratigraphic horizons. This discovery brought the realization that local basinal bathymetry could not be the entire explanation for many black shale units. Some workers concluded that certain widespread black shales were deposited globally and synchronously, perhaps even within relatively narrow (<1 m.y.) time envelopes. These time envelopes, during which the global ocean conditions were propitious for the deposition of OC-rich sediments (but not implying global total anoxia of deep-water masses), were termed "Oceanic Anoxic Events" (OAEs). Ocean drilling and the consequent OAE concept, whether correct or not, renewed interest in black shales in general and promoted considerable debate as to the mechanisms that control black shale deposition.

Controversy continues in regard to the origin of modern OC-rich sediments as well as the origin of ancient widespread black shales. Disagreement on the origin of modern OC-rich sediments centers on the relative importance of primary productivity (OC flux) vs possible preservational mechanisms—water-column anoxia and sedimentation rate in particular. The continuing debate about ancient black shales stems from the "no-analogue" global environment that exists at present and the insufficient stratigraphic, sedimentological, and geochemical data for black shales for resolving accumulation rates and definitive indicators of productivity and anoxia, among other parameters.

In this paper we outline the salient features and inferred depositional regimes for black shales in the stratigraphic record and provide guidelines for further study. Rather than review all of the important studies that have contributed to our knowledge of black shales, we have instead attempted to synthesize data and concepts while providing examples for further reference. We have included a discussion of OC preservation in modern marine environments as a prelude to interpretation of the ancient record and a discussion of proposed biotic and geochemical indicators of degree of oxygenation in the water column.

FUNDAMENTAL PROPERTIES OF BLACK SHALES

Composition

Over the past century the term "black shale" has been applied to a considerable range of related facies interpreted to represent quite varied depositional settings and paleoenvironmental conditions. Thus, it is useful to first establish a clear definition of "black shale" in terms of sedimentologic and geochemical parameters before addressing depositional mechanisms.

In this paper we confine our discussion of black shale to marine mudrocks (see Ingram 1953, Dunbar & Rogers 1957, Blatt et al 1972, Lundegard & Samuels 1980, Spears 1980 for definitions of mudrocks) that are rich in organic matter (as defined below); true color can vary from medium gray to olive-brown to black. Although 10% lamination is a defining characteristic of most black shales, there are numerous OC-rich rocks that fall under the classification of (nonlaminated) mudstones or claystones. In addition to the dominant sediment type (mud or clay), mudrocks may contain various constituents, including calcium carbonate ($CaCO_3$) and biogenic SiO_2, and there are a number of appropriate classification schemes (e.g. Hattin 1975, Pratt 1984, Dean et al 1984). But perhaps the most important constituent that differentiates black shales from other mudrocks is organic matter content, which is the main cause of the dark color in ancient black shale facies (e.g. Trask & Patnode 1942).

Coloration of shales is influenced by amount, type, and maturity of OC, all of which can vary significantly. Sediments containing more than a few percent of immature amorphous organic matter are typically more brown than black; strata containing only 1–2% highly oxidized or thermally mature OC are commonly black. $CaCO_3$ contents in excess of 25% can also impart a lighter color to relatively OC-rich mudrocks. The black coloration of shales also is attributed at times to the presence of very fine-grained iron monosulfide minerals (griegite and/or mackinawite; Berner 1984), but these monosulfides are not stable over long (geologic) periods of time and the black coloration in monosulfide-bearing sediments fades rapidly (within hours) as they are exposed to oxygen. Huyck (1990) has discussed some of the problems and variations in defining "black shale" on the basis of OC contents. Most problems revolve around determining the OC content of "average" shale, which has commonly been taken as 0.65% OC (Vine & Tourtelot 1970). As an operational definition marine "black shales" should contain at least 1% organic matter (e.g. Huyck 1990)—typically a mixed assemblage including material derived from higher plants, algae, and phytoplankton (Cook et al 1981). Proximity to terrestrial sources of organic matter and marine productivity are the primary controls on the type of organic matter that becomes incorporated in black shale deposits as outlined below. This enhanced OC content sets these mudrocks apart from others; however, one must also keep in mind that many ancient black shales that once contained >1% OC may now have <1% OC because of OC loss during thermal maturation or metamorphism (e.g. Raiswell & Berner 1987). Thus the definition of "black shale" by organic content will always be somewhat arbitrary.

A number of workers have referred to "black shales" as "sapropels." It is thus useful to examine briefly the genetic implication of "sapropel" because, in many cases, the term has been misapplied (e.g. Tyson 1986). Many workers apply the operational definition used by Kidd et al (1978) for Mediterranean strata which set an arbitrary lower limit of >2% OC for sapropel and use "sapropelic" to refer to sediments with OC contents between 0.5–1.9%. Anastasakis & Stanley (1984) applied the term "episapropel" to those units that could be determined to have been redeposited on the basis of sedimentary structures. However, although some workers used "sapropel" as a modifier for OC-rich sediments (e.g. Grabau 1913), the original definition of the term includes the inference that "sapropel" is a sediment rich in algal organic matter that was formed under reducing conditions in a "stagnant" water body (e.g. Pontonié 1937, Wasmund 1930). Following the recommendation of Tyson (1986), the term "sapropel" should be reserved only for descriptions of strata that meet the latter criteria; otherwise the term has no more utility than "black shale (clay or mud)."

Lamination and Sedimentary Structures

Lamination is a fundamental distinguishing characteristic of black shales and has long been a topic of interest (e.g. Bradley 1931). Lamination in mudrocks is defined as a parallel arrangement of layers <10 mm thick that results from a regular alternation in fabric, grain size, and/or color (Lundegard & Samuels 1980). Although many classic black shales are finely laminated, not all are, either because original laminae were disturbed by bioturbation or by other physical processes, such as winnowing and redeposition.

Laminae couplets typify sediments in modern environments of formation of OC-rich muds where oxygen concentrations inhibit benthic metazoan activity and where there is pronounced seasonality in productivity or sediment supply. Examples are the Black Sea (e.g. Degens & Ross 1974) and the Gulf of California (e.g. Calvert 1964). One of the most common types of lamination found in Cretaceous and younger rocks are light-dark alternations composed of laminae rich in calcareous material (commonly coccoliths) and clay/organic matter, respectively. Although a number of recent studies have suggested that such laminations result from seasonal phytoplankton blooms (Ross et al 1970, Thomsen 1989, Hattin 1975), evidence from other ancient black shale deposits indicates that there are a wide range of different types of laminae that appear macroscopically similar, but do not reflect a seasonal depositional cycle (Sageman 1991), possibly because sedimentation rates are too slow. In cases such as the Cretaceous Greenhorn Formation of the U. S. Western Interior Basin it appears that individual light-colored laminae represent the amalgamation of numerous annual productivity events, separated by darker clay and organic-rich laminae that probably reflect more rapidly and intermittently deposited mud. Fine lamination in other Cretaceous calcareous black shales of the western North Atlantic has been attributed to bottom-current activity, albeit in an oxygen-deficient basin (e.g. Robertson 1984, Dean & Arthur 1989). Williams & Reimers (1989) attributed lamination in the Miocene Monterey Formation of California, and by inference other OC-rich deposits of upwelling zones associated with oxygen-minimum zones, to periodic (perhaps seasonal) colonization of the sediment/water interface by sulfide-oxidizing bacterial mats.

Other features common to epicontinental black shales include a variety of event beds and horizons that indicate significant condensation and/or erosion. Whereas examples from the Western Interior Cretaceous include skeletal limestone beds and lenses composed of winnowed biogenic debris (Pratt 1984, Hattin 1986, Sageman 1991), the Devonian black shales of New York State contain similarly interpreted pyritic lag deposits (Baird & Brett 1986). Black shale sequences may also be typified by redeposited beds relatively rich in organic matter as suggested for many "episapropels" of Quaternary age in the Mediterranean (e.g. Anastasakis & Stanley 1984) and Cretaceous black mudstones of the deeper marginal North Atlantic (Arthur et al 1984, Degens et al 1986). Many of these redeposited OC-rich beds have sharp basal contacts and are homogeneous to coarsely laminated.

HISTORICAL PERSPECTIVE

In Pettijohn's (1957, pp. 622–26), review of the "Black Shale Facies" (euxinic) he suggested, "That the pyritic black shales were deposited under anaerobic conditions is unquestioned. Whether, however, the basin of accumulation was shallow or deep and whether it was landlocked or freely connected with the sea or even a stagnant area in the open sea has been much debated." With this statement he essentially outlined the major sources of disagreement as to the origin of black shales then and now. His cogent summary of the black shale facies followed years of speculation and assertion by a number of influential workers. For example, Grabau (1913) argued that widely distributed marine black shales (his "sapropelargillyte') were deposited in coastal lagoons or marginal epicontinental seas on extensive mud flats exposed at low tide. Grabau knew of the Black Sea because Pompeckj (1901) had interpreted the Jurassic Posidonienshiefer (Europe) and the Devonian Portage (eastern U.S.) black shales, respectively, as having originated in Black Sea–like environments (see also Schuchert 1915), but he viewed this analogy as doubtful, suggesting only that this model might apply to the Permian Kupferschiefer of Europe. Subsequently, Strøm (1939) summarized his work on organic matter deposition in Norwegian fjords, which probably unintentionally reinforced the barred basin concept. Fleming & Revelle (1939), on the other hand, presented a thorough and insightful treatment of the role of physical oceanographic processes on dissolved oxygen distribution in the water column and even discussed the origin of mid-water oxygen-minimum zones (OMZ); they outlined causes of stable stratification, including the susceptibility of tropical environments to thermal stratification, but, again, particularly emphasized the role of sill depth or basin thresholds in controlling the rate of renewal of bottom water, citing California borderland basins and the Black Sea as examples where stable stratification and/or sills limited exchange of bottom waters and promoted oxygen deficiency. In contrast, Twenhofel (1939) apparently saw a more open-marine origin for "bituminous black shales" stating that ". . . bituminous black shales accumulated, in general, in less stagnant and more open water, but waters that did not have good circulation." Nonetheless, he favored a "Black Sea type" model for Paleozoic laminated black shale deposits of northwest Europe and the U.S. Appalachian basins. He suggested water depths between 300 and 600 feet, however, for the euxinic water masses, arguing that deposition above wave base would have been difficult as a result of oxygen mixing downward periodically.

At the time of publication of Pettijohn's treatise there was little appreciation of the possible role of primary productivity, and the emphasis clearly focused on reducing environments. However, this subject, a source of disagreement among present workers (e.g. Pedersen & Calvert 1990 vs Demaison & Moore 1980), was not completely ignored. For example, Goldman (1924) explicitly suggested the importance of rate of supply of organic matter with or without oxygen depletion in stating that, "In shallow bays that have no threshold, but which are rich in vegetation and have only a feeble deep water circulation, the bottom deposits often consist of black muds high in hydrogen sulfide and containing no animal life. The waters themselves, however, will usually contain dissolved oxygen. The moderate stagnation in these cases is due to the large supply of decomposable organic matter rather than the lack of water exchange." This concept was echoed by Fleming & Revelle (1939) and was implicit in Trask's (1932) discussion of sedimentary organic matter contents, but really was not entertained as a major theme in the origin of black shales until more recently (e.g. Parrish 1982, Parrish & Curtis 1982, Waples 1983). More complete models for black shale deposition, such as that for mid-continent U.S. Pennsylvanian black shales outlined by Heckel (1977), suggested that both high primary productivity (upwelling) and oxygen depletion in deeper water masses (preservation) played a role in producing relatively OC-rich strata.

Like Ruedemann (1935, also Hard 1931), who argued for "toxic" bottom-water conditions and against the concept of small embayments or lagoons as primary sites for black shale deposition in their work on Devonian black shales of the Appalachian region, Pettijohn (1957) also summarized black shale facies as being widespread at certain times and characterized by a slow rate of accumulation. He advocated the idea that many black shales represent deposition in "starved" basin phases (following the terminology of Adams et al 1951), which now would be recognized as relatively condensed zones related to transgressive episodes. Nonetheless, Pettijohn followed the prevailing view that OC-rich strata were the result of more local factors, such as deposition in "barred" or tectonically isolated basins (e.g. Trask 1932, Woolnough 1937, Caspers 1957), and suggested the Black Sea as the "type euxinic basin," thus acknowledging a possible role for density stratification of water masses and dissolved oxygen concentration ("strong reducing environment") in OC preservation, but in a restricted basin context.

That variations in global oceanic circulation likely were involved in the formation of the black shales was recognized by van der Gracht (1931) who argued that during non-glacial periods the equatorward circulation of deep, cold oxygen-rich water would cease and that the bottom water of large oceanic basins would become anoxic. This concept built on the ideas of Chamberlin (1906), who suggested reversal of circulation and warmer, more saline deep waters during warm climatic epochs. Stroøm (1936) also recognized that more widespread episodes of "black sediment" deposition might require more than local causal factors. Thus, long before the 1970s it was conceived that the "Black Shale Facies" represented rather unique but widespread paleoceanographic conditions that recurred episodically. One common theme of nearly all black shale studies links black shales to marine transgression (e.g. Hallam & Bradshaw 1979, Loutit et al 1988)—a cause which has been invoked for modern occurrences of OC-rich sediments in silled basins as well (e.g. Middelburg et al 1991).

A minor revolution in how the community viewed the origin of black shales came as a result of the Deep Sea Drilling Project (DSDP). Cretaceous black shales were recovered in a number of DSDP sites in all ocean basins (Ryan & Cita 1977, Thiede & van Andel 1977, Fischer & Arthur 1977). The recognition that these organic-rich units were equivalent to black shales exposed on a number of continents led to the "OAE" concept, which was originally defined by Schlanger & Jenkyns (1976) to signify time envelopes during which black shale deposition was particularly prevalent. It has become apparent that this terminology may be misleading in the sense that it does not distinguish between events with long and those with short durations. The term "anoxic event" subsequently has been used very loosely in describing almost any relatively OC-rich bed, including, for example, the Quaternary sapropels of the Mediterranean Sea. With respect to the Cretaceous, it is quite clear that the Cenomanian-Turonian OAE was, geologically speaking, a very brief episode (<1 m.y.) recorded nearly globally. By comparison, the Aptian-Albian OAE, as defined by Schlanger & Jenkyns (1976) encompasses nearly 20 m.y. An additional problem is that the term OAE can be construed as indicating that the global ocean was anoxic (widespread "euxinic" conditions), but this sort of genetic implication should be avoided because of the multitude of possible ways in which black shales may have formed, even during one discrete episode.

MODERN ENVIRONMENTS AND ORGANIC CARBON BURIAL

Modern Analogues for Ancient Environments

Interpretation of ancient black shale deposits has been strongly influenced by observations of modern environments where OC-rich muds presently accumulate. Depositional analogues for marine black shales most commonly cited include large landlocked or silled basins, such as the Black Sea and the Baltic Sea, as well as open-ocean environments such as coastal upwelling zones (e.g. Peru, Namibia, NW margin of the Indian Ocean, California borderland basins, and the Gulf of California) affected by impingement of the midwater oxygen-minimum zone. More localized environments of organic matter burial include tropical embayments, coastal swamps, large estuaries, fjords, and brine-stratified slope basins. The physical characteristics of these different environments have been extensively reviewed by Rhoads & Morse (1971), Demaison & Moore (1980), and Arthur et al (1984) among others, and are summarized in Figure 1.

Modern analogues for black shale–forming environments are classified based on relative water depth, and the nature and position of the redox boundary and its affect on benthic communities (Figure 1). The catagories include: 1. deep, enclosed, strongly stratified basins, characterized by a

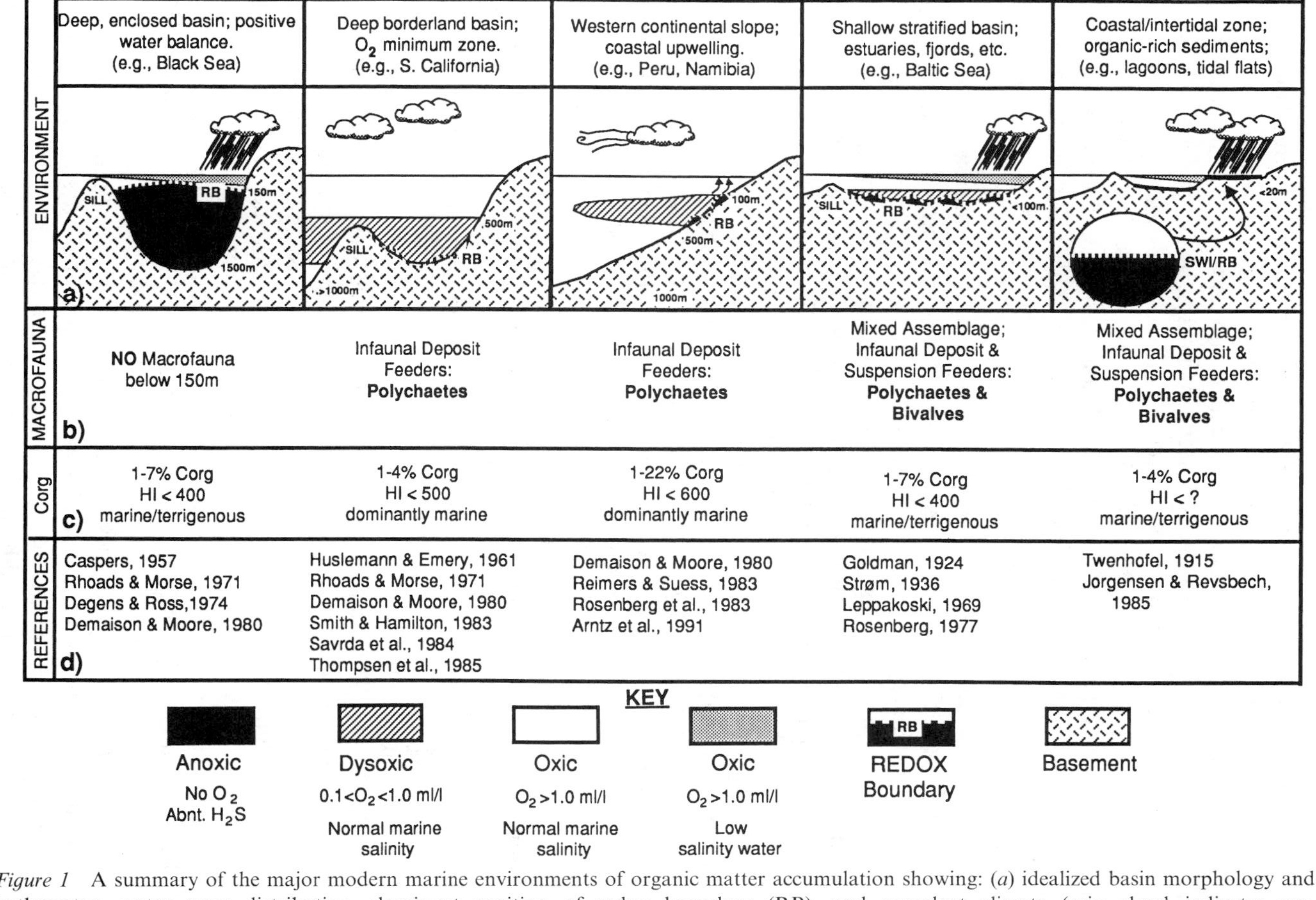

Figure 1 A summary of the major modern marine environments of organic matter accumulation showing: (*a*) idealized basin morphology and bathymetry, water mass distribution, dominant position of redox boundary (RB), and prevalent climate (rain cloud indicates precipitation > evaporation; cloud streamers indicate offshore winds); (*b*) predominant macrofauna reported from oxygen-deficient parts of each environment; (*c*) characteristics of organic matter associated with each environment; and (*d*) key references for each environment.

redox boundary high in the water column and long-term anoxia throughout much of the basin; 2. deep, unenclosed basins in which dysoxic conditions predominate and the redox boundary normally remains at some depth within the substrate, but may fluctuate up and down; 3. shallow, stratified basins that are subject to frequent mixing by storms in which dysoxic to oxic conditions predominate and the redox boundary normally remains at some depth within the substrate, but may fluctuate above and below the sediment/water interface (SWI); and 4. shallow-water areas in which the water column is dysoxic to fully oxic, but the sediments are highly OC-rich and reducing. The redox boundary is commonly coincident with the SWI.

The Black Sea is the best example of a large, deep, stratified basin characterized by an anoxic marine water column and finely laminated, OC-rich sediments (Figure 1). It has played an influential role in the interpretation of ancient black shales (i.e. Rhoads & Morse 1971). However, conditions in the Black Sea are unique and represent one extreme in a spectrum of oxygen-deficient environments. Black Sea waters below about 150 meters are completely anoxic and contain dissolved H_2S; they are devoid of metazoan life. Yet, recent studies of a wide range of ancient black shales have revealed a common history of repeated benthic colonization events, suggesting relatively dynamic, and largely dysoxic benthic environments (e.g. Kauffman 1981, Savrda & Bottjer 1986, Sageman 1989, Wignall 1990). Thus, the late Holocene Black Sea is best viewed as an analogue for sedimentary and geochemical processes under an anoxic water column (euxinic conditions), which, through geologic history, probably occurred more commonly in the benthic zone, than throughout the water column of relatively deep basins.

Oceanic OMZs result from the oxidation of sinking pelagic organic matter and the long residence time and slow rate of exchange between intermediate-depth waters and better oxygenated surface waters (Demaison & Moore 1980). They are particularly intense and well-developed on the margins of ocean basins, where wind-driven upwelling and higher nutrient levels enhance surface-water productivity. In some cases, OMZs impinge upon structurally complex continental margins, such as on the California borderland, where a series of graben-type basins preserve significant thicknesses of OC-rich sediment (Figure 1). These basins are silled below about 500 m depth and contain predominantly dysoxic waters with the main redox boundary at shallow depth within the substrate (Savrda et al 1984). Biotic characteristics of such basins reveal surprisingly high benthic community diversity and biomass from within the dysoxic zone, with up to ca. 300 species reported in some cases (Jumars 1976). The highest species diversity and biomass are among polychaete worms (up to

150 species; Hartman & Barnard 1958), which can bioturbate the sediment at oxygen levels as low as 0.3 ml/l (Thompson et al 1985). In addition, a wide range of taxa comprise less abundant elements in these communities, including arthropods, echinoderms, coelenterates, infaunal protobranch bivalves, nematodes, and benthic foraminifera (Hartman & Barnard 1958, Hülsemann & Emery 1961, Calvert 1964, Nichols 1976, Jumars 1976, Smith & Hamilton 1983, Savrda et al 1984, Thompson et al 1985). Though relatively OC-rich sediments are commonly bioturbated, laminated intervals with up to 4% relatively hydrogen-rich OC occur (Hulsemann & Emery 1961, Summerhayes 1987), suggesting significant spatial and/or temporal variability in the position of the redox boundary.

In upwelling zones the export of abundant pelagic organic matter to the benthic zone, due to high productivity, results in extensive metazoan and bacterial decomposition and consumption of dissolved oxygen (dysoxic to anoxic). The Peru outer shelf to upper slope region is characterized by a well-developed OMZ (e.g. Thiede & Suess 1983, Suess, von Huene, et al 1990, Froelich et al 1988). Sediment OC contents are high relative to the California borderland, due to enhanced productivity [up to 22% TOC (Total Organic Carbon) with HI (Hydrogen index—mg hydrocarbons per g OC) to 600] and lower sedimentation rates; however, the biotic characteristics of these basins are almost identical to those of the California borderland (Rosenberg et al 1983, Arntz et al 1991). Despite evidence of pervasive colonization of the sediment by infauna (polycheates dominant; Arntz et al 1991), intervals of undisturbed laminae described from cores suggest spatial and/or temporal variability in the position of the redox boundary. Mats of sulfide-oxidizing bacteria (e.g. *Thioploca*) that occur in patchy distributions on the surface of these OC-rich sediments (Gallardo 1977, Arntz et al 1991) may play a role in determining redox boundary dynamics.

Oxygen deficiency and accumulation of OC-rich mud has been described from large estuaries (e.g. Chesapeake Bay), shallow basins with high freshwater input (e.g. the Baltic Sea), and fjords, attributed mainly to seasonal intensification of salinity stratification (Goldman 1924, Strøm 1936, Leppakoski 1969, Theede 1973), or in tropical bays (e.g. Kau Bay, Jellyfish Lake, Palau, Cariaco Trench, Venezuela) as the result of thermal and/or salinity stratification (Orem et al 1991, Middelburg et al 1991, Richards & Redfield 1954). At such times, bottom waters become dysoxic, and the substrate is characterized by black, anoxic muds saturated with H_2S, particularly in topographically low areas. Organic content of the sediment typically varies from less than 1 to 7% TOC (but Jellyfish Lake, Palau has up to 46% TOC; Orem et al 1991) and organic matter may be mostly terrestrial in origin, resulting in low hydrogen indices. The benthic fauna is typically dominated by polycheates, although a number of low-oxygen tolerant bivalves also have been described (Leppakoski 1969, Theede 1973). Such environments are quite dynamic, with frequent variations in the position of the redox boundary due to seasonal or longer-term changes in biologic production, stratification, and benthic oxygen levels. Another rather local environment for OC preservation occurs in small, restricted slope basins associated with regions of salt tectonics and salt outcrop (e.g. Orca Basin, Gulf of Mexico, Mediteranean basins; see Cita 1991 for review). These small basins are not likely to produce major preserved black shales in the record but are interesting geochemically.

Certain coastal areas where fine-grained sediments are deposited, such as large bays, lagoons, tidal flats, and intertidal swamps, are also sites of deposition of black, H_2S-containing, OC-rich muds (Twenhofel 1915, Jorgensen & Revsbech 1985, Rosenberg 1977, Martens & Klump 1984). The distribution of organic-rich sediment in such areas is extremely patchy. The OC content of the sediment is highly variable and likely to be dominated by terrestrial components, resulting in low hydrogen richness. Prevalence of mixing in these shallow areas results in mostly oxic to slightly dysoxic conditions in the water column. However, a sharp redox boundary commonly develops at the SWI due to the rapid uptake of oxygen by bacteria in the OC-rich sediment. In some cases, development of a diffusive boundary layer creates a physical microboundary to oxygen exchange; such a boundary may or may not be associated with a conspicuous microbial mat of sulfur-oxidizing bacteria (such as Beggiatoa) on the sediment surface (e.g. Jorgensen & Revsbech 1985). In general, the major faunal constituents of shallow-water OC-rich muds are quite similar to other modern low-oxygen environments (where polycheates and infaunal deposit-feeding bivalves dominate). However, these muds also contain a unique group of highly specialized bivalves that survive within toxic, H_2S-containing muds by virtue of chemosymbioisis with sulfur-oxidizing bacteria. These taxa, including species of the families Lucinidae, Thyasiridae, and Solemyidae, evolved from infaunal suspension-feeding bivalves and developed morphologic and metabolic adaptations for survival in oxygen-deficient reducing environments (Cavanaugh 1985, Dando & Southward 1986).

In summary, most modern environments of OC burial are dominantly dysoxic, but relatively dynamic, with pronounced variability of water column dissolved oxygen concentrations, and at least periodic excursions of the anoxic zone above the SWI. This temporal and spatial heterogeneity in the position of the redox boundary exerts a strong control on the development of benthic communities (soft-bodied infauna vs shelly epifauna), and thus the preservation of primary laminae. Infaunal deposit feeding is the dominant trophic strategy in modern low-oxygen environments while epifaunal suspension-feeding organisms are rare (i.e. Rhoads & Young 1970).

Sulfide-oxidizing bacterial mats, common in modern low-oxygen environments with OC-rich substrates (Spies & Davis 1979, Jannasch 1984, Tuttle 1985), may help to stabilize the redox boundary at the SWI, and thus probably play a significant role in the ecology of most low-oxygen benthic communities. Under conditions of a relatively stable redox boundary at the SWI specialized epifaunal or infaunal suspension feeders have come to dominate the benthic community; where the boundary is more dynamic the associated macrofauna tend to be opportunist colonizers.

Because modern environments of organic-rich sedimentation include examples of either relatively deep basins (types 1 and 2, Figure 1) or relatively shallow environments (types 3 and 4, Figure 1), it is not suprising that one of the longest running geological debates has focused on the relative depth of black shale deposition. However, many epicontinental black shales—the typical subjects of deep vs shallow debates (e.g. Twenhofel 1915, Ruedemann 1935, Trask 1951, Conant & Swanson 1961)—have characteristics that are inconsistent with those of most modern low-oxygen environments. For example: 1. many black shale units are widely distributed and temporally long ranging within epicontinental basins; 2. many intervals are predominantly laminated, rather than predominantly bioturbated; 3. their constituent faunas are dominated by low diversity assemblages of epifaunal suspension feeders, rather than diverse infaunal deposit feeders; and 4. they are characterized by common event communities indicating a dynamic benthic environment (e.g. Kauffman & Sageman 1990, Sageman et al 1991). Increasingly geologists are seeking to understand the *processes* responsible for accumulation of OC-rich sediments in modern marine environments, independent of factors such as basin depth, in an effort to develop more actualistic depositional models for ancient black shales.

Processes: Productivity vs Preservation

Because OC flux to the seafloor is an important determinant in the origin of black shales, we outline briefly the major controls on OC production and initial accumulation on the seafloor. Organic carbon is usually a minor constituent (<0.5 wt.%) of marine sediments; however, its burial over vast areas of the seafloor plays a very important role in biogeochemical cycles. The main sources of OC to marine sediments are terrestrial organic matter transported to the ocean via rivers (e.g. Hedges & Parker 1976, Hedges et al 1986, Meybeck 1982, Berner 1982), and to a lesser extent by winds (e.g. Cachier et al 1986), and photosynthetic production of marine phytoplanktic organic matter with a minor contribution from organisms higher in the food chain. The amount of marine OC fixed in near-surface waters by photosynthesis is dependent, in part, on nutrient availability in surface waters. These principles have been reviewed in Berger et al (1989) and Jahnke (1990) and are summarized briefly here. In central pelagic regions of large oceanic gyres, where primary productivity and surface water nutrient concentrations are generally low (so-called oligotrophic regions), regenerated production (i.e. using nutrients largely recycled from the surface biomass) accounts for most of the primary productivity, and only a very small portion of the OC is exported to deep waters. In highly productive regions, new production (i.e. nutrients supplied from the base of the thermocline) accounts for a more significant fraction of the primary productivity, and a larger portion of the OC is transported from the photic zone. This flux is concentrated near the continental margins and is reflected in sediment OC concentrations (e.g. Premuzic et al 1982, Jahnke et al 1990, Reimers et al 1992).

A global average of less than 20% of the organic matter produced in oceanic surface waters escapes consumption/oxidation within the photic zone (e.g. Müller & Suess 1979, Berger et al 1989) to fall through the main oceanic thermocline. Of the material that escapes the upper 100 m of the water column, 75 to 85% is decomposed within the upper 500 to 1000 m and only 3 to 5% of the total primary productivity is transported below about 1000 m. In addition, greater than 90% of this organic matter that reaches the seafloor is decomposed, primarily by oxygen-based metabolism of organisms (e.g. Emerson & Hedges 1988), so that sediment OC burial fluxes provide minimum estimates of the original organic particle flux to the deep sea. Thus, a small fraction ($<0.5\%$ globally of the primary production) of the OC reaching the sediments is buried (e.g. Berner 1982). Previous studies (e.g. Suess 1980) suggest that, on a regional and local basis, the rate of OC burial reflects primary productivity, in spite of the many processes that diminish OC fluxes to sediments. Such a relationship between sedimentary OC accumulation and primary productivity might allow inferences about surface productivity in the past from sediment accumulation (Müller & Suess 1979, Sarnthein, et al 1988, Stein 1986).

While primary production (photosynthesis) in surface waters and river input control the amount of OC initially supplied to the seafloor, many factors modify the concentration of OC buried in the sediments (e.g. Jahnke et al 1990). For example, because OC may be modified by microbial oxidation and metazoan consumption in the water column and at the SWI (e.g. Aller & Mackin 1984), the OC fluxes (and OC concentrations, see Figure 2) may depend on depth-dependent oxygen concentrations. In areas where the midwater OMZ intersects the seafloor along continental

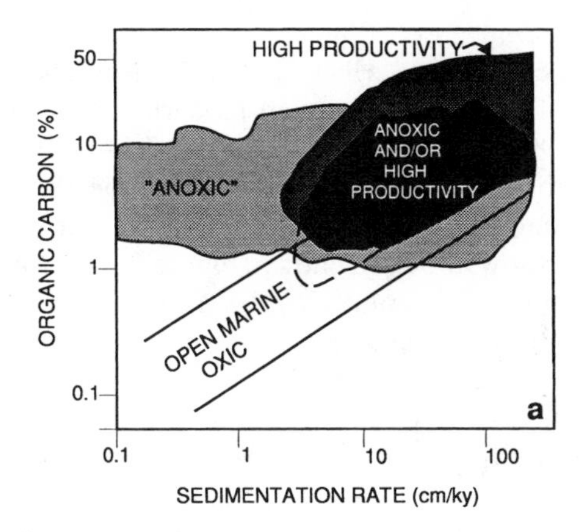

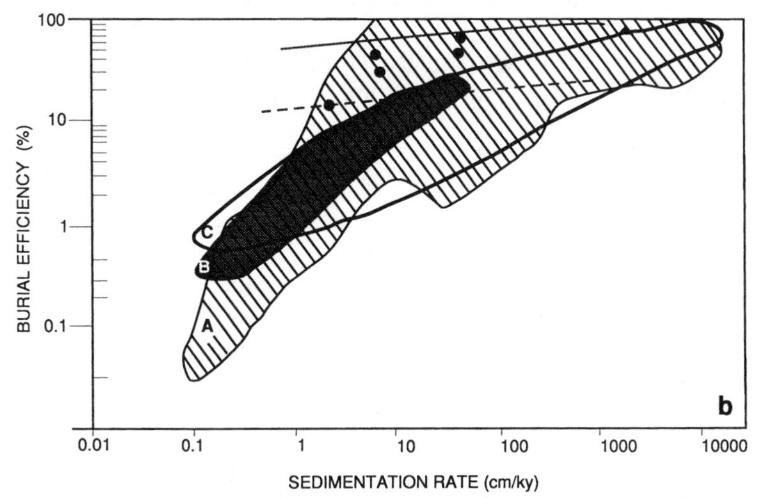

Figure 2 (*a*) Relationship between organic carbon concentration in marine sediments and linear sedimentation rate with inferred depositional conditions (after Stein 1986). (*b*) Burial efficiency (% of OC reaching the sediment/water interface that is ultimately buried) vs linear sedimentation rate from three sources (A—Betts & Holland 1991, B—Stein 1986, C—Henrichs & Reeburgh 1987). Trends B and C represent data from normal oxic marine environments, whereas A is for all available data including data in common with B and C. The black dots and area bounded by a dashed line below and solid line above represent trends from euxinic environments as indicated by Stein (1986) and Canfield (1989).

margins, many workers have argued that OC is preferentially preserved in surface sediments because of the low dissolved-oxygen concentration (e.g. Richards & Redfield 1954, Demaison & Moore 1980, but see Calvert et al 1992a for opposite view).

The rate of OC burial and sedimentation rate covary in all marine sediments (Toth & Lerman 1977, Heath et al 1977, Müller & Suess 1979). A recent summary of available data from modern marine environments (Betts & Holland 1991) attributes increasing OC burial efficiency (the percentage of OC that reaches the SWI, survives degradation, and is ultimately buried) almost entirely to a sedimentation rate effect at linear sedimentation rates below about 60 cm/ky (Figure 2; see also Henrichs & Reeburgh 1987). Higher sedimentation rates apparently reduce the residence time of OC at the SWI and thus remove it from the zone of oxic benthic metabolism, particularly in moderate- to well-oxygenated regions where large benthic metazoans (see below) churn the sediment. In addition, pumping of dissolved oxygen below the SWI which could be utilized to further degrade OC and/or mitigate buildup of inhibitory metabolites (e.g. Lee 1992) is eliminated by the exclusion of large infaunal organisms in dysoxic to anoxic settings (e.g. Aller 1982, Berner & Westrich 1985).

OC distribution may also be affected by the grain size and mineralogy of the accompanying sediments. Some organic compounds are sorbed onto clays (Morris & Calvert 1977) and/or onto calcium carbonate grain surfaces (Müller & Suess 1979), thereby increasing OC preservation in the fine fraction. Fine-grained sediments have been found to be higher in OC than other lithologies in similar environments (e.g. Emery & Uchupi 1972, Bordovskiy 1965). Of course, this relationship may be the result of depositional energy and higher OC fluxes to the seafloor over the continental slope where finer grained sediments preferentially accumulate. However, recent work (Keil et al 1993) suggests that organic matter forms a coating which has a fairly uniform thickness on clastic grains, resulting in a strong relationship between decreasing grain size, increasing grain surface area, and OC concentration. Nonetheless, this association cannot be the main control on black shale occurrence because low-OC clay-rich strata dominate the record.

Substantial debate centers on whether marine sediments and sedimentary rocks containing substantial amounts of organic matter originated because of unusual chemical conditions in the water column overlying the seafloor (enhanced preservation of organic matter as the result of oxygen depletion) or because of an unusually high flux of OC to the seafloor with rate of burial exceeding rate of oxidation. In the latter case the high OC fluxes are commonly viewed as resulting from an increase in available nutrients at the sea surface due to upwelling or other circulation phenomena. Bulk sedimentation rate clearly is a factor in preservation of organic matter. However, the effects of variations in primary productivity (increased carbon flux) and perhaps of enhanced preservation in oxygen-deficient environments on OC accumulation in sediments are probably also important (e.g. Emerson 1985, Emerson & Hedges 1988), yet these have not been calibrated satisfactorily in modern depositional environments. On the basis of data in Figure 2, it appears that a distinction between anoxic (euxinic) and oxic depositional environments at low to moderate primary productivity (<150 g C m^{-2} y^{-1}) is possible at sedimentation rates below 5 cm/ky (Stein 1986). At sedimentation rates above 10 cm/ky, it is difficult to distinguish the relative effects of oxygen concentration from productivity. However, other global data syntheses have found no significant effect of anoxic conditions (at or above the SWI) on OC preservation (Figure 2; e.g. Henrichs & Reeburgh 1987, Canfield 1989, Betts & Holland 1991). For example, burial efficiencies in oxic Dabob Bay (Washington) and nearby euxinic Saanich Inlet (British Columbia) are indistinguishable and the SWI is an important site of degradation in both settings (Cowie & Hedges 1992). Lee (1992) also found that the intrinsic rates of oxic and anoxic degradation are not significantly different (see also Canfield 1989); she proposed, however, that the higher content of lipid-rich OC observed for many sediments deposited in euxinic environments could be the result of buildup of anaerobic bacterial mass that would normally be grazed in oxic environments (see Harvey et al 1986).

The lack of difference in OC degradation rates in oxic vs anoxic settings appears to be confirmed by recent studies of sediments deposited in a high-productivity OMZ settings. In the Gulf of California (Calvert et al 1992) there appears to be no significant difference in OC concentration and pyrolysis hydrogen index between laminated and bioturbated Recent muds; likewise, there is no evidence of enhanced OC preservation within a modern well-developed OMZ on the Oman margin (Pedersen et al 1992). OC accumulation rate data for the Black Sea (Calvert et al 1991, Arthur et al 1993) also indicate no strong difference between productivity-normalized Holocene OC accumulation rates in that deep euxinic basin as compared to oxic open-ocean settings characterized by similar depths and sedimentation rates, although the Black Sea data fall at the high end of the range. In fact, it is inadvisable to use accumulation rates from the Black Sea as a comparison because of apparent focusing of organic matter sedimentation in deep water relative to the margins (Arthur et al 1993). These results are vexing in light of the many workers who have attributed relatively high OC concentrations and evidence of lack of degradation to the effects of low dissolved oxygen concentrations in the depositional milieu (Demaison & Moore 1980). Are productivity and sedimentation rates the only control on OC accumulation in marine environments?

SELECTED TECHNIQUES IN BLACK SHALE STUDIES

Chemical Analyses of Organic Carbon-Rich Sediments: Clues to Origin?

Here we present approaches, using geochemical analyses of OC-rich strata and paleobiological analysis of black shale faunas, to the study of ancient black shales. In each case these should be grounded within detailed stratigraphic and sedimentologic frameworks. We believe that the integration of these different approaches holds great promise for addressing many of the questions facing black shale workers today. Geochemical studies of black shale sequences may provide important information on the causes and consequences of enhanced organic matter burial in ancient environments. Yet none of the potential techniques proposed to date constitute foolproof means of inferring absolute dissolved oxygen concentrations or primary productivities. For example, several workers have suggested that concentrations (and accumulation rates) of Ba in sediments signal the influence of biologic productivity in determining sedimentary organic carbon fluxes (e.g. Brumsack 1986). Although highly correlated with OC flux, the barium is most likely tied up in carbonate phases, highly insoluble barite crystals, and nodules. However, because of subsequent bacterial sulfate reduction and decreasing pore-water sulfate concentrations, barite can be solubilized to migrate within the sediment column. Because of this process, barite in modern OC-rich sediments tends to reprecipitate at or near interfaces representing reduced sedimentation rates or oxygenation events (Brumsack 1986). Thus, drawing conclusions from Ba concentration patterns alone might be misleading. Below we outline several commonly cited geochemical indicators of conditions promoting black shale deposition and briefly discuss their potential and the problems with their interpretation. These are: 1. simple organic geochemical indicators, such as the pyrolysis hydrogen index and applications of carbon isotope data (e.g. Hollander et al 1993), that provide evidence of the degree and conditions of preservation of organic matter; 2. the reduced sulfur to organic carbon (S/C) ratio (e.g. Leventhal 1983, Berner & Raiswell 1983, Berner 1984) and "degree of pyritization" (e.g. Raiswell et al 1988) that have been suggested as a means for discriminating between euxinic conditions as opposed to pore-water anoxia; and 3. ratios of concentrations of transition metals such as V and Ni that may provide evidence for anoxia in the overlying water column (euxinic basin) (e.g. Lewan & Maynard 1982) and the utility of trace metals in general.

PYROLYSIS HYDROGEN INDEX AND SIMPLE INDICES OF PRESERVATION/TYPE

In addition to the problems in discriminating oxic from anoxic depositional environments using OC concentrations or accumulation rates, one of the major differences between anoxic and oxic depositional environments is the apparent better preservation of lipid-rich organic matter in anoxic settings. Although a detailed discussion of the plethora of organic geochemical techniques available is beyond the scope of this review, below we examine one of the more readily applied techniques: the pyrolysis of organic matter by Rock-Eval pyrolysis, which rapidly provides estimates of hydrocarbon type, hydrocarbon-generating potential, and degree of preservation of organic matter in sediments and sedimentary rocks (Espitalié et al 1985). This technique has commonly been applied to the study of modern and ancient OC-rich strata, and, with care, can be relatively easily interpreted in terms of both preservation and source of organic matter (Peters 1986, Katz 1983).

For example, laminated Unit II sediments (2–8 ka) in deep-water sites in the Black Sea are characterized by HI values >400 mgHC/gOC (Figure 3*a*). Laminated Unit II sediments from shallow-water, nearshore cores are characterized by more variable HI values of 150 to 500 mgHC/gOC. Values of HI < 150 mgHC/gOC might indicate either highly oxidized organic matter of planktonic origin or primarily terrestrially derived higher plant material. Mixing of terrestrial material (HI of approximately 100 mgHC/gOC) and well-preserved algal organic matter (HI > 600 mgHC/gOC) in various proportions will lead to HI values between these end-member values. Likewise, partial oxidation of marine authocthonous organic matter can produce HI values lower than 600 as well as higher oxygen indices. In modern marine environments it is unusual to find HI values >400 mgHC/gOC in primarily oxygenated environments, even under highly productive regimes with somewhat oxygen-depleted bottom waters such as the Peru margin (Figure 3*b*) or in the Gulf of California (Calvert et al 1992), where values in surface sediments underlying dysoxic waters are entirely below 300 mgHC/gOC. Note that there is no clear correspondence between OC concentration and HI or OI for Peru sediments in Figure 3*b*. Consistently high HI values of 400–600 are found mainly in environments where organic matter is preserved under euxinic conditions and where there is no significant reworked material. In the upper part of Unit I in the deepest Black Sea, deposited under euxinic conditions, HI values range between 250 and 550 (Arthur et al 1993); thus, HI values of <400 mgHC/gOC do not necessarily indicate deposition under oxic conditions. Pyrolysis data are sensitive to both source and preservation, and interpretation should be aided by other types of information on kerogen type, such as visual kerogen analysis and/or carbon isotopic data on organic carbon, among others.

Hollander et al (1993) proposed a method for discriminating between

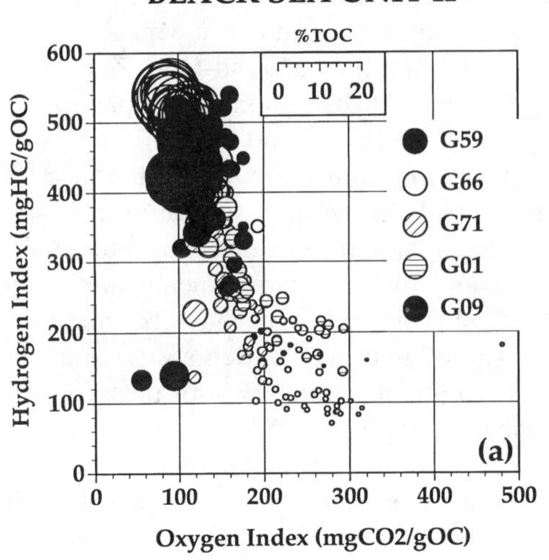

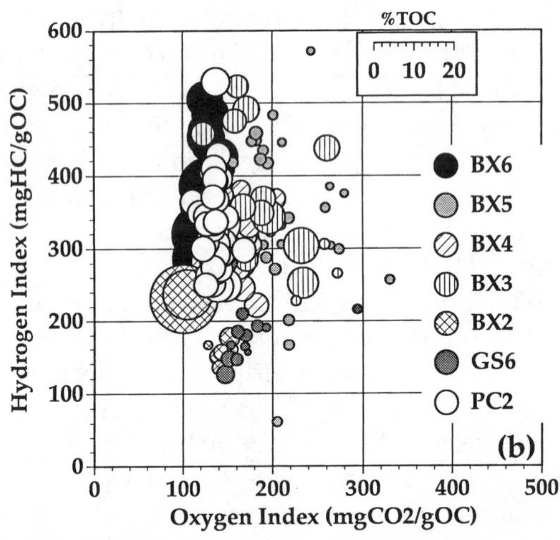

Figure 3 Rock-Eval pyrolysis data for: (*a*) the Black Sea gravity cores sampling "sapropels" of Unit II and equivalents (Arthur & Dean, in preparation), and (*b*) Peru margin box cores (Arthur & Dean, in preparation). Size of dots indicates the %TOC in the sample. Black Sea cores G01 and G71 are shallow slope cores; the remainder are from the deep basin.

laminated black shales produced by high productivity, with concomitant euxinic conditions, and those formed in low-productivity euxinic environments. This method utilizes the difference ($\Delta\delta^{13}C$) between $\delta^{13}Coc$ (organic C) and $\delta^{13}Ccc$ (carbonate C), with the constraint that the OC be entirely marine derived, plotted against the pyrolysis HI, which is assumed to reflect degree of preservation of the marine OC as a function of dissolved oxygen availability. The method relies on the dependency of δ^{13}Coc on the concentration of dissolved, undissociated CO_2 [CO_2aq] in surface waters, which has been shown to increase with increasing productivity and decreasing [CO_2aq] in modern lacustrine (e.g. Hollander & McKenzie 1991) and marine (e.g. Freeman & Hayes 1992) environments. Normalization to carbonate-C isotopic values provides a correction for possible variations in the carbon isotopic compostion of total dissolved carbon.

When applied to the Upper Jurassic Kimmeridge Clay (UK) (Hollander et al 1993) $\Delta\delta^{13}$C decreased with increasing HI and this trend was interpreted as indicating that higher biologic productivity was the main factor in producing the most OC-rich, lipid-rich units. The Lower Jurassic Toarcian Shale of France exhibited the opposite trend—one which suggested that euxinic conditions were most responsible for increasing preservation of OC.

One must apply caution when using this approach for several reasons. First, the carbonate carbon isotope values used as a baseline are commonly altered diagenetically in black shales (e.g. Irwin et al 1977), rendering them more depleted in δ^{13}Ccc and therefore reducing the apparent $\Delta\delta^{13}$C. Secondly, changes in the dominant marine phytoplankton occur across black shale cycles (e.g. Prauss & Riegel 1989, for a Toarcian example), and sufficient data have not yet been collected to constrain possible species effects on δ^{13}Coc.

SULFUR TO CARBON RATIOS AND DEGREE OF PYRITIZATION Following the proposal of Leventhal (1983), the reduced S/C ratio in sediments has commonly been used to infer euxinic conditions (e.g. Beier & Hayes 1989). Leventhal (1983) originally suggested that a positive intercept on the sulfur axis of a sulfur-carbon crossplot and S/C ratios higher than the average of about 0.4 for oxic marine environments (Goldhaber & Kaplan 1974, Berner & Raiswell 1984) would denote deposition in an euxinic environment. Some subsequent workers have stretched this definition to include any set of values with higher S/C ratios. However, care must be taken in interpreting such data because of the role that availability of iron plays in fixing reduced sulfur in sediments (e.g. Raiswell & Berner 1985, Calvert & Karlin 1991). Many black shales are reactive-iron limited, i.e. Fe that can be reduced, solubilized, and made available for pyrite precipitation, particularly if they contain relatively high amounts of OC, calcium carbonate, or biogenic silica (Raiswell et al 1988, Dean & Arthur 1989). Thus, total reduced sulfur need not be high in sediment deposited in euxinic environments. Likewise, sediments deposited in an oxic environment which are rich in reactive iron and have sufficient but low concentrations of reactive OC can have a high S/C ratio. Raiswell et al (1988) suggested that the "degree of pyritization" (DOP) is a more reliable index of euxinic versus anoxic deposition; DOP is the ratio of iron in sulfide minerals to total reactive iron (Fe-sulfides + remaining reactive Fe as determined by boiling with concentrated HCl). High proportions of sulfidized iron (high DOP) tend to occur in sediments deposited in euxinic environments, such as the Black Sea (e.g. Calvert & Karlin 1991). Raiswell et al (1988) suggested the following guidelines: DOP < 0.45 indicates oxic bottom water, for $0.45 < \text{DOP} < 0.75$ one has "restricted" bottom-water conditions, and DOP > 0.75 results from "inhospitable" (euxinic) bottom-water conditions. They also suggest that for the sediment under analysis to be useful, it should contain $<65\%$ skeletal material.

Figures 4*b* and 4*c* show Fe-S-C ternary diagrams (see Figure 4*a* and Dean & Arthur 1989, for construction and principles of interpretation) for the Black Sea and Peru margin, respectively. Peru margin samples (Figure 4*c*) come from either oxic sites (group 1), Fe-poor phosphorites (group 2), or dysoxic sites (all others along *dashed line*-3). The oxic (bioturbated with 1–9% OC) samples fall along the line respresenting S/C of 0.4, typical of oxic marine environments, whereas the dysoxic samples fall in an array along a line representing relatively constant S/Fe, regardless of OC content. The intercept on the Fe-S axis is at an S/Fe ratio of about 0.27. Because the theoretical S/Fe ratio is 0.54 for a sample where all Fe and S reside in pyrite, the plot indicates that only about 50% of the Fe is sulfidized. Although not a direct indicator of DOP, this rapid plotting technique has good utility in environmental determination. For example, DOP averaging 50% was determined for other Peru margin samples by Mossman et al (1992). The highest DOP values for Peru margin sediments would rate an assignment to the "restricted" facies using the above criteria. A Black Sea data set (Figure 4*b*; Calvert & Karlin 1991) allows discrimination between oxic shelf/slope (group 1) and euxinic marginal to deep-basin (group 2) environments. In this case, the intercept for the euxinic samples along a relatively constant S/Fe ratio represented by the dashed line is about 0.38 or about 70% of the Fe sulfidized. Calvert & Karlin (1991) independently determined a mean DOP of about 73% (from a range of 59–85%) for these samples. Using the criteria of Raiswell et al (1988), these samples would fall on the transition between "restricted" and "inhospitable." The oxic samples had DOP values between 3 and 55%. Figure 4*d* shows an application to ancient black shales using data for the Pennsylvanian Stark Shale (Desborough et al 1990). Marginally oxic (group 1) vs anoxic (euxinic; group 2) clusters, based on other geochemical and fabric criteria, are clearly distinguished. For the samples thought to be deposited

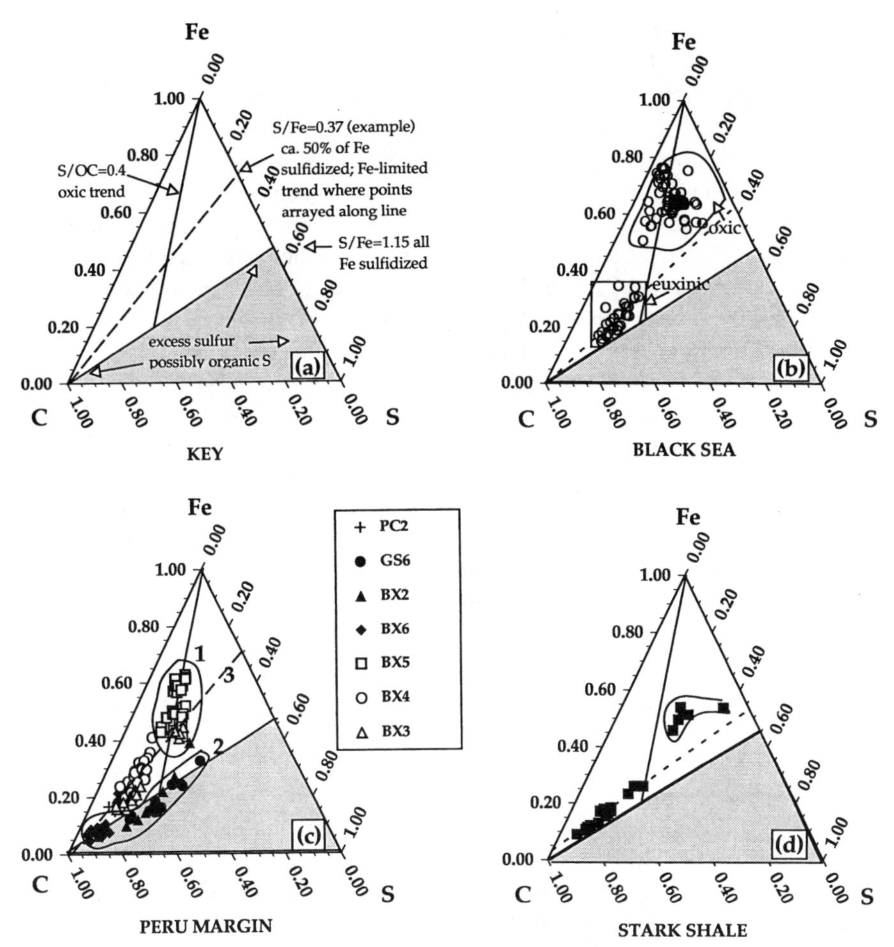

Figure 4 Fe-S-C ternary diagrams (Dean & Arthur 1989). (*a*) Key to interpretation. (*b*) The late Holocene (Unit I) sediments of the Black Sea (data from Calvert & Karlin 1991) recovered in box cores (the box inscribes samples from euxinic setting; an irregular field drawn around samples from oxic settings). (*c*) Holocene Peru margin sediments (Arthur & Dean, in preparation). (*d*) The Pennsylvanian Stark Shale Member of the Dennis Limestone, Kansas (data from Desborough et al 1990).

under euxinic conditions, the degree of Fe sulfidization and apparent DOP is high (about 87%). Therefore, although DOP is certainly one potential indicator of depositional conditions, it is not foolproof, and should not be used solely to differentiate between euxinic and marginally oxic conditions.

TRACE METALS Many ancient black shales are characterized by enrichments in certain trace metals, particularly Cd, Ag, Mo, Zn, V, Cu, Ni, and U (e.g. Vine & Tourtelot 1970, Brumsack 1986, Holland 1979). Such enrichments are quite variable in magnitude and a number of possible causes for this enrichment exist. For example, it is often assumed for sequences of intercalated OC-rich and OC-poor beds that mobilization and migration of redox-sensitive trace elements from relatively OC-poor beds to OC-rich beds can account for enrichments. For many elements (Ag, As, Cd, Cr, Mo, Sb, U, V, and Zn) a sedimentary sequence more than 10 times as thick as the black shale layer has to be leached of its trace metal content with complete transfer to the black shale bed in order to account for enrichments (Brumsack 1986); in general, this mechanism does not explain relatively thick sequences of trace-metal enriched black shale. Other mechanisms, outlined below, include (*a*) bioconcentration and transfer with particulate flux to the seafloor and/or sorption on biogenic and other particles in an oxic water column; (*b*) diffusion from bottom waters into sediments and fixation in authigenic phases associated with organic matter degradation by anaerobic bacteria; and (*c*) particulate flux associated with a water-column oxic/anoxic interface. The latter two mechanisms are explored briefly below as they bear on environmental interpretation of black shales.

In addition to the "direct" contribution from the particulate flux, OC-rich sediments receive trace metals during or shortly after deposition. Oxygen is consumed and disappears at or near the SWI, and ensuing bacterial sulfate reduction causes the generation of hydrogen sulfide. Metals like As, Mo, U, and V are sensitive to changes in redox potential and the presence of hydrogen sulfide. These elements may diffuse from seawater (particularly euxinic water column; see below) into the sediments where they are bound as sulfides (As, Mo) or possibly bound to organic matter (U, V) in their reduced states (Veeh 1967, Brumsack 1986). Because primarily molecular diffusion is involved, the magnitude of enrichment of such elements is likely inversely proportional to the sediment accumulation rate (Anderson et al 1989).

Under euxinic conditions, such as in the present Black Sea or Cariaco Trench, the nutrient/trace metal patterns differ from those observed in oxic waters. Many trace metals undergo a dramatic solubility decrease at the O_2/H_2S–boundary (e.g. Spencer et al 1972, Jacobs et al 1985 1987) and are removed from the water column, usually by precipitation as sulfides, and finally buried in the sediments. Again, this trace element enrichment may be more noticeable when sedimentation rates and corresponding dilution by terrigenous detritus are low. Therefore, sediments deposited under an anoxic water column, with low terrigenous input, and consequent high-OC preservation should exhibit a strong trace metal signal. By contrast the trace metal content of sediments that were deposited in an environment with enhanced surface-water biological productivity and concomittant high sedimentation rates should be comparable to those of modern upwelling sediments, regardless of the presence of an anoxic water column.

It appears then that trace element enrichment in black shales can occur in a variety of ways, and that there is no perfect discriminator of the conditions under which a given black shale was deposited. However, at least two trace-element-based techniques may be of use, when combined with other data: Th/U and V/(V+Ni).

The Th/U ratio is sensitive to the relative inputs of Th-bearing clay minerals and U that is fixed in sediments during early diagenesis in euxinic environments. Zelt (1985), Wignall & Myers (1988), and Doveton (1991) have applied this technique in studies of sedimentary sequences using spectral gamma logs (either hand-held or wireline), and suggest that low Th/U reflects low redox conditions. Soluble uranium in its oxidized state must be reduced and fixed as tetravalent U in the sediments of euxinic basins as opposed to its unreactive behavior in oxic marine basins (Anderson et al 1989). The precipitation of U in the sediments drives diffusion of U into pore waters, thus further enriching U in these OC-rich sediments. Because Th is scavenged by particles and supplied in the clastic fraction, high rates of sedimentation, even in a euxinic basin, will tend to cause high Th/U ratios. Only at relatively low sedimentation rates will the Th/U ratio be a sensitive indicator of a euxinic water column. For example, Mangini & Dominik (1979) demonstrated that the U/OC ratio is inversely proportional to the sedimentation rate in Mediterranean sapropels. This pattern has been confirmed for deep basin sediments of late Holocene age in the Black Sea (Anderson et al 1989). Again, some care must be exercised in applying this technique.

Vanadium is apparently more effectively fixed in sediments in association with organic compounds (e.g. porphyrin compounds) when competing metals, particularly Ni, are removed from anoxic pore waters as sulfide complexes (Lewan & Maynard 1982). Vanadium enrichments in OC-rich sediments appear to be enhanced by high concentrations of dissolved sulfide, especially in the overlying water column (causing V reduction), and slow sedimentation rates, which allow sufficient diffusion of V (reduced by sulfide) into sediment pore waters (e.g. Breit & Wanty 1991) where it can bind with porphyrin complexes. Increasing V enrichment over Ni may therefore indicate prevalence of sulfidic (euxinic) conditions and variations in the V/(V+Ni) ratio would indicate relative changes in oxygenation in the water column. The efficacy of this technique, however, is reduced where high sedimentation rates prevail (both limiting V diffusion and diluting trace metals overall) and/or where there is no source of sufficiently well preserved organic matter with attendant geoporphyrins, regardless of the presence or absence of euxinic conditions.

Examples from two modern OC-rich environments are shown in Figure 5. Holocene Black Sea sapropels (Unit II; Arthur et al 1994) are enriched in trace metals (e.g. Calvert 1990, Brumsack 1989), including V and Ni. Figure 5*a* is a plot of V/(V+Ni) vs %TOC for Unit II samples of laminated sediments from gravity cores taken from shallow to deep water depths (200–2250 m) in the Black Sea, all from below the oxic/anoxic interface. In these sediments, OC concentrations increase toward deeper water and lower sedimentation rates. V/(V+Ni) values are moderately high for all samples (because all were from euxinic environment) and increase slightly with increasing TOC, probably because of more effective diffusion into sediments at slower sedimentation rates (the black squares are from a site anomalously enriched in Ni and therefore do not lie on the same trend). The Peru margin data (Figure 5*b*) show no trend with increasing TOC, and the highest as well as lowest values of V/(V+Ni) occur in oxic slope sites (Bx2, 5, and GS6 are oxic, the remainder dysoxic; Froelich et al 1988); it is likely that V is sorbed onto oxyhydroxides in the surface sediments there. Because there are no true euxinic conditions in the Peru OMZ, there is little opportunity for V reduction in the water column. Nonetheless, samples from Bx6 have TOC and V/(V+Ni) values in the middle range of euxinic Black Sea sediments. V/(V+Ni) values for ancient black shales also exhibit little relationship to TOC, as in the case of the Pennsylvanian Stark Shale (Figure 6*b*), and it is likely that this is because TOC concentrations are not strictly a function of euxinic conditions. On the other hand, there is a strong correspondence between V/(V+Ni) values and DOP in the Stark Shale (Figure 6*a*), suggesting that higher V/(V+Ni) values signal more strongly euxinic conditions of deposition. Again, some care must be taken in applying this technique.

Biofacies Analysis and Paleo-Oxygenation History of Black Shales

Due to excellent preservation, which is a common characteristic of fossil assemblages within organic-rich facies (Canfield & Raiswell 1991a,b), there is a robust record of paleobiologic data from the study of black shales. As a result of early investigations of Devonian, Pennsylvanian, Jurassic, and Cretaceous deposits in the United States and Europe, for example, extensive species lists were compiled (e.g. Ruedemann 1935, Hauf 1921, Einsele & Mosebach 1955). In most cases, however, fossil occurrences were not recorded in detailed stratigraphic sequence, and little taphonomic information is avaiable. Moreover, due to the prevailing interpretation of black shale environments as "stagnant" anoxic seas, black shale faunas were largely believed to have been transported into these environments from shallower areas or from life habits as epi- or pseudoplankton. Yet, throughout the literature there are references to the persistent and widespread occurrence of small, thin-shelled taxa that are normally benthonic, as well as to shell horizons containing abundant specimens of a single or few species that suggest brief episodes of opportunistic benthic colonization.

Following a review of trends in faunal assemblages and associated biogenic sediment fabrics of modern low-oxygen environments, Rhoads & Morse (1971) suggested a model for the interpretation of oxygen-deficient biofacies. In this model, three distinct biofacies were recognized: anaerobic, dysaerobic, and aerobic. Each biofacies was defined by specific faunas and characteristic sediment fabrics that were correlated to specific ranges of dissolved oxygen (Figure 7*a*). This scheme was the first to provide a means of reconstructing paleo-oxygenation histories in OC-rich facies, and has persisted as the central framework for interpretation of oxygen-restricted black shale faunas. Refinement of the model, resulting from increasingly detailed studies of both modern and ancient low-oxygen faunas (e.g. Byers 1977; Morris 1980; Savrda et al 1984, 1991; Thompsen et al 1985; Savrda & Bottjer 1986, 1987; Wignall & Myers 1988; Brett et al 1989; Sageman 1989, 1991; Kauffman & Sageman 1990; Wignall 1990; Sageman et al 1991; Wignall & Hallam 1991) has included the recognition of 1. finer subdivisions of dysaerobic biofacies, 2. patterns of dominance among infaunal vs epifaunal taxa, and 3. trends in low-oxygen community structure, including trophic group dominance, species richness, and community equitability. These improvements have made possible even more detailed paleoenvironmental reconstructions of OC-rich facies.

Figure 7*b* shows an example of a current black shale biofacies model, derived mainly from detailed study of the Late Cenomanian Hartland Shale Member of the Greenhorn Formation in the Western Interior basin (Sageman 1989, 1991; Kauffman & Sageman 1990; Sageman et al 1991). It illustrates a number of the revisions that have been made to the Rhoads & Morse (1971) scheme. In the Hartland Shale Model:

1. Two main biofacies categories are identified—one dominantly infaunal, the other dominantly epifaunal.
2. Seven subdivisions (biofacies levels) are recognized, spanning the range from anaerobic to aerobic.
3. Each biofacies is defined by characteristic taxa with specific life habits and trophic strategies, characteristic sediment fabrics related to the level of infaunal colonization, relatively well-constrained levels of species richness and equitability, and in some cases, recurring ecological relationships between component taxa, such as commensalistic host-epibiont interactions.

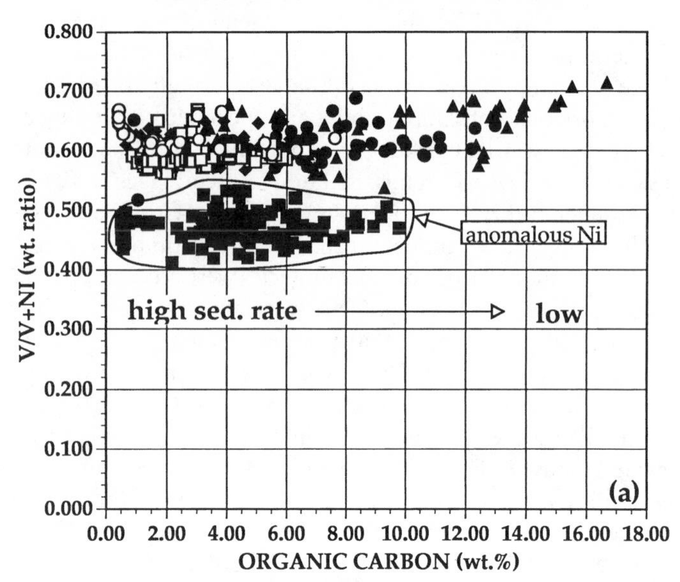
BLACK SEA UNIT II SAPROPELS
V/V+NI (wt. ratio)
0.800
0.700
0.600
0.500
0.400
0.300
0.200
0.100
0.000
anomalous Ni
high sed. rate
low
(a)
0.00 2.00 4.00 6.00 8.00 10.00 12.00 14.00 16.00 18.00
ORGANIC CARBON (wt.%)

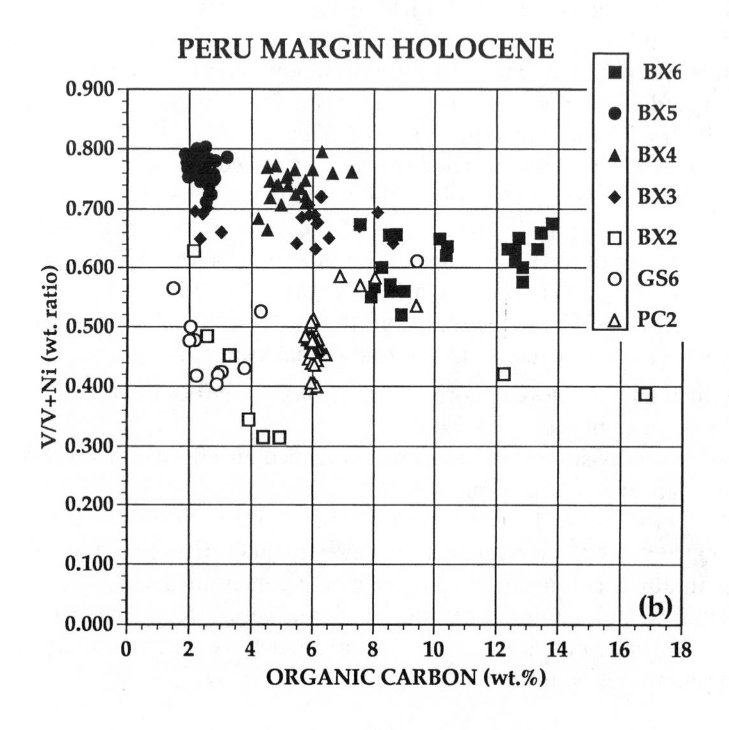
PERU MARGIN HOLOCENE
BX6
BX5
BX4
BX3
BX2
GS6
PC2
V/V+Ni (wt. ratio)
0.900
0.800
0.700
0.600
0.500
0.400
0.300
0.200
0.100
0.000
(b)
0 2 4 6 8 10 12 14 16 18
ORGANIC CARBON (wt.%)

4. The biofacies levels are generally cumulative upward with respect to both taxa and trophic strategies (specific taxa, or members of the same trophic group in any given level may occur at all higher levels), reflecting increasingly favorable conditions for benthic habitation.

INFAUNAL BIOFACIES Predominantly infaunal biofacies are described from strata with greater than 90% of the total fauna represented by sediment feeding or dwelling ichnotaxa, and/or body fossils of organisims that lived infaunally. Infaunal biofacies level 1 (Figure 7*b*) reflects anoxic conditions both above and below the SWI, with laminated sediments, the highest levels of preserved TOC, and no evidence of benthic metazoan life. At biofacies level 2, microburrowing—described from the Greenhorn Formation by Pratt (1984) as the slight disruption of primary laminae by burrows less than 1 mm in diameter—probably indicates the presence of meiofaunal organisms (such as nematodes) in the uppermost layers of the sediment. Level 3 refers to small deposit-feeding pioneer ichnotaxa, such as *Planolites* or *Chondrites*, which are widely recognized as the traces of the most low-oxygen tolerant macroinvertebrate infauna (e.g. Bromley & Ekdale 1984, Savrda & Bottjer 1986, Ekdale & Mason 1988). Level 4 is characterized by an increase in the the size and density of *Planolites* and *Chondrites* burrows, but no additional ichnotaxa, or by the appearance of deposit-feeding infaunal bivalves, such as *Nucula* or *Lucina*. At level 5 more complex feeding traces, such as *Teichichnus* and *Zoophycos* may be found, as well as the body fossils of infaunal grazers such as the anchurid gastropod *Drepanochilus*. Level 6 is indicated by the appearance of sediment-dwelling ichnotaxa ("domichnia"; Seilacher 1964), such as *Thalassinoides*, which indicate the establishment of domiciles within the substrate (see Figure 9). Above level 6 all groups may be included to produce high-diversity, high-equitability assemblages, indicating normal aerobic biofacies.

Although the transition from infaunal biofacies levels 1 to 7 (Figure 7*b*) represents a progressive increase in community diversity and equitability, the extensive reworking of sediment that occurs at higher biofacies levels may reduce the number of taxa ultimately preserved, thus making evaluation of true diversity trends quite difficult. For example, in cases where later episodes of burrowing completely overprint earlier records, and only a single ichnotaxon may be preserved, diversity and equitability estimates

Figure 5 Plots of V/(V + Ni) on a weight percent basis vs organic carbon concentration for: (*a*) Black Sea Holocene "sapropels" of Unit II in gravity cores (Dean & Arthur, in preparation); and (*b*) Holocene Peru margin sediments (Arthur & Dean, in preparation). See text for explanation.

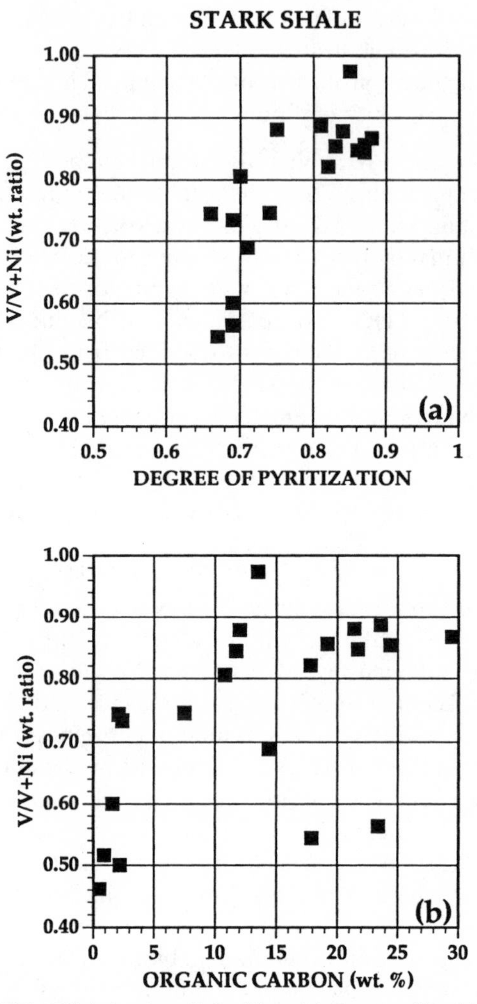

Figure 6 Plot of V/(V+Ni) index (wt. % basis) vs (*a*) Degree of Pyritization (DOP, see text) and (*b*) organic carbon concentration for the Stark Shale data set of Desborough et al (1990).

cannot be made. In cases where burrowing is extensive, careful evaluation of gradations of ichnofabric (e.g. Droser & Bottjer 1986) may be the best proxy for trends in the infaunal community.

EPIFAUNAL BIOFACIES The predominantly epifaunal biofacies series (Figure 7*b*) is described from strata with greater than 90% of the total resident fauna composed of epifaunal taxa. As with the infaunal biofacies series, epifaunal biofacies level 1 reflects anoxic conditions above and below the SWI, as indicated by laminated sediments, the highest levels of preserved TOC, and no evidence of in situ benthic habitation. Level 2 is characterized by laminated sediments containing only ammonites. These taxa are believed to have been nektobenthonic scavengers and predators that were mobile within the water column, and thus could feed on dead or dying organisms at or near the toxic SWI. Level 3 is characterized by pioneer suspension feeders, a group of organisms refered to broadly as "flat clams," which are ubiquitous inhabitants of Phanerozoic organic-rich facies (Kauffman 1988a). These taxa, which include mainly epibyssate suspension-feeding bivalves (e.g. *Inoceramus*) or free-living to pediculate brachiopods (e.g. *Leiorhyncus*) characterized by extremely thin shells, broad/flat shapes, and reclining to slightly erect life habits, were highly adapted to life at the boundary between anoxic and dysoxic to oxic environments. As indicated in Figure 7*b*, the lower levels of dysaerobic biofacies in the epifaunal sequence are characterized by a redox boundary coincident with the SWI. The flat clams which characterize these "benthic boundary biofacies" had special adaptations to low oxygen levels and H_2S, which may have included chemosymbiotic relationships with sulfur-oxidizing bacteria (Kauffman 1988a, Kauffman & Sageman 1989, Sageman et al 1991). As a result, they were the most common epifauna to colonize low-oxygen, OC-rich substrates. At biofacies level 4, pioneer flat clams are joined by successor suspension feeders, taxa that lived as epibionts on the larger flat clam "shell islands" (Kauffman 1981). Level 5 is indicated by the first appearance of opportunistic epifaunal colonizers, such as small bivalves and ostracodes, indicating that the substrate surface was habitable to taxa other than the highly tolerant flat clams. Level 6 is marked by a significant increase in diversity with the addition of various surficial grazers, scavengers, and carnivores, and indicates the initial development of an ecologically complex community. At level 7, all epifaunal groups are included, resulting in the high diversity and equitability levels characteristic of normal aerobic biofacies.

BIOFACIES SUMMARY Application of the biofacies model described above to sections of the Hartland Shale Member that have been measured and collected in great detail (see example, Figure 9) highlights several points concerning black shale biofacies. First, taxa occur either as resident components of benthic communities (they comprise $>90\%$ of an assemblage, are distributed throughout a sample interval, and/or show evidence of long-term in situ habitation), or as event components (they occur in low number, comprise $<10\%$ of an assemblage, occur on single bedding

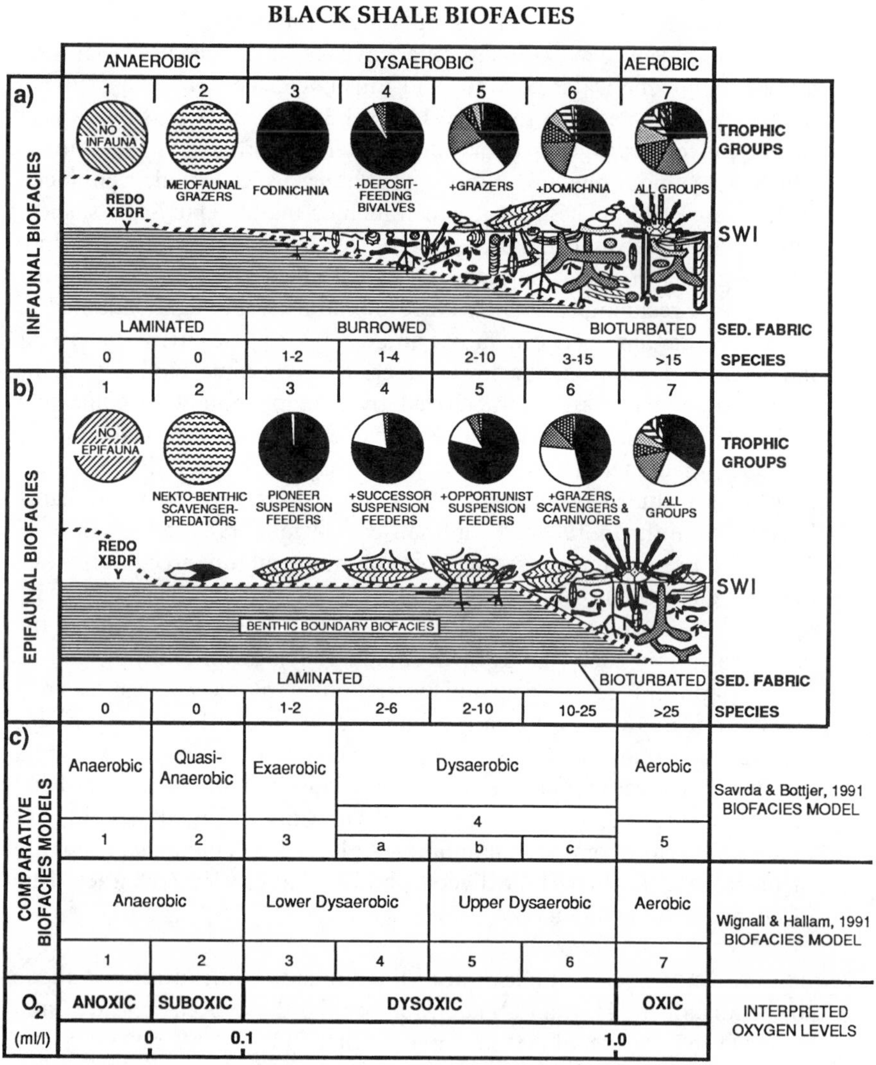

BLACK SHALE BIOFACIES

ANAEROBIC | DYSAEROBIC | AEROBIC

a) INFAUNAL BIOFACIES

1 2 3 4 5 6 7

NO INFAUNA

TROPHIC GROUPS

MEIOFAUNAL GRAZERS
FODINICHNIA
+DEPOSIT-FEEDING BIVALVES
+GRAZERS
+DOMICHNIA
ALL GROUPS

REDOXBDRY

SWI

SED. FABRIC	LAMINATED		BURROWED			BIOTURBATED	
SPECIES	0	0	1-2	1-4	2-10	3-15	>15

b) EPIFAUNAL BIOFACIES

1 2 3 4 5 6 7

NO EPIFAUNA

TROPHIC GROUPS

NEKTO-BENTHIC SCAVENGER-PREDATORS
PIONEER SUSPENSION FEEDERS
+SUCCESSOR SUSPENSION FEEDERS
+OPPORTUNIST SUSPENSION FEEDERS
+GRAZERS, SCAVENGERS & CARNIVORES
ALL GROUPS

REDOXBDRY

SWI

BENTHIC BOUNDARY BIOFACIES

SED. FABRIC	LAMINATED						BIOTURBATED
SPECIES	0	0	1-2	2-6	2-10	10-25	>25

c) COMPARATIVE BIOFACIES MODELS

Savrda & Bottjer, 1991 BIOFACIES MODEL	Anaerobic	Quasi-Anaerobic	Exaerobic	Dysaerobic			Aerobic
				4			
	1	2	3	a	b	c	5
Wignall & Hallam, 1991 BIOFACIES MODEL	Anaerobic		Lower Dysaerobic		Upper Dysaerobic		Aerobic
	1	2	3	4	5	6	7
O_2 INTERPRETED OXYGEN LEVELS	ANOXIC	SUBOXIC	DYSOXIC				OXIC

O_2 (ml/l): 0 — 0.1 — 1.0

planes, and/or show evidence of benthic colonization followed by abrupt mass mortality). The high frequency of event biofacies in many black shales indicates the dynamic nature of low-oxygen benthic environments. Secondly, at higher biofacies levels of the dysaerobic range (levels 5 and 6), fossil assemblages are typically dominated by mixed infaunal-epifaunal biofacies. "Pure" biofacies only occur in the lower part of the dysaerobic range. Finally, whereas changes in biofacies levels have been assumed to reflect mainly changes in benthic oxygen and the position of the redox boundary, major shifts in biofacies type are commonly associated with changes in lithofacies and indicate that the nature of the substrate may play a major role in the colonization of low-oxygen tolerant taxa. Ultimately, biofacies models can be seen as a tool not only to reconstruct paleo-oxygenation trends, but also to determine paleoecological factors that control community development under conditions of oxygen deficiency.

The biofacies model described herein is only one of several including that of Savrda & Bottjer (1991)—who developed a similar scheme based mainly on analyses of ichnotaxa in Miocene and Cretaceous organic-rich facies, and that of Wignall & Hallam (1991)—who describe a biofacies model for British Jurassic black shales, focused mainly on megafossils. Although there are numerous differences between these three models, reflecting both different paleoecological approaches, as well as data bases widely separated in space and time, there is also a remarkable convergence among the three upon a common framework (see Figure 7*c*). All three models address the complexities of infaunal and epifaunal life habit groups and the importance of short-term event communities.

Figure 7 Major features of black shale biofacies models. Characteristics of the Rhoads & Morse (1971) biofacies are shown relative to current terminology for correlative states of benthic oxygenation (based on Tyson & Pearson 1991). Biofacies model based on studies of the Cretaceous Greenhorn Formation, Western Interior basin (modified from Sageman 1991) illustrates revision of the Rhoads & Morse (1971) scheme—including recognition of variation in the position of the redox boundary and the development of (*a*) predominantly infaunal and (*b*) predominantly epifaunal biofacies, finer subdivision of the anaerobic to aerobic transition (dysaerobic biofacies), and detailed paleoecological characterization of each biofacies based on trends in trophic specialization, species diversity, and sediment fabric. Trophic pie charts and faunal examples represent averages of all samples analyzed in study of the Hartland Shale Member (Sageman 1991). [For further details see Sageman (1989), Kauffman & Sageman (1990), and Sageman et al (1991).] (*c*) Comparison of the Greenhorn-based model to biofacies models developed by Savrda & Bottjer (1986, 1987), Savrda et al (1991), and Wignall & Hallam (1991; see also Wignall & Myers 1988, Wignall 1990) illustrating consistent recognition of similar oxygen-related biofacies in strata of Jurassic, Cretaceous, and Miocene age, from both North America and Europe.

Application of Biofacies and Geochemical Techniques

To illustrate how the biofacies models can be combined with geochemical data to interpret depositional environments of ancient black shales we show two examples: 1. the Holocene Black Sea (Figure 8); and 2. the Upper Cenomanian-Lower Turonian basinal marine sequence from the U.S. Western Interior Seaway (Figure 9). The data in Figure 8 illustrate the use of geochemical indicators to detect possible changes in oxygenation for a case in which the sediments are entirely finely laminated and devoid of benthic fauna. These so-called sapropels were deposited during an interval from about 8–2 ka in the deepest part of the Black Sea. Although the deep-water mass has been presumed to have been euxinic throughout this period, Calvert (1990; see also Pedersen & Calvert 1990) suggested that much of the Black Sea Unit II sapropels were deposited under oxic conditions (with high productivity). The data shown in Figure 8 may argue for a brief episode of oxygenation that would have caused the decrease in V/(V+Ni), increase in Mn/Al, and decrease in TOC and pyrolysis HI at about 40 cm in the core. Note the transition from oxic Unit III to inferred euxinic Unit II at the base of the core. At present, we do not understand the origin of the mid-Unit II apparent oxygenation event. However, it does appear that geochemical data are sensitive indicators of changes in oxygen concentrations at very low levels.

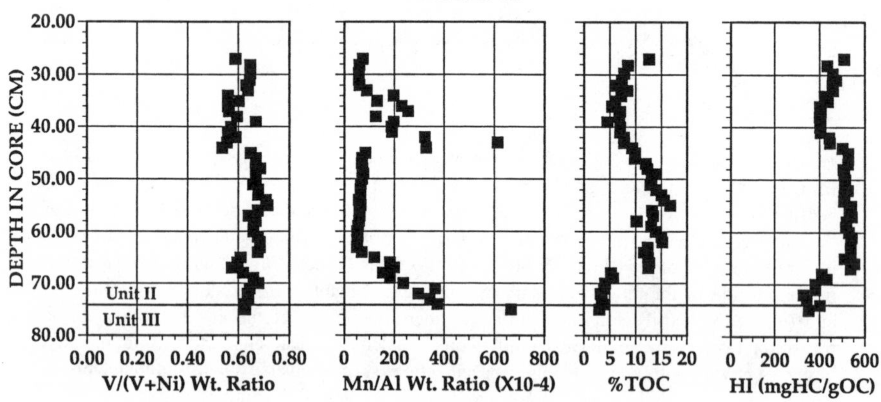

Figure 8 Selected geochemical indices —V/(V+Ni), (Mn/Al) wt. ratio, % organic carbon (TOC), and pyrolysis hydrogen index (HI)—for laminated, deep-basin Black Sea Unit II sapropel (in giant gravity core 66; Dean & Arthur, in preparation). Note that TOC, HI, and V/(V+Ni) all decrease centered at about 40 cm depth whereas Mn/Al increases to a peak. Such changes in these indices might record a brief interval of oxygenation in the deep Black Sea at that time (perhaps 4000 ka).

Shown in Figure 9 are our data for a 23-meter section of the Hartland Shale and the lower part of the Bridge Creek Limestone Member of the Greenhorn Formation. The Hartland Shale is dominantly laminated to sparsely bioturbated with relatively uniform OC concentrations ranging upwards to 4.5% and HI values of about 400 mgHC/gOC. Bioturbation and biofacies analyses suggest deposition primarily under dysoxic conditions that allowed an epifaunal community to survive throughout deposition of most of the unit. Note that species richness and abundance do not necessarily indicate the severity of oxygen depletion suggested by other indicators. The redox boundary is interpreted to have been mainly at or just below the SWI. Sedimentation rates are relatively slow (pelagic) throughout and the V/(V+Ni) values are uniformly high (>0.75). Such high values might suggest deposition under euxinic conditions, but DOP values (not shown) of 0.4 to 0.7 suggest "restricted" but not inhospitable bottoms, in agreement with the biofacies and textural evidence. Mn/Fe values are extremely low, with the exception of one limestone bed, supporting the interpretation of low redox conditions at the SWI. The Hartland Shale is interpreted as a fully marine black (marly) shale unit deposited in an overall transgressive epicontinental sea that was oxygen-depleted but not euxinic. Productivity and high OC fluxes may have played a role in maintaining the redox boundary close to the SWI, but low dissolved oxygen concentrations in the bottom-water mass, perhaps derived by transgression of an external OMZ into the basin from Tethys, may have been important.

A significant change in depositional environment occurs at the contact between the Hartland Shale and Bridge Creek Limestone. The lower part of the Bridge Creek Limestone (S. *gracile* zone) is characterized by more intense bioturbation, biofacies indicating amelioration of redox conditions, low TOC, low HI values, a decrease in V/(V+Ni) values, and fluctuating Mn/Fe. All of these parameters indicate fairly oxic depositional conditions. As these conditions occurred during a transgressive episode, some aspect of watermass stratification must have changed significantly. It is difficult to invoke decreasing productivity in surface waters, and, more likely, the transgression altered the water balance of the basin and led to production of oxygenated "deep" waters within the Western Interior Seaway.

Strongly cyclic alternations between well-oxygenated and dysoxic bottom conditions prevailed for the remainder of the Bridge Creek Limestone as depicted by large variations in bioturbation index, biofacies composition, carbonate and OC concentrations, and V/(V+Ni) and Mn/Fe values. The cyclicity may reflect modulation of climate and watermass conditions by orbital insolation variations (Milankovitch cycles; e.g. Pratt

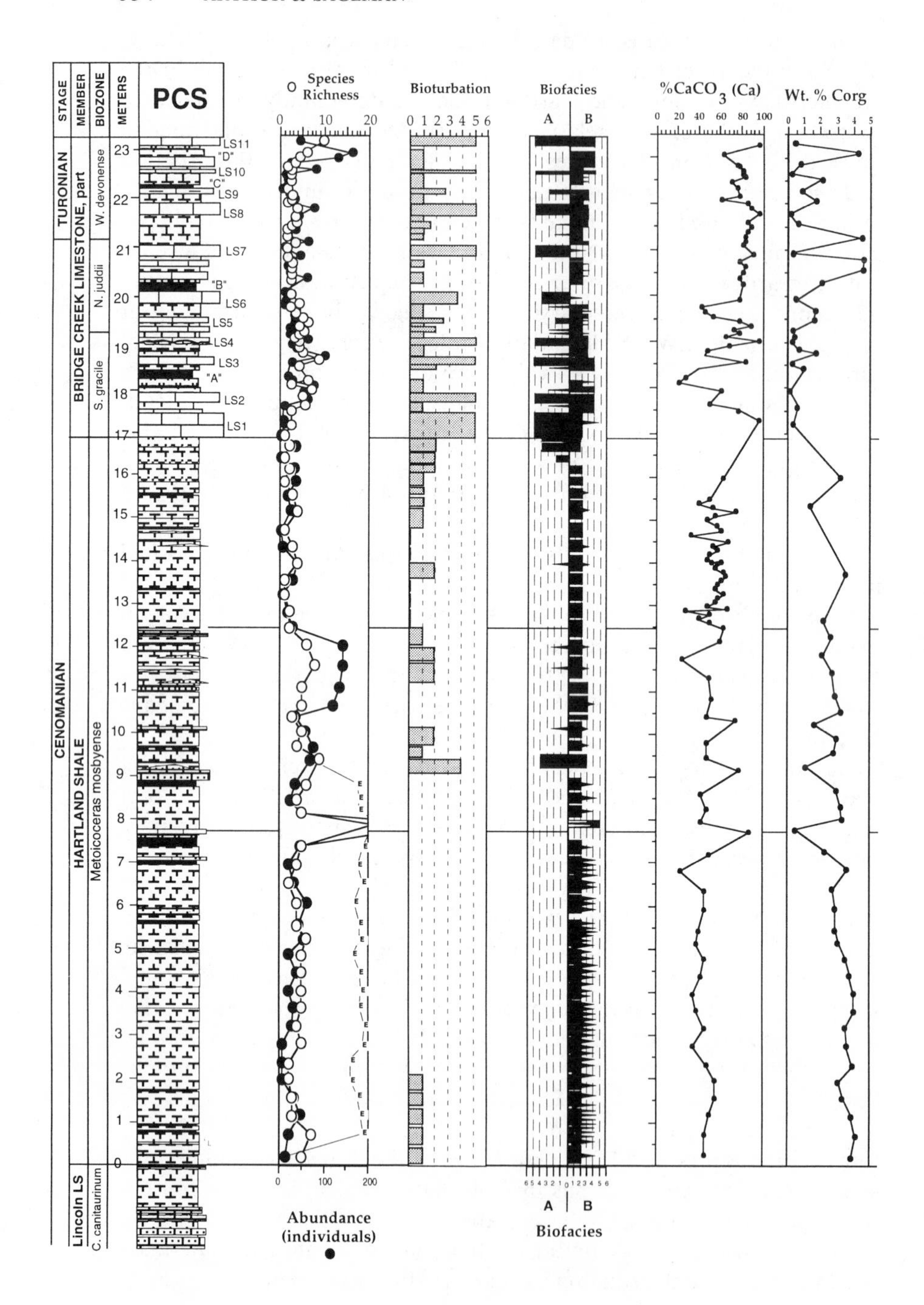
STAGE
MEMBER
BIOZONE
METERS
PCS
TURONIAN
CENOMANIAN
BRIDGE CREEK LIMESTONE, part
HARTLAND SHALE
Lincoln LS
W. devonense
N. juddii
S. gracile
Metoiceras mosbyense
C. canitaurinum
LS11
"D"
LS10
"C"
LS9
LS8
LS7
"B"
LS6
LS5
LS4
LS3
"A"
LS2
LS1
Species Richness
Abundance (individuals)
Bioturbation
Biofacies
A
B
%CaCO3 (Ca)
Wt. % Corg

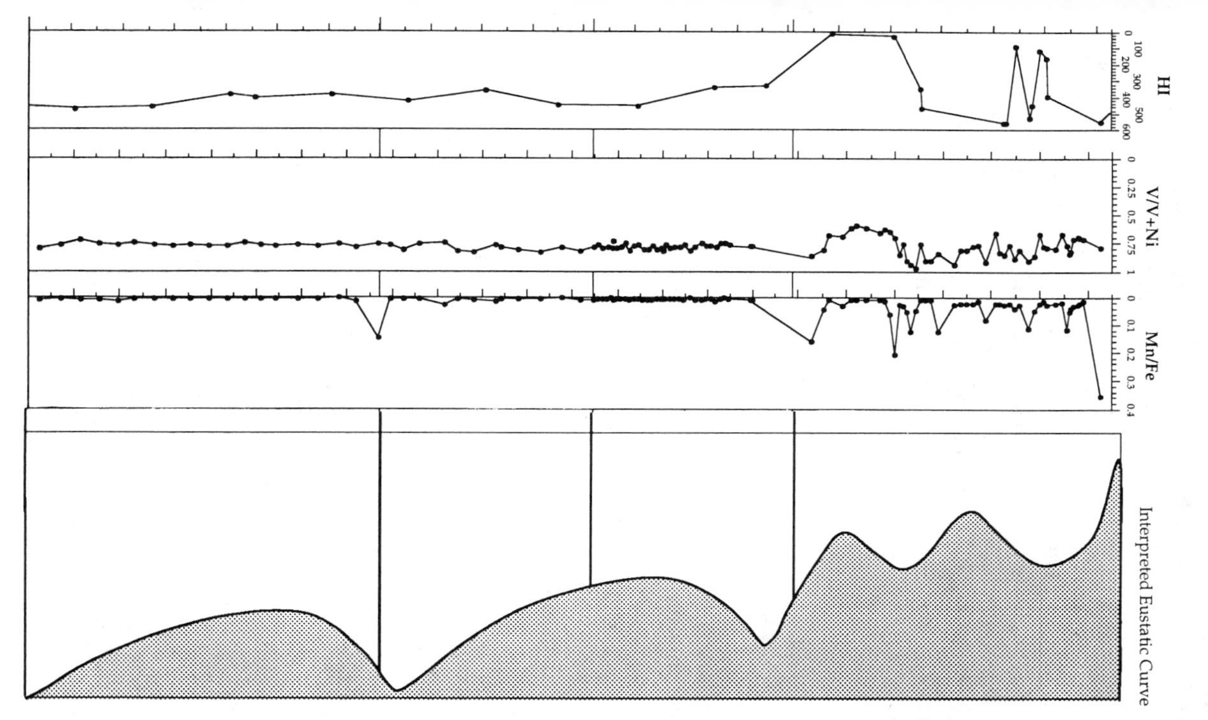

Figure 9 Faunal and geochemical data for the Hartland Shale Member and lower part of the Bridge Creek Limestone Member, Greenhorn Formation. The faunal data include species richness and abundance of macrofauna (the separate abundance curve in the lower Hartland Shale, shown with "e" symbols, reflects the high abundance of tiny epifaunal bivalves of the genus *Entolium* that occur on discrete bedding planes of laminated shales). Bioturbation levels represent a gradation from completely laminated (0) to fully bioturbated with diverse taxa (6). Biofacies follow the model in Figure 2 (infaunal or "A" biofacies plotted to the left; epifaunal or "B" biofacies plotted to the right). Biofacies levels 1–7, described in the text, mainly reflect changes in paleo-oxygenation, whereas major shifts in biofacies types (predominantly epifaunal to infaunal or mixed) reflect changes in additional factors (such as substrate). The geochemical data include weight percent $CaCO_3$ and C_{org}, hydrogen index (from Rock-Eval pyrolysis), the V/(V+Ni) index (see text), and the Mn/Fe ratio. Eustatic curve is interpreted from stratigraphic and faunal data (Sageman et al, in preparation).

1984, Barron et al 1985, Kauffman 1988b). The Pratt model suggests that periodic increases in runoff from highlands to the west of the seaway on 20–100 ky timescales caused surface salinity decreases and increased stratification with concomitant oxygen depletion in bottom waters. The variations in carbonate content appear to reflect dilution cycles (Pratt 1984, Arthur & Dean 1991). However, even these cycles are debated, and Eicher & Diner (1989), for example, prefer productivity variations as a cause, with the carbonate-rich units representing higher productivity. Deposition of this part of the Bridge Creek Limestone corresponds to the early stages of the global Cenomanian/Turonian OAE discussed later.

PHANEROZOIC BLACK SHALE DEPOSITION: TRENDS AND MODELS

Studies of black shales are legion. Yet few deposits are known in detail and definitive paleoenvironmental models are lacking for most deposits. The debates that characterize interpretation of Cretaceous black shales apply to Paleozoic examples as well. However, for the Paleozoic, it is even more difficult to establish organic carbon accumulation rates and other parameters needed to quantify carbon fluxes and establish the relative role of biologic productivity. As discussed earlier, oxygen depletion, either in silled basins, OMZs, or globally, has been interpreted as playing a major role in the formation of black shale units, but there are a number of examples attributed to higher productivity associated with upwelling zones as well.

There apparently were times in the past during which black shales were more widespread. The occurences of major units has been synthesized by North (1979, 1980), Bois et al (1982), and Ulmishek & Klemme (1990); most of the events recognized are referred to in terms of hydrocarbon generation. Leggett (1980) was one of the earliest to point out the apparent coincidence of black shales across parts of Europe in relatively narrow time intervals in the early Paleozoic. Many of the major Phanerozoic episodes of black shale deposition correspond to positive carbon isotope excursions (Figures 10 and 11) which may indicate increased burial of isotopically light organic matter during these black shale depositional intervals. Note also that the most widespread events appear to correspond with overall higher sea level stands. Because, in contrast to the late Mesozoic, there is no substantial preserved deep-sea record, we cannot adequately assess the possible global extent of oxygen-depleted waters or OC-rich facies. However, it does appear that most black shales were deposited in somewhat deeper-water shelf settings even though continental flooding meant that epicontinental seas were more widespread at those times.

Although black shale deposits typify the Cambrian through Lower Silurian in general, data compiled by Ulmishek & Klemme (1990) indicate that Silurian black shales were particularly effective hydrocarbon source beds. For example, they suggest that OC-rich, graptolitic strata of Silurian age covered approximately 42%—a huge proportion—of the total Silurian depositional area. The organic matter was entirely of marine origin (type II), most likely because terrestrial vascular plants were only just evolving (see Thickpenny 1984 and Leventhal 1991 for examples and models). Upper Devonian–Early Mississippian black shales are perhaps the most heralded of Paleozoic units. Ulmishek & Klemme (1990) claim that OC-rich strata extended over about 21% of the marine depositional area in latitudinal positions generally lower than 45°. Again, the organic matter was dominantly marine (Type II), despite the more widespread vascular plants (for detailed examples see Cluff 1980, Ettensohn 1985, 1992). The Late Pennsylvanian–Early Permian is another time of widespread OC burial (Ulmishek & Klemme 1990). The dominant type of organic matter is terrestrial plant material (type III). This is consistent with the known distribution of coal deposits which peaks in this interval. In fact, by comparing sulfur isotope curves with carbon isotope curves and other data, Berner & Raiswell (1983) argue that the major positive carbon isotope excursion (Figure 10) resulted from burial of terrestrial organic matter in coals and associated black shales [for details of deposits and contrasting models see Heckel (1977, 1986), Boardman et al (1984), Zangerl & Richardson (1963), Wenger & Baker (1986), Cecil (1990), and Coveney et al (1991)].

The next major phase of widespread OC burial appears to have occured in the Upper Jurassic following the major regressive period of the Permian through early Jurassic (Figure 10). Ulmishek & Klemme (1990) suggest that, on average, 27% of the marine Jurassic depositional area was typified by OC-rich strata, that the organic matter was dominantly Type II, and that the latitudinal extent of OC-rich facies was much greater than in the Paleozoic. Two major widespread Jurassic episodes have been cited (Jenkyns 1985, 1988; Jenkyns & Clayton 1986; Aigner 1980; Tyson 1986; Wignall 1990; see also Duff 1975, Miller 1990, Hollander et al 1993, Herbin et al 1991 for examples and models). The mid-Cretaceous (Aptian-Turonian) is also considered a major source-bed depositional interval, with black shales deposited over about 27% of the marine area (Ulmishek & Klemme 1991). The record can be resolved into several discrete episodes of widespread OC deposition (Figure 11; see Schlanger & Jenkyns 1976,

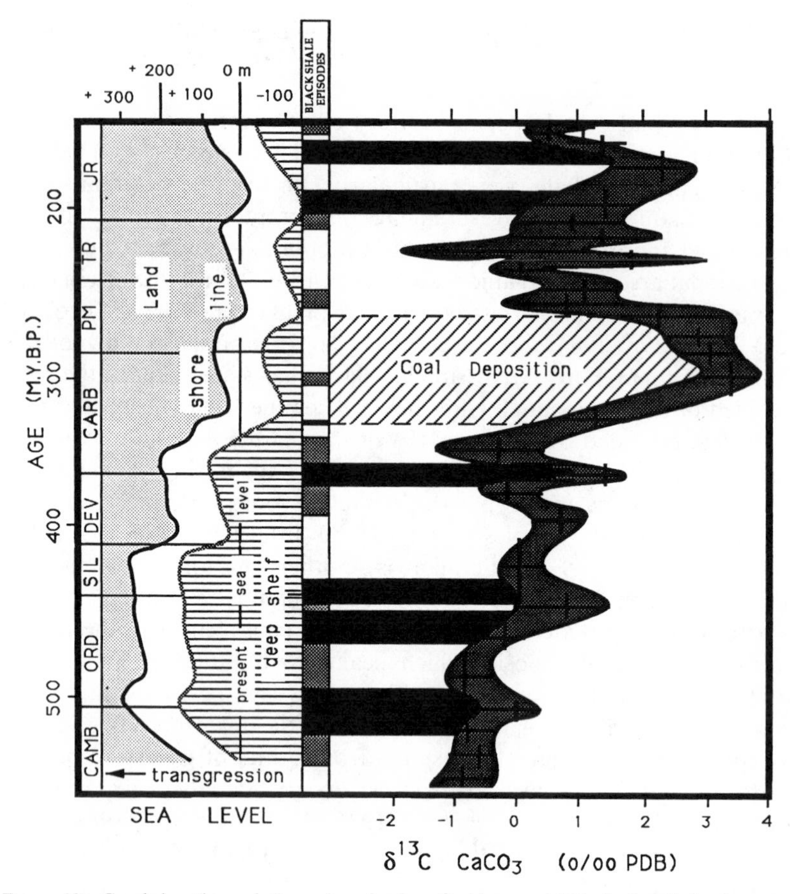

Figure 10 Cambrian through Jurassic episodes of widespread "black shale" development. The major episodes are shown in black and project to the carbon isotope curve. Note the correspondence between more positive carbon isotope excursions and these episodes. Gray shading in the black shale column indicates intervals during which black shales are a typical lithotype in marine sequences but have not reached the proportion of the more global events. Note that the major positive Carboniferous-Permian carbon isotope excursion corresponds mainly to deposition of major coal deposits. Sea-level curve is second-order curve of Vail et al (1977). Carbon isotope curve is period or sub-period average of "bulk" marine carbonates compiled by Lindh (1982).

Arthur & Schlanger 1979, Arthur et al 1990), most of which correspond to global carbon isotope excursions (for detailed examples and models see Jenkyns 1980, Arthur et al 1984, de Graciansky et al 1984, Herbin et al 1986, Schlanger et al 1987, Arthur et al 1987).

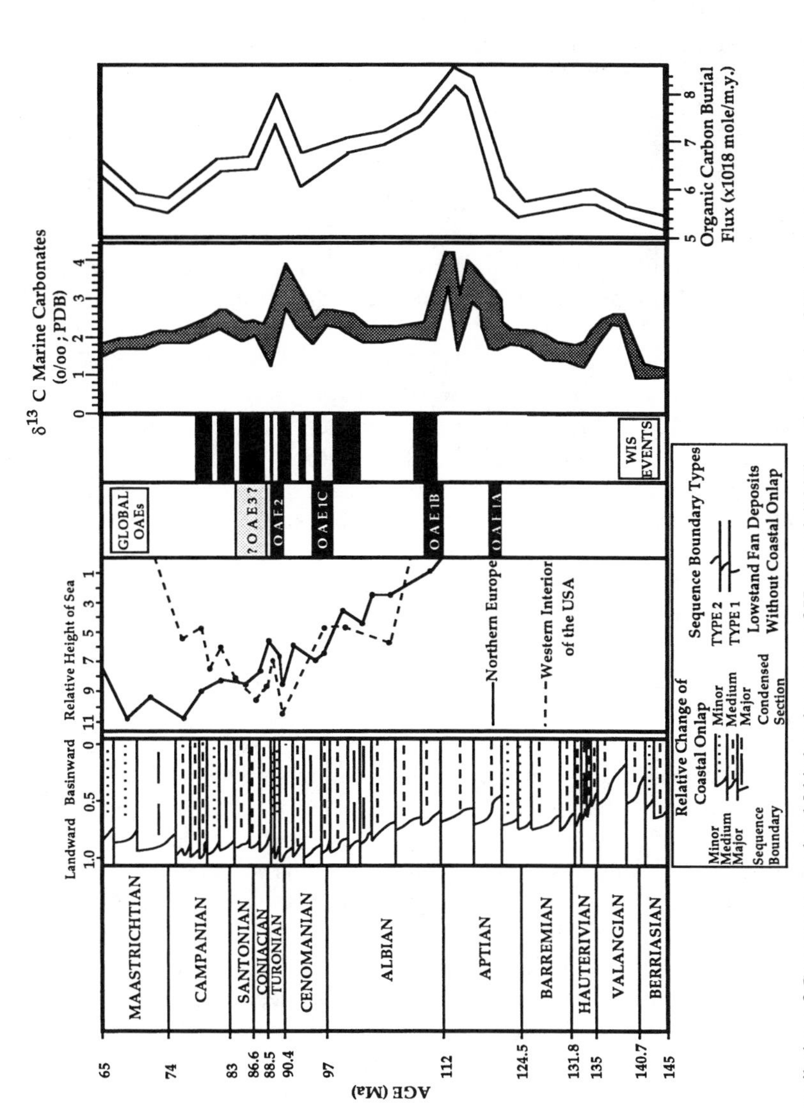

Figure 11 Compilation of Cretaceous sea level [third-order curve of Haq et al (1987) to time scale of Harland et al (1990)], relative sea level for epicontinental seas of the U. S. Western Interior and Northern Europe (after Hancock & Kauffman 1979), timing of global "oceanic anoxic events" (timing after Arthur et al 1990), intervals of organic-carbon rich sediment deposition in the U.S. Western Interior Seaway (WIS events), marine carbonate carbon isotope values (M. Arthur, unpublished), and modeled global organic-carbon burial fluxes (L. Kump & M. Arthur, unpublished).

As for the general causes of black shale episodes, simple and more complex models of climate and ocean circulation can be used to effectively test assumptions about regions of upwelling and higher productivity, higher precipitation/runoff and consequent salinity stratification, areas of deep-water formation, changes in rate or mode of deep-water circulation, and the effects of changing temperature and oxygen solubility or nutrient concentrations on carbon flux, deep-water oxygen concentration, and organic carbon burial rates. Each of these parameters is potentially important in promoting widespread organic matter burial. For example, several studies have emphasized changing wind patterns and coastal upwelling to explain the distribution of black shales (e.g. Parrish & Curtis 1982, Scotese & Summerhayes 1986); numerical general circulation models have been employed to examine patterns of coastal upwelling expected from mid-Cretaceous (e.g. Kruijs & Barron 1990, Barron 1985) and Silurian (Wenlockian; Moore et al 1993) changes in paleogeography.

Another side of the issue is the role of deep-water circulation and anoxia. For example, Wilde & Berry (1982) developed conceptual oceanographic models involving deep circulation changes for the overall pattern of black shale development in the Paleozoic. They suggested that the early Paleozoic ocean was dominated by largely oxygen-deficient deep waters whereas the later Paleozoic deep ocean became more oxic because of more efficient deep circulation associated with glacial climates, particularly in the Pennsylvanian-Permian. Episodes of heightened productivity and black shale deposition have been attributed to a variety of possible causes, including increased deep-water mass formation forced by rising sea level, impact-induced or volcanic-induced upwelling (e.g. Vogt 1989), and/or opening of a deep-water gateway. Somewhat more sophisticated box models can also be manipulated in order to examine the influence of different deep-circulation styles and rates on average deep-water properties. If one employs present-day parameters for exchange rates, preformed nutrients, river input, etc, but allows warm, saline bottom water (WSBW; Brass et al 1982) to form in low latitudes or by other means (mixing at oceanic fronts, Hay et al 1993) instead of sinking of cooler high-latitude water masses, then the average oxygen content of deep-water masses falls to about 36 μmol O_2/kg instead of the present average of about 160 μmol/kg (Sarmiento et al 1988). An average value of dissolved O_2 of 36 μmol/kg suggests that much of the bottom-water mass is anoxic. Ocean deep-water masses can also be made to go anoxic, using a model modern thermohaline circulation, by increasing deep-water residence times by a factor of 2–10 (see also Wilde & Berry 1982; Bralower & Thierstein 1984, 1987), depending on the turnover rate of the intermediate water mass.

Thus, widespread deeper-water anoxia can result from average surface-ocean productivities lower than the present with a warmer deep ocean and lower initial oxygen solubility. Such changes in deep-water turnover rate in either low or high latitudes may be cyclic and related to climatic changes induced by periodic changes in solar insolation, leading to the Milankovitch-like cycles in many Cretaceous black shales (e.g. Barron et al 1985, Herbert et al 1986).

CONCLUSIONS—PROBLEMS AND PROSPECT

Ancient black shales provide a fertile ground for future research because of their importance as hydrocarbon source rocks and thier usefulness as environmental indicators in furnishing clues to secular changes in atmospheric and ocean chemistry and ocean circulation. It is a struggle, however, to uniquely define the full conditions under which a given black shale was deposited. No single parameter can be used to delimit the extent to which black shales reflect deposition under euxinic conditions, with or without high biotic productivity. However, there is good potential for interdisciplinary studies using a combination of benthic biofacies and geochemical techniques, at least for defining oxygen gradients during deposition.

We do not yet fully understand the relative roles of organic carbon flux, dissolved oxygen (anoxia), and rate of sediment accumulation in producing relatively OC-rich sediments in modern environments, nor can we satisfactorily explain why preservation should be enhanced under anaerobic conditions, if it indeed is. Further careful studies of suitable modern environments are required to resolve these questions before ancient black shales can be satisfactorily interpreted in terms of the importance of productivity vs preservational phenomena. Better stratigraphic resolution and constraints on accumulation rates are required for ancient black shales.

There are clearly a number of temporal intervals during which marine black shale facies are much more widespread. These episodes are of interest because they implicate one or more global factors controlling organic matter production and burial. Widespread black shale deposition may provide evidence that oceanic nutrient concentrations were significantly higher than at present and/or that dissolved oxygen was profoundly depleted in oceanic deep waters as the result of any of several factors, including sluggish circulation, carbon loading, or low solubility of dissolved O_2 because of warm deep-water temperatures or low atmospheric pO_2. Considerably more investigations of the stratigraphic significance of black shale occurrence are needed. Likewise, these episodes are of interest because of the geochemical cycle linkages and the relationship between black shales with ore deposits and biotic extinctions. In all of this, we should keep in mind the admonition of Degens & Stoffers (1976) that "The well-mixed oxygenated ocean of today seems not to be the model for the past 600 m.y."

ACKNOWLEDGMENTS

We are deeply indebted to several individuals who have had a profound influence on our thinking about black shales and who have collaborated extensively with us in our studies of modern and ancient black shales. We thank W. E. Dean, A. G. Fischer, E. G. Kauffman, and S. O. Schlanger for their support, guidance, and encouragement, not to mention their generosity with their ideas and data. Although we take the blame for any problems with content, this paper would not have been written without their contributions. Grants from the National Science Foundation (EAR-9117398 and OCE 9014801) and Department of Energy (DOE/DE-FG02-92ER14251) provided support for aspects of this project. We thank David Couzens for his profound patience and careful editing.

Literature Cited

Adams JE, Freznel HN, Rhodes ML, Johnson DP. 1951. Starved Pennsylvanian Midland basin. *Am. Assoc. Petrol. Geol. Bull.* 35: 2600–6

Aigner T. 1980. Biofabrics and Stratinomy of the Lower Kimmeridge Clay (U. Jurassic, Dorset, England). *N. Jahrb. Palaontol. Abh.* 159: 324–38

Aller RC. 1982. The effects of macrobenthos on chemical properties of marine sediment and overlying water. In *Animal-Sediment Relations*, ed. PL McCall, MJS Tevesz, pp. 53–102. New York: Plenum

Aller RC, Mackin JE. 1984. Preservation of reactive organic matter in marine sediments. *Earth Planet. Sci. Lett.* 70: 260–66

Anastasakis GC, Stanley DJ. 1984. Sapropels and organic-rich variants in the Mediterranean: sequence development and classification. In *Fine-Grained Sediments: Processes and Products*, ed. DAV Stow, D Piper, pp. 497–510. *Geol. Soc. London Spec. Publ.*

Anderson RF, Fleisher MQ, LeHuray AP. 1989. Concentration, oxidation state, and particulate flux of uranium in the Black Sea. *Geochim. Cosmochim. Acta* 53: 2215–24

Arntz WE, Tarazona J, Gallardo VA, Flores LA, Salzwedel H. 1991. Benthos communities in oxygen-deficient shelf and upper slope areas of the Peruvian and Chilean Pacific coast, and changes caused by El Niño. In *Modern and Ancient Continental Shelf Anoxia*, ed. RV Tyson, TH Pearson, pp. 131–54. London: Geol. Soc.

Arthur MA, Dean WE. 1991. A holistic geochemical approach to cyclomania: examples from Cretaceous pelagic limestone sequences. In *Cycles and Events in Stratigraphy*, ed. G Einsele, W Ricken, pp. 126–66. Berlin: Springer-Verlag

Arthur MA, Dean WE, Neff ED, Hay BJ, Jones GA, King J. 1994. Late Holocene (0–2000 yBP) organic carbon accumulation in the Black Sea. *Global Biogeochem. Cycles* In press

Arthur MA, Dean WE, Stow DAV. 1984. Models for the deposition of Mesozoic-Cenozoic fine-grained organic-carbon-rich sediments in the deep sea. In *Fine-Grained Sediments: Processes and Products*, ed. DAV Stow, D Piper, pp. 527–62. *Geol. Soc. London Spec. Publ.*

Arthur MA, Jenkyns HC, Brumsack, H-J, Schlanger SO. 1990. Stratigraphy, geochemistry and paleoceanography of organic-carbon rich Cretaceous sequences. In *Cretaceous Resources, Events and Rhythms*, ed. RN Ginsburg, B Beaudoin, pp. 75–119. Dordrecht: Kluwer

Arthur MA, Schlanger SO. 1979. Cretaceous "oceanic anoxic events" as causal factors in development of reef-reservoired giant oil fields. *Bull. Am. Assoc. Petrol. Geol.* 63: 870–85

Arthur MA, Schlanger SO, Jenkyns HC. 1987. The Cenomanian-Turonian oceanic anoxic event, II. Paleoceanographic controls on organic matter production and preservation. See Brooks & Fleet 1987, pp. 401–20

Baird GC, Brett CE. 1986. Erosion on an anaerobic seafloor: significance of reworked pyrite deposits from the Devonian of New York State. *Palaeogeogr. Palaeoclimatol. Palaeoecol.* 57: 157–93

Barron EJ. 1985. Numerical climate modeling: an exploration frontier in petroleum source rock prediction. *Am. Assoc. Petrol. Geol. Bull.* 69: 448–56

Barron EJ, Arthur MA, Kauffman EG. 1985. Cretaceous rhythmic bedding sequences: a plausible link between orbital variations and climate. *Earth Planet. Sci. Lett.* 72: 327–40

Beier JA, Hayes JM. 1989. Geochemical and isotopic evidence for paleoredox conditions during deposition of the Devonian-Mississippian New Albany Shale, southern Indiana. *Geol. Soc. Am. Bull.* 101: 774–82

Berger WH, Smetacek VS, Wefer G. 1989. *Productivity of the Oceans: Present and Past*. New York: Wiley Interscience. 471 pp.

Berner RA. 1982. Burial of organic carbon and pyrite sulfur in the modern ocean: its geochemical and environmental significance. *Am. J. Sci.* 282: 451–73

Berner RA. 1984. Sedimentary pyrite formation: an update. *Geochim. Cosmochim. Acta* 48: 605–15

Berner RA, Raiswell R. 1983. Burial of organic carbon and pyrite sulfur in sediments over Phanerozoic time: a new theory. *Geochim. Cosmochim. Acta* 47: 855–62

Berner RA, Raiswell R. 1984. C/S method for distinguishing freshwater from marine sedimentary rocks. *Geology* 365–68

Berner RA, Westrich JT. 1985. Bioturbation and the early diagenesis of carbon and sulfur. *Am. J. Sci.* 285: 193–206

Betts JN, Holland HD. 1991. The oxygen content of ocean bottom waters, the burial efficiency of organic carbon, and the regulation of atmospheric oxygen. *Global Planet. Change* 5: 5–18

Blatt, H, Middleton GV, Murray RC. 1972. *Origin of Sedimentary Rocks*. Englewood Cliffs, N.J.: Prentice-Hall. 634 pp.

Boardman DRI, Mapes RH, Yancey TE, Malinky JM. 1984. A new model for the depth-related allogenic community succession within North American Pennsylvanian cyclothems and implications on the black shale problem. In *Limestones on the Mid-Continent*, ed. NJ Hyne, pp. 141–82. Tulsa: Tulsa Geol. Soc.

Bois C, Bouche P, Pelet R. 1982. Global geologic history and distribution of hydrocarbon reserves. *Am. Assoc. Petrol. Geol. Bull.* 66: 1248–70

Bordovskiy OK. 1965. Sources of organic matter in marine basins. *Mar. Geol.* 3: 5–31

Bradley WH. 1931. Non-glacial marine varves. *Am. J. Sci.* 22: 318–30

Bralower TJ, Thierstein HR. 1984. Low productivity and slow deep-water circulation in mid-Cretaceous oceans. *Geology* 12: 614–18

Bralower TJ, Thierstein HR. 1987. Organic carbon and metal accumulation rates in Holocene and mid-Cretacous marine sediments: paleoceanographic significance. See Brooks & Fleet 1987, pp. 345–69

Brass GW, Southam JR, Peterson WH. 1982. Warm saline bottom water in the ancient ocean. *Nature* 296: 620–23

Breit GN, Wanty RB. 1991. Vanadium accumulation in carbonaceous rocks: a review of geochemical controls during deposition and diagenesis. *Chem. Geol.* 91: 83–97

Brett CE, Dick VB, Baird GC. 1989. Comparative taphonomy and paleoecology of Middle Devonian dark gray and black shales from Western New York. In Dynamic Stratigraphy and Depositional Environments of the Hamilton Group (Middle Devonian) of New York State, Part 2, ed. E Landing, C Brett, New York: NY State Museum. 32 pp.

Bromley RG, Ekdale AA. 1984. Chondrites: a trace fossil indicator of anoxia in sediments. *Science* 224: 872–74

Brooks J, Fleet A, eds. 1987. *Marine Petroleum Source Rocks*. London: Geol. Soc.

Brumsack HJ. 1986. The inorganic geochemistry of Cretaceous black shales (DSDP Leg 41) in comparison to modern upwelling sediments from the Gulf of California. See Summerhayes & Shackleton, pp. 447–62

Brumsack, H-J. 1989. Geochemistry of recent TOC-rich sediments from the Gulf of California and the Black Sea. *Geol. Rundsch.* 78: 851–82

Byers CW. 1977. Biofacies patterns in euxinic basins: a general model. In *Deep Water Carbonate Environments*, ed. HE Cook, P Enos, pp. 5–17. Tulsa: Soc. Econ. Paleontol. Mineral.

Cachier H, Buat-Menard P, Fontugne M, Chesselet R. 1986. Long-range transport of continentally-derived particulate carbon in the marine atmosphere: evidence from stable carbon isotope studies. *Tellus* 38B: 161–77

Calvert SE. 1964. Factors affecting distribution of laminated diatomaceous sediments in Gulf of California. In *Marine Geology of the Gulf of California*, ed. TH Van Andel, JJ Shor Jr, pp. 311–30. Tulsa: Am. Assoc. Petrol. Geol.

Calvert SE. 1990. Geochemistry and origin of the Holocene sapropel in the Black Sea. In *Facets of Modern Biogeochemistry*, ed. V Ittekot, S Kempe, W Michaelis, A Spitzy, pp. 326–52. Berlin: Springer-Verlag

Calvert SE, Bustin RM, Pederson TF. 1992. Lack of enhanced preservation of sedimentary organic-matter in the oxygen minimum of the Gulf of California. *Geology* 20: 757–60

Calvert SE, Karlin RE. 1991. Relationships between sulphur, organic carbon, and iron in the modern sediments of the Black Sea. *Geochim. Cosmochim. Acta* 55: 2483–90

Calvert SE, Karlin RE, Toolin LJ, Donahue DJ, Southon JR, Vogel JS. 1991. Low organic carbon accumulation rates in Black Sea sediments. *Nature* 350: 692–95

Canfield DE. 1989. Sulfate reduction and oxic respiration in marine sediments: implications for organic carbon preservation in euxinic environments. *Deep-Sea Res.* 36: 121–38

Canfield DE, Raiswell R. 1991a. Chapter 7: Pyrite formation and fossil preservation. In *Topics in Geobiology*, ed. PA Allison, DEG Briggs, pp. 337–87. New York: Plenum

Canfield DE, Raiswell R. 1991b. Chapter 9: Carbonate precipitation and dissolution, Its relevance to fossil preservation. In *Topics in Geobiology*, ed. PA Allison, DEG Briggs, pp. 411–53. New York: Plenum

Caspers H. 1957. Black Sea and Sea of Azov. In *Treatise of Marine Ecology and Paleoecology*, ed. J Hedgepeth, pp. 801–90. Boulder: Geol. Soc. Am.

Cavanaugh CM. 1985. Symbiosis of chemoautotrophic bacteria and marine invertebrates from hydrothermal vents and reducing sediments. *Bull. Biol. Soc. Washington* 6: 373–88

Cecil CB. 1990. Paleoclimate controls on stratigraphic repetitions of chemical and siliciclastic rocks. *Geology* 18: 533–36

Chamberlin TC. 1906. On a possible reversal of deep sea circulation and its influence on geologic climates. *J. Geol.* 14: 1363–73

Cita MB. 1991. Anoxic Basins of the Eastern Mediterranean: an Overview. *Paleoceanography* 6: 133

Cluff RM. 1980. Paleoenvironment of the New Albany Shale Group (Devonian–Mississippian) of Illinois. *J. Sediment. Petrol.* 50: 767–80

Collier RW, Edmond J. 1984. The trace element geochemistry of marine biogenic particulate matter. *Prog. Oceanogr.* 13: 113–99

Conant LC, Swanson VE. 1961. *Chattanooga shale and related rocks of central Tennessee and nearby areas*. Washington, DC: US Geol. Surv. 91 pp.

Cook AC, Hutton AC, Sherwood NR. 1981. Classification of oil shales. *Bull. C. R. Explor.-Prod. Elf-Aquitaine* 5: 353–81

Coveney RM Jr, Watney WL, Maples CG. 1991. Contrasting depositional models for Pennsylvanian black shale discerned from molybdenum abundances. *Geology* 19: 147

Cowie GL, Hedges JI. 1992. The role of anoxia in organic matter preservation in coastal sediments: relative stabilities of the major biochemicals under oxic and anoxic depositional conditions. *Org. Geochem.* 19: 229

Dando PR, Southward AJ. 1986. Chemoautotrophy in bivalve molluscs of the genus *Thyasira*. *J. Mar. Biol. Assoc. U.K.* 66: 915–29

Dean WE, Arthur MA. 1986. Inorganic and organic geochemistry of Eocene to Cretaceous strata recovered from the lower continental rise, North American basin, Site 603, DSDP Leg 93. In *Init. Rep. DSDP* 93, ed. JE van Hinte, SW Wise Jr., et al, pp. 1093–137. Washington, DC: US Govt. Print. Off.

Dean WE, Arthur MA. 1989. Iron-sulfur-carbon relationships in organic-carbon-rich sequences I: Cretaceous Western Interior Seaway. *Am J. Sci* 289: 708–43

Dean WE, Leinen M, Stow DAV. 1984. Classification of deep sea fine-grained sediments. *J. Sediment. Petrol.* 55: 250–56

Degens ET, Emeis K-C, Mycke B, Wiesner MG. 1986. Turbidites, the principal mechanism yielding black shales in the early deep Atlantic Ocean. See Summerhayes & Shackleton 1986, pp. 361–76

Degens ET, Ross DA, eds. 1974. The Black Sea: Geology, Chemistry, Biology. *Am. Assoc. Petrol. Geol. Mem.* 20: 183–99

Degens ET, Stoffers P. 1976. Stratified waters as a key to the past. *Nature* 263: 22–26 de Graciansky PC, Deroo G. 1984. A stagnation event of ocean-wide extent in the upper Cretaceous. *Nature* 308: 346–49

Demaison GJ, Moore GT. 1980. Anoxic environments and oil source bed genesis. *Am. Assoc. Petrol. Geol. Bull.* 64: 1179–209

Desborough GA, Hatch JR, Leventhal JS. 1990. Geochemical and mineralogical comparison of the Upper Pennsylvanian Stark Shale Member of the Dennis Limestone, East-Central Kansas, with the Middle Pennsylvanian Mecca Quarry Shale Member of the Carbondale Formation in Illinois and of the Linton Formation in Indiana. In *Metalliferous Black Shales and Related Ore Deposits*, ed. RI Grauch, HLO Huyck, pp. 85. Washington, DC: US Geol. Survey

Doveton JH. 1991. Lithofacies and geochemical facies profiles from nuclear wireline logs: new subsurface templates for sedimentary modeling. In *Sedimentary Modeling: Computer Simulations and Methods for Improved Parameter Definition*, ed. EK Franseen, WL Watney, CGSC Kendall, W Ross, pp. 101–10. Lawrence: Kansas Geol. Surv.

Droser ML, Bottjer DJ. 1986. A semi-quantitative field classification of ichnofabric. *J. Sediment. Petrol.* 56: 558–59

Duff KL. 1975. Paleoecology of a bituminous shale—The lower Oxford Clay of central England. *Palaeontology* 18: 443–82

Dunbar CD, Rogers J. 1957. *Principles of Stratigraphy*. New York: Wiley. 356 pp.

Eicher DL, Diner R. 1989. Origin of the Cretaceous Bridge Creek cycles in the Western Interior, United States. *Palaeogeogr. Palaeoclimatol. Palaeoecol.* 74: 127–46

Einsele G, Mosebach R. 1955. Zur Petrographie, Fossilerhaltung und Entstehung des Posidonienschiefers im Schwabischen Jura. *N. Jahrb. Geol. Palaontol. Abh.* 101: 319–430

Ekdale AA, Mason T. 1988. Characteristic trace-fossil associations in oxygen-poor sedimentary environments. *Geology* 16: 720–23

Emerson S. 1985. Organic carbon preservation in marine sediments. In *The Carbon Cycle and Atmsoperic CO_2: Natural Variations Archean to Present*, ed. ET Sundquist, WS Broecker, pp. 78–87. Washington DC: Am. Geophys. Union, Geophys. Monogr. Ser.

Emerson S, Hedges JI. 1988. Processes controlling the organic carbon content of open ocean sediments. *Paleoceanography* 3: 621–34

Emery KO, Uchupi E. 1972. Western North Atlantic Ocean. *Mem. Am. Assoc. Petrol. Geol.* 17: 6–532.

Espitalié J, Deroo G, Marquis F. 1985. Rock Eval pyrolysis and its applications. *Inst. Francais Petrole, Dir. Recherche. No.* 27299. 72 pp.

Ettensohn FR. 1985. Controls on development of Catskill Delta complex basin facies. In *Geol. Soc. Am. Spec. Pap.* 201, ed. DW Woodrow, WD Sevon, pp. 65–77

Ettensohn FR. 1992. Controls on the origin of the Devonian-Mississippian oil and gas shales, east-central United States. *Fuel* 71: 1487–92

Fischer AG, Arthur MA. 1977. Secular variations in the pelagic realm. In *Deep Water Carbonate Environments*, ed. HE Cook, P Enos, pp. 19–50. Tulsa: Soc. Econ. Paleontol. Mineral.

Fleming RH, Revelle R. 1939. Physical processes in the ocean. In *Recent Marine Sediments*, ed. PD Trask, pp. 48–141. Tulsa: Am. Assoc. Petrol. Geol.

Freeman LH, Hayes JM. 1992. Fractionation of carbon isotopes by phytoplankton and estimates of ancient pCO_2. *Global Biogeochem. Cycles* 6: 185–98

Froelich PN, Arthur MA, Burnett WC, Deakin M, Hensley V, et al. 1988. Early diagenesis of organic-matter in Peru Continental Margin sediments: phosphorite precipitation. *Mar. Geol.* 80: 309–43

Gallardo VA. 1977. Large benthic microbial communities in sulphide biota under the Peru-Chile Subsurface Countercurrent. *Nature* 268: 331–32

Goldhaber MB, Kaplan IR. 1974. The sulfur cycle. In *The Sea, Marine Chemistry*, ed. ED Goldberg, pp. 569–655. New York: Wiley

Goldman MI. 1924. "Black shale" formation in and about Chesapeake Bay. *Am. Assoc. Petrol. Geol. Bull.* 8: 195–201

Grabau AW. 1913. *Principles of Stratigraphy*. New York: Seiler. 1185 pp.

Hallam A, Bradshaw A. 1979. Bituminous shales and oolitic ironstones as indicators of transgressions and regressions. *J. Geol. Soc. London* 136: 157–64

Hancock JM, Kauffman EG. 1979. The great transgressions of the Late Cretaceous. *J. Geol. Soc. London* 136: 175–86

Haq B, Hardenbol J, Vail PR. 1987. The chronology of fluctuating sea level since the Triassic. *Science* 235: 1156–67

Hard EW. 1931. Black shale deposition in Central New York. *Am. Assoc. Petrol. Geol. Bull.* 15: 165–81

Harland WB, Armstrong RL, Smith AG, Smith DG. 1990. *A Geologic Time Scale*. Cambridge: Press Syndicate, Univ. Cambridge. 2nd ed. 131 pp.

Hartman O, Barnard JL. 1958. *The benthic fauna of the deep basins off Southern California, Part I and II.* Allan Hancock Pacific Expedition 22, Univ. S. Calif. 297 pp.

Harvey RH, Fallon RD, Patton JS. 1986. The effect of organic matter and oxygen on the degradation of bacterial membrane lipids in marine sediments. *Geochim. Cosmochim. Acta* 50: 795–804

Hattin DE. 1975. Petrology and origin of fecal pellets in Upper Cretaceous strata of Kansas and Saskatchewan. *J. Sediment. Petrol.* 45: 686–96

Hattin DE. 1986. Carbonate substrates of the Late Cretaceous Sea, central Great Plains and southern Rocky Mountains. *Palaios* 1: 347–67

Hauf B. 1921. Untersuchung der Fossilfundstätten von Holzmaden im Posidonienschiefer des Oberen Lias Wurttembergs. *Palaontographica* 64: 1–42

Hay WW, Eicher DL, Diner R. 1993. Paleoceanography of the Cretaceous Western Interior Seaway. In *Sedimentation, Climate and Oceanography of the Cretaceous Western Interior Seaway of North America*, ed. WE Caldwell, EG Kauffman. Calgary: Geol. Assoc. Canada. In press

Heath GR, Moore TC, Dauphin JP. 1977. Organic carbon in deep-sea sediments. In *The Fate of Fossil Fuel CO_2 in the Oceans*, ed. NR Anderson, A Malahoff, pp. 605–26. New York: Plenum

Heckel PH. 1977. Origin of phosphatic black shale facies in Pennsylvanian cyclothems of mid-continent North America. *Am. Assoc. Petrol. Geol. Bull.* 61: 1045–68

Heckel PH. 1986. Sea-level curve for Pennsylvanian eustatic marine transgressive-regressive depositional cycles along midcontinent outcrop belt, North America. *Geology* 14: 330–34

Hedges JI, Clark WA, Quay PD, Richey JE, Devol AH, Santos, U de M. 1986. Compositions and fluxes of particulate organic material in the Amazon River. *Limnol. Oceanogr.* 31: 717–38

Hedges JI, Parker PL. 1976. Land-derived organic matter in surface sediments from the Gulf of Mexico. *Geochim. Cosmochim. Acta* 40: 1019–29

Henrichs SM, Reeburg WS. 1987. Anaerobic mineralization of marine sediment organic matter: rates and the role of anerobic processes in the oceanic carbon economy. *Geomicrobiol. J.* 5: 191–237

Herbert TD, Stallard RF, Fischer AG. 1986. Anoxic events, productivity rhythms, and the orbital signature in a mid-Cretaceous pelagic core. *Paleoceanography* 1: 495–506

Herbin JP, Montadert L, Müller C, Gomez R, Thurow J, Wiedmann J. 1986. Organic-rich sedimentation at the Cenomanian-Turonian boundary in oceanic and coastal basins in the North Atlantic and Tethys. See Summerhayes & Shackleton 1986, pp. 389–422

Herbin J-P, Müller C, Geyssant JR, Mélieres F, Penn IE. 1991. Hétérogénéité quantitative et qualitative de la matière organique dans les argiles du Kimméridgien du Val du Pickering (Yorkshire UK). *Rev. Inst. Fr. Pétrole* 46: 675–712

Holland HD. 1979. Metals in black shales—a reassesment. *Econ. Geol.* 74: 1676–79

Hollander DJ, McKenzie JA. 1991. CO_2 control on carbon isotope fractionation during aqueous photosynthesis: a paleo-pCO_2 barometer. *Geology* 19: 929–32

Hollander DJ, McKenzie JA, Hsu KJ, Huc AY. 1993. Application of an eutrophic lake model to the origin of ancient organic-carbon-rich sediments. *Global Biogeochem. Cycles* 7: 157–79

Hülsemann J, Emery KO. 1961. Stratification in recent sediments of Santa Barbara Basin as controlled by organisms and water character. *J. Geol.* 69: 279–90

Huyck HLO. 1990. When is a metalliferous black shale not a black shale?. In *Metalliferous Black Shales and Related Ore Deposits—Proc.* 1989 *US Working Group Meet., Int. Geol. Correlation Prog. Proj.* 254, ed. RI Grauch, HLO Huyck, pp. 42–56. Washington, DC: US Geol. Survey

Ingram RL. 1953. Fissility of mudrocks. *Geol. Soc. Am. Bull.* 65: 869–78

Irwin H, Curtis CD, Coleman M. 1977. Isotopic evidence for sources of diagenetic carbonates formed during burial of organic-rich sediments. *Nature* 269: 209–13

Jacobs L, Emerson S, Huested SS. 1987. Trace metal geochemistry in the Cariaco Trench. *Deep-Sea Res.* 34: 965–81

Jacobs L, Emerson S, Skei J. 1985. Partitioning and transport of metals across the O_2/H_2S interface in a permanently anoxic fjord, Framvaren Fjord, Norway. *Geochim. Cosmochim. Acta* 49: 1433–44

Jahnke RA. 1990. Ocean flux studies : a status report. *Rev. Geophys.* 28: 381–98

Jahnke RA, Reimers CE, Craven DB. 1990. Intensification of recycling of organic matter at the sea floor near ocean margins. *Nature* 348: 50–54

Jannasch HW. 1984. Chemosynthesis: the nutritional basis for life at deep-sea vents. *Oceanus* 27: 73–78

Jenkyns HC. 1980. Cretaceous anoxic events: from continents to oceans. *J. Geol. Soc. London* 137: 171–88

Jenkyns HC. 1985. The Early Toarcian and Cenomanian-Turonian anoxic events in Europe: comparisons and contrasts. *Geol. Rundsch.* 74: 505–18

Jenkyns HC. 1988. The Early Toarcian (Jurassic) Anoxic Event: stratigraphic, sedimentary and geochemical evidence. *Am. J. Sci.* 288: 101–51

Jenkyns HC, Clayton JC. 1986. Black shales and carbon isotopes in pelagic sediments from the Tethyan lower Jurassic. *Sedimentology* 33: 87–106

Jorgensen BB, Revsbech NP. 1985. Diffusive boundary layers and the oxygen uptake of sediments and detritus. *Limnol. Oceanogr.* 30: 111–22

Jumars PA. 1976. Deep sea diversity. *J. Mar. Res.* 34: 217–46

Katz BJ. 1983. Limitations of Rock-Eval pyrolysis for typing organic matter. *Org. Geochem.* 4: 195–99

Kauffman EG. 1981. Ecological reappraisal of the German Posidonienschiefer (Toarcian) and the stagnant basin model. In *Communities of the Past*, ed. J Gray, AJ Boucot, WBN Berry, pp. 311–81. Stroudsburg: Hutchinson Ross

Kauffman EG. 1988a. The case of the missing community: low-oxygen adapted Paleozoic and Mesozoic bivalves ("flat clams") and bacterial symbioses in typical Phanerozoic seas. *Geol. Soc. Am., Centen. Meet.*, 1988, *Denver CO, Abstr. with Programs.* A48

Kauffman EG. 1988b. Concepts and methods of high-resolution event stratigraphy. *Annu. Rev. Earth Planet. Sci.* 16: 605–54

Kauffman EG, Sageman BB. 1990. Biological sensing of benthic environments in dark shales and related oxygen-restricted facies. In *Cretaceous Resources, Events and Rhythms: Background and Plans for Research (Proc. of the 1st Global Sedimentary Geol. Prog. Meet.), Digne, France*, ed. RN Ginsberg, B Beaudoin, pp. 121–38. NATO ASI Ser. C: Math. Phys. Sci.

Keil RG, Tsamakis E, Bor Fuh C, Giddings JC, Hedges JI. 1993. Mineralogic controls on the concentration and elemental composition of organic-matter in marine sediments: hydrodynamic separation using Splitt-fractionation. Unpubl.

Kidd RB, Cita MB, Ryan WBF. 1978. The stratigraphy of Eastern Mediterranean sapropel sequences recovered by DSDP Leg 42A and their paleoenvironmental significance. In *Init. Rep. DSDP*, ed. KJ Hsü, L Montadert, et al, pp. 421–23. Washington, DC: US Govt. Print. Off.

Kruijs E, Barron E. 1990. Climate model prediction of paleoproductivity and potential source-rock distribution. In *Deposition of Organic Facies*, ed. AY Huc, pp. 195–216. Tulsa: Am. Assoc. Petrol. Geol.

Lee C. 1992. Controls on organic carbon preservation: the use of stratified water bodies to compare intrinsic rates of decomposition in oxic and anoxic systems. *Geochim. Cosmochim. Acta* 56: 3323–35

Leggett JK. 1980. British Lower Paleozoic black shales and their palaeo-oceanographic significance. *J. Geol. Soc. London* 137: 139–56

Leppakoski E. 1969. Transitory return of the benthic fauna of Bornholm Basin after extermination by oxygen insufficiency. *Cah. Biol. Mar. Tome* X: 163–72

Leventhal JS. 1983. An interpretation of carbon and sulfur relationships in Black Sea sediments as indicators of environments of deposition. *Geochim. Cosmochim. Acta* 47: 133–38

Leventhal JS. 1991. Comparison of organic geochemistry and metal enrichment in two black shales: Cambrian Alum Shale of Sweden and Devonian Chattanooga Shale of United States. *Mineralium deposita* 26: 104

Lewan MD, Maynard JB. 1982. Factors controlling enrichment of vanadium and nickel in the bitumen of organic sedimentary rocks. *Geochim. Cosmochim. Acta* 46: 2547–60

Lindh TB. 1983. *Temporal variation in ^{13}C, ^{34}S and global sedimentation during the Phanerozoic.* Masters thesis. Univ. Miami, Miami, Fla.

Loutit TS, Hardenbol J, Vail PR, Baum GR. 1988. Condensed sections: the key to age determination and correlation of continental margin sequences. In *Sea-Level Changes: An Integrated Approach*, ed. CK Wilgus, BS Hastings, CG St G Kendall, HW Posamentier, CA Ross, JC Van Wagoner, pp. 183–213. Tulsa: Soc. Econ. Paleontol. Mineral.

Lundegard PD, Samuels ND. 1980. Field classification of fine-grained sedimentary rocks. *J. Sediment. Petrol.* 50: 781–86

Mangini A, Dominik J. 1979. Late Quaternary sapropel on the Mediterranean Ridge: U-budget and evidence for low sedimentation rates. *Sediment. Geol.* 23: 113–25

Martens CS, Klump JV. 1984. Biogeochemical cycling in an organic-rich coastal marine basin 4. An organic carbon budget for sediments dominatd by sulfate reduction and methanogenesis. *Geochim. Cosmochim. Acta* 48: 1987–2004

Meybeck M. 1982. Carbon, nitrogen and phosphorus transport by world rivers. *Am. J. Sci.* 282: 401–50

Middelburg JJ, Calvert SE, Karlin R. 1991. Organic-rich transitional facies in silled basins: response to sea-level change. *Geology* 19: 679–82

Miller RG. 1990. A paleoceanographic

approach to the Kimmeridge Clay Formation. In *Deposition of Organic Facies, Am. Assoc. Petrol. Geol. Studies in Geol.* #30, ed. AY Huc, pp. 13–26. Tulsa: Am. Assoc. Petrol. Geol.

Moore GT, Hayashida DN, Ross CA. 1993. Late Early Silurian (Wenlockian) general circulation model-generated upwelling, graptolitic black shales, and organic-rich source rocks—An accident of plate tectonics? *Geology* 21: 17–20

Morris KA. 1980. Comparison of major sequences of organic-rich mud deposition in the British Jurassic. *J. Geol. Soc. London* 137: 157–70

Morris RJ, Calvert SE. 1977. Geochemical studies of organic-rich sediments from the Namibian Shelf, I. The organic fractions. In *A Voyage of Discovery*, ed. M Angel, pp. 647–65. Oxford: Pergamon

Mossman, J-R, Aplin AC, Curtis CD, Coleman ML. 1992. Geochemistry of inorganic and organic sulfu in organic-rich sediments from the Peru Margin. *Geochim. Cosmochim. Acta* 55: 3581–95

Müller PJ, Suess E. 1979. Productivity, sedimentation rate, and sedimentary organic matter in the oceans, I, Organic carbon preservation. *Deep-Sea Res.* 26: 1347–62

Nichols JA. 1976. The effect of stable dissolved-oxygen stress on marine benthic invertebrate community diversity. *Int. Rev. Gesamten Hydrobiol.* 61: 747–60

North FK. 1979. Episodes of source-sediment deposition. *J. Petrol. Geol.* 2: 199–218

North FK. 1980. Episodes of source-sediment deposition: the episodes in individual close-up. *J. Petrol. Geol.* 3: 323–38

Orem WH, Burnett WC, Landing WM. 1991. Jellyfish Lake, Palau: early diagenesis of organic matter in sediments of an anoxic marine lake. *Limnol. Oceanogr.* 36: 526

Parrish JT. 1982. Upwelling and petroleum source beds, with reference to the Paleozoic. *Am. Assoc. Petrol. Geol. Bull.* 66: 750–74

Parrish JT, Curtis RL. 1982. Atmospheric circulation, upwelling and organic-rich rocks in the Mesozoic and Cenozoic eras. *Palaeogeogr. Palaeoclimatol. Palaeoecol.* 40: 67–101

Pedersen TF, Calvert SE. 1990. Anoxia vs. productivity: What controls the formation of organic-rich sediments and sedimentary rocks? *Am. Assoc. Petrol. Geol. Bull.* 74: 454–66

Pederson TF, Shimmield GB, Price NB. 1992. Lack of enhanced preservation of organic-matter in sediments under the oxygen minimum on the Oman Margin. *Geochim. Cosmochim. Acta* 56: 545–51

Peters KE. 1986. Guidelines for evaluating petroleum source rock using programmed pyrolysis. *Am. Assoc. Pet. Geol. Bull.* 70: 318–29

Pettijohn F. 1957. *Sedimentary Rocks*. New York: Harper. 718 pp.

Pompeckj JF. 1901. Die Juraablagerungen zwischen Regensburg und Regenstauf. *Geogn. Jahrb.* 14: 139–220

Pontonié R. 1937. Die nomenklature der unterwasserablagerungen. *Jahrb. Preuss. Geol. Landesanst. Bergakad.* 58: 426–58

Pratt LM. 1984. Influence of paleoenvironmental factors on the preservation of organic matter in middle Cretaceous Greenhorn Formation near Pueblo, Colorado. *Am. Assoc. Petrol. Geol. Bull.* 68: 1146–59

Prauss M, Riegel W. 1989. Evidence from phytoplankton associations for causes of black shale formation in epicontinental seas. *N. Jahrb. Geol. Palaontol.* 11: 671

Premuzic ET, Benkovitz CM, Gaffrey JS, Walsh JJ. 1982. The nature and distribution of organic matter in the surface sediments of world oceans and seas. *Org. Geochem.* 4: 63–77

Raiswell R, Berner RA. 1985. Pyrite formation in euxinic and semi-euxinic sediments. *Am. J. Sci.* 285: 710–24

Raiswell R, Berner RA. 1987. Organic carbon losses during burial and thermal maturation of normal marine shales. *Geology* 15: 853–56

Raiswell R, Buckley F, Berner RA, Anderson TF. 1988. Degree of pyritization of iron as a paleoenvironmental indicator of bottom-water oxygenation. *J. Sediment. Petrol.* 58: 812–19

Reimers CE, Jahnke RA, McCorkle DC. 1992. Carbon fluxes and burial rates over the continental slope and rise off central California with implications for the global carbon cycle. *Global Biogeochem. Cycles* 6: 199–224

Reimers CE, Suess E. 1983. Spatial and temporal patterns of organic matter accumulation on the Peru continental margin. In *Coastal Upwelling: Its Sediment Record, Part B: Sedimentary Records of Ancient Coastal Upwelling*, ed. J Thiede, E Suess, pp. 311–46. New York: Plenum

Rhoads DC, Morse JM. 1971. Evolutionary and ecologic significance of oxygen-deficient marine basins. *Lethaia* 4: 413–28

Rhoads DC, Young DK. 1970. The influence of deposit-feeding organisms on sediment stability and community trophic structure. *J. Mar. Res.* 28: 150–78

Richards FA, Redfield AC. 1954. A correlation between the oxygen content of seawater and organic content of marine sediments. *Deep-Sea Res.* 1: 279–81

Robertson AHF. 1984. Origin of varve-type lamination, graded claystones and limestone-shale "couplets" in the lower Cretaceous of the western North Atlantic. In *Fine-grained Sediments: Deep-Water Processes and Facies*, *Geol. Soc. Spec. Publ. No.* 15, ed. DAV Stow, DJW Piper, pp. 436–52. Oxford: Blackwell

Rosenberg R. 1977. Benthic macrofaunal dynamics, production, and dispersion in an oxygen-deficient Estuary of Wets Sweden. *J. Exp. Mar. Biol. Ecol.* 26: 107–33

Rosenberg R, Arntz WE, de Flores EC, Flores LA, Carbajal G, et al. 1983. Benthos biomass and oxygen deficiency in the upwelling system off Peru. *J. Mar. Res.* 41: 263–79

Ross DA, Degens ET, Macilvaine J. 1970. The Black Sea: recent sedimentary history. *Science* 170: 163–65

Ruedemann R. 1935. Ecology of black mud shales of eastern New York. *J. Paleontol.* 9: 79–91

Ryan WBF, Cita MB. 1977. Ignorance concerning episodes of ocean-wide stagnation. *Mar. Geol.* 23: 197–215

Sageman BB. 1989. The benthic boundary biofacies model: Hartland Shale Member, Greenhorn Formation (Cenomanian), Western Interior, North America. *Palaeogeogr. Palaeoclimatol. Palaeoecol.* 74: 87–110

Sageman BB. 1991. *High-resolution event stratigraphy, carbon geochemistry and paleobiology of the Upper Cenomanian Hartland Shale Member (Cretaceous), Greenhorn Formation, Western Interior, U.S.* PhD Thesis. Univ. Colo., Boulder

Sageman BB, Wignall PB, Kauffman EG. 1991. Biofacies models for organic-rich facies: tool for paleoenvironmental analysis. In *Cycles and Events in Stratigraphy*, ed. G Einsele, W Ricken, A Seilacher, pp. 542–64. Berlin: Springer-Verlag

Sarmiento JL, Herbert TD, Toggweiler JR. 1988. Causes of anoxia in the world ocean. *Global Biogeochem. Cycles* 2: 115–28

Sarnthein, M, Winn K, Duplessy, J-C, Fontugne MR. 1988. Global variations of surface ocean productivity in low and mid latitudes: influence on CO_2 reservoirs of the deep ocean and atmosphere during the last 21,000 years. *Paleoceanography* 3: 361–99

Savrda CE, Bottjer DJ. 1986. Trace-fossil model for reconstruction of paleooxygenation in bottom waters. *Geology* 14: 3–6

Savrda CE, Bottjer DJ. 1987. The exaerobic zone, a new oxygen-deficient marine biofacies. *Nature* 327: 54–56

Savrda CE, Bottjer DJ, Gorsline DS. 1984. Development of a comprehensive oxygen-deficient marine biofacies model: evidence from Santa Monica, San Pedro, and Santa Barbara basins, California continental borderland. *Am. Assoc. Petrol. Geol. Bull.* 68: 1179–92

Savrda CE, Bottjer DJ, Seilacher A. 1991. Redox-related benthic events. In *Cycles and Events in Stratigraphy*, ed. G Einsele, A Seilacher, W Ricken, pp.524–41. Berlin: Springer-Verlag

Schlanger SO, Arthur MA, Jenkyns HC, Scolle PA. 1987. The Cenomanian-Turonian Oceanic Anoxic Event, I. Stratigraphy and distribution of organic carbon-rich beds and the marine $\delta^{13}C$ excursion. See Brooks & Fleet 1987, pp. 371–99

Schlanger SO, Jenkyns HC. 1976. Cretaceous oceanic anoxic events: causes and consequences. *Geol. Mijnbouw.* 55: 179–84

Schuchert C. 1915. The conditions of black shale deposition as illustrated by Kupferschiefer and Lias of Germany. *Proc. Am. Phil. Soc.* 54: 259–69

Scotese CR, Summerhayes CP. 1986. Computer model of paleoclimate predicts coastal upwelling in the Mesozoic and Cenozoic. *Geobyte* 28–42

Seilacher A. 1964. Biogenic sedimentary structures. In *Approaches to Paleoecology*, ed. J Imbrie, N Newell, pp. 296–316. New York: Wiley

Smith CR, Hamilton SC. 1983. Epibenthic megafauna of a bathyal basin off southern California: patterns of abundance, biomass, and dispersion. *Deep-Sea Res.* 30: 907–28

Spears DA. 1980. Towards a classification of shales. *J. Geol. Soc. London* 137: 123–30

Spencer DW, Brewer PG, Sachs PL. 1972. Aspects of the distribution and trace element composition of suspended matter in the Black Sea. *Geochim. Cosmochim. Acta* 36: 71–86

Spies RB, Davis PH. 1979. The infaunal benthos of a natural oil seep in the Santa Barbara Channel. *Mar. Biol.* 50: 227–37

Stein R. 1986. Surface-water paleo-productivity as inferred from sediments deposited in oxic and anoxic deep-water environments of the Mesozoic Atlantic Ocean. In *Biogeochemistry of Black Shales*, ed. ET Degens, PA Meyers, SC Brassell, pp. 55–70. Hamburg: SCOPE/UNEP, Mittel. Geol.-Paläontol. Inst. Univ. Hamburg

Strøm KM. 1936. Land-Locked Waters. *Det. Norske Videnskaps-Akademi i Oslo, Nat. Naturv. Klasse* 7: 70–81

Strøm KM. 1939. Land-locked waters and the deposition of black muds. In *Recent Marine Sediments*, ed. PD Trask, pp. 356–72. Tulsa: Am. Assoc. Petrol. Geol.

Suess E. 1980. Particulate organic carbon flux in the oceans—surface productivity and oxygen utilization. *Nature* 288: 260–63

Suess E, von Huene R, et al. 1990. *Initial Reports of the Ocean Drilling Program, Leg* 112, *Part II, Sci. Results*. Washington, DC: US Govt. Print. Off. 738 pp.

Summerhayes CP. 1987. Organic-rich Cretaceous sediments from the Atlantic. See Brooks & Fleet 1987, pp. 301–16

Summerhayes CP, Shackleton NJ, eds. 1986. *North Atlantic Paleoceanography*. London: Geol. Soc. Spec. Publ.

Theede H. 1973. Comparative studies on the influence of oxygen deficiency and hydrogen sulphide on marine invertebrates. *Neth. J. Sea Res.* 7: 244–52

Thickpenny A. 1984. The sedimentology of the Swedish Alum Shales. In *Fine-Grained Sediments: Processes and Products*, ed. DAV Stow, D Piper, pp. 511–25. *Geol. Soc. London Spec. Publ.*

Thiede J, van Andel TH. 1977. The paleoenvironment of anaerobic sediments in the late Mesozoic South Atlantic Ocean. *Earth Planet. Sci. Lett.* 33: 301–9

Thompson JB, Mullins HT, Newton CR, Vercoutere TL. 1985. Alternative biofacies model for dysaerobic communities. *Lethaia* 18: 167–79

Thomsen E. 1989. Seasonal variability in the production of Lower Cretaceous calcareous nannoplankton. *Geology* 17: 715–17

Toth DJ, Lerman A. 1977. Organic matter reactivity and sedimentation rate in the ocean. *Am. J. Sci.* 277: 465–85

Trask PD. 1932. *Origin and Environment of Source Sediments of Petroleum*. Houston: Gulf. 67 pp.

Trask PD. 1951. Depositional environment of black shale. *Am. Assoc. Petrol. Geol. Abstr. with Programs*, 1951 Annu. Meet., St. Loius, Mo., p. 44

Trask PD, Patnode HW. 1942. *Source Beds of Petroleum*. Tulsa: Am. Assoc. Petrol. Geol. 561 pp.

Tuttle JH. 1985. The role of sulfur-oxidizing bacteria at deep-sea hydrothermal vents. *Bull. Biol. Soc. Washington* 6: 335–43

Twenhofel WH. 1915. Notes on black shale in the making. *Am. J. Sci.* 4: 272–80

Twenhofel WH. 1939. Environments of origin of black shales. *Am. Assoc. Petrol. Geol. Bull.* 23: 1178–98

Tyson RV. 1987. The genesis and palynofacies characteristics of marine petroleum source rocks. See Brooks & Fleet 1987, pp. 47–67

Tyson RV, Pearson TH. 1991. Modern and ancient continental shelf anoxia: an overview. In *Modern and Ancient Continental Shelf Anoxia*, ed. RV Tyson, TH Pearson, pp. 1–24. London: Geol. Soc.

Ulmishek GF, Klemme HD. 1990. *Depositional Controls, Distribution, and Effectiveness of World's Petroleum Source Rocks*. Washington, DC: US Geol. Survey. 59 pp.

Vail PR, Mitchum RMJ, Thompson SI. 1977. Seismic stratigraphy and global changes of sea level, Part 4: Global cycles of relative changes of sea level. In *Seismic Stratigraphy—Applications to Hydrocarbon Exploration*, ed. CE Payton, pp. 83–98. Tulsa: Am. Assoc. Petrol. Geol.

van der Gracht WAJMvW. 1931. Permocarboniferous orogeny in the south-central United States. *Am. Assoc. Petrol. Geol. Bull.* 15: 991–1057

Veeh HH. 1967. Deposition of uranium from the ocean. *Earth Planet. Sci. Lett.* 3: 145–50

Vine JD, Tourtelot EB. 1970. Geochemistry of black shales—a summary report. *Econ. Geol.* 65: 253–72

Vogt PR. 1989. Volcanogenic upwelling of anoxic, nutrient-rich water: a possible factor in carbonate-bank/reef demise and benthic fauna extinctions. *Geol. Soc. Am. Bull.* 101: 1225

Waples DW. 1983. Reappraisal of anoxia and organic richness, with emphasis on Cretaceous of North Atlantic. *Am. Assoc. Petrol. Geol. Bull.* 67: 963–78

Wasmund E. 1930. Bitumen, sapropel and gyttja. *Geol. Fören. Stockholm Förh* 52: 219–38

Wenger LM, Baker DR. 1986. Variations in organic geochemistry of anoxic-oxic black shale-carbonate sequences in the Pennsylvanian of the Midcontinent, U.S.A. *Org. Geochem.* 10: 85–92

Wignall PB. 1990. *Benthic paleoecology of the Late Jurassic Kimmeridge Clay of England*. London: The Palaeontol. Soc. 74 pp.

Wignall PB, Hallam A. 1991. Biofacies, stratigraphic distribution and depositional models of British onshore Jurassic black shales. In *Modern and Ancient Continenal Shelf Anoxia*, ed. RV Tyson, TH Pearson, pp. 291–310. London: Geol. Soc.

Wignall PB, Myers KJ. 1988. Interpreting benthic oxygen levels in mudrocks: a new approach. *Geology* 16: 452–55

Wilde, P, Berry WBN. 1982. Progressive ventilation of the oceans—potential for return to anoxic conditions in the post-Paleozoic. In *Nature and Origin of Cretaceous Carbon-Rich Facies*, ed. SO Schlanger, MB Cita, pp. 209–24. New York: Academic

Williams LA, Reimers CE. 1983. Role of bacterial mats in oxygen deficient marine basins and coastal upwelling regimes: preliminary report. *Geology* 11: 267–69

Woolnough WG. 1937. Sedimentation in barred basins, and source rocks of petroleum. *Am. Assoc. Petrol. Geol. Bull.* 21: 1101–57

Zangerl R, Richardson ES Jr. 1963. The paleoecological history of two Pennsylvanian black shales. *Fieldiana-Geol. Mem.* 4: 1–158

Zelt FB. 1985. Paleoceanographic events and lithologic/geochemical facies of the Greenhorn marine cycle (Upper Cretaceous) examined using natural gamma-ray spectrometry. In *Field Trip Guidebook No.* 4, ed. LM Pratt, FB Zelt, EG Kauffman, pp. 49–59. Tulsa: Soc. Econ. Paleontol. Mineral.

Annu. Rev. Earth Planet. Sci. 1994. 22:553–95

PHYSICS OF ZODIACAL DUST

Bo Å. S. Gustafson

Department of Astronomy, University of Florida, Gainesville, Florida 32611

KEY WORDS: meteoroids, radiation pressure, Poynting-Robertson drag, solar wind drag, Yarkovsky effect, Lorentz force, dust collisions, light scattering, thermal emission, dust temperature

INTRODUCTION

This review of the physics of zodiacal dust departs from earlier treatments in that it seeks to present a consistent framework to model the dust as nonspherical and inhomogeneous chondritic aggregates ranging from compact to highly porous structures. The objective is to address all aspects needed to model the optical, thermal, and dynamic properties of the zodiacal cloud. Because the use of nonspherical and/or aggregate dust models often is prohibitively cumbersome or solutions nonexistent, laboratory data and approximate solutions are used in comparisons with the spherical case.

A chronological model relating zodiacal dust to solar nebula dust and interstellar grains gives a theoretical basis for the chondritic aggregate model of interplanetary matter and provides a framework to relate specific models to their origin. There is direct evidence that a large fraction of the dust complex is of the chondritic aggregate kind. Relevant modeling parameters are based directly on observations and do not depend on the chronological model.

The zodiacal light is difficult to measure accurately from the ground due to air-glow contamination and a background of unresolved stellar objects. Observers usually report brightness averaged over several days, months, or years of observations. Space-based, optical observations are usually also averaged and presented in a compact form where fine structure and changes with the observer's vantage point are lost in the averaging process. However, extended sets of data on the infrared sky are now available. The

0084–6597/94/0515–0553$05.00

Infrared Astronomical Satellite (*IRAS*) mapped the infrared sky in great detail during its 1983 mission and zodiacal data from the *Cosmic Background Explorer* (*COBE*) will soon be available. The great detail of zodiacal features seen by *IRAS* and the changing perspective over the lifetime of the mission provide mission great impetus for analysis and modeling of the cloud.

Empirical models developed before *IRAS* of the spatial distribution of dust in the solar system were reviewed in a classic article by Giese et al (1986). These models are descriptive in nature and do not account for the laws of celestial mechanics. Dust orbital motion subject to non-gravitational forces give shape to real clouds (Gustafson et al 1987, Dermott et al 1992). Dynamical cloud models simulate the release of dust from a postulated source and use orbital mechanics to produce a spatial distribution at any specified epoch. Such models allow Dermott et al (1992, 1993a,b) to identify the asteroidal origin of the solar system dust bands, and likely contributions from six or more distinct Hirayama asteroid families. Contributions from several sources can be overlapped until the complex structure seen in data sets is reproduced, including seasonal variations and dependence on viewing direction. An understanding of the dust motion is the basis for this modeling. Dust dynamics is also the means by which source probabilities can be assigned to individual dust particles to be collected in Earth orbit.

Heliocentric orbits precess due to planetary perturbations. In the restricted three-body problem, orbits precess along the orbital plane of the massive bodies. The orbit of a "massless" test particle oscillates about a constant inclination as the line of nodes where the orbital planes intersect rotates. With more than 99.8% of the mass of the solar system concentrated in the Sun and 0.1% in Jupiter, the restricted three-body formulation gives a good qualitative approximation of orbital evolution in interplanetary space.

A cloud of particles develops rotational symmetry as orbits precess along Jupiter's orbital plane. This applies to interplanetary dust, comets, and asteroids, and is the basis for assuming rotational symmetry of the zodical dust cloud with the Sun at the center. Other planets have similar influence on interplanetary orbits. Each planet tends to precess orbits along its orbital plane. The resultant plane along which orbits precess depends on the distance to each perturbing planet. The planets also significantly displace the zodiacal cloud's axis of symmetry from the Sun. This can be seen in the *IRAS* data and is reproduced by modeling of particle dynamics (Dermott et al 1992, Xu et al 1993). This illustrates the need for dynamic cloud models.

Differential precession rates result from a finite spread over heliocentric distances and spreads the orbits of Hirayama asteroid families (created from the breakup of larger asteroids) along resultant planes described above (hereafter called local precession planes). An imprint of the distribution of the Hirayama family asteroids is inherited by newly created dust particles as the family asteroids grind down in collisions with each other, background asteroids, and meteoroids. This stochastic production (see modeling by Durda et al 1992), forms a cloud with a spatial distribution that depends only on the fraction of nongravitational forces on the dust (Xu et al 1993). The spatial distribution evolves as dust particles lose orbital momentum and migrate toward the Sun. As the orbits continue to shrink, the local precession plane changes and orbits do not have time to fully spread out along the plane corresponding to a given heliocentric distance before the plane changes. This gives each cloud component, defined as particles with a given descent rate starting from a given source region, a specific and calculable shape. I therefore review the forces acting on zodiacal dust, and the optical and thermal dust properties necessary to this and other modeling efforts. Emphasis is on the underlying physics and on uncertainties involved in estimating the various physical processes on real particles. A model of zodiacal dust particle morphologies is a prerequisite.

DUST MODELS

While it is generally believed that some interplanetary dust particles contain nearly pristine material out of which the solar system formed, the meteoroid complex and the zodiacal cloud are not direct remnants of the presolar nebula. Nebula dust was stored in comets and asteroids, where it has been processed to varying degree before reemerging as part of a continuous size distribution of interplanetary objects. Sizes span from asteroids and cometary nuclei in the km range through meteoroids that are too small to be seen as individual objects, to dust particles that are seen collectively as they produce the zodiacal light through scattering of sunlight. A size range of 1 to 100 μm is usually implied by the word "dust."

Pristine Matter

A simple framework to model interplanetary matter and interpret observations emerges based on a simple chronology carried back to the formation of interstellar grains. Protosolar nebula dust is thought to consist primarily of ancient ice mantles condensed on silicate cores and evolved in interstellar space with a freshly condensed ice mantle from nebula gases (Greenberg 1988).

Silicate spheres grow by condensation in the outflow from cool super giant stars. Aggregates develop from dust-dust collisions. The existence of interstellar polarization indicates that the classical grains responsible for most of the extinction at optical wavelengths are elongated and aligned. Aspect ratios of 1:2 to 1:3, i.e. 2 to 3 sphere arrays, agree with the degree of polarization; the wavelength dependence of extinction (interstellar reddening) is indicative of 10^{-1} μm sizes (Greenberg & Hage 1990). Some molecules condense and consolidate the structure through formation of an ice mantle. Billions of years of ultraviolet photoprocessing in interstellar space and cosmic-ray bombardment changes the ice into a refractory carbon-rich, oxygen-poor material. The grains are thought to evaporate and recondense from passing supernova shocks (Seab 1987); these might reemerge as a homogenized grain population by the time they become part of a star-forming nebula.

Temperatures in the protosolar nebula are thought to have ranged from over 1000 K near Mercury's orbit to the order of 50 K or less between Uranus and Neptune or beyond, where comets formed. Suess (1987) describes how matter in the innermost portions vaporize under these circumstances, allowing the elements contained locally in gas and dust to mix. Excess gas condenses on essentially pristine protosolar dust in the outer parts. Water ice with inclusions of 10^{-2} μm grains of polycyclic aromatic hydrocarbon (PAH) dominates in this outer mantle condensed in the solar nebula. Based on observed quantities and cosmic abundances, Greenberg & Hage (1990) adopted mass fractions in the outer parts of the nebula where comets form of 0.20, 0.19, 0.55, and 0.06 for the silicate, organic refractory, volatile ice, and PHA, respectively. The corresponding densities are 3.5, 1.8, 1.2, and 2 g cm^{-3}. This may be the pristine matter out of which comets and asteroids formed—the grandparents of today's meteoroids and zodiacal dust.

Chondritic Aggregates

The "Bird's-Nest" model of cometary dust originally proposed by Greenberg & Gustafson (1981) represents the dust as pieces of bulk cometary material of aggregated nebular dust from which water ice has sublimated. Assuming comet nuclei bulk densities of 0.6 g cm^{-3} (Sagdeev et al 1988, Rickman 1989), the mixture of solar nebula materials leads to a packing factor $p \approx 0.33$ in comets. The packing factor is defined as one minus the porosity or the fraction of volume occupied by material. In the absence of any compacting process, the depletion of volatiles leaves behind a $p \approx 0.15$ loosely packed tangle of refractory materials. The aggregation process was both computer simulated and mechanically simulated to build models used in analog light scattering measurements (Figure 1). These are updated versions of the original "Bird's-Nest" models of cometary dust. Each model consists of either 250 or 500 spheres with index of refraction

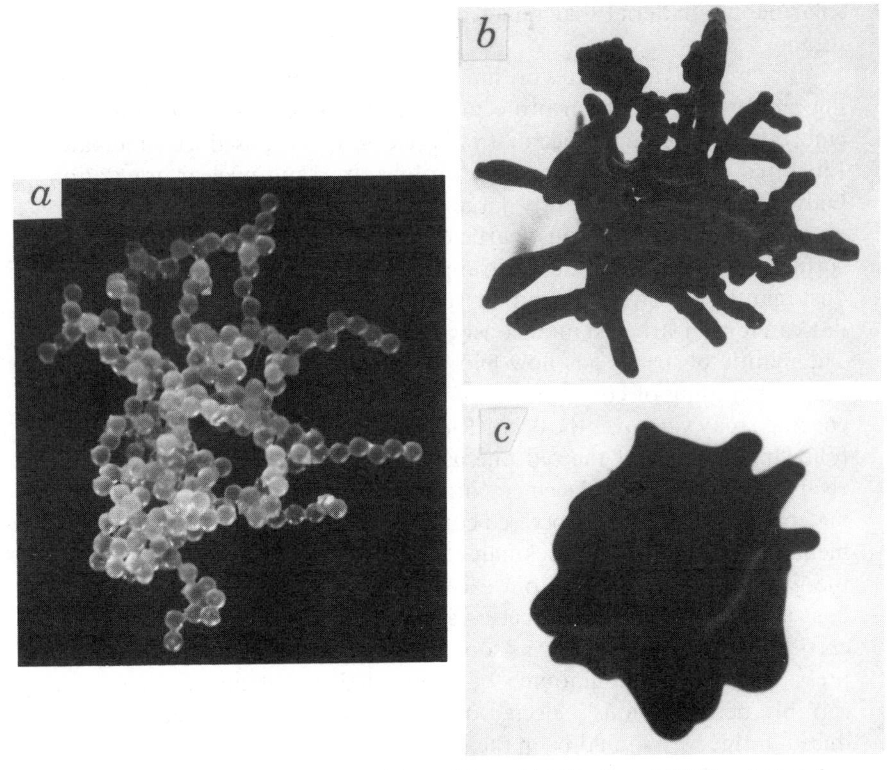

Figure 1 Examples of aggregate particle models used by Zerull et al (1993) and Gustafson et al (in preparation) in microwave analog light scattering measurements. (*a*) Each sphere represents a silicate grain grown by nucleation in the outflow from red giant stars. (*b*) The aggregates were coated by an absorbing compound to represent organic refractories grown in interstellar space. This is a chondritic aggregate of the "Bird's-Nest" type representing fresh comet dust. (*c*) Some aggregates were coated by a thicker mantle to represent a densely packed chondritic aggregate.

$n = 1.735 \pm 0.008 - 0.007i \pm 0.003i$ (Figure 1*a*). This material is representative of many silicates at visual wavelengths. The mantle coating the aggregate in Figure 1*b* has $n = 1.86 \pm 0.10 - 0.12i \pm 0.04i$ representing the organic refractory material. The model shown in Figure 1*c* has a thicker mantle of the same refractive index.

"Bird's-Nests" are aggregates of pristine interstellar grains that are characterized by a narrow size distribution. The aggregates derive from bulk material so that their dimension is three on the average—a whole number. These models are therefore not fractals (of fractal dimensions) despite their appearance. "Bird's-Nest" models represent the most pristine solar nebula material that might be expected to reach Earth by natural means.

"Bird's-Nest" aggregates presumably originate from active areas exposing matter that has been protected in comet interiors. Refractory dust embedded in the porous ice matrix is released as exposed ice sublimates. Liberated dust particles are subject to gas drag and nuclear gravitation. Only pieces with sufficiently large effective surface-to-mass ratios are entrained by the gas. As dust particles with lower ratio accumulate on the surface, newly liberated dust is trapped in the interstices. A consolidated dust mantle of a structure and density largely dependent on the local gas flux can form nearly instantaneously (Shul'man 1972, Rickman et al 1990). The mantle obstructs gas flow and shields the ices from direct sunlight. Partial mantling of comet nuclei is now thought to be common and the coverage may vary over the orbit (Rickman et al 1990). New mantles build following ejection of the old one as the perihelion decreases (Rickman et al 1991). Mantle replacement or changing mantle coverage produces meteoroids and dust of processed cometary matter as ejected mantle fragments. In the models by Rickman et al (1990), dust mantles prevail until the gas pressure at the bottom exceeds the weight of the mantle. Some dust layers were found to eject as soon as they form, i.e. the mantle is carried away by the gas flow as soon as the interstices clog up and the gas pressure rises. Large amounts of compacted comet material can conceivably be continuously ejected in this process. Large meteoroids produced in this way should be in the shape of thin flakes.

Asteroid material appears to range from nearly comet-like assemblages of ice and organics among the Trojan D-type asteroids and Hilda P-types near and beyond the outer edge of the main belt to less pristine C and S types inside the asteroid belt (Gradie et al 1989). It is generally believed that this gradation with heliocentric distance reflects primordial properties of the solar nebula, with incomplete condensation of low-temperature ices in the main asteroid belt. About two thirds of the C-type asteroids appear hydrated, suggesting that water melted and soaked their minerals. This is seen in many carbonaceous meteorites, carbon-rich meteoroids that have survived atmospheric flight and are found on the Earth's surface. S-type asteroids may include assemblages of iron- and magnesium-bearing silicates mixed with metallic nickel-iron.

Dermott et al (1984, 1993b) find that the Themis and Koronis families produce the central solar system dust bands seen in the *IRAS* data and that the Eos family may be responsible for the 10° bands. Together, these families may produce as much as 10% of the zodiacal dust complex, with the whole asteroid belt producing 30% or more (Dermott et al 1993b). The Eos and the Koronis Hirayama families of asteroids have members with spectra ranging from the S-type to intermediary between S and C. The Themis family contains C-type asteroids. Spectral similarities within families suggest that the parent bodies were not fully differentiated [see, for example, the article by Chapman et al (1989) for a review on asteroid families and Lipschutz et al (1989) for the relation between asteroids and meteorites]. Chondritic breccia material found in collisionally highly evolved meteorite materials with similar spectra are interpreted as representative of an extension to these Hirayama families. The breccias have essentially solar abundances of chemical elements. They are assemblies of chondrules, remnants of shocked interstellar grains that have been partially melted and depleted of volatiles. Repeated collisions compacted and partially fused the material into dense refractory assemblies. The compacted grains are typically a few micrometers across or less.

Given the average composition and high density of chondritic breccia (~ 3.5 g cm^{-3}), they may be modeled as a compact mixture of the same refractories as cometary dust. Average bulk densities of 2.1 ± 0.2 g cm^{-3} for S-type asteroids, given by Standish & Hellings (1989) as a preliminary value based on asteroid perturbation on the orbit of Mars, can allow for 15% by mass of water ice or a $p \approx 0.8$ packing. The average bulk density obtained for C-type asteroids, 1.7 ± 0.5 g cm^{-3}, leaves room for all the volatile ice in the primordial mixture or a packing factor as low as 0.7.

There is additional support for the general correctness of these Chondritic Aggregate models (hereafter CA-models) from chondritic interplanetary dust particles (IDPs) collected in the Earth's stratosphere. IDPs are separated from aerosols of terrestrial origin based primarily on their D/H isotopic anomaly. Most ($\sim 60\%$) of the collected IDPs are chondritic with approximately solar bulk composition of the rock-forming elements and are typically 5 to 50 μm across. Recent advances in sample preparation use ultramicrotomy to slice 3 to 10 μm thin sections of IDPs. Analytical electron microscopes (AEMs) allow the direct study of IDP morphology at resolutions approaching 0.01 μm (Bradley 1991). AEM studies confirm that chondritic IDPs can be divided into the olivine, pyroxene, and layer silicate groups proposed by Sanford & Walker (1985) based on infrared measurements, and show that all three classes of chondritic IDPs have an aggregate structure.

Pyroxene IDPs (Figure 2) are highly porous with a packing factor and overall structure approaching that of the "Bird's-Nest" model. However, the aggregate structure of rocky material prevails down to the resolution limit (Bradley 1991). Aggregation apparently took precedence over nucleation during their formation. If pyroxenes are unmodified "Bird's-Nests," then "classical" interstellar grains are aggregates. These could conceivably form in the turbulent wake of supernova shocks. Bradley

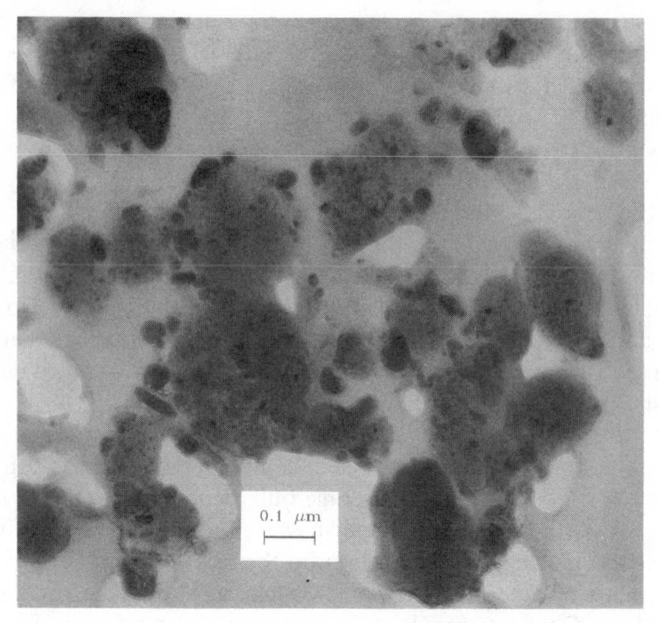

Figure 2 Electron micrograph of an ~80 nm thick section of a pyroxene-rich chondritic aggregate IDP collected in the stratosphere. Reproduced from Bradley (1991), by permission.

showed that only pyroxenes match the Fe to Mg distribution obtained by the PIA and PUMA mass spectrometers at Halley's comet. Analysis of IDP light elements including those in the CHON-particles discovered at Halley (Krueger et al 1991) remains elusive due to use of an epoxy organic polymer substrate to fix the fragile IDP material. The D/H ratio and existence of solar flare tracks show that the pyroxenes were not strongly heated during atmospheric entry. This is believed to be the most primitive class of collected IDP material.

Scanning Electron Microscope (SEM) analysis can distinguish two morphological types of chondritic dust (Flynn & Sutton 1990). Their characteristics are comparable to those inferred for our models. The densities are near 0.6 g cm^{-3}, presumably corresponding to pyroxenes, and 1.9 g cm^{-3}, which is close to the density of C- and S-class asteroids. However, only 15 low density particles and 10 of the higher density were analyzed and strong selection effects during collection could bias the data.

In addition, collected IDPs are a biased sample of the zodiacal dust complex. Love & Brownlee (1991) found that spherical chondritic particles must be smaller than 50 μm in diameter to avoid melting even at 12 km s^{-1}, close to the lowest possible entry velocity at free fall (11.2 km s^{-1}). Most IDPs larger than 20 μm melt at entry velocities of 20 km s^{-1} and evaporate much of their initial material. Since the entry velocity of an average zodiacal particle is thought to be $\sim$30 km s^{-1}, most zodiacal dust particles disintegrate upon atmospheric entry. Samples retrieved from the stratosphere may be both orbit- and material-biased subsets of matter that intersects the Earth's orbit. It is usually assumed that IDPs are biased in favor of asteroidal dust because asteroids are typically on lower inclination, lower eccentricity orbits than comets, which leads to low entry velocities and in addition they may contain tougher material.

Other dust properties may also be biased. Using *IRAS* data, Dermott et al (1993c) found evidence for a configuration of dust trapped in outer resonance with the Earth as suggested from numerical integrations (Gustafson & Misconi 1986). Dust in low eccentricity and inclination orbits is preferentially trapped in these resonances and eventually breaks out of resonance following a close approach to the Earth. Dermott et al suggest that a substantial amount of dust enters the Earth's atmosphere as it breaks away from the resonance and may be a source of low-velocity particles of primarily asteroidal origin. The residence time in the resonances might explain long cosmic-ray exposure age of IDPs and meteorites (Thiel et al 1991) if these ages are confirmed. Major collisions throw particles out of resonance. Only particles that by chance avoid major collisions, and thus have long exposure ages, make it to the close Earth passage with increased probability of becoming IDPs and meteorites.

While the shape of collected chondritic IDPs is complex, they usually are equidimensional although there is some evidence for a shorter dimension, i.e. IDPs may rest flat on the filter (G. J. Flynn, private communication). Particles with elongation ratios greater than 4 would create elongated craters on lunar rocks according to Ashworth (1978), but such craters are not seen in large numbers. In conclusion, dust and most meteoroids are roughly equidimensional with the only exception being the hypothesized evolved comet-mantle meteoroids. Most zodiacal dust has an aggregate structure. The packing varies from $p \approx 0.1$ for comet dust ejected from free-sublimating regions to $p \approx 1$ for compact asteroid chondritic matter. Intermediary packing is expected for some asteroidal matter and comet ejecta from low activity regions. All have approximately solar composition. Although stony and nickel-iron IDPs account for approximately one third of all collected IDPs, this morphology probably corresponds to a smaller fraction of the total asteroid dust population and an even smaller fraction of the total zodiacal dust population. Dust particles and meteroids are usually not homogeneous and spherical.

FORCES

Equations describing motion under solar gravity and planetary perturbations are found in standard celestial mechanics textbooks (i.e. Moulton 1970). Forces due to sunlight or the solar wind also are important on meteoroids and dust particles. Let a linear dimension s define the size of an arbitrary particle. While gravity is proportional to the volume (s^3), pressure forces are in most cases proportional to the surface (s^2), and electromagnetic Lorentz forces to s.

The primary force in each category—solar gravity, sunlight radiation pressure force, and the electric component of the Lorentz force—all are approximately equal on a 10^{-1} μm particle at 1 AU. Second-order gravity and pressure forces—planetary perturbations and Poynting-Robertson light drag—are comparable on particles in the 10^1 to 10^2 μm size range. Both must be accounted for in interpretations of the shape, orientation, and extent of the zodiacal cloud (Gustafson & Misconi 1986, Dermott et al 1992). Only the Lorentz force can usually be safely ignored on micrometer sized or larger zodiacal particles, except outside Jupiter's orbit where any zodiacal light is weak and particle number densities are low.

Solar gravitational attraction dominates on interplanetary particles with dimensions larger than ~ 1 μm. It is often practical to treat other forces as perturbations on the central force field from a solar point mass M acting on the point mass m at heliocentric distance r,

$$\mathbf{F}_g = -\frac{GMm}{r^2}\hat{\mathbf{r}}, \tag{1}$$

where G is the gravitational constant and $\hat{\mathbf{r}}$ the unit heliocentric radius vector.

Radiation Forces

The force due to sunlight radiation pressure is usually the second strongest. Insolation on a stationary surface area, g, perpendicular to the solar direction produces the force

$$\mathbf{F}_r = \frac{S_o g Q_{pr}}{r^2 c}\hat{\mathbf{r}}, \tag{2}$$

where S_o is the solar constant or radiation flux density at unit distance, c is the velocity of light, and Q_{pr} is the efficiency factor for radiation pressure (van de Hulst 1957) weighted by the solar spectrum. Radiation pressure varies with heliocentric distance as the flux density of sunlight, which has an inverse-square dependence.

Because the gravitational attraction to the sun also has an inverse-square dependance on the heliocentric distance, it is customary to eliminate r by introducing the dimensionless quantity

$$\beta = -\frac{F_r}{F_g} = \frac{(S_o/r^2)(Q_{pr}/c)}{GM/r^2} \cdot \frac{g}{m} = C_r Q_{pr}(g/m), \tag{3}$$

where $C_r = 7.6 \times 10^{-5}$ g cm^{-2}. When the geometric cross-section-to-mass ratio of a density ρ (g cm^{-3}) and radius s (cm) sphere, $g/m = 3/4s\rho$, is substituted in Equation (3), it is equivalent to the expression given by Burns et al (1979).

The sun radiates nearly all of its energy in a narrow wave band around 0.6 μm so that the transition from geometric optics to Rayleigh scattering takes place in the micrometer size range. The efficiency Q_{pr} is usually calculated from Mie theory for homogeneous spheres (van de Hulst 1957). Figure 3 compares the resulting β with values for CA-structures. The solid

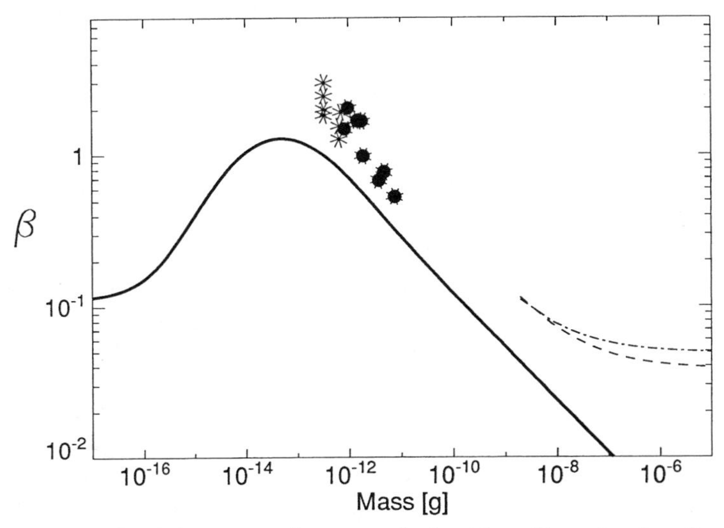

Figure 3 Ratio of radiation pressure force to gravitational force (β) as a function of particle mass. Mie calculations for homogeneous spheres (*solid curve*) consistently underestimates β values for aggregated dust models by a factor 2. Open symbols are for microwave analog measurements using aggregates of the type shown in Figure 1*a* and solid symbols are for those in Figure 1*b* and 1*c*. The dashed curve is for 10 μm thick flakes averaged over random orientation; dash-dot curve represents averages over the most likely spin orientation.

line shows the dependence of β on the mass of spheres using a bulk density $\rho = 2.5$ g cm^{-3} and optical constants for interstellar grain materials as given by Draine & Lee (1984) for their "astronomical silicates." The radiation pressure force is proportional to the geometric cross section at the large particle limit where geometric optics apply. The $\beta \propto s^{-1}$ slope shows that this is a good approximation for spheres above radius $s \approx 10$ μm, corresponding to $m \approx 10^{-11}$ g in the figure. The radiation pressure enhancement in the resonance scattering region below 10^{-11} g is discernible but not prominent in the log-log plot. The curve peaks in the 10^{-14} to 10^{-13} g or micrometer size range where the radiation force exceeds gravity on this particular material and thus the resultant force field is repelling. While β initially drops sharply at smaller sizes, β for natural materials never plummets as it does in some materials in plots by Burns et al (1979) where β for some materials appears to vanish. Instead, β must level off and approach a finite asymptotic value in the Rayleigh scattering region. This is because the efficiency for radiation pressure equals the efficiencies for absorption plus scattering: $Q_{pr} = Q_{abs} + Q_{sca}$. The scattering efficiency is proportional to s^4 and is responsible for the initial decrease. But natural materials have finite absorption and absorption is proportional to volume in Rayleigh scattering so that asymptotically $Q_{pr} = Q_{abs} \propto s$. This makes radiation pressure proportional to the volume (mass) so that β is independent of the size of small particles. The asymptotic dependencies

$$\begin{aligned} &\beta \propto g/m && \text{at large } s \\ &\beta = \text{constant} && \text{at small } s \end{aligned} \tag{4}$$

hold for all materials, particle shapes, and structures as long as only particle size s is varied.

Open symbols in Figure 3 are from microwave analog measurements (Gustafson et al, in preparation) using models like the one shown in Figure 1*a* adopting $\rho = 3.5$ g cm^{-3}. Solid symbols are for the same aggregates coated in a lossy mantle of density 1.8 g cm^{-3} representing an organic refractory compound (Figures 1*b*,*c*). The β values are approximately two times higher than for compact spheres of the same mass. The expected asymptotic behavior for large "Bird's-Nest" structures is that Q_{pr} is independent of size; β then varies as s^{-1} because the dimension is three and not fractal. The dashed curve in Figure 3 is for randomly oriented 10 μm thick cylindrical flakes representing the ejecta from low activity regions on comets. The smallest size corresponds to 10 μm diameter. Surface area-to-mass ratio and β decrease slowly with mass and β soon approaches the large size asymptotic value 0.039. The dash-dot curve is for the same flakes in the most likely nonrandom spin alignment: spin axis perpendicular to the axis of symmetry and the solar direction (see the discussion on alignment below). The same material properties were used in the calculations for flakes as for the spheres. This illustrates that massive particles do not necessarily have small β values [see also the discussion of radiation pressure on nonspherical particles by Gustafson (1989)].

The β values for spherical shapes are the lowest as may be expected from the fact that g/m is at a minimum in this geometry. A widespread use of compact spherical models to represent interplanetary particles systematically underestimates β by a factor close to two, and more for large mantle-fragments. The only exception is at the smallest particle sizes (not shown) where the volume-integrated polarizability alone determines the radiation pressure force. In the Rayleigh scattering region $\beta \approx 0.11$ holds independently of the structure of particles made from "astronomical silicate" as long as the polarizability is isotropic.

Given the particle mass and β, the resulting force field acting on a stationary particle is

$$\mathbf{F} = -\frac{G(1-\beta)Mm}{r^2}\hat{\mathbf{r}}. \tag{5}$$

This can be equated to the gravitational field surrounding a mass $M(1-\beta)$ central star. A nonradial force component usually also is present on an arbitrarily oriented surface, but averages out on a sphere or any other body with rotational symmetry about $\hat{\mathbf{r}}$. In the section on particle spin, we see that random alignment is likely and that the preferred alignment also leads to vanishing nonradial components when averaged over the spin period, so that the form of Equation (5) remains valid.

In the unperturbed central force field, potential energy is constantly exchanged for kinetic energy, total energy and orbital momentum is conserved, and bodies can remain in solar orbit indefinitely. However, interaction with sunlight also generates a drag on moving bodies. Although the drag force is weak compared to the velocity-independent radial pressure force component, it dissipates energy and momentum thereby causing particles to eventually spiral into the Sun.

While there appears to be a consensus about the form of the equation of motion under radiation forces when applied to spherical solar system dust particles, its derivation using a relativistic formulation and the interpretation of the low-order velocity terms in classical (nonrelativistic) physics are still debated (see Klačka 1992 for a recent contribution). The same equation can be used to describe the motion of CA models.

Poynting-Robertson drag is due to the orbital motion around the Sun at velocity v. Solar system orbital velocities remain small compared with the velocity of light and Poynting-Robertson drag can be thought of as arising from an aberration of the sunlight as seen from the particle and a Doppler-shift induced change in momentum. To the first order in v/c, the radiation force acting on a spherical particle is

$$|F_g|\beta[(1-2\dot{r}/c)\hat{\mathbf{r}}-(r\dot{\Theta}/c)\hat{\mathbf{\Theta}}], \tag{6}$$

where the unit vector $\hat{\mathbf{\Theta}}$ is normal to $\hat{\mathbf{r}}$ in the orbital plane (Burns et al 1979, Klačka 1992). The velocity-independent radial term representing force due to radiation pressure is given by Equation (2)—although not a pressure, it is often referred to simply as "radiation pressure." The second term along $\mathbf{r}$ is due to Doppler shift. With the transverse last term, this velocity-dependent part of Equation (6) is the Poynting-Robertson (PR) drag. Doppler shift enters twice: once due to the rate at which energy is received and a second time due to the reradiated and scattered radiation. Because aberration and Doppler shift apply independently of the particle morphology, radiation forces acting on the CA-models also can be expressed through the same formula with the proper β. With the addition of any nonradial pressure terms, Equation (6) applies to arbitrary particles.

PR drag dissipates the orbital angular momentum and energy of meter-size particles and smaller in the inner solar system on time scales of the age of the solar system. A particle with $\beta < 1$ starting in a circular orbit at the heliocentric distance r (AU), which does not experience net transverse pressure terms, spirals into the Sun in 400 r^2/β years. Particles with higher β values escape from the solar system.

Solar Wind Corpuscular Forces

Except for the efficiency factor Q_{pr}, Equation (6) does not depend on the wave nature of light and has also been derived in a corpuscular formulation (Klačka 1992). Corpuscular forces due to collision with solar wind protons are therefore analogous to radiation forces and can be represented by an equation of the same form

$$|F_g|\beta_{sw}[(1-2\dot{r}/v_{sw})\hat{\mathbf{r}}-(r\dot{\Theta}/v_{sw})\hat{\mathbf{\Theta}}], \tag{7}$$

where v_{sw} is the solar wind speed. The ratio of proton pressure to gravity is

$$\beta_{sw} = -\frac{F_{sw}}{F_g} = \frac{(M_{p,o}/r^2)(C_D/2)}{GM/r^2}\cdot\frac{g}{m} = C_p(g/m), \tag{8}$$

where C_D is the free molecular drag coefficient and $C_p \approx 3.6\times10^{-8}\,\mathrm{g\,cm^{-2}}$. The proton momentum flux density at 1 AU, $M_{p,o} \approx 2.15\times10^{-8}\,\mathrm{dyn\,cm^{-2}}$, is nearly the same under fast and slow solar wind conditions. It is one of the most stable solar wind quantities when averaged over a solar rotation (Steinitz & Eyni 1980, Schwenn 1990). It also appears invariant up to 30° heliomagnetic latitudes (Bruno et al 1986), the largest latitudes at which data are available at the time of writing (*Ulysses* will extend these data to higher latitudes). Any heliocentric dependence of the solar wind speed to 1 AU and beyond is neglected in Equation (8) as the possibility of an accelerating flow remains to be settled (Schwenn 1990).

The free molecular flow drag coefficient C_D, due to protons of mass m_p in a Maxwellian velocity distribution of temperature T_p impinging on a sphere from which the fraction ε is specularly reflected, is

$$C_D = \frac{2S^2+1}{\pi^{1/2}S^3}e^{-S^2} + \frac{4S^4+4S^2-1}{2S^4}\operatorname{erf}(S) + \frac{2(1-\varepsilon)\pi^{1/2}}{3S}\left(\frac{T_d}{T_p}\right)^{1/2}\left(1+Y_p\frac{m_s}{m_p}\right), \tag{9}$$

where $S = [m_p/(2kT_p)]^{1/2}|u|$, k is Boltzmann's constant, and u the relative velocity. Equation (9) simplifies considerably as S ranges from approximately 10 under high speed solar wind conditions to 15 at low speed conditions when the plasma is cooler. In this range the error function $\operatorname{erf}(S) = 1$, the first term in Equation (9) is negligible, and the second term is close to 2. The third term includes the contribution from reemitted protons at an average energy given by the dust temperature T_d. The term $Y_p(m_s/m_p)$ was introduced by Mukai & Yamamoto (1982) to account for sputtering of molecules of mass m_s at the yield Y_p. No sputtering corresponds to $Y_p = 0$ and results in $C_d = 2$ when proton velocity dispersions are neglected. This corresponds to complete momentum transfer from all protons that hit the dust particle. At solar wind proton temperatures, this remains a good approximation for most zodiacal dust particles. Even at the high dust temperatures reached around 0.1 AU from the Sun, the efficiency grows by less than 1% when reflections are specular and 1.5% when diffuse. Sputtering of magnetite increases C_D by less than 2.5% while the effect is even smaller on obsidian. These figures are obtained by inserting the same material parameters as Mukai & Yamamoto (1982) in Equation (9). Although water ice leads to slightly higher values, we shall see that volatile ice is not an important dust material inside the asteroid belt where nearly all the zodiacal light is produced.

The pressure due to other solar wind ions can be similarly calculated, but their contribution is less important. Mukai & Yamamoto (1982) showed that the term due to sputtering by helium nuclei is an exception and can increase the effective C_D by 35% for magnetite and 5% for obsidian. It can be shown that randomly oriented particles of any convex shape have the same averge C_D as a sphere by realizing that the distribution of orientation of each surface element is the same as surface elements on a sphere when averaged over all orientations (van de Hulst 1957). Concave and aggregate particles also lead to C_D values close to 2 as multiple scattering leads to more complete transfer of momentum.

The ratio of proton to light pressure is $(C_p \cdot C_D)/(C_r \cdot Q_{pr}) \approx 3.6 \times 10^{-8}/7.6 \times 10^{-5} \approx 4.7 \times 10^{-4}$ (assuming $Q_{pr} = 1$ and $C_D = 2$), thus solar wind corpuscular pressure is negligible compared to sunlight radiation pressure. However, the drag ratio is a factor c/v_{sw} larger because of the greater aberration angle and Doppler shift for protons than for light. The largest variations in the proton drag are caused by the solar wind speed and its dependence on heliocentric latitude and solar distance, not C_D. Average values during conditions of slow wind (<400 km s^{-1}) measured from *Helios 1* and *2* and *IMP 7/8* (Schwenn 1990) give the ratios 0.41 (*Helios*) and 0.43 (*IMP*) decreasing to 0.21 (*Helios*) and 0.20 (*IMP*) in the fast wind (>600 km s^{-1}). Average values over the missions give 0.29 (*Helios*) and 0.30 (*IMP*). *IMP* made measurements at 1 AU while the *Helios* data are normalized to 1 AU. At 0.3 AU *Helios 1* data indicate a 10.5% slower wind speed and *Helios 2* wind speeds were slower by 4.4% indicating that the proton drag to PR drag ratio may increase close to the Sun. The dependence on the latitude angle Θ above the current sheet in 1974 and 1985 is given in Figure 4 based on the velocity dependence derived by Kojima & Kakinuma (1990) using interplanetary scintillation

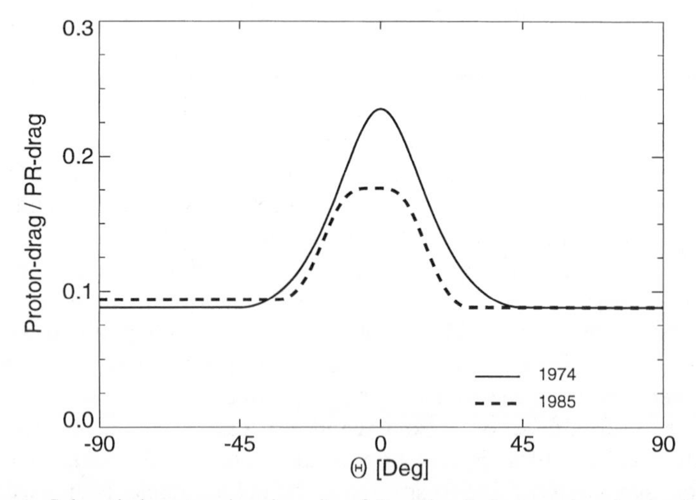

Figure 4 Solar wind proton drag in units of Poynting-Robertson drag in circular orbit plotted as a function of angular distance from the current sheet. The ratio is based on a constant proton momentum flux and interplanetary scintillation measurements (Kojima & Kakinuma 1990) of the solar wind speed in 1974 (*solid curve*) and 1985 (*dashed curve*).

data (IPS). High wind speed and low proton drag is typical at high latitudes.

Mukai & Yamamoto (1982) accounted for an expected net west flow of the solar wind based primarily on theoretical work by Weber & Davis (1967). Observers have been trying to detect this nonradial flow but the results are controversial and a radial flow remains a good approximation when averages are made over a solar rotation (Mariani & Neubauer 1990). The best estimates from *Helios* data is a 0.4° deviation—1/4 of the predicted value (Schwenn 1990). The resulting west pressure component is 0.7% of the radial proton pressure while proton drag is nominally of the order of 5% to 10% at 1 AU. The west pressure force is smaller than the uncertainty in the drag and can usually be neglected.

Solar Wind Lorentz Force

Particles in interplanetary space are charged and therefore couple to the interplanetary magnetic field moving with the solar wind. Photo-emission of electrons from the absorption of solar ultraviolet radiation dominates under normal conditions over the sticking of solar wind electrons so that dust particles have a positive potential $U \approx 5$ V (Goertz 1989). This corresponds to a $Q = 4\pi\varepsilon_o Us$ charge on a spherical particle where $\varepsilon_o = 8.859 \times 10^{-12}$ C V m^{-1} is the permeability of vacuum. The interplanetary magnetic field exerts a force

$$\mathbf{F}_L = Q\mathbf{v} \times \mathbf{B} \tag{10}$$

on the charge, where the velocity $\mathbf{v}$ is relative to the field. It is customary to write Equation (10) in terms of components due to a particle's heliocentric velocity $\mathbf{v}_g$ and the solar wind velocity $\mathbf{v}_{sw}$;

$$\mathbf{F}_L = Q(\mathbf{v}_g \times \mathbf{B} + \mathbf{v}_{sw} \times \mathbf{B}). \tag{11}$$

The second term is independent of the particle motion and can be viewed as resulting from an induced electric field. Equation (11) is then equivalent to Lorentz's equation and the force is referred to as the Lorentz force. The electric component is proportional to the azimuthal component of the magnetic field and always acts along the latitude or $\hat{\mathbf{\Theta}}$-direction in Parker's (1958) model of the interplanetary magnetic field. The electric component dominates everywhere, except at the poles where the azimuthal magnetic field vanishes. It is noteworthy that the electrical term is independent of solar wind speed. The Archimedean spiral becomes steeper with increasing wind speed, which decreases the azimuthal magnetic field component and keeps $\mathbf{v}_{sw} \times \mathbf{B}$ constant (Gustafson & Misconi 1979); wind speed variations have negligible effects on the Lorentz force.

Parker (1958) based his model on an expanding solar corona where the magnetic field from sources on the Sun is "frozen" into the radially expanding solar wind plasma and drawn into an Archimedean spiral as the Sun rotates. At $r_o = 1$ AU, the average radial and azimuthal magnetic field components are $B_{r,o} \approx B_{\Phi,o} \approx 3$ nT, while the normal component $B_\Theta \approx 0$. At the heliocentric distance r and latitude Θ from the solar equatorial plane, the fields $B_r = B_{r,o} r^{-2}$, $B_\Phi = B_{\Phi,o} \cos(\Phi) r^{-1}$, $B_\Theta = 0$, are obtained assuming a constant solar wind velocity. The $\hat{\Theta}$-component vanishes assuming radial expansion and a radial field at the source. Parker's model is a good approximation of the time-averaged observed magnetic field (Mariani & Neubauer 1990).

Over a solar rotation, both the radial and azimuthal components may partially or fully average out in the interplanetary magnetic field sector structure, depending on latitude and the phase of the solar cycle. In the "ballerina model" by Alfvén (1977), a current sheet separates plasma from either hemisphere carrying fields of opposite polarity. A warped or tilted current sheet rotating with the Sun allows plasma from alternating hemispheres to reach the solar equatorial plane. Around solar activity minima the current sheet is nearly flat and aligned with the equator. *Pioneer 11* detected almost no inward-directed field lines over a whole solar rotation as it reached 16° northern heliographic latitudes in early 1976 (Smith et al 1978). This configuration is followed by a phase around the solar maximum when the heliopolar magnetic field undergoes polarity reversal. According to Saito (1988) the current sheet may be pictured as passing the pole and flipping during the reversal so that field directions are reversed at the next solar minimum as observed. Between these phases, a pseudo-aligned phase occurs when the magnetic quadrupole component is 10 to 20% of the strength of the dipole component and the current sheet is highly warped. This is followed by the excursion phase in which the quadrupole component diminishes and the tilt of the current sheet is intermediate between the reversing phase and the aligned phase. In calculating the Lorentz force, a simplified model may be used to describe the magnetic field strength averaged over a solar rotation. In this approximation, the current sheet changes its tilt angle ε at a constant angular velocity of $(1/11)\,\pi$ per year so that the averaged field at latitude Θ is a fraction $2 \arcsin(\tan\Theta/\tan\varepsilon)/\pi$ of the unipolar field.

Because solar wind velocities are high compared to orbital velocities, particle motion relative to the magnetic field is always in the solar direction. Positively charged dust particles always accelerate north when the magnetic field is directed outward and south when it is inward because the magnetic field Archimedean spiral curves in the direction opposite the solar rotation. The dominant electric part in Equation (11) is proportional to B_Φ and therefore to r^{-1}. This can make the instantaneous Lorentz force particularly important on small grains at large heliocentric distances. The electrical component controls the flow of sub-micrometer interstellar dust grains through the heliopause and modulates the penetration deep into the solar system as a function of the solar cycle (Gustafson & Misconi 1979, Grün et al 1993b). As the Lorentz force acts on charges, the particle morphology enters only in the charging process. Charge measurements on individual interplanetary dust particles attempted from *Helios* indicated either extremely low particle densities or surface potentials of the order of 100 V (Leinert & Grün 1990) although $\sim$10 V is often assumed. More work is needed to understand the charge on aggregate particles.

Forces on Rotating Bodies

An additional set of forces arises on rotating bodies but their magnitude and even direction is controversial because these depend on several material properties and the state of rotation. Controversy is compounded by a range of competing effects that can cause spin and lack of direct evidence of the spin state. Some periodic comet nuclei, however, exhibit significant nongravitational forces and are known to have spin. The magnitude of the resulting accelerations is a complex function of the perturbed body's properties (Marsden 1976, Rickman et al 1991). As volatiles sublimate creating the coma and tails, a reaction force results. The nucleus may gain or lose orbital energy depending on its state of rotation as the evening side is warmer and undergoes heavier sublimation than the morning side. The situation is complicated by mantling and the development of active and inactive areas (Rickman et al 1991). Volatile dust particles leaving a comet should be similarly affected but only for a short time as volatiles soon are exhausted and their "jet engines" run out of fuel. Volatiles on dust particles might cause significant random motion and destroy dynamical information on their origin. While sublimation can have a large momentary effect, it will not act for long on small particles. Consequences of this phenomenon remain unexplored in the context of dust and meteoroids.

A related effect is due to asymmetry of the much weaker reaction force from thermal emission and may have a larger effect on rotating meteoroids and dust particles in the long run. The Yarkovsky effect is due to a hot evening hemisphere radiating more thermal energy than the cooler morning side. Like the sublimation force, this effect is hard to estimate because of its dependence on a detailed thermal dust model. However, the force on a large homogeneous sphere can be estimated following Burns et al (1979) by assuming that all energy is deposited at the surface, neglecting heat conduction to the dark side, and averaging the temperature over each hemisphere. Burns et al assumed that this "large particle" approximation applies whenever the surface layer in which the temperature varies by a factor e^{-1} is thinner than the particle radius s. The force decreases rapidly with decreasing size below this limit and is soon negligible.

In this approximation, the Yarkovsky force F_Y on a particle with heat capacity C and thermal conductivity K is

$$\frac{F_Y}{F_{PR}} = \frac{W}{\sqrt{K\rho C}} \cdot \frac{\sqrt{P}}{r} \sin(\xi) \tag{12}$$

when expressed in units of the Poynting-Robertson drag in circular orbit F_{PR}. The spin of period P is about an axis making the angle ξ to the solar direction and the coefficient W is 1.31×10^7 ergs AU sec^{-1} K^{-1} cm^{-2} (Burns et al 1979). Efficiencies for absorption, emission, and Q_{pr} averaged over all wavelengths were assumed to be unity. The possibility of $\xi \neq 90°$ was not considered by Burns et al so the factor $\sin \xi$ is missing in their formula. The component along the orbit pole is $F_Y \sin \zeta$ and that opposing PR drag $F_Y \cos \zeta$, where ζ is the spin axis angle to the orbit pole.

Figure 5 shows a range of possible values of F_Y/F_{PR} as a function of

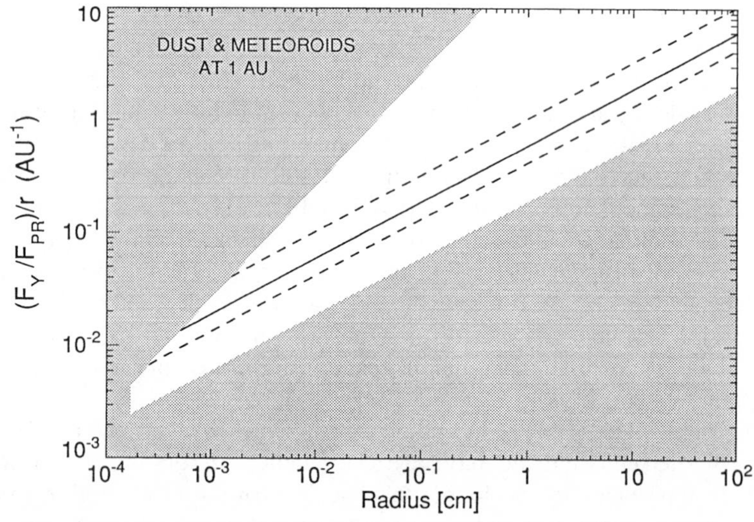

Figure 5 The Yarkovsky force to Poynting-Robertson drag ratio scales with heliocentric distance as r^{-1} and is given as a function of the radius of a spinning sphere. Collision experiments (Fujiwara 1987) gave average fragment spin rates that scale along the solid line. Dashed lines indicate the approximate range in spin rate. Thermal conductivity prevents particles from entering the shaded area to the left where the Yarkovsky force plummets in the diagram. Particles of a given spin rate are along horizontal lines and burst from internal stress before reaching the boundary to the shaded area on the right, which corresponds to the tensile strength of Basalt.

radius for a meteoroid or dust particle at $r = 1$ AU using $C = 10^{-7}$ ergs K^{-1} g^{-1}, $K = 3.5 \times 10^5$ ergs K^{-1} cm^{-1} sec^{-1}, and $\rho = 2.5$ g cm^{-3}. Since F_Y is proportional to the geometric cross section in this approximation as is F_{PR}, a given spin corresponds to a horizontal line in the figure. Thermal conductivity prevents particles from entering the shaded area in the upper left where the "large particle" approximation breaks down and the Yarkovsky force plummets. Centrifugal force-induced nonhydrostatic stresses vary with the size and spin as $\rho s^2(2\pi/P)^2$ (Weidenschilling 1981). Stony spheres cannot survive in the lower shaded area where stresses exceed the tensile strength of basalt (3×10^7 dyn cm^{-2}) (Fujiwara 1987) and particles burst. The incident radiation is attenuated by a factor e^{-1} at the depth $1/\gamma \approx 1.7$ μm, where $\gamma = 4\pi n'/\lambda$ is the absorption coefficient and the imaginary part of the refractive index $n' \approx 0.0294$ is for "astronomical silicate" (Draine & Lee 1984) at the wavelength $\lambda = 0.63$ μm where the solar spectrum peaks. The "large particle" approximation fails below $s \approx 1.7$ μm radius when solar radiation penetrates and deposits heat at a depth comparable to the particle size.

The region between dashed lines in Figure 5 corresponds to rotation periods for fragments from a basalt body. Fujiwara (1987) was able to record the rotation state of fragments produced in a central impact of a 0.37 g polycarbonate projectile on a 367 g basalt sphere at 2.5 km s^{-1}. These sizes correspond to fireballs although the experiment was primarily intended as a scale simulation of asteroid collisions and scales as shown by the lines (Fujiwara et al 1989). Experiments reported by Fujiwara et al also show that fragments have axial ratios near $2{:}\sqrt{2}{:}1$. Thus, given the approximations that went into Equation (12), they can be approximated by a sphere of the average radius.

The Yarkovsky force can be directed anywhere in the tangential plane depending on the spin orientation ζ. While the spin axis of fresh collision fragments has preferred alignments with respect to the impact direction (Fujiwara et al 1989), their spin alignment with respect to the orbital plane ζ is assumed to be random. This approximation may break down at high inclinations.

Competing mechanisms tend to align the spin axis of elongated particles. The solar wind tends to align the spin perpendicular to the sun-meteoroid direction, $\xi = 90°$. The condition of flow relative to a tenuous medium of impactors was also used by Gold (1952) in his discussion of the alignment of interstellar dust. Gold showed that prolate grains acquire spin as a result of randomly occurring collisions with gas molecules as the dust drifts through the interstellar gas. While the angular momentum produced by individual collisions can be in any direction, collisions producing angular momentum perpendicular to both the direction of gas flow and the long axis are the most probable. In this process, Gold suggested that interstellar grains preferentially align their spin perpendicular to the gas flow direction and the long dimension. The same reasoning, if applied to impact fragments or to any prolate body, leads to a preferred spin axis perpendicular to the gas flow and the long axis or the axis of symmetry for oblate cylinders (flakes). Solar radiation forces have the same effect as the solar wind (Radzievskii 1952). While it is possible to conceive of particle geometries that tend to acquire spin alignment along the solar direction (Paddack & Rhee 1976), the radial direction changes continually forcing the particles to align on short time scales compared to the orbital period for the process to be effective. Only particles with a spin axis perpendicular to the orbital plane expose the same geometry all along the trajectory. The most probable systematic spin alignment of nonspherical particles is therefore perpendicular to the orbital plane.

The significance of this spin alignment is that the Yarkovsky force either adds to the Poynting-Robertson effect or counters it. Particles with spin parallel to the orbital momentum axis may remain in orbit while those with anti-parallel spin reach the Sun early. Under conditions where radiation-induced spin is significant, a spin axis could be stabilized by the temperature dependence of the albedo. This effect accelerates spin (Antyukh et al 1973) and maintains the axis orientation. Spin is balanced by rotational damping from a counterpart to the Poynting-Robertson drag (Jones 1990) so that a stable spin state may develop. Spin-aligned particles stay in orbit longer on the average, and elongated particles which acquire spin easily could accumulate in the cloud. However, this does not appear to be the case in the zodiacal cloud.

The degree of polarization from a cloud of randomly oriented particles vanishes in the back-scattering direction for symmetry reasons while small spin-aligned particles cause polarization. Polarization is not seen in the anti-solar, or gegenschein, direction (Weinberg 1985, Levasseur-Regourd et al 1991). Small aligned elongated particles evidently do not contribute a major portion of the zodiacal light. *COBE* measured linear polarization at around right angles to the Sun at 1.2, 2.2, and 3.5 μm. The 3.5 μm data are a particularly powerful indicator of spin-alignment (unless contaminated by scattering) as the neutral color of the zodiacal light suggests that the particles are not small and may be less effective polarizers at shorter wavelengths. Emission measurements can be in any direction and the polarization from a particle in the preferred alignment should be along its orbital plane. If no polarization is found, the likely reason is that the time scale for alignment is longer than the mean time between destructive collisions. Alternatively nondestructive collisions may randomize the spin.

Interestingly, either one precludes alignment of larger but otherwise similar particles by the same mechanisms. Like the Poynting-Robertson and solar wind drag times, the time scale for alignment is proportional to s, as it depends on the intercepted momentum per unit time and particle mass which is proportional to the cross section per unit mass. Probabilities for destructive collisions discussed below do not decrease fast enough with size to compensate for the longer time scale.

COLLISIONS

There are just a few dust particls per km^3 near the Earth's orbit. The particle number density decreases approximately inversely with the distance from the Sun and is even more tenuous away from the ecliptic plane. Collisions are not an important destruction mechanism for particles of micrometer size and smaller because their Poynting-Robertson lifetime is short compared to the mean time between catastrophic collisions. Collisions nevertheless greatly affect the evolution of large zodiacal dust particles and meteoroids. Experiments (Fujiwara et al 1989) confirm that at least in the size range of interest, the mass of the smallest impactor to cause a catastrophic collision is proportional to the target mass. The number of impactors N of mass greater than m is usually represented by a power law of the form $N(m) \propto m^{1-q}$ or $N(s) \propto s^{3-3q}$. The power-law representation is adopted for illustrative purposes; more complex size distributions develop under the combined effect of collisions and PR drag (Gustafson et al 1992). The mean time between catastrophic collisions $\tau_{CC}(s)$ is proportional to $N(s)$, while the Poynting-Robertson time $\tau_{PR}(s)$ at sufficiently large s is proportional to s. The ratio $(\tau_{CC}/\tau_{PR}) \propto s^{2-3q}$. Whenever $q > 2/3$, there is a transition to larger particles whose evolution is dominated by collisions rather than the Poynting-Robertson drag and radiation-induced spin. Most natural size distributions have much larger q values. Assuming that there are no external forces, particles collide and break up and eventually reach an equilibrium size distribution at $q = 11/6$, where newly produced fragments replace destroyed particles (Dohnanyi 1978). The transition to the collision-dominated regime is then quite sharp. This is illustrated in a numerical simulation using the empirical interplanetary particle flux derived by Grün et al (1985) to calculate the probability that a body in a 3 AU circular orbit in the ecliptic survives catastrophic collisions and reaches the Earth's orbit intact. Figure 6 shows that nearly all particles larger than 100 μm are destroyed and cannot reach Earth directly from the asteroid belt—unless they are perturbed into eccentric orbits by planetary perturbations (see Wetherill 1985).

Although Figure 6 shows that most 10 μm and smaller particles created

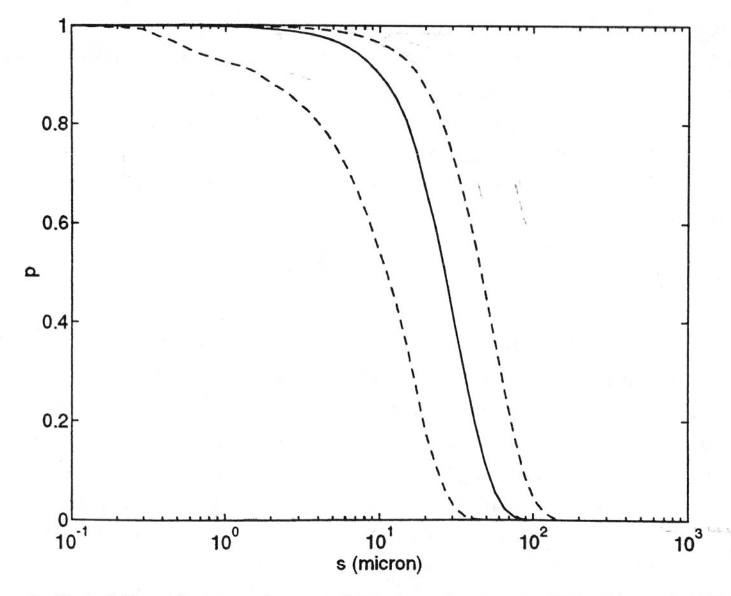

Figure 6 Probability (p) that a dust particle released on a circular orbit at 3 AU, and spiraling toward the Sun under Poynting-Robertson drag, survives catastrophic collisions and reaches the orbit of the Earth at 1 AU. Spheres of 2.5 g cm^{-3} density were exposed to a background of projectiles given by the empirical interplanetary particle flux (Grün et al 1985). Dashed curves indicate the estimated uncertainty.

in the asteroid belt reach the Earth intact, most particles in that size range are collision products from 100 μm or larger particles—unless waves of dust are produced in recent collisions such as modeled by Durda et al 1992. Practically all of the mass starts in the large particles; these break up creating a large amount of debris that also reaches the Earth as 10 μm or smaller fragments. The transition from PR-dominated particle sizes to the collision regime depends on the adopted catastrophic collision criterion, the impact strength, and the mean impact speed. Using reasonable assumptions of solar system conditions, the transition occurs within the 1 to 100 μm size range.

LIGHT SCATTERING

Observed optical properties of the zodiacal light are reviewed elsewhere (Weinberg & Sparrow 1978, Weinberg 1985, Leinert & Grün 1990, Levasseur-Regourd et al 1991). The implied optical properties of zodiacal dust, depending on the assumed particle distribution, are also discussed in these reviews, references therein, and by Lamy & Perrin (1991).

The derivation of an empirical angular scattering function for zodiacal dust from inversion of observed brightness and polarization requires knowledge of the particle number density distribution, although a few observing geometries (Levasseur-Regourd et al 1991) might be less sensitive to the assumptions. Levasseur-Regourd et al give evidence that dust properties vary with location but some general features are invariant: The scattering angle θ ($180^\circ -$phase angle) dependence of the degree of linear polarization above $\theta \approx 70^\circ$ is reminiscent of Rayleigh scattering except that the maximum polarization ($\theta \approx 90^\circ$) is in the 20% to 30% range (or less depending on the distance from the Sun), as opposed to 100%. Another difference is the negative degree of polarization at $\theta > 170^\circ$. This phenomenon is also observed in the scattering from many atmosphereless solar system bodies and in comets. Polarization vanishes at opposition as expected from particles in random orientation. An increased scattering intensity at large scattering angles—the opposition effect—is well established. The opposition effect cannot exclusively be due to a particle enhancement near the Earth, as the enhancement is also seen from deep space craft. Overall, the color of the zodiacal light normalized to the solar spectrum is neutral. The tendency over most of the optical range is toward the red and toward the blue below 2200 Å.

Modeling Angular Distribution

Giese (1973) found that key features of the zodiacal light can be fitted using Mie theory for spheres; this theory is still used by some investigators in the modeling of scattering by zodiacal dust particles. In many cases Mie theory is still the only practical way to integrate over particle sizes and wavelengths in the resonance region. The question addressed here is if Mie theory or any other (relatively) simple scattering theory can reproduce the scattering by CA-structures.

The aggregates shown in Figure 1 are used to test this hypothesis, although they probably are significantly smaller than the dust particles that produce most of the zodiacal light. Figure 3 shows that these models have high β-values and would leave the solar system on hyperbolic trajectories, if released from a large parent in circular orbit. A laboratory is under construction at the University of Florida to simulate and measure the scattering by larger aggregates.

Figure 7*a* shows the angular distribution of scattered optical light intensity (brightness) from an aggregate of 250 silicate spheres (Figure 1*a*) from microwave analog measurements averaged over azimuth rotation (Zerull et al 1993). Also shown are scattering functions based on some simple

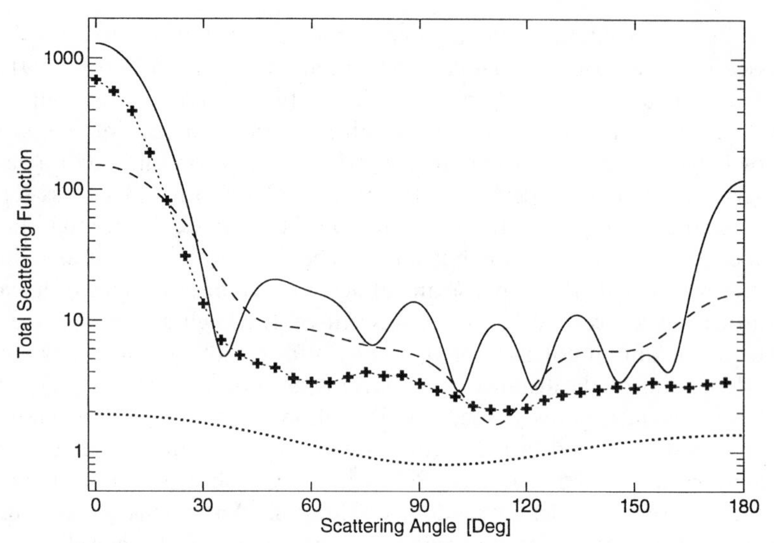

Figure 7 (*a*) Measured angular dependence of total scattered intensity from a 250-sphere aggregate without a mantle (*crosses*) and averaged over azimuthal rotation about three axes (from Zerull et al 1993). (*Solid curve*) Scattering by a sphere of the same material with a geometric cross section equal to the average cross section of the aggregate. (*Dashed curve*) Scattering by a sphere of volume equal to that of all the material in the aggregate. Independent and incoherent scattering by "unit" spheres is given by the dotted curve. None of these curves satisfactorily reproduces the measurements.

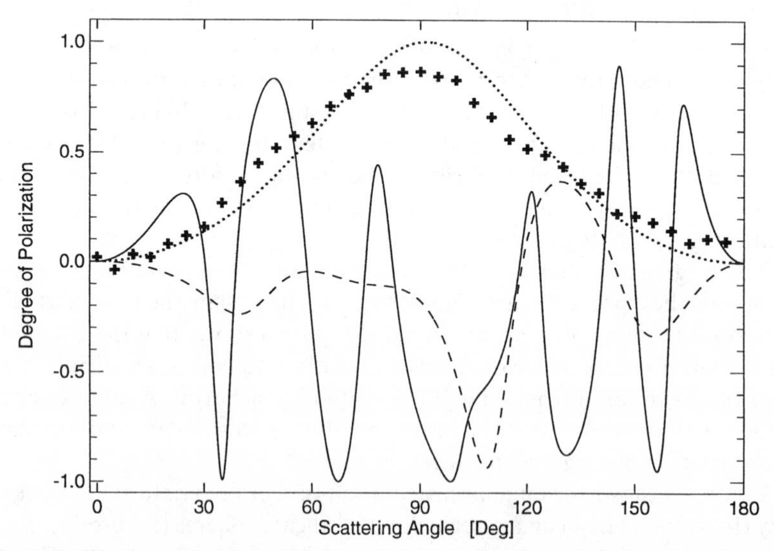

Figure 7 (*b*) Measured degree of linear polarization of radiation scattered from the 250-sphere aggregate as a function of scattering angle (*crosses*) averaged over all available orientations (rotation about three axes) from Zerull et al (1993). The polarization by independent "unit" spheres (*dotted curve*) is a good approximation at all angles, supporting the coherent scattering interpretation. The calculated polarization from both the equal cross section sphere (*solid curve*) and the equal volume sphere (*dashed curve*) are poor approximations.

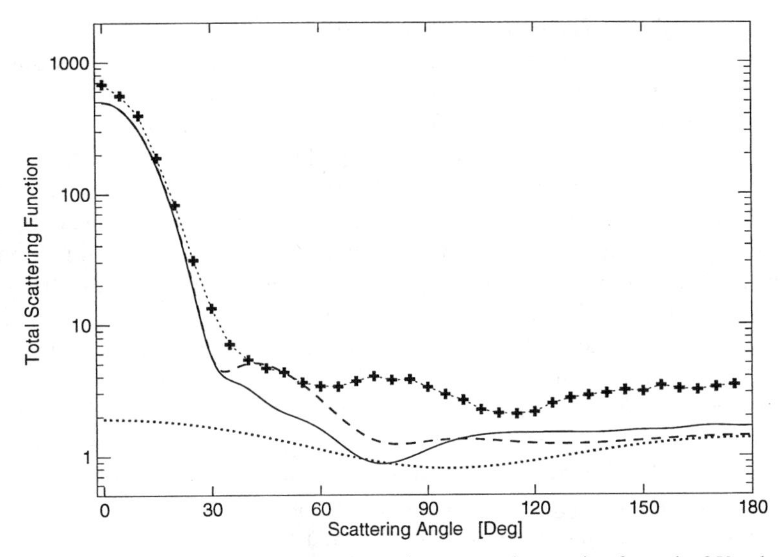

Figure 7 (*c*) Coherent scattering reproduces the measured scattering from the 250-sphere aggregate (*crosses*) better than any of the other approximations. Coherent scattering was computed from a computer reconstruction of the coordinates of each sphere within the aggregate from photographs. Coordinates were obtained for 249 of the 250 particles. (*Solid curve*) Average over rotation about the same axis of rotation as the measurements. (*Dashed curve*) Average over random orientations. (*Dotted curve*) Incoherent scattering. Reproduced from Zerull et al (1993).

model calculations. The dotted curve is for independent scattering—as the 250 spheres scatter light when dispersed in a tenuous cloud, their scattering adds incoherently at random phase. The dashed curve is for a single compact sphere in which all the material is consolidated. Its volume is the same as the compound volume of the 250 spheres in the aggregate without interstices or pores. The solid curve is for a larger sphere with geometric cross section equal to the average cross section of the aggregate. Both comparison spheres have the same index of refraction ($n = 1.735 - 0.007i$) as the individual spheres making up the aggregate.

None of these Mie calculations are satisfactory approximations. However, the Rayleigh-like degree of polarization from independent scattering by "unit" spheres is close to the measured values (Figure 7*b*) while the other calculations fail to reproduce the measurements. The dotted curve also represents polarization in the "coherent scattering" approximation. As used in this article, "coherent scattering" closely follows the Rayleigh-Gans scattering formulation (van de Hulst 1957) but uses summations over discrete scattering centers instead of integrals over a continuous medium. The phase shift suffered by contributions from different parts of the aggregate depends on the optical path length, which is assumed to be independent of the direction of polarization. The coherent scattering solution (solid curve in Figure 7*c*) is also the best approximation to the scattered intensity. Light contributed from all parts of the aggregate is in phase at $\theta = 0$ (forward scattering) but phase differences increase with increasing scattering angle. Phase coherence creates the forward scattering peak by constructive interference. Partial coherence at other directions and coupling between the units in the aggregate lead to increased scattering intensity over the incoherent scattering from a dispersed cloud (dotted curve). The angular scattering can be divided into three zones. Zone I is the forward scattering peak before the first minimum, where constructive interference dominates. The angular extent of this zone reflects the overall dimensions of the aggregate. This zone usually precedes a set of oscillations in zone II that contain information about the detailed distribution of matter within the aggregate. Oscillations dampen at higher scattering angles as phases randomize and any information about the internal structure other than from the increase in scattering intensity due to coupling appears to be lost in zone III which extends to $\theta = 180°$. The zones systematically shift to smaller angles as aggregates grow larger.

The dashed curve is the coherent scattering approximation using an analytic expression to average over orientations and happens by chance to fall between the solid curve and the measurements over much of zone II. The difference between the solid and dashed curves illustrates that averages over one azimuthal rotation (rotation axis perpendicular to the scattering plane) are expected to deviate from averages taken over uniform distributions of orientation, used to simulate randomness. Deviations can be particularly large in zone II. The enhanced measured scattering over the coherent scattering in most directions can largely be explained in terms of coupling between spheres (dependent scattering). The measurements can be reproduced within their uncertainties using a version of the Purcell & Pennypacker (1973) coupled dipole method (Gustafson et al, in preparation).

The coherent scattering approximation works best for aggregates without a coating while an "equivalent spheres" approximation based on the geometric cross section of the aggregate and an average index of refraction works better for coated aggregates. The geometric cross section is obtained by digitizing a photograph using a black and white scanner and counting the number of exposed pixels. This is repeated at twelve orientations. In the "equivalent spheres" approximation, scattering at each orientation is approximated using a sphere of equal geometric cross section and of an index of refraction intended to simulate even distribution of the material over the volume of the sphere (Bruggeman 1935). Any equal cross section sphere approximates the shape of the forward scattering cone where most of the scattered intensity is confined. With the use of an averaged index of refraction, the magnitude also comes close to measured values. The "equivalent spheres" approximation in Figure 8 is the averaged scattering by the set of twelve spheres corresponding to the orientations at which the geometric cross section is known. Zerull et al (1993) show that the approximation grossly underestimates scattering outside the diffraction cone (zone I) by the tenuous aggregates without a mantle.

The most striking effect of applying the $n \approx 1.86 - 0.12i$ absorbing coat is that coherent scattering using core-mantle Mie theory for individual scattering centers grossly overestimates the magnitude of the scattering while the set of "equivalent spheres" becomes a better approximation to the brightness (Figure 8). Mie theory for homogeneous spheres might be a decent approximation for the scattering by aggregate structures when applied with great care, but usually is not. Coherent scattering usually is a decent approximation for the angular scattering by tenuous CA, while the equivalent spheres approximation appears promising for use with compact aggregates.

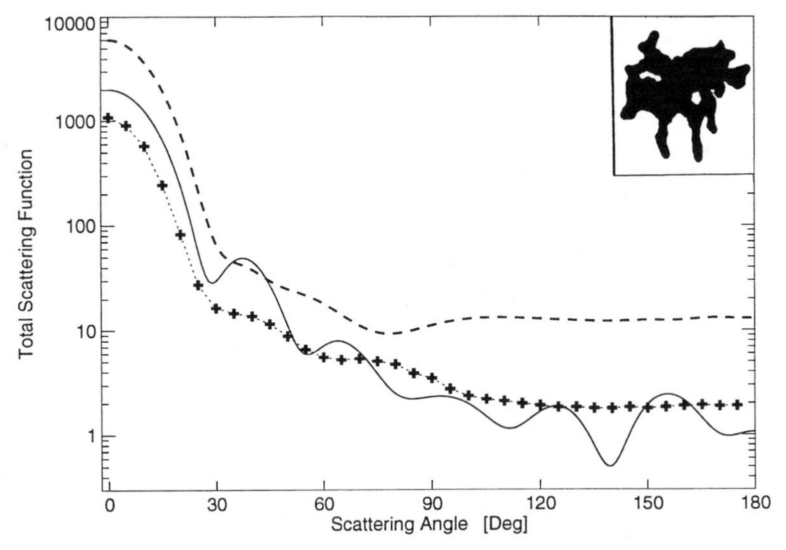

Figure 8 The measured total scattering function (brightness) for the 250-sphere aggregate (*insert*) coated by an absorbing mantle (*crosses*) compared to the "equivalent spheres" approximation (*solid curve*) and to coherent scattering (*dashed curve*). The set of "equivalent spheres" becomes a better approximation as a mantle is added.

Optical and Infrared Cross Section Efficiencies

The efficiency factor for scattering Q_{sca} is obtained by integration over the angular distribution discussed above. Particular care should be taken to fit the forward scattering cone, zone I, where most of the intensity is confined. The radiation pressure efficiency Q_{pr} was discussed in the section on radiation forces. The extinction efficiency is $Q_{ext} = Q_{sca} + Q_{abs}$, where the absorption efficiency Q_{abs} is discussed below.

The asymptotic solution for large spheres with finite absorptivity can be used to calculate $Q_{abs} = 1 - w$ for large convex particles in random orientation from the albedo w (van de Hulst 1957). That the solution for spheres is also valid for large randomly oriented objects of any concave shape is realized by considering that the sum of efficiencies for absorption and scattering by large spheres equals two. This is known as the extinction paradox (van de Hulst 1957). Half the energy goes into diffraction and in practice all refracted light is absorbed, i.e. any scattered light that is not diffracted comes from reflection and $Q_{abs} + Q_{ref} = 1$ where Q_{ref} is the efficiency for reflection. But, van de Hulst (1957, Section 8.42) showed that the scattering pattern caused by reflection from large, randomly oriented, convex particles is identical with the scattering pattern caused by reflection on a large sphere of the same material and surface condition. Logically, Q_{ref} is also the same for any other convex particle satisfying the same conditions. Therefore $Q_{abs} = 1 - Q_{ref}$ also is the same. In particular, Q_{abs} is the same for large randomly oriented plates, cylinders, ellipsoids, spheres, or any large convex particle of a given material.

Figure 9*a* illustrates the wavelength dependence of absorptivity Q_{abs} (=emissivity) for the interstellar grain material "astronomical silicate" (Draine & Lee 1984) at varying packing factors of concave randomly oriented objects at the asymptotic limit for large particle dimensions. Refractive indexes of the porous material were calculated assuming that matter is finely divided on scales of the wavelength so that the Maxwell-Garnett mixing rule applies. The condition is that the circumference to wavelength ratio of individual inclusions is much smaller than 1 (Maxwell Garnett 1904). This condition is fulfilled at infrared wavelengths when the inclusions are individual interstellar grains. To fulfill this condition at shorter wavelengths the inclusions have to be more finely divided, as expected in solar nebula processed matter. This is also a possible outcome of repeated collisions. The Q_{abs} values can be used to calculate the energy balance on concave bodies that are large compared with the longest wavelength at which they emit appreciable thermal energy. This condition requires particle dimensions larger than approximately $r^{1/2}$ centimeter, where r is in AU, but deviations from the dust temperature calculated

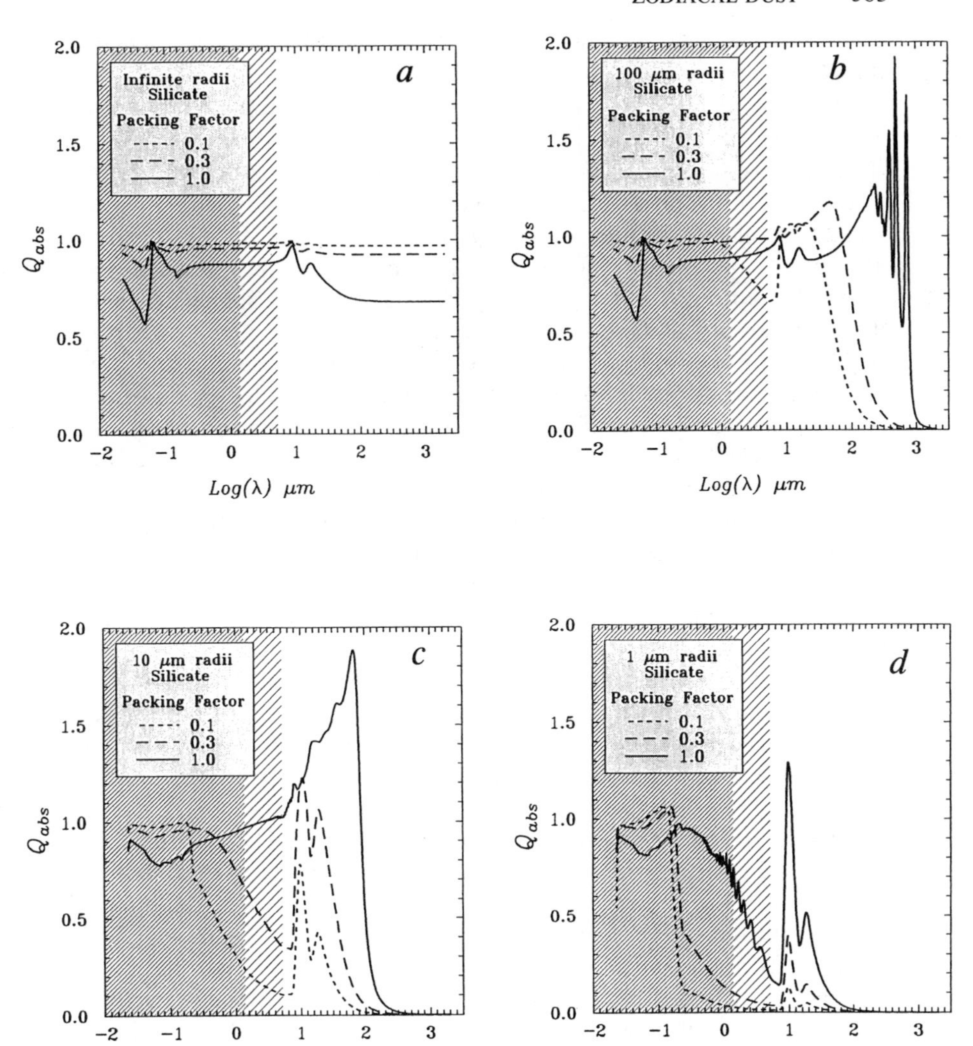

Figure 9 Absorption (and emission) efficiency $Q_{abs} = 1 -$albedo for "astronomical silicate" by Draine & Lee (1984) as a function of wavelength at packing factors 0.1 (*dotted curve*), 0.3 (*dashed curve*), and 1.0 (*solid curve*). The Maxwell-Garnett mixing rule breaks down in the visual range (*shaded*) when unit grains making up the aggregate are $\sim 10^{-1}\mu$m (interstellar grain size) or larger. (*a*) Asymptotic solution for large particle radii. (*b*) For 100 μm radius spheres. The efficiency peaks in a highly structured spectrum beyond 100 μm for compact materials while the porous silicates have low efficiencies. (*c*) For 10 μm radius spheres. The 10 and 20 μm silicate features are seen in the porous materials. (*d*) For 1 μm radius. The silicate feature is seen at all packing factors.

using the asymptotic solution are negligible for particles as small as a tenth this size.

Absorptivity is nearly wavelength independent through the optical region ($\lambda = 0.4$ to 0.7 μm) where the Sun emits most of its energy though there are spectral features in the ultraviolet, and in the 1 to 100 μm region where most of the dust's thermal radiation is emitted. The well-known 10 and 20 μm silicate features are strongest in compact silicate. The drop in absorptivity at longer wavelengths practically disappears when the material is porous as more radiation penetrates inside the aggregate where it can be absorbed. Only a few percent escape absorption ($<4\%$) as opposed to over 10% for compact silicate at optical wavelengths and over 30% in the far infrared. Spectral features in the ultraviolet region of the spectrum have no practical importance to the energy balance.

For particle sizes in the resonance region, Q_{abs} is calculated using Mie theory. The absorptivity of 100 μm radii compact interstellar material shown in Figure 9*b* remains practically indistinguishable from the asymptotic solution through the ultraviolet, optical, and the infrared region down to about 10 μm but is dramatically enhanced at larger wavelengths. At the resonance absorption peak near 500 μm, the particles absorb over twice the energy falling upon them. This is impossible in geometric optics. The peak should best be seen in emission from 10 K or cooler particles that emit an appreciable fraction of their thermal radiation in that part of the spectrum. The resonance peak disappears completely for porous silicates for which that whole part of the spectrum is suppressed. A depression also develops in the 1–10 μm region at the lowest packing factor and the region around the 10 and 20 μm stands out as a broad feature. These features get sharper as the particle size decreases further. At 10 μm radius they become prominent as the absorptivity steadily decreases throughout the optical region until the silicate features are reached at 10 μm (Figure 9*c*). The absorptivity in the optical region for $p = 0.1$ aggregates is half that for a compact particle or of a 100 μm radius aggregate of the same packing factor. Compact silicates do not show distinctive silicate features until the particle size decreases further. Figure 9*d* shows the 10 and 20 μm peaks for a 1 μm radius particle, a sharp drop in absorptivity in the ultraviolet, and suppressed infrared peaks for porous grains.

TEMPERATURE

Reported zodiacal dust color temperatures near 1 AU are in the 255 K to 300 K range and decrease with heliocentric distance as $r^{-\nu}$ where $0.32 \lesssim \nu \lesssim 0.36$ (see the review by Hanner 1991 and references therein). Calibration problems with the *IRAS* data make the higher temperature values more likely but we also should keep in mind that these color temperatures are based on emission near the 10 and 20 μm silicate features and could be significantly different from the actual grain temperature. The heliocentric gradient is less sensitive to calibration errors but is subject to uncertainties from the inversion process. If the gradient v really is significantly flatter than the expected 0.5 for large particles and flatter than 0.4 for small particles, this can be a significant constraint on the nature of the dust.

A large area-to-mass ratio allows dust particles and meteoroids to quickly reach their equilibrium temperature in interplanetary space so that the total energy gain equals the net loss. The energy balance of an isothermal body at heliocentric distance r with average cross section g presented to the Sun and total area G, is given by

$$\frac{1}{4\pi r^2} g \int_0^\infty L_\odot(\lambda) Q_{abs}(\lambda)\, d\lambda = G \int_0^\infty B(\lambda, T) Q_{abs}(\lambda)\, d\lambda + GH(T)Z(T), \quad (13)$$

where the left side represents the energy gain from sunlight of power $L_\odot(\lambda)$ at wavelength λ. The efficiency factor for absorption, $Q_{abs}(\lambda)$, is also the efficiency for emissivity as a consequence of Kirchoff's law or time-reversal symmetry. The right side represents thermal emission across the Planck function $B(\lambda, T)$ at the dust particle temperature T and loss to sublimation, where $H(T)$ is the latent heat of evaporation and $Z(T)$ the sublimation flux.

Equation (13) is usually solved for the case of ideal black spheres with $Q_{abs} \equiv 1$, $G = 4g$, and $Z(T) = 0$. The resulting equilibrium "Black-Sphere" temperature in Kelvin at r AU is $T_{B\text{-}S} = 280\ r^{-1/2}$. It is often convenient to write the radiative equilibrium temperature of an arbitrary particle in terms of $T_{B\text{-}S}$ as

$$T = [(4g/G)(\langle Q_{abs}\rangle_{opt}/\langle Q_{abs}\rangle_{ir})]^{1/4} T_{B\text{-}S}, \quad (14)$$

where the average absorption efficiencies $\langle Q_{abs}\rangle_{opt}$ and $\langle Q_{abs}\rangle_{ir}$ are weighted by the Planck function $B(\lambda, T)$ at the solar color temperature and the dust temperature respectively, and $\langle Q_{abs}\rangle_{ir}$ is an average over all orientations.

Large Gray Particles

The equilibrium "Black-Sphere" temperature $T_{B\text{-}S}$ is a good approximation for a variety of particles larger than $\sim 10^4\ \mu$m. It is seen directly in Equation (14) that the black-sphere radiative equilibrium temperature holds for spheres with the same absorption efficiency in the thermal infrared as in optical part of the spectrum. The condition of spherical shape is also unnecessarily restrictive as the average geometric cross section of any convex particle with random orientation is precisely one quarter of the total area. This fact was rediscovered many times and a proof is given by van de Hulst (1957). This is particularly useful at large particle sizes where it was shown above that geometrical optics gives the same Q_{abs} for any convex particle of a given material when averaged over random orientations.

Temperatures of aligned convex particles also remain close to $T_{B\text{-}S}$. The ratio of the average area exposed to sunlight to the emitting area of a circular cylinder of radius s and length d spinning about the most probable axis is

$$\frac{g}{G} = \frac{2s^2 + 4sd/\pi}{2\pi s^2 + 2\pi sd}, \tag{15}$$

where the denominator comes from Equation (5) by Gustafson (1989). Large spinning flakes or plates that are $\sim 10\ \mu m$ thick (so that geometric optics applies) have a temperature excess over spheres of the same material of less than 6%. The equilibrium temperature of a thin flake $T_{B\text{-}S}(4g/G)^{1/4}$ is obtained using $\lim_{d \to 0}(g/G) = 1/\pi$. Thin needles are less than 5% cooler since $\lim_{s \to 0}(g/G) = 2/\pi^2$. Concave particles can be modeled similarly by adopting higher Q_{abs}-values to account for radiation trapping in cavities.

In conclusion, the radiative equilibrium temperature of a large black or gray particle at distance r (AU) is in the approximate range (Kelvin):

$$T_{B\text{-}B} = (280\, r^{-0.5})^{+6\%}_{-5\%}, \tag{16}$$

assuming that the particle is convex and of a material that has the same absorption efficiency in the infrared as in the optical part of the spectrum. The total radiated power is proportional to T^4 and G; the cylinder emits $g/\pi s^2 = 2/\pi + 4/\pi^2\, d/s$ times the power radiated by a sphere of radius s submerged in the same radiation field. This makes the radiative equilibrium temperature nearly independent of shape of randomly oriented large convex particles.

Small Gray Particles

Particles with dimensions $10^{-2}\ \mu m$ or less are in the Rayleigh region over practically the whole solar spectrum and also radiate their thermal energy as Rayleigh particles. In this limit, the absorption efficiency can be written $Q_{abs} = (8\pi^2/g\lambda)\,\mathrm{Re}(i\alpha)$. The polarizability α depends on the index of refraction n and the particle shape. As long as α is isotropic, it is independent of the distribution of matter within the tiny particle and proportional to the volume (van de Hulst 1957). Integration over the Planck function and substitution into Equation (14) gives $T_{B\text{-}R} = (T_{\odot}/T_{B\text{-}R})^{1/4}T_{B\text{-}S}$, where $T_{\odot} = 5785$ K is the solar color temperature. The small size asymptotic temperature is then

$$T_{B\text{-}R} = 513\,r^{-0.4}. \tag{17}$$

The slower temperature decrease with r is due to decreasing average emissivity efficiency as the particle cools. A black or gray Rayleigh particle is 233 K hotter than a large particle at 1 AU. In reality matter is not black or gray, most materials increase in absorptivity in the infrared, and small particles are sensitive to these variations. However, it is instructive to consider material and size effects separately.

Gray Particles of Arbitrary Size

The dust temperature strongly depends on the scattering regime that contributes most to the optical and infrared averaged absorption efficiencies. The dashed curve in Figure 10 illustrates the effect of changing the size of a sphere made of an ideal "gray" material ($n = 1.41 - 0.1i$). As

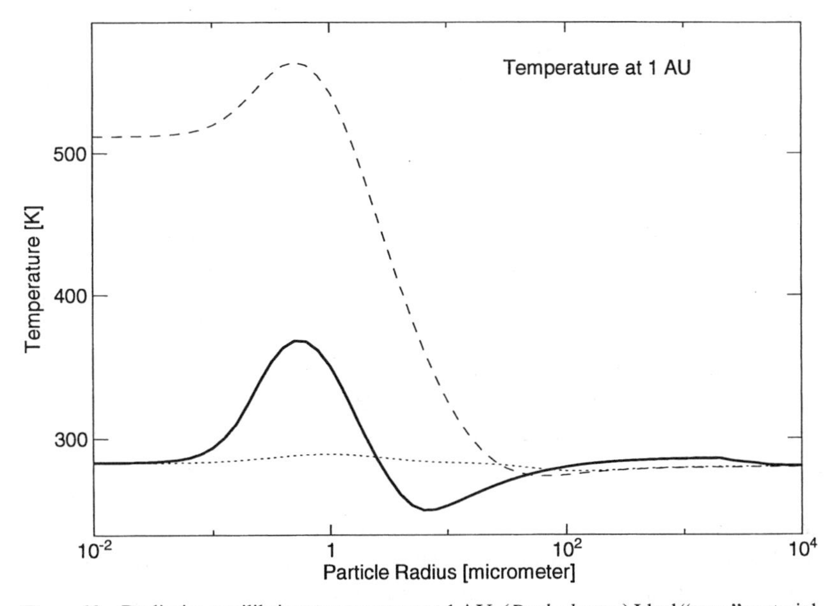

Figure 10 Radiative equilibrium temperature at 1 AU. (*Dashed curve*) Ideal "gray" material, $n = 1.41 - 0.1i$ at all wavelengths. (*Solid curve*) "Astronomical silicate" by Draine & Lee (1984). (*Dotted curve*) A $p = 0.1$ packing of the same material. The heliocentric temperature dependence ν is 0.5 for large particles and 0.4 for small sizes. The temperature difference between large and small particles of a given material is large at most heliocentric distances.

the size of the large sphere decreases, thermal emission enters the resonance region of increased emission causing the dust temperature to drop slightly. The temperature increases as the radius shrinks further and thermal emission enters the Rayleigh region. Peak temperature is reached when the solar spectrum is in the particle's resonance region. At smaller sizes, the solar spectrum also enters the Rayleigh region and the grain reaches its asymptotic Rayleigh temperature. Nonspherical particle shapes lead to qualitatively similar transitions. Equation (17) suggests that small particles in the solar system are always hotter than large particles of a similar material as the crossover point is inside the Sun at 0.5 solar radii. However, natural materials are not black and small particles are not always hotter than large ones of the same material.

Chondritic Aggregate Material

The solid curve in Figure 10 is for a compact sphere using the same material (by Draine & Lee) as used to generate Figure 9. The qualitative features from the ideal black material can be recognized but the net temperature increase towards small sizes is compensated for by increased material-dependent emission in the infrared (the 10 and 20 μm features); this compensates for the Rayleigh $Q_{\mathrm{abs}} \propto \lambda^{-1}$ dependence and brings the asymptotic temperature back to $T_{\mathrm{B\text{-}S}}$. While this qualitative behavior is common to many natural materials it is unusually strong in silicates because of the 10 and 20 μm emission features. The temperature dependence on heliocentric distance is still flatter than $r^{-0.5}$. The size dependence of an ideal "gray" material is approached at large heliocentric distances. The dotted curve is for $p = 0.1$ and the temperature of these porous structures of finely divided matter remains within 10 K of $T_{\mathrm{B\text{-}S}}$. The temperature of "Bird's-Nest" structures is not represented by any of these curves as individual grains in the aggregate are not small compared to the optical wavelengths where most solar energy is emitted. If the aggregate is sufficiently porous, the absorption cross section in the visual range approaches that of the individual grains. However, the aggregate may act as an ensemble at the longer wavelength of thermal emission and porous particles enter the Rayleigh regime at larger dimensions than compact particles, thus the Rayleigh $v = 0.4$ heliocentric temperature dependence may apply. As the porosity increases, grains decouple and the temperature asymptotically approaches that of the individual constituent grains in free space. Figure 10 shows that if this model applies, compact zodiacal particles in the 1 to 100 μm interval should be cooler than $T_{\mathrm{B\text{-}S}}$; sub-micrometer grains should be hotter. Individual or loosely aggregated grains of the size of interstellar grains and many grains found in chondrites are hot. Using similar calculations, Greenberg & Hage (1990) and Hage & Greenberg

(1990) showed that, if the emitting particles in comet Halley's coma can be represented by this kind of model, they must either be smaller than ~ 1 μm or porous with a packing less than $p = 0.2$ to fit the temperature and the emission features at 9.7 and 3.5 μm.

SUBLIMATION

Absorption of sunlight and thermal emission dominates the energy balance of most particles while sublimation can be a large heat sink on dust with volatile materials for a short time. Small particles rapidly run out of volatile coolant because of their high surface-to-volume ratio. The sublimation rate of water ice at temperature T is

$$Z(T) = \frac{b}{\sqrt{T}} e^{-H(T)/RT}, \tag{18}$$

where $b \approx 2 \times 10^{32}$ molecules cm^{-2} sec^{-1} K$^{1/2}$ or 6×10^{9} g cm^{-2} sec^{-1} K$^{1/2}$ of water and R is the gas constant. Division by the ice density $\rho = 0.92$ g cm^{-3} gives the surface recession rate. This rate is very temperature sensitive and pure water ice is nearly transparent at optical wavelengths, i.e. it is a notoriously poor absorber of sunlight while it radiates effectively in the infrared. The resulting temperature is low enough to make water ice meteoroids stable against sublimation throughout much of the outer solar system (Patashnik & Rupprecht 1975). But as little as 1% by volume of the "astronomical silicate" distributed in the ice brings the absorptivity close to unity and the radiative equilibrium temperature close to the black-body temperature. This phenomenon was also pointed out by Hanner (1981) and Lichtenegger & Kömle (1991). Figure 11 shows the estimated surface recession rate of water ice in a permeable isothermal mixture representing large fresh comet dust particles of the "Bird's-Nest" kind. The recession rate is tens of micrometers or more per year throughout the main asteroid belt. Sublimation becomes an efficient coolant inside 2.2 AU where the particle reaches 160 K, but is inconsequential further out.

If water ice fills the interstices in a centimeter-sized "Bird's-Nest" dust particle and sublimates at a controlled rate, a 1% temperature decrease could only be sustained for about a week at 1 AU. This is negligible compared with the dynamical life time. If the ice sublimates freely into vacuum the temperature drop is larger but the ice sublimates in seconds. Figure 12 shows the heliocentric distance where dust reaches 160 K using the same materials as in Figure 9. This is a conservative limit for the existence of water ice on dust particles. A large black sphere reaches $T_{\text{B-S}} =$ 160 K near 3 AU; this is also the asymptotic large size limit both for the

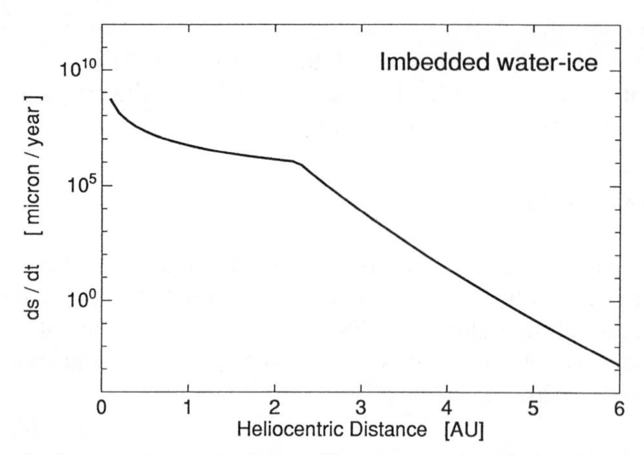

Figure 11 Surface recession rate for freely sublimating water ice imbedded in a large "Bird's-Nest" structure, as a function of heliocentric distance. The refractory materials increase the overall absorptivity so that most of the incident sunlight is absorbed. The equilibrium temperature is dominated by radiative equilibrium outside 2.2 AU where the sublimation rate is too slow to provide efficient cooling. Sublimation rapidly becomes an effective coolant inside this distance.

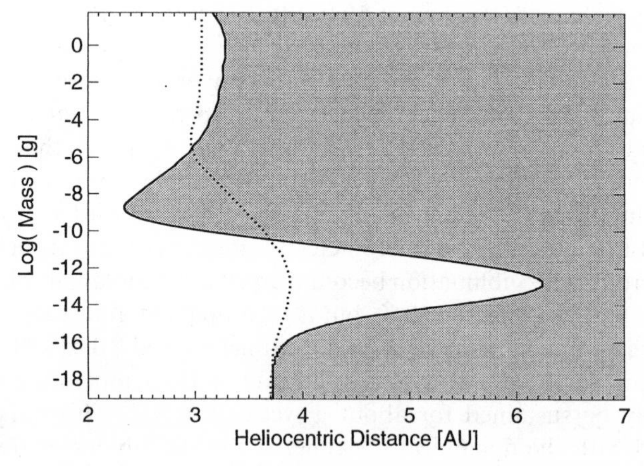

Figure 12 Isotherm of 160 K for dust particles in radiative thermal equilibrium. This is a conservative limit inside which water ice can not exist on dust particles. Interstellar dust grains in the 10^{-13} or 10^{-14}g range are efficient absorbers of sunlight and Rayleigh scatterers in emission so that they become substantially hotter than large particles and should lose their water ice near Jupiter. (*Solid curve*) "Astronomical silicate" by Draine & Lee (1984). (*Dashed curve*) Prous "astronomical silicate" with a packing factor of $p = 0.1$.

compact (*solid curve*) and the porous (*dotted*) models. The curves reflect the same phenomena seen in Figure 10 as thermal emission transits from the geometric optics region at large sizes to the Rayleigh region at the smallest sizes. The Rayleigh region isotherm is clearly shifted outside the large size ice limit due to the $\nu < 0.5$ heliocentric temperature dependence.

The optical constants used in the "compact material" calculations were derived from observed properties of interstellar grains (and inverted using Mie theory; Draine & Lee 1984). Interstellar grains with masses $\sim 5 \times 10^{-13}$ g were recently detected by *Ulysses* (Grün et al 1993). If the local interstellar grains are similar to the ones used to obtain the optical constants, any water ice should be lost near Jupiter's orbit where the grains were detected. Lower isotherms corresponding to more volatile materials have even greater size dependence.

CLOSING REMARKS

At a given wavelength, a dust particle intercepts an amount of energy that is proportional to $gQ_{\text{ext}}r^{-2}$. Each particle absorbs a fraction of this energy and emits it as thermal radiation in the infrared; the remaining energy is scattered and dispersed over scattering angles. This gives each dust particle a wide range of options to handle its energy budget. Large white particles scatter most of the intercepted light close to the original direction. If such particles are present, they would hardly be seen unless we look close to the Sun where observations are difficult. Smaller white particles scatter light over a wider angular range and observations away from the Sun are increasingly biased toward smaller particles. Around the polarization maximum near 70° from the Sun, large and intermediate size particles contribute most of the polarized brightness along the scattering plane where small particles cannot scatter, while the small particles may provide most of the perpendicularly polarized radiation. White particles have spent their energy budget in scattering and have little left over for thermal emission. Black particles, on the other hand, absorb sunlight to emit their own spectrum almost isotropically in the infrared. Particles close to the Sun intercept more energy and therefore have a larger budget. We should therefore not expect that the same average dust properties will fit both optical and infrared data, or optical data at different elongation angles from the Sun. Besides these optical biases there are also orbital dynamic biases. Particles experiencing a given drag in low inclination orbits accommodate better to the local precession plane than similar particles in high inclination, slowly precessing orbits. Their symmetry planes therefore are different as is seen in the *IRAS* data (Dermott et al 1992). It is therefore no surprise that the zodiacal dust cloud has the appearance of a heterogeneous mixture of dust with average optical and thermal properties that depend on the position in the solar system. If we learn to use these biases to our advantage, this might allow us to selectively study components of known origin and of some known particle characteristics.

There is evidence of real differences between dust populations other than what can be expected from bias (Levasseur-Regoud et al 1991, Hanner 1991). The challenge is to identify components of the dust cloud and thereby enable the study of interplanetary dust to be put in a proper context. The contribution from six Hirayama families has been identified by Dermott et al (1993b) and separated from the background. Their contribution is $\sim 10\%$ of the total brightness at 25 μm and the total contribution from the main asteroid belt is estimated to be 1/3 of the total and could be larger. Short period comets are seen to eject material into interplanetary space but a large amount of the visible dust immediately leaves the solar system. Radiation pressure elevates dust particles' potential energy over that of the parent body and the total energy (kinetic plus potential) often leads to unbound trajectories. This is particularly common for release near perihelion where dust production normally peaks. However, many meteor streams are associated with comets and as meteoroids gradually grind down, dust-size comet matter may have a larger chance of reaching a bound orbit.

Recent advances in modeling capabilities and comparisons with observations have led to the realization of new methods to distinguish dust based on its origin and average dynamic drag. This may be the breakthrough that will finally allow us to answer the age-old question of the origin and evolution of the interplanetary dust cloud. The issue has been hotly debated at least since the time of the First World War when Fesenkov (1914) suggested that disintegrating periodic comets feed the interplanetary dust cloud. It is my hope that this review of the physics on which the dynamic modeling of the zodiacal dust cloud depends, and of the optical and thermal dust properties that affect the bias, will contribute to this progress.

Acknowledgments

I am grateful to my family, friends, and colleagues, who each in some way influenced or inspired my work. I am particularly indebted to Prof. S. F. Dermott (Univ. of Florida, USA), the late Prof. R. H. Giese (Ruhr Univ., FRG), Prof. J. M. Greenberg (Univ. Leiden, Netherlands), and Prof. E. Grün (Max-Planck-Institut für Kernphysik, FRG). Graduate students L. G. Adolfsson, D. D. Durda, S. Jayaraman, and J. C. Liou reviewed the manuscript. This work was supported by NASA through grants No. NAGW-2775 and NAGW-2482.

Literature Cited

Alfvén H. 1977. Electric currents in cosmic plasmas. *Rev. Geophys. Space Phys.* 15(3): 271–84

Antyukh EV, Braginskii VB, Manukin AB. 1973. Astrophysical consequences of pondermotive effects of electromagnetic radiation. *Sov. Astron. AJ* 16(5): 893–95

Ashworth DG. 1978. Lunar and planetary impact erosion. In *Cosmic Dust*, ed. JAM McDonnell, pp. 427–526. Chichester: Wiley

Binzel RP, Gehrels T, Matthews MS, eds. 1989. *Asteroids II*. Tucson: Univ. Ariz. Press

Bradley JP. 1991. Newly developed techniques for the analysis of micrometer-sized interplanetary dust particles and comet grains. *Space Sci. Rev.* 56: 131–38

Bruggeman DAG. 1935. Berechnung verschiedener physikalischer Konstanten von heterogenen Substanzen 1. Dielektrizitätskonstanten und Leitfähigkeiten der Mischkörper aus isotropen Substanzen. *Ann. Phys. (Leipzig)* 24: 636–79

Bruno R, Villante U, Bavassano B, Schwenn R, Mariani F. 1986. In-situ observations of the latitudinal gradients of the solar wind parameters during 1976 and 1977. *Sol. Phys.* 104: 431–45

Burns JA, Lamy PL, Soter S. 1979. Radiation forces on small particles in the solar system. *Icarus* 40: 1–48

Chapman CR, Paolicchi P, Zappal V, Binzel RP, Bell JF. 1989. Asteroid families: Physical properties and evolution. See Binzel et al 1989, pp. 386–415

Dermott SF, Durda DD, Gustafson BÅS, Jayaraman S, Xu YL, et al. 1993b. *Proc. IAU Symp.* 160 In press

Dermott SF, Durda DD, Gomes RS, Gustafson BÅS, Jayaraman S, et al. 1993a. Origin of the IRAS dustbands. In *Meteoroids and their Parent Bodies*, ed. J Štohl, IP Williams, pp. 357–66. Bratislava: Astron. Inst., Slovak Acad. Sci.

Dermott SF, Gomes RS, Durda DD, Gustafson BÅS, Jayaraman S, et al. 1992. Dynamics of the Zodiacal cloud. In *Chaos, Resonance, and Collective Dynamical Phenomena in the Solar System*, ed. S Ferraz-Mello, pp. 333–47. Dordrecht: Kluwer

Dermott SF, Jayaraman S, Xu YL, Liou JC. 1993c. *IRAS* observations show that the Earth is embedded in a solar ring of asteroidal dust particles in resonant lock with the planet. *Bull. Am. Astron. Assoc.* 25(3): 1116

Dermott SF, Nicholson PD, Burns JA, Houck JR. 1984. Origin of the Solar System dust bands discovered by IRAS. *Nature* 312: 505–9

Dohnanyi JS. 1978. Particle dynamics. In *Cosmic Dust*, ed. JAM McDonnell, pp. 527–605. Chichester: Wiley

Draine BT, Lee HM. 1984. Optical properties of interstellar graphite and silicate grains. *Astrophys. J.* 285: 89–108

Durda DD, Dermott SF, Gustafson BÅS. 1992. Modeling of asteroidal dust production rates. In *Asteroids, Comets, Meteors* 1991, ed. AW Harris, ELG Bowell, pp. 161–64. Houston: Lunar Planet. Inst.

Fessenkov VG. 1914. Sur l'origine de la lumière zodiacale. *Astron. Nachr.* 198(4752): 465–71

Flynn GJ, Sutton SR. 1990. Evidence for a bimodal distribution of cosmic dust densities. *Proc. Lunar Sci. Conf. XXI*, pp. 375–76

Fujiwara A. 1987. Energy partition into translational and rotational motion of fragments in catastrophic disruption by impact: an experiment and asteroid cases. *Icarus* 70: 536–45

Fujiwara A, Cerroni P, Davis D, Ryan E, Di Martino M, et al. 1989. Experiments and scaling laws for catastrophic collisions. See Binzel et al 1989, pp. 240–65

Giese RH. 1973. Optical properties of single-component zodiacal light models. *Planet. Space Sci.* 21: 513–21

Giese RH, Kneissel B, Rittich U. 1986. Three-dimensional zodiacal dust cloud: a comparative study. *Icarus* 68: 395–411

Goertz CK. 1989. Dusty plasmas in the solar system. *Rev. Geophys.* 27: 271–92

Gold T. 1952. The alignment of galactic dust. *Mon. Not. R Astron. Soc.* 112: 215–18

Gradie JC, Chapman CR, Tedesco EF. 1989. Distribution of taxonomic classes and the compositional structure of the asteroid belt. See Binzel et al 1989, pp. 316–35

Greenberg JM. 1988. The interstellar dust model of comets: post Halley. In *Dust in the Universe*, ed. M Bailey, DA Williams, pp. 121–43. Cambridge: Cambridge Univ. Press

Greenberg JM, Gustafson BÅS. 1981. A comet fragment model for Zodiacal light particles. *Astron. Astrophys.* 93: 35–42

Greenberg JM, Hage JI. 1990. From interstellar dust to comets: a unification of observational constraints. *Astrophys. J.* 361: 260–74

Grün E, Gustafson BÅS, Mann I, Baguhl M, Morfill GE, et al 1993b. Interstellar dust in the heliosphere. *Astron. Astrophys.* Submitted

Grün E, Zook HA, Baguhl M, Balogh A, Bame SJ, et al. 1993. Discovery of Jovian dust streams and interstellar grains by the *Ulysses* spacecraft. *Nature* 362: 428–30

Grün E, Zook HA, Fechtig H, Giese RH. 1985. Collisional balance of the meteoritic complex. *Icarus* 62: 244–72

Gustafson BÅS. 1989. Comet ejection and dynamics of nonspherical dust particles and meteoroids. *Astrophys. J.* 337: 945–49

Gustafson BÅS, Misconi NY. 1979. Streaming of interstellar grains in the solar system. *Nature* 282: 276–78

Gustafson BÅS, Misconi NY. 1986. Interplanetary dust dynamics I. Long-term gravitational effects of the inner planets on zodiacal dust. *Icarus* 66: 280–87

Gustafson BÅS, Misconi NY, Rusk ET. 1987. Interplanetary dust dynamics III. Dust released from P/Encke: distribution with respect to the zodiacal cloud. *Icarus* 72: 582–92

Gustafson BÅS, Grün E, Dermott SF, Durda DD. 1992. Collisional and dynamic evolution of dust from the asteroid belt. In *Asteroids, Comets, Meteors* 1991, ed. AW Harris, ELG Bowell, pp. 223–26. Houston: Lunar Planet. Inst.

Hage JI, Greenberg JM. 1990. A model for the optical properties of porous grains. *Astrophys. J.* 361: 251–59

Hanner MS. 1981. On the detectability of icy grains in the comae of comets. *Icarus* 47: 342–50

Hanner MS. 1991. The infrared zodiacal light. In *Origin and Evolution of Interplanetary Dust*, ed. AC Levasseur-Regourd, H Hasegawa, pp. 171–78. Tokyo: Kluwer

Jones W. 1990. Rotational damping of small interplanetary particles. *Mon. Not. R. Astron. Soc.* 247: 257–59

Klačka J. 1992. Poynting-Robertson effect I. Equation of motion. *Earth, Moon and Planets* 59: 41–59

Kojima M, Kakinuma T. 1990. Solar cycle dependence of global distribution of solar wind speed. *Space Sci. Rev.* 53: 173–222

Krueger FR, Korth A, Kissel J. 1991. The organic matter of comet Halley as inferred by joint gas phase and solid phase analyses. *Space Sci. Rev.* 56: 167–75

Lamy PL, Perrin JM. 1991. The optical properties of interplanetary dust. In *Origin and Evolution of Interplanetary Dust*, ed. AC. Levasseur-Regourd, H. Hasegawa, pp. 163–70. Tokyo: Kluwer

Lichtenegger HIM, Kömle NI. 1991. Heating and evaporation of icy particles in the vicinity of comets. *Icarus* 90: 319–25

Leinert C, Grün E. 1990. Interplanetary dust. In *Physics of the Inner Heliosphere I*, ed. R Schwenn, E Marsch, 207–75. Heidelberg: Springer-Verlag

Levasseur-Regourd AC, Renard JB, Dumont R. 1991. The zodiacal cloud complex. In *Origin and Evolution of Interplanetary Dust*, ed. AC Levasseur-Regourd, H Hasegawa, pp. 131–38. Tokyo: Kluwer

Lipschutz ME, Gaffey MJ, Pellas P. 1989. Meteoritic parent bodies: nature, number, size and relation to present-day asteroids. See Binzel et al 1989, pp. 740–77

Love SG, Brownlee DE. 1991. Heating and thermal transformation of micrometeoroids entering the Earth's atmosphere. *Icarus* 89: 26–43

Mariani F, Neubauer FM. 1990. The interplanetary magnetic field. In *Physics of the Inner Heliosphere I*, ed. R Schwenn, E Marsch, pp. 183–206. Heidelberg: Springer-Verlag

Marsden BG, 1976. Nongravitational forces on comets. In *The Study of Comets*, ed. B Donn, M Jackson, M A'Hearn, R Harrington, pp. 465–89. *NASA SP*-393

Maxwell Garnett JC. 1904. Colors in metal glasses and in metallic films. *Phil. Trans. R. Soc.* A203: 385–420

Moulton FR. 1970. *An Introduction to Celestial Mechanics*. New York: Dover

Mukai T, Yamamoto T. 1982. Solar wind pressure on interplanetary dust. *Astron. Astrophys.* 107: 97–100

Paddack SJ, Rhee JW. 1976. Rotational bursting of interplanetary dust particles. In *Interplanetary Dust and Zodiacal Light*, ed. H Elsässer, H. Fechtig, pp. 453–57. Berlin: Springer-Verlag

Parker EN. 1958. Dynamics of the interplanetary gas and magnetic fields. *Astrophys. J.* 128: 664–75

Patashnik H, Rupprecht G. 1975. The lifetime of ice particles in the solar system. *Icarus* 30: 402–12

Purcell EM, Pennypacker CR. 1973. Scattering and absorption of light by nonspherical dielectric grains. *Astrophys. J.* 186: 705–14

Radzievskii VV. 1952. On the influence of an anisotropic re-emission of solar radiation on the orbital motion of asteroids and meteorites. *Astron. Zh.* 29: 162–70. US Dept. Commerce Transl. 62–10653

Rickman H. 1989. The nucleus of comet Halley: surface structure, mean density, gas and dust production. *Adv. Space Res.* 9(3): 59–71

Rickman H, Fernández JA, Gustafson BÅS. 1990. Formation of stable dust mantles on short-period comet nuclei. *Astron. Astrophys.* 237(2): 524–35

Rickman H, Kamél L, Froeschlé C, Festou MC. 1991. Nongravitational effects and the aging of periodic comets. *Astron. J.* 102(4): 1446–63

Sagdeev RZ, Elyasberg PE, Moroz VI. 1988. Is the nucleus of comet Halley a low density body? *Nature* 331: 240

Saito T. 1988. Solar cycle variation of solar, interplanetary, and terrestrial phenomena. In *Laboratory and Space Plasmas*, ed. H Kikuchi, pp. 473–528. New York: Springer-Verlag

Sanford SA, Walker RM. 1985. Laboratory infrared transmission spectra of individual interplanetary dust particles from 2.5 to 25 microns. *Astrophys. J.* 291: 838–51

Schwenn R. 1990. Large-scale structure of the interplanetary medium. In *Physics of the Inner Heliosphere I*, ed. R Schwenn, E Marsch, pp. 99–181. Heidelberg: Springer-Verlag

Seab CG. 1987. Grain destruction, formation, and evolution. In *Interstellar Processes*, ed. DJ Hollenbach, HA Thronson Jr, pp. 491–512. Dordrecht: Reidel

Shul'man LM. 1972. The evolution of cometary nuclei. In *The Motion, Evolution of Orbits, and Origin of Comets*, ed. GA Chebotarev, EI Kazimirchak-Polonskaya, BG Marsden, pp. 271–76. Dordrecht: Reidel

Smith EJ, Tsurutani BT, Rosenberg LR. 1978. Observations of the interplanetary sector structure up to heliographic latitudes of 16, Pioneer 11. *J. Geophys. Res.* 83: 717–24

Standish EM Jr, Hellings RW. 1989. A determination of the masses of Ceres, Pallas, and Vesta from their perturbations upon the orbit of Mars. *Icarus* 80: 326–33

Steinitz R, Eyni M. 1980. Global properties of the solar wind. I The invarience of the momentum flux density. *Astrophys. J.* 241: 417–24

Suess HE. 1987. *Chemistry of the Solar System*. New York: Wiley

Thiel K, Bradley JP, Spohr R. 1991. Investigation of solar flare tracks in IDPs: some recent results. *Nucl. Tracks Rad. Measurements* 19: 709–16

van de Hulst HC. 1957. *Light Scattering by Small Particles*. New York: Wiley

Weber EJ, Davis L. 1967. The angular momentum of the solar wind. *Astrophys. J.* 148: 217–27

Weidenschilling SJ. 1981. How fast can an asteroid spin? *Icarus* 46: 124–26

Weinberg JL. 1985. Zodiacal light and interplanetary dust. In *Properties and Interactions of Interplanetary Dust*, ed. RH Giese, P Lamy, pp. 1–6. Dordrecht: Reidel

Weinberg JL, Sparrow JG. 1978. Zodiacal light as an indicator of interplanetary dust. In *Cosmic Dust*, ed. JAM McDonnell, pp. 75–122. Chichester: Wiley

Wetherill GW. 1985. Asteroidal source of ordinary chondrites. *Meteoritics* 20: 1–22

Xu YL, Dermott SF, Durda DD, Gustafson BÅS, Jayaraman S, Liou JC. 1993. The Zodiacal Cloud. In *The Ceremony of the 70th Anniversary of the Chinese Astronomical Society*. Beijing: Chinese Acad. Sci. In press

Zerull RH, Gustafson BÅS, Schulz K, Thiele-Corbach E. 1993. Scattering by aggregates with and without an absorbing mantle; microwave analog experiments. *Appl. Opt.* 32(21): 4088–100

Annu. Rev. Earth Planet. Sci. 1994. 22:597–654

TECTONIC AND MAGMATIC EVOLUTION OF VENUS

Roger J. Phillips

McDonnell Center for the Space Sciences and Department of Earth and Planetary Sciences, Washington University, St. Louis, Missouri 63130

Vicki L. Hansen

Department of Geological Sciences, Southern Methodist University, Dallas, Texas 75275

KEY WORDS: *Magellan*, lithosphere, convection, boundary layers

1. INTRODUCTION

Modern Venus exploration dates from the first geochemical measurements of the surface in the early 1970s by Soviet landers (Vinogradov et al 1973) and from the acquisition by the *Pioneer Venus Orbiter*, beginning in 1978, of nearly-global topographic and gravity data (Pettengill et al 1980, Sjogren et al 1980). Venus was revealed geologically as a planet with both strong similarities to and strong differences from Earth. Over the past decade, a wealth of new data has become available from continued surface exploration by the Soviet Union, from improved Earth-based radar astronomy by the United States (e.g. Campbell et al 1991), and from orbiting radar missions by both countries (Barsukov et al 1986, Saunders et al 1992). The tectonic and magmatic style of Venus has come into sharper focus and numerous, conflicting models for the evolution and present state of the planet have been advocated.

Knowledge of Venus increased dramatically with the high-resolution radar mapping of the surface by the *Magellan* spacecraft, commencing in September of 1990. Image resolution to order 100 m and altimetry with vertical and horizontal resolutions of about 80 m and 10 km, respectively,

0084–6597/94/0515–0597$05.00

have allowed a mapping that enables us to consider both local detail and regional to global scale features in constructing evolutionary models.

The purpose of this paper is to provide a review of models of mantle and crustal dynamics and to attempt to tie those models to hypotheses about the magmatic and tectonic evolution and present state of the planet, including the nature of major surface features. We ask the following questions, among others:

1. What are the origins of highlands on Venus and are the different kinds of upland terrain related through evolution or are they manifestations of distinct processes?
2. What are the origins of the forces required for tectonic deformation?
3. What is the importance of lower crustal flow in crustal dynamics?
4. What is the role of melt residuum in the tectonic evolution of the planet?
5. What controls magmatic access to the surface? How much is magmatic style simply a passive response to lithospheric strain?
6. What is the overall style of global tectonics and magmatism?
7. What is the origin of coronae?
8. What role do upgoing mantle plumes play in lithospheric tectonics?
9. What is the behavior of tectonism and magmatism over the past billion years ? Has it been steady, has it undergone a secular decline, or have there been episodes of increased activity?

We begin with a brief review of Venus geology. Our attempt here is to acquaint the reader with the physiography of the planet and provide a limited description of its major tectonic and magmatic features. For an in-depth survey of the planet, the reader is referred to papers in two dedicated issues of the *Journal of Geophysical Research* published in 1992 (Issues E8 and E10, Volume 97, 1992).

2. REVIEW OF VENUS GEOLOGY

2.1 *A Snapshot of Venus*

The hypsometry of Venus is unimodal, in contrast to the bimodal hypsometry of Earth. We distinguish three topographic domains (Figure 1). *Lowlands*, or *plains*, generally lie below the mean planetary radius (MPR); *uplands*, or *highlands*, are greater than ~2 km above MPR; and the region in between (~0 to 2 km above MPR) is here termed the *mesolands* [which correspond approximately to the "rolling plains province" of Masursky et al (1980)]. The highlands comprise <20% of the surface, and the plains and mesolands represent, roughly, equal parts of the remainder. Areas assigned to the mesolands include the chasmata region lying between Thetis and Atla regiones, much of the region included in the triangle whose

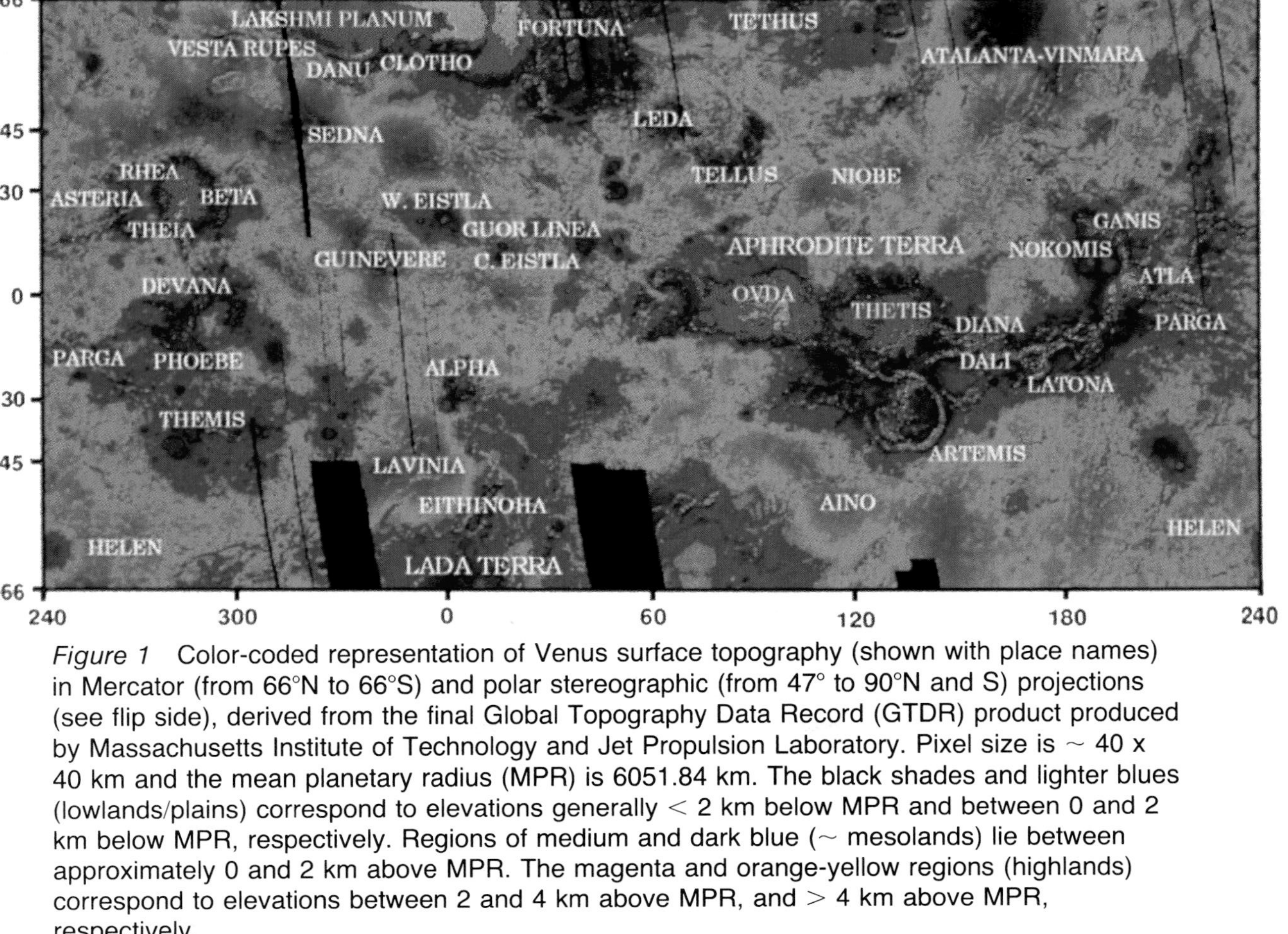

Figure 1 Color-coded representation of Venus surface topography (shown with place names) in Mercator (from 66°N to 66°S) and polar stereographic (from 47° to 90°N and S) projections (see flip side), derived from the final Global Topography Data Record (GTDR) product produced by Massachusetts Institute of Technology and Jet Propulsion Laboratory. Pixel size is ~ 40 x 40 km and the mean planetary radius (MPR) is 6051.84 km. The black shades and lighter blues (lowlands/plains) correspond to elevations generally $<$ 2 km below MPR and between 0 and 2 km below MPR, respectively. Regions of medium and dark blue (~ mesolands) lie between approximately 0 and 2 km above MPR. The magenta and orange-yellow regions (highlands) correspond to elevations between 2 and 4 km above MPR, and $>$ 4 km above MPR, respectively.

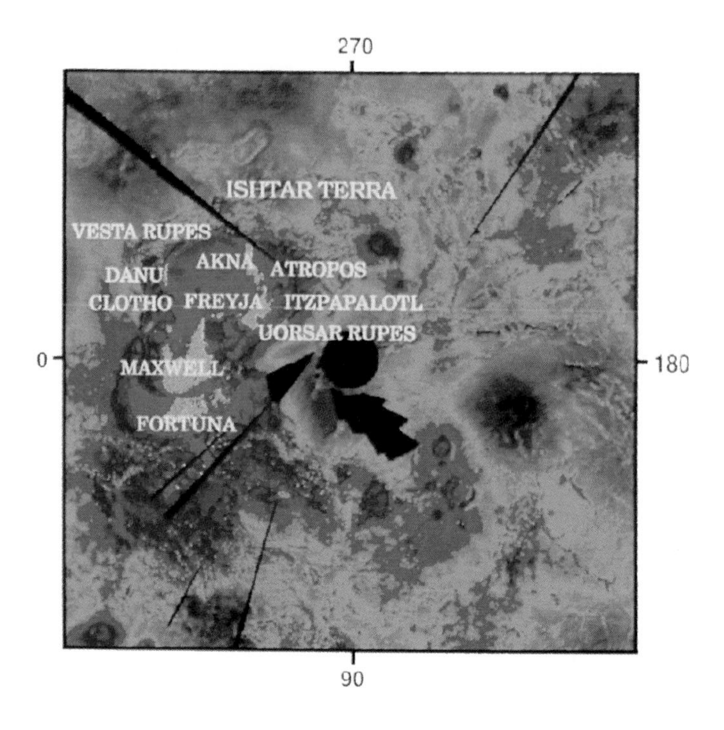
270
ISHTAR TERRA
VESTA RUPES
AKNA
ATROPOS
DANU
CLOTHO
FREYJA
ITZPAPALOTL
UORSAR RUPES
0
180
MAXWELL
FORTUNA
90

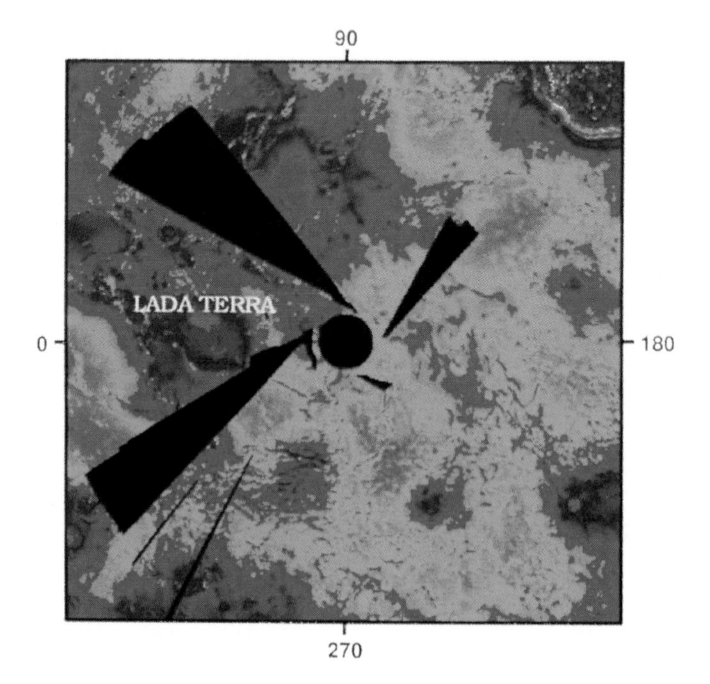
90
LADA TERRA
0
180
270

vertices are defined by Beta, Atla, and Themis regiones (called the "BAT region" by Head et al 1992), the area including Artemis Corona and elevated regions to the west, and parts of the "Southern Highlands," including Lada Terra. Geographically, the uplands include Ishtar Terra, parts of Aphrodite Terra that lie along the equator from $\sim 80°$E to 140°E (Ovda and Thetis regiones), Atla Regio, and the higher portions of Beta, Phoebe, and Themis regiones, which lie along the 285° meridian (Figure 1). Highstanding topographic areas in the equatorial regions, including Aphrodite Terra, and Beta and Eistla regiones are also known collectively as the "Equatorial Highlands" (Phillips et al 1981). Basins surrounding the highlands include: Atalanta, Vinmara, Sedna, Niobe, Leda, and Guinevere planitiae in the north, and Aino, Helen, and Lavinia planitiae in the south. Small highlands, Alpha, Eistla, Bell, and Tellus regiones, lie within the lowlands. Ishtar Terra, the highest region on Venus, is comprised of an interior plateau, Lakshmi Planum (3–4 km above MPR), surrounded by interior mountain belts ranging from 4–11 km above MPR, and outlying tesserae (1–5.5 km above MPR). Coronae (quasi-circular features) and chasmata (deep troughs) lie mainly within the mesolands; chasmata are further restricted to low latitudes. Tessera terrain [now also burdened with the name "complex ridged terrain" (CRT)] was originally described as terrain that ". . . is cut by two or more sets of ridges and grooves . . ." (Barsukov et al 1986); it occurs as a dominant tectonic fabric in some highland regions as well as in isolated single features and isolated patches largely in the plains.

2.2 *Structure*

Plains, or planitiae, lie below MPR, and are defined largely by radar-smooth (dark) fields, interpreted as regions characterized by flood or plains volcanism with locally preserved lava channels. Geologically, the highlands are divisible into Ishtar Terra, crustal plateaus, and volcanic rises. Crustal plateaus include Alpha, Tellus, Ovda, Thetis, and Phoebe regiones, which consist of steep-sided regions dominated by complex tectonic fabrics. These features correlate with small gravity/geoid anomalies, implying small (<100 km) apparent depths of compensation (ADC). Volcanic rises, which include Atla, Beta, Bell, and Eistla regiones, form domical, circular to elongate regions characterized by shield volcanoes, flows, and rifts. Large gravity anomalies, implying large ADCs (100–300 km), coincide with these features. The mesolands host topographic highs of crustal plateaus and volcanic rises and are characterized by linear to arcuate chasmata and quasi-circular coronae. For a summary of Venusian tectonism, see Solomon et al (1991, 1992).

2.2.1 PLAINS The plains record several episodes of volcanism and deformation; superposition demonstrates a continual interplay in time and space. Structural features are dominantly linear in plan view, and include, from largest to smallest: ridge and fracture belts, wrinkle-ridges, and extensional fractures, as well as "parallel lineations." Coherent patterns of these features over large areas (thousands of km^2) imply that they record a crustal response to mantle dynamics. Wrinkle ridges are probably the most common structure on Venus. They are similar to features preserved on the Moon and other terrestrial planets including Earth and Mars (Plesica & Golombek 1986, Watters 1988). Wrinkle ridges are long and sinuous, 1–2 km wide, and tens to hundreds of km long. Their continuity in orientation and spacing (tens of km) over broad regions indicates small amounts (1–5%) of regional crustal shortening normal to their trend (e.g. Suppe & Connors 1992, Bilotti & Suppe 1992, McGill 1992).

Extensional fractures—straighter than their wrinkle ridge counterparts—are also distributed over broad regions, and record crustal extension normal to their trend. Ridge and fracture belts comprise elevated zones of intense deformation characterized by parallel, closely spaced (5–10 km) ridges interpreted as contractional folds, and closely spaced normal faults, locally paired forming graben, respectively. These belts, best developed in Lavinia, and Atalanta-Vinmara planitiae, range in width from tens to a few hundred km and reach over 1000 km length. Ridge and fracture belts commonly occur together, either as distinct domains with different trends (e.g. Lavinia) or within single composite belts (e.g. Atalanta-Vinmara). These belts rise 500 m above the adjacent undeformed plains, and individual belts lie 100–300 km apart. The width of the deformed belts is comparable to the width of the relatively undeformed plain domains between belts. In Atalanta-Vinmara Planitiae, deformation belts trend north merging and bifurcating along strike to form an anastomosing pattern: ridge belts become hybrid fracture belts locally along strike; hybrid belts preserve fold ridges, although fractures, faults, and graben dominate the morphology (Solomon et al 1992). In Lavinia Planitia, fold belts trend NNE and fracture belts trend SSE, together expressing an orthogonal regional fabric. Locally, orthogonal contractional folds and extensional faults are present in a single deformed belt. Deformed belts surround relatively undeformed blocks of plains. Wrinkle-ridges and extensional fractures in the undeformed blocks parallel the NNE- and SSE-trending belts, respectively. The pattern of wrinkle ridges and extension fractures suggests that ridge belts and fracture belts record crustal shortening and extension, respectively, normal to their trends (e.g. Squyres et al 1992b). The coherence of strain patterns over thousands of km^2, with parallelism of principle strain axes inferred for both the deformed belts and adjacent relatively undeformed plains, indicates that the deformed belts record coherent strain over hundreds of thousands of km^2. Coherence in surface strain over such great distances may require that deformation was transmitted to the crust from the mantle below (e.g. Bilotti & Suppe 1992, McGill 1992, Squyres et al 1992b). This contrasts sharply with Earth, where most deformation is concentrated at plate boundaries.

"Parallel lineations," newly recognized in *Magellan* images, are sets of short, thin, straight, parallel fractures whose microwave reflectivity does not depend on the direction of radar illumination (Banerdt & Sammis 1992). These fractures may occur over hundreds of km^2 with little change in their spacing (1–2.5 km $\pm$ 1/3 km) despite diverse settings; therefore, the fractures must either always extend to a depth of 1 km, or the spacing must be independent of layer thickness. Banerdt & Sammis (1992) propose that fracture spacing relates to mechanical properties of the surface basalt layer (<1 km thick) above a horizontal detachment; during lithospheric extension, the horizontal detachment allows frictionally resisted lateral displacement of the surface layer resulting in a fracture spacing that is independent of layer thickness, and dependent on mechanical properties of the layer. Study of parallel lineations may, therefore, provide clues to mechanical properties of the surface materials, and they may also serve as relative temporal markers for deformational sequences.

2.2.2 MESOLANDS Topographic troughs and quasi-circular features dominate the mesolands. Circular to quasi-circular features include: coronae, ranging from 100–2600 km in diameter with a strong mode at 200–300 km; and novae, which have diameters less than 100 km (Stofan et al 1992). Chasmata, marked by topographic troughs and parallel lineaments, are divisible into symmetric and asymmetric varieties.

2.2.2.1 *Coronae* Coronae are unique to Venus, and are divisible into features that display radial, and concentric fractures, or both (Stofan et al 1992, Squyres et al 1992a). Novae generally display radial fractures and may represent an early stage of corona formation (Stofan et al 1991, Janes et al 1992). Coronae typically display a central region higher than surroundings, a raised rim, and a moat. Concentric fractures parallel the rim-moat transition. Radial fractures may be preserved only within the coronae, or may extend beyond the outermost concentric fracture set that marks the typically narrow (generally <100 km) tectonic annuli. Annuli commonly parallel the rim of the coronae just inboard of the moat. Moderate to large amounts of volcanism accompany corona formation. Lava reaches the surface locally through both radial and concentric fractures. Double-ring concentric coronae show a sequence of inner concentric faulting followed by volcanism, followed in turn by faulting and volcanism within an outer ring, indicating that coronae can grow outward by deformation associated with magmatic pulses (Hansen & Phillips 1993a). Extensional faults characterize the annuli, although contractional ridge belts and small-scale strike-slip structures are present locally along parts of individual corona and formed in response to local stresses that added constructionally or destructively with stresses involved in corona formation (e.g. Cyr & Melosh 1993). Radial fractures likely result from crustal doming early in corona evolution (Janes et al 1992, Squyres et al 1992a) which may continue with further development (Hansen & Phillips 1993a). Concentric structures form as the coronae expands, and during crustal relaxation. Intrusive magma is probably responsible for the thickened crust of the interior plateau.

Coronae are common on Venus: Over 360 are recorded across the planet (Stofan et al 1992). Stofan et al (1992) stated that coronae are "randomly" distributed in the plains, display an affinity with low latitudes, and cluster along Parga and Hecate chasmata, and within the Beta-Atla-Themis region. We suggest, however, that the distribution of coronae is sensitive to elevation. Coronae avoid both the plains and the uplands, and are concentrated in the mesolands; they are commonly located near chasmata or rifts. Herrick & Phillips (1992) showed that coronae cluster about MPR, and that they are uncorrelated with both strong positive and strong negative geoid anomalies. Their study suggests to us that if coronae are related to mantle convection, they are not directly linked to either upwellings or downwellings.

2.2.2.2 *Chasmata* Chasmata are long linear to arcuate troughs marked by lineaments that are probably fractures and faults. Chasmata are divisible into two groups—symmetric and asymmetric—depending on their topographic profiles normal to their trend (Hansen 1993). Asymmetric chasmata, the deepest chasmata, are bounded on one side by a ridge that is as high as the trough is deep. The trough-to-ridge differential can be as much as 7 km, over a 30 km distance. Fractures generally parallel the trough. Asymmetric chasmata are spatially associated with coronae, and overlap the coronal troughs with coronae residing on the ridge side (topographic high). Symmetric chasmata are not spatially associated with individual coronae, although they may be associated with corona clusters. These chasmata are significantly shallower than asymmetric chasmata, and trend outward from volcanic rises. They are marked by single, linear topographic troughs, or a chain of troughs, that range in width from 50 to 150 km, and extend for hundreds to thousands of km. On the basis of their structures and the presence of extensive syn-tectonic volcanism, symmetric chasmata are interpreted as rifts (Schaber 1982, McGill et al 1981). The amount of crustal extension across each rift zone is small.

2.2.3 HIGHLANDS Geologically, the uplands are divisible into volcanic rises, crustal plateaus, and Ishtar Terra. We discuss each in turn.

2.2.3.1 *Volcanic rises* Volcanic rises, defined both topographically and geologically, form a class of structures interpreted as the surface expressions of mantle upwellings on the basis of geological and geophysical relations (e.g. Phillips & Malin 1984). The upwelling hypothesis for volcanic rises was originally based on the strong analogy of volcanic rises to terrestrial hotspot terrains (McGill et al 1981), quasi-radial rifting suggesting buoyant uplifting from below, and large ADC values (e.g. Smrekar & Phillips 1991). Volcanic rises include Atla, Beta, Western Eistla, and Bell regiones. We describe these regions below, and then summarize the characteristics that must be accommodated by evolutionary models.

Atla Regio—a broad highland dome marked by numerous volcanic centers and rifts, shield volcanoes, and fractured terrains—occupies easternmost Aphrodite Terra. Atla comprises several large shield volcanoes. Ozza Mons lies at the intersection of five chasmata, each interpreted as a rift (Bindschadler et al 1992b, Senske et al 1992). Volcanic flows fill graben, and are in turn cut by faults, demonstrating temporal and spatial interplay between tectonism and volcanism. Local tesserae comprise the oldest unit. Atla has an ADC of 200 km.

Beta Regio is composed of Rhea and Theia Montes defined by tesserae and constructional volcanoes, respectively (Senske et al 1991, 1992). The three arms of Devana Chasma, marked by numerous normal faults and interpreted as rifts (McGill et al 1981, Schaber, 1982, Campbell et al 1984), intersect at Theia. Rift structures clearly cut across the tesserae of Rhea Mons; rifting and volcanism occurred concurrently in Theia Mons (Senske et al 1992). Beta has an ADC of 270–400 km.

Western and Central Eistla define broad rises with large volcanic constructs and rift zones correlated with strong positive gravity anomalies and ADCs of 100–200 km (Grimm & Phillips 1992). Twin peaks, the volcanoes Sif and Gula Montes, dominate Western Eistla, and Sappho Patera dominates Central Eistla. Radial fractures mark the topographic rise and the individual volcanoes. The largest rift, Guor Linea, extends 1000 km southeast of Gula Mons and reaches a width of 50–75 km; it may record minor clockwise rotation and bulk NNE-SSW crustal extension normal to its trend.

Bell Regio comprises local tesserae and the shield volcano Tepev Mons and several smaller volcanic centers. A north-trending fracture zone, once interpreted as a rift (Janle et al 1987), cuts the region, but unlike the rift zones of Beta and Eistla, there is little evidence of extension in *Magellan* data (Solomon et al 1992). Tesserae form the oldest terrain type. Bell Regio correlates with a 100–200 km ADC (Smrekar & Phillips 1991).

These volcanic rises share the following characteristics: (*a*) they have broad topographic rises; (*b*) they are generally composed of several shield volcanoes; (*c*) they are located at the convergence of rifts; (*d*) they exhibit widespread volcanism synchronous with rift formation; (*e*) their tesserae, when present, constitute the oldest terrain type; and (*f*) they coincide with large ADCs, which taken together with the high topography, probably indicates that a significant component of the topography is compensated by mantle convection. These features are all consistent with the interpretation that volcanic rises represent surface manifestations of mantle upwellings.

2.2.3.2 *Crustal plateaus* Crustal plateaus include Ovda, Thetis, Phoebe, Alpha, and Tellus regiones (Figure 1). Ovda (3000 × 2000 km) and Thetis (1500 × 1500 km) are topographically higher than Phoebe (1700 × 1000 km), Alpha (1500 × 1300 km in extent), and Tellus (2000 × 1600 km). They all share the following characteristics:

1. They are steep-sided regions reaching elevations of 2–4 km above the surrounding plains.
2. The highest elevations are at the margins.
3. They display rough topography at a horizontal scale of tens of km.
4. They are dominated by tesserae or CRT that exhibit structures best described as basin and dome morphology, suggestive of polyphase strain histories. We term this fabric, which is best preserved within the interior of the plateaus, "amorphous terrain" because of its lack of structural coherence. Amorphous terrain commonly hosts small (20–50 km wide and 50–150 km long) elliptical radar-dark and radar-smooth basins, which have been called "intertesserated plains" (Bindschadler & Head 1991); we use the term intratesseral plains, which more appropriately defines the basins as within a single tessera rather than between tesserae. Intratesseral plains show no preferred orientation or spatial distribution within the plateau interior.
5. The boundary of deformation with undeformed plains is extremely irregular.
6. The most coherent structural fabrics are fold belts preserved locally along parts of the plateau margins. Although Bindschadler et al (1992b) state that marginal fold belts parallel their respective boundaries, fold belts do not totally encircle plateaus; fold belts may parallel, or they may trend at a high angle into the surrounding regions. In addition, the boundaries of some plateaus are locally dominated by extensional structures representing the youngest deformation on these plateaus (Solomon et al 1992).
7. Contractional and extensional fabrics are variably developed through-

out plateaus; crosscutting relations and local lava-flooded regions indicate that contractions and extensions took place at different times in different areas.

8. Plateaus have small gravity/geoid anomalies, and ADCs are consistently < 100 km, and often < 50 km.

2.2.3.3 *Ishtar Terra* Ishtar Terra includes Lakshmi Planum (4 km above MPR) and the surrounding deformed belts which include the interior mountain belts—Maxwell, Danu, Akna, and Freyja Montes (3.5–11 km above MPR), as well as the outlying tesserae—Fortuna, Clotho, Atropos, and Itzpapalotl (1–4.5 km above MPR) (Figure 1). Uorsar Rupes marks the steep slope along the northern boundary of Itzpapalotl, and Vesta Rupes marks the steep slope that bounds Danu to the south. Maxwell Montes hosts the highest point on Venus, and Fortuna Tessera comprises the largest tessera terrain. We use Fortuna Tessera for only that part of Fortuna east of Maxwell and west of 30°E; farther east the region has not been imaged by *Magellan* because of a superior conjunction data acquisition gap. Ishtar Terra is notably free of coronae.

Lakshmi Planum is relatively undeformed. Colette and Sacajawea calderas are preserved within it, and flooded tesserae are exposed at the highest elevations (Roberts & Head 1990a,b; Kaula et al 1992). The high plain of Lakshmi is relatively structureless, except along the periphery where wrinkle ridges deform lava flows near the mountains, indicating that some deformation postdates plateau volcanism.

Danu Montes and outboard Clotho Tessera are different from their Ishtar counterparts. Danu is a long, narrow mountain belt that reaches only 1.5 km above Lakshmi; Clotho is lower than other Ishtar tesserae (1–1.5 km above MPR) and is dominated by linear troughs of probable extensional origin.

In contrast, the structural pattern of the other mountain belts and tesserae consists of tightly spaced (< 3 km) "penetrative" lineaments ["penetrative fabric" of Keep & Hansen (1993b), "fine-scale structures" of Bindschadler et al (1992a)] that are deformed by the linear fold ridges and valleys, which dominate the structural fabric. The penetrative lineaments may represent parallel lineations (Banerdt & Sammis 1992) contracted during crustal shortening. Fold ridges—the result of crustal contraction (Campbell et al 1983; Barsukov et al 1986; Crumpler et al 1986; Kaula et al 1992; Solomon et al 1991, 1992)—extend for hundreds of km. Small-scale extensional fractures trend normal to the fold ridges, indicating that local extension parallel to ridge trends accompanied contraction (Hansen & Phillips 1993b,c; Keep & Hansen 1993a,b). Locally, structural valleys are filled with lava floods that formed late in the deformational history (e.g. Akna and Atropos); few vents and channels are identified. In Maxwell and Fortuna, and in Akna and Atropos, parallelism of structures defines a coherent tectonic pattern that extends from Lakshmi wrinkle ridges through peripheral mountain belts and well out (>500 km) into adjacent tesserae; spacing of fold ridges (3–15 km range, ~6 km average) does not change from mountain belt to tessera (Hansen & Phillips 1993b, Keep & Hansen 1993b). Topography, and not structure, marks the major difference between mountain belts and neighboring tesserae. Extensional structures are identified locally in Freyja and Danu, and within the northern and southern slopes of Maxwell (Kaula et al 1992, Smrekar & Solomon 1992), but the highest elevations of the mountain belts and the tesserae display no evidence of gravitational collapse.

2.2.3.4 *Summary* The highlands are divisible into volcanic rises, crustal plateaus, and Ishtar Terra. Although most workers agree that volcanic rises represent hotspots, crustal plateaus and Ishtar Terra are variably interpreted as surface expressions of major subsolidus crustal flow in response to mantle downwelling, and crustal thickening by magmatism related to late-stage evolution of mantle upwelling. Models for the formation of crustal plateaus and Ishtar Terra must account for these important differences: (*a*) Ishtar is structurally much more coherent; dominant fold ridges parallel the orientation of their host mountain belt with respect to Lakshmi; (*b*) elevated regions (mountain belts) are located inside Ishtar Terra, against Lakshmi, rather than against the exterior plains boundary, as is the case for the crustal plateaus; (*c*) the interior of Ishtar, Lakshmi Planum, is a high radar-smooth and radar-dark plateau that is essentially free of deformation; in contrast, crustal plateau interiors record spatially complex strain histories; (*d*) Ishtar tesserae are structurally more coherent than the amorphous terrain that dominates crustal plateaus; and (*e*) the spatial distribution of polyphase versus coherent single-phase deformation is the opposite for Ishtar and crustal plateaus.

2.3 *Volcanism*

Evidence for volcanism is globally widespread on Venus (for a review see Head et al 1992). Volcanic features include shield volcanoes and calderas (20 to >100 km diameter), and are not concentrated in linear zones as they are on Earth where they mark convergent and divergent plate boundaries. On Venus volcanoes are distributed throughout the mesolands and occur locally in the uplands with a concentration of volcanic centers (2–4 times the global average) within the Beta-Atla-Themis region (Head et al 1992). Evidence for volcanism is ubiquitous on the planet. The plains generally host flood volcanism and sinuous channels; obvious volcanic centers are mostly lacking; volcanic rises are dominated by shield volcanoes and rifts; crustal plateaus display evidence of probable early volcanism; and lava–flooded structural valleys are present throughout the Ishtar deformed belts. The majority of volcanic land forms are consistent with a basaltic composition with possible exceptions including steep-sided domes that may represent more siliceous compositions and sinuous rilles possibly indicative of ultramafic lava (Head et al 1991, 1992; Pavri et al 1992; McKenzie et al 1992b; Baker et al 1992). The scale of individual features is typically small—much less than 125,000 km^2 (Head et al 1992).

2.4 *What Impact Craters Tell Us about Resurfacing History*

The distribution and degradation states of impact craters on planetary surfaces have long been used as a guide to both resurfacing history and surface age. Venus has provided a special challenge amongst the terrestrial planets in understanding the spatial disposition and preservation state of impact craters; interpretations of the cratering record have led to wildly different scenarios for the resurfacing history of the planet. In part this is because there are only about 900 impact craters on the surface, so spatial statistics are subject to large uncertainty on a regional basis.

The surface age of Venus has been estimated to be about 400–500 Ma. This value is obtained by dividing the observed number of craters larger than ~ 30 km diameter by an estimate of impact crater formation rate for that size range. Formal uncertainties in regression fits to the crater size–frequency distribution, as well as uncertainties in the meteoroid flux rate, lead to a 90% confidence interval no better than 300–600 Ma (Phillips et al 1992). An age estimate thus obtained has a variety of interpretations because: (*a*) it is a global average, and says nothing about age variations from place to place; (*b*) it gives the "production age" of a surface (or surfaces) under an assumption that once a fresh surface is formed, impact craters subsequently accumulate without disturbance; and (*c*) it is subject to an alternative interpretation that does not give the age of a surface but rather the average lifetime of a crater before it is destroyed (the "retention age").

Phillips et al (1992) have found that: (*a*) The crater distribution cannot be distinguished from a spatially random population. (*b*) Most craters ($\sim 80\%$) appear "fresh" in that the crater and its continuous ejecta blanket are not modified by faulting ("tectonization") or volcanic embayment. (*c*) Modified (embayed and/or tectonized) craters have a statistically significant negative spatial association with unmodified craters. That is, modified craters occur in or near the borders of regions of low overall spatial crater density. (*d*) There is an inverse correlation between radar cross section, σ_o, and spatial crater density.

A useful way to interpret the distribution of surface ages on Venus is to consider an age spectrum, defined as the fractional surface area as a function of age (Phillips 1993). In this regard, points (*a*) and (*b*) have been used to argue that Venus underwent a catastrophic resurfacing (Schaber et al 1992), i.e. the age spectrum is a "spike" at $\sim$300–500 Ma ago. Points (*b*), (*c*), and (*d*) have been used to argue that the age spectrum is much broader, i.e. surface modification is much more widely distributed in time (Phillips et al 1992). Point (*c*) has been used to argue that craters have been removed from the surface, i.e. many low crater density regions exist because craters have been removed by geological processes and are not low density by Poissonian chance as is implicit in the catastrophic resurfacing model. N. Izenberg (personal communication) has shown, by analyzing the degradation states of the complete set of crater ejecta deposits, that crater degradation and removal has proceeded more rapidly in areas of low crater density. Phillips (1993) proposed that the Venus surface can be divided into at least three distinct regions by crater age: 1. parts of the planet ($\sim$25%) that have not been resurfaced for $\sim$0.6–1 Ga, termed "High Crater Density" (HCD) areas, and which are largely associated with volcanic plains; 2. parts of the planet ($\sim$25%) with younger than average surface ages ($\lesssim$0.2 Ga), termed "Low Crater Density" (LCD) areas, and which are correlated strongly with the mesolands; and 3. the remainder of the planet ($\sim$50%), associated with surface retention ages nearer to the planetary mean.

Additional crater observations bear on the tectonic and magmatic history of the planet (Phillips et al 1992). In general, the oldest 90% of impact craters have dark (low values of specific radar cross section, σ_o) floors and the youngest 10% do not. Dark floor material is widely interpreted as volcanic in origin, and the implication is that magma has been widely available in space and time in the interior. Secondly, many of the regions of low crater density follow distinct extensional tectonic trends, and craters that have been volcanically altered have an unusually high association with craters that have been tectonically altered, more than three times the number expected if the processes act independently. The implication is that volcanism follows tectonism in a passive manner. That is, combining both observations, magmas, or at least incipient partial melts, have existed commonly in the subsurface, and where tensional environments occurred in the lithosphere, volcanism resulted.

Ivanov & Basilevsky (1993) measured crater density on tesserae and compared this to nontessera regions. While the overall crater density for the two classes is the same, tesserae display a deficiency (relative to the planetary mean) in craters with diameters of $<$16 km and an excess of craters for diameters greater than this. There is apparently an observational bias for craters of < 16 km diameter that tends to underestimate the number of these craters in tesserae. The excess of larger craters implies that the average age of tessera surfaces exceeds the mean surface age of the planet. These results do not mean that all tesserae are older than all nontessera areas (e.g. Thetis Regio is relatively young), but the overall results indicate that, on average, tesserae are amongst the oldest units on the surface, approximately equivalent in age to HCD areas. This view is supported by the observation that 73% of tessera borders are characterized by onlapping (and hence younger) volcanic units (Ivanov & Head 1993).

3. MANTLE DYNAMICS

On Earth, mantle convection controls tectonism and magmatism, which are largely confined to the vicinity of plate boundaries. The bulk of magma is produced at the midocean spreading centers by pressure release partial melting. Lesser amounts are produced near subduction zones and as intraplate magmatism associated with hotspots. Tectonism is largely associated with the interaction of plates (e.g. plate collision, transform boundaries) and in a secondary way by gravity-driven mechanisms related to topography resulting from plate interaction. The key point for our purposes here is that on Earth the styles of tectonism and volcanism relate directly to the fact that the oceanic lithosphere participates in mantle convection. It is the notion that this phenomenon does not take place on Venus that presages a very different style of magmatism and tectonism for that planet. A very brief review of mantle convection will set the stage for this consideration.

3.1 *Mantle Convection*

The heat passed to a planetary mantle from a core and that generated within the mantle by the decay of radiogenic isotopes must escape across the outer surface of the planet. If the forces associated with thermal buoyancy exceed the viscous resistance of mantle rock, then the heat will be passed outward dominantly by convection. Formally, the ratio of these two effects is the thermal Rayleigh number, which must exceed some critical value for convection to proceed. Interfaces that are a barrier to convection form thermal boundary layers in which heat is transferred conductively. The outer boundary of mantle convection in a terrestrial planet is always a thermal boundary layer marking the transition from mantle convective heat transfer to lithospheric conductive heat transfer.

If horizontal velocity is not impeded at the surface of a planet (a "free" boundary condition in which shear stress vanishes), then convection can be very efficient because cold temperatures in the thermal boundary layer can be advected downward and large quantities of negative buoyancy are available to drive the flow. This is the situation for Earth's ocean basins in which seafloor spreading is the top of the mantle convection system. If, on the other hand, horizontal velocity is zero at the surface (a "rigid" boundary condition), then a velocity boundary layer will form. At least the upper part of the thermal boundary layer will not be easily advected downward and convection will be less efficient. Thus the critical Rayleigh number for a rigid boundary condition will be higher than that for a free boundary condition—all other factors being equal—and the convecting region will be hotter and cool more slowly. This is easily demonstrated in parameterized convection models, and is a basic result of calculations using more sophisticated two- and three-dimensional finite element convection codes (Schubert et al 1990, Leitch & Yuen 1991). That portion of the thermal boundary that is not advected in the main mantle flow is still buoyantly unstable and will spawn cold downgoing plumes that will be superimposed on the general circulation pattern of the mantle.

3.2 *Application to Venus*

3.2.1 LONG-WAVELENGTH GRAVITY AND TOPOGRAPHY CONSTRAINTS

Because of their high correlation, the relationship between long-wavelength gravity and topography on Venus is remarkably different from that on Earth (Phillips & Lambeck 1980, Sjogren et al 1983). On Earth, long-wavelength gravity anomalies arise from convective flow in the mantle and from subducting slabs and have little to do with long-wavelength topography (Hager & Clayton 1989). On Venus, the conventional wisdom is that because there is no asthenosphere (see below), long-wavelength topography is a first-order reflection of mantle convection. The idea that long-wavelength topography is "dynamically supported" ("code words" for support by some type of convective flow in the mantle) is rooted in early estimates of the ADCs of large-scale topographic features (Phillips et al 1981).

A simplified version of the argument that large ADC values imply dynamic support of topography goes like this (Phillips et al 1981, Phillips & Malin 1983): If topography is not supported dynamically, it is supported isostatically. The very term "isostatic" means that the compensating mass stays put, i.e. it is static. The most common picture we have of isostasy is the compensation of mountains by crustal roots (depression of Moho) in Earth's lithosphere. At some depth in any terrestrial planet, the idea of static support becomes implausible simply because the temperature is high enough for lateral density contrasts inherent in a compensation mechanism to be unstable against dispersal by creep on geological time scales. The demonstration of implausibility comes from finding an upper bound on temperature to maintain in place a compensating mass at the given depth, and then showing why such a low temperature is unrealistic. The magical depth on Venus has been around 100 km, but this is not a rigorous bound on the maximum depth of static support.

More recent work (Smrekar & Phillips 1991, Simons & Solomon 1993) has shown that large-scale features range over nearly an order of magnitude in ADC ($\sim$30–300 km) and that multiple compensation mechanisms are probably operative under many features. However, geoid-to-topography ratios (GTR) cluster into two statistically significant groups (Smrekar & Phillips 1991). Generally, the larger group is associated with volcanic rises and the smaller group with crustal plateaus, as discussed in Section 2. This has reinforced the idea that the former features are linked to mantle plumes, whereas the latter have a large component of crustal and lithospheric thermal compensation. Inversion of long-wavelength gravity and topography data for the relative contributions of compensation mechanisms shows a dominant dynamic component beneath such rises as Atla and Beta regiones and a dominant lithospheric component beneath such crustal plateaus as Ovda and Thetis regiones (Phillips et al 1991, Herrick & Phillips 1992). These gravity analyses have underscored the inherent differences between volcanic rises and crustal plateaus. High resolution gravity data from the *Magellan* mission have shown, however, the possible importance of flexurally supported volcanic constructs to the gravity signals associated with volcanic rises, and this may diminish the contribution of the modeled dynamic components to their associated free-air gravity anomalies.

3.2.2 MANTLE VISCOSITY STRUCTURE Kaula (1990a) has proposed that a relative lack of volatiles in the mantle of Venus compared to Earth may lead to a higher viscosity for the Venusian mantle. However, because of the self-regulation of convection, an increase in activation energy (higher solidus temperature) will be offset partially by an increase in mantle temperature. Given also the effect of a rigid upper surface boundary condition (see below) and higher surface temperature, we expect that overall the viscosity of the convecting Venusian mantle will be little different than that of Earth (see also Turcotte 1993).

Phillips (1986) argued that the large ADC values for some topographic features provide evidence against the presence of a low-viscosity zone or asthenosphere beneath the Venusian lithosphere. If an asthenosphere existed, then ADC values would become quite small, the exact value depending on the magnitude of the viscosity contrast between asthenosphere and the rest of the mantle (Robinson et al 1987). Conversely, the presence of an asthenosphere would place actual compensation depths implausibly deep. Phillips (1990) showed with analytical models how observed ADCs on Venus could not be satisfied in the presence of a low-viscosity zone. The lack of an asthenosphere of Venus might be anticipated because of the lack of water or possibly other volatiles in the interior (Phillips 1986). A good review of the Venus interior water issue can be found in Kiefer & Hager (1991a).

Kiefer et al (1986) argued for a uniform-viscosity Venusian mantle based on matching the predictions of viscous flow models to the observed spectral admittances—the spherical harmonic spectral ratios of gravitational potential to topography. Kiefer & Hager (1991a) showed that topographic features in the Equatorial Highlands (Atla, Beta, Ovda, Thetis regiones) could be explained by models with no asthenosphere and upper mantle/lower mantle viscosity ratios ranging from 0.1 to 1.0. These results depend on the assumption that the topography and geoid anomalies of all of these features are caused solely by upwelling mantle plumes. These suppositions are controversial, as we shall see later. Smrekar & Phillips (1991) argued, however, that the GTRs for volcanic rises alone require the absence of an asthenosphere.

A potentially more robust result is that the observed admittance spectrum is positive up to spherical harmonic degree 18 (Bills et al 1987; see also McNamee et al 1993). The viscosity structure that best reproduces Earth's long-wavelength geoid has the following depth range/relative viscosity structure: 0–100 km/1.0; 100–400 km/0.032; 400–670 km/1; 670–2900 km/10.0 (Hager & Clayton 1989). The depth range from 100 to 400 km represents an asthenosphere. The result of a slightly different model (relative viscosity in the 0 to 100 km range is 10.0) produces a negative admittance spectrum for degrees 4–9 (Kiefer et al 1986, Kiefer & Hager 1991a). Because the low-viscosity region tends to uncouple flow stresses from the surface, over this degree range the geoid contribution from the interior outweighs the contribution from the dynamically deformed surface. The admittance calculations were for a perturbing density distribution that is constant throughout the mantle. This cannot be correct, but it is expected that more realistic density distributions will not change these results qualitatively.

For Earth, it is not clear that this exercise would achieve the expected result—a negative portion of the observed admittance spectrum—owing to the presence of an asthenosphere. Variations in lithospheric and crustal structure are mostly responsible for topographic fluctuations (Hager & Clayton 1989), so even in the presence of an asthenosphere, negative admittance values may not be produced. Thus it could be argued that the presence of a non-negative admittance spectrum for Venus is not necessarily an argument against the presence of a low-viscosity zone or asthenosphere.

However, because the long-wavelength topography and geoid are so well correlated on Venus, this likelihood is diminished. The positive correlation itself is a reasonable piece of evidence that Venus does not possess an asthenosphere.

3.2.3 BOUNDARY LAYER PLUMES We consider both the vigor of boundary layer plumes and their transit time through the mantle.

3.2.3.1 *Vigor of plumes* We expect that the ratio of heat passing across the core-mantle boundary to that generated by isotopic decay within the mantle is less for Venus than for Earth (Phillips et al 1991). Venus does not have an internal dipole magnetic field and this is the basis of the argument; the magnetic dipole moment of Venus is no more than 1% of that of Earth even after adjusting for the different rotation rates of the two planets. The most likely explanations for Earth's geodynamo are necessarily accompanied by an outward heat flux, the bulk of which is associated with the freezing of an inner core. The absence of a magnetic field suggests that a freezing inner core does not exist on Venus, and this is borne out by parameterized convection models (Stevenson et al 1983). A lower core heat flux for Venus would imply more infrequent plumes formed by Rayleigh-Taylor instabilities that develop in the core-mantle boundary layer (Phillips et al 1991).

Interior phase transitions and chemical discontinuities may also form barriers to convection and generate plumes. Steinbach & Yuen (1992) considered the relative abilities of terrestrial and Venusian phase transitions to resist penetration by convection. They found that the greater depths for equivalent phase transitions and the probable presence of a conducting lid increases the likelihood that Venusian convection can flow through phase boundaries. Nevertheless, there is no way to decide whether or not interior phase boundaries in Venus are a barrier, or if they are, whether there is still episodic flow through these boundaries. The issue of phase (and chemical) boundaries is important because it bears on the importance of plumes both in removing heat from the interior and in the tectonic and magmatic history of the lithosphere.

3.2.3.2 *Mantle transit times* How long will it take a plume to traverse the mantle? This question has important implications for the origin of coronae. Stofan et al (1992) suppose that those plumes that form coronae come from the core-mantle boundary. The majority of coronae fall in the diameter range 200–300 km; conservatively, we take the diameter of the proposed causative plume as one-half the corona diameter, so as plumes reach the base of the lithosphere, they must have diameters in the range 100–150 km. The trick is to launch a plume from the core-mantle boundary and have it arrive at the base of the lithosphere with the right diameter—and then see how long it takes for the plume to form and traverse the mantle.

A formulation of this problem is given by Whitehead & Luther (1975), who describe a growing spherical plume trailed by a cylindrical conduit. To produce plumes of the correct size, the transit times over the viscosity interval 10^{21} to 10^{22} Pa s range from a few hundred million years up to nearly four billion years. This result is based on a dry olivine flow law (Goetze 1978), a strain rate of $10^{-14}\,s^{-1}$, and a plume temperature contrast, ΔT, of 300 K. It would seem implausible that such small plumes could traverse the entire mantle, given also that the mantle will reorganize its flow regime on a time scale of a few 100 Ma or less (e.g. Steinbach & Yuen 1992). Increasing the plume-mantle temperature contrast, ΔT, which decreases the plume viscosity and density, has only a modest effect on the transit time. Reasonable transit times (<100 Ma) are obtained only for large plumes, the type that might be associated with volcanic rises. This same time interval can also be obtained for the smaller plumes if they are launched from the upper mantle/lower mantle boundary (spinel-perovskite phase transition), assuming a thermal boundary layer exists there.

3.2.4 UPPER BOUNDARY OF CONVECTION *Magellan* radar and altimetry confirmed earlier views that Earth-like plate tectonics does not exist presently on Venus [see Solomon et al (1992) for a thorough discussion]. There are only a few examples of large-scale strike-slip faulting, and there is no evidence of lithospheric spreading such as is found at oceanic spreading centers. By topographic analogy to terrestrial subduction zones, there is evidence for subduction at the edges of some large coronae (McKenzie et al 1992a, Sandwell & Schubert 1992a,b). This hypothesis, discussed elsewhere here, is controversial. Our view is that horizontal lithospheric motion operates presently on a restricted, regional scale (Phillips et al 1991) of order 100 km (Solomon et al 1992).

The absence of large-scale lithospheric motion on Venus implies that a rigid boundary condition at the top of the convecting mantle may be appropriate as long as there is not a weak region (a low-viscosity zone or asthenosphere) separating the convecting region from the lithosphere. We presented several arguments above as to why Venus does not have such a weak zone. However, a rigid boundary condition can only be an approximation for the behavior of flow in the uppermost portion of the mantle.

Flow velocities will decrease upward toward the crust-mantle boundary (Moho) if the surface is immobile and also because the temperature will decrease in the thermal boundary layer. We expect that the upper mantle will flow at ever-decreasing velocity upward, so the imposition of a rigid boundary necessarily must be accompanied by a time scale of interest because over a sufficiently long time interval, even the uppermost mantle will have undergone significant motion. It is possible that the lower crust itself is the zone of decoupling of convective flow from the surface (Buck 1992), and that the lower crust then acts as a velocity boundary layer. Regardless of whether or not the uppermost mantle participates in large-scale flow on geologically interesting time scales, it is buoyantly unstable (unless possibly it contains a magmatically depleted layer; see below) and will initiate cold, downgoing plumes.

The existence of a velocity boundary layer implies a vertical gradient in horizontal velocity, and thus a shear stress acting on the lithosphere in addition to the normal forces associated with buoyancy in the mantle flow regime. In general, normal forces will create dynamic topography and the resulting stresses in the lithosphere relate directly to the gravitational body forces associated with topographic variations. The transfer of shear forces (as well as normal forces) to the overlying lithosphere is the basis of the concept of "convective stress coupling," which is considered to have been a significant element in the tectonic deformation of the Venusian lithosphere (Phillips 1986, 1990). While the flow shear stresses themselves may not be large, they give rise to strong normal stresses acting on vertical planes in the overlying lithosphere. However, flow will be more effectively driven in lithospheric regions, such as the lower crust, if velocity, in addition to stress, is coupled (Ramberg 1968). In the presence of a strong viscosity contrast at the Moho (see Section 4.2.2), neither shear stress nor velocity will be well coupled into the crust. Because viscosity is non-Newtonian, shear stresses will in fact act to maintain crust-mantle decoupling.

3.2.5 THE ROLE OF A MELT RESIDUUM IN THE UPPER MANTLE Phillips & Grimm (1990) proposed that the relative thinness of the Venusian crust could be attributed to the existence of a layer of buoyant melt residuum that has accumulated beneath the lithosphere. If upwelling mantle was unable to penetrate this residuum, then pressure-release partial melting would be inhibited, eventually shutting down completely if the residuum reached a critical thickness. Kaula (1990b) argued that mantle convection should easily penetrate the melt residuum, so that a thick layer would never build up. Work now in progress (Parmentier et al 1993, Smrekar & Parmentier 1993) is examining the conditions for plume penetration of a residuum layer (see also Ogawa 1993). The stability of the residuum depends on the ratio, R, of the chemical density contrast of the residuum (to normal mantle) to the thermal density contrast of the plume (to normal mantle), as well as the ratio, μ, of residuum viscosity to that of underlying mantle. There is a minimum value of R, R_{min}, depending on μ, below which entrainment and mixing of residuum will occur (Parmentier et al 1993). R_{min} will decrease with increasing μ. Because the residuum is colder and more magnesium-rich than the underlying mantle, μ will be greater than unity. It may be that the lithosphere is stable over geological time scales and that only very large plumes, which may be rare, cause disruption. However, Tackley & Stevenson (1993b) find through numerical convection simulations that large-scale convection is able to entrain residuum in downwellings, and they suggest that an equilibrium may exist between generation of residuum by partial melting and its return to the deeper mantle.

The existence of a stable residuum layer has other implications besides limiting crustal growth. Large ADC values for some topographic features could be indicative of plumes trapped at the base of a residuum layer (Smrekar & Parmentier 1993). Also, the residuum layer may become buoyantly unstable and its overturn may have tectonic and magmatic consequences for the lithosphere. Under a very general set of conditions, Parmentier & Hess (1992) showed that a residuum layer in the upper mantle would become buoyantly unstable and that a fraction of it at temperatures above a critical temperature for flow, T_f, would remix with the mantle. In an evolutionary calculation, the planet eventually cools to the point that a stable residuum layer will accumulate. The initial mixing takes place when that part of the layer with temperatures greater than T_f becomes unstable because of the decrease in compositional buoyancy that results when melting is pushed to greater depths as the residuum layer and crust thicken. Episodicity (near periodicity) is a fairly robust feature of the solutions ($\sim$300–500 Ma), but the time of onset depends strongly on initial temperature, viscosity, and heat production rate and exceeds the age of the planet in some cases. Subsequently, we will present a model for the formation of Ishtar Terra that uses this instability mechanism.

3.2.6 OTHER SOURCES OF TIME-DEPENDENT BEHAVIOR We separate our discussion of the time-dependence of mantle flow into episodic/chaotic behavior and secular variation.

3.2.6.1 *Episodic/chaotic behavior* Time-dependent, chaotic behavior is a basic feature of high Rayleigh number convection (e.g. Schubert et al 1990, Leitch & Yuen 1991) and would lead to variable conditions for tectonism and magmatism at several temporal scales over the history of Venus. We have concluded that the viscosity of the Venusian mantle is no higher, and therefore the Rayleigh number is no lower, than that of Earth. We expect time-dependent behavior to have been an important feature of mantle convection over the lifetime of Venus. Arkani-Hamed et al (1993) have argued that the proposed global resurfacing event $\sim$500 Ma ago marks the transition from high Rayleigh number chaotic convection to low Rayleigh number steady, sluggish convection. The argument is based on the supposition that Venus has cooled more rapidly than Earth because the high surface temperature of Venus had allowed the lithosphere prior to 500 Ma ago to fully participate in mantle convection. This assumes, of course, that the atmospheric greenhouse has been in place during most of Venus history. This free boundary condition allowed Venus to cool more rapidly and drop below the transitional Rayleigh number for chaotic behavior. Curiously, these authors argue that despite seafloor spreading and subduction, Earth has operated with a rigid boundary condition and thus has not cooled sufficiently to remove chaotic, time-dependent convection.

Other sources of episodicity include transitions between layered convection controlled by phase boundaries and whole mantle convection. Steinbach & Yuen (1992) speculate that a transition to whole mantle convection, which would be accompanied by a large vertical mass flux in the mantle, could lead to (significant) partial melting and large-scale resurfacing.

3.2.6.2 *Secular behavior* Is Venus "running down" in any sort of secular fashion? Certainly the rate of heat production in the interior is decreasing, as it is for Earth (and all planets). Solomon (1993) proposed that in the lower crust of Venus, the exponential dependence of strain rate on temperature would lead to a rapid cessation of tectonism as the planet went through gradual secular cooling. This shutdown in tectonism would be delayed in the highlands, which are hotter and/or have a thicker crust. This hypothesis was formulated to explain the putative pristine nature of craters in the plains and supposes that prior to approximately 500 Ma ago tectonic strain removed craters very rapidly. It is in contradistinction to models that require episodic events (e.g. Parmentier & Hess 1992, Arkani-Hamed et al 1993) to provide a catastrophe, the latest (or only) one occurring 500 Ma ago. The Solomon model proposes that the critical temperature in the crust required to rapidly decrease the strain rate was reached $\sim$500 Ma ago beneath the plains. Subsequently, the tectonically deformed plains were covered by volcanism to hide the evidence. This volcanism must have been fairly rapid ($\sim$50 Ma) to match the relatively small number of embayed craters; thus the model does not seem to completely avoid a volcanic catastrophe that might still require some profound event in the mantle.

Grimm (1993), following the wisdom of Solomon's basic idea, performed a specific calculation using the secular decrease in strain rate in a convecting mantle. The reciprocal of the present time scale for surface retention ($\sim$500 Ma) provides an upper bound of the presently operative strain rate, and is $O(10^{-16}\ s^{-1})$. Because only a fraction of the craters are even partially faulted, the actual strain rate is at least an order of magnitude less than this. If craters are to have been removed rapidly during an earlier episode, then the strain rate must be of order one-tenth the retention age (see Section 2.4) or $10^{-15}\ s^{-1}$. Therefore the strain rate must have decreased by a factor of 100 or more over some characteristic interval, say 1 Ga. Using parameterized convection techniques, Grimm shows how the net cooling of the mantle attributable to a decline in heat production leads to a corresponding decrease in strain rate by more than a factor of 100 over 1 Ga. However, the crustal strain rate change is diminished sharply, relative to the mantle, owing to its lower creep activation energy, and a mechanism must be available to transmit upper mantle strain rate to the upper crust.

Parameterized convection calculations based on the approach of Phillips & Malin (1983) yield a mantle strain rate change of about a factor of 50 at the crust-mantle boundary over the past 1 Ga. This particular solution produces the Earth-scaled value of surface heat flow of 74 mW m^{-2} (Solomon & Head 1982a). Here, the Moho is within the thermal lithosphere, whose base is defined by the 1500 K isotherm. Over 1 Ga, the Moho temperature decreases from 1176 K to 1061 K. By equating resurfacing rate to strain rate, it can be shown that the present number of craters is consistent with a model in which the resurfacing rate has been declining exponentially, and prior to 1 Ga ago craters were rapidly destroyed.

Grimm (1993) proposes that the corresponding crustal tectonic effect does not take place on Earth because of the uncoupling of mantle convection from the lithosphere resulting from the presence of a low-viscosity zone or asthenosphere (see Phillips et al 1991). Conversely, we argue that convective flow is strongly coupled to the lithosphere on Venus (Phillips 1986, 1990) and its effects would have been greater in the past because of greater heat production in the mantle. Indeed, the inference that tesserae could underlie much of the plains (Ivanov & Head 1993) is consistent with the concept of a decline in surface tectonic activity directly associated with a secular decrease in strain rate, as proposed by Solomon (1993). A decline in tectonism also implies a decline in volcanism because of a decrease in the number of extensional pathways for magma access to the surface.

3.3 *Summary*

We summarize an interior model for Venus. Our confidence level varies for different aspects of the model. We state with reasonable conviction that:

- If both Earth and Venus have the same heat source concentrations, then the interior convecting temperatures and mantle viscosities for these planets are about the same.

- Venus lacks an asthenosphere beneath its lithosphere and shear coupling of mantle convection to the overlying lithosphere is important, which is not the case for Earth.
- Like Earth, convection in the Venusian mantle is both time-dependent and chaotic.
- The strain rate in the mantle has decreased with time; this is reflected at the surface as a secular decrease in tectonism and volcanism.

With relatively less certainty we state that:

- A thermal boundary exists at the core-mantle boundary that initiates large, hot plumes infrequently.
- A residuum layer has existed from time to time in the upper mantle and has had a strong effect on the tectonic and magmatic history of the planet.

At the level of speculation we suggest that:

- Interior phase boundaries act as thermal boundary layers and can initiate plumes, which at times have carried upward a significant amount of the interior heat of the planet. These phase boundaries at times may have stratified flow in the mantle and at other times may have given way to whole-mantle convection.

4. LITHOSPHERIC DYNAMICS

4.1 *Rheological Considerations*

Models for the mechanical behavior of the interior structure of Venus rely on topography and gravity as boundary conditions and plausible postulates about rheology and temperature (unless the latter is considered a free variable). For lithospheric studies, rock strength is assumed to be controlled approximately by the weaker of frictional sliding or viscous (ductile) flow mechanisms (Brace & Kohlstedt 1980); strength as a function of depth can be characterized by a yield strength envelope (YSE). The ductile portion of the YSE defines the stress at which the rock will flow at the adopted strain rate (for a given temperature). The YSE extends to the base of the mechanical lithosphere (BML), which is defined somewhat arbitrarily (e.g. when the yield stress reaches 50 MPa). While frictional strength on preexisting fractures is largely independent of composition (Byerlee 1978), viscosity is not. Under an assumption of a basaltic crustal composition, the viscosity contrast between the crust and mantle (assumed olivine rheology) portions of the lithosphere is large, so lithospheric strength is strongly dependent on estimates of crustal thickness, as well as temperature gradient.

Our understanding of the tectonics of Venus depends strongly on our estimates of lithospheric structure and strength. Hypotheses regarding the support of large mountains, the detachment of the upper crust, subduction, and episodic plate tectonics depend implicitly, if not explicitly, on assumptions regarding the mechanical state of the lithosphere.

4.1.1 CRUSTAL THICKNESS AND TEMPERATURE GRADIENT Estimates of crustal thicknesses on Venus have come primarily from two sources: (*a*) the spacing of tectonic features, and (*b*) the depth of impact craters. These methods yield crustal thicknesses in the range of 20 km or less. Grimm & Solomon (1988) found an upper bound for crustal thickness of 10 to 20 km based on the application of linear viscous relaxation models to the depths of impact craters measured altimetrically by the *Venera* 15/16 spacecrafts. Because crust is most likely weaker than the underlying mantle and because impact craters show little evidence of viscous relaxation, a relatively thin crust is required by the modeling. However, consideration of non-Newtonian rheology and a viscoelastic lithosphere may enable thicker crusts to accommodate the crater depth observations (e.g. Hillgren & Melosh 1989).

Zuber & Parmentier (1990; see also Zuber 1987) applied models of uniform horizontal compression and tension of rheologically stratified lithospheres to examine instability growth as a function of wavelength. Wavelengths with the fastest growth times are related to spacing of tectonic features to constrain model parameters. Rift zones and ridge belts display two horizontal scales of deformation (e.g. for ridge belts, the wavelengths are defined by belt spacing together with the wavelength of structures within the deformed belts) indicating that both a strong upper crust and a strong upper mantle control deformation (as required to produce two maxima in the growth-rate spectra). Results for these features indicate crustal thicknesses of less than 20 km and temperature gradients, dT/dz, of less than 25 K/km. In Ishtar Terra, the mountain belts surrounding Lakshmi Planum have only a single, short (<20 km) wavelength of deformation indicating a crust greater than approximately 20 km in thickness and, because a long-wavelength instability has not developed, a mantle that is weak. This follows logically because the presence of a thick crust elevates the temperature at the Moho. Banerdt & Golombek (1988) also examined rift zone and ridge belt horizontal scales of deformation using classical elastic plate theory modified by more realistic considerations of rheology as described by strength envelopes. They find crustal thicknesses ranging from 5 to 15 km and dT/dz ranging from 5 to 15 K/km. In the Ishtar Terra mountain belts, they also find that the crust may be "considerably thicker" than 15 km.

These last two studies by necessity did not cover all regions of Venus, but converge to similar answers for crustal thickness: less than 20 km. Further, they support the notion that a "jelly sandwich" rheology exists in places, consisting of a strong upper crust overlying a weak lower crust overlying a strong upper mantle. The analyses also suggest that Ishtar Terra may be a region of thicker crust than the plains. Because in general there is a single wavelength of deformation in both the mountain belts (e.g. Akna, Freyja, Maxwell Montes) and the outboard tesserae (e.g. Atropos, Itzpapalotl, Fortuna tesserae) of Ishtar Terra, instability analyses (Zuber & Parmentier 1990) are not able to resolve crustal thickness variations across this region.

The upper bound on temperature gradients found from these studies is not as interesting as the crustal thickness estimates because it includes the average temperature gradient for Earth and, as we shall see below, a more interesting result would have been the lower bound for this parameter.

4.1.2 LITHOSPHERIC THICKNESS AND TEMPERATURE GRADIENT Considerations of both elastic and elastic-plastic rheology of the lithosphere yield estimates of lithospheric thickness and temperature gradient.

4.1.2.1 *Elastic thickness estimates* A planetary lithosphere loaded from above, below, or within will deform in response to the applied load. The resulting topographic profile can be used to constrain the rheological properties of the lithosphere. In the most basic of studies, the lithosphere is treated as a bending elastic plate of thickness T_e and parameters are adjusted until the modeled plate flexure matches the observed topography. Estimates are obtained of the magnitude of the applied loads and of the flexural rigidity of the elastic lithosphere given by $D = ET_e^3/[12(1-\nu^2)]$, where E and ν are Young's modulus and Poisson's ratio, respectively.

Head (1990) proposed that the Freyja Montes mountain belt and the region to its north is a site of ". . . large-scale crustal convergence and imbrication involving the underthrusting of the Northern Polar Plains beneath Ishtar Terra." In this region, Solomon & Head (1990) interpreted the topographic rise north of the linear depression flanking Uorsar Rupes as flexural in origin. Fitting topographic profiles with a broken-plate model, they obtained values of T_e in the range 11–18 km by adopting values of E and ν of 60 GPa and 0.25, respectively. In another study, Sandwell & Schubert (1992b) fit flexural profiles to the outboard topographic highs of four coronal trenches and obtained elastic thickness estimates of 10 to 60 km.

Cyr & Melosh (1993) predicted the deformation patterns around several coronae by combining regional, uniform stress fields with the local stress fields of early- and late-stage models for coronae. They obtained elastic thickness estimates of 10 ± 5 km (and 2 km in one case) and regional stress field magnitudes of approximately 10 to 60 MPa. This thickness result is consistent with the flexural solution obtained in northern Ishtar Terra, but less by factors of two to four than thickness estimates for Artemis, Latona, and Heng-O coronae.

4.1.2.2 *Elastic-plastic models* The meaning of an elastic thickness is not easily understood in terms of a stratified rheology, even in terms of an elastic core. A preferable alternative to simple elastic formulations is to solve the bending problem by treating the lithosphere as an elastic-plastic plate with plastic yielding defined by a yield strength envelope (Bodine et al 1981, Phillips 1990); the solutions then depend on dT/dz. This more complicated approach is possibly circumvented by equating the elastic and elastic-plastic bending moments for a given plate curvature (McNutt 1984). The latter moment is obtained by integrating the stress-distance product from the neutral plane using the lesser magnitude of either the yield stress or the stresses in the elastic cores of the crust and mantle.

Both of the flexure studies cited above ignored the crustal contribution and adopted the moment-matching technique (at the first zero-crossing "seaward" of the trench axis) to obtain temperature gradients. For the Freyja region, Solomon & Head (1990) and Sandwell & Schubert (1992b) estimated gradients of 14–23 K/km and 9.5–26 K/km, respectively. For the coronae Eithinoha, Heng-O, Artemis, and Latona, ranges of 6.8–24 K/km, 5.8–8.9 K/km, 3.2–4.8 K/km, and 3.5–7.7 K/km, respectively, were found by Sandwell & Schubert (1992b).

Plate curvatures for Eithinoha, Latona, and particularly Artemis are large, indicating that lithospheric bending is close to or at plastic saturation; here the moment-matching approximation breaks down. Fully elastic-plastic solutions for Artemis and Latona yield values of about 2 and 4 K/km, respectively. These are upper bounds neglecting the weak crust. Small temperature gradients (i.e. thick mechanical lithospheres) are required to support the large bending moments of these features in the face of large plate curvatures.

Johnson & Sandwell (1993) updated the earlier work of Sandwell & Schubert (1992b), expanding the number of coronae studied to twelve. Seven coronae yielded mechanical thickness estimates of less than 30 km using the McNutt technique. This is in good agreement with the mechanical thickness estimate obtained by Phillips (1990) for a dry olivine rheology, strain rate of 10^{-14} s^{-1}, and linear temperature gradient of 15 K/km (consistent with Earth-scaled expectations of dT/dz). Four of the remaining five coronae are too close to plastic saturation to obtain estimates using the moment-matching scheme.

What should we make of estimates of elastic lithosphere thicknesses (T_e) and dT/dz? First, we see a wide range in estimates of temperature gradients. While we should expect spatial variation in dT/dz, values as low as 2 K/km are certainly suspect (if not guilty of being obviously wrong). For those coronae where curvature indicates a large fraction of saturation in an elastic-plastic rheology, flexural interpretations should be questioned, and it seems worthwhile to consider alternative mechanisms. Relaxation of uncompensated coronal loads is one possibility (Janes et al 1992, Smrekar & Solomon 1992), and the formation of topography by mantle plume–head waves is another (Bercovici & Lin 1993). Generally, temperature gradients inferred at other locales appear to fall within our expectations of Earth-scaled values. However, given models for the origin of coronae (see Section 5.2.1), such regions of the lithosphere must have had anomalously high temperature gradients over all or part of their history.

4.1.3 ABSOLUTE CRUSTAL VISCOSITY There are several models for Venusian topography and tectonics that assume the lower crust can flow large distances on geologically reasonable time scales (e.g. Buck 1992, Bindschadler & Parmentier 1990). Our ability to understand the mechanical behavior of the crust is frustrated by the absence of a reliable flow law for proposed Venusian crustal material. Compositional measurements in the plains of Venus by *Venera* and *Vega* landings indicate a basaltic composition (Surkov et al 1984, 1987) and for that reason the Maryland diabase stress-strain rate measurements of Caristan (1982) have been used often for a flow law for the Venusian crust. There is reason to believe that this flow law provides, at best, a lower bound on the ductile strength of the presumably dry crust (Smrekar & Solomon 1992). For Maryland diabase, an increase in activation energy of 60% would remove the ductile regime in a 20-km-thick crust (temperature gradient 15 K/km; strain rate 10^{-14} s^{-1}). Zuber's work shows, however, that in the plains the lower crust can be weak (relative to the upper crust), and thus it's deformation must be controlled by its ductile strength and not its brittle strength. This constrains a tradeoff (for a given crustal thickness and strain rate) between the allowable bounds of activation energy and temperature gradient, two pairs of which are given above for Maryland diabase.

4.1.4 SUMMARY Our understanding of the rheological and thermal properties of the Venusian lithosphere is meager. For constructing models of tectonics we adopt the most likely set of parameters based on analyses to date:

1. Crustal thickness: $\lesssim$20 km in the plains.
2. Layered rheology: Strong upper crust, weak lower crust, strong upper mantle.
3. Rheology of Ishtar mountain belts (and outboard tesserae): Thick crust overlying weak upper mantle.
4. Nominal temperature gradient: 10 K/km $\lesssim dT/dz \lesssim$ 20 K/km.
5. The flow law applicable to the Venusian crust might be much stronger than Maryland diabase.

4.2 *The Role of Lower Crustal Flow*

Lower crustal flow can be induced in two ways: (*a*) by the pressures associated with topographic gradients (Weertman 1979, Smrekar & Phillips 1988, Smrekar & Solomon 1992), and (*b*) by direct coupling of velocity or shear stress associated with mantle convection (Bindschadler & Parmentier 1990; Bindschadler et al 1990, 1992b; Kiefer & Hager 1991b; Buck 1992).

4.2.1 TOPOGRAPHICALLY DRIVEN FLOW The most straightforward model of crustal dynamics is flow driven by topographic variations. An application of this problem to Venus was carried out by Smrekar & Solomon (1992) in their study of gravitational spreading of high terrain in Ishtar Terra. They used a finite element code ("TECTON," developed by H.J. Melosh and A. Raefsky) that allows analysis of deformation with a visco-elastic non-Newtonian rheology. Relaxation of a crustal plateau was considered by parameterizing the flow law, plateau boundary slope, crustal base boundary condition, crustal thickness, and temperature gradient. Their models indicate the importance of plateau relaxation by lower crustal flow, accompanied by only minor strain at the surface. Using a Maryland diabase flow law and a 15 K/km temperature gradient, Smrekar & Solomon found that for crustal thickness greater than 10 km, failure in a plateau (3° boundary slope, rigid basal boundary condition) higher than 3 km occurred in less one million years; relaxation of the plateau by 25% occurred in less than a few hundred million years. Relaxation times decreased markedly with increasing crustal thickness because the lower part of the crust became hotter. The results obtained are broadly consistent with the observed relaxation of some parts of the Ishtar mountain belts (e.g. Freyja), as inferred from the development of normal faults. More interestingly, they offer the choice that orogeny is recent, or that the crust is a great deal stronger than Maryland diabase, or both. For example, with a websterite flow law (Ave Lallement 1978), 25% relaxation had not been achieved by 1 Ga for the model parameters described above.

4.2.2 HOW WELL DOES CONVECTIVE FLOW COUPLE INTO THE LOWER

CRUST? Bindschadler & Parmentier (1990) analyzed the deformation of a rheologically stratified lithosphere subjected to flow stresses imposed by a density-driven flow representing the effects of sublithospheric convection. The formalism for this problem had been developed earlier (e.g. Ramberg 1968). The results show the importance of an interplay of buoyancy-driven stresses related to dynamic topography with forces introduced by flow and/or shear stress related to convective coupling. In particular, emphasis was given to the concept of a two-stage process for lithospheric response to convective stresses. Considering a downwelling or downgoing cold plume, the stages are: 1. A relatively rapid depression of the surface takes place until the lithosphere and the mantle flow are mutually compensated; this takes place with a characteristic time constant τ_2. 2. The crust then thickens by inward flow with a characteristic time constant τ_1. In isostatic calculations, the time constants τ_2 and τ_1 correspond, respectively, to the characteristic time it takes a topographic disturbance to become compensated and the characteristic time it takes for the compensated system to relax gravitationally. The net result is positive topography over a downwelling (see also Grimm & Phillips 1991). This has led to the idea that some (Bindschadler et al 1992b) to all (Buck 1992) Venusian highlands are over mantle downwellings. Bindschadler in particular has promoted the origin of crustal plateaus by crustal thickening over downwellings, emphasizing that this process is consistent with interpretations of early contractional features observed at crustal plateaus. Earlier we argued that crustal plateaus are altered by complex, polyphase deformation recording many extensional and contractional events.

The question is not whether crust is able to thicken over downwellings (or thin over upwellings), but rather how long it will take. In lithospheric response to perturbations from below, the crust acts as a filter, and most of the topographic energy lies in wavelengths of order 100 times the crustal thickness or greater. For a crustal thickness of 20 km, Bindschadler & Parmentier (1990) show that the long-wavelength response time (one-half steady-state amplitude) for crustal thickening over a downwelling exceeds one billion years for a constant viscosity model of 10^{21} Pa s. These authors also considered more realistic stratified rheologies; for a strong upper crust—weak lower crust—strong upper mantle, they found that thickening occurs only at very long wavelengths and at very long times. For example, for a 20-km-thick crust and a mantle viscosity beneath the strong mantle layer of 10^{21} Pa s, the wavelength that gives one-half the amplitude of the long wavelength limit is 1.26×10^5 km (obviously not very realistic!) and corresponds to a characteristic time τ_1 in excess of two billion years. These large values result because the upper mantle acts as a quasi-rigid boundary and flow is not efficiently coupled into the crust.

4.2.3 LARGE-SCALE FLOW The work of Zuber (1987) demonstrated that the lower crust of Venus has behaved in a ductile manner, deforming locally into boudinage structures. A number of models (e.g. Bindschadler et al 1992b), however, call for large-scale crustal flow driven by mantle convection. By large-scale, we mean that flow is able to take place over a horizontal scale that is many times the thickness of the crust. We summarize arguments against such a mechanism.

1. Arguments for crustal thickening invoking observations of radially-limited azimuthal contractional features associated with crustal plateaus or Ishtar Terra must also account for the full space-time history of strain, which will lead to many styles of faulting.
2. Building crustal plateaus by lower crustal flow requires transport distances that are exceedingly large. For example, material required to thicken the crust at Ovda Regio must travel in the lower crust (~ 10 km thick) at least 4000 km on either side of the Ovda structure. This must happen without surface evidence of the process.
3. An extremely weak lower crust would act as an asthenosphere; buoyant stresses from beneath the crust would be attenuated, leading to weakened surface uplift. By the same arguments we presented above regarding the ADCs estimated from gravity data, it is unlikely that a crustal asthenosphere exists. The lithosphere responds as an essentially coherent unit to buoyancy forces from below.
4. Models that require second-phase thickening or thinning of crust must have surface elevations that pass through zero during this stage. The corresponding ADC (or GTR) will pass through a singularity and then be negative for up to a billion years or more, eventually recovering to a small positive value (Bindschadler & Parmentier 1990, Simons et al 1992). Negative ADC or GTR values are not observed anywhere on Venus (Smrekar & Phillips 1991; M. Simons, personal communication).
5. The persistence of crustal plateaus over long periods of geological time, as inferred from the tessera study of Ivanov & Basilevsky (1993), argues that crustal compensation also persists for long periods, i.e. τ_1 is large.

4.3 *Is Crust Recycled?*

Is crust recycled on Venus? If it is not, why is the typical crustal thickness on Venus no greater than about 20 km? There would seem to be several ways for Venus to recycle crust. A tried-and-true method is to subduct the entire lithosphere, which works quite well for Earth but only in the case of oceanic crust. Continental crust seems quite happy to stay put for billions of years, although the upper part is removed by erosion and eventually recycled and some of the lower crust may be removed in places by delamination. For Venus, lithospheric subduction has been proposed around large coronae, as discussed earlier; we present arguments against this hypothesis in Section 5.2.2. If the crust is thick enough it will enter the garnet granulite or eclogite stability fields and become buoyantly unstable; however, 20 km is not thick enough under a plausible range of temperature gradients. The base of crustal plateaus, if they are compensated by deep roots, may contain these higher density phases. This would not seem to be an efficient way to recycle crust, given the limited number of crustal plateaus in existence today. Convection models have shown that the sinking of overthickened boundary layers beneath crustal plateaus could possibly entrain crust (Lenardic et al 1991, 1993). This phenomenon could take place independently of the origin of crustal plateaus (subsolidus or supersolidus flow); all that is required to overthicken the boundary layer is the thermal blanketing effect of thicker crust and cooperating lithospheric viscosities (which may be unlikely).

In Section 3 we reviewed several models that have been put forward to provide an explanation for the catastrophic resurfacing hypothesis (Schaber et al 1992). While we doubt that a catastrophic resurfacing event took place 500 Ma ago (see Section 2.4), the models are in and of themselves interesting as potential initiators of episodic tectonic and magmatic phenomena. Most of these models have implicit if not explicit recycling of crust. The alternative, of course, is that Venus does not make very much crust and does not return very much crust to the mantle.

5. ORIGIN OF TECTONIC AND MAGMATIC FEATURES

5.1 *Origin of Highlands*

Ishtar, crustal plateaus, and volcanic rises are the main highland types on Venus. Here we review ideas for their origins, and possible linkage. Ishtar Terra is unique on Venus, and possibly requires that some unique aspects come into play during its evolution [the "special pleading" of Solomon et al (1992)].

5.1.1 VOLCANIC RISES As noted in Section 2.2.3.1, it has long been argued that volcanic rises on Venus mark the sites of upwelling mantle plumes, i.e. volcanic rises are hotspots. We see nothing in particular that would replace this notion with a stronger, alternative hypothesis (e.g. all highlands on Venus are associated with downwelling and crustal thickening (Buck 1992)).

Despite the name, volcanism is limited for the most part in these regions and is expressed largely as shield volcanoes (and their attendant flows) associated with rifting (Senske et al 1992). Beta Regio is particularly striking in that it is perhaps the only extensively high area on the planet with a greater than average impact crater density (Phillips et al 1992). Along with the presence of tesserae, the distinct impression is of uplift and rifting of an ancient surface. Other volcanic rises display relatively less tesserae and more shield volcanism (Senske et al 1992), and have crater densities at or less than the planetary average. All in all, the implication is that buoyant uplift from below is responsible for volcanic rises but that partial melting is taking place at relatively great depths, producing large volcanic constructs but otherwise being relatively feeble. Herrick & Phillips (1990) and Phillips et al (1991) proposed that the plumes beneath volcanic rises had not reached the base of the lithosphere to enter a stage of massive partial melting. The most volcano-free of the rises, Beta Regio, also has the largest ADC.

In support of the plume hypothesis, Smrekar & Phillips (1991) showed that the most likely mode of compensation for volcanic rises is dynamic. A detailed study yielded the same result for western and central Eistla Regio, and was also consistent with observed tectonics (Grimm & Phillips 1992). Kiefer & Hager (1991a) showed that even the very large ADC value for Beta Regio ($\sim$300 km) could be fit by a convection model with a high viscosity lid.

What are the remaining issues in the hypothesis that plumes are responsible for volcanic rises? First, we must still assume that there is a hot thermal boundary layer in the mantle to generate plumes. Volcanic rises, of order 2000 km in diameter, number less than a dozen and are of uncertain age. This is consistent with our earlier contention (Section 3.2.3) that core-mantle boundary layer plumes are large and infrequent. Transit time through a 10^{23} Pa s viscosity mantle for a 1000-km-diameter plume (to make a 2000-km-diameter volcanic rise) is 65 Ma. The interpretation that plumes are generated at the upper mantle–lower mantle phase boundary seems less likely because—coronae notwithstanding (see Section 5.2)—hotspots would be much more numerous (and smaller).

Secondly, the contention that volcanic rises are (ironically) relatively magmatically dry and have large ADCs begs for an interpretation beyond the idea that all attendant plumes have been caught in various stages of not quite reaching the base of the lithosphere. While convection modeling shows (nonuniquely) that steady state plumes can account for large ADC values (Kiefer & Hager 1991a), the lack of magmatism is unexplained. The possibility that plumes are trapped at the base of a residuum layer (Phillips & Grimm 1990, Smrekar & Parmentier 1993) might provide a reason for this observation.

Finally, we note that the higher resolution and higher signal-to-noise ratio of *Magellan* gravity data over *Pioneer Venus* gravity data should introduce some caution into earlier interpretations of free-air anomalies in terms of large depths of compensation. In particular, it is becoming increasingly clear that the gravity signals over volcanic rises could contain a near surface component from flexurally supported volcanic constructs; this could compromise simple interpretations of large ADC values.

5.1.2 CRUSTAL PLATEAUS Here we consider how crustal plateaus are formed, and how they are deformed.

5.1.2.1 *Building crustal plateaus* Crustal plateaus are variably interpreted as surface expressions of major subsolidus flow in the crust in response to mantle downwellings (e.g. Bindschadler & Parmentier 1990, Bindschadler et al 1992b) or crustal thickening by magmatism related to late-stage evolution of mantle upwellings (Herrick & Phillips 1990, Phillips et al 1991). This disagreement has developed into the "hotspot-coldspot controversy" for the origin of crustal plateaus, which is a variation of the argument regarding the time scale of significant lower crust flow. Each of these end-member models has serious shortcomings.

Our objections to the downwelling or coldspot model have essentially been given in Section 4.2. The sine qua non of the model requires extremely large amounts of crustal flow inward over a cold mantle downwelling (hence coldspot) to create a plateau. Our main arguments against this are: (*a*) the extremely long time scales required, (*b*) the observed polyphase deformation that neither coldspot or hotspot models explain, (*c*) the lack of negative GTRs, (*d*) the extremely long distances that lower crust must be transported to build the larger crustal plateaus, and (*e*) gravitational evidence against the presence of a crustal asthenosphere.

In the hotspot model, volcanic rises evolve into crustal plateaus because of massive partial melting of the plume head—a process proposed for the formation of basaltic plateaus (oceanic and continental) on Earth (e.g. Richards et al 1989). Objections to this model include: (*a*) the aforementioned problem of generating polyphase deformation, (*b*) the need to do away with large shield volcanoes (because they are generally not associated with crustal plateaus), and (*c*) the lack of evidence for transitional forms between volcanic rises and crustal plateaus. This last point is not only evident from geological observations but also from the clustering of GTRs into two statistically distinct families associated with the two feature types (see Section 3.2.1).

The hotspot and coldspot models do agree that crustal plateaus are indeed regions of thickened crust but formed by supersolidus and subsolidus flow, respectively. The plateau-shaped physiography suggests that the plateau crust is thicker and/or less dense than plains crust, and that plateaus stand above the Venusian plains much as continents stand above ocean basins on Earth. ADC/GTR values for crustal plateaus suggest a large component of lithospheric compensation (Smrekar & Phillips 1991, Simons & Solomon 1993). Additionally, both models also propose that smaller crustal plateaus such as Alpha, Tellus, and Phoebe regiones, which lie 1–2 km above adjacent plains, evolve from larger crustal plateaus such as Ovda and Thetis regiones, which lie 4 km above adjacent plains. The smaller features represent a state of quiescence in which the plateau is no longer being actively built and so is collapsing gravitationally into the plains. Indeed, patches of tesserae observed in the plains would be the death rattle of crustal plateaus, which ultimately become entombed in plains volcanism.

It is not clear that this view is correct. The conclusions of Ivanov & Basilevsky (1993) that crustal plateaus are old and have undergone little gravitational relaxation over the retention age of their crater population of $\sim$700 Ma would suggest that crustal plateaus stay put. If much of the surface of Venus was actively deforming prior to $\sim$1 Ga ago, as outlined in Section 3.2.6.2, then exposed, low-lying patches of tesserae may simply have been low to begin with. Under this view, there is nothing special about the tectonic deformation of crustal plateaus compared to the plains other than their height, which makes it more likely that they would not be covered by volcanism today. However, patches of tessera in the plains often strongly resemble the high-standing, arcuate boundaries of crustal plateaus (Head & Ivanov 1993). Either these tessera patches represent relaxed crustal plateaus, or large-scale dynamical lowering of the surface in places has allowed the embayment of these features. Two other mechanisms could lead to the lowering of crustal plateaus: Simple cooling (following a magmatic origin) which would decrease elevation isostatically in analogy to Earth's oceanic lithosphere, or the development of denser phases in the root of a crustal plateau which would also lead to an isostatic decrease in elevation.

Our hypothesis for crustal plateaus is that 1. they are built magmatically, and 2. that the described sequence from large to small plateaus to tessera patches is essentially correct, but the time scale for this process can be exceedingly long. Our adoption of the former point is the natural outcome of our rejection of large-scale subsolidus crustal flow. However, we do not subscribe to the view that crustal plateaus are always associated with strong upwelling plumes that at first lead to volcanic rises, especially because of objection (*c*) to the hotspot model. Rather, we suggest that volcanic rises and crustal plateaus result from distinctly different magmatic styles.

In the simplest view, we propose that crustal plateaus result from the massive partial melting that can take place in a hot Venusian mantle at relatively shallow depths. McKenzie & Bickle (1988, Figure 7) show how a column of melt ranging from several to many tens of km thick could be produced (at the convecting temperatures of the Venusian mantle) by melting garnet peridotite in a mantle up to a minimum depth ranging from 100 km to 0 km, respectively.

There are two mechanisms that might lead to shallow upwelling of mantle, and thus extensive partial melting. First, crustal plateaus could be caused by plumes, but the difference could be that residuum blocks the plume at depth in the case of a volcanic rise and limits the amount of partial melting. Why should there be a difference? We speculate that either residuum is not pervasive in the upper mantle and volcanic rises, and crustal plateaus occur where residuum is and is not, respectively, or that crustal plateaus are associated with only those plumes hot enough to penetrate residuum.

The second mechanism recognizes that most of the outward convective heat transfer in the Venusian mantle is caused by internal heating and thus takes the form of diffuse upwellings whose temperature modestly exceeds the average convecting temperature of the mantle. Lithospheric stretching following buoyant uplift allows diffuse upwellings to rise to relatively shallow depths. The resulting partial melting adds to the thickness of crust, and the accompanying edifice stresses maintain the lithosphere in tension, which continues the upwelling and melting processes. This is essentially the hypothesis proposed by Solomon & Head (1982b) for the creation of the mammoth Tharsis plateau on Mars by magmatic construction. Once established, this is a passive phenomenon that allows melting to continue. The occurrence of extensional structures throughout Aphrodite Terra (and forming the southern boundaries of Ovda and Thetis regiones) supports the view that crustal plateaus are formed by the mechanism described here and may be linked to mesolands formation.

5.1.2.2 *Deforming crustal plateaus* The complex, polyphase deformation observed for crustal plateaus suggests a long history of deformation resulting from distinct episodes of contractional and extensional strain in the lithosphere. Solomon et al (1991, 1992) proposed that crustal plateaus may undergo deformation from strain supplied exogenically. Part of the evidence lies in the observation that for some crustal plateaus it is clear that boundary deformation can be traced into the surrounding plains. Once a crustal plateau is created, by whatever means, it may represent a relatively weaker portion of the lithosphere because it has a thicker than average crust (as implied by small ADC values). Because of their inherent strength boundaries, crustal plateaus may deform significantly in response to large-scale regional strain fields that may wax and wane on Venus over geological time. A crustal plateau may act as a "strain magnet" (Grimm & Phillips 1990), recording the complex history of strain in any given region. Whatever the level of deformation in the original formation of a crustal plateau, it has been overprinted significantly by other events under this hypothesis. Thus the tectonics of a crustal plateau would not be a guide to its origin. This exogenic deformation may not have been constant over time for the presently observed population of crustal plateaus. Rather, most of the deformation we see may have taken place early, marking the end of a prolonged period of high strain rate in the crust (e.g. Solomon 1993, Grimm 1993, Ivanov & Basilevsky 1993).

Under the concept of a strain magnet, the boundaries of crustal plateaus should exhibit the most recent episodes of deformation because they mark the transition in lithospheric strength. Observationally, this is borne out. The polyphase deformed interiors (amorphous terrain) and the more structurally coherent folds belts located along portions of the margins of these regions suggest that crustal plateaus have deformed progressively outward in time (including modest accretion of surrounding crust). The highest topography marking the outer limits of the crustal plateaus may represent the youngest deformation, or the most recently deformed crust. Intratesseral plains may preserve undisturbed regions along the boundary that were trapped between successive deformed belts. Minor episodes of volcanism are likely associated with extensional strain related to episodes of gravitational relaxation.

5.1.3 ISHTAR TERRA

5.1.3.1 *Existing models* Ishtar Terra is unique on Venus. Any model for its formation must acknowledge and address: (*a*) its irregular planform and complex topographic profile including high interior plateau, peripheral mountain belts, and surrounding high tesserae; (*b*) coherent contractional strain patterns across thousands of kilometers with no strain gradient despite the several kilometer change in elevation; (*c*) dominant fold ridges which parallel the orientation of their host mountain belt (and tesserae) with Lakshmi, which itself is generally undeformed at the surface; (*d*) a lack of evidence for extensional collapse at the highest elevations; (*e*) a mean crater retention age near the planetary mean; and (*f*) volcanic plains contemporaneous with tectonism locally. Previous models fall short in addressing these observations.

Pre-*Magellan* convergent plate-tectonic models (Crumpler et al 1986, Head 1990, Roberts & Head 1990b) set out to explain the general contractional nature of Ishtar structures and high topography; however the topographic and structural "details" observed in *Magellan* data (*a*)–(*d*) create structural and kinematic complications. These models imply subduction all around Lakshmi, yet, Uorsar Rupes and Vesta Rupes form the only slopes similar to terrestrial subduction trenches. The distribution and orientation of Ishtar structures are not consistent with subduction along either of these boundaries; southward subduction along Uorsar Rupes (Head 1990, Solomon & Head 1990) requires that Akna and Maxwell are left-lateral and right-lateral shear zones, respectively; northward subduction along Vesta Rupes would require the opposite shear sense in each belt. Subduction along the other Ishtar margins would require radial contraction in the subducting lower plate, or radial extension in the upper plate; evidence for the structural and kinematic framework implied by these models has not been identified, and the observations are counter to model implications.

Regional crustal shortening preserved in Ishtar deformed belts is variably interpreted as the result of mantle downwelling (Bindschadler et al 1992b) or mantle upwelling (Grimm & Phillips 1991). Neither model is sufficiently precise to predict detailed surface strain or topography. We discuss downwelling and upwelling models, and then present a model we believe addresses the observations summarized above.

The mantle downwelling model proposes that sinking mantle beneath Lakshmi Planum pulls crustal material inward and downward; subsequently, the crust thickens resulting in high surface elevation. When downwelling ceases, the thickened crust spreads gravitationally, decreasing in elevation. Many of the problems with this model are the same as those presented for crustal plateau formation by downwelling. Theoretical time scales for crustal flow are not geologically reasonable, and gravitational evidence argues against the presence of an inferred crustal asthenosphere. In addition, the model calls on huge volumes of lower crust to thicken Ishtar Terra, and therefore requires means by which to produce and transport large volumes of lower crust; evidence for these processes is not identified despite available high-resolution *Magellan* imagery, and the observed surface strains are inconsistent with thickening by lower crustal flow. Furthermore, if Ishtar elevation results from thickened crust, the highest regions would be thickest, and therefore these regions should display evidence of extensional collapse; yet the main part of Maxwell Montes, residing greater than 7.5 km above MPR, is void of extensional structures.

The upwelling model proposes that Ishtar Terra comprises a surface expression of a mantle plume that produces topography dynamically and by volcanic construction, and holds that the mountain belts result from incipient mantle return flow (Grimm & Phillips 1991, Phillips et al 1991). This model supposes that an upwelling plume impinged on a preexisting crustal plateau, which would make the lithosphere in this region weaker than its surroundings (Phillips et al 1991). Following earlier suggestions (Pronin 1986, Basilevsky 1986), the hypothesis is that outward crustal flow is attenuated as the strength barrier is reached at the plateau margin; this results in the accumulation of crust to form mountain belts. Because this is a region of enhanced temperature gradients and thick crust, lower crustal flow presumably takes place on geologically reasonable time scales [$O(10^8)$ Ma or less]. Thus, as in downwelling, mountain belts result from thickened crust; therefore the highest elevations, underlain by the thickest crust, should be loci of gravitation collapse, which is not observed. In addition, this model does not account for outboard tesserae with structures parallel to those of adjacent mountain belts; further, it is not clear that attenuation of outward crustal flow would result in steep slopes leading away from a plateau free of evidence of synchronous deformation.

5.1.3.2 *A mantle imbrication model* We discuss a model for the formation of Ishtar Terra (Hansen & Phillips 1993b,c) that begins with geologic and structural observations and is developed within the context of the theoretical and geophysical constraints outlined previously. Observations (*a*)–(*d*) require a means to deform huge expanses of crust without developing strain gradients, a means to support topography without greatly thickening crust, and specific relations between surface strains and topography. These observations can be accounted for by a model in which Lakshmi Planum represents an old crustal plateau that acts as a buttress to upper mantle residuum at depth, causing residuum to stack up locally around its perimeter during lithospheric contraction resulting from mantle downwelling (Figure 2). The greatly thickened upper mantle and mantle downwelling could be the result of an instability of the type described by Parmentier & Hess (1992). The upper part of the residuum, at temperatures less than T_f, is buoyant, and imbricates in a quasi-brittle fashion in response to shear forces associated with the instability developed in the lower, negatively buoyant, part of the residuum layer. This lower part flows downward in a ductile fashion, eventually becoming entrained in the general mantle circulation pattern. As uppermost upper mantle (buoyant residuum) translates at depth, a décollement forms between the upper crust and upper mantle. Shear stress is transmitted to the base of the upper crust, which responds by forming coherent ridge and valley folds with axes generally perpendicular to the direction of relative displacement; these structures form everywhere that the upper crust and upper mantle are displaced relative to one another. As the uppermost mantle is translated, it thickens locally, deforming in a brittle fashion because of its relative strength. The increase in residuum thickness supports surface topography.

Ishtar's irregular planform and topography reflect the shape and thickness, respectively, of residuum stacked at depth.

Lakshmi Planum forms a crustal nucleus for Ishtar deformation. The buoyant, thickened crust of Lakshmi cannot be drawn downward by mantle downwelling, and it acts as a buttress to the uppermost mantle at depth. We postulate that the "keel" of upper mantle that extends below Ishtar Terra is similar to the mantle keel than extends beneath continents on Earth (Jordon 1975, 1981). Topography of Ishtar deformed belts are supported by variably thickened residuum, and ridge and valleys within deformed belts result from folded, modestly thickened crust. The crust of Ishtar deformed belts, supported by residuum, not by thickened crust, is mechanically stable even at the highest elevations. Ishtar Terra would also be free of coronae, assuming coronae form as a result of melt instabilities within regions of tension. Flood-type volcanism can occur in regions of local lithospheric extension that allow upward movement of residuum and renewed, but modest, pressure-release partial melting. The crater density at Ishtar, close to the planetary average, attests to the tectonic and volcanic stability of the region. Quantitative tests of this model would include verifying that the high-resolution gravity constraints provided by the *Magellan* near-circular orbit can be satisfied, and examining the stability of the thickened residuum and the longevity of the downwelling.

5.2 *Coronae and Mesolands*

Here we discuss the nature of plumes that give rise to coronae, the hypothesis for subduction at the margin of some large coronae, and the origin of the mesolands because they are strongly linked to coronae.

5.2.1 MODELS FOR CORONAE ORIGIN There is widespread agreement that coronae result from buoyant diapirs or plumes impinging on the base of the lithosphere (Basilevsky et al 1986; Schubert et al 1989, 1990; Pronin & Stofan 1990; Stofan & Head 1990; Stofan et al 1991, 1992; Squyres et al 1992a; Janes et al 1992; Hansen & Phillips 1993a). *Magellan* images provided the strongest observations in support of this notion—the oldest tectonic features of coronae are radial fractures; this is the deformation that results from pushing up an elastic plate from underneath (Squyres et al 1992a). However, there is disagreement as to whether the diapirs are thermally driven plumes arising from the core-mantle boundary, and are genetically related to volcanic rises (e.g. Stofan et al 1992, Janes et al 1992, Squyres et al 1992a), or if they are compositional diapirs resulting from melt instabilities within the upper mantle (e.g. Tackley & Stevenson 1991, Phillips & Hansen 1993). Any model for diapir genesis must explain (Stofan et al 1992): (*a*) the widespread global, but not spatially random, dis-

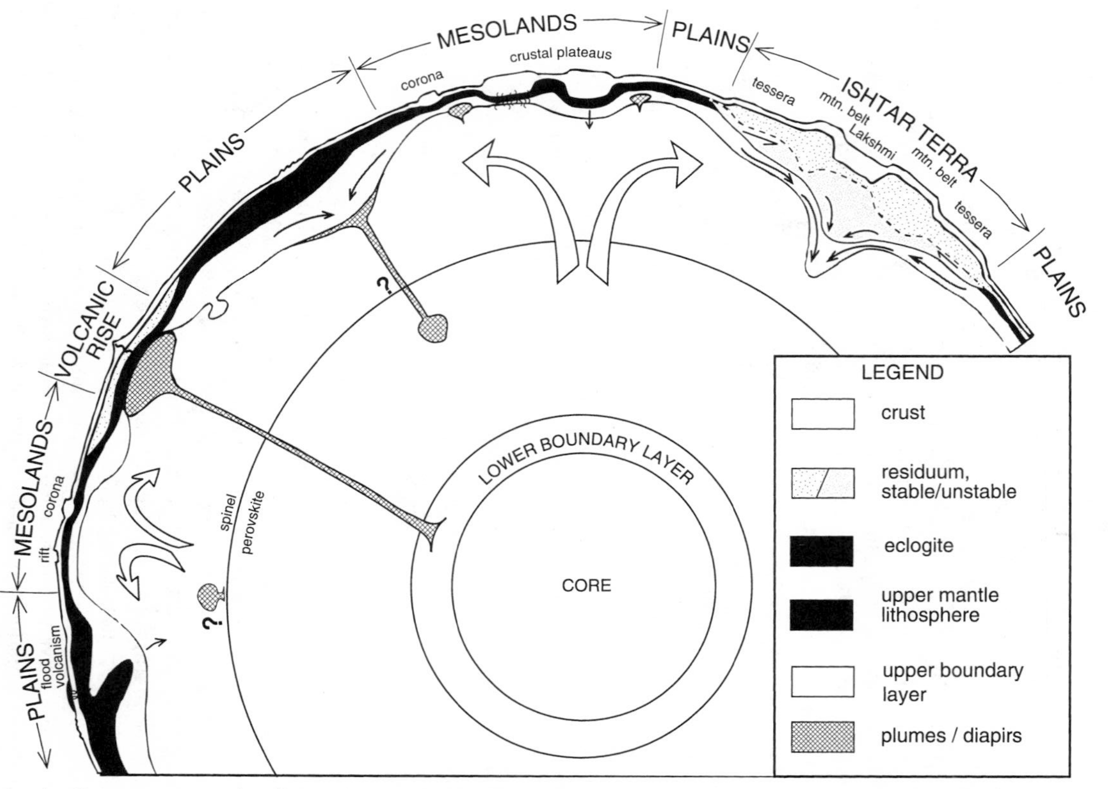

Figure 2 Cartoon cross section illustrating our model of Venus tectonics. Illustration is not to scale and is highly schematic.

tribution of coronae; (*b*) the local concentration of coronae within rift zones, or regions of crustal extension, and the negative association of coronae with both volcanic rises and planitae (Herrick & Phillips 1992); and (*c*) the wide range in corona sizes (100–2600 km) and mode in the size distribution of 200–300 km.

In Section 3.2.3.2 we argued that very long plume transit times through the mantle make it unlikely that the diapirs responsible for coronae origin are thermal plumes arising from the core-mantle boundary. Tackley & Stevenson (1993a) discuss a mechanism for spontaneous and self-perpetuating magmatism that could account for coronae (Tackley & Stevenson 1991, 1993b). In this model, regions of partial melt, or at least incipient partial melt, exist in the upper mantle where the geotherm approaches the solidus temperature. A Rayleigh-Taylor-like instability will develop if an infinitesimal upward velocity perturbation is applied to some portion of the partial melt zone. This is so because the perturbed element of rock will undergo an increment of pressure-release partial melting and thus become more buoyant. The buoyancy will increase the upward velocity leading to still more partial melting and a rising, growing compositional plume will develop until the growth rate is balanced by outward percolation of melt. Lithospheric extension could easily supply the required velocity perturbations, and the strong correlation of coronae with extensional tectonics (e.g. Parga Chasma) provides support for this view. Magee Roberts & Head (1993) have noted the tendency for those coronae with large volcanic flow fields to be associated with lithospheric extension. This is interpreted by them to be the result of increased partial melting of thermal plumes that reached shallower depths than normal because of lithospheric stretching. The Tackley-Stevenson mechanism would also yield a positive correlation between the amount of stretching and the amount of melt produced.

Since not all coronae are clearly associated with extensional tectonics, the Tackley & Stevenson mechanism would not necessarily explain the global distribution of coronae. It is possible that coronae are the result of thermal plumes generated at an interior thermal boundary layer associated with a phase transition—e.g. the spinel-perovskite phase boundary separating the upper and lower mantle at ~700 km depth—where the plume transit times might be 100 Ma or less. If there is a barrier to convection in the interior, then much of the heat from beneath the barrier will be removed in the form of hot plumes. How many plumes would be required to remove this heat if the plumes are also required to make coronae? Following the analysis of Whitehead & Luther (1975), several thousand 300-km-diameter coronae are required to be active at any one time if we assume that plumes carry most of the basal heat (Bercovici et al 1989).

Compared to the observed population of ~ 400, coronae could not possibly represent the total heat output of the lower mantle and core. Either (*a*) such an internal thermal boundary layer does not exist on any permanent basis (e.g. Steinbach & Yuen 1992) and cannot be used to explain the origin of coronae, or (*b*) plumes associated with volcanic rises must represent the majority of heat removal from this interface, and corona–producing plumes play a minor role (with the possible exception of Artemis).

We cannot rule out completely the origin of coronae from thermal plumes associated with relatively shallow thermal boundary layers in the mantle, so it is possible that the global corona population is a mix of a spatially random population (perhaps the largest coronae) associated with this mechanism plus a second, regionally biased, population associated with extensional tectonism and caused by melt instabilities in the uppermost mantle. We note, however, that the absence of associated extensional tectonism with some coronae does not rule out a melt instability origin, and our conclusion is that this is the dominant mechanism in the formation of coronae. The intimate association of magmatism with coronae, including multiple phases of volcanism of apparently different composition (Hansen & Phillips 1993a), also suggests a compositional origin for the causative plume, although pressure release-partial melting associated with a purely thermal plume cannot be ruled out.

5.2.2 SUBDUCTION AT CORONA MARGINS? McKenzie et al (1992a) and Sandwell & Schubert (1992a,b) proposed that the peripheral troughs associated with some large coronae (Artemis, Latona, Eithinoha) and two chasmata (Dali and Diana) bordering corona-like structures are the sites of lithospheric subduction (see Figure 1 for locations). The strongest argument for this hypothesis rests in topographic analogy to terrestrial subduction zones. In particular, these troughs are arcuate in planform and bordered by parallel but asymmetric highstanding topography, the higher of the two border features on the concave side of the trough. Stofan & Bindschadler (1993) have noted that asymmetric topography is a common feature of all coronae with troughs regardless of size. The convex high is reminiscent of the flexural highs observed seaward of terrestrial trenches. Sandwell & Schubert (1992b) matched flexural models to the outboard topographic highs as discussed in Section 4.1.2; attempts to test the subduction hypothesis by comparing the bending moment supplied by inboard topographic loads to the model bending moments obtained by matching topographic profiles proved inconclusive (Sandwell & Schubert 1992a).

At Artemis, Latona, and Eithinoha, the troughs subtend an arc greater than 180°, so underthrusting would lead to strong contraction in the outboard lithosphere, which is not observed. Instead, Sandwell & Schubert

(1992a,b) propose that the outboard lithosphere is rolling back in response to the onslaught of a spreading corona. The subducted lithosphere would fail in tension; the most one might observe at the surface are radial fractures propagating outward from the trench site, for which there is some evidence (Stofan et al 1992).

Through study of high-resolution *Magellan* images, Hansen & Phillips (1993a) documented the presence of pre- to syn-trough tectonic fabrics that trend across a proposed convergent plate boundary at Latona Corona; these structures are not consistent with a subduction interpretation. Studies of this type at other proposed plate boundaries are less conclusive (G. Sharer, personal communication). Hansen & Phillips also documented crosscutting relations that indicate east- and northeast-trending fractures that parallel, and locally coincide with, Dali (and Diana) chasma postdate formation of radial and concentric fractures associated with proposed "upper-plate" coronae. These relations are inconsistent with a subduction interpretation, but they are consistent with Diana and Dali chasmata formation largely postdating spatially associated coronae.

Mechanically, subduction on Venus would appear to be difficult because of the expected large positive buoyancy of the lithosphere resulting mainly from the high surface temperature (Phillips & Malin 1983). Detachment of almost the entire crust would be required to impart negative buoyancy to the remaining lithosphere (Phillips et al 1991). Alternatively, if the lithosphere could be underthrust to depths of 50–100 km, the presumed basaltic crust would undergo a phase change to garnet granulite and eclogite, and resistance to subduction would be diminished (Phillips et al 1991, Sandwell & Schubert 1992a).

The subduction hypothesis remains as a tantalizing idea. The acquisition of new high-resolution gravity data in the latter part of 1993 and into 1994 resulting from the circularization of *Magellan*'s orbit, along with detailed structural mapping, should allow significant progress on this question.

5.2.3 MESOLANDS ORIGIN Morgan & Phillips (1983) proposed that topography on Venus less than about 1 km above the MPR could be supported solely by thermal compensation in the lithosphere, the upper bound being dependent on the assumption of a mean thickness of thermal lithosphere of 100 km. Areas of excess heat flow (compared to the planetary average) will thin the lithosphere, replacing colder material with hotter, less dense material that will support elevated topography. Elevations in excess of the estimated upper bound would be supported by crustal thickness variations. We suggest that the mesolands fit the model put forward by Morgan & Phillips, and this view is supported by the generally positive but modest geoid anomalies associated with mesolands (Bills et al 1987, McNamee et al 1993, Nerem et al 1993). We associate the excess heat flow to broad, diffuse convective upwellings that characterize internally heated convection. In addition to cooling linked to major downwellings, these upwellings must be the dominant heat transfer mechanism out of the mantle.

By virtue of their excess heat flux, mesolands would be characterized by large amounts of partial (or incipient partial) melt in the upper mantle and thus are fertile ground for generating coronae by the Tackley & Stevenson (1991, 1993a) mechanism. Further, rifting and the development of chasmata are natural by-products of thermal uplift, and create a tensional environment to initiate partial melt diapirs that will form coronae. In addition, the crustal plateaus believed to be the youngest (Ovda and Thetis regiones) have a close spatial association with mesolands and may have formed from runaway partial melting events as described in Section 5.1.2.1

Anderson et al (1992) used the level of correlation of terrestrial hotspots with regionally broad, low-velocity anomalies in the upper mantle to argue that hotspots, as well as massive volcanism that forms basaltic plateaus, are the result of a juxtaposition of tensional portions of the lithosphere with broad areas of high temperature in the mantle. We believe this is a good analogy for the formation of mesolands and, in the extreme, the formation of crustal plateaus on Venus.

5.3 *Origins of Planitae*

The most likely interpretation of planitae on Venus is that, at least in their large-scale aspects, they are supported dynamically by sheet-like and possibly cylindrical downwellings associated with cold return flow in the mantle (Bindschadler et al 1990, 1992b; Phillips et al 1991). Large apparent depths of compensation for Guinevere/Sedna Planitae of 100–150 km and for Niobe Planitae of 180 km (Phillips et al 1991) support this view. Herrick & Phillips (1992) showed that inversions of long-wavelength gravity and topography data for a dynamic mantle component and a static lithospheric component produced sheet-like downwellings beneath the planitae. Zuber (1990) argued that the most likely origin of ridge belts is compressive stresses associated with large-scale convective downwellings. An alternative view is that planitae mark the sites of hotspot upwellings where the crust has thinned substantially in response to initial dynamic uplift (Buck 1992).

The simple view that planitae and their attendant ridge belts represent a contractional response to convective mantle downwellings (e.g. Squyres et al 1992b) must be tempered with observations of heterogeneous tectonism and volcanism. An excellent review of this issue is given in Solomon et al (1992). It is clear that there have been a number of horizontal scales of deformation that have operated, indicating a range of lithospheric responses—from deformation of the uppermost few hundred meters of the crust, to the entire strong upper crust, to the crust and strong upper mantle. Three particularly challenging questions to the convective downwelling hypothesis are: 1. What is the origin of tensional tectonism? 2. What is the origin of volcanism? 3. Why aren't deformation belts distributed uniformly in the planitae? Unfortunately, regional-scale variations in lithospheric properties, crustal thickness, and heat flow will probably lead to spatial and temporal overprinting that will be difficult to decipher in light of models with extremely general predictive capabilities. Thus "proof" of one interpretation over another may prove exceedingly difficult. At a geological level, models for Venusian deformed belts in the planitae must explain: (*a*) the elevated nature (0.5–1 km) of deformed belts relative to adjacent plains; (*b*) the similar orientation of principle strain in deformed belts and adjacent "undeformed" plains; (*c*) local orthogonal contraction and extension in ridge and fracture belts (e.g. Lavinia Planitia); and (*d*) local along-strike morphological changes from ridge belts, marked by fold ridges, to hybrid fracture belts which preserve ridges, but are dominated by fractures, faults, and local paired graben (e.g. Atalanta-Vinmara Planitiae).

Underlying planitae may be the vestiges of a period of intense deformation associated with high crustal (Solomon 1993) and/or mantle (Grimm 1993) strain rates. Following an episode of lithospheric extension, there has obviously been a complex interplay of deformation and volcanism in the planitae, and in many places ridge belt formation is older than most or all of the volcanic plains units (Solomon et al 1992). Thus a model that assigns convective downwelling to planitae must allow partial melt to be generated in this environment. We suggest that the most likely possibility for this is that portions of the downwellings are manifested as delaminations of the upper part of the mantle, possibly including separation at the Moho (e.g. Bird 1988). These detachments are replaced by hot mantle material from below that will undergo partial melting because it is emplaced closer to the surface. Local extensional environments in the lithosphere then allow access of magma to the surface. The relatively shallow depth of the magma source would lead, for the most part, to the formation of volcanic plains rather than constructs, although other factors, affecting magma viscosity, may be involved. The emplacement of this hot mantle material should lead to elevated topography, but unlike our proposal for the mesolands, this is a transient event with smaller levels of partial melting.

It is clear that ridge belts are genetically related to widespread distributions of wrinkle ridges and fracture belts are similarly related to extensional grooves (Squyres et al 1992b). Thus the deformation belts appear to be one manifestation of lithospheric response to large-scale regional stress fields. Those cases where extensional features (fractures, graben, fracture belts) are orthogonal to contractional features (wrinkle ridges, ridge belts) are easily understood in terms of principal strain axes in a contractional environment—extensional features will form in accordance with the axis of least contractional strain. However, in many cases contractional and extensional features are not orthogonal, and in a number of instances contractional features change to extensional features along strike. Solomon et al (1992) offer two possible explanations for this: (*a*) Fracture belts are extensional structures that occur in response to local igneous intrusion or lithospheric heating, and (*b*) fracture belts occur as a result of stretching over the crest of contractional structures. These authors favor the former explanation because of a close association between fracture belts and volcanism, but there is a chicken-and-egg issue here in that the occurrence of a tensional environment may simply favor magmas reaching the surface.

Our model for deformed belts follows from our model of Ishtar Terra. Downwelling in the mantle is responsible for broad-scale regional lithospheric contraction normal to the trend of ridge belts. The uppermost mantle deforms in a heterogeneous fashion, as evidenced by the spacing between belts; however, because planitae lack a nucleating crustal plateau, the uppermost mantle is not buttressed at depth, and therefore does not imbricate as it does under Ishtar Terra. Orthogonal ridge and fracture belts may result from crustal shortening due to lithospheric contraction. The basic spacing of ridge belts and the deformational scale within individual belts result from instabilities developed in a jelly-sandwich rheology. Ridge belt elevation results chiefly from crust thickened by minor lower crustal flow. Along-strike changes from ridge belt to hybrid fracture belts record gravitational collapse following crustal thickening beyond the mechanical limit of the upper crust.

5.4 *What is Holding Topography Up (and Down)?*

The enigma of how large-scale topography is maintained on Venus in the face of high surface temperature remains and is punctuated by the presence of the circular Cleopatra impact crater sitting happily undeformed on the steep slope of Maxwell Montes.

Turcotte (1993) proposed that topography is supported by the strong lithosphere that results from a low temperature gradient. The low temperature gradient exists because heat flux from the convecting mantle is subdued while the mantle is rewarming following an episode of extremely active plate tectonics and vigorous subduction cooling ~500 Ma ago, as required by a model of catastrophic resurfacing (Schaber et al 1992). The attendant lithosphere supports topography by thermal isostasy, but is able to achieve a greater maximum elevation than that found by Morgan & Phillips (1983) because of the much thicker lithosphere: $\sim$300 km vs 100 km. Additionally, the stronger crust is able to locally support mountain ranges, such as Maxwell Montes, by Airy compensation.

Thermal isostasy with a 300-km-thick lithosphere will also produce the large GTR values that are ascribed by others to dynamic support (e.g. Kiefer & Hager 1991a). Turcotte's proposal is interesting but speculative in that it requires the initiation of a global and essentially spontaneous subduction event to substantially cool the mantle. Further, Turcotte (1993) does not explain how the topography is created in the first place. If the topography is generated during the plate tectonics episode (presumably the most likely time), the lithosphere is thin and won't support topography by Turcotte's own arguments. If too much time has passed, then the lithosphere is thick and static and it would seem to be difficult to generate the observed highland features. There would have to be a window intermediate in the 500 Ma cycle in which there is enough tectonic activity in the lithosphere to create the highlands, their attendant mountain belts, and the planitae, but the lithosphere would at the same time have to be strong enough so that these features would stick around.

Our view is that topography is supported by a combination of dynamic, thermal, and Airy crustal compensation. At Ishtar Terra we have proposed that thickened mantle residuum is supporting topography. Because we have argued that the time scale for lower crustal flow most likely exceeds 1 Ga, once topography is compensated within the crust, it is relatively stable; although there may be an early episode of partial collapse, relaxation times are very long. The probability that the viscosity of the crust may be a good deal higher than has been assumed in modeling studies strengthens this argument. Finally, we point out that in those regions of the upper mantle in which stable partial melt residuum exists, the mantle layer that transfers heat by conduction can be substantially increased, leading to low temperature gradients and a strong lithosphere (Parmentier & Hess 1992).

5.5 *Is there Horizontal Lithospheric Motion?*

The mountain belts of Ishtar Terra and the northern ridge belts are regions of maximum crustal shortening. However, strong evidence for lithospheric-scale subduction and recycling has not yet been recognized in the geological record of Venus. In Section 5.2.2 we presented arguments against subduction at coronae boundaries. A subduction interpretation outboard of the Ishtar Terra mountain belts may be consistent with broken-plate elastic models representing underthrusting (Solomon & Head

1990), but it is not consistent with surface strain patterns (Section 5.1.3.1).

Earlier predictions that much of Aphrodite Terra is a site of lithospheric divergence (Head & Crumpler 1987, 1990) are invalidated by *Magellan* data (Solomon et al 1992). Evidence for lithospheric extension is limited to two classes of structures: quasi-circular coronae and associated chasmata of the mesolands, and volcanic rises with symmetric rift zones. Although both sites display evidence for volcanic flux, horizontal displacements in each case are probably limited to a few tens of km. In the case of eastern Aphrodite Terra, early-formed coronae, which are generally circular in planform, are cut by fractures of Diana and Dali chasmata, interpreted as rifts (e.g. Schaber 1982); however, the amount of crustal extension must be very small, less than 10% across the entire width of the zone as the coronae remain circular, and are not elongate in a direction normal to the trace of the rift (Hansen & Phillips 1993a). In addition, symmetric rifts that radiate outward from the broad topographic domes of volcanic rises show limited local extension (<30%) normal to strike. Along strike away from the topographic high, the rift trough and associated structures broaden and splay outwards, recording progressively less crustal extension. It is evident that these rifts are simply a response to crustal doming. Not surprisingly, few large-offset strike-slip faults are recognized planetwide (Solomon et al 1992), although evidence for limited local horizontal shear is observed in regions of bulk crustal shortening and extension (e.g. Squyres et al 1992b).

6. THE REALLY BIG PICTURE—GLOBAL TECTONISM AND MAGMATISM

After a brief review of Venus geology, we provided an essentially theoretical view of mantle and lithospheric dynamics that was then used as a framework to provide models for the origin of specific physiographic features on Venus. We continue in this vein, now attempting to put together a model for global tectonism and magmatism. A cartoon of our Venus can be found in Figure 2.

Our view of Venus is simply this: Because Venusian mantle convection is driven almost entirely by internal heating, it is dominated by strong, cold downwellings from the upper thermal boundary layer and by broad, diffuse upwellings; these are manifested at the surface by the planitae and mesolands, respectively. Everything else is nothing more than minor blemishes superposed on this major system. There is but one Ishtar and but two crustal plateaus that by their size may have been the most recently active (Ovda and Thetis regiones). These are the result of unusual events caused, respectively, by (*a*) detachment of residuum and downwelling in the vicinity of a preexisting crustal plateau, and (*b*) runaway partial melting events associated with broad upwellings creating mesolands. There are but a handful of volcanic rises, which are probably linked to upwelling, hot plumes from a sluggish lower thermal boundary layer. We emphasize that this view is based on the Venus that we see today from images, altimetry, and gravity data, and does not necessarily apply to a Venus more than 1 Ga in the past.

While large-scale deformation in the lithosphere is still controlled by mantle convection, crustal tectonics linked directly to convective stress coupling passed through a critical strain rate perhaps 1 Ga ago, after which the rate of tessera production greatly diminished. Contemporary tessera formation may be taking place only at Ishtar Terra and in the mesolands.

This view of Venus is built on the following concepts developed earlier in this paper:

1. Mantle convection in Venus operates at about the same viscosity and Rayleigh number as on Earth.
2. The upper interface of convection acts as a rigid boundary, at least on a time scale of less than 1 Ga. Lithospheric recycling to the interior is a minor process with the possible exception of upper mantle detachment beneath planitae.
3. Venus lacks an asthenosphere; convective stresses couple directly into the lithosphere with a vigor that has decreased with time.
4. Only a modest amount of heat crosses the core-mantle boundary, leading to infrequent, large plumes.
5. The outermost part of the upper thermal boundary layer is relatively static, but it will initiate cold downgoing plumes and/or detach on occasion.
6. A residuum layer exists at various places in the upper mantle. At times it is disrupted, mixing with normal mantle, although a disruption may take the form of a detachment and behave in a semi-rigid manner in the early stages of an event.
7. The crustal thickness is nominally $\lesssim 20$ km in the plains but has achieved greater thickness beneath crustal plateaus, beneath Ishtar Terra, and beneath the mesolands.
8. In the plains there is a stratified rheology of strong upper crust, weak lower crust, and strong upper mantle.
9. Venus has a nominal temperature gradient that lies between 10 K/km and 20 K/km; local variations may fall outside of this range.
10. The lower crust is able to deform in a ductile manner in response to horizontal forces. Tectonic features have resulted up to the scale of ridge and fracture belts.

11. Large-scale lower crustal flow has not taken place and thus does not offer an explanation for high-standing topography resulting from convective downwellings and subsequent crustal thickening.
12. Crustal recycling is minor and possibly occurs only in the deep roots of crustal plateaus that have transformed to denser phases such as eclogite. Large-scale underthrusting may also allow return of crust to the mantle below, but we find no convincing evidence for this.

Crust on Venus today is being generated dominantly in the mesolands. In this sense the mesolands are analogous to oceanic spreading centers in that crust is generated in response to diffuse mantle upwellings and partial melting. On Venus this is manifested at the surface as coronae and flood and constructional volcanism. The mesolands differ from terrestrial spreading centers in that the crust accumulates vertically rather than horizontally (as results from seafloor spreading). Crust may also be formed in the intermediate stages of plume evolution (Phillips et al 1991), but the evidence today is that most volcanic rises are relatively magmatically dry. More likely, crustal plateaus are an exuberant form of mesolands development. Over the history of Venus it would seem that significant crust could be developed. Either crustal development is inhibited by the presence of residuum in the upper mantle (Phillips & Grimm 1990), or significant amounts of crust are being recycled in contradiction to concept 12. above. It is possible that the crust thickens enough beneath mesolands to spawn phase changes to buoyantly unstable garnet granulite or eclogite, or that mesolands eventually evolve into crustal plateaus where such phase changes take place deep in plateau roots. These denser portions of crust may detach and eventually recycle in the mantle. The mesolands are presently the most tectonically/magmatically active areas of the planet, a conclusion strongly supported by crater data (Phillips et al 1992).

We suggest that a residuum layer has had a profound effect on the evolution of Venus. Jordan (1975, 1981) proposed, mainly on the basis of seismic evidence, that beneath the continents on Earth, a layer ("tectosphere") that is depleted in basalt and enriched in magnesium, i.e. a residuum, extends to several hundred km depth. This layer is held to be buoyantly neutral with respect to the surrounding mantle because positive chemical buoyancy is offset by colder temperatures. The tectosphere could be the fate of some of the residuum produced by partial melting in the oceanic mantle, being swept into the continents by processes associated with subduction (see Kaula 1990b). We suggest that residuum is also swept around in the mantle of Venus; lacking the strong, focused lithospheric subduction of Earth, its distribution may be more widespread on Venus. Accumulations may nucleate beneath crustal plateaus, where the mantle may be hotter because of thermal blanketing effects (e.g. Lenardic et al 1991). Buoyant residuum may form stable keels against mantle convection, may compensate large-scale topography, act in places as a barrier to mantle ascent (e.g. plumes), inhibit partial melting, and may occasionally undergo large-scale detachment with dramatic tectonic consequences.

The planitae mark the sites of mantle downwellings. These may be part of the normal mantle flow pattern, or may initiate in the upper part of the thermal boundary layer as plumes or semi-rigid detachments. Leitch & Yuen (1991) showed that formation of cold downgoing plumes is chaotic in variable-α convection (in which the coefficient of thermal expansion α decreases strongly with increasing pressure), and the imposition of a rigid boundary condition increased the number of instabilities in the upper thermal boundary layer. Because the upper thermal boundary layer in internally heated convection is able to store a great deal of potential energy, instability generation, including detachment, has provided the major source of convective stress coupling into the Venusian lithosphere. By the chaotic nature of this activity, we expect that strain in the overlying lithosphere to have been both spatially and temporally chaotic. The net result is that the lithosphere will be constantly subjected to a complex pattern of contractional and extensional strain. The level and rate of this strain have decreased over time as expected from the secular decrease of heat production, and the prolonged period of extensive tesserae formation ended about 1 Ga ago (Solomon 1993, Grimm 1993).

In Section 2.4 we argued from crater observations that at one time or another partial melts have been available on a regional basis everywhere on the planet; the coincidence of an extensional strain environment has allowed volcanism to occur as a passive phenomenon. The local style of volcanism will depend, inter alia, on the specific stress state of the lithosphere, the local state of the upper boundary layer, and the thickness of residuum in the upper mantle. These occurrences of volcanism will be maximized in those regions of hot upper mantle, i.e. the mesolands. Further, volcanism has secularly declined over the history of the planet and not just directly from the reduction in heat production. As the strain rate and corresponding deformation decreased in the crust, the opportunities for magma to reach the surface have also diminished; correspondingly, the ratio of intrusive to extrusive igneous activity has increased.

The chaotic nature of strain in the lithosphere also led to polyphase deformation and tessera formation in both crustal plateaus and lower-lying plains, as discussed in Section 5.1.2. As pointed out in that section, Ovda and Thetis regiones may be the most recently active crustal plateaus in the sense of their formation by crustal thickening. Once this active process ceases, crustal plateaus undergo a slow gravitational collapse governed by a time constant (τ_1) of 1 Ga or greater. These features slowly lose relief, but, acting as strain magnets, their areal extent may be increased by accretion of surrounding crust. Eventually these plateaus are reduced to patches of old tesserae embayed by volcanism. The widespread distribution of low relief tesserae in the plains attests to the importance of both crustal plateau generation and relaxation as well as in situ deformation of the plains themselves over time periods prior to 1 Ga ago.

Volcanic rises are but an energetically minor phase of heat transfer out of the interior, perhaps accounting for the relatively feeble heat produced in a cooling core. They are distinct in both their large gravity anomalies and styles of volcanism, both pointing to relatively deep sources in the upper mantle and possibly interaction of thermal plumes with a residuum layer.

The complex and chaotic nature of mantle downwellings, the spatial variability of residuum, and the relatively well-defined regions of broad upwellings suggest that at any given time the upper mantle of Venus can be divided into a finite number of domains, each characterized by one type of a limited number of characteristic behaviors. On Earth, domains might be defined by upper mantle beneath continents, beneath young oceans (e.g. the Atlantic) and beneath mature subducting oceans (e.g. the Pacific). Venus is different because it does not have plate tectonics and is the epitome of a planet with vertical tectonics. The crust is not recycled, but instead acts like a rug that locally rips and crumples as a result of relative displacement of domains within the mantle below; in a secularly decreasing manner, it has fractured and folded, domed upward and downward; it has been locally stretched by magmatic diapirs, and responded by fracturing, folding, and locally faulting when driven by convective stresses within these domains. The local rips and rumples in the crustal rug of Venus are never far from where they formed, and never far from where the rock that comprises it differentiated at depth. The surface is replenished or "repainted" with volcanism occurring today mostly in the mesolands and at volcanic rises.

Our view of Venus will not please everyone; indeed it may not please anyone. It is based on a number of theoretical arguments regarding mantle and lithospheric dynamics, and on an attempt to reconcile what we see on the surface and infer from gravity data with these theoretical considerations. It is a picture more complex than many earlier views (e.g. Phillips et al 1981, 1991; Phillips & Malin 1983, 1984; Kaula 1990a; Bindschadler et al 1992b), yet does not consider all possibilities [e.g. the tectonic and magmatic consequences of conversion from layered to whole-mantle convection (Steinbach & Yuen 1992)].

This paper is being written as the *Magellan* orbit is being circularized to acquire a high-resolution gravity field. This new data set, along with continued analyses of images and topographic information, will help a great deal in testing the model of Venus proposed here. As with previous attempts to describe Venus, parts of this model will be wrong, but hopefully other parts can be carried forward.

ACKNOWLEDGMENTS

This work was supported by NASA grants NAGW-3024 and NAGW-3701 to Washington University and NAGW-2915 to Southern Methodist University. We thank Rosanna Ridings for her cheerful assistance and Kevin Burke for giving a thorough reading to our manuscript and offering numerous useful suggestions.

Literature Cited

Anderson DL, Tanimoto T, Zhang Y. 1992. Plate tectonics and hotspots: the third dimension. *Science* 256: 1645–51

Arkani-Hamed J, Schaber GG, Strom RG. 1993. Constraints on the thermal evolution of Venus inferred from Magellan data. *J. Geophys. Res.* 98: 5309–15

Avé Lallemant HG. 1978. Experimental deformation of diopside and websterite. *Tectonophysics* 48: 1–27

Baker VR, Komatsu G, Parker TJ, Gulick VC, Kargel JS, Lewis JS. 1992. Channels and valleys on Venus: preliminary analysis of Magellan data. *J. Geophys. Res.* 97: 13,421–44

Banerdt WB, Golombek MP. 1988. Deformational models of rifting and folding on Venus. *J. Geophys. Res.* 93: 4759–72

Banerdt WB, Sammis CG. 1992. Small-scale fracture patterns on the volcanic plains of Venus. *J. Geophys. Res.* 97: 16,149–66

Barsukov VL, Basilevsky AT, Burba GA, Bobinna NN, Kryuchkov VP, et al. 1986. The geology and geomorphology of the Venus surface as revealed by the radar images obtained by Veneras 15 and 16. *Proc. Lunar Planet. Sci. Conf.* 16*th*, Part 2. *J. Geophys. Res.* 91: D378–98 (Suppl.)

Basilevsky AT. 1986. Structure of central and eastern areas of Ishtar Terra and some problems of Venusian tectonics. *Geotectonics* 20: 282–88

Basilevsky AT, Pronin AA, Ronca LB, Kryuchkov VP, Sukhanov AL, Markov MS. 1986. Styles of tectonic deformations on Venus: analysis of Venera 15 and 16 data. *Proc. Lunar Planet. Sci. Conf.* 16*th*, Part 2. *J. Geophys. Res.* 91: D399–411 (Suppl.)

Bercovici D, Lin J. 1993. Mantle plume-head waves: a possible mechanism for the formation of the outer ridges of Venus coronae. *Eos, Trans. Am. Geophys. Union (Suppl.)* 74(16): 189 (Abstr.)

Bercovici D, Schubert G, Glatzmaier GA. 1989. Influence of heating mode on three-dimensional mantle convection. *Geophys. Res. Lett.* 16: 617–20

Bills BG, Kiefer WS, Jones RL. 1987. Venus gravity: a harmonic analysis. *J. Geophys. Res.* 92: 10,335–51

Bilotti F, Suppe J. 1992. Wrinkle ridges and topography on Venus. *Geol. Soc. Am. Abstr. with Programs* 24: A195 (Abstr.)

Bindschadler DL, de Charon A, Beratan KK, Smrekar SE, Head JW. 1992a. Magellan observations of Alpha Regio: implications for formation of complex ridged terrains on Venus. *J. Geophys. Res.* 97: 13,563–77

Bindschadler DL, Head JW. 1991. Tessera terrain, Venus: characterization and models for origin and evolution. *J. Geophys. Res.* 96: 5889–907

Bindschadler DL, Parmentier EM. 1990. Mantle flow tectonics: the influence of a ductile lower crust and implications for the formation of topographic uplands on Venus. *J. Geophys. Res.* 95: 21,329–44

Bindschadler DL, Schubert G, Kaula WM. 1990. Mantle flow tectonics and the origin of Ishtar Terra, Venus. *Geophys. Res. Lett.* 17: 1345–48

Bindschadler DL, Schubert G, Kaula WM. 1992b. Coldspots and hotspots: global tectonics and mantle dynamics of Venus. *J. Geophys. Res.* 97: 13,495–532

Bird P. 1988. Formation of the Rocky Mountains, western United States: a computer continuum model. *Science* 239: 1501–7

Bodine JH, Steckler MS, Watts AB. 1981. Observations of flexure and rheology of the oceanic lithosphere. *J. Geophys. Res.* 86: 3695–707

Brace WF, Kohlstedt DL. 1980. Limits on lithospheric stress imposed by laboratory experiments. *J. Geophys. Res.* 85: 6248–52

Buck WR. 1992. Global decoupling of crust and mantle: implications for topography, geoid and mantle viscosity on Venus. *Geophys. Res. Lett.* 19: 2111–14

Byerlee JD. 1978. Friction of rocks. *Pure Appl. Geophys.* 116: 615–26

Campbell DB, Head JW, Harmon JK, Hine AA. 1984. Venus: volcanism and rift formation in Beta Regio. *Science* 226: 167–70

Campbell DB, Head JW, Hine AA. 1983. Venus: identification of banded terrain in the mountains of Ishtar Terra. *Science* 221: 644–47

Campbell DB, Senske DA, Head JW, Hine AA, Fisher PC. 1991. Venus southern hemisphere: geologic character and age of terrains in the Themis-Alpha-Lada region. *Science* 251: 180–83

Caristan Y. 1982. The transition from high temperature creep to fracture in Maryland diabase. *J. Geophys. Res.* 87: 6781–90

Crumpler LS, Head JW, Campbell DB. 1986. Orogenic belts on Venus. *Geology* 14: 1031–34

Cyr KE, Melosh HJ. 1993. Tectonic patterns and regional stresses near Venusian coronae. *Icarus* 102: 175–84

Goetze C. 1978. The mechanisms of creep in olivine. *Phil Trans. R. Soc. London Ser. A* 288: 99–119

Grimm RE. 1993. Impact crater strains on Venus: implications for tectonic resurfacing rates and lithospheric strength. *J. Geophys. Res.* Submitted

Grimm RE, Phillips RJ. 1990. Tectonics of Lakshmi Planum, Venus: tests for Magellan. 1990. *Geophys. Res. Lett.* 17: 1349–52

Grimm RE, Phillips RJ. 1991. Gravity anomalies, compensation mechanisms, and the geodynamics of western Ishtar Terra, Venus. *J. Geophys. Res.* 96: 8305–24

Grimm RE, Phillips RJ. 1992. Anatomy of a Venusian hot spot: geology, gravity, and mantle dynamics of Eistla Regio. *J. Geophys. Res.* 97: 16,035–54

Grimm RE, Solomon SC. 1988. Viscous relaxation of impact crater relief on Venus: constraints on crustal thickness and thermal gradient. *J. Geophys. Res.* 93: 11,911–29

Hager BH, Clayton RW. 1989. Constraints on the structure of mantle convection using seismic observations, flow models, and the geoid. In *Mantle Convection: Plate Tectonics and Global Dynamics*, ed. WR Peltier, pp. 657–763. New York: Gordon and Breach. 881 pp.

Hansen VL. 1993. Asymmetric Venusian rifts, arguments against subduction. *Eos, Trans. Am. Geophys. Union* (*Suppl.*) 74(43): 377 (Abstr.)

Hansen VL, Phillips RJ. 1993a. Tectonics and volcanism of eastern Aphrodite Terra, Venus: no subduction, no spreading. *Science* 260: 526–30

Hansen VL, Phillips RJ. 1993b. Ishtar deformed belts: evidence for deformation from below? *Lunar Planet. Sci.* 24: 603–4 (Abstr.)

Hansen VL, Phillips RJ. 1993c. Formation of Ishtar Terra, Venus, by mantle imbrication. *Geophys. Res. Lett.* Submitted

Head JW. 1990. Formation of mountain belts on Venus: evidence for large-scale convergence, underthrusting, and crustal imbrication in Freyja Montes, Ishtar Terra. *Geology* 18: 99–102

Head JW, Campbell DB, Elachi C, Guest JE, McKenzie D, et al. 1991. Venus volcanism: initial analysis from Magellan data. *Science* 252: 276–99

Head JW, Crumpler LS. 1987. Evidence for divergent plate boundary characteristics and crustal spreading on Venus. *Science* 238: 1380–85

Head JW, Crumpler LS. 1990. Venus geology and tectonics: hotspot and crustal spreading models and questions for the Magellan mission. *Nature* 346: 525–33

Head JW, Crumpler LS, Aubele JC, Guest JE, Saunders RS. 1992. Venus volcanism: classification of volcanic features and structures, associations, and global distribution from Magellan data. *J. Geophys. Res.* 97: 13,153–97

Head JW, Ivanov M. 1993. Tessera terrain on Venus: implications of tessera flooding models and boundary characteristics for global distribution and mode of formation. *Lunar Planet Sci.* 24: 619–20 (Abstr.)

Herrick RR, Phillips RJ. 1990. Blob tectonics: a prediction for western Aphrodite Terra, Venus. *Geophys. Res. Lett.* 17: 2129–32

Herrick RR, Phillips RJ. 1992. Geological correlations with the interior density structure of Venus. *J. Geophys. Res.* 97: 16,017–34

Hillgren VJ, Melosh HJ. 1989. Crater relaxation on Ganymede: implications for ice rheology. *Geophys. Res. Lett.* 16: 1339–42

Ivanov MA, Basilevsky AT. 1993. Density and morphology of impact craters on tessera terrain, Venus. *Geophys. Res. Lett.* Submitted

Ivanov MA, Head JW. 1993. Tessera terrain: global characterization from Magellan data. *Lunar Planet Sci.* 24: 691–92 (Abstr.)

Janes DM, Squyres SW, Bindschadler DL, Baer G, Schubert G, et al. 1992. Geophysical models for the formation and evolution of coronae on Venus. *J. Geophys. Res.* 97: 16,055–67

Janle P, Jannsen D, Basilevsky AT. 1987. Morphologic and gravimetric investigations of Bell and Eisla regions on Venus. *Earth Moon Planets* 39: 251–73

Johnson C, Sandwell DT. 1993. Lithospheric flexure on Venus. *Geophys. J.* Submitted

Jordan TH. 1975. The continental tectosphere. *Rev. Geophys. Space Phys.* 13: 1–12

Jordan TH. 1981. Continents as a chemical boundary layer. *Phil. Trans. R. Soc. London Ser. A* 301: 359–73

Kaula WM. 1990a. Venus: a contrast in evolution to Earth. *Science* 247: 1191–96

Kaula WM. 1990b. Mantle convection and crustal evolution on Venus. *Geophys. Res. Lett.* 17: 1401–3

Kaula WM, Bindschadler DL, Grimm RE, Hansen VL, Roberts KM, Smrekar, SE. 1992. Styles of deformation in Ishtar Terra and their implications. *J. Geophys. Res.* 97: 16,085–120

Keep M, Hansen VL. 1993a. Structural mapping of Maxwell Montes, Venus. *Lunar Planet Sci.* 24: 775-76. (Abstr.)

Keep M, Hansen VL. 1993b. Structural evolution of Maxwell Montes, Venus: implications for Venusian mountain "belt" formation. *J. Geophys. Res.* Submitted

Kiefer WS, Hager BH. 1991a. A mantle plume model for the equatorial highlands of Venus. *J. Geophys. Res.* 96: 20,947–66

Kiefer WS, Hager BH. 1991b. Mantle downwelling and crustal convergence: a model for Ishtar Terra, Venus. *J. Geophys. Res.* 96: 20,967–80

Kiefer WS, Richards MA, Hager BH, Bills BG. 1986. A dynamic model of Venus's gravity field. *Geophys. Res. Lett.* 13: 14–17

Leitch AM, Yuen DA. 1991. Compressible convection in a viscous Venusian mantle. *J. Geophys. Res.* 96: 15,551–62

Lenardic A, Kaula WM, Bindschadler DL. 1991. The tectonic evolution of western Ishtar Terra, Venus. *Geophys. Res. Lett.* 18: 2209–12

Lenardic A, Kaula WM, Bindschadler DL. 1993. A mechanism for crustal recycling on Venus. *J. Geophys. Res.* 98: 18,697–705

Magee Roberts K, Head JW. 1993. Large-scale volcanism associated with coronae on Venus: implications for formation and evolution. *Geophys. Res. Lett.* 20: 1111–14

Masursky H, Eliason E, Ford PG, McGill GE, Pettengill GH. 1980. Pioneer Venus radar results: geology from images and altimetry. *J. Geophys. Res.* 85: 8232–60

McGill GE. 1992. Wrinkle ridges on Venusian plains: indicators of shallow crustal stress orientations at local and regional scales *Int. Colloq. on Venus*, pp. 67–68. Houston: Lunar Planet. Inst. (Abstr.)

McGill GE, Steenstrup SJ, Barton C, Ford PG. 1981. Continental rifting and the origin of Beta Regio, Venus. *Geophys. Res. Lett.* 8: 737–40

McKenzie D, Bickle MJ. 1988. The volume and composition of melt generated by extension of the lithosphere. *J. Petrol.* 29: 625–79

McKenzie D, Ford PG, Johnson C, Parsons B, Sandwell D, et al. 1992a. Features on Venus generated by plate boundary processes. *J. Geophys. Res.* 97: 13,533–44

McKenzie D, Ford PG, Liu F, Pettengill GH. 1992b. Pancakelike domes on Venus. *J. Geophys. Res.* 97: 15,967–76

McNamee JB, Borderies NJ, Sjogren WL. 1993. Venus: global gravity and topography. *J. Geophys. Res.* 98: 9113–28

McNutt MK. 1984. Lithospheric flexure and thermal anomalies. *J. Geophys. Res.* 89: 11,180–94

Morgan P, Phillips RJ. 1983. Hot spot heat transfer: its application to Venus and implications to Venus and Earth. *J. Geophys. Res.* 88: 8305–17

Nerem RS, Bills BG, McNamee JB. 1993. A high resolution gravity model for Venus: GVM-1. *Geophys. Res. Lett.* 20: 599–602

Ogawa M. 1993. A numerical model of a coupled magmatism-mantle convection system in Venus and the Earth's mantle beneath Archean continental crusts. *Icarus* 102: 40–61

Parmentier EM, Hess PC. 1992. Chemical differentiation of a convecting planetary interior: consequences for a one plate planet such as Venus. *Geophys. Res. Lett.* 19: 2015–18

Parmentier EM, Hess PC, Sotin C., 1993. Mechanisms for episodic large-scale mantle overturn with application to catastrophic resurfacing of Venus. *Eos, Trans. Am. Geophys. Union* (*Suppl.*) 74(16): 188 (Abstr.)

Pavri B, Head JW III, Klose KB, Wilson L. 1992. Steep-sided domes on Venus: characteristics, geologic setting, and eruption conditions from Magellan data. *J. Geophys. Res.* 97: 13,445–78

Pettengill GH, Eliason E, Ford PG, Loriot GB, Masursky H, McGill GE. 1980. Pioneer Venus radar results: altimetry and surface properties. *J. Geophys. Res.* 85: 8261–70

Phillips RJ. 1986. A mechanism for tectonic deformation on Venus. *Geophys. Res. Lett.* 13: 1141–44

Phillips RJ. 1990. Convection-driven tectonics on Venus. *J. Geophys. Res.* 95: 1301–16

Phillips RJ. 1993. The age spectrum of the Venusian surface. *Eos, Trans. Am. Geophys. Union* (*Suppl.*) 74(16): 187 (Abstr.)

Phillips RJ, Grimm RE. 1990. Generation of basaltic crust on Venus. *Lunar Planet. Sci.* 21: 1065–66 (Abstr.)

Phillips RJ, Grimm RE, Malin MC. 1991. Hot-spot evolution and the global tectonics of Venus. *Science* 252: 651–58

Phillips RJ, Hansen VL. 1993. Venus magmatic and tectonic evolution. *Lunar Planet Sci.* 24: 1135–36 (Abstr.)

Phillips RJ, Kaula WM, McGill GE, Malin MC. 1981. Tectonics and evolution of Venus. *Science* 212: 879–87

Phillips RJ, Lambeck K. 1980. Gravity fields of the terrestrial planets: long wavelength anomalies and tectonics. *Rev. Geophys. Space Phys.* 18: 27–76

Phillips RJ, Malin MC. 1983. The interior of Venus and tectonic implications. In *Venus*, ed. DM Hunten, L Colin, TM Donahue, VI Moroz, pp. 159–214. Tucson: Univ. Ariz. Press

Phillips RJ, Malin MC. 1984. Tectonics of Venus. *Annu. Rev. Earth Planet. Sci.* 12: 411–43

Phillips RJ, Raubertas RF, Arvidson RE, Sarkar IC, Herrick RR, et al. 1992. Impact craters and Venus resurfacing history. *J. Geophys. Res.* 97: 15,923–48

Plescia JB, Golombek MP. 1986. Origin of planetary wrinkle ridges based on the study of terrestrial analogs. *Geol. Soc. Am. Bull.* 97: 1289–99

Pronin AA. 1986. The structure of Lakshmi Planum, an indication of horizontal asthenospheric flow on Venus. *Geotectonics* 20: 271–80

Pronin AA, Stofan ER. 1990. Coronae on Venus: morphology, classification, and distribution. *Icarus* 87: 452–74

Ramberg H. 1968. Fluid dynamics of layered systems in the field of gravity, a theoretical basis for certain global structures and isostatic adjustment. *Phys. Earth. Planet. Inter.* 1: 63–87

Richards MA, Duncan RA, Courtillot VE. 1989. Flood basalts and hotspot tracks: plume heads and tails. *Science* 246: 103–7

Roberts KM, Head JW. 1990a. Lakshmi Planum, Venus: characteristics and models of origin. *Earth Moon Planets* 50/51: 193–249

Roberts KM, Head JW. 1990b. Western Ishtar Terra and Lakshmi Planum, Venus: models of formation and evolution. *Geophys. Res. Lett.* 17: 1341–44

Robinson EM, Parsons B, Daly SF. 1987. The effect of a shallow low-viscosity zone on the apparent compensation of mid-plate swells. *Earth Planet. Sci. Lett.* 82: 335–48

Sandwell DT, Schubert G. 1992a. Evidence for retrograde lithospheric subduction on Venus. *Science* 257: 766–70

Sandwell DT, Schubert G. 1992b. Flexural ridges, trenches, and outer rises around coronae on Venus. *J. Geophys. Res.* 97: 16,069–83

Saunders RS, Spear AJ, Allin PC, Austin AL, Berman RC, et al. 1992. Magellan mission summary. *J. Geophys. Res.* 97: 13,067–90

Schaber GG. 1982. Venus: limited extension and volcanism along zones of lithospheric weakness. *Geophys. Res. Lett.* 9: 499–502

Schaber GG, Strom RG, Moore HJ, Soderblom LA, Kirk RL, et al. 1992. Geology and distribution of impact craters on Venus: what are they telling us? *J. Geophys. Res.* 97: 13,257–301

Schubert G, Bercovici D, Glatzmaier GA. 1990. Mantle dynamics in Mars and Venus: influence of an immobile lithosphere on three-dimensional mantle convection. *J. Geophys. Res.* 95: 14,105–29

Schubert G, Bercovici D, Thomas PJ, Campbell DB. 1989. Venus coronae: formation by mantle plumes. *Lunar Planet. Sci.* 20: 968–69 (Abstr.)

Senske DA, Head JW, Stofan ER, Campbell DB. 1991. Geology and structure of Beta Regio, Venus: results from Arecibo radar imaging. *Geophys. Res. Lett.* 18: 1159–62

Senske DA, Schaber GG, Stofan ER. 1992. Regional topographic rises on Venus: geology of Western Eistla Regio and comparison to Beta Regio and Atla Regio. *J. Geophys. Res.* 97: 13,395–420

Simons M, Hager BH, Solomon SC. 1992. Geoid, topography, and convection-driven crustal deformation on Venus. *Int. Colloq. on Venus*, pp. 110–12. Houston: Lunar Planet. Inst. (Abstr.)

Simons M, Solomon MC. 1993. Geoid-to-topography ratios on Venus: a global per-

spective. *Eos, Trans. Am. Geophys. Union* (*Suppl.*) 74(16): 191 (Abstr.)
Sjogren WL, Bills BG, Birkeland PW, Esposito PB, Konopliv AR, et al. 1983. Venus gravity anomalies and their correlations with topography. *J. Geophys. Res.* 88: 1119–28
Sjogren WL, Phillips RJ, Birkeland PW, Wimberly RN. 1980. Gravity anomalies on Venus. *J. Geophys. Res.* 85: 8295–302
Smrekar SE, Parmentier EM. 1993. Mantle upwelling beneath a depleted residuum layer: implications for the resurfacing history of Venus. *Eos, Trans. Am. Geophys. Union* (*Suppl.*) 74(16): 188 (Abstr.)
Smrekar S, Phillips RJ. 1988. Gravity-driven deformation of the crust on Venus. *Geophys. Res. Lett.* 15: 693–96
Smrekar SE, Phillips RJ. 1991. Venusian highlands: geoid to topography ratios and their implications. *Earth Planet. Sci. Lett.* 107: 582–97
Smrekar SE, Solomon SC. 1992. Gravitational spreading of high terrain in Ishtar Terra, Venus. *J. Geophys. Res.* 97: 16,121–48
Solomon SC. 1993. A tectonic resurfacing model for Venus. *Lunar Planet Sci.* 24: 1331–32 (Abstr.)
Solomon SC, Head JW. 1982a. Mechanisms for lithospheric heat transport on Venus: implications for tectonic style and volcanism. *J. Geophys. Res.* 87: 9236–46
Solomon SC, Head JW. 1982b. Evolution of the Tharsis Province of Mars: the importance of heterogeneous lithospheric thickness and volcanic construction. *J. Geophys. Res.* 87: 9755–74
Solomon SC, Head JW. 1990. Lithospheric flexure beneath the Freyja Montes Foredeep, Venus: Constraints on lithospheric thermal gradient and heat flow. *Geophys. Res. Lett.* 17: 1393–96
Solomon SC, Head JW, Kaula WM, McKenzie D, Parsons B, et al. 1991. Venus tectonics: initial analysis from Magellan. *Science* 252: 297–312
Solomon SC, Smrekar SE, Bindschadler DL, Grimm RE, Kaula WM, et al. 1992. Venus tectonics: an overview of Magellan observations. *J. Geophys. Res.* 97: 13,199–255
Squyres SW, Janes DM, Baer G, Bindschadler DL, Schubert G, et al. 1992a. The morphology and evolution of coronae on Venus. *J. Geophys. Res.* 97: 13,611–34
Squyres SW, Jankowski DG, Simons M, Solomon SC, Hager BH, McGill GE. 1992b. Plains tectonism on Venus: the deformation belts of Lavinia Planitia. *J. Geophys. Res.* 97: 13,579–99
Steinbach V, Yuen DA. 1992. The effects of multiple phase transitions on Venusian mantle convection. *Geophys. Res. Lett.* 19: 2243–46
Stevenson DJ, Spohn T, Schubert G. 1983. Magnetism and thermal evolution of the terrestrial planets. *Icarus* 54: 466–89
Stofan ER, Bindschadler DL. 1993. Corona annuli: characteristics and modes of origin. *Eos, Trans. Am. Geophys. Union* (*Suppl.*) 74(43): 377 (Abstr.)
Stofan ER, Bindschadler DL, Head JW, Parmentier EM. 1991. Coronae Structures on Venus: models of origin. *J. Geophys. Res.* 96: 20,933–46
Stofan ER, Head JW. 1990. Coronae of Mnemosyne Regio: morphology and origin. *Icarus* 83: 216–43
Stofan ER, Sharpton VL, Schubert G, Baer G, Bindschadler DL, et al. 1992. Global distribution and characteristics of coronae and related features on Venus: implications for origin and relations to mantle processes. *J. Geophys. Res.* 97: 13,347–78
Suppe J, Connors C. 1992. Critical taper wedge mechanics of fold-and-thrust belts on Venus: initial results from Magellan. *J. Geophys. Res.* 97: 13,545–61
Surkov Yu A, Barsukov VL, Moskalyeva LP, Kharyukova VP, Kemurdzhian AL. 1984. New data on the composition, structure, and properties of Venus rock obtained by Venera 13 and Venera 14. *Proc. Lunar Planet. Sci. Conf.* 14*th*, Part 2. *J. Geophys. Res.* 89: B393–402
Surkov Yu A, Kirnozov FF, Glazov VN, Dunchenko AG, Tatsy LP, Sobornov OP. 1987. Uranium, thorium, and potassium in the Venusian rocks at the landings sites of Vega 1 and 2. *Proc. Lunar Planet. Sci. Conf.* 17*th*, Part 2. *J. Geophys. Res.* 92: E537–40 (Suppl.)
Tackley PJ, Stevenson DJ. 1991. The production of small Venusian coronae by Rayleigh-Taylor instabilities in the uppermost mantle. *Eos, Trans. Am. Geophys. Union* (*Suppl.*) 72(44): 287 (Abstr.)
Tackley PJ, Stevenson DJ. 1993a. A mechanism for spontaneous self-perpetuating volcanism on the terrestrial planets. In *Flow and Creep in the Solar System: Observations, Modeling and Theory*, ed. DB Stone, SK Runcorn, pp. 307–21. Dordrecht: Kluwer
Tackley PJ, Stevenson DJ. 1993b. Volcanism without plumes: melt-driven instabilities, buoyant residuum and global implications. *Eos, Trans. Am. Geophys. Union* (*Suppl.*) 74(16): 188 (Abstr.)
Turcotte DL. 1993. An episodic hypothesis for Venusian tectonics. *J. Geophys. Res.* 98: 17,061–68
Vinogradov AP, Surkov YA, Kirnozov FF. 1973. The content of uranium, thorium and potassium in the rocks of Venus as measured by Venera 8. *Icarus* 20: 253–59

Watters TR. 1988. Wrinkle ridge assemblages on the terrestrial planets. *J. Geophys. Res.* 93: 10,236–54
Weertman J. 1979. Height of mountains on Venus and the creep properties of rock. *Phys. Earth Planet. Inter.* 19: 197–207
Whitehead JA, Luther DS. 1975. Dynamics of laboratory diapir and plume models. *J. Geophys. Res.* 80: 705–17
Zuber MT. 1987. Constraints on the lithospheric structure of Venus from mechanical models and tectonic surface features. *Proc. Lunar Planet. Sci. Conf. 17th*, Part 2, *J. Geophys. Res.* 92: E541–51 (Suppl.)
Zuber MT. 1990. Ridge belts: evidence for regional- and local-scale deformation on the surface of Venus. *Geophys. Res. Lett.* 17: 1369–72
Zuber MT, Parmentier EM. 1990. On the relationship between isostatic elevation and the wavelengths of tectonic surface features on Venus. *Icarus* 85: 290–308

SUBJECT INDEX

D

F

K

L

M

CUMULATIVE INDEXES

CONTRIBUTING AUTHORS, VOLUMES 1–22

CHAPTER TITLES, VOLUMES 1–22

GEOPHYSICS AND PLANETARY SCIENCE

OCEANOGRAPHY, METEOROLOGY, AND PALEOCLIMATOLOGY

PALEONTOLOGY, STRATIGRAPHY, AND SEDIMENTOLOGY

TECTONOPHYSICS AND REGIONAL GEOLOGY

S0-BJP-312
BURIED TREASURES
uthwest
ARKANSAS R.
CIMARRON R.
CANADIAN R.
KANSAS
Wichita
Ft. Scott
Springfield
Joplin
MISSOURI
CUT-THROAT GAP TREASURE
CHEROKEE COW-MAN'S GOLD
OKLAHOMA
RIVER
$2,000,000 JAMES LOOT
Oklahoma City
DEVIL'S CANYON
WICHITA MTS.
Ft. Sill
Lawton
3 BURRO LOADS SPANISH GOLD
buried West side of Twin Mts.
9 CARTLOADS of GOLD and SILVER
buried near 7 Springs on Ft. Sill Rd. between Cache and Beaver Creeks
ARKANSAS RIVER
ZINC
SILVER
OLD SPANISH MINE WORKED BY DE SOTO
Little Rock
ARKANSAS
COPPER ORE
MOUNTAIN OF IRON
GOVERNMENT TREASURE lost in Cache Creek
RED RIVER
LEAD VEIN
BRAZOS
SPIDER ROCK
INDIAN PLATINUM MINE
COPPER
POTHOLE of NUGGETS
STEINHEIMERS MILLIONS
ZINC ORE (LOST CISTERN)
TRINITY RIVER
TREASURE STUFFED CANNON
CADDO INDIAN LEAD MINES
Shreveport
LOUISIANA
MISSISSIPPI RIVER
WHERE DE SOTO SEARCHED FOR GOLD
Ft. Concho (San Angelo)
$70,000
San Luis de las Amarillas (1757)
MISSION SAN SABA
SAN SABA RIVER
PACKSADDLE MOUNTAIN
COPPER-PLATES
7 JACK-LOADS
Nacogdoches
MEXICAN ARMY PAY-CHEST
NECHES RIVER
SABINE R.
Bowie's Fight
LLANO RIVER
(Rio de las Chanas)
YELLOW-WOLF'S GOLD
40 MULE-LOADS
POPE'S GHOST TREASURE
75 JACK-LOADS
Austin
POT OF GOLD IN CRACK
SANTA ANA'S PAY CHESTS
Houston
San Jacinto Battle Ground
LAFITTE'S CHEST
Where Clawson saw "the horrors of hell"
OLD SPANISH TRAIL
LOST RANGERS GOLD
FRIO CANYON
COLORADO RIVER
LAFITTE'S BOOTY
La Porte
San Antonio
RATTLESNAKE CAVE
Del Rio
Presidio
SANTA ANA'S GOLD
GEN. COS' MONEY
DUTCHMAN'S LEAD
Ft. St. Luis
Galveston Island
LAFITTE'S HEADQUARTERS IN TEXAS
BARATARIA BAY
PIRATES RETREAT
ESPANTOSA LAKE
NUECES R.
FT. MERRILL
ROCK PENS
PASO VALENA
CASA BLANCA
FT. RAMIREZ
SAN CAJO MT.
LIPANTITLAN
Corpus Christi
Matagorda Bay
CAILLOU ISLAND $20,000
Laredo
TREASURE-LADEN SPANISH SHIP lost by Lafitte
7 CART-LOADS ARMY MONEY
Palo Alto Battlefield
MEXICAN PAY-CHESTS at Resaca de la Palma
SKELETON and GOLD IN WELL
BURIED by TAYLOR'S MEN
Padre Island
GULF OF MEXICO
N
E
W
S
Monterey
Brownsville
Matamoras

"*Más allá* (farther on)," the Apache told Don Miranda, "are more and richer mines."

CORONADO'S CHILDREN

Tales of Lost Mines and Buried Treasures of the Southwest

BY

J. FRANK DOBIE

ILLUSTRATED BY

BEN CARLTON MEAD

1931

THE LITERARY GUILD OF AMERICA

NEW YORK

Printed in U. S. A.

PRESS OF
BRAUNWORTH & CO., INC.
BOOK MANUFACTURERS
BROOKLYN, NEW YORK

To

MY MOTHER

ELLA BYLER DOBIE

WHO HAS SO OFTEN DELIGHTED ME WITH
CONVERSATIONAL SKETCHES OF SUCH
CHARACTERS AS ENTER
THIS BOOK

AND

TO THE MEMORY OF MY FATHER

R. J. DOBIE

A CLEAN COWMAN OF THE
TEXAS SOIL

IN THE BEGINNING

THESE tales are not creations of mine. They belong to the soil and to the people of the soil. Like all things that *belong*, they have their roots deep in the place of their being, deep too in the past. They are an outgrowth; they embody the geniuses of divergent races and peoples who even while fiercely opposing each other blended their traditions. However all this may be, the tales are just tales. As tales I have listened to them in camps under stars and on ranch galleries out in the brush. As tales, without any ethnological palaver, I have tried to set them down. So it is with something of an apology that I make even a brief explanation before plunging into a veritable Iliad of adventures.

In March, 1536, four hundred years ago tomorrow, a party of Spaniards scouting in the wild lands against the Gulf of California came suddenly upon a spectacle more strange and unexpected than the footprints which greeted Robinson Crusoe's eyes on the desert island. The spectacle was a white man all but naked, his nakedness partly concealed by an uncouth tangle of long hair and beard, his nationality revealed only by wild and whirling words. He was accompanied by a Moor called Estévanico, or Stephen, and eleven Indians. He gave his name as Cabeza de Vaca. He told how he had been shipwrecked on the coast of Florida a continent away and how, passing from tribe to tribe, his life often threatened by starvation, thirst, and savage enemies, he had for eight long years been beating west. He had seen much and heard more. Somewhere to the north, he said, was a galaxy of cities the inhabitants of which wore civilized raiment, lived in palaces ornamented with sapphires and turquoises, and possessed gold without end—the Seven Cities of Cibola.

As the great scholar Adolph F. Bandelier has pointed out in *The Gilded Man,* the story of seven rich cities in an unknown land was old before Columbus discovered America. Despite that, Cabeza de Vaca's story was new—new in a new land where men with fresh hearts were looking for the fabled Fountain of Youth, the giant Amazons, the anthropophagi with heads growing out of their breasts, and many another marvel.

Search for the wondrous Cities was inevitable. The expedition led by Francisco Vasquez Coronado was the first made by white men into what is now known as the Southwest and it was surely the most amazing of all expeditions ever made on the North American continent. Fitted out at an immense cost, it included 300 Spaniards, 1000 Indians, 1000 extra horses, herds of swine and sheep, six swivel guns, and a temperament as superbly sanguine as young men are capable of enjoying.

Ahead of him Coronado sent a Franciscan monk known as Fray Marcos de Nizza, and ahead of Fray Marcos scouted Black Stephen, the Moor who had been with Cabeza de Vaca. If the gorgeous reports about the riches of Cibola turned out to be only partly true, Black Stephen was to send back a small cross "two handfuls long"; if the reports had not been exaggerated, and the Seven Cities really contained more wealth than the palaces of the Montezumas, he should send back "a larger cross." Four days after he had ridden out, the Moor sent back a cross "as high as a man," and the priest relayed the message to Coronado.

Black Stephen was killed; the name of Nizza became a term for contempt; instead of cities with gates of gold and houses with sapphire-studded doors that flashed in the sun, Coronado found among the Zuñis of what is now Arizona walls of mud and naked burrows in barren cliffs. The Seven Cities of Cibola were a dream. but *más allá*—on beyond—so an Indian known as the Turk said, was the Gran Quivira. It was a place "where ordinary dishes were made of wrought plate and the jugs and bowls were of gold." Coronado rode on—*más allá*. He rode a thousand miles across unknown mountains and parched plains—and he found the Gran Quivira to be a handful of naked savages squatting under wind-whisked thatches or skulking at the heels of drifting buffaloes.

At the very time Cibola and Quivira seized the Spaniards of North America, expeditions fitted out by Spanish princes, Dutch bankers, and English sea-dogs were beginning to scour South America in search of El Dorado. El Dorado, the Gilded King, so the story went, had attendants to anoint his body with oil every morning and then to blow gold dust upon him until he was decently clad. At evening he washed himself clean in a lake—a lake somewhere in the uncharted out-yonder. Presumably he slept in undress, but always at sunrise he was freshly robed in golden plating. Thus for uncounted generations kings by the lake had dressed and undressed. It might well be supposed that the lake was paved with golden sands. It might well be supposed that the people who gave their king a fresh suit of gold every day possessed immense stores of it. For more than a hundred years *conquistadores* marched and counter-marched from one extremity of South America to the other, spending the lives of thousands upon thousands of men and the wealth of prodigal treasuries, enduring starvation, fever, cold, thirst, the pests of swamps, and the pitiless heat of deserts—in search of El Dorado. El Dorado was a man, a lake, a city, a country, a people, a name, a dream—a dream at once absurd and sublime—an allegory of every phantom that the high heart of man has ever pursued.

El Dorado and the Seven Cities of Cibola but represented a multitude of tales pointing the buoyant way to untold wealth. La Ciudad Encantada de los Cesares; the fabled palace of Cubanacan in Cuba; the golden mirage of Manoa, in quest of which Raleigh at the age of sixty-three came out of prison to fare forth a second time up the Orinoco; the nebulous treasures of a Casa del Sol, of the Gran Moxo, and of the Golden Temple of Dabaiba, all in South America; a gilded rainbow somewhere in the northern Pacific called the Straits of Annian; the Laguna de Oro in New Mexico hard by the Peak of Gold in the same region; the Seven Hills of the Aijados in Texas, where gold was so plentiful that the natives, "not knowing any of the other metals," tipped arrows and lances with it—all these were but duplications of one theme. Far up the bleak New England coast French voyagers sought "the northern El Dorado" under the name of Norembega—

> And Norembega proved again
> A shadow and a dream.

But despite the madness and fantasticality of such rumors, the Spaniards had a basis for hope and belief. How much the wealth of Montezuma amounted to has never been known, but Cortez plundered enough of it to inflame the imagination of the Old World. In Peru the inca called Atahualpa heaped up for Pizarro, as the price of ransom, golden vessels almost sufficient to fill a room twenty-two by seventeen feet square to a height of nine feet. In a courtyard of Bogotá, Quesada piled up golden booty so high "that a rider on horseback might hide behind it."

When, in 1542, Coronado returned and told the truth about his barren search, men refused to believe him. One who did believe him, Castañeda, chronicler of the expedition, set down an exquisite point of view in these words: "Granted that they did not find the riches of which they had been told, they found a place in which to search for them." The opportunity that Coronado thus opened has never since his time been neglected; the dream he dreamed has never died. For thousands of happy folk the mirage that lured him on has never faded, and today all over the wide, wide lands where *conquistadores* trailed and *padres* built their simple missions—and in yet more places never glimpsed by Spanish eyes—tradition has marked rock and river and ruin with illimitable treasure. The human imagination abhors failure. Hope and credulity are universal among the sons of earth, and so when English-speaking men took over the *sitios* and *porciones* of Spanish lands in the Southwest, they acquired not only the land that Spanish pioneers had surveyed but the traditions they had somehow made an ingredient of the soil itself.

The fact of California gold, which stampeded a nation, and the fact of Nevada silver, which stampeded California, added immensely to the tradition of Spanish wealth and gave it a flavor and coloring characteristic of the American frontier. As a result, the legends of lost mines and buried treasures that pass current today all over the Southwest and West are a blend. An amazing thing about them is that they seem to be increasing rather than diminishing in number; they are incredibly, astoundingly numerous—as are the people who tell them and halfway, at least, believe them.

These people, no matter what language they speak, are truly Coronado's inheritors. I have called them Coronado's children. They follow Spanish trails, buffalo trails, cow trails; they dig where there are no trails; but oftener than they dig or prospect they just sit and tell stories of lost mines, of buried bullion by the jack load, of ghostly *patrones* that guard treasure, and of a thousand other impediments, generally not ghostly at all, that have kept them away from the wealth they are so sure of.

Coronado's children still have the precious ability to wonder. And you who are so sophisticated, unless you can feel at home in the camps of Coronado's children, never imagine that you understand bold Drake or eager Raleigh or stout Cortez or any of the other flaming figures of the spacious, zestful, and wondering times of great Elizabeth and splendid Philip. They are of imagination all compact.

As readers of *The Alhambra* know, generations of Spaniards preceding those who came to America to hunt gold had been bred on tales of wealth hidden by the Moors. Yet the legends of the Old World that have persisted with most vitality and have called most powerfully to the imaginations of men are legends of women: Venus, too supernally beautiful for earth, too lush of flesh for heaven; Helen with a face that burnt the topless towers of Ilium and a kiss that was fair exchange for a man's soul; sacrificial Iphigenia and love-lorn Philomela; Semiramis and fair Rosamond; Egyptian Cleopatra, with Roman Antony against her fragrant bosom, and majestical Lucretia, who made Roman womanhood a synonym for virtue; tragic Dido of Carthage and burning Thisbe of Babylon; wronged Brunhilde and Queen Guinevere, more potent than the whole Round Table of knights; the Maid of Orleans leading armies by the sound of voices and confounding judges by the simplicity of truth; waiting Hero beyond the Hellespont and unfading Deirdre of the Sorrows. To be sure, the Old World has begot and transmitted many fine legends other than those about women; the Old World is very, very old and it has experienced everything. It has imagined legends of strong men like Hercules, Samson, and the Norse giants; of adventurers like Ulysses; of outlaws like Robin Hood and Rob Roy; of treasures like the Golden Fleece and of lost mines like those of King Solomon. But these legends are minor compared with those in which woman has prefigured for all races and times man's conception of loveliness, and even into these minor legends woman is always entering. A woman shears Samson's locks; among women Ulysses makes his most adventurous wanderings. Jason finds Medea more precious than the Golden Fleece; the golden apples are but an incident in the race for marble-limbed Atalanta.

The New World has been a world of men neither lured nor restrained by women. It has been a world of men exploring unknown continents, subduing wildernesses and savage tribes, felling forests, butchering buffaloes, trailing millions of longhorned cattle wilder than buffaloes, digging gold out of mountains, and pumping oil out of hot earth beneath the plains. It has been a world in which men expected, fought for, and took riches beyond computation—a world, indeed, if not of men without women, then of men into whose imaginings woman has hardly entered. The brawny subduers of this New World have conceived legends about a Gargantuan laborer and constructor, Paul Bunyan; about a supreme range rider, Pecos Bill; about matchless mustangs, lost canyons, and mocking mirages of the desert. But above all, their idealizations—their legends—have been about great wealth to be found, the wealth of secret mines and hidden treasures, a wealth that is substantive and has nothing to do with loveliness and beauty. The representative legends of America are the legends of Coronado's children.

In the tales that follow and in the notes herded into the back of the book, I have acknowledged my indebtedness to many helpers. As will be seen, I have drawn freely from *Legends of Texas,* now out of print, which I compiled and edited for the Texas Folk-Lore Society in 1924. Especially would I here thank Mrs. Mattie Austin Hatcher, archivist, and Miss Winnie Allen, assistant archivist, of the University of Texas, for directing me to many a curious item of Southwestern lore. Who preside over the genial branches of the Grasshoppers' Library in the sunshine of the Pecos, beside the elms and oaks on Waller Creek, down the mesquite flats of the Nueces River, up the canyons of the Rio Grande, under the blue haze of the Guadalupes, deep in the soft Wichitas, over the hills of the San Saba, and in many another happily remembered place where I have pursued "scholarly enquiries," I cannot name. I wish I could, for in the wide-spreading Grasshoppers' Library I have learned the most valuable things I know.

The illustrator of this volume, Ben Carlton Mead, of San Antonio, has taken such pains to be faithful to the times, land,

and characters of the tales and has so graciously cooperated in all ways that I cannot refrain from expressing my obligations to him.

Some of Coronado's children and their tales have appeared in *The Country Gentleman*, *American Mercury*, *Yale Review*, *Southwest Review*, *Holland's Magazine*, *Texaco Star*, and *The Alcalde*.

One person has helped me so much and so continuously that in all justice her name should appear on the title page—Bertha McKee Dobie.

J. FRANK DOBIE

University of Texas
Austin, Texas
June 19, 1930

CONTENTS

Contents

LIST OF ILLUSTRATIONS

MAPS AND CHARTS

At the end of each chapter is a pen-and-ink sketch.

CORONADO'S CHILDREN

CHAPTER I

THE LOST SAN SABA MINE

O brave new world
That has such people in't!—*The Tempest.*

WHAT the Golden Fleece was to the Greeks or what El Dorado—the Gilded Man—has been to South America, the lost mines on the San Saba and Llano rivers in Texas have been to all that part of the United States once owned by Spain. The story of these mines is a cycle made up of a thousand cantos. Housed mechanics, preachers, teachers, doctors, lawyers, earth-treading farmers and home-staying women, as well as roaming cowboys, rangers, outlaws and miners, have told the strange story—and believed it. It is a story of yesterday, as obsolete as the claiming of continents by priority in flag-hoisting; it is a story of today, as realistic as the salt of the earth; it is also a story of tomorrow, as fantastic and romantic as the hopes of man. Through it history walks unabashed and in it fancy sets no limit to extravagance.

Sometimes the name of the fabled source of wealth is Los Almagres; sometimes, Las Amarillas; again, La Mina de las Iguanas, or Lizard Mine, from the fact that the ore is said to have been found in chunks called *iguanas* (lizards); oftener the name is simply the Lost San Saba Mine or the Lost Bowie Mine. In seeking it, generations of men have disemboweled mountains, drained lakes, and turned rivers out of their courses. It has been found—and lost—in many places under many conditions. It is here; it is there; it is nowhere. Generally it is silver; sometimes it is gold. Sometimes it is in a cave; sometimes in water; again on top of a mountain. Now it is not a mine at all but an immense storage of bullion. It changes its place like will-o'-the-wisp and it has more shapes than Jupiter assumed in playing lover.

Only the land that hides it does not change. Except that it is brushier, groomed down in a few places by little fields, and cut across by fences, it is today essentially as the Spaniards found it. A soil that cannot be plowed under keeps its traditions—and its secrets. Wherever the mine may be, however it may appear, it has lured, it lures, and it will lure men on. It is bright Glamour and it is dark and thwarting Fate. It is the Wealth of the Indies; it is the Wealth into which Colonel Mulberry Sellers so gloriously transmuted water and turnips.

The preface to this cycle of a thousand cantos goes back to a day of the seventeenth century when a Spanish *conquistador* set out from Nueva Viscaya "to discover a rumored Silver Hill (Cerro de la Plata) somewhere to the north." [1] At a later date La Salle's Frenchmen wandering forth from Saint Louis Bay on the Texas coast listened to Indians tell of "rivers where silver mines are found." Like most great legends, the legend of the San Saba Mine is a magnification of historical fact. The chief fact was Miranda.

Miranda's Report

In February, 1756, Don Bernardo de Miranda, lieutenant-general of the province of Texas, with sixteen soldiers, five citizens, an Indian interpreter, and several peons, rode out from the village of San Fernando (San Antonio) with orders from the governor to investigate thoroughly the mineral riches so long rumored.[2] After traveling eight days towards the northwest, he pitched camp on the Arroyo San Miguel (now called Honey Creek), a southern tributary of the Rio de las Chanas (Llano River). Only one-fourth of a league beyond he reached the Cerro del Almagre (Almagre Hill), so called on account of its color, *almagre* meaning *red hematite,* or *red ochre.* Opening into the hill Miranda found a cave, which he commanded to be called the Cave of Saint Joseph of Alcasar. He prospected both cave and mountain—with results that brought forth the most sanguine and fulsome predictions.

"The mines which are in the Cerro del Almagre," he reported, "are so numerous that I guarantee to give to every settler of the province of Texas a full claim. . . . The principal vein is more than two varas in width and in its westward lead appears to be of immeasurable thickness." Pasturage for stock, wood and water for mining operations, irrigable soil—all the natural requirements for a settlement of workers—were, Miranda added, at hand. The five citizens with him denounced ten mining claims.

On the way back to San Antonio Miranda met a well-known and trusted Apache Indian, who informed him that "many more and better mines" were at "Los Dos Almagres near the source of the Colorado River." Here the Apache people were accustomed to get silver for their own use—not ore, but solid silver, "soft like the buckles of shoes." Miranda offered the bearer of these tidings a red blanket and a butcher knife to lead him to Los Dos Almagres, but the Apache said that the Comanches out there were too numerous and hostile. However, he promised to guide the Spaniards thither later on—*mañana.*

After having been away only three weeks, Miranda reentered San Fernando. He at once dispatched to the viceroy in Mexico City a statement of his findings, together with recommendations. He declared that no mining could be carried on at El Almagre unless a presidio of "at least thirty men" were established near by as a protection against hostile Indians. And since "an abundance of silver and gold was the principal foundation upon which the kingdoms of Spain rested," Miranda urged the establishment of a presidio and the commencement of mining operations.

As evidence of his rich findings, Miranda turned over three pounds of ore to be assayed. This ran at the rate of about ten ounces of silver to the hundredweight—a good showing. But, as Manuel de Aldaco, a rich mine owner of Mexico, pointed out to the viceroy, three pounds of handpicked ore could not be relied upon to represent any extensive location. Moreover, some silver had been present in the reagent used for assaying the three pounds. Aldaco recommended that before Miranda's glowing report was acted upon "at least thirty mule loads of the ore" be carried for reduction to Camp Mazapil, seven or eight hundred miles away from the veins collectively called Los Almagres. Ten of the thirty loads were to be from the surface, ten from a depth of one fathom, and ten from a depth of at least two fathoms. "It is not prudent," concluded Aldaco, "to be exposed to the danger of deception in a matter so grave and important." Finally, to Aldaco it seemed but fair that Miranda and the citizens who had denounced mining claims should pay the cost of transporting the thirty *cargas*.

Miranda had followed his report to Mexico City, where he engaged an attorney to push his enterprise. He strongly objected that the citizens associated with him could not bear the expense of transporting so much ore, for at twenty pesos per *carga* the total cost would amount to more money than all five of the citizens together possessed. However, Miranda at length agreed to pay the cost himself on condition that he be placed in command of a presidio at Los Almagres; such an office would bring valuable perquisites. He professed to have no doubt that, once the ore was assayed, extensive mining would result and a presidio would be required to protect the miners. The haggling went on. At last, on November 23, 1757—more than twenty months after Miranda had made his "discovery"—the viceroy acceded to his proposition.

Meantime Captain Miranda had been dispatched on a mission to the eastern part of Texas. And at this point, so far as the mines are concerned, approved history stops short, drops the subject without one word of explanation. Did Miranda or anyone else ever so much as load the thirty *cargas* of ore on mules to be carried nearly a thousand miles away for assaying? Did Spanish miners then swarm out to the Almagres veins and extract fortunes from the earth?[3] If authenticated documents cannot finish out the half-told story, other kinds of documents,[4] as we shall see, can—and plenty of things have happened in Texas that the records say nothing about.

In the absence of positive testimony we may be sure that no presidio was established on the Llano for the protection of Almagres silver, but even while Miranda was proposing one, the Spanish government had actually planted a fort called San Luis de las Amarillas on the north bank of the San Saba River sixty miles to the northwest. At the same time a mission, for the conversion of Apaches, had been set up three miles below the fort on the south side of the river. This military establishment on the San Saba, though history may regard it as a buffer against Comanches, was according to tradition designed to protect vast mining operations. Thus hunters for the lost Spanish mine—the lost Almagres Mine, the Lost San Saba Mine, or whatever they happen to call it—look for it oftener on the San Saba than on the Llano. The mountain —of silver—went to Mahomet.

It is necessary to trace out the fate of the San Saba enterprise. Captain Diego Ortiz de Parrilla was placed in command, and on June 30, 1757—even before the stockade about his quarters was completed—he asked the viceroy to permit him to move his garrison of one hundred men to the Llano River. The Almagres minerals there should, he said, be protected, for, "*if* worked," they would be "a great credit to the viceroy and of much benefit to the royal treasury." The viceroy evidently thought otherwise, for the move was not allowed. In March, 1758, the mission, three miles away from military assistance, was besieged by two thousand Comanche warriors, who so thoroughly burned and killed that it was never reestablished.

Following this disaster, it was again officially proposed that the presidio be moved south to Los Almagres, "for protection and defense of the work on some rich veins of silver, which, it is claimed, have been discovered by intelligent men who know such things." Nothing came of the proposal. The presidio, always poorly manned and almost constantly terrorized, held out for twelve years and was then (1769) abandoned forever.[5]

The ruins of San Luis de las Amarillas and of a rock wall enclosing three or four acres of ground are still visible about a mile above the present town of Menard. Various old citizens of the region assert that in early days they saw signs of a smelter just outside the stockade, though these signs have been obliterated. Marvin Hunter remembers that his father, a pioneer editor of West Texas, picked up at the smelter a piece of slag weighing about fifteen pounds and containing silver. For years it was used as a door prop in the office of the Menard *Record*. The Hunters had eleven silver bullets, too, found in and around the old presidio—but many years ago Marvin shot them at a wild goose on the Llano River. From the smelter, so the oldest old-timers assert, a clear trail led to what is yet called Silver Mine on Silver Creek to the northwest. Of this creek more later.

In 1847 Doctor Ferdinand Roemer, a German geologist, traveled over Texas gathering material for a book that was printed in Germany two years later. In this book he describes the San Saba ruins and says that, although he looked for a smelter and for slag, he found no sign of either. It is possible that Roemer overlooked the smelter as he most certainly overlooked the old irrigation ditch, remains of which can still be seen. It is also possible that he was too sophisticated to take burnt rocks about an Indian kitchen midden for a smelter—a mistake often made by ranchmen, farmers, and treasure seekers.

Two or three aged men living in Menard recall that when they were boys swimming in the water hole below the old fort they used to stand on a submerged cannon barrel and stick their toes in the muzzle. In 1927 the town authorities diverted the river through an irrigation ditch and drained the water hole in an attempt to find the cannon. No cannon was revealed. Years before this municipal investigation, W. T. Burnum spent fifteen hundred dollars pumping out a cave on the divide north of the old presidio. Failing to find the mine there, he moved his machinery to a small lake just above Menard and pumped it dry. The Spaniards had, before evacuating the region, created the lake and diverted the river into it, thus most effectually concealing their rich workings.

Another persistent rumor has it that a great bell to be hung in the mission was cast within sight of it, gold and silver from near-by mines having been molten into the metal to give it tone. On account of the massacre, however, the bell was never hung, and so to this day certain people versed in recondite history disturb the soil about the mission site looking for it.

The Filibusters

The first Americans who came to Texas came for adventure: Philip Nolan to catch mustangs, Doctor Long to set up a republic, and at least one man in Magee's extraordinary expedition to dig a fortune out of the ground. In 1865, more than fifty years after the remnant of Magee's followers were dispersed, this man, whom history forgets to mention, appeared in the Llano country. He gave his name as Harp Perry, and he told a circumstantial story to explain his presence.

Following the last battle in which the Magee forces took part, he said, he and a fellow adventurer, together with thirty-five Mexicans, engaged in mining on the Little Llano River. Here they had a rich vein of both gold and silver, a vein that bore evidence of Spanish exploitation. It was their custom to take out enough ore at one time to keep their smelter, or "furnace," as he called it, busy for a month. It was some distance from the mine, and ore was carried to it in rawhide kiaks loaded on burros. Always, Perry said, after taking out a supply of ore he and his associates concealed the entrance to the mine. At the smelter they had no regular moulds for running the refined metal into but poured it into hollow canes; the bars, or rods, thus moulded were buried.

The miners had to be constantly on guard against Indians. In the year 1834 a numerous band of Comanches swooped down upon their camp at the smelter, killing everybody but the two Americans and a Mexican girl. The three survivors made their way to Mexico City, where Perry's partner married the girl. Many things postponed their return to Texas—and to wealth. In 1865, however, Perry, an old man now, was back on the Little Llano. He was not looking for the mine. He was looking for twelve hundred pounds of gold and silver that he had helped mould in hollow canes and bury on a high hill a half mile due north of the smelter.

He was utterly unable to orient himself. Brush had encroached on the prairies; gullies had cut up the hillsides. A few frontiersmen were out on the Llano daring the Comanches, who still terrorized the country. Perry offered a reward of $500 to any one of them who would lead him to the old furnace, the key landmark. He said that it was near a spring and that seventy-five steps from it, in a direct line towards the storage of gold and silver, a pin oak should be found with a rock driven into a knot-hole on the east side. But no furnace or stone-marked pin oak could be found. Perry then announced that he was going to Saint Louis and attempt to find his old partner, whom he believed to be living there. He left the Llano. It was afterwards learned that he threw in with a trail outfit going north from Williamson County and that while he was mounting his horse one morning his six-shooter went off accidentally, killing him instantly.

No further attempt was made to locate the furnace until 1878, when a man by the name of Medlin, his ambition having been aroused by Harp Perry's story, engaged to herd sheep for a ranchman on the Llano. Every day while herding he prosecuted his search, and within the year he found the ruins of the old furnace, the tree with a stone fixed in the knot-hole, and the high hill half a mile due north. Medlin's excavations were as wide as the Poor Parson's parish, but the high hill, where Harp Perry had said over and over the sticks of silver and gold were buried, presented such an indefinite kind of mark that the sheep herder did most of his work about the old furnace. Beneath the ruins themselves he unearthed the skeleton of a man, by its side a "miner's spoon," which was made of burnt soapstone and which showed plainly that it had been used for stirring quicksilver into other metals. Shortly after this find, without waiting to dig up anything else, Medlin left his sheep and the Llano hills for South America.

In Galveston, where he had to wait a few days to take ship, a newspaper writer chanced to find him and took down verbatim his story of Harp Perry's unsuccessful and his own successful search for the furnace. In evidence of his veracity he showed the "miner's spoon" he had dug up. Thus, with little chance for error or exaggeration, has been preserved a record of probably the first American—Harp Perry of Magee's expedition—to be lured to Los Almagres.[6]

Bowie's Secret

Flaming above all the other searchers is the figure of James Bowie. It is a great pity that we have no biography of him such as we have of Davy Crockett. This biography would tell—often with only legend for authority—how he rode alligators in Louisiana; how, like Plains Indians chasing the buffalo, he speared wild cattle; how, with the deadly bowie knife, he fought fearful duels in dark rooms; how he trafficked for black ivory with the pirate Laffite on Galveston

Island; and then how he came to San Antonio and married the lovely Ursula de Veramendi, daughter of the vice-governor of Texas.

Bowie was a master of men and a slave to fortune. He was willing to pawn his life for a chance at a chimerical mine, and he asked no odds. Out on the Nueces and Frio rivers, far beyond the last outpost of settlement, he prospected for gold and silver. In his burning quest for the fabled Spanish mines on the San Saba he engaged in one of the most sanguinary and brilliant fights of frontier history. Four years later, at San Antonio, he mistook some bundles of hay loaded on Mexican mules for bags of silver, and led in the so-called Grass Fight.[7] Then on March 6, 1836, leaving not one "messenger of defeat," he and one hundred eighty-odd other Texans died in the Alamo. Thousands of men have believed and yet believe that he died knowing the location of untold riches. At any rate, dying there in the Alamo he carried with him a secret as potent to keep his memory fresh in the minds of the common people as his brave part in achieving the independence of Texas. Thenceforth the mine he sought and that many believe he found took his name.

In the accounts dealing with Bowie's search history and legend freely mingle. I tell the story as frontiersmen and hunters for the Bowie Mine have handed it down.

When Bowie came to western Texas, about 1830, a band of Lipan Indians, a branch of the Apaches, were roaming the Llano region. Their chief was named Xolic, and for a long time Xolic had been in the habit of leading his people down to San Antonio once or twice a year to barter. They always brought with them some silver bullion. They did not bring much at a time, however, for their wants were simple. The Spaniards and Mexicans thought that the Lipan ore had been chipped off some rich vein; there was a touch of gold in it. Of course they tried to learn the secret of such wealth, but the Indians had a tribal understanding that if any man of their number revealed the source of the mineral he should be tortured to death. At length the people of San Antonio grew accustomed to the silver-bearing Lipans and ceased to pry into their secret. Then came the curious Americans.

Bowie laid his plans carefully. He at once began to cultivate the friendship of the Lipans. He sent back east for a fine rifle plated with silver. When it came, he presented it to Chief Xolic. A powwow was held and at San Pedro Springs Bowie was adopted into the tribe. Now followed months of life with the savages. Bowie was expert at shooting the buffalo; he was foremost in fighting against enemies of the Lipans. He became such a good Indian and was so useful a warrior that his adopted brothers finally showed him what he had joined them to see.

He had expected much, but he had hardly expected to be dazzled by such millions as greeted his eyes. Whether it was natural veins of ore that he beheld or a great storage of smelted bullion, legend has not determined. Anyway, it was "Spanish stuff." The sight seemed to overthrow all caution and judgment. Almost immediately after learning their secret, Bowie deserted the Lipans and sped to San Antonio to raise a force for seizing the wealth.

He was between two fires. He did not want too large a body of men to share with; at the same time he must have a considerable number in order to overcome the guarding Indians. It took some time to arrange the campaign. Meanwhile old Chief Xolic died, and a young warrior named Tres Manos (Three Hands) succeeded to his position. Soon after coming to power, Tres Manos visited San Antonio. There he saw Bowie, accused him of treachery, and came near being killed for his effrontery.

The story of Bowie's adventures with the Indians thus far has no support from history. What follows is of record. On November 2, 1831, Bowie set out to find the Spanish mine. His brother, Rezin P. Bowie, was in the company and was perhaps the leading spirit. It has been claimed that he had made a previous trip of exploration into the San Saba country. Both of the brothers were remarkable men and both of them left accounts of the expedition.[8] With them were nine other men, the name of one of whom, Cephas (or Caiaphas) K. Ham, will weave into odd patterns through the long Bowie Mine story.

If James Bowie knew exactly where he was going, he coursed in a strange manner. In fact, he took so much time in "examining the nature of the country," to use his own words, that three weeks after setting out from San Antonio he had not yet arrived at the abandoned presidio on the San Saba only a hundred and fifty miles away. Yet the San Saba fort was a chief, if not the chief, objective of the expedition, for the Bowies were certain that it had protected the Spaniards "while working the silver mines, which are a mile distant." Why then did the Bowies not go directly to the fort and the mine? Did Jim Bowie know—from a Lipan's confidence—of some other place? Where had he spent the three weeks in scouting before he was stopped? *Quien sabe?*

On the nineteenth of November a friendly Comanche warned him that hostile Indians were out. Whether Tres Manos was among them is not recorded; they were mostly Caddos, Wacos, and Tehuacanas. About daylight on the twenty-first one hundred and sixty-four hostiles—fifteen against one—swooped down upon the Bowie camp. The Texans were not unready. They had the advantage of a thicket and of being near water in a creek. The fight lasted all day. One man was killed; three others were wounded. The Indians had fifty dead and thirty-five wounded. With a comrade named Buchanan shot so in the leg that he could not ride and with most of their horses killed or crippled, the mine hunters remained in camp for eight days. They were not provided with surgical instruments or with medicines of any kind, "not even a dose of salts." They "boiled some live oak bark very strong, and thickened it with pounded charcoal and Indian meal, made a poultice of it, and tied it around Buchanan's leg." Then they sewed a piece of buffalo skin around the bandage. The wound healed rapidly.

While waiting for the disabled to recover sufficiently to travel, some of the Bowie party found a cave near camp. This is a point to remember. Ten days were required for the hobbling journey back to San Antonio.

It is generally said that the fight was on what is now known as Calf Creek in McCulloch County, twenty-five miles or so east of the San Saba fort. At any rate, the remains of a barricade called Bowie's Fort are yet visible on Calf Creek, though "the hand of the impious treasure seeker" long since scattered the stones. Rezin P. Bowie said that the fight took place six miles east of the San Saba fort, and there is good reason for accepting his word. Some six or seven miles east of the old presidio a strip of brush growing on what is known to moderns as Jackson's Creek, a tributary of the San Saba, hides a collection of rocks that looked to many frontiersmen like a hastily arranged fortification. Not a great distance from this place is a cave. Jackson's Creek is dry now, but before the country was grazed off it usually furnished water during several months of the year. Thus it affords a site corresponding to Rezin P. Bowie's description. Exact location of the battle ground would be interesting to some seekers of mine and treasure, for they say that the cave near the Bowie camp held "something."

When Doctor Roemer visited the San Saba ruins in 1847, he observed among other carvings on the stone gateposts near the northwest corner of the stockade the name of Bowie and the date 1829. Those gateposts have been shamefully mutilated, but on one of them this legend, neatly carved, is yet visible:

BOWIE MINE 1832

Whether James Bowie carved his name with either of those discrepant dates can never be determined.

Without exception, one might say, the men of that highly individualized class who called themselves Texians knew about the Bowie Mine. Most of them who left any kind of chronicle make mention of it. The unpublished *Memoirs* [9] of Colonel "Rip" (John S.) Ford, border ranger, journalist, and a Texian among Texians—aye, a *Texican*—contains a sequel to the Bowie expedition. This sequel came to "Old Rip" from Cephas K. Ham, who survived "the Calf Creek Fight" for many, many years and became a veritable high priest to the Bowie Mine tradition.

According to Ham's story, he (and not Bowie) was adopted by the Indians and was—*almost*—shown the mine. His warrior brothers were a band of Comanches under the leadership of Chief Incorroy. In 1831 he was wandering around with these Comanches, trading for horses and catching mustangs in order to make up a bunch to drive to Louisiana. One pint of powder, eight balls of lead, one plug of tobacco, one butcher knife, and two brass rings made the price of a good horse. "A certain fat warrior," Ham narrates, "was frequently my hunting companion. One day he pointed to a hill and said: 'There is plenty of silver on the other side. We will go out by ourselves, and I will show it to you. If the other Indians find I have done so, they will kill both of us.' " But camp was hurriedly moved next day and the fat warrior never fulfilled his promise.

Not long afterward Bowie sent a message to Ham advising him that, as the Mexicans were about to make war on the Comanches, he had better cut loose from them. He came into

San Antonio, only to find that Bowie's real motive in sending a warning was to get him to join an expedition in search of the San Saba Mine.

Rezin P. Bowie, Ham's story goes on, had already visited the mine. "It was not far from the fort. The shaft was about eight feet deep." Rezin P. Bowie went down to the bottom of it "by means of steps cut in a live oak log" and hacked off some ore "with his tomahawk." He carried the ore to New Orleans and had it assayed. "It panned out rich." He came back to San Antonio. The results of his next move have already been recounted.

Here Rezin P. Bowie drops out of the story, but Jim Bowie did not give up the quest. Ham and other like authorities agree that he raised a second expedition of thirty men. This time, according to Ham, Bowie reached the San Saba but could not find the shaft, as it had been filled up either by rains or by Indians. Others say that about the time Bowie got ready to exploit the mineral riches he had located, the Texan war for independence broke out. Among many Texans the legend is persistent that Bowie's chief motive in searching for the San Saba treasure was to secure means for financing the Texas army[10]—a view hardly tenable by anyone who knows anything about the real Bowie.

Thus Bowie's name lives on. Wes Burton, a lost mine hunter who has been very successful in telling of his hunts, says that the Lipan Indians never showed Bowie a mine but merely five hundred jack loads of pure silver stored in a cave. The Spaniards mined the silver and moulded it into bars faster than they could transport it. Consequently, when the Indians forced them to abandon their workings they left behind an immense store of bullion. Burton also knew a man who paid $500 to a Mexican in San Antonio for a document purporting to have been taken off Bowie's dead body in the Alamo by one of Santa Anna's officers. The Mexican who sold it claimed to be a descendant of the officer. It gave full and explicit directions to the Bowie Mine; yet somehow the purchaser could never follow them.

Well, James Bowie set out for the San Saba Mine. Therefore he must have known where it was. Miranda found the Almagre vein south of the Llano and powers who did not listen to him established the presidio of San Luis de las Amarillas on the north bank of the San Saba twenty leagues away. Rezin P. Bowie asserted that *the Spanish mine* was only a mile from the presidio. But the Lost San Saba, or Bowie, Mine envelops both these locations as well as many others. Sometimes it is as far east as the Colorado and sometimes it is as far west as the Nueces.

In the Burned Cedar Brake

The early settlers of the San Saba and Llano country found an old road leading south from the presidio of San Luis de las Amarillas. As it was their belief that the Spaniards had hauled bullion over it to San Antonio and Mexico, they called it Silver Trail and they traveled it themselves until the country was fenced. Like other roads laid out by men who must beware of ambuscade, it kept as much as possible to high and open ground. The land it traversed on the North Fork of the Llano came to be known as Lechuza Ranch.

In 1881 the Lechuza came into the possession of a young Scotchman, Captain George Keith Gordon, who, after having hunted slavers on the East Coast of Africa and mapped many of its harbors, had lately retired from the British Navy. Nearly fifty years have passed now since Captain Gordon became interested in the San Saba Mine; he is still interested and has a trail to hunt out. This is his story.

Twenty-five miles or so northwest of the Lechuza Ranch was Fort McKavett, occupied during the seventies for the purpose of frontier defense. After it was abandoned, the camp sutler remained in the country. He was not an uncommunicative sort of being, and he was not tardy in letting the newly arrived owner of the Lechuza know that the hillsides and valleys of his estate contained something more valuable than the eye-delighting mesquite grass.

One day while scouting out in the vicinity of a large cedar brake on the Lechuza range, the sutler, so he said, saw three Indians. He himself was hidden on a hill above them, and he watched. Presently they disappeared in a very queer manner—vanished as if into the earth—then reappeared and left. Curious to see what they had been up to, the sutler rode down to the spot. He found a hole in the ground about thirty inches in diameter. Looking down into it, he could distinguish nothing clearly and so became more curious. He dragged up a small log, tied his lariat to it, and lowered himself. Something over twenty feet down he struck bottom.

He was in a concave about fifteen feet across. Against the wall on one side, the disheveled skeleton of a man sprawled over a heap of silver bars. The bars were so heavy that the sutler could not take even one with him, for it would be all he could do to pull his own weight out of the hole. He would return to the fort, he told himself, and make immediate preparations for securing the silver bars and hauling them away. After clambering to the surface, he marked the spot carefully and left. But a man attached to the army even in the loosest way is very often not his own master. For reasons not necessary to delineate here, it was a full two years before the sutler got back to haul out the silver.

When he did get back, he found that a fire had swept the cedar brake, obliterating all surface markings. The hole thirty inches in diameter was lost.

This account reminded Captain Gordon of a cavity he himself had observed in a cedar brake in his pasture but had not investigated. He was thinking of investigating it when one morning a stranger drove up to the Lechuza headquarters and, after the usual beating about the bush, asked permission to hunt silver on the ranch. He was willing to give the owner half of whatever he should find. Telling him nothing, Captain Gordon hitched up his buggy and drove the prospector to a spot on the old Silver Trail where some irons from a burnt wagon had given rise to a tale about Spanish treasure.

"It was a hot July day," Captain Gordon narrates. "I was not feeling very well; so I sat in the shade and let the stranger have his way. He produced a divining rod and followed its pull into a dense cedar brake—directly away from the wagon irons but towards the cave I knew about. After an hour's struggle with the heat and cedar limbs, he returned, claiming that the farther he went the harder the rod pulled. 'There must be a wagon load of the stuff at least,' he said. I now told him the sutler's story.

"The next morning he, two young Englishmen who were staying with me, and I, all well provided with ropes, picks, crowbars, and shovels, got into a wagon and headed for the cave. On our arrival the rod in the stranger's hand at once told us that the silver was still in the hole. We let him down and I entered also. The rod pulled towards one side. The cave was not so deep as the sutler had described his as being and it appeared to have been filled with loose rocks. My theory was that the cedar fire had ignited bat guano in the cave and that the heat had caused rocks to crack and fall. The young Englishmen flew in to moving the debris and within three hours were down to solid rock bottom—but not to silver. The mineral rod still pointed to the spot, but we were disgusted and quit."

So much for the silver cache. The mine out of which the Spaniards took silver affords the problem that Captain Gordon has really been interested in. Some time after the experience in the cave, he was over on the Nueces River and there met General John R. Baylor, whose exploits as a mineralogist will be told of in the next chapter of this book. Baylor showed Captain Gordon an outcropping of curious sandstone from which he had assayed a showing of silver. The Captain began tracing that formation. He found that, cropping out in various places, it led in a north-northeast direction, across his own ranch, and on straight towards the old presidio at Menard. He and his brother assayed some of the rock found on the Lechuza and got a good percentage of silver.

The way to find the San Saba Mine is to trace this sandstone to the vicinity of the old Spanish fort. Captain Gordon never tracked the outcroppings to the inevitable shaft on the San Saba. Few people realize how tied down the average ranchman is, how little time he has for taking up asides. But the old captain, living now in San Antonio—his house a veritable museum of objects of art and aboriginal artifacts collected from many lands—is at last free of ranch bondage. At the very moment of this writing he is preparing to follow the sandstone trail to its end—the long-hidden Spanish silver mine on the San Saba.

Yellow Wolf: "Three Suns West"

Even if there were no "Spanish charts" to the mineral wealth on the Llano and the San Saba, no Silver Trail, no rocks pointing always *más allá,* there would yet remain as a guide to Coronado's trustful children the tales of the aboriginal red men. In the beginning it was Indians who inspired the search for the Hill of Silver. An Apache told Miranda of the Dos Almagres, where solid silver, "soft like the buckles of shoes," could be found. A Comanche pointed out, somewhere over the hill, a mine to Cephas K. Ham. A legendary Lipan gave the secret to Bowie himself. Years after Bowie's death another Lipan led, for the modest reward of $300, some Austin citizens to "the old Spanish mine" near the San Saba mission. It was out in a "bald-open prairie." According to the latest available report, in the *Telegraph and Texas Register,* Houston, June 22, 1842, the ore from this Lipan's mine "has not yet been accurately analysed."

Oftener than either Spaniards or Bowie, the Comanches appear as witnesses in tales of the great lode. A century and a quarter ago these fierce and extraordinary people boasted their nation to be the peer of the United States, whose citizens they at that time respected and treated as friends. Towards Spanish power, however, never did they express any other feelings than contempt and hatred. They learned to outride the race of horsemen who had introduced the horse to America; they ran Spaniards and Mexicans down, roped them, and dragged them to death. They were particularly jealous of the Llano-San Saba territory, from which they evicted the Apaches; they prevented the Spaniards from getting a firm foothold in the region; they fought against Anglo-American settlement of it until the very end.

Living at Liberty Hill is an old mustanger and trail driver named Andy Mather. The mountains of the West and canyons of the Plains are mapped in his grizzled features. I see him now, a great hat on his head, wearing a vest but no coat, "all booted and spurred and ready to ride," but sitting with monarchial repose in an ample rawhide-bottomed chair on his shady gallery. He was born in 1852 on the North Fork of the San Gabriel River in Williamson County, where his father owned a wheat mill and blacksmith shop. Two miles above the Mather place a band of Comanches under Chief Yellow Wolf made their headquarters. They were peaceable at the time and were friendly with the few settlers.

"In 1851," says Andy Mather, "Yellow Wolf brought some silver ore to my father to be hammered into ornaments. Of course I was not yet born, but one of my earliest and clearest recollections is of my father's telling me about the silver. He knew a good deal about metals and said that the ore was almost pure silver. Yellow Wolf told him that he got it from a place 'three suns to the west.' He described the deposit as being under a bluff near the junction of two streams. He offered to show it, but my father would not at that time think of leaving his family in order to go prospecting into the wild country beyond.

"How far 'three suns west' would be I do not know. A Comanche warrior could go a far piece in a day's time. From our place west to the old San Saba fort it was close to a hundred miles—just about three days' travel as we used to ride. Some people have figured that Yellow Wolf must have got his silver in the vicinity of the fort. I have known plenty of men to lose what little money they had in looking for silver in the hills west of here. I never hunted for the mines myself. All I know is what my father told me. I know that he was a calm man and that he told nothing but the truth. He knew Yellow Wolf well. He hammered Yellow Wolf's ore on his own anvil."

Captive Witnesses

As if to corroborate the Indians, Indian captives have added their tales to the ever-increasing cycle. Captain Jess Billingsley, who won his fame in the Texas Revolution, was the "old rock" itself. Such was his prowess that his followers as long as they lived called themselves and were proud to be called "the Billingsley men." In the speech to his men that preceded the charge at San Jacinto Billingsley used a phrase that became a nation's battle cry and that seems destined to live as long as the name of Texas lives: "Remember the Alamo! Remember Goliad!" [11]

Along in the forties, while San Jacinto and the Alamo were yet freshly remembered, Captain Billingsley heard a Baptist preacher tell how for years he had been held captive by the Comanches and how through them he had obtained knowledge of the Bowie Mine. As proof of his knowledge of the mine he offered to lead the way to it. Billingsley organized a small expedition and with the preacher as guide set out for the San Saba. Up in the hills somewhere the Comanches attacked the party and killed their escaped captive. The survivors of the fight nearly starved to death before they got back to a country in which they were not afraid to shoot a deer—and that was the last of that search.

One time in the early fifties a *Comanchero* (a Mexican trader with the Comanches) while out in the hills north of San Antonio bought a Mexican woman from a band of Comanches, with whom she had been for several years. He took her to San Antonio and released her. Naturally many people questioned her as to her experiences with *los Indios broncos*. Among other things she told of how she had helped them gather silver ore up on the San Saba and beat it into ornaments.

Now, about this time a bachelor named Grumble was ranching on the San Saba below the mouth of Brady Creek—a good seventy-five miles east of the old fort. He had fought Comanches a plenty; he had seen silver on their buckskin trappings; he had even picked up spent silver bullets from their guns. When echoes of the Mexican woman's story reached his ears, he determined to visit her in San Antonio and find out more concerning the source of Comanche silver. This was a year or two after the captive's liberation, she having married in the meantime. She talked to Grumble without reservation.

"The Comanches often camped," she said, "at the old presidio, and right there they made bracelets and conchas and other ornaments out of silver. I have helped to bring the ore into camp, but I will not deceive you by telling you that I saw where it was taken out of the ground. I was a slave and had to obey others. I tell you only what I myself have seen.

"The Indians would leave their camp at the fort and cross the San Saba River. Then they would go on south for about two miles, following up Los Moros Creek. Then the men would leave the squaws and captives at a regular stopping place. Sometimes they would be gone a long time; sometimes not over an hour. I do not think the mine was over half a mile away. When the men got back to where we were waiting, they always gave us their ore to carry on in. It was well understood that the mine was kept hidden so that no stranger could find it."

After hearing this account, Grumble asked the woman if she would go with him to the San Saba and guide him as far as she could towards the Comanche mine.

"All I could do," she answered, "would be to lead you to our waiting ground on Los Moros Creek. At this place I could only point out the brush into which I saw the warriors enter empty-handed and from which I saw them return carrying silver."

All that Grumble asked was to be "put on the right track." The woman and her husband were willing to undertake the trip under his protection. Arrangements were made for immediate departure.

Grumble was a race horse man and gambler. He had formerly lived in New Braunfels, and he chose now to pass through that place and several other towns that offered chances for profitable games and races. Riding at his own convenience, he was sometimes ahead of the Mexicans and sometimes behind them. The arrangement with them was kept secret. Late one evening all three arrived in the village of San Saba, where Grumble went to the hotel while the Mexicans made camp.

As soon as he arose the next morning, the American entered a saloon to get a drink. At the bar he saw another gambler and racer named Sinnet Musset with whom he had recently had "a difficulty." Each man "reached." Musset drew a fraction the quicker, and Grumble was a dead man. This was in 1857.

The Mexicans were naturally frightened. The woman had no desire to fall again into the hands of the Comanches. They went back to Lampasas, through which they had passed on their way up, got work there, and for five years did not tell a soul of their secret engagement with Grumble. When, finally, they did tell, no man was enterprising enough to take up the trail that Grumble had so precipitately quitted.[12]

Perhaps there is some connection between the silver up Los Moros Creek and Mullins' chart. Mullins is a bachelor. He lives at Menard in a feed store and makes money trading real estate. He uses some of his money to keep two or three men digging for San Saba treasure. He is not particular where they dig, provided they dig the holes deep enough. Sometimes they dig above the old presidio, sometimes below it; sometimes on one side of the river, sometimes on the other. The laborers get their wages every Saturday night and are satisfied; so is Mullins.

His prize chart is drawn in blue ink on the scraped hide of a javelina. He bought it from a young man who confessed to have stolen it from an old trapper in East Texas. I doubt if Mullins would sell it for a thousand dollars. It shows the San Saba River and Los Moros Creek. A cross represents the presidio. Near it an Indian with a long scalp-lock is drawing a bow. An owl looks down from a tree. A half dozen irregular stars sprinkled conspicuously over the map indicate where silver is to be found. The largest of the stars is up Los Moros Creek. A line projected southward from the old presidio up Los Moros Creek would very nearly coincide with the line of silver-bearing sandstone that Captain George Keith Gordon traced to the Lechuza Ranch.

The silver ledge on the Frio River must be too many suns west for the location that Yellow Wolf alluded to, and it could not be the location on Los Moros Creek that the Mexican woman told Grumble of. Yet it belongs to the tradition of captives who ranged with their captors over the San Saba territory. A "little rancher" named Whitley out in McMullen County told me about it. Could I reproduce the starlight on his beard, as we sat in front of his house, and the far-away barking of coyotes that mingled with his tones, the story would be a thousand times more real.

"When I was a young man in Refugio County," Whitley said, "I got to knowing an old, old Mexican named Benito who had been raised by the Comanches. They had captured him down on the Rio Grande as a boy and they kept him until he was grown. Whenever the Indians came to San Antonio to trade, he said they always put him and other captives under the supervision of squaws, who stayed hid out.

"The main thing these Indians had to trade was ore. They made bullets out of it too. It was mostly silver. For a long time Benito didn't know where they got it, but finally they trusted him far enough to let him see. There was a big ledge of it up towards the head of the Frio.

"Well, Benito finally slipped away from the Indians and took in a couple of Mexican pardners to go with him for a lot of the ore. The Indians got on their trail and killed both his *compañeros,* and he barely escaped. After that he never tried to go back to the Frio.

"When I knew him he was over a hundred years old, and he would often tell me about the rich silver vein. I wanted to go in search of it, and he thought he could make the trip in spite of his feebleness if we fixed it so he could ride in a hack. He knew he could find the ledge if he ever got up the Frio Canyon, but he would not go unless a good-sized party went. He said that he would pick six Mexicans to go and I could pick six white men.

"Well, we got everything about ready, wagon, provisions, and so forth, when the man in our party who was bearing most of the fitting-out expense up and took down sick. So we naturally had to put the trip off. The man got well and a while after that we got ready to go again. But luck seemed to be against us. This time the old Mexican guide was taken down.

It was out of the question for him to go. He was dying. He gave us the clearest directions he could and thought we could follow them. From what he said, the vein of silver could not be got to on horseback. It is in the south bank of one of three arroyos that run into the Frio close together. At it the arroyo makes a sharp turn, and a man would have to get down and go afoot along the bank. No doubt it was concealed, for the Indians always covered it up well after they had hacked off what they wanted. Benito said that if he could get just one sight of the lay of the land, he could tell which one of the three arroyos the vein was in. But he never got that sight; so he gave the best way-bill he could and died.

"The treasure hunting party broke up and things rocked along for years without me doing anything. Meanwhile a brother-in-law of mine had moved into the upper Frio country. I decided to pay him and my sister a visit and to find the ore at the same time. I took my dogs along, and the first thing we struck the very first morning we rode out to find those three arroyos was a bear. Well, sir, I got to hunting bear, and we never looked for that silver at all. But I know good and well if I had left my dogs at home, I'd 'a' had it.

"I say I know, because my brother-in-law found it after I left. I gave him the directions and he agreed to notify me if he made the find. Well, he made it and was leaving his place to come down the country to tell me, when he was murdered in cold blood. But that is another matter. He had confided in his wife, and of course she told me; but as he hadn't explained to her where he located the vein, that didn't do much good.

"You see, I have known two living witnesses to that silver. It wasn't hearsay with them. If I just had time, I believe I could go up there yet and find it myself."

Living in Fort Worth is an old-time Texas frontiersman named W. A. McDaniel. He appears to have swung a wide loop and to have heard the owl hoot in all sorts of places. The following is one of many tales that he tells.[13]

"Soon after the Civil War, while I was just a kid, my father went out to Coleman County to work for the Stiles and Coggin outfit. The Indians were so bad that at times only about half the cowboys worked on the range, the others keeping guard over the horses and the ranch quarters. My mother cooked on the fireplace for a couple dozen men, but the work and anxiety were so hard on her that father said he wouldn't let her stay in such a country any longer. So one day he put her and us children in a wagon and drove us to Burnet County to stay with an uncle and aunt. The Comanches were raiding there too, but not so bad, and there were more settlers to afford protection.

"Not long after we got there, two or three neighbor families came to see us. Of course, we boys must go swimming and fishing in the creek. The grown folks were afraid for us to go alone—afraid of Indians; so a kind of picnic party was made up. After we got to the creek another boy and I slipped across and ran to a hole that was hid by a bend.

"We were pulling fish out and bragging about our luck when twelve Comanche warriors rode down upon us out of the bushes. Two of them dragged the other boy and myself up on their horses behind them. As soon as they got off a little distance they stopped and blindfolded us. They did not torture us.

"It was about eleven o'clock when they captured us, and all that afternoon and into the night we rode like the devil beating tan bark. We could not see a wink, but we knew by the coolness of the air and the sounds of insects and coyotes when night came. We also knew that we were traveling over a hilly country. At last the Indians stopped, pulled us off, undid our bandages, and told us to lie down and go to sleep. We slept.

"When we awoke next morning, we found ourselves in a

shallow cave. I noticed some of the Comanches picking up what looked to be gravel. They had a fire, a little iron pot, and a bullet mould. They were melting these pebbles and running them into bullets. I picked up four or five of the pebbles and put them in my pocket. We stayed in the cave all that day, all the next night, and until late the third night. Of course both of us boys were looking for a chance to escape, but we were afraid to make any move. We were not tied but we were guarded.

"On the second afternoon of our captivity, the Comanches brought in a jug of fire water—regular old tarantula juice—from somewhere and they all got as drunk as a covey of biled owls. By good dark the warrior guarding us was as drunk as the other Indians. He let the fire go out and keeled over dead to the world. We were in a kind of pen made by the cave wall on one side and the sleeping Comanches on the other. Now was our time.

"We slipped out and found a horse tied in the hollow. We both got on him and headed him southeast. He kept a general course except when we misguided him, which we frequently did; it later turned out that he had been stolen from a settler on the Lampasas. After riding the night out and then, with a few stops, until nearly sundown next day, we struck the settlement from which we had been stolen. While we were telling the story, I pulled the pebbles from my pocket. They proved, upon examination by a man who knew, to contain gold and silver as well as lead.

"I have tried many a time to ride back to the cave. It's been like looking for the white cow with a black face. I went to that cave in darkness; I left it in darkness; it is still in darkness. It can hardly be more than fifty miles from Burnet. When the Comanches saw that the white men were going to take the country for good, they doubtless filled up the entrance to the cave. Some day—perhaps it may be a hundred years from now—the cave and the mine will be found."

Whether this cave paved with pebbles of silver is the same as the one hung with icicles of silver, I cannot say. Living in Sweetwater, Texas, until recently was a blind man named Johnson, very old. Back in the fifties he was a ranger and Indian fighter. One time the band of rangers to which he belonged struck a party of Indians west of the Colorado River and, after killing two or three of them, scattered in pursuit. It was seldom that rangers bothered with Indian prisoners, but the captain of this company took his man alive. The prisoner turned out to be a Mexican.

The Indians, he said, had stolen him as a child, reared him, and shared with him a great secret. If his life were now spared, he would show his captor *mucha plata.* The ranger agreed, and the Mexican led him back into the hills for a few miles until they arrived at a very thick motte of hog, or Mexican, persimmons. Crawling into this, they came upon a broad rock slab. They lifted it back. Underneath was a slanting hole. Peering in, the ranger saw what appeared to be a myriad "icicles of silver." They hung glistening from the roof of the cavity like stalactites.

The captain now regarded himself a wealthy man, but what with chasing Comanche raiders, fighting Mexican bandits, and quelling domestic outlaws, he had no time to realize immediately on his wealth. In fact, he saw the icicles of silver but the one time. He confided the secret to Johnson, and was shortly afterwards killed. The years passed, and by the time Johnson got ready to search for the wonderful hole amid the hog persimmons he was blind. He knew the hill country, every canyon of it; he knew where he and his fellow rangers had met the Indians and in what direction his captain had followed the warrior captive who surrendered and paid such a wonderful ransom for his life; he had the captain's directions in minute detail. But a blind man cannot make his way through dense thickets and into box canyons; a blind man cannot lead the blind.

Beasley's Cavern

Beasley has been dead more than ten years now, and he was perhaps eighty-five years old when he died.[14] As a youth he came to live in the settlement about Lampasas. Following one of the Comanche raids he went with a little band of settlers in pursuit. The trail was plain, and they rode fast towards the west. They crossed the Colorado and veered northward in the direction of the San Saba. The moonlight was too dim to reveal tracks and so at dark the trailers camped to await dawn.

When Beasley went out early next morning to get his horse, he found that a coyote had chewed the picket rope, a rawhide reata greased with fresh tallow, in two, and that his horse had "made tracks." It was well after sunup before he found the animal, on the side of a rocky draw. He caught him and was making a *bozal*—a nose-hitch, used in place of a bridle—preparatory to mounting bareback, when his eye caught the mouth of a cave near at hand.

There is something about a cave that draws all natural men. A cave may conceal anything. This cave faced east, and the rising sun was shining directly into it. Beasley led his horse over and peered within. For fifteen or twenty feet the hole sloped down at a steep angle and then seemed to become a horizontal tunnel. The walls of the opening were reflecting and refracting the sun's rays like a chamber of mirrors. Beasley forgot at once both murderous Indians and his impatient comrades. He made a pair of hobbles out of a bandana, thus securing his mount, and then with the free rope lowered himself into the cave. Now that he could examine the walls near at hand, he saw that they were lined, plated, cased with ore almost pure—ore that was undoubtedly silver.

However, he did not tarry. As quickly as possible he climbed out of the cavern, unhobbled his horse, mounted, and galloped to camp, where he found all hands awaiting him. He privately told the leader of the party what he had found. The leader chided him for having delayed the pursuit of the Indians and at once took up the trail. It was a long trail, a twisting trail, and at the end of it there was blood. When the frontiersmen turned back towards the Lampasas country, they left their leader behind them.

Beasley had communicated his secret to no one else. He alone now possessed it. There was only one other person in the world with whom he would share it. Almost immediately he went east to marry the girl who had been waiting for him to make a stake. He told his bride that their fortune was found, though not yet gathered. Daring the privations and hazards of the then utmost frontier, the couple settled a few miles below the mouth of the San Saba River. Beasley picked the site as being near his silver cavern. He remembered exactly, he thought, where his party had camped that night, where he had found his horse, and where he had seen the metal plating flash against the early sun. He was a good man in the ways of camp, trail, and unfenced range.

Yet after he had established a home—a base of operations—and set out to work his mine, he could not find it. He was so disturbed that for a long time he would not tell his wife of his failure. When he told her, she encouraged him by going out with him. The country settled up, and he went on looking for his cavern. In time he told neighbors of his quest—of his rich find and of his loss. The years by fives and tens shuttled by; he farmed and ran a few cattle; he worked hard; he raised a family; but he never entirely gave up the search. When death came he was still hopeful of some day recovering the fortune that one bright morning in his youth, when the land was youthful like himself, had gleamed before his eyes. To this good day, however, Beasley's cavern, like the tomb of Moses, remains an unseen monument.

Perchance the bald, old eagle
On gray *Packsaddle's* height
Out of his rocky eyrie
Looks on the wondrous sight;
Perchance the *panther* stalking
Still *knows* that *hunted* spot;
For beast and bird have seen and heard
That which man knoweth not.

Pebbles of Gold

Some—many—there are who hold that Beasley's cavern, the cave in which drunken Comanches held McDaniel prisoner, and likewise the cave with icicles of silver are all identical with the Bowie Mine, hasty revelations of it under extraordinary circumstances. However that may be, while the hunters are looking for cavern walls of shining metal, roof of argent stalactites, and floor strewn with pebbles of rich alloy, they had as well look also for pebbles of gold in a brook that flows somewhere in the same country.[15]

About the time that Captain Ben McCulloch of the Texas rangers was introducing the six-shooter to the horseback world, he detailed two of his men to scout for Indian signs west of the upper Colorado River. One morning they arose from their pallets out in that lonely and unsettled region to find their horses gone. A dense fog enveloped the hills and valleys so that they could see nothing. Nevertheless, they struck out to find the horses. The fog held on for hours. The best of woodsmen can become lost in such a fog. When this one lifted, the rangers discovered that they were lost. They had not found the horses and now they could not find the way back to their saddles and canteens. They wandered all day. It was hot summer, in a time of drouth, and they were in a region utterly devoid of water. By the time night came they were so thirsty they could not sleep. The lead, taken from cartridges, that they chewed afforded no relief. Their whole object now was to find water; as soon as they could see to travel, they pushed on.

At length, from the summit of a low range of hills, they saw below them far to the west a winding line of green. What stream it marked they knew not, for they had never seen it before. Its waters were life, and they were so clear that the sun danced on the pebbles at the bottom.

As one of the rangers, after the burnings of thirst were quenched, lay looking into the sun-lit pool from which he had drunk, he was startled to discover that the bottom of it was strewn with minute particles shining like gold. Calling to his companion, he said: "We have lost our horses, our saddles, and our guns, but we have found something better than all of them. Here is gold, gold, world without end!"

The shining particles, some of them as large as coarsely ground corn grits, were so thick among the sand and gravel that they had the appearance of having been sown by the handful. The rangers waded into the water and gathered them until each had a pocketful. Then one of them crept up on a turkey gobbler that was watering and shot it with a six-shooter. That night they feasted.

On their way out they stopped to rest high up on the shoulder of a long, rough hill. They were sprawled out on the ground, neither talking, when suddenly their attention was arrested by that peculiar cry of a hawk so resembling the call of a young wild turkey. Looking towards the hawk, which was alit in a stunted, half-dead post oak, they noticed something sticking out from a crotch of the tree.

It proved to be an ancient rust-eaten pick, its handle gone and one point encased so deeply that it could not be removed. The other point stuck out toward the head of the little stream they had just left. Then the rangers realized that they were not the first to have discovered gold in the region. They went on, leaving the unknown prospector's signboard still pointing. Late in the afternoon they saw Packsaddle Mountain looming in the distance. From this well-known landmark they got their bearings.

A few weeks later they exhibited their gathering of nuggets to an expert on minerals. He pronounced the stuff to be what miners call "drift gold," gold that has washed downstream from a mother lode. The mother lode, he added, might be miles away, but wherever it was it must be exceedingly rich. On many a long ride in after years the rangers sought the golden pool, but though they were Ben McCulloch's own men—plainsmen and woodsmen right—they never found it again. It may be that the mute finger of the old pick on the shoulder of a long, rough mountain still points to the source of the drift gold. The granite hills of the Llano guard well their secrets.

The Magic Circle on Packsaddle

Had McCulloch's rangers on their way to camp climbed Packsaddle Mountain instead of using it merely as a landmark, they might have had an additional story to tell. Every fifteen or twenty years some "prospector" comes along, finds an old shaft on the mountain, imagines that he has discovered the original Spanish mine, and cleans it out, digging it a little deeper down. After he has spent his last two-bits, he goes away, and within a few years the abandoned shaft has the appearance of being as old as the hills—just the appearance to delight searchers for Spanish mines. But never doubt that there is a real mine on or in or under or somewhere around Packsaddle; the shafts have just not been sunk at the right places.

A Spaniard by the name of Blanco, they say, found the right place; hence the object of search is often called the Blanco Mine. Whether it is the same as the Bowie Mine, I shall not undertake to determine. Men have all but shot each other in arguments over this very question. Certainly a plat to it in the form of a "magic circle" is sometimes applied to the Bowie Mine as well.

Among the earliest settlers in Llano County was a man by the name of Larimore. One day while he was hunting javelinas (peccaries) on Packsaddle Mountain his dogs ran three of them into a den. During the tedious process of

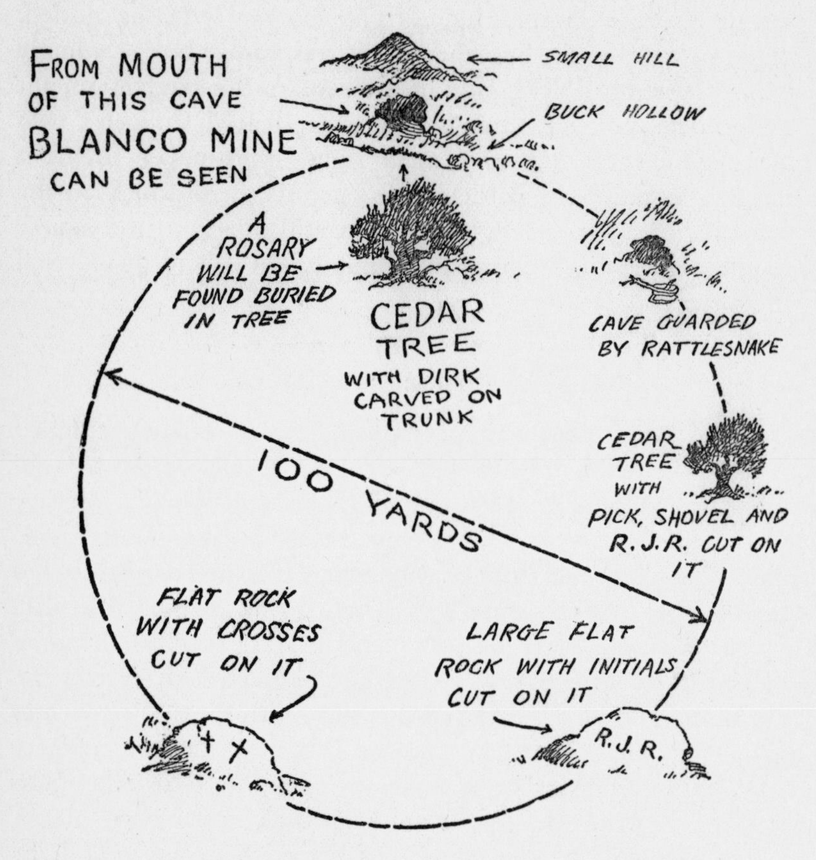

The magic circle. Landmarks within its circumference are supposed to afford a key to the Lost Blanco Mine. (Courtesy Miss Julia Estill.)

smoking them out, Larimore discovered that the den was an old mine—maybe Blanco's—maybe Bowie's. Back in it a profusion of mineral was still exposed. The mineral proved to be lead with a strong percentage of silver. It made excellent bullets; and bullets, with powder, were the most urgent need that Larimore had.

Along in the fall of 1860 Jim Rowland went with Larimore into the mine and helped bring out about a hundred pounds of the ore. This they spent several days moulding into bullets. Larimore was preparing to leave the country. "I don't mind fighting for life," he said to Jim Rowland, "but I am tired of living just to fight. When I leave I'm going to leave that mine so well hidden that no other man can find it." In carrying out his word, Larimore is said to have turned a mountain gully so that it filled the valuable javelina den with silt and then to have covered the entrance with a large flat stone, over which he strewed soil. Grass and weeds soon sealed the last trace of the secret.

Such is the account given in two letters written towards the beginning of the present century by Jim Rowland himself and now in my possession. He knew a good deal more about the Blanco mine than he ever set down on paper.[16] He had even cut his initials, R. J. R., on the flat rock with which Larimore sealed up the entrance. One time a fellow named Chaney who was in the country looking for the mine offered Rowland a thousand dollars to show him the famous rock. Rowland agreed to the offer and took Chaney to Packsaddle Mountain, where he proceeded to lead him about in a manner rather devious for one who knew exactly where to go. Finally Chaney, becoming impatient, told his guide he was tired of "fooling."

"I'm tired too," Rowland replied. "Show me that thousand dollars and I'll show you the mine."

Chaney, however, refused to pay in advance, whereupon Rowland turned on his heel and strode indignantly down the hill. In telling, some time afterward, what "a damned fool" Chaney had showed himself to be, Rowland concluded: "And just to think, at the very minute I told Chaney to hand over them thousand dollars I was a-standing right on top of that mine."

The Lost San Saba Mine

Aurelio's Trunk

It is a long way from Packsaddle Mountain to Calf Creek, and the faith that J. T. Merchant places in that location is based on no ambiguous magic circle. Mr. Merchant lives at Breckenridge, Texas. He used to be a school teacher; he is night watchman for an oil company now. He reads *Current History, Scientific American, The National Geographic Magazine,* and anything he can find bearing on geology and mineralogy. His interest in anthropology and paleontology has led him on long excursions following the tracks of prehistoric animals in rocks and getting evidence on "the Long-Headed Tribe." He has a strange story about a chart to the San Saba Mine.

"While I was a boy attending high school," he says, "I lived at New Boston, near Texarkana, with my sister and her husband, R. H. Jones, who was judge of Bowie County. In April, 1896, legal business took him to San Antonio. There he met a young Mexican, Aurelio Góndora by name, who was looking for a job tutoring Spanish. The young man was a graduate of the University of Mexico and had studied English in New York. As Judge Jones had four children of his own as well as myself to educate and as his experience with Spanish laws and land grants operating in Texas had caused him to value highly a knowledge of the Spanish language, he engaged the tutor and brought him home."

Aurelio's history was a sad one. At the death of his father, who had been quite wealthy, his mother married her dead husband's brother, a drunkard and a gambler. He had quickly run through with most of the estate and had made the life of Aurelio—like Hamlet, "a little more than kin and less than kind"—so wretched that he left home determined never to return. Shy, sensitive, lonely, the young tutor soon became passionately devoted to the member of the Jones household who was nearest his own age and most sympathetic towards him. This was Tom Merchant. The youths roomed together for two years, "often talking for hours on Spanish history."

Towards the end of the second year of this friendship, Tom fell in love. Aurelio as a result showed himself miserable and jealous. Every time the lovers stepped out, they found Aurelio dogging them. He generally took a position about fifteen or twenty steps to the rear. The town talked and joked. The lonely Mexican lad shrank more into himself and craved more the possession of his friend. The girl protested. Tom protested also—and Aurelio took a position a little more to the rear of the lovers. The girl now said that Merchant must either give up her company or force Aurelio to quit following them. Merchant, his patience exhausted, stormed at his friend, nagged him, tried to anger him. The break which finally came brought about a most singular revelation.

"I have always been sorry for my words," says Mr. Merchant. "At the same time it seems to me that an inscrutable fate made me say what I did. Aurelio and I were together in our room. I had been nagging him about the way he followed me around. He was so innocent and kind that it was hard to abuse him. I was looking for something that I could accuse him of in such a way as to shake him. Presently my eye fell on my trunk. It was open, the tray out on the floor, several rumpled garments exposed to view. The idea at once occurred to me that Aurelio was never so untidy. Then another idea came.

" 'Aurelio,' I said, 'there is my trunk open so that you can see everything in it. If you wanted to take anything you could. My trunk is always unlocked. But look at that damned trunk of yours. I have never seen it when it wasn't locked. I have never seen you open it. You follow me around like a sick kitten and yet you do not even trust me with your trunk.'

"I shall never forget the expression that this speech produced on Aurelio's face. For three days we slept together and ate together without exchanging a word. I managed to keep away from him most of the waking hours. On the fourth day one of those dreary, almost interminable drizzles that

East Texas is noted for kept us inside. I went into our room, where I found Aurelio. I had broken him from following me; now I wanted to regain his friendship. I tried to start a conversation, but we could not talk.

"Presently Aurelio got up from his chair and closed the wooden shutters to each of the three windows in the room. Then he pulled down the windows and the shades over them. He even adjusted the lace curtains. Next he shut and locked the two doors opening into the room. A kerosene lamp gave us light.

" 'Now, Tom,' he said, 'your words about my trunk have hurt me so much that I can hardly sleep. I am going to show you why I keep it locked.'

"I remonstrated, telling him I had no desire to see what was in the trunk. 'Let us forget our differences and be the friends that we have been,' I said.

" 'No,' he replied, 'I must put before your eyes what has made me keep the trunk shut. First, though, I must explain some things to you.'

"Then he drew a passionate picture of his idealized father, a Castilian gentleman, highly intelligent, well educated, wealthy. 'When I was a mere child,' Aurelio went on, 'my father took me into his confidence and told me that he possessed the key to a fortune that would be mine even if everything else he had were lost. He repeated over and over the history of this key, specifying how I should use it. He put the details into writing. When he died, this instrument, along with other papers, came into the hands of my mother. After she married my wasteful and debauched uncle and I found I could endure home no longer, I told her that in leaving I wanted but one thing from the estate, that I would never claim anything else —the document my father had made for me. She protested that her husband knew about this document and that he would be terribly angry when he missed it. I took it anyhow. I came to the United States, for here I am to recover the fortune which it describes.'

"Aurelio declared that his uncle would, if he had the chance, murder him in order to possess the document. 'You will see now,' he concluded, 'why I keep my trunk locked.'

"With these words he took the document from his trunk. He showed me a series of plats relating to mines and treasures in Texas, Oklahoma, and New Mexico. There was a good deal of writing, and to this Aurelio added many details that he had heard from his father. I did not, as I shall presently explain, see all the plats. I shall tell of two. While the account of the first has nothing to do with the San Saba Mine, it will establish the validity of Aurelio's information on that subject.

"After finishing school, I parted from my friend and went to live in what is now Haskell County, Oklahoma. One day a man invited me to go fishing at Standing Rock on the Canadian River.

"Standing Rock on the Canadian! That was the key to one of Aurelio's plats. I asked the man if a hatchet was cut into the rock fifty feet up on the east side. He looked at me queerly and said that he had seen that very hatchet. Then I told him that there were nineteen mule loads buried near the rock.

"We went to Standing Rock on the Canadian. Yes, there fifty feet up, on the east side of the rock, just as Aurelio and the plat had described it, was the figure of a hatchet. Six hundred varas away at an angle of seventy degrees east of north should be a cedar tree with a turtle cut on it. We found it. The shaft we sank in the river and all that makes another story."

The plat to the San Saba Mine was so precious to Aurelio that he would not allow a close examination of it. He merely remarked that the Spaniards a short while before leaving the San Saba lost their entire stock of horses to the Comanches and as a result had to abandon all but one mine. It, "the main mine," he added, "was near the junction of a creek with the San Saba River to the east of the presidio. Here the Spaniards had a kind of sub-fort."

This creek Mr. Merchant has, by means of certain rock ruins, identified as Calf Creek, in McCulloch County. "Nevertheless," he concludes, "doubt seems to be the most destroying element to any argument. The contents of Aurelio's trunk would, I am confident, remove all errors and doubts. But many years ago Aurelio was stabbed to death—by an unknown man—in the Indian Territory and at the same time his trunk mysteriously disappeared."

The Relic

It is a persistent belief among treasure seekers, despite the fact that they do not live up to it, that treasure appears only when unsought and always in the most unexpected form. Late one evening some twenty years ago a man in a rickety wagon, hitched to a pair of "crow-bait" horses, drove up to a farm-house in the hill country. He called, "Hello," got out, and stepped into the yard. A widow and her grown son lived on the place, and both of them were at home.

"I'm out buying relics," the stranger said. "Have you got anything to sell?"

"No, I guess not," laughed the widow, "leastwise nothing that you'd carry off."

"Well, just any kind of relic," went on the stranger. "Maybe an old picture, an old gun of some kind, locked deer-horns, queer rocks, Indian spear points. I buy anything."

"But we don't have a thing," repeated the widow. "You can see for yourself that this ain't much like a museum, even if the plows is all wore out and the stove so caved in that I have to bake bread in the yard."

"What's this?" interrupted the stranger, as he walked over and kicked a leaden ball that he had been eyeing most of the time.

"Oh, that's just an old lead ball that one of the boys plowed up in the field years ago," mildly explained the woman. "You can see how the plow grazed it. The boys used to roll it about as a plaything. It's a wonder it wasn't lost long ago."

"Now, that's what I call a relic," exclaimed the stranger. "What'll you take for it?"

"It's not worth anything, I know," slowly answered the woman, "but I'd rather not part with it. It was such a plaything for the boys."

"I'll give you twenty-five dollars," popped out the stranger.

Such a price almost overwhelmed the widow, and she accepted it immediately. The stranger paid down the twenty-five dollars, and then walked towards the wood pile.

"Loan me your axe," he said.

The young man got the axe. Selecting a clean, hard spot of ground, the stranger put the ball between two logs so that it could not roll. "I've been looking for this particular relic a good while," he explained.

Then he cut the ball in two. It was full of gold nuggets. If the plow had gone a little deeper, it would have cut into them.

The stranger was not stingy or mean. Gathering up the spilled nuggets before the speechless widow and son, he mounded them into two equal piles.

"By rights," he said, "at least half is yours. Take it." And a few days later that half was disposed of for exactly $7000.

"This is just one of four lead balls," went on the stranger. "I've been looking for them for fifteen years. According to my directions, the stuff in them come from one of the San Saba mines and the Spaniards planted them. The four balls formed the corners of an exact square four hundred yards to the side. Just by accident the other day while I was up in the Santa Anna Mountains I heard of this old relic in your yard, and

I come a hundred miles as fast as my old plugs could travel."

A prolonged attempt was made to find the other three balls. But the exact spot in which the one ball was plowed up had long since been forgotten. So not one of the four corners could be established. The only definite clue was that the ball had been turned up along one of several washes that seamed the field. If any of the other balls is ever exposed, it will probably be by some plowhand as ignorant of its history as were the widow and her sons.

The Pictured Copper Plates

There were two Kirkpatrick brothers, Moses and James. They lived in the sober and sequestered village of Mullin, where Moses combined banking with his mercantile business and was the stout pillar, both moral and financial, that the Presbyterian church rested upon. He seldom glimpsed the wild flowers growing along the path of dalliance which his brother James so softly trod. James had been educated to medicine in Kentucky, their native state, and riding with his "pill bags" over the far-stretched hills of the Colorado River satisfied his ambition. The people in the country called him "Doctor Jim." He knew a little about rocks and he often brought them home in his saddle pockets or tied to his saddle horn. Home was with his good Scotch parents. Sometimes he fancied a rock too big to carry on horseback; he did not mind spending a day going after it in a buggy. He had a fondness for art. He was a loiterer with the tastes of an adventurer. When he was close to fifty he married a woman twenty-five years his junior. He was a great hand at story-telling, and however idle the heroes of his narratives may have appeared to Brother Moses, they were life and life's romance to nephews and nieces. He never had any children of his own. Thus lived

> Twa Duries in Durrisdeer,
> Ane to tie and ane to ride.

In 1900 "a foreigner" came to Mullin. He was either a Frenchman or a Spaniard—it could not be, or was not, determined which. He was very non-communicative, but as he was in quest of information he had to make human contacts. It was natural that he should "take" to Doctor Kirkpatrick, the easiest-going and the most tolerant man in the country.

He was looking, so he told the doctor, for a wrought-iron spike in an oak tree that should be somewhere north and a little east of the junction of the San Saba River with the Colorado. If he could find the spike, then he could find "the treasure of the banking mission"—the San Saba. The San Saba, according to the foreigner, handled the funds for all the Spanish missions to the northwest and was also a concentration point for bullion from mines in New Mexico and Colorado. It was a kind of clearing house for everything costly freighted between Santa Fé and New Orleans and naturally it had a storage vault for so much wealth. Rather unnaturally, however, this vault was far distant from the mission itself. In view of the danger of Indian attacks, a diagram to it was sent to "headquarters."

Where headquarters were, the foreigner did not say. Perhaps they were in New Orleans, perhaps in Mexico, perhaps in Spain. At any rate, after the San Saba massacre, headquarters alone knew where the wealth was stored; and for a century and a half headquarters took more pains to conceal than to utilize the information. Then, somehow, the foreigner came into possession of it. He went so far as to let Doctor Kirkpatrick get a hurried glimpse at his chart—but not to copy it. After having spent several weeks in a vain search for the wrought-iron spike, he left. No one in Mullin ever heard of him again.

Doctor Jim soon spread the foreigner's story. Consequently, when one morning during the following winter a resident of the village found a wrought-iron spike in the ashes he was cleaning out of his fireplace, he naturally mentioned the matter to the doctor.

Doctor Jim at once traced down the man who had hauled in his neighbor's wood. Then he got the woodhauler to show him where he had cut up a dead live oak tree. It is worth mentioning that a good deal of random digging had been going on in that very vicinity. Doctor Jim now entered into a loose oral understanding with the owner of the land and began operations. He interested Bob Urbach to the extent that Bob nearly broke his back and his credit digging and furnishing grub for a camp of hearty eaters. At last Doctor Jim had an occupation that was a passion.

A great deal of the exploration was through matted shinnery and cedar that a man could hardly penetrate without an axe. The first thing found was a flat limestone rock, perhaps four feet long, near the stump of the tree from which the spike had been taken. One side of it was covered with a picture, partly etched and partly painted, of an extraordinary pageant. The paint seemed to be red and yellow ochre, such as Indians commonly used. What the numbers and certain geometrical signs on this rock were has been forgotten; but many people are alive to testify to the train of ten pack burros that trailed across the stone as if to enter the mouth of a cave. On each of three packs and over the mouth of the cave was pictured a small yellow half-moon—the sign of treasure.

Doctor Kirkpatrick was as fond of deciphering codes as was Edgar Allan Poe of inventing them; he dawdled with mathematics. The figures on the stone fitted into memories of the foreigner's chart. Following Doctor Jim's directions, the laborers dug into what seemed to be the masonry of an ancient altar.

Here, between rocks that had protected them from becoming in the least tarnished, were found two "raw-beaten" copper plates, roughly circular in shape, each about twelve inches in diameter and about a quarter of an inch thick. The plates were covered with crude but clearly marked engravings. The first plate showed a trail winding down a hill, crossing a ravine, and then twisting up a hill to the east. Near the crossing, over what appeared to be a mining shaft, was the half-moon—a sign as magnetic to treasure seekers as the bright star was to the Magi. The second plate showed a setting sun near the mouth of a cave. A man standing by this cave looked across a valley towards another man who approached leading a pack burro. Between the two men three trees formed a triangle. The waiting man was dressed in a long coat of antiquated cut that reached below his knees—a coat too long to be a Prince Albert and too short to be a cassock. He wore on his head something between a derby and a helmet. He had a French appearance. Some of the details were remarkably well finished. A series of bars along one side of the plate meant something to Doctor Kirkpatrick. Taken together, the two plates seemed to suggest the dual character of the San Saba wealth: mines and stored treasure.

Doctor Kirkpatrick now located a triangle of trees a half mile east of the place where the plates were found. At the root of one of these trees, hardly two feet down, Bob Urbach and his fellow laborers dug up a hand-hammered copper box, not more than ten or twelve inches long. The lid of it was etched with the same string of pack burros, headed for the mouth of a cave, that had been found painted on the flat rock. In addition appeared the name Padre Lopez over the date 1762. Within the box were a crucifix set with pearls and two rosaries, one of ivory and the other of exquisitely carved rosewood.

The intermittent hunting that resulted in the finding of these various objects extended over a period of two or three years. As they were unearthed, the public was allowed to look at them, but only Doctor Kirkpatrick understood the ciphers. He translated them into orders.

The Lost San Saba Mine

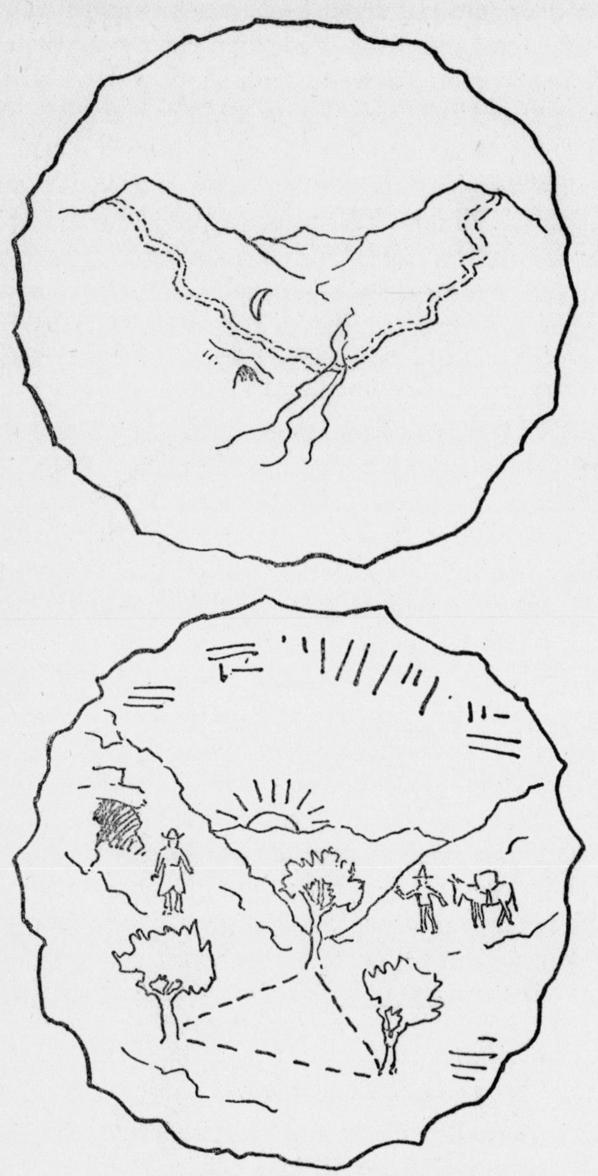

Doctor Jim's pictured copper plates.

At last, when he was ready to open the treasure vault, a work that would require a vast amount of excavation, he considered it wise to draw up a written agreement with the landowner. The landowner, who had in the beginning been contemptuous of the whole business, now demanded a lion's share of the treasure; he would not sign a contract for less. His Scotch stubbornness aroused and his sense of justice outraged, the doctor refused to agree to any such division. He would wait.

His health had been bad for a long time. He and his thrifty brother were not always in accordance, but at this juncture he went to Moses Kirkpatrick and detailed to him the whole story of the search.

"I now have all the information I need," he concluded. "I know where to find the treasure. But I am not going to dig it out until this hog who owns the land becomes more reasonable. I'll wait if I have to wait until I die—and that may be at any time. You are my brother, and I want to tell you where the treasure is so that if anything happens to me the secret will not be lost."

"No," retorted the austere Presbyterian elder and banker, "you will not tell me. I have already listened to too much idle talk. You neglect your practice, you neglect your wife. I have no patience with all this Spanish treasure foolishness."

Shortly after this conversation the doctor became critically ill. For two days he suffered and then he died. This was in 1904.

Following the death of the doctor two things important to the sequel happened: the Moses Kirkpatrick home in Mullin burned down; at the annual fair in Dallas Moses Kirkpatrick for the first time in his life entered the booth of a palmist.

The palmist pored over the deep wrinkles in his hand. "Your house has recently burned down." She described the house. He believed in her clairvoyance.

"You are on the brink of great wealth," she continued.

"How am I going to get the wealth?" the Scotchman asked.

"That is not clear. Your riches are in the ground. You or someone near you has been very close to them already."

Years after this incident Moses Kirkpatrick used to lament in his scriptural manner, "My heart was turned away so that I would not hear. My ears were dull of hearing and my eyes they were closed." He had stood on the brink of wealth—and scorned the hand that beckoned him to knock and enter. That hand now, alas, was invisible beyond the chasm of death.

Not long after Doctor Jim's death, his brother surrendered the copper plates, the stone, and the copper box with its contents to the doctor's young widow.[17] She sold the crucifix, the exquisite rosaries, and the curious box to a dealer in Dallas. She gave the stone to some stranger. The copper plates she took with her to California, where she married again and died. By expending time and money a good detective might trace down those extraordinary objects, but they seem to be lost forever. Even if they were recovered, however, it is very doubtful if any person could be found with Doctor Jim's skill and knowledge to decipher them. The doctor had made copies of the stone and plates on some wooden lids of old-fashioned candy buckets, but when the Kirkpatrick home burned they burned also.

The natives are still digging sporadically and blindly out in the hills where Doctor Jim dug up some very interesting copper objects—quite untarnished.

An Innocent Old Liar

Of all the men who have searched for the Lost Bowie Mine I have heard of only one who in the end admitted that the mine might not exist or that his account of it was not gospel truth.[18]

In the fall of 1876, Mr. J. T. Estill, a pioneer lawyer of

"the hill country," and a fellow barrister, D. Y. Portis, were driving in a two-horse buggy from Mason to attend court in Menard. Portis was at that time perhaps seventy years old, a typical plantation gentleman as well as lawyer of the old school. Rich in anecdote and repartee, he made a fine traveling companion for the equally genial Mr. Estill.

Some fifteen miles west of Mason the soil suddenly changes from a light color to a deep red, and as the travelers approached this divide, Mr. Estill remarked: "Well, we are getting into the *Almagres*—the Red Hills of the San Saba. We must be in the neighborhood of the great Bowie Mine."

"The Bowie Mine," retorted Portis with unusual animation, "is a myth. In my time I have been personally acquainted with a man who accompanied Bowie into the San Saba hills. One night down in Brazoria County a crowd of us young fellows were smoking pipes and telling yarns around a camp fire at which this old Bowie adventurer was present. He told us all about his marvelous experiences with Bowie, and he ended by swearing that he himself had hacked off pure silver from the Bowie vein with a hatchet. Well, sir, such a tale set us wild, and we all agreed right there to fit out an expedition to find the lost mine. We got together teams, wagons, and enough supplies to last us several weeks; we even hired some men to go along to help fight the Indians.

"Our party came through San Antonio and on up into this very country. It hasn't changed much. Bowie's old right bower would tell us where to camp but he stayed with us very little. He was out scouting for the lost mine. Sometimes he would be gone all day, returning only to tell us to move camp.

"By the time the old man had acted in this manner for four or five days, we came to the conclusion that he either knew nothing about the Bowie Mine or was holding back on us, refusing to live up to his contract. The leaders took him aside and in plain terms told him of our suspicions. If he had really hacked ore off a silver vein, they announced, he must tell of its location then and there or else be hanged.

"At this the old guide broke down and actually cried. 'There is no Bowie Mine,' he said. 'It is true that I was with Bowie, but the only thing we found was Indians. They turned us back, for we were only a handful. In after years, with Bowie and the other members of the hunting party dead, I began telling about the mine. I told the story so often and so long that I came to believe it. I am a liar, but I have told you the truth for once. Gentlemen, hang me if you will.'

"The innocent old liar was not hanged, and if the facts were known," concluded Portis, "a half dozen fellows who threatened to hang him have since appropriated his yarns and told them as personal experiences. When I have the right kind of audience, I myself never admit having failed to glimpse the silver of the San Saba."

The Broken Metate

It would be erroneous to conclude that most of the hunters for the Lost Bowie Mine who tell their stories, however they may excel in oral narration, are merely artistic liars. The majority of these men have bought their experiences—the subject of their tales—at the price of years of grinding labor and that habitual abstemiousness which devotion to any single purpose exacts. Many of them have suffered jibes and ridicule only, like other martyrs, to have their patience and confidence strengthened.

We were in a Mexican restaurant in San Antonio. Dishes shoved to one side of him, Longworth [19] was indenting lines in the tablecloth with the point of his knife, occasionally dipping it into *enchilada* sauce for a drop of red to indicate where a hill stood or where a copper peg had been dug up. He was a tall, spare man, perhaps nearer fifty than sixty, not a white streak in his mop of coarse black hair.

"But where did you get this map?" I asked him. "It must have a history."

"Yes," he replied, "it has a history, and God knows in trying to follow it I have had a history too."

The story was told with many backings and windings, and I did not get the whole of it until nearly a year after our first talk, when I found Longworth one day in a little room out in the yard behind the boarding house his wife keeps. He was working on an electrical device for locating minerals that he calls the "radio sleuth." After explaining its mechanism, he brought out a thick pile of documents. From them and from his own testimony I have put Longworth's story together. Like all good stories, it begins far away and long ago.

About 1830, while Texas was still a part of Mexico, a man by the name of Dixon settled on the San Marcos River near the present town of San Marcos. He was a poor man, but not too poor to keep an Indian in his hire. This Indian seems to have been a kind of outcast from his tribe. One day an aged Mexican appeared at Dixon's cabin inquiring for rocks marked in a certain way. Dixon knew of some rocks on his own land that bore the signs described. He, his Indian, and the old Mexican went to them, and in a short time the Mexican dug up a small *olla* of silver coins, which he divided with the landowner.

Not long after this episode the Indian asked Dixon if it was his desire to possess a great deal of silver. Dixon admitted that it was.

"Then," replied the Indian, "I will lead you to it. Yonder to the west in Summer Valley is a cave full of it. The mouth of the cave is stopped up with rocks, but I have been inside. I can go inside again."

It turned out that by Summer Valley the Indian meant the valley of the San Saba River. Thither he and Dixon set forth. When they got into the hills, they saw Indian sign everywhere; the Comanches and the Apaches were at war. Dixon's guide was afraid of both war parties, but from a concealed position he pointed out the vicinity of the cave. Dixon took a good look at the features of the country so that he could recognize the place when he should return at a safer time. Then the two men went back to the San Marcos. The next year the Indian died. This was not long after the close of the Mexican War.

Dixon now took into his confidence three neighboring settlers: Sam Fleming, G. B. Ezell, and Wiley Stroud. The four partners felt that they needed something more definite and reliable than the Indian's tale to guide them. The truth is that the Indian had not on other occasions distinguished himself for veracity. Therefore, they decided to send Dixon at common expense to Monclova, Mexico, once the capital of the united provinces of Texas and Coahuila, to find out what he could from the archives there.

Upon arrival in Monclova, Dixon found that the archives were in the custody of the Catholic Church and that outside examination of them was, for the time being, prohibited. He was on the verge of returning home when he met a Spaniard whom he had known in Texas. He frankly told him his difficulty.

Now it happened that the Spaniard had a daughter, named Carlota, who was engaged in some minor capacity by the priests in charge of the archives. Her father thought she might help in getting the desired information. She readily assented to the plan. She declared that some of the reports in the archives bore on the San Saba mines, though she had no idea of their contents. They were kept secret. She could copy them, she added, only at risk to her life. She would probably have to wait a long time for a chance to get full information. With the understanding that he would hear when it was obtained, Dixon came home.

Months went by without a word, then years—so many years that Dixon and his partners had almost given up hope of ever learning anything from the Monclova documents. Then one day in 1858, the north-bound stage through San Marcos brought a letter. It was from Carlota. She was in San Antonio and she had, she said, information too valuable to trust to the mail. She wanted Dixon to come to her immediately. He went.

In the interview that followed, Carlota was at once open and secretive, definite and indefinite. She was willing to tell only part of what she knew, though that part she disclosed without reservation. She had dug out, she said, documentary evidence concerning fourteen mines located around the old San Saba fort, some of them more than six leagues distant from it, the richest being Las Iguanas. In the bottom of the shaft leading to the Iguanas Mine reposed two thousand bars of silver weighing fifty pounds to the bar. To this vast storage she agreed to furnish a detailed *derrotero* (chart), with the understanding that a fifth of the silver should go to her. Once the silver was secured, it could be utilized in reopening and working the great mines from which it had been extracted before hostile Indians ran the Spanish miners out of the country.

"Dig up this silver," she said, "and then you shall have all the other charts."

Dixon could not read Spanish, but he brought home the *derrotero* to the two thousand bars of silver, and Wiley Stroud, the only "Spanish scholar" in the company, read it. It stated that the shaft was filled up with rocks. The partners agreed that it and the old Indian's "cave" were one. They were on fire with enthusiasm and confidence.

But it always takes lost mine hunters a good while to organize their expeditions. The greater the wealth at stake and the poorer the searchers, the longer the time required. The San Saba wealth was immense; Dixon and his company were as poor as Job's turkey. It took them over two years to get ready to go to the San Saba. Just as they were at last about to set forth, the Civil War began. All four entered the Confederate Army; when the war was over, all four were still alive. Meanwhile Carlota had gone back across the Rio Grande. The Texans lost all trace of her.

In 1868, exactly ten years after Carlota's appearance in San Antonio, and twenty years after the Indian's guidance to-

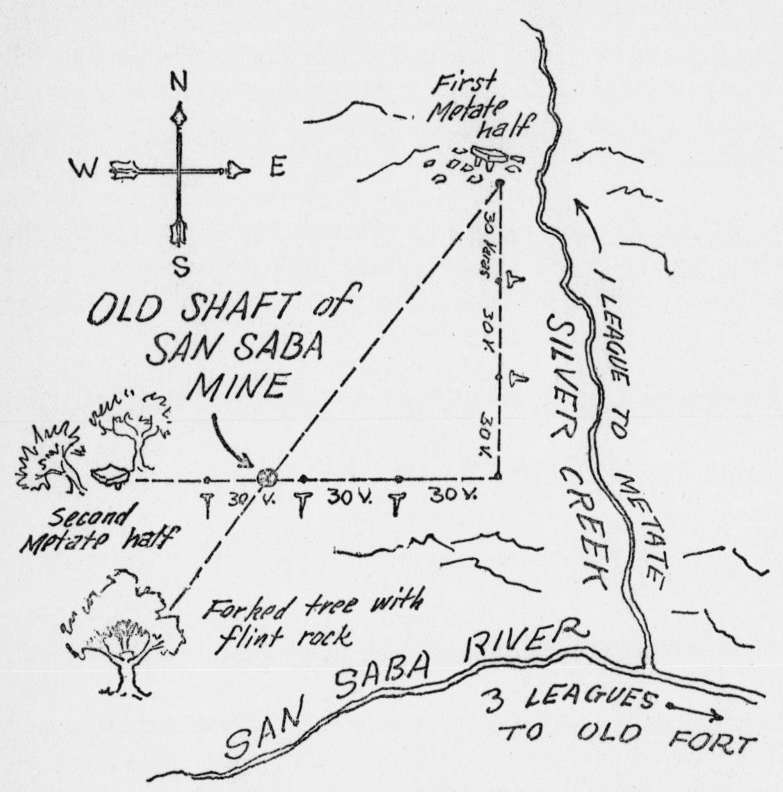

The chart that Carlota brought from the archives of Monclova, Mexico.

wards a cave in "Summer Valley," the San Marcos adventurers, with several grown sons accompanying them, reached the San Saba. They carried along a few spades, axes, and grubbing hoes—and a keg of homemade corn whiskey.

The *derrotero,* as Wiley Stroud spelled it out, directed that they go three leagues towards the west up the San Saba River from the old fort and then turn north and follow up Silver Creek one league. The searchers had no trouble in finding the ruins near Menard. Approximately three leagues (about eight miles) to the west they came to the mouth of a creek called Silver and proceeded another league northward.

The *derrotero* now called for a mound of stones on a hill-side. They found a mound of stones. Under the stones, buried shallow, should be the half of a Mexican *metate* (a stone used to grind corn on). They found half of a *metate.* Now they were to measure off thirty varas due south and dig; there they should find a copper peg. They found it. Another thirty varas to the south should be another copper peg. It was there. Still another thirty varas to the south they should go and then turn west. At intervals of thirty varas each they should on this east-to-west line find three more copper pegs. They found them all. Next, going on west for an unnoted distance, they should come to two mesquite trees growing close together; in the ground between these twin trees they should dig up another half of a *metate.*

As the men ran their lines and dug up copper peg after copper peg, their excitement, as may be imagined, was intense. They worked in a trot. Finding the two mesquites with the piece of *metate* at their roots proved a tedious business. Finally, however, the stump of one tree was located, and, surely enough, excavation around it brought to light the half of a *metate.* This half fitted exactly with the other. Across the gray surface of the rejoined halves the letters of one word showed plainly. The word was EXCAVAD. Dig where?

The chart directed that a tree with three prongs would be found to the south of the last half of the *metate,* and that fixed between these three prongs should be a flint rock "about the size of a turkey egg." It was only after three or four trees had been chopped down and the branches cut out that the flint was found, deeply embedded.

The next step was to sight from the flint in a northeast direction to the initial mound of rocks. The intersection of this line with the east-and-west line was the place to dig into the old shaft. The point of intersection proved to be almost in the bed of Silver Creek, which is generally dry. According to directions, the shaft must be opened for sixty feet straight down; then a complicated tunnel that twisted in various directions on several levels must be cleaned out before the "store room" of two thousand silver bars could be broken into. Dixon and his men began tearing into the earth. As they worked, it became evident to them that they were clearing out a big hole that had been dug through solid rock and then later filled. It did not take them long to realize that they could never remove all the rocks without hoisting machinery and the expenditure of an immense amount of labor. Some of the men became discouraged and skeptical. "If it is necessary to tunnel this whole damned hill out to get the silver," said one, "I don't want none of it." Another man swore that "the whole business was a cheat" and that "not even a Spaniard would go to so much work to hide his stuff."

During the discussion several of the malcontents had frequent recourse to the whiskey keg. Finally G. B. Ezell's son grabbed the *derrotero* out of the hands of Wiley Stroud, who was for the fortieth time interpreting some item, and threw it into the fire. It was immediately consumed. Fortunately, Stroud had spelled out the way-bill aloud so many times that its contents were indelibly impressed upon his own mind and the minds of his partners.

Discouraged and quarreling, the San Marcos men gave up the search, never to renew it. The first rise in Silver Creek after their departure filled up with rocks and gravel the shallow excavation they had made to mark the site of the original Iguanas shaft. The six copper pegs were scattered, and now, so far as is known, not one remains. The pieces of *metate* were carried home by Sam Fleming, who cemented them together. Later he moved to Benton, in Atascosa County, where he ran a blacksmith shop, and it is remembered that his wife used the repaired *metate* as a clabber trough for her chickens. After her death he went to San Antonio, where for fifteen years he lived with his niece and her husband, a man by the name of W. J. Parker.

Now Parker, likewise a Confederate veteran, was for years keeper of the West End Pavilion. Here in 1902, W. M. Longworth met him and was taught the history of the San Saba Mine as it has thus far been recorded.

"Night after night," Parker assured Longworth, "I have heard old man Fleming tell how Dixon was guided by the Indian, how he represented his partners in Monclova, how the Spanish lady, Carlota, brought the *derrotero* to San Antonio, and then how the hunt for the two thousand bars of silver came to a disastrous end. Fleming never varied in a single detail. His story had to be true."

It took Longworth a long time to act. Before he ventured greatly he wanted to make sure of the alleged chain of fourteen mines from which the great storage of silver bars had been extracted. He wanted to see those documents that Carlota claimed to have found in the archives of Monclova but would not show to Dixon.

So, about ten years after becoming interested, he raised five hundred dollars and turned it over to a trusted Mexican, instructing him to go to Monclova and "use the money right." The Mexican seems to have "used the money right." He came back with a copy of a very long document purporting to have been written by one Pablo Bernalles. This document recites how two thousand miners under the direction of a certain José de la Amelgamese worked "fourteen rich mines" on the San Saba in the eighteenth century.

"The San Saba hills are bald hills of limestone," so the document quotes José de la Amelgamese. "They are without surface indications of minerals. But we prospected in caves and found a lead to much silver."

True, the directions to the "fourteen mines" seem a little vague. Yet Longworth felt recompensed for the five hundred dollars he had spent to get them. He went up Silver Creek to a place that he verified as being where Dixon and company had made their location. There he set to work. When he had cleaned out the shaft for sixty feet straight down, he struck a wall that appeared to have been cemented. He broke through it into "a kind of natural cave"—the Spanish tunnel. But the tunnel was also filled with rocks and dirt, *"mostly surface material."* For months he and his helpers carted rocks in wheelbarrows and hoisted them up the shaft. Then he struck a powerful vein of water.

His funds were exhausted, but he managed to buy a good pump. It proved to be inadequate. About this time the United States entered the World War. Longworth made good money working on the cantonments around San Antonio. He bought another pump. It would not do the work. Thus alternating between cantonment jobs and pumps, he spent two years.

Meanwhile he became acquainted with a major in the United States Army who owns an island off the coast of Florida on which pirate loot is buried. As soon as the war was over, this major sent from his home in Kansas City a new kind of "radio machine" guaranteed to indicate any mineral in its vicinity. When the machine got near the San Saba diggings, it "squalled," "bellered," "roared," simply "cut up" beyond all precedent. Longworth felt more certain than ever that he was over a room full of silver bars.

But there was the water. He could find no pump to control it. He was at his row's end, it seemed. Then he heard that a San Antonio lawyer owned some powerful pumps that had proved very successful in a Mexican mine. He went to see the lawyer.

"When I told him what I wanted with pumps and where I was working," says Longworth, "he jumped straight up out of his chair. 'Man,' he exclaimed, 'I've been looking for that location myself. Undoubtedly you are at the right place.' "

In short, for a half interest in the project, the lawyer agreed to finance the search. He secured long leases on the land around the workings. He installed adequate pumps. He put a crew of Mexicans to tunneling. For weeks they carried out from twelve to twenty tons of rock a day.

To describe the labyrinth of holes that Longworth and his moneyed partner now made would be tedious. The farther they dug, the more ambiguous became some of the directions inherited from the San Marcos adventurers. One night, however, Longworth had a dream that set them straight again. Then a series of floods in Silver Creek refilled the tunnel with silt. It had to be reopened.

The lawyer is still spending money. "It may be tomorrow," he says; "it may be a week from now; it may be a year. But some day we are going to break into what is perhaps the greatest treasure chamber known to history. It will create a sensation equalled only by the tomb of King Tut."

This is the dream that never dies.

Always on beyond and beyond! Oh, Spanish *conquistadores* who rode your last ride three centuries back to find the Cerro de la Plata, what think you now of the beyond? Oh, Captain Bernardo de Miranda with your report of silver varas—and nothing else—to keep your memory green while you have been to most unsilvern dust this long while, could you have loaded thirty mules with *cargas* as rich as the viceroy was promised? And you, brave, bloody, rough, romantic, real, enigmatical Jim Bowie, who fought for nebulous treasure as hardily as for a nation's liberty, what besides hostiles did you find in the wild San Saba hills? And you, Captain Jess Billingsley, your cry "Remember the Alamo! Remember Goliad!" with its antiphony, "Me no Alamo! Me no Goliad!" echoing down the vaults of time, what is your judgment at last on the captive preacher's tale? Say, Pioneer Dixon, have Carlota and the secretive priests yet revealed what for so many years you waited for and dreamed of beside the clear San Marcos? Doctor Jim Kirkpatrick, brother to the sober elder of Scotch Presbyterianism, could you really have finished the tale you left half told of pictured copper plates and rosewood rosary in plain Mills County? Shades of old Ben McCulloch's rangers, could you find your way back now to that brook paved with golden pebbles and marked by a dead man's pick? Shade of patient Beasley, leading at last, one may hope, your faithful wife in a land that has fulfilled its promises, have you seen there any light to reillumine the walls of the silver cavern that dazzled your youth? Shades of dead dreamers all, is there somewhere on the Llanos a Hill of Silver, somewhere three suns west of the San Gabriel a ledge of Comanche ore, somewhere about the shattered and silent ruins of the Mission San Saba a shaft down to thousands of bars of Almagres bullion once marked by fragments of a peon's *metate*—is there somewhere a Lost Bowie Mine?

CHAPTER II

DOWN THE NUECES

> Those old credulities, to Nature dear,
> Shall they no longer bloom upon the stock
> Of history?
>
> —Wordsworth.

In 1762 a detachment of the San Saba presidio, together with some priests, moved over to the Nueces and at a site marked by the present village of Camp Wood established a fort and mission called San Lorenzo de la Santa Cruz. The Spaniards often referred to this establishment simply as El Cañon. Like that on the San Saba, it was almost constantly harassed by the Comanches, who on one raid alone drove a thousand horses away from it. It was never well garrisoned and was abandoned only seven years after its founding. Contemporaneous with it, another mission of little significance known as Nuestra Señora de la Candelaria was set up some five leagues downstream, at what is now Montell; it was abandoned in 1766.[1]

No doubt the Spaniards prospected the broken lands up the Nueces. What they found is unknown. Maps of the thirties and forties showing a "Silver Mine" on the San Saba indicate another on the high divide between the Nueces and Frio canyons. The adventurous Bowie prospected in the Frio-Nueces country even before he led the famous expedition to the San Saba.[2] Wherever the Spaniards went they left their sign; the clank of their spurs and the thump of their picks still echo down El Cañon.

Human tracks and human blood will not wash out of a soil, although cement may hide them. The region between the Nueces and the Rio Grande is not cemented over; comparatively little of it will ever be cemented over; it will always be a land with a past. While the Apaches and Lipans were making here their most desperate stand against the encroaching Comanches, a few Spanish rancheros built fortified homes and began stocking the country with horses and cattle, which before long ran as wild and unclaimed as the deer. The bands of maverick horses sometimes numbered thousands, and "Mustangs" was the word that on early maps described the wide blank. Nevertheless, a few roads across it were traversed by all traffic from Mexico to the interior of Texas and by the occasional trains of carts or pack animals to New Orleans and Saint Louis. Counterclaims by the Republic of Texas and Mexico made the strip a kind of no man's land. Long after the claim of Texas was established, the Nueces was called "the dead line for sheriffs." Below it bandits of two languages raided and rendezvoused, and the sparse ranchers who survived to possess the land, where it was possessed at all, were as hardy a breed as ever justified the law of the fittest. Their descendants who hold now the vast thickets of thorned brush, prickly pear, and dagger preserve the frontier temper. They cling to the traditions of their soil. For them the tracks made in that soil by Spanish fortune seekers of the eighteenth century have no more been washed away than the outlaw steer's tracks of yesterday.

General Baylor's Assay

Along about the time of the Civil War, so the story runs, General John R. Baylor was talking with a shaggy trapper who had just come out of the Nueces Canyon.

"Is there any gold up there?" Baylor asked.

"Gold!" exclaimed the trapper. "You can pick up gold nuggets along the gravel bar as big as a cow's liver!"

"Well, why didn't you get some of it?" asked the blunt frontiersman.

"Huh," the trapper explained, "a man has to run too fast in that country to pick up anything. Indians won't let him stop."

Now John R. Baylor was a frontiersman if ever Texas produced one. He was high-handed, high-headed, and high-spirited. He had been sub-agent on the Comanche Reservation on the Clear Fork of the Brazos, but upon being relieved of his office became perhaps the deadliest killer of Comanches known to the frontier. Once, in company with four or five other rough-and-ready pioneers, he attacked a band of Indians, took nine scalps, in return for one of a white woman that dangled from a warrior's belt, brought them to Weatherford, and stretched them on a rope for jubilant citizens to dance around. The fervid orgy of vengeance lasted all night and, except that the women "stomped" as freely as the men, was essentially a scalp dance. Then Baylor exhibited the trophies in other towns and, to arouse yet further rage against the Indians, aided in establishing a fire-eating newspaper called *The White Man*.

When the Civil War broke out, he dashed into New Mexico with a detachment of cavalry and proclaimed himself governor of a territory that embraced about half of New Mexico and Arizona. He was also busy with schemes for bringing Chihuahua and Sonora into the Confederacy. After he had poisoned about half a hundred Apaches like so many coyotes and issued orders that all grown Indians should be killed and the children taken prisoners and sold for slaves "to defray the expense [chiefly for whiskey] of killing," the Confederate authorities relieved him of the governorship. Then he fought as a private gunner at Galveston, was a Confederate congressman, and while the War was ending made plans for some sort of expedition that he should lead into Mexico.[3] General Baylor had dreams of empire—but this is the story of a dream-path to gold.

That remark of the trapper's about nuggets up the Nueces "as big as a cow's liver," coupled with a report he is said to have received from the noted colonizer, Henry Castro, was to cost General Baylor a fortune. Castro, it seems, had just been to Mexico City and there dug out accounts of a Spanish smelter at El Cañon. In detailing to Baylor the results of his researches Castro described how the Spanish miners up the Nueces operated; how, instead of "drifting in" and cross-cutting, they simply followed the vein; how, if it pinched to a narrow width, they made a hole only big enough to crawl through and followed on; how for ladders they used long poles with foot-notches cut into them; how they carried ore out in *zurrones* of javelina skins hung from their heads and shoulders; and how by the use of cheap Indian labor they could work very low-grade ore.

When the Civil War was over, General Baylor filed on some land up the Nueces Canyon—where he could continue to get plenty of action as well as hunt for the Spanish gold. He prospected high and low, in the brush and through rocks. He knew as little about mining as a hog knows about Sunday, but he had his own crucibles and made his own assays.

One day while he was running a test, a neighbor named Stockley, unseen by the General, whittled into the crucible some filings off a perfectly good twenty-dollar gold piece. When the test showed a high percentage of gold, Baylor jumped into the air, yelled, and declared that he had found the famous Spanish lode. Stockley had intended to explain his joke, but now that he saw how happy the General was and remembered what a temper he had, he was afraid to explain. The General's sons knew of the trick, but they, too, were afraid to explain. So the intrepid frontiersman went on spending his time and hundreds of dollars trying to get another test of gold. An Englishman who claimed to be a mineralogist happened by, and at $150 a month he was engaged to apply his knowledge. He put down a shaft known as the John Bull Mine, and to keep the General's hopes up and his own salary forthcoming, he occasionally rubbed a gold ring against some of the samples turned over for assaying. He kept on working until his ring was worn out.

The gold mines of the Nueces are yet unexploited; when, full of years and experience, Baylor died, he still believed in them. He was buried in the Montell cemetery and his grave was mounded over with the specimens of rock he had spent so much time in accumulating. It was a fitting monument, and it has inspired successive prospectors. Chief among them has been Henry Yelvington, journalist, oil speculator, and charming gentleman.

"When just a boy," said Yelvington, "I went with my father and some other men on a hunt up the Nueces, and was enthralled by the stories of lost mines and buried treasure they nightly told around the camp fire. I resolved then to return when I should be 'a big man' and prospect. I was still not very 'big' when, in 1906, as reporter on the San Antonio *Express,* I received a tip that the records in San Fernando cathedral contained information on Spanish mines of the Nueces. I did not tarry to examine the records; I made for the mines. Throwing in with a trapper, I spent eight months prospecting. I did not find a thing; but as deer, turkeys, javelinas, fish, bee trees, foxes, bob cats, squirrels and ringtails were plentiful and an occasional bear could still be found, I did not miss the gold.

"Then in 1915 I got in touch with George Baylor, a son of General Baylor's, and persuaded him to explore with me. The ruins of the walls and smelter of the San Lorenzo Mission, at what is now Camp Wood, appeared to the casual eye to be nothing but a mound, perhaps an acre in extent, with great trees growing on it. About the mound were many old holes, and one of them revealed a stone-arched door of Spanish masonry; another, a bit of the smelter. Near by I picked up a piece of stray ore that assayed $116 per ton in silver and lead.

"On the Baylor homestead Mr. George Baylor and I undertook to explore what they called 'the round cave.' As we climbed the mountain towards it, I noticed a lead running up. Right on this lead we found the 'cave.' I recognized it at once as an entrance to a Spanish mine—though General Baylor had always regarded the hole as natural. We went inside with candles. The shaft dropped down about ten feet, wall rock on one side and loose formation on the other. After following the drift for about fifty feet, we came to a cave-in that almost blocked our way. But, feet foremost, we crawled through and then entered another fairly open tunnel. As is usually the case in old Spanish mines, the circulation of air was perfect. On the way back we wandered into a branch tunnel, the candles burned out, and we were lost. But three hours later we got back to daylight.

"I took some ore from the lead and had it assayed. It ran from $2.70 to $4.75 gold per ton, which is a good showing for a lead in limestone formation; paying ore in limestone is generally very deep. Although we worked at the old mine in a haphazard way for several months, we never did get back to where the Spaniards had left off. There were too many cave-ins, and reopening the tunnels would have cost too much money. Of course it is possible that the vein may have been worked out.

"Above the San Lorenzo, I found another old Spanish shaft, but all I got from it was a crude pick badly rusted. A cave-in near the entrance of the tunnel blocked us. I expect to go back some day and explore that tunnel to its end. I know that the Spaniards worked mines on the Nueces."

It remains but to say that about two years ago a man hunting sheep in a pasture near the ruins of El Cañon found an antique bell weighing ninety pounds and that the find has caused a fresh outburst of digging.

Espantosa Lake

Nearly a hundred miles down the Nueces from the ruins of El Cañon is Espantosa Lake. *Espantosa* means *haunted,* or *horrible,* and things have been seen and heard at Espantosa Lake that well warrant the name. The *camino real* between Presidio Rio Grande and San Antonio used to skirt the lower end of the lake, and here was a favorite camping place. They say that the lake used to harbor a peculiar kind of mermen who sometimes seized young women dipping up water for their camps and carried them away.

One night, they say, some freighters with a wagon of money and other valuables camped on the edge of the lake, watered and hobbled their teams, and went to sleep. Suddenly the ground on which they lay sank, their cargo was engulfed, and they themselves were drowned. For generations after that people used to hear in the darkness a strange wagon that left no tracks rumbling down towards the lake from the hills.

Once a cattleman named Cleary and a hired hand were camped beside the old road. They had run out of tobacco and so had the little store near Presidio Crossing, just east of the lake. A wagon load of supplies was expected daily by the storekeeper. One night the two campers heard a wagon clattering down the hill.

"Well," said Cleary, "we'll get some tobacco in the morning sure."

The next morning early he galloped down to the store. "Gimme some tobacco," he demanded as he alighted from his horse.

"Tobacco!" grumbled the storekeeper. "You know I ain't had any for ten days, and if that freighter don't come right away I'll be out of coffee too."

"Now, looky here!" Cleary replied, "I'm as starved fer

tobacco as a hay-fed horse fer green grass in April. It's no time fer a joke. We heared that freight waggin pass our camp last night jest about the time we was turnin' in. I had a good mind to step out and hold it up and take some tobacco right then. So shell out *pronto.*"

"You mean you dreamed you heard a waggin," the storekeeper rejoined. "There ain't been any more waggin come to this store than a ghost."

Thus the argument went on for some time. Finally Cleary led the storekeeper out to the road, which was sandy at that place, to show him the irrefutable evidence of a wagon—tracks. There were no tracks. He scratched his head and then loped back up the road to his camp. Not a sign of a track was there on the whole stretch.

A certain man in the country known as Mocho—from the fact that his ear had been gotched, or cropped, by a bullet—was one night camped with a "cow crowd" near the place where Cleary had camped. Soon after supper someone called attention to the rumble of a wagon.

"Maybe it's the ghost waggin goin' down to Espantosa Lake," a man suggested.

"Ghost, the devil," Mocho retorted. "I'll make a ghost of that driver if he don't stop tonight."

The rumble of the wagon was by now a distinct clatter. Mocho with drawn six-shooter walked out into the middle of the road. The night was fairly dark, but the sounds were so distinct that one did not need to depend upon sight. As the wagon approached Mocho, the waiting men heard him bellow out, "Halt!" and then in Spanish, *"Parate!"*

For perhaps a minute, not longer, the rolling of the wheels and the tread of hoofs ceased; then they were resumed. They ceased just above Mocho; they began just below him. He stood in the middle of the road; the road was lined with impenetrable brush. Nothing sensible had passed him. The sound of a shot that he directed down the road was not more distinct than the receding clatter of the unhurried wagon. In the morning not a track was to be seen.

No one ever heard the wagon leaving Espantosa Lake; it was always going towards it. In time Peg Leg Tumlinson with much labor put down a shaft at the foot of the lake, where the campers had met their sudden fate, but mud and water prevented anything like a thorough exploration. The precious cargo is still there.

Witching for Silver

On the south bank of the Nueces, seventy-five miles below Espantosa Lake, is the site of Fort Ewell. Nothing remains of it now but a trace of adobe foundation and a graveyard overgrown with prickly pear. A year or so ago fortune hunters stepped over the sagging barbed wire about the enclosure and tore open the last clearly marked grave. Fort Ewell used to be a stage stand, a county seat, and a post office. While O. Henry was at the Dull Ranch, fifteen miles away, he used to ride to Fort Ewell for the mail; on those rides he got something that he later put into "Law and Order" and other stories of the Nueces country. But long ago the stage stopped running, the storekeeper and his post office moved away, the wooden bridge across the Nueces rotted down, and the road became a cow trail. Fort Ewell is a place of memories.

Silence and space are kind to memories. The relics of Fort Ewell moulder in a pasture of 60,000 acres. Across the river is the Rancho de la Mota, with 40,000 acres. Joining the Mota on down the river a few miles is Rancho de los Olmos, containing 56,000 acres. Until recently the owner of the Olmos controlled 200,000 acres of uninhabited pasture lands adjacent to his own estate. Thus the country stretches out—a "big country" of no fields, of few people, and of many memories.

Before "bob wire played hell with Texas," as Big Foot Wallace used to say, a hard "layout" of Mexicans hung around El Fortin (Little Fort). Once they waylaid an Englishman and killed him for his horse. When they were not stealing horses or waylaying travelers, they occasionally looked for a legendary iron safe, full of valuables, that Comanches once took from a *carreta* and dragged out into the tall *sacaguista* grass.

Peg Leg Tumlinson, who sank the shaft in Espantosa Lake, was a "big hombre" among the Fort Ewell Mexicans. One time while he was down on a little *pasear,* a rich old don from below the Rio Grande happened to be visiting in the settlement also. He took a great liking to Peg Leg and one day guided him over to a landmark on the Olmos Ranch known as Estambél Hill, where signs of a mesquite-picket corral are yet visible. At a spot near this old corral the don halted and said:

"I have a *hacienda* with cattle, horses, sheep, and peons a plenty. I need no more. I wish no more. But I wish you to have plenty also. *Por amigo,* I tell you how to get it. Dig here and you will find one burro load of gold money. I know it is here."

Then the old don told Peg Leg Tumlinson specifically what he would find in the hole as he progressed downwards. Peg Leg marked the spot well and went on back up the river to his "stomping ground" around Carrizo Springs. It was twenty years before he returned with a spade and with Doctor Hargus for a partner. If you ask why he delayed so long, Doctor Hargus' explanation will have to suffice: "He was a frontiersman, and those old frontiersmen were peculiar."

Now, according to the directions that the don had given, Peg Leg was to dig down two feet and there he should find some charcoal; two feet under that he should find a saddle blanket; two feet under the saddle blanket, burro bones; two feet under the burro bones, he should come to the bones of a man. And then *"dos pies más abajo y ahí está el oro"* ("two feet on down and there is the gold").

Peg Leg and his partner marked off a hole about ten feet in diameter and began excavation. Two feet down they struck some charcoal. Two feet under that they came to some "dusty dirt" that looked as if it might once have been a saddle blanket. The soil was fairly loose, and a little farther down the diggers struck a pocket of earth, "about the diameter of a barrel," that appeared to have been dug up long before. The loose soil soon petered out, and then they came upon the all but decayed bones of some kind of animal. Two feet under the bones they dug up the knuckle and wrist bones of a man.

"And," Doctor Hargus here interrupted his story, "I ought to know what a human bone is. I drove up the Chisholm Trail before it had a name. When I began practicing medicine down there in the brush below Fort Ewell I was led blindfolded more than once into a thicket to doctor a wounded outlaw. I was with McNelly's rangers while they were cleaning up the border country. I have treated the gunshot wounds of two hundred and fifty-nine men and have seen one hundred and eight men with bullet holes in them that didn't need treating. Those were certainly human bones that Peg Leg and I dug up."

At this stage, the charcoal, the dust of a saddle blanket, animal bones, human bones, all found as the Spanish don said they would be found—the depth only varying a little from his directions—the diggers felt absolutely sure of their burro load of gold. Their pit was now so deep that they could not work in it without a ladder, and they made one by dragging up two poles and lashing short sticks, for rungs, to them with deer skin. That country still abounds in deer.

They dug on down two feet, four feet, six feet. Perhaps the gold had sunk. At the depth of twenty-two feet they crawled out one evening to eat supper. That night the wagon bed caught on fire from a coal that was whiffed into it by the wind. The fire burned up all their provisions. Worst of all, it burned up Peg Leg Tumlinson's vest, which held a pocketful of way-bills to other treasures.

"One of the bills," Tumlinson said, "was without a doubt worth $25,000."

"We intended to come back," Doctor Hargus concluded, "but not long after that old Tumlinson died. I don't know what became of all his charts. He had a lot that were not burned in his vest. He was a strange character. I have decided that when the Spanish don told him to dig *dos pies más abajo* (two feet on down), he probably meant to dig farther down the slope rather than farther down in the hole that contained the charcoal and bones. And instead of *dos pies* he may have said *dos pasos*. A *paso* is a double step, two yards; *dos pasos* would be four yards. I mean to go back down there some fall, when the weather is cool and the bucks are fat, hire me a couple of Tonks, and dig a new hole four yards down the slope from where we dug before."

But Doctor Hargus has never seen the enormous hole that envelops the one he helped to put down. Two or three years after he and Peg Leg were burned out, six strangers drove up to Olmos headquarters and asked the owner's permission to dig for a burro load of gold buried on Estambél Hill.

When told that a hole had already been sunk on the site, they responded that the hole was in the right place but that it was not deep enough. Then they set forth the theory that gold sinks in the ground at the rate of six inches every ten years.

"Look," one of the men jubilantly explained, "how far under ground you sometimes find an ordinary Indian arrowhead that you know must have originally been on the surface. How much deeper would a pack load of heavy gold sink in a hundred years!"

The crew enlarged the hole and dug down to a depth of thirty-four feet. Then they struck a seep vein of water and quit.

The hunt for this treasure on Estambél Hill that I took part in a number of years later was perhaps the most successful I have ever made. One day in the summer of 1924 while I was helping the Olmos outfit brand some calves, the boss suddenly said to me:

"Do you know what old George Ray is doing out here?"

"No," I replied, "I do not. I notice that he and Uncle Jim [Dobie] have been riding around a good deal. I suppose they are on a cattle trade of some kind."

"No," he responded, "old George is locating oil wells."

I had known Mr. Ray a long time. He is a cowman of the old school, except that he is not in debt, owns a fine ranch, is president of a bank, and has large interests in a rich silver mine in Durango.

When I got in to the ranch that night, he, his son Haggard, and Uncle Jim were sitting on the gallery.

"Pancho," Uncle Jim said to me, "George has finished locating some oil pools and now he wants to locate those thirty-one mule loads of silver bullion on Estambél Hill."

"Why," I replied, concealing my joy as best I could, "the thirty-one mule loads are down in McMullen County."

"Well, there are thirty-one mule loads in La Salle County too," and Uncle Jim gave me a wink.

Then I learned about the oil pools. The Rays had arrived at the ranch a day or two ahead of me, each provided with a peach switch and an elongated, thin rubber sack. Also, they had a bottle of *fresh* crude oil. Their method of using this paraphernalia was to fill the rubber sack about two-thirds full of oil, tie the open end of it securely around the stub of a forked peach limb, and then to "witch." Thus equipped, they had been riding over the ranch in an automobile locating places to put down oil wells. And they had, to quote George Ray's own words, "located two of the deepest and strongest pools in Texas." As we rode along horseback next morning to Estambél Hill, he told me all about his "gift."

"When I was a boy," he explained, "an old Scotchman, who was the best man to handle horses that I have ever known, showed me how to witch for water. He said that the same principle could be used in witching for minerals. I never seriously applied my knowledge—or gift, as I call it—until two years ago. I went down there to our mine in Durango and arrived just in time to find that the miners had lost a vein of silver and were having trouble in picking it up again. It had 'pinched out,' as they say. I took a regulation switch, fixed a Mexican dollar in a slit at the end of it, and in five minutes' time located the vein. This experience made me realize the value of my gift.

"Oh, I can't explain it. I feel something in me like electricity. When the switch gets over what I am looking for, I simply can't hold it. It will twist itself in two or go down. I can figure out the depths of some veins of water, but I can't tell the depth of oil or metal because I haven't practiced enough. I've been knowing Jim Dobie for forty years. He's needing money and I have come to help him out. I'm mighty glad that I got the oil pools located, and if we can turn up a little ready silver it will be all the better."

We crossed the river six or seven miles above the ranch, plugged through a mile of the hog-wallow *sacaguista* flats, pulled up on top of a small hill, and halted beside the remains of the old Estambél corral.

"This is the place," announced Uncle Jim.

George Ray cautiously slipped down from his horse, Haggard Ray jumped down, and in no time each of them had a half dollar fixed into his switch and was at work.

The divining rod, water switch, caducean wand, or whatever one wishes to call it, and the manner of using it have been described so often and so thoroughly that little explanation is here necessary.[4] Suffice it to say that the switch is Y-shaped, that the prongs of the Y are usually held in the manner shown by the drawing at the end of this chapter, and that the main stem of the Y, held inverted, "pulls" down towards the object attracting it. If the switch is to be attracted by mineral, it must be "loaded" with mineral; a half dollar is sufficient to make a switch work for any amount of silver.

The two divinators on Estambél Hill were attracted at once—certainly against their desires—into the middle of a very thorny patch of *granjeno* and black chaparral. Finally, after a *granjeno* had torn off most of the elder magician's shirt tail—it had a habit of working out—they got an oblong "box" about the size of a grave marked off on the ground and delivered the joint opinion that the thirty-one mule loads had been buried right there.

Uncle Jim, however, suggested that several of the mule loads had probably been buried separately and that it might be a good idea to explore down the slope. They knew nothing of the great hole which Peg Leg Tumlinson and then the six strangers had put down. A discussion arose as to the space a *carga* of silver bars would occupy. George Ray settled the matter by asserting that in the old days a mule load of bullion in Mexico consisted of six bars, three on each side of the pack, each bar about two and a half feet long and six inches thick. Consequently an entire *carga* could be buried in small space.

A half dozen locations, each over a mule load of silver, were soon made.

"Now mark the strongest deposit and mark it well," Uncle Jim directed.

The Rays both applied the most delicate witchery of their gift. The location was beside a rat's den. To fix the spot beyond any mistake we punched a Spanish dagger stalk into the ground and heaped rocks around it. That afternoon Mr. Ray had to leave to attend a meeting of bank directors. However, he promised to be back in three days with a drop auger, which would greatly facilitate digging. As it turned out, he was detained, but Haggard brought the auger.

The location was not accessible to a wagon, much less to a car, but with the aid of three Mexicans we managed to lug over the auger, block and tackle, some water cans, and other equipment. Three fifteen-foot ash poles, which we "snaked up" from the river at the horn of a saddle and tied together tepee fashion, made a kind of derrick to drop the auger from. Before "spudding in" for the bullion, Haggard switched again and declared that the pull was unmistakable. Discovering that I too possessed a gift, I had learned to hold the switch, and the pull seemed all right to me also. By noon we were ten feet down and the thermometer was about a hundred and ten degrees up.

We rested, poured water in the hole, and dug on. Every once in a while Haggard would manoeuvre his wand, and it never failed to dip straight into the hole. When we were eighteen feet down, Haggard declared that he did not believe it was of any use to dig deeper after the silver bars; something else, he thought, must be attracting his switch. We quit digging.

During the day I had actually spent very little time at the hole but had entertained myself with picking up Indian arrowheads off the hillside. Also I picked up two odd rocks: one that looked to be a petrified oak ball, another that had in it a precipitate of iron. With these specimens I placed three pebbles dug out of the well.

Two evenings later George Ray drove into the ranch posthaste.

"What did you find?" were his first words.

When informed, he became greatly excited.

"Why, there may be radium or anything in that dirt!" he exclaimed. "It pulled so powerfully I know there must be something. Somebody had better ride over and get enough of the dirt for me to test out."

But George Ray never lets his enthusiasm interfere with his rest, and the next morning, after he had had a good night's sleep and was talking again about getting a test of the dirt, I saved somebody a long ride. I pulled out my five rocks. Mr. Ray got his switch, put a half dollar in it, and, squatting on the ground, began testing. Not one of the specimens taken from the well attracted the switch, but when the petrified oak ball came under it, down it went. Then he examined the object more closely.

"Why, that is some sort of old Spanish button," he announced. "The metal in it attracted my switch. It is a great relief to know what it was. I was beginning to doubt my gift. Of course, a man can't listen to all these Mexican yarns about buried treasure. Those people are too durned superstitious."

It was too late for me to explain that the "Spanish button" had come from the hillside a hundred yards away from the well.

The other day an oil company brought in a well on the Ray ranch. It was bored at a spot where Mr. Ray's switch indicated oil. The president of the company, one of the best known and most successful oil men in Texas, has engaged Mr. Ray to make some locations for him in distant sections where he is wildcatting.

> There are more things in heaven and earth, Horatio,
> Than are dreamt of in your philosophy.

THE GOLD THAT TURNED TO *CARBÓN*

After our search for the burro load of gold or the thirty-one mule loads of silver bullion—whichever it might be—the ranch Mexicans naturally unlocked their hoards of treasure talk.

One hot afternoon while Uncle Jim and I were sitting on the wide east gallery of the Olmos ranch house, watching whirlwinds down in the *sacaguista* flat and listening to locusts sizz-z-z in the huisache tree—two very bad dry weather signs—Jacinto de los Santos came up. He is about half Comanche, is nearly as black as a negro, and is as good a hand at building dirt tanks and stringing barbed wire as ever tilted a scraper or drove a fence staple.

"Jacinto," I said, "the Madama Burks has been telling me that one morning a long time ago she found under the trees across the lake from her house an empty wooden box beside a fresh hole. Two strangers had camped there the evening before. They left in the night. You have been around here a long time and you used to work at the Mota. What do you know about such matters?"

"Well," responded Jacinto, "I am just a poor ignorant Mexican, and you know very well how little I know, but I will tell you something.

"One night while I was yet *un mediano* [a young fellow], I was coming with the Madama Burks to the Mota from her sheep camp over on the Quintanilla. We were traveling the old Laredo road and it had been raining. Well, when we got on top of that hill about three miles north of the Nueces, we saw a fire burning off to one side. The name of the hill is Loma de Sauce [Willow Hill]. The Madama asked me why travelers should camp in a place so far from water. I told her that the fire was not made by travelers. We drove on into the ranch.

"The next day I had talk with a Mexican named Justo. He was then *caporal* for the Mota outfit. He did not have a good heart. He asked me if I could go to the place where the fire flickered. I told him I could. Then Justo told me that two priests—*pasajeros* on the old road—long time ago were murdered on the hill and buried in a hole with the $25,000 they carried. As every one knows, fire on the ground is a sure sign of buried gold, and Justo said we should all get rich.

"For my knowledge I was to have half the treasure, and Justo and two vaqueros were to divide the other half. We took some tools and I led them straight to the chaparral where

I had seen the fire. I knew it was the right place. We set to work and right away we dug up *como un costal de carbón* [about a sack of charcoal]. Yes, sir, *carbón.* Then I knew it was of no use to dig farther; nevertheless, we went on down seven feet.

"That *carbón* was the gold itself! The other men had envy of me, and for that reason God changed it into something else. Had God not done so, the envious men would have killed me for my part of the gold. God protected me. It is possible that the *carbón* later turned back into gold. I do not know. I never went back to see. As I said, I was just *un mediano* then, and as the good God had saved my life once I did not wish to risk putting him to that trouble again.

"Another time some men with a *derrotero* came looking for money on that same old road. The Madama Burks told me if I knew anything I was not to tell them. I told them nothing. They had a kind of machine made out of two iron rods. One man held a rod in each hand and when the points came together it was a sign to dig. They dug by a *coma* tree on the Sauce Creek, but that was the wrong place. I will tell you why I think so.

"Not long after they were gone I was running a wild cow down that Sauce flat, and there, right in the middle of a clump of low mesquites, I saw an iron stake with a brass ring in the end of it. I went back later and tried to find it, but cattle had stamped out the tracks of my horse and I could not trail to the stake. I have not seen it since. The treasure was not meant for me. It must have been for the iron stake that the two men with the machine were looking. They were half a mile away from it."

Mysteries of the San Casimiro

It is over on the San Casimiro, however, that signs and mysteries have since the mind of Mexicans runs not to the contrary indicated some enormous treasure. The San Casimiro is an old Spanish grant to five leagues of land now fenced into one of the lonesomest and most formidably thorned pastures to be found anywhere south of the Nueces River. It is incorporated in the Olmos holdings.

A "queer" Mexican *pastor* named Toribio used to pen his goats at the bleak corrals by the *jacal* on San Casimiro Creek. He was always after the *mayordomo* to look for the money that nightly apparitions indicated to him. Sometimes the apparition was a light that flickered back and forth over a hill and then always stopped at a certain place. Sometimes it was *una bruja blanca*—the white ghost of a witch—that with long arms lifted high beckoned him to follow a trail he dared not set foot in. Oftener it was sounds of horrible yelling *como de los indios*—as of Indians. When this yelling set up, the coyotes would cease their howling and listen too. Then there would be groans, and from the mingled sounds one might realize how savages once murdered the rich Spanish rancheros on the San Casimiro—and how the wealth of the *ricos* yet lies there concealed.

Toribio did not like that camp, but Salomé Esparza has never had any uneasiness there, and as pasture man he has lived on the San Casimiro for years. The first night Salomé's bride slept in the *jacal,* she woke him up along about three o'clock in the morning to listen to a strange *tick-tick-tick* of a clock that they did not possess. He could hear nothing. Many nights since then his wife has awakened him to hear the sounds, but always he explains to her that it is the wind or a wire stretched against the house or something like that. She knows better. She knows that the clock has been numbering the hours ever since the treasure was buried, maybe a hundred years ago, and that it will go on numbering them, *tick-tick-tick,* until the treasure is found.

One day I rode over to chat with Salomé's wife. When I arrived, she was turning their little bunch of goats out. She ran into the house, shut the door, presently appeared in a fresh dress, and proffered the inevitable *cafecito.* While we talked and sipped the coffee, it developed that down in Mexico, where she once lived, she had actually located *mucho dinero* by certain strange sounds and sights.

"There," she said, "I used to hear chains rattling and cries coming out of a *chaparro*—a clump of bushes—and at the same time see a light burning. I told a poor family about the matter and very soon they set up a store and had plenty of money. But they never gave me one cent for making them rich."

The most earnest witness of the mysteries of the San Casimiro has been my friend Santos Cortez. During the Madero revolution Santos was a *rural* in Tamaulipas and killed a man. Oxen and log chains could not drag him back across the Rio Grande. I have seldom known him to be without a rifle or pistol; frequently he has both. He is the best hunter I have ever known; indeed, he used to keep a "crowd" of Olmos hands supplied with deer and wild hog meat all through the fall and winter months. He has been a *pastor* too and knows solitude. Many a profitable and happy hour have I listened to Santos tell of wild animals, of his experiences in the revolution, and of the strange things that he has sensed in darkness.

"Oh, it was a lonesome life there on the San Casimiro," he would say, "*una vida muy triste*. I never say anything about some matters except to you. You are my friend, and you will understand.

"I used to sleep with the door well fastened, and sometimes something strange would come and try to shake the door down. I would shout and maybe the thing would go away; if it did not, I would shoot into the door and then it would most certainly leave.

"In the direction where the sun sets, a *corrida* of horsemen used to go galloping by just about the time I was getting to sleep. I could hear the horses trampling and the men seemed to be talking. They were desperate about something, but I could never make out their words. The voices seemed far away and all I could hear was *mum-mum-mum-um-blub-blub-*

blub, like that, until they passed. In the morning I would go out to look at the ground but never a track did I see, not even after a rain.

"One time while I was running some cattle over in the direction where the horsemen always went, I saw three long rocks piled together. They were hewn rocks and appeared very peculiar. I was in a hurry and my horse was jumping through the *chaparro* when I saw them, but I marked the place well. The next day I went back to examine the rocks closely. They were gone. Many times I searched for them, always in vain. This is *una seña muy buena*—a most excellent sign of buried treasure.

"You know how the *bandidos* at El Fortin used to waylay *pasajeros* and rob them. Well, one late afternoon, a long time ago, a ranchero who had sold a big herd of steers in Kansas came to the ferry to cross. He had his gold with him in a belt and he wanted to keep away from the *bandidos,* but the river was up so high that he was afraid to swim it. He was alone and he rode a fine horse. After he crossed the river he took the trail to the east. When he got close to the San Casimiro hills, he saw that he was being followed and so he hid his money quickly and went on. The *bandidos* kept following him and before dark they killed him. When they saw that he no longer wore the money belt they had seen at El Fortin, they knew he had hidden it. They tried to find it, but it was too well secured. Nobody has found it yet. It is still out in the San Casimiro hills."

The Martineño is to the south of old Fort Ewell. Santos Cortez used to be *vaciero* on this ranch under the direction of Don Antonio Salinas, and here he had his most intimate experience with creatures of the darkness. A *vaciero* is a kind of overseer and supply agent for sheep or goat camps. One day when Santos went to a certain camp, he found the *pastor* gone and the sheep shut in the pen.

"Well," said Santos, "I had to tend to the sheep, and that

night I made my pallet down in front of the *jacal*. After I had been asleep a while, I felt something heavy on my stomach and chest. I tried to raise up and throw it off, but the weight was so heavy I could not lift my shoulders. The thing felt like a great round keg. I could see it, too, and it was white.

"I tried to yell, to speak, but all I could do was pant. My tongue would not move and my teeth were locked. I had a pistol on one side of me and a *carabina* on the other. I strained my hand—for the arm was pinned down—to reach the pistol, and then I touched something icy cold that felt like the hand of a dead man. I worked to breathe like a wind-broken horse—*ha-ha-ha-ha-ha,* fast and hard that way.

"Ah Chihuahua! There I was pinned down and the white *bulto* on me never budged. I had touched its hand, and yet it seemed to be round without any hands or feet. Then I began to remember how an evil thing against a man will go away if the man puts good thoughts in his mind. So I began to think about El Dios and the Holy Mary. I thought a long time and in myself I spoke to El Dios. Then the weight began to lift slow, slow, and in the time it takes a *paisano* bird to drink water it was gone and seemed to be in the brush. I jumped up with my pistol and shot towards it. I called it many *maldiciones,* but it made no response, and when daylight came I could find not one sign of it.

"Then I became very sick all in my head and stomach. A vaquero came to the camp and took my place and I went to see a *curandero* [a Mexican medicine man]. He asked me what was the cause of my sickness, but I had shame lest he think me a liar, and so I told him I did not know.

"After that Don Antonio Salinas asked me to tell him what was the matter. I was still sick. Again I had shame, but I saw that Don Antonio was an intelligent man and friendly, and so at last I told him of all that had passed. At once I got well. Then Don Antonio took me with him to the camp, and right there across the old road from it he showed me the graves of three *pasajeros* who had been murdered for their money.

"By this time the missing *pastor* had come in. He told of seeing a black dog at his camp with a head as big as a bull's and fiery eyes. Also I learned that something like a skeleton once jumped out of a tree beside the camp and caught on behind a vaquero. It grabbed him about the waist. His horse pitched and screamed like a panther, but the thing held on until the vaquero ran into the corrals at the ranch. Oh, it was a terrible place! They moved the sheep camp away from it. Nobody would stay there any longer."

El Tigre

The Tigre Ranch is on one of the long, alligator-gar infested lakes of the Nueces flats about five miles below the Olmos headquarters. For many years now the only inhabitants of the dilapidated house have been a family of white owls. The man who built it and was master of the range about it was Zack Hargus—uncle of the Doctor Hargus who dug at Estambél Hill for gold under human bones—and was as odd a character as ever enjoyed his own oddities. He was something of a reader and very much of a tobacco chewer. His favorite position for reading and chewing was on the gallery overlooking the lake, his chair tilted back against the wall on the left-hand side of a window that opened into the dining room. Here by the hour he would sit, reading and chewing. When he wanted to spit, he merely turned his head slightly to the right and spat inside the window. That took less energy than leaning forward to spit off the gallery. He made it a habit to work harder on Sunday than on any other day. Once he built a chimney to his house, working on it only on Sundays. After the country was fenced up, he invariably left the gates open when he went to Cotulla. He was not much of a lover of horses. Sometimes he kept them tied up for a day or two without water. Year in and year out he kept a horse staked to an old wagon axle driven into the ground out in the *sacaguista* grass; naturally there was not much grass around this stake pin. He had odd names for his horses, too; one of them he called Jesus. One time he told a boy who was working for him to drive the team across the river for some posts. When the horses got into the water, one of them drank himself to death. The boy came back to report the loss. "Oh, Mr. Hargus," he cried, "old Jesus has done went and drank water till he died."

I don't know that Zack Hargus has anything to do with this story, but the man who had most to do with it made his headquarters for a while at the Hargus ranch. His name was Musgraves. One day after having sold a considerable amount of stock, for which he received Mexican gold in payment, Musgraves left the Tigre with $12,000 to deposit in San Antonio. Somewhere between home and Dog Town he ran into a supply of *tequila* and after a night's carouse came back to the Tigre still drunk—and without any money. When he sobered up, he remembered having buried the gold, but to save his life he could not tell where, though he had an idea it might be on a high hill near what is now the line between the Tigre and the Coma pastures. In time he died or left the country.

Years later a Tigre Mexican came into Cotulla and reported to a man—nameless here—that he had found enough Mexican gold "to make them all rich." Shortly thereafter the man gave it out that the Mexican had discovered Musgraves' money and "gone *allá*," or, in other words, had pulled his freight for the tall tules. Now, never before had the man shown any evidence of possessing property, but all at once he appeared to be prosperous. Some people flatly said that the Mexican had found the money all right and had left the country, too, but had left it over a route that no man wants to travel or ever returns by. Maybe this is "just a tale."

It reminds me of a story about a Brownsville man. This man started towards San Antonio with $40,000 in gold. His route would have taken him to Beeville, and somewhere between Brownsville and Beeville—legend has lost all but one particular regarding the place—he discovered that bandits were on his trail. He was in camp at the time; the bandits had to ride across a wide brushy flat before they could see him. He simply pulled the coals of his fire aside, made a little hole into which he inserted the gold, covered it up again, and built a roaring fire over the place, thus obliterating signs of excavation. The fire was between two medium-sized oak trees on the west side of the road.

Leaving it still burning, the man saddled his horse and made tracks. But the bandits overtook him. They tried to make him tell what he had done with the money; he would not tell. They took him back to the Rio Grande and crossed into Mexico with him, swearing that they would hold him prisoner until he should reveal the gold. Still he held out and still he was kept prisoner. He remained a prisoner for long years. When, finally, he got back to Texas, barbed wire had been stretched across the country, brush had covered many of the former prairies, roads had been changed, his own memory had probably been shaken. He set out to go to the two oaks on the west side of the old Brownsville road somewhere south of Beeville. He died before he could locate them, or at least before he could locate $40,000 in gold between any two of the various oaks.

The Rock Pens

Austin Texas
April 17th 1873

About six or seven miles below the Laredo Crossing on the south side of the Nueces River near the hills there is or was a tree in the prairie. due west from that tree at the foot of the hills at the mouth of a ravine there is a large rock under the rock there was a small spring of water coming from under the rock, due east from that rock there is a rock pen or rocks laid around like a pen and due east a few yards there is another pen of rock In that pen is the spoils of thirty one mule loads

[Signed] *DANIEL DUNHAM*

When Daniel Dunham dictated this way-bill he was on his death bed, sole survivor of the little band of men who put the thirty-one mule loads of silver bullion, together with various fine images and other precious articles, under ground. Just who these men were and how they came to have so much wealth is a little hazy, but it is generally believed that they were Texas bandits, that they had raided a Mexican mine and church, and had got within sight of the Nueces on their way north when they became aware of a cloud of Indians or Mexicans, perhaps both Indians and Mexicans, following their trail.

The account from this point on is clear. The guards of the pack train picked the best place within reach to make their stand. It was by a small ravine down which flowed water from a wet weather spring. Here they threw up some crude breastworks in the form of low-walled rock pens. In one of the pens they buried the bullion, and then, in order to hide all sign of their secret work, ran the mules around and around over the disturbed earth. The fight soon followed; only Daniel Dunham got out alive. Why he never went back himself to claim what was in the pens, I shall not attempt to say.

It is a fact that the Rock Pens have been sought for more widely and by more men than any other lode between the San Saba and the Rio Grande. When the search began it would be hard to say; I cannot remember when I did not know about the Rock Pens. The way-bill just quoted was in the possession of an old Live Oak County settler named Matt Kivlin, who died along in the nineties. He had shown it to his sons a few times, but there was an accompanying paper that he had never shown. This accompanying paper he destroyed shortly before his death, or else his wife destroyed it immediately thereafter. Even to his own sons—and he never mentioned the matter to anyone else, except his wife—Kivlin's attitude towards the way-bill and the treasure and his connection with Daniel Dunham appeared profoundly mysterious. Certainly Matt Kivlin knew Dunham and certainly he held the way-bill to be veracious; certainly, too, he had an inexplicable aversion to following out its directions. Other charts to the Rock Pens cache seem to be in existence also, for various independent parties were searching for it even before the Kivlin document came into circulation.

Those who know anything about the matter never take the "Laredo Crossing" mentioned in Daniel Dunham's testament to be the crossing at Fort Ewell. A very old road of some kind crossed the Nueces on the Henry Shiner ranch—a ranch of forty or fifty thousand acres, in McMullen County; and this crossing is generally conceded to be the one Dunham had in mind. The country around is hilly, densely thorned, and without springs of any kind. Many a tale tells of early-day travelers perishing of thirst in this drouth-stricken land. Until the advent of the automobile it was indeed a remote country.

So far as is known, no one looking for the Rock Pens has ever found them. Yet they have been sighted time and again—always by men who were not looking for them and who were at the moment ignorant of their significance; when these men, upon being enlightened, attempted to go back to the Pens, they invariably found themselves unable to locate them. The rocks were not piled high in the beginning; very likely they have been scattered.

Along about 1866 Pate McNeill was coming from Dog Town—now called Tilden—down to Lagarto with his young wife. They were in a buggy and were leading a horse, saddled. Somewhere in the Shiner country they glimpsed a fine-looking maverick heifer. McNeill jumped out of the buggy, "forked" his horse, and took after her. After he had roped and tied her, he looked around and saw that he was in a kind of rock pen, the walls of which were low and broken. At that time he did not know that great riches appertained to rock pens in McMullen County; so he calmly ran his famous brand of **PATE** on the heifer's ribs, turned her loose, and went on down the country. Years later when old man Kivlin died and Daniel Dunham's way-bill became common talk, McNeill went back up the Nueces and tried to locate the Rock Pens; but the country had changed so much that he could never find the ground he was looking for.

A deer hunter stumbled into the Pens one time and thought he was in a deserted goat camp. Like Pate McNeill and others, he did not at the time realize how close he was to millions. Wiley Williams, a rancher noted for his truthfulness, used to recall how he once ran across the Rock Pens while trailing some horse thieves. He was considered one of the best trailers and woodsmen of the country; yet he could never go back to the Pens.

Pete Staples, an old negro trail cook, told me that one time while he was hunting turkeys with Judge Marcellus Lowe, they came upon some curiously placed rocks.

"Huh, what's this?" Pete asked. "Looks mighty funny to me for rocks in this place. Where'd they all come from and how come this away? Ain't no other rocks like thesen for a mile out."

"Natural rocks all right," said Judge Lowe, "but this is an old pen."

Years later Judge Lowe tried in vain to go back to the place. Had he got Pete Staples to guide him, he might have succeeded. Pete, however, has a firm conviction that it is dangerous to "monkey" with money that some man now dead buried, and he declares that, although he *could* find the Rock Pens, he "ain't a-gwine to." The Pens, according to him, are in the Guidan Pasture, which joins the Shiner ranch and contains twenty or thirty thousand acres of land.

It does look, as Pete expresses it, as if those thirty-one mule loads of bullion "ain't meant" for any of the people who have searched for it. When the man comes along for whom it is "meant," he will "jes' natchly find it without even trying." Nevertheless, some people are still trying. The inspiring thing about looking for the Rock Pens is that even though the search for them be fruitless, one may at almost any time happen upon some other treasure.

Where Parallel Lines Intersect

The southwestern part of McMullen County is quite rough, various elevations rising, in popular speech, to the dignity of mountains. Of these the most noted is San Cajo. Tradition has it that the mountain used to be called Sin Caja (Without Box, or Without Coffin). Certainly enough human bodies have *sin caja* returned to their native element on or around the hill to validate the name.

Along about 1875 a battered old Mexican came in great distress to John Fogg, who ran a livery stable in Corpus Christi. He said that when he was just a boy he enlisted as guard for a *conducta* of silver going from the San Saba mines to Mexico City. In the San Cajo country they discovered that Indians were upon them. They hastily unpacked, cast the silver between some rocks, and prepared to defend themselves. Meanwhile the boy ran down a ravine to see if he could locate water. The enemy cut off his return. Unseen, he witnessed the annihilation of his companions, and then made his way to Mexico, where he related the circumstances of his escape. Before long he was charged with desertion. Authorities seized him and put him in prison, and there for thirty-seven years he was kept. Years after his release he had wandered back to Texas. His one idea was to return to the cache of silver. He needed help.

After hearing the tale, John Fogg got together a party of three or four men to take the old Mexican up to the San Cajo and make search. On the evening of the third day's travel they camped in sight of the mountain. *"Mañana,"* said the Mexican, "I shall show you." Deer were plentiful, and for supper the elated treasure hunters prepared some freshly killed venison. Their guide ate more than his feeble nature could stand. He was seized with cramps; before daylight he was dead.

While Joe Newberry was bossing a ranch "down in the Sands" along about 1898, an old Mexican who was headed north to look for the Rock Pens gave him a chart to nine jack loads of silver bullion buried squarely on top of the San Cajo. According to the chart, the bullion was put down under a big rock near a *chapote* (Mexican persimmon) tree. It was being transported from mines up the Nueces Canyon when the escort lost their way. A terrible drouth was on the country; the Nueces River for miles at a stretch was as dry as a bone; the Spaniards had missed the lakes that parallel the river; they and their animals were perishing of thirst. They found a sup of water in some rock *tinajas,* but only enough to tantalize them. They chewed prickly pear, but the slimy juice of it tantalized them even more. Behind them lay no hope of water; between them and the Rio Grande stretched seventy miles of parched desert. To reach this far-away water they must cast aside all burdens. They ascended the San Cajo to get the best bearings possible, left the bullion there, and headed southwest. Only one man endured to drink from the Great River. His experience had been so horrible that he never returned to claim the wealth. Of course, though, he made out a chart to it—the chart that Joe Newberry took.

Yet if one knew how to find the water, there was enough, at times at least, to do a band of brigands who lived in a cave under the San Cajo and preyed on traffic passing over "the lower trail" between San Antonio and Laredo. These brigands used an apartment of the cave to stable their horses in. Back of the stable was their treasure room—*el apartado del tesoro.* Here they stored heaps of bullion, Spanish doubloons, golden candlesticks, bridle bits and spurs of precious workmanship, plated firearms, and all manner of other costly plunder taken from grandees and cathedrals. To these riches they added bullion from a mine—legendary to be sure—in the Lechuzas Mountains not far away.

A more prominent landmark than the San Cajo is Loma Alta. At the foot of this hill an early settler named Drummond had a squat. One time a Mexican—and, true to character, he was as old as the hills—came to Drummond looking for bullion that some of his ancestors had left in the vicinity. His *plata* called for a mesquite tree on the southeast slope of Loma Alta marked by a turtle cut into the bark. It called next for a line of smooth, irregularly oblong rocks bearing a resemblance to *manos*—stones used for grinding corn on the *metate.* These rocks had been culled from the hillside and so laid as to point to the hidden bullion. Drummond and the Mexican found the marked tree but rode around for a whole day without being able to find the rocks. They decided that generations of horses and cattle had scattered them so that they could no longer be recognized as forming a line, and gave up the search.

The Mexican left, Drummond died, years passed. Then one day while John Murphy, who ranches near Loma Alta, was holding down a wormy calf out in his pasture to doctor it, he raised his eyes and saw three or four of the *mano*-like rocks lined up in a clump of thick chaparral. He thought of the tale that Drummond had told him, and, looking about further, he found, badly scattered, yet preserving a kind of line, other such rocks. But he could never "settle on a place to dig and so the stuff must still be there."

So many rumors of treasure in one vicinity are enough to inspire almost anyone, and Snowden was the kind of man to catch inspiration. When one of the two or three negro inhabitants of the region showed him a singular boulder on a little *llano* between the Nueces and the San Cajo, he resolved to act. Before beginning excavation, however, Snowden went to San Antonio to consult a fortune teller.

"I see," said the fortune teller, closing his eyes, "a small plain surrounded by brush, north of it rough hills. I see a curious boulder almost in the middle of this plain. Not far from it and six feet underground I see a kind of chest. Some one will unearth this chest by drawing two parallel lines from the northwest and the southwest corners of the boulder and digging at the intersection of the lines."

The fee was twenty-five dollars. Snowden allowed three friends to join him in digging up the chest. But when they got out on the ground and drew a line from the southwest corner of the boulder and then, according to directions, another parallel to it from the northwest corner, they discovered that the lines would not intersect! Snowden sent the chart back to San Antonio for correction. No corrections came. He was for digging anyway, and, refusing to be played upon any longer by a fortune teller, he and his partners made a second trip to the boulder. They drew the lines now so that, while "not exactly parallel," they would *have to intersect.* They dug and dug. Finally one man flung his grubbing hoe as far as he could fling it and swore that he would sell out all his interest in the treasure for "two-bits' worth of Duke's Mixture." Snowden took him up and presently had bought out each of his other partners on terms equally high.

Casa Blanca

A chest or something is buried on the west bank of the Nueces at Puente Piedra, where the *camino real* from Laredo to Goliad used to cross. Fifteen miles or so on down two paltry jack loads of bullion moulder under the great oak trees that mark the ruins of Fort Merrill.[5] At Paso Veleño below that an enormous jag awaits someone. One time a ranchero started from the Veleño to Matamoros with a morral of gold;

at the Arroyo Colorado he staked his horse, buried the morral under one of many mesquite trees about, and went in swimming. The water made him go blind, so that he could not find the gold. It has never been recovered, and very few Mexicans have since its loss bathed in the Arroya Colorado.

Every old ranch in the Nueces country, every old road, every hollow and water hole has its story of treasure. But between the Rock Pens—wherever they are—and Corpus Christi Bay the most famous resort of the treasure seeker is Casa Blanca.

The land records show that in 1807 the Spanish government granted sixteen leagues of land, known as the Casa Blanca grant, to Juan José de la Garza Montemayor and sons. Before this date, presumably, Casa Blanca (White House), which served as both fort and residence and from which the ranch took its name, had been built. By 1810, according to a rather dubious source,[6] the Montemayors numbered their herds at 60,000 sheep, 24,000 cattle, and 14,000 horses; but in that year Indians overpowered them and razed the walls of their stronghold. The tumbled-down rocks can yet be seen, though prickly pear and mesquite have covered all the land about. Along in the forties, Casa Blanca became, on account of the grassy prairies around and abundant water, a kind of rehabilitation headquarters for great trains of Chihuahua carts freighting between Corpus and points in northern Mexico.

Oh, reader,

> Forget six counties overhung with smoke,
> Forget the snorting steam and piston stroke,
> Forget the spreading of the hideous town;
> Think rather of the pack-horse on the down.

In imagination those old freight wagons and Chihuahua carts which rumbled off into silence so long ago, the ruts that they cut long since deepened into gullies, yet creep across the chaparral lands from the Nueces to the Rio Grande. Some of them are drawn by twelve spans of mules and some of them by oxen. The mule trains follow the route leading by the best grass. The ox trains take the road that leads through prickly pear, off which drivers can singe the thorns and thus enable their slow-grazing beasts to get a fill without losing too much time. Many of the wagons are loaded with lumber, unshipped at old Powderhorn or Corpus. Occasionally a Mexican driver gets sick and dies or is killed. His contract has stipulated that in such event his body shall be brought back to Mexico. But embalming fluid is unknown and cremation would be sacrilege. Yonder raised platform of lumber with shallow plank siding and open top is for drying out the body. Buzzards soon strip the flesh, and, unlike coyotes, do not carry off bones. When the skeleton is clean, dry, and bleached, it will be delivered to relatives in Mexico as per contract and the lumber in the platform will also be delivered to its proper owner.

While Casa Blanca was yet a camping place for Chihuahua carts, William Mann took over the ranch. It had already become legendary; people said that the Spanish had here "a kind of sub-mission attached to La Bahia and Goliad." Inevitably all sorts of stories ascribing treasure to the "mission" and to the rich ranchero as well were told, and they are yet believed.

"At a picnic held at Casa Blanca in 1878," Bill Adams, who landed at Corpus Christi in 1852, likes to tell, "I and several other men were seated on some rocks between two mesquite trees near the old *casa*. Before long the talk turned on buried treasure; most of us hooted at the idea of there being anything to the tales, though, as usual, somebody was for digging right there. Well, a man named Brandes was living at the Casa Blanca Ranch, and one morning about a year after our picnic discussion he noticed that the rocks we had sat on between the two trees were all torn up. The signs showed that a wagon had been there in the night, that a hole had been dug, and that a box had been lifted out of the hole. The print of iron bands around the box remained visible in the hole for some time. I saw them myself. What the box contained will never be known."

Had Bill Adams talked to a certain Mexican whose tongue and chart Ed Dubose loosened with *tequila,* he might have been less doubtful about the contents of the box. As the Mexican had the facts, old Montemayor, the ranchero of Casa Blanca, at length sold out his vast herds, acquiring hard cash in payment. He was preparing to transport it and his family to a more civilized place when some Mexican bandits under the leadership of one Carbajal captured him and tortured him until he told where the cash was hidden. Then they killed him. At this juncture they found that they were being spied on by a second band of bandits. Under cover of night they hid their booty in a rock pen adjacent to the Casa Blanca stronghold, burying the body of the murdered ranchero on top of it so that his spirit would act as *patrón,* or guard.

At daybreak the battle between the two bands began. The besiegers far outnumbered the besieged, and in desperation Carbajal and those of his followers who were not killed scattered into the brush. There Carbajal, the chief himself, was shot down, and as he fell he saw his slayer bending over him. That slayer was his own brother. He understood the mistake and, dying, told the brother where the loot was hidden. Even as he told, the last of his companions bit the dust.

But the victorious desperadoes were never to reap the golden benefits of victory. Before they could tend their own dead and wounded, the terrible Texas rangers were upon them. As a matter of fact, Montemayor died or left Casa Blanca many years before there were any Texas rangers—but I follow the story. Ere the bandits could reach the Rio Grande all but one or two of them had been overtaken and, to use a ranger euphemism, been "naturalized."

Just a few miles below Casa Blanca is Lipantitlan, the history of which can be traced back to 1786, but which is now nothing but a thicket of retama and huisache, with an occasional hole at their roots, beside a lake. The story of its treasure, if certain names connected therewith were printed, would certainly provoke a lawsuit. So Lipantitlan will have to remain one of the many, many untold stories down the Nueces.

CHAPTER III

THE FACTS ABOUT FORT RAMIREZ

I was riding in the country,
My companion said to me:
"A captain killed a bandit
Underneath that tree."

I was riding in the country,
My companion said to me:
"A bandit killed a captain
Underneath that tree."

—CAMPOAMOR, *Tradition.*

FORT RAMIREZ—or, more properly, Rancho del Ojo de Agua Ramireña—also belongs to the Nueces country. As a boy I knew it well, for it was situated on our ranch. Frequently I rode by it, and sometimes with legs stiff in leather *chivarras* climbed its walls, there to gaze long at the serpentine winding of Ramireña Creek below and the oak-fringed hills beyond. Often I listened to tales the Mexicans and ranch people told.

The walls that I used to stand on are all down now; treasure hunters are responsible for that, and insensate workmen for a pipeline company recently hauled away most of the rocks. Lines of thorned *granjeno* bushes—those markers sown by birds along old fences all over Southwest Texas—made irregular by more than a hundred years of drouths and sproutings, yet delineate the quadrangular picket corral; but only a native eye can discern those lines. In another generation Fort Ramirez will hardly be more than a name, and treasure hunters may even debate on what hill to sink their holes. Let it be recorded that it is the hill in the southeast corner of what is known as the Primm Pasture overlooking Ramireña Creek to the north and Ramirez Hollow to the west.

The land records in the county seat recite a history that gives to the Ramirez legend a fitting background. Undoubtedly the ancient landmark was the first of any permanence to be erected within the confines of what is now Live Oak County, though when it was built no record tells. Undoubtedly it served for fortification as well as for residence. In 1829 the heirs of Don José Antonio Ramirez and of Don José Victoriano Ramirez entered, from the state of Tamaulipas, a plea to the Mexican government for title to eight leagues of grazing land "known as the Rancho de los Jaboncillos but more commonly as Ojo de Agua Ramireña." On these *sitios,* according to the plea, the Ramirez brothers had cleared land for fields, built a tanyard, erected corrals and ranch houses, made other improvements, and were living in peaceable possession of the estate, though for some reason they had not yet received a grant to it, when in 1813, as a result of the Mexican uprising against Spain, all frontier troops (*presidiales*) were withdrawn. This withdrawal released hordes of Indians to prey on the few scattered rancheros. The Ramirez people were forced to leave their ranch in such haste *"que nada se saco de la casa, ni de los muchos bienes de campo que había"* ("that nothing was taken from the house nor of the extensive herds on the range").

The Texan war of independence found the Ramirez suit still unsettled. For years after the war the claim was prosecuted, but neither the republic nor the state of Texas ever confirmed title to the land, the Señores Ramirez, however just may have been their pretension, having failed to procure a clear grant from either the Spanish or the Mexican government.

While the country was still "all open," Tol McNeill settled on the Ramireña a few miles above the Ramirez stronghold, and in time the talk was that he had dug up $40,000 of the money buried there. It was not until a year or so before his death (in 1927) that I mustered sufficient courage to ask him about the treasure. He told me something, but, so far as interest goes, his story can in no way compete with the man himself, for he was "a character," during forty full years of his own lifetime the theme of crescent speculation and yarn.

He used to be pointed out as a man who had killed "two or three white men and no telling how many Mexicans," and there were other tales—all utterly without proof so far as I know—about the "wide loop" he had swung back in the early days. His right arm terminated in a stub about halfway between wrist and elbow, and on account of the defect he was often referred to as "El Mocho." All sorts of explanations were current as to how he lost the hand. Despite the absence of it, he could roll Bull Durham cigarettes as facilely as anybody, and during his years of activity he was an expert roper both on horseback and in the pen, where he cast the loop with his foot. How he could manage a team of fiery buggy horses! He had a fine ear for music—hymns, dance tunes, Mexican waltzes—and it was a joy to see him at the piano dashing from treble to bass and keeping both going with that one hand.

His cattle were about the wildest in the whole brush country, and many a time I have ridden down Ramireña Creek through his pasture without sighting an animal, though I could hear *ladino* cows popping the bushes a half mile away. He believed in "natural water," but the water holes on the creek always went dry along in the hottest part of the summer and at the same time his two or three little tin windmills quit pumping; then some of his cattle would die of thirst, but the wilder ones survived on prickly pear. A cow that had to be supplied with water by a gasoline pump and couldn't "chaw" it out of prickly pear wasn't worthy of an old-time Texian cowman anyhow.

I remember going over to the Buena Vista—that was the

name of his ranch—one time while his vaqueros were gathering to ship. They had been working for a month and had caught only about a car load and a half of cattle. Most of these were in a big corral built of pickets ten feet high. They were existing on singed prickly pear with a modicum of cotton-seed cake mixed in. Every animal in the pen had been roped and led in necked to an old brindle ox. They looked "as gant as a bunch of gutted snow birds." I did not see them when they were finally driven to the shipping pens, but I heard they were "a sight." Some of the steers were necked together; some had heads tied down to a front foot; some were belled; one or two were hamstrung. They had to be "fixed" somehow so that they could not get away in the brush. Tol McNeill never owed anything on his land or stock, never bought on credit, and he did not have to raise many cattle in order to keep himself and family supplied with such necessities as Bull Durham, frijoles, salt pork, kerosene oil, calico, and a good buggy.

When "Old Man Tol" was converted at a camp-meeting, his conversion was counted as about the greatest victory for the Lord that the Ramireña country had ever witnessed. It was not long, however, before he "backslid." But whatever his spiritual state might be, he always had the blessing asked at his table, and those people who understood him knew that he had a heart "as big as an ox."

The last time I saw him, which was at a meeting of trail drivers in San Antonio, I asked him if oil-leasing was very lively down in his country.

"A fellow was over to my ranch not long ago," he replied, "wanting to lease my land."

"Did you close with him?"

"Well, I don't know. He asked me about an abstract. I just pointed to my old forty-four Winchester over in the corner and told him that was my abstract. 'It has protected me on this land for forty-six years,' I says, 'and nobody ain't ever questioned the abstract yet. It's the only one I'm going to furnish.' The fellow agreed that my abstract must be good, but we haven't signed any papers yet."

While we were talking, a ranger captain walked by.

"I know that man," old Tol remarked, "but he don't seem to know me. If they don't know me, Pancho, I never bother 'em. I never say anything. I'm that way.

"Now, about that old Ramirez Mission down there on your pa's land, I'll tell you all I know. When I first saw it, the walls were all standing and everything about it was good except the roof. They say it was one of a line of Spanish missions extending all the way from Corpus Christi to San Antonio.

"Of course all these old ranch missions and the like were on Spanish grants. I can remember when some Mexkins down in Tamaulipas were trying to establish a title to my own land and all the other land between the Lagarto and Ramireña creeks. They claimed eleven leagues out of the Ramirez grant, but the acreage they wanted to take in would have amounted to twenty leagues. You know how those old grants were surveyed, I guess. They didn't bother with stakes and chains. Two men got on horseback with a forty-foot reata tied from stirrup to stirrup, and they rode in a trot all day counting the rope lengths, throwing in plenty of extra ones so as to have full measure. The man ahead was supposed to halt and hold his position until the other man came up beside him, their stirrups touching, but oftener than not they were both riding at the same time, the second rider merely marking with his eye the starting point of the first. When they came to a creek, they crossed it without marking anything, and it was the same way when they had to go through a bad thicket.

"Well, along after the Civil War brother Pate and I heard of an old Mexkin woman who was said to know all about the Ramirez Mission. People were talking about the treasure buried there and some were already digging. We were interested and we went to see this woman. She lived on Captain Kennedy's Lapara Ranch down on the coast. She was old, old, maybe ninety or a hundred years old.

"Yes, she said she'd been at the Ramirez before it was abandoned. That was when she was a little girl, hardly more'n big enough to carry water from the creek. One day she was just starting up the hill with an olla of water on her head, she told us, when a vaquero galloped up and yelled for her to get on behind him, that the Indians were coming. He had sighted them at the big Ramirez water hole, which, you know, is only a mile or so above the old Ramirez house, and about the time he saw them they took after him. I guess he figgered there were too many Indians for the people at the Ramirez ranch to stand off. Anyway, he stopped just long enough for the girl to jump behind him. She rode astraddle, hanging on to him, until they crossed the Lagarto. Then they went on down to Casa Blanca. It was deserted and half destroyed. The Indians had started their raid at Corpus and come on up the country, making a clean sweep of everything in front of them. They killed every soul on Ramireña Creek.

"Well, after telling us all this, the old woman claimed she didn't know a thing about the money in the Ramirez house, but she said the *dueño* was *muy rico*. She said also that somewhere between Casa Blanca and the Ramireña the Indians had wiped out a wagon train hauling some pretty valuable freight.

"I guess it must of been about a year after picking up all this ancient history that I was running wild horses over south of the Picachos on Lagarto Creek. I was going lickety-split, hell-bent for breakfast, trying to head off a gotch-eared brown stallion and his bunch when all of a sudden I ran into a lot of human bones. I stopped right there to examine them. I knew the other boys would get the horses. There was heads and arms and legs scattered all about. They were as white as bones can bleach. That night I told Oliver Dix about what

I had seen and we rode back next day to take another look at the bones. He had been in the country longer'n I had and he just knew those bones were the remains of the Mexkin freighters who'd been killed by the Indians on their big clean-up.

"One time out on that long hogback running south from the Ramireña water hole I found a wheel belonging to an old-time Mexkin *carreta.* Somebody had been digging around it; it was claimed that the Spaniards had buried a store of money there.

"Of course no such digging went on at this place as has always been going on about the Ramirez Mission itself. It was along in the early seventies, I guess, that I hitched my bridle reins to a *granjeno* bush one day and stepped into the Ramirez building. Right there on the ground were pieces of a *jarro* [an earthen pot] that had been dug out of one corner. The print of some of the coins was still fresh and plain on the caked earth sticking to pieces of the *jarro.* I don't know how much was found. If the Ramirez outfit didn't put all their money in one place, some of it is still there to find. I know damned well I never got any of it."

There is a sequel to this story of the *jarro.* George Givens told it to me thus:

"In 1874 I went to Abilene with a herd of steers, and that same year, I remember distinctly, a saddle, blanket, and bridle were found in a hole pawed up by bulls on the prairie east of the old mission on Ramireña Creek. A little later a man's body was found half buried at the head of the hollow just above the mission. Strange lights had been seen about the old house, and some claimed that a Mexican *jarro* had been dug up inside it. People talked about the saddle and the dead man and the *jarro* a lot. The general supposition was that two men had dug up the treasure and that then one of them killed his pardner and hogged the whole find."

East of the Ramirez ruins there was and is a big field. The

Mexicans who cultivated this field lived in a *jacal* from which at night they were always seeing mysterious lights flickering and flitting between the rock walls and the creek, but never around the fort itself. Many people have held that a tunnel once ran from the fort to the creek; the lights indicated treasure hid in the tunnel. Several shafts have been sunk in attempts to probe the tunnel, but none has succeeded.

One Mexican used to tell how as a child he came with his parents to the Ramireña. They had a little money, and, as land was then very cheap and as the fort was still in tolerable condition, all that it needed to make it habitable being a thatch of beargrass, they considered buying it. One day while they were approaching it to examine it more closely, a white panther leaped out; then when they got inside, they saw many curious coins on the walls and dirt floor. But they were afraid to touch the coins and abandoned the idea of purchasing the place. The white panther was the soul of the dead *dueño* of the treasure, there to watch it.

Yet taking the money would very likely have brought no harm, for a white object—*un bulto blanco*—usually signifies a good spirit; a white cat, a white calf, a white dog, a white mule, a woman dressed in white, or any other form of whiteness may at times appear for the purpose of leading people to buried treasure. It is *un bulto negro* that one had better beware of.

One time a man from East Texas drove up to our ranch in a buckboard. He had a Mexican with him. He asked permission to dig at the fort, and the permission was readily granted. This Mexican claimed to have been digging at the south wall once ten years before when all of a sudden, just as he was sure his *talache* had struck the lid of a chest, he heard an unearthly yell and the rattle of trace chains behind him. It was night. He had enough presence of mind to kick a few clods back into the hole, which was a small one; but he was so frightened that he left in a run and had never been back. He had never even told of his experience until he found this *amo,* whom he could trust. I went along on horseback to guide the treasure hunters. When the Mexican got to the fort he appeared to be a stranger to it. He and his *amo* moved very little dirt.

Not long after this we ourselves had an experience in digging. While looking for the Casa Blanca treasure, Ed Dubose and Stonewall Jackson Wright met with a Mexican sage who gave them "the true facts" about Fort Ramirez. They must look for the treasure in "a secret cell." They possessed a "gold monkey"—a mineral rod—and this instrument they took to the fort; it oscillated towards the west and made two locations. After excavation proved futile at both places, Stonewall Jackson Wright quit, but Ed Dubose kept on. Fort Ramirez "just looked too good" to abandon.

The next step was to consult a noted mulatto fortune teller at Victoria. He described Fort Ramirez satisfactorily and said that for $500 he could and would locate a chest of money buried near it. The agreement was made and one night Dubose drove the mulatto to the fort. He stepped off a few paces, planted his foot down, and said: "Here it is. Dig a round hole here ten feet in diameter." When the two got back to Wade's Switch about daylight, the mulatto demanded his $500. Dubose told him that he would have to wait until the money was dug up and offered to allow him to be present at the ceremony, but he refused to stay. He declared that unless he was paid his fee at once, "spirits would move the box," and it would be useless for anyone to try to find it.

He was not paid at once, but despite the threats of malignant spirits, Ed Dubose persuaded my father to help him dig. I believe that my father had as little superstition and was as little given to extravagant fancies as any man I have ever known; yet there is something about the lure of buried treasure that will cause almost any man to "take one trial." It happened that work was slack about this time, and so one morning in a wagon loaded with tools, bedding, chuck, a lantern, a gun or two, and three Mexican laborers, we all set out for Fort Ramirez. A big hole was to be dug and it was to be guarded until completed.

When the diggers got down six or seven feet, they came upon some loose soil that was quite different in color and substance from the contiguous earth. It appeared to be "the filling" of some old hole. Hopes became feverish, but after about a barrel of the extraneous earth had been removed it petered out. At the depth of twelve feet the men quit digging. Evidently the spirits had moved the box. I saw the old sink the other day. On four or five acres of ground around it are many other holes, some of them freshly dug.

"If I just knew where the big door to the fort was," said the Mexican with me, "I could send for a friend of mine who has a map and get plenty of *dinero*."

All I regret now is that the stones of Ramirez Fort have been carried away. I should like to stand on them once more in April and gaze across the winding Ramireña upon the oak-fringed hills beyond. Yet the hills could hardly be so lush with buffalo-clover—as we used to call the bluebonnet—and red bunch grass, so soft and lovely, as they are in the eyes of memory.

CHAPTER IV

THE CIRCUMSTANCE OF WAR

I have remarked that the stories of treasure buried by the Moors which prevail throughout Spain are most current among the poorest people. It is thus kind nature consoles with shadows for want of substantials.—IRVING, *The Alhambra.*

A POLITICIAN "hunts not the trail of policy so sure" as the seeker after buried treasure hunts chests of "pay money," stuffed cannon, and other valuables hastily secreted by marching armies. A cannon that the Hessians are supposed to have blocked full of gold and pitched into the Delaware River during the Revolutionary War is still looked for.[1] In Sweden, farmers plow with the hope of turning up some of the tribute money paid by Anglo-Saxon kings to a Danish army a thousand years ago. More than any other military people, however, the Spaniards and Mexicans have been given to prodigal abandonment of riches along the path of glorious war. The route pursued by many an armament over the Southwest is kept in mind nowadays solely by accounts of treasure left behind.

RELICS OF DE SOTO

In the year 1539 Hernando De Soto embarked from Cuba with two thousand followers to explore for gold and silver in Florida. For three years the yellow *ignis fatuus* lured him north and west, he and his men killing, pillaging, and baptizing with Christian thoroughness the savage heathen of the wilderness. In 1541 he discovered the Mississippi, and the next year he was buried beneath its waters. "He had crossed a large part of the continent in search of gold, and found nothing so remarkable as his burial place."

One find, however, that has helped immensely the cause of legend was "three hundred and fifty weight of pearls, and figures of babies and birds made from iridescent shells." The quotation is from one of the Spanish chroniclers who accompanied the expedition. The place was in Georgia.

On the long march that followed, one of De Soto's foot soldiers, Juan Terron by name, grew so weary that he drew from his wallet a share of the pearls and offered them to a cavalryman if he would carry them. They weighed six pounds and were in a linen bag. The cavalryman told his comrade to keep them. Then Juan Terron opened the bag, whirled it about his head, and scattered the pearls in every direction. The chronicler estimates that in Spain they would have been worth over $12,000. "There are no pearls for Juan Terron," became a Spanish proverb, meaning that a fool makes no profits. But little did Juan Terron care for pearls or proverbs when, finally, he and three hundred other survivors out of the original two thousand crawled, naked, starving, broken, empty-handed, into Tampico, Mexico.

Such was the fate of De Soto's search for an El Dorado to the north. And now through the enchantment of distance people who inhabit lands that De Soto by the remotest possibility may have traversed go out with sanguine hopes after a plethora of riches that he is supposed to have buried. It cannot be proved that he did not pass through Searcy County, Arkansas. A few years ago an Oklahoma man who had some business respecting a zinc deposit in that county engaged the services of a local surveyor. After the required lines were run, the surveyor said:

"Come over here, and I will show you the oldest mark of civilized man to be found in America." Then the surveyor led the way to a mound that had been dug out so as to give it the appearance of a crater.

"In 1900," he continued, "a Mexican priest came here from a monastery in Mexico. With him he had a chart that he claimed had been made by De Soto and preserved in his own monastery all these centuries. The chart called for a silver mine in a cave, and the location of it was given in degrees and minutes both of latitude and longitude.

"The priest offered me half the silver if I would locate it for him, and the agreement was made. The chart was dated 1541. So I figured to a fine point what the declination of the needle would be for the time elapsed since that date. Then I made my location.

"It was on top of this mound and right in a big oak tree. Well, when the priest saw that I was running into an oak hundreds of years old, he got disgusted. But there had been plenty of time for an oak to grow, and my own theory is that the Spanish set out a sapling as a kind of marker. Anyway, we got a couple of hands, grubbed up the oak, and began to sink a hole.

"When we got down twelve feet we came to a slab of limestone rock. We dug this away and then we found ourselves at the mouth of a kind of cave. And there as fresh as if they had been left yesterday were two or three sticks of zinc and some old clay retorts. De Soto's men made retorts out of native clay in order to control the distillate. I don't know why they called it a silver mine, but they did. Another thing we found was a pearl, scorched like the Indians used to scorch them when they burnt them out of mussels."

The Stuffed Cannon of the Neches

Any party planning to hunt or fish down the Neches River, in East Texas, will do well to secure the services of "Uncle Jimmie" Clanton. He is pretty old but no older than he was twenty years back. As a cook he is not noted for cleanliness and as a handy man he has a remarkable aptitude for letting George do it, but as a recounter of what has happened in the piney woods he is without peer.[2]

"Boys," he began one night, after he had comfortably propped his back against a tree, lighted his pipe, and dismissed from his mind the supper dishes which the "boys" were paying him to wash—"Boys, I want to tell you that my paw helped old Sam Houston whop Santa Anna back in 1836.

"Well now, after the treaty of peace was signed at San Jacinto, or maybe it was just afore, Paw went under Colonel Burleson to smoke the Mexicans out of the fort at Nacogdoches. You all have likely read about this in history, but I'm going to tell you something that never gets into history at all.

"It was early one morning when Burleson's bunch made the charge. Not a soul was inside the walls. They nosed around a little, found the Mexicans' trail leading due south, and struck out to follow it. It was fresh and as the Mexes were traveling with waggins, to haul all their plunder in, Burleson's men figgered they'd probably catch them by dark.

"Well, they rode all day without seeing the enemy, but the trail kept getting fresher. When they camped at dark they was fifty miles from Nacogdoches. About ten o'clock next morning they came to a couple of abandoned waggins. The Mexes was getting scared and cutting loose from their plunder. They pushed on without stopping for dinner, and about three o'clock sighted some Mexes at Boone's Ferry. That ferry wasn't two miles up the river from where we are camped.

"As soon as the Texians saw the Mexes, they made a dash, hoping to cut them off. Just as they got in comfortable shooting distance, the ferry-boat landed on the opposite side with a waggin and three men. The waggin drove on off, but the Texians were too busy to foller it just then. Some of the Mexes took to the timber on the near side and fired back and there was a right lively little scrap. Paw used to tell me all about it, and he said he was laying kinder behind a log firing away when he looked up and noticed three men rolling a cannon toward the river. They rolled it to a high bluff and shoved it right off into the deepest hole in ten miles. He said he wondered what the idear was, but about that time he was too interested in number one to investigate cannons.

"Well, about fifteen of the Mexes were killed and the rest captured. That is, they was all captured except the three that got across the river with the waggin. A detachment took after them. Not more'n a mile and a half from the river they found the waggin, as empty as a last year's bird's nest with the bottom punched out, and one dead Mexican, but the other two got away. Burleson reported with his prisoners to General Houston, and that was the end of the expedition.

"Well, the things that happened in the next few years wouldn't interest you any. Then in the spring of 1875 a slick-haired young Mex showed up in our neighborhood. He didn't appear to have any pertic'lar business—just seemed to be nosing around. After he'd been here for a month or so, he comes to me one day and says that as I seem to be pretty well posted he'd like to make a proposition. We went behind the pen where nobody could see us. There he pulled out an old map. It was in Spanish, but it wasn't no trouble for me to make out the lay of the land. It showed a road beginning at Nacogdoches and running due south. It showed the Neches River and it showed Boone's Ferry. Just below Boone's Ferry—a mile and a half, he said—the road stopped. Well, I was right interested.

"Then he asked me if I knowed where Boone's Ferry used to be.

" 'Shore,' I says.

"Then he opened up. His grandfather, so he says, was at Boone's Ferry when Burleson arrived with the Texians. He was one of two men that got away. The young Mexican wants to know if I am acquainted with the details of the battle.

" 'I certainly am,' I says. 'My paw was in the fight and has told me about it many a time.'

" 'Did he ever tell you about seeing a cannon shoved off into the river?'

" 'Many a time,' says I.

" 'Well, Mr. Clanton,' he says—for he was perlite as a basket of chips— 'did it ever occur to you to wonder why that cannon was shoved into the river?'

" 'Why, yes,' says I, 'I've wondered about that lots of times.'

" 'I'll tell you why,' he says, and he was getting excited, I could see mighty plain, but he didn't let his voice get up. 'That cannon was filled with gold from breech to muzzle.'

" 'Gold!' I whistled. 'So that's it.'

" 'Yes, that's it,' he says. 'Not only that, but I have in my pocket another map giving the exact location of more gold, just on beyond the ferry. You see, the waggin that crossed the river carried a chest of money. The three men with it went on till they got well out of sight; then they buried it. They had a quarrel over it, and one Mexican shot another to shut him up. Then he and my grandfather took down some landmarks on a piece of paper, and pulled for Mexico. On the way the other Mexican died, leaving my grandfather with the map. He died before he could come back and get the money. My own father was killed by bandits; so I was left with the one and original map of the buried treasure.'

"If ever I saw a man that believed he was telling the truth, that young Mexican was him. He didn't ask me for a penny. He just came out flat and said he'd give me half of everything we found providing I'd help him.

"You all can guess I jumped at such an offer. We looked for the cannon first, because I knowed exactly where it ought to be. I still know, for that matter. We dredged and dredged and fished and fished for the blamed thing, but never could locate it. You see, it took about a forty-foot jump off into the river and it had had about forty years to settle in.

"Well, we finally stopped fishing for it and went after

"That cannon was filled with gold from breech to muzzle."

the chest. The map called for a triangle with the chest in the center. Without any trouble at all we found the first of the three corners of the triangle. It was a big rock, kinder queer shaped. The other two landmarks were supposed to be pine trees. But all the big trees in the country had been cut off and rafted down the river. There just wa'n't no trees to sight from. We sighted off triangles from every old stump in the neighborhood, but I guess we didn't hit the right ones. Anyhow, we hunted gold for two months and never dug up a penny. The Mexican finally got discouraged and went back home. But he gave me a copy of his maps and I've been looking for money off and on ever since he left.

"That's about all there is to the history connected with Boone's Ferry, I guess. If any of you all want to see where the cannon was rolled off into the river, we'll go up to Boone's Ferry in the morning."

Santa Anna's Chests

How many chests of "pay money" designed for Spanish and Mexican troops were dropped into Texas streams and left there, it would be impossible to say. Reliable testimony discloses at least one.[3] According to the venerable chronicler, Noah Smithwick, a Mexican officer who was taking pay money to the garrison at Nacogdoches found the San Bernard River so high that he had to cross it in a dugout. The dugout tipped over, and the box of money—"a large sum"—sank to the bottom. After trying in vain to recover it, the Mexican left. During a low stage of water, some time later, a negro who belonged to a settler living near the ford found the box and delivered it to his master. He, as rarely honest as his slave, promptly notified authorities of the Mexican government and turned over the property.

This chest was dropped before the time of the Texas Revolution, but it set an example. At what is known as the Rock Crossing on the Nueces River a detachment of Santa Anna's invading army, so tradition runs, was fording the river in great haste when the "pay cart" broke down, toppling a chest into the channel. It was too heavy to pull out; the cart could not well be repaired to transport it farther even if it were pulled out; the river was turbulent; the commanding officer was in a hurry to annihilate the Texans at Goliad. He decided to leave the chest in the river, where, after speedy subjugation of the revolutionists, the Mexicans could get it on their way home. Fortunately it had a chain attached to it—a log chain. A diver brought up the loose end of the chain and fastened it around the trunk of a tree growing on the bank. As everybody knows, those of Santa Anna's forces who made their way back to Mexico did not tarry anywhere in Texas long enough to haul out a heavy chest. Forty-seven years after it was anchored, a cowman noticed a piece of chain around the trunk of an elm growing on the bank at Rock Crossing—a crossing even then abandoned. When a long time afterwards he learned the purpose of the chain, he took a team of horses to hitch to it and drag out the chest. The tree had washed away.

Now, it is a matter of history that when the Texans captured the commander-in-chief of the Mexican army at San Jacinto in 1836, they found $11,000 in his military chest, not counting the "finery and silver" which were auctioned off for $800. Was that all the treasure Santa Anna had in Texas?

Neal Russell and two fellow "brush poppers" were down in the Nueces country catching wild cattle. They had plenty of horses, plenty of supplies, and a pack outfit. I do not know when this was; it was years ago. Their camp was beside water.

One morning while they were "chousing" a bunch of outlaw cattle, Russell caught a hurried glimpse of two very old Mexicans. In those days any stranger on the range was a curiosity, but there was no time for parley now. However, after the cowhands had "fought brush" for three hours, roped and tied down a *ladino* apiece, and then ridden to the river to water their exhausted mounts, they came again upon the Mexicans. They looked shy and acted rather peculiarly, but they were so thin and worn that Neal Russell had nothing but pity for them. After passing a few remarks, he told them where camp was and invited them to go up and get a fill of beef. The only baggage the old men had was in a couple of frayed serapes carried by a burro.

When, after the usual afternoon hunt, the "white men" rode into camp, the Mexicans were already there, standing respectfully to one side and sucking at some cigarettes that had considerably less tobacco than corn shuck in their "makings."

The Texans talked "Mexican," or "cowpen Spanish," as fluently as though they were natives, and after supper the ancient strangers, evidently feeling at ease, asked if anyone present could tell where the old "War Crossing" on the Nueces was.

"Why," Neal Russell replied, "it's not a mile down the river from our camp. I'll show it to you in the morning."

The Mexicans now seemed to think that they had as well take the Texans into confidence, and one of them began a long explanation.

"I was here on the Nueces," he said, "in the spring of 1836. I was with thirty-eight other soldiers of the Mexican army. We were taking a cart load of money to San Antonio to pay off soldiers. One day's ride north of the Nueces a courier met us and told of Santa Anna's defeat at San Jacinto. We knew that it was foolish to go on and so turned back. We expected to be overtaken at any time by the Texans. Just before we reached the Nueces on our back trail, the axle of the pay-cart broke square in two. The only thing to do was to cut a tree down and hew a new axle. We dragged out of the road a little way and set to work.

"As I told you, we were expecting the Texans at any

moment. The captain was a discreet man. He made us dig a trench beside the cart; he spread two cowhides in the bottom of it, and then emptied the money on the hides. We had to unload the cart anyhow. Then we cut down the straightest mesquite we could find and carried it to the cart. Barely done was all this when we heard a galloping in the distance as if a whole troop of cavalry was coming. *Pronto, pronto,* we threw the log in on top of the coin, spread another cowhide or two over that, and then threw in dirt. It took but a minute to turn the cart over on top of the fresh earth and set fire to it. It blazed up. We mounted our horses and spurred south. I believe now that what we heard running was mustangs instead of cavalry. Anyway, we left the signs of our digging concealed by the fire.

"So far as I know, I am the only survivor of that escort of Mexicans. I do not believe that one of them has ever been back to get the money. Poor fellows, most of them were killed in the war with the United States that came a few years later. I am come now with my old *compadre* to claim what I helped save. You see how we are. We started out poorly prepared. Now we are almost afoot and have no provisions left. If you will help us, we will share with you."

The next morning Neal Russell, the other two cow hunters, and the Mexicans started out with axe and spade. They went to the War Crossing, and then out a short distance on the north side. The elder Mexican led them to three little mounds —knolls common to that part of the country. Beyond these three knolls was a stump, and beyond the stump another knoll.

"That is the place," half panted the Mexican, while they were yet fifty or sixty yards away from it.

The cowboys rode slowly, for the old men on foot almost stumbled at every step. When the party got fairly around the knoll they saw a pile of fresh dirt. Pitched across it was an old log. Mesquite lasts a long time beneath the soil of Southwest Texas. The men looked down into the hole. It was not very deep and had evidently been dug about a week before. A few tags of rotten rawhide lay about.

The Mexicans did not say a word. They were a week too late. The last Neal Russell saw of them they were tottering back towards the Rio Grande, their burro loaded with jerked beef and what other supplies the cow hunters could spare them.

Palo Alto and Resaca de la Palma

The battles of Palo Alto and Resaca de la Palma, which began the war between Mexico and the United States, were fought on Texas soil May 8 and 9, respectively, 1846. A *resaca* is a kind of marshy lake; the Resaca de la Palma is about three miles out from Brownsville (old Fort Brown). The Palo Alto battlefield is something like twice that distance. After a demoralizing defeat at Palo Alto, some of the retreating Mexicans made a stand at Resaca de la Palma, where they were completely routed.

Many years after the war was over, a Mexican named Ramón used to ferry passengers across the Rio Grande. He claimed to have been a ferryman when the Mexican troops crossed to halt General Taylor's army. This is the way he used to describe his experience:

"It took me three days to ferry all the Mexican army over, crossing and crossing back, day and night, night and day. And, oh, I had much desire to go with the troops. There was *música,* oh so lively, and there were the *banderas* all flying bright in the air, and the men were all happy and singing. But I did not go, and in three days more here they were back, but without any *música* or *banderas* and not needing any ferryboat. They came in flocks, running and crawling like *tortugas,* and they fell into the water flat on all fours like *tortugas* and never stopped till they were in the brush of the *república Mejicana.*

"They had been at the fight of what we call Resaca de la Palma, and I was very glad that I had not been with them. They did not have time even to bring back the *señor* general's chest of money or any of the silver that he ate out of. They left it all where three tall palms make a triangle; in the middle of that triangle it is buried. They dug a hole and put the chest and the silverware and a golden cross in it, and then they filled up the hole and made a great fire on top of it so it would look as if some military stores had been burned. And then they came back here like so many *tortugas*, and the gringos were so *bravos* that no one of those who helped hide away the treasure ever would recross the Rio Grande to get it. Besides, most of them were killed at Monterrey."

It was at the Palo Alto battlefield that Santiago's father helped bury army money. Taylor's men were pressing the Mexicans. In order to escape with their lives, the Mexicans must lighten their baggage. The officer in command of the pay-carts received orders to bury the money and retreat. He told off a detail and put them to opening a trench. When the trench was made, he commanded that the money be dumped into it. The place was well hidden by brush. While the last cart load was being dumped, Santiago's father, who had been throwing dirt out like a badger, slipped into the mesquites. He well knew that the poor peons who were doing the work would go into the trench next—for the dead tell no tales. No sooner was he out of sight in the brush than the sound of shots told him how prudent he had been in leaving. He never went back to the place, but he liked to tell about the fine treasure, and his loquacious son has guided several parties of men to the Palo Alto battlefield.[4]

In connection with treasures of the Mexican War a veteran [5] used to tell this tale. After the capture of Monterrey two American soldiers slipped down the San Juan River and robbed an old priest of $10,000. General Taylor discovered the crime—but not the money—and sent the soldiers to New Orleans to be tried. One of them died in prison; the other made his way back to the home of Wash Secrets on the Colorado River in Texas. There he became very ill and on his death bed told "Old Wash" that he and his accomplice had buried the ten thousand dollars in gold at the root of a pecan tree thirty steps from Walnut Springs, where Taylor's army was camped. After the fellow was buried, Wash Secrets went to Mexico to dig up the money. He found Walnut Springs all right, but as Taylor's men had for military purposes cut down all the trees, he had no pecan to go by. He searched for many days among the stumps without finding the gold. "It is there yet."

Steinheimer's Millions

One of the flashy figures of the filibustering era in Texas —vivid like a darting cardinal but too swift to permit of analysis—was Louis Aury, slave smuggler, privateer, and for a time ruler of Galveston Island. History has neglected to delineate Aury's vari-colored accomplices, but concerning one of them at least legend is explicit, even to the date of his birth.[6]

Karl Steinheimer, according to the most knowing of the many men who have dug for his gold, was born near Speyer, Germany, in 1793. At the age of eleven he ran away from home, became a sailor, and by the time he was grown could speak ten languages and had become a renowned captain of pirates. He threw in with Aury and for a while throve in the traffic of "black ivory"; but the days of Aury's prosperity were numbered, the two quarreled, and then Steinheimer withdrew into the mountains of Mexico, where he carried on extensive mining operations. Disappointment in an early love affair made such retirement congenial.

He had mined for nearly ten years when, in 1827, he learned that Hayden Edwards, the noted Texas empresario, had revolted against Mexico and established the Republic of Fredonia. Steinheimer decided at once to make his way to the Edwards' forces and offer his assistance in person and money. He considered himself powerful enough to become ruler of the new country. However, when he reached Monterrey, he learned that the revolt had been put down. Steinheimer returned to his mines and prospered for ten more years. Then he learned something that caused him to close out his business and head north.

He learned that the woman he had wooed but not won or forgotten was living in Saint Louis, still unmarried. His fortune, after he had sold out all his interests and paid all his debts, amounted to ten jack loads of silver and gold. He evidently had little doubt of his ability now to win the one woman. At any rate, he determined to take the fortune with him. He picked two Mexicans to assist in the transportation.

Meantime Texas had become a republic, though Mexico was still claiming it. When Steinheimer arrived in Matamoros, he found a Mexican officer named Manuel Flores with a few followers preparing to invade Texas for the purpose of instigating an Indian uprising. At this time he had no interest in Texas events, but knowing how hostile the Apaches and Comanches were along the San Antonio road, he threw in with the Flores expedition as a means of protection. A few hours before a band of Texans engaged with the Mexicans near the Colorado River, Steinheimer separated himself and his loaded burros from the military party and cut across the country.

His situation was now perilous indeed, for he had not only Indians to avoid but, as a Mexican citizen, the hostile Texans. In order to travel safely he must travel light. He decided to bury all his riches but one package of gold, which would suffice for immediate uses. Accordingly, in the hill country sixty or seventy miles north of Austin, near a place "where three streams intersect and combine into one," he unpacked his ten burros and concealed their freight. The only mark made to designate the spot of concealment was a large brass spike driven into an oak tree some forty or sixty feet away, the spike being of that type formerly used in the construction of boats. The burros were liberated to shift for themselves; then with his two trusted men Steinheimer took a southeastern direction. That was an illogical direction for him to take, it must be admitted, but he took it.

When they had traveled, as he judged, some twelve or fourteen miles, they came to what he termed "a bunch of knobs on a prairie," from the tops of which they could see a valley skirted with trees several miles to the west. At these knobs they were attacked by Indians, the two aides were killed, and Steinheimer himself escaped with a bad wound. He hid for a while in the brush on the central knob of the group, and there he buried the package of gold he had retained from the packs. He had lost his horse and all his supplies, but he still had a gun and some ammunition.

Choosing a northern direction, he now set out afoot. By the time he got far enough away that he was not afraid of being betrayed by the sound of his gun, he was almost starving. His wound had become very painful and gangrene had set in. At this juncture he fell into the hands of some travelers.

They were evidently well disposed and evidently respected the trust that Steinheimer placed in them. Realizing the threat of death, he made a crude map of the region about his buried millions, directed it to his former sweetheart, and sealed it with a letter in which he gave a concise account of his activities during their years of separation down to the moment of writing. He explained that the strangers to whom he was entrusting this message knew nothing of his name or history and that they would get nothing of his but a few coins on his person. Finally, he requested that she keep his message secret for three months after receiving it. If he recovered, he would, he said, reach Saint Louis by the expiration of that time; if he did not arrive, she was to understand that he was dead and that his fortune was hers. These are the last tidings of Steinheimer.

The letter was delivered. A number of years passed before relatives of the lady addressed felt that conditions in Texas would permit of a thorough search for the treasure. They came, and then after months of scouting concluded that the three small streams at the confluence of which the ten burro loads of gold and silver were buried must be the Salado, the Lampasas, and the Leon, which unite below the present town of Belton to form what is called Little River. Between twelve and fifteen miles southeast of this junction a prominent "bunch" of hills known as the Knobs overlooks the tortuous valley of Little River; here amid these Knobs it may be supposed that Steinheimer buried his package of gold and received his death wound.

The men whom the Saint Louis woman sent to secure her strange inheritance found nothing. They were careful not to reveal her name, but, while making their prolonged search, they were naturally obliged to reveal their motives. Thus the story of Steinheimer and his millions became known to the inhabitants of the country. For more than three quarters of a century now the quest has been prosecuted.

CHAPTER V

TALES OF THE COW CAMP

Did you ever notice how much better a pipe smokes when it's lit by a coal out of a camp fire?—*Brushy Joe's Reminiscences.*

The Rider of Loma Escondida [1]

"Boys, ever' time I camp at this crossing I think of the way Jeff Cassidy got waylaid and murdered between here and Loma Escondida."

The speaker was Captain Crouch. He was with his outfit at the old Presidio Crossing on the Leona River near the border. Supper was over.

"Yes," Captain Crouch went on, "that was a long time ago, and tough hombres in this country were thicker'n fiddlers in hell. Horse thieves was so bad that a man couldn't hardly keep a gentle saddle horse without hiding him in a thicket after dark and then sleeping by him. Money was safer'n horses, but a man with a good horse and money both had better be keerful.

"Now, about Jeff Cassidy. He was representative for the 7 D's, a big outfit. Well, he left San 'tonio one day for the old H-Triangle Ranch below here to receive a herd of steers and start 'em up the trail. Like the custom was at the time, he carried money to pay for the stuff in his saddle pockets. He had five thousand pesos, gold, the balance to be paid when the cattle were sold in Kansas and the money brung back.

"The second night out Jeff slept in Friotown. He left there at daylight next morning, coming this way. He was alone, and was riding a little creamy dun pony that I'll tell you about d'reckly. He'd got into Friotown after dark and somehow had failed to learn that the sheriff was going down to inspect brands on the herd, or he might have had company. The sheriff left Frio 'long latish in the morning. When he got about halfway between Loma Escondida and this crossing where we're camped, he found Jeff and his dun horse in the middle of the road both deader'n thunder.

"He hit the trail in a high lope, struck a couple of cowboys that joined him, and together they overtook the murderers somewhere this side of Eagle Pass. Everybody knowed them Newton boys wasn't no 'count and was rustling cattle, but nobody would have suspicioned them of killing a man in cold blood like they did Jeff Cassidy. The case was plain as daylight and the Newtons owned up. Still, they didn't have any of the gold with them, though the pockets had been stripped off Jeff's saddle. Furthermore, neither Jim nor Tom Newton could be forced to tell where the stuff was. They was hung, of course, and——"

But at this point the Captain's narrative was interrupted by an unceremonious announcement from the cook, Alfredo, that the camp was out of coffee.

Alfredo was a new *cocinero,* and in terms free and vigorous Captain Crouch expressed his opinion of a cook who would leave headquarters without enough coffee and not say anything about the shortage until it was nearly time to cook breakfast. The fact that Alfredo had carefully put his coffee in a big can and then loaded on in its place a similar can containing frijoles did not remedy the mistake.

"Well," concluded the Captain, "there ain't but one thing to do, and that's for somebody to ride to Charlie Trebes's commissary and get the coffee."

It was ten miles to the Trebes ranch. Twenty miles on horseback will not leave the rider much of a summer's night for sleeping. None of the hands said anything. It was not for them to say.

Otis Coggins, who had been working on the Crouch ranch

since boyhood, was patching a stirrup-leather. He was a tall, swarthy man under thirty.

"Otis," the Captain went on, "I reckon you're the man to go for the coffee. You know old man Trebes better'n anybody else, and if you get back here by breakfast time you can stay in camp and sleep all morning. Better get about ten pounds, I guess."

"All right," said Otis.

Then, addressing a Mexican, he ordered, "Go get me that Trigeño horse. It's lucky I staked him."

"Guess I taught you never to be caught afoot, even if your back was broke and you couldn't fork a pillow," remarked Captain Crouch. "That reminds me——"

"But, Captain," Otis broke in, still mending on his saddle, "don't you reckon them Newton boys must 'a' buried that gold around somewhere clost to where they killed Jeff Cassidy? They shore aimed to make use of it some day."

"Yes, I guess they did hide it somewhere," Captain Crouch agreed, his mind brought back to the story he was telling when the *cocinero* interrupted. "I don't know as anybody's ever looked for it much. There never was anything that hurt me more'n Jeff's death. He and me had been side pardners—reg'lar yoke mates, you might say—for years. I guess we broke enough horses together to furnish a dozen outfits going up the trail. There was still some mustangs in the country, and one day we laid into a little bunch that had their bellies too full of water to run much and roped one apiece. Mine was a purty black filly; I broke her neck a-trying to get the saddle on her. Jeff's was a creamy dun stallion with a black stripe down his back about the width of your hand—a *bayo coyote*, as the Mexicans call a horse of that color. He was all life and bottom, and Jeff got so he didn't want to ride anything else. He was riding him when he was killed. I always figgered one of them Newton boys made a misshot and killed the horse accidentally. They'd 'a' been mighty glad to take him as well as the gold."

The Mexican had arrived with Otis' horse. The saddle was thrown on him, and a sack for the coffee was tied on the saddle. With an *adios* the rider was off.

Otis Coggins was a silent kind of man. He liked to be alone. His route lay through a vast flat country of mesquite, catclaw, *huajilla, chapote,* prickly pear, and other kinds of brush common to Southwest Texas. The only break on the plain that his brush-lined road traversed was a solitary hill known far and wide as Loma Escondida. It is *escondida*—hidden—because the brush is so high and thick around it that a rider cannot see it until he is almost upon it.

The moon was full; as usual the sky was flawlessly clean; one might have read ordinary book print in the moonlight. As he approached Loma Escondida, not a sound but the pad of his horse's hoofs and the squeak of saddle leather breaking the silence, Otis Coggins noticed an object, or, rather, two objects, in the road some distance ahead of him. They were partly in the shadow of high brush, and at first he took them to be a cow and calf. Cattle in a brush country often bed down in open roads. But presently Otis saw that the larger of the objects was a horse and that the lesser was a man apparently humped over on the ground. The man soon straightened, mounted, and started up the hill ahead of Otis.

Now, the road that Captain Crouch's most trusted hand was traveling was a private ranch road. It was his business to know who was riding around in his employer's pasture at night. He hailed the rider. There was no response. He set spurs to his horse to overtake the trespasser, and, without hard running, was soon close upon him. He saw—saw plainly—that the horse ahead was dun-colored with a black stripe down his back. "He's a *bayo coyote,* like Jeff Cassidy's mustang," Otis reflected.

Near the top of the hill he was actually by the stranger's side; neither spoke a word and Otis could not get a view of the rider's face. Then suddenly the stranger pulled out of the road into some half-open ground and headed straight for a stark, dead mesquite tree, the trunk of which was exceptionally large. Otis knew that tree; he recalled how late one evening while he was driving a bunch of saddle horses over Loma Escondida, something—he did not see what—jumped out of this very tree and stampeded the remuda so that they ran two miles before he could get around them.

He saw something now that he did not believe. He saw horse and rider head into the tree—straight into it—and disappear. The ground beyond was open enough that anything traversing it would have been plainly visible.

He pulled up his horse and looked, just looked. Like most other range men, he had a prosaic head that absolutely refused to harbor ghosts and other such superstitions common among the Mexicans. He guessed maybe he had been half asleep and did not know it. Yet he could not have been mistaken about seeing the man down by his horse back in the road, the man mounting and riding off, the dun horse with a black stripe down his back. All the details were clear.

He rode on now to get the coffee, got it, answered a few questions about the cow work, and started back.

He rode absorbed in contemplation of the strange rider and horse. He determined to investigate the mesquite tree into which they had vanished. Once he turned in his saddle to look behind him, and down a straight stretch of the road saw a coyote following. Off in the brush he heard now and then the lonely wail of other coyotes; he liked to hear them.

True to its name, Loma Escondida did not reveal itself until he was at it, and then Otis saw something that made him know he had not dreamed the rider on the *bayo coyote*. There he was, trotting his horse leisurely up the hill, going in the opposite direction from what he had been going when seen earlier in the night. Otis reined up. He heard distinctly the sound of the dun's hoofs on the dry, hard ground. Those very material sounds made him realize that some trick had been played on him, and he resolved to end the mystery. He knew that his Trigeño horse could put up a hot race, even if the race led clear to camp on the Leona; and if it came to running in the brush, he wore his regular protection of ducking jacket and leather leggins and he considered himself second to no man in skill as a "brush popper."

Vámonos! The stranger was a full hundred yards ahead, up the hill. For a minute he continued the leisurely trot, but when Otis had lessened the interval by half he saw the man ahead of him strike into a run. Now the squat top of the hill was reached; Otis was gaining. He was not more than twenty yards behind. They were coming to the half-opening into which the mysterious rider had before turned out of the road. Here he suddenly turned again. Otis was not surprised. In less time than it takes to draw a six-shooter he had the horn string around his rope loose and a loop shaken out.

"Ojala!" he yelled. "Damn you, if I can't catch you, I'll rope you." But he was too late. Just as he prepared to cast his loop, the pursued rider and horse disappeared into the dead mesquite. There was no sound of impact. Otis himself dodged under a low-hanging limb of the old tree so as to be upon the tricky stranger on the other side. On the other side was nothing but bush, moonlight, vacancy, and silence.

Otis got down and led his horse up to the old mesquite. The first episode had been a surprise. The second, with every sense of the observer alert and expectant, was overwhelming.

"Have I eaten some *raíz diabólica* [2] to be seeing things?" Otis asked himself aloud, for he was determined not to be awed. Then he answered the question with "No I ain't, and I haven't been smoking *marihuana* weed either." There is a vague belief among some of the more ignorant Mexicans that after death their souls enter the mesquite—the mesquite that feeds them such good beans during life. As he rolled a cigarette, Otis Coggins remembered this superstition with contempt. The mesquite had never been sacred to him; he had cursed its thorns a thousand times.

He determined to give the tree a thorough examination. First he bent over to look at the ground around it. The moon, still high, revealed nothing but his own horse's tracks. The tree itself, though dead, was far from rotten. However, his eye caught a long groove in it. He felt the groove with his hands. It had been made with an axe long ago.

With face averted, he was slowly circling the old trunk when his boot caught on a flat rock. It was in shadow and he lit a match to examine more closely for other marks on that side of the tree. Glancing down, he noticed a dull glint of metal. He picked it up, lit another match; it was a twenty-dollar gold piece.

If there was one coin, there must be several. In order to clear the ground for closer inspection, Otis lifted the rock out of place. He hardly needed the light of a match to see what now lay exposed. It was a pair of old saddle pockets. They were so rotten that they almost fell apart when he picked them up. Several coins fell out.

"Jeff Cassidy's gold right where the Newton boys hid it!" Otis exclaimed aloud.

He was not excited. He never got excited. Calmly he went to his saddle, took down the sack, tied the coffee into one end of it, and then secured his find in the other end. Counting it could wait.

When he got into camp on the Leona, Captain Crouch and the *cocinero* were already up.

"I've shore been wanting that coffee," the Captain greeted him. "The water's been bilin' half an hour."

"Well, here it is," Otis sang out, at the same time dismounting and untying the sack, from which he brought to view only a part of the contents. "Hurry up, *cocinero,* I'd like a sip of coffee myself."

Bumblebees and Skilitons

The vaqueros had unrolled their blankets, but with backs propped up against a comfortable old log over by the simmering bean-pot, Pete Staples and I were thoroughly awake.

His tones were strictly confidential. Pete's "people" brought him to Texas from Mississippi before the Civil War. He was raised in the border country, and he used to cook and rustle horses for outfits going up the trail. For a while he lived down in Mexico, where he married a native woman.

"One time," Pete pursued the subject under discussion, "there was a white man who got wind of a lot of Mexican dollars buried down below Roma. He had the place all located, and was so sure of hisself that he brung in an outfit of mules and scrapers to dig away the dirt. He was making a reg'lar tank digging down to that money when a Mexican what I've knowed all my life comed along.

"This Mexican stopped a minute under a mesquite tree to sorter cool off, and right there he seen a hoe laying on the ground half covered up in the dirt. He retched down to pick it up and then he seen a whole *maleta* of coins. A *maleta,* you know, is a kind of bag made out of hide. This one was old and rotten, and when he turned it over with the hoe it broke open and the gold money jes' rolled out in the dirt.

"D'reckly the Mexican went over to where the white man was bossing the scrapers, and he asked him what he was doing.

" 'Oh,' says the white man, 'I'm jes' digging up some buried money.'

" 'Well, you's digging where it ain't no use to dig,' says the Mexican. 'The money ain't there; hit's over here. If you want to see it, come along and I'll show it to you.'

"The white man laughed like he didn't believe what the Mexican was telling him, but he come along. When they got to the mesquite there wa'n't no money in sight, but there was a hole down at the root of the tree kinder like a badger hole, and bumblebees was going in and out making a roaring sound, and the dirt was fairly alive with great big bugs, humming and making a sizzling noise and working round awful like.

" 'Huh, is this what you call money?' says the white man, stamping down on the tumblebugs.

" 'That's all right,' says the Mexican. 'There was dollars right here—dollars of gold and silver too. But there ain't now, I admit, 'cause them dollars is evidently not intinded for you. White man didn't hide that money and it ain't meant for white man to find it. No matter how much you dig or where, you won't find nothing.'

"Shore enough, the man kept on digging and he didn't get nothing. One time I asked the Mexican why he didn't go back and take out the money.

" 'I didn't want none of it,' he says. 'I never put it in the ground. 'Twa'n't mine any more'n that white man's.'

"A few days after he saw the money, though, he went back and scratched around in the dirt a little and picked up an old Mexican square dollar. He brung it to Roma and bought some flour and some coffee and some candy, and give some of the candy to my wife. She was living down there and knowed the man well and she's told me many a time how she et some of the candy the Mexican bought with that old square dollar. I always have thought the money was intinded for him, but you know how some people are, and I can't say as I blame him for not teching what he hadn't a right to. If buried money like that is intinded for a human, he'll come by it jes' easy and nach'al. If it's not intinded, he won't come by it, no matter how much he hunts. Even if he did find it and it wa'n't intinded for him, it ud prove a curse. I'd be afraid of it myself.

"Lemme tell you something. Down there som'er's below Realitos in one of them sand mottes is an old dug well with six jack loads of Mexican silver in it. Nobody ain't a-going to get it neither. When the bandits captured it, they found the rangers was right on top of 'em; so they jes' pitched silver, dead burros, dead men, and all right down in the well. Then they got cleaned up theyselves.

"Well, one of the rangers a long time afterwards got on to the stuff somehow. So he went and bought the land what the well was on and set a bunch of Mexicans to clean it out.

After a while they struck bones. They hollered to the white man, who was bossing on top, and he hollered for them to come on up and let him down. They wa'n't a bit slow coming up.

"When the white man got down there, the first thing he done was to grab holt of a corner of an old *maleta* what he seen sticking out among the bones. He jerked it out and it had the dollars in it all right. Then he looked up and yelled to the Mexicans to pull. He hadn't more'n got the words outen his mouth when he seen a tall skiliton standing alongside the wall of that well. Its toes was right next to hisen and it must of been forty feet tall. It retched clear up to the top, and its face away up there was a-looking down at the white man. He couldn't take his eyes offen it, and all the way up while them Mexicans was a-pulling him slow and jerky he had to look that skiliton in the face. He forgot all about the *maleta* of money and dropped it back, and when he clumb out he was so weak they had to help him on his horse. They managed to get him home and put him to bed, and that night he died. Ain't nobody what I know of undertook to get them six jack loads since."

Pete arose to tinker with the coals around his bean-pot and put a little more water in it before going to bed.

"None of these new-fangled shiny pots cain't make frijoles black and sweet like an old iron kittle," he said.

"That's right, Pete. Hear that owl?"

"Owl in August, rain shore. Gonna be some juicy calf ribs this fall."

The Measure of a Wagon Rod

Belas Carter has a right to certain shady spots in Wichita Falls, where nowadays he can be found. Seven times through sun and storm he drove up the trail; then for years he was a hand on the three million acre X I T Ranch. Before he quit riding, he had this story to exchange around the chuck wagon.[3]

"While I was working for the X I T's in eighty-six, another feller named Jackson and me were sent to rep [4] with the L F D roundup over on the Pecos. With eight good horses apiece and a pack mule, we left Spring Lake before sunup, camped out that night, and the next afternoon about two o'clock came in sight of the Pecos River. We were making for a crossing about twenty miles above Roswell and expected to find the L F D wagon somewhere on the other side.

"It was all prairie country and we could see for a long piece. There was a crowd of horsemen down clost to the river, and directly we saw two of them light out in a neck-and-neck race. Then we saw them go back to where the others were. Next come a shot, one of the riders fell, and the rest of the outfit clumb the hurrikin decks of their ponies and took out. I never saw nor heard of them again.

"When Jackson and I got to the place, there lay a man shot through the neck and bleeding something turrible. We carried him over into the shade of a little abandoned 'dobe hut clost by, and then Jackson roped out the best horse in our string and headed for Roswell to fetch a doctor. The way he lit out I could see there wasn't any grass going to grow under his horse's feet.

"Meantime I'd clamped my thumb and finger on the artery where the blood was flowing from the wounded man. He begun to cough and jerked the artery loose, and the blood spurted so bad I thought he'd die the next minute. He thought so, too, and as soon as I clamped the artery back begun to talk about himself. Said as how he'd been a pretty bad man, that the jig was up now, and he wanted me to have his horses and guns and what money he had. He seemed awful grateful to me. He was a lone wolf, and, though I didn't know his name or he mine, I was the only friend he had in the world.

"He told me that he and four other men had robbed the bank at Monterrey, in Mexico. Before they could get across the Rio Grande, two of them were shot by *rurales,* and after they crossed over, some Texas officers got two more of them. That left him with a pack horse loaded with eighteen thousand dollars. Most of it was in gold doubloons worth about twenty dollars apiece. He decided that the safest place for him was Montana.

"But he was a hunted man, and traveling that way with a pack of gold was ticklish business. He had to keep dodging and couldn't course straight. When he got up on the Brazos about the mouth of the Clear Fork, he decided he'd better hide the money and get it later. So in the night he goes to a ranch house and stole the bean-pot and wagon rod from a chuck wagon. Then he buried the money.

"It took him a long time to tell me all this, for he kept having coughing spells and was awful tired. He got weaker and weaker, until about sundown I saw he was going. All the time I was a-holding his neck trying to stop the blood. Then he pulled me down and whispered: 'Cut open the handle of my war bag and you'll find a written description of where the money is buried. It's sure there and it's all yours.' Then he died.

"Well, before he got cold and while I was still sorter straightening him out, I heard a horse a-galloping. But it wasn't the doctor. The man introduced himself as the sheriff of Lincoln County. Lincoln County at that time took in about a fourth of New Mexico. It turned out that when Jackson got to Roswell and told that a man was shot, this sheriff got the word quicker than the doctor. In them days a man with a bullet hole in him generally didn't need a doctor nohow.

"It was about dark; so I pulled the saddles off mine and the dead man's horses, hobbled them, and turned them loose to graze. I laid the saddles beside the door of the old 'dobe house. Then after we'd et supper I told the sheriff the man had given his things to me.

" 'No,' says the sheriff, 'you ain't got no witnesses to that deal, and the law will have to take charge of the dead man's property.'

"I didn't say anything, but I was determined to have the description of the buried money. I'd already noticed that the dead man's war bag was something unusual with a fancy handle. We generally used just an old sack to carry our duffle in, you know. Along in the night I slipped up, cut the handle of the bag open, and pulled out a piece of paper that was rolled into a tight wad. I just stuck it down in my breeches to wait for daylight to read by.

"The doctor and Jackson rode in during the night. The next morning we got a spade from the L F D wagon, which was about five miles away, and buried the dead outlaw. Then the sheriff took his property and left.

"This is what the paper read: 'Buried one mile of Fort Belknap. 256 steps North of a little creek and 86 steps West of a prickly pear marked swallow-fork. $18,000 is in a bean-pot, the depth of a wagon rod with half the ring sticking out of the ground and three small rocks piled against the ring.'

"Fort Belknap, on the Brazos, was at that time abandoned. A wagon rod is forty-two inches long. So the money was buried pretty deep. I figgered the ear-mark in the prickly pear leaf would be easy to find. There was another description.

"It read: '2½ miles up that branch on the West bank under a large mesquite tree with a limb projecting West, buried 2½ feet deep in the crevice of a rock and wrapped in a Mexican blanket is a Winchester, a six-shooter, and 50 dollars Mexican gold. The crevice is filled up with loose gravel.'

"Well, this last wad of stuff I found. I had to look over a good many prickly pear bushes before I found a leaf marked swallow-fork. The ring on the end of the wagon rod was barely exposed above the surface of the ground. I had actually kicked away the three little rocks piled against it and was about to dig, when I saw a horseman. He saw me too. I gave the ring a stamp with my boot heel so as to drive it out of sight, and then moved to one side. The stranger riding up claimed to be the owner of the land. He did not want any trespassing and he followed me clear out of his range.

"I figgered that I'd go back in a day or two when he was off guard and dig up the money. The first time I went back was at night and I couldn't locate the wagon rod ring. The next time was in broad daylight, but I still couldn't locate it. I played hob when I kicked those rocks away and stamped it down. Still, I thought I'd get the bearing easy enough from that swallow-fork in the pear leaf. The leaf was gone. I guess some cow ate it up, for that was a hard year and cattle were chewing pear lots. You can figger out for yourself how many years back this was. I've been trying off and on ever since to locate the place. I made investigation and found out that the Monterrey bank was really robbed in the way the dying outlaw told me it had been. Within a mile of old Fort Belknap the $18,000 is still down in the ground at the end of a wagon rod waiting to be dug up.

"As the outlaw said when he gave it to me, it's all yours."

CHAPTER VI

POST HOLE BANKS

I used to hear the early settlers tell how a cow and calf was legal tender for ten dollars. Even if a man paid in gold—say forty dollars for a horse—he might say he gave "four cows and calves" for it. So the banks were logically just cowpens.—*Brushy Joe's Reminiscences.*

"THE methods of business were in keeping with the primitive conditions of society," says a chronicler of the open range of southern Texas.[1] "There were no banks in the country. Consequently every ranch home was the depository of more or less money. The coin, if of considerable amount, was put in saddle bags, morrals, etc., and secreted in remote corners of the house or up under the roof or it was buried; it could be brought forth from its hiding place as occasion demanded. . . . In buying stock the ranchmen brought the money in gold and silver to where the animals were to be received and there paid it out dollar by dollar. They generally carried the gold in leather belts buckled around their waists, but the silver, being more bulky, was carried in ducking sacks on a pack horse or mule. . . . It was a matter of current knowledge that one thousand dollars in silver weighed sixty-two and one-half pounds. . . .

"One time a rancher near the line between Karnes and Goliad counties decided to bury a considerable amount of money that he had on hand. Choosing an especially dark night, he went down to the cowpen and, after removing one of the fence posts, dropped his bag of gold in the post hole. He then replaced the post and went to bed satisfied that he had put his treasure where moth and rust could not corrupt nor thieves break through and steal. After a year or two had gone by, he needed the money and went to get it. He had failed to mark the particular post under which it was buried and time had obliterated all trace of his work. There was but one thing for him to do and he did it. He dug up post after post until he came to the right one, and by that time half his pen was torn down."

I don't know whether old Tolbert forgot his post hole or not. He ranched on the Frio, and, as the saying goes, was "stingy enough to skin a flea for its hide and taller." He would never kill a maverick no matter how hungry he was for meat, but would always brand it. He never bought sugar or molasses; "sow bosom," even of the saltiest variety, was a rare luxury; he and his men made out on "poor doe"—often jerked—javelina meat, and frijoles. When he "worked" and had an outfit to feed, he always instructed the *cocinero* to cook the bread early so that it would be cold and hard, and thus go further by the time the hands got to it. He distrusted banks, and during a good part of his life there were no banks to trust. The practice of keeping money on the premises suited him finely.

When he died, none of his money could be found. So, even till this day, people dig for it around the old ranch house. One evening about twenty years ago a man who was working on the place saw two strangers in a wagon go down a ravine that runs near the ranch. He thought they were deer hunters; but when they passed him on their way out next morning, he noted that one of them had a shotgun across his knees and that they avoided conversation. While riding down the ravine a few days later, the ranch hand found that the wagon tracks led from a fresh hole under a live oak tree and that near the hole were pieces of rusted steel hinges with marks of a cold chisel on them. However, not many people believe that the two strangers got Tolbert's money.

Berry got that—and he never hunted for it either. Years ago Berry bought the Tolbert ranch and went to live on it. One day when he had nothing else for his Mexican, Pedro, to do, he told him to put some new posts in the old corral fence. Pedro worked along digging holes and putting in new posts until near ten o'clock. Then at the third post to the east from the south gate he struck something so hard that it turned the edge of his spade. He was used to digging post holes in rocky soil with a crowbar to loosen it and a tin can to dip it out, and so he went to a mesquite tree where the tools were kept and got the crowbar.

But the crowbar would no more dig into the hard substance than the spade would. The sun was mighty hot anyhow; so the Mexican went up to the house where *el señor* Berry was whittling sticks on the gallery and told him that he couldn't dig any more. "Why, *señor,*" he said, "in that third hole from the south gate the devil has humped himself into a rock that nothing can get through."

Berry snorted around considerably at first, but directly he seemed to think of something and told his man, very well, not to dig any more but to saddle up and go out and bring in the main remuda. Now, only the day before they had had the main remuda in the pen and had caught out fresh mounts to keep in the little horse pasture. By this time the released horses would be scattered clear away on the back side of the pasture. The Mexican wondered why his *amo* wanted the remuda again. But it was none of his business. Well, the ride would take him all the rest of the day, and at least he would not have to dig any more post holes before *mañana.*

After Pedro had saddled his horse and drunk a *cafecita* for lunch and fooled away half an hour putting in new stirrup-leather strings and finally had got out of sight, Berry slouched down to the pens. He came back to his shade on the gallery and whittled for an hour or so longer until everything around the *jacal,* even the road-runners and Pedro's wife, was taking a siesta. Then he pulled off his spurs, which always dragged with a big clink when he walked, and went down to the pen again. The spade and the crowbar were where the Mexican had let them fall. Berry punched the crowbar down into the half-made hole. It almost bounced out of his hand, and he heard a kind of metallic thud. No, it was not flint-rock that had stopped the digging.

Berry went around back of the water trough to the huisache where his horse was tied and led him into the pen. Then he started to work. He began digging two or three feet out to one side of the hole. The dry ground was packed from the tramp of thousands of cattle and horses. He had to use the crowbar to loosen the soil. But it was no great task to remove a patch of earth two or three feet square and eighteen or twenty inches deep. Berry knew what he was about, and as he scraped the loosened earth out with his spade he could feel a flat metal surface that seemed to have rivets in it.

It was the lid of a chest. When he had uncovered it, Berry placed one of the new posts so that he could use it as a fulcrum for the crowbar. With that he levered up the end of the chest. As he suspected, it was too heavy and too tightly wedged in the soil for him to lift. He worked a chunk under the raised end of the chest and then looped a stout rope over it. Next, he mounted his horse and dallied the free end of the rope around the horn of his saddle. He had dragged cows out of the bog on that horse, and he knew that the chest was not so heavy as a cow. He had but fifty yards to drag it before he was in the brush, where undetected he could pry the lid off.

When the Mexican got back that night his *mujer* told him that Señor Berry had gone to San Antonio in the buckboard and that he had left word for the remuda to be turned back into the big pasture and for the repair of the corrals to be continued.

"They say" that the deposit Berry made at the Frost National Bank was a clean $17,000, nearly all in silver.

Will McNabb, of Matagorda, on the Gulf coast, while digging post holes a few years ago had a chance at a fortune probably greater than Berry's, but he did not realize it until too late—until he had lost the right post hole. In a certain hole about two feet deep he uncovered eighteen Spanish dollars dated from 1752 to 1814, all neatly stacked one on top of the other and held together by rust from what had probably been an iron brace. The land where he was running the fence is marshy, and six inches below the dollars McNabb hit soil spongy enough "to bog a saddle blanket." He knows now that he found only a few coins out of a heavy, iron-bound chest that sank long ago; at the time he found them, however, the idea of a whole chest of dollars far beneath them and the possibility of sinking a well to recover the fortune did not occur to him. When the idea did occur, he could not even guess what post marked the right place to dig. Putting a well down at every post in a fence line across a marsh would be altogether too expensive a gamble against even a chest—perhaps a pirate chest—of money.

"Uncle Jack" Wilkerson came to Texas from Mississippi in the early days and began ranching in Menard County. Before long he and his partner, Wat Key, were raising cattle by the thousands and were trailing them to New Orleans, Kansas, and other places where they could find a market. Each had a family and they lived close together. Wilkerson was banker for the firm. He would have nothing but gold; he kept it buried in an abandoned smokehouse made of adobe.

When in 1872 he left for New Orleans with a trail herd, Wat Key having gone to Abilene, Kansas, with another herd, Mrs. Wilkerson remained in charge of the ranch. One day a cow dog looking for a cool place in which to take a nap entered the old smokehouse and scratched up the money, scattering it over the floor. When Mrs. Wilkerson discovered the mischief, she called Mrs. Key and together they picked up the coins. It took them some time to do this; the gold amounted to about $10,000. After it was all gathered, they put it in a fruit jar; then, saying that she would go bury it in a place where it would not be disturbed, Mrs. Wilkerson went off alone towards the corrals.

Shortly after this Mrs. Wilkerson became very ill—so ill that she remained most of the time in a state of coma. She rapidly grew worse. Several times, however, she roused into a kind of dim consciousness, at which times she said to Mrs. Key, her constant attendant, "It seems there is something I want to tell you." That was as far as she ever got in telling whatever was on her mind. In a few days she died.

It was weeks later that Jack Wilkerson received word of his wife's death, and then he was nearly home. All Mrs. Key could tell him about the $10,000 in gold was that his wife had carried it in a fruit jar towards the corrals. First and last he dug up most of the posts, but only a few years ago he stated that, so far as he knew, the money had never been found. Other people have been searching for it ever since he moved away from the old ranch quarters.

"One day," relates a young ranchman of the Menard country, "I passed the Wilkerson ranch site while I was gathering some steers. The yard in which the house once stood looked as if it had been plowed. Every inch of dirt inside the ruined foundation had been spaded up. Some distance from the house I came upon a trench about twenty feet long. The upper end of it was shallow, but at the lower end it was perhaps three feet deep. There I noticed a thin flat rock lying on the floor of the trench. I got down off my horse and picked it up.

Under it was a hole about the size and shape that a half-gallon fruit jar would occupy. When I told Uncle Jack about it, he said that it was too far away from the house to have been the place where his wife buried the money. That's all I know; there are people in the country who talk as if they knew more."

CHAPTER VII

MIDAS ON A GOATSKIN

> High on a throne of royal state, which far
> Outshone the wealth of Ormus and of Ind.
>
> —*Paradise Lost.*

"HE's the second sorriest white man in Sabinal," my host said. "The sorriest white man keeps a Mexican woman without marrying her, but Dee Davis lawfully wedded his *pelada*. He's town scavenger, works at night, and sleeps most of the day. He'll probably be awake 'long about four o'clock this evening and more than ready to tell you the kind of yarns you want to hear."

We found Dee Davis just awaking from his siesta. He occupied a one-roomed shack and sat on a goatskin in the door, on the shady side of the house.

"I'm a great hand for goatskins," he said. "They make good settin' and they make good pallets."

I sat in a board-bottomed chair out on the hard, swept ground, shaded by an umbrella-China tree as well as by the wall. The shack was set back in a yard fenced with barbed wire. Within the same enclosure but farther towards the front was a little frame house occupied by Dee Davis's Mexican wife and their three or four half-breed children. The yard, or patio, was gay with red and orange zinnias and blue morning-glories. Out in a ramshackle picket corral to the rear a boy was playing with a burro.

"No, mister," went on Dee Davis, who had got strung out in no time, "I don't reckon anything ever would have come of my dad's picking up those silver bars if it hadn't of been for a surveyor over in Del Rio.

"You see, Dad and Uncle Ben were frontiersmen of the

old style and while they'd had a lot of experiences—yes, mister, a lot of experiences—they didn't know a thing about minerals. Well, along back in the eighties they took up some state land on Mud Creek and begun trying to farm a little. Mud Creek's east of Del Rio. The old Spanish crossing on Mud was worn deep and always washed, but it was still used a little. Well, one day not long after an awful rain, a reg'lar gully-washer and fence-lifter, Dad and Uncle Ben started to town. They were going down into the creek when, by heifers, what should show up right square in the old trail but the corner of some sort of metal bar. They got down out of their buggy and pried the bar out and then three other bars. The stuff was so heavy that after they put it in the buggy they had to walk and lead the horse. Instead of going on into town with it, they went back home. Well, they turned it over to Ma and then more or less forgot all about it, I guess—just went on struggling for a living.

"At that time I was still a kid and was away from home working for the San Antonio Land and Cattle Company, but I happened to ride in just a few days after the find. The Old Man and Uncle Ben never mentioned it, but Ma was so proud she was nearly busting, and as soon as I got inside the house she said she wanted to show me something. In one of the rooms was a bed with an old-timey covering on it that came down to the floor. She carried me to this bed, pulled up part of the cover that draped over to the floor, and told me to look. I looked, and, by heifers, there was bars as big as hogs. Yes, mister, as big as hogs.

"Nothing was done, however. We were a long ways from any kind of buying center and never saw anybody. As I said in the beginning, I don't know how long those bars might have stayed right there under that bed if it hadn't been for the surveyor. I won't call his name, because he's still alive and enjoying the fruits of his visit. My dad was a mighty interesting talker, and this surveyor used to come to see him just to hear him talk. Well, on one of these visits he stayed all night and slept on the bed that hid the bars. One of his shoes got under the bed, and next morning in stooping down to get it he saw the bars. At least that's the explanation he gave. Then, of course, he got the whole story as to how the bars came to be there and where they were dug up.

" 'What you going to do with 'em?' he asked Dad.

" 'Oh, I don't know,' Dad says to him. 'Nothing much, I guess. Ma here figgers the stuff might be silver, but I don't know what it is. More'n likely it's not anything worth having.'

" 'Well,' says the surveyor, 'you'd better let me get it assayed. I'm going down to Piedras Negras in my waggin next week and can take it along as well as not.'

"The upshot was that he took all the bars. Two or three months later when Dad saw him and asked him how the assay turned out, he kinder laughed and says, 'Aw pshaw, 'twan't nothing but babbitting.' Then he went on to explain how he'd left the whole caboodle down there to Piedras Negras because it wasn't worth hauling back.

"Well, it wasn't but a short time before we noticed this surveyor, who had been dog poor, was building a good house and buying land. He always seemed to have money and went right up. Also, he quit coming round to visit his old friend. Yes, mister, quit coming round.

"Some years went by and Dad died. The country had been consider'bly fenced up, though it's nothing but a ranch country yet, and the roads were changed. I was still follering cows, over in Old Mexico a good part of the time. Nobody was left out on Mud Creek. Uncle Ben had moved to Del Rio. One day when I was in there I asked him if he could go back to the old trail crossing on Mud. The idea of them bars and of there being more where they come from seemed to stick in my head.

" 'Sure, I can go to the crossing,' says Uncle Ben. 'It's right on the old Spanish Trail. Furthermore, it's plainly marked by the ruins of an old house on the east bank.'

" 'Well,' says I, 'we'll go over there sometime when we have a day to spare.'

"Finally, two or three years later, we got off. First we went up to the ruins of the house. About all left of it was a tumble-down stick-and-mud chimney.

"Uncle Ben and Dad, you understand, found the bars right down the bank from this place. Just across the creek, on the side next to Del Rio, was a motte of *palo blanco* [hackberry] trees. The day was awfully hot and we crossed back over there to eat our dinner under the shade and rest up a little before we dug any. About the time we got our horses staked, I noticed a little cloud in the northwest. In less than an hour it was raining pitchforks and bob-tailed heifer yearlings, and Mud Creek was tearing down with enough water to swim a steamboat. There was nothing for us to do but go back to Del Rio.

"I've never been back to hunt those bars since. That was close to forty years ago. A good part of that time I've been raising a family, but my youngest boy—the one out there fooling with the burro—is nine years old now. As soon as he's twelve and able to shift for himself a little, I'm going back into that country and make several investigations."

Old Dee shifted his position on the goatskin.

"My eyes won't stand much light," he explained. "I have worked so long at night that I can see better in the darkness than in the daylight."

I noticed that his eyes were weak, but they had a strange light in them. It was very pleasant as we sat there in the shade, by the bright zinnias and the soft morning-glories. Pretty soon Dee Davis would have to milk his cow and then in the dark do his work as scavenger for the town. Still there was no hurry. Dee Davis's mind was far away from scavenger filth. He went on.

"You see, the old Spanish trail crossed over into Texas from Mexico at the mouth of the Pecos River, came on east, circling Seminole Hill just west of Devil's River, on across Mud Creek, and then finally to San Antonio. From there it went to New Orleans. It was the route used by the *antiguas* for carrying their gold and silver out of Mexico to New Orleans. The country was full of Indians; it's still full of dead Spaniards and of bullion and bags of money that the Indians captured and buried or caused the original owners to bury.

"Seminole Hill hides a lot of that treasure. They say that a big jag of Quantrill's loot is located about Seminole too, but I never took much stock in this guerrilla treasure. But listen, mister, and I'll tell you about something that I do take consider'ble stock in.

"Last winter an old Mexican *pastor* named Santiago was staying here in Sabinal with some of his *parientes*. He's a little bit kin to my wife. Now, about nine-tenths of the time a sheepherder don't have a thing to do but explore every cave and examine every rock his sheep get close to. Santiago had a dog that did most of the actual herding. Well, two years ago this fall he was herding sheep about Seminole Hill.

"According to his story—and I don't doubt his word—he went pirooting into a cave one day and stepped right on top of more money than he'd ever seen before all put together. It was just laying there on the floor, some of it stacked up and some of it scattered around every which way. He begun to gather some of it up and had put three pieces in his *hato*—a kind of wallet, you know, that *pastores* carry their provisions in—when he heard the terriblest noise behind him he had ever heard in all his born days. He said it was like the sounds of trace-chains rattling, and dried cowhides being drug at the end of a rope, and panther yells, and the groans of a dying man all mixed up. He was scared half out of his skin. He got out of the cave as fast as his legs would carry him.

"An hour or so later, when he'd kinder collected his wits, he discovered three of the coins still in his *hato.* They were old square 'dobe dollars like the Spanish used to make. As soon as he got a chance, he took them to Villa Acuna across the river from Del Rio, and there a barkeeper traded him three bottles of beer and three silver dollars, American, for them.

"Well, you know how superstitious Mexicans are. Wild horses couldn't drag old Santiago back inside that cave, but he promised to take me out there and show me the mouth of it. We were just waiting for milder weather when somebody sent in here and got him to herd sheep. Maybe he'll be back this winter. If he is, we'll go out to the cave. It won't take but a day."

Dee Davis rolled another cigarette from his supply of Black Horse leaf tobacco and corn shucks. His Mexican wife, plump and easy-going, came out into the yard and began watering the flowers from a tin can. He hardly noticed her, though as he glanced in her direction he seemed to inhale his smoke with a trifle more of deliberation. He was a spare man, and gray moustaches that drooped in Western sheriff style hid only partly a certain nervousness of the facial muscles; yet his few gestures and low voice were as deliberate—and as natural—as the flop of a burro's ears.

"What I'd rather get at than Santiago's cave," he resumed, "is that old smelter across the Rio Grande in Mexico just below the mouth of the Pecos. That smelter wasn't put there to grind corn on, or to boil frijoles in, or to roast goat ribs over, or anything like that. No, mister, not for anything like that.

"It's kinder under a bluff that fronts the river. I know one ranchman who had an expert mining engineer with him, and they spent a whole week exploring up and down the bluff and back in the mountains. I could of told them in a minute that the mine was not above the mouth of the Pecos. If it had of been above, the trails made by miners carrying *parihuelas* could still be seen. I've peered over every foot of that ground and not a *parihuela* trace is there. You don't know what a *parihuela* is? Well, it's a kind of hod, shaped like a stretcher, with a pair of handles in front and a pair behind so two men can carry it. That's what the slave Indians carried ore on.

"No, sir, the mine that supplied that smelter—and it was a big mine—was below the mouth of the Pecos. It's covered up now by a bed of gravel that has probably washed in there during the last eighty or ninety years. All a man has to do to uncover the shaft is to take a few teams and scrapers and clear out the gravel. The mouth of the shaft will then be as plain as daylight. That will take a little capital. You ought to do this. I wish you would. All I want is a third for my information.

"Now, there is an old lost mine away back in the Santa Rosa Mountains that the Mexicans called El Lipano. The story goes that the Lipan Indians used to work it. It was gold and as rich as twenty-dollar gold pieces. El Lipano didn't have no smelter. The Lipans didn't need one.

"And I want to tell you that those Lipan Indians could smell gold as far as a hungry coyote can smell fresh liver. Yes, mister, they could smell it. One time out there in the Big Bend an old-timey Lipan came to D. C. Bourland's ranch and says to him, 'Show me the *tinaja* I'm looking for and I'll show you the gold.' He got down on his hands and knees and showed how his people used to pound out gold ornaments in the rock *tinajas* across the Rio Grande from Reagan Canyon.

"Now that long bluff overlooking the lost mine in the gravel I was just speaking about hides something worth while. I guess maybe you never met old Uncle Dick Sanders. I met him the first time while I was driving through the Indian Territory up the trail to Dodge. He was government interpreter for the Comanche Indians at Fort Sill and was a great hombre among them.

"Well, several years ago an old, old Comanche who was dying sent for Uncle Dick.

" 'I'm dying,' the Comanche says. 'I want nothing more on this earth. You can do nothing for me. But you have been a true friend to me and my people. Before I leave, I want to do you a favor.'

"Then the old Indian, as Uncle Dick Sanders reported the facts to me, went on to tell how when he was a young buck he was with a party raiding horses below the Rio Grande. He said that while they were on a long bluff just south of the river they saw a Spanish cart train winding among the mountains. The soldiers to guard it were riding ahead, and while they were going down into a canyon out of sight, the Comanches made a dash, cut off three *carretas,* and killed the drivers.

"There wasn't a thing in the *carretas* but rawhide bags full of gold and silver coins. Well, this disgusted the Comanches mightily. Yes, mister, disgusted them. They might make an ornament out of a coin now and then, but they didn't know how to trade with money. They traded with buffalo robes and horses.

"So what they did now with the rawhide sacks was to cut them open and pour the gold and silver into some deep cracks they happened to notice in the long bluff. Two or three of the sacks, though, they brought over to this side of the Rio Grande and hid in a hole. Then they piled rocks over the hole. This place was between two forks, the old Comanche said, one a running river walled with rock and the other a deep, dry canyon. Not far below where the canyon emptied into the river, the river itself emptied into the Rio Grande.

"After the Comanche got through explaining all this to Uncle Dick Sanders, he asked for a lump of charcoal and a dressed deerskin. Then he drew on the skin a sketch of the Rio Grande, the bluffs to the south, a stream with a west prong coming in from the north, and the place of the buried coins. Of course he didn't put names on the map. The only name he knew was Rio Grande del Norte. When Sanders came down here looking for the Comanche stuff, of course he brought the map with him and he showed it to me. The charcoal lines had splotched until you could hardly trace them, but Sanders had got an Indian to trace them over with a kind of greenish paint.

"Uncle Dick had some sort of theory that the Comanche had mistook the Frio River for the Rio Grande. Naturally he hadn't got very far in locating the ground, much less the money. He was disgusted with the whole business. Told me I could use his information and have whatever I found. I'm satisfied that Devil's River and Painted Cave Canyon are the forks that the Indians hid the *maletas* of money between, and the long bluff on the south side of the Rio Grande where they poured coins into the chinks is the same bluff I've been talking about."

Dee Davis got up, reached for a stick, squatted on the ground, and outlined the deerskin map that Uncle Dick Sanders had shown him. Then he sat down again on the goatskin and contemplated the map in silence.

It was wonderfully pleasant sitting there in the shade, the shadows growing longer and the evening growing cooler, listening—whether to Dee Davis or to a hummingbird in the morning-glories. I did not want the tales to stop. I remarked that I had just been out in the Big Bend country and had camped on Reagan Canyon, famed for its relation to the Lost Nigger Mine. I expected that Dee Davis would know something about this. He did.

"Now listen," he interposed in his soft voice, "I don't expect you to tell me all you know about the Lost Nigger Mine, and I know some things I can't tell you. You'll understand that. You see I was *vaciero* for a string of *pastores* in that very country and got a good deal farther into the mountains, I guess, than any of the Reagans ever got. You may not believe me, but I'll swear on a stack of Bibles as high as your head that I can lead you straight to the nigger who found the mine. Of course I can't tell you where he is. You'll understand that. It was this away.

"One morning the Reagans sent Bill Kelley—that's the nigger's name—to hunt a horse that had got away with the saddle on. A few hours later Jim Reagan rode up on the nigger and asked him if he had found the horse.

" 'No, sah,' the nigger says, 'but jes' looky here, Mister Jim, I'se foun' a gold mine.'

" 'Damn your soul,' says Jim Reagan, 'we're not paying you to hunt gold mines. Pull your freight and bring in that horse.'

"Yes, mister, that's the way Jim Reagan took the news of the greatest gold mine that's ever been found in the Southwest—but he repented a million times afterwards.

"Well, as you've no doubt heard, the nigger got wind of how he was going to be pitched into the Rio Grande and so that night he lit a shuck on one of the Reagan horses. Then a good while afterwards when the Reagans found out how they'd played the wilds in running off, you might say, the goose that laid the golden egg, they started in to trail him down. No telling how many thousands of dollars they did spend trying to locate Nigger Bill—the only man who could put his hand on the gold.

"I've knowed a lot of the men who looked for the Lost Nigger Mine. Not one of them has gone to the right place. One other thing I'll tell you. Go to that round mountain down in the *vegas* on the Mexican side just opposite the old Reagan camp. They call this mountain El Diablo, also Niggerhead; some calls it El Capitan. Well, about half way up it is a kind of shelf, or mesa, maybe two acres wide. On this shelf close back against the mountain wall is a *chapote* bush. Look under that *chapote* and you'll see a hole about the size of an old-timey dug well. Look down this hole and you'll see an old ladder—the kind made without nails, rungs being tied on the poles with rawhide and the fibre of Spanish dagger. Well, right by that hole, back a little and sorter hid behind the *chapote,* I once upon a time found a *macapal.* I guess you want me to tell you what that is. It's a kind of basket in which Mexican miners used to carry up their ore. It's fastened on the head and shoulders.

"Now, I never heard of a *macapal* being used to haul water up in. And I didn't see any water in that hole. No, mister, I didn't see any water.

"As I said, as soon as my boy gets to be twelve years old—he's nine now—I'm going out in that country and use some of the knowledge I've been accumulating."

Dee Davis leaned over and began lacing the brogan shoes on his stockingless feet. It was about time for him to begin work. But I was loath to leave. How pleasant it was there! Maybe Dee Davis is "the second sorriest white man in Sabinal." I don't know, but it seemed to me then, and it seems to me still, that there are many ways of living worse than the way of this village scavenger with a soft goatskin to sit on, and aromatic Black Horse tobacco to inhale leisurely through a clean white shuck, and bright zinnias and blue morning-glories in the door-yard, and long siestas while the shadows of evening lengthen to soften the light of day, and an easy-going Mexican wife, and playing around a patient burro out in the corral an urchin that will be twelve *mañana,* as it were, and then——. Then silver bars out of Mud Creek as big as hogs—and heaps of old square 'dobe dollars in Santiago's cave on Seminole Hill—and Uncle Dick Sanders' gold in the chinks of the long bluff across the Rio Grande—and somewhere in the gravel down under the bluff a rich mine that a few mules and scrapers might uncover in a day—and, maybe so, the golden Lipano out in the Santa Rosas beyond—and, certainly and above all, the great Lost Nigger Mine of free gold far up the Rio Bravo in the solitude of the Big Bend.

Dee Davis is just one of Coronado's children.

CHAPTER VIII

THE LOST NIGGER MINE

Gold is where you find it.—*Prospectors' Proverb.*

AFTER twisting southward all the way from its source high up in the Rocky Mountains of Colorado, the Rio Grande del Norte veers southeast at El Paso as if to shoot straight for the Gulf of Mexico. For three hundred miles—measuring as the crow flies and not as the stream winds—it keeps this general course; then it turns sharply to cut northeast for more than a hundred miles before it resumes its gulfward trend. The great bow thus made, its string the Southern Pacific Railroad and its arc the Rio Grande, is known as the Big Bend country of Texas.

It is a land of mountain, canyon, and mesa grown over with greasewood, coarse chino grass, dagger, and thorned brush. There are, too, stretches of fine grazing land. Here ranching still is—and forever will be—undisturbed by the plow; here land is measured by the section rather than by the acre, and anything less than forty sections makes "a little ranch." Although a majority of the pastures are abundantly watered, certain far-stretched areas are devoid both of surface springs and of underground veins within tapping distance of the drill. A solitary *tinaja* called Pata de Venado, at which a man and his horse can barely get a drink, is famed two hundred miles away.

Like all soil that has not been transformed by so-called improvements, the Big Bend is a land of traditions. Outlandish pictures painted on the sides of caves by aborigines tell a story that no white man can now decipher; rings of blackened rocks show where Apaches roasted sotol for food. Somewhere in this jagged and gashed land legend has placed a lost canyon, its broad floor carpeted with grass that is always green and watered by gushing springs, its palisaded walls imprisoning a herd of buffaloes—the only buffaloes in all probability that ever entered that rough, thirsty region. Somewhere in this land credulity has fixed a petrified forest with tree trunks "seven hundred feet long." Down in the most remote rincon of the Bend, the Chisos (Phantom) Mountains hide a Spanish mine, the entrance to which was once visible at sunrise from the ancient church of San Vicente across the river. The Big Bend is a strange country where strange characters have lived and yet live—characters like Alice Stillwell Henderson, who single-handed dared fifty Mexican cow thieves, and like "Don Militón" Favor, whose ranch on the Chihuahua Trail was a fort and who alone made treaties with savage tribes. It is a country where strange things have happened and where anything may yet happen.

Across the awful gorge that bounds the Big Bend, the wild, broken country belongs to the deer, the javelina, the panther, and the bear. A great stretch of it a hundred miles long and sixty miles broad has not since the Madero Revolution twenty years ago desolated northern Mexico contained enough cattle or horses to tempt even the bandits. Vacant and silent, it, too, is a land where many strange things have happened and where anything may happen—a land of stories.

And the greatest story of all the Big Bend country and of all the Mexican wasteland across the river is the story of the Lost Nigger Mine. It is not an old story like that of the San Saba Mine; some of the chief contributors to it are still living. The mine is supposed to be either above the mouth of Reagan Canyon, in Brewster County, on the Texas side, or else across the river in the Ladrones (Robber) Mountains, the odds favoring Mexico. From the railroad it is seventy-five miles, with only one house to pass, to the mouth of Reagan Canyon.

I have led a pack mule down the trail that follows Reagan Canyon and I have camped where the dark waters of the Rio

Bravo swishing under the stars make that part of the world seem as remote as it was when Cabeza de Vaca wandered across Texas.

"How far is it up the river to your next neighbor?" I asked Rol Rutledge, who had guided me down to his camp.

"Well," he answered, "it's about fifteen miles on up to where Juan Español lives. He killed his goat-herder not long ago because the herder wouldn't tell him where he had picked up a strange rock."

"How far is it down the river to your next neighbor?"

"There ain't no neighbors down the river. The way a man has to ride, you go on down for fifty miles without finding a crossing. Then you come to where three trails just pitch off the bluff—Las Veredas Coloradas, the Mexicans call them. This is the famous Shafter Crossing."

Along about 1884 the four Reagan brothers, John, Jim, Frank, and Lee, moved their cattle from the Pecos down San Francisco Creek, through Bullis Gap, and against the Rio Grande. The country was all open and unclaimed. The canyon near which they made headquarters camp soon became known as Reagan Canyon. Here the river could be forded, and here it was forded quite frequently. There was stock to drive into Mexico from Texas and there was stock to be driven into Texas from Mexico—and there were no inspectors. Occasionally the Reagans used also the *paso* at the mouth of Maravillas Canyon fifteen miles up the river. The Southern Pacific Railroad connecting San Antonio with El Paso had just been completed. Sometimes the Reagans went to Sanderson, sometimes to Dryden; either place afforded shipping pens, groceries, and liquor.

On one trip to Dryden, one of the Reagan boys saw a Seminole negro who wanted a job and hired him. He had been raised in the Santa Rosa Mountains below Del Rio; he spoke only a few words of English, and his Spanish was very corrupt. His name was Bill Kelley, but he passed variously as Seminole Bill, Nigger Bill, and Santa Rosa. This was in 1887.

One night, while the outfit was sitting around the camp fire, Seminole Bill announced that he had that day found a gold mine. The announcement produced nothing but scornful jeers from the white men. Probably not one of them had ever in his life given a moment's thought to mining.

The next morning Lee Reagan and the Seminole went across the river to hunt a couple of horses that had got away on that side. The horses were valuable, they had been missing several weeks, and the Reagans had already put in considerable time trying to find them. The two men separated in order to cut for sign and did not see anything of each other until along in the shank of the afternoon. Then they pulled up together on a little ridge a few miles out from the ford, dismounting to shift saddles and exchange observations. Neither had found the horses.

"But, Mr. Lee," Seminole Bill said with animation, "we's right heah clost to that gold mine. It's not more'n half a mile ovah yander. Please lemme show it to you." And he gave a sweep with his arm towards the east that Lee Reagan many a time afterwards wished he could recall more accurately. Here the story prongs.

Lee Reagan's own version [1] is that he asked the nigger why he did not bring some of the gold with him.

The reply was, " 'Cause I'se always heared that if you's out by yerself and finds gold and takes some of it, you's sure to die befo' you gets in."

The more common version of the story—said to be the nigger's version—is that the nigger reached into a morral (a fibre bag commonly carried on the horn of the saddle by horsemen of the Southwest), drew out a rock, and handed it to Reagan.

"Damn your gold," Reagan is reported to have growled, hurling the rock as far as he could. "We're not feeding you to hunt mines. What you're to hunt is horses."

At any rate, Reagan did not go with the nigger to see the mine and nothing more was said concerning it. Also, as developments will reveal, the nigger evidently defied the superstition of death by carrying away a piece of ore.

A few days later, when the Reagan outfit was in Dryden with a bunch of cattle, Seminole Bill hopped a freight train for San Antonio. He knew Lock Campbell, conductor on the passenger train between San Antonio and Sanderson, for he had once given Campbell a deer ham, and had found out that he was interested in minerals. In San Antonio he tried to find Campbell but missed him. However, he met a friend of Campbell's and turned over to him a piece of ore to give the conductor.

Then Bill returned to Dryden. There he found the Reagans, who had got back from delivering their cattle to the north. The outfit went on down into the Big Bend. In a week or two somebody rode up to Sanderson to get the mail. He brought back a letter addressed to Bill Kelley. It bore a San Antonio postmark. Here again the story prongs.

According to a widespread belief, one of the Reagans opened the letter during Bill's absence and read that a sample of ore furnished by him was very rich. All that the nigger had to do in order to become immensely wealthy was to reveal the mine. As the story goes on, the Reagans, now thoroughly aroused and very much averse to any stranger's "horning in" on a gold mine in their territory, shot the nigger that night and pitched his body into the Rio Grande. They felt sure, it is claimed, that they could go to the mine, Lee Reagan having been very near it and having seen the direction in which the nigger pointed. But this oft-told tale is not true—not all of it at least. About this time a dead negro was pitched into the Rio Grande of the Big Bend all right, but that was because the Flint gang, for whom he was working, were preparing to forsake the business of smuggling stolen cattle in order to rob a train. They intercepted a letter showing that their negro was planning to give them away, and so they "pecosed" him in the Rio Bravo.

No, the Reagans most assuredly did not kill Nigger Bill, but one night, about the time the letter came, he took the first horse he could get hold of, a plug staked out to wrangle the remuda on, and "pulled his freight for the tules." Why he left so secretly and suddenly has never been satisfactorily explained. Lee Reagan's story is that his brother Jim read the letter to Seminole Bill and then joined the other white men in laughing at him. "The idea of there being any gold in that God-forsaken country and the idea of Bill's knowing anything about gold even if he saw it!" And it was the morning after this laughing bout, according to Lee Reagan, that the nigger turned up missing.

A more commonly believed explanation is that the Mexican *cocinero* (cook), a friend of Bill's, warned him that the Reagans had read a letter about gold and that he had overheard them planning to have the nigger show them the mine and then kill him. This hardly jibes with the surprised manner of the Reagans at a later date when Lock Campbell told them what he knew. Yet another common explanation is that the Reagans had a quantity of "wet" stock on their range, that they knew Nigger Bill was going to "roam" pretty soon, and that in order to rid themselves of a dangerous witness they planned to put him out of the way, with the result that the *cocinero* overheard them and warned the intended victim.

At any rate, the nigger left suddenly on a "borrowed" horse and left in the night.

"When Bill did not show up with the remuda after big daylight," says Jim Reagan in a letter, "we figured he had made a run at a horse and maybe got crippled by his own pony's falling with him. So we all got out and caught some horses and began looking for Bill. The other hands scattered and I took the trail. To my surprise I soon saw that the nigger was headed for Dryden. I trailed him to what we called Dinner Creek. There the tracks turned off down a draw towards the Shafter Crossing on the Rio Grande. I figured that this was just a blind on the nigger's part to throw somebody off his trail and so rode on to Dryden to catch him. But no Bill came there."

The Stillwells were at that time ranching at the Huerfanito in northern Coahuila about a day's ride south of the Shafter Crossing. According to a story they later told, Seminole Bill came to their ranch afoot soon after he had left the Reagan camp. He had a morral full of rocks, which he showed, explaining that he had picked them up out of something like an enormous "ant bed." The elder Stillwell had been a mining engineer and at once recognized in the rocks gold ore of a very high grade. The nigger described in a rather vague way where the outcropping was to be found and then went on towards his people in the Santa Rosa Mountains. The Stillwells were interested enough to take several looks for the mine. Indeed, as will appear in its place, one of them may have found it many years later.

The real story of the hunt for the Lost Nigger Mine, however, will always be associated with the name of Lock Campbell, the railroad conductor. He alone of all the hunters seems to have known what he was about; he alone was both persistent and consistent. He was an extraordinary character. I am glad that I got his story from his own lips before he died, in 1926, less than a year after I visited him.

I found Lock Campbell living in a great house out on Broadway, San Antonio, set deep back from the street with fine trees in front and fine pictures within. He had collected a number of original paintings and his interest in art was genuine—but it was not a passion as was the Lost Nigger Mine. I derived from his beautifully correct speech that he was foreign-bred, and upon question he told me that he was born of Scotch parents and educated in Canada. In the course of our conversation I learned that he had made a neat profit on some Florida lands; that he owned some oil leases in South Texas—"right where the biggest field in Texas is due to come in"; and that he possessed a lot of stock in a dormant copper mine near Lordsburg, New Mexico—a mine "as rich as the famous Number 85 in the same vicinity." Mr. Campbell had passed his seventieth birthday and was retired from railroad duty, but neither age nor retirement had withered his sanguine nature.

"And yet," Campbell went on, "it looks as if something fatalistic hounds every man who follows after these mines. Take that Lordsburg mine. The only reason our company never struck the vein is that we did not go deep enough. Finally one man proposed to advance enough money to sink the shafts on down—and the day after the proposition had been agreed upon, a woman killed him in Mexico. Years passed; another man took up the work and operations were about to be resumed. Then a saddle galled the inside of the man's leg and in a week's time he was dead from blood poisoning.

"A dozen times in my life I have been on the verge of making an immense strike only to have fate step in and thwart me. In 1877 I was mining in Nevada. One day an old-time friend hobbled into camp and said to me: 'Down on the Sitka River in Alaska the Indians are carrying out free gold by the bagful and trading with it. Wait until I get well and we'll go up there together and get rich in one month.' I knew that he could never get well and that there was no use of my waiting. I struck north, was held up at Victoria for months, and then in disgust came back to the States. Ten years after that I picked up the paper in San Francisco one morning and read a headline story about how free gold had just been discovered on the Sitka.

"Even while I was looking for the Lost Nigger Mine, I

missed getting rich by a hair's breadth. Another man and I had hired a prospector to look for the mine. The prospector took cold feet about going across the Rio Grande—was afraid of bandits—and in consequence spent all his time on the Texas side, going as far up the river as Terlingua—clean out of the Reagan nigger's range. I did not see him for six months, but my partner did.

"Well, one day years afterwards, when the Terlingua quicksilver mines were making their owners rich, this former partner casually remarked to me: 'You and I could have had those quicksilver mines if we had had sense enough to take them.'

" 'How's that?' I asked.

" 'Why,' he answered, 'didn't that old prospector ever tell you about finding cinnabar around Terlingua?' Then he went on to tell me how he had met our prospector right after he had found all kinds of cinnabar and had instructed him to leave it alone. I asked why in the world he had let such an opportunity go by. His explanation was that one day in a Denver hotel he had overheard three strangers agree that cinnabar was no good. That is the way chance works."

But to get to Campbell's fatalistic search for the Lost Nigger Mine itself. In the first place, when Seminole Bill left a piece of ore in San Antonio to be delivered to Campbell, he explicitly stated that he had found it on the Mexico side of the Rio Grande. He described the ore as being in a canyon among rocks sticking up so close to each other that if a man tried to ride to it he would break his horse's leg. He was hunting horses, he said, when he saw the sun from the west shining upon it. A shower had fallen not long before and there was some water in the canyon. Some of the gold was shining under the water, which was very shallow.

After Campbell got the rock, he had a piece of it assayed. It ran "about $80,000 to the ton." When it is remembered that ore running $20 to the ton is considered good pay, one will realize what even a small quantity of the nigger's rock would be worth. At the time I talked with Campbell, he still had a piece of the rock. It was iron-stained quartz closely sprinkled with fine gold.

For two years after he had the assay made Campbell kept his counsel while he searched. The Reagans, it seems, had about quit the Big Bend. Campbell himself went down into the Maravillas and Reagan canyons. He grubstaked prospectors and kept them looking. Above all, he tried to trail Seminole Bill down. Then one day he saw Jim Reagan on the train going to the cattlemen's convention at San Antonio. He told Jim how he had secured the ore from the nigger, how it had assayed, and how for two years he had been prospecting on the quiet. He went to his grip and pulled out the piece of ore for Jim to see. Jim Reagan was so excited that he could hardly wait to catch the next train back west. He was desperately afraid that somebody would find the mine before he could reach the Big Bend.

"Why, I never believed before that Bill had a thing!" he explained. "I can go to it in an hour's time after we cross the Rio Grande. I know where we were camped when the nigger found the mine. I know that Lee can go straight to the hill he stood on when the nigger tried to show him the mine and pointed towards it. I know the canyons of that country as well as I know the feel of my own boots."

Reagan offered to take Campbell along on the hunt, but the conductor could not get off, and, besides, horseback riding "tore him all to pieces." It was agreed, however, that he should have an interest in the findings.

As soon as the train pulled into San Antonio, Jim Reagan went to the telegraph office and wired his brother Frank at Del Rio to ship eight good horses to Dryden and have an outfit ready. Jim did not tarry even long enough to bet on the horse races out at the fair grounds. Two days later Lee, John, and Frank Reagan, together with Henry Ware, a kind of jack-leg prospector, were riding towards their old stamping ground in the Big Bend.

When it came to mining, not one of these men knew split beans from coffee, but they were all armed with hammers, and for a day and a half they went about hammering on the boulders (float) that litter the gravel ridges east of where the nigger met Lee Reagan on that memorable afternoon. It seems strange that they should have looked anywhere except in canyons, but there was no canyon due east and within half a mile of the hill that Lee designated as the place where he talked with Bill Kelley. Perhaps Lee had forgotten. Anyhow, the hunters decided that the only sensible thing for them to do was to get back out and try to find Bill Kelley.

They got out. And now began a concerted search for the gold-finding nigger that Lock Campbell joined in and kept up almost as long as he lived and that the three living Reagan brothers have not yet abandoned. Two of the Reagans went to the Seminole settlement in Coahuila, learned that Bill's mother was a hundred miles away, and went there. The old woman said that her son had come home with some rocks rich in gold, that the rocks had become lost, for the family had moved around two or three times, and that soon after his return Bill had left again. She did not know where he had gone. After he went away, she had never heard a word either from or about him.

Meantime Lock Campbell advertised far and wide, offering a reward for accurate knowledge of Seminole Bill's whereabouts. He had several letters about a negro living with the Creek Indians in Oklahoma and claiming to be Bill Kelley; Campbell offered to pay this negro good money if he would come to San Antonio and prove that he was Bill Kelley. The negro never appeared. Another negro asserted that for fifty dollars he would produce Bill Kelley in Brackettville—where some of the Seminoles live now. Campbell told him to produce Kelley and he would pay a hundred dollars. The negro never came to claim the money. He did not have to produce Kelley in order to get money; several men advanced him various sums on the mere promise to produce the one living being who knew the whereabouts of the Lost Nigger Mine. Campbell and the Reagans were by no means alone in their hunt for the nigger.

One day a strange negro brought some gold nuggets to a man named Dickey living in Eagle Pass and for pay offered to show him where he had secured them. Dickey had just lost about a thousand dollars looking for some wildcat mine and he came near kicking the negro out of his place. Later he felt sure that the negro was Bill Kelley—but he could not find him.

While Wes Burton was in Monterrey years ago, he came to know a negro who went by the name of Pablo; and in Burton's mind Pablo was undoubtedly Seminole Bill. He had a peculiar scar on his face, was a desperate drinker and gambler, and, whooping and shooting, frequently rode a fine horse down the street at full speed. He always had plenty of money, and it was well known that he got it by bringing in about every three months a pack horse loaded with gold ore.

On the other hand, it is asserted that when the Spanish-American War broke out Bill Kelley enlisted in the United States Army and was eventually killed in the Philippine Islands. Somebody trailed Bill Kelley to Louisiana and found that he had died there. I have a letter, written in April, 1930, from a man in Marietta, Ohio, assuring me that he has Bill Kelley located in Arizona. A month after this letter came to hand, a man appeared in my office with a communication from a Spanish lady in Durango reciting how a Mexican who worked on her hacienda before the Madero Revolution swore that he had killed Bill Kelley near Del Rio for attempting to steal a horse—after the negro had shown him the mine, on the Texas side. Bill Kelley has become as much a legend as the gold he found, and his vanishing is as uncertain as the vanishing of his mine.

One of the prospectors that Campbell grubstaked was known as "Old Missouri." He is remembered principally for the number of burros he let stray off, and people got to saying that Campbell had stocked the Big Bend with burros. Old Missouri had instructions to scout out from the Reagan camp site only as far as a burro could go and come in a day. Solitary, often not seeing a human being for weeks and months at a time, Old Missouri examined ledge after ledge, canyon after canyon. He was looking for a kind of blue stone, he said. Then one day he found it. He put some of it on his pack burro, loaded on his blankets and a little grub, and started for Sanderson.

On the road he became very sick. He was feeble anyhow. When he reached Sanderson he was too sick to talk. What the "blue" stone in his pack meant or where he had found it, nobody ever ascertained. Old Missouri died before he could explain.

The years passed. Jim Reagan went out into Arizona, "where there are sure enough prospectors," and engaged the services of an expert who "knew rocks from A to izzard." However, when this expert prospector got down into the Big Bend, he swore that he would be ashamed for any mining man who knew straight up to catch him looking for gold in such a country. His liquor played out before he reached the Rio Grande; he was very fond of liquor; and he refused absolutely to prospect.

But "gold is where you find it." And, after all, there are granitic outcrops in the Reagan Canyon country, and the Chisos Mountains are well mineralized. If ever there was any doubt concerning Seminole Bill's gold, old Finky put that doubt to rest.

On July 19, 1899, five men signed a written agreement to hunt and develop the Lost Nigger Mine. The signers were Lock Campbell, railroad conductor; J. G. (Jim) Reagan, who

One of the prospectors was known as Old Missouri.

at that time ran a saloon in Sanderson; D. C. Bourland, who was ranching down in the Reagan Canyon country; O. L. Mueller, a German who had taken out naturalization papers in Mexico and was ranching across from Bourland; and John Finky, prospector. Finky was to be grubstaked by the other four men while he looked for the mine and was to have a fifth interest in it.

And Finky found the mine. He made his headquarters with the Bourlands, who lived close to the river. He hunted consistently across the river in the Ladrones Mountains, which at that time were occasionally visited by Mexican bandits. Every evening, the Bourlands said, Finky would come in from his search dead tired, absolutely worn out. Then one evening, earlier than usual, he came in with his face all lit up, a joyous spring in his step, and a rusty old rifle in his hand. He said that he had found the rifle by the side of a dried-up skeleton of a man and that in less than an hour, not three hundred yards away from the skeleton, he had found gold. He said that he had taken about ten pounds of the ore in his morral and started back with it but buried it before he reached the river. He did not tell why he buried it. That he had reason enough for not producing it will come out later. No phrenologist or archaeologist was by to examine the skull of the skeleton, but in the opinion of Finky it was a negro's skull.

It happened that about the time Finky arrived at ranch quarters that afternoon, Bourland rode in to get a fresh horse. A drouth was on the country, his cattle were dying, and he was working desperately night and day to save them. Finky pleaded with him to go right then to see the mine he had found. "And we had better take along an extra hand and all the artillery we can raise," he said.

But Bourland simply could not leave his cattle. The mine had waited there ten thousand years, he figured, and it could wait a day or two longer. He often declared later that had he known how things were going to turn out he would have let all his cattle die rather than have delayed an hour in going across the river with Finky.

The rest of the story can best be told in Lock Campbell's own words.

"The next morning Finky's face was swollen as if it had been poisoned. Mrs. Bourland doctored it with potato poultice, prickly pear poultice, bread and milk poultice, applications of kerosene oil, and every other remedy she could think of, but it got no better. They brought Finky to Sanderson. When my train ran into the station, Jim Reagan was there to meet me.

" 'Old Finky has found the mine and is going to do us,' was the first thing he said. 'He's here. You can go see him and talk to him yourself.'

"I went over to the rooming-house to see him. Swollen and painful as his face was, he had nevertheless the happiest look I have ever seen on a human countenance.

" 'I have found it, Mr. Campbell! I have found it!' he said. 'I could break the rock with my hammer, but after I had broken it I could not pull the pieces apart, they were so woven together with wire gold.'

"Now the nigger's ore was not wire gold and, therefore, it would seem that Finky found an entirely different deposit. One thing is certain: granting there are two deposits, there is no lead to either of them; each is a 'chimney,' or 'spew.'

"Well, Finky would not tell me or anybody else where he had made his location. He impressed me as having a kind of secret fear, but he said openly that he had been beat out of a fortune more than once and that he was not going to be beat out of this one. He particularly distrusted Jim Reagan, and had it not been for Reagan we should have gotten the mine all right. According to the laws of Mexico, no foreigner could denounce or operate a mine in that country nearer than sixty miles to the border. With this law in mind we had taken in Mueller so that, as a Mexican citizen, he could denounce the mine. But no, Jim Reagan got on a horse and rode clear down to Mueller's ranch in order to tell him not to make a move until old Finky should show us the place. Meantime, Finky recovered and went to the upper Bourland ranch in the Haymond country to await developments.

"The game was blocked. Then Finky became restless and declared he would go to Mexico City himself and see President Diaz. He got as far as El Paso. Now old Finky had one great weakness; that was whiskey. In El Paso he got on a high lonesome and told the barkeeper his business. The barkeeper told him to wait a few days and he would go with him to Mexico and help him fix matters up. The barkeeper seems to have turned the saloon over to Finky. In less than two weeks Finky was dead and his secret was buried with him."

The years went on. John Reagan had died; the other Reagans had moved to Arizona; Lock Campbell had about quit grubstaking prospectors. Then one day in 1909 a man from Oklahoma by the name of Wattenberg drifted into the Big Bend country. He made Alpine his headquarters and stayed around there several weeks, learning what he could about the Mexican border land and picking up details regarding the Lost Nigger Mine tradition. Finally he took into his confidence an old-timer named Felix Lowe and told him his tale. Lowe advised him to see John Young, who has been on the frontier all his life and who has generally been "willing to take a chance." Why Finky did not produce a sample of his wire gold is now to be explained.

Wattenberg had with him a map, and he had a story as to how he had come by that map. According to his story, a nephew of his in the Oklahoma penitentiary, condemned to death, had called him in and, without asking for a cent of remuneration, had told him of an immensely rich gold mine in the Ladrones Mountains across in Mexico about opposite the mouth of Reagan Canyon.[2]

This nephew had been one of a gang of horse thieves

operating all the way from Mexico to Oklahoma. While they were on one horse-stealing expedition across the border, they learned that the Mexican *rurales,* a force corresponding to the Texas rangers, were on their trail. The horse thieves were making tracks for the Rio Grande when they came upon an old prospector lugging a rust-eaten gun and a morral full of rocks.

The prospector, as the condemned horse thief described him to Wattenberg, must have surmised that the gang was after him. Consciousness of the wealth he had suddenly become master of no doubt made him suspicious. At any rate, he went down on his knees and before they had even searched him told the fleeing bandits that if they would spare his life he would show them the richest gold mine in Mexico—the mine from which he had secured the samples in his morral. They took the samples. He showed them the mine. Then the bandits swore that if he should ever tell another soul of the location or come back into Mexico from the Texas side, they would "hound his soul to hell." Before they left the country, the bandits drew a rough sketch of the region, marking plainly the site of the mine.

When John Young examined the sketch presented by Wattenberg, he found the landmarks of it all correct—the Ladrones Mountains, Las Vegas de los Ladrones, Ladrones Canyon, which runs into the Rio Grande above and across from the mouth of Reagan Canyon, Maravillas Canyon, and other features. One of the other features was a dot in the Ladrones entitled "MINE." In short, the map from Oklahoma showed that the maker of it was familiar with the country.

Wattenberg's story dovetailed exactly into Finky's incomplete account. It explained the strange reticence of old Finky, his failure to bring in any ore, his refusal to tell where he found the mine, his hesitation in returning to it.

John Young, Felix Lowe, and Wattenberg now entered into

a partnership to explore for the Lost Nigger Mine. As a preliminary step John Young went to the City of Mexico, saw President Porfirio Diaz personally, and secured a permit to operate a gold mine anywhere in northern Mexico, no matter how near the border. Next he went to Muzquiz and made arrangements to denounce certain mining claims, at the same time engaging for $200 the services of a Muzquiz surveyor. This was early in 1910—on the eve of the great Madero Revolution.

The surveyor came to Alpine and the whole party set out. When they got to the mouth of Maravillas Canyon, where they intended to cross, they found the Rio Grande on a big rise. The surveyor and John Young's son Johnny crossed it on a crude raft, but once across they could do little scouting on foot. The country was so immense and the speck in the Ladrones Mountains marked "MINE" was so little that they knew not where to go. It was like looking for a needle in a haystack—but they looked for two days. Getting into the country at that time in a wagon was a trial; getting out was worse. Everybody was disgusted and John Young's partners quit him.

He, however, did not give up. Thirty days after the first attempt to cross the river, he and his son went down again. This time they took a little flatboat on their wagon. The Rio Grande was up higher than ever. In crossing over, Johnny Young got the Wattenberg map so soaked that it came to pieces. Again he was afoot, the river too swift to put horses into it. The revolution was now going on and nobody who knew anything about Mexico wanted to make permanent camp on the south bank of the Rio Grande.

"I came out and quit the search forever," John Young says.

Then, no longer having any reason for secrecy, he told a newspaper man of his hunt. And, beginning with the beginning—that is, with Seminole Bill and Lock Campbell—the newspaperman wrote the story of the Lost Nigger Mine for the San Antonio *Express*. It was printed September 7, 1913.

A miner named Jack Haggard, who had been living in Mexico for many years, saw the article and immediately wrote a letter that is here produced verbatim.

Muzquiz, Coahuila, Sept. 10 de 1913

Sr. John Young
Alpine, Texas

Dear Sir

The article in the S. A. Express of Sept. 7th attracted my attention. Whether it is a coincidence or a lost mine story, I do not know. But, anyway, about twelve years ago I had a Seminole negro working for me at the mine at Las Esparzas, and he told me a tale of a wonderful rich gold mine that he had found in the Big Bend country. But as I had heard such stories before and spent money and time looking for such mines (without results) I paid very little attention to him.

However, one of the mine foremen named Harry Turner did believe him and from the description given by the negro, Turner and I made a rough sketch of the country in which the mine is supposed to be. Then about five or six years later Turner went on a trip to that locality. He brought back some samples with him and after getting assays made of them he was very much excited. He never showed me the assay reports, stating that anyone who saw them would think he had sent away a twenty-dollar piece and had an assay made of it. He began making hasty preparations to wind up his business as soon as possible. At the time he was contractor in the mine at Rosita, making $2,000 per month. He was in a great hurry to finish his contract and tried to persuade me to go with him, saying that in a year's time we could buy all the coal mines in the state of Coahuila. But as I had a contract for two years I could not leave.

About a week before Turner was to finish his contract at Rosita (Mine No. 3) an explosion of gas killed every man down in the mine, Turner with the balance. Having been a personal friend of his for years, I was called to his house and asked to help settle his affairs. In going through his papers I came across the old sketch we had made according to the Seminole negro's description, also a lot of notes in cipher, which I have not been able to read.

I know where the Seminole negro is. He had some trouble and had to leave Mexico but came back during the Madero Revolution. If there is anything to the newspaper report of the Lost Nigger Mine and you wish any further information, you can write me here. I'll have another talk with the negro.

Very respectfully yours,
Jack Haggard.

John Young immediately replied to this letter and considerable correspondence passed. Then Jack Haggard came to Alpine. He wanted the Youngs to go with him. After all, he had not been able to locate the Seminole negro, and he seemed to prefer a copy of the Wattenberg chart to the one he and Turner had made. Neither of the Youngs was willing to make another trip. Haggard went back to Muzquiz. According to one story he later told, he found the negro. At any rate, he made a pack trip into the Big Bend from Muzquiz, hunted around, crossed the River at the old Stillwell Crossing, and came again to Alpine soliciting the aid of the Youngs. The Youngs still refused to go.

When Haggard left, he said, "If you don't hear from me in fifteen days, you may know that something has happened. The Villaists are chasing all over the country and no man's life is now safe in Mexico."

More than fifteen days passed and nothing was heard of Jack Haggard. Then John Young learned that the revolution had run him out of Mexico. The next news was that he had drowned in Medina Lake.

John Clote of the famous Clote mines in Coahuila believed in the Lost Nigger Mine so thoroughly that he offered to put up $10,000 for Lock Campbell to spend in hunting it. Campbell told him that all had been done that could be done and that the money would be wasted. Clote asserted that he was going to spend it anyhow. He went down with the Lusitania.

One day in 1918 Charlie Stillwell, then in California, received a letter from his brother Will, a Texas ranger on border duty. In part the letter read:

> At last I have located the Lost Nigger Mine. The samples I have correspond with those given to Father by Nigger Bill when he was running away from the Reagan camp and passed our ranch. The distance of the deposit from the river corresponds to the distance described by Nigger Bill. As soon as conditions are so we can work, you must come and we will denounce the mine according to the laws of Mexico. A fortune awaits us even if the ore does not extend to any great depth.

Several months elapsed, and Charlie Stillwell was preparing to return to the Big Bend to help work the mine. A few days before he was to leave, he received a message saying that Will had been killed in Mexico while trying to apprehend a Mexican outlaw. Whatever he may have known about the mine died with him.

Thus those who seek the Lost Nigger Mine may expect to be foiled by fire and water and to meet sudden death.

Many people say that there has never been a mine and that the Reagan nigger merely stumbled on a piece of ore salted out by a certain old California prospector with a sense of humor. As Don Pedro Salinas, the sagest sage in Coahuila, has the facts, three Mexican men and one woman who had been stealing high-grade ore from the gold mines south of Las Cienagas fled with some morrals full of it towards the refuge across the Rio Grande. In the Ladrones Mountains they were so hard pressed that they dropped the ore. A little later *rurales* killed them—and the crude gold cast away remained unseen until Seminole Bill found it. Despite such explanations, most border men—men familiar with the character of the Big Bend country—are of the opinion that the nigger found native gold under a cliff and that the cliff soon afterwards caved off, covering all signs so that they can no longer be detected. Another theory, assuming that the nigger found his gold in a canyon, is that gravel has washed over it, hiding it from view. Not many people doubt that it is hid.

The Lost Nigger Mine—what is it? It is the spell of the Big Bend of unsurveyed mountains and the Lost Canyon, the spell of vast solitudes where men with strange tales have lived strange lives and died strange deaths. It is the lure of unknownness. It is Mystery—Romance. And it is Reality—the Imagination of men of the earth.

There will be other stories of the Lost Nigger Mine.[3] The mine will in all probability be found again—and lost again. But my own story, in the gathering of materials for which I have spent so many vivid hours, is done and, having finished it, I am poignantly conscious of extraordinary scenes and extraordinary characters connected with it that can never be realized on paper. Often and often I recall some of them:

A low camp fire down in the willow *vegas* of the Rio Grande at the old Reagan ford and dark waters swishing under the stars while Rol Rutledge of the border and Carl Raht, historian of the Big Bend, argue low and gravely for and against the possibility of gold in that region.

Another camp, at the mouth of Heath Canyon on the Rio Grande, where customs officers hardly ever come and where red Jim Manning, cowman, ranger, borderer, boils *candelilla* into wax and tells of the Seminoles who used to bring ore to Brackettville. While he is talking, a Mexican rides up on the other side of the river and shouts. We go over. The Mexican wants us to visit his *jacal*. There we see panther skins, and he shows us the flesh and head of an outlaw steer, branded before the revolution, that he has just shot after having trailed it for three days out in the Sierra de los Ladrones. The horns are not the biggest I have ever seen but they have the wildest expression that horns can possibly assume.

And then a cow camp by the famed *tinaja* called Pata de

Venado in Bullis Gap and one of Asa Jones's vaqueros telling a long story about an ancient Indian and an ancient *rural* and gold up the Cañon de los Leones, so difficult of access that no gringo prospector ever gets there, though the vaquero draws me a sketch of the whole land.

And then old Dee Davis, seated on a goatskin in his cabin door at Sabinal, unlocking his hoard of lost mine traditions while a hummingbird sucks at a morning-glory near by.

And John Young in his friendly land office in Alpine going into a steel safe and bringing out documents from the Mexican government, Wattenberg's testimony, a yellowing copy of the San Antonio *Express,* and letters from a miner by the name of Jack Haggard who wrote from Muzquiz, Coahuila.

Then Lee Reagan in the corner of a cafe at El Paso telling of hunt after hunt he and his brothers made for the Lost Nigger Mine and protesting against "wild tales" concerning their abuse of the nigger.

But oftenest of all characters recollected is Lock Campbell. His portly frame rests comfortably in a great cushioned chair in his big house that sits back amid fine trees. He goes over to the mantel to get Seminole Bill's sample of ore, and in a ray of sunlight entering through one of the tall windows holds it up for me to see the myriad flecks of color. Lock Campbell is dressed as an old Scotch gentleman should be dressed, and a great gold watch chain rests on his ample front. His talk is of free gold on the Sitka, prospectors in Frisco, assays in Denver, copper in Lordsburg, land in Florida, oil in South Texas; of Seminole Bill with a deer ham at Dryden and then of Seminole Bill in San Antonio with a rock and a description of the western sun shining on millions in gold beneath a canyon cliff; of booted Jim Reagan rushing from the train to telegraph for horses to carry him and his brother into the Big Bend; of blind old Felix Lowe being led up the high steps to the Campbell gallery and begging for a grubstake on "just one more hunt" that is sure to lead to the gold; of "Old Missouri" and his burros; and of Finky with swollen face and joyous voice crying, "I have found it, Mr. Campbell, I have found it."

Lock Campbell is saying that he has "quit," but over and over he repeats, "I know it is there."

"And remember," he concludes, "remember that, no matter what the geologists say, *gold is where you find it.*"

CHAPTER IX

ON WEST

"Begging your pardon, sır," says Joe, "what sort of wonders might them be?"

"Why, all sorts of wonders," says the parson. "Why, in the west," he says, "there's things you wouldn't believe; not till you'd seen them," he says. "There's diamonds growing on the trees. And great, golden glittering pearls as common as pea-straw. . . . Ah, I could tell you of them."—Masefield, *A Mainsail Haul.*

The Engineer's Ledge

When the Southern Pacific Railroad was building west of Devil's River, back in the eighties, one of the construction engines was driven by an engineer named Hughes. Hughes had a mania for picking up samples of rock along the right of way—and the four hundred miles between Del Rio and El Paso afford plenty of rock to pick up. The engineer's work often allowed him a deal of free time, for when his train brought up supplies to the slowly advancing railhead, it might be hours before the cars were unloaded. Then he would get out and look for rocks.

Hughes had a wonderful memory and he depended on it. He did not label a single one of his scores of specimen rocks. He boasted that he could tell offhand where each came from. He knew the ravines, the bends, the mountains, every cut and canyon from Del Rio to El Paso "like a book."

Thus he went on collecting and remembering for three years. Then one day the railroad construction was ended. The East and the West had run into each other. Engineer Hughes got a pass that took him to Denver for a brief vacation. He carried some of his rocks with him.

He was the kind of man who loves rocks and mountains more than money. He was like old Henry Warren, who, after he had spent a lifetime teaching little Mexican schools along the Rio Grande, was offered a farm by his rich brother in Mississippi if he would come back there and live. "You can't offer me a Western sunset," Warren replied, and remained in the Big Bend. Still, when Hughes got to Denver he became quite interested in having his specimens assayed. All but one proved worthless. The exception was so wonderfully rich in gold that the engineer at once regarded himself as a wealthy man.

He remembered in a flash the black ledge over a ravine where he had found the sample. It was located not very far west of Paisano Pass and just south of the railroad; chips from the ledge were scattered out from it for a hundred yards. The picture in his mind was like a photograph: the angle of clustered sotol above; a gnarled piñon in the ravine below; the curve of the railroad at exactly the point where he would walk southwest.

Before he resigned his position, he had plans all made. First, he must get possession of the land. Then he would need water to work the mine. He had heard that there was an Indian spring ten miles up the ravine; he could pipe the water down. The railroad would put in a switch when the time came.

In El Paso he got a light camping outfit and crawled into the engine cab with an old crony who was headed for Del Rio. It would be a good idea, he thought, to stake his claim before he talked. Besides, he wanted to get out among the mountains. His crony promised to let him off at Paisano Pass, and as they puffed up the grade, it pleased Hughes to see how well he had remembered all the details of the location.

He stood looking after the train until it was out of sight; then he walked back. There were the rusted spikes he had remembered. Of course, there could be no mistake now, But what if there were? After he had crawled up out of the deep railroad cut he sat down to get his breath. He hadn't remembered that the air was so light at Paisano Pass.

Then he went on. He looked first for the angle of sotol; there it was. He swept his eye down the ravine; there was the gnarled piñon. He stumbled. His heart certainly was beating hard. He had gone two hundred yards. Where was the black ledge?

He ran. He ran so far that he got out of sight of the sotol and was away below the piñon. That certainly was wrong. He ran back. He came against the railroad cut before he knew it. Then he zigzagged.

Night found him walking aimlessly about. He had forgotten that he was either thirsty or tired. When he sat down, his feet seemed on fire. At least he could sleep. And he did sleep.

When he awoke, his mind was clear. This was evidently the location of another sample, not the gold one. If he could find a rock to match that other sample, he would try to remember what place he had been associating it with. That other place would be the right one. His memory had simply swapped locations with the rocks.

In ten minutes he found a familiar rock. It was in a pothole and was of a gray color speckled with black quartz. The quartz glistened in the sun, worthless. Where now had he remembered this speckled rock as belonging? He hesitated, but only for a minute. He knew. It was a half mile above the crossing on Eagle Nest Canyon. That was a hundred and seventy miles east. He would go there and find the black ledge he had mistakenly remembered as being at Paisano Pass.

Hughes flagged a freight and rode it to Eagle Nest Canyon. He walked up it a half mile. No black ledge was there!

The only thing to do now was to revisit every place from which he had ever taken a rock specimen. East and west, up and down the railroad, he traveled, sometimes on foot, sometimes on a freight. At first the trains would pick him up. Then his old acquaintances dropped away and strangers did not want to bother with a "cracked pebble hunter," even if he had been a trainman.

Years passed. Engineer Hughes told his story often. He told it in great detail. Occasionally some old prospector would join him for a week or so. He has been dead thirty-odd years now, but people out on the S. P. still talk about the Engineer's ledge.

The Lost Padre Mine

The guide tapped the ponderous old bell with the clapper so that I might hear its tones. We were in the church of Nuestra Señora de la Guadalupe in Juarez, across the river from El Paso. His own tones gave the guide—and me also—as much pleasure as those of the bell; he had reason to be proud of his precise English. "Why, sir," he clattered on, "that bell swings from the very same rawhide that was stretched up there when this church was built in 1659."

The rawhide does look weathered. A guide so well informed on Juarez bells must, I thought, know something about the Lost Padre Mine. I was not mistaken.

When he was only a boy, he knew an Indian in Juarez—a "pure quill," as the gringos down in Mexico call an aboriginal of undiluted blood—who, although he never did a lick of work, was always well provided. Once a year, regularly, this Indian would absent himself a few days and then return with a small *carga* of gold and silver ore. It was well known that he got it from the Padre Mine, but he would never tell a soul where the Padre Mine was.

The tradition of this famous mine was hoary long before guides became the chief functionaries of the church at Juarez.[1] By standing either at the front portal of the church or else in the tower—authorities differ—exactly at sunrise and looking to the northeast, one should be able to see the black opening of a tunnel in Franklin Mountain across the Rio Grande in Texas—the entrance to La Mina del Padre. Since the sun rises at a different angle every morning of the year and since it is occasionally obscured by clouds, I doubt if of the hundreds who have strained their eyes towards the northeast from the church a single matutinal watcher has ever given the directions for locating the Padre tunnel a thoroughly scientific test. Anyway, the tunnel was filled up over a hundred and fifty years ago.

In the year 1888 a man by the name of Robinson and an old government packer known as Big Mick began business in a sensible way by starting with the mountain instead of the church. They located the Padre shaft and set to cleaning it out. Backing them was a man who had examined the records at Santa Fé sufficiently to conclude that in addition to working the Padre Mine the Jesuits had secreted three hundred jack loads of silver bullion at the bottom of the shaft. The Jesuits were expelled from Spanish America along about 1780—and when they left the country, they left their mines and their bullion all well hidden. They say, also, that Oñate, founder of the province of New Mexico, walled up in the Padre tunnel 4336 ingots of gold, 5000 bars of silver, nine mule loads of jewels pilfered from Aztec treasure houses, and four codices—smuggled from Spain—each as precious as the Codex Vaticanus itself and, therefore, worth their weight in diamonds.

The prospectors got down into the shaft leading to all this wonderful booty only far enough to clean out about twenty feet of "very peculiar filling." It was of a reddish cast and appeared to be river soil. Certainly it could not have come from the mountain side. Thirty years after this abortive attempt to open the shaft, an old, old Mexican woman told Clabe Robinson, nephew of Big Mick's partner, that before she was born a priest from Ysleta had his people carry river dirt—red river dirt—up the mountain and fill a shaft with it. Singularly enough, about the same time a retired railroad engineer from Mexico produced a chart, presented to him by a Tarahumare centenarian, that recited how a valuable mine in Mount Franklin had been filled at the top with twenty feet of red silt from the Rio Grande.

Clabe Robinson is a logician as well as a plainsman. Incidentally, he has put together the most interesting collection of facts about mustangs that has ever been printed. He is out in the mountains above El Paso now, and news of his project for getting to the bottom of the Padre shaft may be expected at any time. He has the spur of rivalry to urge him on.

Not a great while ago an Italian came to El Paso with a chart as fresh and "hot" as were the contents of the Pardoner's wallet. He elicited the interest of an aviator, and this aviator, fired probably by Lindbergh's success in locating lost Mayan cities, attempted to locate the old mine from the air. He discovered something that calls for a vast amount of work, but whether he will have faith enough to remove a mountain remains to be seen.

Although I have ridden clear across the Sierra Madre on a mule in order to view the Lost Tayopa Mine so dramatically rediscovered by C. B. Ruggles, and although I intend to go in search of the Lost Adams Diggings in New Mexico—for a wonderful story is attached to it—I should not spend a minute looking for the Padre shaft. This is why.[2]

Upon one of the peaks of Mount Franklin there stands out against the perennially clear sky the distinct outline of an Indian's head. "The spinsters and the knitters in the sun" and the mantilla-muffled old women who squat in the plaza with palm outstretched and a mumbled *"pido por Dios"* to every passerby know how that head came there and what it means.

It is the head of Cheetwah, chief of an ancient tribe of Indians. For two hundred years he held the mountain land of his people secure against all invaders. Then came a new enemy, pale of features, pompous of manner, commanding all

Indians to deliver up their gold and silver. The order so incensed Cheetwah that he climbed to the top of the mountain and there cried out a great summons to all the warriors in the spirit world to rally for exterminating the enemy. The warriors of the spirit world heard. History sets the date as 1680, when every Spaniard in New Mexico was either killed or driven south.

After this victory Cheetwah and his people vanished into the mountains, there to keep vigil forever that no alien with pick and shovel should prosper from the mineral wealth of the land. As a symbol of this watch, the peak of the mountain took on the features of Cheetwah's face. The white men returned. They own the country now, but they can never possess its gold and silver. The decree is written in the sky, as whoever looks up to the immutable outlines of that peak of Mount Franklin may read. No, the Lost Padre Mine is lost forever.

CHAPTER X

LOS MUERTOS NO HABLAN

"There is something in a treasure that fastens upon a man's mind. He will pray and blaspheme and still persevere, and will curse the day he ever heard of it, and will let his last hour come upon him unawares, still believing that he missed it only by a foot. He will see it every time he closes his eyes. He will never forget it until he is dead—and even then—Doctor, did you ever hear of the miserable gringos on Azuera, that cannot die? There is no getting away from a treasure that once fastens upon your mind."—Conrad, *Nostromo.*

Red Curly's bandits made a raid
  One night in Monterrey;
They put the loot under granite rock
  And they put it there to stay.

They killed their Mexkin pardners all,
  Next shot a poor stranger;
A mob then hung Red to a joist
  In the town of Shakespeare.

Now, just before they strung him up,
  Red offered, so I'm told,
To show them at El Muerto Springs
  A wagon load of gold.

ACCORDING to Bill Cole, the ballad—and it has twenty verses—should be sung to the tune of "The Drunkard's Hiccoughs." Bill knows, for Bill has not done much the last fifteen years but think about "the Monterrey holdup stuff," as it is called, sing about it to himself, talk about it to anybody who will listen, and dig, dig, dig after it—out by El Muerto Springs in Jeff Davis County on the Texas border. They are digging for it, too, the identical loot, five hundred miles farther west in Skeleton Canyon on the Arizona line. It is the most magnificent booty, not even excepting Pancho Villa's, to have been buried in modern times.

Valentine, in Jeff Davis County, where Bill Cole lives, is noted chiefly for its shipping pens and a lunch room for freight crews. North and east of it lie the Davis Mountains. I went to Valentine to see Bill Cole.

"Just follow them whittlings," one of the railroad men replied in answer to my query, pointing at the same time to the ground beside a shaded section of the depot platform, "and you'll find ole Bill."

I followed the whittlings and found Bill comfortably chewing tobacco in his mother's lodging house. She was a very old woman, and he was, I judged, about fifty-five, with a drooping black moustache and a drawl that drooped in harmony. Few men have ever welcomed me with more warmth than Bill. For two nights and one day we talked, meanwhile visiting his workings at El Muerto Springs, where he has been digging since 1917. As "the wad," according to Bill's calculations, amounts to "a fraction over $800,000," he was naturally very much worried over the income tax laws.

"In 1914," he began, "I was sitting in my pool hall talking with Preacher Bloys. He wasn't too damn good to come in and see a pool hall man or shake hands with a saloon keeper. Well, while we were chatting, a man and a boy who said they were from Oklahoma stepped in to ask me where they could get burros for a trip into the Davis Mountains. In them days I was considerable of a hunter, and somebody had advised them to come to me to get outfitted. They said they wanted to go on a hunt for bear and black-tails.

"I told 'em where to get burros and packs, and after they left, old Preacher Bloys turned to me and said, 'Those men are not hunters. What they are hunting is that pack train money. It's out there in the Davis Mountains all right, but the man who finds it will have to have a true map.'

"That was all he said on the subject. I knew, though, that

Preacher Bloys never talked idly and from that day I began looking for the true map. The Oklahoma man and boy were out ten days and came in without a thing. They were back the next year and then gave up the search for good. I don't know what kind of map they had, but they had one. I began piecing evidence together from first one place and then another. After the state closed my business, I was free to go after the stuff right and I have been after it ever since."

The quest for evidence led Bill Cole to the Pacific coast, down into Mexico, and up into the Rocky Mountains of Arizona and Colorado; and it has caused him to spend many a long day whittling on the depot platform while receiving "authentic tidings of invisible things." The result of all these researches Bill told with many jerks and cutbacks. I have tried to put the story into connected form.

It begins with a band of nineteen Mexican bandits, led by Juan Estrada, who at one time were the terror of the Big Bend. Frequently they dressed themselves up like Indians, and many a foul act of their doing was laid to the innocent Apaches. With feathers in their hair, they robbed the stage coach and chased Big Foot Wallace. Painted like savages, they raided across into Mexico. They were inordinate cattle thieves, and even old Milt Favor, the biggest cowman at that time on the Rio Grande, with a hundred and fifty vaqueros under him, a ranch that was a fort, and an orchard that supplied peach brandy to every traveler over the Chihuahua Trail, was helpless against their depredations.

In 1854 the United States government had established Fort Davis in the Davis Mountains. After the Civil War the fort was rehabilitated. In 1879—the very year that Victorio and his Apaches broke out—a detachment of Fort Davis troops was camped near Lobo in the wide Van Horn valley. The detachment consisted of ninety-nine negroes and a white lieutenant. Their business was to cut toboso grass to be used for hay by the Fort Davis cavalry.

While the negroes were haying and the lieutenant was commanding, four American outlaws came into the country from the west. Their names were Zwing Hunt, Jim Hughes, Red Curly (called also Sandy King), and Doctor Neal. Hughes was the leader. Some of them had been mixed up in the Lincoln County War, wherein Billy the Kid won his fame. Following John Ringo, the gang had raided cattle off the Sonora ranches. They had all helped materially in giving Tombstone the reputation of being the rowdiest and roughest mining town in the West. They were desperate men and they had a desperate plan. It was to throw in with the Estrada bandits, seize a quota of mules and supplies from the United States troops, and, thus outfitted, make a flying swoop upon the riches of Monterrey.

They had no trouble in making an alliance with the Estrada gang. Then one fine morning when the ninety-nine negroes and the white lieutenant fell out unarmed to the picket lines, the twenty Mexican and the four American desperadoes arose from their concealment in the tall grass near at hand and opened fire. Only one negro escaped alive.

"The official report sent to Washington," says Bill Cole, "gave sixteen negroes killed by Indians, but those government reports are always wrong. The officers were afraid to report the cold facts. I myself have seen enough old .44 Winchester shells on the ground of that Lobo camp to fill a sack." [1]

The desperadoes took an ample supply of provisions, ammunition, horseshoes, blankets, saddles, and other goods and picked out twenty-five choice mules—one around and an extra. They headed south for Mexico. At Paso Viejo in the San Antonio, or Lost, Mountains they paused to shoe their animals in a thorough manner.

A day's ride below the Presidio del Norte crossing on the Rio Grande, Estrada of the Mexicans and Hughes of the Americans halted their followers to load the mules with guano from the immense bat caves still to be found in that mountainous region. Posing as traders, they then went on into Monterrey and sold their guano at a hundred dollars per ton.

Their character thus established, they pitched camp near the edge of the city and opened a monte game, in which they at times lost heavily. One night, after the mules were well rested from the trip, Estrada's men were sent out with explicit directions to toll the guards of the mint and smelter to camp. The bait was to be free *tequila* and a game of monte in which the banker was losing thousands. No peon soldier could resist the allurement of monte and *tequila*. But as the truant sentinels slipped into camp, they found instead of a monte game a death trap. Twelve guards in all walked into it. The proverb that guided the greatest ruler Mexico has ever had, Porfirio Diaz, president of the republic at this time, was *Los muertos no hablan*—The dead do not talk. So Hughes and company had good precedent. No one of the twelve guards ever talked after entering the bandit camp.

Meantime a picked crew looted the mint, raided the smelter, and sacked the cathedral. The combined bandits now loaded their jag on the twenty-five government mules and pulled for the Davis Mountains. They knew that they would be pursued and they lost no time. By the map it is three hundred and ninety miles from the rich city of Monterrey in the state of Nuevo León to El Muerto Springs in Jeff Davis County, Texas. The desperate bandits traveled this distance without once taking the packs off their mules!

"I have been disputed on this point," Bill Cole explains. "Three hundred and ninety miles is a long ways for any animal loaded down with a dead weight of gold and silver to travel. But it is possible. Those were picked mules. Tom Bybee rode from Cisco to Fort Stockton, over three hundred miles, without once taking the saddle off his sorrel horse."

Anyway, the raiders made a forced march, and it was well that they did so. As soon as their depredations were discovered, dispatches were sent to the Mexican troops at Monclova.

The Monclova cavalry made for the Presidio del Norte crossing, where they expected to intercept the robbers. The robbers, however, in anticipation of just this move, crossed the Rio Grande at the mouth of Reagan Canyon nearly two hundred miles below. According to international agreement, troops of either country could at that time cross the border between Mexico and the United States, but only *on the trail of marauders*. The Monclova pursuers lost a week riding back down the river to strike the trail.

So far as testimony has come down, only one man on the Texas side saw the mule train as it forged on to its destination. This may well be, for the Big Bend is a wide and lonesome country now and it was a great deal wider and a great deal more lonesome back in 1879. That one witness was a Mexican *pastor,* or shepherd, named Quintana.

As Quintana told the story to Cole, he was herding a little flock of goats near Barrel Springs, seventeen miles east of El Muerto Springs, when he saw the first *mulero* coming up the draw. The bandit train was strung out for nearly a mile, he said. The mules were so worn and tired that the only means by which the *muleros* could keep them going was to jab them in the sides with sotol stalks. Two Americans were riding ahead of the column on mounts that were fairly fresh; each of them was leading two mules. Quintana hid in the brush but pretty soon he was surprised by two other Americans. From his description they must have been Red Curly and Hughes, for these two had turned off to buy some flour in Murphysville (now Alpine). They asked Quintana if he had seen the pack train. He said that he had. He could not understand English, but he understood that the two men now debated killing him. However, the pair rode off without having lived up to their principle: *Los muertos no hablan.* That is all of Quintana's story.

When Hughes and Red Curly joined the other two Ameri-

No peon soldier could resist the allurement of tequila and a monte game.

cans, the lead *muleros* were halted until the straggling rear should catch up. Then with Juan Estrada accompanying them, they galloped ahead, leaving their mules to be brought on by some of the Mexicans. Just how they murdered Estrada is not known, but certainly they murdered with dispatch and in silence. As the Mexicans, "carajoing," hissing their mules, and jabbing sotol stalks into their bellies, turned up the blind pass leading to El Muerto Springs, they knew that their long three hundred and ninety mile march was almost over. Indeed, it was to be over sooner than they expected.

In all their unreadiness the Mexicans were met by a deadly fire from the treacherous gringos. Only one escaped. He happened to be on his mule, the freshest of the outfit, and ran away. About ten years ago, so the tale runs, Beau McCutcheon, a well-known ranchman of the Davis Mountains, found two silver bars weighing over a hundred pounds each on his ranch. He will not talk about these bars, but Bill Cole is sure that they were dropped by the escaped Mexican. Furthermore, Bill has talked with a Texan who while down in Mexico heard from the escaped Mexican's own lips the whole story of the raid in Monterrey, of the almost unparalleled flight, and of the massacre of his confederates by the American outlaws.

During the massacre Zwing Hunt was severely wounded by a machete that one of the Mexicans managed to seize before he was killed. Hunt's companions put him in a cave, transplanted some river grass, which they got at the springs, in front of the cave and watered it copiously. Meantime, in a hole twelve feet deep, put down in plain sight of Hunt's cave, they buried their plunder. Then they did a strange thing. If I were making this story up, I should not allow them to do it—but "facts are stubborn things." After providing Hunt with a supply of food and water, they left him alone while they went back farther into the mountains. There with brush and pickets they made a kind of fence across the mouth of a grassy canyon, wherein they turned their jaded mounts loose to eat, drink, and rest. The mules—that is, those that had not been killed in the mêlée—were released, and some of them eventually drifted back to the hay camp at Lobo. At the end of two weeks Hughes, Red Curly, and Doctor Neal returned to their wounded comrade.

They found him nearly well. He said that eight days after the Estrada gang had been butchered, Mexican cavalry appeared in the valley below his cave. He saw them ride about, take spurs and other equipment off the dead bodies, and then at the end of reatas drag them all to a gully and cover them up with rocks and loose gravel. The cavalry went back towards Mexico. The officer in command, so it was later reported, advised the Mexican government that the bandits had been annihilated by him in a pitched battle, but that the Monterrey loot had been taken over by a fresh force of outlaws and carried on into the interior out of reach.

Jim Hughes and his *compadres* now had little fear of further pursuit. Their cache was safe. They filled their saddle bags and morrals with coin and rode to El Paso, where they spent money right and left. After carousing a spell, they drifted on out into New Mexico. At Hachita, a stage stand, Hughes wrote his mother: "We rounded up the stuff and buried it in a twelve-foot hole." Hughes had the reputation of being a very dutiful and affectionate son.

In the spring of 1881 the gang made three separate robberies—apparently not from need of money but for the love of the game. The boldest of these acts was holding up the train near Tombstone. In this holdup they killed a mail clerk, blew open the express safe, and got off with a sum of gold so vast that the express company would never admit the amount.

Next, in the Chiricahua Mountains, east of Tombstone, the outlaws killed a man and his son for nothing more than a wagon and two horses. They put their plunder in the wagon and proceeded on across the parched world to El Muerto Springs. Nothing had been touched during their absence. They now had new riches to add to their storage, and they decided to make a thorough job of burying it.

Not far away from the springs they found four Mexican miners tunneling into a mountain. This tunnel, incidentally, has been "lost." They engaged the Mexican miners to dig a deep hole, through solid rock, right down beside the twelve-foot hole in which the Monterrey stuff was stored. The Mexicans went down eighty-five feet and then tunneled back under a shelving rock for eighteen feet. In lieu of dynamite they had to use sotol heads, which they heated until the sap exploded in the form of steam. They made buckets out of deer and antelope hides. A log was fixed horizontally above the shaft; then by means of a rope cast over it and tied to the horn of a saddle the loosened rock was hoisted in the hide buckets.

After the great hole was completed, the twenty-five mule loads of Monterrey stuff were taken out of the shallow hole and lowered into it. Then on top of this was let down the wagon load of booty freshly brought from Arizona. In the big hole but somewhat apart from the main storage was carefully placed a tomato can full of precious stones taken from the Monterrey cathedral, among them the jeweled eyes of a priceless image of the Virgin Mary. Of course each of the desperadoes held out all he wished, but these personal reservations hardly made a showing on the main treasure.

When the gold and silver and jewels had all been stowed away, the Mexican miners were ordered to seal the hole so that no prying hand could ever get to it. Surely the bandits must have intended to return at some time and take more of the wealth; yet they now, apparently, locked out themselves as well as all meddlers. There were thousands of antelopes in the country, many of them watering at El Muerto Springs.

Blood is stronger than water. The outlaws shot antelopes and brought the carcasses in. The miners took the blood from them, and, mixing it with granite gravel, made a cement as irrefragable as the Pillars of Hercules. In the end Jim Hughes boasted that he could tell the world where the riches were buried and they would still be as secure from pillage as though they were locked within the strongest vault of the Bank of England.

While the Mexican miners toiled, Hughes, Red Curly, Zwing Hunt, and Doctor Neal sat in the shade under a juniper tree growing upon the mountain-side and there played poker by the hour. The juniper is still growing. I have sat under it myself, gazed down upon the holes that Bill Cole has been so many years digging, heard a mocking-bird quarreling in the limbs overhead, and watched a bumblebee at red phlox flourishing in the stony soil. What a peaceful and far away world it is that one realizes under that juniper tree in the boundless pasture lands surrounding El Muerto Springs! But that quiet and restful juniper tree forms a valuable piece of evidence in locating the site of an immense prize mortared down not only with antelope blood but with the blood of one hundred and forty-two men who met their death because of it.

The outlaws spent five months, from April to September, in the Davis Mountains. While they were here a consumptive in quest of fresh air came by. He offered them six hundred dollars—all he had—if they would let him live in their camp. They made him keep his money and took care of him until he died, which was within a few weeks. Then they gave him a decent burial, stowing away with him the six hundred dollars in a boot.

And now, with the great treasure all cemented down, the outlaws had to decide what they should do with the Mexican miners who had done the work. *Los muertos no hablan.* They put the bodies of the four miners down in the original treasure hole and covered them with twelve feet of earth.

Whether they so purposed or not, the Hughes gang were through with El Muerto forever. Loaded with money, they went back into New Mexico. They were on the dodge now. They made camp in the mountains east of Silver City. Here a fifth outlaw by the name of Russian Bill joined them. One day while they were all "celebrating" in Silver City, one of them asked a quiet, rather retiring young stranger to have a drink. He declined.

"You're too damned nice to live in this country," yelped an insulted member of the gang and forthwith shot the young man dead.

This was the straw that broke the camel's back. A heavy reward was out for the train robbers, and it was pretty well known who the train robbers were. A posse organized to round them up. Doctor Neal was caught in camp and mortally wounded. He died leaning against a juniper tree. The other bandits scattered.

Zwing Hunt was shot, captured, and taken to Tombstone, where he escaped from the hospital only, it is thought, to be killed by Apache Indians. Many people, however, claim that he was not killed but that he eluded the Apaches and on his death bed in San Antonio years later gave a map—a rather baffling map—to the Monterrey loot.

Russian Bill and Red Curly (Sandy King) were overtaken by the Law and Order League in Shakespeare, a mining town near Lordsburg, and in the absence of trees were hanged, as the ballad has it, from the rafters of the Pioneer House dining room. As they were about to be strung up, Red Curly begged to be shot instead of hanged. He was refused; then he offered to show his captors a wagon load of gold if they would grant his request. Nobody believed that he could fulfill his promise, and so, with handcuffs on, he and Russian Bill had to swing.

This was late in the afternoon; the Pioneer House served supper as usual that night. After the bodies were taken down, Jim Engle, a blacksmith and horseshoer, was called upon to file the handcuffs off the dead men. He later moved to West Texas, where Bill Cole has had much talk with him. True, certain chroniclers of the bad man tradition have set it down that when Russian Bill and Sandy King were hanged, the only charges against them were that Bill was a horse thief and Sandy "a damned nuisance"—but even granting thus much, "what," in the language of Bill Cole, "has that to do with the Monterrey holdup stuff?" As for Jim Hughes, it seems that after leaving the San Simon valley he ran a saloon in Lordsburg for a while, but the only certain thing about the rest of his career is that he never came back to El Muerto Springs.

About ten years after Doctor Neal died leaning against a juniper tree, a New Mexico goat man named Stevens visited the old outlaw camp east of Silver City. He was looking for evidence concerning the Monterrey loot. The first thing he noticed about the camp was "a mummified deer," with only one quarter gone, hanging in a tree. He began to dig around. He uncovered a rock slab. Under the slab he found a box, and in the box, along with some letters addressed to the affectionate Jim Hughes by his mother, was a map to the El Muerto Springs cache. A handful of flour in the bottom of the box showed that it had been used for holding provisions.

When, after devious negotiations, Stevens turned the map over to Bill Cole, Bill at once recognized it to be true to the El Muerto Springs topography. He located thirteen points called for by the map and then beyond all question—in his own mind—he located the exact spot under which the treasure was cemented down. Among the points he located were the juniper tree that shaded the outlaws while they played poker and overlooked the toiling Mexican miners; the cave in which the wounded Hunt stayed for two weeks; the ruins of some adóbe corrals that Quintana, the *pastor,* once penned his goats in;

also the remains of an early day stage stand; the trace of an old Indian irrigation ditch; an Indian grave on a peak west of the treasure hole; and a rock called, from its shape, Woman's Head Rock. In the cave Bill found an old gun with some of Hunt's blood still on it. Any Doubting Thomas who wishes to inspect that blood-stained gun may do so.

After making his location, Cole had dug only a few feet before he was convinced beyond the shadow of a doubt that he was following the shaft sunk by the four Mexican miners. Less than ten feet down he struck that extraordinary cement made of granite gravel mixed in antelope blood. The dried blood was still so strong that it attracted blow flies by the tens of thousands; many a hot day Cole had to stop working, the flies were that thick in the hole. On down he found clots of antelope hair. He found a piece of antelope hide, doubtless from one of the hide buckets. The miners had left various rocks jutting out from the walls of the shaft, and in some of these rocks Cole discovered round holes worn by the rope that carried the buckets up and down. On other rocks he found chisel marks. He found a sliver of steel "from some old Spanish tool." Fifty feet down he found a piece of bark from a juniper tree; then a horn button cut out round like a marble with an eye through it for a buckskin string. All along as he went deeper and deeper he found pieces of surface rock.

But a shaft eighty-five feet straight down and then a tunnel eighteen feet back require an enormous amount of digging and lifting for one man's hands. Cole had no funds. The winter after the first season of digging he undertook to enlist capital. He convinced a doctor that there was a fortune at the bottom of the well. The doctor gave up a few dollars to carry on the work, and then the doctor's wife found out what he was doing and put a stop to such extravagance. Meanwhile the Valentine folk and the ranch people around joked, as they still joke, about "the champion money finder of West Texas."

"Talk about the big fight at Muerto Springs," exclaimed

John Z. Means, who came to that country about the time the Apaches left it and who is known as the mildest-mannered gentleman that ever drove a cow, "the only fight that ever occurred there was between two old horned frogs."

But no banter has ever caused Bill Cole to draw a faster breath or swallow a drop of tobacco juice. He went on digging. At last when he was nearly down to the eighty-five foot level and was ready to tunnel back to the booty, he struck water. He could not pump it out. Then he got a well driller, for an interest in the loot, to dig an off-set hole, thinking by it to control the water. It was controlled long enough for Cole, working on the floor of his shaft, to jab his crowbar into solid silver. A bit of the silver stuck to the bar, he says. Then the water rushed in again.

Bill has had other partners, other foilings. He sank a second shaft, but water broke into it. Repeatedly the owner of the land has threatened to run Bill off the premises. But his camp is still out there at El Muerto Springs, and the last letter I had from him expressed the expectation of "lifting the wad out in a couple of weeks now." It also expressed unusual concern over income tax laws.

And this concern brings on a sequel. Without knowing anything of my particular interest in the Monterrey loot, a dentist who lives about six hundred miles away from it apprised me not long ago that he had been out to Valentine and while there had become acquainted with an old blacksmith named Engle—the man who cut the handcuffs off the dead bodies of Red Curly and Russian Bill. Engle was morally outraged and was expressing himself in no ambiguous tones. "A certain ne'er-do-well" of the village had proposed that the blacksmith melt up "some cathedral candlesticks to be excavated in the near future." He had indignantly refused to take part in any such sacrilege. Engle also reported the unnamed ne'er-do-well to be constantly discussing means of getting around the income tax laws and at the same time to be living comfortably without any *visible* means of support.

"All this talk about excavation in the near future," concluded Engle, "is just a blind to cover up what has already been done. I can put two and two together."

The reader can make the same addition.[2]

CHAPTER XI

THE CHALLENGE OF THE DESERT

The Paiute Indians who used to live in the Death Valley region called the valley *Tomesha*—ground afire.—Bourke Lee, *Death Valley*.

THE impulse to reproduce the species, the impulse to protect offspring, the impulse to wrest gold from the earth—I know not which of these three can operate the most madly and fiercely. The last, usually operating far removed from the humdrum of family cares, seems most romantic. It is on the other side of the world from merely acquisitive money-grubbing. A man who lived among the early California gold hunters wrote of them thus:[1]

"I felt as though I had been translated to another planet. There was nothing here that I had ever seen or heard of before. The great forests, the deep cañons with rivers of clear water dashing over the boulders, the azure sky with never a cloud were all new to me, and the country swarmed with game, such as elk, deer, and antelope, with occasionally a grizzly bear, and in the valleys were many waterfowls. Tall bearded men were digging up the ground and washing it in long toms and rockers, and on the banks by their sides was a sheet iron pan in which were various amounts of yellow gold. These men had neither tents nor houses. They camped under lofty pine or spreading oak trees, for it never rained there in the summer time. They were strong and healthy and lived a life as free as the air they breathed. . . .

"Where all the gold came from was a much mooted question, and they pondered deeply over it and finally settled down to the belief that it must have been thrown out by volcanoes, as the country bore evidence of ancient volcanic convulsions. In 1851, a report was started (but no one knew who started it) that high up in the Sierra Nevada was a lake, evidently the crater of a large volcano, and that the shores of this lake were covered with gold so plentiful that there was little sand or gravel there. As soon as the miners heard this rumor, they at once said, 'That's just where we thought all this gold came from, and what is the use of us digging here in the mud and water for a few hundred dollars a day when we can go up there and just shovel it up by the ton?' And incredible as it now seems, several hundred of them abandoned rich claims and went up into the mountains and spent all summer looking for the 'gold lake.' When they came back, ragged and footsore, they found that all their rich claims had been taken by others. I personally knew two men who, in 1851, were working in Rich Gulch on the West Branch of Feather River, taking out three hundred dollars a day apiece. As soon as they heard of Gold Lake, they at once quit their rich claims and spent months and all the money they had searching for this imaginary lake. [It is often called Lingard's Lake now, for Lingard, somebody said, found it, though he was unable to find his way back to it, and it is still an object of search.]. . . .

"I once asked some men who, I knew, had families in the states, 'Why don't you quit drinking and gambling and save your money and go back home to your families?' To which they all answered, 'So I would, but don't you see that at the rate gold is being taken out, by the time I got home with a lot of it, it wouldn't be worth any more than so much copper; therefore, I am going to stay right here and have a hell of a good time while it is worth something.' And so they stayed and had that kind of a time. No such enormous amounts of gold had been found anywhere before, and, as they all believed the supply was inexhaustible, there was some justification for their thinking it would soon lose its value."

Nuggets in the Sand

Nothing in nature is more maddening than a summer sandstorm in the desert. The thermometer mounts to 110, 120, even more degrees in the blazing sun. Then the wind rises and begins shifting the dunes. It moves them fifty feet, five hundred feet. Above the swirling, cutting sand the sun becomes a dim copper disc; then there is no sun. The peaks of arid mountains, generally so clearly defined in the distance, blur out. A man caught in the storm cannot see his own hand. At one place the wind scoops out sand until "bottom" is reached; at another it piles up sand into overwhelming crests. On the grazeable fringes of the desert the sand sometimes plays humorous tricks. It has covered up a windmill. The yarn goes that a cowboy awoke one morning to find his horse standing on top of a mesquite tree instead of under it where he had staked him. But in the deadliest wastes of the desert there are no grazing grounds or windmills. A lizard does well to live there.

Somewhere out in the midst of the desert sands of Arizona—I am unable to give the particular region—is a famed aggregation of gold nuggets—thousands, millions of them. They have been seen twice at least.

One day a Mexican girl was herding goats in the scraggly brush along the eastern fringe of the desert. Some of the goats had strayed considerably out, and it was getting along towards night when the girl saw that a windstorm was rising. She gave the customary call to start the goats homeward, but the strays refused to come. She hurried out to drive them in, but she could not hurry like the wind.

She had walked less than a hundred yards when she became almost blinded by the driving sand. She lost view of the goats. Then she faced the teeth of the wind to make for the shelter of the little Mexican ranch where she lived. Now the sand was cutting like knives. She tried to bear against it, and veered. She zigzagged. She could not see a foot away. She let the wind carry her as it would. She never knew how many hours she stumbled, crawled, cried. She lost all sense of time, direction, purpose.

Finally, though, she realized that she was being swept away like the sand itself and that she had done much more wisely had she not attempted to walk in it. She knew that she was lost, and then she huddled down against the earth to remain there until the storm was over and she could get her bearings. Sometimes such a storm lasts for days, sometimes only for hours. She found herself in what seemed a kind of depression, and there with back against the storm she waited.

The wind grew stronger, if possible, and she felt the sands being carried away from around her. But she remained where she was, settling as the sand settled. After a time, a very long time, she seemed to be at the "bottom," on solid ground. She could feel little rocks about her. There was a difference in light but not in visibility between night and day; the night and half of a day went by, and then the wind lulled.

The shepherd girl could see. She looked at the mute and unfamiliar horizon about her. She looked at the appalling sky. She looked at the ground beneath her feet. The little rocks she had felt were gold nuggets! Far away across the waste of sand she saw the smoke of a train. Home was perhaps nearer than the railroad, but the girl did not know where she was. She had lost all sense of direction and could not recognize a single landmark. She knew that if she could reach the railroad, the next train would stop for her.

She gathered into her lap as many nuggets as she could carry and started towards the railroad. Not a mark on the sand was there to go by, only the memory of the faint line of smoke and the smoky sun. She ran, she walked, she fell, she staggered, she crawled. She reached the railroad.

When a train came, it, according to the custom of the desert, stopped, and the trainmen took the shepherd girl on board. The sight of her nuggets drove them wild and assured her careful attention in the nearest town. She was quickly restored; meantime a train crew that had quit their jobs waited for her to guide them.

They need not have waited. All she knew was the direction she had waded from in the sands to reach the track. She was vague as to the time it had taken her to traverse the distance. Perhaps she had not moved in a straight line. There was not a mark on the horizon to go by; there remained not a footprint in the restless sands.

The search was fruitless, and the winds came again, and the sands shifted back and forth in their way, and the Mexican shepherd girl never saw again the golden rocks that lie below the sands.

However, tradition has it that a cowboy who was once trying to cross the desert came accidentally upon the marvelous display of gold. He gathered enough of the nuggets to fill his saddle pockets and he even removed his saddle blanket so that he could roll the gold up in it. He forgot the fearful price one may pay for gold in the desert. Had he taken the right direction, he might have ridden to water. But he forgot or missed directions. His horse at length sank exhausted, and he shot him to drink his blood. Then the cowboy went on. He threw away everything that he had even to the last nugget. At last, as it seemed by the providence of God, he dragged his perishing frame to a water hole and to human aid.

In the desert the hot winds blow and the sands shift and there is never a traveler's track but is blown away. The skeletons out there buried so deep one day and exposed so bare and naked the next tell no stories, write no epilogues. Only be sure of this: what the sands uncover one day they will cover again the next. And, the desert rat says, what they cover they will also uncover—some day.

The Breyfogle Mine

One day in 1862 while racing across Nevada a horse ridden by Pony Bob Haslam of the Pony Express stumbled to his knees. In recovering his feet, the horse kicked loose a chunk of rock that caught the eye of his rider. He took it to Virginia City, where it was pronounced to be silver ore of extraordinary richness. Hell breaking loose in Georgia was nothing compared with the stampede that California made to the Reese River district. The "excitement" centered in and around the present town of Austin, Nevada.

Staying in Los Angeles at the time the news broke were three men who, although without funds or means of conveyance, determined to get to Reese River. Their names were McLeod, O'Bannion, and Breyfogle. The great silver strike was four hundred miles north across the most desolate, forbidding, and inexorable region of mountain and desert on the North American continent. The stage route led nearly four hundred miles northwest of Los Angeles to Sacramento City, still three hundred miles away from the silver, and then cut east. I am giving air line measurements. Either route twisted like a corkscrew. There was no traveled road of any kind across the desert. All people of sound judgment took the stage route. Some of the forty-niners who had tried the short-cut paid their lives to give one spot it traversed a name—Death Valley. Still, if you are going afoot, it makes a difference whether you are to walk, say, six hundred miles or a thousand. Breyfogle and his partners were going to Nevada silver afoot. They decided to cut straight across.

It was about the first of June, summer in the desert, when they set out, carrying some provisions, a blanket apiece, canteens, and rifles with which they hoped to procure jack rabbit meat along the way. At the San Fernando Mission the hospitable padres tried to persuade them to abandon such a perilous undertaking, but they trudged on. They crossed the Mohave Desert, skirted the southern spurs of the Argus Range, crept across the glittering waste known as the Panamint Valley, and at length began ascending the awful Panamint Mountains, from the heights of which can be seen to the east the weird, unearthly basin of horrors called Death Valley and on beyond it the Funeral Range.

On the eastern slope of the Panamints they came, while following a crude Indian trail, to a rock *tinaja* in which they found water. Here they prepared to spend the night. The ground was so rough that they experienced great difficulty in finding smooth places on which to lie down. McLeod and O'Bannion made their pallet together near the water hole; Breyfogle found a bedding place about two hundred yards down the slope. The men, as was their custom, slept with all their clothing on, removing only their shoes.

That unusual separation of himself from his comrades saved Breyfogle's life. He woke in the night to hear shouts and groans and to realize that Indians were murdering the other sleepers. He jumped from his blanket, grabbed his shoes, and with them and nothing else in his hand fled barefooted to the valley below. Only a crazy man of brute toughness could have run barefooted in darkness over rocks and thorn stubble as Breyfogle ran. Breyfogle was very near the brute both physically and mentally, and now he was utterly crazed with fear.

At daylight he found himself down in the bottom of Death Valley. Fearful lest the Indians might still follow him, he secreted himself for several hours in a fold of gravel and sand before attempting to cross to the eastern side, a distance of about ten miles. His feet were so bruised and torn that he was unable to put on his shoes.

The terrific June sun beat upon his bare head. Thirst became stronger than fear. In the afternoon he began traveling. By some mad chance he came at the eastern edge of the valley to a little geyser-like hole of alkali water. He drank it, the first water he had tasted since the previous evening. It made him deathly sick, but he soon recovered, and, filling his shoes with water—they were big shoes and they were stout—limped on. Not after the experience of the night before would he ever again lie down to sleep near a water hole.

After traveling about an hour into the lower foothills of

the Funeral Range, he halted, heaped up some rocks in the form of a wall to lie behind, and went to sleep. During the night he drank the contents of one of his shoes. At the break of day he drank the water from the other shoe and then set out to gain the top of the range eight or ten miles ahead of him. He was sick. The alkali water whetted more than it allayed thirst.

About half way up the mountain Breyfogle saw off to the south a green spot that he took to be growth marking a spring. He judged it to be about three miles away. He turned towards it. He had covered about half the distance to the green spot when his attention was arrested by float rock of a soft grayish-white cast with free gold showing plainly all through it.

Fearful as he was of Indians, exhausted and battered as he was from the torture he had endured, mad as he was for a swallow of fresh, cool water, he paused at the sight of the gold ore. He picked up several of the richest pieces and tied them in his bandana. He started on again towards the green spot and had taken but a few steps when he came upon the vein itself from which the float had washed. Here the ore was pinkish feldspar, much richer in gold than the float. Breyfogle discarded his first samples and gathered a bandana of the pink ore.

The time spent gathering ore amounted to only a few minutes. Breyfogle skulked, limped on towards the green spot. It proved to be a low, bushy mesquite tree, very green and full of green beans. The man ate so ravenously of them and was so disappointed in not finding water that he collapsed, and, as he afterwards said, lost his mind.

But he apparently never lost his sense of direction. He recovered, though he could never recollect when. The experiences he endured for days following remained ever afterwards absolutely blank to him. Water of some kind he must have somehow found, but how and where he could not remember.

He knew the value of *viznaga* juice. He no doubt ate roots and herbs. The indisputable fact is that he kept walking north, across the Funeral Range, and then across the wide Amargosa Desert. At the clear fresh water of Baxter Springs, fully two hundred and fifty miles—as one must travel—from the point where he had emerged from Death Valley, Breyfogle came to his right mind. After remaining here for two days, drinking water and eating whatever green and edible vegetation he could find, he continued on—bound for the Reese River silver strike. He crossed into Smoky Valley and there saw the first human being he had glimpsed since the murder of his partners.

A man by the name of Wilson was ranching in Smoky Valley. While out one morning cutting for horse tracks, he came upon the prints of a man's bare feet. Astonished at their size and shape, he put spurs to his horse and within a few miles overtook Breyfogle. For many years afterwards his description of the human object before him was a part of a fireside story familiar all over Nevada and eastern California.

Breyfogle, he said, was all but naked. His pants were in shreds, the shreds coming only to his knees, while the tattered remains of a shirt did little more than cover the shoulders. His black hair and beard were long and matted. Breyfogle was a Bavarian and at this time he was about forty years old. He was heavy-boned, thick through the breast, stood all of six feet high, and under normal conditions weighed around two hundred pounds. He was strikingly bow-legged, and, as has already been suggested, had enormous feet. He was naturally of a swarthy complexion. He appeared to Wilson a cadaverous giant parched and seared as if by the fires of hell. He was still carrying his shoes. In one of them was stuffed a bandana tied around some specimens of ore.

The rancher took the wild man of the desert home with him and, aided by his wife, provided him with food and clothing. A few days later he took him to Austin and there turned him over to a mining friend named Jake Gooding, who put Breyfogle to work in a quartz mill.

Breyfogle told Gooding all he could about his mine. The samples of ore he showed told more. Some were almost half gold. The season was too hot for an immediate expedition, but three months later Gooding and Breyfogle, accompanied by five or six other men and well provided with saddle horses, pack mules, water casks, and provisions, set out. Upon reaching the Funeral Range, however, they were met by a war party of Panamint Indians and turned back to Austin for reinforcements.

Meantime authorities in Los Angeles had been notified of the fate of Breyfogle's partners; a search party had gone out and had found the remains of the victims at the place described by the survivor.

During the winter a second expedition made up of about a dozen men set forth to find the gold. They got through the mountains to Death Valley without Indian troubles. Breyfogle led them to the geyser-like hole of alkali water where he had filled his shoes. Without much difficulty, he led them to a low, wall-shaped heap of rocks, where he had spent the night after his partners were murdered. From this he led them on up the Funeral Range a distance; then he turned abruptly south—towards a spot no longer green, though only a few months before so green that it had appeared to mark a spring of water. About three miles from where they started south the party came to a bare, scrubby mesquite tree.

"This," said Breyfogle, "is where I gorged the mesquite beans, fainted, and lost my mind. We ought to have passed the gold on our way here from the north. I picked up the specimens of pink feldspar just over yonder and put them in my bandana."

Of course there were other mesquite shrubs in the country, but Breyfogle was sure of the one. He was sure of the water hole; he was sure of the heap of rocks. But the gold? Breyfogle coursed and recoursed away from and back to the mesquite. He saw another mesquite. He wavered. The men with him searched frantically in every direction. Then some of them jeered him; some cursed him for having led them on a wild goose chase; some were sure that if they could remain in the region a reasonable length of time they could find the gold. But a party full of discord will not persist at anything. The gold hunters packed up and returned to Austin. Breyfogle left the country, and thus ended what promised to be but an easy walk to the mine he gave his name to.

Not all the hunters who have gone out since have got back to explain their failure. But desert rats still search. George Hearst, father of the notorious publisher and one of the most successful mining men of his day, secured a piece of Breyfogle's ore and for two winters kept prospectors in the field looking for the lost vein. He believed in it. Many men still believe in it, though most of them think that while Breyfogle was waiting in Austin for cooler weather before returning to claim his gold a cloudburst swept down the slopes of the Funeral Mountains and covered it up. They are hoping that another cloudburst will uncover it.

This is just one of many stories about Breyfogle, some of them very different. But I put particular credence in this, for the original narrator of it, Donald F. MacCarthy, made, thirty years ago, two visits to Jake Gooding, and thus received his information first-hand. Gooding had talked many times with Wilson, rescuer of Breyfogle; he had himself befriended Breyfogle, seen his golden rocks, listened to him tell his story over and over, and then gone with him to refind the gold. Mr. MacCarthy's own hunts for the Breyfogle Mine at least acquainted him with the terrain.[2]

Yuma's Gold

Some sixty miles to the northwest of Tucson, Arizona, the sand bed of Arivaipa Creek joins with the sand bed of San Pedro River, both sand beds generally being "as dry as a lime-burner's hat." At this junction was Old Camp Grant—to be distinguished from new Camp Grant—and at the time of which we are to tell, the early seventies, the Arivaipa band of Apaches had their main camp amid the Arivaipa Hills about ten miles distant. The route between Camp Grant and Tucson was marked by piles of loose stones indicating places where Apaches had killed their white enemies, the cairns for Mexican victims usually being distinguished by rude crosses of stalks from the mescal and Spanish dagger plants. Those crosses were unutterably lonely. The Arivaipa Indians wanted to keep their homeland.

The one white man who became intimate with them and wormed from them the secret their harsh land concealed, thus unwittingly acquiring a fame that will probably last as long as that so striven for by "the poet of the Sierras," is remembered only by the name of Yuma. A graduate of West Point, he had seen several years of border service when, still a lieutenant, he came to Fort Yuma on the Colorado River as acting-quartermaster of that post. Ocean steamers in those days brought freight through the Gulf of California to the mouth of the Colorado, where it was transferred to river steamers, which carried it on up to Fort Yuma. Here it was discharged to be hauled by wagons far inland. Thus the acting-quartermaster's position gave him supervision over all supplies not only for his own post but for other posts strung across a vast territory. The tonnage he handled was enormous, payment for supplies usually being made by vouchers on the Quartermaster General's office in San Francisco. Such a volume of business in such an isolated region gave opportunity for peculation. The lieutenant fell under the sway of dishonest army contractors, and when official investigation revealed their practices he was court-martialed and discharged from the army.

A man of gentle breeding and an officer in whom military training had inculcated the highest degree of pride, he felt his disgrace keenly. He became a pariah from his own people and took refuge among the Yuma Indians, who lived about the fort and among whom he had made many friends. The chief of these Indians was Pascual, grave and gaunt, with leathery wrinkled cheeks and a prodigious nose, from which hung an ornament made of white bone embellished by swinging pendants. Few chiefs among American Indians have enjoyed such absolute power as Pascual wielded over the Yumas. His people at this time were living in peace and plenty. Their superior physique struck the eye of every traveler. The women were like noble partridges, and one of these old Pascual gave to the exiled lieutenant, who married her and became an adopted member of the tribe. Thenceforth he was known to white men by no other name than Yuma.

Yuma became an Indian trader. His adopted people were at peace with the various divisions of the Apache tribes, and among these, with his bride as a protective talisman, he was free to travel and traffic. Taking pack mules loaded with rifle ammunition and other goods desired by the Apaches, Yuma and his wife penetrated where no white trader before him had dared intrude. He seems to have been genuinely in love, and he was loved. The life was congenial; profits were satisfactory; he confessed himself happier than he had ever been before in his life.

About a year after his marriage he found himself among the Arivaipa Apaches, who received him well. From his wife and from other sources he had heard that the Arivaipas possessed a deposit of gold from which they had been known to barter rich specimens. This deposit they, of course, guarded with fierce secrecy. Presuming now on his own reputation for being "a good Indian," which to all intents and purposes he was, he took the friendly chief apart, and, displaying before him a fine rifle, a beaded belt full of ammunition, and some curious silver spangles, offered them in exchange for a glimpse of the coveted spot.

The chief parried and debated with Yuma—and no doubt with himself—a long time. It was his office as it was that of no other Apache to keep the traditional secret of the gold.

Still, the fact that Yuma was a member of an allied tribe entitled him to certain rights and privileges. Also, the fine rifle, the beautiful belt, and the gaudy spangles were powerful orators. Finally the chief acquiesced.

"I will show you tomorrow morning," he said.

Shortly after daybreak the next morning he and Yuma, unaccompanied, left camp afoot, ostensibly to hunt deer. Traveling in a northerly direction, they ascended a long ridge, on which they kept for about three miles. Then they came to the crest of a low but asperous range of mountains overlooking the San Pedro valley to the east. Still pursuing a route generally northward, they continued in this rough country for about six miles.

They were picking their way along the side of a gulch, keeping well up from the bottom, when the chief stopped. He stood beside an inconspicuous crater-like depression, perhaps six feet in diameter at its shallow bottom and rimmed with rock.

"Here," he said, scanning the horizon.

Yuma got inside the depression, and began digging with his hunting knife and hands. A few inches down he struck ore. It was so compact that he could only with great difficulty break off a handful. But that handful, together with the sight of the marvelous vein from which it had been taken, filled him with such joy that he could scarcely keep from shouting to the hilltops. The ore, wonderfully rich in gold, was a rose quartz. It was very beautiful.

After getting the sample, Yuma, aided by the chief, filled in the hole and carefully smoothed it over. Before leaving, he noted the lay of the ground on every side; he noted that the gulch below him headed only a few hundred yards beyond. The terrain, as he later described it, was so exceedingly broken and rocky that no one who had not seen the spot marking the ore would be likely ever to come upon it. The chief told him that, while he had nothing to fear from himself, if any of his braves ever found him in that vicinity they would kill him on sight.

Now, Yuma knew nothing of mining, but from his description of the place from which he took the ore, it must have been at the top of what is technically known as a chimney. Its position assured it against being obliterated by either cloudburst or landslide. A "chimney" of such ore as Yuma showed samples of might well produce a million dollars in gold in a very short time without much expense.

On the way back to camp Yuma and the chief killed a deer. Then to avoid the suspicion that a hasty departure might breed, the trader remained among the Apaches for several days before leaving for Tucson. There he expected to meet a friend whom he wanted for a partner. He found him.

The man was a young freighter named Crittenden, who had several wagons hauling ore from mining camps about Tucson to Fort Yuma for loading on river steamers. He was from Kentucky, a worthy kinsman of the brilliant statesman, John J. Crittenden.

Keeping the secret of the great find to themselves, Yuma and Crittenden prepared to explore the mine at once, it being arranged that the young Indian wife should stay with Mrs. Crittenden.

Leaving Tucson late one afternoon, the two men rode on horseback all night and early next morning reached Camp Grant. Here they rested until evening and then, without committing themselves, struck northward down the San Pedro. After they had ridden some ten miles, Yuma said that they were about opposite the mine; accordingly, they unsaddled, picketed their horses, and lay down to await daylight. They did not sleep much.

At daylight they began climbing the range to the west. It was so steep and rough that they were obliged to lead their horses most of the time. Hours of climbing brought them to the gulch up which the Arivaipa chief had led Yuma. They had a pick and shovel. They worked two hours, took out about thirty pounds only of the richest ore, and then, after covering the hole over and burying the pick and shovel, they set out for Tucson. Instead of returning to the valley of the San Pedro and going by way of Old Fort Grant, they coursed down the western slope of the mountains and then crossed a trackless basin. They traveled all night and arrived in Tucson about daylight.

Immediately they had the ore crushed—all but a few lumps to be preserved as specimens—and the gold panned out. From less than thirty pounds of ore they recovered $1200 worth of gold. The operation could not well be kept secret; the whole town went wild.

Yuma and Crittenden now decided that it would be wise to allow excitement to subside before attempting development. So Crittenden continued with his freighting business, and Yuma, once more accompanied by his wife, struck out on a trading expedition among the Papago Indians on the Papago Desert, a hundred miles west of Tucson. Never before had this idyllic couple traveled so gleefully, in such gay spirits, with prospects so bright. It was their last journey.

The Papagos were ever a gentle people towards the whites. It was the Papagos who under Father Kino built the mission of San Xavier del Bac on the Santa Cruz, nine miles above Tucson, pronounced by competent critics to be the most beautiful example of mission architecture in America. But gentle as they were towards *Cristianos,* towards the Apaches the Papagos were as fierce and relentless as "the tigers of the desert" themselves, and they regarded the Yuma Indians as Apache allies.

Why Yuma should have taken his wife among them or why she allowed herself to go among them will never be known. She was immediately recognized as belonging to a hostile tribe. The older Indians were consternated. They believed the traders to be spies sent by Apaches to forerun a raid. No details of the fate of the young couple ever reached the outside world, but in Tucson and elsewhere it came to be the general opinion that they had been lured into the fastness of the desert and there destroyed.

Crittenden knew where his partner had gone. He waited long weeks for his return. Finally he gave up hope and prepared to set out alone for the mine in the Arivaipa country, so that he might there post notice of his discovery and thus legally complete his title. As he and Yuma had traveled before, he now rode to Old Camp Grant. He remained there for two days, and this time he seems to have made no secret of his mission. He rode a particularly fine horse, which the soldiers much admired. Telling them that he would return that evening, he left the camp early one morning.

Crittenden did not come back. Three days later some soldiers who scouted out on his trail found his horse entangled in a picket rope and almost starved for water. He had been picketed on the west side of the San Pedro sand bed about ten miles below (north of) the army post and near the foot of a broken range of mountains. Near by were saddle and bridle untouched. The best trailers in camp scoured the surrounding country for days; they found where Crittenden had climbed the mountain afoot. That was all. He must have taken his rifle with him, but the probability is that he had no chance to use it. Quite likely he was near the mine when the bullet from an ambushed Apache put an end to his life.

It may easily be surmised that Apaches had come upon the sign of horses in their hills and had seen where the forbidden earth had been disturbed. Ever alert, they had then awaited the return of prospectors.

Soon after the disappearance of Crittenden a band of about a hundred Papagos, led by a few Americans and aided by some Mexicans, surprised the Arivaipa Apaches, who were at the time in a condition unwontedly docile, and literally butchered man, woman, and child. This horror is known in history as the Camp Grant Massacre, 1871. It assured perpetual muteness to most of the band's secrets.

As to Yuma's Mine, it is still out there in the Arivaipa hills untouched. The million dollars in gold are still to be taken from its "chimney" formation. The beautiful rose quartz that Yuma uncovered and then covered up again is still hidden by the loose shale of the mountain side. And though "the tigers of the desert" no longer guard it, it is guarded by something fiercer and more relentless than all the tigers of the world. It is guarded by the desert itself.

CHAPTER XII

IN THE SUNSHINE OF THE PECOS

For them still
Somewhere on the sunny hill,
Or along the winding stream,
Through the willows, flits a dream;
Flits, but shows a smiling face,
Flees, but with so quaint a grace,
None can choose to stay at home,
All must follow, all must roam.
—Robert Louis Stevenson.

AND nowhere are the tales of Coronado's children quite so bright and fresh as along the upper Pecos. That must be because of the sunshine, in which all day long the children sing:

Tu eres Lupita divina
Como los rayos del sol.

Pecos village, seven thousand feet up in the mountains of New Mexico and thirty miles southeast of antique Santa Fé, was a pueblo of Indians centuries before Columbus dreamed of a sail. On his long march in search of the Gran Quivira, Coronado camped at the village, then called Cicuye; it was at that time the largest and strongest of the pueblos. By 1620 the Spaniards had erected a church at the place, and in the great Pueblo revolt of 1680 the priest of this church was assassinated while fleeing the village. The Pecos Mexicans point to the red earth along the high mesa southwest of the Arroyo Pecos and say that the priest's blood stained it. The Santa Fé Trail from Missouri twisted by Pecos village. In the quadrangle against the Pecos church the fated Texans of the Santa Fé expedition were, in 1841, herded as prisoners before setting out on their two-thousand mile Journey of Death to the prisons of Mexico City. Meanwhile the Pecos Indians had declined until the remnant, only seventeen souls, "abandoned the crumbling ruins of the dwellings that had housed their ancestors for so many centuries" and went to Jemez eighty miles away, leaving the relics for legend to mystify and archaeology to excavate.[1]

The modern Pecos village is down under the hill from the ancient pueblo. Here lives José Vaca. I shall never forget him. First of all, he took me to inspect the deep cave that for four years he had been boring into the crooked seam of a mountain. After we had explored it, we climbed high above, whence we could see the Pecos River winding and shining in the sun and land after land of mountains beyond. Then, high up there in the sunshine, José told me the history of his cave and many another tale out of his inexhaustible supply. I say inexhaustible, for his father and grandfather sought all their lives and he has sought all his life to find the golden treasure that for each succeeding generation shines brighter and more alluring.

With many apologies to José—and to the sunshine on the Pecos—I set down some of the stories that he told, only omitting the most joyous profanity that I have ever heard. Picture him cross-eyed, his mouth twisted into an extraordinary slant, and his whole body gesturing with Latin energy while he talks and talks. Think of him, too, as very honest, for he knows how to say "I do not know." And remember, as you listen to the tales, that New Mexico is a land of dead cities and that the Indian of today is as chary of his secrets as are the Indian skeletons that have lain drying in those dead cities hundreds of years.

Bewitched Sand

"My grandfather live in Pecos down there to be more than a hundred years old," said José Vaca. "When I was young before he die, I hear him say many things, but I was not careful then to listen. He knew Indians that lived here in this Pecos pueblo and he bought a piece of land from one of them. After he pay for the land, and the Indian was leaving to go far away, the Indian he say: 'You have here now more wealth than is in the world elsewhere.'

" 'How?' ask my grandfather. 'Show me.'

"Then the Indian take him and a burro to where was some sand in the creek. They put some sand in sacks and bring it on the burro, and they get twenty-five dollars worth of gold out of that one load of sand.

"That night the Indian disappear, and the next day my grandfather he go with two burros and load both with the sand. He bring it up, and from it he do not get one thing. Nothing, I tell you. That old Indian is gone, but he has his eyes on the sand. Maybe he was *un brujo* [a wizard]. Maybe the sand was *embrujada* [bewitched]. I do not know. I know when the Indian is here the sand has gold. I know when the Indian is gone the gold is all gone too."

El Macho

"El Macho is little way up the river from Pecos. A long time ago one Mexican living there named Epimeño bought a boy from the Indians and raise him to be *pastor*. This boy used to go out with the sheep and stay all day. He was very wise. He was strong wrestler too. Now, this Epimeño has a son named Licencio. One day while Licencio is wrestling with the Indian, he feel something hard against his own chest.

" 'What is that?' Licencio ask.

"But the Indian boy will not tell. Then Licencio t'rew him down and found a piece of pure silver, as big as hen egg, tied to a buckskin string around the Indian's neck.

"The Mexican boy tole his father, and then old Epimeño commence to follow the Indian everywhere. 'Tell me where you find that silver,' he say. 'If you do not tell me I will kill you.'

"The Indian he took Epimeño first to one hill, then to another, but he never show him the silver. Old Epimeño beat him and starve him and crawl behind him in the pasture like a coyote. After a while the Indian boy tole Licencio that he is going to run away and find his own people. 'I cannot stand such hard master,' he say.

" 'You been brought all this trouble on me,' he say to Licencio, 'but you are kind. I remember always the food you steal for me. I will tell you where the silver is. It is on the mountain just west of Epimeño's house. Go there and you will see an old shoe hanging from a tree. By that shoe is enough silver to make you rich forever.'

"The Indian left. Licencio was scared to tell his father how the Indian came to trust him. Maybe he did not believe the story about the silver. He was too lazy to climb the mountain to hunt the old shoe. But one day he tole a friend named Parrilla.

" 'Let's go,' say Parrilla.

"They go. Sure enough they found the old shoe. But they cannot find the silver.

"Perhaps the Indians were watching them."

La Mina Perdida

The big mine of the Pecos country is somewhere far up towards the head of the canyon. It was worked centuries ago by the Spanish, and much of the wealth buried by them down the river, while they were trying to reach Mexico with their precious cargoes, came from it. The mine may have been either gold or silver, but the consensus of opinion is that it was gold. It is called simply La Mina Perdida—The Lost Mine.[2]

"One time," said José, "my father-in-law and I were hunting deer up in that country. Oh, it is far, far from here! We had a deer loaded on one horse when it began to rain. We got under some trees, and while we waited what do you think we found there? We found an ole flume and an anvil. The anvil was more big than any anvil I ever see before. Very certainly some big work was done there at one time. Well, it kept raining and we stay there for two hours, and I look and look at the mountains all about and the trees and at everything so that I can know the place when I come back.

"There was a canyon close by with very pretty water in it, and I want to stay and try for fish, but my father-in-law he keep saying, 'Let's go, let's go.' I know very well there are fine trout in that canyon, but I must wait for another time to catch them. Then we left while it was still raining.

"We come back here to Pecos, and I tole ole Martín Salazar about that flume and anvil, and he say: 'Very surely you have found The Lost Mine.' Ole Man Hughes he say the same thing, and Francisco Ortejo, and many other ole mens also. Those ole mens been live a long time and surely they know some things, but all are dead now.

"I tole my father-in-law if he can go back to the flume and anvil, and he say of course he can. So when the time is come to hunt we go. But, son-of-a-gun, if we can find that place. I never even see that canyon with nice water in it where I know very well the trout are waiting. By holy saints, but to hunt for lost mine and lost money is hard work, very hard. Always something, always something, look like, is against the man. The Indians they watch that lost mine too."

That Ole Man Devil

José's most fatalistic experience, however, is connected not with gold or silver but with what he calls a mica mine. It is above the headwaters of the Pecos. Like La Mina Perdida, it was found by him while he was hunting in company with his father-in-law.

"We come on a whole mountain side of the mica," said José. "The mica was clear to see like that shining water down there in the *acequia*. We stay there a long time and get a lot of the mica and pack it on the burro. The slabs be so big I have to break them up. Oh, there was a quarter, maybe a half mile, of the mica all clear, right there on the ground, uncovered. It was joy to see it.

"After I load the burro, I feel a nail in my shoe, and I take the shoe off and hammer down the nail with a rock. While I am working, a piece of the heel come off and I leave that ole heel there on the ground—right where I get the mica to load on the burro. Then we all feel fine and come back to Pecos.

"Soon afterwards I was working on the new road up the river. Mr. McCarch he was boss, and I have plenty of time to talk to him. It is my business to shoot the dynamite, and often I have to wait long times for the other men to get ready for me. While I wait and talk, I tell Mr. McCarch about this mica mine. He get very excite and ask me to show him samples. I bring him a piece; he send it off, and after he hear from Denver—or maybe it was some other place—he get more excite. I never see a man so eager like that.

" 'We will have to work no longer,' he say. 'There is plenty mica for you and all your children and grandchildren and all your *parientes* and all the children of your *parientes*. Plenty too for me. I will be pardners with you, and we will work the mine.'

"Then we make pardnership and go before a judge and sign paper. I was one; Mr. McCarch was one; my father-in-law he make three, and Francisco he make four. Francisco is brother-in-law. He want to bring another American, but we will not let that, and we talk too much. Then we get ready to go up for the mica. It was the right time of year, in July, when it is not too cold. But Mr. McCarch he cannot leave his work. So he sent a man was named Iglehart in his place.

"It takes us three days to go from the Pecos River into those mountains. And then high, high up we make camp close to the mica mine. When we stop, Mr. Iglehart and Francisco begin to sleep, and I never see such men for sleepy. I do not know what is the matter with that man Iglehart. He sleep till eleven o'clock next day; then he eat some breakfast and dinner and go back to sleep. He sleep all that day and all that night. The second day I say to him if he has come to hunt mica mine or to sleep. 'Oh, I am so sleepy,' he say, and he sleep all that day.

"When third day was come, I say to him, 'Let's go find that mine. What for you come up here? You never going quit sleep?'

"Then he look round and say, 'Oh, see that cloud. Rain coming this day. I do not like to go out.'

"Then he sleep all that day.

"I do not know what to do. I look and I look and I know very well we are in the right place, but there is no mica. I tell you I find that heel of my old shoe which I leave there when I was loading mica on the burro. But there is no mica now. My father-in-law he look too and he say that sure we are in the right place of the mountains. Francisco at last was looking also, but that man Iglehart will do nothing.

" 'I cannot walk,' he say. 'It looks like rain,' he say.

"Nobody lives in that mountains—nobody. After while I go down into a canyon, and there I meet an ole Mexican man. He is looking at the ground and has no gun.

" 'What you doing here?' I ask him.

" 'I hunt bear,' he answer.

" 'You hunt bear here all alone and so ole and have no gun!' I say.

" 'Huh,' he say, 'what you do here? You hunt mica mine. You no can find it. No use for you to hunt mica mine. You never put cross up when you see it first time.'

"I do not know who that ole man is. Maybe he is devil.

I never see such looking man before. But I want to be friends with him. He knows something about these mountains, I think. Perhaps he has been here long time. I want to treat him nice so he will tell me something.

" 'Come to camp,' I say. 'We have plenty to eat. Come and eat with us.'

" 'Oh, I have lots to eat,' he answer. 'I eat up there.' And he point to the sky. I cannot understand what it is.

"Then I go back to camp and tell my father-in-law and Francisco what I have seen. Francisco laughs at me and thinks I am liar. Then next morning over in another canyon, there he see that same ole man, and the ole man talk to him just like he been talk to me. Then Francisco knows I am speaking truth.

"It is no use to hunt for mica any longer, I see very well. When I say, 'All right, let us go back to Pecos,' that man Iglehart jump up.

" 'Oh, I am feeling fine,' he say. 'I can walk. I do not know what has been the matter with me. It is not going to rain now, sure.'

"Son-of-a-gun, damn! I know that he has bad heart, and I was wishing many times on that trip that he was not with us. That Francisco, too, had some bad heart.

"We come back and Mr. McCarch was very sorrowful. 'You and I go all alone sometime,' he say. 'We will not take the others, but we will be honest and give them their part when we have found the mine.'

"We make plan to go the next summer, but when summer is come Mr. McCarch is having trouble with his wife. She is getting divorce from him. He cannot go. I do not know what has become of that man. That was eight, ten year ago. I wish he write me and say, 'Let's go.'

"Sandoval—he lives here in Pecos—has seen that mica too. It is there. If we had it we could live always in pleasure. I never see anything like that mica. I thought I could go to it with my eyes shut; I went with them open and I could not see. What you think about that ole man devil had talk with me?"

The Montezuma of the Pecos

All over northern Mexico and the southwestern part of the United States the fable of Montezuma is believed and told. The word *montezumas* has among Mexicans become a synonym for *ruins.* At Del Rio, four hundred miles down the Rio Grande from the *montezumas* of El Paso, Sugar Loaf Hill hides the fabulous riches of Montezuma secreted there to prevent their falling into the hands of the Spaniards. The famous ruins of Casas Grandes, Chihuahua, are but "the old houses of Montezuma." At the head of an inaccessible canyon in Sonora, Apaches today are guarding Montezuma's millions. Out in Arizona a gigantic sink into which legend has cast Montezuma's wealth goes by the name of Montezuma's Well, and in the same region some ancient cliff-dwellings bear the name of Montezuma's Castle.[3]

Various localities of New Mexico claim to be the birthplace of Montezuma and to be now the repository of his hidden treasures; but the claim of Pecos village is most insistent, most famous. Somewhere in the mountains about Pecos, according to belief among both Indians and Mexicans, a fire is kept constantly burning in a cave awaiting the return of Montezuma from the south, though a giant serpent sometimes devours the pious tenders of the fire. Some of the Pecos Mexicans say that Montezuma wore golden shoes and that he walked in them to Mexico City, where the Spaniards confiscated them—so that he could not walk back. According to other *Pecoseños,* Montezuma never left their village at all.[4]

Be all this as it may, a young Pueblo Indian a good many years ago was put in jail at Las Vegas on the rare charge of raping a girl of his tribe. Now the tribal punishment for rape, so José says, is to strip the guilty man naked, stake him down in an ant bed, and leave him there until the ants have left nothing but the bleached bones. The Indian in Las Vegas jail was less afraid of the courts than of tribal punishment; he knew that even if he were sent to the penitentiary, he would be watched by his tribesmen and put to their terrible punishment the day he was released. But I shall let José tell the story.

"The Indian tole his fear to Don Salamón. He was sheriff to watch the jail. This Salamón has a very kind heart, and he tole the Indian he will let him run away in the night time so he can go far, far from his people and not have to die in ant bed. So he let the Indian loose, and that Indian is so happy and so grateful that he tole Don Salamón the great secret of all the Indians of the Pecos country. Now the Indian could surely never come back, for it was death to tell this secret.

" 'You go,' the Indian say to Don Salamón, 'to the ole church at Pecos pueblo and find the Spanish road to Santa Fé. Right on that hill where the pueblo is, not more than the length of a *cabestro* [hair rope] from the road, you will find a white rock with an old cross on it. The cross is made of wood; it is very little, and it is planted down in the rock so that it will not fall. Right next to that white rock is a black rock. The black rock is to hide a cave. But do not go into the cave. Dig under the white rock. You will not have to dig deep. A very little work will do the business. As soon as the white rock with the cross on it is pulled up, you will find the *patrón;* then only seven feet under the *patrón* you will find the gold.'

"You know," José reminded me, "what *patrón* is. It is the dead man who guards the treasure. All these peoples long time ago who hide great treasure been careful to have *patrón*. The Indian tole Don Salamón this *patrón* was Montezuma. Montezuma was half Spanish and half Indian, but he love his Indian people more, for he grow up with them and is a great chief. And Montezuma have gold—more than any man today can count—and he knows if the Spanish find this gold is among Indian people, it be worse with them than if they keep poor.

"So when Montezuma was dying, he order that all his gold be put in one great hole close to his pueblo. When he see the gold all down in the ground, he order that they fill up the hole but leave room for him at top. Then he make his people put him in the hole so he can sit with his back against the wall. All his people are around him. 'Swear,' say he to them, 'that you will never take the gold away from this place until the Spanish are gone from Indians' land.' And they all swear. Then Montezuma he die in the hole and they leave him there for *patrón* to guard the gold, and over him they place big white rock with cross.

" 'All these centuries,' the Indian tole Don Salamón, 'my people been keep the order of Montezuma, and it is the law with us if any man tell the secret he will die more hard than in the ant bed. But you save my life. I have gratitude. I leave now my people forever. I tell you the secret to thank you, for I have nothing else. Let me tell you when you go to dig, you must go in the night, you must dig quick. Work like lightning, for eyes will be on you. It will be danger to stay there after you dig. Have big car ready to take you away as soon as you get the gold. The *patrón* watches also. You must guard your life.'

"After the Indian tole these things he left, and then Don Salamón he come up here to see me, for I know the country. I do not know where the white rock with the cross is, but I know all ole Spanish roads. Don Salamón say he can find the rock if I show him the right road.

"Then we get axes and spades and picks and everything in his car ready to go. Don Salamón's wife was there and she begin to cry out that her husband do not go. 'You will get killed, you will die,' she cried. She put her arms around his neck and there she cry until Don Salamón is ashamed to go. He must stay home.

"Since that time I have been looking for that white rock with cross on it next to the black rock. I have not found it. Perhaps the cross is gone now, but I know very well no man has got Montezuma's gold. I cannot understand this *patrón*. I am strong man. He is dead, but he keep me off. I wish I know when he sleep."

José Vaca's Cave

All of José Vaca's other labors in search for lost treasure dwindle into insignificance when compared with his *magnum opus*. For years now he has been tunneling into one of the Tecolote Mountains overlooking the Pecos River just below the village of Pecos. During the summer he works for wages, saving what he can; during the fall and winter he follows the business of excavation with an unflagging abandonment that only genius can realize. "I used to have many guns, plenty of horses; now I have nothing. All is spent on the cave," José says, "and I am not through yet. I wish some rich man with interest in this work might come along to help me finish."

José's own "interest" began with his finding in the cave a conch with an Indian design on it—a "spoon," as he calls it. His initial hope and expectation was to uncover a rich hoard of antique Indian pottery. But he had not worked long at clearing out the cave before the sole object of search became Spanish treasure. The fame of his digging brought from every quarter other treasure hunters with their *historias* and *derroteros*. A local woman became interested and ventured a hundred dollars in the enterprise. In the cave today may be seen picks, cans, ropes, spades, lanterns, wheelbarrows, wooden supports, all the paraphernalia of a mighty work uncompleted. Tons and tons of earth substance have been carried out and dumped down the mountain side.

During his years of work José has had various partners. One partner, a man well along in years, came from Mexico, worked a while, and went on. He was to return, but word came that he died in Canada. He was an Englishman, and he was significant for the kind of instrument that he used in locating treasure. It was a *cartucho* (a kind of metal cartridge shell) filled with opium, poison, black rock finely powdered, and a varied assortment of other elements. Leaning over, the man swung the *cartucho* from his forehead and followed the direction in which it oscillated most strongly. When over the treasure, the instrument was supposed to cease its oscillations.

Another man, a Mexican who had made "a big find" at Albuquerque and who had the reputation of being extraordinarily successful as a treasure hunter, came seeking to join in the Pecos Cave venture. But one of José's party was against him, and so the greatest opportunity of all to enlist a man of means and also of "interest" was let pass. This Albuquerque Mexican had an *horqueta* (fork) fashioned from the flat scapula of some animal.[5] It was shaped thus:

The upright branch fitted into a glass knob, on top of which was a hole with threads. An assortment of hollow screws fitted into these threads. One of the screws was filled with gold dust, another with powdered silver, another with lead, another with copper, and so on. The owner used the screw filled with whatever metal he was looking for. To make the *horqueta* work, he grasped the lower prongs in his two hands, thumbs out and palms turned up, and then followed the "pull" if there was any—and there generally was. The *horqueta* would twist and strain in the direction of the sought-for mineral as if drawn by some powerful magnet, and then, when over it, would point straight down in the manner of "switches" that "turn" to water.

In testing out José's cave the owner of this remarkable instrument used both gold and silver screws, and he declared that his *horqueta* had never "pulled" so hard in any other place. He went into the cave and remained there alone for half a day. According to José, he was transported into a kind of ecstasy, lying on his back and holding the instrument over his head so that he might sense the "pullings" more delicately. It certainly is a pity that such a meticulous artist should have been excluded from the treasure hunt.

Once José went to Santa Fé for a famous *brujo* (wizard) and fortune teller named Nicolaso. Nicolaso declared absolutely that treasure is stored in the cave. "Go on," he said to José. "The treasure is hard to find, but if you work long enough you will get it out." Indeed, Nicolaso was himself going to join in the undertaking, but a short time afterwards he located a great treasure in a certain old *gachupin* house at Santa Fé, and as a result he at once bought a fine automobile and left the country.

The most remarkable of all the men who came to José, however, seems to have been Charlie Rose. Rose knew about every lost mine and buried treasure from Mexico to Alaska. He is dead now, but the *derrotero* that he brought from Las Vegas has been the basis for practically all of the work so far done in the cave.

This *derrotero* called for a rock cemented into the ground somewhere down the slope from the cave entrance. On this rock was to be found a small wooden cross. Beside an ancient wood road that led down from the mountain, José and Charlie Rose found a rock that seemed to be cemented into the ground.

The cross was gone, as naturally it would be after so many centuries, but they detected a hole where it had been and some "cement." José showed me the rock. It weighs several hundred pounds and is flat on one side and convex on the other, resembling in shape those old Mexican cart-wheels hewn out of solid tree trunks. I was unable to detect any cement.

The *derrotero* called next for a marked rock at the entrance of the cave. This rock should have certain letters, figures, words on it. It was not hard to find a rock at the mouth of the cave. The one settled on as *the* rock weighed several tons. The treasure hunters believed that they saw signs and figures on one of the exposed surfaces, but they could not make out any particular letter or figure. They dynamited the rock so as to ascertain what might be inside or under it.

The legend on which the *derrotero* is based is that the early Spanish were bringing an immense cargo of pure ore from La Mina Perdida up the Pecos, when they were attacked by Indians. The Spanish retreated with their bullion to the cave, stored it, and cemented some rocks over the entrance in such a way that they should seem left thus by nature. Then followed a battle in which all the Spaniards were killed.

Certainly the rocks of the cave now have the appearance of having been placed there by nature. Some of them weigh thousands of tons and cover acres of ground, and there are gigantic seams that extend probably for miles. Far back in one of these seams José and one of his ill-starred partners have sunk a well thirty feet deep. While they were digging rock out of this well, a violent explosion of stinking gas occurred. This explosion is the most encouraging thing that has ever happened, for, according to all gold hunters, buried gold gives off a gas that is deathly sickening and highly explosive. The problem now is to get at the source of the explosion.

Meantime, the silent watchers of all things in the Pecos country, the Indians, have not been blind. No one has seen them near, but the track of their moccasins has more than once been found by the cave. Sometimes, too, the whole side of the mountain has been enveloped in a vague light rising and falling, the portent of which José and his partners—these modern inheritors of the wisdom and hope of Coronado—are in no doubt of.

José's One Lucky Find

The last partner to appear in the strange procession that has dreamed and delved through José's cave called himself a Frenchman. "I never been lucky only one time in my life," went on José as he told of the Frenchman. "That was right here in Pecos. A man got scared to live in his ole house. He was hearing strange noises and seeing lights. So he move into another house, but he can still see lights in the adobe he left.

"One day he come to my pardner, the Frenchman, and say he think there must be some treasure in that house. He will go with the Frenchman and help get it, he say. But when they are ready to go one night, they see strange lights burning and the ole man is afraid to go. So he give the key to the Frenchman and tole him to look.

"Well, the Frenchman he say for me to go with him. When we come to the house, that Frenchman say for me to stay outside. He go in and I can hear him talking. I do not know who he is talking to. I do not know what language he speak in. It is not Spanish; it is not English; it is not French. Then after while the Frenchman come out. 'All right,' he say, 'we come back tomorrow night at this hour and take out the money.'

"When we got back the next night, the Frenchman unlock the door. We step inside, and while he is holding the lantern, he gives me the pick. 'Hit into the wall over the door,' he say.

I do not want to dig into that wall, but he keep saying, 'Go on, go on.' So I take the pick and give just one hit into the adobe where he say, and out come money rolled up in an old paper. There is one hundred and twenty-five dollars. The Frenchman say it is not necessary to give any to the man who was coward. So he take half and I take half.

"It is not much money for so many years of trying."

CHAPTER XIII

THE PECOS BARRICADE

On the rocky banks of the Pecos
  We laid him down to rest,
With his saddle for a pillow
  And his gun acrost his breast.
—Cowboy ballad.

THE waters, fresh and crystal, of the upper Pecos River, springing out into sunshine that blesses rather than blisters, are not the waters that hurl themselves bitter and murky into the Rio Bravo. The Pecos is a long river, a strange river, a thousand miles of twisting canyon from the pine-clad mountains of New Mexico to the gray, bleak bluffs of the Rio Grande on the Texas border. The traveler from the East upon reaching the drainage of the Pecos can yet say, "I have arrived in the West." And until environment and occupation cease to operate upon the character of men, the sparse ranchers of the Pecos will still be Westerners. The mountainous breaks, the alkali flats, the vast stretches of shifting sands, the treeless plains rolling out to far away hills—the Pecos world, despite narrow strips of irrigated land, will not be plowed up or, save around isolated oil fields, transformed by the structures of population into standardized mediocrity.

Far south of the Santa Fé Trail, three other transcontinental routes crossed the Pecos, converging west of it: the Butterfield (also called the Southern Overland and the Emigrant) Trail running from Saint Louis to California, the Chihuahua Trail from New Orleans and San Antonio west, and between these two the San Antonio-El Paso stage road. Of the three principal crossings, the Fort Lancaster, the Pontoon, and the Horsehead, Horsehead Crossing was the most famous by far.

No stranger ever reached it from either side of the Pecos without having yearned for many hours and without being surprised at the sudden appearance, at his very feet, of the timberless banks; but no traveler was ever tempted to stop and settle at Horsehead Crossing. Early Spaniards, so tradition says, marked it with the bleached skulls of mustangs; hence the name. Bones of cattle used to line the waterless Goodnight-Loving Trail that stretched for ninety-six miles between the head of the Concho River and Horsehead Crossing. Bones and graves of human beings likewise marked the routes traversing the Pecos lands. Indians, thirst, and "bad men" were all responsible for these desolate remainders.

Here history is so stark, so real, so dramatic that no fiction can surpass it in picturesque qualities. "Sunday, July 2," runs an entry in the diary of J. M. Bell, who was helping trail a herd of cattle from San Antonio to California in 1854. "The bones of a man were found [at Comanche Springs west of the Pecos]; on the knee-cap and foot the muscles still remain, although it has been 3 years since he was killed. Some of the clothing is laying about him." At Howard's Well, a spring between Beaver Lake and Fort Lancaster, travelers used to see the charred remains of six wagons, the occupants of which had to the last man been killed by Indians. Such a sight was characteristic of the Chihuahua Trail—and the Pecos. "From El Paso to San Antonio," recorded Edward F. Beale in 1857, while conducting what was probably the strangest train that ever traversed the route—the train of United States Army camels—"is but one long battle ground—a surprise here, robbery of animals there. Every spring and watering place has its history or anecdote connected with Indian violence and bloodshed." [1]

The great trains of wagons passing between Chihuahua City and San Antonio were sometimes veritable argosies of treasure. In five years' time the enterprising August Santleben [2] alone transported over a million dollars in freshly minted Mexican silver, one of his caravans in 1876 carrying 350,000 pesos. "In 1853 and 1854," says Theophilus Noel in his strange autobiography, "the ore from all the mines of North Mexico was hauled to San Antonio, much of it on wooden-wheeled carts, where it was taken by Texas teamsters to Port Lavaca and thence to England for refinement."

Out of such a background legend was inevitable. Legend says that not all the silver and gold that braved the routes crossing the Pecos got through. Some of it is still at Castle Gap.

Maximilian's Gold

In 1867 the Rio Grande was the refuge line for two distinct classes of vanquished men. From the north various bands of Confederate soldiers, distrustful of their fate at the hands of Union carpet-baggers and too proud to submit to dominion of a foe, were crossing the Great River to make their fortunes in Mexico and other Latin countries. They were generally desperate men, indeed. To the south Emperor Maximilian, a puppet placed on the throne of Mexico by Napoleon III, had been overthrown, and various adherents of his were fleeing for the safety that lay beyond the Rio Bravo.

How Maximilian took refuge in Queretaro; how here, after months of miserable indecision and squalid existence, he was captured; and how he was then at the orders of a savage but patriotic Indian, Benito Juarez, shot to death—all this is familiar history. As he stood before the wall waiting for the "lucky bullet," he gave each of the soldiers a piece of gold and told them to aim true. He had brought with him to Mexico a large private fortune, including rich services of plate. Whether those gold coins handed to his executioners were all that remained of the fortune, history can hardly say. Maximilian and the mad, forlorn Empress Carlotta became the object of pity and the theme of story from Yaqui huts in the canyons of Sonora to the court of Austria. One of the stories long current in Texas deals with Maximilian's fortune.[3]

Among the ex-Confederate soldiers who, after a brief trial of carpet-bag rule, swore they would no longer live under the flag of the United States was a band of six Missourians. Their names are unknown. They came to Texas and rode west over the Chihuahua Trail to Presidio del Norte on the Rio Grande. There they met a caravan of wagons coming out of Mexico.

The teams of this train appeared to have been driven very hard. The man in charge of it was an Austrian, but he spoke English. He was exceedingly anxious for information concerning road conditions to San Antonio. The ex-Confederates told him that the route was infested with Indians and lawless characters. At this the foreigner volunteered the information that he had a valuable cargo of flour, and he offered good pay to the exiles if they would turn back with him and act as guard. Time meant nothing to them; they were ready to accept what fortune offered; they turned back.

The travelers with the wagons numbered fifteen, being made up of Austrians, Mexican peons, and a beautiful young girl who appeared to be the daughter of the leader. During the day, as the newly hired guards came to observe, the drivers were constantly in their seats, and at night some one slept in each wagon. The wagons were tightly covered and no opportunity for looking inside was allowed.

Such caution aroused first the curiosity and then the suspicion of the daring men who had engaged to accompany the train. After consultation they selected one of their number to make a close investigation. His report was astounding. The wagons were loaded with gold—heaps of it—Spanish, Austrian, and American coins, all sorts of vessels of gold and silver, bullion too—a fortune fabulous in value. The guards at once began planning to kill their charges and seize the cargo.

By now the caravan was approaching the Pecos River. The ground seemed clear for action.

At Castle Gap, fifteen miles east of Horsehead Crossing, about midnight, the crime was committed. As the six desperadoes themselves constituted the watch, and as all the Austrians and Mexicans were asleep, the surprise was complete. Not even the girl was spared. The bodies were burned in a fire made of wagons, harness, and other properties. Papers taken from a chest revealed that the leader of the dead band was one of Maximilian's followers entrusted with carrying the royal fortune out of Mexico to Galveston. From Galveston it was to be shipped to Austria, where the Empress Carlotta had already gone and whither at the time the instructions were written Maximilian intended to flee.

The robbers were not prepared to take such a tell-tale store of valuables into the settlements. They agreed to take only enough coin to satisfy their immediate desires and then, when they should have made arrangements for a quiet disposition of the vast fortune, to return for it. Consequently they buried it, noted well the landscape of rock, sands, and lake, and rode on east.

The wasted condition of their horses delayed them in crossing the parched land between Castle Gap and the Concho and they suffered greatly. One of them became so ill that he was forced to stop at Fort Concho. It was well for him that he remained behind, for a day's ride on ahead his five confederates were attacked by Indians and all killed. The lone survivor, having recovered, set out for San Antonio, where a rendezvous had been appointed. He knew nothing of the massacre of the men who had preceded him, but on his way he saw their mutilated bodies. He was now sole owner of Maximilian's fortune. He decided to go to Missouri and there secure the help of the James boys in disposing of it.

Near Denton, on his way north, he camped one night, merely by chance, with a party of three or four men who turned out to be horse thieves. A sheriff's posse surrounded the camp about daylight and took every man in it prisoner, the Missourian included. Horse thieves were frequently hanged without delay, but these were not. In the Denton jail the malady that had forced the Missourian to halt at Fort Concho renewed itself. He was now a very sick man. A certain Doctor Black, a resident of Denton, gave him no hope of recovery unless he could secure his freedom. He sent for a lawyer named O'Connor. Before O'Connor could do anything, however, his client was in a dying condition. Realizing his fate, he turned over to Black and O'Connor a plat to the fortune buried at Castle Gap, told them all the circumstances connected with it, and then gave up the ghost.

When Black and O'Connor got out to Castle Gap—a far journey then that required time and preparation—the lake was dry, and the terrific sandstorms common to the region had shifted the landscape. Hence marks called for by the plat could not be identified. All that was found were some wagon irons marked by fire. Sphinx-like in its muteness amid the deep and silent sands, Castle Gap still guards Maximilian's gold.

Rattlesnake Cave on the Pecos

Away to the southeast of Castle Gap, out of the sand country, the Chihuahua Trail followed up Devil's River, left it at Beaver Lake, ran alongside Dry Draw, and then crossed over the Divide to the Pecos. Somewhere along Dry Draw *was*—for it is not now—a cave known as Rattlesnake Cave. I have never heard how much gold ore was stored in this cave or how it came to be there. Maybe some of it was native to the place. Maybe it was the "gold blocks," stuffed hurriedly into hiding by Spaniards, that early-day cowboys used to look for in every cave they glimpsed between the Concho River and Fort Stockton. One of the J M riders years ago discovered just east of the old Pontoon Crossing on the Pecos what he thought was the right cave, but upon exploring it found only stacks—enormous stacks—of buffalo hides. They were all rotten.

One day while a sixteen-year-old Mexican *pastor* was herding sheep out from his camp at Beaver Lake, he discovered a cave. The entrance was just about the size of a barrel, but he wriggled through it. This was along in the spring, and if he thought anything at all about rattlesnakes, he must have thought that they were out, for rattlesnakes do not "lay up" much in caves except in the winter time or when the weather is blazing hot.

The *pastor* found the inside of the cave as dark as a stack of black cats, but he lit a sotol stalk, an excellent torch, and went on. Then all of a sudden he heard a rattle, and right in front of him he made out "the biggest rattler raring up that ever he'd dreamed of." From the way he described it, it must have been reared a yard high, and was "as thick through as a man's leg." Its tail was singing like a buzz-saw, and then a thousand other snakes set up a rattling in every direction.

The Mexican was scared almost into a state of lunacy, but he had sense enough to begin backing towards the mouth of the cave, the big snake following him. About the time he reached daylight, his foot kicked against some rocks. He reached down and picked up a couple and threw one, either killing or stunning the snake.

Then he was out and thirty steps away before he realized that he was still carrying the other rock in his hand. As he was about to cast it away, he noticed how heavy it was. He looked at it. He hurriedly concealed it in his *hato*.

Three or four days later the *vaciero*—the man who supplies and oversees the various camps on a big sheep range—came along in his buckboard to leave some "grub" at the Beaver Lake camp. When he got there, he found the *pastor* all hunkered up, *"muy malo,"* he said, and wanting to go to town for some *medicina.* The Mexican laborer generally gets sick when he wants to quit or lay off. There was nothing for the *vaciero* to do but leave a substitute with the sheep and take the *pastor* to Ozona.

Well, the first thing the Mexican did when he got to town was to go to the bank. The rock he presented proved to be a crude block of gold and silver, mixed. It weighed three pounds and seven ounces. The banker was considerably excited. But not a word of explanation could he get out of the Mexican. The boy simply would not talk. Then the banker sent out and got the ranchman for whom the boy worked and also the leading merchant. Meanwhile he held the Mexican.

All three of the white men could *hablar Español,* and they had no trouble making the boy understand what they wanted. He was naturally a little scared of his boss. The upshot of the interview was that he promised to take the men to the place where he had found the "rock." He told all about how he discovered it, but he did not describe a single landmark. All they knew was that it was within herding distance of the camp at Beaver Lake. That meant that it could not be more than six or seven miles at most in any direction and was probably much closer in.

The boy said that before he left town he wanted to see his mother. That was all right with the white men. Naturally, when he got to his mother's house he told her all about the cave, the snakes, the gold, and his promise to the gringos.

"Oh, my son," the old woman cried, "do not show them that place. Is that your gold to give away? Did you put it there? Don't you understand that the snakes are spirits to guard it? You will be cursed forever if you tell the secret. Go now and tell the *padre* and he will tell you the same."

Then the boy went to the *padre* and told him the whole story. It was just as his mother had said. The *padre* became even more excited than the old woman had been.

"Do not show the cave to anybody," he said. "That gold belongs to the Spanish and the church. The great rattlesnake and all the other snakes in the cave are there to guard it. They are spirits, and if you tell their secret, you will pay with your life and your soul will end in hell."

The boy had been scared all along and this kind of talk decided him. So he hung around Mexican town and kept out of sight as long as he could. Then his *amo* came after him, and it wasn't any use trying to hold out longer. The white men tanked him up on whiskey, and finally headed him towards the ranch. He and the ranchman rode horseback, the banker and the merchant following in a buggy.

What happened after the outfit left town nobody knows. They made camp somewhere on Johnson's Run. The next day the Mexican was brought back to Ozona with a bullet through him, and he died without having told his secret. Some of the circumstances got out, and soon the Mexican consul at San Antonio took the matter up with the governor. The governor ordered Ranger Neal Russell to investigate.

The white men swore that the Mexican killed himself. But the wound was made with a high-powered gun, and the only weapon the Mexican had to shoot with was an old .44 Winchester. The bullet had entered the thigh from the rear and come out at the groin, and men do not shoot themselves in the thigh when they want to quit living.

Neal Russell always held that after the Mexican sobered up he tried to run away and that one of the white men shot at him to scare him and shot too far up. Nothing ever came of the matter in the courts. The men camped for a few days at Beaver Lake and hunted out from it for the cave, but, failing to find it, soon lost interest in the treasure.

In working up the evidence, Russell found that between the time the Mexican entered the cave and the time he went to town, one man had seen him and talked with him. The Mexican told this man a few more details than he had told the banker. They were not too definite at that. From the facts gathered Neal Russell made the best way-bill he could and gave it to Wes Burton. The rest of the story can best be told in Burton's own words.

"I kept the information close for a good while, but just before we got into the war with Germany, I took Preacher Crumley in with me—Neal had died meanwhile—and we started out on an exploring trip.

"When we got into the country Neal had described, we found so many places for caves that I'll confess I didn't have much hope of locating one with an opening no bigger than a man's body.

"Well, while Crumley and I were jogging along up Dry Draw, about thirty miles below Ozona, a mesquite thorn stuck into the front casing. Crumley said he'd patch the tube, and for me to scout around a bit.

"I knew all along that we were close to the old Chihuahua Trail, but I didn't realize how close till I walked right into the gully that marks where it used to run. I follered the gully up a ways until it kinder played out and then I could actually see ruts those old Mexican cart wheels had worn. Then, bang, I came to where it looked as if an explosion of some kind had blowed the whole earth out, and I never seen the like of rattlesnake rattles and bones in my life as was laying all about. I picked up eighteen rattles and put 'em in my pocket to show Crumley.

"By the time I got back to the car I'd killed two rattlesnakes myself, and durned if my pardner hadn't killed another. We were ready to go on, but seeing a man loping over a hill towards us, we waited.

" 'Well, stranger,' says I, 'this seems to be a purty good country for rattlesnakes. I wonder if you know anything of a cave that sometimes goes by the name of Rattlesnake Cave?'

"The feller introduced himself as named Cox, and I could see by his reply that he had us sized up for suckers of some kind.

" 'I guess it is a good country for rattlesnakes,' he says. 'Durned if it don't look like they growed around here on bushes. Look at them bones scattered abouts. They're cattle that died of snake bites. Right up yonder last winter,' and he jerked his arm up the way I'd been, 'we throwed a stick of dynamite into a hole and blowed out a wagon load of the stinking devils. You ought to go up there and see that place. I never heard of it going by any name, but for all I know it might be the Rattlesnake Cave you're looking for. Ain't much of it left now, I guess. It shore was one cave of rattlesnakes.'

"There wasn't much more passed between us. I didn't let on as to why I was interested in the cave, and d'reckly Cox rode on off.

"Well, Crumley and me concluded right there that we were as good as inside the hole. It seemed as if for once luck was just naturally with us, and I don't know as I ever felt better in my life. We were short on grub, though, and had to have a new spade. So we pulled right on in to Ozona.

"When we got there we found the hardware store sold clean out of spades. It was all the same anyhow, for a letter was waiting for Crumley saying his baby was bad sick and for him to come back home.

"No railroad in that country, you know. So all we could do was to come back east in the car. The baby died, and one thing and another kept us tied up for a year. We figgered that nothing wasn't going to bother our cave while we were gone, anyhow.

"At the end of the year, we were back in Ozona. First thing we learned was how a man named Cox had bought a 50,000 acre ranch down Howard's Draw, stocked it, and had money in the bank. Twelve months before and he'd been nothing but an ordinary cowboy. Now he was swaller-forking all over seven counties. There didn't seem to be any secret about him having picked his fortune out of a cave.

"Of course there wasn't any use for us to go further. We never even went down to see what Rattlesnake Cave looked like after it had been cleaned out.

"Looks like something's always coming up that way. If it hadn't been for my blabbing to that man Cox, and if old man Crumley's baby hadn't took sick and died, I'd have a fortune now the size of Major Littlefield's."

The Fateful Opals

No form of ancient folk-lore has survived more persistently than superstitions about precious stones. People in general, as any jeweler will testify, still believe in birth stones and buy them. The imperial jewels of the Romanoffs, which made Lenin's regime cry, "Take the accursed things out of Russia"; the famous stone of the Sultan of Succadana, which in old times brought wars and untold calamities upon that country; the fabulous emerald of the Archipelago, which the Dutch government during the last century officially inquired after and which Joseph Conrad brought into his story of *Lord Jim;* the Sultana Diamond, the presence of which in Morocco herdsmen and fortune-tellers have for centuries regarded as necessary to prevent disaster; the curse laid on the possessors of stolen gems so exemplified in the fate of seven persons who a few years ago stole the Schoellkopf jewels at Buffalo, New York —these are stones and stories of the present as well as of the past.

The opal—symbol of hope—a cure for bad eyes—a stimulant to the heart—a chameleon-hued indicator of the health of its wearer—is, or was, the birth stone for October. During comparatively recent times, however, it has come to be considered a baneful stone, though, according to ancient tradition, the sons and daughters of October escape its malignancy.

> October's child is born for woe,
> And life's vicissitudes must know;
> But lay an opal on her breast
> And hope will lull the woes to rest.

So fearful of the opal nowadays are many folk in quest of October birth stones that organized jewelers have arbitrarily substituted the tourmaline.

From the Kremlin of Moscow or from the Sultan's guarded palace in El Ksar El Kebir to the rocky banks of the Pecos near its mouth in one of the loneliest and bleakest reaches of the Rio Grande is a far cry. But the Pecos, too, hides fateful jewels.

Rena Decker was a native of Brockton, Massachusetts, beautiful and adventuresome. In the spring of 1876 she quit her job as waitress in one of the Brockton hotels and went to Jacksonville, Florida. There she made the acquaintance of an exceedingly attractive Cuban grisette by the name of Montez Veronica Rodriguez. In a short time these two young women decided to seek their fortune—by whatever route it could be most speedily gained—in San Antonio.

San Antonio in 1876 was a place for fortune seekers. The dust raised by thousands of longhorn cattle on their way up the trail to Kansas sifted over the town, and cowboys and cowmen jingled their spurs in the streets, eager to surrender their coin to attractive women. The Jack Harris saloon and variety theatre, where the bad Ben Thompson and the brazen King Fisher were later killed, was probably the most noted resort of its kind between the Gulf of Mexico and Cheyenne. Hard by in the ample gambling emporium over the White Elephant saloon enough gold and silver were stacked on the tables nightly to run a bank. Indian fighters with "vouchers," as they called scalps dangling from their belts, paraded the streets. Great trains of Chihuahua wagons drawn by sixteen mules each and freighted with flour or silver clattered across Main Plaza. Samuel Bell the jeweler was making fine bowie knives. Wild turkeys at fifteen cents a pound were a drug on the market. Dried buffalo, "fresh from the plains," was for sale alongside the turkeys. The military telegraph had just been strung. The Sunset railroad was coming. The big pot was in the little one; the goose was hanging high; the skillet was a-frying. San Antonio was lusty, free, booming, with the sky for the limit and the lid thrown away.

Among the sporting gentry were two men known as Pronto Green and Dirk Pacer. They had plenty of money and they were free with it. That does not mean they were not shrewd. For business reasons as well as *par amour,* they formed an alliance with Rena Decker and Montez Rodriguez. From the arrival of the young women until February, 1877, the two couples were the most conspicuous figures about the amusement places of San Antonio. Then they went to Mexico City.

Here they opened what they denominated The Open Palace—a combination of saloon, café, gambling hall, and extensive suites of "private" bedrooms. For some time The Open Palace paid well, but so much trouble arose among the patrons that the owners saw they would before long be forced to close. They decided to try next the city of Durango. While they were preparing to make this change, Pronto Green by methods very dubious managed to acquire from a Chinese trader then in Mexico City a golden idol with eyes, necklace, and other adornments of rare opals. The opals were removed and two of them—harlequins—were set into rings for Rena and Montez, and the idol itself was melted into common bullion. Rena, however, had such a superstition against opals that she refused to wear her ring, whereupon Montez added it to her own jewelry. Meantime Green, as banker for the company, invested a considerable amount of their funds in other opals, for, at this time, Mexico had only one competitor, Hungary, in the production of opals, and Mexican opals were so cheap as to offer a promising investment.

In Durango the company opened a resort similar to The Open Palace and for a while prospered. Then one night a general row broke out in which a man was killed and several other men were wounded. The adventurers were again forced to close. Now they began roving from city to city, often taking in remote mining camps. They continued to make money, but at the same time they became more debauched and less desirable to keepers of the law.

At length Rena Decker grew so disgusted with the life she was living that she determined to quit it. Therefore she asked for her share of the gains, refusing to take any part of the precious stones, and set out for the Texas border. Her comrades saw her across the Rio Grande, whence she journeyed to Brockton, Massachusetts, and passed from the story.

The remaining three adventurers stayed on in northern Mexico until March, 1881, at which time, prompted by a difficulty that arose at a *fandango,* they considered it wise to quit the country. Procuring pack horses and saddle mounts, they set out for Del Rio, but when about opposite that place they learned that certain Mexican *empleados* were designing to arrest them, and so they turned up the river. At the mouth of the Pecos, unwatched, they crossed with their goods into Texas.

Montez Rodriguez had contracted a fever and as soon as she was on Texas soil became too ill to travel farther. Dirk Pacer and Pronto Green nursed her for a few days as best they could, but despite their attentions she died. With tools borrowed from a sheep camp they dug a grave and buried her with the two opal rings still on her hand. A Mexican *pastor* made the sign of the cross over her grave, and a stone with the initials M. V. R. cut on it was placed at her head. The grave was on the east side of the Pecos only a short distance from its mouth.

Green and Pacer now decided to make their way back to the east, taking care to avoid San Antonio and other places in Texas where they were too well known. In order that they might be free to scout, they considered it prudent to leave the major part of their fortune behind until they were ready to dispose of it. This fortune consisted mostly of cumbersome silver. So, taking an ample supply of gold, they secreted the silver to the amount of $37,500 and with it a bag of Mexican opals.

Upon reaching the Kiamichi Mountains, on the line between the Indian Territory and Arkansas, Dirk Pacer became too ill to proceed; years of dissipation had made him a physical wreck. Pronto Green arranged with a generous-natured mountaineer to care for the sick man and pressed on into Kansas. There he was killed in a railroad wreck. A short time afterwards Pacer himself breathed his last breath—but not before he had learned of his partner's death and imparted to his host both the history that has just been told and the location of the $37,500 in silver secreted with the bag of opals.

The cache, as he described it, is in one of the side canyons of the Pecos not far above the rock on which are carved the initials of Montez Veronica Rodriguez. Several years ago, L. D. Bertillion, a rover in far places, collector of horns, grower of strange plants, teller of strange tales—as witness this, which he told to me—started out with a man from Mississippi to hunt this treasure. At Piedras Negras, across from Eagle Pass, the Mississippian bought some opals, at a bargain, and then the two followed the Rio Grande up to the Pecos. They should have outfitted with packs and water kegs, but they did not. On the hunt, mostly afoot, that followed, the Easterner became crazed with thirst, let his suitcase containing the opals, ammunition, and other goods fall into an inaccessible gulch, and had to be half carried back to the railroad by Bertillion. So far as is known, this is the only hunt that has ever been made for the gamblers' opals on the Pecos, though some *pastor* or range rider must have seen the desolate rock marked M. V. R.

CHAPTER XIV

THE SECRET OF THE GUADALUPES[1]

He ne'er would sleep within a tent,
  No comforts would he know,
But like a brave old Texian
  A-ranging he would go.

Mustang Gray, frontier ballad.

THE tradition of gold in the Guadalupes runs back a long, long way. While governor of New Mexico, General Lew Wallace—at least so he claimed in a written article[2]—dug out of the basement of the Palace at Santa Fé an ancient document reciting how a converted Indian of Tabíra conducted Captain de Gavilán and thirty other Spaniards to a wonderfully rich gold deposit on the eastern spurs of the Guadalupe Mountains. The Spaniards named the place, on account of volcanic evidences, Sierra de Cenizas—Ashes Mountains—and left loaded down with nuggets and ore in the form of both "wires" and "masses." Then came the great uprising of 1680, in which the Pueblos killed every Spaniard who did not flee from New Mexico. About the same time Tabíra, the home of the guide to Sierra de Cenizas, was wiped out. Sierra de Cenizas has for centuries been a lost spot in geography as well as a lost mine.

Since the advent of English-speaking prospectors it has been the Apaches who knew the whereabouts of gold in the Guadalupes. Indians have "the best eyes in the world." The wilder they are, the better they can see. Excepting the Yaquis, who still have most of the gold of Sonora under surveillance, the Apaches were the wildest Indians on the North American continent. Their most famous leader, hard, untamable old Geronimo, used to say that the richest gold mines in the western world lay hidden in the Guadalupes.

The setting is worthy of its traditions. Guadalupe Peak, the highest point in Texas, rises 9500 feet above sea level, just below the New Mexico line. It is a beacon from all sides. The long, narrow chain of mountains above which it towers, extends, with gaps, southward clear to the Rio Grande and northward for nearly a hundred miles. Here in the Guadalupes the only mountain sheep left in Texas and a majority of those left in New Mexico are, under the protection of the law, making their last stand, eagles and panthers molesting them more than man, their haunts so wild, rough, and waterless that only occasionally does a human being intrude thither. Here the Apaches made final retreat, and on the Mescalero Apache Indian Reservation hardly a day's horseback ride from the northwestern spurs of the Guadalupe chain, remnants of that fierce, secretive, and outraged people yet live, their tribal name an inseparable element in the traditions of the whole Southwestern world.

Of all the seekers for the gold of the Guadalupes "Old Ben" Sublett—William Colum Sublett being his correct name—was the most picturesque and has become most famous. Like the much besung Joe Bowers he went west from Missouri—"yes, all the way from Pike"—to prospect in the Rocky Mountains. He saw other men grow rich from virgin gold, but the pay streak never opened under his pick. He went in rags; at times his wife and children went hungry. Life for them must have been fearfully hard. It killed his wife. Then Sublett with two little girls and an infant son turned southeast and, crossing the Guadalupes, made for "civilization."

Civilization was the new Texas and Pacific railroad track—a double line of steel that glittered across hundreds of miles of West Texas land too waterless at that date even for the scant population of ranches. Sublett put up a tent beside a section house where the town of Monahans now stands and where a well had been dug. He got odd jobs from the railroad. Other men were coming in; they contributed to the support of "the children in the tent."

In fact, strangers were so charitable that one day Ben Sublett drove away alone in his rickety old buckboard, pulled by a pair of bony horses and carrying a meager supply of frijoles, flour, and coffee. He did not need much. He could, as the saying goes, live on what a hungry coyote would leave. He again took to prospecting—in the nearest mountains of any size, the Guadalupes. This was long before an oil well was dreamed of in West Texas; it was before the Mescalero Apaches had been securely rounded up. Men who packed lead under their skins and could show the scars of arrow wounds warned Sublett that he had better stay away from the Guadalupes and the Apaches. He laughed at them. Trip after trip he made into the mountains, returning only to work long enough to buy a fresh store of supplies and contribute a little to the direst needs of his children.

He moved them over to Odessa, where there were a few saloons but no churches, where women were scarce, and where the click of six-shooters synchronized with the click of spurs. There the oldest child of the family, a girl, made something by taking in washing. The father was freer than ever to prospect. He knew what he was about. Every time he came in, his return was a surprise to the people of the town; they scoffed at his crazy mode of life. Occasionally he brought in a nugget hardly of enough value to keep him in shoe leather. In vain his children begged him to quit the mountains and settle down to some steady-paying job. He was stubborn; he would take advice from no one. He had a "hunch" that he would some day find the gold in the Guadalupes.

Sometimes he tinkered on the ranch windmills that were dotting the country. Other times he trapped quail and killed antelopes to ship to Chicago. A catch he made of ninety-seven quail together in a net is still remembered. One while, it seems, he trapped in the White Mountains of Arizona. This must have been before he came to Texas. After he had been prospecting for years, he admitted that an Apache whom he met in the White Mountains had told him a story of gold in the Guadalupes.

"Old Ben, the crazy prospector," became the jest of the country. Then one day, after having been gone for an unusually long time, he drove his rickety rig up to the Mollie Williams saloon in Odessa, strode—despite his habitual limp, caused by an old bullet wound—boldly to the bar, in a hearty voice invited everybody present to join him, and called for drinks all around. The bar-keeper hesitated, the men sniggered. But when Old Ben threw a buckskin pouch full of nuggets on the bar, the crowd went wild.

"Boys," he said, "I have been poor, but I ain't poor no longer. I can buy out this town and have plenty left. Drink."

They drank. They cheered. They drank again. Then between drinks Old Ben went out to his buckboard and brought in a small canvas sack filled with gold "so pure that a jeweler could have hammered it out."

"My friends," urged the crazy prospector, "drink all you want. Drink all you can hold. I have at last found the richest gold mine in the world. I can build a palace of California marble and buy up the whole state of Texas as a back yard for my children to play in. Let's celebrate."

Old Ben never built the palace, it seems, or encumbered himself with leagues of land. In reality he had no desire for estates or cushioned halls. His wants were few and elemental. He was not greedy for riches. The golden secret that he bore in his breast—like the hidden light of the "Lantern Bearers"—and the notoriety that the secret brought meant more to him than any amount of taxable properties. He had a kind of hunger for fame. Human chicanery and the mad grasping for property perplexed him, and thus he came to distrust all "prosperity friends." After he struck it rich, he was never known to work at all. Every few months he would slip out to the mountains alone, "and he generally brung back around a thousand dollars' worth of gold." The chief pleasure he derived from it seemed to be in displaying it.

As may be imagined, many men tried to get Sublett to show them the location of his gold. "If anybody wants my mine," he would say, "let him go out and hunt for it like I did. People have laughed at me and called me a fool. The plains of the Pecos and the peaks of the Guadalupes have been my only friends. They are my home. When I die, I want to be buried with the Guadalupes in sight of my grave on one side and the Pecos on the other. I am going to carry this secret with me so that for years and years after I am gone people will remember me and talk about 'the rich gold mine old man Sublett found.' I will leave something behind me to talk about."

Sublett was trailed, spied upon, "laid for," but no lobo wolf was ever more wily in avoiding traps than was Sublett in avoiding detection. His habit was to leave town at some unexpected time, camp on the Pecos a day or two, and then strike out from camp during the night. He might be gone only a few days; he might be gone for months. It is said that he at one time kept his money in W. E. Connell's bank at Midland, which is about thirty miles east of Odessa. Where Sublett turned his raw gold into cash nobody has explained, but the banker came to observe that when the mysterious old prospector's deposit ran low he invariably made a trip and not long after returning invariably banked "hard money." Of course, there are people who say that Sublett never owned a check book in his life; some people will talk. Anyhow, as the best of the talkers tell it, banker Connell and a cowman by the name of George Gray offered Sublett ten thousand dollars if he would show them the source of his cash.

When Old Ben Sublett emptied a buckskin pouch of nuggets on the bar, the crowd went wild.

Sublett just laughed at them. "Why," he replied, "I could go out and dig up that much in less than a week's time."

After this conversation Gray and Connell engaged Jim Flannigan to follow Sublett on the next trip. Sublett's funds in the bank were running low and he was due to "pull out" any hour. For two weeks Lee Driver, who was then keeping a livery stable in Midland, fed a horse for Flannigan. Then one day word came that Sublett had left Odessa in a hack pulled by two burros. Flannigan followed his tracks through the sand for fifty miles west along the railroad to Pecos on the Pecos River, and then for twenty-five miles on up the river. There the trail played out—stopped—quit—just disappeared. How any West Texan could lose the plain trail of a hack in soft soil uncut by other tracks is almost inconceivable, but lose the trail Flannigan did. He was not the first or the last man to lose it.

He was still riding around trying to pick it up when he happened to meet a man who had just seen Sublett traveling down the river towards Pecos. He turned back, but before he reached Odessa the cunning old prospector had already arrived. He had been gone from town "only four days," and had in that time traveled at least a hundred and fifty miles. Very good traveling for a pair of burros pulling a hack through sand dunes! Evidently Sublett had not got even into the foothills of the Guadalupes on this trip. He must have had a cache on the Pecos, for, as usual, he brought in a sack of gold.

But, despite his secretive ways, Sublett occasionally relented, and before he died took several people more or less into his confidence. Once when he was coming out of the Guadalupes he met an old crony named Mike Wilson; he must have been feeling almost insanely generous, for he gave his friend such minute directions for reaching the mine that Wilson actually got to it. There he emptied provisions out of a tow-sack and crammed into it as much ore as he could carry home. The trip wore him out, and as soon as he reached town he went on a spree that lasted for three weeks. When he sobered up and tried to go to the mine a second time, he found himself utterly bewildered. Old Sublett just laughed at him and refused to direct him again. "If anybody wants that mine," he said, "let him go out and hunt for it like I did." Years ago Mike Wilson died in a hut within sight of the Guadalupes, trying vainly until the end to recall the way to Sublett's lost gold.

Another time, some men at Pecos finally, after much persuasion, "ribbed up" Sublett to show them the mine. They felt so gay and prosperous that they loaded a big assortment of fancy canned goods into their chuck wagon to supplement the regular camp supplies. The first night out a tin of pineapple gave Sublett a case of ptomaine poisoning. He was probably already sick from having promised to give away his secret. At any rate, he claimed that someone had tried to poison him, became as stubborn as a government mule, and refused to go a step farther.

Perhaps, though, unknown to Sublett, there was an independent sharer of his secret. Or maybe Sublett discovered the sharer and acted as his jealousy might have prompted him to act out in the wild loneliness of the Guadalupes. Every man is entitled to his own conclusion from the testimony offered by F. H. Hardesty, who used to ranch in El Paso County.

One evening along in the eighties a fellow by the name of Lucius Arthur, known better as Frenchy, rode up to Hardesty's ranch, watered, and accepted the invitation to unsaddle and stay all night. While the two men were talking after supper, Frenchy confided to his host that he was trailing two Mexicans who had left Ysleta, on the Rio Grande, the preceding night. He said that he had started to follow them once before but that his grub and water had played out. He knew that they were bound for a gold mine somewhere in the Guadalupes to the east.

Frenchy had been keeping his eye on these Mexicans for a long time. One of them, according to him, belonged to a wealthy old family of rancheros down in Mexico. Perhaps, as he suggested, some Mexican had found out about the gold back in the days when *gente* from below the Rio Grande used to come up and get salt from the great beds west of the Guadalupe Mountains. A gringo's attempt to control this salt resulted in what is still referred to as the Salt War. Anyway, a member of the rancher's family made a trip to the Guadalupe gold mine each year and brought out a supply of ore. The Mexican now after it had come to Ysleta to meet his brother-in-law and together they had left that place in the dead of night.

"After hearing all this," Hardesty related, "I told Frenchy he ought to go better equipped. I told him he might have to stay out for weeks trailing the Mexicans and waiting for them to clear out from the gold mine before he could get into it. Then I offered to stake him with everything he needed. Well, we went in pardners, and when he left my place he had as good an outfit as any man could want and was carrying enough supplies to last two months. Six weeks later he was back. He had gold quartz to show.

"According to his story, he had trailed the Mexicans and from a place of concealment had watched them climb a rope ladder into a chasm. He saw them haul up sacks of ore and water for their horses, which were staked on the rim. But he himself had to depend on water so far away that he couldn't keep regular watch. After he had hung around several days, the Mexicans left and then he made a closer inspection. The chasm, from the way he described it, must have varied in width from forty to a hundred feet and was all of sixty feet deep. Down at the bottom he could see the entrance to a cave with freshly broken rock in front of it. He claimed that he didn't go down into the chasm because he was short on rope for a ladder. I thought he might have been a little more resourceful, but I said nothing. The chunks of quartz he brought in had been dropped by the Mexicans, so he said.

"Frenchy rested up a few days, took a fresh pack of supplies, including enough rope to picket out a whole *caballada,* and left again for the Guadalupes. He never came back. I have never heard of him since. That's all I know about the gold of the Guadalupes."

As we shall see presently, Frenchy's description of the Mexican mine jibes perfectly with that which has come down of the place where Sublett resorted. Let us get back to Sublett.

Something over thirty-five years ago a jack-leg carpenter by the name of Stewart was putting a tin roof on a house for Judge J. J. Walker, of Barstow. He could many a tale unfold, and one very hot day while he rested in the shade he unfolded this one to his employer.

Along in the late eighties several officials of the Texas and Pacific Railroad engaged Stewart to guide them into the Pecos country on a hunt. Camp was made in some trackless hills east of the river. A rumor racing over the range had it that the Apaches had broken out and were back on their old stamping grounds. Stewart was naturally uneasy lest they foray down from the mountains and either kill some of his party or drive off their horses. He had his son, a mere child, with him.

One evening about sundown he saw a wagon coming towards camp. It was a light spring wagon drawn by a single horse—a very large horse. When it reached camp, the driver alighted. He was Ben Sublett and he was alone. Stewart had known him for years. Of course he was invited to stay all night, and he unhitched. After the hunters and Stewart's boy had settled to sleep, Sublett told Stewart that he was going to his gold mine *at the point of the Guadalupes.* He said that while riding along that day he had realized as never before how old he was and that he had decided positively never to make another trip after gold. "I have always declared that the secret would die with me," he said, "but now that I have met up with you out here I somehow want to take you with me and show you the mine."

Stewart replied that he would not think of leaving the men who were depending on him for guidance in that wild country and that even if he were willing to leave them he would never take his own child on into Apache range. In reply to this argument Sublett remarked that no man accompanying him would ever be in danger from Indians. Nevertheless, Stewart did not go.

When Sublett set out next morning, however, Stewart did accompany him "as far as the top of a blue mound towards the west." Here Sublett halted and, while Stewart looked through a long spy glass, tried to show him where the mine was located, asserting at the same time that such long-range directions would never be of any use. He said that he would be back in three days.

The third day, just after dark, he drove into camp. As soon as supper was over and the hunters had bedded down, Stewart asked, "What luck did you have?"

For answer Sublett picked up a dried deer hide and put it, flesh side up, on the ground where the low fire cast a light over it and also where some boxes of provisions hid it from the eyes of any man who might be awake on his pallet. Then he poured on it a Bull Durham tobacco sack, of the fifteen cent size, as full of gold nuggets as it would hold. Stewart ran his hand through them and scattered them over the hide.

"You do not seem to have any small nuggets," he observed.

"What," Sublett rejoined, "would be the use of picking up a small nugget when with one more rake in the gravel I could bring up a big one?"

In the morning Sublett left. He had gathered his last nuggets. The next that Stewart heard of him he was dead. Stewart soon afterwards attempted to find the mine—but failed. What has become of him or what "point of the Guadalupes" he gazed at through the long spy glass while from an unidentified "blue mound" Sublett pointed towards the gold, are unknown. It turned out as Sublett had predicted.

"Come with me and I will show you the gold," he had said, "but if you go alone, even after I have pointed out its general location, you will never be able to find it."

So far as is known, Sublett never wavered again in his determination to hold fast the secret. When he was dying, his son-in-law, Sid Pitts, of Roswell, New Mexico, tried to persuade him to hand over the golden key he had clutched so long. Apparently the old man—he was eighty—started to tell him how to go to the mine. "First," he began, "you cross the Pecos at. . . ." Then he broke off with, "Hell, it ain't no use. They'd beat you out of it even if you found it." Evidently it was not the philanthropic desire to save his kin from the worries of wealth but his tenacious determination to keep "the damned human race" from sharing it and from learning what he had spent so many happily contrary years in withholding that caused old Sublett to keep silent to the end.

Sublett's death occurred in 1892. He was buried in Odessa, a little too far away from the Guadalupes to realize his wish for a grave within sight of them. Certainly, however, he left behind him "something to talk about." In the wide, wide lands of the Pecos, from its mouth far down on the Rio Grande to old Fort Sumner in New Mexico, there is hardly a town, a squatter's cabin, or a rancher's home in which the story of Sublett's Mine has not been told. Prospectors by the score have looked for it, and prospectors as well as many men who are not prospectors are still looking for it.

Among the most constant seekers has been Sublett's son Ross. In fact, Ross has been as constant as grubstakings from strangers would allow him to be; he has even on occasions grubstaked himself. This constancy is logical, for Ross Sublett probably knows more concerning the whereabouts of the mine than any other living man. When he was a little past his ninth birthday his father took him to it. That was the only time he saw it, but see it he did. Five years later when the secretive old prospector lay on his death-bed, Ross, of sufficient age by that time to feel responsibility, tried to get him to describe the way to the gold. The dying man —a little gentler to him than he was to the son-in-law— merely mumbled: "It's too late. Any description would be useless. You'll just have to go out and hunt it down like I did."

An accommodating disposition on the part of Ross Sublett to recount his childhood memories has not dimmed them. Like the annular rings of a tree, his memories increase in both number and compass. He lives at Carlsbad, New Mexico, is easy of access, and should you interview him, he would respond in this wise: "Yes, I have a distinct recollection of how the mine looked. The last stage to it, going west from the Pecos, was always made on horseback or with pack burros. It was down in a crevice, and the only way to get to it was by a rope ladder that my father always removed as soon as he came up with the gold. I played around while he got the ore out of a kind of cave. I seem to remember, too, that pieces of ore were in plain sight right in front of the cave. I am confident the mine is within six miles of a spring in the Rustler Hills."

The Rustler Hills [3] are a good forty miles east of Guadalupe Peak, but more than one tradition has made them the site of Sublett's gold. Not long after Sublett died, a New Mexico sheriff familiarly known as Cicero, while prospecting in these hills, met a cowboy called Grizzly Bill.

"You'd as well pull in your horns, Cicero," bawled out Grizzly as soon as the two men, both on horseback, came within hailing distance of each other. "I've done found Sublett's gold and I'm on my way to spend it."

Grizzly went on to Pecos, Texas. He rode into town shooting his six-shooter at the sky and yelling, "Hide out, little ones, yer daddy's come home." He got on a high lonesome, displayed his gold, and, while trying to "show off" on a wild horse, was thrown in such a way as to have his neck broken. Nobody who knows anything about the matter, however, has ever supposed that Grizzly Bill found the Lost Sublett Mine. In fact, all he found was a nugget near a spring—where more than likely Sublett had camped and lost it.

CHAPTER XV

NOT ONLY GOLD AND SILVER

He had bought a large map representing the sea,
  Without the least vestige of land:
And the crew were much pleased when they found it to be
  A map they could all understand.

—LEWIS CARROLL, *The Hunting of the Snark.*

THE FOXES HAVE HOLES

"GOLD is where you find it." So is lead, so is zinc, so are quicksilver and copper.[1] The story of mineral wealth is the story of accident. In 1545 an Indian hunter named Diego Hualca was chasing a goat up one of the most formidable mountains of Bolivia. In order to pull himself over a ledge he caught hold of a bush. The bush came out by the roots. Clinging to them and strewing the torn ground, masses of silver glittered in the eyes of the hunter. Thus were discovered what for centuries have been the world-famous silver mines of Potosi.

The badger, the prairie dog, and the gopher are the mascots of the mining West. Coyotes and burros seem to have directed more prospectors to "pay" than all the practical geologists combined. In 1859 "Pancake" Comstock, a half-witted prospector, saw "some queer-looking stuff" in a gopher hole. He ran his hand down and scooped out dirt sprinkled with gold and silver. The great Comstock lode was located.

In the fall of 1857 a mob of army teamsters, freshly discharged by General Albert Sidney Johnston, struck out from Fort Bridger, Wyoming, for somewhere. In the vicinity of what is now Denver they noticed fresh signs of another party of white men. They followed the tracks, and pretty soon came upon two newly-made graves. Something about the graves, however, appeared suspicious. In those days traders cached their goods in the ground. The Fort Bridger men investigated the mounds and without much trouble dug up two casks of Kentucky whiskey. In no time they were hilariously drunk and had matched a horse race. The course across the prairie was indented with prairie dog holes; that made no difference to the frolicsome riders. A fellow by the name of Nathaniel Ax, called Than for short, led the race, but just before he reached the line his horse plunged into one of the holes, throwing Than so that his head was rammed up to the shoulders in another hole. The lick slightly sobered him, and as he extricated himself he took note of the gravel into which he had burrowed.

"Boys," he yelled, "this looks like Hang Town grit."

Hang Town was the most famous mining camp in California. Pick and pan were soon out; sluice boxes were hollowed from cottonwood logs. Within six months the hills and mountains around the prairie dog town were alive with gold diggers.[2]

About 1880 ten postal clerks in Washington put up a hundred dollars apiece and sent the money to Alexander Topence at Bellevue, Idaho, to stake a prospector. When Topence received the money, he went out on the street and spied Dan Scribner.

"Dan," he asked, "what are you doing now?"

"Nothing."

"How would you like to do a little prospecting?"

"Fine."

Dan outfitted, struck for Wood River, and pitched his camp in a shady place near water and grass. He hobbled one horse, put a bell on the other, and turned them loose. The next morning he could neither see nor hear anything of his horses. The often quoted saying of an old prospector comes to mind. "How long have you been prospecting in these mountains?" somebody asked him. "Nigh unto forty years," he answered. Then he caught himself. "I haven't exactly been prospecting all that time, though. About thirty of them years I've been looking for burros; just ten were left for actual prospecting."

Dan Scribner began climbing a hill so that he could get a view of the country and thus perhaps locate his horses. Halfway up he saw a badger hole and stopped to look at the rock that the badger had thrown out. He found a fine piece of galena. He went back to his tent and got a pick. Before noon he had uncovered three feet of almost solid galena ore running high in silver. He let the horses hunt him up. What became of the badger is not told. The mine was named the "Minnie Moore" and it sold for half a million.[3]

At Tonopah, Nevada, Jim Butler's mule kicked off a shallow cap of rock that hid one of the richest bodies of ore discovered in recent times. In 1927 a desert rat by the name of Horton chanced upon a badger hole out in the malpais north of Tonopah; the dirt from it assayed thousands of dollars per ton—and the Weepah stampede was on.

That ground squirrel on the Sabinal River in Texas did not burrow deep enough maybe to establish a mining camp, but it exposed enough quicksilver to start a story that has already lasted longer than most mines last. Along about 1870 a company of rangers were camped four miles above Sabinal town. They often practiced shooting at prickly pear leaves and ground squirrels. Then suddenly, in accordance with ranger custom, these protectors of the frontier were ordered to some other place.

After thirty years or more had passed, one of the veteran rangers appeared in Sabinal and went up the river. He was gone a day or two, took no one into his confidence, and quietly left the country. Within a few weeks he returned with another member of his all but forgotten company. They secured the help of some of the oldest settlers and definitely located the site of the ranger camp. Next, the former rangers drew up with the owner of the land a contract allowing them to mine quicksilver. Then they told their story.

According to it, while they were camped on the Sabinal a generation back, one of the rangers shot a ground squirrel on the edge of its hole. In stooping over to pick up the dead creature, he happened to glance into the hole. The sun was shining at just the right angle to throw light down it, and there the ranger glimpsed some free globules of quicksilver. He got a can, dipped up some of it, and passed it around for his comrades to examine. Some of them rubbed their guns with it. None of them knew anything about minerals, although they knew that quicksilver was worth something. It was a day or two after this episode that they were ordered on.

The two veterans who, after so long a time, returned to the site, set up camp and began work. They dug many trenches, but there was no friendly ground squirrel to direct them now, and, to be brief, they were unable to find even a trace of what they were after. The ground where they dug is near a great fault that has exposed millions of tons of igneous rock. Quicksilver, it is said, sometimes occurs under just such conditions; but to this day the elusive mineral once glimpsed by the Texas rangers—down in the hole of a ground squirrel—has not been found, and their story has passed into the tradition of the Sabinal country.[4]

Precious Lead

No story of the frontiers is more common than that about savages extravagantly shooting bullets of silver. Felix Aubry, who, among other exploits, rode eight hundred miles from Santa Fé, New Mexico, to Independence, Missouri, in five days and sixteen hours, claimed to have found Indians on the Gila River who were actually shooting gold bullets. But, after all, lead moulds better into bullets than either silver or gold. A ball of lead was often worth more to a frontiersman than all the gold and silver in the world. Lead for bullets meant life; lack of it meant death. The pioneers melted bulk lead and moulded their own balls. A bullet mould was as necessary as a rifle. It is not strange that even in territory lacking almost altogether in evidences of mineralization, stories of hidden lead mines should have grown up.

Caddo Indians, so the story is told all along the upper Louisiana-Texas line,[5] used to bring lead to Shreve's Landing (now Shreveport) to trade. They would never tell where they got it, nor would they allow any white man to accompany them to the source of supply. Secret spying on their movements, however, led some woodsmen to conclude that the lead mines must be in what is now Cass County, Texas. Years later a fox hunter while following his hounds one morning up John's Creek, a tributary to Caddo Lake, noticed as he broke with difficulty through a strange mixture of debris at the mouth of a crude shaft some pieces of galena lead and broken timbers. He was just then too much interested in getting a fox to care about galena ore, and after the hunt was over he was too busy to investigate it, but eventually he became thoroughly convinced that he had seen the lost lead mine of the Caddos.

The resourceful pioneers had more lead mines, now lost, scattered over the country than Indians ever had. James Goacher, so history relates,[6] came to Texas from Alabama in 1835 and settled on Rabb's Creek, near the present town of Giddings, in Lee County. On the way up from Austin's coastal settlement to his wilderness home he blazed a way that was long known as Goacher's Trace, but his name is kept alive now by his supposed connection with a deposit of lead, which he not only used but sold and traded to others.

When settlers came to Goacher's home to buy lead, so legend remembers, he would, if he did not have a sufficient supply on hand, insist upon their staying at the house while he went for some. He always reappeared in a quarter different from that in which he had gone out. Sometimes he would be gone for hours; again, only a short while. He guarded the secret of his leaden ore as jealously as though it had been a trove of precious gems. Only three souls shared with him knowledge of the whereabouts of the mine; they were his two sons and a son-in-law named Crawford. One day about two years after the family had settled on Rabb's Creek, all four of the men met death while rushing to the house, unarmed, to repel a horde of savages. Those savages, so it is claimed, knew the whereabouts of the lead, knew the fatal worth of lead bullets, and in annihilating the Goacher men and carrying off the women and children they did not kill, were but following a plan to render the colonists less formidable. Perhaps, after the massacre, they covered up all trace of the lead mine; so remote was the Goacher homestead that other settlers knew nothing of the havoc until several days later. Certainly the mine has never been found, though more than one sign pointing to it has been glimpsed, always, however, in such a manner as to foil arrival at the goal.

One fine spring morning thirty-five years ago Doctor John M. Johnson and his newly wedded wife drove out of Giddings in their buggy to spend the day on Rabb's Creek. They stopped at Two-Mile Crossing, fished until noon, and ate lunch. Then, gun in hand, the doctor struck afoot down the creek after squirrels, leaving Mrs. Johnson to drive the horse. There was no hurry, the road was rocky, Mrs. Johnson had a fancy for pretty rocks, and several times she stopped to pick some up and put them in the buggy. At Five-Mile Crossing the couple met and drove back home.

A few days later the young housewife arranged the rocks to border a flower bed—all but one. She thought that it would be suitable for a door stop; so she took it inside to use for that purpose, and for many years it kept doors from slamming. Then the Johnson family—there were children now—moved to a new home. A few days later Mrs. Johnson found that her old door stop had been lost in the move; she was regretful, but as it was only a rock, no particular search was made for it and it was soon forgotten.

More years passed. Then one morning Doctor Johnson was called in to see the town jeweler. The jeweler and his family were living in the house formerly occupied by the Johnsons. After prescribing and talking for a while, the doctor picked up his case and was leaving the room when his glance fell on a rock beside the door.

"Why," he exclaimed, a tone of friendly pleasure in his voice, "there is Mrs. Johnson's long-lost door stop."

The sick man, failing to grasp the meaning of the remark, asked:

"What door stop?"

The doctor pointed to the rock by the door and said: "Mrs. Johnson and I began married life in this house, and she used that rock for a door stop just as your wife is using it." Then he went to his office.

He had scarcely arrived when again he was summoned to the jeweler's home. Fearing that his patient had suffered a sudden attack, he hurried into the sick room without knocking.

"Where," the jeweler eagerly greeted him, "did that chunk of lead come from?"

"What chunk of lead?"

"That door stop you called a rock. After you left just now I cut into it and found it to be pure lead. Here, see for yourself."

The doctor looked. The familiar door stop was indeed lead ore, almost pure.

But the doctor had forgotten where it had come from. He took it home to his wife. Of course she recalled instantly how she and her lover had brought it in from a honeymoon excursion to Rabb's Creek so many years back. The couple drove again to Two-Mile Crossing and then slowly, examining many rocks, on down to Five-Mile Crossing. But Mrs. Johnson could not recall at what particular place she had picked up the heavy rock, and no other rock resembling it could be found. This is just one of the stories that hinge on Goacher's lead mine.[7]

Along in the fifties an old Dutchman by the name of Frank Vanlitsen, who lived near Wallace Bridge on the Lavaca River down towards the Texas coast, began supplying his neighbors with lead in a pure form. No one knew where he got it; certainly he did not receive it off ships that then regularly put into Lavaca and Matagorda bays. As may be imagined, the source of his supply became a subject for conjecture and imaginative play.[8] But he never told. In fact, he seldom told anything, and no one ever heard him refer to his past. A great part of the country along Lavaca River was, as it is today, heavily timbered and sparsely settled; the goings and comings of Vanlitsen could not well be spied upon. About three years after he had begun selling lead he was found dead in his cabin with a bullet hole through his head and a pistol clutched in his hand. When discovered, he had been dead for several days, and his neighbors, judging that he had committed suicide, buried him without making any particular investigation. The Civil War came on, the country was disrupted, the demand for lead was stronger than ever. The "Dutchman's lead mine" was not forgotten.

About 1867 a noted character of the Lavaca country, nameless here, for it is necessary to say that he bears the reputation of being an extraordinary liar—also, he has killed several men—noised it about that he had at last discovered Vanlitsen's lead. One afternoon while riding after cattle in the Lavaca bottom he became so tired, he said, that he dismounted and threw himself down on a well-shaded sand bed in the bottom of a gully and went to sleep. When he awoke, his eye fell on some chunks of lead sticking out through sandstone that the gully had cut through. He broke some of it off, tied it on his saddle, and rode home. Darkness was coming on when he left the gully; he made no mark to guide him back to the place; he had no reason to doubt his ability as a woodsman to return and open up the mine. That night a heavy rain fell, washing out all signs of his tracks. As soon as the water had run off the country, he set out with pick and sacks to gather more of the lead. He could not find it. He tried many times to relocate it, always in vain. To prove the veracity of his claim, he used to offer for inspection about half a bucket of poorly smelted lead. Many men have looked for the lead-bearing gully, but no one has ever found it.

But there is no doubt that solitary old Frank Vanlitsen had plenty of lead and that for three years he sold lead to settlers of the region. Geologists have declared that the country shows absolutely no indications of lead or associated minerals. Where did the Dutchman get his supply? Some people—for it is human nature to demand explanations—claim that Spanish pack trains following the old Atascosita Road, which traversed the Lavaca country, used to carry bullion, generally silver, from Mexican mines to New Orleans. They tell how one of these trains was robbed near Chicolete Creek and opine that Vanlitsen, who came to the country very early, helped rob it, found that the bullion was lead, took his share of it, years later began selling it, and then was murdered by a confederate, who arranged that the deed should appear suicidal. Anyway, the Dutchman had lead.

Yet another early day provider of lead from a secret mine was a rancher by the name of Hoffman who lived north of Sabinal.[9] He had come from California—and so knew all about minerals. Scattering settlers used to go to his cabin to buy lead. One day while Will and High Thompson, brothers, were helping Hoffman brand calves at his ranch pens, they said something about needing lead for bullets. Hoffman replied that if they would keep on working he would get them all the lead they could use. He had plenty of it, he said. The Thompson boys kept on branding; Hoffman rode away, and in about two hours returned with the lead. He said that he had got it out of his mine and that as soon as he could sell his cattle he was going to work the mine on an extensive basis. Within a short time he sold out his stock of cattle, but almost immediately thereafter he was killed by Indians.

The Thompson brothers then began to hunt for the mine. One day while they were searching for sign of it, High called out to Will to come and see "this great big blue cow chip." The cow chip proved to be lead. They were at the mine—and didn't know it. A very short time after this Will, who was always leader, was killed either by Indians or by bandits. The mine was now forgotten, or neglected, for a period of years, and meanwhile the land on which it is supposed to be located passed into the hands of a man who would not allow any but his own kin to prospect on it.

Finally a brother-in-law of the land-owner resumed the search and got High Thompson to help him. They sank several shafts near the place where High, as he remembered, had seen the "big blue cow chip," but they never found any lead. The mine is still a lost mine, talked about by many and even today searched for by more than a few.

In 1887 Thomas Longest and Luke Callaway, who kept a stable at Dalton, Georgia, came to Texas to buy horses. They bought five carloads on the upper Brazos, and while Callaway went back to Georgia with the shipment Longest remained to look over the country.

One day while riding with a cowboy near the intersection of the Salt Fork and Double Mountain Fork of the Brazos,

Longest saw a steer with such wide-spreading horns that he expressed a desire to have the animal killed. He wanted to send the horns to a friend in New York, his former home. The cowboy, however, declared that the horns on this steer were short compared with the horns on a certain brand of steers ranging a day's ride to the northwest. Longest promptly set out to find the horns, the cowboy going with him only far enough to show him a crossing on the Salt Fork free from quicksands and telling him the general direction of trails to Croton Creek.

After Longest had ridden for several hours, a storm forced him to take shelter under a canyon bluff in a very rough and desolate-looking country. Here, while waiting for the weather to clear, he noticed on the ground a rusty piece of iron. Upon closer examination, he found it to be an old pick. With it he poked around in the dirt and uncovered the remains of a shovel. He kept on investigating and presently discovered a ledge of ore. From it he broke off a piece weighing four or five pounds. He was sure that it was silver. He lost all interest in cow horns and at once returned to Georgia, where he dispatched the ore to an assayer in New York. To his great disappointment, it was pronounced lead, but seventy per cent pure—a valuable find.

Longest now set about trying to interest a mining company, and by the spring of 1888 arranged to show its representative the mine. However, during the trip to Texas the year before he had contracted a severe cold and cough, which developed into tuberculosis. Hoping to get better, he put off the trip, and affairs were at this stage when L. D. Bertillion—the man who knows where the Mexican opals are buried on the Pecos—met Longest in Georgia.

"In a few months," says Bertillion,[10] "Longest was dead. I don't think that he ever made a way-bill to the mine. Somebody had evidently worked it before Longest saw it. Thus for a second time became lost what is perhaps one of the richest lead veins in America. From the description of the country given by Longest, I judge that the ore is located in either Stonewall or King County, more likely in the latter."

Copper on the Brazos

The tradition of copper worked by the Spanish up the Brazos River—high up where the water runs clear and where cliffy eagle perches look down upon it—goes far back. And along with the tales of copper that used to stir settlers plowing in the flat muddy bottoms many hundreds of miles downstream were tales too of "inexhaustible" mines of gold and silver, "a mountain of iron," and a monstrous "lump of platina" that Indians worshiped and fashioned arrow spikes from.[11] As frontiersmen pushed west and north into the Cross Timbers along the Brazos they came to fix the location of the Spanish copper workings somewhere near the junction of the Salt and Double Mountain forks and within sighting distance of Kiowa Peak—the region in which Thomas Longest of Georgia found such a fine specimen of lead.

One very pretentious expedition to find the Brazos copper was organized in Washington by a group of men who called themselves "the Washington and Texas Land and Copper Company."[12] The personnel of this expedition when it set out from Fort Richardson, at Jacksboro, Texas, in 1872, consisted of about sixty men, among them a Virginia congressman of ante-bellum dignity, an orientalist, an "official artist," a chemist, the "sometime State Geologist of Texas" named Roessler, and various other "characters." They did little but travel leisurely and "locate" ten or twelve sections of land about Kiowa Peak.

Yet let no one scoff at the idea of copper in Texas, and let not the prospectors who follow legend lag in their sanguine pursuit. Geologists have actually conceded the presence of copper on Los Brazos de Dios.[13] The most extraordinary hunt into these copper realms, however, seems to have been chiefly concerned with gold. Mr. R. E. Sherrill has supplied an account of it. His narrative follows.[14]

"About 1908 a large old gentleman, whose name I cannot now recall, suddenly appeared in the sleepy little town of Haskell from somewhere on the Mexican border and began inquiring about the topography of the country and the tradition of Spanish treasure. Having learned what he could, he took into his confidence a few select men and explained to them that he had gathered from reliable Mexicans definite information regarding a gold storage on the Brazos and that he proposed to search for 'the key' to this hidden wealth.

"Adding his own information to what he heard from the natives, the stranger gradually let out a tale that ran somewhat as follows. At an early date, when Spanish miners were gathering great quantities of gold in Mexico, a company of them in search of further treasure wandered far to the northwest, taking along, for some reason, a large store of the precious metal. Directed by an Indian or by their own keen instinct for such things, they located the copper mines on the Brazos and proceeded to work them. Here they aroused the hostility of the native Indians and were in danger of massacre. So they hastily hid their treasure, made a plat of the country, and fled. The Indians continued so hostile that they never dared return to take away their gold. Amidst the turmoil and dangers of Mexico at that time, the plat was delivered for safe-keeping to a faithful mestizo who was attached to the Spanish party. It remained in his hands until the old man, approaching death, delivered it to a member of his family. Thus the plat passed along for two or three generations until Texas fell into the hands of the hated gringos and it became certain that no poor Mexican could ever get possession of the treasure. Finally, for some small favors and a little money, a Mexican turned the plat over to the portly American who had now come with it and its tale to Haskell County.

"The company he organized to assist him in digging up the treasure kept the plat a secret, though it and the oral directions pertaining to it were so complicated that not even the wisest of them pretended not to be confused. The map covered a large territory, including the two branches of the Brazos, Kiowa Peak, and numerous minor features of the vicinity. It called for many specified rocks and many marked trees. The rocks had been covered with soil or the markings on them had been weathered away. Most of the trees had perished in fires long years past. An explanation was given to some of the signs, but the meaning of more had to be guessed at.

"The search was thorough and long continued, and a deal of money was spent in digging. Most of the prospecting was along the river, and a Mexican who was herding sheep in the neighborhood began to enter into counsels of the treasure hunters. To use his information, they made him a partner. He announced that if they could produce a rock bearing certain letters and symbols, the picture of which he drew, he could find the gold. Only a few days after this, the party did uncover, about eight or ten inches under the surface of the soil, a rock that they called the 'Spider Rock.'

"The rock had many curious markings on it, among them the letter H, in old Spanish chirography, as the Mexican had called for. Presuming to interpret the markings, he said that the little hill on which the Spider Rock was found was underlaid with the 'base rock'; that underneath the 'base rock' were buried a great many bodies; and that buried nineteen steps west of the skeletons would be found a large bone of some prehistoric animal. He said that, in excavating, the diggers would find a kind of wall, as if a trench had been dug and then filled in with a much harder substance.

"Fired with hope, the treasure hunters set to digging for

the 'base rock.' They came against a wall-like wedge of very firm substance, wider at the top and narrower at the base, as if a trench had been filled in with it. When they had got down some fifteen feet, they were met by such a stench that they could hardly work. They found a great many decayed bodies and many relics of various kinds. Furthermore, at the specified distance, they found the bone of a prehistoric animal. It was of about the thickness of a man's body and very porous.

"The Mexican now directed that the diggers go to the bluff a little farther to the west. He said that there they would uncover a great bone like the first and other things buried by the Spaniards. The bone was found, and with it were an old-fashioned sword, some copper ornaments thought to be epaulets, some silver ornaments also, forty-two gold buttons, and a great number of beads.

"But here ended the findings. A majority of the relics were placed in Doctor Terrell's drug store at Haskell, where they remained until a fire, about 1909, destroyed them. After turning up more than an acre of ground, the depth of the excavations varying from a few inches to twenty feet, the diggers dispersed to their farms, the large man from the border left, and the Mexican disappeared. Many men think he knew more than he would tell. Not long after he vanished, a skeleton was found several miles to the east across the river, in the opposite direction from that in which the Mexican had led the Americans. Near the skeleton were two small, heavy copper pots, one shaped somewhat in the form of a canoe, the other round and of the capacity of a gallon and a half, built much stronger than any vessel now made for commerce and capable of holding molten metal. The popular conclusion is that the Mexican took from these copper vessels at least a part of the vast Spanish treasure.

"Nearly every man of that searching party was a friend of mine. I wish to give an illustration of their sanguine nature. At one time the party believed that they were within a foot or two of their treasure, but they feared to uncover it before they had made arrangements to take care of it. They were afraid, so one of them confided to me, to put much of the money in local banks, lest the banks be robbed, but wished to entrust it to our private vault, where no one would suspect its presence. I agreed to take care of the money and was to be notified a little after midnight. The amount to be deposited was $60,000 in gold. I was never called to open the vault."

CHAPTER XVI

SARTIN FOR SURE

I cannot say how the truth may be;
I say the tale as 'twas told to me.
—SCOTT, *Lay of the Last Minstrel.*

MORO'S GOLD

ABOUT a hundred years ago some daring young *caballeros* who lived south of the Rio Grande in the state of Tamaulipas organized for protection against Indians and desperadoes. As a badge of office they rode garbed in the picturesque robes of the Moors and were for that reason called Moros. One of them, Ramón Berrera by name, owned a ranch in Texas, and on a certain occasion when he was returning home with money for which he had sold a herd of cattle, he was murdered and robbed by his servant. The fellow Moros of the dead man spurred far and long in search of the servant but could never capture him. However, years later some of them heard that he was in a distant part of Texas living under the name of Moro and was a very rich man.

What connection, if any, may exist between this tradition and another that has for more than three generations been famous all over southern Texas, the reader must conclude for himself. Fannie Ratchford, a descendant of the chief participant—other than Moro himself—has thus told the story.[1]

"Before the Civil War, my grandfather, Preston R. Rose, lived on a large plantation, called Buena Vista, lying along the Guadalupe River, seven miles from Victoria. Late one afternoon in the summer of 1859 he was sitting on the porch reading, when my mother, then a little girl, called his attention to the unusual sight of a stranger coming across the field from the direction of the river. The stranger was of small stature and dark complexion, evidently a Spaniard. When he reached the porch, he addressed my grandfather in the easy, courteous manner of a gentleman and an equal, and requested hospitality for the night, explaining that his pack mule had gotten away from him and that he had exhausted himself in a fruitless search.

"His request was granted without question, and Moro took up his residence at Buena Vista, which on one pretext or another lasted for several months, in spite of the suspicious and disquieting circumstances that soon arose. The first of these was a report brought in by the negroes the next morning after Moro's arrival, that a mule, with a pistol shot through his head, had been found partly buried in the river bottom. Another was the fact that Moro was never seen without a glove on his right hand, not even at meal time. The negro boy who waited on him in his room reported that he once saw him without the glove when he was washing his hand, and described a strange device on his wrist that was probably a tattooed figure. But the most disturbing circumstance connected with Moro was his eagerness to get rid of money. He distributed gold coins (of what coinage, I never heard) among the household servants like copper pennies, until Grandfather rather sharply requested him to stop.

"Though there was not much to be bought in Victoria, Moro never came back from a trip to the little town without the most expensive presents that could be bought for all the family in spite of the fact that they were invariably refused. My mother seems to have been particularly impressed by a large oil painting which he once bought from a local artist at an exorbitant price, as a present for my grandmother. When she refused to accept it, he asked permission to hang it in the library, and there it hung as long as the house was in possession of the family.

"Frequently Moro proposed the most extravagant things. Once he urged Grandfather to allow him to build a stone mansion of feudal magnificence to replace the colonial frame house in which he lived. Again he proposed that he take the entire family to Europe at his expense and leave the girls there to receive an elaborate education.

"One day as Moro was walking about the plantation with Grandfather, the question of plantation debts came up, and Moro remarked in a significant tone that Grandfather was at that minute standing within fifty feet of enough gold to enable him to pay all the debts of the plantation and still be a rich man, even if he did not own an acre of land or a negro slave. Grandfather's prideful indignation prevented his continuing the disclosure that he was evidently eager to make. The only landmark of any kind near was a large fig tree fifty feet away.

"In the meantime the negroes had caught the idea of buried treasures, and many were the tales they told of seeing Moro digging about the place at night.

"A guest staying in the house one night reported that he had been drawn to the door of his room by an unusual noise, and had seen Moro painfully heaving a small chest up the stairway, step at a time.

"My grandfather was a man in whom the spirit of adventure was strong. He had left his plantation to the direction of his wife while he went adventuring into the California gold fields in '49. Consequently Moro was able to catch his interest by the story of buried treasures down on the Rio Grande, and Grandfather consented to go if he were allowed to make up his own party. The party as finally organized consisted of friends and neighbors, most of whom were well-to-do planters, but one man included was somewhat out of the social class of the others, though well known and trusted throughout the neighborhood. To this man Moro objected strenuously, saying that he would either prevent their finding the treasure or, if it were found, would murder them all to get the whole for himself. Grandfather insisted, and the man went.

"Moro was nervous and sulky from the start, and so aroused the suspicions of the party that by the time they reached the Rio Grande, he was not allowed out of sight. But despite the close watch kept upon him, he finally made his escape by diving from one of the boats in which the party were crossing the Rio Grande to the point where he said the treasure was to be found. The man whom Moro feared would have shot him as he rose above the surface of the water if Grandfather had not intervened.

"There was nothing left for the party to do but return home, for Moro had given them no map or directions that would enable them to make an independent search. But before setting out on the return, Grandfather foolishly accepted a dare to swim the river in a very wide place, and in doing so caught a severe cold that developed into 'galloping consumption,' from which he died a few months later.

"The rest of the story, so far as there is any, is confused and contradictory. A few weeks before Grandfather's death, some of the negroes on the place came to the house, begging for relief from Moro's ghost, which was seen almost nightly digging at various spots on the plantation, but most often near the big fig tree in the field.

"Grandfather was too ill to make any investigation for himself, but he questioned the negroes closely, and came to the conclusion that all the stories had grown out of one real incident—that Moro had probably come back to recover money he had buried on the place.

"The man whom Moro feared went to Mexico to escape service in the Confederate Army, and his sudden rise to fortune, coupled with a wild story he told on his return of having met with Moro below the Rio Grande, convinced my grandmother that he had in some way come into possession of the treasure.

"The legend of buried money still lingers around the old plantation of Buena Vista. About ten years after the Civil War, my father bought the part of the estate on which the home was situated, and during the years that he lived there was much annoyed by treasure seekers who begged permission to dig for 'Moro's gold,' or who came at night and dug without permission. One day as he was showing a 'free negro' how to run a straight furrow in the field not far from the old fig tree, the horse stumbled and his right foreleg sank in the ground up to the shoulder. Moro's gold seems not to have entered my father's thoughts at the time, but later he remembered it and was convinced that if there had ever been any money buried on the plantation it was in that spot."

The Mystery of the Palo Duro

Lighthouse Canyon is a tributary of that strange and wonderful cleft across the Panhandle Plains of Texas known variously as Prairie Dog Creek, Red River, and the Palo Duro. Coronado doubtless wandered over the region while in quest of the chimerical Gran Quivira. Long after him the *Comancheros*—Spaniards and Mexicans who traded with the Comanches—were familiar with it.

In Lighthouse Canyon itself was born in the year 1850 Jesus Ramón Grachias. He became a man of marked intelligence, fair education, and wide experience. At the age of seventy-six he happened one day while hunting burros to enter the camp of a college professor who was taking his vacation in New Mexico, and there he related one of his experiences.[2]

"I lived in Lighthouse Canyon until my father died in 1854, and then my mother brought me away. My remembrances of those four years are very dim. More than thirty years later I was working on the Fort Worth and Denver Railroad then building into Amarillo. As I was thus again on Palo Duro Canyon and as I had some money saved up,

I decided to visit the place where I was born and where my father was buried. I put my wife, baby, and camp things into a wagon and started west.

"We traveled slow. The country was still wide—ranches, ranches, ranches. I did not know if there was a road into Lighthouse Canyon. I only knew that it was west. After two weeks I came upon a Mexican who was trapping lobos for the J A Ranch. He guided me to within ten miles of the place I wanted to go and then, after telling me how to take the wagon, turned back.

"Many times my mother had described the place to me: the great pillar projecting up from the center of the canyon almost to the level of the plains like a lighthouse; the strange markings on the walls of rock; the fine spring of water; the shaggy old cedars around whose gnarled roots I used to play; the herds of buffaloes and antelopes that came into the breaks for winter; the owls and panthers and lobos that made their cries in the night. Remembrance of all these things helped me locate the spot where my father had built his *jacal* nearly forty years before. The place was still wild and there was not one sign of former habitation, but when I unloaded the wagon and put up a tent, I knew I would sleep that night on the spot where I was born.

"Many times I had promised my mother that some day I would come back here and put a cross over my father's grave. I found two flat stones lying near each other on the ground. When I turned one of them over, I saw the letters J. R. G. They were roughly cut. They stood for my father's name, my name also, Jesus Ramón Grachias. The stones had marked his grave and had very likely been pushed over by cattle or buffaloes. Also, my mother had told me that the grave was fifty feet east of an old cedar. I found only the burned stump of a cedar, fifty feet away in the right direction. Accordingly, I heaped up a hill of earth between the stones, set them in place again, and made a cross.

"The night after I had accomplished this I had a dream. I saw in this dream someone digging near the stump of the burned cedar. As I had myself dug there to get earth to put on the grave, and as I had, moreover, dug in several other places to find fishing worms, I did not regard the dream as very unusual. But the next night the dream came again, and this time I heard very distinctly these words: *'Dig fifteen feet east of old cedar tree.'*

"All the next day I was disturbed, thinking that my mother had perhaps said the grave was fifteen feet east of the cedar instead of fifty feet east and that I had maybe marked the wrong spot. I was much troubled, for now that I was on the ground I could not bear to leave the exact spot of my father's remains unmarked. Then I determined to make another mound fifteen feet east of the cedar tree and put a cross on it also. I measured off the distance and began to work. But the voice of the dream kept ringing in my ears: *'Dig fifteen feet east of the old cedar tree.' 'Dig,'* it said. *'Dig.'*

"If I dig, I thought, I can tell if the earth has ever been disturbed. So I began digging. At first the ground seemed no softer than any other ground. Then very soon my spade struck something hard, something of iron. Quickly I uncovered it. It was the lid of a chest, about eighteen inches wide and thirty inches long. More quickly yet I uncovered one end and prized it up. The chest was about eighteen inches deep. With the help of my wife I now got it out of the ground. We were very excited. I had to break the clasps of the lid. We opened it. It was full of coins. They were all Spanish and were all dated before 1821, the year of Mexico's independence.

"The grave of my father was after all properly marked. We could now leave. I backed my wagon up to the chest and dug little trenches for the hind wheels so that the rear end of the bed would be low. Thus I managed to load the chest. I drove up the canyon to its head and topped out upon the plains. I had no desire to go back east. I made for Santa Fé. It took us ten days to arrive. The banker there was a friend to the family of my mother. He counted the coins and gave me credit for $7600.

"I do not know which was bigger in me, the joy or the mystery. My mother was still alive and I at once traveled to her to tell her all things and to take her money. She could not understand the meaning any more than I could. We talked and talked. She told me much about my father that I had never known or had forgotten. You will see if what I learned explained anything.

"When the Texans whipped Santa Anna, a great many Mexicans living in that state left. My father was among them. He and four other Mexicans set out for Santa Fé over a country they did not know even by report. There were no roads. Winter overtook them on the plains. They camped in a deep canyon—Lighthouse Canyon—well protected. They made friends with the Comanches; game was everywhere. Early in the spring all their horses but two got away with the mustangs. On these horses two men left for Santa Fé to secure aid. Those messengers never returned. My father and his companions stayed, hunting and living on the country. I think they were happy.

"In the late summer of 1841 a body of Texas traders and soldiers—the Santa Fé Expedition—came among them, lost. The Mexicans joined them as guides. It is well known how the Texans were made prisoners and sent to Mexico City. My father was made prisoner also. He was kept for four years. Then he was released and came back to Santa Fé.

"There he married my mother. She was thirty-five years old; this was in 1849. He was twenty years older. As soon almost as they were married, he loaded a wagon with goods and set out for the plains of Texas. He told her that he was going to hunt buffaloes, but she said he seemed to hunt nothing but canyons. At last they came to the place he had been looking for. It was Lighthouse Canyon. 'I once lived here,' he said. 'I will live here again.' He died there, as I have already told you, in 1854.

"For a day before he died he seemed out of his head. He kept talking wildly to my mother of fights, of gold coin, of escapes, of enemies, of friends—of many things that she did not understand, for he explained nothing. He had never explained anything to her. The last hour of his life was calm. She was holding me over him to see if he would take notice when she heard, 'Buried fifty feet east of old cedar tree.' I recall only being lifted up and looking at him.

"So my mother with her own hands dug a grave fifty feet east of the big cedar tree, just as she supposed my father had directed. Then alone she got the horses, which were hobbled in the canyon, hitched them to our wagon, and drove with me to Santa Fé, leaving forever the lonely Palo Duro."

It has been said that the narrator—this finder who was not a searcher—was a fairly well educated man. The activities of the subconscious mind have long interested many people who make no claims to education. Jesus Ramón Grachias was quite averse to being regarded as superstitious.

"I do not believe in spirits," he said. "I do not believe in voices of the dead. Now listen. My mother was even when she married slightly deaf. What my dying father said to her was probably this: 'Dig fifteen feet east of the old cedar tree. It is buried fifteen feet east of old cedar tree.' My mother, failing to catch the first word of his command, thought he was talking of his grave and heard 'fifteen' as 'fifty.' At the same time, although I was too young to attach any meaning to the words or to store them away in my memory, my subconscious mind registered the sounds. More than thirty years later when I returned to Lighthouse Canyon and rested in the shade of the cedars around which I had played as a child and drank from the spring that my little mother as a bride used to dip from and stepped on the soil where my father's feet had worn a trail and where he had died—then old associations stirred the subconscious mind to activity. My nerves were all active, my imagination was alive. It was natural for me to dream a dream in which the subconscious mind brought to the surface those words stored away so long ago: 'Dig fifteen feet east of the old cedar tree.'

"The real mystery to me is how the money came to be there. Was my father at some time a robber—some time before 1821, for no coins were dated later than that year? Why did he live for years beside the money without using it, without telling his wife of it? Was he awaiting the return of some confederate? I do not know. I only wish that all the money were not spent now."

CHAPTER XVII

THE TREASURE OF THE WICHITAS

Glendower: I can call the spirits from the vasty deep.
Hotspur: Why so can I, or so can any man;
But will they come when you call for them?
—*Henry IV, Part I.*

THE metropolis of the Wichita Mountains, in southern Oklahoma, is Lawton. One time while I was traveling thither by train I met a young lieutenant who had been assigned to duty at Fort Sill, which is adjacent to Lawton. He was eager to learn all about "the old Indian country" and to explore the Wichita National Forest. As he did not have to report for duty for several days, he intended spending some of his free time in scouting around.

"Well," said I, "I'm going down into the Wichitas myself to learn something about the country, and I'll tell you what I'll do. I'll bet you a pair of shopmade officers' boots against the best Stetson hat in Lawton that within two hours' time after we register at the hotel I can get you up a hunt for one of the biggest lost treasures in America."

I needed a new hat and the lieutenant did not, apparently, need any more boots.

"Did you ever hunt for it?" he asked.

"No, I have never been in this section before."

"Do you know anybody there?"

"Not a soul, so far as I am aware."

"Have you a hunt planned out?"

"Oh, no, not at all."

"Then what makes you think you can get one up in two hours' time?"

"I've lived in the Southwest a good while. I have read some history. I know that in the seventeenth century Spanish priests came into the Wichitas to convert Indians and that Spanish prospectors tried to mine. When they left, they left a tradition.[1] There's bound to be Spanish gold buried back in the hills and there are bound to be all sorts of lost mines."

The lieutenant was simply wild to get out into the hills and look for Spanish gold. He took up my bet and was afraid he would win.

Twelve minutes after we registered at the hotel we were at the court house. A stenographer told us that she thought the county surveyor was at his home, three or four blocks away. We found him; he made us comfortable on some logs under a shade tree on his grassy lawn. His name is Sam Joyner and he is a most affable gentleman. Within twenty-five minutes after we registered at the hotel we were down to business.

In a Chicken's Craw

"Well," Joyner said, "you've certainly come to the right place for information. I guess there is more stuff hidden right here in the Wichitas than anywhere else in America and I guess I've had more calls for running lines than any other surveyor in Oklahoma.

"For instance, about 1905 an old duffer from up in Kansas dropped into this country with a map. At first he was mighty cautious about letting anybody have a look at it. He began by asking if I knew where a group of seven springs might be. I had to confess that I did not, 'but,' says I, 'if you'll give me your other locations, I might be able to work out the site you're looking for.'

"Finally he showed me the map. The words on it were in English, but the lettering seemed to be German. It called for exactly nine cart loads of gold and silver buried near

Seven Springs thirty miles northwest of Duncan's Store on the old road to Fort Sill and between Cache and Big Beaver creeks.

"The spot called for put us at some springs I was familiar with on Charlie Thomas's place. There were only four of them, but it was easy to figure that time and the tramping of cattle had covered up the other three. We located a mark on a very old cottonwood near the springs and dug around considerably. Then the old duffer got tired and went back to Kansas.

"Years passed and another character floated into the country. He kept hanging around Charlie Thomas's place and harping on what a good piece of land that was by the springs to raise truck on. Finally Charlie told him he'd fence off a field and rent it on shares.

"He put in a crop and worked it fairly well, but every now and then Charlie or one of his boys would catch him punching down into the ground with a steel rod. They never let him know they were watching. He seemed to be making a systematic probing of the whole valley. This went on three or four seasons.

"Then one spring, just as his truck was ready to market, he disappeared without a word, leaving his crop, his wagon, everything belonging to his shack, and a team of horses behind. From that day to this he has not been heard of. Undoubtedly that steel rod of his finally plunked the right spot, though Charlie Thomas was unable to locate the hole. The old fellow was clever enough to cover it up, I guess.

"Now, whatever the Spaniards concealed in the ground they took out of the ground. They had mines somewhere. Yet it is a fact that although I myself have surveyed hundreds of mineral claims in the Wichita Mountains, not a single mine in this country has to my personal knowledge paid. That does not mean a paying mine won't some day be opened.

"Before I came here a prospector who had a dugout on Deep Red found nuggets galore. One time when I was in that vicinity trying to locate the abandoned dugout, a farmer came to me and said he had something he wanted examined. It was a gold nugget polished as slick as a greased pinto bean.

" 'Why, good gracious,' I said, 'where did you get this?'

"Well, his wife had accidently found it in the craw of a chicken she was cleaning.

" 'But where did the chicken get it?' I wanted to know.

" 'I wish I could tell,' the farmer answered me. 'The chicken must have picked it up in a sand bed. The only sand about is some I hauled up for the children to play in. I hauled it from two different creeks and they have both been up since then, so that I don't know exactly where I took it out.'

"Now, down here on Cache Creek . . ."

"Hold on, Mr. Joyner," I cried. "Wait a minute." Then I turned to my companion.

"Lieutenant," I said, "we have been in Lawton over an hour. If we go on listening to Mr. Joyner you'll win your boots and miss the hunt and I'll be out both money and a Stetson hat. The decision is yours to make. Shall we attend to business or get more tales?"

"Then," replied the lieutenant—and I tip my Stetson hat, which I am wearing this minute for an eyeshade, to his gameness—"we'll do both. Mr. Joyner is going to take me to the farm where that chicken swallowed the nugget and we're going to pan sand in both those creeks. You are invited to come along."

I could not go along, for I had business with Pete Givers.

Pete Givers is what Trader Horn would pronounce "convivial." A little, dark, wiry man with a kind of foreign accent to his speech, he combines in himself the ease that characterizes lovers of good tales and the intensity that great dreamers habitually evince. By trade he is a tailor, but his thoughts habitually dwell as far above stitches and ironing boards as the mind of the tinker who wrote *Pilgrim's Progress* dwelt above pots and solder.

"Yes, sir," Pete Givers announced soon after he got launched on his subject, "I studied Hindu philosophy under that great teacher Lorentz. I can take an astral body out of its spiritual relation. That is something! Yet my science is helpless to recover the Spanish treasure in the Wichitas. This is why.

"Clairvoyancy is simply opening up the subconscious mind. The subconscious mind, though, can open up only while the conscious mind is closed. That is why the subject must be put to sleep. To the sleeper space is nothing. If there is anywhere in the world a mind that consciously or subconsciously knows the fact sought for, a good subject can realize it. In Wisconsin, for instance, a man once came to me to locate some money that had been stolen from him. I procured the services of an excellent subject, put her to sleep, and she revealed the money as being hidden in a barn. Within twenty-four hours we had secured it. A living person, the thief, knew where that money was.

"On the other hand, not for a hundred years perhaps has a human being lived who knew the hiding place of the Spanish treasures here in the Wichitas. Concerning that stuff no mind exists for a subject to work upon. However, one treasure hereabouts does have a chance of being revealed. It is modern enough that some human being may know subconsciously where it is buried. That is the James' loot."

The James Boys' Loot

The best book about any American outlaw that has been written, *The Rise and Fall of Jesse James* by Robertus Love, has a great deal in it about Frank James, but it has not one word about the two million dollar loot he helped bury in the Wichitas and then could not find. To get that incident in the history of the famous Missouri brothers one must go to Pete Givers.

Just how and where the James gang came by the $2,000,000 is not quite clear. Some say they took it from a Mexican transport crossing Oklahoma on the way to Saint Louis; others say that it was an accumulation from various bank and train robberies. The James gang robbed not to live; they lived to rob, and they robbed on a magnificent scale. Anyway, they stored $2,000,000 in the earth somewhere along the old road between Fort Sill and the Keeche Hills to the northeast, intending to leave it there until peaceful days should come in which to spend it.

The peaceful days never came. After having been hunted for nearly twenty years, "poor Jesse" was laid in his grave, as the ballad runs, by a "dirty little coward" named Bob Ford. All his companions but Frank were in their graves too or behind bars. Frank was still on the dodge. But Frank "came in," gave himself up, was tried, and acquitted.

Thereupon, as the Wichitas have the history, he set out to recover the long buried loot. He had left the region a hunting ground for Indians; he found it homesteaded, fenced, plowed. He could not locate the spot. He knew that it was alongside the old Western Cattle Trail leading to Dodge City; he knew in a general way where it was; but despite his tenacious memory and his falcon eye, the bit of earth concealing $2,000,000 looked no different from ten thousand other bits of earth.

Frank James bought a little farm so that he might have a strategic base from which to search. He would ride daily from his farm to Cache Creek over the piece of trail he once traversed at the head of the most daring band of robbers known to American history. He always covered the distance in a flying gallop, hoping that when he got to the scene of the secreted treasure the doors of memory would flash open. But daily the doors of memory remained locked. Riding thus, he wore out six horses before he galloped away from the Wichitas forever.

Thus the treasures of the Wichitas unfold themselves in legend. There is the great government treasure lost to view in the sands and waters of Cache Creek. There is the treasure of Cut Throat Gap—a name that was not bestowed without reason. And always there are characters and characters.

Devil's Canyon

For tradition and digging the cynosure of the Wichitas is Devil's Canyon. It seems to be a fact that one of the homesteaders of the region, in plowing a rocky patch of soil, turned up nearly as many human bones as boulders. Here at first was a mystery, but soon it became known that those bones belonged to treasure-bearing Spaniards who had been overwhelmed by Indians.

"Well, it was this away," began "Old Man" Sloan after I had ribbed him up to tell his story. "When that Wichita country was opened for settlement, I took up a squat that enclosed a good part of Twin Mountains, seven or eight miles from Devil's Canyon. A few years later an old Indian woman came in there with some of her people and began poking around. She claimed to be one hundred and five years old and she looked every day of it.

"According to her story, when she was a girl her people had a big fight with some Spaniards bringing a pack train out of Devil's Canyon and killed off the whole band. After the battle, she and two warriors captured three of the Spanish burros. They were loaded with gold ore. They buried it at the foot of Twin Mountains. As the old woman had been gone eighty years or so, it wasn't any wonder she couldn't find it now.

"She got me interested. I knew about that field of Spanish bones, but these were the first definite details I had learned. I began to investigate.

"There was a young woman fortune teller in Snyder named Dolly who'd been successful in finding a good many things, and I took the matter up with her. She looked down at the ground a while with her eyes shet, like she always done when she was a-studying, and then she says, 'Yes, the gold is there.'

"Dolly wasn't very definite, though, and she seemed afraid all along that somebody would prosecute her for telling fortunes. She asked me if there wasn't three hardwood trees close to my fence. I told her yes. Next, she asked if there wasn't a big flat rock near by. I told her yes. Then to test her, I asked if there was anything on the rock.

"She kept her eyes shet a long time. 'Yes,' she finally replied, 'there's some letters on it, but I can't make out what they are.'

"That satisfied me, for there were letters on the rock which I'd put there myself in white paint—NO HUNTING ALLOWED—and they'd been weathered off.

"She ended up by saying that if I expected to find anything, I'd have to keep the search absolutely secret and not reveal her part in it. I told her I just had to have somebody help me dig.

" 'Well,' says she, 'Fay can help you, can't he?'

"Fay's my son, you know, and I said, 'That's exactly who I was figgering on getting to help me.'

"I was so anxious about the matter that I couldn't think about anything else. I let out and told Fay even before I'd settled on a place to dig. About this time a nephew came out to visit us. Before I knowed it, Fay had told him and both boys were deviling me to start the search.

"Well, I went back to Dolly to get more specific directions. As soon as I saw her, I knowed something had happened. The first thing she said was: 'Certain people has been told what you're doing.'

"I couldn't deny it.

" 'Mr. Sloan,' says she, 'I can't tell you another thing.'

"I argued, but it didn't do no good.

"Of course, I dug. I dug a plenty—and never found a thing. Nevertheless, the gold's there all right, I guess. There's certainly been plenty of proof as to Dolly's power in finding things.

"For instance, one time a merchant there in Snyder was busy with a customer when a man came in in a big hurry to buy something. I forget now what it was. He helped himself, laid a silver dollar down on the counter, and rushed out. The merchant was occupied for a few minutes. Then when he went to get the dollar, it had disappeared. That night he went to consult Dolly.

" 'Dolly,' says he, 'I've suffered a business loss, and I wish you'd help me. I want to know who it was took my dollar.'

"Dolly looked down in her way for a little while. 'I know who it is,' she says. 'Go to your store in the morning and the third man that comes in, look him square in the eye and say to him, 'I guess you've carried my dollar about long enough.'

"Sure enough, the next morning when the third man came into the store, the merchant looks him in the eye and says, 'I guess you've carried my dollar about long enough.'

"The man pulled out a dollar and laughed and said he'd took it off just for fun.

"I could go on with other instances. You see what sort of vision this woman has. If it hadn't been for my blabbing when Dolly told me not to, that gold might be doing somebody some good right now instead of laying out there rusting to dust."

The Pothole of Nuggets

One time in the early days a band of men who were going across the plains to trade in New Mexico were attacked by Indians somewhere south of the Wichitas, in Texas. They made a corral of their wagons and fought off the Indians as long as they could, but when night came they were so thinned in numbers and the Indians were so strong that the survivors decided to break for their lives. They broke, and all but one man were speedily overtaken, killed, and scalped.

He saved his life by falling into a pothole that lay concealed down a ravine. He found it big enough for him to move around in and remained in it. When daylight came, he detected amid the rounded pebbles and sand at the bottom of the hole something that made him glad he was a prisoner. He was in a hole of gold nuggets. He began sorting them from the gravel, and while digging down with his bare hands found a little water. Thus hiding, living on water, and growing rich, he remained all day. He said afterwards that there must have been a barrel of the nuggets.

Finally, feeling satisfied that the Indians had left and having collected as much gold as he could carry, he set out for a distant fort. The Indians had burned all the supplies with the wagons. A shot at game might mean betrayal and death. Soon he began to realize how weak hunger can make the strongest man—even a man loaded with gold. He left his gun in a wild plum thicket loaded with green fruit that mocked him. His fear of being discovered led him to pursue a very twisting course. The precious gold made him exceedingly cautious. At length he grew so weak that he put a part of his burden in a gopher hole, tamping dirt down on top of it. He grew weaker; he began dropping the gold. When at last, only three specimen nuggets left in his pocket, he staggered into the army walls, he was hardly conscious of his loss.

As his strength and health recovered, the one subject that engrossed his mind day and night was the pothole of nuggets. He knew that he could never retrace his path. He knew that when he and his companions were attacked they were in a pathless region, traveling only by the sun and watershed. But the hills around the besieged wagons and the ravine down which he had stumbled into the pothole were pictured indelibly in his memory. The irons of the burned wagons would verify the exact spot. He began the quest alone. The riches belonged to him; he had bought them fair.

For years and years he wandered. He entered the region from the north; he entered it from the east, from the south, and from the west. But no hill or ravine corresponding to the picture so clear in his brain ever revealed itself. As the country settled up, he, vainly hoping to receive intelligence of the burned wagon irons, told his story to the settlers. It became, one might say, an element of the soil. Years ago he died in Wichita Falls, leaving to descendants three specimen nuggets that yet bear testimony to the truth of his often told tale.

CHAPTER XVIII

LAFFITE AND PIRATE BOOTY

Or if thou hast uphoarded in thy life
Extorted treasure in the womb of earth,
For which, they say, you spirits oft walk in death,
Speak of it.

. . . .

It faded on the crowing of the cock.—*Hamlet.*

THE MAN OF MYSTERY

FOR more than a hundred years Laffite's treasure has been the El Dorado of the Gulf coast. The search for it today is as fresh and eager as it was when Jackson fought the British at New Orleans. Probably a majority of the people in New Orleans could give a more extensive account of Jean Laffite than of Andrew Jackson. In Galveston schoolboys cherish a bit of what is purported to have been Laffite's jacket, and a fine new hotel—perhaps with unintended irony—bears his name. A recent dispatch from Yucatan said that his seal had there been dug up. His strange career, his fabled hoard, and his uneasy ghost will not let his name die. Yet, despite a considerable body of undisputed facts about Laffite's political machinations, the man himself remains veiled, enigmatical.

Various popular historians have glorified him as patriot; others have denounced him as pirate; hardly one who has written of Louisiana or Texas has neglected him. Regardless of all this interest, however, only one scholar, Mr. Stanley Faye, has studied him with any thoroughness, and Mr. Faye's extraordinary revelations remain yet to be published.[1] Enough novels and pulp paper stories to fill a deep five-foot shelf have made Laffite their theme. Folk whose only knowledge of history consists of inherited tradition tell of his daring adventures and look for his legendary millions all the way from the Keys of Florida to Point Isabel at the mouth of the Rio Grande. They call him "the Pirate of the Gulf."

Maybe he was not a pirate. Maybe, as he always claimed, he was a gentleman smuggler and privateer. His birthplace has been variously fixed as St. Malo, as a village on the Garonne, as Bayonne, as Marseilles. Mr. Stanley Faye, relying on documentary evidence not wholly satisfying, thinks he was born in Orduña, a valley in the Basque provinces of Spain. He could pass for either Spaniard or Frenchman. A Spanish agent in New Orleans reported that his friendship for Spaniards was equaled only by his detestation of English and Americans; indisputably he spent a good part of his life preying on Spain. Biographers with imaginary gift have said that his family were Bourbon aristocrats; also, that they were mere peasants. His very name has been in dispute; he signed it—two or three times at least—as Laffite, but traditionally it has been spelled Lafitte.

His whole life was a series of contradictions. In New Orleans he owned a blacksmith shop; and he lived in the manner of a cavalier, winning the sobriquet of "Gentleman" Laffite. Defying American law, he pitted his own cannon and cruisers against the American Navy; then at the battle of New Orleans, he, by fighting with Jackson against the British, won pardon for all past offenses and glamour for future memorialization. The government placed a price on his head; then the government placed batteries of defense under his hand. He was gallant to women, but whether he was ever in love is doubtful. Legend says that he married more than once; it also says that he had a low caste Creole mistress in Louisiana who gave birth to a son. At Galveston he was accompanied by a luscious quadroon. A journalist who saw this quadroon and asked Laffite for "the story of his life," received—so he reported—the following account. At the opening of the nineteenth century Laffite was a rich merchant in San Domingo, where he married a rich and beautiful wife. Soon after marrying, he sold out his business with the intention of going to Europe to live. He bought a ship and loaded it with goods and specie. At sea he was captured by a Spanish man-of-war. "They took everything—goods, specie, even his wife's jewels." Then they landed the Laffites on a barren sand key with just enough provisions to keep them alive a few days. An American schooner rescued them and took them to New Orleans, where the wife contracted fever and died.

It is asserted that from boyhood Laffite "loved to play with old ocean's locks" and that once he recklessly dared a West Indian hurricane by driving his fleet straight across water-covered Galveston Island. On the other hand, it is asserted that he was so subject to seasickness that he seldom boarded a vessel and that he "did not know enough of the art of navigation to manage a jolly boat." In truth, he was not so much a seaman as he was a boss of seamen. He was a brilliant conversationalist, but in conversing he was careful to avoid the secrets and duplicities that characterized nearly his whole existence. It seems safe to assert that he died quietly in a bed in Yucatan, in 1826; nevertheless, the story has come down that he died in a dare-devil engagement with a British war-sloop, his buccaneers cheering around him, his locks "matted with blood," the dagger in his swarthy hand streaming red.

Patriotic in one act, yet not a patriot; piratical through a lifetime of activities, yet not a pirate, he was—he is—Legend, Paradox, Mystery. His whole life, as Montaigne defined death, was *un grand Peutêtre*—a great Perhaps. He must have been a puzzle to himself. Truth is precious; so is an interesting story, even though the facts therein be overshadowed by fable. Legend, ever contemptuous of history, is still expanding and spinning the story of Laffite. I tell it as I have gleaned it from sources dubious as well as authentic.

There may have been three Laffite brothers, but only two, Jean and Pierre, enter the story. One tradition has it that Jean was an adopted brother. Whether brothers by blood or adoption, never were two men more devoted to each other than Jean and Pierre. They appeared in New Orleans about the time of the Louisiana Purchase (1803). New Orleans in those days was French to the backbone, and—despite any Spanish blood that may have flowed in their veins—the Laffites found themselves among their own kind. What they did upon arrival we know not. They apparently had money. Ere long they were the proprietors of a mighty blacksmith shop between Bourbon and Dauphine streets. It seems, however, that they never worked at the smithy themselves but had the work done by a corps of efficient slaves. Their real business was over on Barataria. Some old maps mark Barataria as "Smugglers' Retreat."

From New Orleans the Mississippi River sprangles out into the Gulf through a maze of bayous and interlocking lakes. To the west the great Bayou La Fourche sprawls gulfward through another maze of marsh and sluggish, twisting currents. Between the two is Barataria Bay. Curtaining off Barataria Bay from the Gulf is a sliver of an island called Grand Terre, sometimes Barataria. From Grand Terre to the coast the Bay is sprinkled with islands. A pass from the Gulf into the Bay gives entrance to a fine harbor on the main island.

Even today, says Albert Phelps, historian of Louisiana, this land of delta and island is "a desolate waste of salt marsh, jungle, and forests of cypress and water oak. A sad land with a sombre beauty of its own, these wide acres are still the haunt of wild things, untraversed for the most part save by the pirogue of the pot-hunter or the negro moss-gatherer. The waters of the almost currentless bayous are alive with garfish; turtles, snakes, and alligators bask on many a log, and herons, cranes, flamingoes, kingfishers, and pelicans hold a monopoly of the fisheries. More dense, more intricately water-threaded than all is the district lying about Barataria Bay."

The men who plied their piratical trade in this region, the Baratarians, as they were called, were a motley crew—Portuguese, French, Italian, Malay, adventurers from every nation. For a full and vivid picture of these highly interesting people there is only one place to go, though there are many places where one might go—a chapter upon them in *New Orleans, the Place and the People,* by Miss Grace King, who has, I think, blended literature and history more effectively than any other writer of the land where Laffite's memory dwells. The Baratarians were not exactly pirates in the manner of the ship-scuttling and throat-slitting Blackbeard, Morgan, and L'Olonoise crews who more than a hundred years before had given to the Mexican waters their tradition of piracy. The days of the great pirates had waned. British warships had chased their successors into marshy holes of refuge. The Baratarians were privateers—licensed pirates.

The times were propitious to privateers. South America, Central America, Mexico were all seething with the yeast of revolt against Spain. The republic of Cartagena, a mere seaport of Colombia, was glad to issue letters of marque against Spanish shipping. France and, for a time, the United States were authorizing privateers to prey on English commerce. England, embroiled in the Napoleonic wars, had only limited forces to police the western seas. The naval power of America was a farce. Armed with their letters of marque and also with brass cannon and steel cutlasses, the Baratarians could sally forth from their snug refuge and thrive off Spanish merchantmen, with now and then a prize of some other nationality thrown in for lagniappe.

But the Baratarians needed a market for their plunder.

Next door to them was New Orleans. Perhaps thirty thousand consuming human beings made up its population; it was the gateway to the commerce of the Mississippi valley. Most of the goods that the Baratarians had for sale could not be declared. It was not their manner to pay duty anyhow, and they sold cheap, as buyers from Memphis and Saint Louis as well as from New Orleans soon learned. The marshes and bayous afforded approaches that no revenue officer could ever follow. The citizens, almost without exception, were as friendly to the smugglers as was the secret land.

With plunder in hand and a market at their door, the Baratarians required an agent and banker. Jean Laffite became that agent. He was an energetic and efficient business man. He spoke English, French, Spanish, and Italian fluently, if not correctly. He had a gift for making phrases. He had a conscience as elastic as any politician could wish for. Nature seemed to have designed him for agent to the Baratarians.

From agent to chieftain is only a step. Laffite insisted upon two things: strict obedience and that word "privateer." He avoided the term "pirate" as a "mortician" avoids "undertaker." Once, according to an old story, a certain Grambo, who had known rougher days, hooted at the name, boldly declaring himself a pirate and calling upon his comrades to put down this genteel privateer who had come to rule over them. Laffite pulled his pistol and shot Grambo through the heart. Thereafter his rule and his choice of diction were undisputed.

Early in 1813 certain American merchants and bankers of New Orleans became so alarmed over the loss of their legitimate business to the smugglers that they called on the naval authorities for help. The naval authorities sent out two minor expeditions that were successively put to flight. Then the merchants and bankers called on the state legislature. The legislature debated and declared that they had no funds. It was clear that Laffite had friends. While the legislature was debating, he held at Barataria one public auction of 450 negroes.

Meantime Governor Claiborne of Louisiana had proclaimed the Laffite brothers to be "banditti and pirates," and had offered a reward of $500 for the arrest and delivery of Jean. Jean retaliated by offering a reward of $15,000 for the arrest and delivery of Claiborne to him! He continued to visit New Orleans when he pleased. He even dallied on the streets, laughing and chatting with his friends while he leaned on a wall that placarded the governor's proclamation.

One day Claiborne's men captured Pierre and clapped him into jail. He soon escaped, but the federal grand jury brought an indictment. To fight the case Jean engaged the two best lawyers in Louisiana at a stipulated fee of $20,000 each. One of the lawyers was the district attorney, John R. Grymes, who resigned his office to enter Laffite's services. After the trial, which amounted to nothing, he was invited out to the "Pirates' Lair" to receive his fee. He stayed a week amid feast and revelry, and then "in a superb yawl laden with boxes of Spanish gold and silver" was returned to the mainland. "What a cruel misnomer it is," he declared upon returning, "to call the most honest and polished gentlemen the world ever produced bandits and pirates!" Only on the stage he had elected, at the time nature had destined him, could Laffite have gestured so magnificently.

In 1812 the United States declared war on England. The British prepared to lay siege to New Orleans. In September, 1814, two English officers landed on Grand Terre and offered Jean Laffite $30,000 in cash, a captaincy in the British navy, and a chance at enlistment for all of his men provided he would aid in the proposed capture of New Orleans.

The British did not know Laffite. He played for time. Then he informed the United States officials of what was brewing and offered his services in defense of the city. "Though proscribed by my adopted country," he wrote, "I never let slip any occasion of serving her or of proving that she has never ceased to be dear to me."

As a reply to such friendly advances Commodore Patterson of the United States Navy made an attack on the privateer stronghold and captured a large quantity of booty. Laffite himself was away at the time; most of his men escaped and fortified themselves on Last Island.

Hickory-tough old Andrew Jackson was at Mobile when he heard of the British proposal and of Laffite's offer. Forthwith he issued a thundering proclamation to the Louisianians in which he bitterly denounced the British for attempting to form an alliance with "hellish banditti." "The undersigned," he concluded with a flourish, "calls not upon pirates and robbers to join him in the glorious cause."

When he reached New Orleans, however, and saw the desperate need for more men, he came down off his high horse. These Baratarians were men of his own mettle. He placed Laffite and his "hellish banditti" in charge of two important batteries. The battle was won, and in a general army order Old Hickory praised Laffite and his captains as "gentlemen of courage and fidelity." President Madison issued a full pardon to all Baratarians who had taken part in the battle.

What Laffite next did legend has been profuse in explaining. One story has him going to Washington and squandering $60,000 in gaudy living. A persistent story has him returning to Europe and in his own ship carrying Napoleon from Elba to France—and the Hundred Days' War that ended with Waterloo. The story goes on that Laffite had even made arrangements to bring Napoleon to America and that he did bring a vast treasure belonging to the fallen emperor, which, of course, he properly buried. Impossible fictions!

Laffite had become associated with Toledo, Herrera, Gutierrez, Peter Ellis Bean, Perry, and other adventurers who were seeking to overthrow Spanish rule in Mexico and establish an independent state in Texas. He now secretly engaged with Spanish agents to act as spy upon such filibusters and insurgents. Such an engagement, however, in nowise checked his privateering upon Spanish shipping. The record of his double-dealings from the time he landed on Galveston Island until he was driven away from it would make a steel windmill giddy.

It was along in 1816 when he began making a new Barataria at Galveston and, with Pierre to aid, resumed the old business of distributing "purchases"—goods and negroes—to Louisiana buyers. To his American audience he announced that he had selected Galveston as headquarters, first, in order to be near the United States should that dear country again need his services; secondly, in order to further the cause of liberty in Mexico. To his Spanish masters he announced that he was going to collect in one place—Galveston—all the privateers infesting the Gulf so that they could be captured at one fell swoop. He seems to have collected them all right.

Galveston Island already had a history. At the time Laffite arrived, Louis Aury, a Mexican "republican" soon to vacate, was using it as a base for smuggling slaves and pilfering ships. Here with a thousand men of mongrel breed under him, making and unmaking captains, Laffite lived in his Maison Rouge like a lord in feudal splendor. An old French legend has it that the devil built Maison Rouge in a single night. In contracting with the devil for its erection, Laffite agreed to give him the life and soul of the first creature he cast his eyes upon in the morning. Laffite then contrived to have a dog pitched into his tent about daylight; so all the devil got out of the deal was a dog.

They called him "the Lord of Galveston Island." He was at this time about forty years old, and is described as being exceedingly handsome, even noble in appearance. He had magnetism, charm, suavity, every quality necessary for one who would run innocently with the hare and at the same time bay lustily with the hounds. He seldom smiled, but he cultivated in a rare manner the art of being agreeable. He set an orderly table with abundance of plate, linen, and choice wines. Generally he went unarmed, but, with a nose that sniffed the lightest wind of adversity, he could be depended upon to appear at the right moment provided with a brace of pistols and a "boarding sword." When aroused, he was a desperate man indeed, and he was both an expert swordsman and an unerring shot.

On his lonely island, a wilderness of wild land behind it, a world of silent waters before it, "the Pirate of the Gulf" played host to a train of strange characters. Here—if report be true—came Peter Ellis Bean, who had mustanged in Texas with the filibuster Nolan, who had for six years somehow existed in a solitary cell of a Spanish prison, his only companion a pet lizard, and who had then secured his liberty in time to fight beside Laffite at New Orleans. Here came "Old Ben" Milam, "war-born," who had also fought at New Orleans, who was to help Mexico throw off the Spanish yoke, and who was to meet his death leading the Texans into San Antonio. Doctor Long, who at the head of three hundred men had declared Texas a republic—this was years before Austin settled Texas with Americans—came also, seeking Laffite's aid in his enterprise. Laffite was generous in giving "good wishes"—and at the same time reported him to Spanish authorities. Here, too, came half a thousand French refugees seeking an asylum, and Laffite sent them up the Trinity River, where they established the short-lived and tragic Champ d'Asile—happily ignorant of their benefactor's plot to annihilate them. The savage Carankawas came to wonder and barter. Their visit ended in blood.

"Spanish doubloons," said a frontiersman whom Maison Rouge entertained, "were as plentiful as biscuits." Jim Campbell, one of Laffite's lieutenants, who remained on to become a citizen of the Republic of Texas after his master had sailed away, used to tell how Galveston Bay, preceding any dangerous expedition, "was covered with boats seeking select places to bury treasures." Once from a rich haul, so the story goes, Laffite took for his own share—though he usually received a "royal fifth"—only a delicate gold chain and seal that had been removed from the neck of a Spanish bishop on his way to Rome. He gave the chain to Rezin Bowie, brother of the famous James Bowie. The Bowies must have been visitors more than once, for we hear of their buying negroes from Laffite at a dollar a pound to smuggle into Louisiana. Another man who was to win a name in Texas, L. D. Lafferty, in his old age recalled clearly how in urging him to enlist as a buccaneer Laffite "frankly confessed that he had enough silver and gold on the island to freight a ship."

Of all Laffite's men at Galveston, Jim Campbell was the most famous. There was much that he never told, but he told plenty. That fellow had seen service, and he was not niggardly in telling about it. He had been with the notorious Captain Rapp on the *Hotspur,* privateer, when two Spanish ships engaged her. The fight lasted from dawn till dusk. In the end the *Hotspur* got away with most of her crew dead and only fourteen unwounded. One of the Spaniards was so shot up that she could not pursue; the deck of the other, the *Consalada,* at last sight "was literally covered with the dead—they had to walk on the bodies." Then Campbell had served with Aury and Mina. On one cruise of six weeks under Laffite he had captured five prizes with cargoes valued at $200,000; another time he "took a Guinea-man" with 308 slaves.

In time an American warship put in to call on the Lord of Galveston. His letters of marque, furnished now by Mexican revolutionists, authorized him to prey, as usual, upon Spanish shipping; but his men frequently made no distinction in flags. To show his sentiment and patriotism, Laffite had an offending pirate hanged on the seashore, and the warship departed. But the offense against American traders was soon repeated. Lieutenant Kearney, in command of a United States man-of-war, appeared one day in 1820 with polite orders that Laffite abandon Galveston. Laffite left forever. The rest is legend—mostly about pirate treasure.

The Legends

Just before he sailed away, so one oft-told tale runs, some of Kearney's men saw him walking to and fro, apparently in great distress, and heard him muttering words about "my treasure" and "the three trees." The three trees were a well-known location on Galveston Island. The eavesdroppers stole thither and began digging in ground that had evidently been disturbed only a short time before. The earth was loose. They made fast time. Soon they struck a box. They dug it up. They tore it open. It contained the body of a beautiful girl, Laffite's bride—his "treasure." Henry Ford was not the first man to regard history as "bunk."

For a thousand miles along the Gulf coast every inlet and island has its Laffite treasure. Somebody is always searching for this treasure, and nearly always there is a legend of great detail and realistic circumstance to back up the search. Once or twice a year—of recent years more frequently perhaps than in the past—a newspaper item from some town in Texas or Louisiana reports a hunt for Laffite treasure; but the most ardent hunts are made in secret and the best tales never get into print. Sometimes, of course, sheer pertinacity will betray the golden dream of a secret hunter.

Newell, a New Orleans printer, was one of these solitary and silent seekers. For twenty years he lived with and for his aureate vision. Along about the middle of the last century Newell's father befriended a battered old sea-faring derelict who was soon to make the final voyage. The derelict was grateful and bequeathed to the elder Newell a chart to a vast treasure purporting to have been buried by Laffite's men on a little island in Lake Borgne. The father turned the chart over to his son, the printer.

Printer Newell had all confidence in the chart and at once began hoarding his wages to fit out an expedition. He bought an old smack, a camping outfit, and tools, and disappeared from the printing establishment. For many weeks he cruised among the islands along the coast line, sometimes digging for days in a barren sand bar. The winds and tides were so constantly shifting these islands that he could never be sure which one his chart called for. Yet as he searched he was all the more certain the great swag of doubloons was there—if he could only find it. His supply of food ran out, his clothes wore out, and he returned to set type a while longer and save another stake.

He lived apart from the other printers, seldom speaking, never treating, his mind in another world. He became a marked man. It must have been a great relief to him when he could escape for another expedition. These expeditions went on at irregular intervals for years. Finally Newell took in a partner; but increasing eagerness and intentness made him suspicious of the partner, and again he wandered alone.

Years and years passed—five, ten, fifteen, twenty of them, the golden dream more luminous with each lustrum. Among the unnumbered islands outside the Rigolets, Newell spaded up thousands of tons of sand. Sometimes he dug holes so deep that "you could have buried a fishing smack in one of them." His persistence was sheer genius. Youth turned to middle age, and Newell's hair was white. In the summer of 1871 some coast men saw him scudding out to sea in the teeth of a fearful hurricane. The next day a lumber vessel limped into Pearl River harbor towing a little boat that was recognized as Newell's. Then his body was found washed up on a drift. Presumably the storm had swept him off the deck of his smack and he had drowned.

Laffite and Pirate Booty

As uncertain as the island in search of which Newell spent and lost his life was the Spanish dagger (or yucca plant) on Padre Island west of Corpus Christi, Texas, that a band of adventurers came looking for a generation ago. They engaged old Charlie Blutcher, surveyor of Nueces County, to run lines for them, and all went to Padre. Their orienting points were to be one Spanish dagger and three brass spikes. They found hundreds of daggers but not one brass spike. So far as anybody knows, the Spanish ship with its hold full of plate still rests against Padre Island, where Laffite's men lost it during a storm and where the welling sands covered it up.

When Laffite was run away from Galveston Island, so more than one legend avers, he did not leave the Texas coast as he was supposed to do. He released three of his four vessels to shift for themselves, but first he picked from his followers a crew to sail his own craft, the *Pride*. Then he slipped down to the mouth of Lavaca River, to hide when necessary and to sally out and privateer when opportunity offered.

Port Lavaca consisted of treacherous sand bars through which meandered a narrow channel known only to Laffite's men. The channel twisted its way up the river to a landing well out of view of any ships that might pass on the sea.

One day while cruising for a prize, Laffite came in view of a United States revenue cutter. He fled for the Lavaca refuge, getting inside the sand bars just in time to keep from being overhauled. The captain of the cruiser was as sly as Laffite. He sailed away as if he had given up the chase, but as soon as he was out of sight he put into a cove to await the reappearance of his quarry. In a few days Laffite slipped out. Again the cutter bore down upon him, and again he headed for his hole. But the cutter was so near him this time that it could follow his path through the bars, at the same time raking his ship with shell. Laffite was bottled up.

He ran as far as he could up the river, then nosed the

Pride, already sinking, into the bank. In the few minutes that were left to him he divided a vast treasure among his men and told them to scatter and make out the best they could. Two men only remained with him.

When they went ashore with Laffite, they carried a chest containing a million dollars' worth of gold and jewels. It was very heavy. They could not lug it far. So when they had gone about a quarter of a mile east from the Lavaca River and were well hidden in a salt grass flat, they buried it, at Laffite's orders.

After the chest was in the ground and covered up, the pirate captain took a Jacob's staff—a brass rod used by surveyors instead of tripod—which he had brought along, and set it up immediately over the chest. Then he fixed his compass in the socket on top of the Jacob's staff and took bearings on two mottes of trees within sight. These mottes are well known today as the Kentucky Motte and the Mauldin Motte. What the readings were only Laffite knew; his men observed that he did not make notes on paper. Finally he drove the Jacob's staff down in the ground until the socket on the end of it stuck out only a foot or so.

After leaving the chest, Laffite and his two companions traveled for three days without food. It was winter time, and they were cold and almost starved when they came to the cabin of a settler. There they were fed and cared for; Laffite arranged for three horses, though no saddles were to be procured, and when he got ready to leave, he gave his hosts a thousand dollars. He now told the two men, whom he had retained thus far, that he was going among the Indians north of Red River and that they had better go to New Orleans, their old home, and live honest lives. He said that if they or their representatives came back at any time after the expiration of three years and found the chest unremoved, they might have it. He rode off towards the north, and that was the last his trusted mates ever heard of him.

A few months later one of these men, while dying in a New Orleans saloon, confided his secret to an Irish bartender. The other married, had two sons, and when they were mature told them facts that corroborated the bartender's story. Many years ago the two sons came into the Lavaca country searching for Laffite's chest. They found nothing and departed, but they left behind them a tradition that yet abides.

In time a ranchman named Hill acquired the land down the Lavaca River and stocked it with horses. The custom on the open range was to put a stallion and about twenty-five mares into a *manada* and loose-herd them until they were "broke in" to staying together. The *manada* was penned every night. Now, Hill had one negro herder who was particularly sleepy-headed. Every day this negro would take his mares to grass, turn them loose, stake his saddle horse, go to sleep, and wake up only in time to reach the pens by sundown.

One day he took his mares down to a flat along the river. He looked around for something to stake his horse to, but could see nothing but salt grass. He was kicking through it, hoping to find a stick that water or wood-rats had left, when he barked his shin against something hard and solid. It was a brass rod, fast in the earth, with a kind of knob or socket on the end of it—the very thing for a stake-pin.

The horse was not well broken to being staked, however, and in the course of the afternoon he got tangled up in the rope and "set back." When the negro, aroused from his nap, unwound him, he found the brass rod pulled far out of the ground. He pulled it on up and as a curiosity took it to the ranch, arriving considerably ahead of his usual time.

The minute Hill saw the rod he knew that it was a Jacob's staff, and he knew what a Jacob's staff stuck down in a salt grass flat near the Lavaca River meant. He at once ordered the negro to lead him back to the place where the rod had been found. Perhaps the negro was lazy and did not want to go so far; perhaps he was so ignorant of woodcraft that he could not find the place. At any rate, he did not lead Hill to it. He did, however, give an accurate description of the location with reference to the two famous mottes. Hill took the best trailer he had and spent all the next day and the next searching for the spot where the brass rod had been found. But the whole country was covered with coarse, thick grass tracked over everywhere by horses. Finally he quit the search.

Other searchers came and went. Then J. C. Wise began his search. This was only two or three years ago. Wise lives in San Antonio—that rendezvous for centuries of lost fortune seekers and filibustering adventurers. He had spent years in gathering evidence, then more years in perfecting a machine to locate minerals under the ground. In the very beginning he and his two associates made the mistake of hunting for a week on the Colorado River instead of the Lavaca. When they got near the Lavaca site, a ranchman refused to let them through his pasture. They went around the bay, procured a motor boat, and came up the river.

"The first thing we had to do," says Wise, "was to locate Laffite's old ship. Of course it had long since rotted, but we thought that by diving we might find some of the bulwarks. I think all the fish in the bay had a playground right where we wanted to work—and they weren't all red fish or mackerel either. Finally we located what was left of the ship and then took our measurements. With the wreck and the two mottes as marks, we felt reasonably sure of getting close enough to the treasure chest for our mineral machine to operate.

"It had rained and the salt grass flat was knee deep in mud. We slushed about for five days, and that machine of ours never did indicate a thing. I know that the chest is there, though, and some day when I have more time I'm going back after it."

Not all authorities agree that *all* of Laffite's treasure on the Lavaca River is in a chest. Here is a letter from an old Texas ranger who lives five hundred miles upland.

"I was living in Lavaca County in 1870," the letter reads, "and while I was there my brother hired to a farmer named Bundick. One rainy day Bundick went turkey hunting, and just as he was crawling up on a flock of turkeys near the river, he struck his knee against something very hard. The pain almost sickened him, and when he looked down he saw what he took to be a pile of bricks, most of them half covered in the dirt. They were peculiar-looking bricks, and he wondered how they came to be out there. He picked up one, put it in his shot pouch, and when he got home showed it to my brother. Brother knew at once what it was. It was silver bullion hid there by some of Laffite's men. If Bundick had not been in such pain when he picked the thing up, he certainly would have noticed that it was silver or lead.

"Well, he and my brother went right back to get the whole pile of bricks. They found a moulted turkey feather that Bundick had noticed near the place, but they simply could not locate the bricks. There Bundick's tracks were too. Since that time, fifty-odd years ago, that pile of brick-looking silver has been stumbled on by two different persons ignorant of their worth, and in each instance the person when put wise has been unable to go back to the location. I myself have spent many days looking for the bullion, and while I am sure that I was often within a hundred steps of it, I never had the good fortune to stump my toe against it."

The mouth of the Colorado River is—or was—a better place than the Lavaca to look for pirate treasure. Along in the twenties William Selkirk, a prosperous citizen of New York, befriended a very sick sailor named Robinson. The sailor was exceedingly grateful. He was about to die and could, therefore, well afford to give away all he had to his benefactor. What he had, amounted to the story of his life—and a map.

The sailor had been a pirate under one of Laffite's captains, he said. Once while cruising for a prize in the Gulf of Mexico, the brig he was in awoke one morning to find herself under the guns of a towering Spaniard. She ran for shore and about sundown slipped over a bar into the mouth of the Colorado River. However, she could not proceed upstream; she fully expected that her deep-draughted pursuer would be standing outside the bar at daylight ready to shell.

"There's a bare chance of getting away," said the captain. "If we don't, then we'll just fight back the best we can."

He was a desperate devil. Darkness came on. Three sailors, one of them Robinson, were sent ashore with the ship's treasure chest. It was very heavy. They buried it on a promontory, and then, returning, delivered to the captain a true and detailed description of the place—henceforth to be called Gold Point.

On the morrow, sure enough, the Spaniard beat up within range and opened fire. The captain was killed; so were the two sailors who had accompanied Robinson ashore with the chest. Thus he alone was left with exact knowledge of the treasure. He escaped the Spaniard, but he had no opportunity to take a sou with him. The winds of fortune, strong and in cross-currents, blew him for the next several years here and there over the sailing world. Now, at last, the treasure chest still unrecovered, he was dying, an object of charity, in New York.

William Selkirk had so much faith in the sailor's recital that he soon pulled up stakes, came to Texas, and purchased from the Mexican government 6000 acres of land at the mouth of the Colorado River—including Gold Point. Of his attempts to locate the pirate chest there is nothing unusual to tell.

The Selkirks have kept that land and paid taxes on it all these years. For the last forty years or so they have had a negro by the name of George Ellis watching certain marshy tracts of the land to keep off squatters—and treasure hunters. Not long ago some men came to Gold Point and, instead of asking George Ellis for permission to dig after a chest of gold, hired him to help them. George figured that by working with the intruders he could watch them and thus protect the Selkirk interests. He knew that if he scared them off, they would come back secretly and dig anyhow.

After they had dug a while, George felt his spade scrape against metal that he knew was the lid of a box.

"Boss," he said, "I jes' don' know what's come ovah me. There's a pow'ful mis'ry in my back; my hands is gettin' whicherly-like and my haid's all globble-globble. 'Deed, boss, I'se jes' bound to quit. You-all ain't goin' find nuthin' 'round 'mongst this sand nohow. You jes' as well quit too."

While he was making this speech the faithful George got right down over the box lid that his spade had scraped, so as to protect it against the discovery of another spade. He was mightily surprised when his employers declared that they were tired and disgusted and ready to quit also. They quit on the spot, and set out toward Bay City.

If it had not been for ha'nts, George the faithful would most certainly have returned under cover of dusk to finish unearthing what he had discovered, but "ha'nts is turrible, jes' turrible, about these yere old treasures." It was really after sunup before George got back to Gold Point. A rectangular hole in the earth showed where a chest had been lifted out during the night. One of the hunters, presumably, had struck it with his spade and recognized the *plumb* about the same time that George made his discovery. The negro's getting out of the way was just what the white men wanted.

Probably the earth was never visited by a more extravagant or a more uneasy ghost than the ghost of Laffite. Sometimes this ghost strains with all the agony of a purgatoried soul to get its treasure removed and put to uses of virtue. Again, the ghost—or perhaps it is the ghost of some man slain and buried over the treasure—repulses the most daring and godless prospectors as they come near the object of their search. Back in the days of Reconstruction one of the most hard-headed and upright lawyers in Texas used to get messages from Laffite's spirit trying to explain to him where a whole shipload of precious plunder was secreted on Galveston Island. The lack of coherence in the messages revealed plainly enough what torture the spirit was suffering.

Over on the Louisiana-Texas line a man by the name of Marion Meredith has, or used to have, a remarkable chart.[2] He got it from a neighboring widow whose husband had secured it from a Mexican woman. This chart was a kind of widow-maker.

It called for a tree with a chain about it, somewhere near the mouth of the Neches River. According to the history that went with the chart, some of Laffite's men had their ship cabled to the tree and had just finished planting a fine "wad" of loot near by when they were jumped by a Spanish galleon. They cut their hawser and got away only to be bombarded to the bottom of the ocean. One man alone survived, the man that handed down the chart.

Marion Meredith's neighbor, who finally came into possession of this chart, felt so sure of finding the treasure that he would trust no one to go with him in search of it. Without trouble he found the tree and the chain. Then he stepped directly to the spot at which he was to dig. After he had shoveled down a few feet, some unseen power seized him. At least his subsequent gesticulations were interpreted to indicate as much. When he reached home his organs of speech were paralyzed and in less than a week he was dead.

As soon as Meredith got hold of the chart—from the widow of the late searcher—he took in as partner a rough old character by the name of Clawson. The two men found the tree with the chain about it—a rather rusty chain. At the spot where the chart called for the treasure to be, they found a hole already started. Near at hand were some decayed tools.

They had not dug down very far before they came to a skeleton. They carefully removed it and laid it out on the bank. As the hole got deeper, only one man at a time could work in it. It was Clawson's turn to dig, and Meredith was peering down from above, expecting every minute to see a shovelful of doubloons pitched up, when all at once Clawson gave a wild leap for the surface. His face was haggard; his eyes had the look of the haunted; when he spoke his voice was terror itself.

"Come! For God's sake, let's get away from this place," he half whispered, half shrieked, clutching Meredith's arm in a viselike grip.

"What's the matter? What have you seen?" demanded Meredith, who had seen nothing.

"I have seen hell and its horrors. Come away from here, I tell you."

And Clawson pulled so powerfully and his terror was so contagious that they left without even taking their tools. Clawson would never make explanation. He only begged Meredith, as he valued his life, never to dig at that place again. In time Meredith returned to get his tools. He found the skeleton back inside the hole out of which he and Clawson had so carefully taken it. He covered it up, shoveling in sand and shell until the sink was level. Then he came away, never to go back. He had absolute confidence in Clawson's judgment. Long afterwards he met Clawson in Beaumont and again asked him for an explanation.

"For God's sake," Clawson replied, "never ask me about that matter again. It has haunted me all these years."

At La Porte, on the Gulf, not far from Houston, stands an abandoned house beneath which the spirit of Laffite, harried and desperate, keeps guard over his blood-marked booty. It leads any occasional sleeper in the house to a great mound of yellow coins, jeweled watches, bracelets, diamond rings, and strings of pearls. Then in an attempt to win absolution for sins committed on earth, sometimes in distress and sometimes in anger, it begs the visitant to take those things and dispose of them *without spending one penny evilly or selfishly.* No one has ever taken the treasure. Coast dwellers used to gather around firesides on northery winter nights, and while the rich juice of sweet potatoes roasting in the ashes oozed through the jackets, tell tales of the old house and its ghostly tenant. One such tale, written by Julia Beazley[3] under the title of *The Uneasy Ghost of Lafitte,* follows.

One time Lafitte and his buccaneering crew sailed up to what is now Bay Ridge (which is opposite the haunted house of La Porte). He anchored his schooner off shore, and rowed to the beach with two trusted lieutenants and the heavy chest which none dared touch except at his orders. When the skiff grounded, the watchers on the schooner saw their chief blindfold his helpers; then they saw the three disappear with the chest behind a screen of grapevine-laden trees. Two hours later Lafitte returned alone. He was in a black mood and no one had the temerity to question him. It was supposed that he had caught one of his helpers trying to mark the location of his cache, and had killed them both. Some say that he led them back to the pit they had dug and filled up, made them reopen and enlarge it, and while they were bent down digging, shot them dead. Soon afterwards Laffite and his followers went down together in a West Indian hurricane, and his crime-stained treasure still lies buried in its secret hiding place.

Yet to many, as I have intimated, that place has not been secret. It is under the old house. As faithfully as I can follow the tale, I shall relate an experience connected with that old house as it was told me by a Confederate veteran who has now passed on. I shall call him Major Walcart, though that was not his real name.

"It was on a February night back in the eighties," the Major used to say. "The early darkness of a murky day had overtaken me, and I was dead tired. I do not think mud ever lay deeper along the shore of Galveston Bay or that an east wind ever blew more bleakly. When I came to a small stream, I rode out into the open water, as the custom then was, to find shallow passage. A full moon was rising out of the bay. Heavy clouds stretched just above it, and I remember the unearthly aspect of the blustering breakers in its cheerless light. The immensity and unfriendliness of the scene made me feel lonesome, and I think the horse shared my mood. By common consent we turned across before we had gone far enough from shore, and fell into the trench cut by the stream in the bottom of the bay.

"We were wretchedly wet as we scrambled up a clayey slope and gained the top of the bluff. A thin cry which I had not been sure was real when I first heard it now became insistent. It was like the wail of a child in mortal pain, and I confess that it reminded me of tales I had heard of the werewolf, which lures unwary travelers to their doom by imitating the cry of a human infant. By the uncertain light of the moon, which the next moment was cut off entirely, I saw that I had reached a kind of stable that crowned the bluff, and from this structure the uncanny summons seemed to come.

"The sounds were growing fainter, and I hesitated but a moment. Dismounting, I led my horse through the doorless entrance, and now the mystery was explained. Huddled together for warmth lay a flock of sleeping goats. A kid had rashly squeezed itself into the middle of the heap, and the insensate brutes were crushing its life out. I found the perishing little creature, and its flattened body came back to the full tide of life in my arms. Its warmth was grateful to my cold fingers, and I fondled it a moment before setting it down on the dry dirt floor.

"I tied my horse to a post that upheld the roof of the stable, and with saddle and blanket on my arm started toward the house, which I could make out in its quadrangle of oaks, not many yards distant. The horse whinnied protestingly as I left him, and when the moaning of the wind in the eaves smote my ears I was half in mind to turn back and bunk with the goats. It was a more forbidding sound than the hostile roar of the breakers had been in the bay.

"I called, but only the muddy waves incessantly tearing at the bluff made answer. I had scarcely hoped to hear the sound of a human voice. The great double doors leading in from the front porch were barred, but the first window I tried yielded entrance. Striking a match, I found myself in a room that gave promise of comfort. Fat pine kindling lay beside the big fireplace, and dry chunks of solid oak were waiting to glow for me the whole night through.

"I was vaguely conscious that the brave fire I soon had going did not drive the chill from the air so promptly as it should, but my head was too heavy with sleep to be bothered. I spread my horse blanket quite close to the cheerful blaze, and with saddle for pillow and slicker for cover I abandoned myself to the luxury of rest.

"I do not know how long I had slept when I became aware of a steady gaze fixed on my face. A man was looking down on me, and no living creature ever stood so still. There was imperious command in the unblinking eyes, and yet I saw a sort of profound entreaty also.

"It was plain that the visitant had business with me. I arose, and together we left the room, passed its neighbor, and entered a third, a barren little apartment through the cracks of which the wind came mercilessly. I think it was I who had opened the doors. My companion did not seem to move. He was merely present all the time.

" 'It is here,' he said, as I halted in the middle of the bare floor, 'that more gold lies buried than is good for any man. You have but to dig, and it is yours. You can use it; I cannot. However, it must be applied only to purposes of highest beneficence. Not one penny may be evilly or selfishly

"The treasure is mine to give," said the ghost of Laffite. "I paid for it with the substance of my soul."

spent. On this point you must keep faith and beware of any failing. Do you accept?'

"I answered, 'Yes,' and the visitant was gone, and I was shivering with cold. I groped my way back to my fire, bumping into obstructions I had not found in my journey away from it. I piled on wood with a generous hand, and the flames leaped high. I watched the unaccountable shadows dance on the whitewashed walls, and marked how firebeams flickered across the warpings of the boards in the floor. Then I dozed off.

"I do not know how long I had been asleep when I felt the presence of the visitant again. The still reproach of his fixed eyes was worse than wrath. 'I need your help more than you can know,' he said, 'and you would fail me. The treasure is mine to give. I paid for it with the substance of my soul. I want you to have it. With it you can balance somewhat the burden of guilt I carry for its sake.'

"Again we made the journey to the spot where the treasure was buried, and this time he showed it to me. There were yellow coins, jeweled watches, women's bracelets, diamond rings, and strings of pearls. It was just such a trove as I had dreamed of when as a boy I had planned to dig for Lafitte's treasure, except that the quantity of it was greater. With the admonition, 'Do not force me to come again,' my companion was gone, and once more I made my way back to the fire.

"This time I took up my saddle and blanket and went out to the company of my horse. The wind and the waves were wailing together, but I thought I saw a promise of light across the chilly bay, and never was the prospect of dawn more welcome. As I saddled up and rode off, the doleful boom of the muddy water at the foot of the bluff came to me like an echoed anguish."

But Lafitte does not appear to every one who spends a night in the house, and any person seeking the treasure from purely selfish motives is likely to rue his pains. A story is told of an acquisitive and enterprising man who came hundreds of miles with the purpose of helping himself to the chance of finding pirate gold, but who abruptly changed his mind after spending a night in the house. As Lafitte steadily pursues his object of finding a fit recipient for his dangerous gift, never succeeding, his disappointment is sometimes terrible, so they say, and some simple folk believe that when there is a particularly dolorous moan in the wash of the waves, it is the despair of the pirate finding voice.

Like the man Laffite, his treasures are uncertain, elusive, mysterious. Doubtless, too, most of the legends concerning them are, like many of those concerning Laffite, without foundation. Yet there are few monuments so potent to make a name remembered as its association with a great lost treasure. As long as Grand Terre and Galveston Island are above water, "Laffite's Treasure" is likely to keep the name of Jean Laffite green.

More than one writer has drawn a parallel between Laffite and Byron's Corsair. At least Byron's closing couplet seems proper to the Patriot and Pirate of the Mexican Gulf:

> He left a Corsair's name to other times,
> Linked with one virtue and a thousand crimes.

CHAPTER XIX

SHADOWS AND SYMBOLS

And after he went to the ship's board, and wrote there other letters which said: Thou man that wilt enter within me, beware that thou be full within the faith, for I ne am but Faith and Belief.

—MALORY, *Le Morte d'Arthur.*

NOT long ago an Irishman from Indiana came to me and asked, "Did you ever hear of the Montezuma treasure buried in Sugar Loaf Mound at Del Rio?"

"Yes," I replied.

"Well," the Irishman responded, "that Montezuma treasure is buried all right, but not in Sugar Loaf Mound. I'm going to tell you something.

"When I was a boy, my father was foreman of a section gang working on the Southern Pacific Railroad. We lived for a time in a camp several miles east of Del Rio. Along about five o'clock one summer afternoon my brother and I were wandering around in the hills between the railroad and the Rio Grande. Looking up towards a rocky ridge, I saw a giant arrow pointing southward. We were traveling in that direction and before long I saw another enormous arrow, pointing down hill. We went on, and before long we saw a roughly formed ring of rocks, maybe fifteen steps in diameter. It was down in a swag.

"When we got back to camp, I told an old-timer about what we had seen. He said that the arrows pointed towards treasure and that the treasure was undoubtedly buried in the ring of rocks. The next morning my brother and I and this old fellow went back to examine the signs more closely. Well, we couldn't find those arrows at all and somehow missed the circle of rocks. I made a dozen other attempts to sight the arrows—all unsuccessful.

"We left the country. Years afterwards I learned that in giving secret directions the Spaniards sometimes made use of indirect lighting. They arranged rocks or timbers so that the shadows cast at a certain time of day would point to the treasure they had concealed. Any man looking for the signs would have to be on the ground at a particular hour. Now, neither I nor my brother after the first and only glimpse of the arrows again looked for them at the exact hour in which we saw them—between two and three hours by sun on a summer afternoon. I have just returned from a trip to Del Rio, but the man I was with had little faith in my theory and our time was limited. I want to go back with some one really interested."

I could not go. I thought of how Jim Bridger once told a prospector that there was a diamond on top of a mountain in the Yellowstone country which could be sighted fifty miles away "if a man got the right range on it when the sun was right." The prospector offered the old scout a fine horse and a new rifle if he would show it to him.

Notwithstanding the advantages of shadows, buriers of treasure have from time immemorial generally marked their deposits with signs that are directly visible, though usually quite enigmatical. The Spaniards, since they buried most of the treasure in this country, developed, of course, the most elaborate code of symbols. A register of these signs, collected and deciphered for the most part by a lawyer who refuses to allow the disclosure of his name, is now—for the first time, I believe—printed.

X — On line to the treasure. X is also a common designation on landmarks.

✝ — Cross and other rich objects pertaining to the Church are buried here.

The cross might mean many things. So potent was its symbolism that the very sign of it might protect a man's possessions as well as himself. It often said: "I have been here"; "A Christian has passed this way." When Coronado went east in search of the Gran Quivira, he gave instructions that he was to be trailed by means of wooden crosses which he would erect from time to time along his route. Again when Fray Marcos de Nizza set out in 1539 to hunt the Seven Cities of Cibola, he had instructions, should he find himself on the coast of El Mar de Sur (the Pacific Ocean), "to bury written reports at the foot of a tree distinguished by its size, and to cut a cross in the bark of the tree, so that in case a ship was sent along the coast, its crew might know how to identify it by that mark."

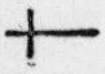

Horizontal cross. The long part of the upright points towards the treasure.

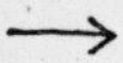

Horizontal arrow without heft pointing towards treasure; sometimes towards water.

Arrow without heft inclined upward pointing to other signs farther on.

Arrow without heft pointing downward to treasure.

Two or more arrows so connected indicate that treasure has been divided into as many parcels and buried in the directions pointed to.

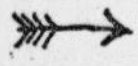

Arrow with feathered heft flying away from mine or treasure.

Bowie knife pointing to treasure.

Mule shoe lying horizontal: En route to treasure; keep traveling.

Mule shoe with toe down: Treasure is below.

Treasure directly underneath this sign.

Spanish gourd. On way to spring of water.

Turtle, or dry land terrapin, with head pointing towards treasure. The turtle also means death, defeat, destruction, and the burial of possessions somewhere in the vicinity.

Snake going up tree. Treasure on opposite side of tree. Travel on to next sign.

Snake coming down tree. A snake or turtle coming down a tree means that treasure is on that side of it. Measure distance from the tip of the reptile's tail to the ground. Step off ten times that distance straight out. At the termination of the distance stepped, one should find either the treasure or another sign.

Snake in striking position with head pointed towards treasure.

Snake coiled on tree or rock indicates presence of treasure directly beneath.

A straight line indicates a certain number of varas to be measured off, the vara being 33 1/3 inches; the number of varas called for usually ranges between 50 and 100.

Two straight lines indicate double the distance of one line.

Flight of steps. This sign indicates that the treasure is down in a cave or shaft.

Treasure is to be found within a triangle formed by trees or rocks.

Over deposit, which is located within a triangle made by trees or rocks.

Triangle formed by trees or rocks enclosing treasure.

Triangle formed by trees or rocks with treasure in the middle.

This sign indicates that while the deposit is marked by a triangle of trees or rocks, it is to be found to one side of the triangle.

Deposit is around a bend or curve away from triangle formed by trees or rocks.

Treasure buried in box or chest.

Peace pipe. Friendly Indians.

Sombrero, or hat. The number of sombreros shown indicates how many people were in the party that buried the treasure. The sombreros may also indicate the number of men killed by an enemy.

Mines close by. Any representation of the sun indicates proximity of mineral wealth.

ORO *Oro* (gold) is short distance away.

G Gold short distance away.

A tunnel.

OR Stop; change direction.

OR Perhaps variant signs of the cross.

Greyhound. As to the meaning, there is some doubt.

Whoever has read this book through will have found many other significations of treasure; as, fox fire, a white animal, ghosts, etc. Indeed, riches hoarded in the ground no more need a sign to signify their presence than good wine needs a bush. Like murder, they will out—but not out of the ground.

In addition to attracting divining rods, switches, and various other objects, buried treasure is supposed to have a particularly powerful attraction for mercury. I know of one man who put mercury in a bottle and went out to look for treasure. At a certain place the bottom fell out of his bottle. The treasure was so big and pulled so hard that it drew the mercury through the glass!

Treasure hunters often refer to the so-called "Sixth and Seventh Books of Moses." My own copy of this curious and absurd farrago of matter pertaining to black magic and other superstitions was printed in New York, 1880. The title page reads: *The Sixth and Seventh Books of Moses; or, Moses' Magical Spirit Art, Known as the Wonderful Arts of the Old Wise Hebrews, taken from the Mosaic Books of the Cabala and the Talmud, for the good of mankind. Translated from the German, word for word, according to old writings.* Instead of tabulating signs that indicate the way to treasure, the work affords directions for recovering it. Thus, Mephistopheles is an effective agent to "bring treasure from the earth and from the deep very quickly." The best time to enlist the aid of angels in the business of recovering treasure is in the sign of the Scorpion, for then "they rule over legacies and riches." If devils and angels are lacking, certain cabalistic configurations will draw treasure upward.

I have in my possession an odd form of candle secured from a Mexican in San Antonio, who, with three companions, used it in an attempt to locate treasure. This candle is a lump of wax with four wicks protruding at right angles to each other. It is supposed to be taken on a windy night to a place near treasure, the *exact* whereabouts of which is unknown. Each of four men grasps a wick in his hand, against the mould of wax. Then the wicks, well greased, are lighted. The one that burns longest in the wind shows in which direction to proceed towards the treasure. By the method of trial and error the spot sought for may thus be arrived at—provided the wicks burn correctly.

But suppose a treasure hunter has no signs or shadows to go by, possesses no chart; suppose he does not believe in occultism and has no faith in any of the various instruments to be made or purchased. What resort is left to him? Why, dreams—for "dremes ben significaciouns."

This is a story related many years ago by a Confederate veteran.

SIXTH AND SEVENTH BOOKS OF MOSES. 53

THE

Schemhamforas

Which will certainly bring to light the Treasures of Earth, if buried in the Treasure-Earth.

From the Arcan Bible of Moses.

FROM

P. HOFFMAN, Jesuit.

Composed ad Proxim.

L. MISCHINSKY, at RAOL, MDCCXLVI.

"After the war I got back to Texas, broke like everybody else. But I bought a farm in Leon County—on credit—married, and began to make a home. I had the farm about paid off and was getting along very well when one summer I had a dream, or vision. At the time it appeared to me I was sleeping on a pallet on the gallery, my wife and two small children occupying the bed just inside the door.

"I saw a woman come into the yard through the gate, a strange-looking woman with strange headgear and queer dress, and I marveled that my fierce watch dogs did not attack her. She came to the side of the gallery and said in a clear voice: 'Dig in your little pasture and you will find treasure.'

"I sat up and watched her go out of the gate, just as she had come, and could hardly persuade myself that what I saw was a dream. The next morning I told my wife of the dream—and then forgot it. Now, the little pasture was a few fenced acres near the house where we kept our milk calves. It was drouth stricken; the soil was hard and dry and had no growth except a few brambles.

"Not many nights later, while I lay as before, the same woman came again. I saw her plainly in the moonlight. She spoke, very quietly but distinctly, the same words: 'Dig in your little pasture. Dig beneath the white rose.'

"I knew positively that there was no growth in the little pasture excepting the few brambles I have mentioned. But on my telling my wife of seeing the woman again in a dream, she said, 'Come on; let's look for roses.' And catching my hand, she laughingly dragged me to the pasture. There, as sure as I am a Reb, we found a rose bush with two white flowers on it. Then we got busy, but, after digging down about two feet, I struck a large rock and quit.

"The story got out and I became the butt of many jokes. A few months afterward my brother-in-law offered me a fancy price for the place and I quit farming. Later on in the year I noticed that the little pasture had been plowed—the only improvement noticeable. About the same time I noticed my brother-in-law buying property, including a fine family carriage, sending his daughter to boarding-school, and getting himself elected to the state legislature. Maybe there was something under the roses."

And maybe there was. This is the dream that never dies.

NOTES

CHAPTER I

[1] Bolton, H. E., *Spanish Explorations of the Southwest,* pp. 283–284.

[2] The account of this expedition is based on a Spanish transcript in the archives of the University of Texas entitled "Expedition to Los Almagres and Plans for Developing the Mines, 1755–1756." It is generally referred to as "The Miranda Report," and to it is attached a considerable amount of correspondence. It has never been published, and I am deeply indebted to Mrs. Margaret Kenney Kress, instructor of Spanish in the University of Texas, for the use of her careful translation.

[3] In 1907 Doctor Herbert E. Bolton, "With Miranda's Report in hand, . . . beyond question identified the mine opened by Miranda with the Boyd Shaft, near Honey Creek Cove [in Llano County]. . . . On the basis of [this] identification the Los Almagres Mining Company was formed and [it] purchased about seventeen hundred acres of land round about Boyd Shaft."—Bolton, *Texas in the Middle Eighteenth Century,* page 83. I have been unable to learn that the Los Almagres Mining Company has spent any money in developing mines.

[4] One of the ubiquitous "authorities" relied on by various exploring parties has been a certain Don Ignacio Obregon, called *Inspector Real de las Minas.* According to John Warren Hunter, who printed at Mason, Texas, in 1905, an interesting but confusing pamphlet entitled *Rise and Fall of the Mission San Saba,* this Don Ignacio reported to the Mexican government, in 1812, that the Almagres ore tested out 1680 pesos to the ton. In a feature article, "The Lost Gold Mines of Texas May be Found Again," by W. D. Hornaday, Dallas *News,* January 7, 1923, Don Ignacio bobs up again with even more sanguine news.

Much credence has been placed in certain widely circulated letters purporting to have been transcribed from the archives at Monterrey, Mexico. See the *Texas Almanac* for 1868; also *History of San Antonio and the Early Days of Texas,* compiled by Robert Sturmberg, San Antonio, 1920, Chapter III.

[5] The proposal to move the presidio south "to some rich veins of sil-

ver" is found in *Algunas Cartas á Don Pedro Romero de Terreros, . . . 1758–1759* (Garcia Library, University of Texas).

For a succinct history of the San Saba mission and presidio, see William E. Dunn, "The Apache Mission of the San Saba River," *Southwestern Historical Quarterly,* Vol. XVII, pp. 379–414; also H. E. Bolton, *Texas in the Middle Eighteenth Century,* pp. 79–93.

[6] The story was printed in the Galveston *News,* date unknown, and was contributed in abbreviated form, by E. G. Littlejohn to *Legends of Texas,* pp. 22–23.

A word here concerning newspaper exploitation of the Lost Almagres Mine theme will not be out of place. In 1853, while the California gold strike was fresh in the minds of everybody, Texas newspapers gave it out that the "gold veins enclosed by the Colorado on the east, the San Saba on the north, and the Llano on the south are as rich as any in California." Men rushed into this region by the hundreds, two hundred gold hunters from New Orleans alone landing at Lavaca Bay. The newspapers have never again succeeded in bringing about a Llano-San Saba gold rush, but the story never grows stale to feature writers. Year after year, they reiterate it, generally in a very diluted form.

For accounts of the gold rush, see the Galveston *Weekly Journal,* May 13, June 6, and June 16, 1853.

[7] "Several days previous to the fight [November 26, 1835] it was currently reported in camp that there was a quantity of silver coming from Mexico on pack mules to pay off the soldiers of General Cos. Our scouts kept a close watch, to give the news as soon as the convoy should be espied, so that we might intercept the treasure. On the morning of the 26th, Colonel Bowie was out in the direction of the Medina, with a company, and discovered some mules with packs approaching. Supposing this to be the expected train, he sent a messenger for reinforcements."—Baker, D. W. C., *Texas Scrap Book,* p. 92.

[8] Mary Austin Holley, first historian of Texas, in *Texas,* Lexington, Ky., 1836, pp. 161–173, quotes the Rezin P. Bowie account of the Indian fight, said to have been printed first in a Philadelphia periodical. It has been reprinted dozens of times. Less known, and differing little in essential facts, is the account of the battle purporting to have been written by James Bowie; it is to be found in J. C. F. Kyger's *Texas Gems* (Denison, Texas, 1885, pp. 130–134) and also in John Henry Brown's *History of Texas* (Vol. I, pp. 170–175).

[9] Archives, University of Texas. For further account of Ham's experiences, see De Shields, James T., *Border Wars of Texas*, Tioga, Texas, 1912, pp. 74–76.

[10] This is the theme of a Texan novel in which Bowie is the hero, William O. Stoddard's *The Lost Gold of the Montezumas: A Story of the Alamo.* Matt Bradley, editor and publisher of *Border Wars of Texas,* said in a signed article printed in the Dallas *News,* January 28, 1923, that only three months before the fall of the Alamo Bowie was trying to reach the riches of which he alone among white men knew the secret.

Some notice of the recognition of the San Saba Mine in the literature of Europe as well as of America may here prove interesting. Henri Fournel, *Coup d'oeil . . . sur le Texas,* 1841, page 23, speaks *"des richesses metalliques depuis longtemps signalées par les Espagnoles."* I am unable now to verify the reference, but I am sure that Gustave Aimard introduces the subject in one of his romances, probably *The Freebooters.*

An English fabrication published in 1843 has this sentence: "The Comanches have a great profusion of gold, which they obtain from the neighborhood of the San Seba [*sic*] hills, and work it themselves into bracelets, armlets, diadems, as well as bits for their horses, and ornaments to their saddles."—Marryat, Captain, *Monsieur Violet,* etc., p. 175.

Citation of all references to the San Saba mines in early Texasana would be both tedious and useless. Elias R. Wrightman, surveyor for Stephen F. Austin, in locating "the first three hundred" Texas colonists, made in 1828 a map of Texas that shows the general location of the San Saba Mine; in notes for a history of Texas written about the same time he describes the mineral riches in most extravagant phrases.—Helm, Mary S., *Scraps of Early Texas,* Austin, 1884, p. 136 ff.

So with most early histories of Texas, notably Kennedy's. In early fiction dealing with Texas use of the San Saba riches was freely made. "As soon as the Comanches, giving away, will permit the whites to work the mines north of San Antonio, the immense profits will prove an irresistible attraction," says Anthony Ganilk (pseudonym for A. T. Myrthe) in *Ambrosio de Letinez, or The First Texian Novel,* N. Y., 1842, Vol. II, p. 63.

The popular writer, Charles W. Webber, who had been ranger with Captain Jack Hays, brought out in 1848 and 1849, respectively, two novels, *Old Hicks the Guide* and *The Gold Mines of the Gila,* that make effective use of tales about the San Saba Mine.

Later fictional uses of the same material are to be found in *The Three Adventurers,* a novel by J. S. (K. Lamity) Bonner, Austin, no date; and in "The Llano Treasure Cave," a short story by T. B. Baldwin (Dick Naylor), *The Texas Magazine,* Vol. III, pp. 195–204, reprinted in the Dallas *Semi-Weekly Farm News,* July 11 and July 14, 1922.

[11] A son, Jeptha Billingsley, wrote two letters from Elgin, Texas, Oct. and Nov., 1922, that afford the basis for this sketch. They are in the archives of the University of Texas. Walter Billingsley, of San Antonio, a nephew of "Captain Jess," knew many of the San Jacinto veterans, and, according to their story, the Billingsley men were so relentless during the rout at San Jacinto that Houston, seeing the Mexicans wanted to surrender, sent word to Captain Billingsley to "slow down." "Present my compliments to General Houston," Billingsley is said to have replied to the commander-in-chief's messenger, "and tell him to go to hell."

[12] Thus, three score years ago and more, B. F. Gholson, who was born on the Texas frontier in 1842, who became one of the most noted Indian fighters, cowmen, and all around frontiersmen of the Southwest, and who gave his name to Gholson Gap, heard the story. Thus he told it to me at a reunion of pioneer Texas rangers held at San Saba just seventy-two years after Sinnet Musset's bullet halted Grumble and the ransomed captive at that place.

[13] See "W. A. McDaniel, Cowboy," etc., by Cora Melton Cross, the *Semi-Weekly Farm News,* Dallas, Texas, June 22, 1928.

[14] For this story and also for the one about the icicles of silver I am indebted to Clyde Smith, a lath plasterer of Beaumont.

[15] The story that follows was clipped many years ago by E. G. Littlejohn, of Galveston, from the Galveston *News* and contributed by him to *Legends of Texas.*

[16] See "A Legend of the Blanco Mine," by Julia Estill, *Legends of Texas,* pp. 24–26.

[17] Dr. S. E. Chandler, president of Daniel Baker College, at Brownwood, Texas, frequently visited the home of Moses Kirkpatrick. He saw the copper plates and other objects. I have interviewed various other people who saw them also. But nobody else can tell the whole story so well as Dr. Chandler tells it. I thank him most emphatically.

[18] The story of this exceptional man comes through Miss Julia Estill, of Fredericksburg, who wrote it for *Legends of Texas.*

[19] W. M. Longworth thinks, justly, that his street address should be given so that people who know of hidden treasures may write to him and secure his assistance with the "radio sleuth" in finding such wealth. His address is 1110 Denver Boulevard, San Antonio, Texas.

Notes

CHAPTER II

[1] For history of the establishments consult Bolton's *Texas in the Middle Eighteenth Century,* using index.

[2] See Sowell, A. J., *Early Settlers and Indian Fighters of Southwest Texas,* Austin, 1900, pp. 405–408.

[3] It is a pity that we have no unvarnished biography of Lieutenant-Colonel Baylor. ("General" seems to have been a title of courtesy.) For the account of his search for Spanish gold on the Nueces I am indebted to his son, Henry Baylor, and to Henry Yelvington, both of San Antonio. *The War of the Rebellion: A Compilation of the Official Records of the Union and Confederate Armies* has scattered through its many and well-indexed volumes numerous facts concerning John R. Baylor.

[4] An appalling amount of literature has been printed on the subject. Water Supply Paper 416, U. S. Geological Survey, Department of the Interior, 1917, entitled *The Divining Rod: A History of Water Witching,* by Arthur J. Ellis, contains a bibliography of nearly thirty pages listing books and pamphlets from 1532 to 1917. Men have claimed to be able to locate not only water and minerals but also criminals with the divining rod. The Orient and the Occident have both been familiar with it for centuries. Learned doctors and societies have repeatedly investigated it—and pronounced it a sham. Persons claiming to be expert with it have at times drawn to themselves the attention of whole nations. The wideness of the subject can not here be even suggested.

George Ray located, just outside the front gate of the Olmos ranch house, a vein of water that he said was between 25 and 29 feet underground. His manner of computing the depth was thus: After locating one vein, he stepped forward and 9 paces beyond located a second vein; 8 paces still farther on he located a third vein. 8 and 9 make 17; a pace is 3 feet; 3 times 17 equals 51. There were 2 interstices between the three veins; 51 divided by 2 gives 25½, say 26 feet—the approximate depth necessary to go for water. Fortunately a lake affords a never-failing supply of water for Olmos headquarters; so no test of the vein that Mr. Ray located was made.

[5] "Old Rip" Ford's men had a strange engagement at Fort Merrill. (See A. J. Sowell's *Early Settlers and Indian Fighters of Southwest Texas,* Austin, 1900, pages 819–820.) It was located on land patented to the ambitious John McMullen. The McGloins came into a share of it, and just before Live Oak County was organized, in 1856, the McGloin heirs agreed "that the houses and other improvements" situ-

ated upon the land should "be sold at public outcry." Barlow's Ferry, later replaced by the bridge near what is now Dinero, several miles down the river, diverted the public road on which the Fort Merrill crossing was once famous.

[6] Corpus Christi *Star,* Dec. 23, 1848.

CHAPTER IV

[1] Cannon were in early days not infrequently buried, thus giving a certain basis of fact for legend.

In 1759 Parrilla marched from San Antonio with a force of six hundred men and attacked the Taovayas villages on Red River. He found the Indians "intrenched behind a strong stockade of breastworks, flying a French flag, and skilfully using French weapons and tactics." After a sanguinary battle resulting in heavy losses to both sides, the Spanish withdrew, leaving "two cannon and extra baggage" concealed behind. Seventeen years later the cannon were recovered; inhabitants of Cooke and Montague counties, Texas, are still looking for something buried about "Old Spanish Fort."

Traders journeying over the Santa Fé Trail used to cache cannon on the Big Arkansas River, for on returning east they left the "dangerous country" at that line and would not need the cannon until they got back with another west-bound cargo.

During the Civil War Confederate forces buried two cannon near Santa Fé, which were exhumed a generation later—thus giving courage to a host of treasure hunters.

[2] For the story that follows see "The Treasure Cannon on the Neches," by Roscoe Martin, *Legends of Texas,* pp. 84–89.

[3] Smithwick, Noah, *The Evolution of a State, or Recollections of Old Texas Days,* Austin, 1900, p. 83.

[4] Five years after these stories of battlefield treasures appeared in *Legends of Texas,* Mr. Bob Nutt, of Beeville, who is responsible for Ramón's tale of Resaca de la Palma, received a letter from one Harley Johnson, written at Kerrville, as follows:

"My father Jim Johnson of Goliad, now dead went up the Kansas trail with cattle for Dillard Fant for four year hand running 1872–1875 and on last trip befriended the remuda Mexican who was very sick. Said Mexican told him how he and his troop buried this seven cart loads on Palo Alto Battle field (but the stuff was hurried thrown on 50 pack jacks and taken to hole). The Mexican took my father and some other men down to Palo Alto and they dug and uncovered it but it sank into quick sand of unknown depth when ten men jumped into the hole to roll one of the *maletas* out. I know the spot; have made 4 trips down there but could never get to it for quick sand. The original hole was 9 ft sq. x 6 ft deep. The stuff is some where at bottom. If you will pay all expenses to get it out I'll give you one tenth $25,000 of the $250,000. I taught Berclair school on Indian creek and knew your father also brothers. Have sick wife and need money."

[5] Robert Hall. See *Life of Robert Hall,* etc., by Brazos (pseudonym), Austin, 1898, p. 85.

[6] This account of Steinheimer came to me from L. D. Bertillion and was printed in *Legends of Texas,* pp. 91–95.

CHAPTER V

[1] The original version of this tale was supplied by a ranch boy of Frio County named Charles A. Beever, through Leon Denny Moses, School of Mines, El Paso.

[2] Peyote, "dry whiskey."

[3] For allowing me to make use of this tale I am indebted to Mrs. G. B. Smedley, of Wichita Falls and Fort Worth. It will be found incorporated in her paper, "Legends of Wichita County," *Publications* No. VIII of the Texas Folk-Lore Society.

[4] A "rep" is a cowboy who represents his brand on an "outside" range in order to bring in cattle that have strayed away.

CHAPTER VI

[1] Luther A. Lawhon, in *Trail Drivers of Texas,* San Antonio, 1920, pp. 176–177.

CHAPTER VIII

[1] See El Paso *Herald,* "Week-end Edition," January 22, 1927.

[2] Nothing is commoner in traditions of buried treasure than death cell maps. In his vivid autobiography, *Beating Back* (New York, 1914, pp. 44–45), Al Jennings tells of a hunt he once made in Oklahoma after treasure that a man about to be hanged gave directions to. Ben Hughes was an outlaw in the federal jail at Fort Smith, Arkansas. He had a cell mate named Jim Cash, who was charged with murder.

"The case against him was so strong that he never stood a chance, and he went to the gallows. [Before he went, however, he] told Ben

Hughes where he had buried seven thousand dollars in stolen money—gave him a map and full directions. The next spring a party of us—all wanted by the law—went hunting for this buried treasure. We camped, leaving a negro to guard our horses. The map was perfectly plain; we found the place according to directions, and began to dig under a flat rock. Before we had gone a foot we heard a shot. We grabbed our guns, crawled to the bushes, and surveyed the camp. It looked peaceable, and we rushed in for our horses. We found the negro rolling and moaning with pain. He had got to monkeying with a loaded revolver, and, not understanding guns, had shot himself in the foot. We were patching him up when we spied a body of marshals looking for gentlemen answering to our description. When the subsequent episode was finished we found ourselves miles from the treasure and afraid to go back. If that seven thousand dollars ever existed, it may be there yet. And there it will stay, for all of me."

"I understand," Jennings continues, "that people are digging all over Creek Nation for the buried treasure of the Jennings gang. I buried my treasure all right, but not that way. It used to run through my fingers like water."

Later he repented of his rash disbelief in outlaw treasure and, according to an article in *The Texas Monthly,* Dallas, June, 1929, went, in 1925, in search of a vast fortune purported to be buried at Nevas, Durango, by one of Pancho Villa's henchmen.

[3] Following the appearance of an account of the Lost Nigger Mine which I wrote for *Holland's Magazine* (March and April, 1930), people from a dozen states wrote me seeking further guidance to the mine!

CHAPTER IX

[1] At a time when the principal town, Paso del Norte, was on the Mexican side of the river and what is now the City of El Paso was the village of Franklin, the Galveston *Daily News* (June 22, 1873) printed this dispatch:

"Franklin, June 4, 1873.

"A short time ago two old shafts were discovered in a mountain about a mile from town, and a company formed to clear them out and prospect them. This they have done, and have found a well-defined silver lead several feet in thickness, variously estimated at from $30 to $75 per ton. One of the shafts is 90 feet deep and the other over 100. There is an old Mexican tradition of a very rich mine about two and a half miles from the cathedral tower of El Paso [Juarez], and this mine, the locators at least seem to think, fills the bill."

[2] See "The Legend of Cheetwah," by Edith C. Lane, *Legends of Texas,* pp. 130–132.

CHAPTER X

[1] As a matter of fact, the army records show that only two negroes were killed and buried in the vicinity of Lobo—but history has no more business interfering with legend than legend has interfering with history.

[2] Should anyone wish to know more about Jim Hughes, Red Curly (Sandy King), Zwing Hunt, and Russian Bill, he can do no better than read three books that have recently been published around the outlaws of the Southwest: *Helldorado,* by William M. Breakenridge; *Tombstone's Yesterday,* by Lorenzo D. Walters; and *Tombstone,* by Walter Noble Burns. Mr. Burns devotes a whole chapter to the Skeleton Canyon treasure, as Bill Cole's phantom is called in Arizona.

CHAPTER XI

[1] The extract is reprinted by permission of the publishers, The Arthur H. Clark Company, from Granville Stuart's *Forty Years on the Frontier,* Cleveland, Ohio, 1925, Vol. I, pages 60–62.

[2] Accounts—all poor—of the Gunsight, Pegleg, Breyfogle, and other lost mines of the Death Valley area have been printed in many places. I have dozens of newspaper stories on the subject. In 1924–1925 the *Western Story Magazine* ran a series of brief tales on lost mines of the West, written by Roderick O'Hargan. A flippant account of the Breyfogle Mine is to be found in Chapter LI of *Adventures in the Apache Country,* by J. Ross Browne, New York, 1869. In 1926 P. C. Kullman and Co., a brokerage firm of New York, issued an odd pamphlet entitled *Lost Mines in the Western Mountains* that contains brief legendary tales about several desert mines. A recent book, *Death Valley,* by Bourke Lee (1930), has an interesting chapter on "Mines and Miners" in which Breyfogle is slightly touched on.

I list these sources merely to show the persistence of the traditions; I owe them nothing. For the stories of both the Breyfogle and the Yuma mines I owe everything to Donald F. MacCarthy, of Montrose, California, who has looked for them and who knows what makes a good story. For the story of the Mexican shepherd girl who found nuggets in the desert, I am indebted to my friend, W. W. Burton, of Austin, now (1930) eighty-six years old.

CHAPTER XII

[1] Bandelier began the study in 1880. Since 1915 it has been continued by Dr. A. V. Kidder, whose hospitable camp beside the Pecos ruins, which he has been years studying, has been my abode on more than one occasion and whose "History of Pecos" (in *An Introduction to the Study of Southwestern Archæology,* Yale University Press, 1924) has afforded the basis for this outline of the Pecos background.

[2] A man like José Vaca does not require documentary proofs of *antigua* mines. Why should he? Other lost mine hunters do, however. For generations they have haunted the archives of Santa Fé. Some of the contradictions that they may there find to inspire their hopes follow.

"In like manner we discovered in said land [New Mexico] eleven silver mines of very rich veins, the ore of three of which was brought to this city [Mexico] and given to his excellency. . . . The assayer found one sample to run 50% silver . . ."—Report on Chamuscado's expedition to New Mexico, 1581–1582, in *Leading Facts of New Mexico History,* by R. E. Twitchell, Vol. I, p. 258.

"In New Mexico there were no mines [worked by the Spanish] until after 1725, and compulsory labor on the part of the Indians even after that date was limited to service in the missions."—Bandelier, A. F., *Investigations in the Southwest,* Final Report, Part I, p. 194.

Don Pedro Pino, from New Mexico, in reporting to a Spanish congress at Cadiz, 1812, said: "In this province mines have been found closed, some of them with work tools inside; but it is not known at what time they were discovered and worked. There are many mineral veins in the mountains of gold and silver."—Read, Benjamin M., *Illustrated History of New Mexico,* Santa Fé, 1912, p. 516.

"In 1803, in his report, Governor Chacon says that copper is abundant, but no mines are worked. . . . Pike in 1807 refers to a copper mine west of the Rio Grande . . . yielding 20,000 mule loads of metal annually. . . . This must have been the Santa Rita. . . . The metal was transported to the City of Mexico by pack mules and wagons, 100 mules, carrying 300 pounds each, being constantly employed. There is very little of record which shows any mining done in New Mexico during Spanish rule."—Twitchell, R. E., *Leading Facts of New Mexico History,* Vol. I, p. 475.

Notes

[3] For description of these remarkable sites in Arizona see *Mesa, Cañon, and Pueblo,* by Charles F. Lummis, New York, 1925, pp. 294–306.

[4] The Pecos Indians kept the sacred fire for Montezuma burning until 1840, when the remnants of the village moved to Jemez.—Bandelier, A. F., "A Visit to the Aboriginal Ruins in the Valley of the Pecos," *Papers of the Archaeological Institute of America,* Boston, 1881, p. 112. In Willa Cather's recent *Death Comes for the Archbishop,* a noble and beautiful novel, use is made of the secret fire-chamber.

For an account of the Pima Indian belief in Montezuma as their ancestor, see Major William H. Emory's *Report on the United States and Mexican Boundary Survey,* Washington, 1857, Vol. I, p. 117. The Papago Indians applied the name of Montezuma to a character, "Elder Brother," that had existed in their myths probably centuries before they heard the name of Montezuma. See "The Papago Migration Myth," by J. Alden Mason, *Journal of American Folk-Lore,* Vol. 34, pp. 254–268. Compare also with "The Pima Indians," by Frank Russell, 26th *Annual Report of the Bureau of American Ethnology,* p. 225.

The rise of the Montezuma cult in the Southwest has been brilliantly traced by Adolph F. Bandelier in *The American Anthropologist,* Vol. V, October, 1892, pp. 319–326. As Bandelier shows, neither the war chieftain Montezuma nor his ancestors had anything more to do with the Southwest than had Julius Cæsar. The name of Montezuma was hardly known among the aborigines of the Southwest prior to the Mexican War of 1846. At that time, however, officials of the Mexican government had manuscripts circulated in New Mexico representing that Cortez had married "the Malinche," a character familiar to Indians through the "Matachines" dance. As a matter of history, Marina—in the Aztec tongue, Malinche—was mistress to Cortez. The common people hated her. This Malinche, so the government document circulated in New Mexico asserted, was the daughter of a great Indian from the north named Montezuma, and thus she brought to the conqueror of New Spain, as a part of her dowry, the territory of New Mexico. Such a tale was fabricated to inspire loyalty to the Mexican republic among New Mexicans, most of whom detested the *gachupines* and therefore had a sympathy for the symbolic object of *gachupín* tyranny—Montezuma. What such propaganda really inspired was legend.

[5] This, of course, is but a variation of the ordinary "mineral rod,"

"switch," described elsewhere in this book. The Mexicans call it *bara de virtud,* or *bara de San Ignacio.*

CHAPTER XIII

[1] For Beale's Report (35th Congress, 1st Session, House of Representatives, Ex. Doc., No. 124) and the whole story of the camels, which is also the story of the Texas-California route, see *Uncle Sam's Camels,* edited by Lewis Burt Lesley, Harvard University Press, 1929. Quotation is from pp. 154–155.

[2] Santleben was king of the freighters. His book, *A Texas Pioneer,* New York, 1910, is a mine of information. For record of his transportation of money see pages 104, 153, 182, 195, 200.

[3] This legend has been supplied by J. A. Rickard, of O'Donnell, Texas, who heard it from a frontiersman named T. J. Kellis.

CHAPTER XIV

[1] To D. F. MacCarthy, generous gentleman, of Montrose, California, who has looked for the Lost Breyfogle Mine in Death Valley and the Yuma Mine of Arizona as well as for Guadalupe gold; to J. Evetts Haley, whose rich store of information relating to West Texas is partly revealed in his excellent history of *The X I T Ranch;* to Ralph H. Shuffler, newspaper man and good fellow of Odessa, who interviewed various old-timers in my behalf; to my good friend, Clabe Robinson, of Live Oak County; to Green Ussery, New Mexico ranchman; to J. E. Ellison, Marietta, Ohio, once a hunter of the Guadalupe gold; to J. J. Walker, a lawyer of Barstow, Texas; to a newspaper article of unknown origin reprinted in *Hunter's Frontier Magazine,* Melvin, Texas, October, 1916, and then again reprinted in *Legends of Texas;* to a newspaper interview that appeared in the El Paso *Times* in 1912 and was reprinted in *Frontier Times,* March, 1924; and to a summary account under the general title of "Legends of Lost Mines," by Frederick O'Hargan, in *Far West* magazine, July, 1928—to all of these I express my indebtedness for traditional material relating to the lost gold of the Guadalupes—and I wish I were out again with Asa Jones on his Culberson County ranch breathing the air from the Guadalupe Mountains themselves.

[2] I have not seen the article written by Lew Wallace, nor have I been able to determine the magazine in which it appeared, but I have a copy of part of it as Donald F. MacCarthy transcribed it in Helena,

Montana, in 1889. General Wallace was much interested in mines, as witness his article on "The Mines of Santa Eulalia, Chihuahua," *Harper's New Monthly Magazine,* November, 1867, pp. 681–702, and also as witness one of his sanguine mining ventures related by Henry F. Hoyt, *The Frontier Doctor,* Boston, 1929, pp. 160–177. General Wallace likewise was an historical romancer.

[3] Although the Rustler Hills rise 3700 feet above sea level, they are hardly more than 250 feet higher than the adjacent plains. They are capped by massive grey limestone and are not nearly so rough as the Guadalupes proper. An excellent description of the topography and geology of the whole region, with especial attention to mineral resources, is to be found in George Burr Richardson's "Report of a Reconnaissance in Trans-Pecos Texas," University of Texas Mineral Survey Bulletin No. 9, Austin, 1904.

CHAPTER XV

[1] When the great Washoe boom of Nevada was on in 1860 and everybody in Virginia City as well as in San Francisco was a potential millionaire and stocks in the "Wake-up-Jake," the "Root Hog or Die," the "Let Her Rip," the "Gouge-Eye," and hundreds of other "mines" were doubling in value over night, ledges of iridium, platinum, and plumbago were claimed by prospectors. Men tunneled into the granite of Mount Davidson, which towers above Virginia City, in order to tap a lake—an ocean—of coal oil that "spirits," through mediums, had directed them to.—See Charles Howard Shinn, *The Story of the Mine,* New York, 1896, p. 141; also Dan De Quille, *History of the Big Bonanza,* Hartford, Conn., 1877, pp. 100–102.

[2] Bruffey, George A., *Eighty-one Years in the West,* Butte, Montana, 1925, pp. 131–132.

[3] The story of the "Minnie Moore" is taken from a privately printed but exceedingly interesting and valuable book entitled *Reminiscences of Alexander Topence,* Ogden, Utah, 1923, pp. 214–217.

[4] This legend of the rangers' quicksilver was contributed by Edgar B. Kincaid, who ranches near Sabinal, to *Legends of Texas,* p. 62.

[5] My informant is G. T. Bludworth, State Department of Education, Austin.

[6] Wilbarger, J. W., *Indian Depredations in Texas,* Austin, 1889, pp. 15–19; De Shields, James T., *Border Wars of Texas,* Tioga, Texas, 1912, pp. 212–215. The legend of Goacher's mine appeared

under the title of "The Legend of a Lost Lead Mine," by John Knox, in the Houston *Post-Dispatch,* Sunday, Aug. 29, 1926.

[7] It is said that during the Civil War settlers worked a vein of lead to the north of Smithville, which is in the same formation that Rabb's Creek is in. That mine has been lost also. In a letter published in the Brenham, Texas, *Banner-Press,* Feb. 26, 1927, "Prof." Rundus exhorts the farmers to look out for lead, copper, quicksilver, and other minerals likely to occur in the country.

[8] Grateful acknowledgment is here made to Bennett Lay, of Hallettsville, Texas, for furnishing this story of the lost lead of Lavaca County.

[9] This account of Hoffman's mine was written for *Legends of Texas* by Edgar B. Kincaid.

[10] See *Legends of Texas,* pp. 77–78.

[11] In 1774 De Mézières reported Spaniards gone in search of mines that Indians said were "in the direction of the Brazos de Dios." In 1823 Daniel Shipman and two other men went from Austin's settlement on the coast in search of "an inexhaustible silver mine" reported to exist up the Brazos River. In 1836 the Reverend David B. Edward was strong in his belief in "a mountain of iron" on the headwaters of the Brazos. In the remarkable *Narrative* (1839) of her capture by the Indians and subsequent experiences Mrs. Rachel Plummer described "a large lump of platina" near the Brazos in the Cross Timbers. In 1844 the Houston *Telegraph and Texas Register* urged prospectors to go up the Brazos River a hundred miles above the mouth of the Bosque and there locate copper deposits "laid down on all old Spanish maps." In 1867 and 1868 at least three expeditions left Parker County to prospect for copper on the Brazos; some of them pushed on to the Wichita and there found "favorable sign."—Bolton, H. E., *Athanase de Mézières,* Vol. I, p. 104, and Vol. II, pp. 33, 34, 47; Shipman, Daniel, *Frontier Life,* 1879, pp. 23–26; Edward, David B., *History of Texas,* Cincinnati, 1836, pp. 44–45; Plummer, Mrs. Rachel, *Narrative of the Capture and Subsequent Sufferings,* 1839, reprinted at Palestine, Texas, 1926, p. 111; *Telegraph and Texas Register,* Houston, May 1, 1844; *Pioneer Days in the Southwest,* compiled by Emanuel Dubbs, Guthrie, Okla., 1909, pp. 180–187.

[12] See *Five Years a Cavalryman,* by H. H. McConnell, Jacksboro, Texas, 1899, pp. 294–296—a delightful book.

[13] "The existence of copper ores in the Permian measures of Texas has long been known, and these ores have been, from time to time, the object of geological researches and mining developments. . . . The ore appears principally in two zones of Permian rocks, namely, the

Red River zone in the counties of Archer, Wichita, Montague, Hardeman, and Wilbarger, and the Brazos River zone in the counties of Haskell, Baylor, Stonewall, and Knox."—Schmitz, E. J., *Transactions* of the American Institute of Mining Engineers, Vol. 26, p. 97 ff., 1897, quoted in *The Minerals and Mineral Localities of Texas,* by Frederic W. Simonds, University of Texas Mineral Survey Bulletin No. 5, Austin, 1902, p. 23.

[14] *Legends of Texas,* pp. 72–77.

CHAPTER XVI

[1] It is reproduced from *Legends of Texas.* The account, traditional in her family, of the Moros of Tamaulipas comes from Jovita González.

[2] The legend follows closely the telling by E. O. McNew, "Palo Duro May Be Treasure's Hiding Place," in the Fort Worth *Star Telegram,* Sunday, Nov. 25, 1928.

CHAPTER XVII

[1] Dr. John Sibley, U. S. Indian Agent (Thoburn's *A Standard History of Oklahoma,* Vol. I, pp. 20–21), after making a reconnaissance of the Wichita country in 1805, reported a traveler thus: "Amongst these mountains of mines we often heard a noise like the explosion of cannon or distant thunder. The Indians said it was the spirits of the white people [Spaniards] working in their treasure."

In *Astoria,* Chap. XXVI, Washington Irving notes that Lewis and Clarke heard the same kind of mysterious explosions in the Rocky Mountains, attributed by Indians "to the bursting of rich mines of silver." Irving cites similar phenomena over Spanish America.

CHAPTER XVIII

[1] In the Autumn Number of *The Yale Review,* 1928, appeared an article of mine entitled "The Mystery of Lafitte's Treasure." Anyone comparing the chapter on "Laffite and Pirate Booty" with that article would observe that various remarks dealing with the man Laffite have been changed. The changes are due to the work of Mr. Stanley Faye, of Aurora, Illinois, with whom *The Yale Review* article brought me in contact and who with extraordinary generosity placed the manuscript of his rarely scholarly work on *Privateers of the Gulf* at my disposal. I cannot express sufficiently my obligation to Mr. Faye. It is to be hoped that his ripe and revealing work will soon be in book form.

Notes

Of articles on Laffite there has been no dearth, but hardly one of them shows any critical sifting. The best biographical sketch heretofore obtainable is that entitled "Life of Jean Lafitte," by "W. B." (William Bollaert), in *Littell's Living Age,* Vol. XXXII, March 6, 1852, pp. 433–446.

"The Cruise of the Enterprise—A Day with La Fitte," by "T.," in *The United States Democratic Review,* Washington, D. C., Vol. VI, July, 1839, pp. 33–42, affords an interesting picture of the privateer boss's headquarters at Galveston at the time the United States government ordered him to leave.

De Bow's Review, New Orleans, 1851–1853, ran several articles dealing with Laffite. See Vol. XI, pp. 372–387, "Life and Times of Lafitte"; Vol. XIII, pp. 378–383, "Early Life in the Southwest—The Bowies"; and Vol. XV, pp. 572–584, "Early Life in the Southwest."

The Pirates' Own Book, Boston, 1837 (reprinted at Salem, 1924), contains an anonymously written sketch twenty-five pages long entitled "Life of Lafitte, the Pirate of the Gulf."

The *Papers* of Mirabeau Buonaparte Lamar, one time president of the Republic of Texas, edited by Charles A. Gulick, Jr., and Winnie Allen, Austin, 1925, contain some excellent material on Laffite. See *Papers* Nos. 19, 24, and 2492 (old Jim Campbell's story).

Fortier, Phelps, Gayarre, Cable, Grace King, and other Louisiana writers have all treated of Laffite; among them Miss King is best. The Texas historians have copied each other without saying much of anything. However, Moses Austin's deposition in the *Austin Papers,* edited by Barker, is interesting, if erroneous.

The articles in the *American Mercury,* Feb., 1926, and the *Atlantic Monthly,* June, 1903, add nothing, though the latter entertains.

As for newspaper articles, they have been printed by the score, one feature writer rehashing another until the hash has generally become very thin indeed. An exception is an article compiled by Harry Benge Crozier for the Galveston *Daily News,* April 11, 1917. Another article signed by W. L. Bradley in the same newspaper, April 28, 1895, has some good stories in it.

Among fantastic things on Campbell and other characters associated with Laffite are to be mentioned the account, pp. 125–131, in Charles Hooten's *St. Louis Isle, or Texiana,* London, 1847; also articles in the Galveston *News* for Feb. 8, 1878; April 21, 1878; and May 25, 1879.

The bibliography of hearsay dilution might be extended indefinitely.

Notes

Some legendary material on Laffite that appears in *Legends of Texas* has not been used in this chapter. A very large number of weak novels and short stories have been built around Laffite. I want to thank the staff of the Rosenberg Library of Galveston for having placed at my disposal their voluminous Laffite material.

[2] The story of Meredith's chart is appropriated from "Life and Legends of Lafitte the Pirate," by E. G. Littlejohn, *Legends of Texas,* pp. 179–185.

[3] Taken from *Legends of Texas,* pp. 185–189.

GLOSSARY OF MEXICAN AND OTHER LOCALISMS OF THE SOUTHWEST

It would be possible to write of the folk of the Southwest in language free from colloquialism—but not to write them down. Their language, seasoned as it is with Mexicanisms and metaphor peculiar to the range, is as much a part of the folk's lore as are the legendary tales themselves.

Acequia, irrigation ditch.
Adios, goodbye.
Allá, yonder; **más allá,** farther on.
Almagre, red hematite.
Amigo, friend; **por amigo,** for the sake of friendship.
Amo, master.
Antigua, ancient; an old-timer.
Arroyo Colorado, Red Creek.
Artillery, pistols, personal weapons.

Bad man, outlaw.
Bandera, flag.
Bandido, bandit.
Bara de San Ignacio, mineral rod.
Bara de virtud, mineral rod.
Bayo coyote, dun horse with a black stripe down his back.
Black ivory, negro slaves.
Borrowed, euphemism for stolen.
Bozal, nose-hitch.
Bravo, fierce, brave.
Brazos de Dios, Arms of God.
Bronco, wild; **los Indios broncos,** unchristianized Indians.
Brujo, wizard.
Brush popper, brush hand; cowboy expert at running in the brush.
Buena Vista, Fine Prospect.
Bulto, bulk, object.

Caballada, band of horses.
Caballero, horseman, gentleman.
Cabestro, hair rope.

Glossary

Caboodle, lot, aggregation, amount.
Cafecito, cup of coffee.
Camino real, public highway.
Candelilla, a plant from which wax is extracted.
Caporal, boss.
Carabina, carbine, rifle.
Carajoing, shouting "Carajo!"—an exclamation much used by mule drivers, vaqueros, and other outdoor workers.
Carbón, charcoal.
Carga, load.
Carreta, old-fashioned wooden-wheeled Mexican cart.
Cartucho, cartridge shell.
Casa Blanca, White House.
Cerro, hill.
Chaparro, brush.
Chapote, Mexican persimmon.
Chihuahua, name of a state and city in Mexico—used as a harmless expletive.
Chihuahua cart, a heavy wooden cart. See **Carreta**.
"Chimney," mining term for a "blow out" or "shoot" of ore.
Chisos, phantoms.
Chivarras, leggins, chaps.
Chouse, to chase.
Cienaga, marsh.
Cocinero, cook.
Coma, an evergreen, thorny tree; also small berry borne by the tree.
Comancheros, traders with Comanches.
Compadre, (1) co-godfather; (2) close companion. See **Compañero**.
Compañero, partner, companion.
Concha, shell-shaped ornament of silver.
Conducta, a guard or convoy.
Conquistador, conqueror.
Corrida, cow crowd, outfit of cow hands.
Cow chips, dried cow dung.
Cowpen Spanish, illiterate Spanish.
Cristianos, civilized people, Christians; tantamount to "white people."
Crow-bait, a poor, decrepit animal (usually a horse).
Curandero, one who cures; a kind of quack doctor who uses both home remedies and something of the Indian medicine man's magic.
Cut for sign, to examine the ground for tracks and droppings (the two "signs").

Glossary

Dally, to wrap a rope around the horn of a saddle. (From **dar la vuelta**, to give a turn.)
Derrotero, a set of directions, a chart; also a diary.
Difficulty, a quarrel, often resulting in a "killing."
Dinero, money.
Dios, God.
Dueño, owner.
Duffer, codger.
Duffle, personal effects; clothes, etc.

Embrujada, bewitched.
Empleado, officer; name often applied by Mexicans to Texas rangers.
Enchilada, a highly seasoned Mexican dish.
Escondida, hidden.
Espantosa, haunted, horrible.
Excavad, dig (imperative form).

Fandango, a dance; more freely, any gay party.
Fire-water, whiskey.
Frijoles, dried Mexican beans—a staple diet in the ranch country.

Gachupin, upper class Spaniard.
Gallery, porch.
Gente, people.
Granjeno, one of various thorned bushes of the Southwest.
Grubstake (as verb) to furnish provisions; (as noun) provisions.
Gully-washer and fence-lifter, a very hard rain.

Hablar Español, to speak Spanish.
Hacienda, a plantation or ranch; a landed estate.
Hard money, coin.
Hato, wallet.
Hear the owl hoot, to have many and varied experiences.
High lonesome, a big drunk.
Historia, a tale; historical account.
Hombre, man.
Horn in, to intrude.
Horqueta, a fork.
Huajilla, a bush, valuable for browsing, belonging to the catclaw family.

Iguana, lizard; name of an ore formation.

Jacal, hut, cabin.
Jarro, earthern pot. See **Olla**.

Glossary

Javelina (from Spanish **jabalina**), peccary.

Ladrón, robber.
Lagarto, alligator.
Lay for, lie in wait for.
Lechuza, owl, screech owl.
Like the devil beating tan bark, fast and furious.
Llano, prairie; a flat, open plain.
Lobo, "loafer" wolf.
Loose-herd, to herd loosely so as to allow animal to graze.
Loma, hill; **Loma Alta**, High Hill.

Macapal, basket for carrying ore out of shaft.
Machete, cutlass.
Macho, mule.
Madama, Madam.
Maldiciones, bad words.
Maleta, bag made out of rawhide; also, satchel.
Manada, bunch of mares with stallion.
Mano, (1) hand; (2) stone (used in the hand) for grinding corn against the **metate**.
Mantilla, head shawl.
Mañana, tomorrow; sometime.
Maravillas, wonders.
Marihuana, a narcotic.
Mayordomo, overseer, major-domo.
Mediano, half grown.
Medicina, medicine.
Mesa, flat-topped hill; mountain shaped like a table (**mesa**).
Mescal, the agave, or century, plant; liquor made from the plant.
Mestizo, a Mexican of mixed blood.
Metate, a stone on which corn is rubbed to meal.
Mina, mine.
Mocho, gotched; droop-horned.
Montezuma, ruin; name of Aztec ruler.
Moro, Moor; Mexican vigilantes who dressed like **Moors**.
Morral, fiber bag, usually carried on horn of saddle.
Mota, clump of trees; motte.
Motte. See **Mota**.
Mucha, much.
Muerto, dead man. **Los muertos no hablan**, the dead do not talk.
Mujer, woman, wife.

Mulero, mule driver.
Música, music.
Mustang, (as verb) to catch mustangs; (as noun) wild horse.
Muy malo, very ill.

Nueces, pecan, nut.

Ojala, an exclamation of surprise or encouragement.
Olla, pot, jar.
Olmos, elms.

Padre, (1) father; (2) priest.
Paisano, road-runner, chaparral bird.
Palo, tree; Palo Alto, High Tree.
Palo blanco, hackberry.
Palo Duro, (literally, hard wood) a bush common on the upper Texas plains, from which Palo Duro Canyon takes its name.
Parate, halt.
Parientes, relatives.
Parihuela, hod for carrying ore.
Pasajero, traveler.
Pasear, (1) a journey or trip; (2) to travel.
Paso, (1) a pass, a ford; (2) a double-step, six feet.
Pastor, shepherd.
Pata, foot; Pata de Venado, Deer's Foot.
Patrón, (1) owner, employer, patron; (2) ghostly warden.
Pecos, as verb, to throw into the Pecos River; hence, to kill by drowning.
Pecoseños, inhabitants of Pecos.
Pelado(a), low class Mexican.
Peon, laboring class of Mexicans.
Perdida, lost.
Peso, dollar.
Picachos, Peaks.
Pido por Dios, I beg in God's name—a beggar's cry.
Pinto, (1) spotted horse; (2) spotted bean.
Piñon, dwarf pine.
Pirooting, meandering, "fooling around," probably from pirouetting.
Plata, (1) silver, (2) chart.
Play out, cease to be.
Plug, a broken down horse.
Poor doe, lean, tough venison of any kind.

Porción, an allotment, or portion, of land.
Presidio, fort; Presidio del Norte, Fort of the North.
Pronto, quickly, soon.
Pueblo, town.
Puente, bridge; Puente Piedra, Stone Bridge—a rocky ford.
Pull freight for the tules, take to the wilds, or "tall timber."
Pull in your horns, stop.

Quien sabe, who knows?
Quill, a pure blooded Indian of Mexico.

Raíz diabólica (literally, devil root), peyote or "mescal button," a drug.
Ranchero, ranchman.
Reach, to make a motion as if to draw a pistol.
Reata, rope; more especially a rawhide rope.
Remuda, bunch of saddle horses.
Rep, (as noun) a cowboy who represents his brand at outside ranches; (as verb) to represent.
República Mejicana, Mexican Republic.
Resaca, marsh; Resaca de la Palma, Palm Marsh.
Rib up, persuade.
Rico, rich, a rich man.
Rural, Mexican peace officer corresponding to Texas ranger.
Rustle, (1) to wrangle, or herd, horses; (2) to steal.

Sacaguista, a kind of coarse, salt grass.
Señor, sir.
Serape, shawl, blanket.
Set back, to pull back.
Shank of the afternoon, late afternoon.
Shuck, (1) cigarette made with corn shuck for wrapping; (2) slang name for Mexican.
Siesta, afternoon nap.
Sitio, grant of land.
Sotol, a plant of the yucca family having long, stiff, saw-edged leaves.
Sow bosom, salt pork.
Spanish dagger, yucca plant.
Squat, a bit of land, a claim.
Stake, (1) as verb, to furnish supplies or money for an enterprise; (2) "to pull up stakes," to leave; (3) "to make a stake," to make a beginner's fortune.

Stake-pin, a pin or peg used for tying a stake-rope to.
Stamping ground, home range.
Stick-and-mud chimney, constructed of adobe (mud) and sticks.
Stomp, stamp; stomp dance, war dance.
String up, hang.
Stuff, buried treasure.
Swag, (1) quantity, load; (2) a low place, coulee.
Swallow fork, (1) an ear-mark; (2) used as a verb, to "sashay," to travel carelessly.
Swing a wide loop, live a free life.

Talache, grubbing-hoe.
Tarantula juice, whiskey.
Tequila, alcoholic drink made from the maguey plant.
Terlingua, a name meaning Three Tongues.
Tigers of the desert, Apaches.
Tinaja, a rock water hole.
Toboso, a coarse grass.
Tomahawk, often used for hatchet.
Tortuga, turtle, tortoise.
Tracks, to make, to leave, travel.
Trigueño, a brown horse.

Vaciero, supply agent for several sheep (or goat) camps; also, a kind of boss over **pastores.**
Vámonos, let us go.
Vaquero, cowboy.
Vereda, trail; Veredas Coloradas, Red Trails.
Vega, a meadow; a stretch of low, flat country.
Viznaga, a species of cactus.
Voucher, Indian scalp.

War bag, a sack or bag for personal belongings.
Wet stock, cattle or horses smuggled across the Rio Grande.
White man, (1) English-speaking Caucasian as distinguished from Mexican; (2) a decent human being.
Wrangle, to herd horses; to drive them.

Zurrón, a bag fastened to the head for carrying ore in.